ALGEBRA

Quadratic Formula

The solutions of the quadratic equation $ax^2 + bx + c = 0$ are given by

$$x = \frac{-b \pm \sqrt{b^2 - 4ac}}{2a}.$$

Factorial notation

For each positive integer n,

$$n! = n(n-1)(n-2)\cdots 3\cdot 2\cdot 1;$$

by definition, $0! = 1$.

Radicals

$$\sqrt[n]{x^m} = \left(\sqrt[n]{x}\right)^m = x^{m/n}$$

Exponents

$$(ab)^r = a^r b^r \qquad a^r a^s = a^{r+s} \qquad x^{-n} = \frac{1}{x^n}$$

$$(a^r)^s = a^{rs} \qquad \frac{a^r}{a^s} = a^{r-s}$$

Binomial Formula

$$(x+y)^2 = x^2 + 2xy + y^2$$
$$(x+y)^3 = x^3 + 3x^2y + 3xy^2 + y^3$$
$$(x+y)^4 = x^4 + 4x^3y + 6x^2y^2 + 4xy^3 + y^4$$

In general, $(x+y)^n = x^n + \binom{n}{1}x^{n-1}y + \binom{n}{2}x^{n-2}y^2 + \cdots + \binom{n}{k}x^{n-k}y^k + \cdots + \binom{n}{n-1}xy^{n-1} + y^n$,

where the binomial coefficient $\binom{n}{m}$ is the integer $\dfrac{n!}{m!(n-m)!}$.

Factoring

If n is a positive integer, then

$$x^n - y^n = (x-y)(x^{n-1} + x^{n-2}y + x^{n-3}y^2 + \cdots + x^{n-k-1}y^k + \cdots + xy^{n-2} + y^{n-1}).$$

If n is an *odd* positive integer, then

$$x^n + y^n = (x+y)(x^{n-1} - x^{n-2}y + x^{n-3}y^2 - \cdots \pm x^{n-k-1}y^k \mp \cdots - xy^{n-2} + y^{n-1}).$$

GEOMETRY

Distance Formulas

Distance on the real number line:

$$d = |a - b|$$

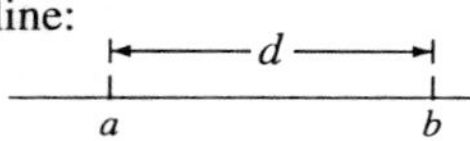

Distance in the coordinate plane:

$$d = \sqrt{(x_1 - x_2)^2 + (y_1 - y_2)^2}$$

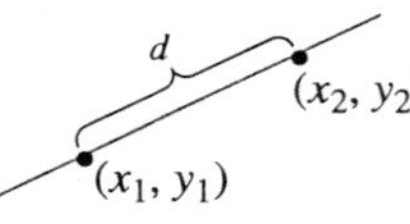

Equations of Lines and Circles

Slope-intercept equation:

$$y = mx + b$$

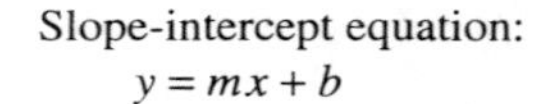

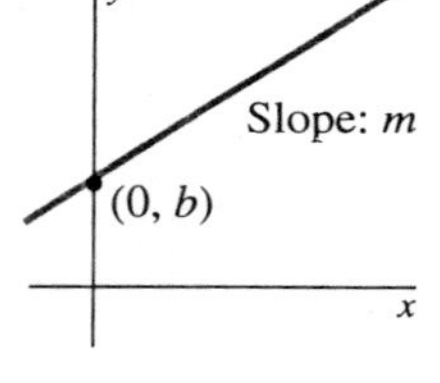

Point-slope equation:

$$y - y_1 = m(x - x_1)$$

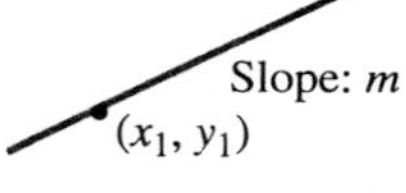

Circle with center (h, k) and radius r:

$$(x-h)^2 + (y-k)^2 = r^2$$

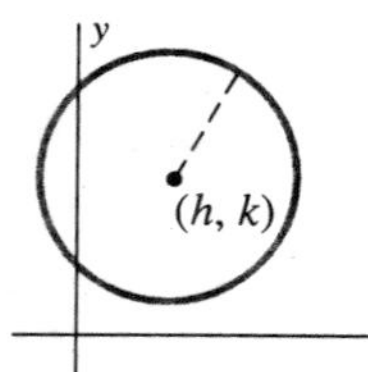

Triangle area: $A = \frac{1}{2}bh$

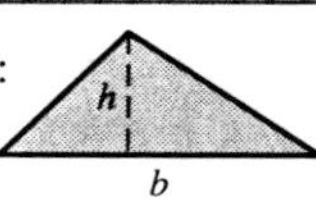

Rectangle area: $A = bh$

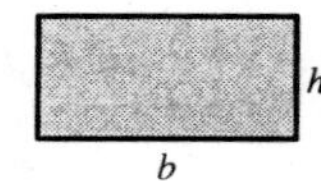

Trapezoid area: $A = \dfrac{b_1 + b_2}{2}h$

Circle area: $A = \pi r^2$

Circumference: $C = 2\pi r$

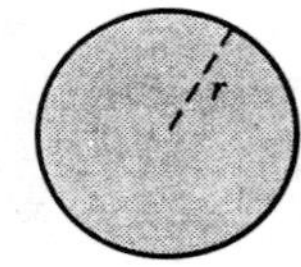

Sphere volume: $V = \frac{4}{3}\pi r^3$

Surface area: $A = 4\pi r^2$

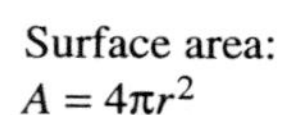

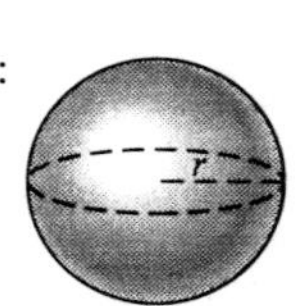

Cylinder volume: $V = \pi r^2 h$

Curved surface area: $A = 2\pi rh$

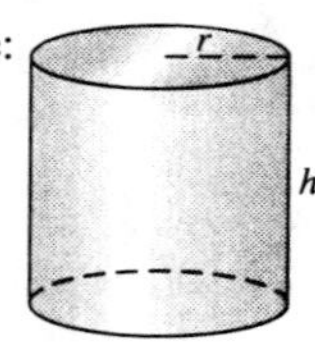

Cone volume: $V = \frac{1}{3}\pi r^2 h$

Curved surface area: $A = \pi r\sqrt{r^2 + h^2}$

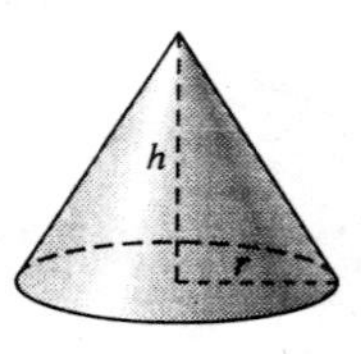

TRIGONOMETRY

$$\sin^2 A + \cos^2 A = 1 \quad \text{(the \textit{fundamental identity})}$$
$$\tan^2 A + 1 = \sec^2 A$$

$$\cos 2A = \cos^2 A - \sin^2 A = 1 - 2\sin^2 A = 2\cos^2 A - 1$$
$$\sin 2A = 2\sin A\cos A$$

$$\cos(A+B) = \cos A\cos B - \sin A\sin B$$
$$\cos(A-B) = \cos A\cos B + \sin A\sin B$$
$$\sin(A+B) = \sin A\cos B + \cos A\sin B$$
$$\sin(A-B) = \sin A\cos B - \cos A\sin B$$

$$\cos^2 A = \frac{1 + \cos 2A}{2} \qquad \sin^2 A = \frac{1 - \cos 2A}{2}$$

See the Appendices for more reference formulas.

PROJECTS

Note: The icon indicates projects that are supported by additional Maple/Mathematica/MATLAB/graphing-calculator resources on the CD-ROM that accompanies this textbook.

MULTIVARIABLE CALCULUS

MULTIVARIABLE CALCULUS

Sixth Edition

C. HENRY EDWARDS
The University of Georgia, Athens

DAVID E. PENNEY
The University of Georgia, Athens

"THIS TITLE WAS ORIGINALLY PUBLISHED CONTAINING SOFTWARE THAT IS NO LONGER AVAILABLE."

Prentice Hall, Upper Saddle River, New Jersey 07458

Editor in Chief: Sally Yagan
Acquisition Editor: George Lobell
Vice President/Director of Production and Manufacturing: David W. Riccardi
Executive Managing Editor: Kathleen Schiaparelli
Senior Managing Editor: Linda Mihatov Behrens
Assistant Managing Editor: Bayani Mendoza de Leon
Production Editor: Jeanne Audino
Assistant Managing Editor, Math Media Production: John Matthews
Media Production Editors: Donna Crilly, Wendy Perez
Manufacturing Buyer: Alan Fischer
Manufacturing Manager: Trudy Pisciotti
Marketing Manager: Angela Battle
Marketing Assistant: Rachel Beckman
Development Editor: Ed Millman
Editor in Chief, Development: Carol Trueheart
Assistant Editor of Media: Vince Jansen
Editorial Assistant/Supplements Editor: Melanie Van Benthuysen
Art Director: Jonathan Boylan
Assistant to the Art Director: John Christiana
Interior Designer: Donna Young
Cover Designer: Bruce Kenselaar
Art Editor: Thomas Benfatti
Managing Editor, Audio/Video Assets: Grace Hazeldine
Creative Director: Carole Anson
Director of Creative Services: Paul Belfanti
Photo Researcher: Karen Pugliano
Photo Editor: Beth Boyd
Cover Photo: Antonio Martinelli/Arche De La Defense, Sunset View–Paris, Arch. J. O. Spreckelsen-P. Andreu.
David Cardelus/Esto. All rights reserved.
Back Cover Photo: City of Art and Sciences, Valencia, España. Barbara Burg/Oliver Schuh/Palladium Photodesign, Co, Germany.
Art Studio: Network Graphics

Upper Saddle River, New Jersey 07458

Printed in the United States of America
10 9 8 7 6 5 4 3

ISBN 0-13-033967-9

Pearson Education LTD.
Pearson Education Australia PTY, Limited
Pearson Education Singapore, Pte. Ltd.
Pearson Education North Asia Ltd.
Pearson Education Canada, Ltd.
Pearson Educacion de Mexico, S.A. de C.V.
Pearson Education–Japan
Pearson Education Malaysia, Pte. Ltd.

CONTENTS

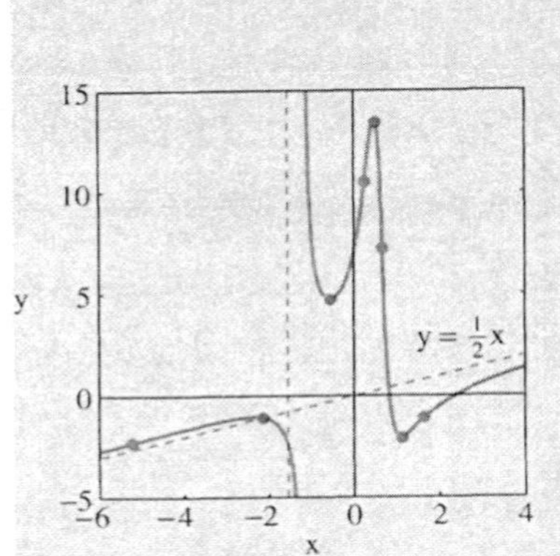

15
10
5
0
−5
y
$y = \frac{1}{2}x$
−6 −4 −2 0 2 4
x

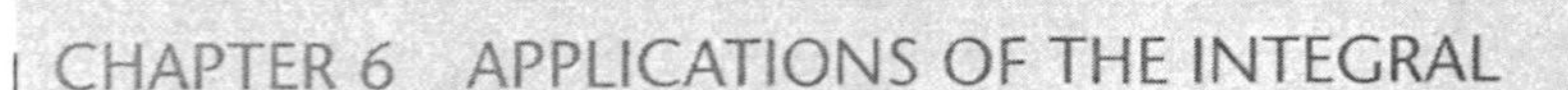

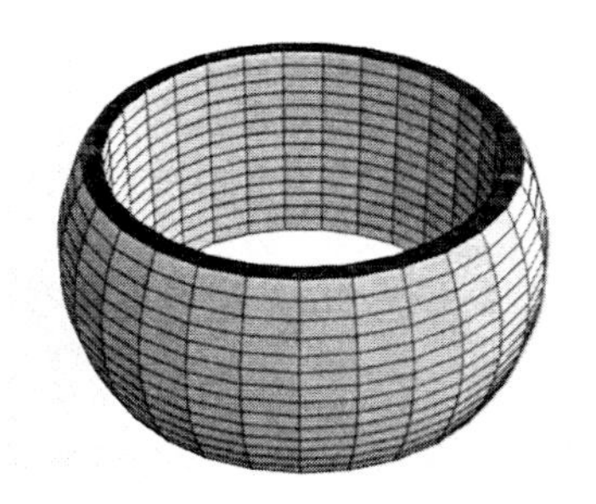

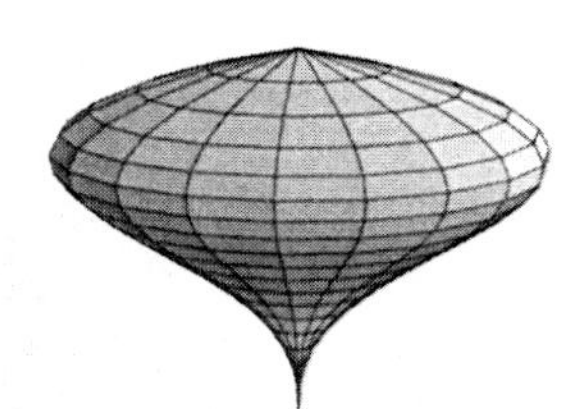
y
$y = 1/x$
1
2
$\frac{1}{2}$
x

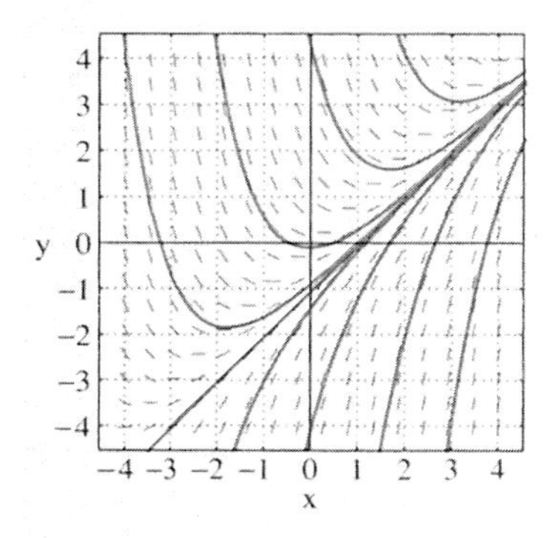
4
3
2
1
y 0
−1
−2
−3
−4
−4 −3 −2 −1 0 1 2 3 4
x

CHAPTER 10 POLAR COORDINATES AND PARAMETRIC CURVES 623

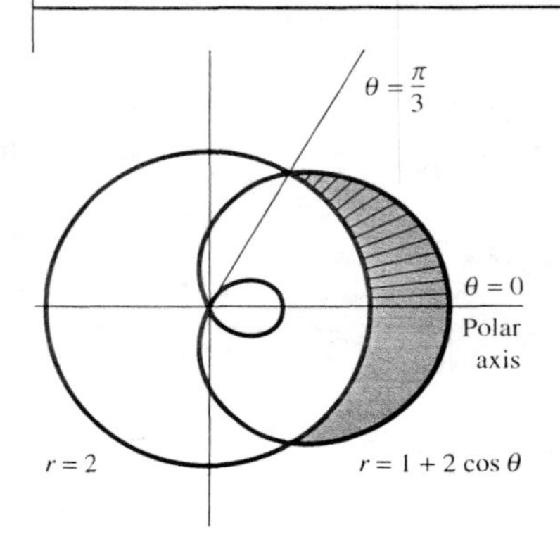

CHAPTER 11 INFINITE SERIES 681

CHAPTER 12 VECTORS, CURVES, AND SURFACES IN SPACE 771

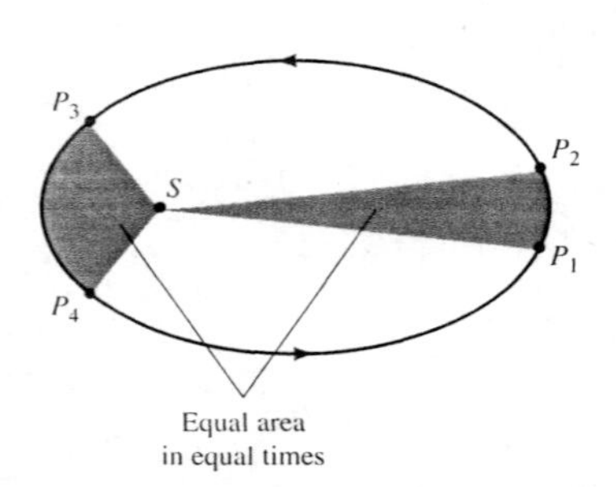

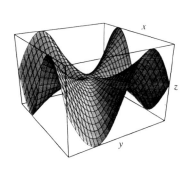

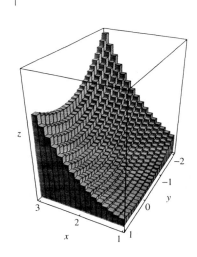

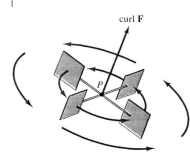

PHOTO CREDITS p. 623 (top left) Stock Montage, Inc./Historical Pictures Collection; (bottom right) Stephen Gerard/Science Service/Photo Researchers, Inc. p. 681 (top left) University of Cambridge; (bottom right) Image by David E. Penney p. 804 Robert Garvey/ Black Star p. 849 Courtesy of the Library of Congress p. 939 (top left) The Granger Collection, New York; (bottom right) Jeff Greenberg/PhotoEdit p. 1013 (top left) CORBIS; (bottom right) Chandra X-Ray Center/A. Hobart

ABOUT THE AUTHORS

C. Henry Edwards is emeritus professor of mathematics at the University of Georgia. He earned his Ph.D. at the University of Tennessee in 1960, and recently retired after 40 years of classroom teaching (including calculus or differential equations almost every term) at the universities of Tennessee, Wisconsin, and Georgia, with a brief interlude at the Institute for Advanced Study (Princeton) as an Alfred P. Sloan Research Fellow. He has received numerous teaching awards, including the University of Georgia's *honoratus* medal in 1983 (for sustained excellence in honors teaching), its Josiah Meigs award in 1991 (the institution's highest award for teaching), and the 1997 state-wide Georgia Regents award for research university faculty teaching excellence. His scholarly career has ranged from research and dissertation direction in topology to the history of mathematics to computing and technology in the teaching and applications of mathematics. In addition to being author or co-author of calculus, advanced calculus, linear algebra, and differential equations textbooks, he is well-known to calculus instructors as author of *The Historical Development of the Calculus* (Springer-Verlag, 1979). During the 1990s he served as a principal investigator on three NSF-supported projects: (1) A school mathematics project including Maple for beginning algebra students, (2) A Calculus-with-*Mathematica* program, and (3) A MATLAB-based computer lab project for numerical analysis and differential equations students.

David E. Penney, University of Georgia, completed his Ph.D. at Tulane University in 1965 (under the direction of Prof. L. Bruce Treybig) while teaching at the University of New Orleans. Earlier he had worked in experimental biophysics at Tulane University and the Veteran's Administration Hospital in New Orleans under the direction of Robert Dixon McAfee, where Dr. McAfee's research team's primary focus was on the active transport of sodium ions by biological membranes. Penney's primary contribution here was the development of a mathematical model (using simultaneous ordinary differential equations) for the metabolic phenomena regulating such transport, with potential future applications in kidney physiology, management of hypertension, and treatment of congestive heart failure. He also designed and constructed servomechanisms for the accurate monitoring of ion transport, a phenomenon involving the measurement of potentials in microvolts at impedances of millions of megohms. Penney began teaching calculus at Tulane in 1957 and taught that course almost every term with enthusiasm and distinction until his retirement at the end of the last millennium. During his tenure at the University of Georgia he received numerous University-wide teaching awards as well as directing several doctoral dissertations and seven undergraduate research projects. He is the author of research papers in number theory and topology and is the author or co-author of textbooks on calculus, computer programming, differential equations, linear algebra, and liberal arts mathematics.

PREFACE

Contemporary calculus instructors and students face traditional challenges as well as new ones that result from changes in the role and practice of mathematics by scientists and engineers in the world at large. As a consequence, this sixth edition of our calculus textbook is its most extensive revision since the first edition appeared in 1982.

Two chapters of the fifth edition have been combined in a single more tightly organized one. An entirely new chapter now appears in the table of contents, and most of the remaining chapters have been extensively rewritten. About 125 of the book's over 750 worked examples are new for this edition and the 1825 figures in the text include 225 new computer-generated graphics. About 600 of its over 7000 problems are new, and these are augmented by 320 new conceptual discussion questions that now precede the problem sets. Moreover, 1050 new true/false questions are included in the Study Guides on the new CD-ROM that accompanies this edition. In summary, almost 2000 of these 8400-plus problems and questions are new, and the text discussion and explanations have undergone corresponding alteration and improvement.

PRINCIPAL NEW FEATURES

The current revision of the text features

- More unified treatment of **transcendental functions** in Chapter 7, and
- A new Chapter 9 on **differential equations** and applications.

The new chapter on differential equations now appears immediately after Chapter 8 on techniques of integration. It includes both direction fields and Euler's method together with the more elementary symbolic methods (which exploit techniques from Chapter 8) and interesting applications of both first- and second-order equations. Chapter 11 (Infinite Series) now ends with a new section on power series solutions of differential equations, thus bringing full circle a unifying focus of second-semester calculus on elementary differential equations.

NEW LEARNING RESOURCES

Conceptual Discussion Questions The set of problems that concludes each section is now preceded by a brief **Concepts: Questions and Discussion** set consisting of several open-ended conceptual questions that can be used for either individual study or classroom discussion.

The Text CD-ROM The content of the new CD-ROM that accompanies this text is fully integrated with the textbook material, and is designed specifically for use hand-in-hand with study of the book itself. This CD-ROM features the following resources to support learning and teaching:

- **Interactive True/False Study Guides** that reinforce and encourage student reading of the text. Ten author-written questions for each section carefully guide students through the section, and students can request individual hints suggesting where in the section to look for needed information.
- **Live Examples** feature dynamic multimedia and computer algebra presentations—many accompanied by audio explanations—which enhance student intuition and understanding. These interactive examples expand upon many of the textbook's principal examples; students can change input data and conditions and then observe the resulting changes in step-by-step solutions and accompanying graphs and figures. **Walkthrough videos** demonstrate how students can interact with these live examples.
- **Homework Starters** for the principal types of computational problems in each textbook section, featuring both interactive presentations similar to the live examples and (Web-linked) voice-narrated videos of pencil-and-paper investigations illustrating typical initial steps in the solution of selected textbook problems.
- **Computing Project Resources** support most of the almost three dozen projects that follow key sections in the text. For each such project marked in the text by a CD-ROM icon, more extended discussions illustrating Maple, *Mathematica,* MATLAB, and graphing calculator investigations are provided. Computer algebra system commands can be copied and pasted for interactive execution.
- **Hyperlinked Maple Worksheets** contributed by Harald Pleym of Telemark University College (Norway) constitute an interactive version of essentially the whole textbook. Students and faculty using Maple can change input data and conditions in most of the text examples to investigate the resulting changes in step-by-step solutions and accompanying graphs and figures.
- **PowerPoint Presentations** provide classroom projection versions of about 350 of the figures in the text that would be least convenient to reproduce on a blackboard.
- **Web Site** The contents of the CD-ROM together with additional learning and teaching resources are maintained and updated at the textbook Web site **www.prenhall.com/edwards,** which includes a Comments and Suggestions center where we invite response from both students and instructors.

PH Grade Assist (Computerized Homework Grading System)

About 2000 of the textbook problems are incorporated in an automated grading system that is now available. Each problem solution in the system is structured algorithmically so that students can work in a computer lab setting to submit homework assignments for automatic grading. (There is a small annual fee per participating student.)

New Solutions Manuals

The entirely new 1810-page **Instructor's Solutions Manual** (available in three volumes) includes a detailed solution for every problem in the book. These solutions were written exclusively by the authors and have been checked independently by others.

The entirely new 930-page **Student Solutions Manual** (available in two volumes) includes a detailed solution for every odd-numbered problem in the text. The answers (alone) to most of these odd-numbered problems are included in the answers section at the back of this book.

New Technology manuals

Each of the following manuals is available shrink-wrapped with any version of the text for half the normal price of the manual (all of which are inexpensive):

- Jensen, *Using MATLAB in Calculus* (0-13-027268-X)
- Freese/Stegenga, *Calculus Concepts Using Derive* (0-13-085152-3)
- Gresser, *TI Graphing Calculator Approach, 2e* (0-13-092017-7)
- Gresser, *A Mathematica Approach, 2e* (0-13-092015-0)
- Gresser, *A Maple Approach, 2e* (0-13-092014-2)

THE TEXT IN MORE DETAIL . . .

In preparing this edition, we have taken advantage of many valuable comments and suggestions from users of the first five editions. This revision was so pervasive that the individual changes are too numerous to be detailed in a preface, but the following paragraphs summarize those that may be of widest interest.

▼ **New Problems** Most of the 600 new problems lie in the intermediate range of difficulty, neither highly theoretical nor computationally routine. Many of them have a new technology flavor, suggesting (if not requiring) the use of technology ranging from a graphing calculator to a computer algebra system.

▼ **Discussion Questions and Study Guides** We hope the 320 conceptual discussion questions and 1050 true/false study-guide questions constitute a useful addition to the traditional fare of student exercises and problems. The True/False Study Guide for each section provides a focus on the key ideas of the section, with the single goal of motivating guided student reading of the section.

▼ **Examples and Explanations** About one-sixth of the book's worked examples are either new or significantly revised, together with a similar percentage of the text discussion and explanations. Additional computational detail has been inserted in worked examples where students have experienced difficulty, together with additional sentences and paragraphs in similar spots in text discussions.

▼ **Project Material** Many of the text's 33 projects are new for this edition. These appear following the problem sets at the ends of key sections throughout the text. Most (but not all) of these projects employ some aspect of modern computational technology to illustrate the principal ideas of the preceding section, and many contain additional problems intended for solution with the use of a graphing calculator or computer algebra system. Where appropriate, project discussions are significantly expanded in the CD-ROM versions of the projects.

▼ **Historical Material** Historical and biographical chapter openings offer students a sense of the development of our subject by real human beings. Indeed, our exposition of calculus frequently reflects the historical development of the subject—from ancient times to the ages of Newton and Leibniz and Euler to our own era of new computational power and technology.

TEXT ORGANIZATION

▼ **Introductory Chapters** Instead of a routine review of precalculus topics, Chapter 1 concentrates specifically on functions and graphs for use in mathematical modeling. It includes a section cataloging informally the elementary transcendental functions of calculus, as background to their more formal treatment using calculus itself. Chapter 1 concludes with a section addressing the question "What *is* calculus?" Chapter 2 on limits begins with a section on tangent lines to motivate the official introduction of limits in Section 2.2. Trigonometric limits are treated throughout Chapter 2 in order to encourage a richer and more visual introduction to the limit concept.

▼ **Differentiation Chapters** The sequence of topics in Chapters 3 and 4 differs a bit from the most traditional order. We attempt to build student confidence by introducing topics more nearly in order of increasing difficulty. The chain rule appears quite early (in Section 3.3) and we cover the basic techniques for differentiating algebraic functions before discussing maxima and minima in Sections 3.5 and 3.6 (in order to illustrate early some significant applications of the derivative). Section 3.7 treats the derivatives of all six trigonometric functions. The authors' fondness for Newton's method (Section 3.8) will be apparent.

The mean value theorem and its applications are deferred to Chapter 4 (following implicit differentiation and related rates in Section 4.1, and differentials and linear approximation in Section 4.2). In addition, a dominant theme of Chapter 4 is the use of calculus both to construct graphs of functions and to explain and interpret graphs that have been constructed by a calculator or computer. This theme is developed in Sections 4.4 on the first derivative test and 4.6 on higher derivatives and concavity.

▼ **Integration Chapters** Chapter 5 begins with a section on antiderivatives—which could logically be included in the preceding chapter, but benefits from the use of integral notation. When the definite integral is introduced in Sections 5.3 and 5.4, we emphasize endpoint and midpoint sums rather than upper and lower and more general Riemann sums. This concrete emphasis carries through the chapter to its final section on numerical integration. Chapter 6 begins with a largely new section on Riemann sum approximations, with new examples centering on fluid flow and medical applications. Section 6.6 is a new treatment of centroids of plane regions and curves.

Chapter 8 (Techniques of Integration) is organized to accommodate those instructors who feel that methods of formal integration now require less emphasis, in view of modern techniques for both numerical and symbolic integration. Integration by parts (Section 8.3) precedes trigonometric integrals (Section 8.4). The method of partial fractions appears in Section 8.5, and trigonometric substitutions and integrals involving quadratic polynomials follow in Sections 8.6 and 8.7. Improper integrals appear in Section 8.8, with new and substantial subsections on special functions and probability and random sampling. This rearrangement of Chapter 8 makes it more convenient to stop wherever the instructor desires.

▼ **Calculus of Transcendental Functions** Section 7.1 (much strengthened for this edition) introduces the exponential and logarithmic functions from a fairly intuitive viewpoint; the approach based on the natural logarithm as an integral appears in Section 7.4. Sections 7.2 and 7.3 introduce l'Hôpital's rule and apply it to round out the calculus of exponential and logarithmic functions. Sections 7.5 and 7.6 cover both derivatives of and integrals involving inverse trigonometric functions and hyperbolic functions.

▼ **Differential Equations** This entirely new chapter begins with the most elementary differential equations and applications (Section 9.1) and then proceeds to introduce both graphical (slope field) and numerical (Euler) methods in Section 9.2. Subsequent sections of the chapter treat separable and linear first-order differential equations and (in more depth than usual in a calculus course) applications such as population growth (including logistic and predator-prey populations) and motion with resistance. The final two sections of Chapter 9 treat second-order linear equations and applications to mechanical vibrations. Instructors desiring still more coverage of differential equations can arrange with the publisher to bundle and use appropriate sections of Edwards and Penney, **Differential Equations: Computing and Modeling 2/e** (Prentice-Hall, 2000).

▼ **Parametric Curves and Polar Coordinates** The principal change in Chapter 10 is the replacement of three separate sections in the 5th edition on parabolas, ellipses, and hyperbolas with a single Section 10.6 that provides a unified treatment of all the conic sections.

▼ **Infinite Series** After the usual introduction to convergence of infinite sequences and series in Sections 11.2 and 11.3, a combined treatment of Taylor polynomials and Taylor series appears in Section 11.4. This makes it possible for the instructor to experiment with a briefer treatment of infinite series, but still offer exposure to the Taylor series that are so important for applications. The principal change in Chapter 11 is the addition of a new final section on power series methods and their use to introduce new transcendental functions, thereby concluding the middle third of the book with a return to differential equations.

▼ **Multivariable Calculus** The treatment of calculus of more than a single variable is rather traditional, beginning with vectors, curves, and surfaces in Chapter 12. Chapter 13 features a strong treatment of multivariable maximum-minimum problems in Sections 13.5 (initial approach to these problems), 13.9 (Lagrange multipliers), and 13.10 (critical points of functions of two variables). Chapters 13 (Partial Differentiation), 14 (Multiple Integrals), and 15 (Vector Calculus) have been significantly rewritten for this edition.

OPTIONS IN TEACHING CALCULUS

The Calculus Sequence The present version of the text is accompanied by a less traditional version that treats transcendental functions earlier in single-variable calculus and includes matrices for use in multivariable calculus. Both versions of the complete text are also available in two-volume split editions. By appropriate selection of first and second volumes, the instructor can therefore construct a complete text for a calculus sequence with

- Early transcendentals in single-variable calculus and matrices in multivariable calculus;
- Early transcendentals in single-variable calculus but traditional coverage of multivariable calculus;
- Transcendental functions delayed until after the integral in single-variable calculus, but matrices used in multivariable calculus;
- Neither early transcendentals in single-variable calculus nor matrices in multivariable calculus.

Maximum-Minimum Problems The text includes first coverage of maximum-minimum problems in Chapter 3 (Sections 3.5 and 3.6) to provide early motivation in the form of concrete applications of the derivative, and then returns with the first- and second-derivative tests of Sections 4.4 and 4.6. However, some instructors may prefer to treat these applications later—following trigonometric derivatives and related rates, and along with the bulk of the associated material in Chapter 4. The modular character of the pertinent sections in these two differential calculus chapters permits such desired rearrangements of the material within the typical pair of instructional units (each likely followed by its own hour test). For instance, Sections 3.5 and 3.6 on max-min problems can be deferred and used to begin the second unit on differential calculus, after appending Sections 4.1 (Implicit Differentiation and Related Rates) and 4.2 (Increments, Differentials, and Linear Approximation) to the first unit. Thus the material in Chapters 3 and 4 would be covered in the following order.

Unit I: Differentiation

3.1 The Derivative and Rates of Change
3.2 Basic Differentiation Rules

3.3 The Chain Rule
3.4 Derivatives of Algebraic Functions
3.7 Derivatives of Trigonometric Functions (skipping Example 13 and Problems 78–86)
3.8 Successive Approximations and Newton's Method
4.1 Implicit Functions and Related Rates
4.2 Increments, Differentials, and Linear Approximation

Unit II: Applications of the Derivative

3.5 Maxima and Minima of Functions on Closed Intervals
3.6 Applied Optimization Problems
4.3 Increasing and Decreasing Functions and the Mean Value Theorem
4.4 The First Derivative Test and Applications
4.5 Simple Curve Sketching
4.6 Higher Derivatives and Concavity
4.7 Curve Sketching and Asymptotes

ACKNOWLEDGMENTS

All experienced textbook authors know the value of critical reviewing during the preparation and revision of a manuscript. In our work on this edition we have profited greatly from the unusually detailed and constructive advice of the following very able reviewers:

- Kenzu Abdella—Trent University
- Martina Bode—Northwestern University
- David Caraballo—Georgetown University
- Tom Cassidy—Bucknell University
- Lucille Croom—Hunter College
- Yuanan Diao—University of North Carolina at Charlotte
- Victor Elias—University of Western Ontario
- Haitao Fan—Georgetown University
- James J. Faran, V—The State University of New York at Buffalo
- K. N. Gowrisankaran—McGill University
- Qing Han—University of Notre Dame
- Melvin D. Lax—California State University, Long Beach
- Robert H. Lewis—Fordham University
- Allan B. MacIssac—University of Western Ontario
- Rudolph M. Najar—California State University, Fresno
- George Pletsch—Albuquerque Technical and Vocational Institute
- Nancy Rallis—Boston College
- Robert C. Reilly—University of California, Irvine
- James A. Reneke—Clemson University
- Alexander Retakh—Yale University
- Carl Riehm—McMaster University
- Ira Sharenow—University of Wisconsin, Madison
- Kay Strangman—University of Wisconsin, Madison
- Sophie Tryphonas—University of Toronto at Scarborough
- Clifford E. Weil—Michigan State University
- Kamran Vakili—Princeton University
- Cathleen M. Zucco-Teveloff—Trinity College

Many of the best improvements that have been made must be credited to colleagues and users of the previous five editions throughout the United States, Canada, and abroad. We are grateful to all those, especially students, who have written to us, and hope they will continue to do so. We thank the accuracy checkers of M. and N. Toscano, who verified the solution of every worked-out example and odd-numbered answer, as well as all of the solutions in the Instructor's and Student Solutions Manuals. We believe that the appearance and quality of the finished book is clear testimony to the skill, diligence, and talent of an exceptional staff at Prentice-Hall. We owe special thanks to George Lobell, our mathematics editor, whose advice and criticism guided and shaped this revision in many significant and tangible ways, as did the constructive comments and suggestions of Ed Millman, our developmental editor. We also thank Gale Epps and Melanie Van Benthuysen for their highly varied and detailed services in aid of editors and authors throughout the work of revision. The visual graphics of this text have been widely praised in previous editions, and it is time for us to thank Ron Weichart of Network Graphics, who has worked with us through the past three editions. Jeanne Audino, our production editor, expertly and smoothly managed the whole process of book production. Our art director, Jonathan Boylan, supervised and coordinated the attractive design and layout of the text and the cover for this edition. Vince Jansen coordinated the production of the CD-ROM, for which we thank especially Robert Curtis and Lee Wayand for their interactive examples and Harald Pleym for his Maple worksheets. Finally, we again are unable to thank Alice Fitzgerald Edwards and Carol Wilson Penney adequately for their unrelenting assistance, encouragement, support, and patience extending through six editions and over two decades of work on this textbook.

C. Henry Edwards
hedwards@math.uga.edu

David E. Penney
dpenney@math.uga.edu

MULTIVARIABLE CALCULUS

POLAR COORDINATES AND PARAMETRIC CURVES

10

Pierre de Fermat (1601–1665)

Pierre de Fermat exemplifies the distinguished tradition of great amateurs in mathematics. Like his contemporary René Descartes, he was educated as a lawyer. But unlike Descartes, Fermat actually practiced law as his profession and served in the regional parliament. His ample leisure time was, however, devoted to mathematics and to other intellectual pursuits, such as the study of ancient Greek manuscripts.

In a margin of one such manuscript (by the Greek mathematician Diophantus) was found a handwritten note that has remained an enigma ever since. Fermat asserts that for *no* integer $n > 2$ do positive integers x, y, and z exist such that $x^n + y^n = z^n$. For instance, although $15^2 + 8^2 = 17^2$, the sum of two (positive integer) cubes cannot be a cube. "I have found an admirable proof of this," Fermat wrote, "but this margin is too narrow to contain it." Despite the publication of many incorrect proofs, "Fermat's last theorem" remained unproved for three and one-half centuries. But in a June 1993 lecture, the British mathematician Andrew Wiles of Princeton University announced a long and complex proof of Fermat's last theorem. Although the proof as originally proposed contained some gaps, these have been repaired, and experts in the field agree that Fermat's last *conjecture* is, finally, a *theorem*.

Descartes and Fermat shared in the discovery of analytic geometry. But whereas Descartes typically used geometrical methods to solve algebraic equations (see the Chapter 1 opening), Fermat concentrated on the investigation of geometric curves defined by algebraic equations. For instance, he introduced the translation methods of this chapter together with rotation methods to show that the graph of an equation of the form $Ax^2 + Bxy + Cy^2 + Dx + Ey + F = 0$ is generally a conic section. Most of his mathematical work remained unpublished during his lifetime, but it contains numerous tangent line (derivative) and area (integral) computations.

The brilliantly colored left-hand photograph is a twentieth-century example of a geometric object defined by means of algebraic operations. Starting with the point $P(a, b)$ in the xy-plane, we interpret P as the complex number $c = a + bi$ and define the sequence $\{z_n\}$ of points of the complex plane iteratively (as in Section 3.8) by the equations

$$z_0 = c, \qquad z_{n+1} = z_n^2 + c \quad (\text{for } n \geqq 0).$$

If this sequence of points remains inside the circle $x^2 + y^2 = 4$ for all n, then the original point $P(a, b)$ is colored black. Otherwise, the color assigned to P is determined by the speed with which this sequence "escapes" that circular disk. The set of all black points is the famous *Mandelbrot set,* discovered in 1980 by the French mathematician Benoit Mandelbrot.

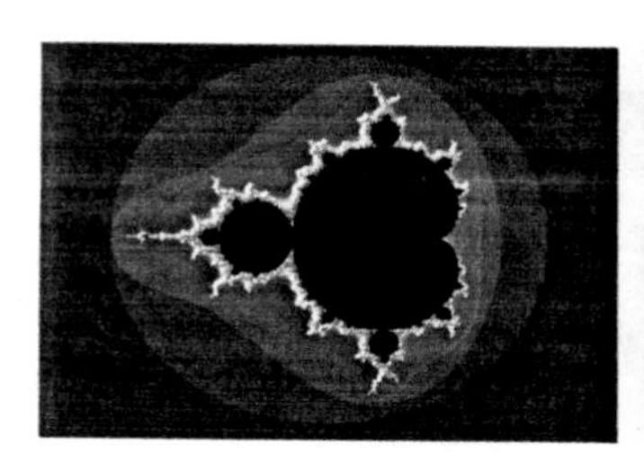

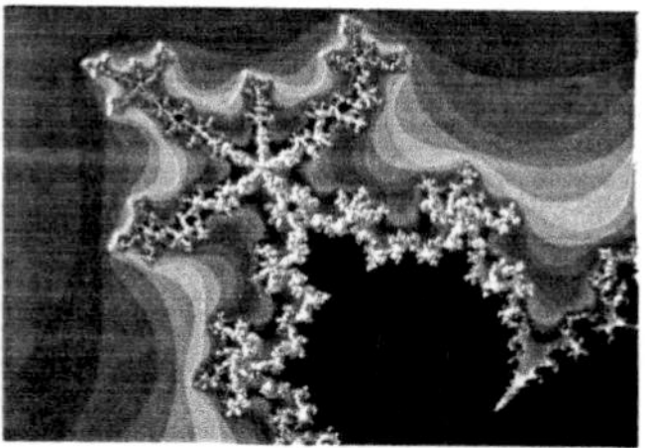

The object in the right-hand figure is a subset of that in the left-hand figure.

10.1 ANALYTIC GEOMETRY AND THE CONIC SECTIONS

Plane analytic geometry, a central topic of this chapter, is the use of algebra and calculus to study the properties of curves in the xy-plane. The ancient Greeks used deductive reasoning and the methods of axiomatic Euclidean geometry to study lines, circles, and the **conic sections** (parabolas, ellipses, and hyperbolas). The properties of conic sections have played an important role in diverse scientific applications since the seventeenth century, when Kepler discovered—and Newton explained—the fact that the orbits of planets and other bodies in the Solar System are conic sections.

The French mathematicians Descartes and Fermat, working almost independently of one another, initiated analytic geometry in 1637. The central idea of analytic geometry is the correspondence between an equation $F(x, y) = 0$ and its **locus** (typically, a curve), the set of all those points (x, y) in the plane with coordinates that satisfy this equation.

A central idea of analytic geometry is this: Given a geometric locus or curve, its properties can be derived algebraically or analytically from its defining equation $F(x, y) = 0$. For example, suppose that the equation of a given curve turns out to be the linear equation

$$Ax + By = C, \tag{1}$$

where A, B, and C are constants with $B \neq 0$. This equation may be written in the form

$$y = mx + b, \tag{2}$$

where $m = -A/B$ and $b = C/B$. But Eq. (2) is the slope-intercept equation of the straight line with slope m and y-intercept b. Hence the given curve is this straight line. We use this approach in Example 1 to show that a specific geometrically described locus is a particular straight line.

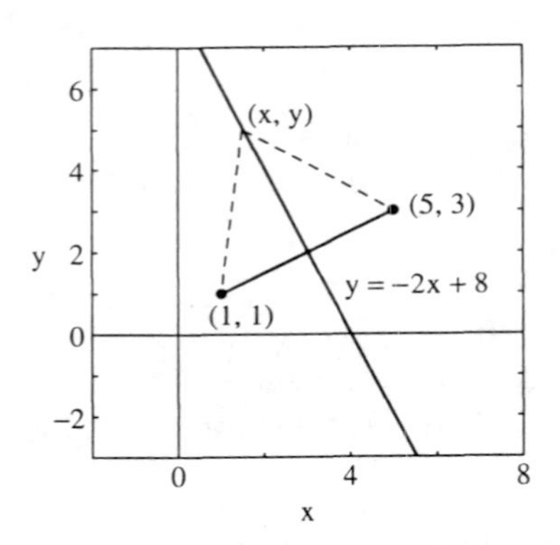

FIGURE 10.1.1 The perpendicular bisector of Example 1.

EXAMPLE 1 Prove that the set of all points equidistant from the points (1, 1) and (5, 3) is the perpendicular bisector of the line segment that joins these two points.

Solution The typical point $P(x, y)$ in Fig. 10.1.1 is equally distant from (1, 1) and (5, 3) if and only if

$$(x-1)^2 + (y-1)^2 = (x-5)^2 + (y-3)^2;$$
$$x^2 - 2x + 1 + y^2 - 2y + 1 = x^2 - 10x + 25 + y^2 - 6y + 9;$$
$$2x + y = 8;$$
$$y = -2x + 8. \tag{3}$$

Thus the given locus is the straight line in Eq. (3) whose slope is −2. The straight line through (1, 1) and (5, 3) has equation

$$y - 1 = \tfrac{1}{2}(x - 1) \tag{4}$$

and thus has slope $\frac{1}{2}$. Because the product of the slopes of these two lines is −1, it follows (from Theorem 2 in Appendix B) that these lines are perpendicular. If we solve Eqs. (3) and (4) simultaneously, we find that the intersection of these lines is, indeed, the midpoint (3, 2) of the given line segment. Thus the locus described is the perpendicular bisector of this line segment. ◆

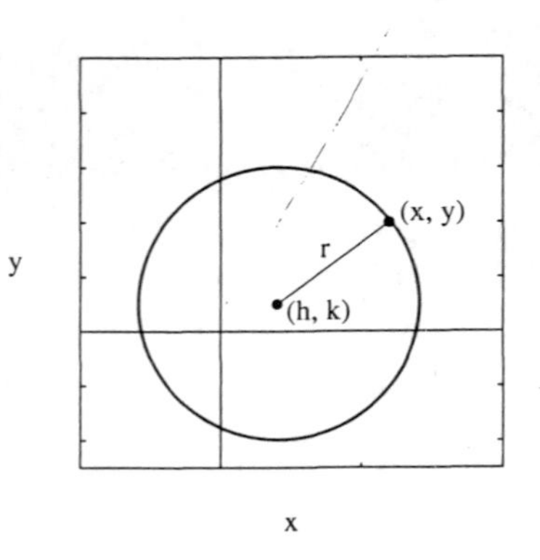

FIGURE 10.1.2 The circle with center (h, k) and radius r.

The circle shown in Fig. 10.1.2 has center (h, k) and radius r. It may be described geometrically as the set or locus of all points $P(x, y)$ whose distance from (h, k) is r.

The distance formula then gives

$$(x-h)^2+(y-k)^2=r^2 \tag{5}$$

as the equation of this circle. In particular, if $h=k=0$, then Eq. (5) takes the simple form

$$x^2+y^2=r^2. \tag{6}$$

We can see directly from this equation, without further reference to the definition of *circle,* that a circle centered at the origin has the following symmetry properties:

- *Symmetry around the x-axis:* The equation of the curve is unchanged when y is replaced with $-y$.
- *Symmetry around the y-axis:* The equation of the curve is unchanged when x is replaced with $-x$.
- *Symmetry with respect to the origin:* The equation of the curve is unchanged when x is replaced with $-x$ and y is replaced with $-y$.
- *Symmetry around the* 45° *line* $y=x$*:* The equation is unchanged when x and y are interchanged.

The relationship between Eqs. (5) and (6) is an illustration of the *translation principle* stated informally in Section 1.2. Imagine a translation (or "slide") of the plane that moves each point (x, y) to the new position $(x+h, y+k)$. Under such a translation, a curve C is moved to a new position. The equation of the new translated curve is easy to obtain from the old equation—we simply replace x with $x-h$ and y with $y-k$. Conversely, we can recognize a translated circle from its equation: Any equation of the form

$$x^2+y^2+Ax+By+C=0 \tag{7}$$

can be rewritten in the form

$$(x-h)^2+(y-k)^2=p$$

by completing squares, as in Example 2 of Section 1.2. Thus the graph of Eq. (7) is either a circle (if $p>0$), a single point (if $p=0$), or no points at all (if $p<0$). We use this approach in Example 2 to discover that the locus described is a particular circle.

EXAMPLE 2 Determine the locus of a point $P(x, y)$ if its distance $|AP|$ from $A(7, 1)$ is twice its distance $|BP|$ from $B(1, 4)$.

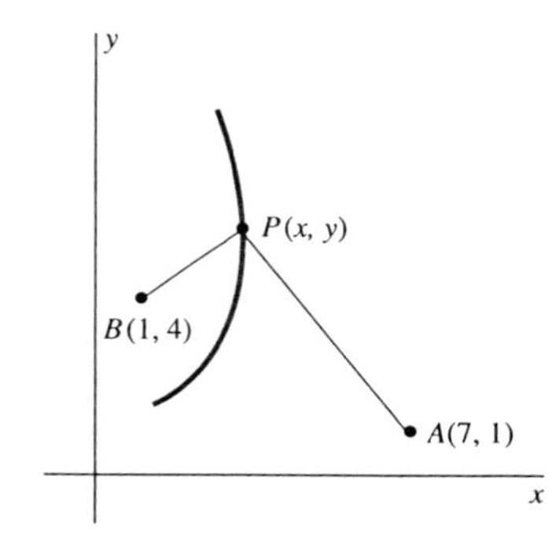

FIGURE 10.1.3 The locus of Example 2.

Solution The points A, B, and P appear in Fig. 10.1.3, along with a curve through P that represents the given locus. From

$$|AP|^2=4|BP|^2 \quad (\text{because } |AP|=2|BP|),$$

we get the equation

$$(x-7)^2+(y-1)^2=4[(x-1)^2+(y-4)^2].$$

Hence

$$3x^2+3y^2+6x-30y+18=0;$$
$$x^2+y^2+2x-10y=-6;$$
$$(x+1)^2+(y-5)^2=20.$$

Thus the locus is a circle with center $(-1, 5)$ and radius $r=\sqrt{20}=2\sqrt{5}$. ◆

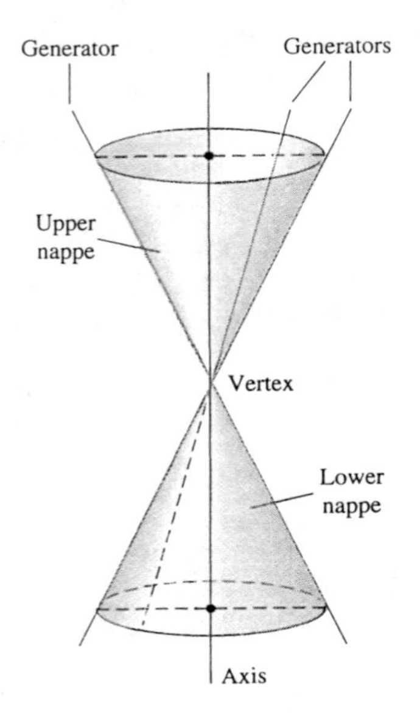

FIGURE 10.1.4 A cone with two nappes.

Conic Sections

Conic sections are so named because they are the curves formed by a plane intersecting a cone. The cone used is a right circular cone with two *nappes* extending infinitely far in both directions (Fig. 10.1.4). There are three types of conic sections, as illustrated in Fig. 10.1.5. If the cutting plane is parallel to some generator of the cone (a line that, when revolved around an axis, forms the cone), then the curve of intersection is a *parabola*. If the plane is not parallel to a generator, then the curve of intersection is either a single closed curve—an *ellipse*—or a *hyperbola* with two *branches*.

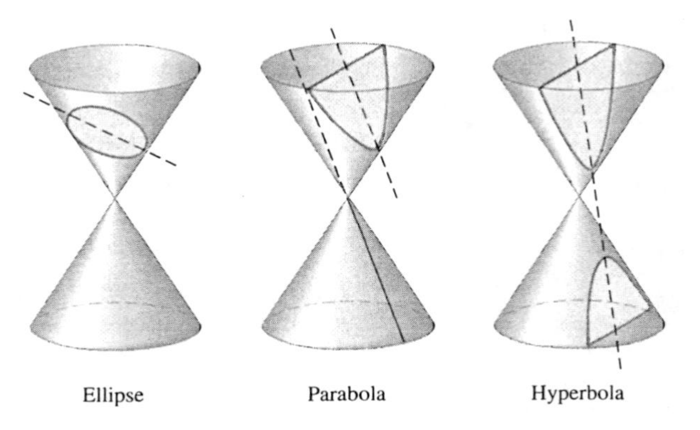

FIGURE 10.1.5 The conic sections.

In Appendix J we use the methods of three-dimensional analytic geometry to show that if an appropriate xy-coordinate system is set up in the intersecting plane, then the equations of the three conic sections take the following forms:

$$\text{Parabola:} \qquad y^2 = kx; \tag{8}$$

$$\text{Ellipse:} \qquad \frac{x^2}{a^2} + \frac{y^2}{b^2} = 1; \tag{9}$$

$$\text{Hyperbola:} \qquad \frac{x^2}{a^2} - \frac{y^2}{b^2} = 1. \tag{10}$$

In Section 10.6 we discuss these conic sections on the basis of definitions that are two-dimensional—they do not require the three-dimensional setting of a cone and an intersecting plane. Example 3 illustrates one such approach to the conic sections.

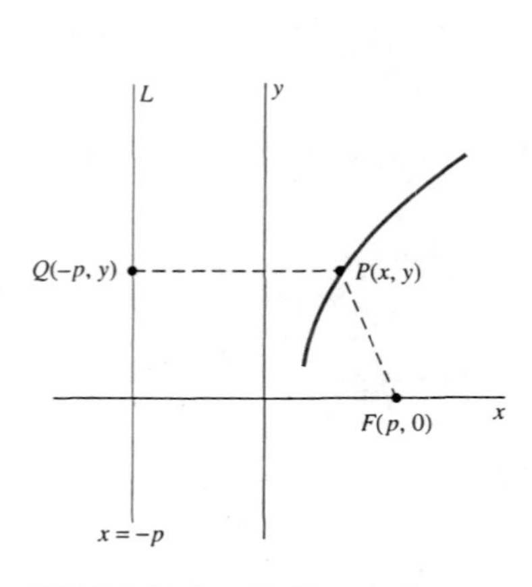

FIGURE 10.1.6 The locus of Example 3.

EXAMPLE 3 Let e be a given positive number (*not* to be confused with the natural logarithm base; in the context of conic sections, e stands for *eccentricity*). Determine the locus of a point $P(x, y)$ if its distance from the fixed point $F(p, 0)$ is e times its distance from the vertical line L whose equation is $x = -p$ (Fig. 10.1.6).

Solution Let PQ be the perpendicular from P to the line L. Then the condition

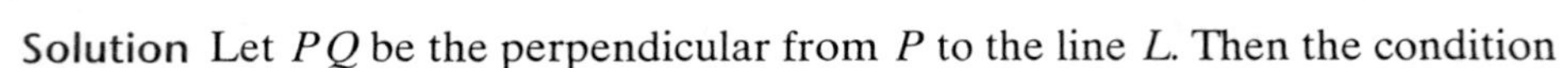

$$|PF| = e|PQ|$$

takes the analytic form

$$\sqrt{(x-p)^2 + y^2} = e|x - (-p)|.$$

That is,

$$(x^2 - 2px + p^2) + y^2 = e^2(x^2 + 2px + p^2),$$

so

$$x^2(1 - e^2) - 2p(1 + e^2)x + y^2 = -p^2(1 - e^2). \tag{11}$$

- *Case 1:* $e = 1$. Then Eq. (11) reduces to

$$y^2 = 4px. \tag{12}$$

We see upon comparison with Eq. (8) that the locus of P is a *parabola* if $e = 1$.

- *Case 2:* $e < 1$. Dividing both sides of Eq. (11) by $1 - e^2$, we get

$$x^2 - 2p \cdot \frac{1+e^2}{1-e^2}x + \frac{y^2}{1-e^2} = -p^2.$$

We now complete the square in x. The result is

$$\left(x - p \cdot \frac{1+e^2}{1-e^2}\right)^2 + \frac{y^2}{1-e^2} = p^2\left[\left(\frac{1+e^2}{1-e^2}\right)^2 - 1\right] = a^2.$$

This equation has the form

$$\frac{(x-h)^2}{a^2} + \frac{y^2}{b^2} = 1, \tag{13}$$

where

$$h = +p \cdot \frac{1+e^2}{1-e^2} \quad \text{and} \quad b^2 = a^2(1-e^2). \tag{14}$$

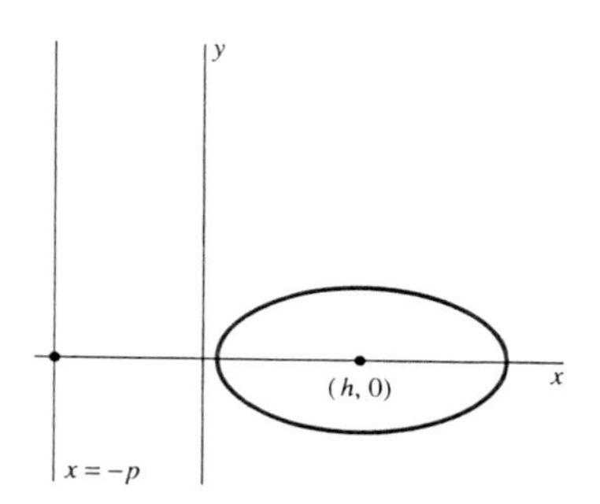

FIGURE 10.1.7 An ellipse: $e < 1$ (Example 3).

When we compare Eqs. (9) and (13), we see that if $e < 1$, then the locus of P is an *ellipse* with (0, 0) translated to $(h, 0)$, as illustrated in Fig. 10.1.7.

- *Case 3:* $e > 1$. In this case, Eq. (11) reduces to a translated version of Eq. (10), so the locus of P is a *hyperbola*. The details, which are similar to those in Case 2, are left for Problem 35.

Thus the locus in Example 3 is a *parabola* if $e = 1$, an *ellipse* if $e < 1$, and a *hyperbola* if $e > 1$. The number e is called the **eccentricity** of the conic section. The point $F(p, 0)$ is commonly called its **focus** in the parabolic case. Figure 10.1.8 shows the parabola of Case 1; Fig. 10.1.9 illustrates the hyperbola of Case 3. ◆

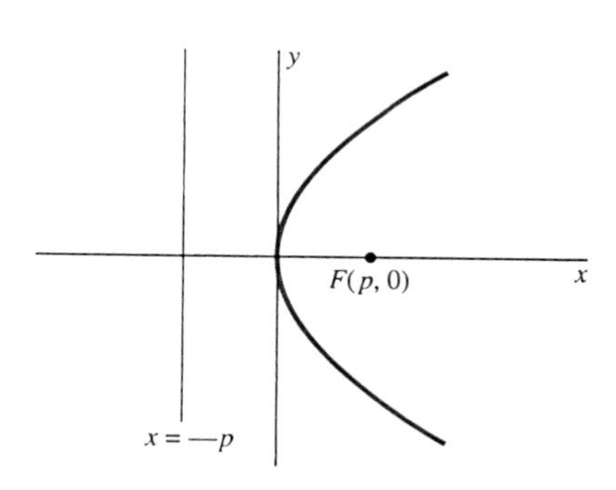

FIGURE 10.1.8 A parabola: $e = 1$ (Example 3).

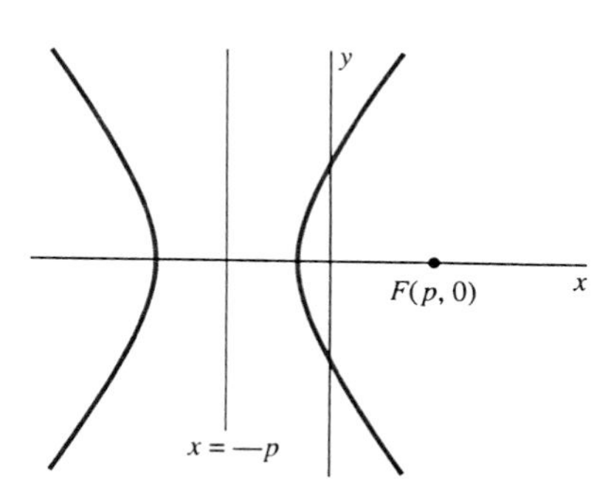

FIGURE 10.1.9 A hyperbola: $e > 1$ (Example 3).

If we begin with Eqs. (8) through (10), we can derive the general characteristics of the three conic sections shown in Figs. 10.1.7 through 10.1.9. For example, in the case of the parabola of Eq. (8) with $k < 0$, the curve passes through the origin, $x \geqq 0$ at each of the curve's points, $y \to \pm\infty$ as $x \to \infty$, and the graph is symmetric around the x-axis (because the curve is unchanged when y is replaced with $-y$).

In the case of the ellipse of Eq. (9), the graph must be symmetric around both coordinate axes. At each point (x, y) of the graph, we must have $|x| \leqq a$ and $|y| \leqq b$. The graph intersects the axes at the four points $(\pm a, 0)$ and $(0, \pm b)$.

Finally, the hyperbola of Eq. (10)—or its alternative form

$$y = \pm\frac{b}{a}\sqrt{x^2 - a^2}$$

—is symmetric around both coordinate axes. Its meets the x-axis at the two points $(\pm a, 0)$ and has one branch consisting of points with $x \geqq a$ and has another branch where $x \leqq -a$. Also, $|y| \to \infty$ as $|x| \to \infty$.

10.1 TRUE/FALSE STUDY GUIDE

10.1 CONCEPTS: QUESTIONS AND DISCUSSION

You may want to use the implicit plotting facility of a computer algebra system to investigate the following questions.

1. The graph of the equation $x^2 - y^2 = 0$ consists of the two lines $x - y = 0$ and $x + y = 0$ through the origin. What is the graph of the equation $x^n - y^n = 0$? Does it depend on whether the positive integer n is even or odd? Explain your answers.
2. How do the graphs of the equations $x^3 + y^3 = 1$ and $x^4 + y^4 = 1$ differ from the unit circle $x^2 + y^2 = 1$ (and from each other)? How does the graph of the equation $x^n + y^n = 1$ change as the positive integer n gets larger and larger? Discuss the possibility of a "limiting set" as $n \to +\infty$. Do these questions depend on whether n is even or odd?
3. The graph of the equation $x^2 - y^2 = 1$ is a hyperbola. Discuss (as in Question 2) the graph of the equation $x^n - y^n = 1$.

10.1 PROBLEMS

In Problems 1 through 6, write an equation of the specified straight line.

1. The line through the point $(1, -2)$ that is parallel to the line with equation $x + 2y = 5$
2. The line through the point $(-3, 2)$ that is perpendicular to the line with equation $3x - 4y = 7$
3. The line that is tangent to the circle $x^2 + y^2 = 25$ at the point $(3, -4)$
4. The line that is tangent to the curve $y^2 = x + 3$ at the point $(6, -3)$
5. The line that is perpendicular to the curve $x^2 + 2y^2 = 6$ at the point $(2, -1)$
6. The perpendicular bisector of the line segment with endpoints $(-3, 2)$ and $(5, -4)$

In Problems 7 through 16, find the center and radius of the circle described in the given equation.

7. $x^2 + 2x + y^2 = 4$
8. $x^2 + y^2 - 4y = 5$
9. $x^2 + y^2 - 4x + 6y = 3$
10. $x^2 + y^2 + 8x - 6y = 0$
11. $4x^2 + 4y^2 - 4x = 3$
12. $4x^2 + 4y^2 + 12y = 7$
13. $2x^2 + 2y^2 - 2x + 6y = 13$
14. $9x^2 + 9y^2 - 12x = 5$
15. $9x^2 + 9y^2 + 6x - 24y = 19$
16. $36x^2 + 36y^2 - 48x - 108y = 47$

In Problems 17 through 20, show that the graph of the given equation consists either of a single point or of no points.

17. $x^2 + y^2 - 6x - 4y + 13 = 0$
18. $2x^2 + 2y^2 + 6x + 2y + 5 = 0$
19. $x^2 + y^2 - 6x - 10y + 84 = 0$
20. $9x^2 + 9y^2 - 6x - 6y + 11 = 0$

In Problems 21 through 24, write the equation of the specified circle.

21. The circle with center $(-1, -2)$ that passes through the point $(2, 3)$
22. The circle with center $(2, -2)$ that is tangent to the line $y = x + 4$
23. The circle with center $(6, 6)$ that is tangent to the line $y = 2x - 4$
24. The circle that passes through the points $(4, 6)$, $(-2, -2)$, and $(5, -1)$

In Problems 25 through 30, derive the equation of the set of all points $P(x, y)$ that satisfy the given condition. Then sketch the graph of the equation.

25. The point $P(x, y)$ is equally distant from the two points $(3, 2)$ and $(7, 4)$.
26. The distance from P to the point $(-2, 1)$ is half the distance from P to the point $(4, -2)$.
27. The point P is three times as far from the point $(-3, 2)$ as it is from the point $(5, 10)$.
28. The distance from P to the line $x = -3$ is equal to its distance from the point $(3, 0)$.

29. The sum of the distances from P to the points $(4, 0)$ and $(-4, 0)$ is 10.

30. The sum of the distances from P to the points $(0, 3)$ and $(0, -3)$ is 10.

31. Find all the lines through the point $(2, 1)$ that are tangent to the parabola $y = x^2$.

32. Find all lines through the point $(-1, 2)$ that are normal to the parabola $y = x^2$.

33. Find all lines that are normal to the curve $xy = 4$ and simultaneously are parallel to the line $y = 4x$.

34. Find all lines that are tangent to the curve $y = x^3$ and are also parallel to the line $3x - y = 5$.

35. Suppose that $e > 1$. Show that Eq. (11) of this section can be written in the form

$$\frac{(x-h)^2}{a^2} - \frac{y^2}{b^2} = 1,$$

thus showing that its graph is a hyperbola. Find a, b, and h in terms of p and e.

10.2 POLAR COORDINATES

A familiar way to locate a point in the coordinate plane is by specifying its rectangular coordinates (x, y)—that is, by giving its abscissa x and ordinate y relative to given perpendicular axes. In some problems it is more convenient to locate a point by means of its *polar coordinates.* The polar coordinates give its position relative to a fixed reference point O (the **pole**) and to a given ray (the **polar axis**) beginning at O.

For convenience, we begin with a given xy-coordinate system and then take the origin as the pole and the nonnegative x-axis as the polar axis. Given the pole O and the polar axis, the point P with **polar coordinates** r and θ, written as the ordered pair (r, θ), is located as follows. First find the terminal side of the angle θ, given in radians, where θ is measured counterclockwise (if $\theta > 0$) from the x-axis (the polar axis) as its initial side. If $r \geqq 0$, then P is on the terminal side of this angle at the distance r from the origin. If $r < 0$, then P lies on the ray opposite the terminal side at the distance $|r| = -r > 0$ from the pole (Fig. 10.2.1). The **radial coordinate** r can be described as the *directed* distance of P from the pole along the terminal side of the angle θ. Thus, if r is positive, the point P lies in the same quadrant as θ, whereas if r is negative, then P lies in the opposite quadrant. If $r = 0$, the angle θ does not matter; the polar coordinates $(0, \theta)$ represent the origin whatever the **angular coordinate** θ might be. The origin, or pole, is the only point for which $r = 0$.

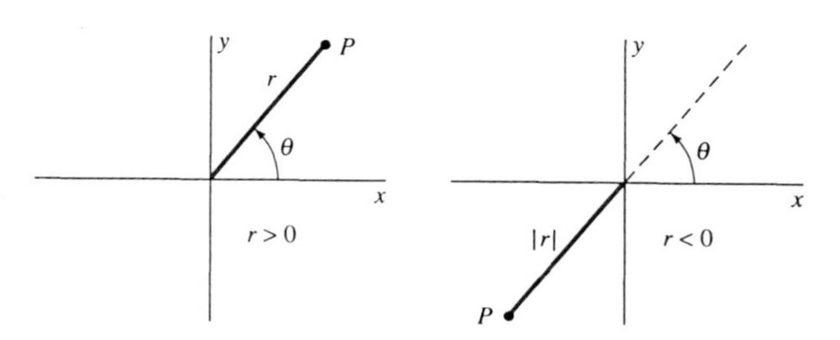

FIGURE 10.2.1 The difference between the two cases $r > 0$ and $r < 0$.

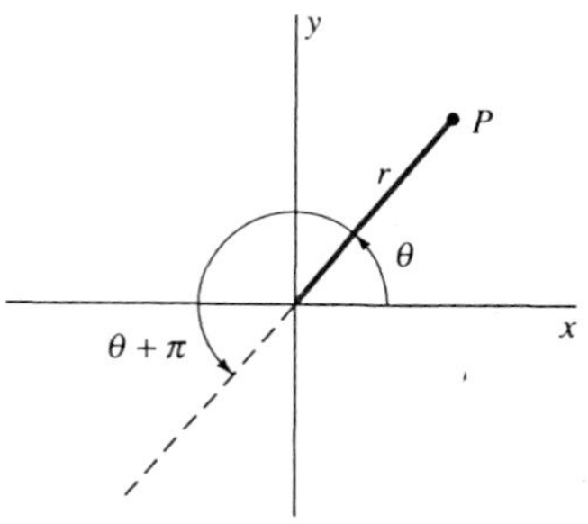

FIGURE 10.2.2 The polar coordinates (r, θ) and $(-r, \theta + \pi)$ represent the same point P (Example 1).

EXAMPLE 1 Polar coordinates differ from rectangular coordinates in that any point has more than one representation in polar coordinates. For example, the polar coordinates (r, θ) and $(-r, \theta + \pi)$ represent the same point P, as shown in Fig. 10.2.2. More generally, this point P has the polar coordinates $(r, \theta + n\pi)$ for any even integer n *and* the coordinates $(-r, \theta + n\pi)$ for any odd integer n. Thus the polar-coordinate pairs

$$\left(2, \frac{\pi}{3}\right), \quad \left(-2, \frac{4\pi}{3}\right), \quad \left(2, \frac{7\pi}{3}\right), \quad \text{and} \quad \left(-2, -\frac{2\pi}{3}\right)$$

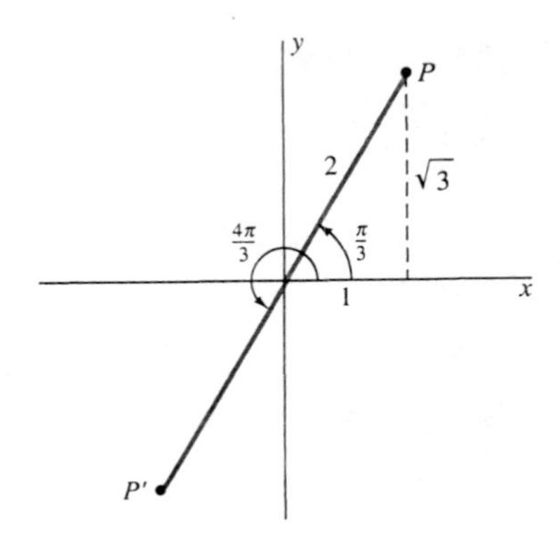

FIGURE 10.2.3 The point P of Example 1 can be described in many different ways using polar coordinates.

all represent the same point P in Fig. 10.2.3. (The rectangular coordinates of P are $(1, \sqrt{3})$.) ◆

To convert polar coordinates into rectangular coordinates, we use the basic relations

$$x = r\cos\theta, \quad y = r\sin\theta \tag{1}$$

that we read from the right triangle in Fig. 10.2.4. Converting in the opposite direction, we have

$$r^2 = x^2 + y^2, \qquad \tan\theta = \frac{y}{x} \quad \text{if } x \neq 0. \tag{2}$$

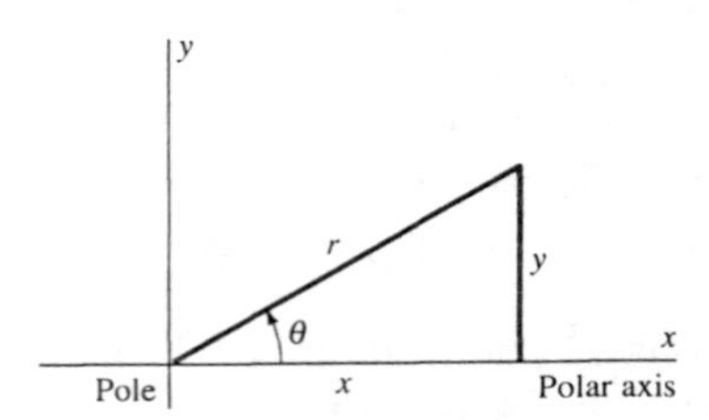

FIGURE 10.2.4 Read Eqs. (1) and (2)—conversions between polar and rectangular coordinates—from this figure.

Some care is required in making the correct choice of θ in the formula $\tan\theta = y/x$. If $x > 0$, then (x, y) lies in either the first or fourth quadrant, so $-\pi/2 < \theta < \pi/2$, which is the range of the inverse tangent function. Hence if $x > 0$, then $\theta = \arctan(y/x)$. But if $x < 0$, then (x, y) lies in the second or third quadrant. In this case a proper choice for the angle is $\theta = \pi + \arctan(y/x)$. In any event, the signs of x and y in Eq. (1) with $r > 0$ indicate the quadrant in which θ lies.

Polar Coordinate Equations

Some curves have simpler equations in polar coordinates than in rectangular coordinates; this is an important reason for the usefulness of polar coordinates. The **graph** of an equation in the polar-coordinate variables r and θ is the set of all those points P such that P has some pair of polar coordinates (r, θ) that satisfy the given equation. The graph of a polar equation $r = f(\theta)$ can be constructed by computing a table of values of r against θ and then plotting the corresponding points (r, θ) on polar-coordinate graph paper.

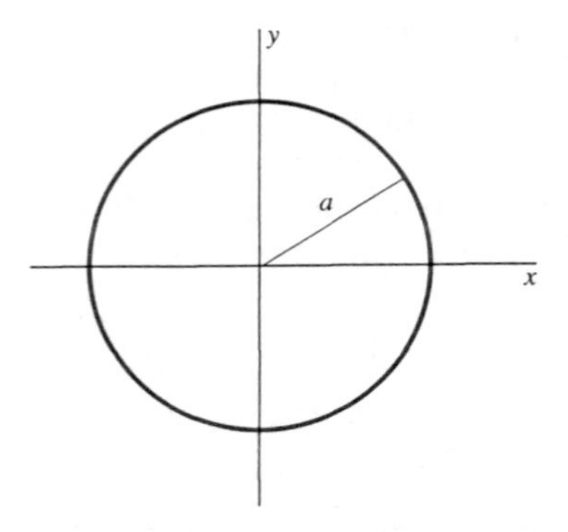

FIGURE 10.2.5 The circle $r = a$ centered at the origin (Example 2).

EXAMPLE 2 One reason for the importance of polar coordinates is that many real-world problems involve circles, and the polar-coordinate equation (or *polar equation*) of the circle with center $(0, 0)$ and radius $a > 0$ (Fig. 10.2.5) is very simple:

$$r = a. \tag{3}$$

Note that if we begin with the rectangular-coordinates equation $x^2 + y^2 = a^2$ of this circle and transform it using the first relation in (2), we get the polar-coordinate equation $r^2 = a^2$. Then Eq. (3) results upon taking positive square roots. ◆

EXAMPLE 3 Construct the polar-coordinate graph of the equation $r = 2\sin\theta$.

Solution Figure 10.2.6 shows a table of values of r as a function of θ. The corresponding points (r, θ) are plotted in Fig. 10.2.7, using the rays at multiples of $\pi/6$ and the circles (centered at the pole) of radii 1 and 2 to locate these points. A visual inspection of the smooth curve connecting these points suggests that it is a circle of radius 1. Let us assume for the moment that this is so. Note then that the point $P(r, \theta)$ moves *once around this circle counterclockwise* as θ increases from 0 to π and then moves around this circle a *second time* as θ increases from π to 2π. This is because the negative values of r for θ between π and 2π give—in this example—the same geometric points as do the positive values of r for θ between 0 and π. (Why?) ◆

The verification that the graph of $r = 2\sin\theta$ is the indicated circle illustrates the general procedure for transferring back and forth between polar and rectangular coordinates, using the relations in (1) and (2).

θ	r
0	0.00
$\pi/6$	1.00
$\pi/3$	1.73
$\pi/2$	2.00
$2\pi/3$	1.73
$5\pi/6$	1.00
π	0.00
$7\pi/6$	−1.00
$4\pi/3$	−1.73
$3\pi/2$	−2.00
$5\pi/3$	−1.73
$11\pi/6$	−1.00
2π	0.00
	(data rounded)

FIGURE 10.2.6 Values of $r = 2\sin\theta$ (Example 3).

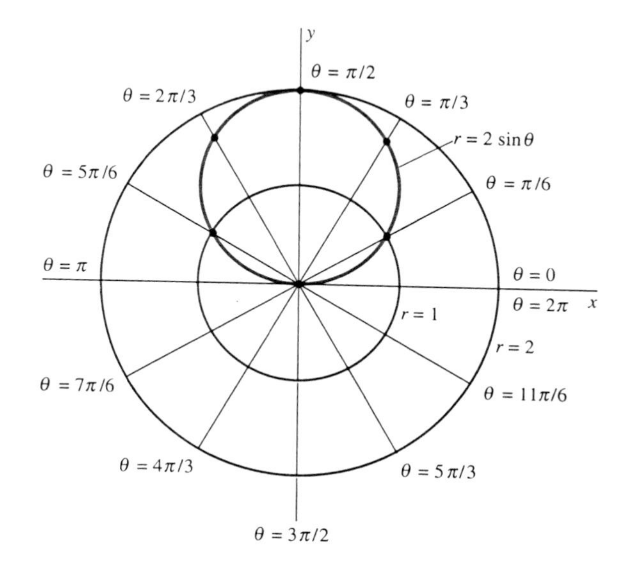

FIGURE 10.2.7 The graph of the polar equation $r = 2\sin\theta$ (Example 3).

EXAMPLE 4 To transform the equation $r = 2\sin\theta$ of Example 3 into rectangular coordinates, we first multiply both sides by r to get

$$r^2 = 2r\sin\theta.$$

Equations (1) and (2) now give

$$x^2 + y^2 = 2y.$$

Finally, after we complete the square in y, we have

$$x^2 + (y-1)^2 = 1,$$

the rectangular-coordinate equation (or *rectangular equation*) of a circle whose center is $(0, 1)$ and whose radius is 1. ◆

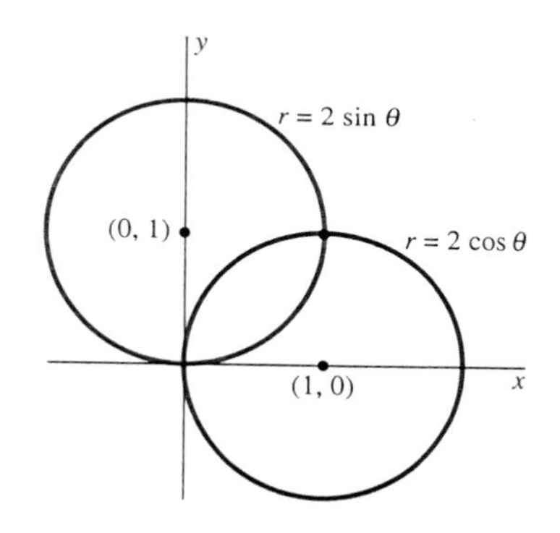

FIGURE 10.2.8 The graphs of the circles whose equations appear in Eq. (4) with $a = 1$.

More generally, the graphs of the equations

$$r = 2a\sin\theta \quad \text{and} \quad r = 2a\cos\theta \tag{4}$$

are circles of radius a centered, respectively, at the points $(0, a)$ and $(a, 0)$. This is illustrated (with $a = 1$) in Fig. 10.2.8.

By substituting the equations given in (1), we can transform the rectangular equation $ax + by = c$ of a straight line into

$$ar\cos\theta + br\sin\theta = c.$$

Let us take $a = 1$ and $b = 0$. Then we see that the polar equation of the vertical line $x = c$ is $r = c\sec\theta$, as we can deduce directly from Fig. 10.2.9.

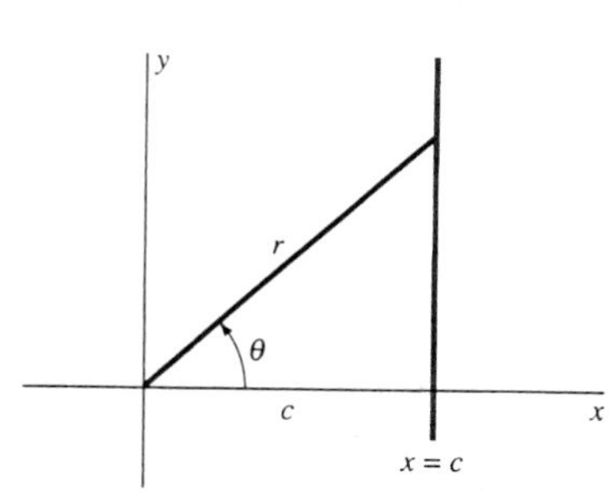

FIGURE 10.2.9 Finding the polar equation of the vertical line $x = c$.

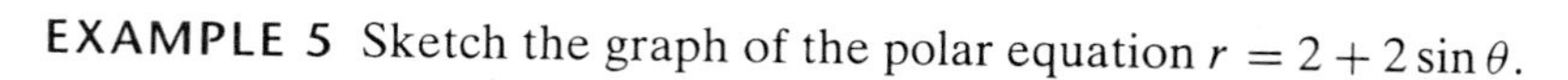

EXAMPLE 5 Sketch the graph of the polar equation $r = 2 + 2\sin\theta$.

Solution If we scan the second column of the table in Fig. 10.2.6, mentally adding 2 to each entry for r, we see that

- r increases from 2 to 4 as θ increases from 0 to $\pi/2$;
- r decreases from 4 to 2 as θ increases from $\pi/2$ to π;
- r decreases from 2 to 0 as θ increases from π to $3\pi/2$;
- r increases from 0 to 2 as θ increases from $3\pi/2$ to 2π.

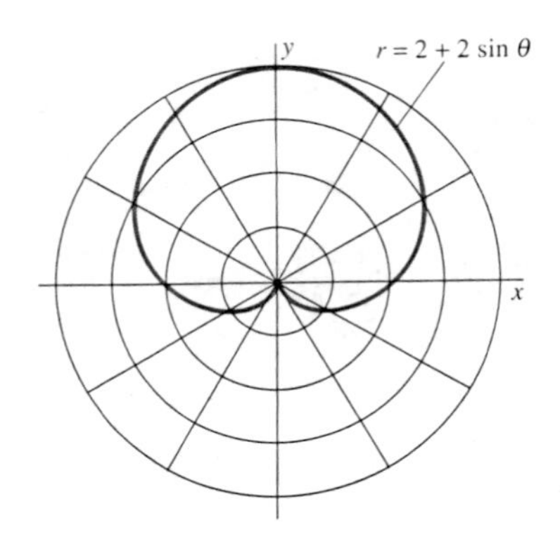

FIGURE 10.2.10 A cardioid (Example 5).

This information tells us that the graph resembles the curve shown in Fig. 10.2.10. This heart-shaped graph is called a **cardioid.** The graphs of the equations

$$r = a(1 \pm \sin\theta) \quad \text{and} \quad r = a(1 \pm \cos\theta)$$

are all cardioids, differing only in size (determined by a), axis of symmetry (horizontal or vertical), and the direction in which the cusp at the pole points. ◆

EXAMPLE 6 Sketch the graph of the equation $r = 2\cos 2\theta$.

Solution Rather than constructing a table of values of r as a function of θ and then plotting individual points, let us begin with a *rectangular-coordinate graph* of r as a function of θ. In Fig. 10.2.11, we see that $r = 0$ if θ is an odd integral multiple of $\pi/4$, and that r is alternately positive and negative on successive intervals of length $\pi/2$ from one odd integral multiple of $\pi/4$ to the next.

Now let's think about how r changes as θ increases, beginning at $\theta = 0$. As θ increases from 0 to $\pi/4$, r decreases in value from 2 to 0, and so we draw the first portion (labeled "1") of the polar curve in Fig. 10.2.12. As θ increases from $\pi/4$ to $3\pi/4$, r first decreases from 0 to -2 and then increases from -2 to 0. Because r is now negative, we draw the second and third portions (labeled "2" and "3") of the polar curve in the third and fourth quadrants (rather than in the first and second quadrants) in Fig. 10.2.12. Continuing in this fashion, we draw the fourth through eighth portions of the polar curve, with those portions where r is negative in the quadrants opposite those in which θ lies. The arrows on the resulting polar curve in Fig. 10.2.12 indicate the direction of motion of the point $P(r, \theta)$ along the curve as θ increases. The whole graph consists of four loops, each of which begins and ends at the pole. ◆

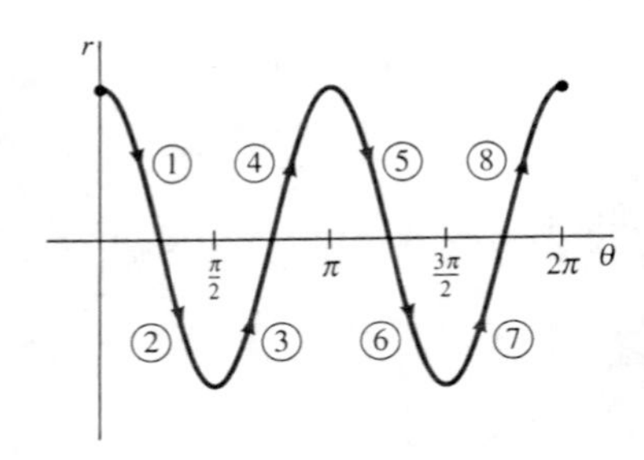

FIGURE 10.2.11 The rectangular coordinates graph of $r = 2\cos 2\theta$ as a function of θ. Numbered portions of the graph correspond to numbered portions of the polar-coordinates graph in Fig. 10.2.12.

The curve in Example 6 is called a *four-leaved rose.* The equations $r = a\cos n\theta$ and $r = a\sin n\theta$ represent "roses" with $2n$ "leaves," or loops, if n is even and $n \geqq 2$ but with n loops if n is odd and $n \geqq 3$.

The four-leaved rose exhibits several types of symmetry. The following are some *sufficient* conditions for symmetry in polar coordinates:

- *For symmetry around the x-axis:* The equation is unchanged when θ is replaced with $-\theta$.
- *For symmetry around the y-axis:* The equation is unchanged when θ is replaced with $\pi - \theta$.
- *For symmetry with respect to the origin:* The equation is unchanged when r is replaced with $-r$.

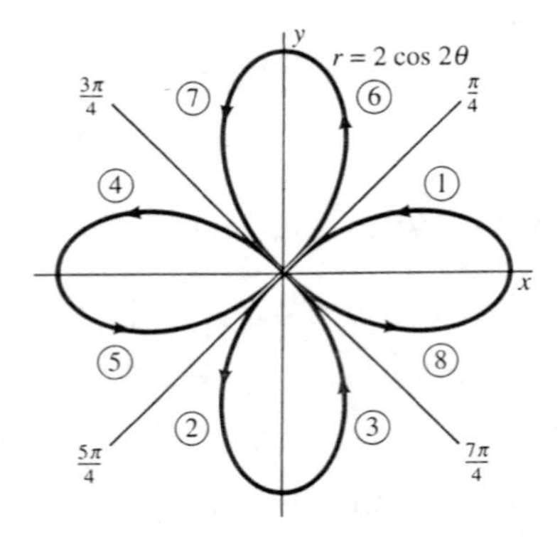

FIGURE 10.2.12 A four-leaved rose (Example 6).

Because $\cos 2\theta = \cos(-2\theta) = \cos 2(\pi - \theta)$, the equation $r = 2\cos 2\theta$ of the four-leaved rose satisfies the first two symmetry conditions, and therefore its graph is symmetric around both the x-axis and the y-axis. Thus it is also symmetric around the origin. Nevertheless, this equation does *not* satisfy the third condition, the one for symmetry around the origin. This illustrates that although the symmetry conditions given are *sufficient* for the symmetries described, they are not *necessary* conditions.

EXAMPLE 7 Figure 10.2.13 shows the lemniscate with equation

$$r^2 = -4\sin 2\theta.$$

To see why it has loops only in the second and fourth quadrants, we examine a table of signs of values of $-4\sin 2\theta$.

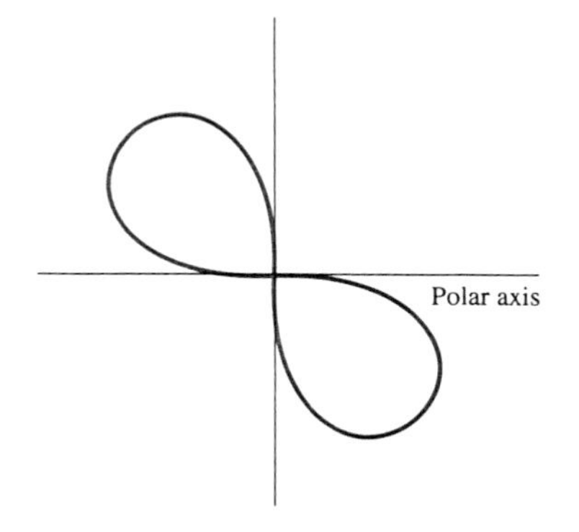

FIGURE 10.2.13 The lemniscate $r^2 = -4\sin 2\theta$ (Example 7).

θ	2θ	$-4\sin 2\theta$
$0 < \theta < \frac{1}{2}\pi$	$0 < 2\theta < \pi$	Negative
$\frac{1}{2}\pi < \theta < \pi$	$\pi < 2\theta < 2\pi$	Positive
$\pi < \theta < \frac{3}{2}\pi$	$2\pi < 2\theta < 3\pi$	Negative
$\frac{3}{2}\pi < \theta < 2\pi$	$3\pi < 2\theta < 4\pi$	Positive

When θ lies in the first or the third quadrant, the quantity $-4\sin 2\theta$ is negative, so the equation $r^2 = -4\sin 2\theta$ cannot be satisfied for any real values of r. ◆

Example 6 illustrates a peculiarity of graphs of polar equations, caused by the fact that a single point has multiple representations in polar coordinates. The point with polar coordinates $(2, \pi/2)$ clearly lies on the four-leaved rose, but these coordinates do *not* satisfy the equation $r = 2\cos 2\theta$. This means that a point may have one pair of polar coordinates that satisfy a given equation and others that do not. Hence we must be careful to understand this: The graph of a polar equation consists of all those points with *at least one* polar-coordinate representation that satisfies the given equation.

Another result of the multiplicity of polar coordinates is that the simultaneous solution of two polar equations does not always give all the points of intersection of their graphs. For instance, consider the circles $r = 2\sin\theta$ and $r = 2\cos\theta$ shown in Fig. 10.2.8. The origin is clearly a point of intersection of these two circles. Its polar representation $(0, \pi)$ satisfies the equation $r = 2\sin\theta$, and its representation $(0, \pi/2)$ satisfies the other equation, $r = 2\cos\theta$. But the origin has no *single* polar representation that satisfies both equations simultaneously! If we think of θ as increasing uniformly with time, then the corresponding moving points on the two circles pass through the origin at different times. Hence the origin cannot be discovered as a point of intersection of the two circles merely by solving their equations $r = 2\sin\theta$ and $r = 2\cos\theta$ simultaneously in a straightforward manner. But one fail-safe way to find *all* points of intersection of two polar-coordinate curves is to graph both curves.

EXAMPLE 8 Find all points of intersection of the graphs of the equations $r = 1 + \sin\theta$ and $r^2 = 4\sin\theta$.

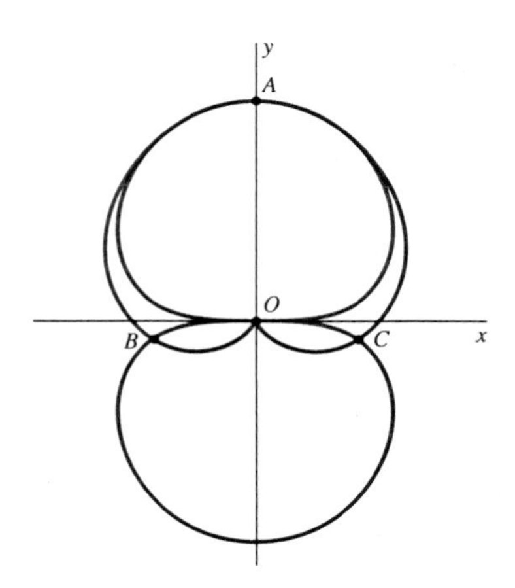

FIGURE 10.2.14 The cardioid $r = 1 + \sin\theta$ and the figure eight $r^2 = 4\sin\theta$ meet in four points (Example 8).

Solution The graph of $r = 1 + \sin\theta$ is a scaled-down version of the cardioid of Example 5. In Problem 52 we ask you to show that the graph of $r^2 = 4\sin\theta$ is the figure-eight curve shown with the cardioid in Fig. 10.2.14. The figure shows four points of intersection: A, B, C, and O. Can we find all four using algebra?

Given the two equations, we begin by eliminating r. Because

$$(1 + \sin\theta)^2 = r^2 = 4\sin\theta,$$

it follows that

$$\sin^2\theta - 2\sin\theta + 1 = 0;$$

$$(\sin\theta - 1)^2 = 0;$$

and thus that $\sin\theta = 1$. So θ must be an angle of the form $\frac{1}{2}\pi + 2n\pi$ where n is an integer. All points on the cardioid and all points on the figure-eight curve are produced by letting θ range from 0 to 2π, so $\theta = \pi/2$ will produce all the solutions that we can obtain by simple algebraic elimination. The only such point is $A(2, \pi/2)$, and the other three points of intersection are detected only when the two equations are graphed. ◆

Calculator/Computer-Generated Polar Curves

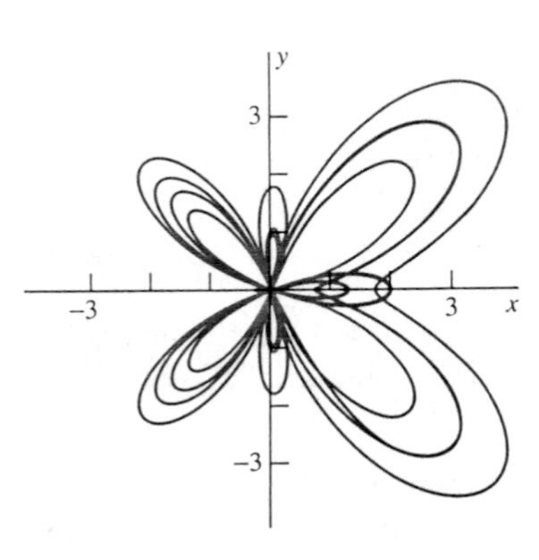

FIGURE 10.2.15 $r = e^{\cos\theta} - 2\cos 4\theta + \sin^3(\theta/4)$.

It might take you quite a while to construct by hand the "butterfly curve" shown in Fig. 10.2.15. But most graphing calculators and computer algebra systems have facilities for plotting polar curves. For instance, with a TI calculator set in "polar graph mode," one need only enter and graph the equation

```
r = e∧(cos(θ)) - 2*cos(4θ) + sin(θ/4)∧3
```

on the interval $0 \leqq \theta \leqq 8\pi$. With *Maple* and *Mathematica* the graphics package commands

```
polarplot(exp(cos(t)) - 2*cos(4*t) + sin(t/4)∧3, t=0..8*Pi);
```

and

```
PolarPlot[ Exp[Cos[t]] - 2*Cos[4*t] + Sin[t/4]∧3, { t, 0, 8*Pi } ];
```

(respectively) give the same result (with t in place of θ).

Because of the presence of the term $\sin^3(\theta/4)$, the more usual interval $0 \leqq \theta \leqq 2\pi$ gives only a part of the curve shown in Fig. 10.2.15. (Try it to see for yourself.) But

$$\sin^3\left(\frac{\theta + 8\pi}{4}\right) = \sin^3\left(\frac{\theta}{4} + 2\pi\right) = \sin^3\left(\frac{\theta}{4}\right),$$

so values of $\sin^3(\theta/4)$ repeat themselves when θ exceeds 8π. Therefore the interval $0 \leqq \theta \leqq 8\pi$ suffices to give the entire butterfly curve. You might try plotting a butterfly curve with the term $\sin^3(\theta/4)$ replaced with $\sin^5(\theta/12)$—as originally recommended by Temple H. Fay in his article "The Butterfly Curve" (*American Mathematical Monthly,* May 1989, p. 442). What range of values of θ will now be required to obtain the whole butterfly?

10.2 TRUE/FALSE STUDY GUIDE

10.2 CONCEPTS: QUESTIONS AND DISCUSSION

1. Figures 10.2.16 through 10.2.18 illustrate the polar curve $r = a + b\cos\theta$ for various values of a and b. What determines whether the curve exhibits a cusp (Fig. 10.2.16), a loop (Fig. 10.2.17), or neither (Fig. 10.2.18)? Does your answer apply also to polar curves of the form $r = a + b\sin\theta$? Given a and b, what is the difference between the curves $r = a + b\cos\theta$ and $r = a + b\sin\theta$?

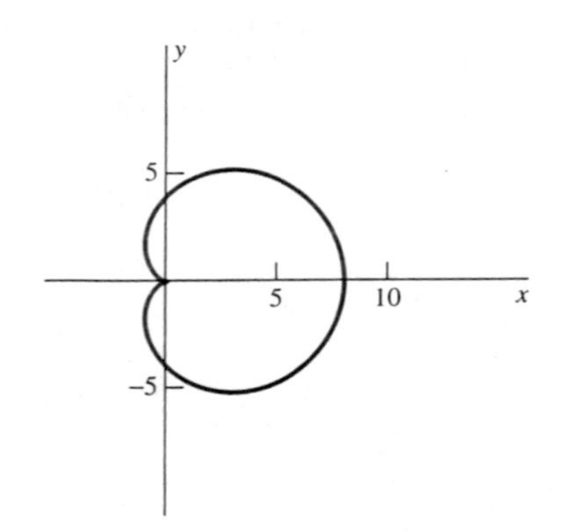

FIGURE 10.2.16 $r = 4 + 4\cos\theta$.

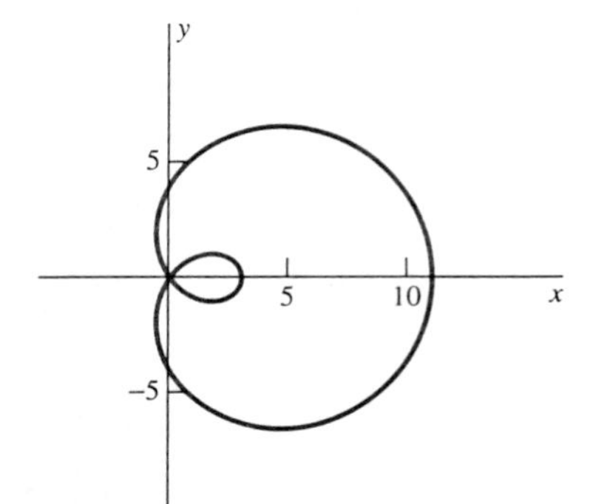

FIGURE 10.2.17 $r = 4 + 7\cos\theta$.

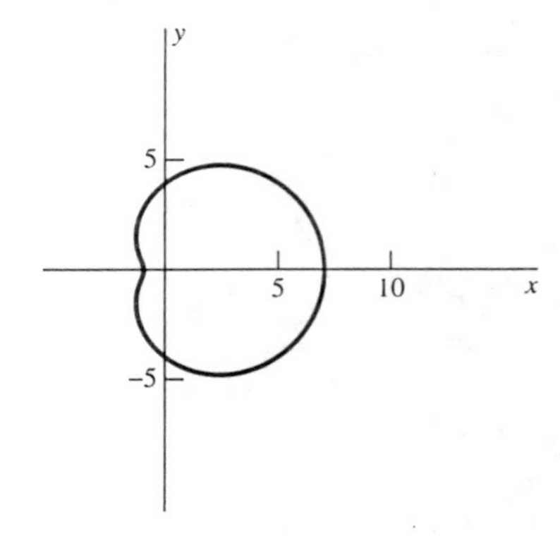

FIGURE 10.2.18 $r = 4 + 3\cos\theta$.

2. Figures 10.2.19 and 10.2.20 show the graphs of the equations $r = \cos 3\theta$ and $r = \sin 4\theta$. Given a positive integer n, what is the difference between the "rose graphs" $r = \cos n\theta$ and $r = \sin n\theta$? Explain precisely how the number of leaves in the complete graph depends on n. What determines whether $0 \leqq \theta \leqq \pi$ or $0 \leqq \theta \leqq 2\pi$ gives all the leaves?

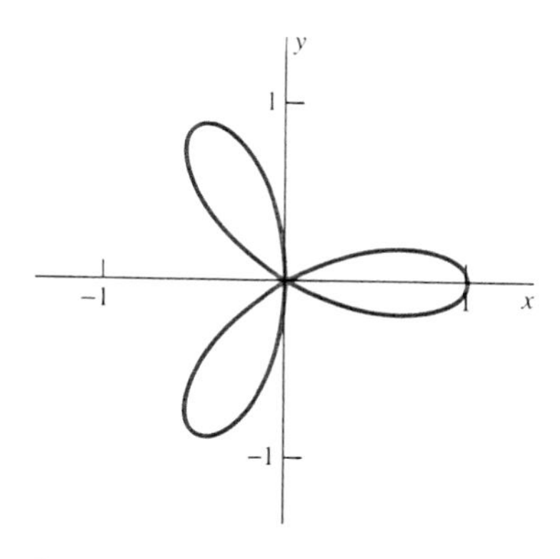

FIGURE 10.2.19 $r = \cos 3\theta$.

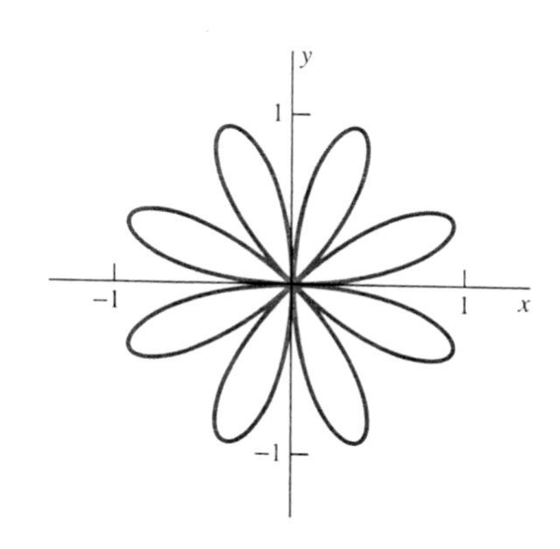

FIGURE 10.2.20 $r = \sin 4\theta$.

10.2 PROBLEMS

1. Plot the points with the given polar coordinates, and then find the rectangular coordinates of each.

(a) $(1, \pi/4)$ (b) $(-2, 2\pi/3)$ (c) $(1, -\pi/3)$
(d) $(3, 3\pi/2)$ (e) $(2, -\pi/4)$ (f) $(-2, -7\pi/6)$
(g) $(2, 5\pi/6)$

2. Find two polar-coordinate representations, one with $r > 0$ and the other with $r < 0$, for the points with the given rectangular coordinates.

(a) $(-1, -1)$ (b) $(\sqrt{3}, -1)$ (c) $(2, 2)$
(d) $(-1, \sqrt{3})$ (e) $(\sqrt{2}, -\sqrt{2})$ (f) $(-3, \sqrt{3})$

In Problems 3 through 10, express the given rectangular equation in polar form.

3. $x = 4$
4. $y = 6$
5. $x = 3y$
6. $x^2 + y^2 = 25$
7. $xy = 1$
8. $x^2 - y^2 = 1$
9. $y = x^2$
10. $x + y = 4$

In Problems 11 through 18, express the given polar equation in rectangular form.

11. $r = 3$
12. $\theta = 3\pi/4$
13. $r = -5\cos\theta$
14. $r = \sin 2\theta$
15. $r = 1 - \cos 2\theta$
16. $r = 2 + \sin\theta$
17. $r = 3\sec\theta$
18. $r^2 = \cos 2\theta$

For the curves described in Problems 19 through 28, write equations in both rectangular and polar form.

19. The vertical line through $(2, 0)$

20. The horizontal line through $(1, 3)$

21. The line with slope -1 through $(2, -1)$

22. The line with slope 1 through $(4, 2)$

23. The line through the points $(1, 3)$ and $(3, 5)$

24. The circle with center $(3, 0)$ that passes through the origin

25. The circle with center $(0, -4)$ that passes through the origin

26. The circle with center $(3, 4)$ and radius 5

27. The circle with center $(1, 1)$ that passes through the origin

28. The circle with center $(5, -2)$ that passes through the point $(1, 1)$

In Problems 29 through 32, transform the given polar-coordinate equation into a rectangular-coordinate equation, then match the equation with its graph among those in Figs. 10.2.21 through 10.2.24.

29. $r = -4\cos\theta$
30. $r = 5\cos\theta + 5\sin\theta$
31. $r = -4\cos\theta + 3\sin\theta$
32. $r = 8\cos\theta - 15\sin\theta$

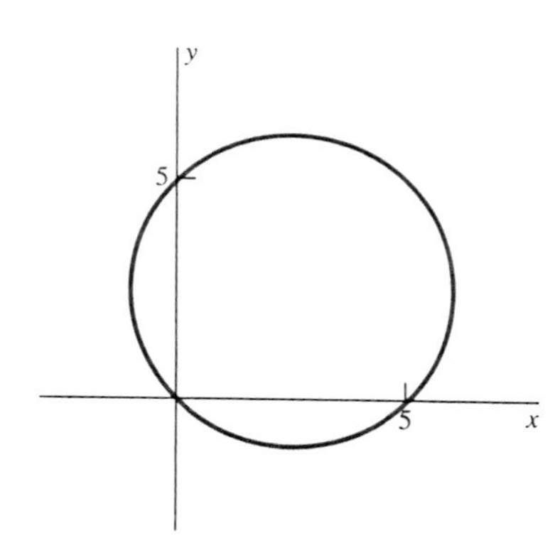

FIGURE 10.2.21

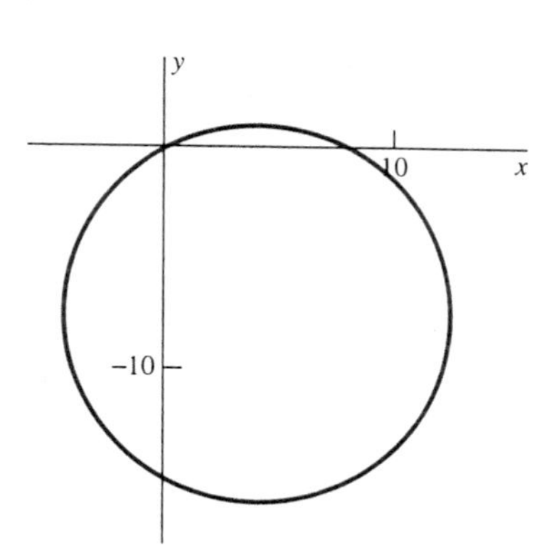

FIGURE 10.2.22

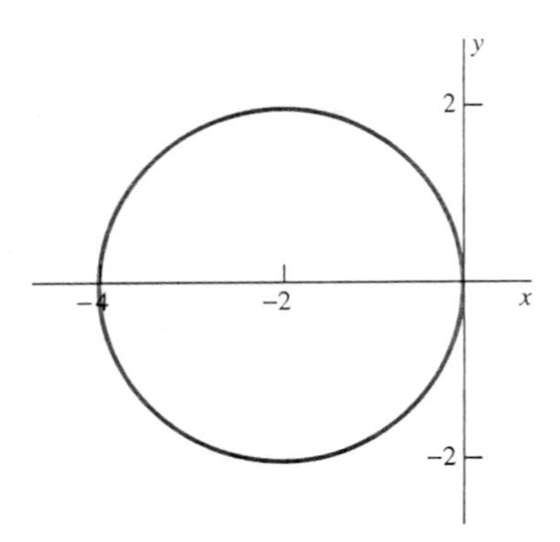

FIGURE 10.2.23

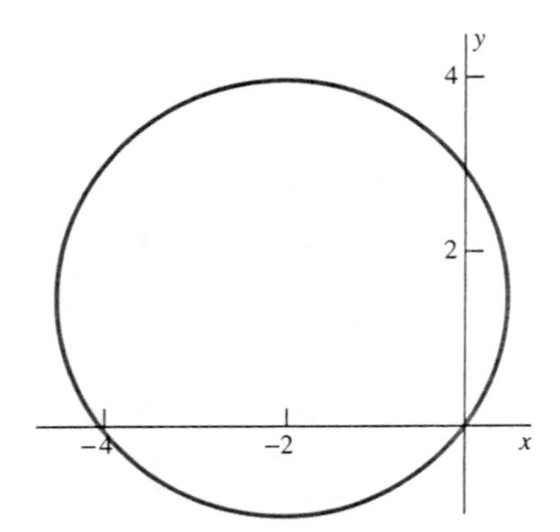

FIGURE 10.2.24

The graph of a polar equation of the form $r = a + b\cos\theta$ *(or* $r = a + b\sin\theta$*) is called a* limaçon *(from the French word for* snail*). In Problems 33 through 36, match the given polar-coordinate equation with its graph among the limaçons in Figs. 10.2.25 through 10.2.28.*

33. $r = 8 + 6\cos\theta$

34. $r = 7 + 7\cos\theta$

35. $r = 5 + 9\cos\theta$

36. $r = 3 + 11\cos\theta$

37. Show that the graph of the polar equation $r = a\cos\theta + b\sin\theta$ is a circle if $a^2 + b^2 \neq 0$. Express the center (h, k) and radius r of this circle in terms of a and b.

38. Show that if $0 < a < b$, then the limaçon with polar equation $r = a + b\cos\theta$ has an inner loop (as in Figs. 10.2.25 and 10.2.27). In this case, find (in terms of a and b) the range of values of θ that correspond to points of the inner loop.

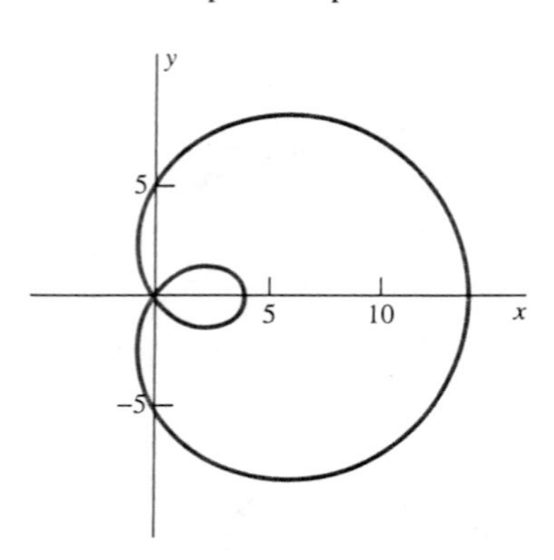

FIGURE 10.2.25

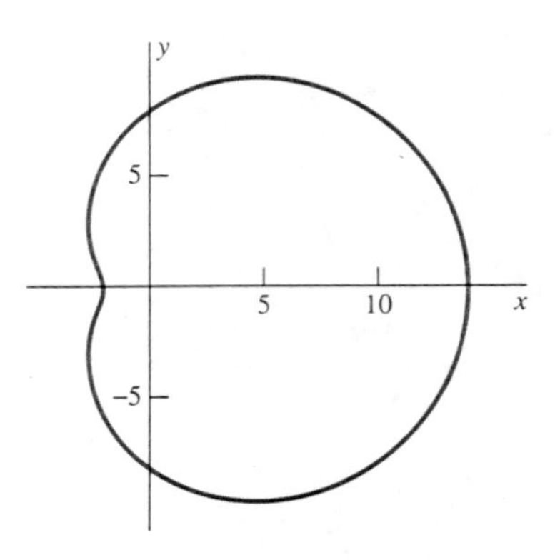

FIGURE 10.2.26

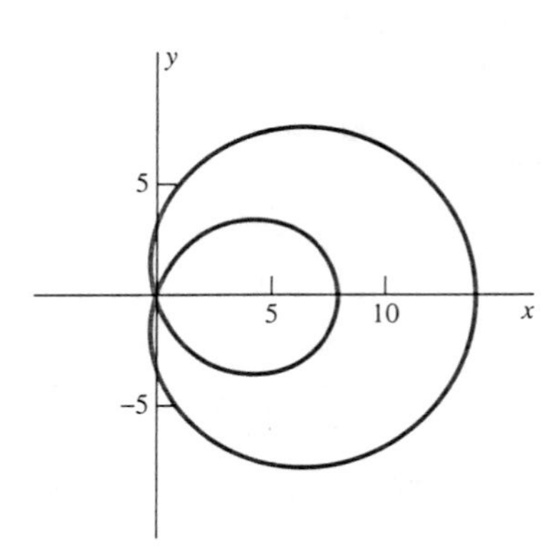

FIGURE 10.2.27

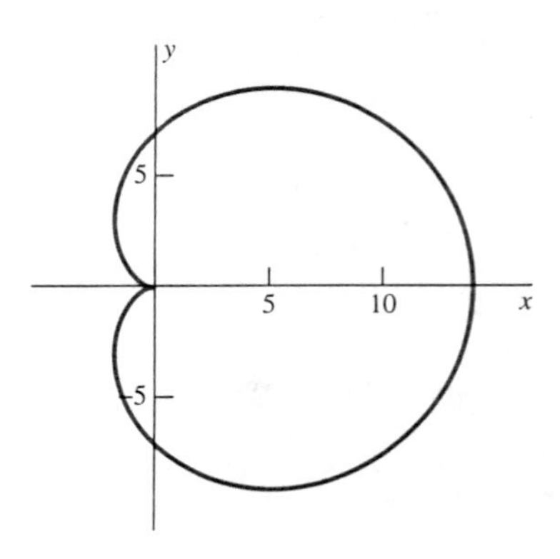

FIGURE 10.2.28

Sketch the graphs of the polar equations in Problems 39 through 52. Indicate any symmetries around either coordinate axis or the origin.

39. $r = 2\cos\theta$ (circle)

40. $r = 2\sin\theta + 2\cos\theta$ (circle)

41. $r = 1 + \cos\theta$ (cardioid)

42. $r = 1 - \sin\theta$ (cardioid)

43. $r = 2 + 4\sin\theta$ (limaçon)

44. $r = 4 + 2\cos\theta$ (limaçon)

45. $r^2 = 4\sin 2\theta$ (lemniscate)

46. $r^2 = 4\cos 2\theta$ (lemniscate)

47. $r = 2\sin 2\theta$ (four-leaved rose)

48. $r = 3\sin 3\theta$ (three-leaved rose)

49. $r = 3\cos 3\theta$ (three-leaved rose)

50. $r = 3\theta$ (spiral of Archimedes)

51. $r = 2\sin 5\theta$ (five-leaved rose)

52. $r^2 = 4\sin\theta$ (figure eight)

In Problems 53 through 58, find all points of intersection of the curves with the given polar equations.

53. $r = 1, \quad r = \cos\theta$

54. $r = \sin\theta, \quad r^2 = 3\cos^2\theta$

55. $r = \sin\theta, \quad r = \cos 2\theta$

56. $r = 1 + \cos\theta, \quad r = 1 - \sin\theta$

57. $r = 1 - \cos\theta, \quad r^2 = 4\cos\theta$

58. $r^2 = 4\sin\theta, \quad r^2 = 4\cos\theta$

59. (a) The straight line L passes through the point with polar coordinates (p, α) and is perpendicular to the line segment joining the pole and the point (p, α). Write the polar-coordinate equation of L. (b) Show that the rectangular-coordinate equation of L is

$$x\cos\alpha + y\sin\alpha = p.$$

60. Find a rectangular-coordinate equation of the cardioid with polar equation $r = 1 - \cos\theta$.

61. Use polar coordinates to identify the graph of the rectangular-coordinate equation $a^2(x^2 + y^2) = (x^2 + y^2 - by)^2$.

62. Plot the polar equations

$$r = 1 + \cos\theta \quad \text{and} \quad r = -1 + \cos\theta$$

on the same coordinate plane. Comment on the results.

63. Figures 10.2.29 and 10.2.30 show the graphs of the equations $r = \cos(5\theta/3)$ and $r = \cos(5\theta/2)$. Why does one have five

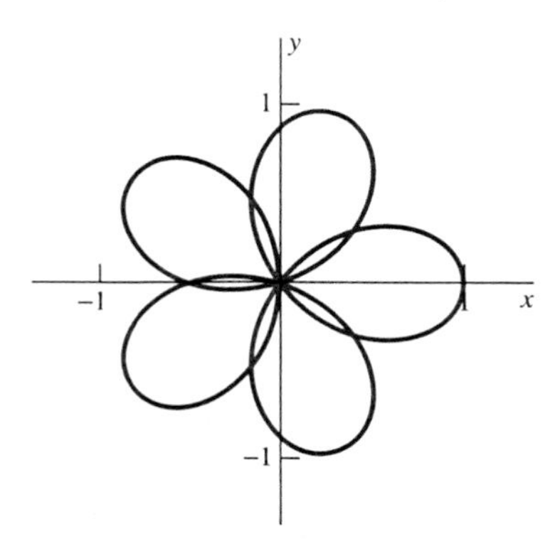

FIGURE 10.2.29 $r = \cos\left(\dfrac{5\theta}{3}\right)$.

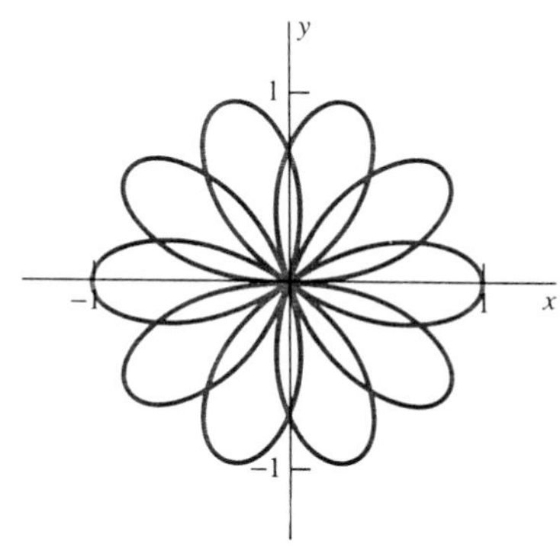

FIGURE 10.2.30 $r = \cos\left(\dfrac{5\theta}{2}\right)$.

(overlapping) loops while the other has ten loops? In each case, what range of values of θ is required to obtain all the loops? In the more general case $r = (\cos p\theta/q)$ where p and q are positive integers, is it p or q (or both) that determine the number of loops and the range of values of θ required to show all the loops in the complete graph?

64. Figures 10.2.31 and 10.2.32 show the graphs of the equations $r = 1 + 4\sin 3\theta$ and $r = 1 + 4\cos 4\theta$. What determines whether a polar curve of the form $r = a + b\sin(n\theta)$—with a and b positive constants and n a positive integer—has both larger and smaller loops? What determines whether the smaller loops are within or outside of the larger ones?

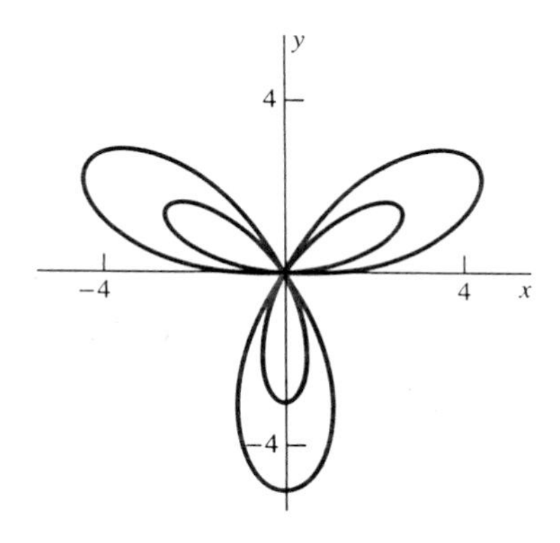

FIGURE 10.2.31 $r = 1 + 4\sin 3\theta$.

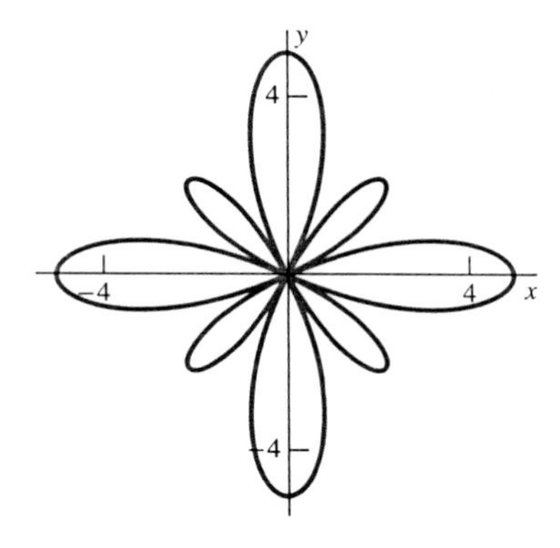

FIGURE 10.2.32 $r = 1 + 4\cos 4\theta$.

10.3 AREA COMPUTATIONS IN POLAR COORDINATES

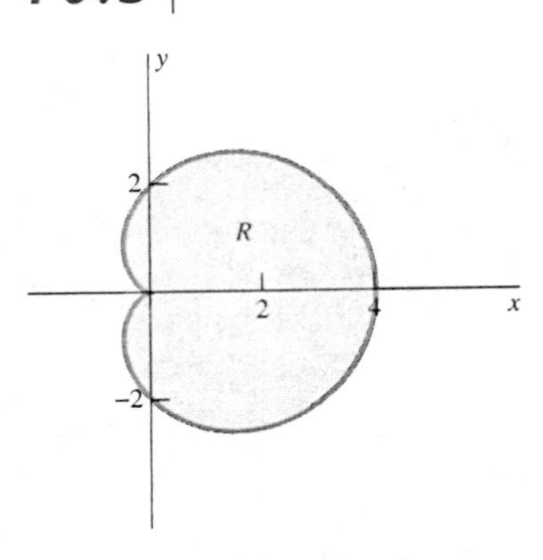

FIGURE 10.3.1 What is the area of the region R bounded by the cardioid $r = 2(1 + \cos\theta)$?

The graph of the polar-coordinate equation $r = f(\theta)$ may bound an area, as does the cardioid $r = 2(1 + \cos\theta)$—see Fig. 10.3.1. To calculate the area of this region, we may find it convenient to work directly with polar coordinates rather than to change to rectangular coordinates.

To see how to set up an area integral using polar coordinates, we consider the region R of Fig. 10.3.2. This region is bounded by the two radial lines $\theta = \alpha$ and $\theta = \beta$ and by the curve $r = f(\theta)$, $\alpha \leqq \theta \leqq \beta$. To approximate the area A of R, we begin with a partition

$$\alpha = \theta_0 < \theta_1 < \theta_2 < \cdots < \theta_n = \beta$$

of the interval $[\alpha, \beta]$ into n subintervals, all with the same length $\Delta\theta = (\beta - \alpha)/n$. We select a point $\theta_i^\star$ in the ith subinterval $[\theta_{i-1}, \theta_i]$ for $i = 1, 2, \ldots, n$.

Let ΔA_i denote the area of the sector bounded by the lines $\theta = \theta_{i-1}$ and $\theta = \theta_i$ and by the curve $r = f(\theta)$. We see from Fig. 10.3.2 that for small values of $\Delta\theta$, ΔA_i is approximately equal to the area of the *circular* sector that has radius $r_i^\star = f(\theta_i^\star)$ and is bounded by the same lines. That is,

$$\Delta A_i \approx \tfrac{1}{2}(r_i^\star)^2\,\Delta\theta = \tfrac{1}{2}\,[f(\theta_i^\star)]^2\,\Delta\theta.$$

We add the areas of these sectors for $i = 1, 2, \ldots, n$ and thereby find that

$$A = \sum_{i=1}^{n} \Delta A_i \approx \sum_{i=1}^{n} \tfrac{1}{2}\,[f(\theta_i^\star)]^2\,\Delta\theta.$$

The right-hand sum is a Riemann sum for the integral

$$\int_\alpha^\beta \tfrac{1}{2}[f(\theta)]^2\,d\theta.$$

Hence, if f is continuous, the value of this integral is the limit, as $\Delta\theta \to 0$, of the preceding sum. We therefore conclude that the *area A of the region R bounded by the lines $\theta = \alpha$ and $\theta = \beta$ and the curve $r = f(\theta)$* is

$$A = \int_\alpha^\beta \tfrac{1}{2}[f(\theta)]^2\,d\theta. \qquad \textbf{(1)}$$

The infinitesimal sector shown in Fig. 10.3.3, with radius r, central angle $d\theta$, and area $dA = \frac{1}{2}r^2\,d\theta$, serves as a useful device for remembering Eq. (1) in the abbreviated

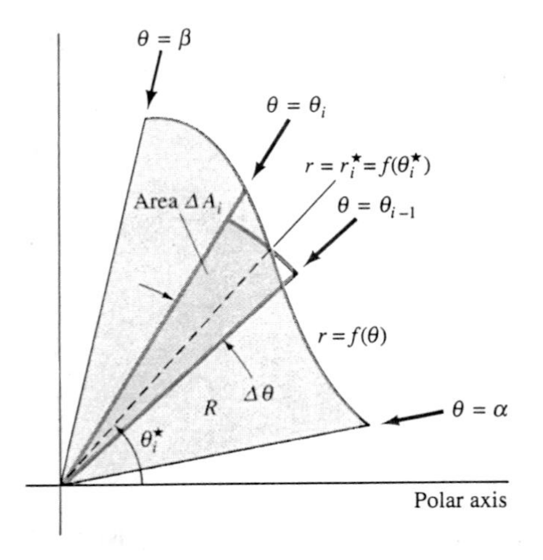

FIGURE 10.3.2 We obtain the area formula from Riemann sums.

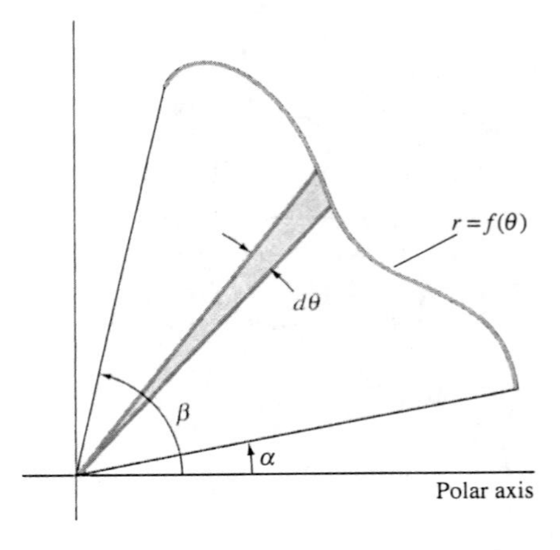

FIGURE 10.3.3 Nonrigorous derivation of the area formula in polar coordinates.

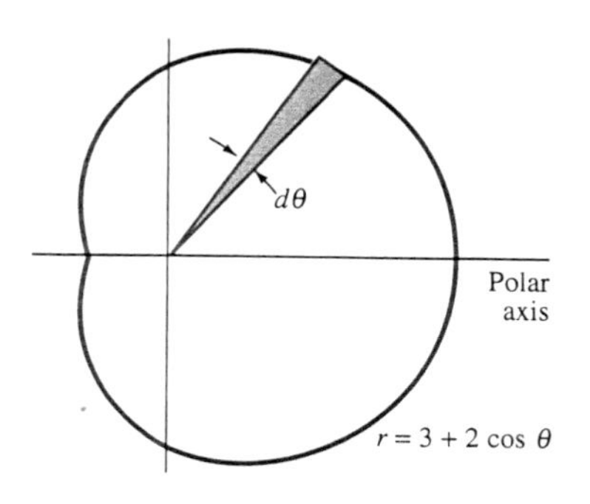

FIGURE 10.3.4 The limaçon of Example 1.

form

$$A = \int_{\alpha}^{\beta} \tfrac{1}{2} r^2 \, d\theta. \tag{2}$$

EXAMPLE 1 Find the area of the region bounded by the limaçon with equation $r = 3 + 2\cos\theta$, $0 \leqq \theta \leqq 2\pi$ (Fig. 10.3.4).

Solution We could apply Eq. (2) with $\alpha = 0$ and $\beta = 2\pi$. Here, instead, we will make use of symmetry. We will calculate the area of the upper half of the region and then double the result. Note that the infinitesimal sector shown in Fig. 10.3.4 sweeps out the upper half of the limaçon as θ increases from 0 to π (Fig. 10.3.5). Hence

$$\begin{aligned} A &= 2\int_{\alpha}^{\beta} \tfrac{1}{2} r^2 \, d\theta = \int_0^{\pi} (3 + 2\cos\theta)^2 \, d\theta \\ &= \int_0^{\pi} (9 + 12\cos\theta + 4\cos^2\theta) \, d\theta. \end{aligned}$$

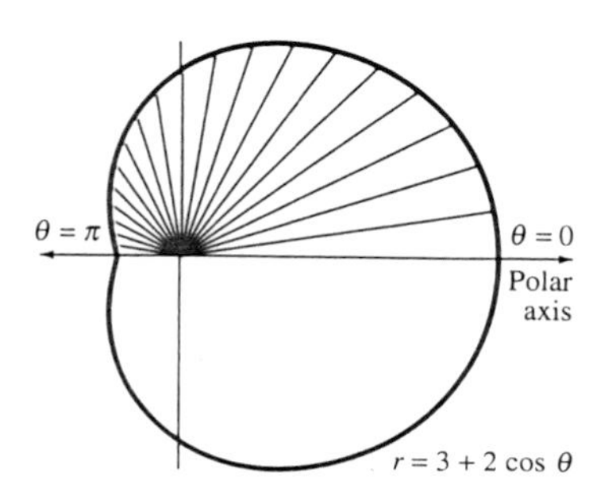

FIGURE 10.3.5 Infinitesimal sectors from $\theta = 0$ to $\theta = \pi$ (Example 1).

Because

$$4\cos^2\theta = 4 \cdot \frac{1 + \cos 2\theta}{2} = 2 + 2\cos 2\theta,$$

we now get

$$\begin{aligned} A &= \int_0^{\pi} (11 + 12\cos\theta + 2\cos 2\theta) \, d\theta \\ &= \Big[11\theta + 12\sin\theta + \sin 2\theta \Big]_0^{\pi} = 11\pi. \end{aligned}$$

◆

EXAMPLE 2 Find the area bounded by each loop of the limaçon with equation $r = 1 + 2\cos\theta$ (Fig. 10.3.6).

Solution The equation $1 + 2\cos\theta = 0$ has two solutions for θ in the interval $[0, 2\pi]$: $\theta = 2\pi/3$ and $\theta = 4\pi/3$. The upper half of the outer loop of the limaçon corresponds to values of θ between 0 and $2\pi/3$, where r is positive. Because the curve is symmetric around the x-axis, we can find the total area A_1 bounded by the outer loop by integrating from 0 to $2\pi/3$ and then doubling. Thus

$$\begin{aligned} A_1 &= 2\int_0^{2\pi/3} \tfrac{1}{2}(1 + 2\cos\theta)^2 \, d\theta = \int_0^{2\pi/3} (1 + 4\cos\theta + 4\cos^2\theta) \, d\theta \\ &= \int_0^{2\pi/3} (3 + 4\cos\theta + 2\cos 2\theta) \, d\theta \\ &= \Big[3\theta + 4\sin\theta + \sin 2\theta \Big]_0^{2\pi/3} = 2\pi + \tfrac{3}{2}\sqrt{3}. \end{aligned}$$

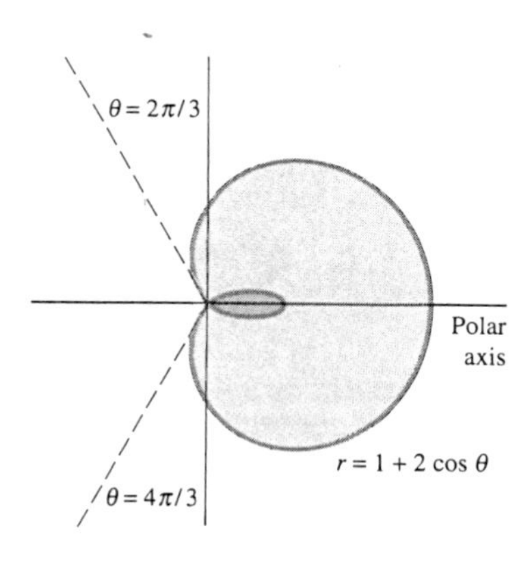

FIGURE 10.3.6 The limaçon of Example 2.

The inner loop of the limaçon corresponds to values of θ between $2\pi/3$ and $4\pi/3$, where r is negative. Hence the area bounded by the inner loop is

$$\begin{aligned} A_2 &= \int_{2\pi/3}^{4\pi/3} \tfrac{1}{2}(1 + 2\cos\theta)^2 \, d\theta \\ &= \tfrac{1}{2}\Big[3\theta + 4\sin\theta + \sin 2\theta \Big]_{2\pi/3}^{4\pi/3} = \pi - \tfrac{3}{2}\sqrt{3}. \end{aligned}$$

The area of the region lying *between* the two loops of the limaçon is then

$$A = A_1 - A_2 = 2\pi + \tfrac{3}{2}\sqrt{3} - \left(\pi - \tfrac{3}{2}\sqrt{3}\right) = \pi + 3\sqrt{3}.$$

◆

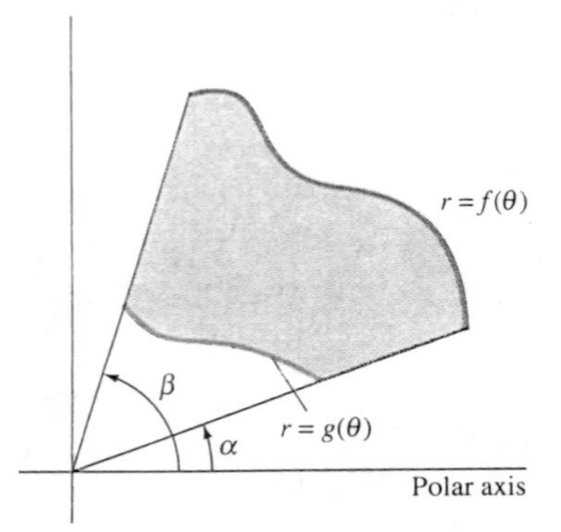

FIGURE 10.3.7 The area between the graphs of f and g.

The Area Between Two Polar Curves

Now consider two curves $r = f(\theta)$ and $r = g(\theta)$, with $f(\theta) \geqq g(\theta) \geqq 0$ for $\alpha \leqq \theta \leqq \beta$. Then we can find the area of the region bounded by these curves and the rays (radial lines) $\theta = \alpha$ and $\theta = \beta$ (Fig. 10.3.7) by subtracting the area bounded by the inner curve from that bounded by the outer curve. That is, the area A between the two curves is given by

$$A = \int_{\alpha}^{\beta} \tfrac{1}{2}[f(\theta)]^2\, d\theta - \int_{\alpha}^{\beta} \tfrac{1}{2}[g(\theta)]^2\, d\theta,$$

so that

$$A = \frac{1}{2}\int_{\alpha}^{\beta} \{[f(\theta)]^2 - [g(\theta)]^2\}\, d\theta. \tag{3}$$

With r_{outer} for the outer curve and r_{inner} for the inner curve, we get the abbreviated formula

$$A = \frac{1}{2}\int_{\alpha}^{\beta} [(r_{\text{outer}})^2 - (r_{\text{inner}})^2]\, d\theta \tag{4}$$

for the area of the region shown in Fig. 10.3.8.

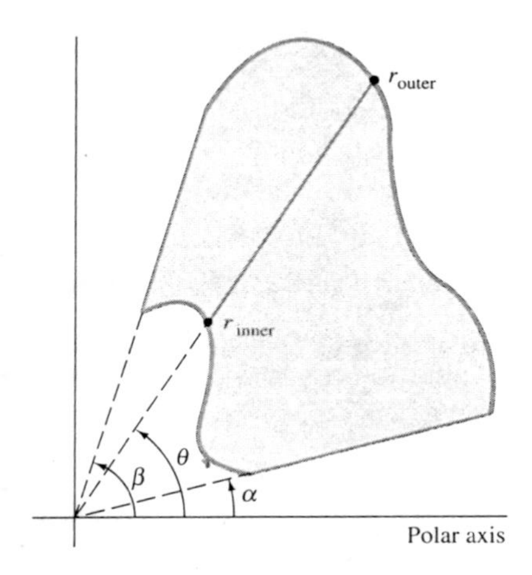

FIGURE 10.3.8 The radial line segment illustrates the radii r_{inner} and r_{outer} of Eq. (4).

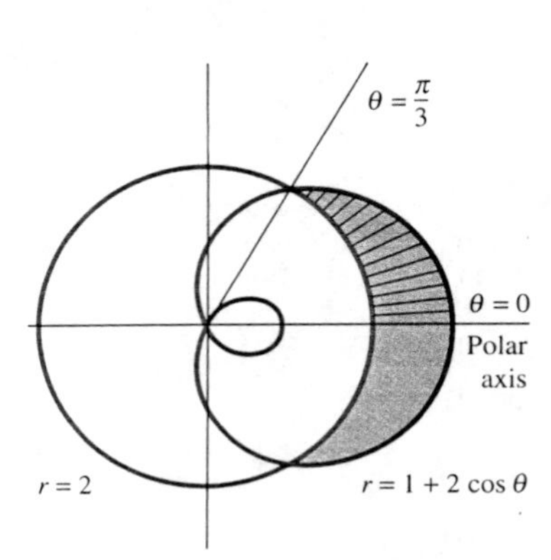

FIGURE 10.3.9 The region of Example 3.

EXAMPLE 3 Find the area A of the region that lies within the limaçon $r = 1 + 2\cos\theta$ and outside the circle $r = 2$.

Solution The circle and limaçon are shown in Fig. 10.3.9, with the area A between them shaded. The points of intersection of the circle and limaçon are given by

$$1 + 2\cos\theta = 2, \quad \text{so} \quad \cos\theta = \tfrac{1}{2},$$

and the figure shows that we should choose the solutions $\theta = \pm\pi/3$. These two values of θ are the needed limits of integration. When we use Eq. (3), we find that

$$\begin{aligned}
A &= \frac{1}{2}\int_{-\pi/3}^{\pi/3} [(1 + 2\cos\theta)^2 - 2^2]\, d\theta \\
&= \int_{0}^{\pi/3} (4\cos\theta + 4\cos^2\theta - 3)\, d\theta \qquad \text{(by symmetry)} \\
&= \int_{0}^{\pi/3} (4\cos\theta + 2\cos 2\theta - 1)\, d\theta \\
&= \Big[4\sin\theta + \sin 2\theta - \theta\Big]_0^{\pi/3} = \frac{15\sqrt{3} - 2\pi}{6}.
\end{aligned}$$

◆

10.3 TRUE/FALSE STUDY GUIDE

10.3 CONCEPTS: QUESTIONS AND DISCUSSION

1. Give an example of a plane region whose area can be calculated both by a rectangular-coordinate integral and a polar-coordinate integral, but the latter is easier to evaluate.
2. Give an example of a plane region whose area can be calculated both by a rectangular-coordinate integral and a polar-coordinate integral, but the former is easier to evaluate.
3. Give an example of an unbounded plane region such that its polar-coordinate area integral is improper but convergent.

10.3 PROBLEMS

In Problems 1 through 6, sketch the plane region bounded by the given polar curve $r = f(\theta)$, $\alpha \leqq \theta \leqq \beta$, and the rays $\theta = \alpha$, $\theta = \beta$.

1. $r = \theta$, $0 \leqq \theta \leqq \pi$ **2.** $r = \theta$, $0 \leqq \theta \leqq 2\pi$

3. $r = 1/\theta$, $\pi \leqq \theta \leqq 3\pi$ **4.** $r = 1/\theta$, $3\pi \leqq \theta \leqq 5\pi$

5. $r = e^{-\theta}$, $0 \leqq \theta \leqq \pi$ **6.** $r = e^{-\theta}$, $\pi/2 \leqq \theta \leqq 3\pi/2$

In Problems 7 through 16, find the area bounded by the given curve.

7. $r = 2\cos\theta$ **8.** $r = 4\sin\theta$

9. $r = 1 + \cos\theta$

10. $r = 2 - 2\sin\theta$ (Fig. 10.3.10) **11.** $r = 2 - \cos\theta$

12. $r = 3 + 2\sin\theta$ (Fig. 10.3.11)

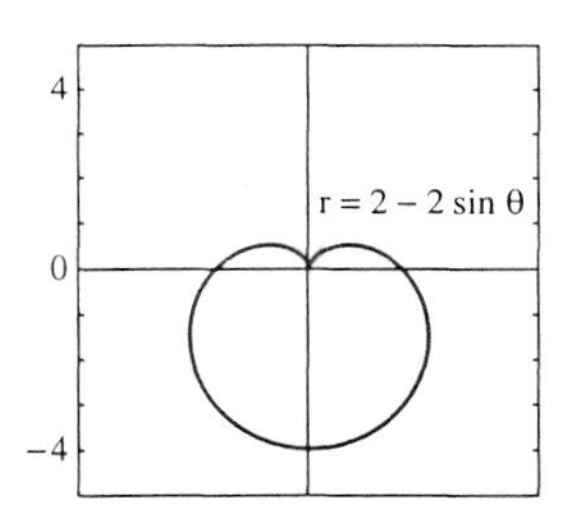

FIGURE 10.3.10 The cardioid of Problem 10.

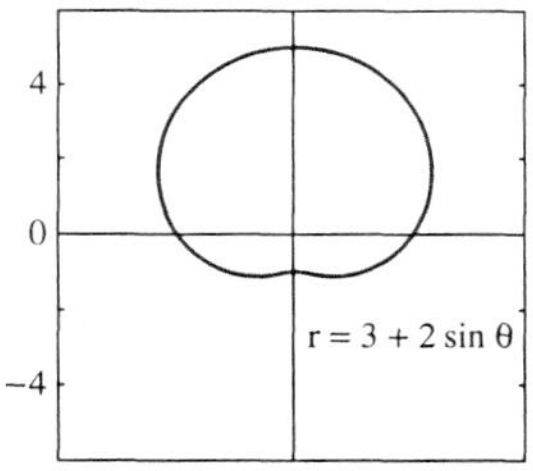

FIGURE 10.3.11 The limaçon of Problem 12.

13. $r = -4\cos\theta$ **14.** $r = 5(1 + \sin\theta)$

15. $r = 3 - \cos\theta$ **16.** $r = 2 + \sin\theta + \cos\theta$

In Problems 17 through 24, find the area bounded by one loop of the given curve.

17. $r = 2\cos 2\theta$ **18.** $r = 3\sin 3\theta$ (Fig. 10.3.12)

19. $r = 2\cos 4\theta$ (Fig. 10.3.13)

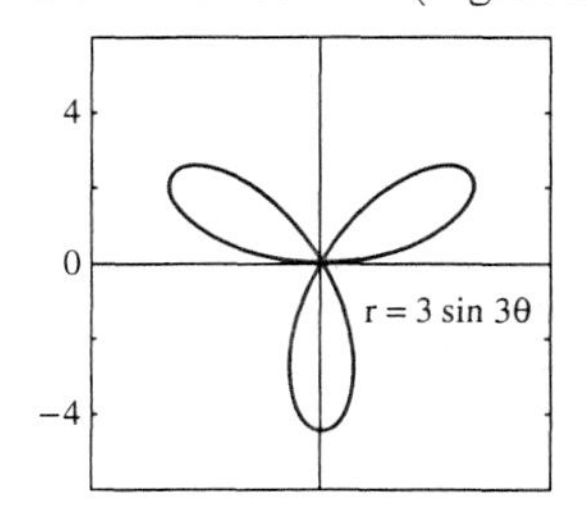

FIGURE 10.3.12 The three-leaved rose of Problem 18.

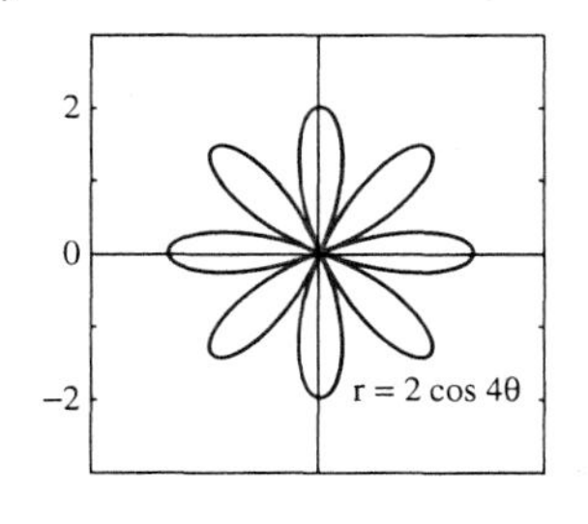

FIGURE 10.3.13 The eight-leaved rose of Problem 19.

20. $r = \sin 5\theta$ (Fig. 10.3.14)

21. $r^2 = 4\sin 2\theta$

22. $r^2 = 4\cos 2\theta$ (Fig. 10.3.15)

23. $r^2 = 4\sin\theta$

24. $r = 6\cos 6\theta$

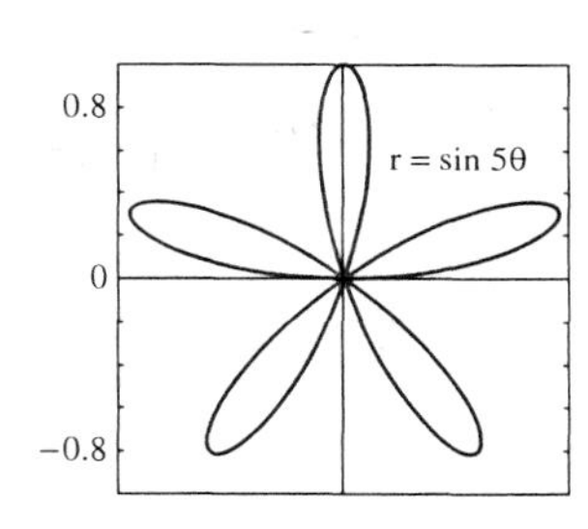

FIGURE 10.3.14 The five-leaved rose of Problem 20.

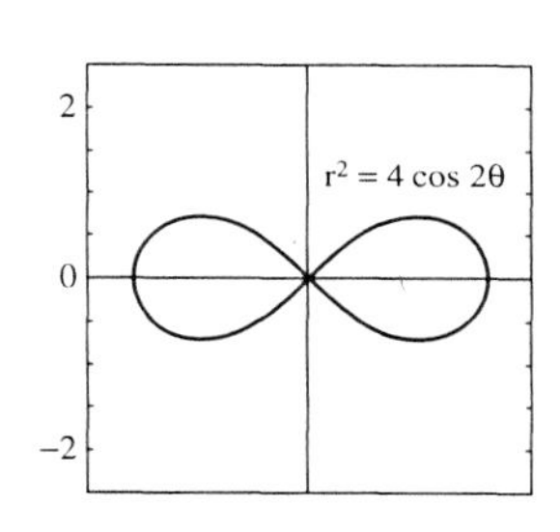

FIGURE 10.3.15 The lemniscate of Problem 22.

In Problems 25 through 36, find the area of the region described.

25. Inside $r = 2\sin\theta$ and outside $r = 1$

26. Inside both $r = 4\cos\theta$ and $r = 2$

27. Inside both $r = \cos\theta$ and $r = \sqrt{3}\sin\theta$

28. Inside $r = 2 + \cos\theta$ and outside $r = 2$

29. Inside $r = 3 + 2\cos\theta$ and outside $r = 4$

30. Inside $r^2 = 2\cos 2\theta$ and outside $r = 1$

31. Inside $r^2 = \cos 2\theta$ and $r^2 = \sin 2\theta$ (Fig. 10.3.16)

32. Inside the large loop and outside the small loop of $r = 1 - 2\sin\theta$ (Fig. 10.3.17)

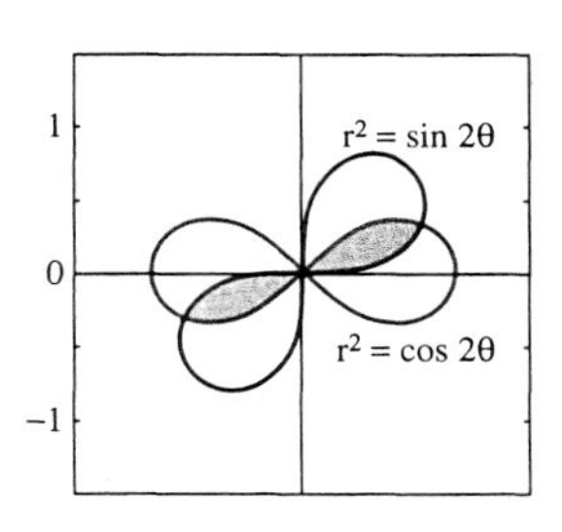

FIGURE 10.3.16 Problem 31.

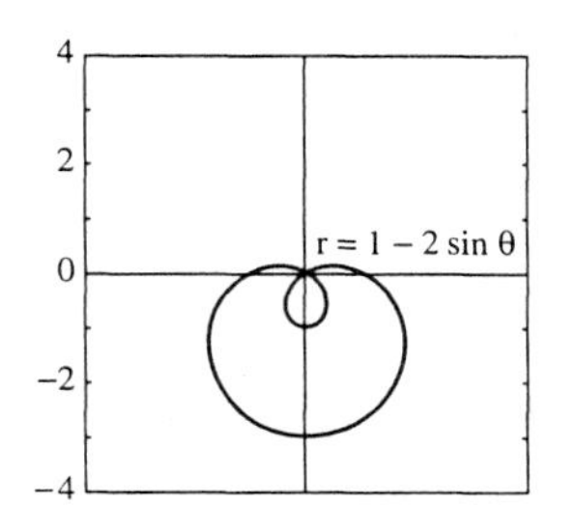

FIGURE 10.3.17 Problem 32.

33. Inside $r = 2(1 + \cos\theta)$ and outside $r = 1$

34. Inside the figure-eight curve $r^2 = 4\cos\theta$ and outside $r = 1 - \cos\theta$

35. Inside both $r = 2\cos\theta$ and $r = 2\sin\theta$

36. Inside $r = 2 + 2\sin\theta$ and outside $r = 2$

37. Find the area of the circle $r = \sin\theta + \cos\theta$ by integration in polar coordinates (Fig. 10.3.18). Check your answer by writing the equation of the circle in rectangular coordinates, finding its radius, and then using the familiar formula for the area of a circle.

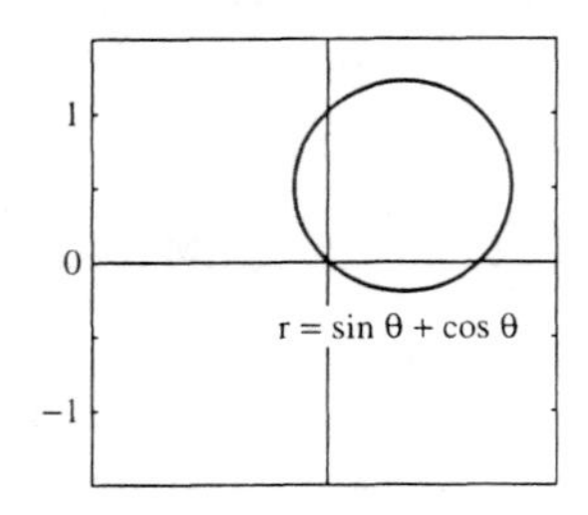

FIGURE 10.3.18 The circle $r = \sin\theta + \cos\theta$ (Problem 37).

38. Find the area of the region that lies interior to all three circles $r = 1$, $r = 2\cos\theta$, and $r = 2\sin\theta$.

39. The *spiral of Archimedes,* shown in Fig. 10.3.19, has the simple equation $r = a\theta$ (a is a constant). Let A_n denote the area bounded by the nth turn of the spiral, where $2(n-1)\pi \leqq \theta \leqq 2n\pi$, and by the portion of the polar axis joining its endpoints. For each $n \geqq 2$, let $R_n = A_n - A_{n-1}$

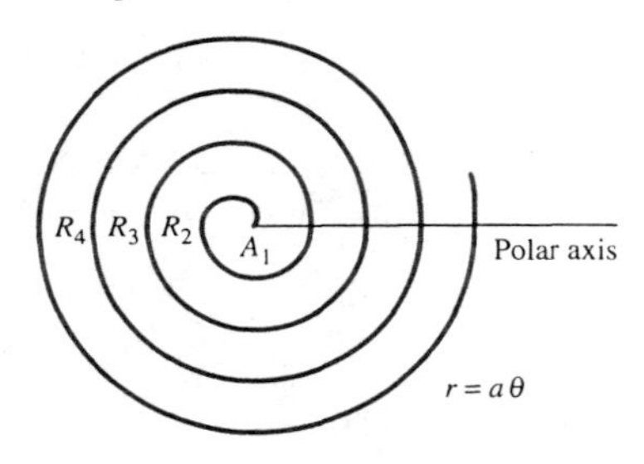

FIGURE 10.3.19 The spiral of Archimedes (Problem 39).

denote the area between the $(n-1)$th and the nth turns. Then derive the following results of Archimedes:

(a) $A_1 = \frac{1}{3}\pi(2\pi a)^2$; (b) $A_2 = \frac{7}{12}\pi(4\pi a)^2$;

(c) $R_2 = 6A_1$; (d) $R_{n+1} = nR_2$ for $n \geqq 2$.

40. Two circles both have radius a, and each circle passes through the center of the other. Find the area of the region that lies within both circles.

41. A polar curve of the form $r = ae^{-k\theta}$ is called a *logarithmic spiral,* and the portion given by $2(n-1)\pi \leqq \theta \leqq 2n\pi$ is called the nth *turn* of this spiral. Figure 10.3.20 shows the first five turns of the logarithmic spiral $r = e^{-\theta/10}$, and the area of the region lying between the second and third turns is shaded. Find:

(a) The area of the region that lies between the first and second turns.

(b) The area of the region that lies between the $(n-1)$th and nth turns for $n > 1$.

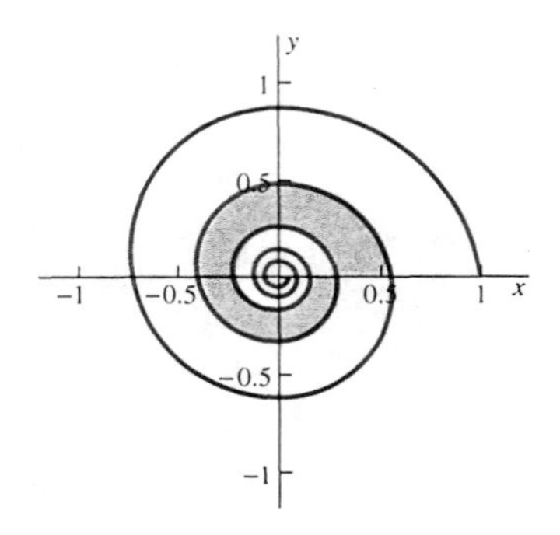

FIGURE 10.3.20 The logarithmic spiral of Problem 41.

42. Figure 10.3.21 shows the first turn of the logarithmic spiral $r = 2e^{-\theta/10}$ together with the two circles, both centered at $(0, 0)$, through the endpoints of the spiral. Find the areas of the two shaded regions and verify that their sum is the area of the annular region between the two circles.

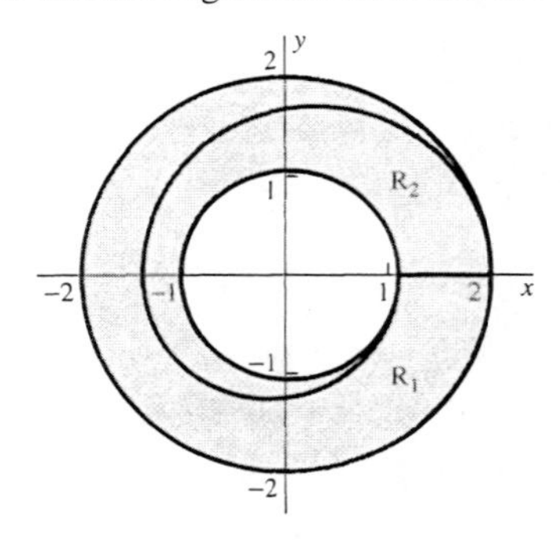

FIGURE 10.3.21 The two regions of Problem 42.

43. The shaded region R in Fig. 10.3.22 is bounded by the cardioid $r = 1 + \cos\theta$, the spiral $r = e^{-\theta/5}$, $0 \leqq \theta \leqq \pi$, and the spiral $r = e^{\theta/5}$, $-\pi \leqq \theta \leqq 0$. Graphically estimate the points of intersection of the cardioid and the spirals, then approximate the area of the region R.

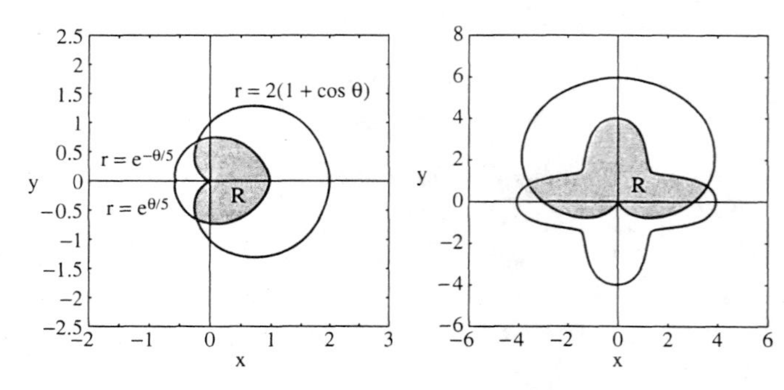

FIGURE 10.3.22 The region of Problem 43.

FIGURE 10.3.23 The region of Problem 44.

44. The shaded region R in Fig. 10.3.23 lies inside both the cardioid $r = 3 + 3\sin\theta$ and the polar curve $r = 3 + \cos 4\theta$. Graphically estimate the points of intersection of the two curves; then approximate the area of the region R.

10.4 PARAMETRIC CURVES

Until now we have encountered *curves* mainly as graphs of equations. An equation of the form $y = f(x)$ or of the form $x = g(y)$ determines a curve by giving one of the coordinate variables explicitly as a function of the other. An equation of the form $F(x, y) = 0$ may also determine a curve, but then each variable is given implicitly as a function of the other.

Another important type of curve is the trajectory of a point moving in the coordinate plane. The motion of the point can be described by giving its position $(x(t), y(t))$ at time t. Such a description involves expressing both the rectangular-coordinate variables x and y as functions of a third variable, or *parameter,* t rather than as functions of one another. In this context a **parameter** is an independent variable (not a constant, as is sometimes meant in popular usage). This approach motivates the following definition.

DEFINITION Parametric Curve
A **parametric curve** C in the plane is a pair of functions

$$x = f(t), \quad y = g(t), \tag{1}$$

that give x and y as continuous functions of the real number t (the parameter) in some interval I.

Each value of the parameter t determines a point $(f(t), g(t))$, and the set of all such points is the **graph** of the curve C. Often the distinction between the curve—the pair of **coordinate functions** f and g—and the graph is not made. Therefore, we may refer interchangeably to the curve and to its graph when the context makes clear the intended meaning. The two equations in (1) are called the **parametric equations** of the curve.

The graph of a parametric curve may be sketched by plotting enough points to indicate its likely shape. In some cases we can eliminate the parameter t and thus obtain an equation in x and y. This equation may give us more information about the shape of the curve.

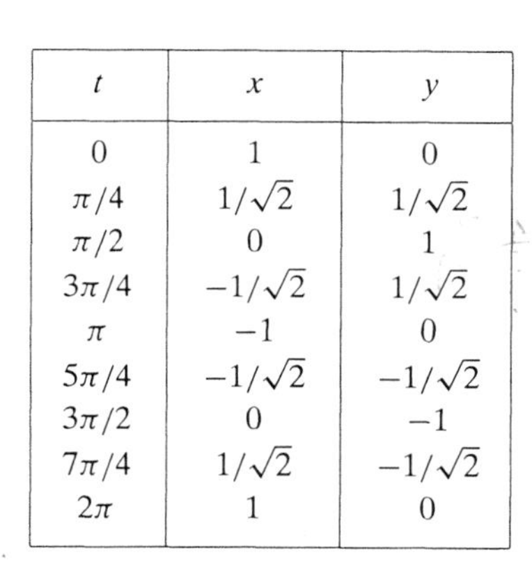

t	x	y
0	1	0
$\pi/4$	$1/\sqrt{2}$	$1/\sqrt{2}$
$\pi/2$	0	1
$3\pi/4$	$-1/\sqrt{2}$	$1/\sqrt{2}$
π	-1	0
$5\pi/4$	$-1/\sqrt{2}$	$-1/\sqrt{2}$
$3\pi/2$	0	-1
$7\pi/4$	$1/\sqrt{2}$	$-1/\sqrt{2}$
2π	1	0

FIGURE 10.4.1 A table of values for Example 1.

EXAMPLE 1 Determine the graph of the curve

$$x = \cos t, \quad y = \sin t, \quad 0 \leqq t \leqq 2\pi. \tag{2}$$

Solution Figure 10.4.1 shows a table of values of x and y that correspond to multiples of $\pi/4$ for the parameter t. These values give the eight points highlighted in Fig. 10.4.2, all of which lie on the unit circle. This suggests that the graph is, in fact, the unit circle. To verify this, we note that the fundamental identity of trigonometry gives

$$x^2 + y^2 = \cos^2 t + \sin^2 t \equiv 1,$$

so every point of the graph lies on the circle with equation $x^2 + y^2 = 1$. Conversely, the point of the circle with angular (polar) coordinate t is the point $(\cos t, \sin t)$ of the graph. Thus the graph is precisely the unit circle. ◆

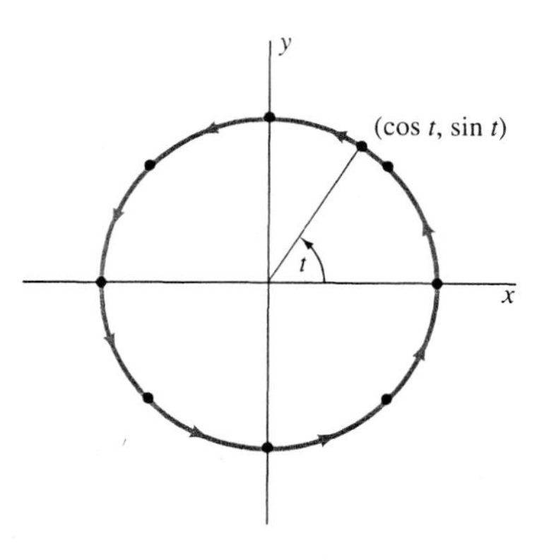

FIGURE 10.4.2 The graph of the parametric functions of Example 1.

What is lost in the process in Example 1 is the information about how the graph is produced as t goes from 0 to 2π. But this is easy to determine by inspection. As t travels from 0 to 2π, the point $(\cos t, \sin t)$ begins at $(1, 0)$ and travels counterclockwise around the circle, ending at $(1, 0)$ when $t = 2\pi$.

A given figure in the plane may be the graph of different curves. To speak more loosely, a given curve may have different **parametrizations.**

EXAMPLE 2 The graph of the parametric curve

$$x = \frac{1 - t^2}{1 + t^2}, \qquad y = \frac{2t}{1 + t^2}, \qquad -\infty < t < +\infty$$

also lies on the unit circle, because we find that $x^2 + y^2 = 1$ here as well. If t begins at 0 and increases, then the point $P(x(t), y(t))$ begins at $(1, 0)$ and travels along the upper half of the circle. If t begins at 0 and decreases, then the point $P(x(t), y(t))$ travels along the lower half of the circle. As t approaches either $+\infty$ or $-\infty$, the point P approaches the point $(-1, 0)$. Thus the graph consists of the unit circle with the single point $(-1, 0)$ deleted. A slight modification of the curve of Example 1,

$$x = \cos t, \qquad y = \sin t, \qquad -\pi < t < \pi,$$

is a different parametrization of this same graph. ◆

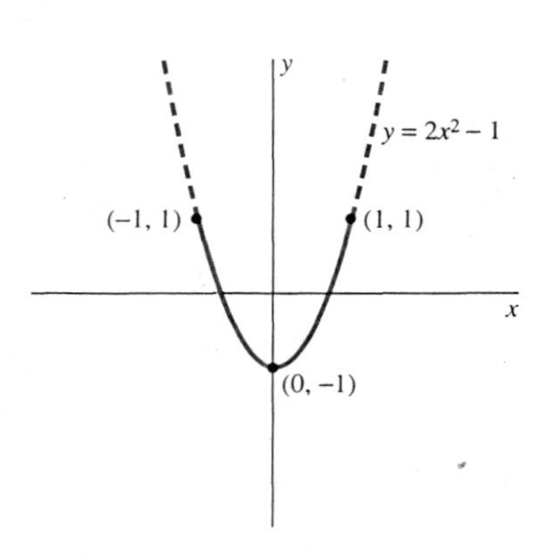

FIGURE 10.4.3 The curve of Example 3 is part of a parabola.

EXAMPLE 3 Eliminate the parameter to determine the graph of the parametric curve

$$x = t - 1, \qquad y = 2t^2 - 4t + 1, \qquad 0 \leqq t \leqq 2.$$

Solution We substitute $t = x + 1$ (from the equation for x) into the equation for y. This yields

$$y = 2(x + 1)^2 - 4(x + 1) + 1 = 2x^2 - 1$$

for $-1 \leqq x \leqq 1$. Thus the graph of the given curve is a portion of the parabola $y = 2x^2 - 1$ (Fig. 10.4.3). As t increases from 0 to 2, the point $(t - 1, 2t^2 - 4t + 1)$ travels along the parabola from $(-1, 1)$ to $(1, 1)$. ◆

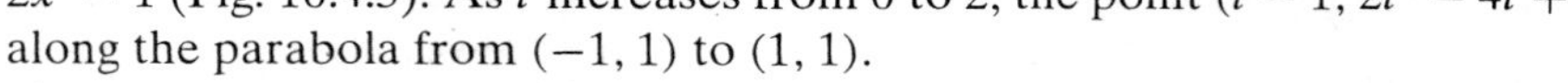

REMARK The parabolic arc of Example 3 can be reparametrized with

$$x = \sin t, \qquad y = 2\sin^2 t - 1.$$

Now, as t increases, the point $(\sin t, 2\sin^2 t - 1)$ travels back and forth along the parabola between the two points $(-1, 1)$ and $(1, 1)$, rather like the bob of a pendulum.

The parametric curve of Example 3 is one in which we can eliminate the parameter and thus obtain an explicit equation $y = f(x)$. Moreover, any explicitly presented curve $y = f(x)$ can be viewed as a parametric curve by writing

$$x = t, \qquad y = f(t),$$

with the parameter t taking on values in the original domain of f. By contrast, the circle of Example 1 illustrates a parametric curve whose graph is not the graph of any single function. (Why not?) Example 4 exhibits another way in which parametric curves can differ from graphs of functions—they can have self-intersections.

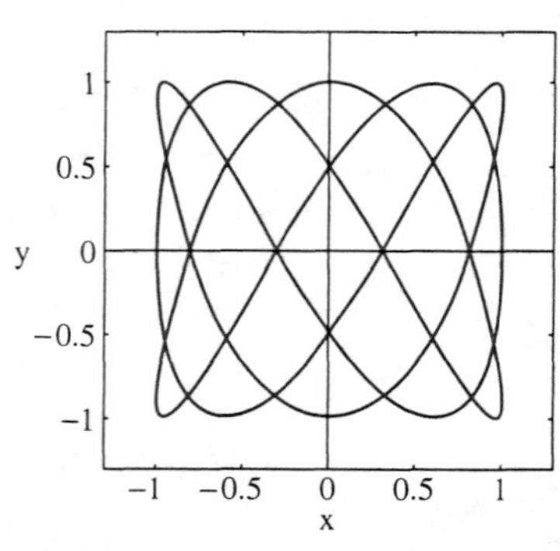

FIGURE 10.4.4 The Lissajous curve with $a = 3$, $b = 5$.

EXAMPLE 4 The parametric equations

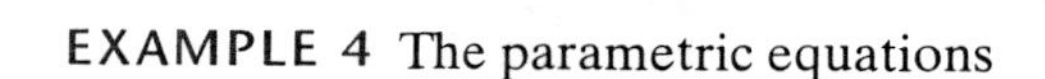

$$x = \cos at, \qquad y = \sin bt$$

(with a and b constant) define the *Lissajous curves* that typically appear on oscilloscopes in physics and electronics laboratories. The Lissajous curve with $a = 3$ and $b = 5$ is shown in Fig. 10.4.4. You probably would not want to calculate and plot by hand enough points to produce a Lissajous curve. Figure 10.4.4 was plotted with a computer program that generated it almost immediately. But it is perhaps more instructive to watch a slower graphing calculator plot a parametric curve like this, because the curve is traced by a point that moves on the screen as the parameter t increases (from 0 to 2π in this case). For instance, with a TI calculator set in "parametric graph mode," one need only enter and graph the equations

```
XT = cos(3T)      YT = sin(5T)
```

on the interval $0 \leqq t \leqq 2\pi$. With *Maple* and *Mathematica* the commands

```
plot([cos(3*t),sin(5*t),t=0..2*pi]);
```

and

```
ParametricPlot[{Cos[3*t],Sin[5*t]},{t,0,2*Pi}];
```

(respectively) give the same figure. ◆

The use of parametric equations $x = x(t)$, $y = y(t)$ is most advantageous when elimination of the parameter is either impossible or would lead to an equation $y = f(x)$ that is considerably more complicated than the original parametric equations. This often happens when the curve is a geometric locus or the path of a point moving under specified conditions.

EXAMPLE 5 The curve traced by a point P on the edge of a rolling circle is called a **cycloid.** The circle rolls along a straight line without slipping or stopping. (You will see a cycloid if you watch a patch of bright paint on the tire of a bicycle that crosses your path.) Find parametric equations for the cycloid if the line along which the circle rolls is the x-axis, the circle is above the x-axis but always tangent to it, and the point P begins at the origin.

Solution Evidently the cycloid consists of a series of arches. We take as parameter t the angle (in radians) through which the circle has turned since it began with P at the origin. This is the angle TCP in Fig. 10.4.5.

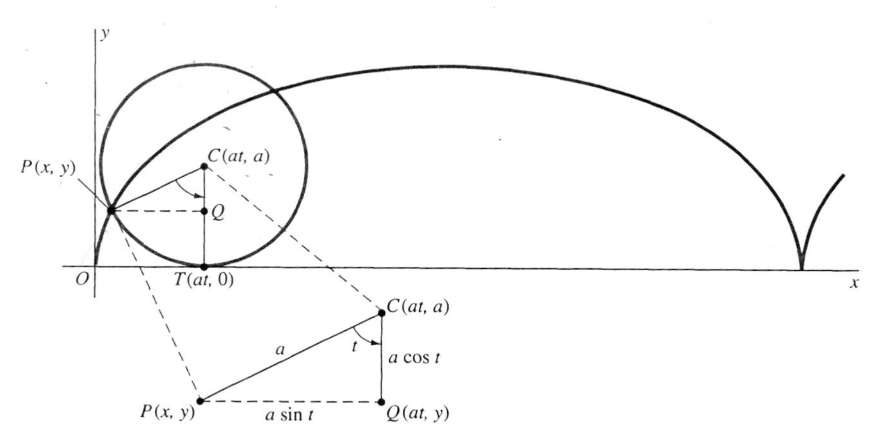

FIGURE 10.4.5 The cycloid and the right triangle CPQ (Example 5).

The distance the circle has rolled is $|OT|$, so this is also the length of the circumference subtended by the angle TCP. Thus $|OT| = at$ if a is the radius of the circle, so the center C of the rolling circle has coordinates (at, a) when the angle TCP is t. The right triangle CPQ in Fig. 10.4.5 provides us with the relations

$$at - x = a \sin t \quad \text{and} \quad a - y = a \cos t.$$

Therefore the cycloid—the path of the moving point P—has parametric equations

$$x = a(t - \sin t), \quad y = a(1 - \cos t). \tag{3}$$

◆

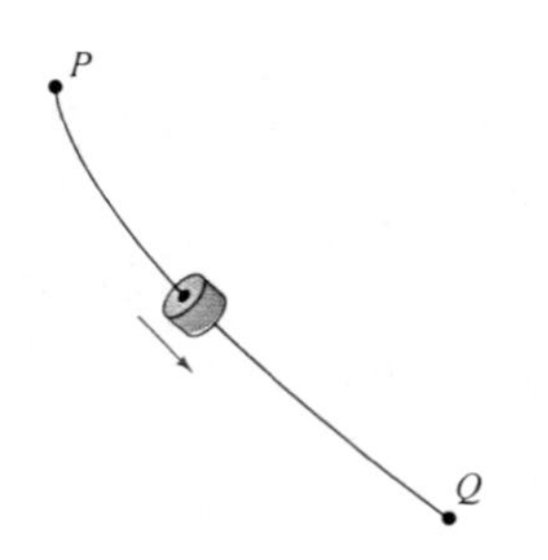

FIGURE 10.4.6 A bead sliding down a wire—the brachistochrone problem.

HISTORICAL NOTE Figure 10.4.6 shows a bead sliding down a frictionless wire from point P to point Q. The *brachistochrone problem* asks what shape the wire should be to minimize the bead's time of descent from P to Q. In June of 1696, John Bernoulli proposed the brachistochrone problem as a public challenge, with a 6-month deadline (later extended to Easter 1697 at Leibniz's request). Isaac Newton, then retired from academic life and serving as Warden of the Mint in London, received Bernoulli's challenge on January 29, 1697. The very next day he communicated his own solution—the curve of minimal descent time is an arc of an inverted cycloid—to the Royal Society of London.

Lines Tangent to Parametric Curves

The parametric curve $x = f(t)$, $y = g(t)$ is called **smooth** if the derivatives $f'(t)$ and $g'(t)$ are continuous and never simultaneously zero. In some neighborhood of each point of its graph, a smooth parametric curve can be described in one or possibly both of the forms $y = F(x)$ and $x = G(y)$. To see why this is so, suppose (for example) that $f'(t) > 0$ on the interval I. Then $f(t)$ is an increasing function on I and therefore has an inverse function $t = \phi(x)$ there. If we substitute $t = \phi(x)$ into the equation $y = g(t)$, then we get

$$y = g(\phi(x)) = F(x).$$

We can use the chain rule to compute the slope dy/dx of the line tangent to a smooth parametric curve at a given point. Differentiating $y = F(x)$ with respect to t yields

$$\frac{dy}{dt} = \frac{dy}{dx} \cdot \frac{dx}{dt},$$

so

$$\frac{dy}{dx} = \frac{dy/dt}{dx/dt} = \frac{g'(t)}{f'(t)} \tag{4}$$

at any point where $f'(t) \neq 0$. The tangent line is vertical at any point where $f'(t) = 0$ but $g'(t) \neq 0$.

Equation (4) gives $y' = dy/dx$ as a function of t. Another differentiation with respect to t, again with the aid of the chain rule, results in the formula

$$\frac{dy'}{dt} = \frac{dy'}{dx} \cdot \frac{dx}{dt},$$

so

$$\frac{d^2y}{dx^2} = \frac{dy'}{dx} = \frac{dy'/dt}{dx/dt}. \tag{5}$$

EXAMPLE 6 Calculate dy/dx and d^2y/dx^2 for the cycloid with the parametric equations in (3).

Solution We begin with

$$x = a(t - \sin t), \quad y = a(1 - \cos t). \tag{3}$$

Then Eq. (4) gives

$$\frac{dy}{dx} = \frac{dy/dt}{dx/dt} = \frac{a \sin t}{a(1 - \cos t)} = \frac{\sin t}{1 - \cos t}. \tag{6}$$

This derivative is zero when y is an odd integral multiple of π, so the tangent line is horizontal at the midpoint of each arch of the cycloid. The endpoints of the arches correspond to even integral multiples of π, where both the numerator and the denominator in Eq. (6) are zero. These are isolated points (called *cusps*) at which the cycloid fails to be a smooth curve. (See Fig. 10.4.7.)

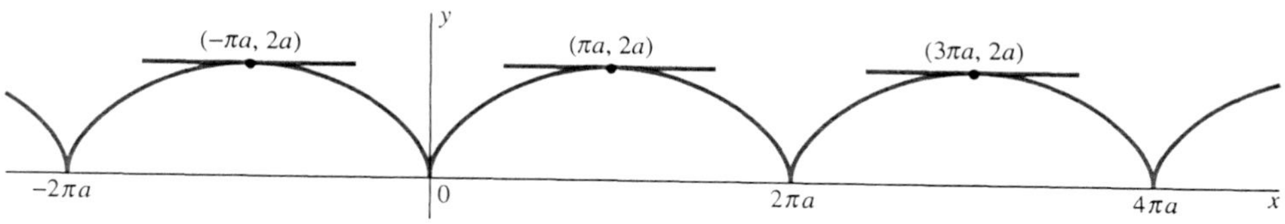

FIGURE 10.4.7 Horizontal tangents and cusps of the cycloid.

Next, Eq. (5) yields

$$\frac{d^2y}{dx^2} = \frac{(\cos t)(1-\cos t) - (\sin t)(\sin t)}{(1-\cos t)^2 \cdot a(1-\cos t)} = -\frac{1}{a(1-\cos t)^2}.$$

Because $d^2y/dx^2 < 0$ for all t (except for the isolated even integral multiples of π), this shows that each arch of the cycloid is concave downward (Fig. 10.4.5). ◆

REMARK In Fig. 10.4.7 it appears that the cycloid has a vertical tangent line at each cusp point $(2n\pi a, 0)$. We can verify this observation by calculating the limit as $t \to 2n\pi$ of the derivative in (6). Using l'Hôpital's rule, we get

$$\lim_{t\to 2n\pi} \frac{dy}{dx} = \lim_{t\to 2n\pi} \frac{\sin t}{1-\cos t} = \lim_{t\to 2n\pi} \frac{\cos t}{\sin t} = \pm\infty,$$

because $\cos t \to 1$ and $\sin t \to 0$ as $t \to 2n\pi$. The limit is $+\infty$ or $-\infty$ according as t approaches $2n\pi$ from the right or the left. In either event, we conclude that the tangent line is, indeed, vertical at the cusp point.

EXAMPLE 7 It would be impractical to attempt to graph the curve

$$x^3 = 2y^6 - 5y^4 + 9y \tag{7}$$

by solving for y as a function of x. But we can parametrize this curve by defining

$$y = t, \quad x = (2t^6 - 5t^4 + 9t)^{1/3}. \tag{8}$$

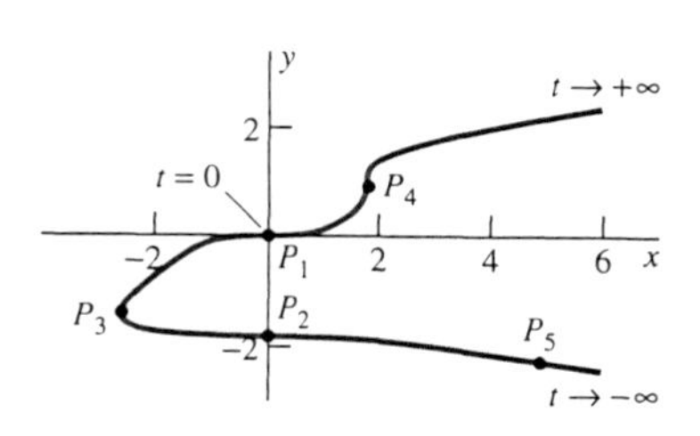

FIGURE 10.4.8 The parametric curve of Example 7.

Figure 10.4.8 shows a computer plot of this parametric curve for $-2.5 \leqq t \leqq 2.5$. We see at least four likely critical and inflection points. It appears that there are horizontal tangent lines at the points P_1 and P_2 on the y-axis, and vertical tangent lines at P_3 and P_4. Let's investigate the character of these points by calculating the pertinent derivatives.

To investigate the possibility of horizontal and vertical tangent lines, we use Eq. (4) to calculate the first derivative

$$\frac{dy}{dx} = \frac{dy/dt}{dx/dt} = \frac{3(2t^6 - 5t^4 + 9t)^{2/3}}{12t^5 - 20t^3 + 9}. \tag{9}$$

Using a computer algebra system, we find that the only real zeros of the polynomial $2t^6 - 5t^4 + 9t$ in the numerator are $t = 0$ and $t \approx -1.8065$. These values of t yield the points $P_1(0, 0)$ and $P_2(-0.00002422, -1.86065)$, respectively, that are shown in the figure. Thus P_2 does not lie precisely on the y-axis, after all.

The denominator polynomial $12t^5 - 20t^3 + 9$ in (9) has only the single real zero $t \approx -2.5587$, which yields the single point $P_3(-2.5587, -1.3941)$ on the curve where the tangent line is vertical. In particular, there is *no* vertical tangent line near the point P_4 indicated in the figure.

To investigate the possibility of possible inflection points, we use Eq. (5) and a computer algebra system to calculate the second derivative

$$\begin{aligned} \frac{d^2y}{dx^2} &= \frac{d}{dt}\left(\frac{dy}{dx}\right) \div \frac{dx}{dt} \\ &= -\frac{6(2t^6 - 5t^4 + 9t)^{1/3}(36t^{10} - 150t^8 + 50t^6 + 594t^5 - 450t^3 - 81)}{(12t^5 - 20t^3 + 9)^3}. \end{aligned} \tag{10}$$

The two trinomials that appear in the numerator and denominator here are the same as those in (9), and correspond to the three critical points already found. Our computer algebra system reports that the tenth-degree numerator polynomial in (10) has only two real zeros: $t \approx 1.0009$ and $t \approx -2.2614$. These two zeros of the second derivative yield the two points $P_4(1.8172, 1.0009)$ and $P_5(4.8820, -2.2614)$ that are shown in the figure. It is visually clear that the concavity of the curve changes at P_4—where $dy/dx \approx 0.9063$ so the tangent line is steep but not vertical—but the character of the remaining point is not so obvious. Nevertheless, you can graph the second derivative in (10) to verify that it is positive to the right and negative to the left of P_5—so this final candidate is, indeed, also an inflection point.

Finally, because our viewing window in Fig. 10.4.8 is large enough to include all the critical points and inflection points on the curve in (7)—and since it is clear from the equations in (8) that $|x|$ and $|y| \to \infty$ as $|t| \to \infty$—we are assured that the figure shows all of the principal features of the curve. ◆

Polar Curves as Parametric Curves

A curve given in polar coordinates by the equation $r = f(\theta)$ can be regarded as a parametric curve with parameter θ. To see this, we recall that the equations $x = r\cos\theta$ and $y = r\sin\theta$ allow us to change from polar to rectangular coordinates. We replace r with $f(\theta)$, and this gives the parametric equations

$$x = f(\theta)\cos\theta, \quad y = f(\theta)\sin\theta, \tag{11}$$

which express x and y in terms of the parameter θ.

EXAMPLE 8 The *spiral of Archimedes* has the polar-coordinate equation $r = a\theta$ (Fig. 10.4.9). The equations in (11) give the spiral the parametrization

$$x = a\theta\cos\theta, \quad y = a\theta\sin\theta.$$ ◆

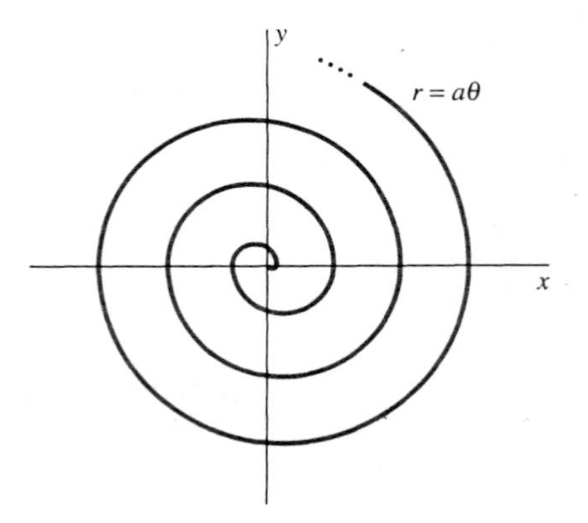

FIGURE 10.4.9 The spiral of Archimedes (Example 8).

The slope dy/dx of a tangent line can be computed in terms of polar coordinates as well as rectangular coordinates. Given a polar-coordinate curve $r = f(\theta)$, we use the parametrization shown in (11). Then Eq. (4), with θ in place of t, gives

$$\frac{dy}{dx} = \frac{dy/d\theta}{dx/d\theta} = \frac{f'(\theta)\sin\theta + f(\theta)\cos\theta}{f'(\theta)\cos\theta - f(\theta)\sin\theta}, \tag{12}$$

or, alternatively, denoting $f'(\theta)$ by r',

$$\frac{dy}{dx} = \frac{r' \sin\theta + r\cos\theta}{r'\cos\theta - r\sin\theta}. \tag{13}$$

Equation (13) has the following useful consequence. Let ψ denote the angle between the tangent line at P and the radius OP (extended) from the origin (Fig. 10.4.10). Then

$$\cot\psi = \frac{1}{r}\cdot\frac{dr}{d\theta} \quad (0 \leqq \psi \leqq \pi). \tag{14}$$

In Problem 32 we indicate how Eq. (14) can be derived from Eq. (13).

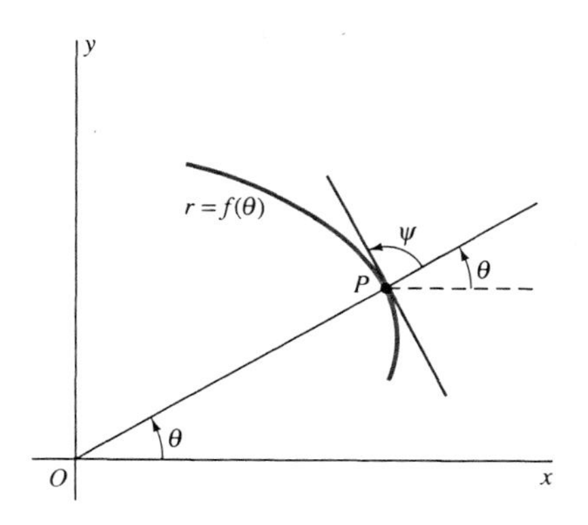

FIGURE 10.4.10 The interpretation of the angle ψ. [See Eq. (14).]

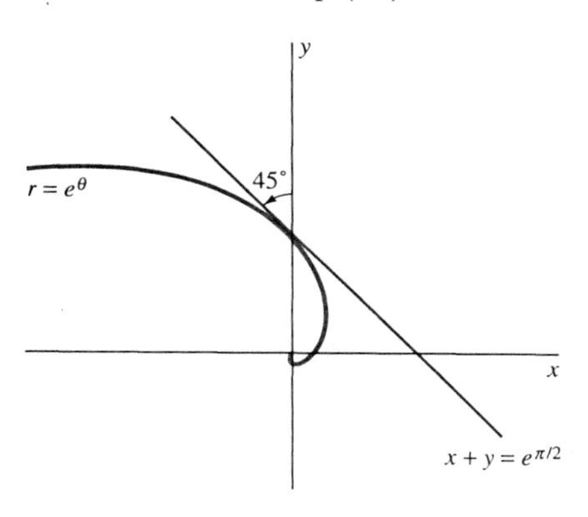

FIGURE 10.4.11 The angle ψ is always 45° for the logarithmic spiral (Example 9).

EXAMPLE 9 Consider the *logarithmic spiral* with polar equation $r = e^{\theta}$. Show that $\psi = \pi/4$ at every point of the spiral, and write an equation of its tangent line at the point $(e^{\pi/2}, \pi/2)$.

Solution Because $dr/d\theta = e^{\theta}$, Eq. (14) tells us that $\cot\psi = e^{\theta}/e^{\theta} = 1$. Thus $\psi = \pi/4$. When $\theta = \pi/2$, Eq. (13) gives

$$\frac{dy}{dx} = \frac{e^{\pi/2}\sin(\pi/2) + e^{\pi/2}\cos(\pi/2)}{e^{\pi/2}\cos(\pi/2) - e^{\pi/2}\sin(\pi/2)} = -1.$$

But when $\theta = \pi/2$, we have $x = 0$ and $y = e^{\pi/2}$. It follows that an equation of the desired tangent line is

$$y - e^{\pi/2} = -x; \quad \text{that is,} \quad x + y = e^{\pi/2}.$$

The line and the spiral appear in Fig. 10.4.11. ◆

10.4 TRUE/FALSE STUDY GUIDE

10.4 CONCEPTS: QUESTIONS AND DISCUSSION

1. Pick two points A and B in the plane. Then define a parametrization $P(t) = (x(t), y(t))$ of the line segment $\overline{AB}$ such that $P(0) = A$ and $P(1) = B$.
2. Pick two points A and B equidistant from the origin. Then define a parametrization of a circular arc AB such that $P(0) = A$ and $P(1) = B$.
3. Pick two points A and B on the parabola $y = x^2$. Then define a parametrization of the parabola such that $P(0) = A$ and $P(1) = B$.
4. Let A and B be two points on a given parametric curve. Is it always possible to define a parametrization of the curve such that $P(0) = A$ and $P(1) = B$?

10.4 PROBLEMS

In Problems 1 through 12, eliminate the parameter and then sketch the curve.

1. $x = t + 1, \quad y = 2t - 1$

2. $x = t^2 + 1, \quad y = 2t^2 - 1$

3. $x = t^2, \quad y = t^3$

4. $x = \sqrt{t}, \quad y = 3t - 2$

5. $x = t + 1, \quad y = 2t^2 - t - 1$

6. $x = t^2 + 3t, \quad y = t - 2$

7. $x = e^t, \quad y = 4e^{2t}$

8. $x = 2e^t, \quad y = 2e^{-t}$

9. $x = 5\cos t, \quad y = 3\sin t$

10. $x = \sinh t, \quad y = \cosh t$

11. $x = 2\cosh t, \quad y = 3\sinh t$

12. $x = \sec t, \quad y = \tan t$

In Problems 13 through 16, first eliminate the parameter and sketch the curve. Then describe the motion of the point $(x(t), y(t))$ as t varies in the given interval.

13. $x = \sin 2\pi t, \quad y = \cos 2\pi t; \quad 0 \leqq t \leqq 1$

14. $x = 3 + 2\cos t, \quad y = 5 - 2\sin t; \quad 0 \leqq t \leqq 2\pi$

15. $x = \sin^2 \pi t, \quad y = \cos^2 \pi t; \quad 0 \leqq t \leqq 2$

16. $x = \cos t, \quad y = \sin^2 t; \quad -\pi \leqq t \leqq \pi$

In Problems 17 through 20, (a) first write the equation of the line tangent to the given parametric curve at the point that corresponds to the given value of t, and (b) then calculate d^2y/dx^2 to determine whether the curve is concave upward or concave downward at this point.

17. $x = 2t^2 + 1, \quad y = 3t^3 + 2; \quad t = 1$

18. $x = \cos^3 t, \quad y = \sin^3 t; \quad t = \pi/4$

19. $x = t\sin t, \quad y = t\cos t; \quad t = \pi/2$

20. $x = e^t, \quad y = e^{-t}; \quad t = 0$

In Problems 21 through 24, find the angle ψ between the radius OP and the tangent line at the point P that corresponds to the given value of θ.

21. $r = \exp(\theta\sqrt{3}), \quad \theta = \pi/2$

22. $r = 1/\theta, \quad \theta = 1$

23. $r = \sin 3\theta, \quad \theta = \pi/6$

24. $r = 1 - \cos\theta, \quad \theta = \pi/3$

In Problems 25 through 28, find

(a) The points on the curve where the tangent line is horizontal.

(b) The slope of each tangent line at any point where the curve intersects the x-axis.

25. $x = t^2, \quad y = t^3 - 3t$ (Fig. 10.4.12)

26. $x = \sin t, \quad y = \sin 2t$ (Fig. 10.4.13)

27. $r = 1 + \cos\theta$

28. $r^2 = 4\cos 2\theta$ (See Fig. 10.3.15.)

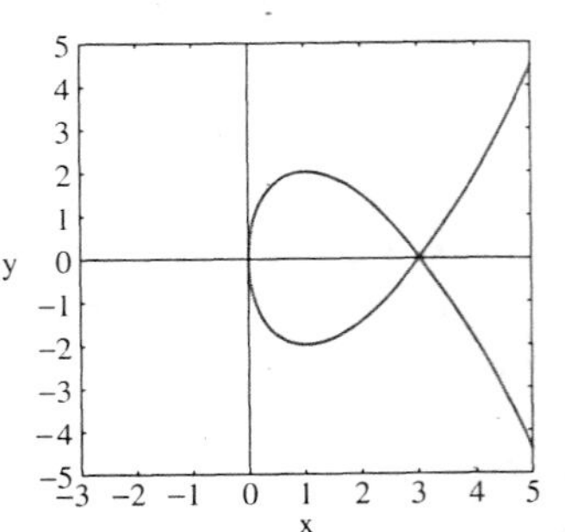

FIGURE 10.4.12 The curve of Problem 25.

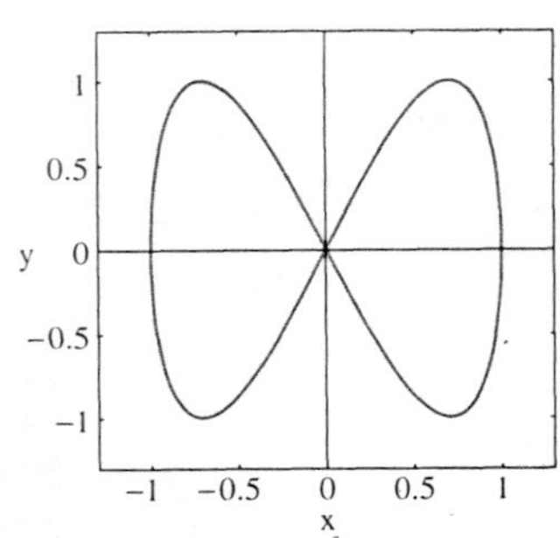

FIGURE 10.4.13 The curve of Problem 26.

29. The curve C is determined by the parametric equations $x = e^{-t}$, $y = e^{2t}$. Calculate dy/dx and d^2y/dx^2 directly from these parametric equations. Conclude that C is concave upward at every point. Then sketch C.

30. The graph of the folium of Descartes with rectangular equation $x^3 + y^3 = 3xy$ appears in Fig. 10.4.14. Parametrize its loop as follows: Let P be the point of intersection of the line $y = tx$ with the loop; then solve for the coordinates x and y of P in terms of t.

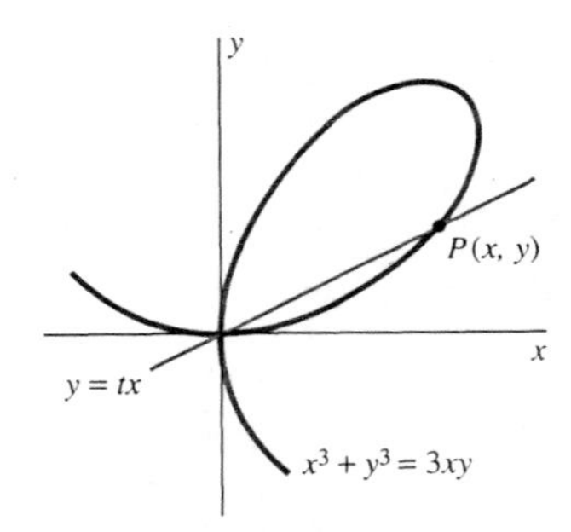

FIGURE 10.4.14 The loop of the folium of Descartes (Problem 30).

31. Parametrize the parabola $y^2 = 4px$ by expressing x and y as functions of the slope m of the tangent line at the point $P(x, y)$ of the parabola.

32. Let P be a point of the curve with polar equation $r = f(\theta)$, and let ψ be the angle between the extended radius OP and the tangent line at P. Let α be the angle of inclination of this tangent line, measured counterclockwise from the horizontal. Then $\psi = \alpha - \theta$. Verify Eq. (14) by substituting $\tan\alpha = dy/dx$ from Eq. (13) and $\tan\theta = y/x = (\sin\theta)/(\cos\theta)$ into the identity

$$\cot\psi = \frac{1}{\tan(\alpha - \theta)} = \frac{1 + \tan\alpha\tan\theta}{\tan\alpha - \tan\theta}.$$

33. Let P_0 be the highest point of the circle of Fig. 10.4.5—the circle that generates the cycloid of Example 5. Show that the line through P_0 and the point P of the cycloid (the point P is shown in Fig. 10.4.5) is tangent to the cycloid at P. This fact gives a geometric construction of the line tangent to the cycloid.

34. A circle of radius b rolls without slipping inside a circle of radius $a > b$. The path of a point fixed on the circumference of the rolling circle is called a *hypocycloid* (Fig. 10.4.15). Let P begin its journey at $A(a, 0)$ and let t be the angle AOC, where O is the center of the large circle and C is the center of the rolling circle. Show that the coordinates of P are given by the parametric equations

$$x = (a - b)\cos t + b\cos\left(\frac{a-b}{b}t\right),$$

$$y = (a - b)\sin t - b\sin\left(\frac{a-b}{b}t\right).$$

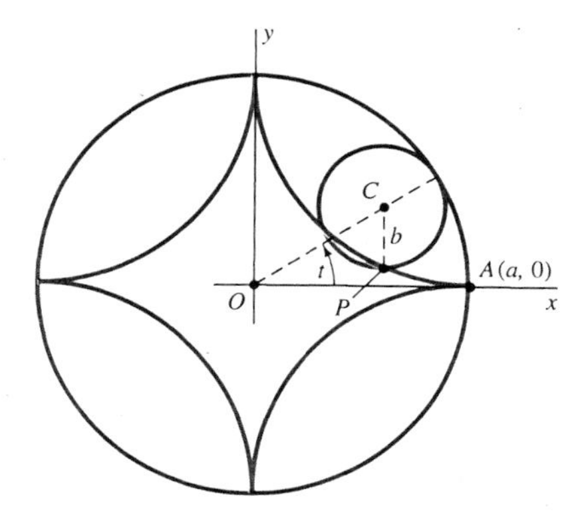

FIGURE 10.4.15 The hypocycloid of Problem 34.

35. If $b = a/4$ in Problem 34, show that the parametric equations of the hypocycloid reduce to

$$x = a\cos^3 t, \qquad y = a\sin^3 t.$$

36. (a) Prove that the hypocycloid of Problem 35 is the graph of the equation

$$x^{2/3} + y^{2/3} = a^{2/3}.$$

(b) Find all points of this hypocycloid where its tangent line is either horizontal or vertical, and find the intervals on which it is concave upward and those on which it is concave downward. (c) Sketch this hypocycloid.

37. Consider a point P on the spiral of Archimedes, the curve shown in Fig. 10.4.16 with polar equation $r = a\theta$.

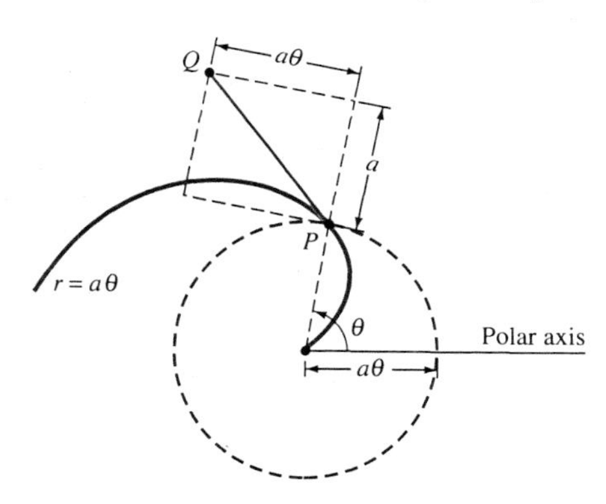

FIGURE 10.4.16 The segment PQ is tangent to the spiral (a result of Archimedes; see Problem 37).

Archimedes viewed the path of P as compounded of two motions, one with speed a directly away from the origin O and another a circular motion with unit angular speed around O. This suggests Archimedes' result that the line PQ in the figure is tangent to the spiral at P. Prove that this is indeed true.

38. (a) Deduce from Eq. (6) that if t is not an integral multiple of 2π, then the slope of the tangent line at the corresponding point of the cycloid is $\cot(t/2)$. (b) Conclude that at the cusp of the cycloid where t is an integral multiple of 2π, the cycloid has a vertical tangent line.

39. A *loxodrome* is a curve $r = f(\theta)$ such that the tangent line at P and the radius OP in Fig. 10.4.10 make a constant angle. Use Eq. (14) to prove that every loxodrome is of the form $r = Ae^{k\theta}$, where A and k are constants. Thus every loxodrome is a logarithmic spiral similar to the one considered in Example 9.

40. Let a curve be described in polar coordinates by $r = f(\theta)$ where f is continuous. If $f(\alpha) = 0$, then the origin is the point of the curve corresponding to $\theta = \alpha$. Deduce from the parametrization $x = f(\theta)\cos\theta$, $y = f(\theta)\sin\theta$ that the line tangent to the curve at this point makes the angle α with the positive x-axis. For example, the cardioid $r = f(\theta) = 1 - \sin\theta$ shown in Fig. 10.4.17 is tangent to the y-axis at the origin. And, indeed, $f(\pi/2) = 0$. The y-axis is the line $\theta = \alpha = \pi/2$.

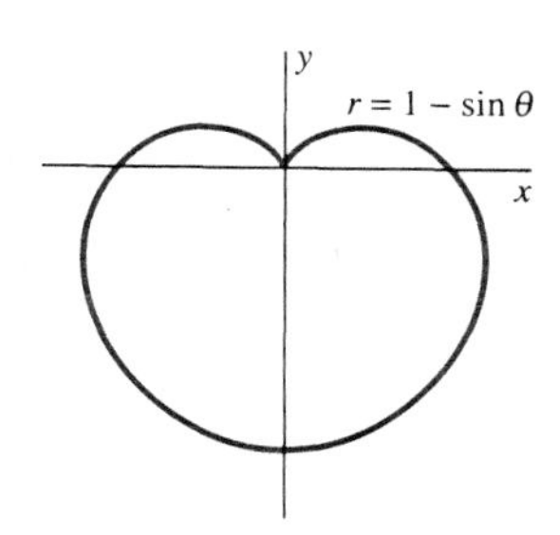

FIGURE 10.4.17 The cardioid of Problem 40.

41. Use the technique of Problem 30 to parametrize the first-quadrant loop of the folium-like curve $x^5 + y^5 = 5x^2y^2$.

42. A line segment of length $2a$ has one endpoint constrained to lie on the x-axis and the other endpoint constrained to lie on the y-axis, but its endpoints are free to move along those axes. As they do so, its midpoint sweeps out a locus in the xy-plane. Obtain a rectangular-coordinate equation of this locus and thereby identify this curve.

In Problems 43–46, investigate (as in Example 7) the given curve and construct a sketch that shows all the critical points and inflection points on it.

43. $x = y^3 - 3y^2 + 1$

44. $x = y^4 - 3y^3 + 5y$

45. $x^3 = y^5 - 5y^3 + 4$

46. $x^5 = 5y^6 - 17y^3 + 13y$

10.4 Project: Trochoid Investigations

A *trochoid* is traced by a point P on a spoke of a wheel of radius a as it rolls along the x-axis. If the distance of P from the center of the rolling wheel is $b > 0$, show that the trochoid is described by the parametric equations

$$x = at - b\sin t, \qquad y = a - b\cos t.$$

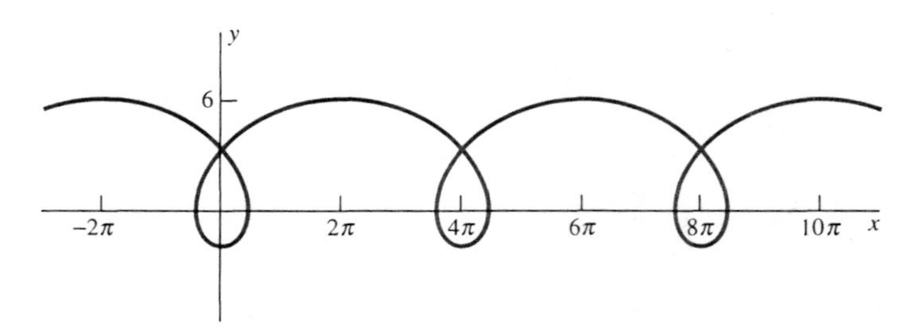

FIGURE 10.4.18 The trochoid with $a = 2$ and $b = 4$.

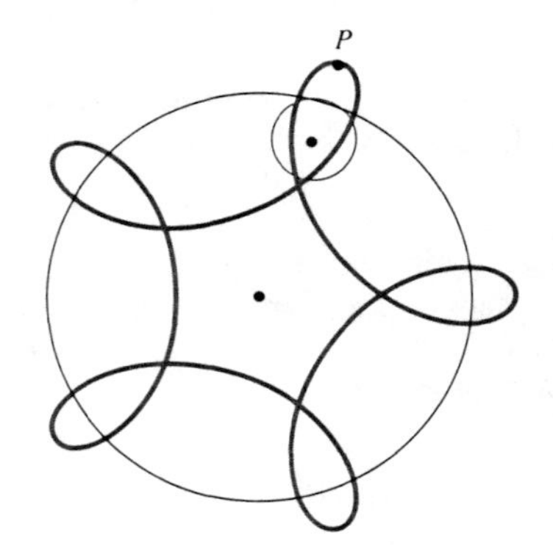

FIGURE 10.4.19 The hypotrochoid with $a = 10$, $b = 2$, $c = 4$.

Note that the trochoid is a familiar cycloid if $b = a$. We allow the possibility that $b > a$. Figure 10.4.18 shows the trochoid with $a = 2$ and $b = 4$. Experiment with different values of a and b. What determines whether the trochoid has loops, cusps, or neither?

Hypotrochoids

A hypotrochoid is to a hypocycloid (Problem 34) as a trochoid is to a cycloid. Thus a hypotrochoid is traced by a point P on a spoke of a wheel of radius b as it rolls around inside a circle of radius a. If the distance of P from the center of the rolling wheel is $c > 0$, show that the hypotrochoid is described by the parametric equations

$$x = (a - b)\cos t + c\cos\left(\frac{a-b}{b}t\right), \qquad y = (a - b)\sin t - c\sin\left(\frac{a-b}{b}t\right).$$

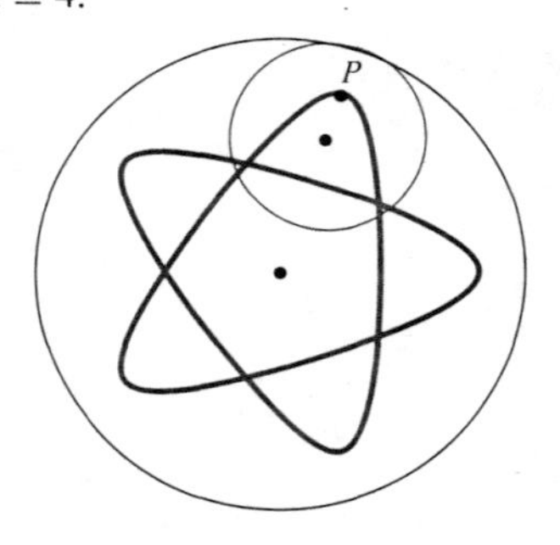

FIGURE 10.4.20 The hypotrochoid with $a = 10$, $b = 4$, $c = 2$.

Note that the hypotrochoid is a hypocycloid if $c = b$. There are a number of different ways a hypotrochoid can look. Figures 10.4.19 and 10.4.20 illustrate two possibilities. Experiment with different values of a, b, and c. What determines whether the trochoid has loops, cusps, or neither? If there are loops, what determines how many there are? Does a hypotrochoid always repeat itself after a finite number of turns around the origin? What happens if a is an integer but b is an irrational number?

Epitrochoids

An *epitrochoid* is generated in the same way as a hypotrochoid, except now the small circle rolls around on the outside of the large circle. With the same notation otherwise, show that the epitrochoid is described by the parametric equations

$$x = (a + b)\cos t - c\cos\left(\frac{a+b}{b}t\right), \qquad y = (a + b)\sin t - c\sin\left(\frac{a+b}{b}t\right).$$

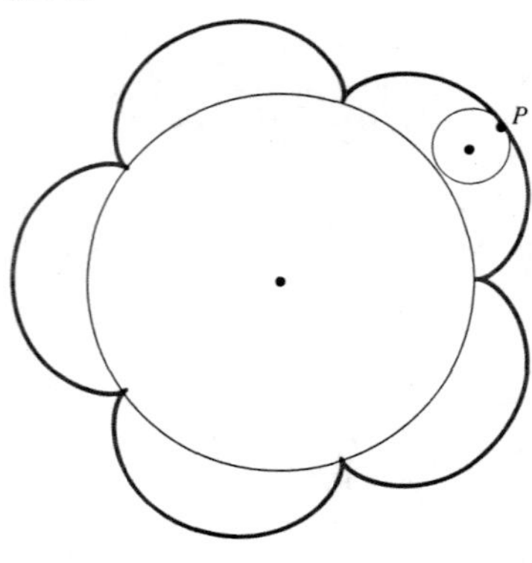

FIGURE 10.4.21 The epitrochoid with $a = 10$, $b = 2$, $c = 2$.

If $b = c$—so the point P lies on the rim of the rolling circle, then the epitrochoid is an *epicycloid* (illustrated in Fig. 10.4.21). Experiment with different values of a, b, and c, and investigate for epitrochoids the same questions posed previously for hypotrochoids.

10.5 INTEGRAL COMPUTATIONS WITH PARAMETRIC CURVES

In Chapter 6 we discussed the computation of a variety of geometric quantities associated with the graph $y = f(x)$ of a nonnegative function on the interval $[a, b]$. These included the following.

- The area under the curve:

$$A = \int_a^b y\,dx. \tag{1}$$

- The volume of revolution around the x-axis:

$$V_x = \int_a^b \pi y^2\,dx. \tag{2a}$$

- The volume of revolution around the y-axis:

$$V_y = \int_a^b 2\pi xy\,dx. \tag{2b}$$

- The arc length of the curve:

$$s = \int_0^s ds = \int_a^b \sqrt{1 + (dy/dx)^2}\,dx. \tag{3}$$

- The area of the surface of revolution around the x-axis:

$$S_x = \int_{x=a}^b 2\pi y\,ds. \tag{4a}$$

- The area of the surface of revolution around the y-axis:

$$S_y = \int_{x=a}^b 2\pi x\,ds. \tag{4b}$$

We substitute $y = f(x)$ into each of these integrals before we integrate from $x = a$ to $x = b$.

We now want to compute these same quantities for a smooth parametric curve

$$x = f(t), \quad y = g(t), \quad \alpha \leqq t \leqq \beta. \tag{5}$$

The area, volume, arc length, and surface integrals in Eqs. (1) through (4) can then be evaluated by making the formal substitutions

$$\begin{aligned} x &= f(t), & y &= g(t), \\ dx &= f'(t)\,dt, & dy &= g'(t)\,dt, \quad \text{and} \\ ds &= \sqrt{[f'(t)]^2 + [g'(t)]^2}\,dt. \end{aligned} \tag{6}$$

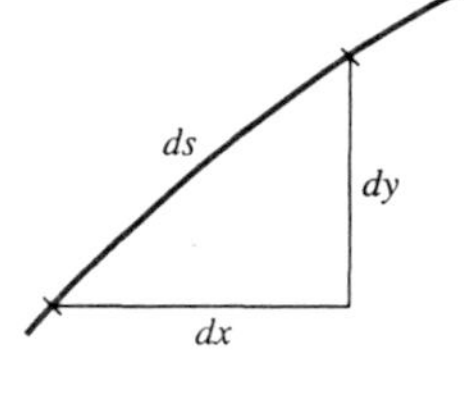

FIGURE 10.5.1 Nearly a right triangle for dx and dy close to zero.

The infinitesimal "right triangle" in Fig. 10.5.1 serves as a convenient device for remembering the latter substitution for ds. The Pythagorean theorem then leads to the symbolic manipulation

$$ds = \sqrt{dx^2 + dy^2} = \sqrt{\left(\frac{dx}{dt}\right)^2 + \left(\frac{dy}{dt}\right)^2}\,dt = \sqrt{[f'(t)]^2 + [g'(t)]^2}\,dt. \tag{7}$$

It simplifies the discussion to assume that the graph of the parametric curve in (5) resembles Fig. 10.5.2, in which $y = g(t) \geqq 0$ and $x = f(t)$ is either increasing on

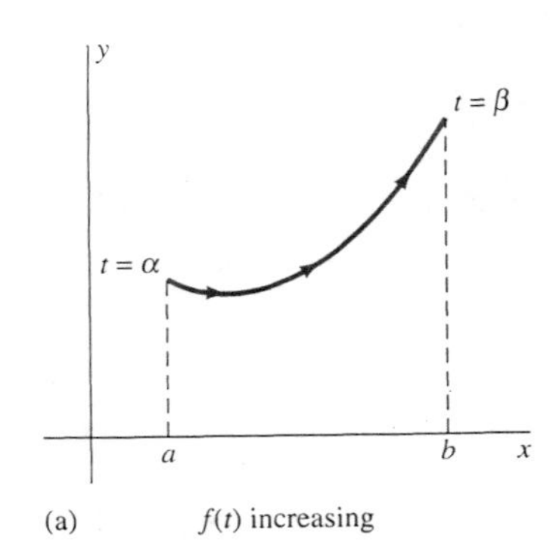

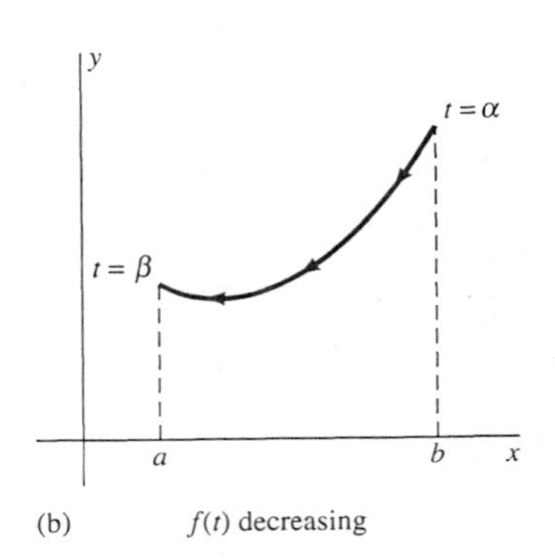

FIGURE 10.5.2 Tracing a parametrized curve: (a) $f(t)$ increasing; (b) $f(t)$ decreasing.

the entire interval $\alpha \leqq t \leqq \beta$ or is decreasing there. The two parts of Fig. 10.5.2 illustrate the two possibilities—whether as t increases the curve is traced in the positive x-direction from left to right, or in the negative x-direction from right to left. How and whether to take this direction of motion into account depends on which integral we are computing.

CASE 1 Area and Volume of Revolution To evaluate the integrals in (1) and (2), which involve dx, we integrate *either* from $t = \alpha$ to $t = \beta$ *or* from $t = \beta$ to $t = \alpha$—the proper choice of limits on t being the one that corresponds to traversing the curve in the positive x-direction *from left to right.* Specifically,

$$A = \int_\alpha^\beta g(t)\, f'(t)\, dt \quad \text{if } f(\alpha) < f(\beta),$$

whereas

$$A = \int_\beta^\alpha g(t)\, f'(t)\, dt \quad \text{if } f(\beta) < f(\alpha).$$

The validity of this method of evaluating the integrals in Eqs. (1) and (2) follows from Theorem 1 of Section 5.7, on integration by substitution.

CASE 2 Arc Length and Surface Area To evaluate the integrals in (3) and (4), which involve ds rather than dx, we integrate from $t = \alpha$ to $t = \beta$ irrespective of the direction of motion along the curve. To see why this is so, recall from Eq. (4) of Section 10.4 that $dy/dx = g'(t)/f'(t)$ if $f'(t) \neq 0$ on $[\alpha, \beta]$. Hence

$$s = \int_a^b \sqrt{1 + \left(\frac{dy}{dx}\right)^2}\, dx = \int_{f^{-1}(a)}^{f^{-1}(b)} \sqrt{1 + \left[\frac{g'(t)}{f'(t)}\right]^2}\, f'(t)\, dt.$$

Assuming that $f'(t) > 0$ if $f(\alpha) = a$ and $f(\beta) = b$, whereas $f'(t) < 0$ if $f(\alpha) = b$ and $f(\beta) = a$, it follows in either event that

$$s = \int_\alpha^\beta \sqrt{1 + \left[\frac{g'(t)}{f'(t)}\right]^2}\, |f'(t)|\, dt,$$

and so

$$s = \int_\alpha^\beta \sqrt{[f'(t)]^2 + [g'(t)]^2}\, dt = \int_\alpha^\beta \sqrt{\left(\frac{dx}{dt}\right)^2 + \left(\frac{dy}{dt}\right)^2}\, dt. \tag{8}$$

This formula, derived under the assumption that $f'(t) \neq 0$ on $[\alpha, \beta]$, may be taken to be the *definition* of arc length for an arbitrary smooth parametric curve. Similarly, the area of a surface of revolution is defined for smooth parametric curves as the result of first substituting Eq. (6) into Eq. (4a) or (4b) and then integrating from $t = \alpha$ to $t = \beta$.

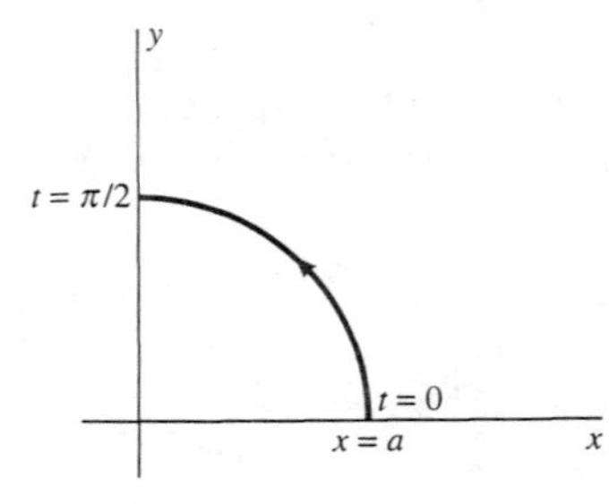

FIGURE 10.5.3 The quarter-circle of Example 1.

EXAMPLE 1 Use the parametrization $x = a\cos t$, $y = a\sin t$ $(0 \leqq t \leqq 2\pi)$ of the circle with center $(0, 0)$ and radius a to find (a) the area A of this circle; (b) the volume V of the sphere obtained by revolving the circle around the x-axis; and (c) the surface area S of this sphere.

Solution (a) The left-to-right direction along the quarter circle shown in Fig. 10.5.3 is from $t = \pi/2$ to $t = 0$, and $dx = -a\sin t\, dt$. Therefore Eq. (1) and multiplication by 4 give

$$\begin{aligned}A &= 4\int_{t=\pi/2}^{0} y\,dx = 4\int_{\pi/2}^{0} (a\sin t)(-a\sin t)\,dt\\ &= 4a^2\int_{0}^{\pi/2} \sin^2 t\,dt = 2a^2\int_{0}^{\pi/2} (1-\cos 2t)\,dt\\ &= 2a^2\left[t - \frac{1}{2}\sin 2t\right]_0^{\pi/2} = 2a^2\cdot\frac{\pi}{2} = \pi a^2\end{aligned}$$

for yet another derivation of the familiar formula $A = \pi a^2$ for the area of a circle of radius a.

(b) To calculate the volume of the sphere, we apply Eq. (2a) and double to get

$$\begin{aligned}V &= 2\int_{t=\pi/2}^{0} \pi y^2\,dx\\ &= 2\int_{\pi/2}^{0} \pi(a\sin t)^2(-a\sin t\,dt) = 2\pi a^3\int_{0}^{\pi/2} (1-\cos^2 t)\sin t\,dt\\ &= 2\pi a^3\left[-\cos t + \frac{1}{3}\cos^3 t\right]_0^{\pi/2} = \frac{4}{3}\pi a^3.\end{aligned}$$

(c) To find the surface area of the sphere, we calculate first the arc-length differential

$$ds = \sqrt{(-a\sin t)^2 + (a\cos t)^2}\,dt = a\,dt$$

of the parametrized curve. Then Eq. (4a) gives

$$\begin{aligned}S &= 2\int_{t=0}^{\pi/2} 2\pi y\,ds = 2\int_{0}^{\pi/2} 2\pi(a\sin t)\cdot a\,dt\\ &= 4\pi a^2\int_{0}^{\pi/2} \sin t\,dt = 4\pi a^2\Big[-\cos t\Big]_0^{\pi/2} = 4\pi a^2.\end{aligned}$$

◆

Of course, the results of Example 1 are familiar. In contrast, Example 2 requires the methods of this section.

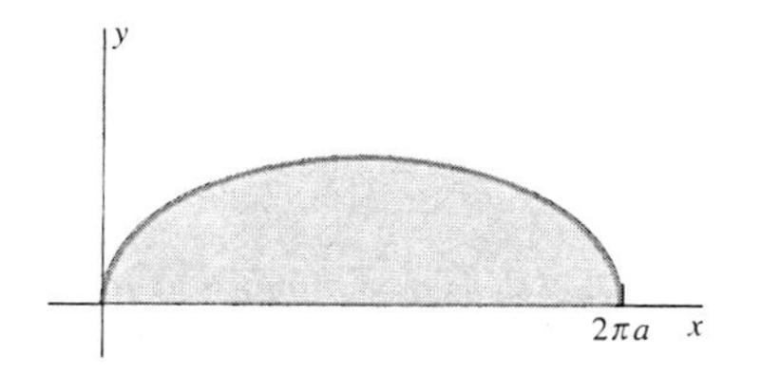

FIGURE 10.5.4 The cycloidal arch of Example 2.

EXAMPLE 2 Find the area under, and the arc length of, the cycloidal arch of Fig. 10.5.4. Its parametric equations are

$$x = a(t - \sin t), \quad y = a(1-\cos t), \quad 0 \leqq t \leqq 2\pi.$$

Solution Because $dx = a(1-\cos t)\,dt$ and the left-to-right direction along the curve is from $t = 0$ to $t = 2\pi$, Eq. (1) gives

$$\begin{aligned}A &= \int_{t=0}^{2\pi} y\,dx\\ &= \int_{0}^{2\pi} a(1-\cos t)\cdot a(1-\cos t)\,dt = a^2\int_{0}^{2\pi} (1-\cos t)^2\,dt\end{aligned}$$

for the area. Now we use the half-angle identity

$$1 - \cos t = 2\sin^2\left(\frac{t}{2}\right)$$

and a consequence of Problem 58 in Section 8.3:

$$\int_{0}^{\pi} \sin^{2n} u\,du = \pi\cdot\frac{1}{2}\cdot\frac{3}{4}\cdot\frac{5}{6}\cdots\frac{2n-1}{2n}.$$

We thereby get

$$A = 4a^2 \int_0^{2\pi} \sin^4 \left(\frac{t}{2}\right) dt = 8a^2 \int_0^{\pi} \sin^4 u \, du \qquad \left(u = \frac{t}{2}\right)$$

$$= 8a^2 \cdot \pi \cdot \frac{1}{2} \cdot \frac{3}{4} = 3\pi a^2$$

for the area under one arch of the cycloid. The arc-length differential is

$$ds = \sqrt{a^2(1 - \cos t)^2 + (a \sin t)^2}\, dt$$

$$= a\sqrt{2(1 - \cos t)}\, dt = 2a \sin \left(\frac{t}{2}\right) dt,$$

so Eq. (3) gives

$$s = \int_0^{2\pi} 2a \sin \frac{t}{2} \, dt = \left[-4a \cos \frac{t}{2}\right]_0^{2\pi} = 8a$$

for the length of one arch of the cycloid. ◆

Parametric Polar Coordinates

Suppose that a parametric curve is determined by giving its polar coordinates

$$r = r(t), \qquad \theta = \theta(t), \qquad \alpha \leqq t \leqq \beta$$

as functions of the parameter t. Then this curve is described in rectangular coordinates by the parametric equations

$$x(t) = r(t) \cos \theta(t), \qquad y(t) = r(t) \sin \theta(t), \quad \alpha \leqq t \leqq \beta,$$

giving x and y as functions of t. The latter parametric equations may then be used in the integral formulas in Eqs. (1) through (4).

To compute ds, we first calculate the derivatives

$$\frac{dx}{dt} = (\cos \theta)\frac{dr}{dt} - (r \sin \theta)\frac{d\theta}{dt}, \qquad \frac{dy}{dt} = (\sin \theta)\frac{dr}{dt} + (r \cos \theta)\frac{d\theta}{dt}.$$

Upon substituting these expressions for dx/dt and dy/dt in Eq. (8) and making algebraic simplifications, we find that the arc-length differential in parametric polar coordinates is

$$ds = \sqrt{\left(\frac{dr}{dt}\right)^2 + \left(r\frac{d\theta}{dt}\right)^2}\, dt. \tag{9}$$

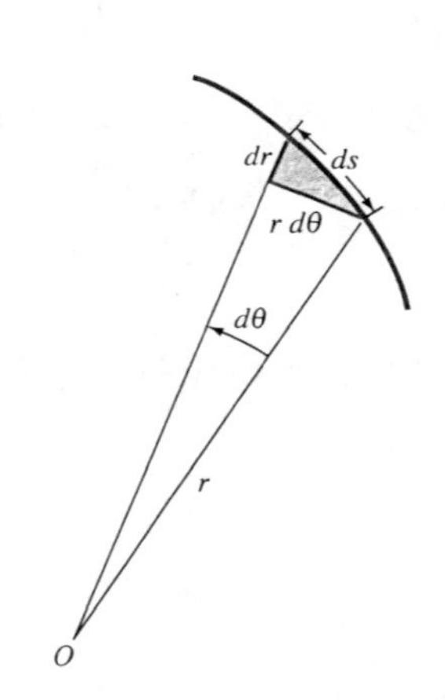

FIGURE 10.5.5 The differential triangle in polar coordinates.

In the case of a curve with the explicit polar-coordinate equation $r = f(\theta)$, we may use θ itself as the parameter. Then Eq. (9) takes the simpler form

$$ds = \sqrt{\left(\frac{dr}{d\theta}\right)^2 + r^2}\, d\theta. \tag{10}$$

The formula $ds = \sqrt{(dr)^2 + (r\, d\theta)^2}$, equivalent to Eq. (9), is easy to remember with the aid of the tiny "almost-triangle" shown in Fig. 10.5.5.

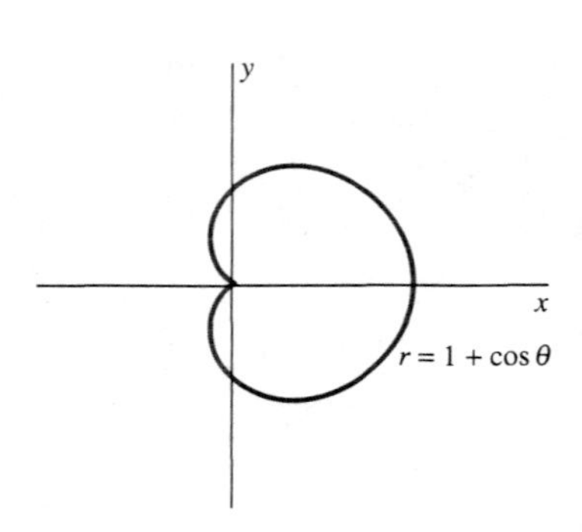

FIGURE 10.5.6 The cardioid of Example 3.

EXAMPLE 3 Find the perimeter (arc length) s of the cardioid with polar equation $r = 1 + \cos \theta$ (Fig. 10.5.6.). Find also the surface area S generated by revolving the cardioid around the x-axis.

Solution Because $dr/d\theta = -\sin \theta$, Eq. (10) and the identity

$$1 + \cos \theta = 2 \cos^2 \left(\frac{\theta}{2}\right) \tag{11}$$

give

$$ds = \sqrt{(-\sin\theta)^2 + (1+\cos\theta)^2}\,d\theta = \sqrt{2(1+\cos\theta)}\,d\theta$$

$$= \sqrt{4\cos^2\left(\frac{\theta}{2}\right)}\,d\theta = \left|2\cos\left(\frac{\theta}{2}\right)\right|\,d\theta.$$

Hence $ds = 2\cos(\theta/2)\,d\theta$ on the upper half of the cardioid, where $0 \leqq \theta \leqq \pi$, and thus $\cos(\theta/2) \geqq 0$. Therefore

$$s = 2\int_0^{\pi} 2\cos\frac{\theta}{2}\,d\theta = 8\left[\sin\frac{\theta}{2}\right]_0^{\pi} = 8.$$

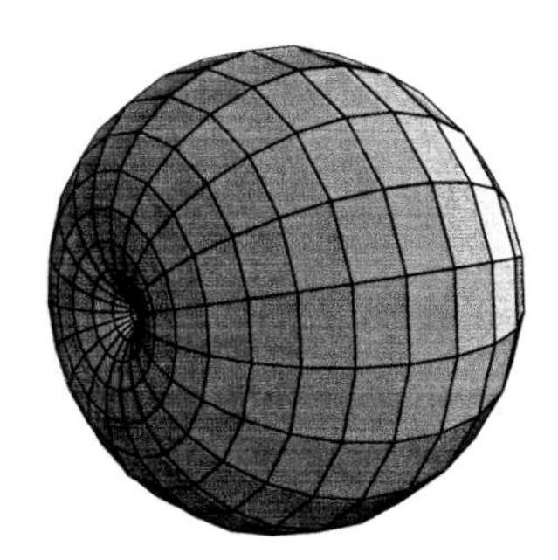

FIGURE 10.5.7 The surface generated by rotating the cardioid around the x-axis.

The surface area of revolution around the x-axis (Fig. 10.5.7) is given by

$$S = \int_{\theta=0}^{\pi} 2\pi y\,ds$$

$$= \int_{\theta=0}^{\pi} 2\pi(r\sin\theta)\,ds = \int_0^{\pi} 2\pi(1+\cos\theta)(\sin\theta)\cdot 2\cos\left(\frac{\theta}{2}\right)d\theta$$

$$= 16\pi\int_0^{\pi}\cos^4\frac{\theta}{2}\sin\frac{\theta}{2}\,d\theta = 16\pi\left[-\frac{2}{5}\cos^5\frac{\theta}{2}\right]_0^{\pi} = \frac{32\pi}{5},$$

using the identity

$$\sin\theta = 2\sin\left(\frac{\theta}{2}\right)\cos\left(\frac{\theta}{2}\right)$$

as well as the identity in Eq. (11). ◆

10.5 TRUE/FALSE STUDY GUIDE

10.5 CONCEPTS: QUESTIONS AND DISCUSSION

1. If the circle of radius a is parametrized by $x = a\cos t$, $y = a\sin t$ as in Example 1, explain carefully why the integral

$$\int_{t=0}^{2\pi} y\,dx = \int_{t=0}^{\pi} y\,dx + \int_{t=\pi}^{2\pi} y\,dx$$

does *not* give the correct area of the circle. Relate the two integrals on the right to the upper and lower halves of the circle.

2. If the circle of radius a is parametrized by $x = a\sin\pi t$, $y = a\cos\pi t$, explain carefully why the integral

$$\int_{t=0}^{2} y\,dx = \int_{t=0}^{1/2} y\,dx + \int_{t=1/2}^{3/2} y\,dx + \int_{t=3/2}^{2} y\,dx$$

does give the correct area. Relate the three integrals on the right to appropriate parts of the circular area.

10.5 PROBLEMS

In Problems 1 through 6, find the area of the region that lies between the given parametric curve and the x-axis.

1. $x = t^3$, $y = 2t^2 + 1$; $-1 \leqq t \leqq 1$

2. $x = e^{3t}$, $y = e^{-t}$; $0 \leqq t \leqq \ln 2$

3. $x = \cos t$, $y = \sin^2 t$; $0 \leqq t \leqq \pi$

4. $x = 2 - 3t$, $y = e^{2t}$; $0 \leqq t \leqq 1$

5. $x = \cos t$, $y = e^t$; $0 \leqq t \leqq \pi$

6. $x = 1 - e^t$, $y = 2t + 1$; $0 \leqq t \leqq 1$

In Problems 7 through 10, find the volume obtained by revolving around the x-axis the region described in the given problem.

7. Problem 1

8. Problem 2

9. Problem 3

10. Problem 5

In Problems 11 through 16, find the arc length of the given curve.

11. $x = 2t$, $y = \frac{2}{3}t^{3/2}$; $5 \leqq t \leqq 12$

12. $x = \frac{1}{2}t^2$, $y = \frac{1}{3}t^3$; $0 \leqq t \leqq 1$

13. $x = \sin t - \cos t,\ y = \sin t + \cos t;\quad \frac{1}{4}\pi \leqq t \leqq \frac{1}{2}\pi$

14. $x = e^t \sin t,\ y = e^t \cos t;\quad 0 \leqq t \leqq \pi$

15. $r = e^{\theta/2};\quad 0 \leqq \theta \leqq 4\pi$

16. $r = \theta;\quad 2\pi \leqq \theta \leqq 4\pi$

In Problems 17 through 22, find the area of the surface of revolution generated by revolving the given curve around the indicated axis.

17. $x = 1 - t,\ y = 2\sqrt{t},\quad 1 \leqq t \leqq 4;\quad$ the x-axis

18. $x = 2t^2 + t^{-1},\ y = 8\sqrt{t},\quad 1 \leqq t \leqq 2;\quad$ the x-axis

19. $x = t^3,\ y = 2t + 3,\quad -1 \leqq t \leqq 1;\quad$ the y-axis

20. $x = 2t + 1,\ y = t^2 + t,\quad 0 \leqq t \leqq 3;\quad$ the y-axis

21. $r = 4\sin\theta,\quad 0 \leqq \theta \leqq \pi;\quad$ the x-axis

22. $r = e^\theta,\quad 0 \leqq \theta \leqq \frac{1}{2}\pi;\quad$ the y-axis

23. Find the volume generated by revolving around the x-axis the region under the cycloidal arch of Example 2.

24. Find the area of the surface generated by revolving around the x-axis the cycloidal arch of Example 2.

25. Use the parametrization $x = a\cos t,\ y = b\sin t$ to find: (a) the area bounded by the ellipse $x^2/a^2 + y^2/b^2 = 1$; (b) the volume of the ellipsoid generated by revolving this ellipse around the x-axis.

26. Find the area bounded by the loop of the parametric curve $x = t^2,\ y = t^3 - 3t$ of Problem 25 in Section 10.4.

27. Use the parametrization $x = t\cos t,\ y = t\sin t$ of the Archimedean spiral to find the arc length of the first full turn of this spiral (corresponding to $0 \leqq t \leqq 2\pi$).

28. The circle $(x - b)^2 + y^2 = a^2$ with radius $a < b$ and center $(b, 0)$ can be parametrized by

$$x = b + a\cos t,\quad y = a\sin t,\quad 0 \leqq t \leqq 2\pi.$$

Find the surface area of the torus obtained by revolving this circle around the y-axis (Fig. 10.5.8).

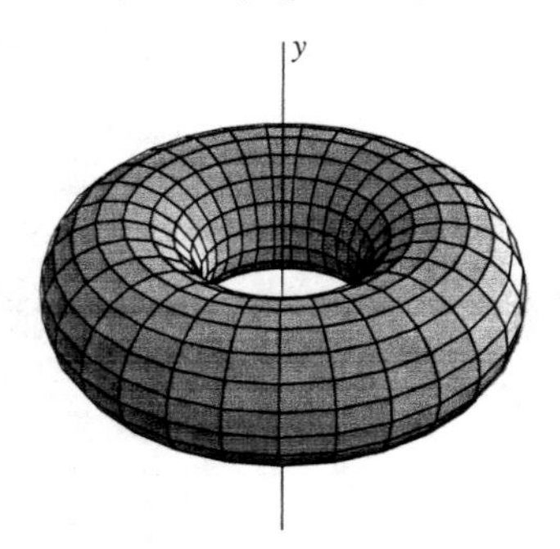

FIGURE 10.5.8 The torus of Problem 28.

29. The *astroid* (four-cusped hypocycloid) has equation $x^{2/3} + y^{2/3} = a^{2/3}$ (Fig. 10.4.15) and the parametrization

$$x = a\cos^3 t,\quad y = a\sin^3 t,\quad 0 \leqq t \leqq 2\pi.$$

Find the area of the region bounded by the astroid.

30. Find the total length of the astroid of Problem 29.

31. Find the area of the surface obtained by revolving the astroid of Problem 29 around the x-axis.

32. Find the area of the surface generated by revolving the lemniscate $r^2 = 2a^2\cos 2\theta$ around the y-axis (Fig. 10.5.9). [*Suggestion:* Use Eq. (10); note that $r\,dr = -2a^2 \sin 2\theta\, d\theta$.]

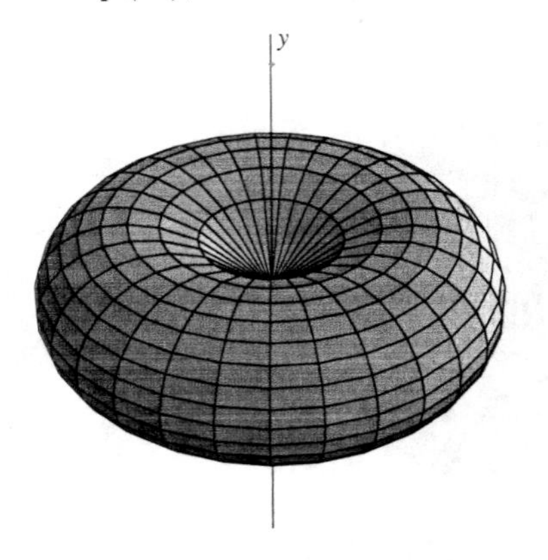

FIGURE 10.5.9 The surface generated by rotating the lemniscate of Problem 32 around the y-axis.

33. Figure 10.5.10 shows the graph of the parametric curve

$$x = t^2\sqrt{3},\quad y = 3t - \tfrac{1}{3}t^3.$$

The shaded region is bounded by the part of the curve for which $-3 \leqq t \leqq 3$. Find its area.

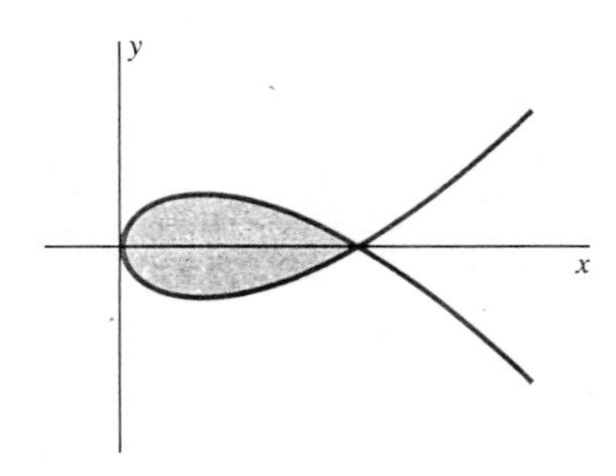

FIGURE 10.5.10 The parametric curve of Problems 33 through 36.

34. Find the arc length of the loop of the curve of Problem 33.

35. Find the volume of the solid obtained by revolving around the x-axis the shaded region in Fig. 10.5.10.

36. Find the surface area of revolution generated by revolving around the x-axis the loop of Fig. 10.5.10.

37. (a) With reference to Problem 30 and Fig. 10.4.14 in Section 10.4, show that the arc length of the first-quadrant loop of the folium of Descartes is

$$s = 6\int_0^1 \frac{\sqrt{1 + 4t^2 - 4t^3 - 4t^5 + 4t^6 + t^8}}{(1 + t^3)^2}\,dt.$$

(b) Use a programmable calculator or a computer to approximate this length.

38. Find the surface area generated by rotating around the y-axis the cycloidal arch of Example 2. [*Suggestion:* $\sqrt{x^2} = x$ only if $x \geqq 0$.]

39. Find the volume generated by rotating around the y-axis the region under the cycloidal arch of Example 2.

40. Suppose that after a string is wound clockwise around a circle of radius a, its free end is at the point $A(a, 0)$. (See Fig. 10.5.11.) Now the string is unwound, always stretched tight so the unwound portion TP is tangent to the circle at T. The locus of the string's free endpoint P is called the **involute** of the circle.

(a) Show that the parametric equations of the involute (in terms of the angle t of Fig. 10.5.11) are

$$x = a(\cos t + t \sin t), \quad y = a(\sin t - t \cos t).$$

(b) Find the length of the involute from $t = 0$ to $t = \pi$.

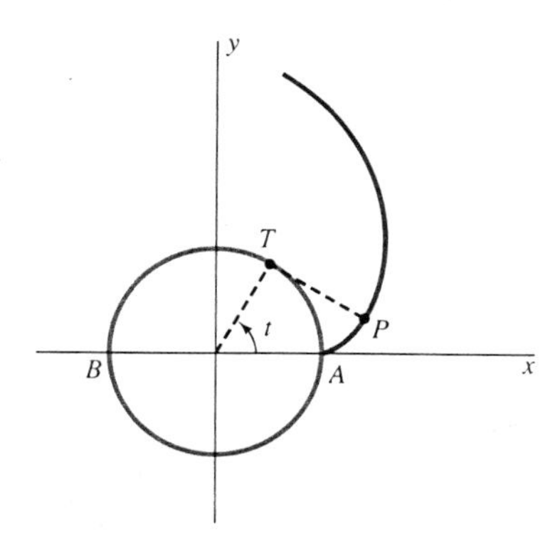

FIGURE 10.5.11 The involute of a circle.

41. Suppose that the circle of Problem 40 is a water tank and the "string" is a rope of length πa. It is anchored at the point B opposite A. Figure 10.5.12 depicts the total area that can be grazed by a cow tied to the free end of the rope. Find this total area. (The three labeled arcs of the curve in the figure represent, respectively, an involute APQ generated as the cow unwinds the rope in the counterclockwise direction, a semicircle QR of radius πa centered at B, and an involute RSA generated as the cow winds the rope around the tank proceeding in the counterclockwise direction from B to A. These three arcs form a closed curve that resembles a cardioid, and the cow can reach every point that lies inside this curve and outside the original circle.)

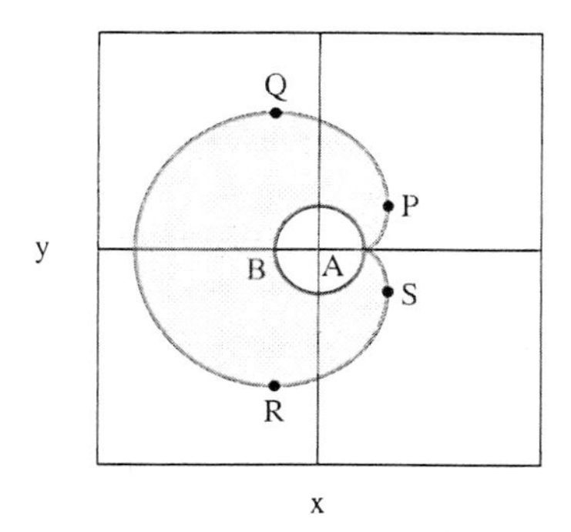

FIGURE 10.5.12 The area that the cow of Problem 41 can graze.

42. Now suppose that the rope of the previous problem has length $2\pi a$ and is anchored at the point A before being wound completely around the tank. Now find the total area that the cow can graze. Figure 10.5.13 shows an involute APQ, a semicircle QR of radius $2\pi a$ centered at A, and an involute RSA. The cow can reach every point that lies inside the outer curve and outside the original circle.

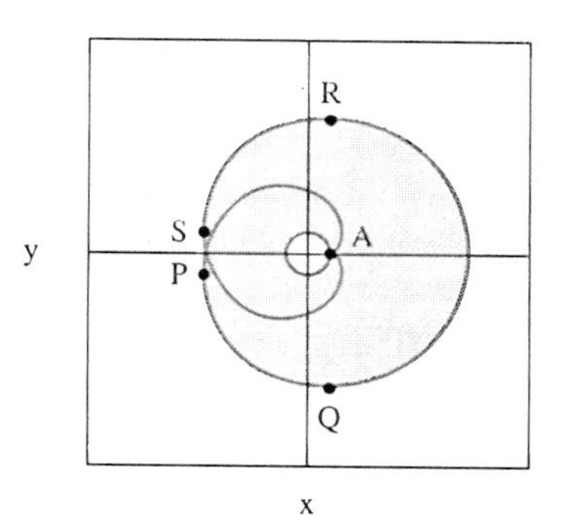

FIGURE 10.5.13 The area that the cow of Problem 42 can graze.

In Problems 43 through 54, use a graphing calculator or computer algebra system as appropriate. Approximate (by integrating numerically) the desired quantity if it cannot be calculated exactly.

43. Find the total arc length of the 3-leaved rose $r = 3 \sin 3\theta$ of Fig. 10.3.12.

44. Find the total surface area generated by rotating around the y-axis the 3-leaved rose of Problem 43.

45. Find the total length of the 4-leaved rose $r = 2 \cos 2\theta$ of Fig. 10.2.12.

46. Find the total surface area generated by revolving around the x-axis the 4-leaved rose of Problem 45.

47. Find the total arc length of the limaçon (both loops) $r = 5 + 9 \cos \theta$ of Fig. 10.2.25.

48. Find the total surface area generated by revolving around the x-axis the limaçon of Problem 47.

49. Find the total arc length (all seven loops) of the polar curve $r = \cos(\frac{7}{3}\theta)$ of Fig. 10.5.14.

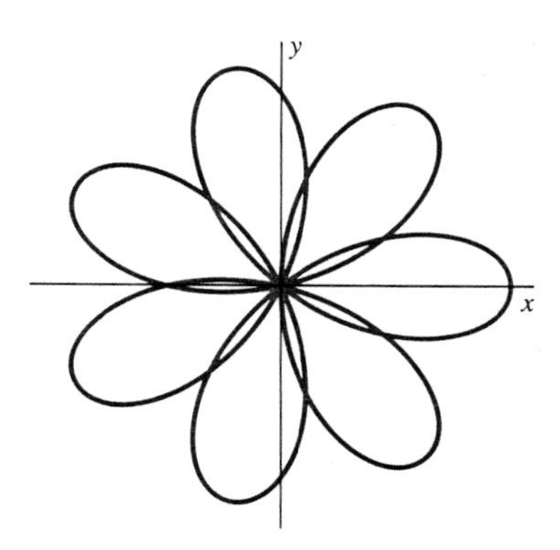

FIGURE 10.5.14 The curve $r = \cos(\frac{7}{3}\theta)$ of Problem 49.

50. Find the total arc length of the figure-8 curve $x = \sin t$, $y = \sin 2t$ of Fig. 10.4.13.

51. Find the total surface area and volume generated by revolving around the x-axis the figure-8 curve of Problem 50.

52. Find the total surface area and volume generated by revolving around the y-axis the figure-8 curve of Problem 50.

53. Find the total arc length of the Lissajous curve $x = \cos 3t$, $y = \sin 5t$ of Fig. 10.4.4.

54. Find the total arc length of the epitrochoid $x = 8\cos t - 5\cos 4t$, $y = 8\sin t - 5\sin 4t$ of Fig. 10.5.15.

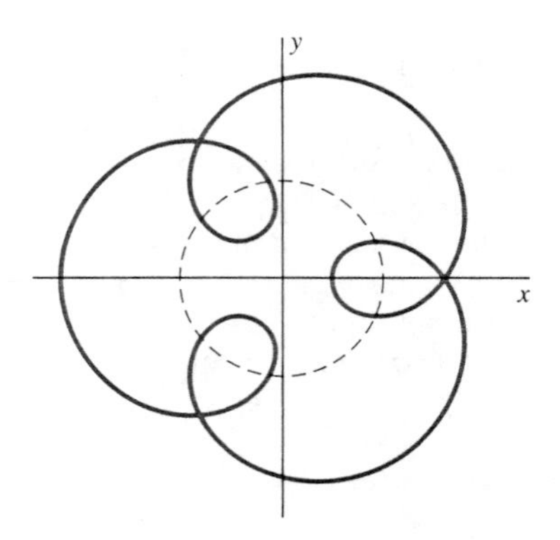

FIGURE 10.5.15 The epitrochoid of Problem 54.

55. Frank A. Farris of Santa Clara University, while designing a computer laboratory exercise for his calculus students, discovered an extremely lovely curve with the parametrization

$$x(t) = \cos t + \tfrac{1}{2}\cos 7t + \tfrac{1}{3}\sin 17t,$$

$$y(t) = \sin t + \tfrac{1}{2}\sin 7t + \tfrac{1}{3}\cos 17t.$$

For information on what these equations represent, see his article "Wheels on Wheels on Wheels—Surprising Symmetry" in the June 1996 issue of *Mathematics Magazine.* Plot these equations so you can enjoy this extraordinary figure, then numerically integrate to approximate the length of its graph. What kind of symmetry does the graph have? Is this predictable from the coefficients of t in the parametric equations?

10.5 Project: Moon Orbits and Race Tracks

The investigations in this project call for the use of numerical integration techniques (using a calculator or computer) to approximate the parametric arc-length integral

$$s = \int_a^b \sqrt{[x'(t)]^2 + [y'(t)]^2}\, dt. \tag{1}$$

Consider the ellipse with equation

$$\frac{x^2}{a^2} + \frac{y^2}{b^2} = 1 \quad (a > b) \tag{2}$$

and *eccentricity* $\epsilon = \sqrt{1 - (b/a)^2}$. Substitute the parametrization

$$x = a\cos t, \quad y = b\sin t \tag{3}$$

into Eq. (1) to show that the perimeter of the ellipse is given by the *elliptic integral*

$$p = 4a\int_0^{\pi/2} \sqrt{1 - \epsilon^2\cos^2 t}\, dt. \tag{4}$$

This integral is known to be nonelementary if $0 < \epsilon < 1$. A common simple approximation to it is

$$p \approx \pi(A + R), \tag{5}$$

where

$$A = \frac{1}{2}(a + b) \quad \text{and} \quad R = \sqrt{\frac{a^2 + b^2}{2}}$$

denote the arithmetic mean and root-square mean, respectively, of the semiaxes a and b of the ellipse.

Investigation A As a warm-up, consider the ellipse whose major and minor semiaxes a and b are, respectively, the largest and smallest nonzero digits of your student I.D. number. For this ellipse, compare the arc-length estimate given by (5) and by numerical evaluation of the integral in (4).

Investigation B If we ignore the perturbing effects of the sun and the planets other than the earth, the orbit of the moon is an almost perfect ellipse with the earth at one focus. Assume that this ellipse has major semiaxis $a = 384{,}403$ km (exactly) and eccentricity $\epsilon = 0.0549$ (exactly). Approximate the perimeter p of this ellipse [using Eq. (4)] to the nearest meter.

Investigation C Suppose that you are designing an elliptical auto racetrack. Choose semiaxes for *your* racetrack so that its perimeter will be somewhere between a half mile and two miles. Your task is to construct a table with *time* and *speed* columns that an observer can use to determine the average speed of a particular car as it circles the track. The times listed in the first column should correspond to speeds up to perhaps 150 mi/h. The observer clocks a car's circuit of the track and locates its time for the lap in the first column of the table. The corresponding figure in the second column then gives the car's average speed (in miles per hour) for that circuit of the track. Your report should include a convenient table to use in this way—so you can successfully sell it to racetrack patrons attending the auto races.

10.6 CONIC SECTIONS AND APPLICATIONS

Here we discuss in more detail the three types of conic sections—parabolas, ellipses, and hyperbolas—that were introduced in Section 10.1.

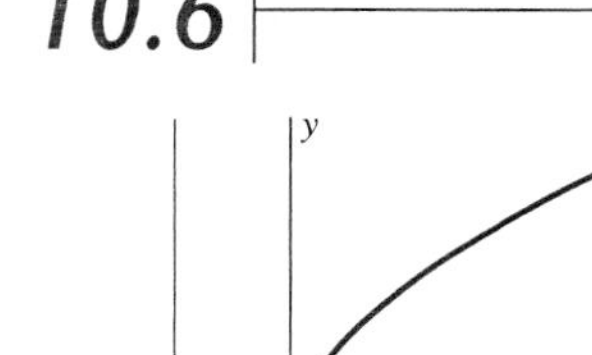

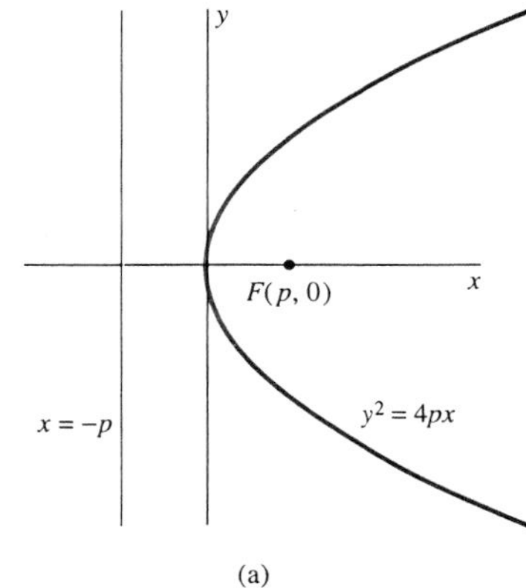

(a)

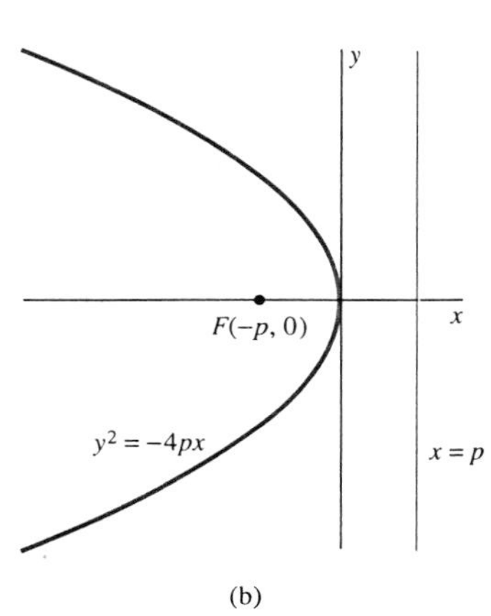

(b)

FIGURE 10.6.1 Two parabolas with vertical directrices.

The Parabola

The case $e = 1$ of Example 3 in Section 10.1 is motivation for this formal definition.

DEFINITION The Parabola
A **parabola** is the set of all points P in the plane that are equidistant from a fixed point F (called the **focus** of the parabola) and a fixed line L (called the parabola's **directrix**) not containing F.

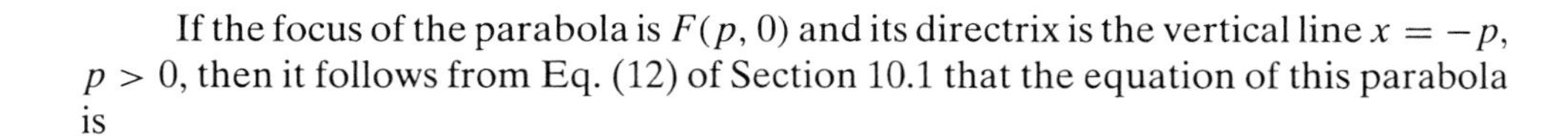

If the focus of the parabola is $F(p, 0)$ and its directrix is the vertical line $x = -p$, $p > 0$, then it follows from Eq. (12) of Section 10.1 that the equation of this parabola is

$$y^2 = 4px. \tag{1}$$

When we replace x with $-x$ both in the equation and in the discussion that precedes it, we get the equation of the parabola whose focus is $(-p, 0)$ and whose directrix is the vertical line $x = p$. The new parabola has equation

$$y^2 = -4px. \tag{2}$$

The old and new parabolas appear in Fig. 10.6.1.

We could also interchange x and y in Eq. (1). This would give the equation of a parabola whose focus is $(0, p)$ and whose directrix is the horizontal line $y = -p$.

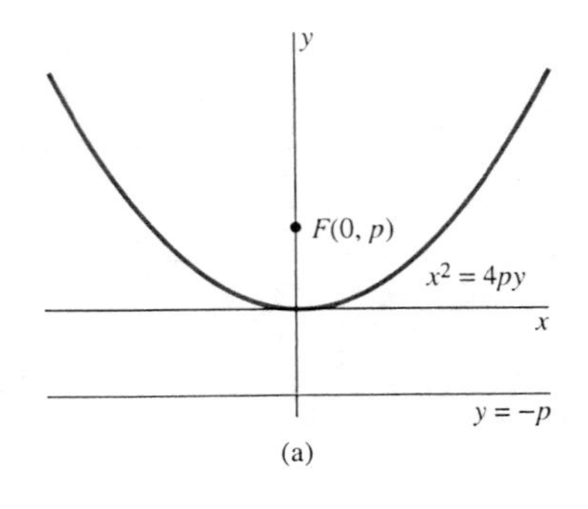

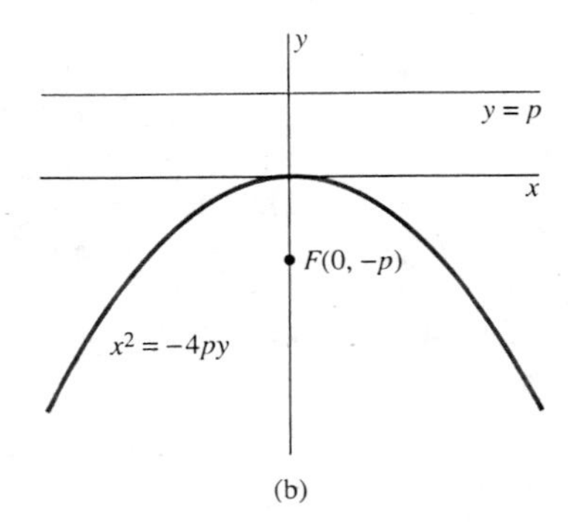

FIGURE 10.6.2 Two parabolas with horizontal directrices: (a) opening upward; (b) opening downward.

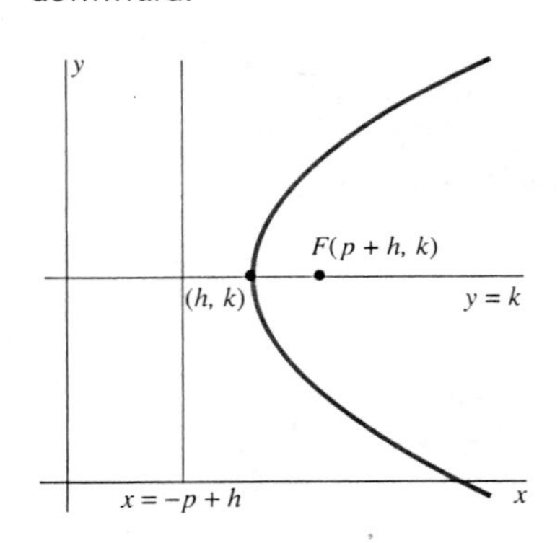

FIGURE 10.6.3 A translation of the parabola $y^2 = 4px$.

This parabola opens upward, as in Fig. 10.6.2(a); its equation is

$$x^2 = 4py. \tag{3}$$

Finally, we replace y with $-y$ in Eq. (3). This gives the equation

$$x^2 = -4py \tag{4}$$

of a parabola opening downward, with focus $(0, -p)$ and with directrix $y = p$, as in Fig. 10.6.2(b).

Each of the parabolas discussed so far is symmetric around one of the coordinate axes. The line around which a parabola is symmetric is called the **axis** of the parabola. The point of a parabola midway between its focus and its directrix is called the **vertex** of the parabola. The vertex of each parabola that we discussed in connection with Eqs. (1) through (4) is the origin $(0, 0)$.

EXAMPLE 1 Determine the focus, directrix, axis, and vertex of the parabola $x^2 = 12y$.

Solution We write the given equation as $x^2 = 4 \cdot (3y)$. In this form it matches Eq. (3) with $p = 3$. Hence the focus of the given parabola is $(0, 3)$ and its directrix is the horizontal line $y = -3$. The y-axis is its axis of symmetry, and the parabola opens upward from its vertex at the origin. ◆

Suppose that we begin with the parabola of Eq. (1) and translate it in such a way that its vertex moves to the point (h, k). Then the translated parabola has equation

$$(y - k)^2 = 4p(x - h). \tag{1a}$$

The new parabola has focus $F(p + h, k)$ and its directrix is the vertical line $x = -p + h$ (Fig. 10.6.3). Its axis is the horizontal line $y = k$.

We can obtain the translates of the other three parabolas in Eqs. (2) through (4) in the same way. If the vertex is moved from the origin to the point (h, k), then the three equations take these forms:

$$(y - k)^2 = -4p(x - h), \tag{2a}$$

$$(x - h)^2 = 4p(y - k), \quad \text{and} \tag{3a}$$

$$(x - h)^2 = -4p(y - k). \tag{4a}$$

Equations (1a) and (2a) both take the general form

$$y^2 + Ax + By + C = 0 \quad (A \neq 0), \tag{5}$$

whereas Eqs. (3a) and (4a) both take the general form

$$x^2 + Ax + By + C = 0 \quad (B \neq 0). \tag{6}$$

What is significant about Eqs. (5) and (6) is what they have in common: Both are linear in one of the coordinate variables and quadratic in the other. In fact, we can reduce *any* such equation to one of the standard forms in Eqs. (1a) through (4a) by completing the square in the coordinate variable that appears quadratically. This means that the graph of any equation of the form of either Eqs. (5) or (6) is a parabola. The features of the parabola can be read from the standard form of its equation, as in Example 2.

EXAMPLE 2 Determine the graph of the equation

$$4y^2 - 8x - 12y + 1 = 0.$$

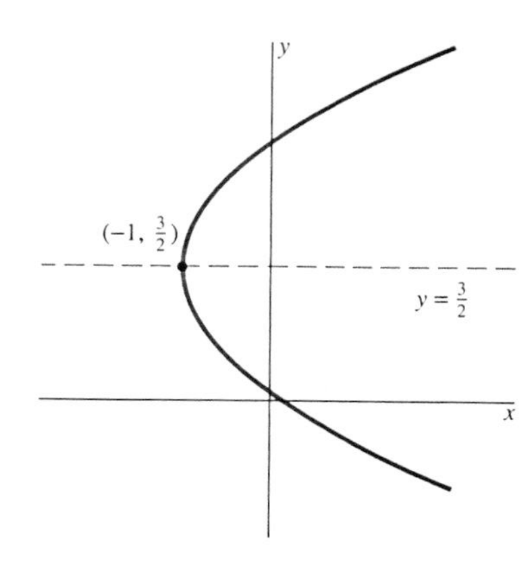

FIGURE 10.6.4 The parabola of Example 2.

Solution This equation is linear in x and quadratic in y. We divide through by the coefficient of y^2 and then collect on one side of the equation all terms that include y:

$$y^2 - 3y = 2x - \tfrac{1}{4}.$$

Then we complete the square in the variable y and thus find that

$$y^2 - 3y + \tfrac{9}{4} = 2x - \tfrac{1}{4} + \tfrac{9}{4} = 2x + 2 = 2(x + 1).$$

The final step is to write in the form $4p(x - h)$ the terms on the right-hand side that include x:

$$\left(y - \tfrac{3}{2}\right)^2 = 4 \cdot \tfrac{1}{2} \cdot (x + 1).$$

This equation has the form of Eq. (1a) with $p = \frac{1}{2}$, $h = -1$, and $k = \frac{3}{2}$. Thus the graph is a parabola that opens to the right from its vertex at $(-1, \frac{3}{2})$. Its focus is at $(-\frac{1}{2}, \frac{3}{2})$, its directrix is the vertical line $x = -\frac{3}{2}$, and its axis is the horizontal line $y = \frac{3}{2}$. It appears in Fig. 10.6.4. ◆

Applications of Parabolas

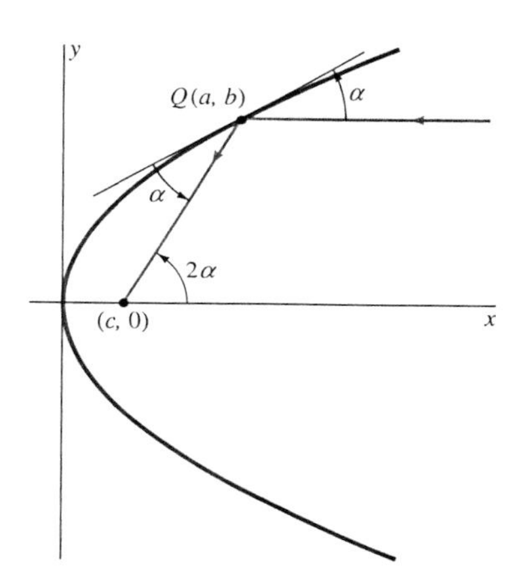

FIGURE 10.6.5 The reflection property of the parabola: $\alpha = \beta$.

The parabola $y^2 = 4px (p > 0)$ is shown in Fig. 10.6.5 along with an incoming ray of light traveling to the left and parallel to the x-axis. This light ray strikes the parabola at the point $Q(a, b)$ and is reflected toward the x-axis, which it meets at the point $(c, 0)$. The light ray's angle of reflection must equal its angle of incidence, which is why both of these angles—measured with respect to the tangent line L at Q—are labeled α in the figure. The angle vertical to the angle of incidence is also equal to α. Hence, because the incoming ray is parallel to the x-axis, the angle the reflected ray makes with the x-axis at $(c, 0)$ is 2α.

Using the points Q and $(c, 0)$ to compute the slope of the reflected light ray, we find that

$$\frac{b}{a - c} = \tan 2\alpha = \frac{2\tan\alpha}{1 - \tan^2\alpha}.$$

(The second equality follows from a trigonometric identity in Problem 64 of Section 7.5.) But the angle α is related to the slope of the tangent line L at Q. To find that slope, we begin with

$$y = 2\sqrt{px} = 2(px)^{1/2}$$

and compute

$$\frac{dy}{dx} = \left(\frac{p}{x}\right)^{1/2}.$$

Hence the slope of L is both $\tan\alpha$ and dy/dx evaluated at (a, b); that is,

$$\tan\alpha = \left(\frac{p}{a}\right)^{1/2}.$$

Therefore,

$$\frac{b}{a - c} = \frac{2\tan\alpha}{1 - \tan^2\alpha} = \frac{2\sqrt{\dfrac{p}{a}}}{1 - \dfrac{p}{a}} = \frac{2\sqrt{pa}}{a - p} = \frac{b}{a - p},$$

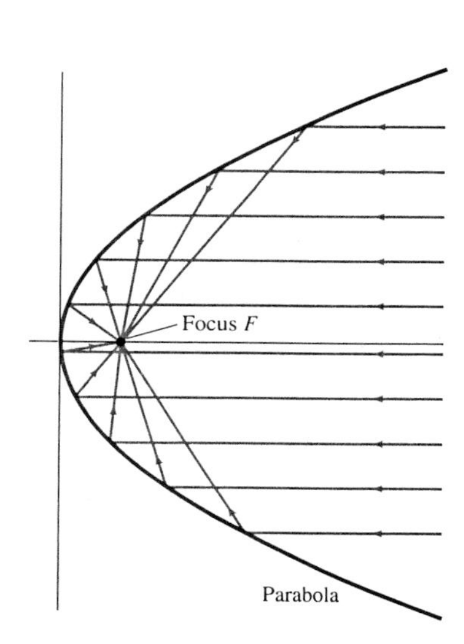

FIGURE 10.6.6 Incident rays parallel to the axis reflect through the focus.

because $b = 2\sqrt{pa}$. Hence $c = p$. The surprise is that c is independent of a and b and depends only on the equation $y^2 = 4px$ of the parabola. Therefore *all* incoming light rays parallel to the x-axis will be reflected to the single point $F(p, 0)$. This is why F is called the *focus* of the parabola.

This **reflection property** of the parabola is exploited in the design of parabolic mirrors. Such a mirror has the shape of the surface obtained by revolving a parabola around its axis of symmetry. Then a beam of incoming light rays parallel to the axis will be focused at F, as shown in Fig. 10.6.6. The reflection property can also be used in reverse—rays emitted at the focus are reflected in a beam parallel to the axis, thus keeping the light beam intense. Moreover, applications are not limited to light rays alone; parabolic mirrors are used in visual and radio telescopes, radar antennas, searchlights, automobile headlights, microphone systems, satellite ground stations, and solar heating devices.

Galileo discovered early in the seventeenth century that the trajectory of a projectile fired from a gun is a parabola (under the assumptions that air resistance can be ignored and that the gravitational acceleration remains constant). Suppose that a projectile is fired with initial velocity v_0 at time $t = 0$ from the origin and at an angle α of inclination from the horizontal x-axis. Then the initial velocity of the projectile splits into the components

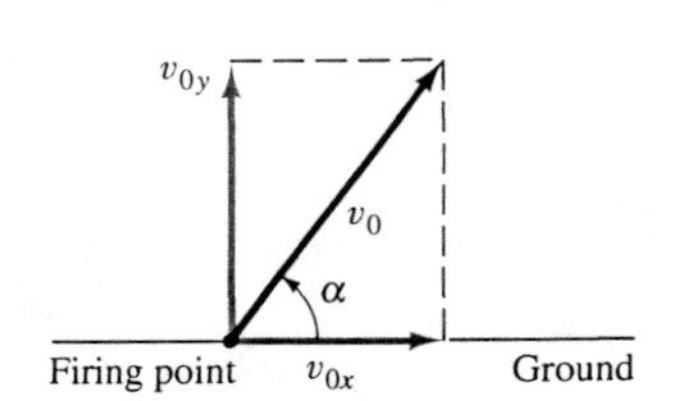

FIGURE 10.6.7 Resolution of the initial velocity v_0 into its horizontal and vertical components.

$$v_{0x} = v_0 \cos\alpha \quad \text{and} \quad v_{0y} = v_0 \sin\alpha,$$

as indicated in Fig. 10.6.7. The fact that the projectile continues to move horizontally with *constant* speed v_{0x}, together with Eq. (34) of Section 5.2, implies that its x- and y-coordinates after t seconds are

$$x = (v_0 \cos\alpha)t, \tag{7}$$

$$y = -\tfrac{1}{2}gt^2 + (v_0 \sin\alpha)t. \tag{8}$$

By substituting $t = x/(v_0 \cos\alpha)$ from Eq. (7) into Eq. (8) and then completing the square, we can derive (as in Problem 70) an equation of the form

$$y - M = -4p\left(x - \tfrac{1}{2}R\right)^2. \tag{9}$$

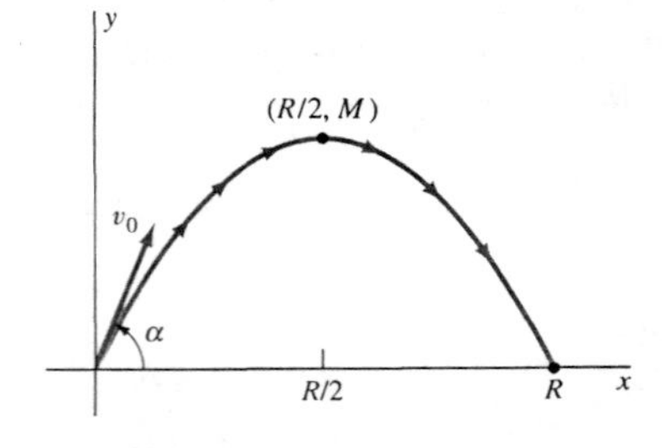

FIGURE 10.6.8 The trajectory of the projectile, showing its maximum altitude M and its range R.

Here,

$$M = \frac{v_0^2 \sin^2\alpha}{2g} \tag{10}$$

is the maximum height attained by the projectile, and

$$R = \frac{v_0^2 \sin 2\alpha}{g} \tag{11}$$

is its **range,** the horizontal distance the projectile will travel before it returns to the ground. Thus its trajectory is the parabola shown in Fig. 10.6.8.

The Ellipse

An ellipse is a conic section with eccentricity e less than 1, as in Example 3 of Section 10.1.

DEFINITION The Ellipse

Suppose that $e < 1$, and let F be a fixed point and L a fixed line not containing F. The **ellipse** with **eccentricity** e, **focus** F, and **directrix** L is the set of all points P such that the distance $|PF|$ is e times the (perpendicular) distance from P to the line L.

The equation of the ellipse is especially simple if F is the point $(c, 0)$ on the x-axis and L is the vertical line $x = c/e^2$. The case $c > 0$ is shown in Fig. 10.6.9. If Q is the point $(c/e^2, y)$, then PQ is the perpendicular from $P(x, y)$ to L. The condition $|PF| = e|PQ|$ then gives

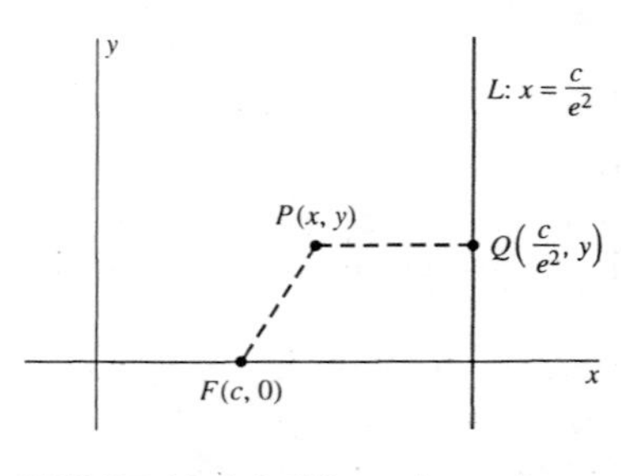

FIGURE 10.6.9 Ellipse: focus F, directrix L, eccentricity e.

$$(x - c)^2 + y^2 = e^2\left(x - \frac{c}{e^2}\right)^2;$$

$$x^2 - 2cx + c^2 + y^2 = e^2x^2 - 2cx + \frac{c^2}{e^2};$$

$$x^2(1 - e^2) + y^2 = c^2\left(\frac{1}{e^2} - 1\right) = \frac{c^2}{e^2}(1 - e^2).$$

Thus

$$x^2(1-e^2)+y^2=a^2(1-e^2),$$

where

$$a=\frac{c}{e}. \tag{12}$$

We divide both sides of the next-to-last equation by $a^2(1-e^2)$ and get

$$\frac{x^2}{a^2}+\frac{y^2}{a^2(1-e^2)}=1.$$

Finally, with the aid of the fact that $e<1$, we may let

$$b^2=a^2(1-e^2)=a^2-c^2. \tag{13}$$

Then the equation of the ellipse with focus $(c, 0)$ and directrix $x=c/e^2=a/e$ takes the simple form

$$\frac{x^2}{a^2}+\frac{y^2}{b^2}=1. \tag{14}$$

We see from Eq. (14) that this ellipse is symmetric around both coordinate axes. Its x-intercepts are $(\pm a, 0)$ and its y-intercepts are $(0, \pm b)$. The points $(\pm a, 0)$ are called the **vertices** of the ellipse, and the line segment joining them is called its **major axis.** The line segment joining $(0, b)$ and $(0, -b)$ is called the **minor axis** [note from Eq. (13) that $b<a$]. The alternative form

$$a^2=b^2+c^2 \tag{15}$$

of Eq. (13) is the Pythagorean relation for the right triangle of Fig. 10.6.10. Indeed, visualization of this triangle is an excellent way to remember Eq. (15). The numbers a and b are the lengths of the major and minor **semiaxes,** respectively.

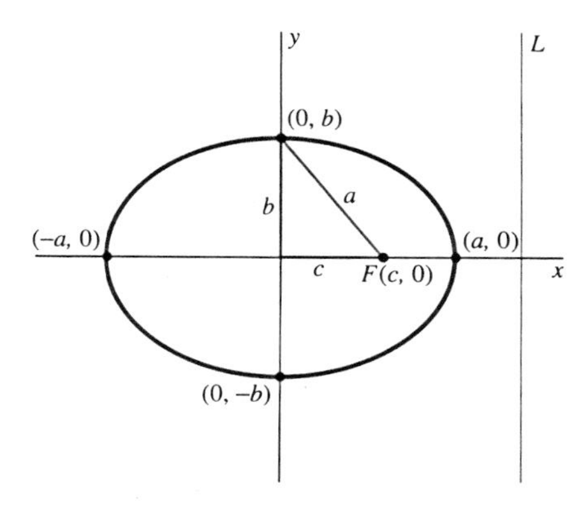

FIGURE 10.6.10 The parts of an ellipse.

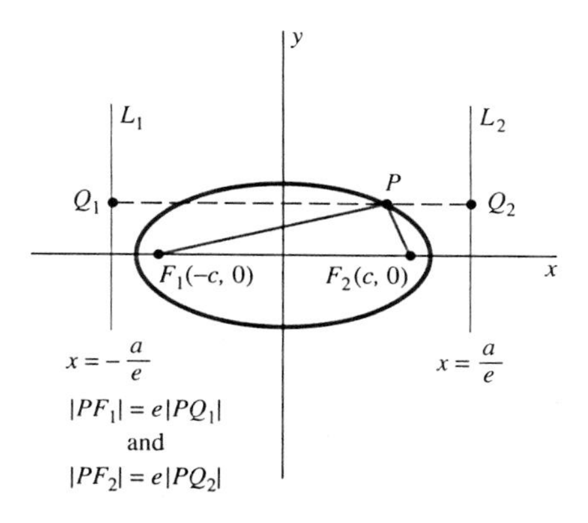

FIGURE 10.6.11 The ellipse as a conic section: two foci, two directrices.

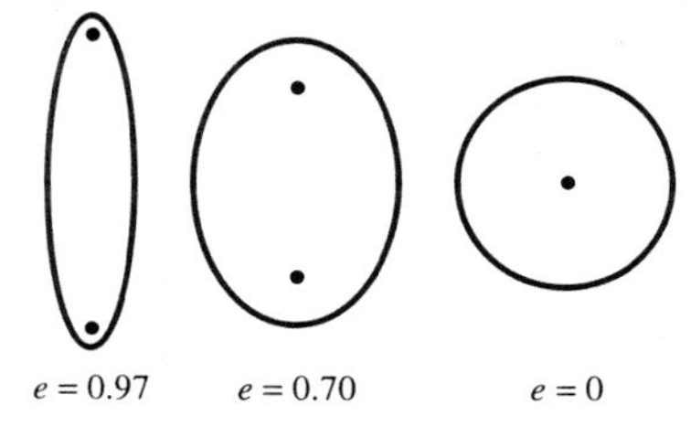

FIGURE 10.6.12 The relation between the eccentricity of an ellipse and its shape.

Because $a=c/e$, the directrix of the ellipse in Eq. (14) is $x=a/e$. If we had begun instead with the focus $(-c, 0)$ and directrix $x=-a/e$, we would still have obtained Eq. (14), because only the squares of a and c are involved in its derivation. Thus the ellipse in Eq. (14) has *two* foci, $(c, 0)$ and $(-c, 0)$, and *two* directrices, $x=a/e$ and $x=-a/e$ (Fig. 10.6.11).

The larger the eccentricity $e<1$, the more elongated the ellipse. (Remember that $e=1$ is the eccentricity of every parabola). But if $e=0$, then Eq. (13) gives $b=a$, so Eq. (14) reduces to the equation of a circle of radius a. Thus a circle is an ellipse of eccentricity zero. Compare the three cases shown in Fig. 10.6.12.

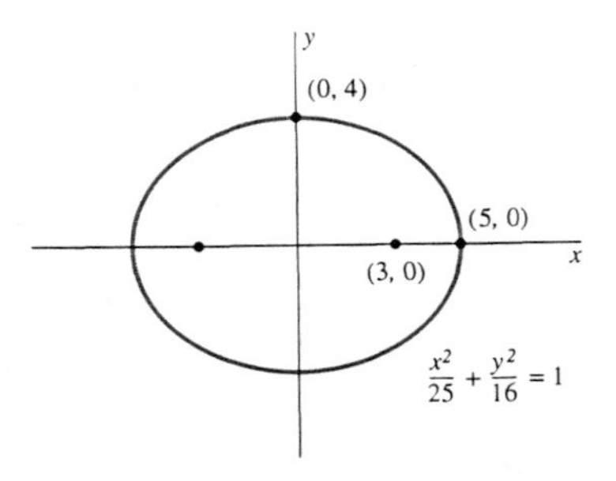

FIGURE 10.6.13 The ellipse of Example 3.

EXAMPLE 3 Find an equation of the ellipse with foci $(\pm 3, 0)$ and vertices $(\pm 5, 0)$.

Solution We are given $c = 3$ and $a = 5$, so Eq. (13) gives $b = 4$. Thus Eq. (14) gives

$$\frac{x^2}{25} + \frac{y^2}{16} = 1$$

for the desired equation. This ellipse is shown in Fig. 10.6.13. ◆

If the two foci of an ellipse are on the y-axis, such as $F_1(0, c)$ and $F_2(0, -c)$, then the equation of the ellipse is

$$\frac{x^2}{b^2} + \frac{y^2}{a^2} = 1, \tag{16}$$

and it is still true that $a^2 = b^2 + c^2$, as in Eq. (15). But now the major axis of length $2a$ is vertical and the minor axis of length $2b$ is horizontal. The derivation of Eq. (16) is similar to that of Eq. (14); see Problem 79. Figure 10.6.14 shows the case of an ellipse whose major axis is vertical. The vertices of such an ellipse are at $(0, \pm a)$; they are always the endpoints of the major axis.

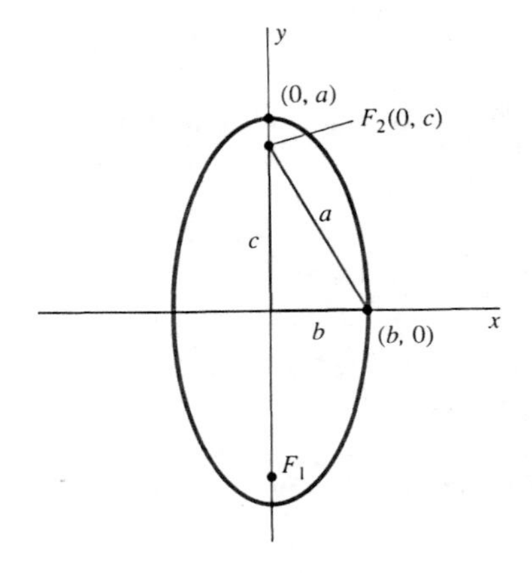

FIGURE 10.6.14 An ellipse with vertical major axis.

In practice there is little chance of confusing Eqs. (14) and (16). The equation or the given data will make it clear whether the major axis of the ellipse is horizontal or vertical. Just use the equation to read the ellipse's intercepts. The two intercepts that are farthest from the origin are the endpoints of the major axis; the other two are the endpoints of the minor axis. The two foci lie on the major axis, each at distance c from the center of the ellipse—which will be the origin if the equation of the ellipse has the form of either Eq. (14) or Eq. (16).

EXAMPLE 4 Sketch the graph of the equation

$$\frac{x^2}{16} + \frac{y^2}{25} = 1.$$

Solution The x-intercepts are $(\pm 4, 0)$; the y-intercepts are $(0, \pm 5)$. So the major axis is vertical. We take $a = 5$ and $b = 4$ in Eq. (15) and find that $c = 3$. The foci are thus at $(0, \pm 3)$. Hence this ellipse has the appearance of the one shown in Fig. 10.6.15. ◆

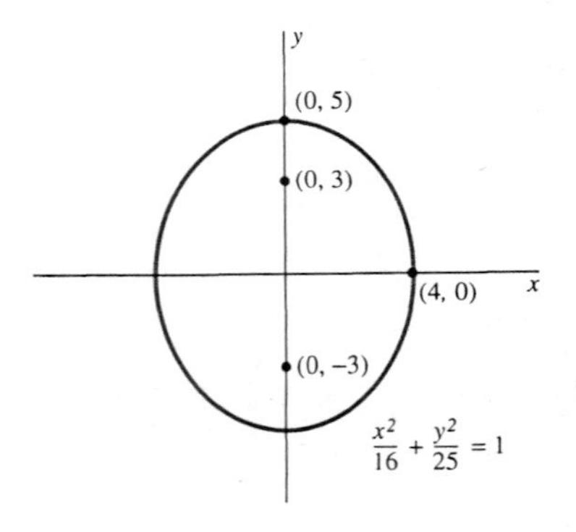

FIGURE 10.6.15 The ellipse of Example 4.

Any equation of the form

$$Ax^2 + Cy^2 + Dx + Ey + F = 0, \tag{17}$$

in which the coefficients A and C of the squared variables are *both nonzero* and *have the same sign,* may be reduced to the form

$$A(x-h)^2 + C(y-k)^2 = G$$

by completing the square in x and y. We may assume that A and C are both positive. Then if $G < 0$, there are no points that satisfy Eq. (17), and the graph is the empty set. If $G = 0$, then there is exactly one point on the locus—the single point (h, k). And if $G > 0$, we can divide both sides of the last equation by G and get an equation that resembles one of these two:

$$\frac{(x-h)^2}{a^2} + \frac{(y-k)^2}{b^2} = 1, \tag{18a}$$

$$\frac{(x-h)^2}{b^2} + \frac{(y-k)^2}{a^2} = 1. \tag{18b}$$

Which equation should you choose? Select the one that is consistent with the condition $a \geqq b > 0$. Finally, note that either of the equations in (18) is the equation of a translated ellipse. Thus, apart from the exceptional cases already noted, the graph of Eq. (17) is an ellipse if $AC > 0$.

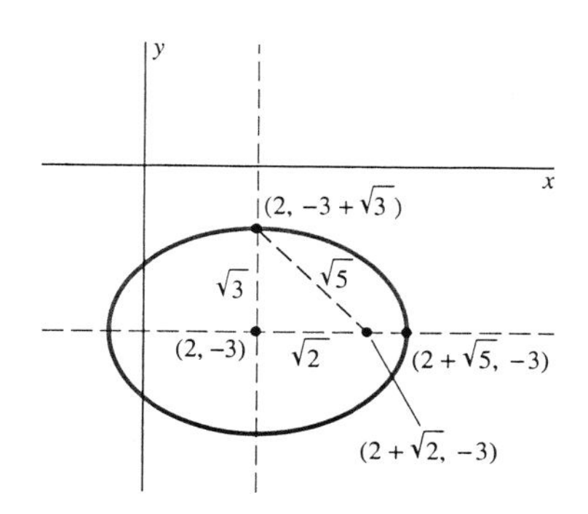

FIGURE 10.6.16 The ellipse of Example 5.

EXAMPLE 5 Determine the graph of the equation

$$3x^2 + 5y^2 - 12x + 30y + 42 = 0.$$

Solution We collect terms containing x, terms containing y, and complete the square in each variable. This gives

$$3(x^2 - 4x) + 5(y^2 + 6y) = -42;$$

$$3(x^2 - 4x + 4) + 5(y^2 + 69 + 9) = 15;$$

$$\frac{(x-2)^2}{5} + \frac{(y+3)^2}{3} = 1.$$

Thus the given equation is that of a translated ellipse with center at $(2, -3)$. Its horizontal major semiaxis has length $a = \sqrt{5}$ and its minor semiaxis has length $b = \sqrt{3}$ (Fig. 10.6.16). The distance from the center to each focus is $c = \sqrt{2}$ and the eccentricity is $e = c/a = \sqrt{2/5}$. ◆

Applications of Ellipses

EXAMPLE 6 The orbit of the earth is an ellipse with the sun at one focus. The planet's maximum distance from the center of the sun is 94.56 million miles and its minimum distance is 91.44 million miles. What are the major and minor semiaxes of the earth's orbit, and what is its eccentricity?

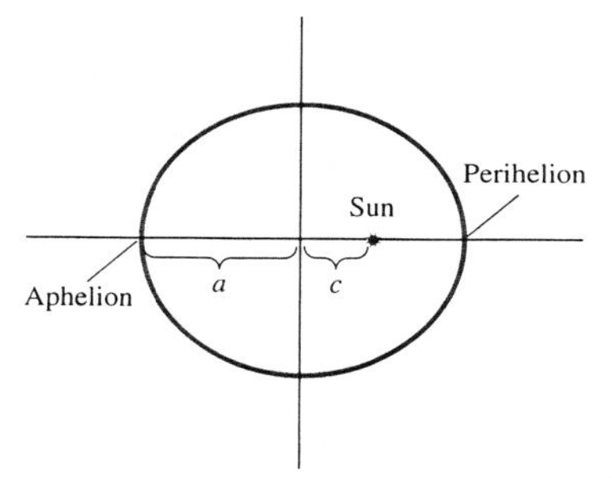

FIGURE 10.6.17 The orbit of the earth with its eccentricity exaggerated (Example 6).

Solution As Fig. 10.6.17 shows, we have

$$a + c = 94.56 \quad \text{and} \quad a - c = 91.44,$$

with units in millions of miles. We conclude from these equations that $a = 93.00$, that $c = 1.56$, and then that

$$b = \sqrt{(93.00)^2 - (1.56)^2} \approx 92.99$$

million miles. Finally,

$$e = \frac{c}{a} = \frac{1.56}{93.00} \approx 0.017,$$

a number relatively close to zero. This means that the earth's orbit is nearly circular. Indeed, the major and minor semiaxes are so nearly equal that, on any usual scale, the earth's orbit would appear to be a perfect circle. But the difference between uniform circular motion and the earth's actual motion has some important aspects, including the facts that the sun is 1.56 million miles off center and that the orbital speed of the earth is not constant. ◆

EXAMPLE 7 One of the most famous comets is Halley's comet, named for Edmund Halley (1656–1742), a disciple of Newton. By studying the records of the paths of earlier comets, Halley deduced that the comet of 1682 was the same one that had been sighted in 1607, in 1531, in 1456, and in 1066 (an omen at the Battle of Hastings). In 1682 Halley predicted that this comet would return in 1759, in 1835, and in 1910; he was correct each time. The period of Halley's comet is about 76 years—it can vary a couple of years in either direction because of perturbations of its orbit by the planet Jupiter. The orbit of Halley's comet is an ellipse with the sun at one focus. In terms of astronomical units (1 AU is the mean distance from the earth to the sun), the major and minor semiaxes of this elliptical orbit are 18.09 AU and 4.56 AU, respectively. What are the maximum and minimum distances from the sun of Halley's comet?

Solution We are given that $a = 18.09$ (all distance measurements are in astronomical units) and that $b = 4.56$, so

$$c = \sqrt{(18.09)^2 - (4.56)^2} \approx 17.51.$$

Hence the maximum distance of the comet from the sun is $a + c \approx 35.60$ AU, and its minimum distance is $a - c \approx 0.58$ AU. The eccentricity of its orbit is

$$e = \frac{c}{a} = \frac{17.51}{18.09} \approx 0.97,$$

a very eccentric orbit (but see Problem 77). ◆

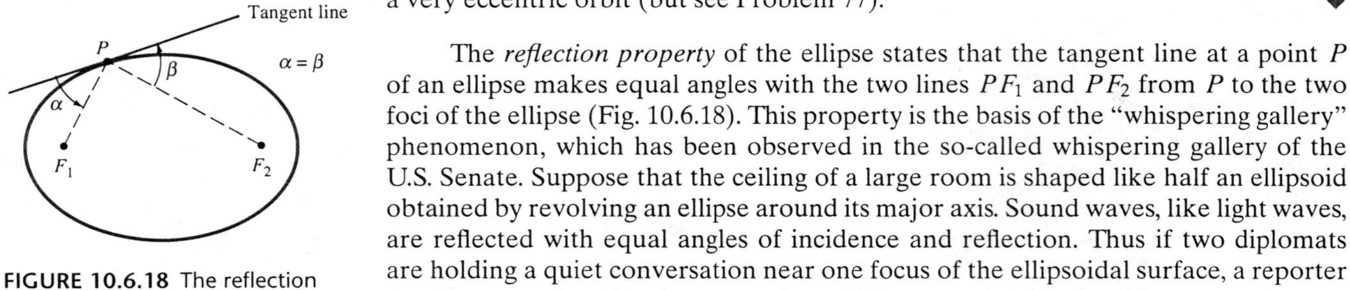

FIGURE 10.6.18 The reflection property: $\alpha = \beta$.

The *reflection property* of the ellipse states that the tangent line at a point P of an ellipse makes equal angles with the two lines PF_1 and PF_2 from P to the two foci of the ellipse (Fig. 10.6.18). This property is the basis of the "whispering gallery" phenomenon, which has been observed in the so-called whispering gallery of the U.S. Senate. Suppose that the ceiling of a large room is shaped like half an ellipsoid obtained by revolving an ellipse around its major axis. Sound waves, like light waves, are reflected with equal angles of incidence and reflection. Thus if two diplomats are holding a quiet conversation near one focus of the ellipsoidal surface, a reporter standing near the other focus—perhaps 50 feet away—might be able to eavesdrop on their conversation even if the conversation were inaudible to others in the same room.

Some billiard tables are manufactured in the shape of an ellipse. The foci of such tables are plainly marked for the convenience of enthusiasts of this unusual game.

A more serious application of the reflection property of ellipses is the nonsurgical kidney-stone treatment called *shockwave lithotripsy.* An ellipsoidal reflector with a transducer (an energy transmitter) at one focus is positioned outside the patient's body so that the offending kidney stone is located at the other focus. The stone then is pulverized by reflected shockwaves emanating from the transducer. (For further details, see the COMAP *Newsletter* **20**, November, 1986.)

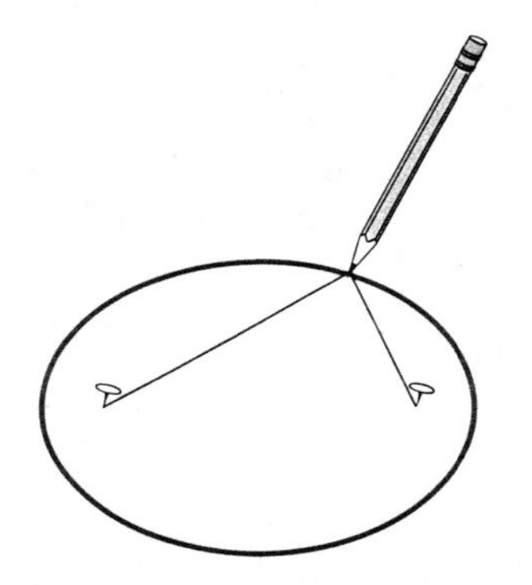
FIGURE 10.6.19 One way to draw an ellipse.

An alternative definition of the ellipse with foci F_1 and F_2 and major axis of length $2a$ is this: It is the locus of a point P such that the sum of the distances $|PF_1|$ and $|PF_2|$ is the constant $2a$. (See Problem 82.) This fact gives us a convenient way to draw the ellipse by using two tacks placed at F_1 and F_2, a string of length $2a$, and a pencil (Fig. 10.6.19).

The Hyperbola

A hyperbola is a conic section defined in the same way as is an ellipse, except that the eccentricity e of a hyperbola is greater than 1.

DEFINITION The Hyperbola

Suppose that $e > 1$, and let F be a fixed point and L a fixed line not containing F. Then the **hyperbola** with **eccentricity** e, **focus** F, and **directrix** L is the set of all points P such that the distance $|PF|$ is e times the (perpendicular) distance from P to the line L.

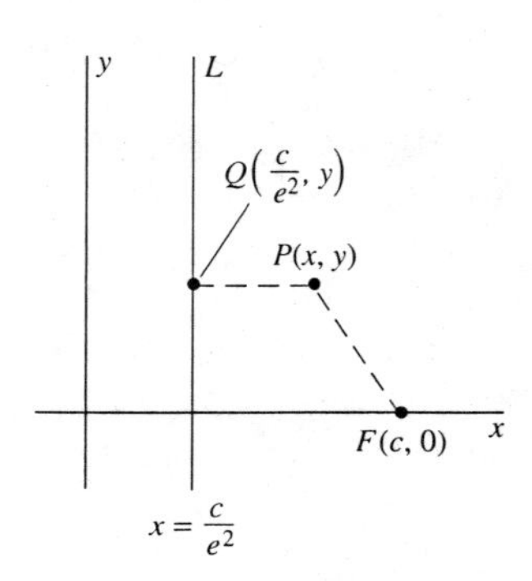

FIGURE 10.6.20 The definition of the hyperbola.

As with the ellipse, the equation of a hyperbola is simplest if F is the point $(c, 0)$ on the x-axis and L is the vertical line $x = c/e^2$. The case $c > 0$ is shown in Fig. 10.6.20. If Q is the point $(c/e^2, y)$, then PQ is the perpendicular from $P(x, y)$ to L. The condition $|PF| = e|PQ|$ gives

$$(x - c)^2 + y^2 = e^2\left(x - \frac{c}{e^2}\right)^2;$$

$$x^2 - 2cx + c^2 + y^2 = e^2x^2 - 2cx + \frac{c^2}{e^2};$$

$$(e^2 - 1)x^2 - y^2 = c^2\left(1 - \frac{1}{e^2}\right) = \frac{c^2}{e^2}(e^2 - 1).$$

Thus

$$(e^2 - 1)x^2 - y^2 = a^2(e^2 - 1),$$

where

$$a = \frac{c}{e}. \quad \textbf{(19)}$$

If we divide both sides of the next-to-last equation by $a^2(e^2 - 1)$, we get

$$\frac{x^2}{a^2} - \frac{y^2}{a^2(e^2 - 1)} = 1.$$

To simplify this equation, we let

$$b^2 = a^2(e^2 - 1) = c^2 - a^2. \quad \textbf{(20)}$$

This is permissible because $e > 1$. So the equation of the hyperbola with focus $(c, 0)$ and directrix $x = c/e^2 = a/e$ takes the form

$$\frac{x^2}{a^2} - \frac{y^2}{b^2} = 1. \quad \textbf{(21)}$$

The minus sign on the left-hand side is the only difference between the equation of a hyperbola and that of an ellipse. Of course, Eq. (20) also differs from the relation

$$b^2 = a^2(1 - e^2) = a^2 - c^2$$

for the case of the ellipse.

The hyperbola of Eq. (21) is clearly symmetric around both coordinate axes and has x-intercepts $(\pm a, 0)$. But it has no y-intercept. If we rewrite Eq. (21) in the form

$$y = \pm\frac{b}{a}\sqrt{x^2 - a^2}, \quad \textbf{(22)}$$

then we see that there are points on the graph only if $|x| \geqq a$. Hence the hyperbola has two **branches,** as shown in Fig. 10.6.21. We also see from Eq. (22) that $|y| \to \infty$ as $|x| \to \infty$.

The x-intercepts $V_1(-a, 0)$ and $V_2(a, 0)$ are the **vertices** of the hyperbola, and the line segment joining them is its **transverse axis** (Fig. 10.6.22). The line segment

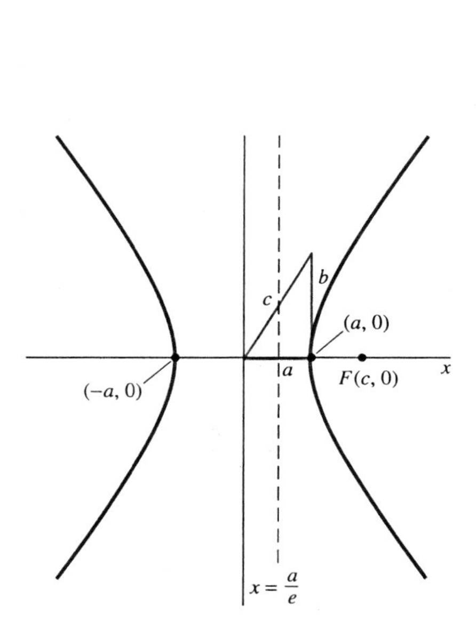

FIGURE 10.6.21 A hyperbola has two *branches.*

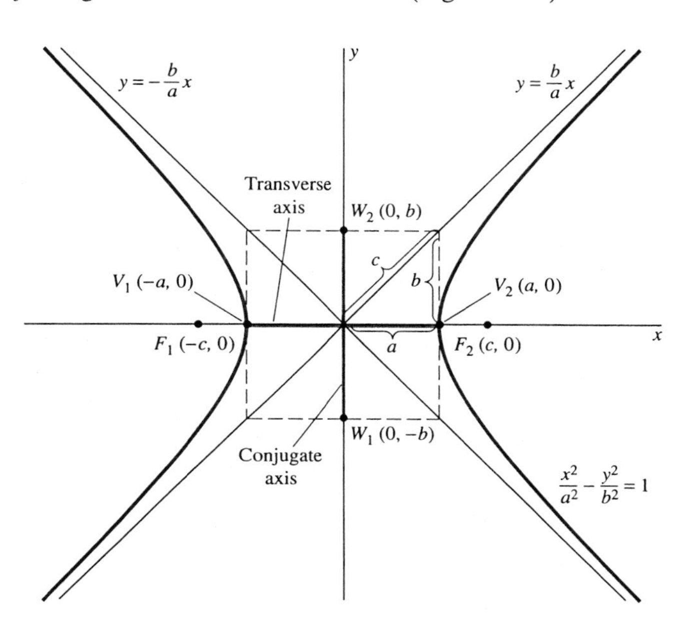

FIGURE 10.6.22 The parts of a hyperbola.

joining $W_1(0, -b)$ and $W_2(0, b)$ is its **conjugate axis.** The alternative form

$$c^2 = a^2 + b^2 \tag{23}$$

of Eq. (20) is the Pythagorean relation for the right triangle shown in Fig. 10.6.22.

The lines $y = \pm bx/a$ that pass through the **center** $(0, 0)$ and the opposite vertices of the rectangle in Fig. 10.6.22 are **asymptotes** of the two branches of the hyperbola in both directions. That is, if

$$y_1 = \frac{bx}{a} \quad \text{and} \quad y_2 = \frac{b}{a}\sqrt{x^2 - a^2},$$

then

$$\lim_{x\to\infty} (y_1 - y_2) = 0 = \lim_{x\to-\infty} (y_1 - (-y_2)). \tag{24}$$

To verify the first limit (for instance), note that

$$\begin{aligned}\lim_{x\to\infty} \frac{b}{a}\left(x - \sqrt{x^2 - a^2}\right) &= \lim_{x\to\infty} \frac{b}{a} \cdot \frac{\left(x - \sqrt{x^2 - a^2}\right)\left(x + \sqrt{x^2 - a^2}\right)}{x + \sqrt{x^2 - a^2}} \\ &= \lim_{x\to\infty} \frac{b}{a} \cdot \frac{a^2}{x + \sqrt{x^2 - a^2}} = 0.\end{aligned}$$

Just as in the case of the ellipse, the hyperbola with focus $(c, 0)$ and directrix $x = a/e$ also has focus $(-c, 0)$ and directrix $x = -a/e$ (Fig. 10.6.23). Because $c = ae$ by Eq. (19), the foci $(\pm ae, 0)$ and the directrices $x = \pm a/e$ take the same forms in terms of a and e for both the hyperbola ($e > 1$) and the ellipse ($e < 1$).

If we interchange x and y in Eq. (21), we obtain

$$\frac{y^2}{a^2} - \frac{x^2}{b^2} = 1. \tag{25}$$

This hyperbola has foci at $(0, \pm c)$. The foci as well as this hyperbola's transverse axis lie on the y-axis. Its asymptotes are $y = \pm ax/b$, and its graph generally resembles the one in Fig. 10.6.23.

When we studied the ellipse, we saw that its orientation—whether the major axis is horizontal or vertical—is determined by the relative sizes of a and b. In the

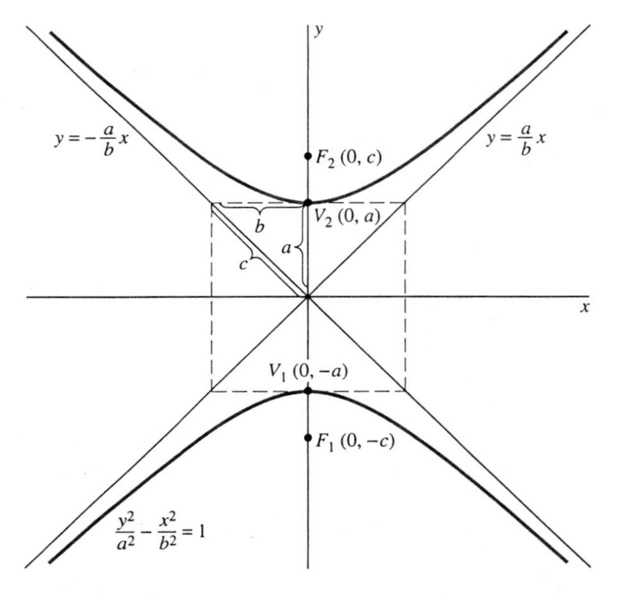

FIGURE 10.6.23 The hyperbola of Eq. (25) has horizontal directrices.

case of the hyperbola, the situation is quite different, because the relative sizes of a and b make no such difference: They affect only the slopes of the asymptotes. The direction in which the hyperbola opens—horizontal as in Fig. 10.6.22 or vertical as in Fig. 10.6.23—is determined by the signs of the terms that contain x^2 and y^2.

EXAMPLE 8 Sketch the graph of the hyperbola with equation

$$\frac{y^2}{9} - \frac{x^2}{16} = 1.$$

Solution This is an equation of the form in Eq. (25), so the hyperbola opens vertically. Because $a = 3$ and $b = 4$, we find that $c = 5$ by using Eq. (23): $c^2 = a^2 + b^2$. Thus the vertices are $(0, \pm 3)$, the foci are the two points $(0, \pm 5)$, and the asymptotes are the two lines $y = \pm 3x/4$. This hyperbola appears in Fig. 10.6.24. ◆

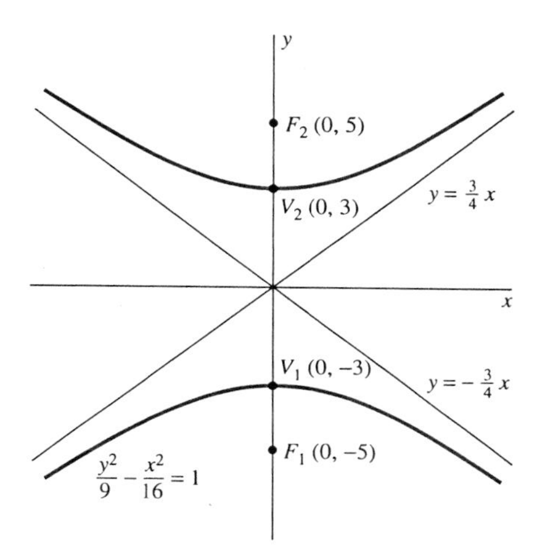

FIGURE 10.6.24 The hyperbola of Example 8.

EXAMPLE 9 Find an equation of the hyperbola with foci $(\pm 10, 0)$ and asymptotes $y = \pm 4x/3$.

Solution Because $c = 10$, we have

$$a^2 + b^2 = 100 \quad \text{and} \quad \frac{b}{a} = \frac{4}{3}.$$

Thus $b = 8$ and $a = 6$, and the standard equation of the hyperbola is

$$\frac{x^2}{36} - \frac{y^2}{64} = 1.$$

◆

As we noted earlier, any equation of the form

$$Ax^2 + Cy^2 + Dx + Ey + F = 0 \tag{26}$$

with both A and C nonzero can be reduced to the form

$$A(x - h)^2 + B(y - k)^2 = G$$

by completing the square in x and y. Now suppose that the coefficients A and C of the quadratic terms have *opposite signs.* For example, suppose that $A = p^2$ and $B = -q^2$. The last equation then becomes

$$p^2(x - h)^2 - q^2(y - k)^2 = G. \tag{27}$$

If $G = 0$, then factorization of the difference of squares on the left-hand side yields the equations

$$p(x - h) + q(y - k) = 0 \quad \text{and} \quad p(x - h) - q(y - k) = 0$$

of two straight lines through (h, k) with slopes $m = \pm p/q$. If $G \neq 0$, then division of Eq. (27) by G gives an equation that looks either like

$$\frac{(x-h)^2}{a^2} - \frac{(y-k)^2}{b^2} = 1 \quad (\text{if } G > 0)$$

or like

$$\frac{(y-k)^2}{a^2} - \frac{(x-h)^2}{b^2} = 1 \quad (\text{if } G < 0).$$

Thus if $AC < 0$ in Eq. (26), the graph is either a pair of intersecting straight lines or a hyperbola.

EXAMPLE 10 Determine the graph of the equation

$$9x^2 - 4y^2 - 36x + 8y = 4.$$

Solution We collect the terms that contain x and those that contain y, and we then complete the square in each variable. We find that

$$9(x - 2)^2 - 4(y - 1)^2 = 36,$$

so

$$\frac{(x-2)^2}{4} - \frac{(y-1)^2}{9} = 1.$$

Hence the graph is a hyperbola with a horizontal transverse axis and center $(2, 1)$. Because $a = 2$ and $b = 3$, we find that $c = \sqrt{13}$. The vertices of the hyperbola are $(0, 1)$ and $(4, 1)$, and its foci are the two points $(2 \pm \sqrt{13}, 1)$. Its asymptotes are the two lines

$$y - 1 = \pm\tfrac{3}{2}(x - 2),$$

translates of the asymptotes $y = \pm 3x/2$ of the hyperbola $\frac{1}{4}x^2 - \frac{1}{9}y^2 = 1$. Figure 10.6.25 shows the graph of the translated hyperbola. ◆

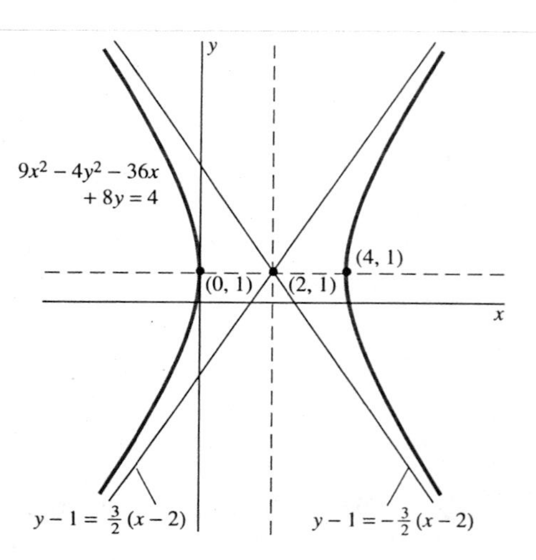

FIGURE 10.6.25 The hyperbola of Example 10, a translate of the hyperbola $x^2/4 - y^2/9 = 1$.

Applications of Hyperbolas

The *reflection property* of the hyperbola takes the same form as that for the ellipse. If P is a point on a hyperbola, then the two lines PF_1 and PF_2 from P to the two foci make equal angles with the tangent line at P. In Fig. 10.6.26 this means that $\alpha = \beta$.

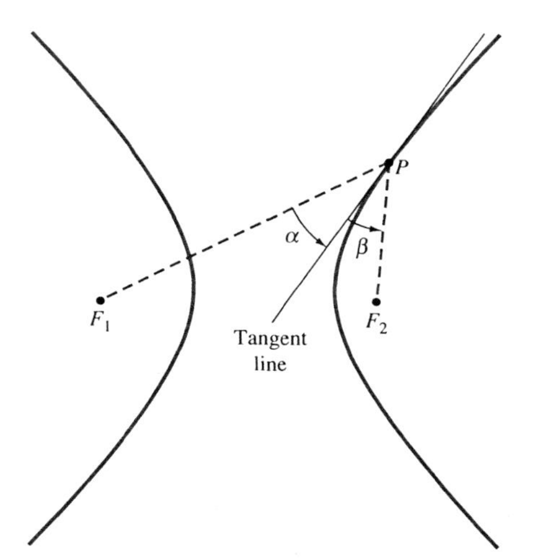

FIGURE 10.6.26 The reflection property of the hyperbola.

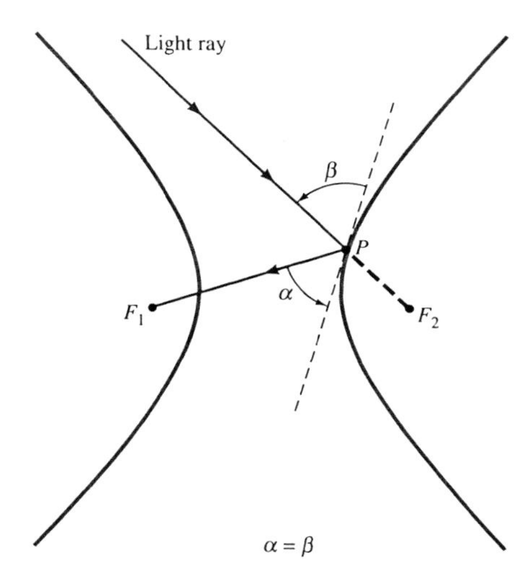

FIGURE 10.6.27 How a hyperbolic mirror reflects a ray aimed at one focus: $\alpha = \beta$ again.

For an important application of this reflection property, consider a mirror that is shaped like one branch of a hyperbola and is reflective on its outer (convex) surface. An incoming light ray aimed toward one focus will be reflected toward the other focus (Fig. 10.6.27). Figure 10.6.28 indicates the design of a reflecting telescope that makes use of the reflection properties of the parabola and the hyperbola. The parallel incoming light rays first are reflected by the parabola toward its focus at F. Then they are intercepted by an auxiliary hyperbolic mirror with foci at E and F and reflected into the eyepiece located at E.

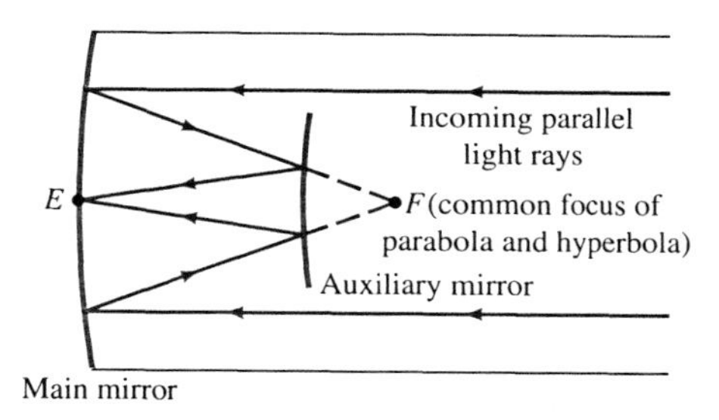

FIGURE 10.6.28 One type of reflecting telescope: main mirror parabolic, auxiliary mirror hyperbolic.

Example 11 illustrates how hyperbolas are used to determine the positions of ships at sea.

EXAMPLE 11 A ship lies in the Labrador Sea due east of Wesleyville, point A, on the long north-south coastline of Newfoundland. Simultaneous radio signals are transmitted by radio stations at A and at St. John's, point B, which is on the coast 200 km due south of A. The ship receives the signal from A 500 microseconds (μs) before it receives the signal from B. Assume that the speed of radio signals is 300 m/μs. How far out at sea is the ship?

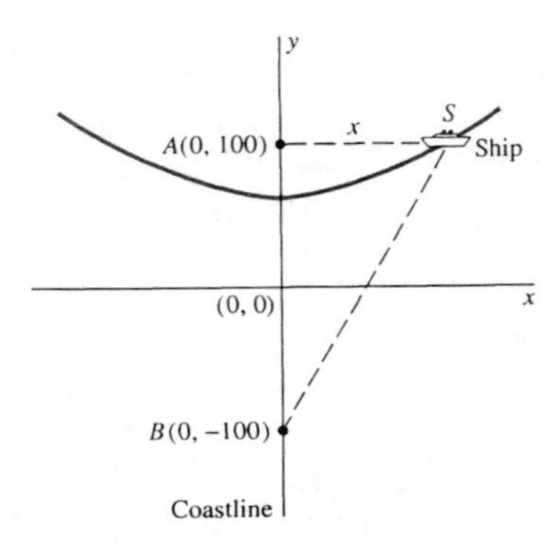

FIGURE 10.6.29 A navigation problem (Example 11).

Solution The situation is diagrammed in Fig. 10.6.29. The difference between the distances of the ship at S from A and B is

$$|SB| - |SA| = 500 \cdot 300 = 150{,}000$$

meters; that is, 150 km. Thus (by Problem 88) the ship lies on a hyperbola with foci A and B. From Fig. 10.6.29 we see that $c = 100$, so $a = \frac{1}{2} \cdot 150 = 75$, and thus

$$b = \sqrt{c^2 - a^2} = \sqrt{100^2 - 75^2} = 25\sqrt{7}.$$

In the coordinate system of Fig. 10.6.29, the hyperbola has equation

$$\frac{y^2}{75^2} - \frac{x^2}{7 \cdot 25^2} = 1.$$

We substitute $y = 100$ because the ship is due east of A. Thus we find that the ship's distance from the coastline is $x = \frac{175}{3} \approx 58.3$ km. ◆

Conics in Polar Coordinates

In order to investigate orbits of satellites—such as planets or comets orbiting the sun or natural or artificial moons orbiting a planet—we need equations of the conic sections in polar coordinates. As a bonus, we find that all three conic sections have the same general equation in polar coordinates.

To derive the polar equation of a conic section, suppose its focus is the origin O and that its directrix is the vertical line $x = -p$ (with $p > 0$). In the notation of Fig. 10.6.30, the fact that $|OP| = e|PQ|$ then tells us that $r = e(p + r\cos\theta)$. Solution of this equation for r yields

$$r = \frac{pe}{1 - e\cos\theta}.$$

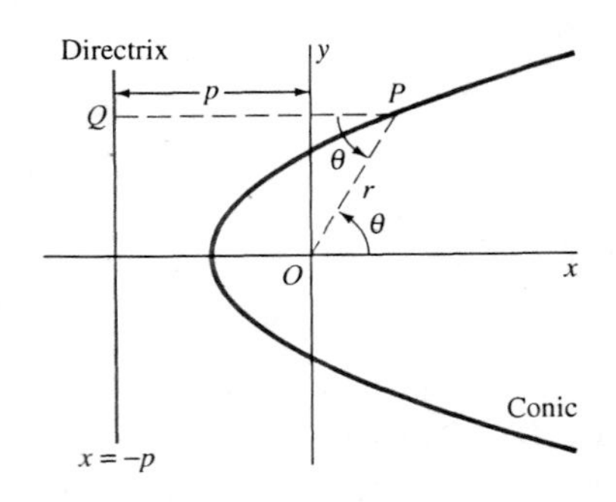

FIGURE 10.6.30 A conic section: $|OP| = e\,|PQ|$.

If the directrix is the vertical line $x = +p > 0$ to the right of the origin, then a similar calculation gives the same result, except with a change of sign in the denominator.

Polar Coordinate Equation of a Conic Section

The polar equation of a conic section with eccentricity e, focus O, and directrix $x = \pm p$ is

$$r = \frac{pe}{1 \pm e\cos\theta}. \tag{28}$$

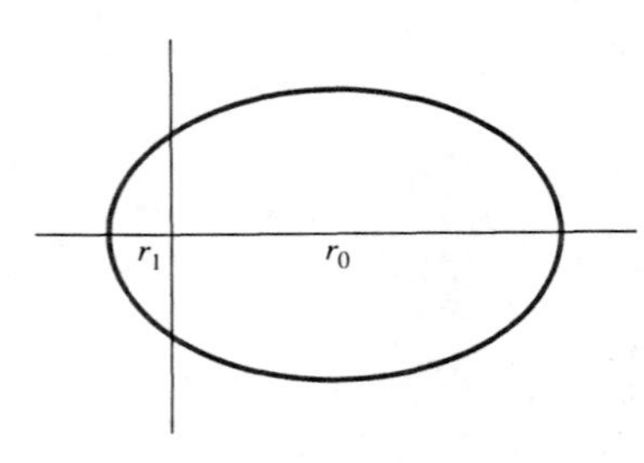

FIGURE 10.6.31 The maximal radius $r_0 = \dfrac{pe}{1-e}$ and the minimal radius $r_1 = \dfrac{pe}{1+e}$ of the ellipse.

Figure 10.6.31 shows an ellipse with eccentricity $e < 1$ and directrix $x = -p$. Its vertices correspond to $\theta = 0$ and $\theta = \pi$, where maximal and minimal radii r_0 and r_1 occur. It follows that the length $2a$ of its major axis is

$$2a = r_0 + r_1 = \frac{pe}{1-e} + \frac{pe}{1+e} = \frac{2pe}{1-e^2}.$$

Cross multiplication gives the relation

$$pe = a(1 - e^2), \tag{29}$$

and substituting in (28) then yields the equation

$$r = \frac{a(1 - e^2)}{1 \pm e\cos\theta} \tag{30}$$

of an ellipse with eccentricity e and major semiaxis a.

EXAMPLE 12 Sketch the graph of the equation

$$r = \frac{16}{5 - 3\cos\theta}.$$

Solution First we divide numerator and denominator by 5 and find that

$$r = \frac{\frac{16}{5}}{1 - \frac{3}{5}\cos\theta}.$$

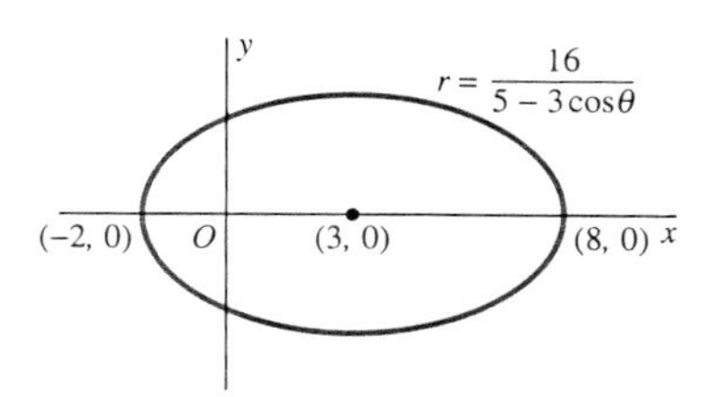

FIGURE 10.6.32 The ellipse of Example 12.

Thus $e = \frac{3}{5}$ and $pe = \frac{16}{5}$. Equation (29) then implies that $a = 5$. Finally, $c = ae = 3$ and

$$b = \sqrt{a^2 - c^2} = 4.$$

So we have here an ellipse with major semiaxis $a = 5$, minor semiaxis $b = 4$, and center at (3, 0) in Cartesian coordinates. The ellipse is shown in Fig. 10.6.32. ◆

REMARK 1 The limiting form of Eq. (30) as $e \to 0$ is the equation $r = a$ of a circle. Because $p \to \infty$ as $e \to 0$ with a fixed in Eq. (29), we may therefore regard any circle as an ellipse with eccentricity zero and with directrix at infinity.

REMARK 2 If we begin with an ellipse with eccentricity $e < 1$ and directrix $x = -p$, then the limiting form of Eq. (30) as $e \to 1^-$ is the equation

$$r = \frac{p}{1 - \cos\theta} \tag{31}$$

of a parabola. For instance, Fig. 10.6.33 shows a parabola and an ellipse of eccentricity $e = 0.99$, both with directrix $p = -1$. Observe that the two curves appear to almost coincide near the origin where $30° < \theta < 330°$. This sort of approximation of an ellipse by a parabola is useful in studying comets with highly eccentric elliptical orbits.

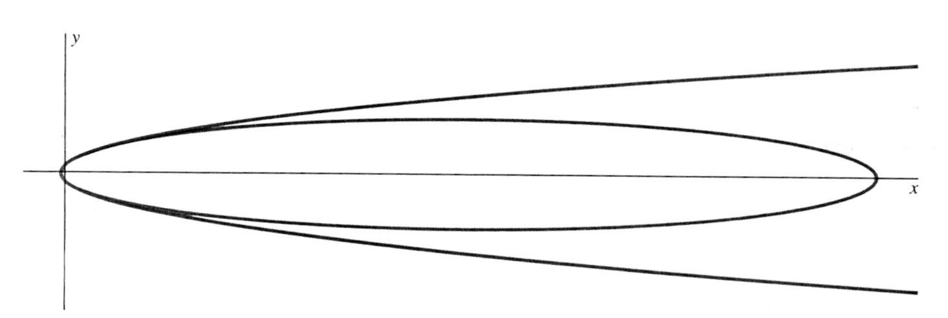

FIGURE 10.6.33 The parabola $r = \dfrac{1}{1 - \cos\theta}$ and the ellipse $r = \dfrac{0.99}{1 - 0.99\cos\theta}$.

EXAMPLE 13 A certain comet is known to have a highly eccentric elliptical orbit with the sun at one focus. Two successive observations as this comet approached the sun gave the measurements $r = 6$ AU when $\theta = 60°$, and $r = 2$ AU when $\theta = 90°$ (relative to a fixed polar-coordinate system). Estimate the position of the comet at its point of closest approach to the sun.

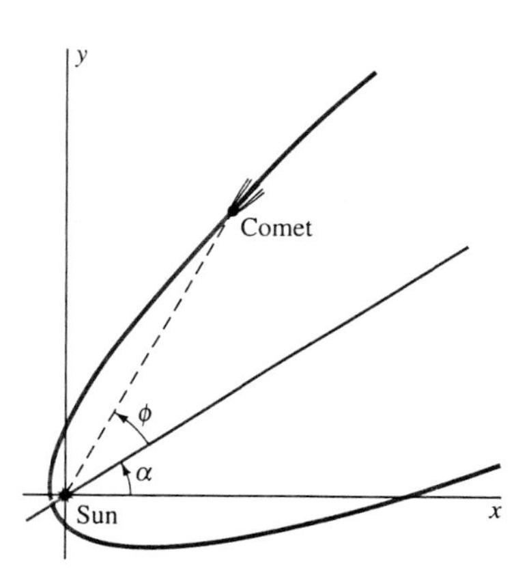

FIGURE 10.6.34 The comet of Example 13.

Solution Because the elliptical orbit is highly eccentric, we assume that near the sun it can be approximated closely by a parabola. The angle $\theta = \alpha$ of the axis is unknown, but a preliminary sketch indicates that α will be less than the initial angle of observation; thus $0 < \alpha < \pi/3$. Using the polar coordinate system with this unknown polar axis and counterclockwise angular variable $\phi = \theta - \alpha$ (Fig. 10.6.34), the equation in (31) of the parabola takes the form

$$r = \frac{p}{1 - \cos\phi} = \frac{p}{1 - \cos(\theta - \alpha)}. \tag{32}$$

The vertex of a parabola is its point closest to its focus (Problem 65), so the minimum distance of the comet from the sun will be $r = p/2$ when $\theta = \pi + \alpha$. Our problem, then, is to determine the values of p and α.

Substituting the given observational data into Eq. (32) yields the two equations

$$6 = \frac{p}{1 - \cos(\pi/3 - \alpha)} \quad \text{and} \quad 2 = \frac{p}{1 - \cos(\pi/2 - \alpha)}. \tag{33}$$

Elimination of p yields

$$6 - 6\cos\left(\frac{\pi}{3} - \alpha\right) = 2 - 2\cos\left(\frac{\pi}{2} - \alpha\right);$$

$$6 - 6\left(\frac{1}{2}\cos\alpha + \frac{\sqrt{3}}{2}\sin\alpha\right) = 2 - 2\sin\alpha.$$

We therefore need to solve the single equation

$$3\cos\alpha + (3\sqrt{3} - 2)\sin\alpha - 4 = 0.$$

A calculator or computer yields the approximate root $\alpha = 0.3956 \approx 22.67°$. Then the second equation in (33) gives $p = 2(1 - \sin\alpha) \approx 1.2293$ (AU). Because 1 astronomical unit is about 93 million miles, the comet's closest approach to the sun is about $\frac{1}{2}p = (0.5)(1.2293)(93) \approx 57.16$ million miles. ◆

10.6 TRUE/FALSE STUDY GUIDE

10.6 CONCEPTS: QUESTIONS AND DISCUSSION

1. Summarize the definitions and alternative constructions of the parabola, ellipse, and hyperbola.
2. Compare the reflection properties of the three types of conic sections.
3. Summarize the applications of the conic sections. You might like to consult an encyclopedia or do a Web search.

10.6 PROBLEMS

In Problems 1 through 5, find the equation and sketch the graph of the parabola with vertex V and focus F.

1. $V(0, 0)$, $F(3, 0)$

2. $V(0, 0)$, $F(0, -2)$

3. $V(2, 3)$, $F(2, 1)$

4. $V(-1, -1)$, $F(-3, -1)$

5. $V(2, 3)$, $F(0, 3)$

In Problems 6 through 10, find the equation and sketch the graph of the parabola with the given focus and directrix.

6. $F(1, 2)$, $x = -1$

7. $F(0, -3)$, $y = 0$

8. $F(1, -1)$, $x = 3$

9. $F(0, 0)$, $y = -2$

10. $F(-2, 1)$, $x = -4$

In Problems 11 through 18, sketch the parabola with the given equation. Show and label its vertex, focus, axis, and directrix.

11. $y^2 = 12x$

12. $x^2 = -8y$

13. $y^2 = -6x$

14. $x^2 = 7y$

15. $x^2 - 4x - 4y = 0$

16. $y^2 - 2x + 6y + 15 = 0$

17. $4x^2 + 4x + 4y + 13 = 0$

18. $4y^2 - 12y + 9x = 0$

In Problems 19 through 33, find an equation of the ellipse specified.

19. Vertices $(\pm 4, 0)$ and $(0, \pm 5)$

20. Foci $(\pm 5, 0)$, major semiaxis 13

21. Foci $(0, \pm 8)$, major semiaxis 17

22. Center $(0, 0)$, vertical major axis 12, minor axis 8

23. Foci $(\pm 3, 0)$, eccentricity $\frac{3}{4}$

24. Foci $(0, \pm 4)$, eccentricity $\frac{2}{3}$

25. Center $(0, 0)$, horizontal major axis 20, eccentricity $\frac{1}{2}$

26. Center $(0, 0)$, horizontal minor axis 10, eccentricity $\frac{1}{2}$

27. Foci $(\pm 2, 0)$, directrices $x = \pm 8$

28. Foci $(0, \pm 4)$, directrices $y = \pm 9$

29. Center $(2, 3)$, horizontal axis 8, vertical axis 4

30. Center $(1, -2)$, horizontal major axis 8, eccentricity $\frac{3}{4}$

31. Foci $(-2, 1)$ and $(4, 1)$, major axis 10

32. Foci $(-3, 0)$ and $(-3, 4)$, minor axis 6

33. Foci $(-2, 2)$ and $(4, 2)$, eccentricity $\frac{1}{3}$

Sketch the graphs of the equations in Problems 34 through 38. Indicate centers, foci, and lengths of axes.

34. $4x^2 + y^2 = 16$

35. $4x^2 + 9y^2 = 144$

36. $4x^2 + 9x^2 = 24x$

37. $9x^2 + 4y^2 - 32y + 28 = 0$

38. $2x^2 + 3y^2 + 12x - 24y + 60 = 0$

In Problems 39 through 52, find an equation of the hyperbola described.

39. Foci $(\pm 4, 0)$, vertices $(\pm 1, 0)$

40. Foci $(0, \pm 3)$, vertices $(0, \pm 2)$

41. Foci $(\pm 5, 0)$, asymptotes $y = \pm 3x/4$

42. Vertices $(\pm 3, 0)$, asymptotes $y = \pm 3x/4$

43. Vertices $(0, \pm 5)$, asymptotes $y = \pm x$

44. Vertices $(\pm 3, 0)$, eccentricity $e = \frac{5}{3}$

45. Foci $(0, \pm 6)$, eccentricity $e = 2$

46. Vertices $(\pm 4, 0)$ and passing through $(8, 3)$

47. Foci $(\pm 4, 0)$, directrices $x = \pm 1$

48. Foci $(0, \pm 9)$, directrices $y = \pm 4$

49. Center $(2, 2)$, horizontal transverse axis of length 6, eccentricity $e = 2$

50. Center $(-1, 3)$, vertices $(-4, 3)$ and $(2, 3)$, foci $(-6, 3)$ and $(4, 3)$

51. Center $(1, -2)$, vertices $(1, 1)$ and $(1, -5)$, asymptotes $3x - 2y = 7$ and $3x + 2y = -1$

52. Focus $(8, -1)$, asymptotes $3x - 4y = 13$ and $3x + 4y = 5$

Sketch the graphs of the equations given in Problems 53 through 58; indicates centers, foci, and asymptotes.

53. $x^2 - y^2 - 2x + 4y = 4$

54. $x^2 - 2y^2 + 4x = 0$

55. $y^2 - 3x^2 - 6y = 0$

56. $x^2 - y^2 - 2x + 6y = 9$

57. $9x^2 - 4y^2 + 18x + 8y = 31$

58. $4y^2 - 9x^2 - 18x - 8y = 41$

In each of Problems 59 through 64, identify and sketch the conic section with the given polar equation.

59. $r = \dfrac{6}{1 + \cos\theta}$

60. $r = \dfrac{6}{1 + 2\cos\theta}$

61. $r = \dfrac{3}{1 - \cos\theta}$

62. $r = \dfrac{8}{8 - 2\cos\theta}$

63. $r = \dfrac{6}{2 - \sin\theta}$

64. $r = \dfrac{12}{3 + 2\cos\theta}$

65. Prove that the point of the parabola $y^2 = 4px$ closest to its focus is its vertex.

66. Find an equation of the parabola that has a vertical axis and passes through the points $(2, 3)$, $(4, 3)$, and $(6, -5)$.

67. Show that an equation of the line tangent to the parabola $y^2 = 4px$ at the point (x_0, y_0) is

$$2px - y_0 y + 2px_0 = 0.$$

Conclude that the tangent line intersects the x-axis at the point $(-x_0, 0)$. This fact provides a quick method for constructing a line tangent to a parabola at a given point.

68. A comet's orbit is a parabola with the sun at its focus. When the comet is $100\sqrt{2}$ million miles from the sun, the line from the sun to the comet makes an angle of $45°$ with the axis of the parabola (Fig. 10.6.35). What will be the minimum distance between the comet and the sun? [*Suggestion:* Write the equation of the parabola with the origin at the focus, then use the result of Problem 65.]

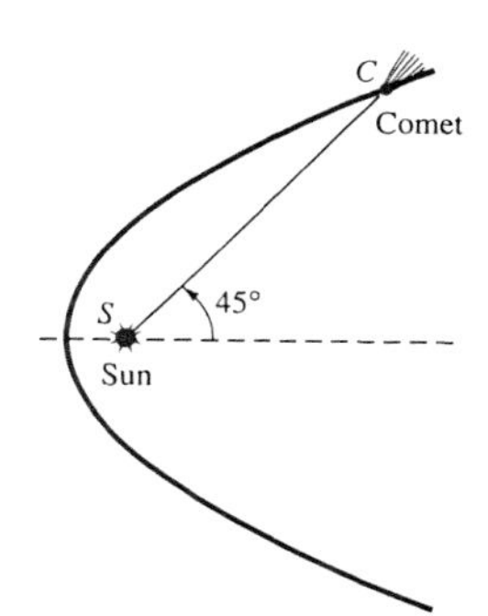

FIGURE 10.6.35 The comet of Problem 68 in parabolic orbit around the sun.

69. Suppose that the angle of Problem 68 increases from $45°$ to $90°$ in 3 days. How much longer will be required for the comet to reach its point of closest approach to the sun? Assume that the line segment from the sun to the comet sweeps out area at a constant rate (Kepler's second law).

70. Use Eqs. (7) and (8) to derive Eq. (9) with the values of M and R given in Eqs. (10) and (11).

71. Deduce from Eq. (11) that, given a fixed initial velocity v_0, the maximum range of the projectile is $R_{\max} = v_0^2/g$ and is attained when $\alpha = 45°$.

In Problems 72 through 74, assume that a projectile is fired with initial velocity $v_0 = 50$ m/s from the origin and at an angle of inclination α. Use $g = 9.8$ m/s².

72. If $\alpha = 45°$, find the range of the projectile and the maximum height it attains.

73. For what value or values of α is the range $R = 125$ m?

74. Find the range of the projectile and the length of time it remains above the ground if (a) $\alpha = 30°$; (b) $\alpha = 60°$.

75. The book *Elements of Differential and Integral Calculus* by William Granville, Percey Smith, and William Longley (Ginn and Company: Boston, 1929) lists a number of "curves for reference"; the curve with equation $\sqrt{x} + \sqrt{y} = \sqrt{a}$ is called a parabola. Verify that the curve in question actually is a parabola, or show that it is not.

76. The 1992 edition of the study guide for the national actuarial examinations has a problem quite similar to this one: Every point on the plane curve K is equally distant from the point $(-1, -1)$ and the line $x + y = 1$, and K has equation

$$x^2 + Bxy + Cy^2 + Dx + Ey + F = 0.$$

Which is the value of D: $-2, 2, 4, 6$, or 8?

77. (a) The orbit of the comet Kahoutek is an ellipse of extreme eccentricity $e = 0.999925$; the sun is at one focus of this ellipse. The minimum distance between the sun and Kahoutek is 0.13 AU. What is the maximum distance between Kahoutek and the sun? (b) The orbit of the comet Hyakutake is an ellipse of extreme eccentricity $e = 0.999643856$; the sun is at one focus of this ellipse. The minimum distance between the sun and Hyakutake is 0.2300232 AU. What is the maximum distance between Hyakutake and the sun?

78. The orbit of the planet Mercury is an ellipse of eccentricity $e = 0.206$. Its maximum and minimum distances from the sun are 0.467 and 0.307 AU, respectively. What are the major and minor semiaxes of the orbit of Mercury? Does "nearly circular" accurately describe the orbit of Mercury?

79. Derive Eq. (16) for an ellipse whose foci lie on the y-axis.

80. Show that the line tangent to the ellipse

$$\frac{x^2}{a^2} + \frac{y^2}{b^2} = 1$$

at the point $P(x_0, y_0)$ of that ellipse has equation

$$\frac{x_0 x}{a^2} + \frac{y_0 y}{b^2} = 1.$$

81. Use the result of Problem 80 to establish the reflection property of the ellipse. [*Suggestion:* Let m be the slope of the line normal to the ellipse at $P(x_0, y_0)$ and let m_1 and m_2 be the slopes of the lines PF_1 and PF_2, respectively, from P to the two foci F_1 and F_2 of the ellipse. Show that

$$\frac{m - m_1}{1 + m_1 m} = \frac{m_2 - m}{1 + m_2 m};$$

then use the identity for $\tan(A - B)$.]

82. Given $F_1(-c, 0)$ and $F_2(c, 0)$ with $a > c > 0$, prove that the ellipse

$$\frac{x^2}{a^2} + \frac{y^2}{b^2} = 1$$

(with $b^2 = a^2 - c^2$) is the locus of those points P such that $|PF_1| + |PF_2| = 2a$.

83. Find an equation of the ellipse with horizontal and vertical axes that passes through the points $(-1, 0)$, $(3, 0)$, $(0, 2)$, and $(0, -2)$.

84. Derive an equation for the ellipse with foci $(3, -3)$ and $(-3, 3)$ and major axis of length 10. Note that the foci of this ellipse lie on neither a vertical line nor a horizontal line.

85. Show that the graph of the equation

$$\frac{x^2}{15 - c} - \frac{y^2}{c - 6} = 1$$

is (a) a hyperbola with foci $(\pm 3, 0)$ if $6 < c < 15$ and (b) an ellipse if $c < 6$. (c) Identify the graph in the case $c > 15$.

86. Establish that the line tangent to the hyperbola

$$\frac{x^2}{a^2} - \frac{y^2}{b^2} = 1$$

at the point $P(x_0, y_0)$ has equation

$$\frac{x_0 x}{a^2} - \frac{y_0 y}{b^2} = 1.$$

87. Use the result of Problem 86 to establish the reflection property of the hyperbola. (See the suggestion for Problem 81.)

88. Suppose that $0 < a < c$, and let $b = \sqrt{c^2 - a^2}$. Show that the hyperbola $x^2/a^2 - y^2/b^2 = 1$ is the locus of a point P such that the *difference* between the distances $|PF_1|$ and $|PF_2|$ is equal to $2a$ (F_1 and F_2 are the foci of the hyperbola).

89. Derive an equation for the hyperbola with vertices $(\pm 3/\sqrt{2}, \pm 3/\sqrt{2})$ and foci $(\pm 5, \pm 5)$. Use the difference definition of a hyperbola implied by Problem 88.

90. Two radio signaling stations at A and B lie on an east-west line, with A 100 mi west of B. A plane is flying west on a line 50 mi north of the line AB. Radio signals are sent (traveling at 980 ft/μs) simultaneously from A and B, and the one sent from B arrives at the plane 400 μs before the one sent from A. Where is the plane?

91. Two radio signaling stations are located as in Problem 90 and transmit radio signals that travel at the same speed. But now we know only that the plane is generally somewhere north of the line AB, that the signal from B arrives 400 μs before the one sent from A, and that the signal sent from A and reflected by the plane takes a total of 600 μs to reach B. Where is the plane?

92. A comet has a parabolic orbit with the sun at one focus. When the comet is 150 million miles from the sun, the sun-comet line makes an angle of 45° with the axis of the parabola. What will be the minimum distance between the comet and the sun?

93. A satellite has an elliptical orbit with the center of the earth (take its radius to be 4000 mi) at one focus. The lowest point of its orbit is 500 mi above the North Pole and the highest point is 5000 mi above the South Pole. What is the height of the satellite above the surface of the earth when the satellite crosses the equatorial plane?

94. Find the closest approach to the sun of a comet as in Example 13 of this section; assume that $r = 2.5$ AU when $\theta = 45°$ and that $r = 1$ AU when $\theta = 90°$.

95. An ellipse has semimajor axis a and semiminor axis b. Use the polar-coordinate equation of an ellipse to derive the formula $A = \pi ab$ for its area.

96. The orbit of a certain comet approaching the sun is the parabola

$$r = \frac{1}{1 - \cos\theta}.$$

The units for r are in astronomical units. Suppose that it takes 15 days for the comet to travel from the position $\theta = 60°$ to the position $\theta = 90°$. How much longer will it require for the comet to reach its point of closest approach to the sun? Assume that the radius from the sun to the comet sweeps out area at a constant rate as the comet moves (Kepler's second law of planetary motion).

CHAPTER 10 REVIEW: CONCEPTS AND DEFINITIONS

Use the following list as a guide to additional concepts that you may need to review.

1. Conic sections
2. The relationship between rectangular and polar coordinates
3. The graph of an equation in polar coordinates
4. The area formula in polar coordinates
5. Definition of a parametric curve and of a smooth parametric curve
6. The slope of the line tangent to a smooth parametric curve (both in rectangular and in polar coordinates)
7. Integral computations with parametric curves [Eqs. (1) through (4) of Section 10.5]
8. Arc length of a parametric curve

REVIEW OF CONIC SECTIONS

The parabola with focus $(p, 0)$ and directrix $x = -p$ has eccentricity $e = 1$ and equation $y^2 = 4px$. The accompanying table compares the properties of an ellipse and a hyperbola, each with foci $(\pm c, 0)$ and major axis of length $2a$.

	Ellipse	Hyperbola
Eccentricity	$e = \frac{c}{a} < 1$	$e = \frac{c}{a} > 1$
a, b, c relation	$a^2 = b^2 + c^2$	$c^2 = a^2 + b^2$
Equation	$\frac{x^2}{a^2} + \frac{y^2}{b^2} = 1$	$\frac{x^2}{a^2} - \frac{y^2}{b^2} = 1$
Vertices	$(\pm a, 0)$	$(\pm a, 0)$
y-intercepts	$(0, \pm b)$	None
Directrices	$x = \pm\frac{a}{e}$	$x = \pm\frac{a}{e}$
Asymptotes	None	$y = \pm\frac{bx}{a}$

CHAPTER 10 MISCELLANEOUS PROBLEMS

Sketch the graphs of the equations in Problems 1 through 30. In Problems 1 through 18, if the graph is a conic section, label its center, foci, and vertices.

1. $x^2 + y^2 - 2x - 2y = 2$
2. $x^2 + y^2 = x + y$
3. $x^2 + y^2 - 6x + 2y + 9 = 0$
4. $y^2 = 4(x + y)$
5. $x^2 = 8x - 2y - 20$
6. $x^2 + 2y^2 - 2x + 8y + 8 = 0$
7. $9x^2 + 4y^2 = 36x$
8. $x^2 - y^2 = 2x - 2y - 1$
9. $y^2 - 2x^2 = 4x + 2y + 3$
10. $9y^2 - 4x^2 = 8x + 18y + 31$
11. $x^2 + 2y^2 = 4x + 4y - 12$
12. $y^2 - 6y + 4x + 5 = 0$
13. $9(x^2 - 2x + 1) = 4(y^2 + 9)$
14. $(x^2 - 4)(y^2 - 1) = 0$
15. $x^2 - 8x + y^2 - 2y + 16 = 0$
16. $(x - 1)^2 + 4(y - 2)^2 = 1$
17. $(x^2 - 4x + y^2 - 4y + 8)(x + y)^2 = 0$
18. $x = y^2 + 4y + 5$
19. $r = -2\cos\theta$
20. $\cos\theta + \sin\theta = 0$
21. $r = \dfrac{1}{\sin\theta + \cos\theta}$
22. $r\sin^2\theta = \cos\theta$
23. $r = 3\csc\theta$
24. $r = 2(\cos\theta - 1)$
25. $r^2 = 4\cos\theta$
26. $r\theta = 1$
27. $r = 3 - 2\sin\theta$
28. $r = \dfrac{1}{1 + \cos\theta}$
29. $r = \dfrac{4}{2 + \cos\theta}$
30. $r = \dfrac{4}{1 - 2\cos\theta}$

In Problems 31 through 38, find the area of the region described.

31. Inside both $r = 2\sin\theta$ and $r = 2\cos\theta$
32. Inside $r^2 = 4\cos\theta$
33. Inside $r = 3 - 2\sin\theta$ and outside $r = 4$
34. Inside $r^2 = 2\sin 2\theta$ and outside $r = 2\sin\theta$
35. Inside $r = 2\sin 2\theta$ and outside $r = \sqrt{2}$
36. Inside $r = 3\cos\theta$ and outside $r = 1 + \cos\theta$
37. Inside $r = 1 + \cos\theta$ and outside $r = \cos\theta$
38. Between the loops of $r = 1 - 2\sin\theta$

In Problems 39 through 43, eliminate the parameter and sketch the curve.

39. $x = 2t^3 - 1, \quad y = 2t^3 + 1$
40. $x = \cosh t, \quad y = \sinh t$
41. $x = 2 + \cos t, \quad y = 1 - \sin t$
42. $x = \cos^4 t, \quad y = \sin^4 t$
43. $x = 1 + t^2, \quad y = t^3$

In Problems 44 through 48, write an equation of the line tangent to the given curve at the indicated point.

44. $x = t^2, \quad y = t^3; \quad t = 1$
45. $x = 3\sin t, \quad y = 4\cos t; \quad t = \pi/4$
46. $x = e^t, \quad y = e^{-t}; \quad t = 0$
47. $r = \theta; \quad \theta = \pi/2$
48. $r = 1 + \sin\theta; \quad \theta = \pi/3$

In Problems 49 through 52, find the area of the region between the given curve and the x-axis.

49. $x = 2t + 1, \quad y = t^2 + 3; \quad -1 \leqq t \leqq 2$
50. $x = e^t, \quad y = e^{-t}; \quad 0 \leqq t \leqq 10$
51. $x = 3\sin t, \quad y = 4\cos t; \quad 0 \leqq t \leqq \pi/2$
52. $x = \cosh t, \quad y = \sinh t; \quad 0 \leqq t \leqq 1$

In Problems 53 through 57, find the arc length of the given curve.

53. $x = t^2, \quad y = t^3; \quad 0 \leqq t \leqq 1$
54. $x = \ln(\cos t), \quad y = t; \quad 0 \leqq t \leqq \pi/4$
55. $x = 2t, \quad y = t^3 + \dfrac{1}{3t}; \quad 1 \leqq t \leqq 2$

56. $r = \sin\theta;\quad 0 \leqq \theta \leqq \pi$

57. $r = \sin^2(\theta/3);\quad 0 \leqq \theta \leqq \pi$

In Problems 58 through 62, find the area of the surface generated by revolving the given curve around the x-axis.

58. $x = t^2 + 1,\quad y = 3t;\quad 0 \leqq t \leqq 2$

59. $x = 4\sqrt{t},\quad y = \dfrac{t^3}{3} + \dfrac{1}{2t^2};\quad 1 \leqq t \leqq 4$

60. $r = \cos\theta$

61. $r = e^{\theta/2};\quad 0 \leqq \theta \leqq \pi$

62. $x = e^t \cos t,\quad y = e^t \sin t;\quad 0 \leqq t \leqq \pi/2$

63. Consider the rolling circle of radius a that was used to generate the cycloid in Example 5 of Section 10.4. Suppose that this circle is the rim of a disk, and let Q be a point of this disk at distance $b < a$ from its center. Find parametric equations for the curve traced by Q as the circle rolls along the x-axis. Assume that Q begins at the point $(0, a - b)$. Sketch this curve, which is called a **trochoid.**

64. If the smaller circle of Problem 34 in Section 10.4 rolls around the *outside* of the larger circle, the path of the point P is called an **epicycloid.** Show that it has parametric equations

$$x = (a + b)\cos t - b\cos\left(\frac{a+b}{b}t\right),$$

$$y = (a + b)\sin t - b\sin\left(\frac{a+b}{b}t\right).$$

65. Suppose that $b = a$ in Problem 64. Show that the epicycloid is then the cardioid $r = 2a(1 - \cos\theta)$ translated a units to the right.

66. Find the area of the surface generated by revolving the lemniscate $r^2 = 2a^2 \cos 2\theta$ around the x-axis.

67. Find the volume generated by revolving around the y-axis the area under the cycloid

$$x = a(t - \sin t),\quad y = a(1 - \cos t),\quad 0 \leqq t \leqq 2\pi.$$

68. Show that the length of one arch of the hypocycloid of Problem 34 in Section 10.4 is $s = 8b(a - b)/a$.

69. Find a polar-coordinate equation of the circle that passes through the origin and is centered at the point with polar coordinates (p, α).

70. Find a simple equation of the parabola whose focus is the origin and whose directrix is the line $y = x + 4$. Recall from Miscellaneous Problem 93 of Chapter 3 that the distance from the point (x_0, y_0) to the line with equation $Ax + By + C = 0$ is

$$\frac{|Ax_0 + By_0 + C|}{\sqrt{A^2 + B^2}}.$$

71. A **diameter** of an ellipse is a chord through its center. Find the maximum and minimum lengths of diameters of the ellipse with equation

$$\frac{x^2}{a^2} + \frac{y^2}{b^2} = 1.$$

72. Use calculus to prove that the ellipse of Problem 71 is normal to the coordinate axes at each of its four vertices.

73. The parabolic arch of a bridge has base width b and height h at its center. Write its equation, choosing the origin on the ground at the left end of the arch.

74. Use methods of calculus to find the points of the ellipse

$$\frac{x^2}{a^2} + \frac{y^2}{b^2} = 1$$

that are nearest to and farthest from (a) the center $(0, 0)$; (b) the focus $(c, 0)$.

75. Consider a line segment QR that contains a point P such that $|QP| = a$ and $|PR| = b$. Suppose that Q is constrained to move on the y-axis, whereas R must remain on the x-axis. Prove that the locus of P is an ellipse.

76. Suppose that $a > 0$ and that F_1 and F_2 are two fixed points in the plane with $|F_1F_2| > 2a$. Imagine a point P that moves in such a way that $|PF_2| = 2a + |PF_1|$. Prove that the locus of P is one branch of a hyperbola with foci F_1 and F_2. Then—as a consequence—explain how to construct points on a hyperbola by drawing appropriate circles centered at its foci.

77. Let Q_1 and Q_2 be two points on the parabola $y^2 = 4px$. Let P be the point of the parabola at which the tangent line is parallel to Q_1Q_2. Prove that the horizontal line through P bisects the segment Q_1Q_2.

78. Determine the locus of a point P such that the product of its distances from the two fixed points $F_1(-a, 0)$ and $F_2(a, 0)$ is a^2.

79. Find the eccentricity of the conic section with equation $3x^2 - y^2 + 12x + 9 = 0$.

80. Find the area bounded by the loop of the *strophoid*

$$r = \sec\theta - 2\cos\theta$$

shown in Fig. 10.MP.1.

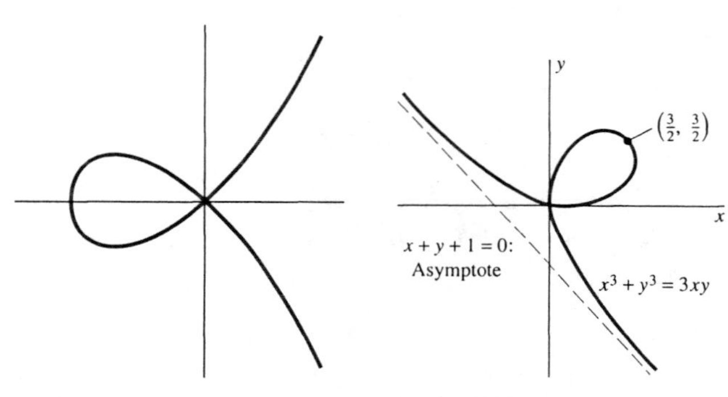

FIGURE 10.MP.1 The strophoid of Problem 80.

FIGURE 10.MP.2 The folium of Descartes $x^3 + y^3 = 3xy$ (Problem 81).

81. Find the area bounded by the loop of the folium of Descartes with equation $x^3 + y^3 = 3xy$ shown in Fig. 10.MP.2. (*Suggestion:* Change to polar coordinates and then substitute $u = \tan\theta$ to evaluate the area integral.)

82. Use the method of Problem 81 to find the area bounded by the first-quadrant loop of the curve $x^5 + y^5 = 5x^2y^2$ (similar to the folium of Problem 81).

83. The graph of a conic section in the xy-plane has intercepts at $(5, 0)$, $(-5, 0)$, $(0, 4)$, and $(0, -4)$. Deduce all the information you can about this conic. Can you determine whether it is a parabola, a hyperbola, or an ellipse? What if you also know that the graph of this conic is normal to the y-axis at the point $(0, 4)$?

11 INFINITE SERIES

Srinivasa Ramanujan
(1887–1920)

On a cold January day in 1913, the eminent Cambridge mathematics professor G. H. Hardy received a letter from an unknown 25-year-old clerk in the accounting department of a government office in Madras, India. Its author, Srinivasa Ramanujan, had no university education, he admitted—he had flunked out—but "after leaving school I have employed the spare time at my disposal to work at Mathematics . . . I have not trodden through the conventional regular course . . . but am striking out a new path for myself." The ten pages that followed listed in neat handwritten script approximately 50 formulas, most dealing with integrals and infinite series that Ramanujan had discovered, and asked Hardy's advice whether they contained anything of value. The formulas were of such exotic and unlikely appearance that Hardy at first suspected a hoax, but he and his colleague J. E. Littlewood soon realized that they were looking at the work of an extraordinary mathematical genius.

Thus began one of the most romantic episodes in the history of mathematics. In April 1914 Ramanujan arrived in England a poor, self-taught Indian mathematical amateur called to collaborate as an equal with the most sophisticated professional mathematicians of the day. For the next three years a steady stream of remarkable discoveries poured forth from his pen. But in 1917 he fell seriously ill, apparently with tuberculosis. The following year he returned to India to attempt to regain his health but never recovered, and he died in 1920 at the age of 32. Up to the very end he worked feverishly to record his final discoveries. He left behind notebooks outlining work whose completion occupied prominent mathematicians throughout the twentieth century.

With the possible exception of Euler, no one before or since has exhibited Ramanujan's virtuosity with infinite series. An example of his discoveries is the infinite series

$$\frac{1}{\pi} = \frac{\sqrt{8}}{9801} \sum_{n=0}^{\infty} \frac{(4n)!}{(n!)^4} \cdot \frac{(1103 + 26390n)}{396^{4n}},$$

whose first term yields the familiar approximation $\pi \approx 3.14159$, and with each additional term giving π to roughly eight more decimal places of accuracy. For instance, just four terms of Ramanujan's series are needed to calculate the 30-place approximation

$$\pi \approx 3.14159\,26535\,89793\,23846\,26433\,83279$$

that suffices for virtually any imaginable "practical" application—if the universe were a sphere with a radius of 10 billion light years, then this value of π would give its circumference accurate to the nearest hundredth of an inch. But in recent years Ramanujan's ideas have been used to calculate the value of π accurate to a *billion* decimal places. Indeed, such gargantuan computations of π are commonly used to check the accuracy of new supercomputers.

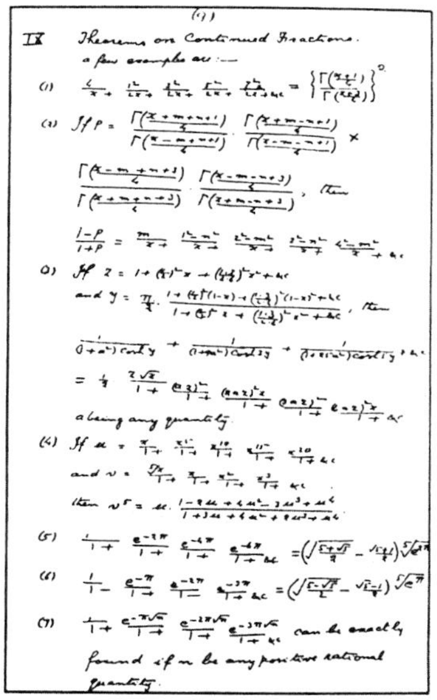

A typical page of Ramanujan's letter to Hardy, listing formulas Ramanujan had discovered, but with no hint of proof or derivation.

11.1 INTRODUCTION

In the fifth century B.C., the Greek philosopher Zeno proposed the following paradox: In order for a runner to travel a given distance, the runner must first travel halfway, then half the remaining distance, then half the distance that yet remains, and so on *ad infinitum.* But, Zeno argued, it is clearly impossible for a runner to accomplish infinitely many such tasks in a finite period of time, so motion from one point to another is impossible.

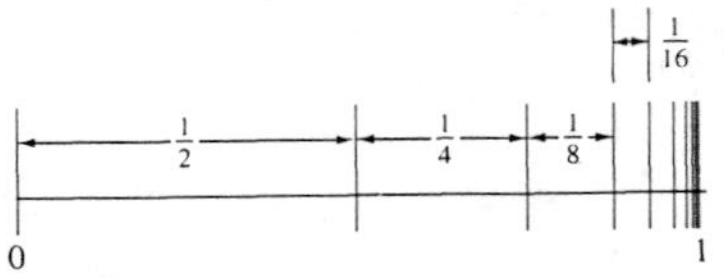

Zeno's paradox suggests the infinite subdivision of $[0, 1]$ indicated in Fig. 11.1.1. There is one subinterval of length $1/2^n$ for each integer $n = 1, 2, 3, \dots$. If the length of the interval is the sum of the lengths of the subintervals into which it is divided, then it would appear that

FIGURE 11.1.1 Subdivision of an interval to illustrate Zeno's paradox.

$$1 = \frac{1}{2} + \frac{1}{4} + \frac{1}{8} + \frac{1}{16} + \cdots + \frac{1}{2^n} + \cdots,$$

with infinitely many terms somehow adding up to 1. But the formal infinite sum

$$1 + 2 + 3 + \cdots + n + \cdots$$

of all the positive integers seems meaningless—it does not appear to add up to *any* (finite) value.

The question is this: What, if anything, do we mean by the sum of an *infinite* collection of numbers? This chapter explores conditions under which an *infinite* sum

$$a_1 + a_2 + a_3 + \cdots + a_n + \cdots,$$

known as an *infinite series,* is meaningful. We discuss methods for computing the sum of an infinite series and applications of the algebra and calculus of infinite series. Infinite series are important in science and mathematics because many functions either arise most naturally in the form of infinite series or have infinite series representations (such as the Taylor series of Section 11.4) that are useful for numerical computations.

11.2 INFINITE SEQUENCES

An **infinite sequence** of real numbers is an ordered, unending list

$$a_1,\ a_2,\ a_3,\ a_4,\ \dots,\ a_n,\ a_{n+1},\ \dots \tag{1}$$

of numbers. That this list is *ordered* implies that it has a first term a_1, a second term a_2, a third term a_3, and so forth. That the sequence is unending, or *infinite,* implies that (for every n) the ***n*th term** a_n has a successor a_{n+1}. Thus, as indicated by the final ellipsis in (1), an infinite sequence never ends and—despite the fact that we write explicitly only a finite number of terms—it actually has an infinite number of terms. Concise notation for the infinite sequence in (1) is

$$\{a_n\}_{n=1}^{\infty}, \qquad \{a_n\}_1^{\infty}, \qquad \text{or simply} \qquad \{a_n\}. \tag{2}$$

Frequently an infinite sequence $\{a_n\}$ of numbers can be described "all at once" by a single function f that gives the successive terms of the sequence as successive values of the function:

$$a_n = f(n) \quad \text{for} \quad n = 1, 2, 3, \dots. \tag{3}$$

Here $a_n = f(n)$ is simply a *formula for the nth term* of the sequence. Conversely, if the sequence $\{a_n\}$ is given in advance, we can regard (3) as the definition of the function f having the set of positive integers as its domain of definition. Ordinarily we will use the subscript notation a_n in preference to the function notation $f(n)$.

EXAMPLE 1 The following table exhibits several particular infinite sequences. Each is described in three ways: in the concise sequential notation $\{a_n\}$ of (2), by writing the formula as in (3) for its nth term, and in extended list notation as in (1). Note that n need not begin with the initial value 1.

$$\left\{\frac{1}{n}\right\}_1^\infty \qquad a_n = \frac{1}{n} \qquad 1, \frac{1}{2}, \frac{1}{3}, \frac{1}{4}, \dots, \frac{1}{n}, \dots$$

$$\left\{\frac{1}{10^n}\right\}_0^\infty \qquad a_n = \frac{1}{10^n} \qquad 1, \frac{1}{10}, \frac{1}{100}, \frac{1}{1000}, \dots, \frac{1}{10^n}, \dots$$

$$\{\sqrt{3n-7}\}_3^\infty \qquad a_n = \sqrt{3n-7} \qquad \sqrt{2}, \sqrt{5}, \sqrt{8}, \sqrt{11}, \dots, \sqrt{3n-7}, \dots$$

$$\left\{\sin\frac{n\pi}{2}\right\}_1^\infty \qquad a_n = \sin\frac{n\pi}{2} \qquad 1, 0, -1, 0, \dots, \sin\frac{n\pi}{2}, \dots$$

$$\{3+(-1)^n\}_1^\infty \qquad a_n = 3+(-1)^n \qquad 2, 4, 2, 4, \dots, 3+(-1)^n, \dots$$

Sometimes it is inconvenient or impossible to give an explicit formula for the nth term of a particular sequence. The following example illustrates how sequences may be defined in other ways. ◆

EXAMPLE 2 Here we give the first ten terms of each sequence.

(a) The sequence of prime integers (those positive integers n having precisely two divisors, 1 and n with $n > 1$),

$$2, 3, 5, 7, 11, 13, 17, 19, 23, 29, \dots$$

(b) The sequence whose nth term is the nth decimal digit of the number

$$\pi = 3.14159265358979323846\dots,$$

$$1, 4, 1, 5, 9, 2, 6, 5, 3, 5, \dots$$

(c) The **Fibonacci sequence** $\{F_n\}$, which may be defined by

$$F_1 = 1, \quad F_2 = 1, \quad \text{and} \quad F_{n+1} = F_n + F_{n-1} \quad \text{for} \quad n \geqq 2.$$

Thus each term after the second is the sum of the preceding two terms:

$$1, 1, 2, 3, 5, 8, 13, 21, 34, 55, \dots$$

This is an example of a *recursively defined* sequence in which each term (after the first few) is given by a formula involving its predecessors. The 13th-century Italian mathematician Fibonacci asked the following question: If we start with a single pair of rabbits that gives birth to a new pair after two months, and each such new pair does the same, how many pairs of rabbits will we have after n months? See Problems 55 and 56.

(d) If the amount $A_0 = 100$ dollars is invested in a savings account that draws 10% interest compounded annually, then the amount A_n in the account at the end of n years is defined (for $n \geqq 1$) by the *iterative formula* $A_n = (1.10)A_{n-1}$ (rounded to the nearest number of cents) in terms of the preceding amount:

$$110.00, 121.00, 133.10, 146.41, 161.05, 177.16,$$
$$194.87, 214.36, 235.79, 259.37, \dots$$

◆

Limits of Sequences

The limit of a sequence is defined in much the same way as the limit of an ordinary function (Section 2.2).

DEFINITION Limit of a Sequence

We say that the sequence $\{a_n\}$ **converges** to the real number L, or has the **limit** L, and we write

$$\lim_{n\to\infty} a_n = L, \tag{4}$$

provided that a_n can be made as close to L as we please merely by choosing n to be sufficiently large. That is, given any number $\epsilon > 0$, there exists an integer N such that

$$|a_n - L| < \epsilon \quad \text{for all } n \geqq N. \tag{5}$$

If the sequence $\{a_n\}$ does *not* converge, then we say that $\{a_n\}$ **diverges.**

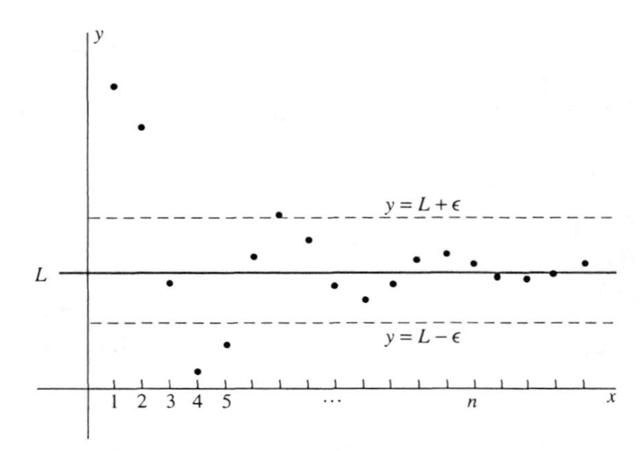

FIGURE 11.2.1 The point (n, a_n) approaches the line $y = L$ as $n \to +\infty$.

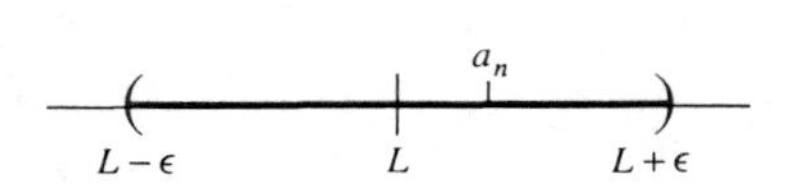

FIGURE 11.2.2 The inequality $|a_n - L| < \epsilon$ means that a_n lies somewhere between $L - \epsilon$ and $L + \epsilon$.

Figure 11.2.1 illustrates geometrically the definition of the limit of a sequence. Because

$$|a_n - L| < \epsilon \quad \text{means that} \quad L - \epsilon < a_n < L + \epsilon,$$

the condition in (5) means that if $n \geqq N$, then the point (n, a_n) lies between the horizontal lines $y = L - \epsilon$ and $y = L + \epsilon$. Alternatively, if $n \geqq N$, then the number a_n lies between the points $L - \epsilon$ and $L + \epsilon$ on the real line (Fig. 11.2.2).

EXAMPLE 3 Suppose that we want to establish rigorously the intuitively evident fact that the sequence $\{1/n\}_1^\infty$ converges to zero,

$$\lim_{n\to\infty} \frac{1}{n} = 0. \tag{6}$$

Because $L = 0$ here, we need only convince ourselves that to each positive number ϵ there corresponds an integer N such that

$$\left|\frac{1}{n}\right| = \frac{1}{n} < \epsilon \quad \text{if} \quad n \geqq N.$$

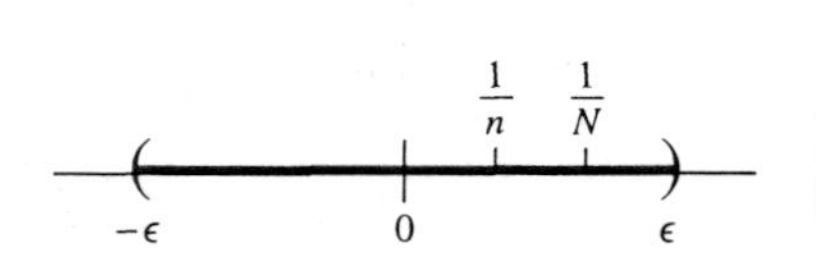

FIGURE 11.2.3 If $N > \dfrac{1}{\epsilon}$ and $n \geqq N$ then $0 < \dfrac{1}{n} \leqq \dfrac{1}{N} < \epsilon$.

But evidently it suffices to choose any fixed integer $N > 1/\epsilon$. Then $n \geqq N$ implies immediately that

$$\frac{1}{n} \leqq \frac{1}{N} < \epsilon,$$

as desired (Fig. 11.2.3). ◆

EXAMPLE 4 (a) The sequence $\{(-1)^n\}$ diverges because its successive terms "oscillate" between the two values $+1$ and -1. Hence $(-1)^n$ cannot approach any single value as $n \to \infty$. (b) The terms of the sequence $\{n^2\}$ increase without bound as $n \to \infty$. Thus the sequence $\{n^2\}$ diverges. In this case, we might also say that $\{n^2\}$ diverges *to infinity.* ◆

Using Limit Laws

The limit laws in Section 2.2 for limits of functions have natural analogues for limits of sequences. Their proofs are based on techniques similar to those used in Appendix D.

THEOREM 1 Limit Laws for Sequences
If the limits

$$\lim_{n\to\infty} a_n = A \quad \text{and} \quad \lim_{n\to\infty} b_n = B$$

exist (so A and B are real numbers), then

1. $\lim_{n\to\infty} ca_n = cA$ (c any real number);

2. $\lim_{n\to\infty} (a_n + b_n) = A + B$;

3. $\lim_{n\to\infty} a_n b_n = AB$;

4. $\lim_{n\to\infty} \dfrac{a_n}{b_n} = \dfrac{A}{B}$.

In part 4 we must assume that $B \neq 0$ (so that $b_n \neq 0$ for all sufficiently large values of n).

THEOREM 2 Substitution Law for Sequences
If $\lim_{n\to\infty} a_n = A$ and the function f is continuous at $x = A$, then

$$\lim_{n\to\infty} f(a_n) = f(A).$$

THEOREM 3 Squeeze Law for Sequences
If $a_n \leqq b_n \leqq c_n$ for all n and

$$\lim_{n\to\infty} a_n = L = \lim_{n\to\infty} c_n,$$

then $\lim_{n\to\infty} b_n = L$ as well.

These theorems can be used to compute limits of many sequences formally, without recourse to the definition. For example, Eq. (6) and the product law of limits yield

$$\lim_{n\to\infty} \frac{1}{n^k} = 0 \tag{7}$$

for every positive integer k.

EXAMPLE 5 Eq. (7) and the limit laws give (after dividing numerator and denominator by the highest power of n that is present)

$$\lim_{n\to\infty} \frac{7n^2}{5n^2 - 3} = \lim_{n\to\infty} \frac{7}{5 - \dfrac{3}{n^2}}$$

$$= \frac{\lim_{n\to\infty} 7}{\left(\lim_{n\to\infty} 5\right) - 3 \cdot \left(\lim_{n\to\infty} \dfrac{1}{n^2}\right)} = \frac{7}{5 - 3 \cdot 0} = \frac{7}{5}.$$ ◆

EXAMPLE 6 Show that $\lim_{n\to\infty} \dfrac{\cos n}{n} = 0.$

Solution This follows from the squeeze law and the fact that $1/n \to 0$ as $n \to \infty$, because

$$-\frac{1}{n} \leq \frac{\cos n}{n} \leq \frac{1}{n}$$

for every positive integer n. ◆

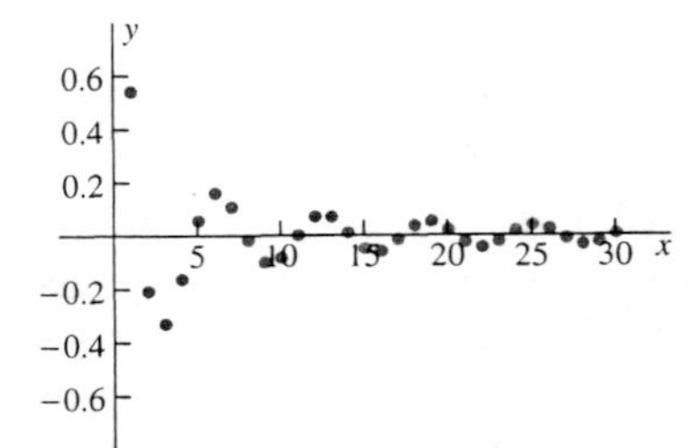

FIGURE 11.2.4 The points $(n, (\cos n)/n)$ for $n = 1, 2, \ldots, 30$.

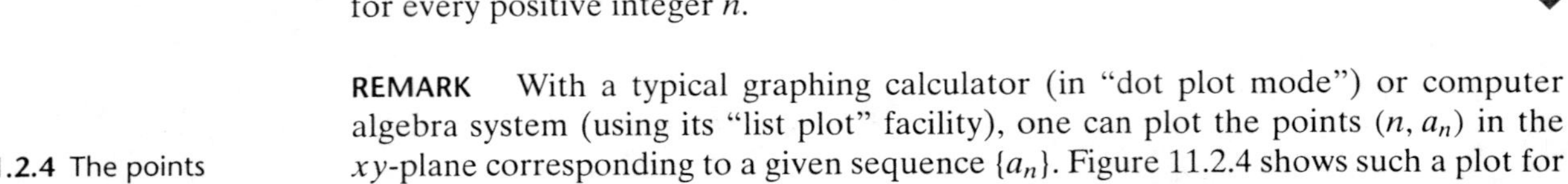

REMARK With a typical graphing calculator (in "dot plot mode") or computer algebra system (using its "list plot" facility), one can plot the points (n, a_n) in the xy-plane corresponding to a given sequence $\{a_n\}$. Figure 11.2.4 shows such a plot for the sequence of Example 6 and provides visual evidence of its convergence to zero.

EXAMPLE 7 Show that if $a > 0$, then $\lim_{n\to\infty} \sqrt[n]{a} = 1$.

Solution We apply the substitution law with $f(x) = a^x$, $a_n = 1/n$, and $A = 0$. Because $1/n \to 0$ as $n \to \infty$ and f is continuous at $x = 0$, this gives

$$\lim_{n\to\infty} a^{1/n} = \lim_{n\to\infty} f(1/n) = f(0) = a^0 = 1.$$ ◆

EXAMPLE 8 The limit laws and the continuity of $f(x) = \sqrt{x}$ at $x = 4$ yield

$$\lim_{n\to\infty} \sqrt{\frac{4n-1}{n+1}} = \left(\lim_{n\to\infty} \frac{4 - \dfrac{1}{n}}{1 + \dfrac{1}{n}} \right)^{1/2} = \sqrt{4} = 2.$$ ◆

EXAMPLE 9 Show that if $|r| < 1$, then $\lim_{n\to\infty} r^n = 0$.

Solution Because $|r^n| = |(-r)^n|$, we may assume that $0 < r < 1$. Then $1/r = 1 + a$ for some number $a > 0$, so the binomial formula yields

$$\frac{1}{r^n} = (1+a)^n = 1 + na + \{\text{positive terms}\} > 1 + na;$$

$$0 < r^n < \frac{1}{1+na}.$$

Now $1/(1+na) \to 0$ as $n \to \infty$. Therefore, the squeeze law implies that $r^n \to 0$ as $n \to \infty$. ◆

Figure 11.2.5 shows the graph of a function f such that $\lim_{x\to\infty} f(x) = L$. If the sequence $\{a_n\}$ is defined by the formula $a_n = f(n)$ for each positive integer n,

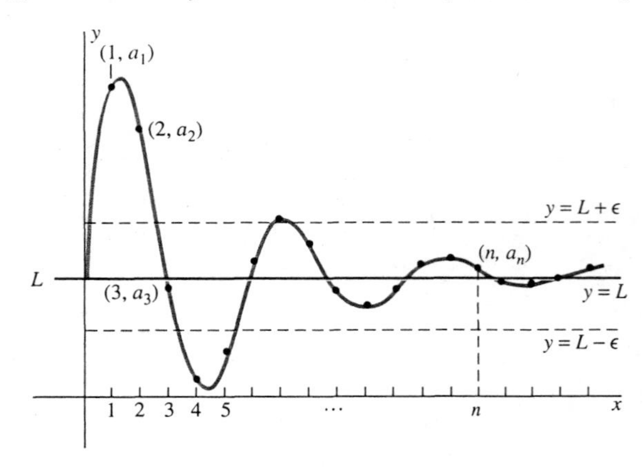

FIGURE 11.2.5 If $\lim_{x\to\infty} f(x) = L$ and $a_n = f(n)$, then $\lim_{n\to\infty} a_n = L$.

then all the points $(n, f(n))$ lie on the graph of $y = f(x)$. It therefore follows from the definition of the limit of a function that $\lim_{n\to\infty} a_n = L$ as well.

THEOREM 4 Limits of Functions and Sequences
If $a_n = f(n)$ for each positive integer n, then

$$\lim_{x\to\infty} f(x) = L \quad \text{implies that} \quad \lim_{n\to\infty} a_n = L. \tag{8}$$

The converse of the statement in (8) is generally false. For example, take $f(x) = \sin \pi x$ and, for each positive integer n, let $a_n = f(n) = \sin n\pi$. Then $\sin n\pi \equiv 0$, but $\sin nx$ oscillates between 1 and -1, so

$$\lim_{n\to\infty} a_n = \lim_{n\to\infty} \sin n\pi = 0, \quad \text{but}$$

$$\lim_{x\to\infty} f(x) = \lim_{x\to\infty} \sin \pi x \quad \text{does not exist.}$$

Because of (8) we can use **l'Hôpital's rule for sequences:** If $a_n = f(n), b_n = g(n)$, and $f(x)/g(x)$ has the indeterminate form ∞/∞ as $x \to \infty$, then

$$\lim_{n\to\infty} \frac{a_n}{b_n} = \lim_{x\to\infty} \frac{f(x)}{g(x)} = \lim_{x\to\infty} \frac{f'(x)}{g'(x)}, \tag{9}$$

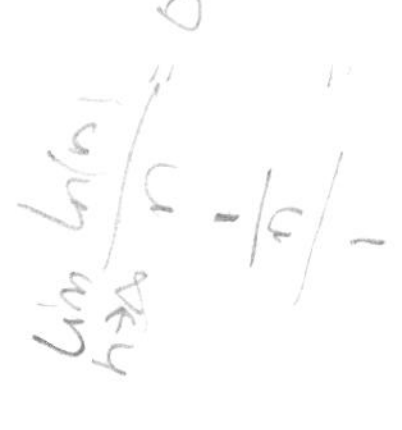

provided that f and g satisfy the other hypotheses of l'Hôpital's rule, including the important assumption that the right-hand limit exists.

EXAMPLE 10 Show that $\lim_{n\to\infty} \dfrac{\ln n}{n} = 0.$

Solution The function $(\ln x)/x$ is defined for all $x \geqq 1$ and agrees with the given sequence $\{(\ln n)/n\}$ when $x = n$, a positive integer. Because $(\ln x)/x$ has the indeterminate form ∞/∞ as $x \to \infty$, l'Hôpital's rule gives

$$\lim_{n\to\infty} \frac{\ln n}{n} = \lim_{x\to\infty} \frac{\ln x}{x} = \lim_{x\to\infty} \frac{\frac{1}{x}}{1} = 0. \quad \blacklozenge$$

EXAMPLE 11 Show that $\lim_{n\to\infty} \sqrt[n]{n} = 1.$

Solution First we note that

$$\ln \sqrt[n]{n} = \ln n^{1/n} = \frac{\ln n}{n} \to 0 \quad \text{as } n \to \infty,$$

by Example 10. By the substitution law with $f(x) = e^x$, this gives

$$\lim_{n\to\infty} n^{1/n} = \lim_{n\to\infty} \exp\left(\ln n^{1/n}\right) = e^0 = 1. \quad \blacklozenge$$

EXAMPLE 12 Find $\lim_{n\to\infty} \dfrac{3n^3}{e^{2n}}.$

Solution We apply l'Hôpital's rule repeatedly, although we must be careful at each intermediate step to verify that we still have an indeterminate form. Thus we find that

$$\lim_{n\to\infty} \frac{3n^3}{e^{2n}} = \lim_{x\to\infty} \frac{3x^3}{e^{2x}} = \lim_{x\to\infty} \frac{9x^2}{2e^{2x}} = \lim_{x\to\infty} \frac{18x}{4e^{2x}} = \lim_{x\to\infty} \frac{18}{8e^{2x}} = 0. \quad \blacklozenge$$

Bounded Monotonic Sequences

The set of all *rational* numbers has by itself all of the most familiar elementary algebraic properties of the entire real number system. To guarantee the existence of irrational numbers, we must assume in addition a "completeness property" of the real numbers. Otherwise, the real line might have "holes" where the irrational numbers ought to be. One way of stating this completeness property is in terms of the convergence of an important type of sequence, a bounded monotonic sequence.

The sequence $\{a_n\}_1^\infty$ is said to be **increasing** if

$$a_1 \leqq a_2 \leqq a_3 \leqq \cdots \leqq a_n \leqq \cdots$$

and **decreasing** if

$$a_1 \geqq a_2 \geqq a_3 \geqq \cdots \geqq a_n \geqq \cdots.$$

The sequence $\{a_n\}$ is **monotonic** if it is either increasing or decreasing. The sequence $\{a_n\}$ is **bounded** if there is a number M such that $|a_n| \leqq M$ for all n. The following assertion may be taken to be an axiom for the real number system.

Bounded Monotonic Sequence Property

Every bounded monotonic infinite sequence converges—that is, has a finite limit.

Suppose, for example, that the increasing sequence $\{a_n\}_1^\infty$ is bounded above by a number M, meaning that $a_n \leqq M$ for all $n \geqq 1$. Because the sequence is also bounded below (by a_1, for instance), the bounded monotonic sequence property implies that

$$\lim_{n\to\infty} a_n = A \quad \text{for some real number } A \leqq M,$$

as in Fig. 11.2.6(a). If the increasing sequence $\{a_n\}$ is not bounded above, then it follows that

$$\lim_{n\to\infty} a_n = +\infty$$

as in Fig. 11.2.6(b). (See Problem 52.) Figure 11.2.7 illustrates the graph of a typical bounded increasing sequence, with the heights of the points (n, a_n) steadily rising toward A.

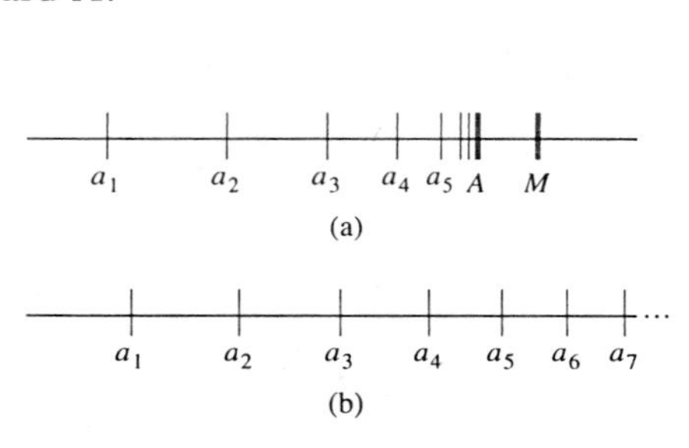

FIGURE 11.2.6 (a) If the increasing sequence $\{a_n\}$ is bounded above by M, then its terms "pile up" at some point $A \leqq M$. (b) If the sequence is unbounded, then its terms "keep going" and diverge to infinity.

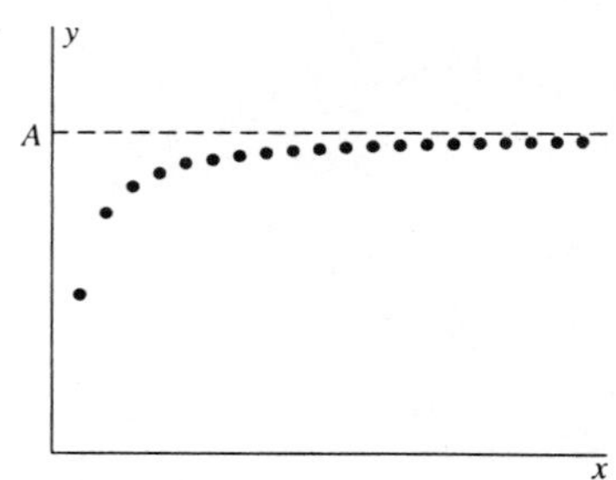

FIGURE 11.2.7 Graph of a bounded increasing sequence with limit A.

EXAMPLE 13 Investigate the sequence $\{a_n\}$ that is defined recursively by

$$a_1 = \sqrt{6}, \qquad a_{n+1} = \sqrt{6 + a_n} \quad \text{for} \quad n \geqq 1. \tag{10}$$

Solution The first four terms of this sequence are

$$\sqrt{6}, \quad \sqrt{6+\sqrt{6}}, \quad \sqrt{6+\sqrt{6+\sqrt{6}}}, \quad \sqrt{6+\sqrt{6+\sqrt{6+\sqrt{6}}}}. \tag{11}$$

If the sequence $\{a_n\}$ converges, then its limit A would seem to be the natural interpretation of the infinite expression

$$\sqrt{6+\sqrt{6+\sqrt{6+\sqrt{6+\cdots}}}}.$$

A calculator gives 2.449, 2.907, 2.984, and 2.997 for the approximate values of the terms in (11). This suggests that the sequence may be bounded above by $M = 3$. Indeed, if we assume that a particular term a_n satisfies the inequality $a_n < 3$, then it follows that

$$a_{n+1} = \sqrt{6 + a_n} < \sqrt{6 + 3} = 3;$$

that is, $a_{n+1} < 3$ as well. Can you see that this implies that *all* terms of the sequence are less than 3? (If there were a first term not less than 3, then its predecessor would be less than 3, and we would have a contradiction. This is a "proof by mathematical induction.")

In order to apply the bounded monotonic sequence property to conclude that the sequence $\{a_n\}$ converges, it remains to show that it is an increasing sequence. But

$$(a_{n+1})^2 - (a_n)^2 = (6 + a_n) - (a_n)^2 = (2 + a_n)(3 - a_n) > 0$$

because $a_n < 3$. Because all terms of the sequence are positive (why?), it therefore follows that $a_{n+1} > a_n$ for all $n \geqq 1$, as desired.

Now that we know that the limit A of the sequence $\{a_n\}$ exists, we can write

$$A = \lim_{n\to\infty} a_{n+1} = \lim_{n\to\infty} \sqrt{6 + a_n} = \sqrt{6 + A},$$

and thus $A^2 = 6 + A$. The roots of this quadratic equation are -2 and 3. Because $A > 0$ (why?), we conclude that $A = \lim_{n\to\infty} a_n = 3$, and so

$$\sqrt{6 + \sqrt{6 + \sqrt{6 + \sqrt{6 + \cdots}}}} = 3. \quad \textbf{(12)}$$

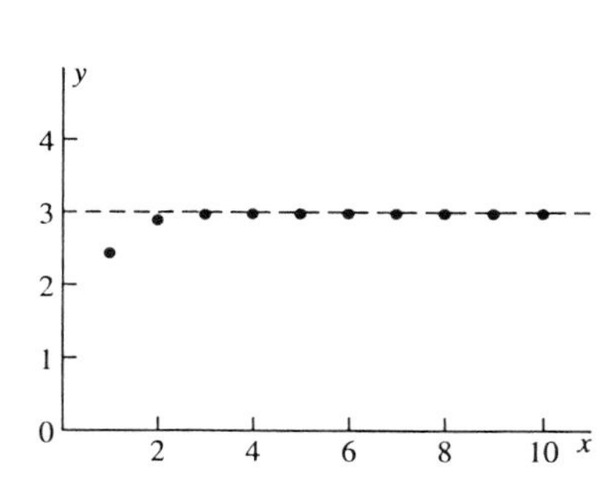

FIGURE 11.2.8 Graph of the sequence of Example 13.

The graph in Fig. 11.2.8 of the first ten terms of the sequence $\{a_n\}$ shows that the convergence to its limit 3 is quite rapid. ◆

To indicate what the bounded monotonic sequence property has to do with the "completeness property" of the real numbers, in Problem 63 we outline a proof, using this property, of the existence of the number $\sqrt{2}$. In Problems 61 and 62, we outline a proof of the equivalence of the bounded monotonic sequence property and another common statement of the completeness of the real number system—the *least upper bound property.*

11.2 TRUE/FALSE STUDY GUIDE

11.2 CONCEPTS: QUESTIONS AND DISCUSSION

1. Can a sequence $\{a_n\}_1^\infty$ converge to two different numbers?
2. Suppose it is known that every open interval containing the point L contains all but finitely many members of the sequence $\{a_n\}_1^\infty$. Does this imply that $\lim_{n\to\infty} a_n = L$?
3. Suppose that the sequence $\{a_n\}_1^\infty$ is obtained by interspersing the members of the two convergent infinite sequences $\{p_n\}_1^\infty$ and $\{q_n\}_1^\infty$. Does it follow that the sequence $\{a_n\}_1^\infty$ also converges?

11.2 PROBLEMS

In Problems 1 through 8, find a pattern in the sequence with given terms a_1, a_2, a_3, a_4 and (assuming that it continues as indicated) write a formula for the general term a_n of the sequence.

1. $1, 4, 9, 16, \dots$ **2.** $2, 7, 12, 17, \dots$

3. $\frac{1}{3}, \frac{1}{9}, \frac{1}{27}, \frac{1}{81}, \dots$ **4.** $1, -\frac{1}{2}, \frac{1}{4}, -\frac{1}{8}, \dots$

5. $\frac{1}{2}, \frac{1}{5}, \frac{1}{8}, \frac{1}{11}, \dots$ **6.** $\frac{1}{2}, \frac{1}{5}, \frac{1}{10}, \frac{1}{17}, \dots$

7. $0, 2, 0, 2, \dots$ **8.** $10, 5, 10, 5, \dots$

In Problems 9 through 42, determine whether or not the sequence $\{a_n\}$ converges, and find its limit if it does converge.

9. $a_n = \dfrac{2n}{5n - 3}$ **10.** $a_n = \dfrac{1 - n^2}{2 + 3n^2}$

11. $a_n = \dfrac{n^2 - n + 7}{2n^3 + n^2}$ **12.** $a_n = \dfrac{n^3}{10n^2 + 1}$

13. $a_n = 1 + \left(\frac{9}{10}\right)^n$ **14.** $a_n = 2 - \left(-\frac{1}{2}\right)^n$

15. $a_n = 1 + (-1)^n$

16. $a_n = \dfrac{1 + (-1)^n}{\sqrt{n}}$

17. $a_n = \dfrac{1 + (-1)^n\sqrt{n}}{\left(\frac{3}{2}\right)^n}$

18. $a_n = \dfrac{\sin n}{3^n}$

19. $a_n = \dfrac{\sin^2 n}{\sqrt{n}}$

20. $a_n = \sqrt{\dfrac{2 + \cos n}{n}}$

21. $a_n = n \sin \pi n$

22. $a_n = n \cos \pi n$

23. $a_n = \pi^{-(\sin n)/n}$

24. $a_n = 2^{\cos \pi n}$

25. $a_n = \dfrac{\ln n}{\sqrt{n}}$

26. $a_n = \dfrac{\ln 2n}{\ln 3n}$

27. $a_n = \dfrac{(\ln n)^2}{n}$

28. $a_n = n \sin\left(\dfrac{1}{n}\right)$

29. $a_n = \dfrac{\tan^{-1} n}{n}$

30. $a_n = \dfrac{n^3}{e^{n/10}}$

31. $a_n = \dfrac{2^n + 1}{e^n}$

32. $a_n = \dfrac{\sinh n}{\cosh n}$

33. $a_n = \left(1 + \dfrac{1}{n}\right)^n$

34. $a_n = (2n + 5)^{1/n}$

35. $a_n = \left(\dfrac{n-1}{n+1}\right)^n$

36. $a_n = (0.001)^{-1/n}$

37. $a_n = \sqrt[n]{2^{n+1}}$

38. $a_n = \left(1 - \dfrac{2}{n^2}\right)^n$

39. $a_n = \left(\dfrac{2}{n}\right)^{3/n}$

40. $a_n = (-1)^n(n^2 + 1)^{1/n}$

41. $a_n = \left(\dfrac{2 - n^2}{3 + n^2}\right)^n$

42. $a_n = \dfrac{\left(\frac{2}{3}\right)^n}{1 - \sqrt[n]{n}}$

In Problems 43 through 50, investigate the given sequence $\{a_n\}$ numerically or graphically. Formulate a reasonable guess for the value of its limit. Then apply limit laws to verify that your guess is correct.

43. $a_n = \dfrac{n - 2}{n + 13}$

44. $a_n = \dfrac{2n + 3}{5n - 17}$

45. $a_n = \sqrt{\dfrac{4n^2 + 7}{n^2 + 3n}}$

46. $a_n = \left(\dfrac{n^3 - 5}{8n^3 + 7n}\right)^{1/3}$

47. $a_n = e^{-1/\sqrt{n}}$

48. $a_n = n \sin \dfrac{2}{n}$

49. $a_n = 4 \tan^{-1} \dfrac{n - 1}{n + 1}$

50. $a_n = 3 \sin^{-1} \sqrt{\dfrac{3n - 1}{4n + 1}}$

51. Prove that if $\lim_{n\to\infty} a_n = A \neq 0$, then the sequence $\{(-1)^n a_n\}$ diverges.

52. Prove that if the increasing sequence $\{a_n\}$ is not bounded, then $\lim_{n\to\infty} a_n = +\infty$. (It's largely a matter of saying precisely what this means.)

53. Suppose that $A > 0$. Given $x_1 \neq 0$ but otherwise arbitrary, define the sequence $\{x_n\}$ recursively by

$$x_{n+1} = \frac{1}{2} \cdot \left(x_n + \frac{A}{x_n}\right) \quad \text{if} \quad n \geqq 1.$$

Prove that if $L = \lim_{n\to\infty} x_n$ exists, then $L = \pm\sqrt{A}$.

54. Suppose that A is a fixed real number. Given $x_1 \neq 0$ but otherwise arbitrary, define the sequence $\{x_n\}$ recursively by

$$x_{n+1} = \frac{1}{3} \cdot \left(2x_n + \frac{A}{(x_n)^2}\right) \quad \text{if} \quad n \geqq 1.$$

Prove that if $L = \lim_{n\to\infty} x_n$ exists, then $L = \sqrt[3]{A}$.

55. (a) Suppose that every newborn pair of rabbits becomes productive after two months, and thereafter gives birth to a new pair of rabbits every month. If we begin with a single newborn pair of rabbits, denote by F_n the total number of pairs of rabbits we have after n months. Explain carefully why $\{F_n\}$ is the Fibonacci sequence of Example 2. (b) If, instead, every newborn pair of rabbits becomes productive after three months, denote by $\{G_n\}$ the number of pairs of rabbits we have after n months. Give a recursive definition of the sequence $\{G_n\}$ and calculate its first ten terms.

56. Let $\{F_n\}$ be the Fibonacci sequence of Example 2, and assume that

$$\tau = \lim_{n\to\infty} \frac{F_{n+1}}{F_n}$$

exists. (It does.) Show that $\tau = \frac{1}{2}(1+\sqrt{5})$. (*Suggestion:* Write $a_n = F_n/F_{n-1}$ and show that $a_{n+1} = 1 + (1/a_n)$.)

57. Let the sequence $\{a_n\}$ be defined recursively as follows:

$$a_1 = 2; \qquad a_{n+1} = \tfrac{1}{2}(a_n + 4) \quad \text{for } n \geqq 1.$$

(a) Prove by induction on n that $a_n < 4$ for each n and that $\{a_n\}$ is an increasing sequence. (b) Find the limit of this sequence.

58. Investigate as in Example 13 the sequence $\{a_n\}$ that is defined recursively by

$$a_1 = \sqrt{2}, \qquad a_{n+1} = \sqrt{2 + a_n} \quad \text{for} \quad n \geqq 1.$$

In particular, show that

$$\sqrt{2 + \sqrt{2 + \sqrt{2 + \sqrt{2 + \cdots}}}} = 2.$$

Verify the results stated in Problems 59 and 60.

59. $\sqrt{20 + \sqrt{20 + \sqrt{20 + \sqrt{20 + \cdots}}}} = 5.$

60. $\sqrt{90 + \sqrt{90 + \sqrt{90 + \sqrt{90 + \cdots}}}} = 10.$

Problems 61 and 62 deal with the least upper bound *property of the real numbers: If the nonempty set S of real numbers has an upper bound, then S has a least upper bound. The number M is an* **upper bound** *for the set S if $x \leqq M$ for all x in S. The upper bound L of S is a* **least upper bound** *for S if no number smaller than L is an upper bound for S. You can easily show that if the set S has least upper bounds L_1 and L_2, then $L_1 = L_2$; in other words, if a least upper bound for a set exists, then it is unique.*

61. Prove that the least upper bound property implies the bounded monotonic sequence property. (*Suggestion:* If $\{a_n\}$ is a bounded increasing sequence and A is the least upper bound of the set $\{a_n : n \geqq 1\}$ of terms of the sequence, you can prove that $A = \lim_{n\to\infty} a_n$.)

62. Prove that the bounded monotonic sequence property implies the least upper bound property. (*Suggestion:* For each positive integer n, let a_n be the least integral multiple of $1/10^n$ that is an upper bound of the set S. Prove that $\{a_n\}$ is a bounded decreasing sequence and then that $A = \lim_{n\to\infty} a_n$ is a least upper bound for S.)

63. For each positive integer n, let a_n be the largest integral multiple of $1/10^n$ such that $a_n^2 \leqq 2$. (a) Prove that $\{a_n\}$ is a bounded increasing sequence, so $A = \lim_{n\to\infty} a_n$ exists. (b) Prove that if $A^2 > 2$, then $a_n^2 > 2$ for n sufficiently large. (c) Prove that if $A^2 < 2$, then $a_n^2 < B$ for some number $B < 2$ and all sufficiently large n. (d) Conclude that $A^2 = 2$.

64. Investigate the sequence $\{a_n\}$, where

$$a_n = \left[\!\left[n + \tfrac{1}{2} + \sqrt{n} \right]\!\right].$$

You may need a computer or programmable calculator to discover what is remarkable about this sequence.

11.2 Project: Nested Radicals and Continued Fractions

This project is an investigation of the relation

$$\sqrt{q + p\sqrt{q + p\sqrt{q + p\sqrt{q + \cdots}}}} = p + \cfrac{q}{p + \cfrac{q}{p + \cfrac{q}{p + \cdots}}} \tag{1}$$

where p and q are positive. We ask not only whether this equation could possibly be true, but also what it means. In the following two numerical explorations, you can (for instance) take p and q to be the last two nonzero digits in your student I.D. number.

Exploration 1 Define the infinite sequence $\{a_n\}$ recursively by

$$a_1 = \sqrt{q} \quad \text{and} \quad a_{n+1} = \sqrt{q + pa_n} \quad \text{for} \quad n \geqq 1. \tag{2}$$

Use a calculator or computer to approximate enough terms of this sequence numerically to determine whether it appears to converge. Assuming that it does, write the first several terms symbolically and conclude that $A = \lim_{n\to\infty} a_n$ is a natural interpretation of the *nested radical* on the left-hand side in (1). Finally, take the limit in the recursion in (2) to show that A is the positive solution of the quadratic equation $x^2 - px - q = 0$. Does the quadratic formula then yield a result consistent with your numerical evidence?

Exploration 2 Define the infinite sequence $\{b_n\}$ recursively by

$$b_1 = p \quad \text{and} \quad b_{n+1} = p + \frac{q}{b_n} \quad \text{for} \quad n \geqq 1. \tag{3}$$

Use a calculator or computer to approximate enough terms of this sequence numerically to determine whether or not it appears to converge. Assuming that it does, write the first several terms symbolically and conclude that $B = \lim_{n\to\infty} b_n$ is a natural interpretation of the *continued fraction* on the right-hand side in (1). Finally, take the limit in the recursion in (3) to show that B is also the positive solution of the quadratic equation $x^2 - px - q = 0$. Conclude thereby that Eq. (1) is indeed true.

11.3 INFINITE SERIES AND CONVERGENCE

An **infinite series** is an expression of the form

$$\sum_{n=1}^{\infty} a_n = a_1 + a_2 + a_3 + \cdots + a_n + \cdots, \tag{1}$$

where $\{a_n\}$ is an infinite sequence of real numbers. The number a_n is called the ***n*th term** of the series. The symbol $\sum_{n=1}^{\infty} a_n$ is simply an abbreviation for the right-hand side of Eq. (1). In this section we discover what is meant by the **sum** of an infinite series.

EXAMPLE 1 Consider the infinite series

$$\sum_{n=1}^{\infty} \frac{1}{2^n} = \frac{1}{2} + \frac{1}{4} + \frac{1}{8} + \frac{1}{16} + \cdots + \frac{1}{2^n} + \cdots \qquad (2)$$

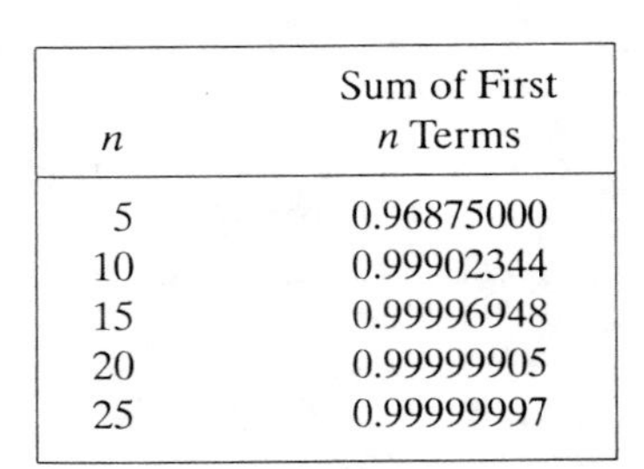

n	Sum of First n Terms
5	0.96875000
10	0.99902344
15	0.99996948
20	0.99999905
25	0.99999997

FIGURE 11.3.1 Sums of terms in the infinite series of Example 1.

that was mentioned in Section 11.1; its nth term is $a_n = 1/2^n$. Although we cannot literally add an infinite number of terms, we can add any finite number of the terms in (2). For instance, the sum of the first five terms is

$$\frac{1}{2} + \frac{1}{4} + \frac{1}{8} + \frac{1}{16} + \frac{1}{32} = \frac{31}{32} = 0.96875.$$

We could add five more terms, then five more, and so forth. The table in Fig. 11.3.1 shows what happens. It appears that the sums get closer and closer to 1 as we add more and more terms. If indeed this is so, then it is natural to say that the sum of the (whole) infinite series in (2) is 1, and hence to write

$$\sum_{n=1}^{\infty} \frac{1}{2^n} = \frac{1}{2} + \frac{1}{4} + \frac{1}{8} + \frac{1}{16} + \cdots + \frac{1}{2^n} + \cdots = 1. \qquad ◆$$

Motivated by Example 1, we introduce the *partial sums* of the general infinite series in (1). The **nth partial sum** S_n of the series is the sum of its first n terms:

$$S_n = a_1 + a_2 + a_3 + \cdots + a_n = \sum_{k=1}^{n} a_k. \qquad (3)$$

Thus each infinite series has not only an infinite sequence of terms, but also an **infinite sequence of partial sums** $S_1, S_2, S_3, \ldots, S_n, \ldots$, where

$$\begin{aligned} S_1 &= a_1, \\ S_2 &= a_1 + a_2, \\ S_3 &= a_1 + a_2 + a_3, \\ &\vdots \\ S_{10} &= a_1 + a_2 + a_3 + a_4 + a_5 + a_6 + a_7 + a_8 + a_9 + a_{10}, \end{aligned}$$

and so forth. We define the sum of the infinite series to be the limit of its sequence of partial sums, provided that this limit exists.

DEFINITION The Sum of an Infinite Series

We say that the infinite series

$$\sum_{n=1}^{\infty} a_n \quad \textbf{converges} \text{ (or is } \textbf{convergent}\text{)}$$

with **sum** S provided that the limit of its sequence of partial sums,

$$S = \lim_{n \to \infty} S_n, \qquad (4)$$

exists (and is finite). Otherwise we say that the series **diverges** (or is **divergent**). If a series diverges, then it has no sum.

Thus the sum of an infinite series is a limit of finite sums,

$$S = \sum_{n=1}^{\infty} a_n = \lim_{N \to \infty} \sum_{n=1}^{N} a_n,$$

provided that this limit exists.

EXAMPLE 1 (continued) Show that the series

$$\sum_{n=1}^{\infty}\left(\frac{1}{2}\right)^n = \frac{1}{2}+\frac{1}{4}+\frac{1}{8}+\frac{1}{16}+\cdots$$

converges. Then find its sum.

Solution The first four partial sums are

$$S_1 = \frac{1}{2},\qquad S_2 = \frac{3}{4},\qquad S_3 = \frac{7}{8},\qquad \text{and} \qquad S_4 = \frac{15}{16}.$$

It seems likely that $S_n = (2^n - 1)/2^n$, and indeed this follows easily by induction, because

$$S_{n+1} = S_n + \frac{1}{2^{n+1}} = \frac{2^n-1}{2^n} + \frac{1}{2^{n+1}} = \frac{2^{n+1}-2+1}{2^{n+1}} = \frac{2^{n+1}-1}{2^{n+1}}.$$

Hence the sum of the given series is

$$S = \lim_{n\to\infty} S_n = \lim_{n\to\infty}\frac{2^n-1}{2^n} = \lim_{n\to\infty}\left(1-\frac{1}{2^n}\right) = 1.$$

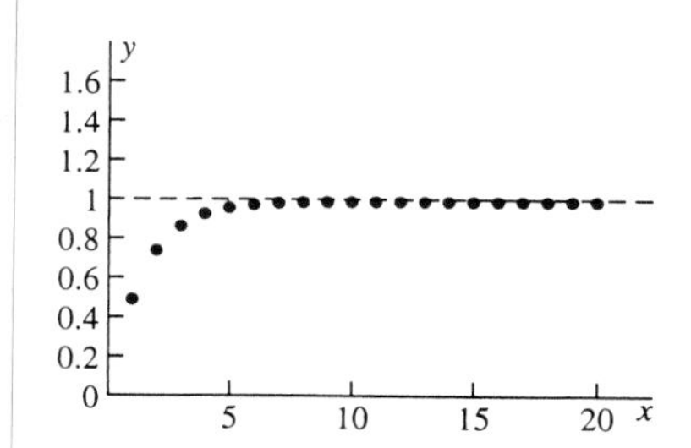

FIGURE 11.3.2 Graph of the first 20 partial sums of the infinite series in Example 1.

The graph in Fig. 11.3.2 illustrates the convergence of the partial sums to the number 1. ◆

EXAMPLE 2 Show that the series

$$\sum_{n=1}^{\infty}(-1)^{n+1} = 1-1+1-1+\cdots$$

diverges.

Solution The sequence of partial sums of this series is

$$1, 0, 1, 0, 1, \ldots,$$

which has no limit. Therefore the series diverges. ◆

EXAMPLE 3 Show that the infinite series

$$\sum_{n=1}^{\infty}\frac{1}{n(n+1)}$$

converges. Then find its sum.

Solution We need a formula for the nth partial sum S_n so that we can evaluate its limit as $n \to \infty$. To find such a formula, we begin with the observation that the nth term of the series is

$$a_n = \frac{1}{n(n+1)} = \frac{1}{n} - \frac{1}{n+1}.$$

(In more complicated cases, such as those in Problems 50 through 55, such a decomposition can be obtained by the method of partial fractions.) It follows that the sum of the first n terms of the given series is

$$\begin{aligned} S_n &= \left(1-\frac{1}{2}\right)+\left(\frac{1}{2}-\frac{1}{3}\right)+\left(\frac{1}{3}-\frac{1}{4}\right) \\ &\quad + \left(\frac{1}{4}-\frac{1}{5}\right)+\cdots+\left(\frac{1}{n}-\frac{1}{n+1}\right) \\ &= 1-\frac{1}{n+1} = \frac{n}{n+1}. \end{aligned}$$

Hence

$$\sum_{n=1}^{\infty} \frac{1}{n(n+1)} = \lim_{n\to\infty} \frac{n}{n+1} = 1. \qquad \blacklozenge$$

The sum for S_n in Example 3, called a *telescoping* sum, provides us with a way to find the sums of certain series. The series in Examples 1 and 2 are examples of a more common and more important type of series, the *geometric series.*

DEFINITION Geometric Series
The series $\sum_{n=0}^{\infty} a_n$ is said to be a **geometric series** if each term after the first is a fixed multiple of the term immediately before it. That is, there is a number r, called the **ratio** of the series, such that

$$a_{n+1} = r a_n \quad \text{for all } n \geqq 0.$$

If we write $a = a_0$ for the initial constant term, then $a_1 = ar, a_2 = ar^2, a_3 = ar^3$, and so forth. Thus every geometric series takes the form

$$a + ar + ar^2 + ar^3 + \cdots = \sum_{n=0}^{\infty} ar^n. \tag{5}$$

Note that the summation begins at $n = 0$ (rather than at $n = 1$). It is therefore convenient to regard the sum

$$S_n = a(1 + r + r^2 + r^3 + \cdots + r^n)$$

of the first $n+1$ terms as the nth partial sum of the series.

EXAMPLE 4 The infinite series

$$\sum_{n=0}^{\infty} \frac{2}{3^n} = 2 + \frac{2}{3} + \frac{2}{9} + \cdots + \frac{2}{3^n} + \cdots$$

is a geometric series whose first term is $a = 2$ and whose ratio is $r = \frac{1}{3}$. $\blacklozenge$

THEOREM 1 The Sum of a Geometric Series
If $|r| < 1$, then the geometric series in Eq. (5) converges, and its sum is

$$S = \sum_{n=0}^{\infty} ar^n = \frac{a}{1-r}. \tag{6}$$

If $|r| \geqq 1$ and $a \neq 0$, then the geometric series diverges.

PROOF If $r = 1$, then $S_n = (n+1)a$, so the series certainly diverges if $a \neq 0$. If $r = -1$ and $a \neq 0$, then the series diverges by an argument like the one in Example 2. So we may suppose that $|r| \neq 1$. Then the elementary identity

$$1 + r + r^2 + r^3 + \cdots + r^n = \frac{1 - r^{n+1}}{1 - r}$$

follows if we multiply each side by $1 - r$. Hence the nth partial sum of the geometric series is

$$S_n = a(1 + r + r^2 + r^3 + \cdots + r^n) = a\left(\frac{1}{1-r} - \frac{r^{n+1}}{1-r}\right).$$

If $|r| < 1$, then $r^{n+1} \to 0$ as $n \to \infty$, by Example 9 in Section 11.2. So in this case the geometric series converges to

$$S = \lim_{n\to\infty} a \cdot \left(\frac{1}{1-r} - \frac{r^{n+1}}{1-r}\right) = \frac{a}{1-r}.$$

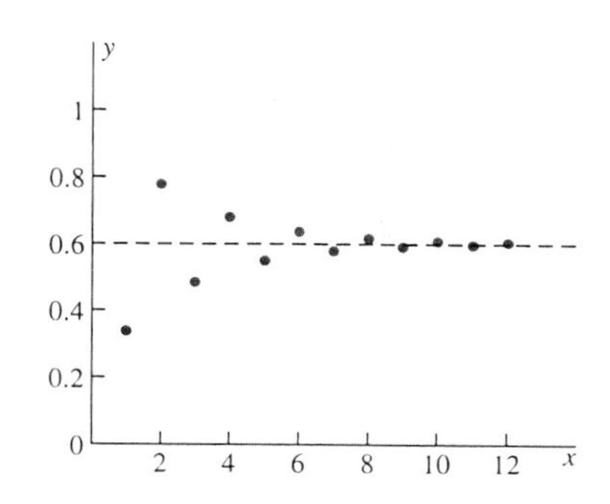

FIGURE 11.3.3 Graph of the first dozen partial sums of the infinite series in Example 5.

But if $|r| > 1$, then $\lim_{n\to\infty} r^{n+1}$ does not exist, so $\lim_{n\to\infty} S_n$ does not exist. This establishes the theorem. ◄

EXAMPLE 5 With $a = 1$ and $r = -\frac{2}{3}$, we find that

$$1 - \frac{2}{3} + \frac{4}{9} - \frac{8}{27} + \cdots = \sum_{n=0}^{\infty}\left(-\frac{2}{3}\right)^n = \frac{1}{1-\left(-\frac{2}{3}\right)} = \frac{3}{5}.$$

The graph in Fig. 11.3.3 shows the partial sums of this series approaching its sum $\frac{3}{5}$ alternately from above and below. ◆

EXAMPLE 6 Determine whether or not the infinite series $\displaystyle\sum_{n=1}^{\infty}\frac{2^{2n-1}}{3^n}$ converges.

Solution If we write this series in the form

$$\sum_{n=1}^{\infty}\frac{2^{2n-1}}{3^n} = \frac{2}{3} + \frac{8}{9} + \frac{32}{27} + \frac{128}{81} + \cdots = \frac{2}{3}\left(1 + \frac{4}{3} + \frac{16}{9} + \frac{64}{27} + \cdots\right),$$

then we recognize it as a geometric series with first term $a = \frac{2}{3}$ and ratio $r = \frac{4}{3}$. Because $r > 1$, the second part of Theorem 1 implies that this series diverges. ◆

Theorem 2 implies that the operations of addition and of multiplication by a constant can be carried out term by term in the case of *convergent* series. Because the sum of an infinite series is the limit of its sequence of partial sums, this theorem follows immediately from the limit laws for sequences (Theorem 1 of Section 11.2).

THEOREM 2 Termwise Addition and Multiplication
If the series $A = \sum a_n$ and $B = \sum b_n$ converge to the indicated sums and c is a constant, then the series $\sum(a_n + b_n)$ and $\sum ca_n$ also converge, with sums

1. $\sum(a_n + b_n) = A + B$;
2. $\sum ca_n = cA$.

The geometric series in Eq. (6) may be used to find the rational number represented by a given infinite repeating decimal.

EXAMPLE 7

$$0.55555\cdots = \frac{5}{10} + \frac{5}{100} + \frac{5}{1000} + \cdots = \frac{5}{10}\left(1 + \frac{1}{10} + \frac{1}{100} + \cdots\right)$$

$$= \sum_{n=0}^{\infty}\frac{5}{10}\left(\frac{1}{10}\right)^n = \frac{\frac{5}{10}}{1-\frac{1}{10}} = \frac{5}{10}\cdot\frac{10}{9} = \frac{5}{9}.$$

In a more complicated situation, we may need to use the termwise algebra of Theorem 2:

$$0.7282828\cdots = \frac{7}{10} + \frac{28}{10^3} + \frac{28}{10^5} + \frac{28}{10^7} + \cdots$$

$$= \frac{7}{10} + \frac{28}{10^3}\left(1 + \frac{1}{10^2} + \frac{1}{10^4} + \cdots\right)$$

$$= \frac{7}{10} + \frac{28}{1000}\sum_{n=0}^{\infty}\left(\frac{1}{100}\right)^n = \frac{7}{10} + \frac{28}{1000}\left(\frac{1}{1-\frac{1}{100}}\right)$$

$$= \frac{7}{10} + \frac{28}{1000}\cdot\frac{100}{99} = \frac{7}{10} + \frac{28}{990} = \frac{721}{990}.$$

This technique can be used to show that every repeating infinite decimal represents a rational number. Consequently, the decimal expansions of irrational numbers such as π, e, and $\sqrt{2}$ must be nonrepeating as well as infinite. Conversely, if p and q are integers with $q \neq 0$, then long division of q into p yields a repeating decimal expansion for the rational number p/q, because such a division can yield at each stage only q possible different remainders. ◆

EXAMPLE 8 Suppose that Paul and Mary toss a fair six-sided die in turn until one of them wins by getting the first "six." If Paul tosses first, calculate the probability that he will win the game.

Solution Because the die is fair, the probability that Paul gets a "six" on the first round is $\frac{1}{6}$. The probability that he gets the game's first "six" on the second round is $\left(\frac{5}{6}\right)^2\left(\frac{1}{6}\right)$—the product of the probability $\left(\frac{5}{6}\right)^2$ that neither Paul nor Mary rolls a "six" in the first round and the probability $\frac{1}{6}$ that Paul rolls a "six" in the second round. Paul's probability p of getting the first "six" in the game is the *sum* of his probabilities of getting it in the first round, in the second round, in the third round, and so on. Hence

$$\begin{aligned} p &= \frac{1}{6} + \left(\frac{5}{6}\right)^2\left(\frac{1}{6}\right) + \left(\frac{5}{6}\right)^2\left(\frac{5}{6}\right)^2\left(\frac{1}{6}\right) + \cdots \\ &= \frac{1}{6}\left[1 + \left(\frac{5}{6}\right)^2 + \left(\frac{5}{6}\right)^4 + \cdots\right] \\ &= \frac{1}{6} \cdot \frac{1}{1 - \left(\frac{5}{6}\right)^2} = \frac{1}{6} \cdot \frac{36}{11} = \frac{6}{11}. \end{aligned}$$

Because he has the advantage of tossing first, Paul has more than the fair probability $\frac{1}{2}$ of getting the first "six" and thus winning the game. ◆

Theorem 3 is often useful in showing that a given series does *not* converge.

THEOREM 3 The nth-Term Test for Divergence

If either

$$\lim_{n\to\infty} a_n \neq 0$$

or this limit does not exist, then the infinite series $\sum a_n$ diverges.

PROOF We want to show under the stated hypothesis that the series $\sum a_n$ diverges. It suffices to show that *if* the series $\sum a_n$ does converge, then $\lim_{n\to\infty} a_n = 0$. So suppose that $\sum a_n$ converges with sum $S = \lim_{n\to\infty} S_n$, where

$$S_n = a_1 + a_2 + a_3 + \cdots + a_n$$

is the nth partial sum of the series. Because $a_n = S_n - S_{n-1}$,

$$\lim_{n\to\infty} a_n = \lim_{n\to\infty} (S_n - S_{n-1}) = \left(\lim_{n\to\infty} S_n\right) - \left(\lim_{n\to\infty} S_{n-1}\right) = S - S = 0.$$

Consequently, if $\lim_{n\to\infty} a_n \neq 0$, then the series $\sum a_n$ diverges. ◄

REMARK It is important to remember also the *contrapositive* of the nth-term divergence test: *If the infinite series* $\sum a_n$ *converges with sum* S*, then its sequence* $\{a_n\}$ *of terms converges to* 0. Thus we have *two* sequences associated with the single infinite series $\sum a_n$: its sequence $\{a_n\}$ of *terms* and its sequence $\{S_n\}$ of *partial sums.* And (assuming that the series converges to S) these two sequences have generally different limits:

$$\lim_{n\to\infty} a_n = 0 \quad \text{and} \quad \lim_{n\to\infty} S_n = S.$$

EXAMPLE 9 The series

$$\sum_{n=1}^{\infty} (-1)^{n-1} n^2 = 1 - 4 + 9 - 16 + 25 - \cdots$$

diverges because $\lim_{n\to\infty} a_n$ does not exist, whereas the series

$$\sum_{n=1}^{\infty} \frac{n}{3n+1} = \frac{1}{4} + \frac{2}{7} + \frac{3}{10} + \frac{4}{13} + \cdots$$

diverges because

$$\lim_{n\to\infty} \frac{n}{3n+1} = \frac{1}{3} \neq 0. \qquad \blacklozenge$$

WARNING The converse of Theorem 3 is *false*! The condition

$$\lim_{n\to\infty} a_n = 0$$

is necessary *but not sufficient* to guarantee convergence of the series

$$\sum_{n=1}^{\infty} a_n.$$

That is, a series may satisfy the condition $a_n \to 0$ as $n \to \infty$ and yet diverge. An important example of a divergent series with terms that approach zero is the **harmonic series**

$$\sum_{n=1}^{\infty} \frac{1}{n} = 1 + \frac{1}{2} + \frac{1}{3} + \frac{1}{4} + \frac{1}{5} + \cdots. \tag{7}$$

THEOREM 4
The harmonic series diverges.

PROOF The nth term of the harmonic series in (7) is $a_n = 1/n$, and Fig. 11.3.4 shows the graph of the related function $f(x) = 1/x$ on the interval $1 \leqq x \leqq n+1$. For each integer k, $1 \leqq k \leqq n$, we have erected on the subinterval $[k, k+1]$ a rectangle with height $f(k) = 1/k$. All of these n rectangles have base length 1, and their respective heights are the successive terms $1, 1/2, 1/3, \ldots, 1/n$ of the harmonic series. Hence the sum of their areas is the nth partial sum

$$S_n = 1 + \frac{1}{2} + \frac{1}{3} + \frac{1}{4} + \cdots + \frac{1}{n}$$

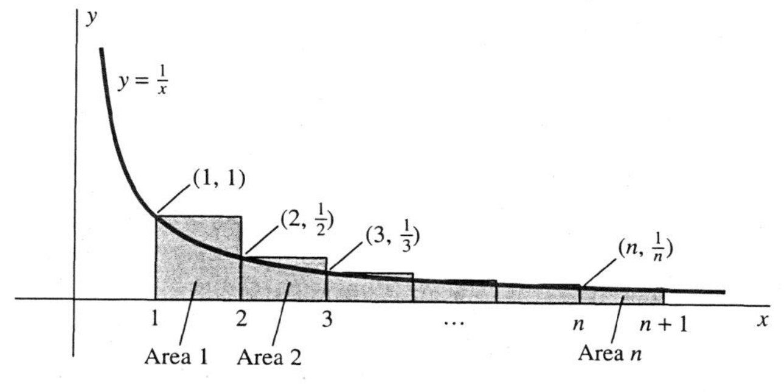

FIGURE 11.3.4 Idea of the proof of Theorem 4.

of the series. Because these rectangles circumscribe the area under the curve $y = 1/x$ from $x = 1$ to $x = n + 1$, we therefore see that S_n must exceed this area. That is,

$$S_n > \int_1^{n+1} \frac{1}{x}\,dx = \Big[\ln x\Big]_1^{n+1} = \ln(n+1).$$

But $\ln(n + 1)$ takes on arbitrarily large positive values with increasing n. Because $S_n > \ln(n + 1)$, it follows that the partial sums of the harmonic series also take on arbitrarily large positive values. Now the terms of the harmonic series are positive, so its sequence of partial sums is increasing. We may therefore conclude that $S_n \to +\infty$ as $n \to +\infty$, and hence that the harmonic series diverges. ◄

If the sequence of partial sums of the series $\sum a_n$ diverges to infinity, then we say that the series **diverges to infinity,** and we write

$$\sum_{n=1}^{\infty} a_n = \infty.$$

The series $\sum(-1)^{n+1}$ of Example 2 is a series that diverges but does not diverge to infinity. In the nineteenth century it was common to say that such a series was divergent by *oscillation;* today we say merely that it diverges.

Our proof of Theorem 4 shows that

$$\sum_{n=1}^{\infty} \frac{1}{n} = \infty.$$

But the partial sums of the harmonic series diverge to infinity very slowly. If N_A denotes the smallest integer such that

$$\sum_{n=1}^{N_A} \frac{1}{n} \geqq A,$$

then with the aid of a programmable calculator you can verify that $N_5 = 83$. With the aid of a computer and refinements of estimates like those in the proof of Theorem 4, one can show that

$$\begin{aligned} N_{10} &= 12367, \\ N_{20} &= 272{,}400{,}600, \\ N_{100} &\approx 1.5 \times 10^{43}, \qquad \text{and} \\ N_{1000} &\approx 1.1 \times 10^{434}. \end{aligned}$$

Thus you would need to add more than a quarter of a billion terms of the harmonic series to get a partial sum that exceeds 20. At this point each of the next few terms would be approximately $0.00000004 = 4 \times 10^{-9}$. The number of terms you'd have to add to reach 1000 is far greater than the estimated number of elementary particles in the entire universe (10^{80}). If you enjoy such large numbers, see the article "Partial sums of infinite series, and how they grow," by R. P. Boas, Jr., in *American Mathematical Monthly* **84** (1977): 237–248.

Theorem 5 says that if two infinite series have the same terms from some point on, then either both series converge or both series diverge. The proof is left for Problem 63.

THEOREM 5 Series that Are Eventually the Same

If there exists a positive integer k such that $a_n = b_n$ for all $n > k$, then the series $\sum a_n$ and $\sum b_n$ either both converge or both diverge.

It follows that a *finite* number of terms can be changed, deleted from, or adjoined to an infinite series without altering its convergence or divergence (although the *sum* of a convergent series will generally be changed by such alterations). In particular, taking $b_n = 0$ for $n \leqq k$ and $b_n = a_n$ for $n > k$, we see that the series

$$\sum_{n=1}^{\infty} a_n = a_1 + a_2 + a_3 + \cdots + a_k + a_{k+1} + \cdots$$

and the series

$$\sum_{n=k+1}^{\infty} a_n = a_{k+1} + a_{k+2} + a_{k+3} + a_{k+4} + \cdots$$

that is obtained by deleting its first k terms either both converge or both diverge.

11.3 TRUE/FALSE STUDY GUIDE

11.3 CONCEPTS: QUESTIONS AND DISCUSSION

1. Can one ever obtain a convergent infinite series by interspersing the terms of two divergent series?
2. Suppose that an infinite series has the property that, given any positive number, all but finitely many terms of the series are positive and less than this number. Does it follows that this series converges? What if it's true that, given any positive number, all but finitely many partial sums of the series are greater than this number? Does it then follow that this series diverges?
3. Can one determine whether a given infinite series converges or diverges merely by computing a sufficiently large number of partial sums?
4. Can one determine the sum—accurate to a given fixed number of decimal places—of a convergent geometric series merely by computing a sufficiently large number of partial sums?

11.3 PROBLEMS

In Problems 1 through 37, determine whether the given infinite series converges or diverges. If it converges, find its sum.

1. $1 + \frac{1}{3} + \frac{1}{9} + \cdots + \frac{1}{3^n} + \cdots$
2. $1 + e^{-1} + e^{-2} + e^{-3} + \cdots + e^{-n} + \cdots$
3. $1 + 3 + 5 + 7 + \cdots + (2n - 1) + \cdots$
4. $\frac{1}{2} + \frac{1}{\sqrt{2}} + \frac{1}{\sqrt[3]{2}} + \cdots + \frac{1}{\sqrt[n]{2}} + \cdots$
5. $1 - 2 + 4 - 8 + 16 - \cdots + (-2)^n + \cdots$
6. $1 - \frac{1}{4} + \frac{1}{16} - \cdots + \left(-\frac{1}{4}\right)^n + \cdots$
7. $4 + \frac{4}{3} + \frac{4}{9} + \frac{4}{27} + \cdots + \frac{4}{3^n} + \cdots$
8. $\frac{1}{3} + \frac{2}{9} + \frac{4}{27} + \frac{8}{81} + \cdots + \frac{2^{n-1}}{3^n} + \cdots$
9. $1 + (1.01) + (1.01)^2 + (1.01)^3 + \cdots + (1.01)^n + \cdots$
10. $1 + \frac{1}{\sqrt{2}} + \frac{1}{\sqrt[3]{3}} + \cdots + \frac{1}{\sqrt[n]{n}} + \cdots$
11. $\sum_{n=0}^{\infty} \frac{(-1)^n n}{n+1}$
12. $\sum_{n=1}^{\infty} \left(\frac{e}{10}\right)^n$
13. $\sum_{n=0}^{\infty} (-1)^n \left(\frac{3}{e}\right)^n$
14. $\sum_{n=0}^{\infty} \frac{3^n - 2^n}{4^n}$
15. $\sum_{n=1}^{\infty} (\sqrt{2})^{1-n}$
16. $\sum_{n=1}^{\infty} \left(\frac{2}{n} - \frac{1}{2^n}\right)$
17. $\sum_{n=1}^{\infty} \frac{n}{10n + 17}$
18. $\sum_{n=1}^{\infty} \frac{\sqrt{n}}{\ln(n+1)}$
19. $\sum_{n=1}^{\infty} (5^{-n} - 7^{-n})$
20. $\sum_{n=0}^{\infty} \frac{1}{1 + \left(\frac{9}{10}\right)^n}$
21. $\sum_{n=1}^{\infty} \left(\frac{e}{\pi}\right)^n$
22. $\sum_{n=1}^{\infty} \left(\frac{\pi}{e}\right)^n$
23. $\sum_{n=0}^{\infty} \left(\frac{100}{99}\right)^n$
24. $\sum_{n=0}^{\infty} \left(\frac{99}{100}\right)^n$
25. $\sum_{n=0}^{\infty} \frac{1 + 2^n + 3^n}{5^n}$
26. $\sum_{n=0}^{\infty} \frac{1 + 2^n + 5^n}{3^n}$
27. $\sum_{n=0}^{\infty} \frac{7 \cdot 5^n + 3 \cdot 11^n}{13^n}$
28. $\sum_{n=1}^{\infty} \sqrt[n]{2}$
29. $\sum_{n=1}^{\infty} \left[\left(\frac{7}{11}\right)^n - \left(\frac{3}{5}\right)^n\right]$
30. $\sum_{n=1}^{\infty} \frac{2n}{\sqrt{4n^2 + 3}}$
31. $\sum_{n=1}^{\infty} \frac{n^2 - 1}{3n^2 + 1}$
32. $\sum_{n=1}^{\infty} \sin^n 1$
33. $\sum_{n=1}^{\infty} \tan^n 1$
34. $\sum_{n=1}^{\infty} (\arcsin 1)^n$
35. $\sum_{n=1}^{\infty} (\arctan 1)^n$
36. $\sum_{n=1}^{\infty} \arctan n$

37. $\sum_{n=2}^{\infty} \frac{1}{n \ln n}$ (*Suggestion:* Mimic the proof of Theorem 4 to show divergence.)

38. Use the method of Example 6 to verify that

(a) $0.666666666\cdots = \frac{2}{3}$; (b) $0.111111111\cdots = \frac{1}{9}$;

(c) $0.249999999\cdots = \frac{1}{4}$; (d) $0.999999999\cdots = 1$.

In Problems 39 through 43, find the rational number represented by the given repeating decimal.

39. 0.4747 4747... **40.** 0.2525 2525...

41. 0.123 123 123... **42.** 0.3377 3377 3377...

43. 3.14159 14159 14159...

In Problems 44 through 49, find the set of all those values of x for which the given series is a convergent geometric series, then express the sum of the series as a function of x.

44. $\sum_{n=1}^{\infty} (2x)^n$ **45.** $\sum_{n=1}^{\infty} \left(\frac{x}{3}\right)^n$

46. $\sum_{n=1}^{\infty} (x-1)^n$ **47.** $\sum_{n=1}^{\infty} \left(\frac{x-2}{3}\right)^n$

48. $\sum_{n=1}^{\infty} \left(\frac{x^2}{x^2+1}\right)^n$ **49.** $\sum_{n=1}^{\infty} \left(\frac{5x^2}{x^2+16}\right)^n$

In Problems 50 through 55, express the nth partial sum of the infinite series as a telescoping sum (as in Example 3) and thereby find the sum of the series if it converges.

50. $\sum_{n=1}^{\infty} \frac{1}{4n^2-1}$ **51.** $\sum_{n=1}^{\infty} \frac{1}{9n^2+3n-2}$

52. $\sum_{n=1}^{\infty} \ln \frac{n+1}{n}$ **53.** $\sum_{n=1}^{\infty} \frac{1}{16n^2-8n-3}$

54. $\sum_{n=1}^{\infty} \frac{1}{n(n+2)}$ **55.** $\sum_{n=2}^{\infty} \frac{1}{n^2-1}$

In Problems 56 through 60, use a computer algebra system to find the partial fraction decomposition of the general term, then apply the method of Problems 50 through 55 to sum the series.

56. $\sum_{n=1}^{\infty} \frac{2n+1}{n^2(n+1)^2}$ **57.** $\sum_{n=1}^{\infty} \frac{6n^2+2n-1}{n(n+1)(4n^2-1)}$

58. $\sum_{n=1}^{\infty} \frac{2}{n(n+1)(n+2)}$ **59.** $\sum_{n=1}^{\infty} \frac{6}{n(n+1)(n+2)(n+3)}$

60. $\sum_{n=3}^{\infty} \frac{6n}{n^4-5n^2+4}$

61. Prove: If $\sum a_n$ diverges and c is a nonzero constant, then $\sum ca_n$ diverges.

62. Suppose that $\sum a_n$ converges and that $\sum b_n$ diverges. Prove that $\sum (a_n + b_n)$ diverges.

63. Let S_n and T_n denote the nth partial sums of $\sum a_n$ and $\sum b_n$, respectively. Suppose that k is a fixed positive integer and that $a_n = b_n$ for all $n \geqq k$. Show that $S_n - T_n = S_k - T_k$ for all $n > k$. Hence prove Theorem 5.

64. A ball has *bounce coefficient* $r < 1$ if, when it is dropped from a height h, it bounces back to a height of rh (Fig. 11.3.5). Suppose that such a ball is dropped from the initial height a and subsequently bounces infinitely many times. Use a geometric series to show that the total up-and-down distance it travels in all its bouncing is

$$D = a \cdot \frac{1+r}{1-r}.$$

Note that D is *finite.*

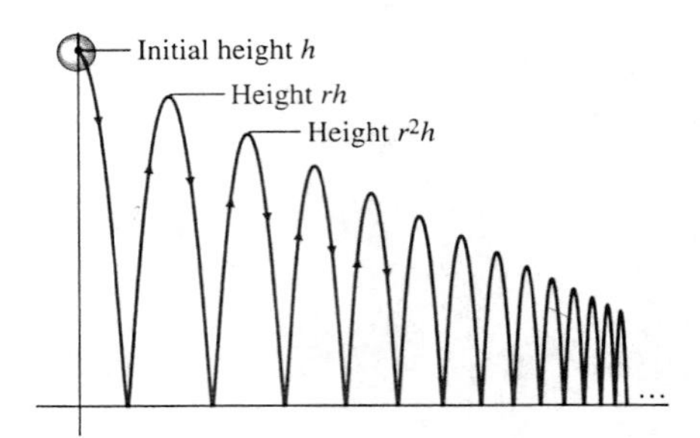

FIGURE 11.3.5 Successive bounces of the ball of Problems 64 and 65.

65. A ball with bounce coefficient $r = 0.64$ (see Problem 64) is dropped from an initial height of $a = 4$ ft. Use a geometric series to compute the total time required for it to complete its infinitely many bounces. The time required for a ball to drop h feet (from rest) is $\sqrt{2h/g}$ seconds, where $g = 32$ ft/s^2.

66. Suppose that the government spends \$1 billion and that each recipient of a fraction of this wealth spends 90% of the dollars that he or she receives. In turn, the secondary recipients spend 90% of the dollars they receive, and so on. How much total spending thereby results from the original injection of \$1 billion into the economy?

67. A tank initially contains a mass M_0 of air. Each stroke of a vacuum pump removes 5% of the air in the container. Compute: (a) The mass M_n of air remaining in the tank after n strokes of the pump; (b) $\lim_{n\to\infty} M_n$.

68. Paul and Mary toss a fair coin in turn until one of them wins the game by getting the first "head." Calculate for each the probability that he or she wins the game.

69. Peter, Paul, and Mary toss a fair coin in turn until one of them wins by getting the first "head." Calculate for each the probability that he or she wins the game. Check your answer by verifying that the sum of the three probabilities is 1.

70. Peter, Paul, and Mary roll a fair die in turn until one of them wins by getting the first "six." Calculate for each the probability that he or she wins the game. Check your answer by verifying that the sum of the three probabilities is 1.

71. A pane of a certain type of glass reflects half the incident light, absorbs one-fourth, and transmits one-fourth. A window is made of two panes of this glass separated by a small space (Fig. 11.3.6). What fraction of the incident light I is transmitted by the double window?

FIGURE 11.3.6 The double-pane window of Problem 71.

72. Criticize the following evaluation of the sum of an infinite series:

Let $x = 1 - 2 + 4 - 8 + 16 - 32 + 64 - \cdots$.

Then $2x = 2 - 4 + 8 - 16 + 32 - 64 + \cdots$.

Add the equations to obtain $3x = 1$. Thus $x = \frac{1}{3}$, and "therefore"

$$1 - 2 + 4 - 8 + 16 - 32 + 64 - \cdots = \tfrac{1}{3}.$$

11.3 Project: Numerical Summation and Geometric Series

With a modern calculator or computer, the computation of partial sums of infinite series—historically a tedious and time-consuming task—is now (ordinarily) a simple matter. Graphing calculators and computer algebra systems typically include one-line command such as

`sum(seq(a,k), k,1,n))`	TI calculator
`sum( a(k), k = 1..n )`	*Maple*
`Sum[ a[k], { k, 1, n } ]`	*Mathematica*

for the calculation of the nth partial sum of the infinite series $\sum_{k=1}^{\infty} a_k$ whose kth term is denoted by $a(k)$. For instance, we can check numerically the fact that

$$\sum_{k=0}^{\infty} \left(\frac{1}{5}\right)^k = \frac{5}{4}$$

by very quickly calculating the first seven partial sums 1.0000, 1.2000, 1.2400, 1.2480, 1.2496, 1.2499, and 1.2500. While not conclusive, this numerical evidence is nevertheless reassuring.

Investigation A Calculate partial sums of the geometric series

$$\sum_{n=0}^{\infty} r^n$$

with $r = 0.2, 0.5, 0.75, 0.9$, and 0.99. For each value of r, calculate the partial sums S_n with $n = 10, 20, 30, \dots$, continuing until two successive results agree to four or five decimal places. (For $r = 0.9$ and 0.99, you may decide to use $n = 100, 200, 300, \dots$.) How does the apparent rate of convergence—as measured by the number of terms required for the desired accuracy—depend on the value of r?

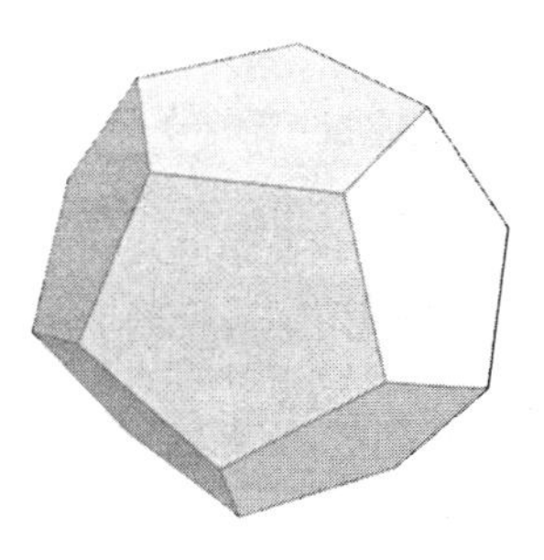

FIGURE 11.3.7 The 12-sided dodecahedron.

Investigation B Archaeological evidence indicates that the ancient (pre-Roman) Etruscans played dice using a dodecahedral die having 12 pentagonal faces numbered 1 through 12 (Fig. 11.3.7). One could simulate such a die by drawing a random card from a deck of 12 cards numbered 1 through 12. Here let's think of a deck having k cards numbered 1 through k. For your own personal value of k, begin with the largest digit in the sum of the digits in your student ID number. This is your value of k unless this digit is less than 5, in which case subtract it from 10 to get your value of k.

(a) John and Mary draw alternately from a shuffled deck of k cards. The first one to draw an ace—the card numbered 1—wins. Assume that John draws first. Use the formula for the sum of a geometric series to calculate (both as a rational number and as a four-place decimal) the probability J that John wins, and similarly the probability M that Mary wins. Check that $J + M = 1$.

(b) Now John, Mary, and Paul draw alternately from the deck of k cards. Calculate separately their respective probabilities of winning, given that John draws first and Mary draws second. Check that $J + M + P = 1$.

11.4 TAYLOR SERIES AND TAYLOR POLYNOMIALS

The infinite series we studied in Section 11.3 have *constant* terms, and the sum of such a series (assuming it converges) is a *number*. In contrast, much of the practical importance of infinite series derives from the fact that many functions have useful representations as infinite series with *variable* terms.

EXAMPLE 1 If we write $r = x$ for the ratio in a geometric series, then Theorem 1 in Section 11.3 gives the infinite series representation

$$\frac{1}{1-x} = \sum_{n=0}^{\infty} x^n = 1 + x + x^2 + x^3 + \cdots \tag{1}$$

of the function $f(x) = 1/(1-x)$. That is, for each fixed number x with $|x| < 1$, the infinite series in (1) converges to the number $1/(1-x)$. The nth partial sum

$$S_n(x) = 1 + x + x^2 + x^3 + \cdots + x^n \tag{2}$$

of the geometric series in (1) is now an nth-degree *polynomial* that approximates the function $f(x) = 1/(1-x)$. The convergence of the infinite series for $|x| < 1$ suggests that the approximation

$$\frac{1}{1-x} \approx 1 + x + x^2 + x^3 + \cdots + x^n \tag{3}$$

should then be accurate if n is sufficiently large. Figure 11.4.1 shows the graphs of $1/(1-x)$ and the three approximations $S_1(x)$, $S_2(x)$, and $S_3(x)$. It appears that the approximations are more accurate when n is larger and when x is closer to zero. ◆

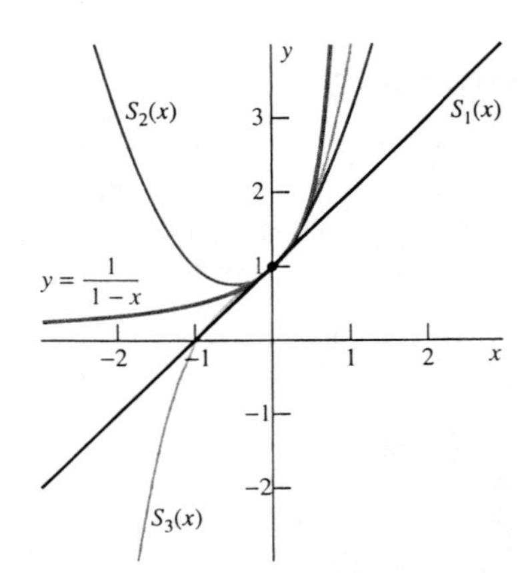

FIGURE 11.4.1 Graphs of the partial sums $S_1(x)$, $S_2(x)$, and $S_3(x)$ of the power series $\sum_{n=0}^{\infty} x^n = \dfrac{1}{1-x}$ of Example 1.

REMARK The approximation in (3) could be used to calculate numerical quotients with a calculator that has only $+$, $-$, $\times$ keys (but no $\div$ key). For instance,

$$\begin{aligned} \frac{329}{73} &= \frac{3.29}{0.73} = 3.29 \times \frac{1}{1-0.27} \\ &\approx (3.29)[1 + (0.27) + (0.27)^2 + \cdots + (0.27)^{10}] \\ &\approx (3.29)(1.36986); \qquad \text{thus} \\ \frac{329}{73} &\approx 4.5068, \end{aligned}$$

accurate to four decimal places. This is a simple illustration of the use of polynomial approximation for numerical computation.

The definitions of the various elementary transcendental functions leave it unclear how to compute their values precisely, except at a few isolated points. For example,

$$\ln x = \int_1^x \frac{1}{t}\,dt \quad (x > 0)$$

by definition, so obviously $\ln 1 = 0$, but no other value of $\ln x$ is obvious. The natural exponential function is the inverse of $\ln x$, so it is clear that $e^0 = 1$, but it is not at all clear how to compute e^x for $x \neq 0$. Indeed, even such an innocent-looking expression as $\sqrt{x}$ is not computable (precisely and in a finite number of steps) unless x happens to be the square of a rational number.

But *any* value of a polynomial

$$P(x) = c_0 + c_1 x + c_2 x^2 + \cdots + c_n x^n$$

with known coefficients $c_0, c_1, c_2, \ldots, c_n$ is easy to calculate—as in the preceding remark, only addition and multiplication are required. One goal of this section is to use the fact that polynomial values are so readily computable to help us calculate approximate values of functions such as $\ln x$ and e^x.

Polynomial Approximations

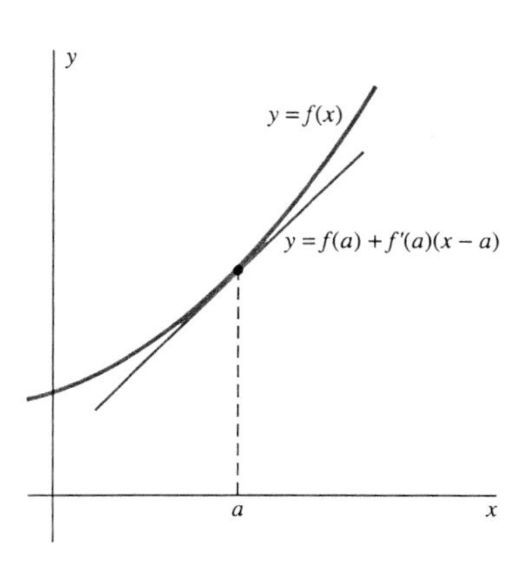

FIGURE 11.4.2 The tangent line at $(a, f(a))$ is the best linear approximation to $y = f(x)$ near a.

Suppose that we want to calculate (or, at least, closely approximate) a specific value $f(x_0)$ of a given function f. It would suffice to find a polynomial $P(x)$ with a graph that is very close to that of f on some interval containing x_0. For then we could use the value $P(x_0)$ as an approximation to the actual value of $f(x_0)$. Once we know how to find such an approximating polynomial $P(x)$, the next question would be how accurately $P(x_0)$ approximates the desired value $f(x_0)$.

The simplest example of polynomial approximation is the linear approximation

$$f(x) \approx f(a) + f'(a)(x - a)$$

obtained by writing $\Delta x = x - a$ in the linear approximation formula, Eq. (3) of Section 4.2. The graph of the first-degree polynomial

$$P_1(x) = f(a) + f'(a)(x - a) \tag{4}$$

is the line tangent to the curve $y = f(x)$ at the point $(a, f(a))$; see Fig. 11.4.2. This first-degree polynomial agrees with f and with its first derivative at $x = a$. That is,

$$P_1(a) = f(a) \quad \text{and} \quad P_1'(a) = f'(a).$$

EXAMPLE 2 Suppose that $f(x) = \ln x$ and that $a = 1$. Then $f(1) = 0$ and $f'(1) = 1$, so $P_1(x) = x - 1$. Hence we expect that $\ln x \approx x - 1$ for x near 1. With $x = 1.1$, we find that

$$P_1(1.1) = 0.1000, \quad \text{whereas} \quad \ln(1.1) \approx 0.0953.$$

The error in this approximation is about 5%.

To better approximate $\ln x$ near $x = 1$, let us find a second-degree polynomial

$$P_2(x) = c_0 + c_1 x + c_2 x^2$$

that not only has the same value and the same first derivative as does f at $x = 1$, but also has the same second derivative there: $P_2''(1) = f''(1) = -1$. To satisfy these conditions, we must have

$$\begin{aligned} P_2(1) &= c_2 + c_1 + c_0 = 0, \\ P_2'(1) &= 2c_2 + c_1 = 1, \qquad \text{and} \\ P_2''(1) &= 2c_2 = -1. \end{aligned}$$

When we solve these equations, we find that $c_0 = -\frac{3}{2}$, $c_1 = 2$, and $c_2 = -\frac{1}{2}$, so

$$P_2(x) = -\tfrac{3}{2} + 2x - \tfrac{1}{2}x^2.$$

With $x = 1.1$ we find that $P_2(1.1) = 0.0950$, which is accurate to three decimal places because $\ln(1.1) \approx 0.0953$. The graph of $y = P_2(x)$ is a parabola through $(1, 0)$ with the same value, slope, *and curvature* there as $y = \ln x$ (Fig. 11.4.3). ◆

The tangent line and the parabola used in the computations of Example 2 illustrate one general approach to polynomial approximation. To approximate the function $f(x)$ near $x = a$, we look for an nth-degree polynomial

$$P_n(x) = c_0 + c_1 x + c_2 x^2 + \cdots + c_n x^n$$

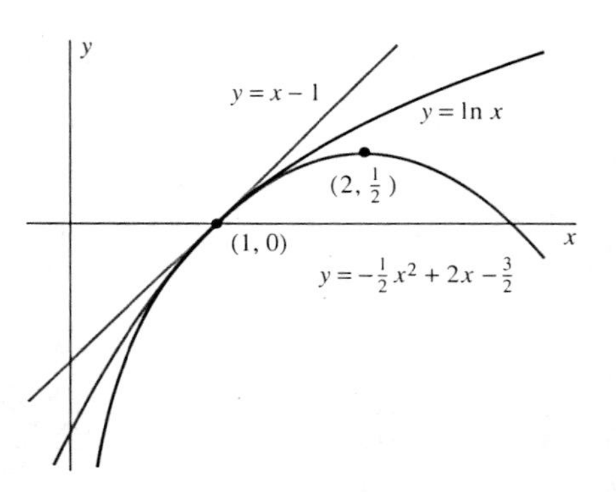

FIGURE 11.4.3 The linear and parabolic approximations to $y = \ln x$ near the point (1, 0) (Example 2).

such that its value at a and the values of its first n derivatives at a agree with the corresponding values of f. That is, we require that

$$\begin{aligned} P_n(a) &= f(a), \\ P_n'(a) &= f'(a), \\ P_n''(a) &= f''(a), \\ &\vdots \\ P_n^{(n)}(a) &= f^{(n)}(a). \end{aligned} \tag{5}$$

We can use these $n+1$ conditions to evaluate the values of the $n+1$ coefficients $c_0, c_1, c_2, \ldots, c_n$.

The algebra involved is much simpler, however, if we begin with $P_n(x)$ expressed as an nth-degree polynomial in powers of $x - a$ rather than in powers of x:

$$P_n(x) = c_0 + c_1(x-a) + c_2(x-a)^2 + \cdots + c_n(x-a)^n. \tag{6}$$

Then substituting $x = a$ in Eq. (6) yields

$$c_0 = P_n(a) = f(a)$$

by the first condition in Eq. (5). Substituting $x = a$ into

$$P_n'(x) = c_1 + 2c_2(x-a) + 3c_3(x-a)^2 + \cdots + nc_n(x-a)^{n-1}$$

yields

$$c_1 = P_n'(a) = f'(a)$$

by the second condition in Eq. (5). Next, substituting $x = a$ into

$$P_n''(x) = 2c_2 + 3 \cdot 2c_3(x-a) + \cdots + n(n-1)c_n(x-a)^{n-2}$$

yields $2c_2 = P_n''(a) = f''(a)$, so

$$c_2 = \tfrac{1}{2}f''(a).$$

We continue this process to find $c_3, c_4, \ldots, c_n$. In general, the constant term in the kth derivative $P_n^{(k)}(x)$ is $k!c_k$, because it is the kth derivative of the kth-degree term $b_k(x-a)^k$ in $P_n(x)$:

$$P_n^{(k)}(x) = k!c_k + \{\text{powers of } x - a\}.$$

(Recall that $k! = 1 \cdot 2 \cdot 3 \cdots (k-1) \cdot k$ denotes the *factorial* of the positive integer k, read "k factorial.") So when we substitute $x = a$ into $P_n^{(k)}(x)$, we find that

$$k!c_k = P_n^{(k)}(a) = f^{(k)}(a)$$

and thus that

$$c_k = \frac{f^{(k)}(a)}{k!} \tag{7}$$

for $k = 1, 2, 3, \dots, n$.

Indeed, Eq. (7) holds also for $k = 0$ if we use the universal convention that $0! = 1$ and agree that the zeroth derivative $g^{(0)}$ of the function g is just g itself. With such conventions, our computations establish the following theorem.

THEOREM 1 The *n*th-Degree Taylor Polynomial

Suppose that the first n derivatives of the function $f(x)$ exist at $x = a$. Let $P_n(x)$ be the nth-degree polynomial

$$\begin{aligned} P_n(x) &= \sum_{k=0}^{n} \frac{f^{(k)}(a)}{k!}(x-a)^k \\ &= f(a) + f'(a)(x-a) + \frac{f''(a)}{2!}(x-a)^2 + \cdots + \frac{f^{(n)}(a)}{n!}(x-a)^n. \end{aligned} \tag{8}$$

Then the values of $P_n(x)$ and its first n derivatives agree, at $x = a$, with the values of f and its first n derivatives there. That is, the equations in (5) all hold.

The polynomial in Eq. (8) is called the **nth-degree Taylor polynomial of the function f at the point $x = a$**. Note that $P_n(x)$ is a polynomial in powers of $x - a$ rather than in powers of x. To use $P_n(x)$ effectively for the approximation of $f(x)$ near a, we must be able to compute the value $f(a)$ and the values of its derivatives $f'(a)$, $f''(a)$, and so on, all the way to $f^{(n)}(a)$.

The line $y = P_1(x)$ is simply the line tangent to the curve $y = f(x)$ at the point $(a, f(a))$. Thus $y = f(x)$ and $y = P_1(x)$ have the same slope at this point. Now recall from Section 4.6 that the second derivative measures the way the curve $y = f(x)$ is bending as it passes through $(a, f(a))$. Therefore, let us call $f''(a)$ the "concavity" of $y = f(x)$ at $(a, f(a))$. Then, because $P_2''(a) = f''(a)$, it follows that $y = P_2(x)$ has the same value, the same slope, *and* the same concavity at $(a, f(a))$ as does $y = f(x)$. Moreover, $P_3(x)$ and $f(x)$ will also have the same rate of change of concavity at $(a, f(a))$. Such observations suggest that the larger n is, the more closely the nth-degree Taylor polynomial will approximate $f(x)$ for x near a.

EXAMPLE 3 Find the nth-degree Taylor polynomial of $f(x) = \ln x$ at $a = 1$.

Solution The first few derivatives of $f(x) = \ln x$ are

$$f'(x) = \frac{1}{x}, \quad f''(x) = -\frac{1}{x^2}, \quad f^{(3)}(x) = \frac{2}{x^3}, \quad f^{(4)}(x) = -\frac{3!}{x^4}, \quad f^{(5)}(x) = \frac{4!}{x^5}.$$

The pattern is clear:

$$f^{(k)}(x) = (-1)^{k-1}\frac{(k-1)!}{x^k} \quad \text{for } k \geqq 1.$$

Hence $f^{(k)}(1) = (-1)^{k-1}(k-1)!$, so Eq. (8) gives

$$P_n(x) = (x-1) - \frac{1}{2}(x-1)^2 + \frac{1}{3}(x-1)^3 - \frac{1}{4}(x-1)^4 + \cdots + \frac{(-1)^{n-1}}{n}(x-1)^n.$$

With $n = 2$ we obtain the quadratic polynomial

$$P_2(x) = (x-1) - \tfrac{1}{2}(x-1)^2 = -\tfrac{1}{2}x^2 + 2x - \tfrac{3}{2},$$

the same as in Example 2. With the third-degree Taylor polynomial

$$P_3(x) = (x-1) - \tfrac{1}{2}(x-1)^2 + \tfrac{1}{3}(x-1)^3$$

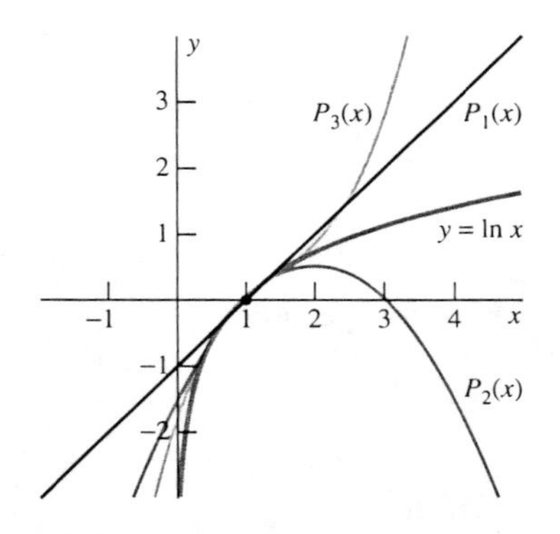

FIGURE 11.4.4 The first three Taylor polynomials approximating $f(x) = \ln x$ near $x = 1$.

we can go a step further in approximating $\ln(1.1) = 0.095310179\ldots \approx 0.0953$. The value

$$P_3(1.1) = (0.1) - \tfrac{1}{2}(0.1)^2 + \tfrac{1}{3}(0.1)^3 \approx 0.095333 \approx 0.0953$$

is accurate to four decimal places (rounded). In Fig. 11.4.4 we see that, the higher the degree and the closer x is to 1, the more accurate the approximation $\ln x \approx P_n(x)$ appears to be. ◆

In the common case $a = 0$, the nth-degree Taylor polynomial in Eq. (8) reduces to

$$P_n(x) = f(0) + f'(0)x + \frac{f''(0)}{2!}x^2 + \cdots + \frac{f^{(n)}(0)}{n!}x^n. \tag{9}$$

EXAMPLE 4 Find the nth-degree Taylor polynomial for $f(x) = e^x$ at $a = 0$.

Solution This is the easiest of all Taylor polynomials to compute, because $f^{(k)}(x) = e^x$ for all $k \geqq 0$. Hence $f^{(k)}(0) = 1$ for all $k \geqq 0$, so Eq. (9) yields

$$P_n(x) = 1 + x + \frac{x^2}{2!} + \frac{x^3}{3!} + \cdots + \frac{x^n}{n!}.$$ ◆

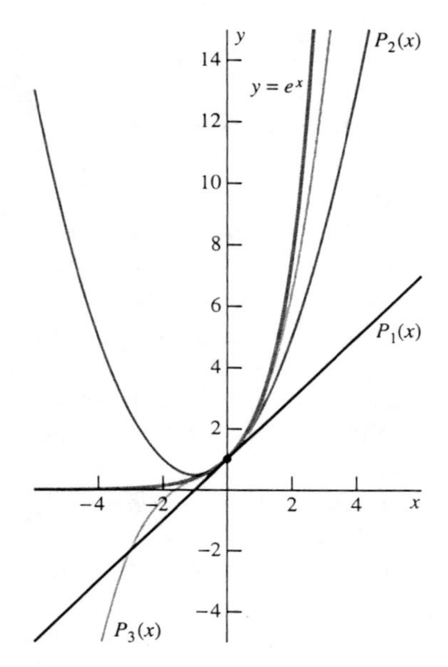

FIGURE 11.4.5 The first three Taylor polynomials approximating $f(x) = e^x$ near $x = 0$.

The first few Taylor polynomials of the natural exponential function at $a = 0$ are, therefore,

$$P_0(x) = 1,$$
$$P_1(x) = 1 + x,$$
$$P_2(x) = 1 + x + \tfrac{1}{2}x^2,$$
$$P_3(x) = 1 + x + \tfrac{1}{2}x^2 + \tfrac{1}{6}x^3,$$
$$P_4(x) = 1 + x + \tfrac{1}{2}x^2 + \tfrac{1}{6}x^3 + \tfrac{1}{24}x^4,$$
$$P_5(x) = 1 + x + \tfrac{1}{2}x^2 + \tfrac{1}{6}x^3 + \tfrac{1}{24}x^4 + \tfrac{1}{120}x^5.$$

Figure 11.4.5 shows the graphs of $P_1(x)$, $P_2(x)$, and $P_3(x)$. The table in Fig. 11.4.6 shows how these polynomials approximate $f(x) = e^x$ for $x = 0.1$ and for $x = 0.5$. At least for these two values of x, the closer x is to $a = 0$, the more rapidly $P_n(x)$ appears to approach $f(x)$ as n increases.

$x = 0.1$

n	$P_n(x)$	e^x	$e^x - P_n(x)$
0	1.00000	1.10517	0.10517
1	1.10000	1.10517	0.00517
2	1.10500	1.10517	0.00017
3	1.10517	1.10517	0.00000
4	1.10517	1.10517	0.00000

$x = 0.5$

n	$P_n(x)$	e^x	$e^x - P_n(x)$
0	1.00000	1.64872	0.64872
1	1.50000	1.64872	0.14872
2	1.62500	1.64872	0.02372
3	1.64583	1.64872	0.00289
4	1.64844	1.64872	0.00028
5	1.64879	1.64872	0.00002

FIGURE 11.4.6 Approximating $y = e^x$ with Taylor polynomials at $a = 0$.

Taylor's Formula

The closeness with which the polynomial $P_n(x)$ approximates the function $f(x)$ is measured by the difference

$$R_n(x) = f(x) - P_n(x),$$

for which

$$f(x) = P_n(x) + R_n(x). \tag{10}$$

This difference $R_n(x)$ is called the **nth-degree remainder for $f(x)$ at $x = a$**. It is the *error* made if the value $f(x)$ is replaced with the approximation $P_n(x)$.

The theorem that lets us estimate the error, or remainder, $R_n(x)$ is called **Taylor's formula,** after Brook Taylor (1685–1731), a follower of Newton who introduced Taylor polynomials in an article published in 1715. The particular expression for $R_n(x)$ that we give next is called the *Lagrange form* for the remainder because it first appeared in 1797 in a book written by the French mathematician Joseph Louis Lagrange (1736–1813).

THEOREM 2 Taylor's Formula

Suppose that the $(n + 1)$th derivative of the function f exists on an interval containing the points a and b. Then

$$\begin{aligned} f(b) = f(a) + f'(a)(b-a) + \frac{f''(a)}{2!}(b-a)^2 \\ + \frac{f^{(3)}(a)}{3!}(b-a)^3 + \cdots + \frac{f^{(n)}(a)}{n!}(b-a)^n + \frac{f^{(n+1)}(z)}{(n+1)!}(b-a)^{n+1} \end{aligned} \tag{11}$$

for some number z between a and b.

REMARK With $n = 0$, Eq. (11) reduces to the equation

$$f(b) = f(a) + f'(z)(b-a),$$

the conclusion of the mean value theorem (Section 4.3). Thus Taylor's formula is a far-reaching generalization of the mean value theorem of differential calculus.

A proof of Taylor's formula is given in Appendix I. If we replace b with x in Eq. (11), we get the **nth-degree Taylor formula with remainder at $x = a$**,

$$\begin{aligned} f(x) = f(a) + f'(a)(x-a) + \frac{f''(a)}{2!}(x-a)^2 + \frac{f^{(3)}(a)}{3!}(x-a)^3 \\ + \cdots + \frac{f^{(n)}(a)}{n!}(x-a)^n + \frac{f^{(n+1)}(z)}{(n+1)!}(x-a)^{n+1}, \end{aligned} \tag{12}$$

where z is some number between a and x. Thus the nth-degree remainder term is

$$R_n(x) = \frac{f^{(n+1)}(z)}{(n+1)!}(x-a)^{n+1}, \tag{13}$$

which is easy to remember—it's the same as the *last* term of $P_{n+1}(x)$, except that $f^{(n+1)}(a)$ is replaced with $f^{(n+1)}(z)$.

EXAMPLE 3 (continued) To estimate the accuracy of the approximation

$$\ln 1.1 \approx 0.095333,$$

we substitute $x = 1$ into the formula

$$f^{(k)}(x) = (-1)^{k-1}\frac{(k-1)!}{x^k}$$

for the kth derivative of $f(x) = \ln x$ and get

$$f^{(k)}(1) = (-1)^{k-1}(k-1)!.$$

Hence the third-degree Taylor formula *with remainder* at $x = 1$ is

$$\ln x = (x-1) - \frac{1}{2}(x-1)^2 + \frac{1}{3}(x-1)^3 - \frac{3!}{4!z^4}(x-1)^4$$

with z between $a = 1$ and x. With $x = 1.1$ this gives

$$\ln(1.1) \approx 0.095333 - \frac{(0.1)^4}{4z^4},$$

where $1 < z < 1.1$. The value $z = 1$ gives the largest possible magnitude $(0.1)^4/4 = 0.000025$ of the remainder term. It follows that

$$0.095308 < \ln(1.1) < 0.095334,$$

so we can conclude that $\ln(1.1) = 0.0953$ to four-place accuracy. ◆

Taylor Series

If the function f has derivatives of all orders, then we can write Taylor's formula (Eq. (11)) with any degree n that we please. Ordinarily, the exact value of z in the Taylor remainder term in Eq. (13) is unknown. Nevertheless, we can sometimes use Eq. (13) to show that the remainder approaches zero as $n \to +\infty$:

$$\lim_{n\to\infty} R_n(x) = 0 \qquad \textbf{(14)}$$

for some particular *fixed* value of x. Then Eq. (10) gives

$$f(x) = \lim_{n\to\infty}[P_n(x) + R_n(x)] = \lim_{n\to\infty} P_n(x) = \lim_{n\to\infty}\sum_{k=0}^{n}\frac{f^{(k)}(a)}{k!}(x-a)^k;$$

that is,

$$f(x) = \sum_{k=0}^{\infty}\frac{f^{(k)}(a)}{k!}(x-a)^k. \qquad \textbf{(15)}$$

The infinite series

$$\sum_{n=0}^{\infty}\frac{f^{(n)}(a)}{n!}(x-a)^n = f(a) + f'(a)(x-a) + \frac{f''(a)}{2!}(x-a)^2 + \cdots + \frac{f^{(n)}(a)}{n!}(x-a)^n + \cdots \qquad \textbf{(16)}$$

is called the **Taylor series** of the function f at $x = a$. Its partial sums are the successive Taylor polynomials of f at $x = a$.

We can write the Taylor series of a function f without knowing that it converges. But if the limit in Eq. (14) can be established, then it follows as in Eq. (15) that the Taylor series in Eq. (16) actually converges to $f(x)$. If so, then we can approximate the value of $f(x)$ sufficiently accurately by calculating the value of a Taylor polynomial of f of sufficiently high degree.

EXAMPLE 5 In Example 4 we noted that if $f(x) = e^x$, then $f^{(k)}(x) = e^x$ for all integers $k \geqq 0$. Hence the Taylor formula

$$f(x) = f(0) + f'(0)x + \frac{f''(0)}{2!}x^2 + \cdots + \frac{f^{(n)}(0)}{n!}x^n + \frac{f^{(n+1)}(z)}{(n+1)!}x^{n+1}$$

at $a = 0$ gives

$$e^x = 1 + x + \frac{x^2}{2!} + \frac{x^3}{3!} + \cdots + \frac{x^n}{n!} + \frac{e^z x^{n+1}}{(n+1)!} \tag{17}$$

for some z between 0 and x. If x and hence z are negative then $e^z < 1$, whereas $e^z < e^x$ if both are positive. Thus the remainder term $R_n(x)$ satisfies the inequalities

$$0 < |R_n(x)| < \frac{|x|^{n+1}}{(n+1)!} \quad \text{if } x < 0,$$

$$0 < |R_n(x)| < \frac{e^x x^{n+1}}{(n+1)!} \quad \text{if } x > 0.$$

Therefore, the fact that

$$\lim_{n\to\infty} \frac{x^n}{n!} = 0 \tag{18}$$

for all x (see Problem 55) implies that $\lim_{n\to\infty} R_n(x) = 0$ for all x. This means that the Taylor series for e^x converges to e^x for all x, and we may write

$$e^x = \sum_{n=0}^{\infty} \frac{x^n}{n!} = 1 + x + \frac{x^2}{2!} + \frac{x^3}{3!} + \frac{x^4}{4!} + \cdots. \tag{19}$$

The series in Eq. (19) is the most famous and most important of all Taylor series.

With $x = 1$, Eq. (19) yields a numerical series

$$e = \sum_{n=0}^{\infty} \frac{1}{n!} = 1 + \frac{1}{1!} + \frac{1}{2!} + \frac{1}{3!} + \frac{1}{4!} + \cdots \tag{20}$$

for the number e itself. The 10th and 20th partial sums of this series give the approximations

$$e \approx 1 + \frac{1}{1!} + \frac{1}{2!} + \cdots + \frac{1}{10!} \approx 2.7182818$$

and

$$e \approx 1 + \frac{1}{1!} + \frac{1}{2!} + \cdots + \frac{1}{20!} \approx 2.71828\,18284\,59045\,235,$$

both of which are accurate to the number of decimal places shown. ◆

EXAMPLE 6 To find the Taylor series at $a = 0$ for $f(x) = \cos x$, we first calculate the derivatives

$$\begin{aligned} f(x) &= \cos x, & f'(x) &= -\sin x, \\ f''(x) &= -\cos x, & f^{(3)}(x) &= \sin x, \\ f^{(4)}(x) &= \cos x, & f^{(5)}(x) &= -\sin x, \\ &\vdots & &\vdots \\ f^{(2n)}(x) &= (-1)^n \cos x, & f^{(2n+1)}(x) &= (-1)^{n+1} \sin x. \end{aligned}$$

It follows that

$$f^{(2n)}(0) = (-1)^n \quad \text{but} \quad f^{(2n+1)}(0) = 0,$$

so the Taylor polynomials and Taylor series of $f(x) = \cos x$ include only terms of *even* degree. The Taylor formula of degree $2n$ for $\cos x$ at $a = 0$ is

$$\cos x = 1 - \frac{x^2}{2!} + \frac{x^4}{4!} - \cdots + (-1)^n \frac{x^{2n}}{(2n)!} + (-1)^{n+1} \frac{\cos z}{(2n+2)!} x^{2n+2},$$

where z is between 0 and x. Because $|\cos x| \leqq 1$ for all z, it follows from Eq. (18) that the remainder term approaches zero as $n \to \infty$ *for all* x. Hence the desired Taylor series of $f(x) = \cos x$ at $a = 0$ converges to $\cos x$ for all x, so we may write

$$\cos x = \sum_{n=0}^{\infty} \frac{(-1)^n x^{2n}}{(2n)!} = 1 - \frac{x^2}{2!} + \frac{x^4}{4!} - \frac{x^6}{6!} + \cdots. \tag{21}$$

◆

In Problem 41 we ask you to show similarly that the Taylor series at $a = 0$ of $f(x) = \sin x$ is

$$\sin x = \sum_{n=0}^{\infty} \frac{(-1)^n x^{2n+1}}{(2n+1)!} = x - \frac{x^3}{3!} + \frac{x^5}{5!} - \frac{x^7}{7!} + \cdots. \tag{22}$$

Figures 11.4.7 and 11.4.8 illustrate the increasingly better approximations to $\cos x$ and $\sin x$ that we get by using more and more terms of the series in Eqs. (21) and (22).

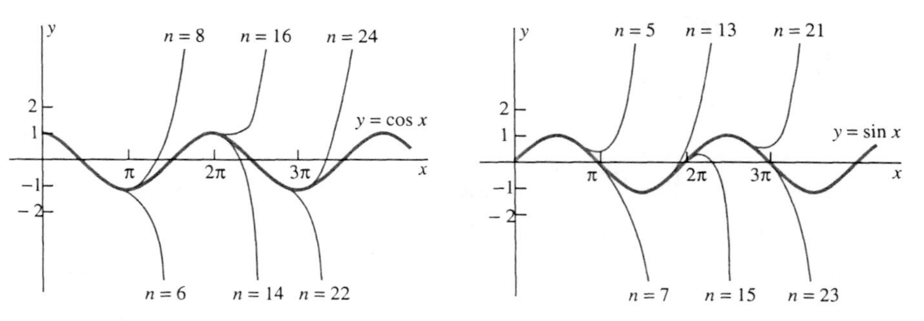

FIGURE 11.4.7 Approximating $\cos x$ with nth-degree Taylor polynomials.

FIGURE 11.4.8 Approximating $\sin x$ with nth-degree Taylor polynomials.

The case $a = 0$ of Taylor's series in (16) is called the **Maclaurin series** of the function $f(x)$,

$$\sum_{n=0}^{\infty} \frac{f^{(n)}(0)}{n!} x^n = f(0) + f'(0)x + \frac{f''(0)}{2!}x^2 + \frac{f^{(3)}(0)}{3!}x^3 + \cdots. \tag{23}$$

Colin Maclaurin (1698–1746) was a Scottish mathematician who used this series as a basic tool in a calculus book he published in 1742. The three Maclaurin series

$$e^x = \sum_{n=0}^{\infty} \frac{x^n}{n!} = 1 + x + \frac{x^2}{2!} + \frac{x^3}{3!} + \frac{x^4}{4!} + \cdots, \tag{19}$$

$$\cos x = \sum_{n=0}^{\infty} \frac{(-1)^n x^{2n}}{(2n)!} = 1 - \frac{x^2}{2!} + \frac{x^4}{4!} - \frac{x^6}{6!} + \cdots, \quad \text{and} \tag{21}$$

$$\sin x = \sum_{n=0}^{\infty} \frac{(-1)^n x^{2n+1}}{(2n+1)!} = x - \frac{x^3}{3!} + \frac{x^5}{5!} - \frac{x^7}{7!} + \cdots \tag{22}$$

(which actually were discovered by Newton) bear careful examination and comparison. Observe that:

- The terms in the *even* cosine series are the *even*-degree terms in the exponential series but with alternating signs.
- The terms in the *odd* sine series are the *odd*-degree terms in the exponential series but with alternating signs.

Equations (19), (21), and (22) are *identities* that hold for all values of x. Consequently, new series can be derived by substitution, as in Examples 7 and 8.

EXAMPLE 7 Substituting $x = -t^2$ into Eq. (19) yields

$$e^{-t^2} = 1 - t^2 + \frac{t^4}{2!} - \frac{t^6}{3!} + \cdots + (-1)^n \frac{t^{2n}}{n!} + \cdots. \qquad \blacklozenge$$

EXAMPLE 8 Substituting $x = 2t$ into Eq. (22) gives

$$\sin 2t = 2t - \frac{4}{3}t^3 + \frac{4}{15}t^5 - \frac{8}{315}t^7 + \cdots. \qquad \blacklozenge$$

Euler's Formula

The sum of an infinite series $\sum c_n$ with complex terms $c_n = a_n + ib_n$ is defined by

$$\sum_{n=1}^{\infty} c_n = \sum_{n=1}^{\infty} a_n + i \sum_{n=1}^{\infty} b_n$$

provided that the two infinite series of real terms on the right-hand side converge, in which case we say that the series of complex terms on the left-hand side converges.

It can be shown that the exponential series in (19) converges whenever the number x is replaced with a complex number $z = x + iy$. Consequently, the exponential function e^z can be *defined* (for complex as well as for real arguments) by means of the series

$$e^z = \sum_{n=0}^{\infty} \frac{z^n}{n!} = 1 + z + \frac{z^2}{2!} + \frac{z^3}{3!} + \frac{z^4}{4!} + \cdots.$$

If we substitute the *pure imaginary* number $z = i\theta$ (with θ real), we get

$$\begin{aligned} e^{i\theta} &= \sum_{n=0}^{\infty} \frac{(i\theta)^n}{n!} = 1 + i\theta + \frac{(i\theta)^2}{2!} + \frac{(i\theta)^3}{3!} + \frac{(i\theta)^4}{4!} + \cdots \\ &= 1 + i\theta - \frac{\theta^2}{2!} - \frac{i\theta^3}{3!} + \frac{\theta^4}{4!} + \frac{i\theta^5}{5!} - \cdots \\ &= \left(1 - \frac{\theta^2}{2!} + \frac{\theta^4}{4!} - \cdots\right) + i\left(\theta - \frac{\theta^3}{3!} + \frac{\theta^5}{5!} - \cdots\right), \end{aligned}$$

using the facts that $i^2 = -1$, $i^3 = -i$, $i^4 = 1$, and so on. We recognize the Maclaurin series for $\cos\theta$ and $\sin\theta$ on the right-hand side and conclude that

$$e^{i\theta} = \cos\theta + i\sin\theta$$

for every real number θ. This is the famous **Euler's formula.** For instance, with $\theta = \pi$ it gives $e^{i\pi} = \cos\pi + i\sin\pi = -1$, and hence the extraordinary relation

$$e^{i\pi} + 1 = 0$$

relating the five most important special numbers in mathematics: 0, 1, i, π, and e.

The Number π

In Section 5.3 we described how Archimedes used polygons inscribed in and circumscribed about the unit circle to show that $3\frac{10}{71} < \pi < 3\frac{1}{7}$. With the aid of electronic computers, π has been calculated to well over a *billion* decimal places. We describe now some of the methods that have been used for such computations. [For a chronicle of humanity's perennial fascination with the number π, see Peter Beckmann, *A History of* π, New York: St. Martin's Press, 1971.]

We begin with the elementary algebraic identity

$$\frac{1}{1+x} = 1 - x + x^2 - x^3 + \cdots + (-1)^{k-1}x^{k-1} + \frac{(-1)^k x^k}{1+x}, \qquad \textbf{(24)}$$

which can be verified by multiplying both sides by $1 + x$. We substitute t^2 for x and

$n+1$ for k and thus find that

$$\frac{1}{1+t^2} = 1 - t^2 + t^4 - t^6 + \cdots + (-1)^n t^{2n} + \frac{(-1)^{n+1} t^{2n+2}}{1+t^2}.$$

Because $D_t \tan^{-1} t = 1/(1+t^2)$, integrating both sides of this last equation from $t = 0$ to $t = x$ gives

$$\tan^{-1} x = x - \frac{x^3}{3} + \frac{x^5}{5} - \frac{x^7}{7} + \cdots + (-1)^n \frac{x^{2n+1}}{2n+1} + R_{2n+1}, \tag{25}$$

where

$$|R_{2n+1}| = \left| \int_0^x \frac{t^{2n+2}}{1+t^2}\, dx \right| \leqq \left| \int_0^x t^{2n+2}\, dx \right| = \frac{|x|^{2n+3}}{2n+3}. \tag{26}$$

This estimate of the error makes it clear that

$$\lim_{n\to\infty} R_n = 0$$

if $|x| \leqq 1$. Hence we obtain the Taylor series for the inverse tangent function:

$$\tan^{-1} x = \sum_{n=0}^{\infty} (-1)^n \frac{x^{2n+1}}{2n+1} = x - \frac{x^3}{3} + \frac{x^5}{5} - \frac{x^7}{7} + \cdots, \tag{27}$$

valid for $-1 \leqq x \leqq 1$.

If we substitute $x = 1$ into Eq. (27), we obtain *Leibniz's series*

$$\frac{\pi}{4} = 1 - \frac{1}{3} + \frac{1}{5} - \frac{1}{7} + \cdots.$$

Although this is a beautiful series, it is not an effective way to compute π. But the error estimate in Eq. (26) shows that we can use Eq. (25) to calculate $\tan^{-1} x$ if $|x|$ is small. For example, if $x = \frac{1}{5}$, then the fact that

$$\frac{1}{9 \cdot 5^9} \approx 0.000000057 < 0.0000001$$

implies that the approximation

$$\tan^{-1}\left(\tfrac{1}{5}\right) \approx \tfrac{1}{5} - \tfrac{1}{3}\left(\tfrac{1}{5}\right)^3 + \tfrac{1}{5}\left(\tfrac{1}{5}\right)^5 - \tfrac{1}{7}\left(\tfrac{1}{5}\right)^7$$

is accurate to six decimal places.

Accurate inverse tangent calculations lead to accurate computations of the number π. For example, we can use the addition formula for the tangent function to show (Problem 52) that

$$\frac{\pi}{4} = 4 \tan^{-1}\left(\frac{1}{5}\right) - \tan^{-1}\left(\frac{1}{239}\right). \tag{28}$$

HISTORICAL NOTE In 1706, John Machin used Eq. (28) to calculate the first 100 decimal places of π. (In Problem 54 we ask you to use it to show that $\pi = 3.14159$ to five decimal places.) In 1844 the lightning-fast mental calculator Zacharias Dase of Germany computed the first 200 decimal places of π, using the related formula

$$\frac{\pi}{4} = \tan^{-1}\left(\frac{1}{2}\right) + \tan^{-1}\left(\frac{1}{5}\right) + \tan^{-1}\left(\frac{1}{8}\right). \tag{29}$$

You might enjoy verifying this formula. (See Problem 53.) A recent computation of 1 million decimal places of π used the formula

$$\frac{\pi}{4} = 12 \tan^{-1}\left(\frac{1}{18}\right) + 8 \tan^{-1}\left(\frac{1}{57}\right) - 5 \tan^{-1}\left(\frac{1}{239}\right).$$

For derivations of this formula and others like it, with further discussion of the computations of the number π, see the article "An algorithm for the calculation of π" by George Miel in the *American Mathematical Monthly* **86** (1979), pp. 694–697. Although few practical applications require more than ten or twelve decimal places of π, these computations provide dramatic evidence of the power of Taylor's formula. Moreover, the number π continues to serve as a challenge both to human ingenuity and to the accuracy and efficiency of modern electronic computers. For an account of how investigations of the Indian mathematical genius Srinivasa Ramanujan (1887–1920) have led recently to the computation of over a billion decimal places of π, see the article "Ramanujan and pi," Jonathan M. Borwein and Peter B. Borwein, *Scientific American* (Feb. 1988), pp. 112–117.

11.4 TRUE/FALSE STUDY GUIDE

11.4 CONCEPTS: QUESTIONS AND DISCUSSION

1. Suppose that we take Euler's formula $e^{i\theta} = \cos\theta + i\sin\theta$ as a starting point and *define* the exponential $e^z = e^{x+iy}$ by writing
$$e^z = e^x e^{iy} = e^x(\cos y + i\sin y).$$
Can you then *prove* on this basis that $e^{z+w} = e^z e^w$ if $z = x+iy$ and $w = u+iv$ are complex numbers?
2. Can you use the definition of e^z in Question 1 to prove that $D_x e^{kx} = ke^{kx}$ if $k = a + bi$ is a complex constant and x is a real variable?

11.4 PROBLEMS

In Problems 1 through 10, find Taylor's formula for the given function f at a = 0. Find both the Taylor polynomial $P_n(x)$ of the indicated degree n and the remainder term $R_n(x)$.

1. $f(x) = e^{-x}$, $n = 5$
2. $f(x) = \sin x$, $n = 4$
3. $f(x) = \cos x$, $n = 4$
4. $f(x) = \dfrac{1}{1-x}$, $n = 4$
5. $f(x) = \sqrt{1+x}$, $n = 3$
6. $f(x) = \ln(1+x)$, $n = 4$
7. $f(x) = \tan x$, $n = 3$
8. $f(x) = \arctan x$, $n = 2$
9. $f(x) = \sin^{-1} x$, $n = 2$
10. $f(x) = x^3 - 3x^2 + 5x - 7$, $n = 4$

In Problems 11 through 20, find the Taylor polynomial with remainder by using the given values of a and n.

11. $f(x) = e^x$; $a = 1, n = 4$
12. $f(x) = \cos x$; $a = \pi/4, n = 3$
13. $f(x) = \sin x$; $a = \pi/6, n = 3$
14. $f(x) = \sqrt{x}$; $a = 100, n = 3$
15. $f(x) = \dfrac{1}{(x-4)^2}$; $a = 5, n = 5$
16. $f(x) = \tan x$; $a = \pi/4, n = 4$
17. $f(x) = \cos x$; $a = \pi, n = 4$
18. $f(x) = \sin x$; $a = \pi/2, n = 4$
19. $f(x) = x^{3/2}$; $a = 1, n = 4$
20. $f(x) = \dfrac{1}{\sqrt{1-x}}$; $a = 0, n = 4$

In Problems 21 through 28, find the Maclaurin series of the given function f by substituting in one of the known series in Eqs. (19), (21), and (22).

21. $f(x) = e^{-x}$
22. $f(x) = e^{2x}$
23. $f(x) = e^{-3x}$
24. $f(x) = \exp(x^3)$
25. $f(x) = \sin 2x$
26. $f(x) = \sin \dfrac{x}{2}$
27. $f(x) = \sin x^2$
28. $f(x) = \sin^2 x = \frac{1}{2}(1 - \cos 2x)$

In Problems 29 through 40, find the Taylor series [Eq. (16)] of the given function at the indicated point a.

29. $f(x) = \ln(1+x)$, $a = 0$
30. $f(x) = \dfrac{1}{1-x}$, $a = 0$
31. $f(x) = e^{-x}$, $a = 0$
32. $f(x) = \sin x$, $a = \pi/2$
33. $f(x) = \ln x$, $a = 1$
34. $f(x) = e^{2x}$, $a = 0$
35. $f(x) = \cos x$, $a = \pi/4$
36. $f(x) = \dfrac{1}{(1-x)^2}$, $a = 0$
37. $f(x) = \dfrac{1}{x}$, $a = 1$
38. $f(x) = \cos x$, $a = \pi/2$
39. $f(x) = \sin x$, $a = \pi/4$
40. $f(x) = \sqrt{1+x}$, $a = 0$

41. Derive, as in Example 5, the Taylor series in Eq. (22) of $f(x) = \sin x$ at $a = 0$.

42. Granted that it is valid to differentiate the sine and cosine Taylor series in a term-by-term manner, use these series to verify that $D_x \cos x = -\sin x$ and $D_x \sin x = \cos x$.

43. Use the differentiation formulas $D_x \sinh x = \cosh x$ and $D_x \cosh x = \sinh x$ to derive the Maclaurin series

$$\cosh x = \sum_{n=0}^{\infty} \frac{x^{2n}}{(2n)!} \quad \text{and} \quad \sinh x = \sum_{n=0}^{\infty} \frac{x^{2n+1}}{(2n+1)!}$$

for the hyperbolic sine and cosine functions. What is their relationship to the Maclaurin series of the ordinary sine and cosine functions?

44. Derive the Maclaurin series stated in Problem 43 by substituting the known Maclaurin series for the exponential function in the definitions

$$\cosh x = \frac{e^x + e^{-x}}{2} \quad \text{and} \quad \sinh x = \frac{e^x - e^{-x}}{2}$$

of the hyperbolic functions.

The sum commands listed for several computer algebra systems in the Section 11.3 Project can be used to calculate Taylor polynomials efficiently. For instance, when the TI graphing calculator definitions

```
Y1 = sin(x)

Y2 = sum(seq((-1)∧(N-1)*X∧(2N-1)/
        (2N-1)!,N,1,7))
```

are graphed, the result is Fig. 11.4.9, showing that the 13th-degree Taylor polynomial $P_{13}(x)$ approximates $\sin x$ rather closely if $-3\pi/2 < x < 3\pi/2$ but not outside this range. By plotting several successive Taylor polynomials of a function $f(x)$ simultaneously, we can get a visual sense of the way in which they approximate the function. Do this for each function given in Problems 45 through 50.

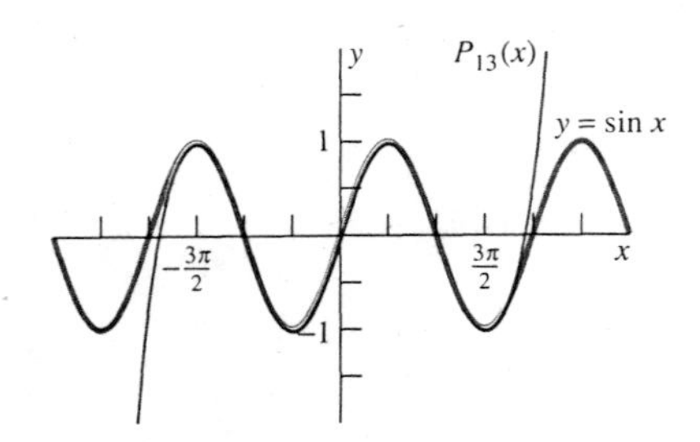

FIGURE 11.4.9 Graphs of $y = \sin x$ and its 13th-degree Taylor polynomial $P_{13}(x)$.

45. $f(x) = e^{-x}$

46. $f(x) = \sin x$

47. $f(x) = \cos x$

48. $f(x) = \ln(1 + x)$

49. $f(x) = \dfrac{1}{1+x}$

50. $f(x) = \dfrac{1}{1-x^2}$

51. Let the function

$$f(x) = \sum_{n=0}^{\infty} \frac{(-1)^n x^n}{(2n)!} = 1 - \frac{x}{2!} + \frac{x^2}{4!} - \frac{x^3}{6!} + \cdots$$

be defined by replacing x with $\sqrt{x}$ in the Maclaurin series for $\cos x$. Plot partial sums of this series to verify graphically that $f(x)$ agrees with the function $g(x)$ defined by

$$g(x) = \begin{cases} \cos \sqrt{x} & \text{if } x \geqq 0, \\ \cosh \sqrt{|x|} & \text{if } x < 0. \end{cases}$$

52. Beginning with $\alpha = \tan^{-1}(\frac{1}{5})$, use the addition formula

$$\tan(A + B) = \frac{\tan A + \tan B}{1 - \tan A \tan B}$$

to show in turn that (a) $\tan 2\alpha = \frac{5}{12}$; (b) $\tan 4\alpha = \frac{120}{119}$; (c) $\tan(\pi/4 - 4\alpha) = -\frac{1}{239}$. Finally, show that part (c) implies Eq. (28).

53. Apply the addition formula for the tangent function to verify Eq. (29).

54. Every young person deserves the thrill, just once, of calculating personally the first several decimal places of the number π. The seemingly random nature of this decimal expansion demands an explanation; how, indeed, are the digits 3.14159 26535 89793 . . . determined? For a partial answer, set your calculator to display nine decimal places. Then add enough terms of the arctangent series in (27) with $x = \frac{1}{5}$ to calculate $\arctan(\frac{1}{5})$ accurate to nine places. Next, calculate the value of $\arctan(\frac{1}{239})$ similarly. Finally, substitute these numerical results in Eq. (28) and solve for π. How many accurate decimal places do you get?

55. Prove that

$$\lim_{n \to \infty} \frac{x^n}{n!} = 0$$

if x is a real number. [*Suggestion:* Choose an integer k such that $k > |2x|$, and let $L = |x|^k/k!$. Then show that

$$\frac{|x|^n}{n!} < \frac{L}{2^{n-k}}$$

if $n > k$.]

56. Suppose that $0 < x \leqq 1$. Integrate both sides of the identity

$$\frac{1}{1+t} = 1 - t + t^2 - t^3 + \cdots + (-1)^n t^n + \frac{(-1)^{n+1} t^{n+1}}{1+t}$$

from $t = 0$ to $t = x$ to show that

$$\ln(1 + x) = x - \frac{x^2}{2} + \frac{x^3}{3} - \cdots + (-1)^n \frac{x^{n+1}}{n+1} + R_n,$$

where $\lim_{n \to \infty} R_n = 0$. Hence conclude that

$$\ln(1 + x) = \sum_{n=1}^{\infty} (-1)^{n+1} \frac{x^n}{n}$$

if $0 < x \leqq 1$.

57. Criticize the following "proof" that $2 = 1$. Substituting $x = 1$ into the result in Problem 56 yields the fact that

$$\ln 2 = 1 - \tfrac{1}{2} + \tfrac{1}{3} - \tfrac{1}{4} + \cdots.$$

If

$$S = 1 + \tfrac{1}{2} + \tfrac{1}{3} + \tfrac{1}{4} + \cdots,$$

then

$$\ln 2 = S - 2 \cdot \left(\tfrac{1}{2} + \tfrac{1}{4} + \tfrac{1}{6} + \tfrac{1}{8} + \cdots\right) = S - S = 0.$$

Hence $2 = e^{\ln 2} = e^0 = 1$.

58. Deduce from the result of Problem 56 first that

$$\ln(1 - x) = -\sum_{n=1}^{\infty} \frac{x^n}{n} = -x - \frac{x^2}{3} - \frac{x^3}{3} - \cdots$$

and then that

$$\ln \frac{1+x}{1-x} = \sum_{n \text{ odd}} \frac{2x^n}{n} = 2\left(x + \frac{x^3}{3} + \frac{x^5}{5} + \cdots\right)$$

if $0 \leqq x \leqq 1$.

59. Approximate the number $\ln 2 \approx 0.69315$ first by substituting $x = 1$ in the Maclaurin series of Problem 56, and then by substituting $x = \frac{1}{3}$ (why?) in the second series of Problem 58. Which approach appears to require the fewest terms to yield the value of $\ln 2$ accurate to a given number of decimal places?

11.4 Project: Calculating Logarithms on a Deserted Island

The problem is that you're stranded for life on a desert island with only a very basic calculator that does not calculate natural logarithms. So to get modern science going on this miserable island, you need to use the infinite series for $\ln[(1 + x)/(1 - x)]$ in Problem 58 to produce a simple table of logarithms (with five-place accuracy, say), giving $\ln x$ at least for the integers $x = 1, 2, 3, \ldots, 9$, and 10.

The most direct way might be to use the series for $\ln[(1+x)/(1-x)]$ to calculate first $\ln 2$, $\ln 3$, $\ln 5$, and $\ln 7$. Then use the law of logarithms $\ln xy = \ln x + \ln y$ to fill in the other entries in the table by simple addition of logarithms already computed. Unfortunately, larger values of x result in series that are more slowly convergent. So you could save yourself time and work by exercising some ingenuity: Calculate from scratch some four *other* logarithms from which you can build up the rest. For example, if you know $\ln 2$ and $\ln 1.25$, then $\ln 10 = \ln 1.25 + 3 \ln 2$. (Why?) Be as ingenious as you wish. Can you complete your table of ten logarithms by calculating directly (using the series) *fewer* than four logarithms to begin with?

For a finale, calculate somehow (from scratch, and accurate to five rounded decimal places) the natural logarithm $\ln(pq.rs)$, where p, q, r, and s denote the last four *nonzero* digits in your student I.D. number.

11.5 THE INTEGRAL TEST

A Taylor series (as in Section 11.4) is a special type of infinite series with *variable* terms. We saw that Taylor's formula can sometimes be used—as in the case of the exponential, sine, and cosine series—to establish the convergence of such a series.

But given an infinite series $\sum a_n$ with *constant* terms, it is the exception rather than the rule when a simple formula for the nth partial sum of that series can be found and used directly to determine whether the series converges or diverges. There are, however, several *convergence tests* that use the *terms* of an infinite series rather than its partial sums. Such a test, when successful, will tell us whether or not the series converges. Once we know that the series $\sum a_n$ does converge, it is then a separate matter to find its sum S. It may be necessary to approximate S by adding sufficiently many terms; in this case we shall need to know how many terms are required for the desired accuracy.

Here and in Section 11.6, we concentrate our attention on **positive-term series**—that is, series with terms that are all positive. If $a_n > 0$ for all n, then

$$S_1 < S_2 < S_3 < \cdots < S_n < \cdots,$$

so the sequence $\{S_n\}$ of partial sums of the series is increasing. Hence there are just two possibilities. If the sequence $\{S_n\}$ is *bounded*—there exists a number M such that $S_n < M$ for all n—then the bounded monotonic sequence property (Section 11.2) implies that $S = \lim_{n\to\infty} S_n$ exists, so the series $\sum a_n$ *converges.* Otherwise, it diverges to infinity (by Problem 52 in Section 11.2).

A similar alternative holds for improper integrals. Suppose that the function f is continuous and positive-valued for $x \geqq 1$. Then it follows (from Problem 51) that the improper integral

$$\int_1^\infty f(x)\,dx = \lim_{b\to\infty} \int_1^b f(x)\,dx \tag{1}$$

either converges (the limit is a real number) or diverges to infinity (the limit is $+\infty$). This analogy between positive-term series and improper integrals of positive functions is the key to the **integral test.** We compare the behavior of the series $\sum a_n$ with that of the improper integral in Eq. (1), where f is an appropriately chosen function. [Among other things, we require that $f(n) = a_n$ for all n.]

THEOREM 1 The Integral Test

Suppose that $\sum a_n$ is a positive-term series and that f is a positive-valued, decreasing, continuous function for $x \geqq 1$. If $f(n) = a_n$ for all integers $n \geqq 1$, then the series and the improper integral

$$\sum_{n=1}^{\infty} a_n \quad \text{and} \quad \int_1^\infty f(x)\,dx$$

either both converge or both diverge.

PROOF Because f is a decreasing function, the rectangular polygon with area

$$S_n = a_1 + a_2 + a_3 + \cdots + a_n$$

shown in Fig. 11.5.1 contains the region under $y = f(x)$ from $x = 1$ to $x = n + 1$. Hence

$$\int_1^{n+1} f(x)\,dx \leqq S_n. \tag{2}$$

Similarly, the rectangular polygon with area

$$S_n - a_1 = a_2 + a_3 + a_4 + \cdots + a_n$$

shown in Fig. 11.5.2 is contained in the region under $y = f(x)$ from $x = 1$ to $x = n$. Hence

$$S_n - a_1 \leqq \int_1^n f(x)\,dx. \tag{3}$$

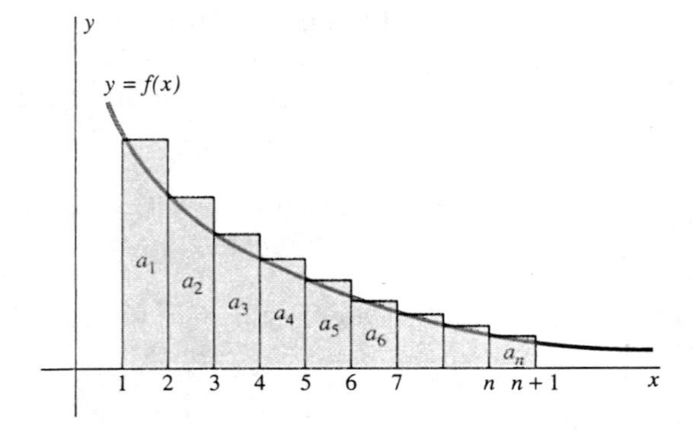

FIGURE 11.5.1 Underestimating the partial sums with an integral.

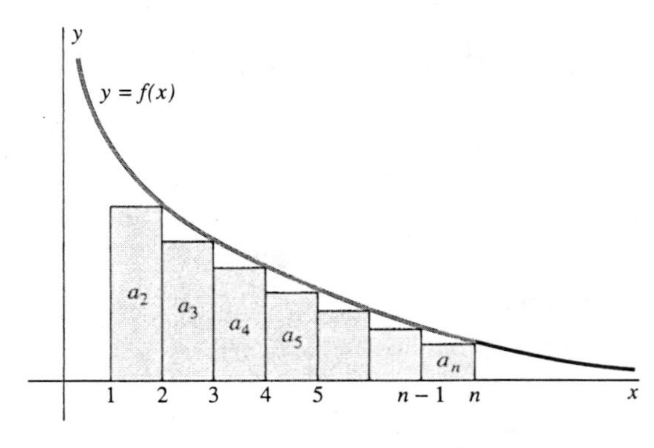

FIGURE 11.5.2 Overestimating the partial sums with an integral.

Suppose first that the improper integral $\int_1^\infty f(x)\,dx$ diverges (necessarily to $+\infty$). Then

$$\lim_{n\to\infty} \int_1^{n+1} f(x)\,dx = +\infty,$$

so it follows from (2) that $\lim_{n\to\infty} S_n = +\infty$ as well, and hence the infinite series $\sum a_n$ likewise diverges.

Now suppose instead that the improper integral $\int_1^\infty f(x)\,dx$ converges and has the (finite) value I. Then (3) implies that

$$S_n \leqq a_1 + \int_1^n f(x)\,dx \leqq a_1 + I,$$

so the increasing sequence $\{S_n\}$ is bounded. Thus the infinite series

$$\sum_{n=1}^{\infty} a_n = \lim_{n\to\infty} S_n$$

converges as well. Hence we have shown that the infinite series and the improper integral either both converge or both diverge. ◄

EXAMPLE 1 We used a version of the integral test to prove in Section 11.3 that the harmonic series

$$\sum_{n=1}^{\infty} \frac{1}{n} = 1 + \frac{1}{2} + \frac{1}{3} + \frac{1}{4} + \cdots$$

diverges. Using the test as stated in Theorem 1 is a little simpler: We note that $f(x) = 1/x$ is positive, continuous, and decreasing for $x \geqq 1$ and that $f(n) = 1/n$ for each positive integer n. Now

$$\int_1^\infty \frac{1}{x}\,dx = \lim_{b\to\infty} \int_1^b \frac{1}{x}\,dx = \lim_{b\to\infty} \Big[\ln x\Big]_1^b = \lim_{b\to\infty} (\ln b - \ln 1) = +\infty.$$

Thus the improper integral diverges and, therefore, so does the harmonic series. ◆

The harmonic series is the case $p = 1$ of the ***p*-series**

$$\sum_{n=1}^{\infty} \frac{1}{n^p} = 1 + \frac{1}{2^p} + \frac{1}{3^p} + \cdots + \frac{1}{n^p} + \cdots. \tag{4}$$

Whether the p-series converges or diverges depends on the value of p.

EXAMPLE 2 Show that the p-series converges if $p > 1$ but diverges if $0 < p \leqq 1$.

Solution The case $p = 1$ has already been settled in Example 1. If $p > 0$ but $p \neq 1$, then the function $f(x) = 1/x^p$ satisfies the conditions of the integral test, and

$$\int_1^\infty \frac{1}{x^p}\,dx = \lim_{b\to\infty} \int_1^b \frac{1}{x^p}\,dx = \lim_{b\to\infty} \left[-\frac{1}{(p-1)x^{p-1}}\right]_1^b$$
$$= \lim_{b\to\infty} \frac{1}{p-1}\left(1 - \frac{1}{b^{p-1}}\right).$$

If $p > 1$, then

$$\int_1^\infty \frac{1}{x^p}\,dx = \frac{1}{p-1} < \infty,$$

so the integral and the series both converge. But if $0 < p < 1$, then

$$\int_1^\infty \frac{1}{x^p}\,dx = \lim_{b\to\infty} \frac{1}{1-p}(b^{1-p} - 1) = \infty,$$

and in this case the integral and the series both diverge. ◆

As specific examples, the series

$$\sum_{n=1}^{\infty} \frac{1}{n^2} = 1 + \frac{1}{2^2} + \frac{1}{3^2} + \cdots + \frac{1}{n^2} + \cdots$$

converges ($p = 2 > 1$), whereas the series

$$\sum_{n=1}^{\infty} \frac{1}{\sqrt{n}} = 1 + \frac{1}{\sqrt{2}} + \frac{1}{\sqrt{3}} + \cdots + \frac{1}{\sqrt{n}} + \cdots$$

diverges ($p = \frac{1}{2} \leqq 1$).

Now suppose that the positive-term series $\sum a_n$ converges by the integral test and that we wish to approximate its sum by adding sufficiently many of its initial terms. The difference between the sum S of the series and its nth partial sum S_n is the **remainder**

$$R_n = S - S_n = a_{n+1} + a_{n+2} + a_{n+3} + \cdots. \tag{5}$$

This remainder is the error made when the sum is estimated by using in its place the partial sum S_n.

THEOREM 2 The Integral Test Remainder Estimate

Suppose that the infinite series and improper integral

$$\sum_{n=1}^{\infty} a_n \quad \text{and} \quad \int_1^{\infty} f(x)\,dx$$

satisfy the hypotheses of the integral test, and suppose in addition that both converge. Then

$$\int_{n+1}^{\infty} f(x)\,dx \leqq R_n \leqq \int_n^{\infty} f(x)\,dx, \tag{6}$$

where R_n is the remainder given in Eq. (5).

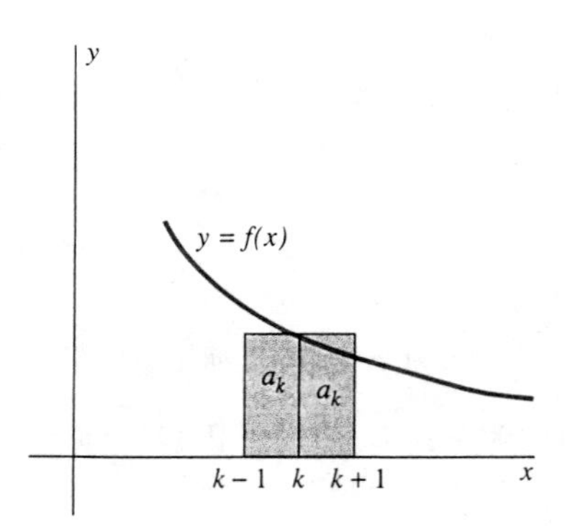

FIGURE 11.5.3 Establishing the integral test remainder estimate.

PROOF We see from Fig. 11.5.3 that

$$\int_k^{k+1} f(x)\,dx \leqq a_k \leqq \int_{k-1}^{k} f(x)\,dx$$

for $k = n+1, n+2, \ldots$. We add these inequalities for all such values of k, and the result is the inequality in (6), because

$$R_n = \sum_{k=n+1}^{\infty} a_k,$$

$$\sum_{k=n+1}^{\infty} \int_k^{k+1} f(x)\,dx = \int_{n+1}^{\infty} f(x)\,dx,$$

and

$$\sum_{k=n+1}^{\infty} \int_{k-1}^{k} f(x)\,dx = \int_n^{\infty} f(x)\,dx.$$

◄

If we substitute $R_n = S - S_n$, then it follows from (6) that the sum S of the series satisfies the inequality

$$S_n + \int_{n+1}^{\infty} f(x)\,dx \leqq S \leqq S_n + \int_n^{\infty} f(x)\,dx. \tag{7}$$

If the nth partial sum S_n is known and the difference

$$\int_n^{n+1} f(x)\,dx$$

between the two integrals is small, then (7) provides an accurate estimate of the sum S of the infinite series.

EXAMPLE 3 We will see in Section 11.8 that the exact sum of the p-series with $p=2$ is $\pi^2/6$, thus giving the beautiful formula

$$\frac{\pi^2}{6} = 1 + \frac{1}{2^2} + \frac{1}{3^2} + \frac{1}{4^2} + \cdots. \tag{8}$$

Use this series to approximate the number π by applying the integral test remainder estimate, first with $n=50$, then with $n=200$.

Solution Obviously we take $f(x)=1/x^2$ in the remainder estimate. Because

$$\int_n^{\infty} \frac{1}{x^2}\,dx = \lim_{b\to\infty}\left[-\frac{1}{x}\right]_n^b = \lim_{b\to\infty}\left(\frac{1}{n} - \frac{1}{b}\right) = \frac{1}{n},$$

Eq. (7) gives

$$S_n + \frac{1}{n+1} \leqq \frac{\pi^2}{6} \leqq S_n + \frac{1}{n}, \tag{9}$$

where

$$S_n = 1 + \frac{1}{2^2} + \frac{1}{3^2} + \cdots + \frac{1}{n^2}$$

is the nth partial sum of the series in (8). Upon multiplying by 6 and taking square roots, (9) gives the inequality

$$\sqrt{6\left(S_n + \frac{1}{n+1}\right)} \leqq \pi \leqq \sqrt{6\left(S_n + \frac{1}{n}\right)}. \tag{10}$$

You could add the first 50 terms in (8) one by one in a few minutes using a simple four-function calculator, but this kind of arithmetic is precisely the task for which a modern calculator or computer algebra system is designed. A one-line instruction such as the calculator command `sum(seq(1/n∧2,n,1,50))` yields

$$S_{50} = \sum_{n=1}^{50} \frac{1}{n^2} \approx 1.625132734.$$

Then, using (9) for illustration rather than (10), we calculate

$$1.62513273 + \frac{1}{51} < \frac{\pi^2}{6} < 1.62513274 + \frac{1}{50};$$

$$1.64474057 < \frac{\pi^2}{6} < 1.64513274;$$

$$3.14140787 < \pi < 3.14178237.$$

Finally, rounding down on the left and up on the right (why?), we conclude that $3.1414 < \pi < 3.1418$. The average of these two bounds is the traditional four-place approximation $\pi \approx 3.1416$.

The 200th partial sum of the series in (8) is

$$S_{200} = \sum_{n=1}^{200} \frac{1}{n^2} \approx 1.639946546.$$

Substituting this sum and $n = 200$ in (10), we get

$$3.14158081 < \pi < 3.14160457.$$

This proves that $\pi \approx 3.1416$ rounded accurate to four decimal places. ◆

EXAMPLE 4 Show that the series

$$\sum_{n=2}^{\infty} \frac{1}{n(\ln n)^2} \tag{11}$$

converges, and determine how many terms you would need to add to find its sum accurate to within 0.01. That is, how large must n be in order that the remainder satisfy the inequality $R_n < 0.01$?

Solution We begin the sum at $n = 2$ because $\ln 1 = 0$. Let $f(x) = 1/[x(\ln x)^2]$. Then

$$\int_n^{\infty} \frac{1}{x(\ln x)^2}\,dx = \lim_{b\to\infty}\left[-\frac{1}{\ln x}\right]_n^b = \lim_{b\to\infty}\left(\frac{1}{\ln n} - \frac{1}{\ln b}\right) = \frac{1}{\ln n}.$$

Substituting $n = 2$ shows that the series in (11) converges (by the integral test). Our calculations and the right-hand inequality in (6) now give $R_n < 1/(\ln n)$, so we need

$$\frac{1}{\ln n} \leqq 0.01; \quad \ln n \geqq 100; \quad n \geqq e^{100} \approx 2.7 \times 10^{43}.$$

A computer that could calculate a billion (10^9) terms per second would require about 8.5×10^{26} years—far longer than the expected lifetime of the universe—to sum this many terms. But you can check that accuracy to only one decimal place—that is, $R_n < 0.05$—would require only about $n = 4.85 \times 10^8$ (fewer than a half billion) terms, well within the range of a powerful desktop computer. ◆

11.5 TRUE/FALSE STUDY GUIDE

11.5 CONCEPTS: QUESTIONS AND DISCUSSION

1. What might it mean to say that one infinite series converges more slowly than another? Perhaps that more terms of one than the other must be added to make the remainder R_n less than a preassigned error? If so, compare the rates at which the series $\sum n^{-2}, \sum n^{-3/2}, \sum n^{-4/3}, \dots, \sum n^{-101/100}, \dots,$ converge.
2. Can you use infinite series such as those listed in Question 1 to illustrate the claim that, however slowly one infinite series converges, there's another one that converges even more slowly?
3. Can you think of a way in which the convergent infinite series in Question 1 seem to resemble more and more closely the divergent harmonic series $\sum n^{-1}$? Discuss the possibility that two infinite series can resemble each other arbitrarily closely, yet one converges and the other diverges.

11.5 PROBLEMS

In Problems 1 through 30, use the integral test to test the given series for convergence.

1. $\sum_{n=1}^{\infty} \frac{n}{n^2+1}$

2. $\sum_{n=1}^{\infty} \frac{n}{e^{n^2}}$

3. $\sum_{n=1}^{\infty} \frac{1}{\sqrt{n+1}}$

4. $\sum_{n=1}^{\infty} \frac{1}{(n+1)^{4/3}}$

5. $\sum_{n=1}^{\infty} \frac{1}{n^2+1}$

6. $\sum_{n=1}^{\infty} \frac{1}{n(n+1)}$

7. $\sum_{n=2}^{\infty} \frac{1}{n \ln n}$

8. $\sum_{n=1}^{\infty} \frac{\ln n}{n}$

9. $\sum_{n=1}^{\infty} \frac{1}{2^n}$

10. $\sum_{n=1}^{\infty} \frac{n}{e^n}$

11. $\sum_{n=1}^{\infty} \frac{n^2}{e^n}$

12. $\sum_{n=1}^{\infty} \frac{1}{17n-13}$

13. $\sum_{n=1}^{\infty} \frac{\ln n}{n^2}$

14. $\sum_{n=1}^{\infty} \frac{n+1}{n^2}$

15. $\sum_{n=1}^{\infty} \frac{n}{n^4+1}$

16. $\sum_{n=1}^{\infty} \frac{1}{n^3+n}$

17. $\sum_{n=1}^{\infty} \frac{2n+5}{n^2+5n+17}$

18. $\sum_{n=1}^{\infty} \ln\left(\frac{n+1}{n}\right)$

19. $\sum_{n=1}^{\infty} \ln\left(1+\frac{1}{n^2}\right)$

20. $\sum_{n=1}^{\infty} \frac{2^{1/n}}{n^2}$

21. $\sum_{n=1}^{\infty} \frac{n}{4n^2+5}$

22. $\sum_{n=1}^{\infty} \frac{n}{(4n^2+5)^{3/2}}$

23. $\sum_{n=2}^{\infty} \frac{1}{n\sqrt{\ln n}}$

24. $\sum_{n=2}^{\infty} \frac{1}{n(\ln n)^3}$

25. $\sum_{n=1}^{\infty} \frac{1}{4n^2+9}$

26. $\sum_{n=1}^{\infty} \frac{n+1}{n+100}$

27. $\sum_{n=1}^{\infty} \frac{n}{n^4+2n^2+1}$

28. $\sum_{n=1}^{\infty} \frac{1}{(n+1)^3}$

29. $\sum_{n=1}^{\infty} \frac{\arctan n}{n^2+1}$

30. $\sum_{n=3}^{\infty} \frac{1}{n(\ln n)[\ln(\ln n)]}$

In Problems 31 through 34, tell why the integral test does not apply to the given series.

31. $\sum_{n=1}^{\infty} \frac{(-1)^n}{n}$

32. $\sum_{n=1}^{\infty} e^{-n} \sin n$

33. $\sum_{n=1}^{\infty} \frac{2+\sin n}{n^2}$

34. $\sum_{n=1}^{\infty} \left(\frac{\sin n}{n}\right)^4$

In Problems 35 through 38, determine the values of p for which the given series converges.

35. $\sum_{n=1}^{\infty} \frac{1}{p^n}$

36. $\sum_{n=1}^{\infty} \frac{n}{(n^2+1)^p}$

37. $\sum_{n=2}^{\infty} \frac{1}{n(\ln n)^p}$

38. $\sum_{n=3}^{\infty} \frac{1}{n(\ln n)\,[\ln(\ln n)]^p}$

In Problems 39 through 42, find the least positive integer n such that the remainder R_n in Theorem 2 is less than E.

39. $\sum_{n=1}^{\infty} \frac{1}{n^2}$; $E = 0.0001$

40. $\sum_{n=1}^{\infty} \frac{1}{n^2}$; $E = 0.00005$

41. $\sum_{n=1}^{\infty} \frac{1}{n^3}$; $E = 0.00005$

42. $\sum_{n=1}^{\infty} \frac{1}{n^6}$; $E = 2 \times 10^{-11}$

In Problems 43 through 46, find the sum of the given series accurate to the indicated number k of decimal places. Begin by finding the smallest value of n such that the remainder satisfies the inequality $R_n < 5 \times 10^{-(k+1)}$. Then use a calculator to compute the partial sum S_n and round off appropriately.

43. $\sum_{n=1}^{\infty} \frac{1}{n^{3/2}}$; $k = 2$

44. $\sum_{n=1}^{\infty} \frac{1}{n^3}$; $k = 3$

45. $\sum_{n=1}^{\infty} \frac{1}{n^5}$; $k = 5$

46. $\sum_{n=1}^{\infty} \frac{1}{n^7}$; $k = 7$

In Problems 47 and 48, use a computer algebra system (if necessary) to determine the values of p for which the given infinite series converges.

47. $\sum_{n=1}^{\infty} \frac{\ln n}{n^p}$

48. $\sum_{n=1}^{\infty} \frac{1}{p^{\ln n}}$

49. Deduce from the inequalities in (2) and (3) with the function $f(x) = 1/x$ that

$$\ln n \leqq 1 + \frac{1}{2} + \frac{1}{3} + \cdots + \frac{1}{n} \leqq 1 + \ln n$$

for $n = 1, 2, 3, \ldots$. If a computer adds 1 millions terms of the harmonic series per second, how long will it take for the partial sum to reach 50?

50. (a) Let

$$c_n = 1 + \frac{1}{2} + \frac{1}{3} + \cdots + \frac{1}{n} - \ln n$$

for $n = 1, 2, 3, \ldots$. Deduce from Problem 49 that $0 \leqq c_n \leqq 1$ for all n. (b) Note that

$$\int_n^{n+1} \frac{1}{x}\,dx \geqq \frac{1}{n+1}.$$

Conclude that the sequence $\{c_n\}$ is decreasing. Therefore the sequence $\{c_n\}$ converges. The number

$$\gamma = \lim_{n\to\infty} c_n = \lim_{n\to\infty} \left(1 + \frac{1}{2} + \frac{1}{3} + \cdots + \frac{1}{n} - \ln n\right) \approx 0.57722$$

is known as **Euler's constant.**

51. Suppose that the function f is continuous and positive-valued for $x \geqq 1$. Let

$$b_n = \int_1^n f(x)\,dx$$

for $n = 1, 2, 3, \ldots$. (a) Suppose that the increasing sequence $\{b_n\}$ is bounded, so that $B = \lim_{n\to\infty} b_n$ exists. Prove that

$$\int_1^{\infty} f(x)\,dx = B.$$

(b) Prove that if the sequence $\{b_n\}$ is not bounded, then

$$\int_1^{\infty} f(x)\,dx = +\infty.$$

11.5 Project: The Number π, Once and for All

When we replace the parameter p in the p-series $\sum 1/n^p$ with the variable x, we get one of the most important transcendental functions in higher mathematics, the **Riemann zeta function**

$$\zeta(x) = \sum_{n=1}^{\infty} \frac{1}{n^x} = 1 + \frac{1}{2^x} + \frac{1}{3^x} + \frac{1}{4^x} + \cdots.$$

REMARK One can substitute a complex number $x = a + bi$ in the zeta function. Now that Fermat's last theorem has been proved, the most famous unsolved conjecture in mathematics is the **Riemann hypothesis**—that $\zeta(a + bi) = 0$ implies that $a = \frac{1}{2}$; that is, that the only complex zeros of the Riemann zeta function have real part $\frac{1}{2}$. (The smallest such example is approximately $\frac{1}{2} + 14.13475i$.) The truth of the Riemann hypothesis would have profound implications in number theory, including information about the distribution of the prime numbers.

In Problems 1 through 4, use the given value of the zeta function and the integral-test remainder estimate (as in Example 3 of this section) with the given value of n to determine how accurately the value of the number π is thereby determined. Knowing that

$$\pi \approx 3.14159\,26535\,89793\,23846,$$

write each final answer in the form $\pi \approx 3.abcde\ldots$, giving precisely those digits that are correct or correctly rounded.

1. $\zeta(2) = \dfrac{\pi^2}{6}$ with $n = 25$.

2. $\zeta(4) = \dfrac{\pi^4}{90}$ with $n = 20$.

3. $\zeta(6) = \dfrac{\pi^6}{945}$ with $n = 15$.

4. $\zeta(8) = \dfrac{\pi^8}{9450}$ with $n = 10$.

5. Finally, use one of the preceding four problems and your own careful choice of n to show that $\pi \approx 3.141592654$ with all digits correct or correctly rounded.

Euler showed that if n is even then $\zeta(n)$ is a rational multiple of π^n (as in the cases $n = 2, 4, 6, 8$ cited above). Because any integral power of π is irrational, it follows that the number $\zeta(n)$ is irrational if n is even. But little was known about $\zeta(n)$ for n odd until 1978, when Roger Apéry proved that $\zeta(3)$ is irrational. In Section 7.7 of Andrews, Askey, and Roy, *Special Functions* (Cambridge Univ. Press: 1999), the authors show that there exist infinite sequences $\{A_n\}$ and $\{B_n\}$ of integers such that

$$0 < |A_n + B_n\zeta(3)| < 3 \cdot \left(\frac{9}{10}\right)^n$$

for each integer $n \geqq 1$. Can you explain why this implies that $\zeta(3)$ is irrational? (Assume, to the contrary, that $\zeta(3) = p/q$ is rational.)

11.6 COMPARISON TESTS FOR POSITIVE-TERM SERIES

With the integral test we attempt to determine whether or not an infinite series converges by comparing it with an improper integral. The methods of this section involves comparing the terms of the *positive-term* series $\sum a_n$ with those of another positive-term series $\sum b_n$ whose convergence or divergence is known. We have already developed two families of *reference series* for the role of the known series $\sum b_n$; these are the geometric series of Section 11.3 and the p-series of Section 11.5. They are well adapted for our new purposes because their convergence or divergence is quite easy to determine. Recall that the geometric series $\sum r^n$ converges if $|r| < 1$ and diverges if $|r| \geqq 1$, and that the p-series $\sum 1/n^p$ converges if $p > 1$ and diverges if $0 < p \leqq 1$.

Let $\sum a_n$ and $\sum b_n$ be positive-term series. Then we say that the series $\sum b_n$ **dominates** the series $\sum a_n$ provided that $a_n \leqq b_n$ for all n. Theorem 1 says that the positive-term series $\sum a_n$ converges if it is dominated by a convergent series and diverges if it dominates a positive-term divergent series.

THEOREM 1 Comparison Test

Suppose that $\sum a_n$ and $\sum b_n$ are positive-term series. Then

1. $\sum a_n$ converges if $\sum b_n$ converges and $a_n \leqq b_n$ for all n;
2. $\sum a_n$ diverges if $\sum b_n$ diverges and $a_n \geqq b_n$ for all n.

PROOF Denote the nth partial sums of the series $\sum a_n$ and $\sum b_n$ by S_n and T_n, respectively. Then $\{S_n\}$ and $\{T_n\}$ are increasing sequences. To prove part (1), suppose that $\sum b_n$ converges, so $T = \lim_{n\to\infty} T_n$ exists (so that T is a real number). Then the fact that $a_n \leqq b_n$ for all n implies that $S_n \leqq T_n \leqq T$ for all n. Thus the sequence $\{S_n\}$ of partial sums of $\sum a_n$ is bounded and increasing and therefore converges. Thus $\sum a_n$ converges.

Part (2) is merely a restatement of part (1). If the series $\sum a_n$ converged, then the fact that $\sum a_n$ dominates $\sum b_n$ would imply—by part (1), with a_n and b_n interchanged—that $\sum b_n$ converged. But $\sum b_n$ diverges, so it follows that $\sum a_n$ must also diverge. ◄

We know by Theorem 5 of Section 11.3 that the convergence or divergence of an infinite series is not affected by the insertion or deletion of a finite number of terms. Consequently, the conditions $a_n \leqq b_n$ and $a_n \geqq b_n$ in the two parts of the comparison test really need to hold only for all $n \geqq k$, where k is some fixed positive integer. Thus we can say that the positive-term series $\sum a_n$ converges if it is "eventually dominated" by the convergent positive-term series $\sum b_n$.

EXAMPLE 1 Because

$$\frac{1}{n(n+1)(n+2)} < \frac{1}{n^3}$$

for all $n \geqq 1$, the series

$$\sum_{n=1}^{\infty} \frac{1}{n(n+1)(n+2)} = \frac{1}{1\cdot 2\cdot 3} + \frac{1}{2\cdot 3\cdot 4} + \frac{1}{3\cdot 4\cdot 5} + \cdots$$

is dominated by the series $\sum 1/n^3$, which is a convergent p-series with $p = 3$. Both are positive-term series, and hence the series $\sum 1/[n(n+1)(n+2)]$ converges by part (1) of the comparison test. ◆

EXAMPLE 2 Because

$$\frac{1}{\sqrt{2n-1}} > \frac{1}{\sqrt{2n}}$$

for all $n \geqq 1$, the positive-term series

$$\sum_{n=1}^{\infty} \frac{1}{\sqrt{2n-1}} = 1 + \frac{1}{\sqrt{3}} + \frac{1}{\sqrt{5}} + \frac{1}{\sqrt{7}} + \cdots$$

dominates the series

$$\sum_{n=1}^{\infty} \frac{1}{\sqrt{2n}} = \frac{1}{\sqrt{2}} \sum_{n=1}^{\infty} \frac{1}{n^{1/2}}.$$

But $\sum 1/n^{1/2}$ is a divergent p-series with $p=\frac{1}{2}$, and a constant nonzero multiple of a divergent series diverges. So part (2) of the comparison test implies that the series $\sum 1/\sqrt{2n-1}$ also diverges. ◆

EXAMPLE 3 Test the series

$$\sum_{n=0}^{\infty} \frac{1}{n!} = 1 + \frac{1}{1!} + \frac{1}{2!} + \frac{1}{3!} + \cdots$$

for convergence.

Solution We note first that if $n \geqq 1$, then

$$\begin{aligned} n! &= n(n-1)(n-2)\cdots 3 \cdot 2 \cdot 1 \\ &\geqq 2 \cdot 2 \cdot 2 \cdots 2 \cdot 2 \cdot 1 \qquad \text{(the same number of factors);} \end{aligned}$$

that is, $n! \geqq 2^{n-1}$ for $n \geqq 1$. Thus

$$\frac{1}{n!} \leqq \frac{1}{2^{n-1}} \quad \text{for } n \geqq 1,$$

so the series

$$\sum_{n=0}^{\infty} \frac{1}{n!} \quad \text{is dominated by the series} \quad 1 + \sum_{n=1}^{\infty} \frac{1}{2^{n-1}} = 1 + \sum_{n=0}^{\infty} \frac{1}{2^n},$$

which is a convergent geometric series (after the first term). Both are positive-term series, so by the comparison test the given series converges. We saw in Section 11.4 that the sum of the series is the number e, so

$$e = 1 + \frac{1}{1!} + \frac{1}{2!} + \frac{1}{3!} + \cdots + \frac{1}{n!} + \cdots.$$

Indeed, this series provides perhaps the simplest way of showing that

$$e \approx 2.71828\,1828\,459045\,23536.$$ ◆

Limit Comparison of Terms

Suppose that $\sum a_n$ is a positive-term series such that $a_n \to 0$ as $n \to +\infty$. Then, in connection with the nth-term divergence test of Section 11.3, the series $\sum a_n$ has at least a *chance* of converging. How do we choose an appropriate positive-term series $\sum b_n$ with which to compare it? A good idea is to express b_n as a *simple* function of n, simpler than a_n but such that a_n and b_n approach zero at the same rate as $n \to +\infty$. If the formula for a_n is a fraction, we can try discarding all but the terms of largest magnitude in its numerator and denominator to form b_n. For example, if

$$a_n = \frac{3n^2 + n}{n^4 + \sqrt{n}},$$

then we reason that n is small in comparison with $3n^2$, and that $\sqrt{n}$ is small in comparison with n^4, when n is quite large. This suggests that we choose $b_n = 3n^2/n^4 = 3/n^2$. The series $\sum 3/n^2$ converges $(p=2)$, but when we attempt to compare $\sum a_n$ and $\sum b_n$, we find that $a_n \geqq b_n$ (rather than $a_n \leqq b_n$). Consequently, the comparison test does not apply immediately—the fact that $\sum a_n$ dominates a convergent series does *not* imply that $\sum a_n$ itself converges. Theorem 2 provides a convenient way of handling such a situation.

THEOREM 2 Limit Comparison Test
Suppose that $\sum a_n$ and $\sum b_n$ are positive-term series. If the limit

$$L = \lim_{n\to\infty} \frac{a_n}{b_n}$$

exists and $0 < L < +\infty$, then either both series converge or both series diverge.

PROOF Choose two fixed positive numbers P and Q such that $P < L < Q$. Then $P < a_n/b_n < Q$ for n sufficiently large, and so

$$Pb_n < a_n < Qb_n$$

for all sufficiently large values of n. If $\sum b_n$ converges, then $\sum a_n$ is eventually dominated by the convergent series $\sum Qb_n = Q\sum b_n$, so part (1) of the comparison test implies that $\sum a_n$ also converges. If $\sum b_n$ diverges, then $\sum a_n$ eventually dominates the divergent series $\sum Pb_n = P\sum b_n$, so part (2) of the comparison test implies that $\sum a_n$ also diverges. Thus the convergence of either series implies the convergence of the other. ◄

EXAMPLE 4 With

$$a_n = \frac{3n^2 + n}{n^4 + \sqrt{n}} \quad \text{and} \quad b_n = \frac{1}{n^2}$$

(motivated by the discussion preceding Theorem 2), we find that

$$\lim_{n\to\infty} \frac{a_n}{b_n} = \lim_{n\to\infty} \frac{3n^4 + n^3}{n^4 + \sqrt{n}} = \lim_{n\to\infty} \frac{3 + \dfrac{1}{n}}{1 + \dfrac{1}{n^{7/2}}} = 3.$$

Because $\sum 1/n^2$ is a convergent p-series $(p = 2)$, the limit comparison test tells us that the series

$$\sum_{n=1}^{\infty} \frac{3n^2 + n}{n^4 + \sqrt{n}}$$

also converges. ◆

EXAMPLE 5 Test for convergence: $\displaystyle\sum_{n=1}^{\infty} \frac{1}{2n + \ln n}$.

Solution Because $\lim_{n\to\infty}(\ln n)/n = 0$ (by l'Hôpital's rule), $\ln n$ is very small in comparison with $2n$ when n is large. We therefore take $a_n = 1/(2n + \ln n)$ and, ignoring the constant coefficient 2, we take $b_n = 1/n$. Then we find that

$$\lim_{n\to\infty} \frac{a_n}{b_n} = \lim_{n\to\infty} \frac{n}{2n + \ln n} = \lim_{n\to\infty} \frac{1}{2 + \dfrac{\ln n}{n}} = \frac{1}{2}.$$

Because the harmonic series $\sum 1/n = \sum b_n$ diverges, it follows that the given series $\sum a_n$ also diverges. ◆

It is important to remember that if $L = \lim_{n\to\infty}(a_n/b_n)$ is either zero or infinite, then the limit comparison test does not apply. (See Problem 52 for a discussion of what conclusions may sometimes be drawn in these cases.) Note, for example, that if $a_n = 1/n^2$ and $b_n = 1/n$, then $\lim_{n\to\infty}(a_n/b_n) = 0$. But in this case $\sum a_n$ converges, whereas $\sum b_n$ diverges.

Estimating Remainders

Suppose that $0 \leqq a_n \leqq b_n$ for all n and we know that $\sum b_n$ converges, so the comparison test implies that $\sum a_n$ converges as well. Let us write $s = \sum a_n$ and $S = \sum b_n$. If a numerical estimate is available for the remainder

$$R_n = S - S_n = b_{n+1} + b_{n+2} + \cdots$$

in the dominating series $\sum b_n$, then we can use it to estimate the remainder

$$r_n = s - s_n = a_{n+1} + a_{n+2} + \cdots$$

in the series $\sum a_n$. The reason is that $0 \leqq a_n \leqq b_n$ (for all n) implies that $0 \leqq r_n \leqq R_n$. We can apply this fact if, for instance, we have used the integral test remainder estimate to calculate an upper bound for R_n—which is, then, an upper bound for r_n as well.

EXAMPLE 6 The series

$$\sum_{n=1}^{\infty} a_n = \sum_{n=1}^{\infty} \frac{1}{n^3 + \sqrt{n}}$$

converges because it is dominated by the convergent p-series

$$\sum_{n=1}^{\infty} b_n = \sum_{n=1}^{\infty} \frac{1}{n^3}.$$

It therefore follows by the integral test remainder estimate (Section 11.5) that

$$0 < r_n \leqq R_n \leqq \int_n^{\infty} \frac{1}{x^3}\,dx = \lim_{b \to \infty} \left[-\frac{1}{2x^2} \right]_n^b = \frac{1}{2n^2}.$$

Now a calculator gives

$$s_{100} = \sum_{n=1}^{100} \frac{1}{n^3 + \sqrt{n}} \approx 0.680284 \quad \text{and} \quad R_{100} \leqq \frac{1}{2 \cdot 100^2} = 0.00005.$$

It follows that $0.680284 \leqq s \leqq 0.680334$. In particular,

$$\sum_{n=1}^{\infty} \frac{1}{n^3 + \sqrt{n}} \approx 0.6803$$

rounded accurate to four decimal places. ◆

Rearrangement and Grouping

We close our discussion of positive-term series with the observation that the sum of a convergent *positive*-term series is not altered by grouping or rearranging its terms. For example, let $\sum a_n$ be a convergent positive-term series and consider

$$\sum_{n=1}^{\infty} b_n = (a_1 + a_2 + a_3) + a_4 + (a_5 + a_6) + \cdots.$$

That is, the new series has terms

$$b_1 = a_1 + a_2 + a_3,$$
$$b_2 = a_4,$$
$$b_3 = a_5 + a_6,$$

and so on. Then every partial sum T_n of $\sum b_n$ is equal to some partial sum $S_{n'}$ of $\sum a_n$. Because $\{S_n\}$ is an increasing sequence with limit $S = \sum a_n$, it follows easily that $\{T_n\}$ is an increasing sequence with the same limit. Thus $\sum b_n = S$ as well. The argument is more subtle if terms of $\sum a_n$ are moved "out of place," as in

$$\sum_{n=1}^{\infty} b_n = a_1 + a_2 + a_4 + a_3 + a_6 + a_8 + a_5 + a_{10} + a_{12} + \cdots,$$

but the same conclusion holds: Any rearrangement of a convergent *positive*-term series also converges, and it converges to the same sum.

Similarly, it is easy to prove that any grouping or rearrangement of a divergent positive-term series also diverges. But these observations all fail in the case of an infinite series with both positive and negative terms. For example, the series $\sum(-1)^n$ diverges, but it has the convergent grouping

$$(-1+1)+(-1+1)+(-1+1)+\cdots = 0+0+0+\cdots = 0.$$

It follows from Problem 56 of Section 11.4 that

$$\ln 2 = 1 - \frac{1}{2} + \frac{1}{3} - \frac{1}{4} + \frac{1}{5} - \cdots,$$

but the rearrangement

$$1 + \frac{1}{3} - \frac{1}{2} + \frac{1}{5} + \frac{1}{7} - \frac{1}{4} + \frac{1}{9} + \frac{1}{11} - \frac{1}{6} + \cdots$$

converges instead to $\frac{3}{2}\ln 2$. This series for $\ln 2$ even has rearrangements that converge to zero and others that diverge to $+\infty$. (See Problem 64 of Section 11.7.)

11.6 TRUE/FALSE STUDY GUIDE

11.6 CONCEPTS: QUESTIONS AND DISCUSSION

1. Can you give examples of a pair of positive-term infinite series $\sum a_n$ and $\sum b_n$ such that $\lim_{n\to\infty}(a_n/b_n) = 0$ and (a) both series converge; (b) both diverge; (c) one converges and the other diverges?
2. Can you give an example of two convergent positive-term infinite series $\sum a_n$ and $\sum b_n$ such that $\lim_{n\to\infty}(a_n/b_n) = 1$ but neither series dominates the other?
3. Can you give an example of two positive-term infinite series $\sum a_n$ and $\sum b_n$ that either both converge or both diverge, but the limit $\lim_{n\to\infty}(a_n/b_n)$ does not exist?

11.6 PROBLEMS

Use comparison tests to determine whether the infinite series in Problems 1 through 36 converge or diverge.

1. $\displaystyle\sum_{n=1}^{\infty} \frac{1}{n^2+n+1}$

2. $\displaystyle\sum_{n=1}^{\infty} \frac{n^3+1}{n^4+2}$

3. $\displaystyle\sum_{n=1}^{\infty} \frac{1}{n+\sqrt{n}}$

4. $\displaystyle\sum_{n=1}^{\infty} \frac{1}{n+n^{3/2}}$

5. $\displaystyle\sum_{n=1}^{\infty} \frac{1}{1+3^n}$

6. $\displaystyle\sum_{n=1}^{\infty} \frac{10n^2}{n^4+1}$

7. $\displaystyle\sum_{n=2}^{\infty} \frac{10n^2}{n^3-1}$

8. $\displaystyle\sum_{n=1}^{\infty} \frac{n^2-n}{n^4+2}$

9. $\displaystyle\sum_{n=1}^{\infty} \frac{1}{\sqrt{37n^3+3}}$

10. $\displaystyle\sum_{n=1}^{\infty} \frac{1}{\sqrt{n^2+1}}$

11. $\displaystyle\sum_{n=1}^{\infty} \frac{\sqrt{n}}{n^2+n}$

12. $\displaystyle\sum_{n=1}^{\infty} \frac{1}{3+5^n}$

13. $\displaystyle\sum_{n=2}^{\infty} \frac{1}{\ln n}$

14. $\displaystyle\sum_{n=1}^{\infty} \frac{1}{n-\ln n}$

15. $\displaystyle\sum_{n=1}^{\infty} \frac{\sin^2 n}{n^2+1}$

16. $\displaystyle\sum_{n=1}^{\infty} \frac{\cos^2 n}{3^n}$

17. $\displaystyle\sum_{n=1}^{\infty} \frac{n+2^n}{n+3^n}$

18. $\displaystyle\sum_{n=1}^{\infty} \frac{1}{2^n+3^n}$

19. $\displaystyle\sum_{n=2}^{\infty} \frac{1}{n^2\ln n}$

20. $\displaystyle\sum_{n=1}^{\infty} \frac{1}{n^{1+\sqrt{n}}}$

21. $\displaystyle\sum_{n=1}^{\infty} \frac{\ln n}{n^2}$

22. $\displaystyle\sum_{n=1}^{\infty} \frac{\arctan n}{n}$

23. $\displaystyle\sum_{n=1}^{\infty} \frac{\sin^2(1/n)}{n^2}$

24. $\displaystyle\sum_{n=1}^{\infty} \frac{e^{1/n}}{n}$

25. $\displaystyle\sum_{n=1}^{\infty} \frac{\ln n}{e^n}$

26. $\displaystyle\sum_{n=1}^{\infty} \frac{n^2+2}{n^3+3n}$

27. $\sum_{n=1}^{\infty} \frac{n^{3/2}}{n^2+4}$

28. $\sum_{n=1}^{\infty} \frac{1}{n \cdot 2^n}$

29. $\sum_{n=1}^{\infty} \frac{3}{4+\sqrt{n}}$

30. $\sum_{n=1}^{\infty} \frac{n^2+1}{e^n(n+1)^2}$

31. $\sum_{n=1}^{\infty} \frac{2n^2-1}{n^2 \cdot 3^n}$

32. $\sum_{n=1}^{\infty} \frac{1}{\sqrt[3]{2n^4+1}}$

33. $\sum_{n=1}^{\infty} \frac{2+\sin n}{n^2}$

34. $\sum_{n=1}^{\infty} \frac{\ln n}{n^3}$

35. $\sum_{n=1}^{\infty} \frac{(n+1)^n}{n^{n+1}}$ $\left[\textit{Suggestion: } \lim_{n\to\infty}\left(1+\frac{1}{n}\right)^n = e.\right]$

36. $\sum_{n=1}^{\infty} \left(\frac{\sin n}{n}\right)^4$

In Problems 37 through 40, calculate the sum of the first ten terms of the series, then estimate the error made in using this partial sum to approximate the sum of the series.

37. $\sum_{n=1}^{\infty} \frac{1}{n^2+1}$

38. $\sum_{n=1}^{\infty} \frac{1}{3^n+1}$

39. $\sum_{n=1}^{\infty} \frac{\cos^2 n}{n^2}$

40. $\sum_{n=2}^{\infty} \frac{1}{(n+1)(\ln n)^2}$

In Problems 41 through 44, first determine the smallest positive integer n such that the remainder satisfies the inequality $R_n < 0.005$. Then use a calculator or computer to approximate the sum of the series accurate to two decimal places.

41. $\sum_{n=1}^{\infty} \frac{1}{n^3+1}$

42. $\sum_{n=1}^{\infty} \frac{n}{(n+1)2^n}$

43. $\sum_{n=1}^{\infty} \frac{\cos^4 n}{n^4}$

44. $\sum_{n=1}^{\infty} \frac{1}{n^{2+(1/n)}}$

45. Show that if $\sum a_n$ is a convergent positive-term series, then the series $\sum \sin(a_n)$ also converges.

46. (a) Prove that $\ln n < n^{1/8}$ for all sufficiently large values of n. (b) Explain why part (a) shows that the series $\sum 1/(\ln n)^8$ diverges.

47. Prove that if $\sum a_n$ is a convergent positive-term series, then $\sum (a_n/n)$ converges.

48. Suppose that $\sum a_n$ is a convergent positive-term series and that $\{c_n\}$ is a sequence of positive numbers with limit zero. Prove that $\sum a_n c_n$ converges.

49. Use the result of Problem 48 to prove that if $\sum a_n$ and $\sum b_n$ are convergent positive-term series, then $\sum a_n b_n$ converges.

50. Prove that the series

$$\sum_{n=1}^{\infty} \frac{1}{1+2+3+\cdots+n}$$

converges.

51. Use the result of Problem 50 in Section 11.5 to prove that the series

$$\sum_{n=1}^{\infty} \frac{1}{1+\frac{1}{2}+\frac{1}{3}+\cdots+\frac{1}{n}}$$

diverges.

52. Adapt the proof of the limit-comparison test to prove the following two results. (a) Suppose that $\sum a_n$ and $\sum b_n$ are positive-term series and that $\sum b_n$ converges. If

$$L = \lim_{n\to\infty} \frac{a_n}{b_n} = 0,$$

then $\sum a_n$ converges. (b) Suppose that $\sum a_n$ and $\sum b_n$ are positive-term series and that $\sum b_n$ diverges. If

$$L = \lim_{n\to\infty} \frac{a_n}{b_n} = +\infty,$$

then $\sum a_n$ diverges.

11.7 ALTERNATING SERIES AND ABSOLUTE CONVERGENCE

In Sections 11.5 and 11.6 we considered only positive-term series. Now we discuss infinite series that have both positive terms and negative terms. An important example is a series with terms that are alternatively positive and negative. An **alternating series** is an infinite series of the form

$$\sum_{n=1}^{\infty} (-1)^{n+1} a_n = a_1 - a_2 + a_3 - a_4 + a_5 - \cdots \tag{1}$$

or of the form $\sum_{n=1}^{\infty} (-1)^n a_n$, where $a_n > 0$ for all n. For example, the *alternating harmonic series*

$$\sum_{n=1}^{\infty} \frac{(-1)^{n+1}}{n} = 1 - \frac{1}{2} + \frac{1}{3} - \frac{1}{4} + \frac{1}{5} - \cdots$$

and the geometric series

$$\sum_{n=0}^{\infty}\left(-\frac{1}{2}\right)^n = 1 - \frac{1}{2} + \frac{1}{4} - \frac{1}{8} + \frac{1}{16} - \cdots$$

are both alternating series. Theorem 1 shows that both these series converge because the sequence of absolute values of their terms is decreasing and has limit zero.

THEOREM 1 Alternating Series Test
If the alternating series in Eq. (1) satisfies the two conditions

1. $a_n \geqq a_{n+1} > 0$ for all n and
2. $\lim_{n\to\infty} a_n = 0$,

then the infinite series converges.

PROOF We first consider the even-numbered partial sums $S_2, S_4, S_6, \ldots, S_{2n}, \ldots$. We may write

$$S_{2n} = (a_1 - a_2) + (a_3 - a_4) + \cdots + (a_{2n-1} - a_{2n}).$$

Because $a_k - a_{k+1} \geqq 0$ for all k, the sequence $\{S_{2n}\}$ is increasing. Also, because

$$S_{2n} = a_1 - (a_2 - a_3) - \cdots - (a_{2n-2} - a_{2n-1}) - a_{2n},$$

$S_{2n} \leqq a_1$ for all n. So the increasing sequence $\{S_{2n}\}$ is bounded above. Hence the limit

$$S = \lim_{n\to\infty} S_{2n}$$

exists by the bounded monotonic sequence property of Section 11.2. It remains only for us to verify that the odd-numbered partial sums $S_1, S_3, S_5, \ldots$ also converge to S. But $S_{2n+1} = S_{2n} + a_{2n+1}$ and $\lim_{n\to\infty} a_{2n+1} = 0$, so

$$\lim_{n\to\infty} S_{2n+1} = \left(\lim_{n\to\infty} S_{2n}\right) + \left(\lim_{n\to\infty} a_{2n+1}\right) = S.$$

Thus $\lim_{n\to\infty} S_n = S$, and therefore the series in Eq. (1) converges. ◀

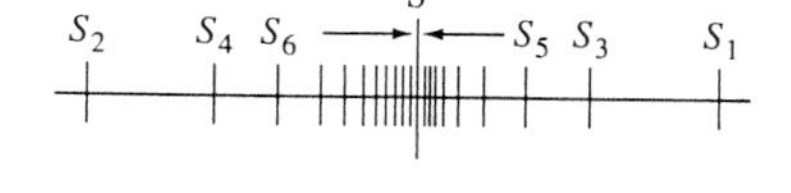

FIGURE 11.7.1 The even partial sums $\{S_{2n}\}$ increase and the odd partial sums $\{S_{2n+1}\}$ decrease.

Figure 11.7.1 illustrates the way in which the partial sums of a convergent alternating series (with positive first term) approximate its sum S, with the even partial sums $\{S_{2n}\}$ approaching S from below and the odd partial sums $\{S_{2n+1}\}$ approaching S from above.

EXAMPLE 1 The series

$$\sum_{n=1}^{\infty}\frac{(-1)^{n+1}}{2n-1} = 1 - \frac{1}{3} + \frac{1}{5} - \frac{1}{7} + \frac{1}{9} - \cdots$$

satisfies the conditions of Theorem 1 and therefore converges. The alternating series test does not tell us the sum of this series, but we saw in Section 11.4 that its sum is $\pi/4$. The graph in Fig. 11.7.2 of the partial sums of this series illustrates the typical convergence of an alternating series, with its partial sums approaching its sum alternately from above and below. ◆

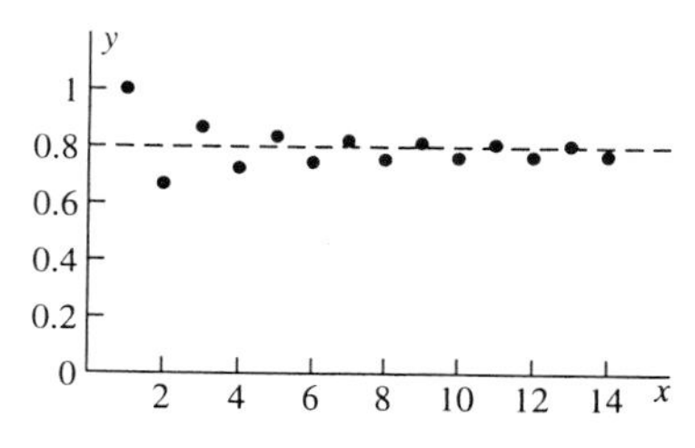

FIGURE 11.7.2 Graph of the first 14 partial sums of the alternating series in Example 1.

EXAMPLE 2 The series

$$\sum_{n=1}^{\infty}\frac{(-1)^{n+1}n}{2n-1} = 1 - \frac{2}{3} + \frac{3}{5} - \frac{4}{7} + \frac{5}{9} - \cdots$$

is an alternating series, and by expanding we verify that $n(2n+1) > (n+1)(2n-1)$, so it follows that

$$a_n = \frac{n}{2n-1} > \frac{n+1}{2n+1} = a_{n+1}$$

for all $n \geqq 1$. But

$$\lim_{n\to\infty} a_n = \frac{1}{2} \neq 0,$$

so the alternating series test *does not apply.* (This fact alone does not imply that the series in question diverges—many series in Sections 11.5 and 11.6 converge even though the alternating series test does not apply. But the series of this example diverges by the nth-term divergence test.) ◆

If a series converges by the alternating series test, then Theorem 2 shows how to approximate its sum with any desired degree of accuracy—*if* you have a computer fast enough to add a large number of its terms.

THEOREM 2 Alternating Series Remainder Estimate

Suppose that the series $\sum(-1)^{n+1}a_n$ satisfies the conditions of the alternating series test and therefore converges. Let S denote the sum of the series. Denote by $R_n = S - S_n$ the error made in replacing S with the nth partial sum S_n of the series. Then this **remainder** R_n has the same sign as the next term $(-1)^{n+2}a_{n+1}$ of the series, and

$$0 \leqq |R_n| < a_{n+1}. \tag{2}$$

In particular, the *sum S of a convergent alternating series lies between any two consecutive partial sums.* This follows from the proof of Theorem 1, where we saw that $\{S_{2n}\}$ is an increasing sequence and that $\{S_{2n+1}\}$ is a decreasing sequence, both converging to S. The resulting inequalities

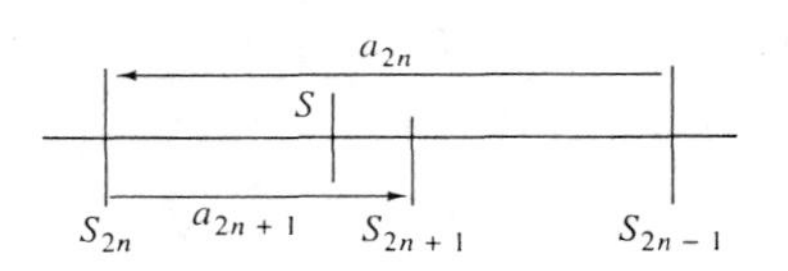

FIGURE 11.7.3 Illustrating the proof of the alternating series remainder estimate.

$$S_{2n-1} > S > S_{2n} = S_{2n-1} - a_{2n}$$

and

$$S_{2n} < S < S_{2n+1} = S_{2n} + a_{2n+1}$$

(see Fig. 11.7.3) imply the inequality in (2).

REMARK The inequality in (2) means the following. Suppose that you are given an *alternating series* that satisfies the conditions of Theorem 2 and has sum S. Then, if S is replaced with a partial sum S_n, the error made is numerically less than the first term a_{n+1} not retained and has the same sign as this first neglected term. **Important:** This error estimate *does not* apply to other types of series.

EXAMPLE 3 We saw in Section 11.4 that

$$e^x = \sum_{n=0}^{\infty} \frac{x^n}{n!}$$

for all x and thus (with $x = -1$) that

$$\frac{1}{e} = e^{-1} = 1 - 1 + \frac{1}{2!} - \frac{1}{3!} + \frac{1}{4!} - \cdots.$$

Use this alternating series to compute e^{-1} accurate to four decimal places.

Solution To attain four-place accuracy, we want the error to be less than a half unit in the fourth place. Thus we want

$$|R_n| < \frac{1}{(n+1)!} \leqq 0.00005.$$

If we use a calculator to compute the reciprocals of the factorials of the first several integers, we find that the least value of n for which this inequality holds is $n = 7$. Then

$$e^{-1} = 1 - \frac{1}{1!} + \frac{1}{2!} - \frac{1}{3!} + \frac{1}{4!} - \frac{1}{5!} + \frac{1}{6!} - \frac{1}{7!} + R_7 \approx 0.367857 + R_7.$$

(Relying on a common "+2 rule of thumb," we are carrying six decimal places because we want four-place accuracy in the final answer.) Now the first neglected term $1/8!$ is positive, so the inequality in (2) gives

$$0 < R_7 < \frac{1}{8!} < 0.000025.$$

Therefore

$$S_7 \approx 0.367857 < e^{-1} < S_7 + 0.000025 \approx 0.367882.$$

The two bounds here both round to $e^{-1} \approx 0.3679$. Although this approximation is accurate to four decimal places, its reciprocal

$$e = 1/e^{-1} \approx 1/(0.3679) \approx 2.7181 \approx 2.718$$

gives the number e accurate to only three decimal places. ◆

Absolute Convergence

The series

$$\sum_{n=1}^{\infty} \frac{(-1)^{n+1}}{n} = 1 - \frac{1}{2} + \frac{1}{3} - \frac{1}{4} + \frac{1}{5} - \cdots$$

converges, but if we simply replace each term with its absolute value, we get the *divergent* series

$$1 + \frac{1}{2} + \frac{1}{3} + \frac{1}{4} + \frac{1}{5} + \cdots.$$

In contrast, the *convergent* series

$$\sum_{n=0}^{\infty} \frac{(-1)^n}{2^n} = 1 - \frac{1}{2} + \frac{1}{4} - \frac{1}{8} + \cdots = \frac{2}{3}$$

has the property that the associated positive-term series

$$1 + \frac{1}{2} + \frac{1}{4} + \frac{1}{8} + \cdots = 2$$

also converges. Theorem 3 tells us that if a series of *positive* terms converges, then we may insert minus signs in front of any of the terms—every other one, for instance—and the resulting series will also converge.

THEOREM 3 Absolute Convergence Implies Convergence

If the series $\sum |a_n|$ converges, then so does the series $\sum a_n$.

PROOF Suppose that the series $\sum |a_n|$ converges. Note that

$$0 \leqq a_n + |a_n| \leqq 2|a_n|$$

for all n. Let $b_n = a_n + |a_n|$. It then follows from the comparison test that the positive-term series $\sum b_n$ converges, because it is dominated by the convergent series $\sum 2|a_n|$. It is easy to verify, too, that the termwise difference of two convergent series also converges. Hence we now see that the series

$$\sum a_n = \sum (b_n - |a_n|) = \sum b_n - \sum |a_n|$$

converges. ◄

Thus we have another convergence test, one not limited to positive-term series nor limited to alternating series: Given the series $\sum a_n$, test the series $\sum |a_n|$ for convergence. If the latter converges, then so does the former. (But the converse is *not* true!) This phenomenon motivates us to make the following definition.

DEFINITION Absolute Convergence

The series $\sum a_n$ is said to **converge absolutely** (and is called **absolutely convergent**) provided that the series

$$\sum |a_n| = |a_1| + |a_2| + |a_3| + \cdots + |a_n| + \cdots$$

converges.

Thus we have explained the title of Theorem 3, and we can rephrase the theorem as follows: *If a series converges absolutely, then it converges.* The two examples preceding Theorem 3 show that a convergent series may either converge absolutely or fail to do so:

$$1 - \frac{1}{2} + \frac{1}{4} - \frac{1}{8} + \frac{1}{16} - \cdots$$

is an absolutely convergent series because

$$1 + \frac{1}{2} + \frac{1}{4} + \frac{1}{8} + \frac{1}{16} + \cdots$$

converges, whereas

$$1 - \frac{1}{2} + \frac{1}{3} - \frac{1}{4} + \frac{1}{5} - \cdots$$

is a series that, though convergent, is *not* absolutely convergent. A series that converges but does not converge absolutely is said to be **conditionally convergent.** Consequently, the terms *absolutely convergent, conditionally convergent,* and *divergent* are simultaneously all inclusive and mutually exclusive: Any given numerical series belongs to exactly one of those three classes.

There is some advantage in the application of Theorem 3, because to apply it we test the *positive*-term series $\sum |a_n|$ for convergence—and we have a variety of tests, such as comparison tests or the integral test, designed for use on positive-term series.

Note also that absolute convergence of the series $\sum a_n$ means that a *different* series $\sum |a_n|$ converges, and the two sums will generally differ. For example, with $a_n = \left(-\frac{1}{3}\right)^n$, the formula for the sum of a geometric series gives

$$\sum_{n=0}^{\infty} a_n = \sum_{n=0}^{\infty} \left(-\frac{1}{3}\right)^n = \frac{1}{1 - \left(-\frac{1}{3}\right)} = \frac{3}{4},$$

whereas

$$\sum_{n=0}^{\infty} |a_n| = \sum_{n=0}^{\infty} \left(\frac{1}{3}\right)^n = \frac{1}{1 - \frac{1}{3}} = \frac{3}{2}.$$

EXAMPLE 4 Discuss the convergence of the series

$$\sum_{n=1}^{\infty} \frac{\cos n}{n^2} = \cos 1 + \frac{\cos 2}{4} + \frac{\cos 3}{9} + \cdots.$$

Solution Let $a_n = (\cos n)/n^2$. Then

$$|a_n| = \frac{|\cos n|}{n^2} \leqq \frac{1}{n^2}$$

for all $n \geqq 1$. Hence the positive-term series $\sum |a_n|$ converges by the comparison test, because it is dominated by the convergent p-series $\sum(1/n^2)$. Thus the given series is absolutely convergent, and it therefore converges by Theorem 3. ◆

One reason for the importance of absolute convergence is the fact (proved in advanced calculus) that the terms of an absolutely convergent series may be regrouped or rearranged without changing the sum of the series. As we suggested at the end of Section 11.6, this is *not* true of conditionally convergent series.

The Ratio Test and the Root Test

Our next two convergence tests involve a way of measuring the rate of growth or decrease of the sequence $\{a_n\}$ of terms of a series to determine whether $\sum a_n$ converges absolutely or diverges.

THEOREM 4 The Ratio Test

Suppose that the limit

$$\rho = \lim_{n\to\infty} \left| \frac{a_{n+1}}{a_n} \right| \tag{3}$$

either exists or is infinite. Then the infinite series $\sum a_n$ of nonzero terms

1. Converges absolutely if $\rho < 1$;
2. Diverges if $\rho > 1$.

If $\rho = 1$, the ratio test is inconclusive.

PROOF If $\rho < 1$, choose a (fixed) number r with $\rho < r < 1$. Then Eq. (3) implies that there exists an integer N such that $|a_{n+1}| \leqq r|a_n|$ for all $n \geqq N$. It follows that

$$\begin{aligned} |a_{N+1}| &\leqq r|a_N|, \\ |a_{N+2}| &\leqq r|a_{N+1}| \leqq r^2|a_N|, \\ |a_{N+3}| &\leqq r|a_{N+2}| \leqq r^3|a_N|, \end{aligned}$$

and in general that

$$|a_{N+k}| \leqq r^k|a_N| \quad \text{for } k \geqq 0.$$

Hence the series

$$|a_N| + |a_{N+1}| + |a_{N+2}| + \cdots$$

is dominated by the geometric series

$$|a_N|(1 + r + r^2 + r^3 + \cdots),$$

and the latter converges because $|r| < 1$. Thus the series $\sum |a_n|$ converges, so the series $\sum a_n$ converges absolutely.

If $\rho > 1$, then Eq. (3) implies that there exists a positive integer N such that $|a_{n+1}| > |a_n|$ for all $n \geqq N$. It follows that $|a_n| > |a_N| > 0$ for all $n > N$. Thus the sequence $\{a_n\}$ cannot approach zero as $n \to +\infty$, and consequently, by the nth-term divergence test, the series $\sum a_n$ diverges. ◄

To see that $\sum a_n$ may either converge or diverge if $\rho = 1$, consider the divergent series $\sum(1/n)$ and the convergent series $\sum(1/n^2)$. You should verify that, for both series, the value of the ratio ρ is 1.

EXAMPLE 5 Consider the series

$$\sum_{n=1}^{\infty} \frac{(-1)^n 2^n}{n!} = -2 + \frac{4}{2!} - \frac{8}{3!} + \frac{16}{4!} - \cdots.$$

Then

$$\rho = \lim_{n\to\infty}\left|\frac{a_{n+1}}{a_n}\right| = \lim_{n\to\infty}\left|\frac{\dfrac{(-1)^{n+1}2^{n+1}}{(n+1)!}}{\dfrac{(-1)^n 2^n}{n!}}\right| = \lim_{n\to\infty}\frac{2}{n+1} = 0.$$

Because $\rho < 1$, the series converges absolutely. ◆

EXAMPLE 6 Test for convergence: $\displaystyle\sum_{n=1}^{\infty}\frac{n}{2^n}$.

Solution We have

$$\rho = \lim_{n\to\infty}\left|\frac{a_{n+1}}{a_n}\right| = \lim_{n\to\infty}\frac{\dfrac{n+1}{2^{n+1}}}{\dfrac{n}{2^n}} = \lim_{n\to\infty}\frac{n+1}{2n} = \frac{1}{2}.$$

Because $\rho < 1$, this series converges (absolutely). ◆

EXAMPLE 7 Test for convergence: $\displaystyle\sum_{n=1}^{\infty}\frac{3^n}{n^2}$.

Solution Here we have

$$\rho = \lim_{n\to\infty}\left|\frac{a_{n+1}}{a_n}\right| = \lim_{n\to\infty}\frac{\dfrac{3^{n+1}}{(n+1)^2}}{\dfrac{3^n}{n^2}} = \lim_{n\to\infty}\frac{3n^2}{(n+1)^2} = 3.$$

In this case $\rho > 1$, so the given series diverges. ◆

THEOREM 5 The Root Test

Suppose that the limit

$$\rho = \lim_{n\to\infty}\sqrt[n]{|a_n|} \tag{4}$$

exists or is infinite. Then the infinite series $\sum a_n$

1. Converges absolutely if $\rho < 1$;
2. Diverges if $\rho > 1$.

If $\rho = 1$, the root test is inconclusive.

PROOF If $\rho < 1$, choose a (fixed) number r such that $\rho < r < 1$. Then $|a_n|^{1/n} < r$, and hence $|a_n| < r^n$, for n sufficiently large. Thus the series $\sum |a_n|$ is eventually dominated by the convergent geometric series $\sum r^n$. Therefore $\sum |a_n|$ converges, and so the series $\sum a_n$ converges absolutely.

If $\rho > 1$, then $|a_n|^{1/n} > 1$, and hence $|a_n| > 1$, for n sufficiently large. Therefore the nth-term test for divergence implies that the series $\sum a_n$ diverges. ◄

The ratio test is generally simpler to apply than the root test, and therefore it is ordinarily the one to try first. But there are certain series for which the root test succeeds and the ratio test fails, as in Example 8.

EXAMPLE 8 Consider the series

$$\sum_{n=0}^{\infty} \frac{1}{2^{n+(-1)^n}} = \frac{1}{2} + \frac{1}{1} + \frac{1}{8} + \frac{1}{4} + \frac{1}{32} + \frac{1}{16} + \cdots.$$

Then $a_{n+1}/a_n = 2$ if n is even, whereas $a_{n+1}/a_n = \frac{1}{8}$ if n is odd. So the limit required for the ratio test does not exist. But

$$\lim_{n\to\infty} |a_n|^{1/n} = \lim_{n\to\infty} \left| \frac{1}{2^{n+(-1)^n}} \right|^{1/n} = \lim_{n\to\infty} \frac{1}{2} \left| \frac{1}{2^{(-1)^n/n}} \right| = \frac{1}{2},$$

so the given series converges by the root test. (Its convergence also follows from the fact that it is a rearrangement of the positive-term convergent geometric series $\sum 1/2^n$.) ◆

11.7 TRUE/FALSE STUDY GUIDE

11.7 CONCEPTS: QUESTIONS AND DISCUSSION

1. Can you give an example of a divergent alternating series $\sum(-1)^{n+1}a_n$ such that $\lim_{n\to\infty} a_n = 0$? In view of the alternating series test stated in Theorem 1 of this section, how is such an example possible?
2. Give your own example of an infinite series such that both $\sum a_n$ and $\sum |a_n|$ converge but have different sums, both of which you can calculate. What can you conclude about the series if $\sum a_n$ and $\sum |a_n|$ have the same sum?
3. Can you give an example of a conditionally convergent positive-term series? Why or why not?

11.7 PROBLEMS

Determine whether or not the alternating series in Problems 1 through 20 converge or diverge.

1. $\sum_{n=1}^{\infty} \frac{(-1)^{n+1}}{n^2}$

2. $\sum_{n=1}^{\infty} \frac{(-1)^{n+1}}{\sqrt{n^2+1}}$

3. $\sum_{n=1}^{\infty} \frac{(-1)^n n}{3n+2}$

4. $\sum_{n=1}^{\infty} \frac{(-1)^n n}{3n^2+2}$

5. $\sum_{n=1}^{\infty} \frac{(-1)^{n+1} n}{\sqrt{n^2+2}}$

6. $\sum_{n=1}^{\infty} \frac{(-1)^{n+1} n^2}{\sqrt{n^5+5}}$

7. $\sum_{n=2}^{\infty} \frac{(-1)^{n+1} n}{\ln n}$

8. $\sum_{n=1}^{\infty} \frac{(-1)^n \ln n}{\sqrt{n}}$

9. $\sum_{n=1}^{\infty} \frac{(-1)^n n}{2^n}$

10. $\sum_{n=1}^{\infty} n \cdot \left(-\frac{2}{3}\right)^{n+1}$

11. $\sum_{n=1}^{\infty} \frac{(-1)^n n}{\sqrt{2^n+1}}$

12. $\sum_{n=1}^{\infty} \left(-\frac{n\pi}{10}\right)^{n+1}$

13. $\sum_{n=1}^{\infty} \frac{1}{n^{2/3}} \sin\left(\frac{n\pi}{2}\right)$

14. $\sum_{n=1}^{\infty} \frac{\cos n\pi}{n^{3/2}}$

15. $\sum_{n=1}^{\infty} (-1)^n \sin\left(\frac{1}{n}\right)$

16. $\sum_{n=1}^{\infty} (-1)^n n \sin\left(\frac{\pi}{n}\right)$

17. $\sum_{n=1}^{\infty} \frac{(-1)^{n+1}}{\sqrt[n]{2}}$

18. $\sum_{n=1}^{\infty} \frac{(-1.01)^{n+1}}{n^4}$

19. $\sum_{n=1}^{\infty} \frac{(-1)^{n+1}}{\sqrt[n]{n}}$

20. $\sum_{n=1}^{\infty} \frac{(-1)^{n+1} n!}{(2n)!}$

Determine whether the series in Problems 21 through 42 converge absolutely, converge conditionally, or diverge.

21. $\sum_{n=1}^{\infty} \frac{(-1)^{n+1}}{2^n}$

22. $\sum_{n=1}^{\infty} \frac{1}{n^2+1}$

23. $\sum_{n=1}^{\infty} \frac{(-1)^n \ln n}{n}$

24. $\sum_{n=1}^{\infty} \frac{1}{n^n}$

25. $\sum_{n=1}^{\infty} \left(\frac{10}{n}\right)^n$

26. $\sum_{n=1}^{\infty} \frac{3^n}{n!n}$

27. $\sum_{n=0}^{\infty} \frac{(-10)^n}{n!}$

28. $\sum_{n=1}^{\infty} \frac{(-1)^{n+1} n!}{n^n}$

29. $\sum_{n=1}^{\infty} (-1)^{n+1} \left(\frac{n}{n+1}\right)^n$

30. $\sum_{n=1}^{\infty} \frac{n!n^2}{(2n)!}$

31. $\sum_{n=1}^{\infty} \left(\frac{\ln n}{n}\right)^n$

32. $\sum_{n=0}^{\infty} \frac{(-1)^n 2^{3n}}{7^n}$

33. $\sum_{n=0}^{\infty} (-1)^n \left(\sqrt{n+1} - \sqrt{n}\right)$

34. $\sum_{n=1}^{\infty} n \cdot \left(\frac{3}{4}\right)^n$

35. $\sum_{n=1}^{\infty} \left[\ln\left(\frac{1}{n}\right)\right]^n$

36. $\sum_{n=0}^{\infty} \frac{(n!)^2}{(2n)!}$

37. $\displaystyle\sum_{n=1}^{\infty} \frac{(-1)^{n+1}3^n}{n(2^n+1)}$

38. $\displaystyle\sum_{n=1}^{\infty} \frac{(-1)^{n+1}\arctan n}{n}$

39. $\displaystyle\sum_{n=1}^{\infty} \frac{(-1)^{n+1}n!}{1\cdot 3\cdot 5\cdots(2n-1)}$

40. $\displaystyle\sum_{n=1}^{\infty} (-1)^{n+1}\frac{1\cdot 3\cdot 5\cdots(2n-1)}{1\cdot 4\cdot 7\cdots(3n-2)}$

41. $\displaystyle\sum_{n=1}^{\infty} \frac{(n+2)!}{3^n(n!)^2}$

42. $\displaystyle\sum_{n=1}^{\infty} \frac{(-1)^{n+1}n^n}{3^{n^2}}$

In Problems 43 through 48, sum the indicated number of initial terms of the given alternating series. Then apply the alternating series remainder estimate to estimate the error in approximating the sum of the series with this partial sum. Finally, approximate the sum of the series, writing precisely the number of decimal places that thereby are guaranteed to be correct (after rounding).

43. $\displaystyle\sum_{n=1}^{\infty} \frac{(-1)^{n+1}}{n^3}$, 5 terms

44. $\displaystyle\sum_{n=1}^{\infty} \frac{(-1)^{n+1}}{3^n}$, 8 terms

45. $\displaystyle\sum_{n=1}^{\infty} \frac{(-1)^{n+1}}{n!}$, 6 terms

46. $\displaystyle\sum_{n=1}^{\infty} \frac{(-1)^{n+1}}{n^n}$, 7 terms

47. $\displaystyle\sum_{n=1}^{\infty} \frac{(-1)^{n+1}}{n}$, 12 terms

48. $\displaystyle\sum_{n=1}^{\infty} \frac{(-1)^{n+1}}{n^2}$, 15 terms

In Problems 49 through 54, sum enough terms (tell how many) to approximate the sum of the series, writing the sum rounded to the indicated number of correct decimal places.

49. $\displaystyle\sum_{n=1}^{\infty} \frac{(-1)^{n+1}}{n^4}$, 3 decimal places

50. $\displaystyle\sum_{n=1}^{\infty} \frac{(-1)^{n+1}}{n^5}$, 4 decimal places

51. $\displaystyle\frac{1}{\sqrt{e}} = \sum_{n=0}^{\infty} \frac{(-1)^n}{n!2^n}$, 4 decimal places

52. $\displaystyle\cos 1 = \sum_{n=0}^{\infty} \frac{(-1)^n}{(2n)!}$, 5 decimal places

53. $\displaystyle\sin 60^\circ = \sum_{n=0}^{\infty} \frac{(-1)^n}{(2n+1)!}\left(\frac{\pi}{3}\right)^{2n+1}$, 5 decimal places

54. $\displaystyle\ln(1.1) = \sum_{n=1}^{\infty} \frac{(-1)^{n+1}}{n\cdot 10^n}$, 7 decimal places

In Problems 55 and 56, show that the indicated alternating series $\sum(-1)^{n+1}a_n$ *satisfies the condition that* $a_n \to 0$ *as* $n \to +\infty$, *but nevertheless diverges. Tell why the alternating series test does not apply. It may be informative to graph the first* 10 *or* 20 *partial sums.*

55. $a_n = \begin{cases} \dfrac{1}{n} & \text{if } n \text{ is odd,} \\[2mm] \dfrac{1}{n^2} & \text{if } n \text{ is even.} \end{cases}$

56. $a_n = \begin{cases} \dfrac{1}{\sqrt{n}} & \text{if } n \text{ is odd,} \\[2mm] \dfrac{1}{n^3} & \text{if } n \text{ is even.} \end{cases}$

57. Give an example of a pair of convergent series $\sum a_n$ and $\sum b_n$ such that $\sum a_n b_n$ diverges.

58. Prove that $\sum |a_n|$ diverges if the series $\sum a_n$ diverges.

59. Prove that

$$\lim_{n\to\infty} \frac{a^n}{n!} = 0$$

(for any real number a) by applying the ratio test to show that the infinite series $\sum a^n/n!$ converges.

60. (a) Suppose that r is a (fixed) number such that $|r| < 1$. Use the ratio test to prove that the series $\sum_{n=0}^{\infty} nr^n$ converges. Let S denote its sum. (b) Show that

$$(1-r)S = \sum_{n=1}^{\infty} r^n.$$

Show how to conclude that

$$\sum_{n=0}^{\infty} nr^n = \frac{r}{(1-r)^2}.$$

61. Let

$$H_n = \sum_{k=1}^{n} \frac{1}{k} \quad \text{and} \quad S_n = \sum_{k=1}^{n} \frac{(-1)^{k+1}}{k}$$

denote the nth partial sums of the harmonic and alternating harmonic series, respectively. (a) Show that $S_{2n} = H_{2n} - H_n$ for all $n \geqq 1$. (b) Problem 50 in Section 11.5 says that

$$\lim_{n\to\infty} (H_n - \ln n) = \gamma$$

(where $\gamma \approx 0.57722$ denotes Euler's constant). Explain why it follows that

$$\lim_{n\to\infty} (H_{2n} - \ln 2n) = \gamma.$$

(c) Conclude from parts (a) and (b) that $\lim_{n\to\infty} S_{2n} = \ln 2$. Thus

$$\ln 2 = 1 - \frac{1}{2} + \frac{1}{3} - \frac{1}{4} + \frac{1}{5} - \frac{1}{6} + \cdots.$$

62. Suppose that $\sum a_n$ is a conditionally convergent infinite series. For each n, let

$$a_n^+ = \frac{a_n + |a_n|}{2} \quad \text{and} \quad a_n^- = \frac{a_n - |a_n|}{2}.$$

(a) Explain why $\sum a_n^+$ consists of the positive terms of $\sum a_n$ and why $\sum a_n^-$ consists of the negative terms of $\sum a_n$. (b) Given a real number r, show that some rearrangement of the conditionally convergent series $\sum a_n$ converges to r. *Suggestion:* If r is positive, for instance, begin with the first partial sum of the positive series $\sum a_n^+$ that exceeds r.

Then add just enough terms of the negative series $\sum a_n^-$ so that the cumulative sum is less than r. Next add just enough terms of the positive series that the cumulative sum is greater than r, and continue in this way to define the desired rearrangement. Why does it follow that this rearranged infinite series converges to r?

63. Use the method of Problem 62 to write the first dozen terms of a rearrangement of the alternating harmonic series (Problem 61) that converges to 1 rather than to ln 2.

64. Describe a way to rearrange the terms of the alternating harmonic series to obtain (a) A rearranged series that converges to -2; (b) A rearranged series that diverges to $+\infty$.

65. Here is another rearrangement of the alternating harmonic series of Problem 61:

$$1 - \frac{1}{2} - \frac{1}{4} - \frac{1}{6} - \frac{1}{8}$$
$$+ \frac{1}{3} - \frac{1}{10} - \frac{1}{12} - \frac{1}{14} - \frac{1}{16}$$
$$+ \frac{1}{5} - \frac{1}{18} - \frac{1}{20} - \frac{1}{22} - \frac{1}{24}$$
$$+ \frac{1}{7} - \frac{1}{26} - \frac{1}{28} - \frac{1}{30} - \frac{1}{32} + \cdots.$$

Use a computer to collect evidence about the value of its sum.

11.8 POWER SERIES

The most important infinite series representations of functions are those whose terms are constant multiples of (successive) integral powers of the independent variable x—that is, series that resemble "infinite polynomials." For example, we discussed in Section 11.4 the geometric series

$$\frac{1}{1-x} = 1 + x + x^2 + x^3 + \cdots \quad (|x| < 1) \tag{1}$$

and the Taylor series

$$e^x = \sum_{n=0}^{\infty} \frac{x^n}{n!} = 1 + x + \frac{x^2}{2!} + \frac{x^3}{3!} + \frac{x^4}{4!} + \cdots, \tag{2}$$

$$\cos x = \sum_{n=0}^{\infty} \frac{(-1)^n x^{2n}}{(2n)!} = 1 - \frac{x^2}{2!} + \frac{x^4}{4!} - \frac{x^6}{6!} + \cdots, \quad \text{and} \tag{3}$$

$$\sin x = \sum_{n=0}^{\infty} \frac{(-1)^n x^{2n+1}}{(2n+1)!} = x - \frac{x^3}{3!} + \frac{x^5}{5!} - \frac{x^7}{7!} + \cdots. \tag{4}$$

There we used Taylor's formula to show that the series in Eqs. (2) through (4) converge, for all x, to the functions e^x, $\cos x$, and $\sin x$, respectively. Here we investigate the convergence of a "power series" without knowing in advance the function (if any) to which it converges.

All the infinite series in Eqs. (1) through (4) have the form

$$\sum_{n=0}^{\infty} a_n x^n = a_0 + a_1 x + a_2 x^2 + \cdots + a_n x^n + \cdots \tag{5}$$

with the constant *coefficients* $a_0, a_1, a_2, \ldots$. An infinite series of this form is called a **power series** in (powers of) x. In order that the initial terms of the two sides of Eq. (5) agree, we adopt here the convention that $x^0 = 1$ even if $x = 0$.

Convergence of Power Series

The partial sums of the power series in (5) are the *polynomials*

$$s_1(x) = a_0 + a_1 x, \qquad s_2(x) = a_0 + a_1 x + a_2 x^2, \qquad s_3(x) = a_0 + a_1 x + a_2 x^2 + a_3 x^3,$$

and so forth. The nth partial sum is an nth-degree polynomial. When we ask *where* the power series converges, we seek those values of x for which the limit

$$s(x) = \lim_{n\to\infty} s_n(x)$$

exists. The sum $s(x)$ of a power series is then a function of x that is defined wherever the series converges.

The power series in (5) obviously converges when $x = 0$. In general, it will converge for some nonzero values of x and diverge for others. Because of the way in which powers of x are involved, the ratio test of Section 11.7 is particularly effective in determining the values of x for which a given power series converges.

Assume that the limit

$$\rho = \lim_{n\to\infty} \left| \frac{a_{n+1}}{a_n} \right| \tag{6}$$

exists. This is the limit that we need if we want to apply the ratio test to the series $\sum a_n$ of constants. To apply the ratio test to the power series in Eq. (5), we write $u_n = a_n x^n$ and compute the limit

$$\lim_{n\to\infty} \left| \frac{u_{n+1}}{u_n} \right| = \lim_{n\to\infty} \left| \frac{a_{n+1}x^{n+1}}{a_n x^n} \right| = \rho\,|x|. \tag{7}$$

If $\rho = 0$, then $\sum a_n x^n$ converges absolutely for all x. If $\rho = +\infty$, then $\sum a_n x^n$ diverges for all $x \neq 0$. If ρ is a positive real number, we see from Eq. (7) that $\sum a_n x^n$ converges absolutely for all x such that $\rho \cdot |x| < 1$—that is, when

$$|x| < R = \frac{1}{\rho} = \lim_{n\to\infty} \left| \frac{a_n}{a_{n+1}} \right|. \tag{8}$$

In this case the ratio test also implies that $\sum a_n x^n$ diverges if $|x| > R$ but is inconclusive when $x = \pm R$. We have therefore proved Theorem 1, under the additional hypothesis that the limit in Eq. (6) exists. In Problems 69 and 70 we outline a proof that does not require this additional hypothesis.

THEOREM 1 Convergence of Power Series

If $\sum a_n x^n$ is a power series, then either

1. The series converges absolutely for all x, or
2. The series converges only when $x = 0$, or
3. There exists a number $R > 0$ such that $\sum a_n x^n$ converges absolutely if $|x| < R$ and diverges if $|x| > R$.

The number R of Case 3 is called the **radius of convergence** of the power series $\sum a_n x^n$. We write $R = \infty$ in Case 1 and $R = 0$ in Case 2. The set of all real numbers x for which the series converges is called its **interval of convergence** (Fig. 11.8.1); note that this set *is* an interval. If $0 < R < \infty$, then the interval of convergence is one of the intervals

$$(-R, R), \quad (-R, R], \quad [-R, R), \quad \text{or} \quad [-R, R].$$

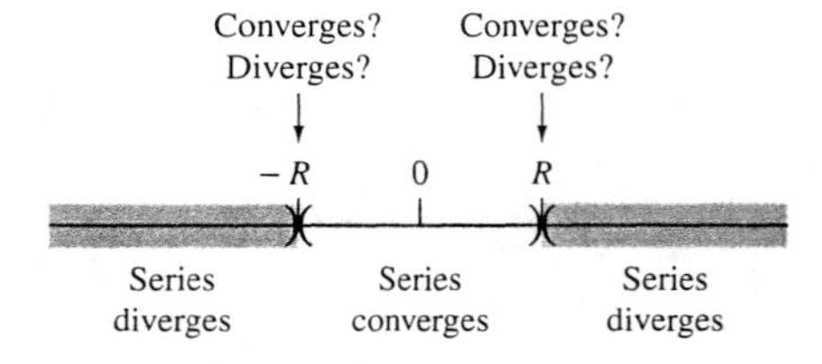

FIGURE 11.8.1 The interval of convergence if $0 < R = \lim_{n\to\infty} \left| \frac{a_n}{a_{n+1}} \right| < \infty$.

When we substitute either of the endpoints $x = \pm R$ into the series $\sum a_n x^n$, we obtain an infinite series with constant terms whose convergence must be determined separately. Because these will be numerical series, the earlier tests of this chapter are appropriate.

EXAMPLE 1 Find the interval of convergence of the series

$$\sum_{n=1}^{\infty} \frac{x^n}{n \cdot 3^n}.$$

Solution With $u_n = x^n/(n \cdot 3^n)$ we find that

$$\lim_{n\to\infty} \left| \frac{u_{n+1}}{u_n} \right| = \lim_{n\to\infty} \left| \frac{\dfrac{x^{n+1}}{(n+1)\cdot 3^{n+1}}}{\dfrac{x^n}{n \cdot 3^n}} \right| = \lim_{n\to\infty} \frac{n\,|x|}{3(n+1)} = \frac{|x|}{3}.$$

Now $|x|/3 < 1$ provided that $|x| < 3$, so the ratio test implies that the given series converges absolutely if $|x| < 3$ and diverges if $|x| > 3$. When $x = 3$, we have the divergent harmonic series $\sum(1/n)$, and when $x = -3$ we have the convergent alternating series $\sum(-1)^n/n$. Thus the interval of convergence of the given power series is $[-3, 3)$. We see dramatically in Fig. 11.8.2 the difference between convergence at $x = -3$ and divergence at $x = +3$. ◆

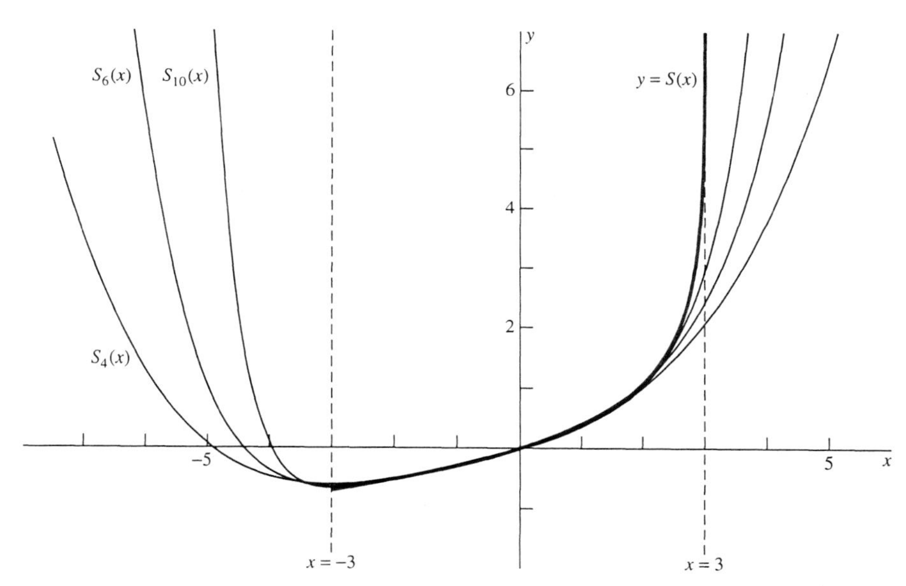

FIGURE 11.8.2 Graphs of the partial sums $S_4(x)$, $S_6(x)$, and $S_{10}(x)$ of the power series $S(x) = \sum_{n=1}^{\infty} \frac{x^n}{n \cdot 3^n}$ of Example 1. We see convergence at $x = -3$, but apparently $S(x) \to \infty$ as x approaches $+3$, where the series diverges harmonically.

EXAMPLE 2 Find the interval of convergence of the power series

$$\sum_{n=0}^{\infty} \frac{(-2)^n x^n}{(2n)!} = 1 - \frac{2x}{2!} + \frac{4x^2}{4!} - \frac{8x^3}{6!} + \frac{16x^4}{8!} - \cdots.$$

Solution With $u_n = (-2)^n x^n/(2n)!$ we find that

$$\lim_{n\to\infty} \left| \frac{u_{n+1}}{u_n} \right| = \lim_{n\to\infty} \left| \frac{\dfrac{(-2)^{n+1}x^{n+1}}{(2n+2)!}}{\dfrac{(-2)^n x^n}{(2n)!}} \right| = \lim_{n\to\infty} \frac{2|x|}{(2n+1)(2n+2)} = 0$$

for all x [using the fact that $(2n+2)! = (2n)!(2n+1)(2n+2)$]. Hence the ratio test implies that the given power series converges for all x, and its interval of convergence is therefore $(-\infty, +\infty)$, the entire real line. ◆

REMARK The power series of Example 2 results upon substituting $\sqrt{2x}$ for x in the Taylor series for $\cos x$ [Eq. (3)]. But only for $x > 0$ does the sum $S(x)$ of the series exhibit the oscillatory character of the function $\cos\sqrt{2x}$ (Fig. 11.8.3). For $x < 0$ the power series converges to the quite different (and nonoscillatory) function $\cosh\sqrt{|2x|}$.

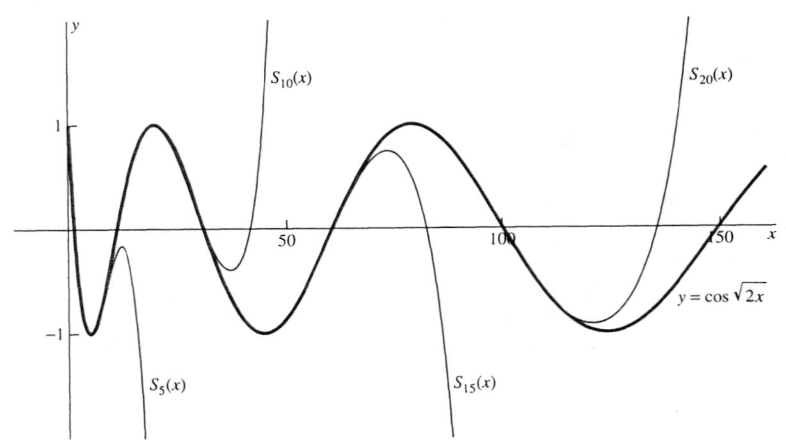

FIGURE 11.8.3 Graphs of the partial sums $S_5(x)$, $S_{10}(x)$, $S_{15}(x)$, and $S_{20}(x)$ of the power series $S(x) = \sum_{n=0}^{\infty} \frac{(-2x)^n}{(2n)!}$ of Example 2, which converges to $\cos\sqrt{2x}$ for $x > 0$.

EXAMPLE 3 Find the interval of convergence of the series $\sum_{n=1}^{\infty} n^n x^n$.

Solution With $u_n = n^n x^n$ we find that

$$\lim_{n\to\infty} \left|\frac{u_{n+1}}{u_n}\right| = \lim_{n\to\infty} \left|\frac{(n+1)^{n+1}x^{n+1}}{n^n x^n}\right| = \lim_{n\to\infty} (n+1)\left(1+\frac{1}{n}\right)^n |x| = +\infty$$

for all $x \neq 0$, because

$$\lim_{n\to\infty} \left(1+\frac{1}{n}\right)^n = e.$$

Thus the given series diverges for all $x \neq 0$, and its interval of convergence consists of the single point $x = 0$. ◆

EXAMPLE 4 Use the ratio test to verify that the Taylor series for $\cos x$ in Eq. (3) converges for all x.

Solution With $u_n = (-1)^n x^{2n}/(2n)!$ we find that

$$\lim_{n\to\infty} \left|\frac{u_{n+1}}{u_n}\right| = \lim_{n\to\infty} \left|\frac{\dfrac{(-1)^{n+1}x^{2n+2}}{(2n+2)!}}{\dfrac{(-1)^n x^{2n}}{(2n)!}}\right| = \lim_{n\to\infty} \frac{x^2}{(2n+1)(2n+2)} = 0$$

for all x, so the series converges for all x. ◆

IMPORTANT In Example 4, the ratio test tells us only that the series for $\cos x$ converges to *some* number, *not* necessarily the particular number $\cos x$. The argument of Section 11.4, using Taylor's formula with remainder, is required to establish that the sum of the series is actually $\cos x$.

Power Series in Powers of $x - c$

An infinite series of the form

$$\sum_{n=0}^{\infty} a_n(x-c)^n = a_0 + a_1(x-c) + a_2(x-c)^2 + \cdots, \tag{9}$$

where c is a constant, is called a **power series in** (powers of) $x - c$. By the same reasoning that led us to Theorem 1, with x^n replaced with $(x-c)^n$ throughout, we conclude that either

1. The series in Eq. (9) converges absolutely for all x, or
2. The series converges only when $x - c = 0$—that is, when $x = c$—or
3. There exists a number $R > 0$ such that the series in Eq. (9) converges absolutely if $|x - c| < R$ and diverges if $|x - c| > R$.

As in the case of a power series with $c = 0$, the number R is called the **radius of convergence** of the series, and the **interval of convergence** of the series $\sum a_n(x-c)^n$ is the set of all numbers x for which it converges (Fig. 11.8.4). As before, when $0 < R < \infty$, the convergence of the series at the endpoints $x = c - R$ and $x = c + R$ of its interval of convergence must be checked separately.

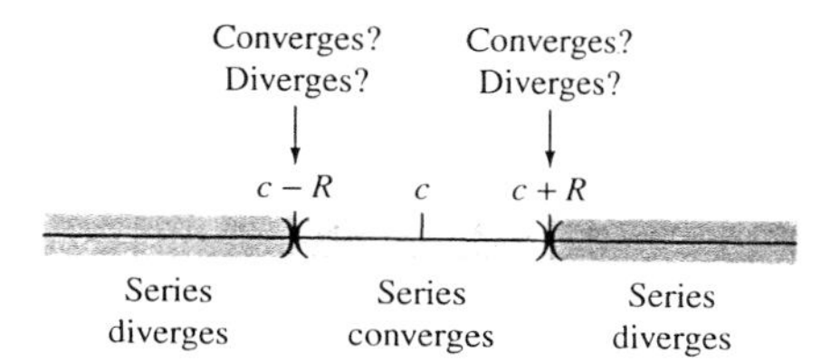

FIGURE 11.8.4 The interval of convergence of $\sum_{n=0}^{\infty} a_n(x-c)^n$.

EXAMPLE 5 Determine the interval of convergence of the series

$$\sum_{n=1}^{\infty} \frac{(-1)^n(x-2)^n}{n \cdot 4^n}.$$

Solution We let $u_n = (-1)^n(x-2)^n/(n \cdot 4^n)$. Then

$$\lim_{n\to\infty}\left|\frac{u_{n+1}}{u_n}\right| = \lim_{n\to\infty}\left|\frac{\dfrac{(-1)^{n+1}(x-2)^{n+1}}{(n+1)\cdot 4^{n+1}}}{\dfrac{(-1)^n(x-2)^n}{n\cdot 4^n}}\right|$$

$$= \lim_{n\to\infty}\frac{|x-2|}{4}\cdot\frac{n}{n+1} = \frac{|x-2|}{4}.$$

Hence the given series converges when $|x - 2| < 4$, so the radius of convergence is $R = 4$. Because $c = 2$, the series converges when $-2 < x < 6$ and diverges if either $x < -2$ or $x > 6$. When $x = -2$, the series reduces to the divergent harmonic series, and when $x = 6$ it reduces to the convergent alternating series $\sum(-1)^n/n$. Thus the interval of convergence of the given power series is $(-2, 6]$. ◆

Power Series Representations of Functions

Power series are important tools for computing (or approximating) values of functions. Suppose that the series $\sum a_n x^n$ converges to the value $f(x)$; that is,

$$f(x) = a_0 + a_1 x + a_2 x^2 + \cdots + a_n x^n + \cdots$$

for each x in the interval of convergence of the power series. Then we call $\sum a_n x^n$ a **power series representation** of $f(x)$. For example, the geometric series $\sum x^n$ in Eq. (1) is a power series representation of the function $f(x) = 1/(1-x)$ on the interval $(-1, 1)$.

We saw in Section 11.4 how Taylor's formula with remainder can often be used to find a power series representation of a given function. Recall that the nth-degree Taylor's formula for $f(x)$ at $x = a$ is

$$f(x) = f(a) + f'(a)(x-a) + \frac{f''(a)}{2!}(x-a)^2 + \frac{f^{(3)}(a)}{3!}(x-a)^3 + \cdots + \frac{f^{(n)}(a)}{n!}(x-a)^n + R_n(x). \quad \textbf{(10)}$$

The remainder $R_n(x)$ is given by

$$R_n(x) = \frac{f^{(n+1)}(z)}{(n+1)!}(x-a)^{n+1},$$

where z is some number between a and x. If we let $n \to +\infty$ in Eq. (10), we obtain Theorem 2.

THEOREM 2 Taylor Series Representations

Suppose that the function f has derivatives of all orders on some interval containing a and also that

$$\lim_{n\to\infty} R_n(x) = 0 \quad \textbf{(11)}$$

for each x in that interval. Then

$$f(x) = \sum_{n=0}^{\infty} \frac{f^{(n)}(a)}{n!}(x-a)^n \quad \textbf{(12)}$$

for each x in the interval.

The power series in Eq. (12) is the **Taylor series** of the function f **at** $x = a$ (or *in powers of* $x - a$, or *with center* a). If $a = 0$, we obtain the power series

$$f(x) = \sum_{n=0}^{\infty} \frac{f^{(n)}(0)}{n!}x^n = f(0) + f'(0)x + \frac{f''(0)}{2!}x^2 + \cdots, \quad \textbf{(13)}$$

commonly called the **Maclaurin series** of f. Thus the power series in Eqs. (2) through (4) are the Maclaurin series of the functions e^x, $\cos x$, and $\sin x$, respectively.

EXAMPLE 6 New power series can be constructed from old ones. For instance, upon replacing x with $-x$ in the Maclaurin series for e^x, we obtain

$$e^{-x} = 1 - x + \frac{x^2}{2!} - \frac{x^3}{3!} + \cdots + (-1)^n\frac{x^n}{n!} + \cdots.$$

Let us now add the series for e^x and e^{-x} and divide by 2. This gives

$$\cosh x = \frac{e^x + e^{-x}}{2} = \frac{1}{2}\left(1 + x + \frac{x^2}{2!} + \frac{x^3}{3!} + \frac{x^4}{4!} + \cdots\right) + \frac{1}{2}\left(1 - x + \frac{x^2}{2!} - \frac{x^3}{3!} + \frac{x^4}{4!} - \cdots\right),$$

so

$$\cosh x = 1 + \frac{x^2}{2!} + \frac{x^4}{4!} + \frac{x^6}{6!} + \cdots.$$

Similarly,

$$\sinh x = x + \frac{x^3}{3!} + \frac{x^5}{5!} + \frac{x^7}{7!} + \cdots.$$

Note the strong resemblance to Eqs. (3) and (4), the series for $\cos x$ and $\sin x$, respectively.

Upon replacing x with $-x^2$ in the series for e^x, we obtain

$$e^{-x^2} = \sum_{n=0}^{\infty} (-1)^n \frac{x^{2n}}{n!} = 1 - x^2 + \frac{x^4}{2!} - \frac{x^6}{3!} + \cdots.$$

Because this power series converges to $\exp(-x^2)$ for all x, it must be the Maclaurin series for $\exp(-x^2)$. (See Problem 66.) Think how tedious it would be to compute the derivatives of $\exp(-x^2)$ needed to write its Maclaurin series directly from Eq. (13). ◆

EXAMPLE 7 Sometimes a function is originally defined by means of a power series. One of the most important "higher transcendental functions" of applied mathematics is the Bessel function $J_0(x)$ of order zero defined by

$$J_0(x) = \sum_{n=0}^{\infty} \frac{(-1)^n x^{2n}}{2^{2n}(n!)^2} = 1 - \frac{x^2}{4} + \frac{x^4}{64} - \frac{x^6}{2304} + \cdots.$$

Only terms of even degree appear, so let us write $u_n = (-1)^n x^{2n}/[2^{2n}(n!)^2]$ for the nth term in this series (not counting its constant term). Then

$$\lim_{n\to\infty} \left| \frac{u_{n+1}}{u_n} \right| = \lim_{n\to\infty} \left| \frac{\dfrac{(-1)^{n+1}x^{2n+2}}{2^{2n+2}[(n+1)!]^2}}{\dfrac{(-1)^n x^{2n}}{2^{2n}(n!)^2}} \right| = \lim_{n\to\infty} \frac{x^2}{4(n+1)^2} = 0$$

for all x, so the ratio test implies that $J_0(x)$ is defined on the whole real line. The series for $J_0(x)$ resembles somewhat the cosine series, but the graph of $J_0(x)$ exhibits *damped* oscillations (Fig. 11.8.5). Bessel functions are important in such applications as the distribution of temperature in a cylindrical steam pipe and distribution of thermal neutrons in a cylindrical reactor. ◆

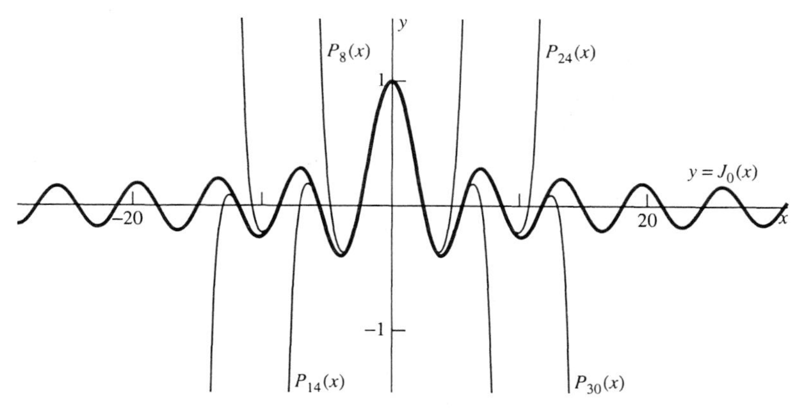

FIGURE 11.8.5 Graphs of the Bessel function $J_0(x)$ and its Taylor polynomials $P_8(x)$, $P_{14}(x)$, $P_{24}(x)$, and $P_{30}(x)$.

The Binomial Series

Example 8 gives one of the most famous and useful of all series, the *binomial series,* which was discovered by Newton in the 1660s. It is the infinite series generalization of the (finite) binomial theorem of elementary algebra.

EXAMPLE 8 Suppose that α is a nonzero real number. Show that the Maclaurin series of $f(x) = (1+x)^\alpha$ is

$$\begin{aligned}(1+x)^\alpha &= 1 + \sum_{n=1}^{\infty} \frac{\alpha(\alpha-1)(\alpha-2)\cdots(\alpha-n+1)}{n!}x^n \\ &= 1 + \alpha x + \frac{\alpha(\alpha-1)}{2!}x^2 + \frac{\alpha(\alpha-1)(\alpha-2)}{3!}x^3 + \cdots. \end{aligned} \quad \textbf{(14)}$$

Also determine the interval of convergence of this **binomial series.**

Solution To derive the series itself, we simply list all the derivatives of $f(x) = (1+x)^\alpha$, including its "zeroth" derivative:

$$\begin{aligned} f(x) &= (1+x)^\alpha \\ f'(x) &= \alpha(1+x)^{\alpha-1} \\ f''(x) &= \alpha(\alpha-1)(1+x)^{\alpha-2} \\ f^{(3)}(x) &= \alpha(\alpha-1)(\alpha-2)(1+x)^{\alpha-3}, \\ &\vdots \\ f^{(n)}(x) &= \alpha(\alpha-1)(\alpha-2)\cdots(\alpha-n+1)(1+x)^{\alpha-n}. \end{aligned}$$

Thus

$$f^{(n)}(0) = \alpha(\alpha-1)(\alpha-2)\cdots(\alpha-n+1).$$

If we substitute this value of $f^{(n)}(0)$ into the Maclaurin series formula in Eq. (13), we get the binomial series in Eq. (14).

To determine the interval of convergence of the binomial series, we let

$$u_n = \frac{\alpha(\alpha-1)(\alpha-2)\cdots(\alpha-n+1)}{n!}x^n.$$

We find that

$$\begin{aligned}\lim_{n\to\infty}\left|\frac{u_{n+1}}{u_n}\right| &= \lim_{n\to\infty}\left|\frac{\dfrac{\alpha(\alpha-1)(\alpha-2)\cdots(\alpha-n)x^{n+1}}{(n+1)!}}{\dfrac{\alpha(\alpha-1)(\alpha-2)\cdots(\alpha-n+1)x^n}{n!}}\right| \\ &= \lim_{n\to\infty}\left|\frac{(\alpha-n)x}{n+1}\right| = |x|.\end{aligned}$$

Hence the ratio test shows that the binomial series converges absolutely if $|x| < 1$ and diverges if $|x| > 1$. Its convergence at the endpoints $x = \pm 1$ depends on the value of α; we shall not pursue this problem. Problem 67 outlines a proof that the sum of the binomial series actually is $(1+x)^\alpha$ if $|x| < 1$. ◆

If $\alpha = k$, a positive integer, then the coefficient of x^n is zero for $n > k$, and the binomial series reduces to the binomial formula

$$(1+x)^k = \sum_{n=0}^{k} \frac{k!}{n!(k-n)!}x^n.$$

Otherwise Eq. (14) is an infinite series. For example, with $\alpha = \frac{1}{2}$, we obtain

$$\begin{aligned}\sqrt{1+x} &= 1 + \frac{\frac{1}{2}}{1!}x + \frac{\left(\frac{1}{2}\right)\left(-\frac{1}{2}\right)}{2!}x^2 + \frac{\left(\frac{1}{2}\right)\left(-\frac{1}{2}\right)\left(-\frac{3}{2}\right)}{3!}x^3 \\ &\quad + \frac{\left(\frac{1}{2}\right)\left(-\frac{1}{2}\right)\left(-\frac{3}{2}\right)\left(-\frac{5}{2}\right)}{4!}x^4 + \cdots \\ &= 1 + \tfrac{1}{2}x - \tfrac{1}{8}x^2 + \tfrac{1}{16}x^3 - \tfrac{5}{128}x^4 + \cdots. \end{aligned} \tag{15}$$

If we replace x with $-x$ and take $\alpha = -\frac{1}{2}$, we get the series

$$\frac{1}{\sqrt{1-x}} = 1 + \frac{-\frac{1}{2}}{1!}(-x) + \frac{\left(-\frac{1}{2}\right)\left(-\frac{3}{2}\right)}{2!}(-x)^2 + \cdots + \frac{1 \cdot 3 \cdot 5 \cdots (2n-1)}{n! \cdot 2^n}x^n + \cdots,$$

which in summation notation takes the form

$$\frac{1}{\sqrt{1-x}} = 1 + \sum_{n=1}^{\infty} \frac{1 \cdot 3 \cdot 5 \cdots (2n-1)}{2 \cdot 4 \cdot 6 \cdots (2n)}x^n. \tag{16}$$

We will find this series quite useful in Example 12 and in Problem 68.

Differentiation and Integration of Power Series

Sometimes it is inconvenient to compute the repeated derivatives of a function in order to find its Taylor series. An alternative method of finding new power series is by the differentiation and integration of known power series.

Suppose that a power series representation of the function $f(x)$ is known. Then Theorem 3 (we leave its proof to advanced calculus) implies that the function $f(x)$ may be differentiated by separately differentiating the individual terms in its power series. That is, the power series obtained by termwise differentiation converges to the derivative $f'(x)$. Similarly, a function can be integrated by termwise integration of its power series.

THEOREM 3 Termwise Differentiation and Integration

Suppose that the function f has a power series representation

$$f(x) = \sum_{n=0}^{\infty} a_n x^n = a_0 + a_1 x + a_2 x^2 + a_3 x^3 + \cdots$$

with nonzero radius of convergence R. Then f is differentiable on $(-R, R)$ and

$$f'(x) = \sum_{n=1}^{\infty} n a_n x^{n-1} = a_1 + 2a_2 x + 3a_3 x^2 + 4a_4 x^3 + \cdots. \tag{17}$$

Also,

$$\int_0^x f(t)\,dt = \sum_{n=0}^{\infty} \frac{a_n x^{n+1}}{n+1} = a_0 x + \tfrac{1}{2}a_1 x^2 + \tfrac{1}{3}a_2 x^3 + \cdots \tag{18}$$

for each x in $(-R, R)$. Moreover, the power series in Eqs. (17) and (18) have the same radius of convergence R.

REMARK 1 Although we omit the proof of Theorem 3, we observe that the radius of convergence of the series in Eq. (17) is

$$R = \lim_{n\to\infty} \left| \frac{n a_n}{(n+1)a_{n+1}} \right| = \left(\lim_{n\to\infty} \frac{n}{n+1} \right) \cdot \left(\lim_{n\to\infty} \left| \frac{a_n}{a_{n+1}} \right| \right) = \lim_{n\to\infty} \left| \frac{a_n}{a_{n+1}} \right|.$$

Thus, by Eq. (8), the power series for $f(x)$ and the power series for $f'(x)$ have the same radius of convergence (under the assumption that the preceding limit exists).

REMARK 2 Theorem 3 has this important consequence: If both the power series $\sum a_n x^n$ and $\sum b_n x^n$ converge and, for all x with $|x| < R$ $(R > 0)$, $\sum a_n x^n = \sum b_n x^n$, then $a_n = b_n$ for all n. In particular, the Taylor series of a function is its unique power series representation (if any). (See Problem 66.)

EXAMPLE 9 Termwise differentiation of the geometric series for

$$f(x) = \frac{1}{1-x}$$

yields

$$\begin{aligned} \frac{1}{(1-x)^2} &= D_x\left(\frac{1}{1-x}\right) = D_x\,(1 + x + x^2 + x^3 + \cdots) \\ &= 1 + 2x + 3x^2 + 4x^3 + \cdots. \end{aligned}$$

Thus

$$\frac{1}{(1-x)^2} = \sum_{n=1}^{\infty} nx^{n-1} = \sum_{n=0}^{\infty}(n+1)x^n.$$

The series converges to $1/(1-x)^2$ if $-1 < x < 1$. ◆

EXAMPLE 10 Replacing x with $-t$ in the geometric series of Example 9 gives

$$\frac{1}{1+t} = 1 - t + t^2 - t^3 + \cdots + (-1)^n t^n + \cdots.$$

Because $D_t \ln(1+t) = 1/(1+t)$, termwise integration from $t = 0$ to $t = x$ now gives

$$\begin{aligned} \ln(1+x) &= \int_0^x \frac{1}{1+t}\,dt \\ &= \int_0^x (1 - t + t^2 - \cdots + (-1)^n t^n + \cdots)\,dt; \end{aligned}$$

$$\ln(1+x) = x - \frac{1}{2}x^2 + \frac{1}{3}x^3 - \frac{1}{4}x^4 + \cdots + \frac{(-1)^{n+1}}{n}x^n + \cdots \tag{19}$$

if $|x| < 1$. ◆

EXAMPLE 11 Find a power series representation for the arctangent function.

Solution Because $D_t \tan^{-1} t = 1/(1+t^2)$, termwise integration of the geometric series

$$\frac{1}{1+t^2} = 1 - t^2 + t^4 - t^6 + t^8 - \cdots$$

gives

$$\tan^{-1} x = \int_0^x \frac{1}{1+t^2}\,dt = \int_0^x (1 - t^2 + t^4 - t^6 + t^8 - \cdots)\,dt$$

if x is in the interval $(-1, 1)$ where the geometric series converges. Therefore

$$\tan^{-1} x = \sum_{n=1}^{\infty} \frac{(-1)^{n+1} x^{2n-1}}{2n-1} = x - \tfrac{1}{3}x^3 + \tfrac{1}{5}x^5 - \tfrac{1}{7}x^7 + \tfrac{1}{9}x^9 - \cdots \tag{20}$$

if $-1 < x < 1$. Figure 11.8.6 illustrates both the convergence of the power series within this interval and the divergence outside it. ◆

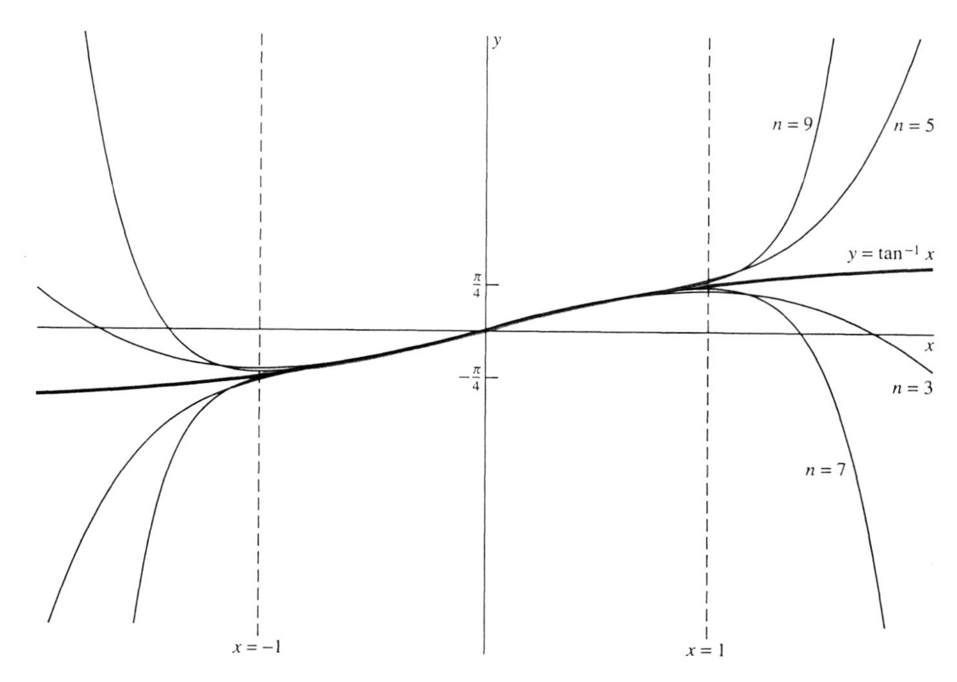

FIGURE 11.8.6 The graphs of the Taylor polynomials of degrees $n = 3, 5, 7$, and 9 illustrate the convergence within the interval $-1 < x < 1$ and divergence outside this interval.

EXAMPLE 12 Find a power series representation for the arcsine function.

Solution First we substitute t^2 for x in Eq. (16). This yields

$$\frac{1}{\sqrt{1-t^2}} = 1 + \sum_{n=1}^{\infty} \frac{1 \cdot 3 \cdot 5 \cdots (2n-1)}{2 \cdot 4 \cdot 6 \cdots (2n)} t^{2n}$$

if $|t| < 1$. Because $D_t \sin^{-1} t = 1/\sqrt{1-t^2}$, termwise integration of this series from $t = 0$ to $t = x$ gives

$$\sin^{-1} x = \int_0^x \frac{1}{\sqrt{1-t^2}}\,dt = x + \sum_{n=1}^{\infty} \frac{1 \cdot 3 \cdot 5 \cdots (2n-1)}{2 \cdot 4 \cdot 6 \cdots (2n)} \cdot \frac{x^{2n+1}}{2n+1} \tag{21}$$

if $|x| < 1$. Problem 68 shows how to use this series to derive the series

$$\frac{\pi^2}{6} = 1 + \frac{1}{2^2} + \frac{1}{3^2} + \frac{1}{4^2} + \cdots + \frac{1}{n^2} + \cdots,$$

which we used in Example 3 of Section 11.5 to approximate the number π. ◆

11.8 TRUE/FALSE STUDY GUIDE

11.8 CONCEPTS: QUESTIONS AND DISCUSSION

1. Suppose that you started with the Maclaurin series of the sine and cosine functions as their definitions. How many of the familiar properties of $\cos x$ and $\sin x$—such as their derivatives and addition formulas—could you establish using only these series?
2. Use the Maclaurin series for the sine and cosine functions and the corresponding hyperbolic series in Example 6 to explore relations between function pairs trig ix and trigh x, where trig denotes one of the cos/sin/tan trigonometric functions, and trigh denotes the corresponding hyperbolic function.

11.8 PROBLEMS

Find the interval of convergence of each power series in Problems 1 through 30.

1. $\sum_{n=1}^{\infty} nx^n$

2. $\sum_{n=1}^{\infty} \frac{x^n}{\sqrt{n}}$

3. $\sum_{n=1}^{\infty} \frac{nx^n}{2^n}$

4. $\sum_{n=1}^{\infty} \frac{(-1)^n x^n}{n^{1/2} 5^n}$

5. $\sum_{n=1}^{\infty} n! x^n$

6. $\sum_{n=1}^{\infty} \frac{(-1)^n x^n}{n^n}$

7. $\sum_{n=1}^{\infty} \frac{3^n x^n}{n^3}$

8. $\sum_{n=1}^{\infty} \frac{(-4)^n x^n}{\sqrt{2n+1}}$

9. $\sum_{n=1}^{\infty} (-1)^n n^{1/2} (2x)^n$

10. $\sum_{n=1}^{\infty} \frac{n^2 x^n}{3n-1}$

11. $\sum_{n=1}^{\infty} \frac{(-1)^n n x^n}{2^n (n+1)^3}$

12. $\sum_{n=1}^{\infty} \frac{n^{10} x^n}{10^n}$

13. $\sum_{n=1}^{\infty} \frac{(\ln n) x^n}{3^n}$

14. $\sum_{n=2}^{\infty} \frac{(-1)^n 4^n x^n}{n \ln n}$

15. $\sum_{n=0}^{\infty} (5x-3)^n$

16. $\sum_{n=1}^{\infty} \frac{(2x-1)^n}{n^4+16}$

17. $\sum_{n=1}^{\infty} \frac{2^n (x-3)^n}{n^2}$

18. $\sum_{n=1}^{\infty} \frac{n!}{n^n} x^n$ (Do not test the endpoints; the series diverges at each.)

19. $\sum_{n=1}^{\infty} \frac{(2n)!}{n!} x^n$

20. $\sum_{n=1}^{\infty} \frac{1 \cdot 3 \cdot 5 \cdots (2n+1)}{n!} x^n$ (Do not test the endpoints; the series diverges at each.)

21. $\sum_{n=1}^{\infty} \frac{n^3 (x+1)^n}{3^n}$

22. $\sum_{n=1}^{\infty} \frac{(-1)^{n+1} (x-2)^n}{n^2}$

23. $\sum_{n=1}^{\infty} \frac{(3-x)^n}{n^3}$

24. $\sum_{n=1}^{\infty} \frac{(-1)^{n+1} 10^n}{n!} (x-10)^n$

25. $\sum_{n=1}^{\infty} \frac{n!}{2^n} (x-5)^n$

26. $\sum_{n=1}^{\infty} \frac{(-1)^{n+1}}{n \cdot 10^n} (x-2)^n$

27. $\sum_{n=0}^{\infty} x^{(2^n)}$

28. $\sum_{n=0}^{\infty} \left(\frac{x^2+1}{5} \right)^n$

29. $\sum_{n=1}^{\infty} \frac{(-1)^n x^n}{1 \cdot 3 \cdot 5 \cdots (2n-1)}$

30. $\sum_{n=1}^{\infty} \frac{1 \cdot 3 \cdot 5 \cdots (2n-1)}{2 \cdot 5 \cdot 8 \cdots (3n-1)} x^n$

In Problems 31 through 42, use power series established in this section to find a power series representation of the given function. Then determine the radius of convergence of the resulting series.

31. $f(x) = \frac{x}{1-x}$

32. $f(x) = \frac{1}{10+x}$

33. $f(x) = x^2 e^{-3x}$

34. $f(x) = \frac{x}{9-x^2}$

35. $f(x) = \sin(x^2)$

36. $f(x) = \cos^2 2x = \frac{1}{2}(1 + \cos 4x)$

37. $f(x) = \sqrt[3]{1-x}$

38. $f(x) = (1+x^2)^{3/2}$

39. $f(x) = (1+x)^{-3}$

40. $f(x) = \frac{1}{\sqrt{9+x^3}}$

41. $f(x) = \frac{\ln(1+x)}{x}$

42. $f(x) = \frac{x - \arctan x}{x^3}$

In Problems 43 through 48, find a power series representation for the given function $f(x)$ by using termwise integration.

43. $f(x) = \int_0^x \sin t^3 \, dt$

44. $f(x) = \int_0^x \frac{\sin t}{t} \, dt$

45. $f(x) = \int_0^x \exp(-t^3) \, dt$

46. $f(x) = \int_0^x \frac{\arctan t}{t} \, dt$

47. $f(x) = \int_0^x \frac{1 - \exp(-t^2)}{t^2} \, dt$

48. $\tanh^{-1} x = \int_0^x \frac{1}{1-t^2} \, dt$

Beginning with the geometric series $\sum_{n=0}^{\infty} x^n$ as in Example 9, differentiate termwise to find the sums (for $|x| < 1$) of the power series in Problems 49 through 51.

49. $\sum_{n=1}^{\infty} nx^n$

50. $\sum_{n=1}^{\infty} n(n-1)x^n$

51. $\sum_{n=1}^{\infty} n^2 x^n$

52. Use the power series of the preceding problems to sum the numerical series

$$\sum_{n=1}^{\infty} \frac{n}{2^n} \quad \text{and} \quad \sum_{n=1}^{\infty} \frac{n^2}{3^n}.$$

53. Verify by termwise differentiation of its Maclaurin series that the exponential function $y = e^x$ satisfies the differential equation $dy/dx = y$. (Thus the exponential series arises naturally as a power series that is its own termwise derivative.)

54. Verify by termwise differentiation of their Maclaurin series that the sine function $y = \sin x$ and the cosine function $y = \cos x$ both satisfy the differential equation

$$\frac{d^2y}{dx^2} + y = 0.$$

55. Verify by termwise differentiation of the hyperbolic sine and cosine series in Example 6 that each of the functions $\cosh x$ and $\sinh x$ is the derivative of the other, and that each satisfies the differential equation $y'' - y = 0$.

56. In elementary mathematics one sees various definitions (some circular!) of the trigonometric functions. One approach to a rigorous foundation for these functions is to begin by defining $\cos x$ and $\sin x$ by means of their Maclaurin series. For instance, never having heard of sine, cosine, or the number π, we might define the function

$$S(x) = \sum_{n=1}^{\infty} \frac{(-1)^{n-1}x^{2n-1}}{(2n-1)!}$$

and verify using the ratio test that this series converges for all x. Use a computer algebra system to plot graphs of high-degree partial sums $s_n(x)$ of this series. Does it appear that the function $S(x)$ appears to have a zero somewhere near the number 3? Solve the equation $s_n(x) = 0$ numerically (for some large values of n) to verify that this least positive zero of the sine function is approximately 3.14159 (and thus the famous number π makes a fresh new appearance).

57. The Bessel function of order 1 is defined by

$$J_1(x) = \sum_{n=0}^{\infty} \frac{(-1)^n x^{2n+1}}{2^{2n+1} n!(n+1)!} = \frac{x}{2} - \frac{x^3}{16} + \frac{x^5}{384} - \cdots.$$

Verify that this series converges for all x and that the derivative of the Bessel function of order zero is given by $J_0'(x) = -J_1(x)$. Are the graphs in Fig. 11.8.7 consistent with this latter fact?

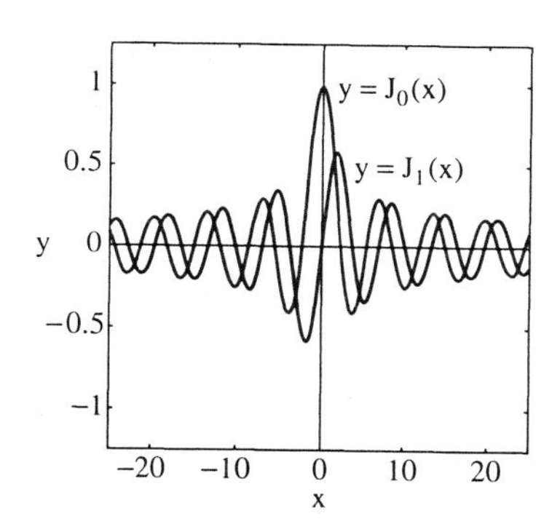

FIGURE 11.8.7 Graphs of the Bessel functions $J_0(x)$ and $J_1(x)$. Note that their zeros are interlaced, like the zeros of the cosine and sine functions.

58. Verify by termwise integration that

$$\int xJ_0(x)\,dx = xJ_1(x) + C.$$

59. Bessel's equation of order n is the second-order differential equation

$$x^2y'' + xy' + (x^2 - n^2)y = 0.$$

Verify by termwise differentiation that $y = J_0(x)$ satisfies Bessel's equation of order zero.

60. Verify that $y = J_1(x)$ satisfies Bessel's equation of order 1 (Problem 59).

61. First use the sine series to find the Taylor series of $f(x) = (\sin x)/x$. Then use a graphing calculator or computer to illustrate the approximation of $f(x)$ by its Taylor polynomials with center $a = 0$.

62. First find the Taylor series of the function

$$g(x) = \int_0^x \frac{\sin t}{t}\,dt.$$

Then determine where this power series converges. Finally, use a graphing calculator or computer to illustrate the approximation of $g(x)$ by its Taylor polynomials with center $a = 0$.

63. Deduce from the arctangent series (Example 11) that

$$\pi = \frac{6}{\sqrt{3}} \sum_{n=0}^{\infty} \frac{(-1)^n}{2n+1}\left(\frac{1}{3}\right)^n.$$

Then use this alternating series to show that $\pi = 3.14$ accurate to two decimal places.

64. Substitute the Maclaurin series for $\sin x$, and then assume the validity of termwise integration of the resulting series, to derive the formula

$$\int_0^{\infty} e^{-t}\sin xt\,dt = \frac{x}{1+x^2} \quad (|x| < 1).$$

Use the fact from Section 8.8 that

$$\int_0^{\infty} t^n e^{-t}\,dt = \Gamma(n+1) = n!.$$

65. (a) Deduce from the Maclaurin series for e^t that

$$\frac{1}{x^x} = \sum_{n=0}^{\infty} \frac{(-1)^n}{n!}(x\ln x)^n.$$

(b) Assuming the validity of termwise integration of the series in part (a), use the integral formula of Problem 53 in Section 8.8 to conclude that

$$\int_0^1 \frac{1}{x^x}\,dx = \sum_{n=1}^{\infty} \frac{1}{n^n}.$$

66. Suppose that $f(x)$ is represented by the power series

$$\sum_{n=0}^{\infty} a_n x^n$$

for all x in some open interval centered at $x = 0$. Show by repeated differentiation of the series, substituting $x = 0$ after each differentiation, that $a_n = f^{(n)}(0)/n!$ for all $n \geqq 0$. Thus the only power series in x that represents a function at and near $x = 0$ is its Maclaurin series.

67. (a) Consider the binomial series

$$f(x) = \sum_{n=0}^{\infty} \frac{\alpha(\alpha-1)(\alpha-2)\cdots(\alpha-n+1)}{n!}x^n,$$

which converges (to *something*) if $|x| < 1$. Compute the derivative $f'(x)$ by termwise differentiation, and show that it satisfies the differential equation $(1+x)f'(x) = \alpha f(x)$. (b) Solve the differential equation in part (a) to obtain $f(x) = C(1+x)^\alpha$ for some constant C. Finally, show that $C = 1$. Thus the binomial series converges to $(1+x)^\alpha$ if $|x| < 1$.

68. (a) Show by direct integration that

$$\int_0^1 \frac{\arcsin x}{\sqrt{1-x^2}}\,dx = \frac{\pi^2}{8}.$$

(b) Use the result of Problem 58 in Section 8.3 to show that

$$\int_0^1 \frac{x^{2n+1}}{\sqrt{1-x^2}}\,dx = \frac{2\cdot 4\cdot 6\cdots(2n)}{1\cdot 3\cdot 5\cdots(2n+1)}.$$

(c) Substitute the series of Example 10 for $\arcsin x$ into the integral of part (a); then use the integral of part (b) to integrate termwise. Conclude that

$$\int_0^1 \frac{\arcsin x}{\sqrt{1-x^2}}\,dx = 1 + \frac{1}{3^2} + \frac{1}{5^2} + \frac{1}{7^2} + \cdots.$$

(d) Note that

$$\sum_{n=1}^{\infty} \frac{1}{n^2} = \sum_{n=1}^{\infty} \frac{1}{(2n-1)^2} + \sum_{n=1}^{\infty} \frac{1}{(2n)^2}.$$

Use this information and parts (a) and (c) to show that

$$\sum_{n=1}^{\infty} \frac{1}{n^2} = \frac{\pi^2}{6}.$$

69. Prove that if the power series $\sum a_n x^n$ converges for some $x = x_0 \neq 0$, then it converges absolutely for all x such that $|x| < |x_0|$. [*Suggestion:* Conclude from the fact that $\lim_{n\to\infty} a_n x_0^n = 0$ that $|a_n x^n| \leqq |x/x_0|^n$ for all n sufficiently large. Thus the series $\sum |a_n x^n|$ is eventually dominated by the geometric series $\sum |x/x_0|^n$, which converges if $|x| < |x_0|$.]

70. Suppose that the power series $\sum a_n x^n$ converges for some but not all nonzero values of x. Let S be the set of real numbers for which the series converges absolutely. (a) Conclude from Problem 69 that the set S is bounded above. (b) Let λ be the least upper bound of the set S. (See Problem 61 of Section 11.2.) Then show that $\sum a_n x^n$ converges absolutely if $|x| < \lambda$ and diverges if $|x| > \lambda$. Explain why this proves Theorem 1 without the additional hypothesis that $\lim_{n\to\infty} |a_{n+1}/a_n|$ exists.

11.9 POWER SERIES COMPUTATIONS

Power series often are used to approximate numerical values of functions and integrals. *Alternating* power series (such as the sine and cosine series) are especially common and useful. Recall the alternating series remainder (or "error") estimate of Theorem 2 in Section 11.7. It applies to a convergent alternating series $\sum(-1)^{n+1}a_n$ whose terms are decreasing (so $a_n > a_{n+1}$ for every n). If we write

$$\sum_{k=1}^{\infty} (-1)^{k+1} a_k = (a_1 - a_2 + a_3 - \cdots \pm a_n) + E, \tag{1}$$

then $E = \mp a_{n+1} \pm a_{n+2} \mp a_{n+3} \pm \cdots$ is the error made when the series is *truncated*—the terms following $(-1)^{n+1}a_n$ are simply chopped off and discarded, and the n-term partial sum is used in place of the actual sum of the whole series. The remainder estimate then says that the error E has the same sign as the first term not retained, and is less in magnitude than this first neglected term; that is, $|E| < a_{n+1}$.

EXAMPLE 1 Use the first four terms of the binomial series

$$\sqrt{1+x} = 1 + \tfrac{1}{2}x - \tfrac{1}{8}x^2 + \tfrac{1}{16}x^3 - \tfrac{5}{128}x^4 + \cdots \tag{2}$$

to estimate the number $\sqrt{105}$ and to estimate the accuracy in the approximation.

Solution If $x > 0$ then the binomial series is, after the first term, an alternating series. In order to match the pattern on the left-hand side in Eq. (2), we first write

$$\sqrt{105} = \sqrt{100+5} = 10\sqrt{1+\tfrac{5}{100}} = 10\sqrt{1+0.05}.$$

Then with $x = 0.05$ the series in (2) gives

$$\begin{aligned}\sqrt{105} &= 10\left[1 + \tfrac{1}{2}(0.05) - \tfrac{1}{8}(0.05)^2 + \tfrac{1}{16}(0.05)^3 + E\right]\\ &= 10\,[1.02469531 + E] = 10.2469531 + 10E.\end{aligned}$$

Note that the error $10E$ in our approximation $\sqrt{105} \approx 10.2469531$ is 10 times the error E in the truncated series itself. It follows from the remainder estimate that E

is negative and that

$$|10E| < 10 \cdot \tfrac{5}{128}(0.05)^4 \approx 0.0000024.$$

Therefore,

$$10.2469531 - 0.0000024 = 10.2469507 < \sqrt{105} < 10.2469531,$$

so it follows that $\sqrt{105} \approx 10.24695$ rounded accurate to five decimal places. ◆

REMARK Suppose that we had been asked in advance to approximate $\sqrt{105}$ accurate to five decimal places. A convenient way to do this is to continue writing terms of the series until it is clear that they have become too small in magnitude to affect the fifth decimal place. A good rule of thumb is to use two more decimal places in the computations than are required in the final answer. Thus we use seven decimal places in this case and get

$$\begin{aligned}\sqrt{105} &= 10 \cdot (1 + 0.05)^{1/2} \\ &\approx 10 \cdot (1 + 0.025 - 0.0003125 + 0.0000078 - 0.0000002 + \cdots) \\ &\approx 10.246951 \approx 10.24695.\end{aligned}$$

EXAMPLE 2 Figure 11.9.1 shows the graph of the function $f(x) = (\sin x)/x$. Approximate (accurate to three decimal places) the area

$$A = \int_{-\pi}^{\pi} \frac{\sin x}{x}\,dx = 2\int_0^{\pi} \frac{\sin x}{x}\,dx \tag{3}$$

of the shaded region lying under the "principal arch" from $x = -\pi$ to π.

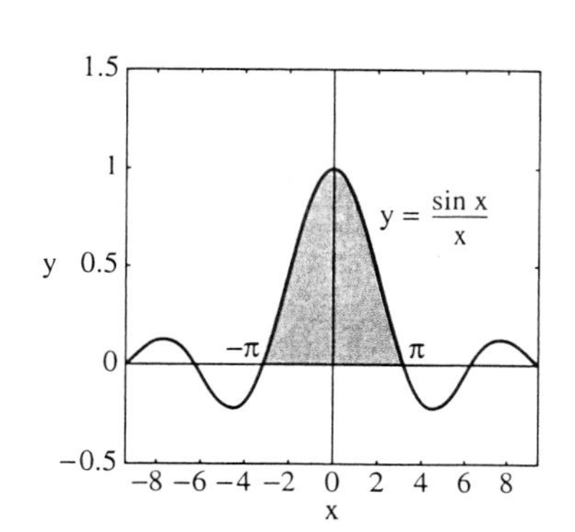

FIGURE 11.9.1 The graph $y = \dfrac{\sin x}{x}$ of Example 2.

Solution When we substitute the Taylor series for $\sin x$ in Eq. (3) and integrate termwise, we get

$$\begin{aligned}A &= 2\int_0^{\pi} \frac{1}{x}\left(x - \frac{x^3}{3!} + \frac{x^5}{5!} - \frac{x^7}{7!} + \cdots\right) dx \\ &= 2\int_0^{\pi}\left(1 - \frac{x^2}{3!} + \frac{x^4}{5!} - \frac{x^6}{7!} + \cdots\right) dx \\ &= 2\left[x - \frac{x^3}{3!3} + \frac{x^5}{5!5} - \frac{x^7}{7!7} + \cdots\right]_0^{\pi},\end{aligned}$$

and thus

$$A = 2\pi - \frac{2\pi^3}{3!3} + \frac{2\pi^5}{5!5} - \frac{2\pi^7}{7!7} + \frac{2\pi^9}{9!9} - \frac{2\pi^{11}}{11!11} + \cdots.$$

Following the "+2 rule of thumb" and retaining five decimal places, we calculate

$$A = 6.28319 - 3.44514 + 1.02007 - 0.17122 + 0.01825 - 0.00134 + 0.00007 - \cdots.$$

The sum of the first six terms gives $A \approx 3.70381$. Because we are summing an alternating series, the error in this approximation is positive and less than the next term 0.00007. Neglecting possible roundoff in the last place, we would conclude that $3.70381 < A < 3.70388$. Thus $A \approx 3.704$ rounded accurate to three decimal places. ◆

The Algebra of Power Series

Theorem 1, which we state without proof, implies that power series may be added and multiplied much like polynomials. The guiding principle is that of collecting coefficients of like powers of x.

THEOREM 1 Adding and Multiplying Power Series
Let $\sum a_n x^n$ and $\sum b_n x^n$ be power series with nonzero radii of convergence. Then

$$\sum_{n=0}^{\infty} a_n x^n + \sum_{n=0}^{\infty} b_n x^n = \sum_{n=0}^{\infty} (a_n + b_n) x^n \tag{4}$$

and

$$\left(\sum_{n=0}^{\infty} a_n x^n\right)\left(\sum_{n=0}^{\infty} b_n x^n\right) = \sum_{n=0}^{\infty} c_n x^n$$
$$= a_0 b_0 + (a_0 b_1 + a_1 b_0)x + (a_0 b_2 + a_1 b_1 + a_2 b_0)x^2 + \cdots, \tag{5}$$

where

$$c_n = a_0 b_n + a_1 b_{n-1} + a_2 b_{n-2} + \cdots + a_{n-1} b_1 + a_n b_0. \tag{6}$$

The series in Eqs. (4) and (5) converge for any x that lies interior to the intervals of convergence of both $\sum a_n x^n$ and $\sum b_n x^n$.

Thus if $\sum a_n x^n$ and $\sum b_n x^n$ are power series representations of the functions $f(x)$ and $g(x)$, respectively, then the product power series $\sum c_n x^n$ found by "ordinary multiplication" and collection of terms is a power series representation of the product function $f(x)g(x)$. This fact can also be used to divide one power series by another, *provided* that the quotient is known to have a power series representation.

EXAMPLE 3 Assume that the tangent function has a power series representation $\tan x = \sum a_n x^n$ (it does). Use the Maclaurin series for $\sin x$ and $\cos x$ to find a_0, a_1, a_2, and a_3.

Solution We multiply series to obtain

$$\begin{aligned}\sin x &= \tan x \cos x\\ &= (a_0 + a_1 x + a_2 x^2 + a_3 x^3 + \cdots)\left(1 - \frac{x^2}{2} + \frac{x^4}{24} - \cdots\right).\end{aligned}$$

If we multiply each term in the first factor by each term in the second, then collect coefficients of like powers, the result is

$$\sin x = a_0 + a_1 x + \left(a_2 - \tfrac{1}{2}a_0\right)x^2 + \left(a_3 - \tfrac{1}{2}a_1\right)x^3 + \cdots.$$

But because

$$\sin x = x - \tfrac{1}{6}x^3 + \tfrac{1}{120}x^5 - \cdots,$$

comparison of coefficients gives the equations

$$\begin{aligned} a_0 & & &= 0,\\ & a_1 & &= 1,\\ -\tfrac{1}{2}a_0 & & + a_2 &= 0,\\ & -\tfrac{1}{2}a_1 & + a_3 &= -\tfrac{1}{6}.\end{aligned}$$

Thus we find that $a_0 = 0$, $a_1 = 1$, $a_2 = 0$, and $a_3 = \frac{1}{3}$. So

$$\tan x = x + \tfrac{1}{3}x^3 + \cdots.$$

Things are not always as they first appear. A computer algebra system gives the continuation

$$\tan x = x + \tfrac{1}{3}x^3 + \tfrac{2}{15}x^5 + \tfrac{17}{315}x^7 + \tfrac{62}{2835}x^9 + \tfrac{1382}{155{,}925}x^{11} + \cdots \tag{7}$$

of the tangent series. For the general form of the nth coefficient, see K. Knopp's *Theory and Application of Infinite Series* (New York: Hafner Press, 1971), p. 204.

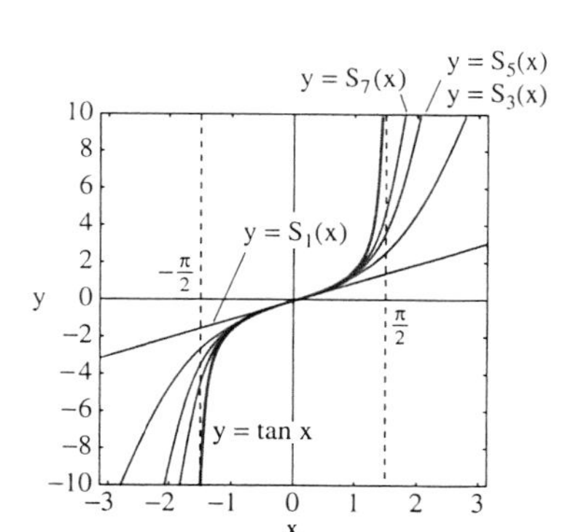

FIGURE 11.9.2 The graphs of $y = \tan x$ and the first four partial sums of the power series in (7).

You may also check that the first few terms agree with the result of ordinary division of the Maclaurin series for $\cos x$ into the Maclaurin series for $\sin x$:

$$1 - \frac{1}{2}x^2 + \frac{1}{24}x^4 - \cdots \overline{\Big) \, x - \frac{1}{6}x^3 + \frac{1}{120}x^5 - \cdots} \qquad \text{quotient: } x + \frac{1}{3}x^3 + \frac{2}{15}x^5 + \cdots$$

Figure 11.9.2 shows the approximation of the tangent function (on $-\pi/2 < x < \pi/2$) by the first four odd-degree polynomial partial sums corresponding to the terms exhibited in Eq. (7). Evidently these polynomial approximations have difficulty "keeping up" with $\tan x$ as it approaches $\pm\infty$ as $x \to \pm\pi/2$. ◆

Power Series and Indeterminate Forms

According to Theorem 3 of Section 11.8, a power series is differentiable and therefore continuous within its interval of convergence. It follows that

$$\lim_{x \to c} \sum_{n=0}^{\infty} a_n (x - c)^n = a_0. \tag{8}$$

Examples 4 and 5 illustrate the use of this simple observation to find the limit of the indeterminate form $f(x)/g(x)$. The technique is to first substitute power series representations for $f(x)$ and $g(x)$.

EXAMPLE 4 Find $\displaystyle\lim_{x \to 0} \frac{\sin x - \arctan x}{x^2 \ln(1 + x)}$.

Solution The power series of Eqs. (4), (19), and (20) in Section 11.8 give

$$\begin{aligned} \sin x - \arctan x &= \left(x - \tfrac{1}{6}x^3 + \tfrac{1}{120}x^5 - \cdots\right) - \left(x - \tfrac{1}{3}x^3 + \tfrac{1}{5}x^5 - \cdots\right) \\ &= \tfrac{1}{6}x^3 - \tfrac{23}{120}x^5 + \cdots \end{aligned}$$

and

$$x^2 \ln(1 + x) = x^2 \cdot \left(x - \tfrac{1}{2}x^2 + \tfrac{1}{3}x^3 + \cdots\right) = x^3 - \tfrac{1}{2}x^4 + \tfrac{1}{3}x^5 - \cdots.$$

Hence

$$\begin{aligned} \lim_{x \to 0} \frac{\sin x - \arctan x}{x^2 \ln(1 + x)} &= \lim_{x \to 0} \frac{\tfrac{1}{6}x^3 - \tfrac{23}{120}x^5 + \cdots}{x^3 - \tfrac{1}{2}x^4 + \cdots} \\ &= \lim_{x \to 0} \frac{\tfrac{1}{6} - \tfrac{23}{120}x^2 + \cdots}{1 - \tfrac{1}{2}x + \cdots} = \frac{1}{6}. \end{aligned}$$ ◆

EXAMPLE 5 Find $\displaystyle\lim_{x \to 1} \frac{\ln x}{x - 1}$.

Solution We first replace x with $x - 1$ in the power series for $\ln(1 + x)$ used in Example 4. [Equation (8) makes it clear that this method requires all series to have center c if we are taking limits as $x \to c$.] This gives us

$$\ln x = (x - 1) - \tfrac{1}{2}(x - 1)^2 + \tfrac{1}{3}(x - 1)^3 - \cdots.$$

Hence

$$\begin{aligned} \lim_{x \to 1} \frac{\ln x}{x - 1} &= \lim_{x \to 1} \frac{(x - 1) - \tfrac{1}{2}(x - 1)^2 + \tfrac{1}{3}(x - 1)^3 - \cdots}{x - 1} \\ &= \lim_{x \to 1} \left[1 - \tfrac{1}{2}(x - 1) + \tfrac{1}{3}(x - 1)^2 - \cdots\right] = 1. \end{aligned}$$ ◆

The method of Examples 4 and 5 provides a useful alternative to l'Hôpital's rule, especially when repeated differentiation of numerator and denominator is inconvenient or too time-consuming. (See Problems 59 and 60.)

Numerical and Graphical Error Estimation

The following examples show how to investigate the accuracy in a power-series partial-sum approximation for a specified interval of values of x. We will take the statement that a given approximation is "accurate to p decimal places" to mean that its error E is numerically less than half a unit in the pth decimal place; that is, that $|E| < 0.5 \times 10^{-p}$. For instance, four-place accuracy means that $|E| < 0.00005$. (Note that $p = 4$ is the number of zeros here.) Nevertheless, we should remember that in some cases a result accurate to within a half unit in the pth place may round "the wrong way," so that the result rounded to p places may still be in error by a unit in the pth decimal place (as in Problem 12).

EXAMPLE 6 Consider the polynomial approximation

$$\sin x \approx x - \frac{x^3}{3!} + \frac{x^5}{5!} - \cdots + (-1)^{n+1}\frac{x^{2n-1}}{(2n-1)!} \tag{9}$$

obtained by truncating the alternating Taylor series of the sine function.

(a) How accurate is the cubic approximation $P_3(x) \approx x - x^3/3!$ for angles from $0°$ to $10°$? Use this approximation to estimate $\sin 10°$.

(b) How many terms in (9) are needed to guarantee six-place accuracy in calculating $\sin x$ for angles from $0°$ to $45°$? Use the corresponding polynomial to approximate $\sin 30°$ and $\sin 40°$.

(c) For what values of x does the fifth-degree approximation yield five-place accuracy?

Solution (a) Of course we must substitute x in radians in (9), so we deal here with values of x in the interval $0 \leqq x \leqq \pi/18$. For any such x, the error E is positive (Why?) and is bounded by the magnitude of the next term:

$$|E| < \frac{x^5}{5!} \leqq \frac{(\pi/18)^5}{5!} \approx 0.00000135 < 0.000005.$$

We count five zeros on the right, and thus we have five-place accuracy. For instance, substituting $x = \pi/18$ in the cubic polynomial $P_3(x)$ gives

$$\sin 10° = \sin\left(\frac{\pi}{18}\right) \approx \frac{\pi}{18} - \frac{1}{3!}\cdot\left(\frac{\pi}{18}\right)^3$$
$$\approx 0.1736468 \approx 0.17365.$$

This five-place approximation $\sin 10° \approx 0.17365$ is correct, because the actual seven-place value of $\sin 10°$ is $0.1736482 \approx 0.17365$.

Solution (b) For any x in the interval $0 \leqq x \leqq \pi/4$, the error E made if we use the polynomial value in (9) in place of the actual value $\sin x$ is bounded by the first neglected term,

$$|E| < \frac{x^{2n+1}}{(2n+1)!} \leqq \frac{(\pi/4)^{2n+1}}{(2n+1)!}.$$

n	$\dfrac{(\pi/4)^{2n+1}}{(2n+1)!}$
1	0.08074551
2	0.00249039
3	0.00003658
4	0.00000031
5	0.00000000

FIGURE 11.9.3 Estimating the error in Example 6(b).

The table in Fig. 11.9.3 shows calculator values for $n = 1, 2, 3, \ldots$ of this maximal error (rounded to eight decimal places). For six-place accuracy we want $|E| < 0.0000005$, so we see that $n = 4$ will suffice. We therefore use the seventh-degree Taylor polynomial

$$P_7(x) = x - \frac{x^3}{3!} + \frac{x^5}{5!} - \frac{x^7}{7!} \tag{10}$$

to approximate $\sin x$ for $0 \leqq x \leqq \pi/4$. With $x = \pi/6$ we get

$$\sin 30° \approx \frac{\pi}{6} - \frac{(\pi/6)^3}{3!} + \frac{(\pi/6)^5}{5!} - \frac{(\pi/6)^7}{7!} \approx 0.49999999 \approx \frac{1}{2},$$

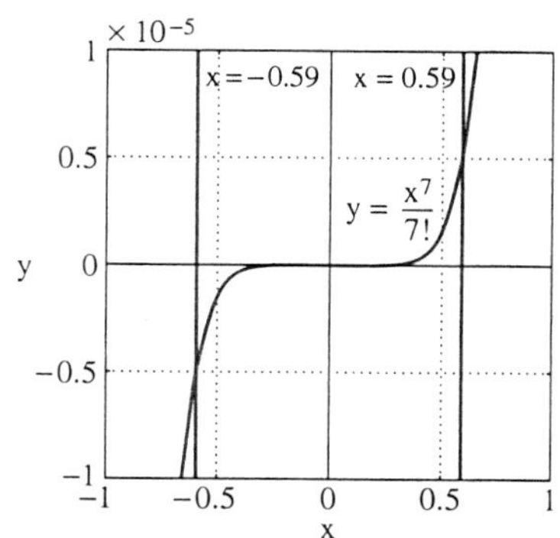

FIGURE 11.9.4 The graph of the maximal error $y = \frac{x^7}{7!}$ in Example 6(c).

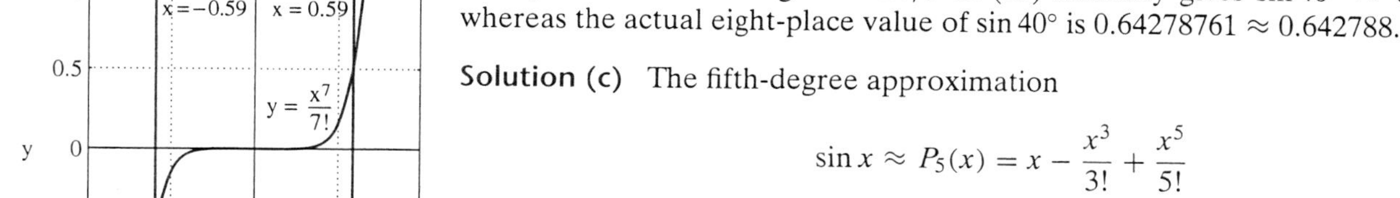

as expected. Substituting $x = 2\pi/9$ in (10) similarly gives $\sin 40° \approx 0.64278750$, whereas the actual eight-place value of $\sin 40°$ is $0.64278761 \approx 0.642788$.

Solution (c) The fifth-degree approximation

$$\sin x \approx P_5(x) = x - \frac{x^3}{3!} + \frac{x^5}{5!} \tag{11}$$

gives five-place accuracy when x is such that the error E satisfies the inequality

$$|E| < \frac{|x|^7}{7!} = \frac{|x|^7}{5040} \leqq 0.000005;$$

that is, when $|x| \leqq [(5040) \cdot (0.000005)]^{1/7} \approx 0.5910$ (radians). In degrees, this corresponds to angles between $-33.86°$ and $+33.86°$. In Fig. 11.9.4 the graph of $y = x^7/7!$ in the viewing window $-1 \leqq x \leqq 1$, $-0.00001 \leqq y \leqq 0.00001$ provides visual corroboration of this analysis—we see clearly that $x^7/7!$ remains between -0.000005 and 0.000005 when x is between -0.59 and 0.59. ◆

EXAMPLE 7 Suppose now that we want to approximate $f(x) = \sin x$ with three-place accuracy on the whole interval from 0° to 90°. Now it makes sense to begin with a Taylor series centered at the midpoint $x = \pi/4$ of the interval. Because the function $f(x)$ and its successive derivatives are $\sin x$, $\cos x$, $-\sin x$, $-\cos x$, and so forth, their values at $x = \pi/4$ are $\frac{1}{2}\sqrt{2}$, $\frac{1}{2}\sqrt{2}$, $-\frac{1}{2}\sqrt{2}$, $-\frac{1}{2}\sqrt{2}$, and so forth. Consequently Taylor's formula with remainder (Section 11.4) for $f(x) = \sin x$ centered at $x = \pi/4$ takes the form

$$\sin x = \frac{\sqrt{2}}{2} \cdot \left[1 + \left(x - \frac{\pi}{4}\right) - \frac{1}{2!}\left(x - \frac{\pi}{4}\right)^2 - \frac{1}{3!}\left(x - \frac{\pi}{4}\right)^3 + \cdots \pm \frac{1}{n!}\left(x - \frac{\pi}{4}\right)^n\right] + E(x) \tag{12}$$

where

$$|E(x)| = \left|\frac{f^{(n+1)}(z)}{(n+1)!}\left(x - \frac{\pi}{4}\right)^{n+1}\right| \leqq \frac{1}{(n+1)!}\left|x - \frac{\pi}{4}\right|^{n+1} \tag{13}$$

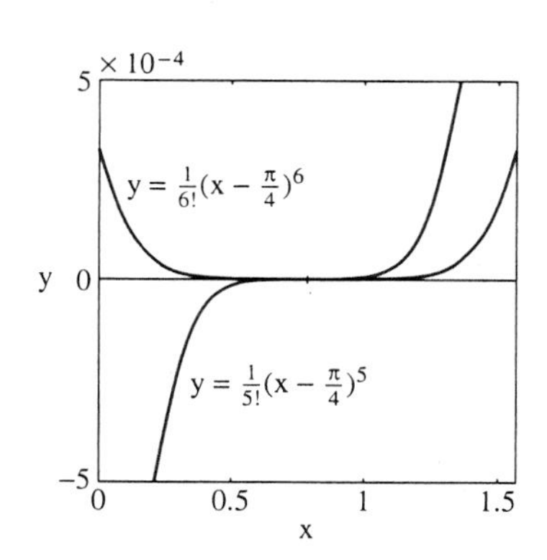

FIGURE 11.9.5 Comparing errors in Example 7.

for some z in the interval $0 \leqq x \leqq \pi/2$. Observe that the corresponding Taylor series is not alternating—if $x > \pi/4$ it has instead a "$++--++--$" pattern of signs—but we can still use the remainder estimate in (13). For three-place accuracy we need to choose n so that $y = E(x)$ remains within the viewing window $-0.0005 \leqq y \leqq 0.0005$ on the whole interval $0 \leqq x \leqq \pi/2$. Looking at the graphs plotted in Fig. 11.9.5, we see that this is so if $n = 5$ but not if $n = 4$. The desired approximation is therefore

$$\sin x \approx \frac{\sqrt{2}}{2} \cdot \left[1 + \left(x - \frac{\pi}{4}\right) - \frac{1}{2!}\left(x - \frac{\pi}{4}\right)^2 - \frac{1}{3!}\left(x - \frac{\pi}{4}\right)^3 + \frac{1}{4!}\left(x - \frac{\pi}{4}\right)^4 + \frac{1}{5!}\left(x - \frac{\pi}{4}\right)^5\right].$$

For instance, substituting $x = 0$ we get $\sin 0° \approx 0.00020 \approx 0.000$ as desired, and $x = \pi/2$ gives $\sin 90° \approx 1.00025 \approx 1.000$. ◆

11.9 TRUE/FALSE STUDY GUIDE

11.9 CONCEPTS: QUESTIONS AND DISCUSSION

1. Outline how you might use the binomial series (as in Example 1) to construct a *table of roots*—perhaps the square roots, cube roots, and fourth roots of the first 100 positive integers.

2. Give your own examples of several integrals for which numerical approximation using series (as in Example 2) would be useful.
3. Give your own examples of several indeterminate forms for which numerical evaluation using series (as in Examples 4 and 5) would be useful.

11.9 PROBLEMS

In Problems 1 through 10, use an infinite series to approximate the indicated number accurate to three decimal places.

1. $\sqrt[3]{65}$
2. $\sqrt[4]{630}$
3. $\sin(0.5)$
4. $e^{-0.2}$
5. $\tan^{-1}(0.5)$
6. $\ln(1.1)$
7. $\sin\left(\frac{\pi}{10}\right)$
8. $\cos\left(\frac{\pi}{20}\right)$
9. $\sin 10^\circ$
10. $\cos 5^\circ$

In Problems 11 through 22, use power series to approximate the value of the given integrals accurate to four decimal places.

11. $\int_0^1 \frac{\sin x}{x}\,dx$
12. $\int_0^1 \frac{\sin x}{\sqrt{x}}\,dx$
13. $\int_0^{1/2} \frac{\arctan x}{x}\,dx$
14. $\int_0^1 \sin x^2\,dx$
15. $\int_0^{1/10} \frac{\ln(1+x)}{x}\,dx$
16. $\int_0^{1/2} \frac{1}{\sqrt{1+x^4}}\,dx$
17. $\int_0^{1/2} \frac{1-e^{-x}}{x}\,dx$
18. $\int_0^{1/2} \sqrt{1+x^3}\,dx$
19. $\int_0^1 e^{-x^2}\,dx$
20. $\int_0^1 \frac{1-\cos x}{x^2}\,dx$
21. $\int_0^{1/2} \sqrt[3]{1+x^2}\,dx$
22. $\int_0^{1/2} \frac{x}{\sqrt{1+x^3}}\,dx$

In Problems 23 through 28, use power series rather than l'Hôpital's rule to evaluate the given limit.

23. $\lim_{x\to 0} \frac{1+x-e^x}{x^2}$
24. $\lim_{x\to 0} \frac{x-\sin x}{x^3\cos x}$
25. $\lim_{x\to 0} \frac{1-\cos x}{x(e^x-1)}$
26. $\lim_{x\to 0} \frac{e^x-e^{-x}-2x}{x-\arctan x}$
27. $\lim_{x\to 0} \left(\frac{1}{x}-\frac{1}{\sin x}\right)$
28. $\lim_{x\to 1} \frac{\ln(x^2)}{x-1}$

In Problems 29 through 32, calculate the indicated number with the required accuracy using Taylor's formula for an appropriate function centered at the given point $x = a$.

29. $\sin 80^\circ$; $a = \pi/4$, four decimal places
30. $\cos 35^\circ$; $a = \pi/4$, four decimal places
31. $\cos 47^\circ$; $a = \pi/4$, six decimal places
32. $\sin 58^\circ$; $a = \pi/3$, six decimal places

In Problems 33 through 36, determine the number of decimal places of accuracy the given appropriate formula yields for $|x| \leqq 0.1$.

33. $e^x \approx 1 + x + \frac{1}{2}x^2 + \frac{1}{6}x^3 + \frac{1}{24}x^4$
34. $\sin x \approx x - \frac{1}{6}x^3 + \frac{1}{120}x^5$
35. $\ln(1+x) \approx x - \frac{1}{2}x^2 + \frac{1}{3}x^3 - \frac{1}{4}x^4$
36. $\sqrt{1+x} \approx 1 + \frac{1}{2}x - \frac{1}{8}x^2$
37. Show that the approximation in Problem 33 gives the value of e^x accurate to within 0.001 if $|x| \leqq 0.5$. Then calculate $\sqrt[3]{e}$ accurate to two decimal places.
38. For what values of x is the approximation $\sin x \approx x - \frac{1}{6}x^3$ accurate to five decimal places?
39. (a) Show that the values of the cosine function for angles between 40° and 50° can be calculated with five-place accuracy using the approximation
$$\cos x \approx \frac{\sqrt{2}}{2}\left[1 - \left(x-\frac{\pi}{4}\right) - \frac{1}{2}\left(x-\frac{\pi}{4}\right)^2 + \frac{1}{6}\left(x-\frac{\pi}{4}\right)^3\right].$$
(b) Show that this approximation yields eight-place accuracy for angles between 44° and 46°.
40. Extend the approximation in Problem 39 to one that yields the values of $\cos x$ accurate to five decimal places for angles between 30° and 60°.

In Problems 41 through 44, use termwise integration of an appropriate power series to approximate the indicated area or volume accurate to two decimal places.

41. Figure 11.9.1 shows the region that lies between the graph of $y = (\sin x)/x$ and the x-axis from $x = -\pi$ to $x = \pi$. Substitute $\sin^2 x = \frac{1}{2}(1 - \cos 2x)$ to approximate the volume of the solid that is generated by revolving this region around the x-axis.
42. Approximate the area of the region that lies between the graph of $y = (1 - \cos x)/x^2$ and the x-axis from $x = -2\pi$ to $x = 2\pi$ (Fig. 11.9.6).

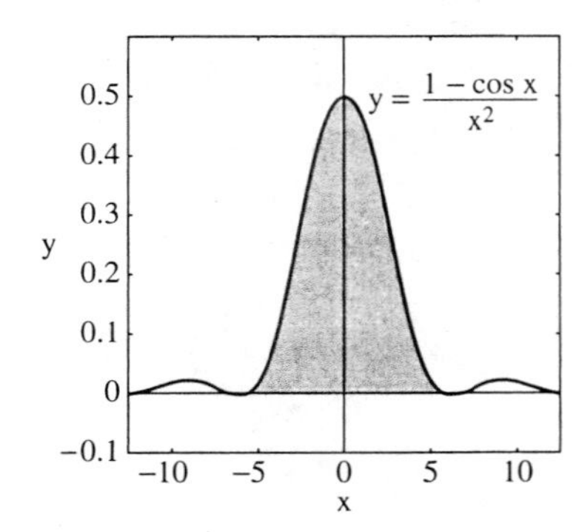

FIGURE 11.9.6 The region of Problem 42.

43. Approximate the volume of the solid generated by rotating the region of Problem 42 around the y-axis.

44. Approximate the volume of the solid generated by rotating the region of Problem 42 around the x-axis.

45. Derive the geometric series by long division of $1-x$ into 1.

46. Derive the series for $\tan x$ listed in Example 3 by long division of the Maclaurin series of $\cos x$ into the Maclaurin series of $\sin x$.

47. Derive the geometric series representation of $1/(1-x)$ by finding $a_0, a_1, a_2, \ldots$ such that

$$(1-x)(a_0 + a_1x + a_2x^2 + a_3x^3 + \cdots) = 1.$$

48. Derive the first five coefficients in the binomial series for $\sqrt{1+x}$ by finding a_0, a_1, a_2, a_3, and a_4 such that

$$(a_0 + a_1x + a_2x^2 + a_3x^3 + a_4x^4 + \cdots)^2 = 1 + x.$$

49. Use the method of Example 3 to find the coefficients a_0, a_1, a_2, a_3, and a_4 in the series

$$\sec x = \frac{1}{\cos x} = \sum_{n=0}^{\infty} a_n x^n.$$

50. Multiply the geometric series for $1/(1-x)$ and the series for $\ln(1-x)$ to show that if $|x| < 1$, then

$$\frac{\ln(1-x)}{1-x} = -x - \left(1 + \tfrac{1}{2}\right)x^2 - \left(1 + \tfrac{1}{2} + \tfrac{1}{3}\right)x^3 - \left(1 + \tfrac{1}{2} + \tfrac{1}{3} + \tfrac{1}{4}\right)x^4 - \cdots.$$

51. Take as known the logarithmic series

$$\ln(1+x) = x - \tfrac{1}{2}x^2 + \tfrac{1}{3}x^3 - \tfrac{1}{4}x^4 + \cdots.$$

Find the first four coefficients in the series for e^x by finding a_0, a_1, a_2, and a_3 such that

$$1 + x = e^{\ln(1+x)} = \sum_{n=0}^{\infty} a_n\left(x - \tfrac{1}{2}x^2 + \tfrac{1}{3}x^3 - \tfrac{1}{4}x^4 + \cdots\right)^n.$$

This is exactly how the power series for e^x was first discovered (by Newton)!

52. Use the method of Example 3 to show that

$$\frac{x}{\sin x} = 1 + \frac{1}{6}x^2 + \frac{7}{360}x^4 + \cdots.$$

53. Show that long division of power series gives

$$\frac{2+x}{1+x+x^2} = 2 - x - x^2 + 2x^3 - x^4 - x^5 + 2x^6 - x^7 - x^8 + 2x^9 - x^{10} - x^{11} + \cdots.$$

Show also that the radius of convergence of this series is $R = 1$.

54. Use the series in Problem 53 to approximate with two-place accuracy the value of the integral

$$\int_0^{1/2} \frac{x+2}{x^2+x+1}\,dx.$$

Compare your estimate with the exact result given by a computer algebra system.

Use the power series in Problem 53 to approximate with two-place accuracy the rather formidable integrals in Problems 55 and 56. Compare your estimates with the exact values given by a computer algebra system.

55. $\displaystyle\int_0^{1/2} \frac{1}{1+x^2+x^4}\,dx$

56. $\displaystyle\int_0^{1/2} \frac{1}{1+x^4+x^8}\,dx$

In Problems 57 and 58, graph the given function and several of its Taylor polynomials of the indicated degrees.

57. $f(x) = \dfrac{\sin x}{x}$; degrees $n = 2, 4, 6, \ldots$.

58. $f(x) = \displaystyle\int_0^x \frac{\sin t}{t}\,dt$; degrees $n = 3, 5, 7, \ldots$.

59. Use known power series to evaluate $\displaystyle\lim_{x\to 0} \frac{\sin x - \tan x}{\sin^{-1} x - \tan^{-1} x}$.

60. Substitute series such as

$$\sin(\tan x) = x + \frac{x^3}{6} - \frac{x^5}{40} - \frac{55x^7}{1008} + \cdots$$

provided by a computer algebra system to evaluate

$$\lim_{x\to 0} \frac{\sin(\tan x) - \tan(\sin x)}{\sin^{-1}(\tan^{-1} x) - \tan^{-1}(\sin^{-1} x)}.$$

61. (a) First use the parametrization $x(t) = a\cos t$, $y(t) = b\sin t$, $0 \leqq t \leqq 2\pi$ of the ellipse $(x/a)^2 + (y/b)^2 = 1$ to show that its perimeter (arc length) p is given by

$$p = 4a\int_0^{\pi/2} \sqrt{1 - \epsilon^2\cos^2 t}\,dt$$

where $\epsilon = \sqrt{1-(b/a)^2}$ is the *eccentricity* of the ellipse. This so-called *elliptic integral* is nonelementary, and so must be approximated numerically. (b) Use the binomial series to expand the integrand in the perimeter formula in part (a). Then integrate termwise—using Formula 113 from the table of integrals inside the back cover—to show that the perimeter of the ellipse is given in terms of its major semiaxis and eccentricity by the power series

$$p = 2\pi a\left(1 - \frac{1}{4}\epsilon^2 - \frac{3}{64}\epsilon^4 - \frac{5}{256}\epsilon^6 - \frac{175}{16384}\epsilon^8 - \cdots\right).$$

62. The *arithmetic mean* of the major and minor semiaxes of the ellipse of Problem 61 is $A = \frac{1}{2}(a+b)$; their *root-square mean* is $R = \sqrt{\frac{1}{2}(a^2+b^2)}$. Substitute $b = a\sqrt{1-\epsilon^2}$ and use the binomial series to derive the expansions

$$A = a\left(1 - \frac{1}{4}\epsilon^2 - \frac{1}{16}\epsilon^4 - \frac{1}{32}\epsilon^6 - \frac{5}{256}\epsilon^8 - \cdots\right)$$

and

$$R = a\left(1 - \frac{1}{4}\epsilon^2 - \frac{1}{32}\epsilon^4 - \frac{1}{128}\epsilon^6 - \frac{5}{2048}\epsilon^8 - \cdots\right).$$

Something wonderful happens when you average these two series; show that

$$\frac{1}{2}(A+R)=a\left(1-\frac{1}{4}\epsilon^2-\frac{3}{64}\epsilon^6-5\frac{5}{256}\epsilon^6-\frac{180}{16384}\epsilon^8-\cdots\right),$$

and then note that the first four terms of the series within the parentheses here are the same as in the ellipse perimeter series of Problem 61(b). Conclude that the perimeter p of the ellipse is given by

$$p=\pi(A+R)+\frac{5\pi a}{8192}\epsilon^8+\cdots. \tag{14}$$

If ϵ is quite small—as in a nearly circular ellipse—then the difference between the exact value of p and the simple approximation

$$p\approx\pi(A+R)=\pi\left(\tfrac{1}{2}(a+b)+\sqrt{\tfrac{1}{2}(a^2+b^2)}\right)$$

is extremely small. For instance, suppose that the orbit of the moon around the earth is an ellipse with major semiaxis a exactly 238,857 miles long and eccentricity ϵ exactly 0.0549. Then use Eq. (14) and a computer algebra system with extended-precision arithmetic to find the perimeter of the moon's orbit accurate *to the nearest inch;* give your answer in miles-feet-inches format.

11.9 Project: Calculating Trigonometric Functions on a Deserted Island

Again (as in the 11.4 Project) you're stranded for life on a desert island with only a very basic calculator that doesn't know about transcendental functions. Now your task is to use the (alternating) sine and cosine series to construct a table presenting (with five-place accuracy) the sines, cosines, and tangents of angles from 0° to 90° in increments of 5°.

To begin with, you can find the sine, cosine, and tangent of an angle of 45° from the familiar 1-1-$\sqrt{2}$ right triangle. Then you can find the values of these functions at an angle of 60° from an equilateral triangle. Once you know all about 45° and 60° angles, you can use the sine and cosine addition formulas

$$\sin(\alpha\pm\beta)=\sin\alpha\cos\beta\pm\cos\alpha\sin\beta$$

and

$$\cos(\alpha\pm\beta)=\cos\alpha\cos\beta\mp\sin\alpha\sin\beta$$

and/or equivalent forms to find the sine, cosine, and tangent of such angles as 15°, 30°, 75°, and 90°.

But algebra and simple trigonometric identities will probably never give you the sine or cosine or an angle of 5°. For this you will need to use the power series for sine and cosine. Sum enough terms (and then some) so you know your result is accurate to nine decimal places. Then fill in all the entries in your table, rounding them to five places. Tell—honestly—whether your entries agree with those your *real* calculator gives.

Finally, explain what strategy you would use to complete a similar table of values of trigonometric functions with angles in increments of 1° rather than 5°.

11.10 SERIES SOLUTIONS OF DIFFERENTIAL EQUATIONS

In Section 9.6 we saw that solving a homogeneous linear differential equation with constant coefficients can be reduced to the algebraic problem of finding the roots of its characteristic equation. There is no simple or similarly routine procedure for solving linear differential equations with *variable* coefficients. Even such a simple-looking equation as $y''-xy=0$ has no solution that can be expressed in terms of the standard elementary functions of calculus. One of the most important applications of power series is their use to solve such differential equations.

The Power Series Method

The *power series method* for solving a differential equation consists of substituting the power series

$$y=\sum_{n=0}^{\infty}c_nx^n=c_0+c_1x+c_2x^2+c_3x^3+\cdots \tag{1}$$

in the differential equation, and then attempting to determine what the values of the coefficients $c_0, c_1, c_2, c_3, \ldots$ must be in order that the series in (1) will actually satisfy the given differential equation. At first glance this might seem to be a formidable problem, because we have infinitely many unknowns $c_0, c_1, c_2, c_3, \ldots$ to find. Nevertheless, we will see that the method frequently succeeds. When it does, we obtain a power series representation of a solution, in contrast to the closed form solutions that result from the solution techniques we saw in Chapter 9.

Before we can substitute the series in (1) in a differential equation, we must first know what to substitute for the derivatives $y', y'', \ldots$ of the unknown function $y(x)$. But recall from Theorem 3 in Section 11.8 that the derivative of a power series can be calculated by termwise differentiation. Hence the first and second derivatives of the series in (1) are given by

$$y' = \sum_{n=1}^{\infty} nc_n x^{n-1} = c_1 + 2c_2 x + 3c_3 x^2 + \cdots \tag{2}$$

and

$$y'' = \sum_{n=2}^{\infty} n(n-1)c_n x^{n-2} = 2c_2 + 6c_3 x + 12c_4 x^2 + \cdots. \tag{3}$$

Also, these two series have the same radius of convergence as the original series in (1).

The process of determining the coefficients $c_0, c_1, c_2, c_3, \ldots$ in the series so that it will satisfy a given differential equation depends also on the following consequence of termwise differentiation: If two power series represent the same function on an open interval, then they are identical series. That is, they are one and the same power series. (See Problem 66 in Section 11.8.) In particular, *if* $\sum a_n x^n \equiv 0$ *on an open interval, then it follows that* $a_n = 0$ *for all* n. This fact is sometimes called the **identity principle** for power series.

EXAMPLE 1 Solve the equation $y' + 2y = 0$.

Solution We substitute the series

$$y = \sum_{n=0}^{\infty} c_n x^n \quad \text{and} \quad y' = \sum_{n=1}^{\infty} nc_n x^{n-1},$$

and obtain

$$\sum_{n=1}^{\infty} nc_n x^{n-1} + 2\sum_{n=0}^{\infty} c_n x^n = 0. \tag{4}$$

To compare coefficients here, we need the general term in each sum to be the term containing x^n. To accomplish this, we shift the index of summation in the first sum. To see how to do this, note that

$$\sum_{n=1}^{\infty} nc_n x^{n-1} = c_1 + 2c_2 x + 3c_3 x^2 + \cdots = \sum_{n=0}^{\infty} (n+1)c_{n+1} x^n.$$

Thus we can replace n with $n+1$ if, at the same time, we start counting one step lower; that is, at $n = 0$ rather than at $n = 1$. This is a shift of $+1$ in the index of summation. The result of making this shift in Eq. (4) is the identity

$$\sum_{n=0}^{\infty} (n+1)c_{n+1} x^n + 2\sum_{n=0}^{\infty} c_n x^n = 0;$$

that is,

$$\sum_{n=0}^{\infty} [(n+1)c_{n+1} + 2c_n]x^n = 0.$$

If this equation holds on some interval, then it follows from the identity principle that $(n+1)c_{n+1} + 2c_n = 0$ for all $n \geqq 0$; consequently,

$$c_{n+1} = -\frac{2c_n}{n+1} \tag{5}$$

for all $n \geqq 0$. Equation (5) is a **recurrence relation** from which we can successively compute $c_1, c_2, c_3, \dots$ in terms of c_0; the latter will turn out to be the arbitrary constant that we expect to find in a general solution of a first-order differential equation.

With $n = 0$, Eq. (5) gives

$$c_1 = -\frac{2c_0}{1}.$$

With $n = 1$, Eq. (5) gives

$$c_2 = -\frac{2c_1}{2} = +\frac{2^2 c_0}{1 \cdot 2} = \frac{2^2 c_0}{2!}.$$

With $n = 2$, Eq. (5) gives

$$c_3 = -\frac{2c_2}{3} = -\frac{2^3 c_0}{1 \cdot 2 \cdot 3} = -\frac{2^3 c_0}{3!}.$$

By now it should be clear that after n such steps, we will have

$$c_n = (-1)^n \frac{2^n c_0}{n!}, \quad n \geqq 1.$$

(This is easy to prove by induction on n.) Consequently, our solution takes the form

$$y(x) = \sum_{n=0}^{\infty} c_n x^n = \sum_{n=0}^{\infty} (-1)^n \frac{2^n c_0}{n!} x^n = c_0 \sum_{n=0}^{\infty} \frac{(-2x)^n}{n!} = c_0 e^{-2x}.$$

In the final step we have used the familiar exponential series to identify our power series solution as the same solution $y(x) = c_0 e^{-2x}$ we could have obtained by the method of separation of variables. ◆

Shift of Index of Summation

In the solution of Example 1 we wrote

$$\sum_{n=1}^{\infty} n c_n x^{n-1} = \sum_{n=0}^{\infty} (n+1) c_{n+1} x^n \tag{6}$$

by shifting the index of summation by $+1$ in the series on the left. That is, we simultaneously *increased* the index of summation by 1 (replacing n with $n+1$, $n \to n+1$) and *decreased* the starting point by 1, from $n = 1$ to $n = 0$, thereby obtaining the series on the right. This procedure is valid because each infinite series in (6) is simply a compact notation for the single series

$$c_1 + 2c_2 x + 3c_3 x^2 + 4c_4 x^3 + \cdots. \tag{7}$$

More generally, we can shift the index of summation by k in an infinite series by simultaneously *increasing* the summation index by k ($n \to n+k$) and *decreasing* the starting point by k. For instance, a shift by $+2$ ($n \to n+2$) yields

$$\sum_{n=3}^{\infty} a_n x^{n-1} = \sum_{n=1}^{\infty} a_{n+2} x^{n+1}.$$

If k is negative we interpret a "decrease by k" as an increase by $-k = |k|$. Thus a shift by -2 ($n \to n-2$) in the index of summation yields

$$\sum_{n=1}^{\infty} nc_n x^{n-1} = \sum_{n=3}^{\infty} (n-2)c_{n-2}x^{n-3};$$

we have *decreased* the index of summation by 2, but *increased* the starting point by 2, from $n = 1$ to $n = 3$. You should check that the summation on the right is merely another representation of the series in (7).

EXAMPLE 2 Solve the equation $(x-3)y' + 2y = 0$.

Solution As before, we substitute

$$y = \sum_{n=0}^{\infty} c_n x^n \quad \text{and} \quad y' = \sum_{n=1}^{\infty} nc_n x^{n-1}$$

to obtain

$$(x-3)\sum_{n=1}^{\infty} nc_n x^{n-1} + 2\sum_{n=0}^{\infty} c_n x^n = 0,$$

so that

$$\sum_{n=1}^{\infty} nc_n x^n - 3\sum_{n=1}^{\infty} nc_n x^{n-1} + 2\sum_{n=0}^{\infty} c_n x^n = 0.$$

In the first sum we can replace $n = 1$ with $n = 0$ with no effect on the sum. In the second sum we shift the index of summation by $+1$. This yields

$$\sum_{n=0}^{\infty} nc_n x^n - 3\sum_{n=0}^{\infty} (n+1)c_{n+1} x^n + 2\sum_{n=0}^{\infty} c_n x^n = 0;$$

that is,

$$\sum_{n=0}^{\infty} \left[nc_n - 3(n+1)c_{n+1} + 2c_n\right] x^n = 0.$$

The identity principle then gives

$$nc_n - 3(n+1)c_{n+1} + 2c_n = 0,$$

from which we obtain the recurrence relation

$$c_{n+1} = \frac{n+2}{3(n+1)}c_n \quad \text{for} \quad n \geqq 0.$$

We apply this formula with $n = 0$, $n = 1$, and $n = 2$ in turn, and find that

$$c_1 = \frac{2}{3}c_0, \quad c_2 = \frac{3}{3 \cdot 2}c_1 = \frac{3}{3^2}c_0, \quad \text{and} \quad c_3 = \frac{4}{3 \cdot 3}c_2 = \frac{4}{3^3}c_0.$$

This is almost enough to make the pattern evident; it is not difficult to show by induction on n that

$$c_n = \frac{n+1}{3^n}c_0 \quad \text{if} \quad n \geqq 1.$$

Hence our proposed power series solution is

$$y(x) = c_0 \sum_{n=0}^{\infty} \frac{n+1}{3^n} x^n. \tag{8}$$

Its radius of convergence is

$$\rho = \lim_{n\to\infty} \left|\frac{c_n}{c_{n+1}}\right| = \lim_{n\to\infty} \frac{3n+3}{n+2} = 3.$$

Thus the series in (8) converges if $-3 < x < 3$ and diverges if $|x| > 3$. In this particular example we can explain why. An elementary solution (obtained by separation of variables) of our differential equation is $y = 1/(3-x)^2$. If we differentiate termwise the geometric series

$$\frac{1}{3-x} = \frac{\frac{1}{3}}{1-\frac{x}{3}} = \frac{1}{3}\sum_{n=0}^{\infty}\frac{x^n}{3^n},$$

we get a constant multiple of the series in (8). Thus this series (with the arbitrary constant c_0 appropriately chosen) represents the solution

$$y(x) = \frac{1}{(3-x)^2}$$

on the interval $-3 < x < 3$, and the singularity at $x = 3$ is the reason why the radius of convergence of the power series solution turned out to be $\rho = 3$. ◆

EXAMPLE 3 Solve the equation $x^2y' = y - x - 1$.

Solution We make the usual substitutions $y = \sum c_n x^n$ and $y' = \sum nc_n x^{n-1}$, which yield

$$x^2\sum_{n=1}^{\infty} nc_n x^{n-1} = -1 - x + \sum_{n=0}^{\infty} c_n x^n,$$

so that

$$\sum_{n=1}^{\infty} nc_n x^{n+1} = -1 - x + \sum_{n=0}^{\infty} c_n x^n.$$

Because of the presence of the two terms -1 and $-x$ on the right-hand side, we need to split off the first two terms, $c_0 + c_1x$, of the series on the right for comparison. If we also shift the index of summation on the left by -1 (replace $n = 1$ with $n = 2$ and n with $n - 1$), we get

$$\sum_{n=2}^{\infty}(n-1)c_{n-1}x^n = -1 - x + c_0 + c_1x + \sum_{n=2}^{\infty} c_n x^n.$$

Because the left-hand side contains neither a constant term nor a term containing x to the first power, the identity principle now yields $c_0 = 1$, $c_1 = 1$, and $c_n = (n-1)c_{n-1}$ for $n \geqq 2$. It follows that

$$c_2 = 1 \cdot c_1 = 1!, \qquad c_3 = 2 \cdot c_2 = 2!, \qquad c_4 = 3 \cdot c_3 = 3!,$$

and, in general, that

$$c_n = (n-1)! \quad \text{for} \quad n \geqq 2.$$

Thus we obtain the power series

$$y(x) = 1 + x + \sum_{n=2}^{\infty}(n-1)!x^n.$$

But the radius of convergence of this series is

$$\rho = \lim_{n\to\infty}\frac{(n-1)!}{n!} = \lim_{n\to\infty}\frac{1}{n} = 0,$$

so the series converges only for $x = 0$. What does this mean? Simply that the given differential equation does not have a [convergent] power series solution of the assumed form $y = \sum c_n x^n$. This example serves as a warning that the simple act of writing $y = \sum c_n x^n$ involves an assumption that may be false. ◆

EXAMPLE 4 Solve the equation $y'' + y = 0$.

Solution If we assume a solution of the form

$$y = \sum_{n=0}^{\infty} c_n x^n,$$

we find that

$$y' = \sum_{n=1}^{\infty} n c_n x^{n-1} \quad \text{and} \quad y'' = \sum_{n=2}^{\infty} n(n-1) c_n x^{n-2}.$$

Substituting for y and y'' in the differential equation then yields

$$\sum_{n=2}^{\infty} n(n-1) c_n x^{n-2} + \sum_{n=0}^{\infty} c_n x^n = 0.$$

We shift the index of summation in the first sum by $+2$ (replace $n = 2$ with $n = 0$ and n with $n + 2$). This gives

$$\sum_{n=0}^{\infty} (n+2)(n+1) c_{n+2} x^n + \sum_{n=0}^{\infty} c_n x^n = 0.$$

The identity $(n+2)(n+1)c_{n+2} + c_n = 0$ now follows from the identity principle, and thus we obtain the recurrence relation

$$c_{n+2} = -\frac{c_n}{(n+1)(n+2)} \tag{9}$$

for $n \geqq 0$. It is evident that this formula will determine the coefficients c_n with even subscript in terms of c_0 and those of odd subscript in terms of c_1; c_0 and c_1 are not predetermined, and thus will be the two arbitrary constants we expect to find in a general solution of a second-order equation.

When we apply the recurrence relation in (9) with $n = 0$, 2, and 4 in turn, we get

$$c_2 = -\frac{c_0}{2!}, \quad c_4 = \frac{c_0}{4!}, \quad \text{and} \quad c_6 = -\frac{c_0}{6!}.$$

Taking $n = 1$, 3, and 5 in turn, we find that

$$c_3 = -\frac{c_1}{3!}, \quad c_5 = \frac{c_1}{5!}, \quad \text{and} \quad c_7 = -\frac{c_1}{7!}.$$

Again, the pattern is clear; we leave it for you to show (by induction) that for $k \geqq 1$,

$$c_{2k} = \frac{(-1)^k c_0}{(2k)!} \quad \text{and} \quad c_{2k+1} = \frac{(-1)^k c_1}{(2k+1)!}.$$

Thus we get the power series solution

$$y(x) = c_0\left(1 - \frac{x^2}{2!} + \frac{x^4}{4!} - \frac{x^6}{6!} + \cdots\right) + c_1\left(x - \frac{x^3}{3!} + \frac{x^5}{5!} - \frac{x^7}{7!} + \cdots\right);$$

that is, $y(x) = c_0 \cos x + c_1 \sin x$. Note that we have no problem with the radius of convergence here; the Taylor series for the sine and cosine functions converge for all x. ◆

Power Series Definitions of Functions

The solution of Example 4 can bear further comment. Suppose that we had never heard of the sine and cosine functions, let alone their Taylor series. We would then have discovered the two power series solutions

$$C(x) = \sum_{n=0}^{\infty} \frac{(-1)^n x^{2n}}{(2n)!} = 1 - \frac{x^2}{2!} + \frac{x^4}{4!} - \cdots \tag{10}$$

and

$$S(x) = \sum_{n=0}^{\infty} \frac{(-1)^n x^{2n+1}}{(2n+1)!} = x - \frac{x^3}{3!} + \frac{x^5}{5!} - \cdots \tag{11}$$

of the differential equation $y'' + y = 0$. It is clear that $C(0) = 1$ and that $S(0) = 0$. After verifying that the two series in (10) and (11) converge for all x, we can differentiate them term by term to find that

$$C'(x) = -S(x) \quad \text{and} \quad S'(x) = C(x). \tag{12}$$

Consequently $C'(0) = 0$ and $S'(0) = 1$. Thus with the aid of the power series method (all the while knowing nothing about the sine and cosine functions), we have discovered that $y = C(x)$ is the unique solution of

$$y'' + y = 0$$

that satisfies the initial conditions $y(0) = 1$ and $y'(0) = 0$, and that $y = S(x)$ is the unique solution that satisfies the initial conditions $y(0) = 0$ and $y'(0) = 1$. It follows that $C(x)$ and $S(x)$ are linearly independent, and—recognizing the importance of the differential equation $y'' + y = 0$—we can agree to call C the *cosine* function and S the *sine* function. Indeed, all the usual properties of these two functions can be established, using only their initial values (at $x = 0$) and the derivatives in (12); there is no need to refer to triangles or even to angles. (Can you use the series in (10) and (11) to show that $[C(x)]^2 + [S(x)]^2 = 1$ for all x?) This demonstrates that the cosine and sine functions are fully determined by the differential equation $y'' + y = 0$ of which they are the natural linearly independent solutions. Figures 11.10.1 and 11.10.2 show how the geometric character of the graphs of $\cos x$ and $\sin x$ is revealed by the graphs of the Taylor polynomial approximations that we get by truncating the infinite series in (10) and (11).

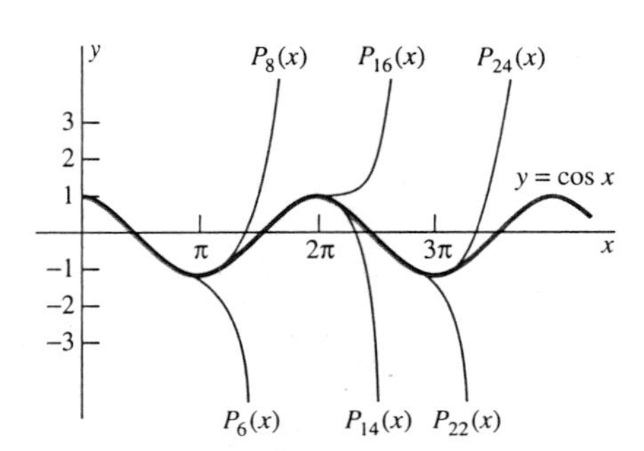

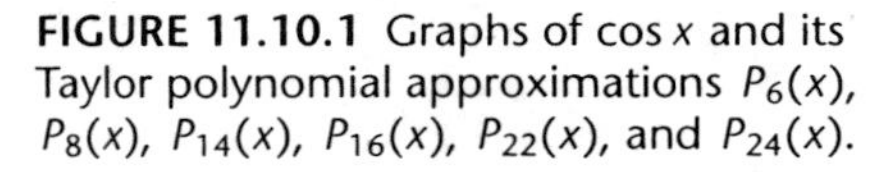
FIGURE 11.10.1 Graphs of $\cos x$ and its Taylor polynomial approximations $P_6(x)$, $P_8(x)$, $P_{14}(x)$, $P_{16}(x)$, $P_{22}(x)$, and $P_{24}(x)$.

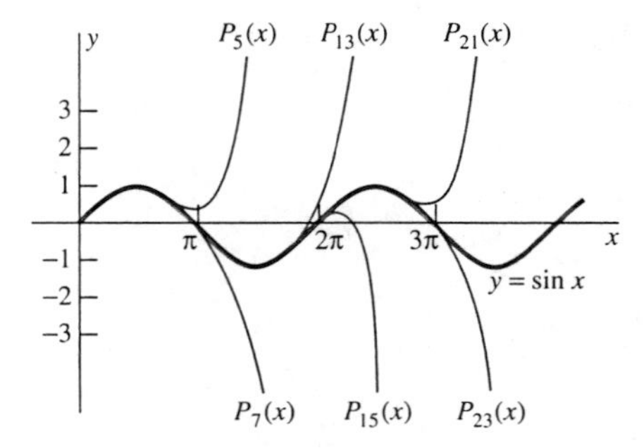

FIGURE 11.10.2 Graphs of $\sin x$ and its Taylor polynomial approximations $P_5(x)$, $P_7(x)$, $P_{13}(x)$, $P_{15}(x)$, $P_{21}(x)$, and $P_{23}(x)$.

This is by no means an uncommon situation. Many important special functions of mathematics occur in the first instance as power series solutions of differential equations, and thus are in practice *defined* by means of these power series. Example 5 introduces in this manner the *Airy functions* that appear in applications ranging from the propagation of radio waves to vibrations in atoms and molecules.

EXAMPLE 5 Solve the Airy equation $y'' - xy = 0$.

Solution Substituting $y = \sum c_n x^n$ and $y'' = \sum n(n-1)c_n x^{n-2}$ as usual yields

$$\sum_{n=2}^{\infty} n(n-1)c_n x^{n-2} - \sum_{n=0}^{\infty} c_n x^{n+1} = 0.$$

A shift of indices—replacing n with $n+2$ in the first sum and with $n-1$ in the second—yields

$$\sum_{n=0}^{\infty} (n+2)(n+1)c_{n+2}x^n - \sum_{n=1}^{\infty} c_{n-1}x^n = 0.$$

Splitting off the term corresponding to $n = 0$ in the first sum and combining the remaining terms, we get

$$2c_2 + \sum_{n=1}^{\infty} \left[(n+2)(n+1)c_{n+2} - c_{n-1}\right] x^n = 0.$$

The identity principle now gives $c_2 = 0$—because there is no other constant term on the left-hand side—and the recurrence relation $(n+2)(n+1)c_{n+2} - c_{n-1} = 0$ for $n \geqq 1$. Replacement of n with $n+1$ gives the recurrence relation

$$c_{n+3} = \frac{c_n}{(n+2)(n+3)} \tag{13}$$

for $n \geqq 0$. Thus each coefficient (after the first three) depends on the third previous one. Hence the fact that $c_2 = 0$ implies that

$$c_2 = c_5 = c_8 = \cdots = 0.$$

Beginning with c_0 as an arbitrary constant, we apply (13) with $n = 0$, $n = 3$, and $n = 6$ in turn and calculate

$$c_3 = \frac{c_0}{2 \cdot 3} = \frac{c_0}{6}, \quad c_6 = \frac{c_3}{5 \cdot 6} = \frac{c_0}{180}, \quad \text{and} \quad c_9 = \frac{c_6}{8 \cdot 9} = \frac{c_0}{12960}.$$

Beginning with c_1 as a second arbitrary constant, we calculate similarly

$$c_4 = \frac{c_1}{3 \cdot 4} = \frac{c_1}{12}, \quad c_7 = \frac{c_4}{6 \cdot 7} = \frac{c_1}{504}, \quad \text{and} \quad c_{10} = \frac{c_7}{9 \cdot 10} = \frac{c_1}{45360}.$$

When we collect the terms that involve c_0 and those that involve c_1, we get the general solution

$$y(x) = c_0\left(1 + \frac{x^3}{6} + \frac{x^6}{180} + \frac{x^9}{12960} + \cdots\right) + c_1\left(x + \frac{x^4}{12} + \frac{x^7}{504} + \frac{x^{10}}{45360} + \cdots\right)$$

of the Airy equation, with arbitrary constants c_0 and c_1. We see here the independent (why?) particular solutions

$$y_1(x) = 1 + \frac{x^3}{6} + \frac{x^6}{180} + \frac{x^9}{12960} + \cdots \quad \text{and} \quad y_2(x) = x + \frac{x^4}{12} + \frac{x^7}{504} + \frac{x^{10}}{45360} + \cdots.$$

Recognizing the pattern of coefficients is not so easy as in Example 4, but you can verify that the terms shown agree with the formulas

$$y_1(x) = 1 + \sum_{k=1}^{\infty} \frac{1 \cdot 4 \cdots (3k-2)}{(3k)!} x^{3k} \quad \text{and} \quad y_2(x) = x + \sum_{k=1}^{\infty} \frac{2 \cdot 5 \cdots (3k-1)}{(3k+1)!} x^{3k+1}.$$

The special combinations

$$\mathrm{Ai}(x) = \frac{y_1(x)}{3^{2/3}\Gamma\left(\frac{2}{3}\right)} - \frac{y_2(x)}{3^{1/3}\Gamma\left(\frac{1}{3}\right)} \quad \text{and} \quad \mathrm{Bi}(x) = \frac{y_1(x)}{3^{1/6}\Gamma\left(\frac{2}{3}\right)} + \frac{y_2(x)}{3^{-1/6}\Gamma\left(\frac{1}{3}\right)}$$

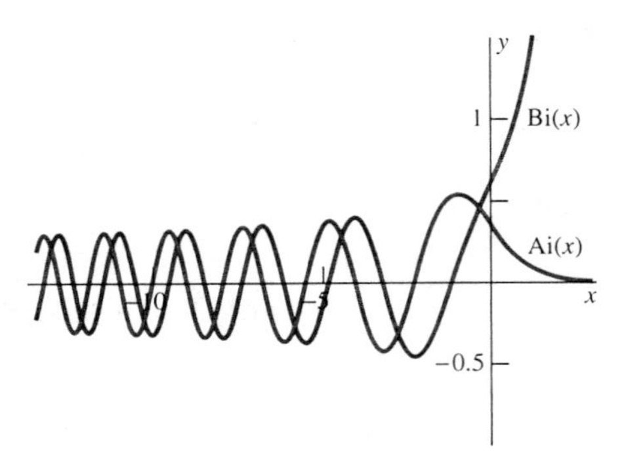

FIGURE 11.10.3 The graphs $y = \text{Ai}(x)$ and $y = \text{Bi}(x)$ of the Airy functions.

—with $\Gamma(x)$ denoting the gamma function defined in Section 8.8—are the standard *Airy functions*, which appear in mathematical tables and computer algebra systems. Their graphs in Fig. 11.10.3 exhibit oscillatory behavior for $x < 0$, but $\text{Ai}(x)$ decreases exponentially and $\text{Bi}(x)$ increases exponentially as $x \to \infty$. ◆

11.10 TRUE/FALSE STUDY GUIDE

11.10 CONCEPTS: QUESTIONS AND DISCUSSION

1. Suppose that the exponential function $E(x) = e^x$ is *defined* as the solution of the initial value problem $y' = y$, $y(0) = 1$. Beginning with this definition, what properties of the function $E(x)$ can be established?
2. Suppose that the hyperbolic functions $\text{Ch}(x) = \cosh x$ and $\text{Sh}(x) = \sinh x$ are *defined* as the solutions of the differential equation $y'' = y$ that satisfy the initial conditions $y(0) = 1$, $y'(0) = 0$ and $y(0) = 0$, $y'(0) = 1$, respectively. Beginning with these definitions, what properties of the functions $\text{Ch}(x)$ and $\text{Sh}(x)$ can be established? Can you discover a connection between these functions and the function $E(x)$ of Question 1?

11.10 PROBLEMS

In Problems 1 through 10, find a power series solution of the given differential equation. Determine the radius of convergence of the resulting series, and use your knowledge of familiar Maclaurin series and the binomial series to identify the series solution in terms of familiar elementary functions. (Of course, no one can prevent you from checking your work by also solving the equations by the methods of Chapter 9!)

1. $y' = y$

2. $y' = 4y$

3. $2y' + 3y = 0$

4. $y' + 2xy = 0$

5. $y' = x^2 y$

6. $(x - 2)y' + y = 0$

7. $(2x - 1)y' + 2y = 0$

8. $2(x + 1)y' = y$

9. $(x - 1)y' + 2y = 0$

10. $2(x - 1)y' = 3y$

In Problems 11 through 14, use the method of Example 4 to find two linearly independent power series solutions of the given differential equation. Determine the radius of convergence of each series, and identify the general solution in terms of familiar elementary functions.

11. $y'' = y$

12. $y'' = 4y$

13. $y'' + 9y = 0$

14. $y'' + y = x$

Show (as in Example 3) that the power series method fails to yield a power series solution of the form $y = \sum c_n x^n$ for the differential equations in Problems 15 through 18.

15. $xy' + y = 0$

16. $2xy' = y$

17. $x^2 y' + y = 0$

18. $x^3 y' = 2y$

In Problems 19 through 22, first derive a recurrence relation giving c_n for $n \geqq 2$ in terms of c_0 or c_1 (or both). Then apply the given initial conditions to find the values of c_0 and c_1. Next determine c_n (in terms of n, as in the text) and, finally, identify the particular solution in terms of familiar elementary functions.

19. $y'' + 4y = 0$; $y(0) = 0$, $y'(0) = 3$

20. $y'' - 4y = 0$; $y(0) = 2$, $y'(0) = 0$

21. $y'' - 2y' + y = 0$; $y(0) = 0$, $y'(0) = 1$

22. $y'' + y' - 2y = 0$; $y(0) = 1$, $y'(0) = -2$

23. Show that the equation

$$x^2 y'' + x^2 y' + y = 0$$

has no power series solution of the form $y = \sum c_n x^n$.

24. Use the power series method to discover the solution

$$J_0(x) = \sum_{k=0}^{\infty} \frac{(-1)^k x^{2k}}{2^{2k}(k!)^2} = 1 - \frac{x^2}{4} + \frac{x^4}{64} - \frac{x^6}{2304} + \cdots$$

of the Bessel equation $xy'' + y' + xy = 0$. Explain why the series method does not yield an independent second solution.

25. (a) Show that the solution of the initial value problem

$$y' = 1 + y^2, \qquad y(0) = 0$$

is $y(x) = \tan x$. (b) Because $y(x) = \tan x$ is an odd function with $y'(0) = 1$, its Taylor series is of the form

$$y = x + c_3x^3 + c_5x^5 + c_7x^7 + \cdots.$$

Substitute this series in $y' = 1 + y^2$ and equate like powers of x to derive the following relations:

$$3c_3 = 1, \qquad 5c_5 = 2c_3,$$
$$7c_7 = 2c_5 + (c_3)^2, \qquad 9c_9 = 2c_7 + 2c_3c_5,$$
$$11c_{11} = 2c_9 + 2c_3c_7 + (c_5)^2.$$

(c) Conclude that

$$\tan x = x + \frac{1}{3}x^3 + \frac{2}{15}x^5 + \frac{17}{315}x^7 + \frac{62}{2835}x^9 + \frac{1382}{155925}x^{11} + \cdots.$$

CHAPTER 11 REVIEW: DEFINITIONS, CONCEPTS, RESULTS

Use the following list as a guide to concepts that you may need to review.

1. Definition of the limit of a sequence
2. The limit laws for sequences
3. The bounded monotonic sequence property
4. Definition of the sum of an infinite series
5. Formula for the sum of a geometric series
6. The nth-term test for divergence
7. Divergence of the harmonic series
8. The nth-degree Taylor polynomial of the function f at the point $x = a$
9. Taylor's formula with remainder
10. The Taylor series of the elementary transcendental functions
11. The integral test
12. Convergence of p-series
13. The comparison and limit comparison tests
14. The alternating series test
15. Absolute convergence: definition *and* the fact that it implies convergence
16. The ratio test
17. The root test
18. Power series; radius of convergence and interval of convergence
19. The binomial series
20. Termwise differentiation and integration of power series
21. The use of power series to approximate values of functions and integrals
22. The sum and product of two power series
23. The use of power series to evaluate indeterminate forms
24. Power series solution of differential equations

CHAPTER 11 MISCELLANEOUS PROBLEMS

In Problems 1 through 15, determine whether or not the sequence $\{a_n\}$ converges, and find its limit if it does converge.

1. $a_n = \dfrac{n^2+1}{n^2+4}$

2. $a_n = \dfrac{8n-7}{7n-8}$

3. $a_n = 10 - (0.99)^n$

4. $a_n = n \sin \pi n$

5. $a_n = \dfrac{1 + (-1)^n\sqrt{n}}{n+1}$

6. $a_n = \sqrt{\dfrac{1 + (-0.5)^n}{n+1}}$

7. $a_n = \dfrac{\sin 2n}{n}$

8. $a_n = 2^{-(\ln n)/n}$

9. $a_n = (-1)^{\sin(n\pi/2)}$

10. $a_n = \dfrac{(\ln n)^3}{n^2}$

11. $a_n = \dfrac{1}{n}\sin\dfrac{1}{n}$

12. $a_n = \dfrac{n - e^n}{n + e^n}$

13. $a_n = \dfrac{\sinh n}{n}$

14. $a_n = \left(1 + \dfrac{2}{n}\right)^{2n}$

15. $a_n = (2n^2 + 1)^{1/n}$

Determine whether each infinite series in Problems 16 through 30 converges or diverges.

16. $\displaystyle\sum_{n=1}^{\infty} \frac{(n^2)!}{n^n}$

17. $\displaystyle\sum_{n=1}^{\infty} \frac{(-1)^{n+1}\ln n}{n}$

18. $\displaystyle\sum_{n=0}^{\infty} \frac{3^n}{2^n + 4^n}$

19. $\displaystyle\sum_{n=0}^{\infty} \frac{n!}{e^{n^2}}$

20. $\displaystyle\sum_{n=1}^{\infty} \frac{1}{n^{3/2}}\sin\frac{1}{n}$

21. $\displaystyle\sum_{n=0}^{\infty} \frac{(-2)^n}{3^n + 1}$

22. $\sum_{n=1}^{\infty} 2^{-(2/n^2)}$

23. $\sum_{n=2}^{\infty} \frac{(-1)^n n}{(\ln n)^3}$

24. $\sum_{n=1}^{\infty} \frac{(-1)^n}{10^{1/n}}$

25. $\sum_{n=1}^{\infty} \frac{\sqrt{n}+\sqrt[3]{n}}{n^2+n^3}$

26. $\sum_{n=1}^{\infty} \frac{(-1)^{n+1}}{n^{[1+(1/n)]}}$

27. $\sum_{n=1}^{\infty} \frac{(-1)^{n+1}\arctan n}{\sqrt{n}}$

28. $\sum_{n=1}^{\infty} n \sin \frac{1}{n}$

29. $\sum_{n=3}^{\infty} \frac{1}{n(\ln n)(\ln \ln n)}$

30. $\sum_{n=3}^{\infty} \frac{1}{n(\ln n)(\ln \ln n)^2}$

Find the interval of convergence of the power series in Problems 31 through 40.

31. $\sum_{n=0}^{\infty} \frac{2^n x^n}{n!}$

32. $\sum_{n=0}^{\infty} \frac{(3x)^n}{2^{n+1}}$

33. $\sum_{n=1}^{\infty} \frac{(x-1)^n}{n \cdot 3^n}$

34. $\sum_{n=0}^{\infty} \frac{(2x-3)^n}{4^n}$

35. $\sum_{n=1}^{\infty} \frac{(-1)^n x^n}{4n^2-1}$

36. $\sum_{n=0}^{\infty} \frac{(2x-1)^n}{n^2+1}$

37. $\sum_{n=0}^{\infty} \frac{n! x^{2n}}{10^n}$

38. $\sum_{n=2}^{\infty} \frac{x^n}{\ln n}$

39. $\sum_{n=0}^{\infty} \frac{1+(-1)^n}{2(n!)} x^n$

40. $\sum_{n=1}^{\infty} \left(1+\frac{1}{n}\right)^n (x-1)^n$

Find the set of all values of x for which the series in Problems 41 through 43 converge.

41. $\sum_{n=1}^{\infty} (x-n)^n$

42. $\sum_{n=1}^{\infty} (\ln x)^n$

43. $\sum_{n=0}^{\infty} \frac{e^{nx}}{n!}$

44. Find the rational number that has repeated decimal expansion 2.7 1828 1828 1828 . . . .

45. Give an example of two convergent numerical series $\sum a_n$ and $\sum b_n$ such that the series $\sum a_n b_n$ diverges.

46. Prove that if $\sum a_n$ is a convergent positive-term series, then $\sum a_n^2$ converges.

47. Let the sequence $\{a_n\}$ be defined recursively as follows:

$$a_1 = 1; \qquad a_{n+1} = 1 + \frac{1}{1+a_n} \quad \text{if } n \geqq 1.$$

The limit of the sequence $\{a_n\}$ is the value of the *continued fraction*

$$1 + \cfrac{1}{2 + \cfrac{1}{2 + \cfrac{1}{2 + \cfrac{1}{2 + \cdots}}}}.$$

Assuming that $A = \lim_{n\to\infty} a_n$ exists, prove that $A = \sqrt{2}$.

48. Let $\{F_n\}_1^\infty$ be the Fibonacci sequence of Example 2 in Section 11.2. (a) Prove that $0 < F_n \leqq 2^n$ for all $n \geqq 1$, and hence conclude that the power series

$$F(x) = \sum_{n=1}^{\infty} F_n x^n$$

converges if $|x| < \frac{1}{2}$. (b) Show that $(1 - x - x^2)F(x) = x$, so

$$F(x) = \frac{x}{1-x-x^2}.$$

49. We say that the *infinite product* indicated by

$$\prod_{n=1}^{\infty} (1+a_n) = (1+a_1)(1+a_2)(1+a_3)\cdots$$

converges provided that the infinite series

$$S = \sum_{n=1}^{\infty} \ln(1+a_n)$$

converges, in which case the value of the infinite product is, by definition, e^S. Use the integral test to prove that

$$\prod_{n=1}^{\infty} \left(1+\frac{1}{n}\right)$$

diverges.

50. Prove that the infinite product (see Problem 49)

$$\prod_{n=1}^{\infty} \left(1+\frac{1}{n^2}\right)$$

converges, and use the integral test remainder estimate to approximate its value. The actual value of this infinite product is known to be

$$\frac{\sinh \pi}{\pi} \approx 3.67607\,79103\,74977\,72069\,56975.$$

In Problems 51 through 55, use infinite series to approximate the indicated number accurate to three decimal places.

51. $\sqrt[5]{1.5}$

52. $\ln(1.2)$

53. $\int_0^{0.5} e^{-x^2}\,dx$

54. $\int_0^{0.5} \sqrt[3]{1+x^4}\,dx$

55. $\int_0^1 \frac{1-e^{-x}}{x}\,dx$

56. Substitute the Maclaurin series for $\sin x$ into that for e^x to obtain

$$e^{\sin x} = 1 + x + \tfrac{1}{2}x^2 - \tfrac{1}{8}x^4 + \cdots.$$

57. Substitute the Maclaurin series for the cosine and then integrate termwise to derive the formula

$$\int_0^\infty e^{-t^2} \cos 2xt\,dt = \frac{\sqrt{\pi}}{2} e^{-x^2}.$$

Use the reduction formula

$$\int_0^\infty t^{2n} e^{-t^2}\,dt = \frac{2n-1}{2} \int_0^\infty t^{2n-2} e^{-t^2}\,dt$$

that follows from the one derived in Problem 50 of Section 8.3. The validity of this improper termwise integration is subject to verification.

58. Prove that

$$\tanh^{-1} x = \int_0^x \frac{1}{1-t^2}\,dt = \sum_{n=0}^{\infty} \frac{x^{2n+1}}{2n+1}$$

if $|x| < 1$.

59. Prove that

$$\sinh^{-1} x = \int_0^x \frac{1}{\sqrt{1+t^2}}\,dt$$

$$= \sum_{n=0}^{\infty} (-1)^n \frac{1\cdot 3\cdot 5\cdots(2n-1)}{2\cdot 4\cdot 6\cdots(2n)} \cdot \frac{x^{2n+1}}{2n+1}$$

if $|x| < 1$.

60. Suppose that $\tan y = \sum a_n y^n$. Determine a_0, a_1, a_2, and a_3 by substituting the inverse tangent series [Eq. (27) of Section 11.4] into the equation

$$x = \tan(\tan^{-1} x) = \sum_{n=0}^{\infty} a_n (\tan^{-1} x)^n.$$

61. According to *Stirling's series*, the value of $n!$ for large n is given to a close approximation by

$$n! \approx \sqrt{2\pi n}\left(\frac{n}{e}\right)^n e^{\mu(n)},$$

where

$$\mu(n) = \frac{1}{12n} - \frac{1}{360n^3} + \frac{1}{1260n^5}.$$

Substitute $\mu(n)$ into Maclaurin's series for e^x to show that

$$e^{\mu(n)} = 1 + \frac{1}{12n} + \frac{1}{288n^2} - \frac{139}{51840n^3} + \cdots.$$

Can you show that the next term in the last series is $-571/(2{,}488{,}320n^4)$?

62. Define

$$T(n) = \int_0^{\pi/4} \tan^n x\,dx$$

for $n \geqq 0$. (a) Show by "reduction" of the integral that

$$T(n+2) = \frac{1}{n+1} - T(n)$$

for $n \geqq 0$. (b) Conclude that $T(n) \to 0$ as $n \to \infty$. (c) Show that $T(0) = \pi/4$ and that $T(1) = \frac{1}{2}\ln 2$. (d) Prove by induction on n that

$$T(2n) = (-1)^{n+1}\left(1 - \frac{1}{3} + \frac{1}{5} - \cdots \pm \frac{1}{2n-1} - \frac{\pi}{4}\right).$$

(e) Conclude from parts (b) and (d) that

$$1 - \frac{1}{3} + \frac{1}{5} - \frac{1}{7} + \cdots = \frac{\pi}{4}.$$

(f) Prove by induction on n that

$$T(2n+1) = \frac{1}{2}(-1)^n\left(1 - \frac{1}{2} + \frac{1}{3} - \cdots \pm \frac{1}{n} - \ln 2\right).$$

(g) Conclude from parts (b) and (f) that

$$1 - \tfrac{1}{2} + \tfrac{1}{3} - \tfrac{1}{4} + \cdots = \ln 2.$$

63. Prove as follows that the number e is irrational. First suppose to the contrary that $e = p/q$, where p and q are positive integers. Note that $q > 1$. Write

$$\frac{p}{q} = e = 1 + \frac{1}{1!} + \frac{1}{2!} + \frac{1}{3!} + \cdots + \frac{1}{q!} + R_q,$$

where $0 < R_q < 3/(q+1)!$. (Why?) Then show that multiplying of both sides of this equation by $q!$ would lead to the contradiction that one side of the result is an integer but the other side is not.

64. Evaluate the infinite product (see Problem 49)

$$\prod_{n=2}^{\infty} \frac{n^2}{n^2-1}$$

by finding an explicit formula for

$$\prod_{n=2}^{k} \frac{n^2}{n^2-1} \quad (k \geqq 2)$$

and then taking the limit as $k \to \infty$.

65. Find a continued fraction representation (see Problem 47)

$$a_0 + \cfrac{1}{a_1 + \cfrac{1}{a_2 + \cfrac{1}{a_3 + \cfrac{1}{a_4 + \cdots}}}}$$

of $\sqrt{5}$.

66. Evaluate

$$1 + \frac{1}{2} - \frac{2}{3} + \frac{1}{4} + \frac{1}{5} - \frac{2}{6} + \frac{1}{7} + \frac{1}{8} - \frac{2}{9} + \frac{1}{10} + \cdots.$$

VECTORS, CURVES, AND SURFACES IN SPACE

12

Johannes Kepler (1571–1630)

Ancient Greek mathematicians and astronomers developed an elaborate mathematical model to account for the complicated motions of the sun, moon, and six planets then known as viewed from the earth. A combination of uniform circular motions was used to describe the motion of each body around the earth—if the earth is arbitrarily placed at the origin of coordinates, then each body *does* orbit the earth.

In this system, it was typical for a planet P to travel uniformly around a small circle (the *epicycle*) with center C, which in turn traveled uniformly around a circle centered at the earth (labeled E in the figure at the lower left). The radii of the circles and the angular speeds of P and C were chosen to match the observed motion of the planet as closely as possible. For greater accuracy, secondary "circles on circles" could be used. In fact, several circles were required for each body in the Greek theory of epicycles, which reached its definitive form in Ptolemy's *Almagest* of the second century A.D.

In 1543, Copernicus altered Ptolemy's approach by placing the center of each primary circle at the sun rather than at the earth. But this change was of greater philosophical than mathematical significance. His *heliocentric system* was still overly complicated, still requiring many secondary circles, and still beset with inaccuracies in representing the motions of the heavenly bodies.

It was Johannes Kepler who finally got rid of all these circles. On the basis of a detailed analysis of planetary observations accumulated by the Danish astronomer Tycho Brahe, Kepler stated his three famous *laws of planetary motion,* which describe elliptical (rather than circular) orbits of planets around the sun (Section 12.6). Ironically, his original goal had been to prove that the placement of Mercury, Venus, Earth, Mars, and Jupiter is determined by the five regular polyhedra as indicated in the figure at the lower right, which appeared in his *Mysterium Cosmographicum* (1596). This model of the solar system shows a cube inscribed in the sphere containing Saturn's orbit, and the sphere of Jupiter's orbit is inscribed in this cube. A tetrahedron (with four triangular faces) is inscribed in Jupiter's sphere, and in this tetrahedron is inscribed the sphere of the orbit of Mars. Continuing in this way, the spheres of the three remaining planets then known were interspersed with the remaining three regular solids—the octahedron (eight triangular faces), the dodecahedron (12 pentagonal faces), and the icosahedron (20 triangular faces). It is said that Kepler always remained prouder of his five solids than of his three laws.

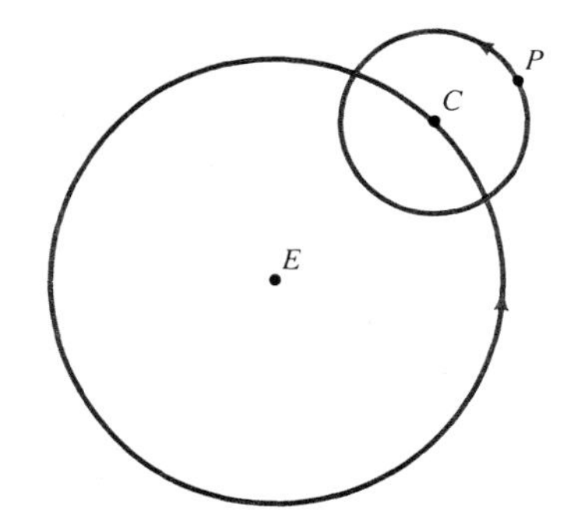

The small circle is the epicycle.

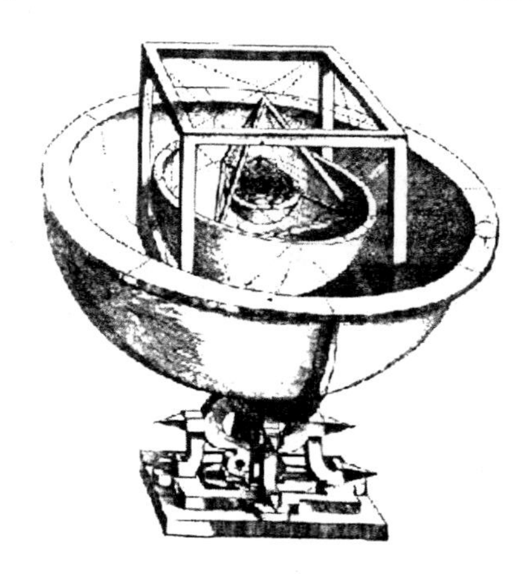

Kepler's regular polyhedron model of the solar system.

In his *Principia Mathematica* (1687), Newton showed that Kepler's laws follow from the basic principles of mechanics ($F = ma$ and so on) and the inverse-square law of gravitational attraction. His success in using mathematics to explain natural phenomena ("I now demonstrate the frame of the System of the World") inspired confidence that the universe could be understood and perhaps even mastered. This new confidence permanently altered humanity's perception of itself and of its place in the scheme of things. Newton employed a powerful but now antiquated form of geometrical calculus in the *Principia*. In Section 12.6 we apply the modern calculus of vector-valued functions to outline the relation between Newton's laws and Kepler's laws.

12.1 VECTORS IN THE PLANE

A physical quantity such as length, temperature, or mass can be specified in terms of a single real number, its *magnitude*. Such a quantity is called a **scalar.** Other physical quantities, such as force and velocity, possess both magnitude and *direction;* these quantities are called **vector quantities,** or simply **vectors.**

For example, to specify the velocity of a moving point in the coordinate plane, we must give both the rate at which it moves (its speed) and the direction of that motion. The *velocity vector* of the moving point incorporates both pieces of information—direction and speed. It is convenient to represent this velocity vector by an arrow, with its initial point located at the current position of the moving point on its trajectory (Fig. 12.1.1).

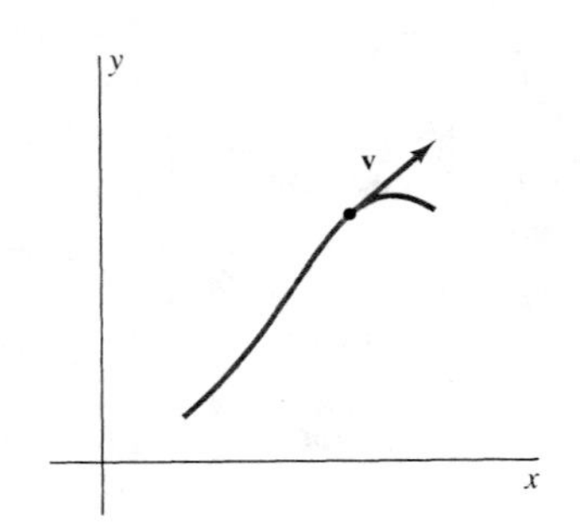

FIGURE 12.1.1 A velocity vector may be represented by an arrow.

Although the arrow, a directed line segment, carries the desired information—both magnitude (the segment's length) and direction—it is a pictorial representation rather than a quantitative object. The following formal definition of a vector captures the essence of magnitude in combination with direction.

DEFINITION Vector

A **vector** $\mathbf{v}$ in the Cartesian plane is an ordered pair of real numbers that has the form $\langle a, b\rangle$. We write $\mathbf{v} = \langle a, b\rangle$ and call a and b the **components** of the vector $\mathbf{v}$.

The directed line segment $\overrightarrow{OP}$ from the origin O to the point $P(a, b)$ is one geometric representation of the vector $\mathbf{v}$. (See Fig. 12.1.2.) For this reason, the vector $\mathbf{v} = \langle a, b\rangle$ is called the **position vector** of the point $P(a, b)$. In fact, the relationship between $\mathbf{v} = \langle a, b\rangle$ and $P(a, b)$ is so close that, in certain contexts, it is convenient to confuse the two deliberately—to regard $\mathbf{v}$ and P as the same mathematical object.

The directed line segment from the point $Q(a_1, b_1)$ to the point $R(a_2, b_2)$ has the same direction and magnitude as the directed line segment from the origin $O(0, 0)$ to the point $P(a, b)$ with $a = a_2 - a_1$ and $b = b_2 - b_1$ (Fig. 12.1.2), and consequently they represent the same vector $\mathbf{v} = \overrightarrow{OP} = \overrightarrow{QR}$. This observation makes it easy to find the components of the vector with arbitrary initial point Q and arbitrary terminal point R.

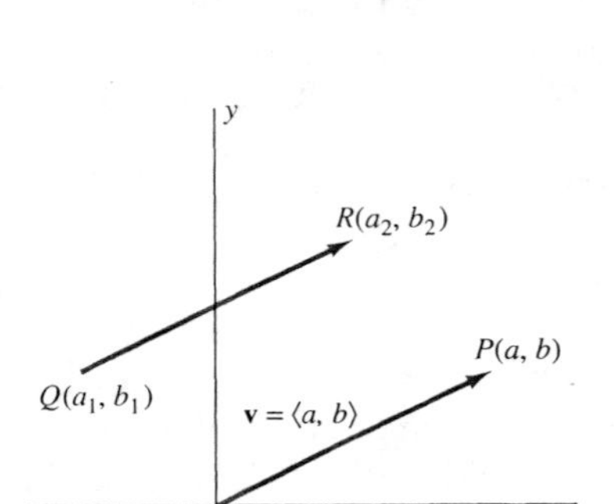

FIGURE 12.1.2 The position vector $\mathbf{v}$ of the point P and another representation $\overrightarrow{QR}$ of $\mathbf{v}$.

REMARK When discussing vectors we often use the term *scalar* to refer to an ordinary numerical quantity, one that is *not* a vector. In printed work we use **bold** type to distinguish the names of vectors from those of other mathematical objects, such as the scalars a and b that are the components of the vector $\mathbf{v} = \langle a, b\rangle$. In handwritten work a suitable alternative is to place an arrow—or just a bar—over every symbol that denotes a vector. Thus you may write $\vec{v} = \langle a, b\rangle$ or $\overline{v} = \langle a, b\rangle$. There is no need for an arrow or a bar over a vector $\langle a, b\rangle$ already identified by angle brackets, so none should be used there.

A directed line segment has both length and direction. The **length** of the vector $\mathbf{v} = \langle a, b\rangle$ is denoted by $v = |\mathbf{v}|$ and is defined as follows:

$$v = |\mathbf{v}| = |\langle a, b\rangle| = \sqrt{a^2 + b^2}. \tag{1}$$

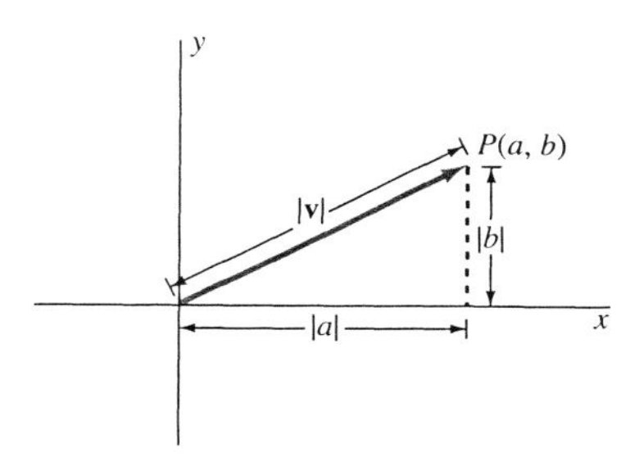

FIGURE 12.1.3 The length $v = |\mathbf{v}|$ of the vector $\mathbf{v}$.

The notation $v = |\mathbf{v}|$ is used because the length of a vector is in many ways analogous to the absolute value of a real number (Fig. 12.1.3).

EXAMPLE 1 The length of the vector $\mathbf{v} = \langle 1, -2 \rangle$ is

$$v = |\langle 1, -2 \rangle| = \sqrt{(1)^2 + (-2)^2} = \sqrt{5}. \quad \blacklozenge$$

The only vector with length zero is the **zero vector** with both components zero, denoted by $\mathbf{0} = \langle 0, 0 \rangle$. The zero vector is unique in that it has no specific direction. Every nonzero vector has a specified direction; the vector represented by the arrow $\overrightarrow{OP}$ from the origin O to another point P in the plane has direction specified (for instance) by the counterclockwise angle from the positive x-axis to $\overrightarrow{OP}$.

What is important about the vector $\mathbf{v} = \langle a, b \rangle$ represented by $\overrightarrow{OP}$ often is not *where* it is, but how long it is and which way it points. If the directed line segment $\overrightarrow{QR}$ with endpoints $Q(a_1, b_1)$ and $R(a_2, b_2)$ has the same length and direction as $\overrightarrow{OP}$, then we say that $\overrightarrow{QR}$ **represents** (or is a **representation** of) the vector $\mathbf{v}$ (Fig. 12.1.2). Thus a single vector has many representatives (Fig. 12.1.4).

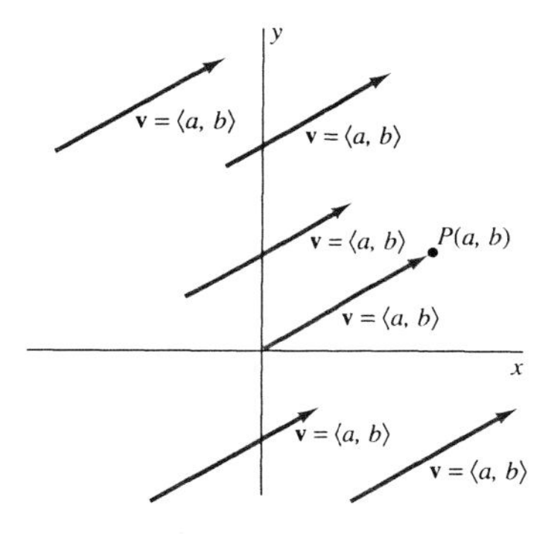

FIGURE 12.1.4 All these arrows represent the same vector $\mathbf{v} = \langle a, b \rangle$.

Algebraic Operations with Vectors

The operations of addition and multiplication of real numbers have analogues for vectors. We shall define each of these operations of *vector algebra* in terms of components of vectors and then give a geometric interpretation in terms of arrows.

DEFINITION Equality of Vectors

The two vectors $\mathbf{u} = \langle u_1, u_2 \rangle$ and $\mathbf{v} = \langle v_1, v_2 \rangle$ are **equal** provided that $u_1 = v_1$ and $u_2 = v_2$.

In other words, two vectors are equal if and only if *corresponding components* are the same. Moreover, two directed line segments $\overrightarrow{PQ}$ and $\overrightarrow{RS}$ represent the same vector provided that they have the same length and direction. This will be the case provided that the segments $\overrightarrow{PQ}$ and $\overrightarrow{RS}$ are opposite sides of a parallelogram (Fig. 12.1.5).

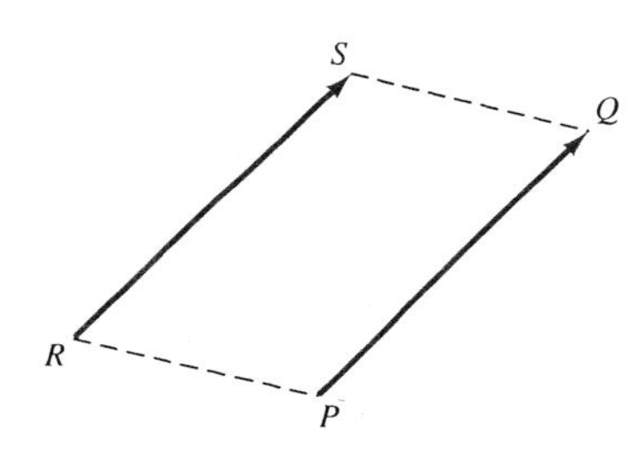

FIGURE 12.1.5 Parallel directed segments representing equal vectors.

DEFINITION Addition of Vectors

The **sum** $\mathbf{u} + \mathbf{v}$ of the two vectors $\mathbf{u} = \langle u_1, u_2 \rangle$ and $\mathbf{v} = \langle v_1, v_2 \rangle$ is the vector

$$\mathbf{u} + \mathbf{v} = \langle u_1 + v_1, u_2 + v_2 \rangle. \quad \textbf{(2)}$$

Thus we add vectors by adding corresponding components—that is, by *componentwise addition.* The geometric interpretation of vector addition is the **triangle law of addition,** illustrated in Fig. 12.1.6, where the labeled lengths indicate why this interpretation is valid. An equivalent interpretation is the **parallelogram law of addition,** illustrated in Fig. 12.1.7.

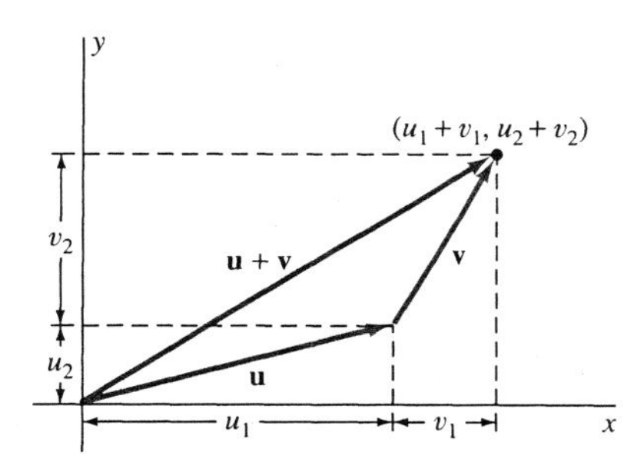

FIGURE 12.1.6 The triangle law is a geometric interpretation of vector addition.

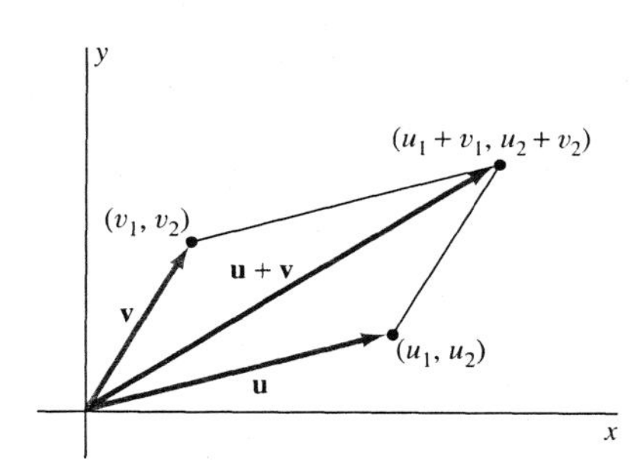

FIGURE 12.1.7 The parallelogram law for vector addition.

EXAMPLE 2 The sum of the vectors $\mathbf{u} = \langle 4, 3 \rangle$ and $\mathbf{v} = \langle -5, 2 \rangle$ is the vector

$$\mathbf{u} + \mathbf{v} = \langle 4, 3 \rangle + \langle -5, 2 \rangle = \langle 4 + (-5), 3 + 2 \rangle = \langle -1, 5 \rangle. \qquad \blacklozenge$$

It is natural to write $2\mathbf{u} = \mathbf{u} + \mathbf{u}$. But if $\mathbf{u} = \langle u_1, u_2 \rangle$, then

$$2\mathbf{u} = \mathbf{u} + \mathbf{u} = \langle u_1, u_2 \rangle + \langle u_1, u_2 \rangle = \langle 2u_1, 2u_2 \rangle.$$

This suggests that multiplication of a vector by a scalar (real number) also is defined in a componentwise manner.

DEFINITION Multiplication of a Vector by a Scalar

If $\mathbf{u} = \langle u_1, u_2 \rangle$ and c is a real number, then the **scalar multiple** $c\mathbf{u}$ is the vector

$$c\mathbf{u} = \langle cu_1, cu_2 \rangle. \tag{3}$$

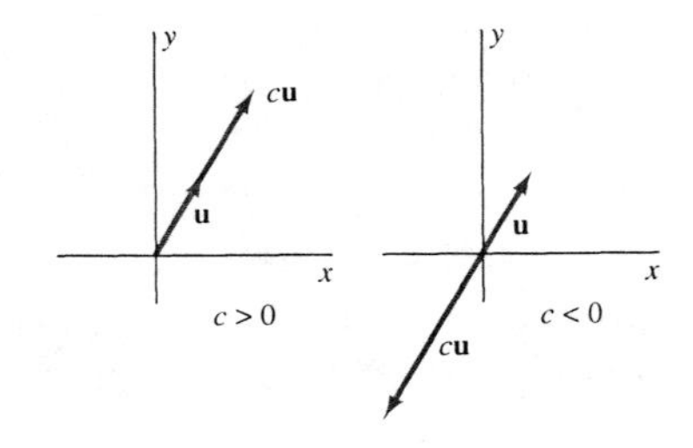

FIGURE 12.1.8 The vector $c\mathbf{u}$ may have the same direction as $\mathbf{u}$ or the opposite direction, depending on the sign of c.

Note that

$$|c\mathbf{u}| = \sqrt{(cu_1)^2 + (cu_2)^2} = |c|\sqrt{(u_1)^2 + (u_2)^2} = |c| \cdot |\mathbf{u}|.$$

Thus the length of $|c\mathbf{u}|$ is $|c|$ times the length of $\mathbf{u}$. The **negative** of the vector $\mathbf{u}$ is the vector

$$-\mathbf{u} = (-1)\mathbf{u} = \langle -u_1, -u_2 \rangle,$$

with the same length as $\mathbf{u}$ but the opposite direction. We say that the two nonzero vectors $\mathbf{u}$ and $\mathbf{v}$ have

- The **same direction** if $\mathbf{u} = c\mathbf{v}$ for some $c > 0$;
- **Opposite directions** if $\mathbf{u} = c\mathbf{v}$ for some $c < 0$.

The geometric interpretation of scalar multiplication is that $c\mathbf{u}$ is the vector with length $|c| \cdot |\mathbf{u}|$, with the same direction as $\mathbf{u}$ if $c > 0$ but with the opposite direction if $c < 0$ (Fig. 12.1.8).

The **difference** $\mathbf{u} - \mathbf{v}$ of the vectors $\mathbf{u} = \langle u_1, u_2 \rangle$ and $\mathbf{v} = \langle v_1, v_2 \rangle$ is defined to be

$$\mathbf{u} - \mathbf{v} = \mathbf{u} + (-\mathbf{v}) = \langle u_1 - v_1, u_2 - v_2 \rangle. \tag{4}$$

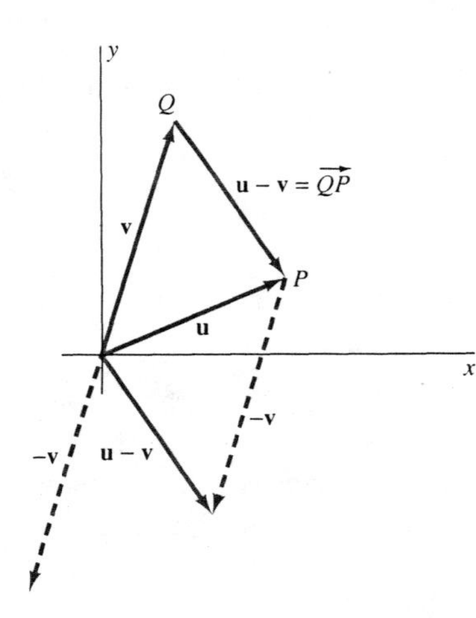

FIGURE 12.1.9 Geometric interpretation of the difference $\mathbf{u} - \mathbf{v}$.

If we think of $\langle u_1, u_2 \rangle$ and $\langle v_1, v_2 \rangle$ as position vectors of the points P and Q, respectively, then $\mathbf{u} - \mathbf{v}$ may be represented by the arrow $\overrightarrow{QP}$ from Q to P. We may therefore write

$$\mathbf{u} - \mathbf{v} = \overrightarrow{OP} - \overrightarrow{OQ} = \overrightarrow{QP},$$

as illustrated in Fig. 12.1.9.

EXAMPLE 3 Suppose that $\mathbf{u} = \langle 4, -3 \rangle$ and $\mathbf{v} = \langle -2, 3 \rangle$. Find $|\mathbf{u}|$ and the vectors $\mathbf{u} + \mathbf{v}$, $\mathbf{u} - \mathbf{v}$, $3\mathbf{u}$, $-2\mathbf{v}$, and $2\mathbf{u} + 4\mathbf{v}$.

Solution

$$|\mathbf{u}| = \sqrt{4^2 + (-3)^2} = \sqrt{25} = 5.$$

$$\mathbf{u} + \mathbf{v} = \langle 4 + (-2), -3 + 3 \rangle = \langle 2, 0 \rangle.$$

$$\mathbf{u} - \mathbf{v} = \langle 4 - (-2), -3 - 3 \rangle = \langle 6, -6 \rangle.$$

$$3\mathbf{u} = \langle 3 \cdot 4, 3 \cdot (-3) \rangle = \langle 12, -9 \rangle.$$

$$-2\mathbf{v} = \langle -2 \cdot (-2), -2 \cdot 3 \rangle = \langle 4, -6 \rangle.$$

$$2\mathbf{u} + 4\mathbf{v} = \langle 2 \cdot 4 + 4 \cdot (-2), 2 \cdot (-3) + 4 \cdot 3 \rangle = \langle 0, 6 \rangle. \qquad \blacklozenge$$

The familiar algebraic properties of real numbers carry over to the following analogous properties of vector addition and scalar multiplication. Let $\mathbf{a}$, $\mathbf{b}$, and $\mathbf{c}$ be vectors and r and s real numbers. Then

$$
\begin{aligned}
&1.\quad \mathbf{a}+\mathbf{b}=\mathbf{b}+\mathbf{a}.\\
&2.\quad \mathbf{a}+(\mathbf{b}+\mathbf{c})=(\mathbf{a}+\mathbf{b})+\mathbf{c}.\\
&3.\quad r(\mathbf{a}+\mathbf{b})=r\mathbf{a}+r\mathbf{b}.\\
&4.\quad (r+s)\mathbf{a}=r\mathbf{a}+s\mathbf{a}.\\
&5.\quad (rs)\mathbf{a}=r(s\mathbf{a})=s(r\mathbf{a}).
\end{aligned}
\tag{5}
$$

You can easily verify these identities by working with components. For example, if $\mathbf{a} = \langle a_1, a_2\rangle$ and $\mathbf{b} = \langle b_1, b_2\rangle$, then

$$
\begin{aligned}
r(\mathbf{a}+\mathbf{b}) &= r\langle a_1+b_1,\ a_2+b_2\rangle = \langle r(a_1+b_1),\ r(a_2+b_2)\rangle\\
&= \langle ra_1+rb_1,\ ra_2+rb_2\rangle = \langle ra_1,\ ra_2\rangle + \langle rb_1,\ rb_2\rangle = r\mathbf{a}+r\mathbf{b}.
\end{aligned}
$$

The proofs of the other four identities in (5) are left as exercises.

The Unit Vectors i and j

A **unit** vector is a vector of length 1. If $\mathbf{a} = \langle a_1, a_2\rangle \neq \mathbf{0}$, then

$$\mathbf{u} = \frac{\mathbf{a}}{|\mathbf{a}|} \tag{6}$$

is the unit vector with the same direction as **a**, because

$$|\mathbf{u}| = \sqrt{\left(\frac{a_1}{|\mathbf{a}|}\right)^2 + \left(\frac{a_2}{|\mathbf{a}|}\right)^2} = \frac{1}{|\mathbf{a}|}\sqrt{a_1^2+a_2^2} = 1.$$

For example, if $\mathbf{a} = \langle 3, -4\rangle$, then $|\mathbf{a}| = 5$. Thus $\langle \frac{3}{5}, -\frac{4}{5}\rangle$ is a unit vector that has the same direction as **a**.

Two particular unit vectors play a special role, the vectors

$$\mathbf{i} = \langle 1, 0\rangle \quad\text{and}\quad \mathbf{j} = \langle 0, 1\rangle.$$

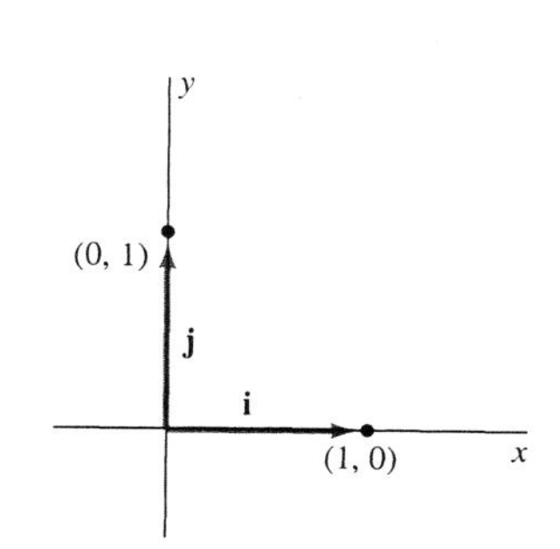

FIGURE 12.1.10 The vectors **i** and **j**.

The first points in the positive x-direction; the second points in the positive y-direction (Fig. 12.1.10). Together they provide a useful alternative notation for vectors. If $\mathbf{a} = \langle a_1, a_2\rangle$, then

$$\mathbf{a} = \langle a_1, 0\rangle + \langle 0, a_2\rangle = a_1\langle 1, 0\rangle + a_2\langle 0, 1\rangle = a_1\mathbf{i} + a_2\mathbf{j}. \tag{7}$$

Thus every vector in the plane is a **linear combination** of **i** and **j**. The usefulness of this notation is based on the fact that such linear combinations of **i** and **j** may be manipulated as if they were ordinary sums. For example, if

$$\mathbf{a} = a_1\mathbf{i} + a_2\mathbf{j} \quad\text{and}\quad \mathbf{b} = b_1\mathbf{i} + b_2\mathbf{j},$$

then

$$\mathbf{a}+\mathbf{b} = (a_1\mathbf{i}+a_2\mathbf{j}) + (b_1\mathbf{i}+b_2\mathbf{j}) = (a_1+b_1)\mathbf{i} + (a_2+b_2)\mathbf{j}.$$

Also,

$$c\mathbf{a} = c(a_1\mathbf{i}+a_2\mathbf{j}) = (ca_1)\mathbf{i} + (ca_2)\mathbf{j}.$$

EXAMPLE 4 Suppose that $\mathbf{a} = 2\mathbf{i} - 3\mathbf{j}$ and $\mathbf{b} = 3\mathbf{i} + 4\mathbf{j}$. Express $5\mathbf{a} - 3\mathbf{b}$ in terms of **i** and **j**.

Solution

$$
\begin{aligned}
5\mathbf{a} - 3\mathbf{b} &= 5\cdot(2\mathbf{i}-3\mathbf{j}) - 3\cdot(3\mathbf{i}+4\mathbf{j})\\
&= (10-9)\mathbf{i} + (-15-12)\mathbf{j} = \mathbf{i} - 27\mathbf{j}.
\end{aligned}
$$

◆

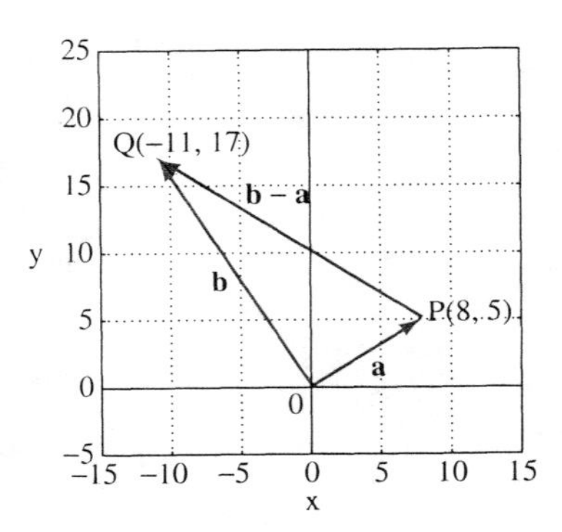

FIGURE 12.1.11 The vectors **a**, **b**, and **b** − **a** of Example 5.

EXAMPLE 5 When the vectors $\mathbf{a} = 8\mathbf{i} + 5\mathbf{j}$ and $\mathbf{b} = -11\mathbf{i} + 17\mathbf{j}$ are plotted carefully (Fig. 12.1.11), they look as though they might be perpendicular. Determine whether or not this is so.

Solution If the vectors **a** and **b** are regarded as position vectors of the points $P(8, 5)$ and $Q(-11, 17)$, then their difference $\mathbf{c} = \mathbf{b} - \mathbf{a} = -19\mathbf{i} + 12\mathbf{j}$ represents the third side $\overrightarrow{PQ}$ of the triangle OPQ (Fig. 12.1.11). According to the Pythagorean theorem, this triangle is a right triangle with hypotenuse PQ if and only if $|\mathbf{c}|^2 = |\mathbf{a}|^2 + |\mathbf{b}|^2$. But

$$|\mathbf{c}|^2 = (-19)^2 + 12^2 = 505 \quad \text{whereas} \quad |\mathbf{a}|^2 + |\mathbf{b}|^2 = [8^2 + 5^2] + [(-11)^2 + 17^2] = 499.$$

It follows that the vectors **a** and **b** are not perpendicular. ◆

Equation (7) expresses the vector $\mathbf{a} = \langle a_1, a_2 \rangle$ as the sum of a horizontal vector $a_1\mathbf{i}$ and a vertical vector $a_2\mathbf{j}$, as Fig. 12.1.12 shows. The decomposition or *resolution* of a vector into its horizontal and vertical components is an important technique in the study of vector quantities. For example, a force **F** may be decomposed into its horizontal and vertical components $F_1\mathbf{i}$ and $F_2\mathbf{j}$, respectively. The physical effect of the single force **F** is the same as the combined effect of the separate forces $F_1\mathbf{i}$ and $F_2\mathbf{j}$. (This is an instance of the empirically verifiable parallelogram law of addition of forces.) Because of this decomposition, many two-dimensional problems can be reduced to one-dimensional problems, the latter solved, and the two results combined (again by vector methods) to give the solution of the original problem.

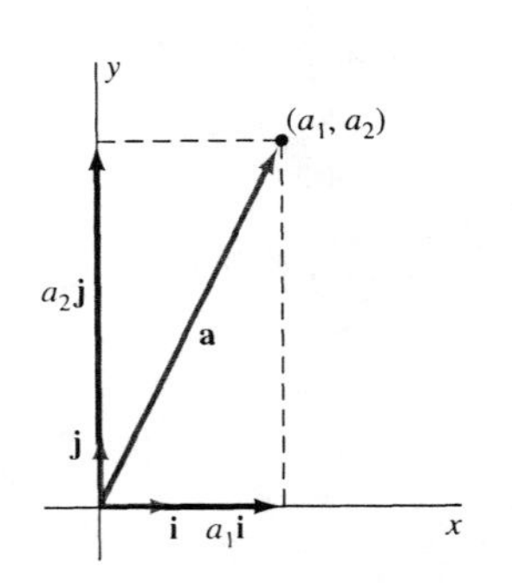

FIGURE 12.1.12 Resolution of $\mathbf{a} = \langle a_1, a_2 \rangle$ into its horizontal and vertical components.

EXAMPLE 6 A 100-lb weight is suspended from the ceiling by means of two perpendicular flexible cables of equal length (Fig. 12.1.13). Find the tension (in pounds) in each cable.

Solution Each cable is inclined at an angle of 45° from the horizontal, so it follows readily upon calculating horizontal and vertical components that the indicated tension force vectors $\mathbf{T}_1$ and $\mathbf{T}_2$ are given by

$$\mathbf{T}_1 = (T_1 \cos 45°)\mathbf{i} + (T_1 \sin 45°)\mathbf{j} \quad \text{and} \quad \mathbf{T}_2 = (-T_2 \cos 45°)\mathbf{i} + (T_2 \sin 45°)\mathbf{j},$$

where $T_1 = |\mathbf{T}_1|$ and $T_2 = |\mathbf{T}_2|$ are the tension forces we seek. The downward force of gravity acting on the weight is given by $\mathbf{F} = -100\mathbf{j}$. In order that the weight hangs motionless, the three forces must "balance," so that $\mathbf{T}_1 + \mathbf{T}_2 + \mathbf{F} = \mathbf{0}$; that is,

$$[(T_1 \cos 45°)\mathbf{i} + (T_1 \sin 45°)\mathbf{j}] + [(-T_2 \cos 45°)\mathbf{i} + (T_2 \sin 45°)\mathbf{j}] = 100\mathbf{j}.$$

When we equate the components of **i** in this equation and separately equate the components of **j**, we get the two scalar equations

$$T_1 \cos 45° - T_2 \cos 45° = 0 \quad \text{and} \quad T_1 \sin 45° + T_2 \sin 45° = 100.$$

The first of these scalar equations implies that $T_1 = T_2 = T$, and then the second yields $T = 100/(2 \sin 45°) = 50\sqrt{2} \approx 70.71$ (pounds) for the tension in each cable. ◆

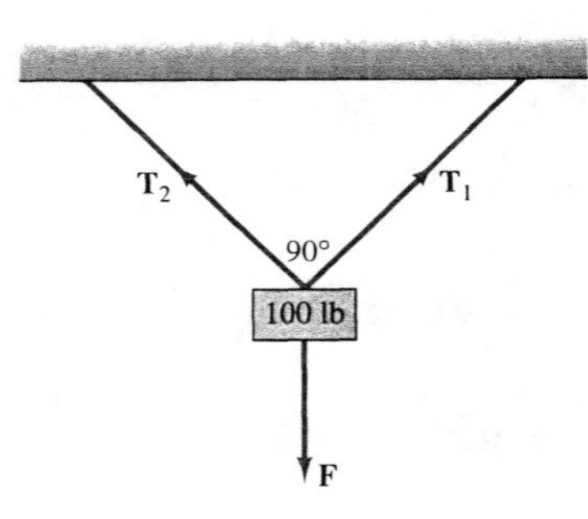

FIGURE 12.1.13 The suspended weight of Example 6.

12.1 TRUE/FALSE STUDY GUIDE

12.1 CONCEPTS: QUESTIONS AND DISCUSSION

1. Discuss the relation between a 2-dimensional vector and a point in the plane.
2. Give several examples of quantities that possess both magnitude and direction. For each, discuss whether and how such quantities might be added.
3. If a person owns stock in two companies, how might the worth of his portfolio be described by a 2-dimensional vector? If several people owning stock in these same two companies form a partnership, is the "worth vector" of the partnership equal to the sum of the worth vectors of the partners?

12.1 PROBLEMS

In Problems 1 through 4, find a vector $\mathbf{v} = \langle a, b \rangle$ that is represented by the directed line segment $\overrightarrow{RS}$. Then sketch both $\overrightarrow{RS}$ and the position vector of the point $P(a, b)$.

1. $R(1, 2)$, $S(3, 5)$

2. $R(-2, -3)$, $S(1, 4)$

3. $R(5, 10)$, $S(-5, -10)$

4. $R(-10, 20)$, $S(15, -25)$

In Problems 5 through 8, find the sum $\mathbf{w} = \mathbf{u} + \mathbf{v}$ and illustrate it geometrically.

5. $\mathbf{u} = \langle 1, -2 \rangle$, $\mathbf{v} = \langle 3, 4 \rangle$

6. $\mathbf{u} = \langle 4, 2 \rangle$, $\mathbf{v} = \langle -2, 5 \rangle$

7. $\mathbf{u} = 3\mathbf{i} + 5\mathbf{j}$, $\mathbf{v} = 2\mathbf{i} - 7\mathbf{j}$

8. $\mathbf{u} = 7\mathbf{i} + 5\mathbf{j}$, $\mathbf{v} = -10\mathbf{i}$

In Problems 9 through 16, find $|\mathbf{a}|$, $|-2\mathbf{b}|$, $|\mathbf{a} - \mathbf{b}|$, $\mathbf{a} + \mathbf{b}$, and $3\mathbf{a} - 2\mathbf{b}$.

9. $\mathbf{a} = \langle 1, -2 \rangle$, $\mathbf{b} = \langle -3, 2 \rangle$

10. $\mathbf{a} = \langle 3, 4 \rangle$, $\mathbf{b} = \langle -4, 3 \rangle$

11. $\mathbf{a} = \langle -2, -2 \rangle$, $\mathbf{b} = \langle -3, -4 \rangle$

12. $\mathbf{a} = -2\langle 4, 7 \rangle$, $\mathbf{b} = -3\langle -4, -2 \rangle$

13. $\mathbf{a} = \mathbf{i} + 3\mathbf{j}$, $\mathbf{b} = 2\mathbf{i} - 5\mathbf{j}$

14. $\mathbf{a} = 2\mathbf{i} - 5\mathbf{j}$, $\mathbf{b} = \mathbf{i} - 6\mathbf{j}$

15. $\mathbf{a} = 4\mathbf{i}$, $\mathbf{b} = -7\mathbf{j}$

16. $\mathbf{a} = -\mathbf{i} - \mathbf{j}$, $\mathbf{b} = 2\mathbf{i} + 2\mathbf{j}$

In Problems 17 through 20, find a unit vector $\mathbf{u}$ with the same direction as the given vector $\mathbf{a}$. Express $\mathbf{u}$ in terms of $\mathbf{i}$ and $\mathbf{j}$. Also find a unit vector $\mathbf{v}$ with the direction opposite that of $\mathbf{a}$.

17. $\mathbf{a} = \langle -3, -4 \rangle$

18. $\mathbf{a} = \langle 5, -12 \rangle$

19. $\mathbf{a} = 8\mathbf{i} + 15\mathbf{j}$

20. $\mathbf{a} = 7\mathbf{i} - 24\mathbf{j}$

In Problems 21 through 24, find the vector $\mathbf{a}$, expressed in terms of $\mathbf{i}$ and $\mathbf{j}$, that is represented by the arrow $\overrightarrow{PQ}$ in the plane.

21. $P = (3, 2)$, $Q = (3, -2)$

22. $P = (-3, 5)$, $Q = (-3, 6)$

23. $P = (-4, 7)$, $Q = (4, -7)$

24. $P = (1, -1)$, $Q = (-4, -1)$

In Problems 25 through 28, determine whether or not the given vectors $\mathbf{a}$ and $\mathbf{b}$ are perpendicular.

25. $\mathbf{a} = \langle 6, 0 \rangle$, $\mathbf{b} = \langle 0, -7 \rangle$

26. $\mathbf{a} = 3\mathbf{j}$, $\mathbf{b} = 3\mathbf{i} - \mathbf{j}$

27. $\mathbf{a} = 2\mathbf{i} - \mathbf{j}$, $\mathbf{b} = 4\mathbf{j} + 8\mathbf{i}$

28. $\mathbf{a} = 8\mathbf{i} + 10\mathbf{j}$, $\mathbf{b} = 15\mathbf{i} - 12\mathbf{j}$

In Problems 29 and 30, express $\mathbf{i}$ and $\mathbf{j}$ in terms of $\mathbf{a}$ and $\mathbf{b}$.

29. $\mathbf{a} = 2\mathbf{i} + 3\mathbf{j}$, $\mathbf{b} = 3\mathbf{i} + 4\mathbf{j}$

30. $\mathbf{a} = 5\mathbf{i} - 9\mathbf{j}$, $\mathbf{b} = 4\mathbf{i} - 7\mathbf{j}$

In Problems 31 and 32, write $\mathbf{c}$ in the form $r\mathbf{a} + s\mathbf{b}$ where r and s are scalars.

31. $\mathbf{a} = \mathbf{i} + \mathbf{j}$, $\mathbf{b} = \mathbf{i} - \mathbf{j}$, $\mathbf{c} = 2\mathbf{i} - 3\mathbf{j}$

32. $\mathbf{a} = 3\mathbf{i} + 2\mathbf{j}$, $\mathbf{b} = 8\mathbf{i} + 5\mathbf{j}$, $\mathbf{c} = 7\mathbf{i} + 9\mathbf{j}$

33. Find a vector that has the same direction as $5\mathbf{i} - 7\mathbf{j}$ and is (a) three times its length; (b) one-third its length.

34. Find a vector that has the opposite direction from $-3\mathbf{i} + 5\mathbf{j}$ and is (a) four times its length; (b) one-fourth its length.

35. Find a vector of length 5 with (a) the same direction as $7\mathbf{i} - 3\mathbf{j}$; (b) the direction opposite that of $8\mathbf{i} + 5\mathbf{j}$.

36. For what numbers c are the vectors $\langle c, 2 \rangle$ and $\langle c, -8 \rangle$ perpendicular?

37. For what numbers c are the vectors $2c\mathbf{i} - 4\mathbf{j}$ and $3\mathbf{i} + c\mathbf{j}$ perpendicular?

38. Given the three points $A(2, 3)$, $B(-5, 7)$, and $C(1, -5)$, verify by direct computation of the vectors and their sum that $\overrightarrow{AB} + \overrightarrow{BC} + \overrightarrow{CA} = \mathbf{0}$.

In Problems 39 through 42, give a componentwise proof of the indicated property of vector algebra. Take $\mathbf{a} = \langle a_1, a_2 \rangle$, $\mathbf{b} = \langle b_1, b_2 \rangle$, and $\mathbf{c} = \langle c_1, c_2 \rangle$ throughout.

39. $\mathbf{a} + (\mathbf{b} + \mathbf{c}) = (\mathbf{a} + \mathbf{b}) + \mathbf{c}$

40. $(r + s)\mathbf{a} = r\mathbf{a} + s\mathbf{a}$

41. $(rs)\mathbf{a} = r(s\mathbf{a})$

42. If $\mathbf{a} + \mathbf{b} = \mathbf{a}$, then $\mathbf{b} = \mathbf{0}$.

43. Find the tension in each cable of Example 6 if the angle between them is 120°.

In Problems 44 through 46, a given weight (in pounds) is suspended by two cables as shown in the figure. Find the tension in each cable.

44.

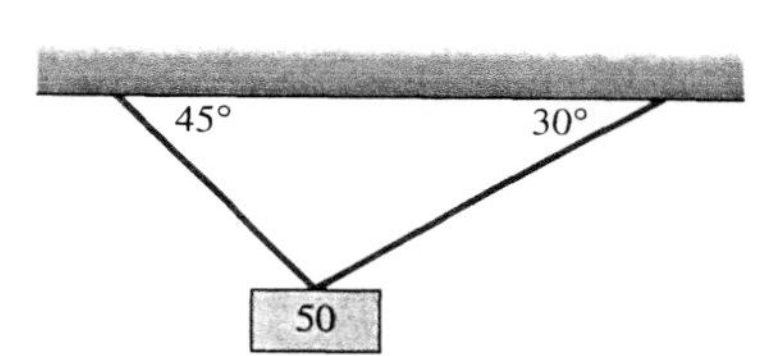

FIGURE 12.1.14

45.

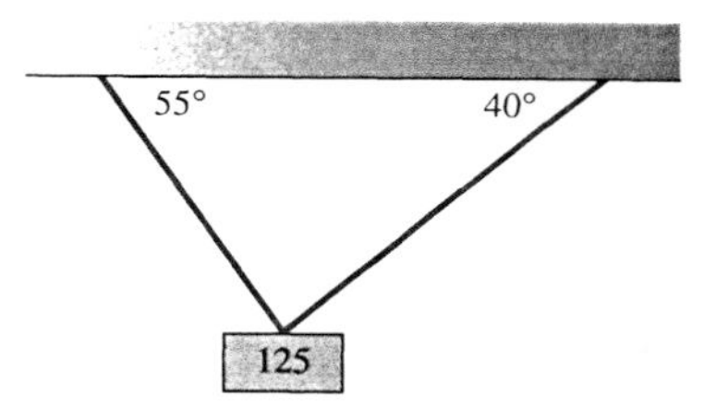

FIGURE 12.1.15

46.

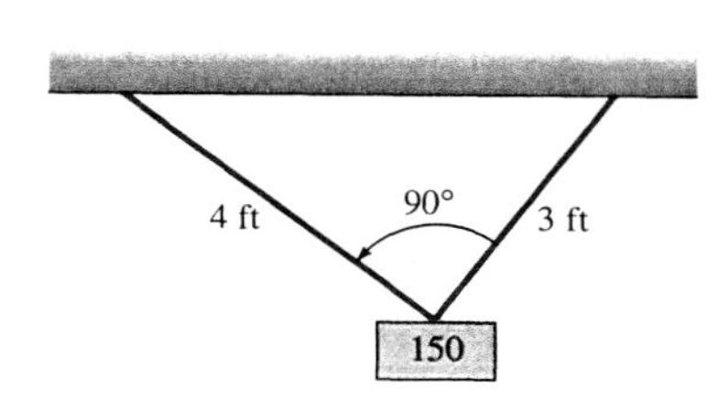

FIGURE 12.1.16

In Problems 47 through 49, assume the following fact: If an airplane flies with velocity vector $\mathbf{v}_a$ *relative to the air and the velocity of the wind is* $\mathbf{w}$, *then the velocity vector of the plane relative to the ground is* $\mathbf{v}_g = \mathbf{v}_a + \mathbf{w}$ *(Fig. 12.1.17). The vector* $\mathbf{v}_a$ *is called the* apparent velocity vector *and the vector* $\mathbf{v}_g$ *is called the* true velocity vector.

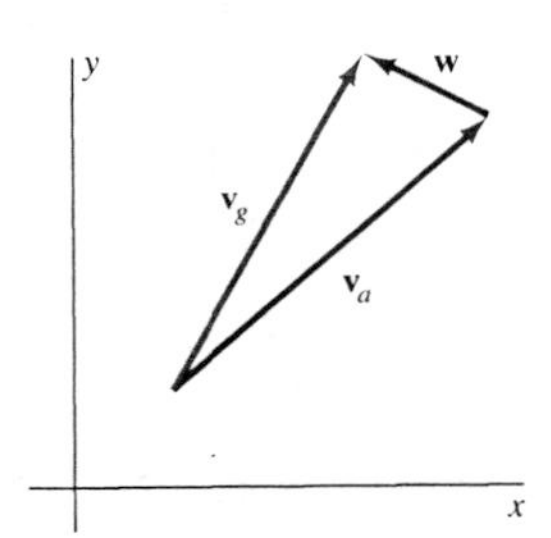

FIGURE 12.1.17 The vectors of Problems 47 through 49
- Apparent velocity: $\mathbf{v}_a$
- Wind velocity: $\mathbf{w}$
- True velocity: $\mathbf{v}_g = \mathbf{v}_a + \mathbf{w}$

47. Suppose that the wind is blowing from the northeast at 50 mi/h and that the pilot wishes to fly due east at 500 mi/h. What should the plane's apparent velocity vector be?

48. Repeat Problem 47 with the phrase *due east* replaced with *due west*.

49. Repeat Problem 47 in the case that the pilot wishes to fly northwest at 500 mi/h.

50. Given any three points A, B, and C in the plane, show that $\overrightarrow{AB} + \overrightarrow{BC} + \overrightarrow{CA} = \mathbf{0}$. [*Suggestion:* Picture the triangle ABC.]

51. If $\mathbf{a}$ and $\mathbf{b}$ are the position vectors of the points P and Q in the plane and M is the point with position vector $\mathbf{v} = \frac{1}{2}(\mathbf{a} + \mathbf{b})$, show that M is the midpoint of the line segment PQ. Is it sufficient to show that the vectors $\overrightarrow{PM}$ and $\overrightarrow{QM}$ are equal and opposite?

52. In the triangle ABC, let M and N be the midpoints of AB and AC, respectively. Show that $\overrightarrow{MN} = \frac{1}{2}\overrightarrow{BC}$. Conclude that the line segment joining the midpoints of two sides of a triangle is parallel to the third side. How are their lengths related?

53. Prove that the diagonals of a parallelogram $ABCD$ bisect each other. [*Suggestion:* If M and N are the midpoints of the diagonals AC and BD, respectively, and O is the origin, show that $\overrightarrow{OM} = \overrightarrow{ON}$.]

54. Use vectors to prove that the midpoints of the four sides of an arbitrary quadrilateral are the vertices of a parallelogram.

55. Figure 12.1.18 shows the vector $\mathbf{a}_\perp$ obtained by rotating the vector $\mathbf{a} = a_1\mathbf{i} + a_2\mathbf{j}$ through a counterclockwise angle of 90°. Show that

$$\mathbf{a}_\perp = -a_2\mathbf{i} + a_1\mathbf{j}.$$

[*Suggestion:* Begin by writing $\mathbf{a} = (r\cos\theta)\mathbf{i} + (r\sin\theta)\mathbf{j}$.]

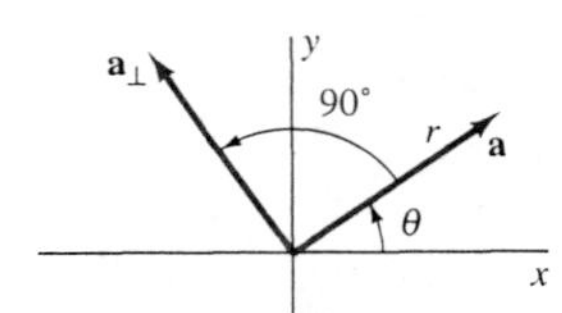

FIGURE 12.1.18 Rotate $\mathbf{a}$ counterclockwise 90° to obtain $\mathbf{a}_\perp$ (Problem 55).

12.2 THREE-DIMENSIONAL VECTORS

In the first eleven chapters we discussed many aspects of the calculus of functions of a *single* variable. The geometry of such functions is two-dimensional, because the graph of a function of a single variable is a curve in the coordinate plane. Most of the remaining chapters deal with the calculus of functions of *several* (two or more) independent variables. The geometry of functions of two variables is three-dimensional because the graphs of such functions are generally surfaces in space.

Rectangular coordinates in the plane may be generalized to rectangular coordinates in space. A point in space is determined by giving its location relative to three mutually perpendicular **coordinate axes** that pass through the origin O. We shall usually draw the x-, y-, and z-axes as shown in Fig. 12.2.1, sometimes with arrows indicating the positive direction along each axis; the positive x-axis will always be labeled x, and similarly for the positive y- and z-axes. With this configuration of axes, our rectangular coordinate system is said to be **right-handed:** If you curl the fingers of your right hand in the direction of a 90° rotation from the positive x-axis to the positive y-axis, then your thumb points in the direction of the positive z-axis. If the x- and y-axes were interchanged, then the coordinate system would be left-handed. These two coordinate systems are different in that it is impossible to bring one into coincidence with the other by means of rotations and translations. This is why the L- and D-alanine molecules shown in Fig. 12.2.2 are different; you can metabolize the left-handed ("levo") version but not the right-handed ("dextro") version. In this book we shall discuss right-handed coordinate systems exclusively and always draw the x-, y-, and z-axes with the right-handed orientation shown in Fig. 12.2.1.

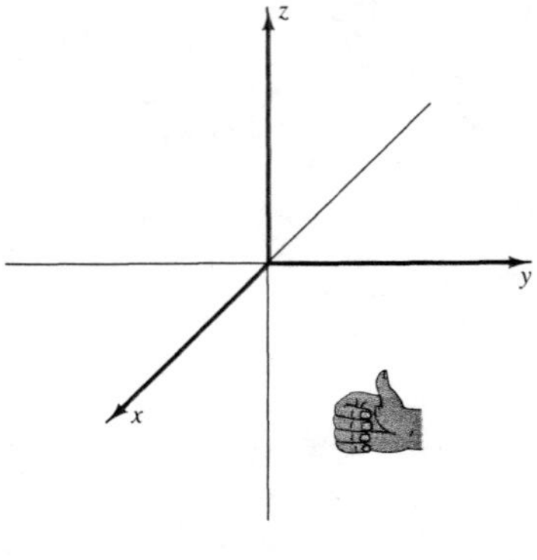

FIGURE 12.2.1 The right-handed coordinate system.

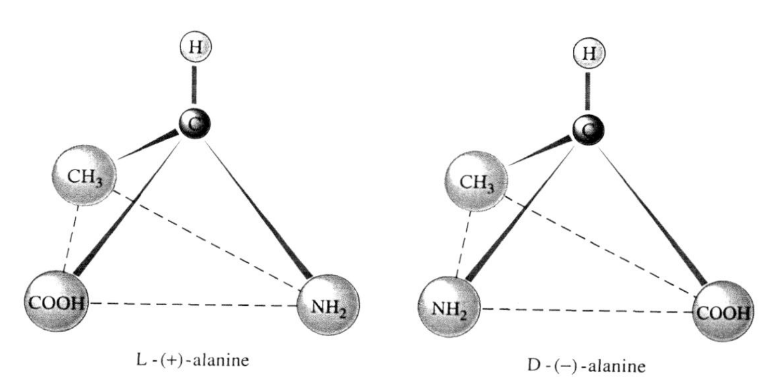

FIGURE 12.2.2 The stereoisomers of the amino acid alanine are physically and biologically different even through they have the same molecular formula.

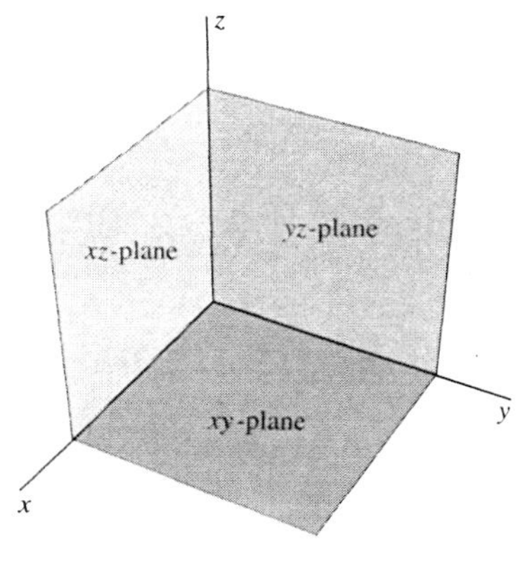

FIGURE 12.2.3 The coordinate planes in space.

The three coordinate axes taken in pairs determine the three **coordinate planes** (Fig. 12.2.3):

- The (horizontal) xy-plane, where $z = 0$;
- The (vertical) yz-plane, where $x = 0$; and
- The (vertical) xz-plane, where $y = 0$.

The point P in space is said to have **rectangular coordinates** (x, y, z) if

- x is its signed distance from the yz-plane,
- y is its signed distance from the xz-plane, and
- z is its signed distance from the xy-plane.

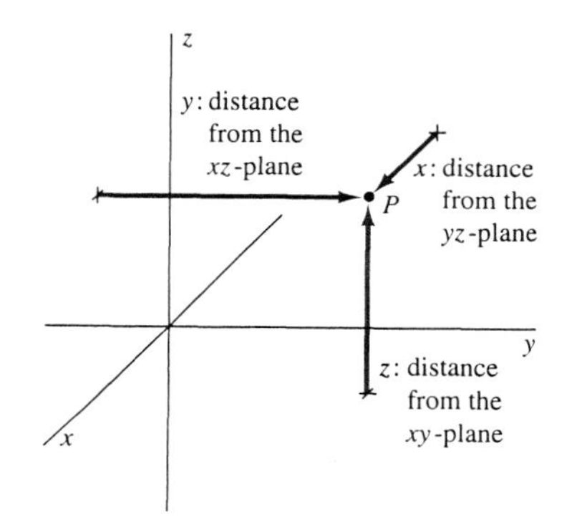

FIGURE 12.2.4 Locating the point P in rectangular coordinates.

(See Fig. 12.2.4.) In this case we may describe the location of P simply by calling it "the point $P(x, y, z)$." There is a natural one-to-one correspondence between ordered triples (x, y, z) of real numbers and points P in space; this correspondence is called a **rectangular coordinate system** in space. In Fig. 12.2.5 the point P is located in the **first octant**—the eighth of space in which all three rectangular coordinates are positive.

If we apply the Pythagorean theorem to the right triangles P_1QR and P_1RP_2 in Fig. 12.2.6, we get

$$\begin{aligned}|P_1P_2|^2 &= |RP_2|^2 + |P_1R|^2 = |RP_2|^2 + |QR|^2 + |P_1Q|^2 \\ &= (x_1 - x_2)^2 + (y_1 - y_2)^2 + (z_1 - z_2)^2.\end{aligned}$$

Thus the **distance formula** for the **distance** $|P_1P_2|$ between the points P_1 and P_2 is

$$|P_1P_2| = \sqrt{(x_1 - x_2)^2 + (y_1 - y_2)^2 + (z_1 - z_2)^2}. \tag{1}$$

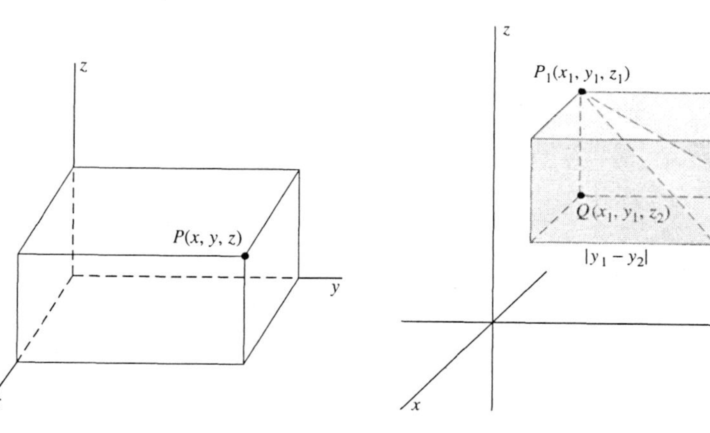

FIGURE 12.2.5 Completing the box to show P with the illusion of the third dimension.

FIGURE 12.2.6 The distance between P_1 and P_2 is the length of the long diagonal of the box.

EXAMPLE 1 The distance between the points $A(1, 3, -2)$ and $B(4, -3, 1)$ is

$$|AB| = \sqrt{(4-1)^2 + (-3-3)^2 + (1+2)^2} = \sqrt{54} \approx 7.348. \quad \blacklozenge$$

You can apply the distance formula in Eq. (1) to show that the **midpoint** M of the line segment joining $P_1(x_1, y_1, z_1)$ and $P_2(x_2, y_2, z_2)$ is

$$M\left(\frac{x_1 + x_2}{2}, \frac{y_1 + y_2}{2}, \frac{z_1 + z_2}{2}\right). \tag{2}$$

(See Problem 63.)

The **graph** of an equation in three variables x, y, and z is the set of all points in space with rectangular coordinates that satisfy that equation. In general, the graph of an equation in three variables is a *two-dimensional surface* in $\boldsymbol{R}^3$ (three-dimensional space with rectangular coordinates).

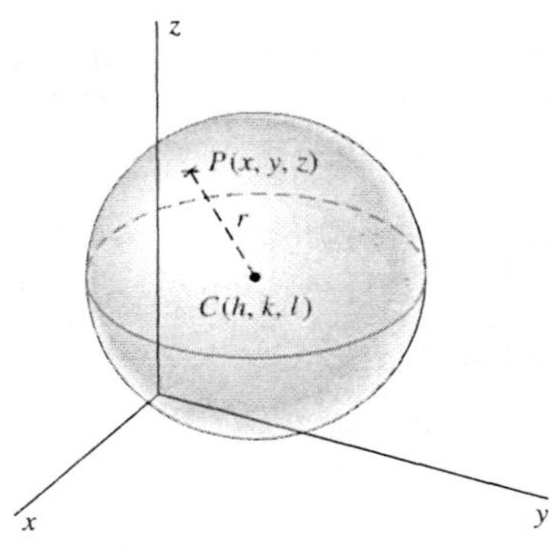

FIGURE 12.2.7 The sphere with center (h, k, l) and radius r.

EXAMPLE 2 Given a fixed point $C(h, k, l)$ and a number $r > 0$, find an equation of the sphere with radius r and center C.

Solution By definition, the sphere is the set of all points $P(x, y, z)$ such that the distance from P to C is r. That is, $|CP| = r$, and thus $|CP|^2 = r^2$. Therefore

$$(x-h)^2 + (y-k)^2 + (z-l)^2 = r^2. \tag{3}$$

$\blacklozenge$

Equation (3) is worth remembering as the equation of the **sphere with radius** r and **center** $C(h, k, l)$ shown in Fig. 12.2.7. Moreover, given an equation of the form

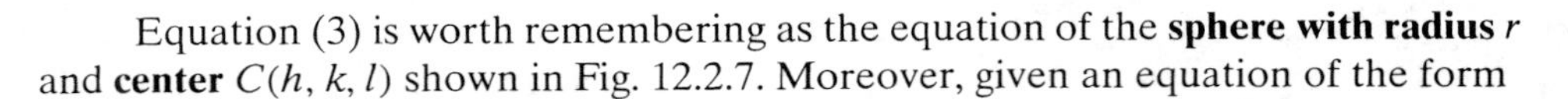

$$x^2 + y^2 + z^2 + Ax + By + Cz + D = 0,$$

we can attempt—by completing the square in each variable—to write it in the form of Eq. (3) and thereby show that its graph is a sphere.

EXAMPLE 3 Determine the graph of the equation

$$x^2 + y^2 + z^2 + 4x + 2y - 6z - 2 = 0.$$

Solution We complete the square in each variable. The equation then takes the form

$$(x^2 + 4x + 4) + (y^2 + 2y + 1) + (z^2 - 6z + 9) = 2 + (4 + 1 + 9) = 16;$$

that is,

$$(x+2)^2 + (y+1)^2 + (z-3)^2 = 4^2.$$

Thus the graph of the given equation is the sphere with radius 4 and center $(-2, -1, 3)$. $\blacklozenge$

Vectors in Space

The discussion of vectors in space parallels the discussion in Section 12.1 of vectors in the plane. The difference is that a vector in space has three components rather than two. The point $P(x, y, z)$ has **position vector** $\mathbf{v} = \overrightarrow{OP} = \langle x, y, z\rangle$, which is represented by the directed line segment (or arrow) $\overrightarrow{OP}$ from the origin O to the point P (as well as by any parallel translate of this arrow—see Fig. 12.2.8). The distance formula in Eq. (1) gives

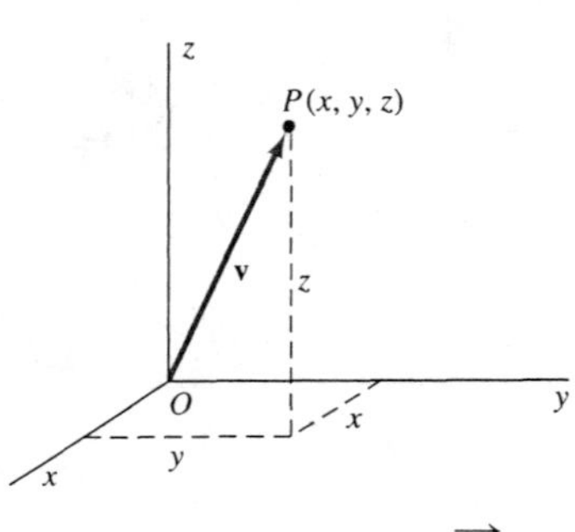

FIGURE 12.2.8 The arrow $\overrightarrow{OP}$ represents the position vector $\mathbf{v} = \langle x, y, z\rangle$.

$$|\mathbf{v}| = \sqrt{x^2 + y^2 + z^2} \tag{4}$$

for the **length** (or **magnitude**) of the vector $\mathbf{v} = \langle x, y, z\rangle$.

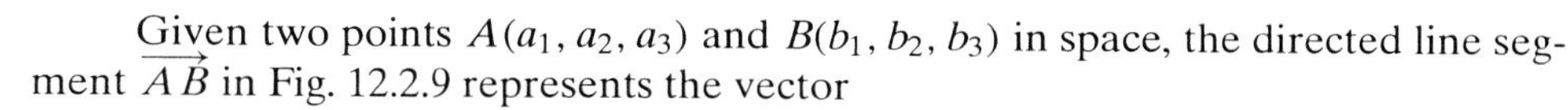

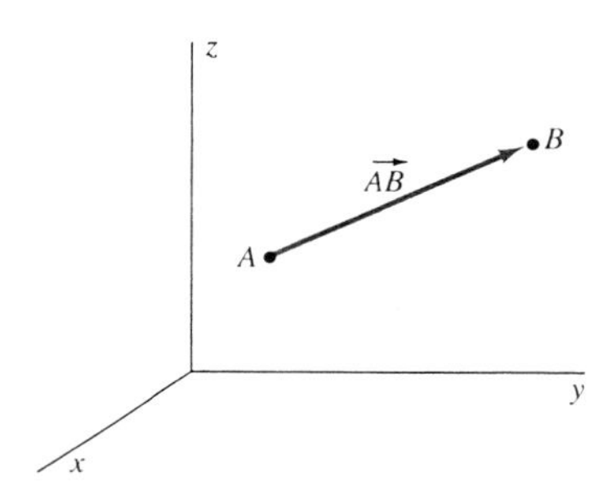

FIGURE 12.2.9 The arrow $\overrightarrow{AB}$ represents the vector $\mathbf{v} = \langle b_1 - a_1, b_2 - a_2, b_3 - a_3 \rangle$.

Given two points $A(a_1, a_2, a_3)$ and $B(b_1, b_2, b_3)$ in space, the directed line segment $\overrightarrow{AB}$ in Fig. 12.2.9 represents the vector

$$\mathbf{v} = \langle b_1 - a_1, b_2 - a_2, b_3 - a_3 \rangle.$$

Its length is the distance between the two points A and B:

$$|\mathbf{v}| = |\overrightarrow{AB}| = \sqrt{(b_1 - a_1)^2 + (b_2 - a_2)^2 + (b_3 - a_3)^2}.$$

What it means for two vectors in space to be equal is essentially the same as in the case of two-dimensional vectors: The vectors $\mathbf{a} = \langle a_1, a_2, a_3 \rangle$ and $\mathbf{b} = \langle b_1, b_2, b_3 \rangle$ are **equal** provided that $a_1 = b_1$, $a_2 = b_2$, and $a_3 = b_3$. That is, two vectors are equal exactly when corresponding components are equal.

We define addition and scalar multiplication of three-dimensional vectors exactly as we did in Section 12.1, taking into account that the vectors now have three components rather than two: The **sum** of the vectors $\mathbf{a} = \langle a_1, a_2, a_3 \rangle$ and $\mathbf{b} = \langle b_1, b_2, b_3 \rangle$ is the vector

$$\mathbf{a} + \mathbf{b} = \langle a_1 + b_1, a_2 + b_2, a_3 + b_3 \rangle. \tag{5}$$

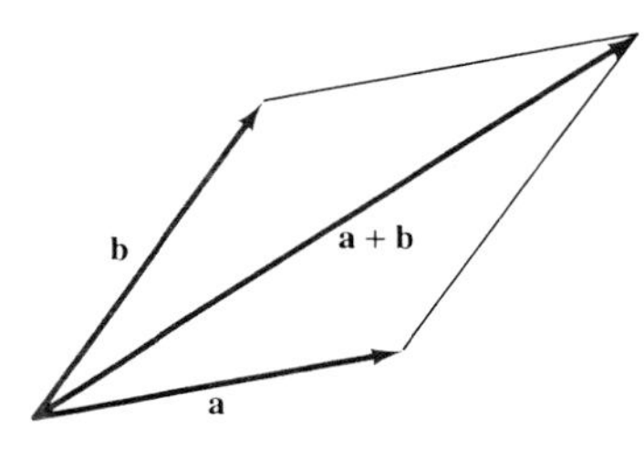

FIGURE 12.2.10 The parallelogram law for addition of vectors.

Because **a** and **b** lie in a plane (although not necessarily the xy-plane) if their initial points coincide, addition of three-dimensional vectors obeys the same **parallelogram law** as in the two-dimensional case (Fig. 12.2.10).

If c is a real number, then the **scalar multiple** $c\mathbf{a}$ is the vector

$$c\mathbf{a} = \langle ca_1, ca_2, ca_3 \rangle. \tag{6}$$

The length of $c\mathbf{a}$ is $|c|$ times the length of **a**, and $c\mathbf{a}$ has the same direction as **a** if $c > 0$ but the opposite direction if $c < 0$. The following algebraic properties of vector addition and scalar multiplication for three-dimensional vectors are easy to establish; they follow from computations with components, exactly as in Section 12.1:

$$\begin{aligned} \mathbf{a} + \mathbf{b} &= \mathbf{b} + \mathbf{a}, \\ \mathbf{a} + (\mathbf{b} + \mathbf{c}) &= (\mathbf{a} + \mathbf{b}) + \mathbf{c}, \\ r(\mathbf{a} + \mathbf{b}) &= r\mathbf{a} + r\mathbf{b}, \\ (r + s)\mathbf{a} &= r\mathbf{a} + s\mathbf{a}, \\ (rs)\mathbf{a} &= r(s\mathbf{a}) = s(r\mathbf{a}). \end{aligned} \tag{7}$$

EXAMPLE 4 If $\mathbf{a} = \langle 3, 4, 12 \rangle$ and $\mathbf{b} = \langle -4, 3, 0 \rangle$, then

$$\begin{aligned} \mathbf{a} + \mathbf{b} &= \langle 3 - 4, 4 + 3, 12 + 0 \rangle = \langle -1, 7, 12 \rangle, \\ |\mathbf{a}| &= \sqrt{3^2 + 4^2 + 12^2} = \sqrt{169} = 13, \\ 2\mathbf{a} &= \langle 2 \cdot 3, 2 \cdot 4, 2 \cdot 12 \rangle = \langle 6, 8, 24 \rangle, \quad \text{and} \\ 2\mathbf{a} - 3\mathbf{b} &= \langle 6 + 12, 8 - 9, 24 - 0 \rangle = \langle 18, -1, 24 \rangle. \end{aligned}$$

◆

A **unit vector** is a vector of length 1. We can express any vector in space (or *space vector*) in terms of the three **basic unit vectors**

$$\mathbf{i} = \langle 1, 0, 0 \rangle, \qquad \mathbf{j} = \langle 0, 1, 0 \rangle, \qquad \mathbf{k} = \langle 0, 0, 1 \rangle.$$

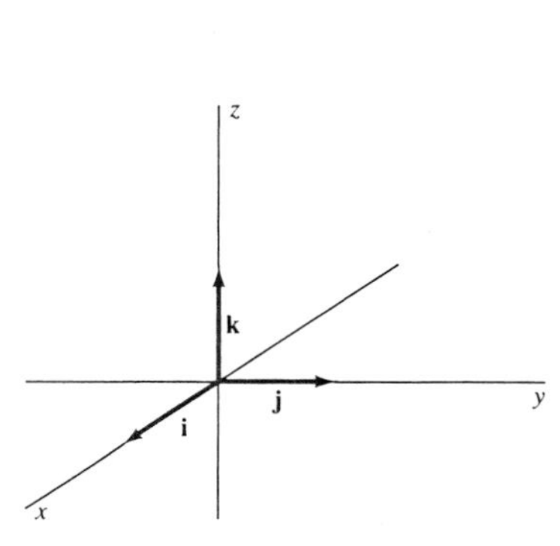

FIGURE 12.2.11 The basic unit vectors **i**, **j**, and **k**.

When located with their initial points at the origin, these basic unit vectors form a right-handed triple of vectors pointing in the positive directions along the three coordinate axes (Fig. 12.2.11).

The space vector $\mathbf{a} = \langle a_1, a_2, a_3 \rangle$ can be written as

$$\mathbf{a} = a_1\mathbf{i} + a_2\mathbf{j} + a_3\mathbf{k},$$

a linear combination of the basic unit vectors. As in the two-dimensional case, the usefulness of this representation is that algebraic operations involving vectors may be carried out simply by collecting coefficients of **i**, **j**, and **k**.

EXAMPLE 5 Given the vectors $\mathbf{a} = \langle 3, -4, 2\rangle$ and $\mathbf{b} = \langle 5, 2, -7\rangle$, we can write

$$\mathbf{a} = 3\mathbf{i} - 4\mathbf{j} + 2\mathbf{k} \quad \text{and} \quad \mathbf{b} = 5\mathbf{i} + 2\mathbf{j} - 7\mathbf{k}$$

in order to calculate

$$\begin{aligned} 7\mathbf{a} + 5\mathbf{b} &= 7 \cdot (3\mathbf{i} - 4\mathbf{j} + 2\mathbf{k}) + 5 \cdot (5\mathbf{i} + 2\mathbf{j} - 7\mathbf{k}) \\ &= (21 + 25)\mathbf{i} + (-28 + 10)\mathbf{j} + (14 - 35)\mathbf{k} \\ &= 46\mathbf{i} - 18\mathbf{j} - 21\mathbf{k} = \langle 46, -18, -21\rangle. \end{aligned}$$

◆

The Dot Product of Two Vectors

The **dot product** of the two vectors

$$\mathbf{a} = a_1\mathbf{i} + a_2\mathbf{j} + a_3\mathbf{k} \quad \text{and} \quad \mathbf{b} = b_1\mathbf{i} + b_2\mathbf{j} + b_3\mathbf{k}$$

is the number obtained when we multiply corresponding components of **a** and **b** and add the results. That is,

$$\mathbf{a} \cdot \mathbf{b} = a_1b_1 + a_2b_2 + a_3b_3. \tag{8}$$

Thus the dot product of two vectors is the *sum of the products of their corresponding components.* In the case of plane vectors $\mathbf{a} = \langle a_1, a_2\rangle$ and $\mathbf{b} = \langle b_1, b_2\rangle$, we simply dispense with third components and write $\mathbf{a} \cdot \mathbf{b} = a_1b_1 + a_2b_2$.

EXAMPLE 6 To apply the definition to calculate the dot product of the two vectors $\mathbf{a} = \langle 3, 4, 12\rangle$ and $\mathbf{b} = \langle -4, 3, 0\rangle$, we simply follow the pattern in Eq. (8):

$$\mathbf{a} \cdot \mathbf{b} = (3)(-4) + (4)(3) + (12)(0) = -12 + 12 + 0 = 0.$$

And if $\mathbf{c} = \langle 4, 5, -3\rangle$, then

$$\mathbf{a} \cdot \mathbf{c} = (3)(4) + (4)(5) + (12)(-3) = 12 + 20 - 36 = -4.$$

◆

IMPORTANT The dot product of two *vectors* is a *scalar*—that is, an ordinary real number. For this reason the dot product is often called the **scalar product.** Example 6 illustrates the fact that the scalar product of two nonzero vectors (with positive lengths) may be zero or even a negative number.

The following **properties of the dot product** show that dot products of vectors behave in many ways in analogy to the ordinary algebra of real numbers.

$$\begin{aligned} \mathbf{a} \cdot \mathbf{a} &= |\mathbf{a}|^2, \\ \mathbf{a} \cdot \mathbf{b} &= \mathbf{b} \cdot \mathbf{a}, \\ \mathbf{a} \cdot (\mathbf{b} + \mathbf{c}) &= \mathbf{a} \cdot \mathbf{b} + \mathbf{a} \cdot \mathbf{c}, \\ (r\mathbf{a}) \cdot \mathbf{b} &= r(\mathbf{a} \cdot \mathbf{b}) = \mathbf{a} \cdot (r\mathbf{b}). \end{aligned} \tag{9}$$

Each of the properties in (9) can be established by working with components of the vectors involved. For instance, to establish the second equation, suppose that $\mathbf{a} = \langle a_1, a_2, a_3\rangle$ and $\mathbf{b} = \langle b_1, b_2, b_3\rangle$. Then

$$\mathbf{a} \cdot \mathbf{b} = a_1b_1 + a_2b_2 + a_3b_3 = b_1a_1 + b_2a_2 + b_3a_3 = \mathbf{b} \cdot \mathbf{a}.$$

This derivation makes it clear that the commutative law for the dot product is a consequence of the commutative law $ab = ba$ for multiplication of ordinary real numbers.

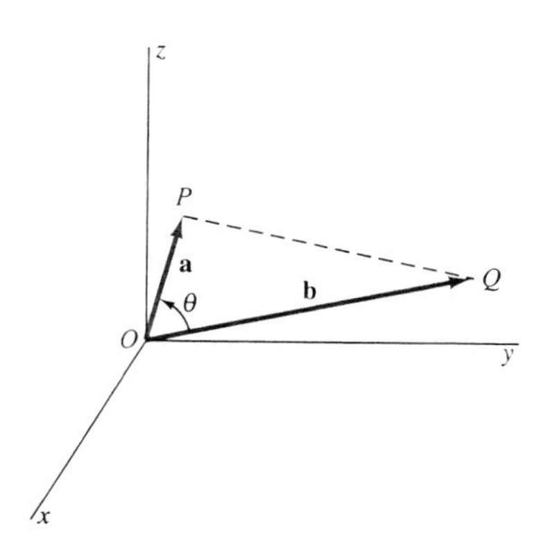

FIGURE 12.2.12 The angle θ between the vectors **a** and **b**.

Example 6 shows that the *algebraic definition* of the dot product is easy to apply in routine calculations. But what does it mean? The significance and **meaning of the dot product** lie in its *geometric interpretation*.

Let the vectors **a** and **b** be represented as position vectors by the directed segments $\overrightarrow{OP}$ and $\overrightarrow{OQ}$, respectively. Then the angle θ between **a** and **b** is the angle at O in triangle OPQ of Fig. 12.2.12. We say that **a** and **b** are **parallel** if $\theta = 0$ or if $\theta = \pi$ and that **a** and **b** are **perpendicular** if $\theta = \pi/2$. For convenience, we regard the zero vector $\mathbf{0} = \langle 0, 0, 0 \rangle$ as both parallel to *and* perpendicular to *every* vector.

THEOREM 1 Interpretation of the Dot Product

If θ is the angle between the vectors **a** and **b**, then

$$\mathbf{a} \cdot \mathbf{b} = |\mathbf{a}|\,|\mathbf{b}| \cos\theta. \tag{10}$$

PROOF If either $\mathbf{a} = \mathbf{0}$ or $\mathbf{b} = \mathbf{0}$, then Eq. (10) follows immediately. If the vectors **a** and **b** are parallel, then $\mathbf{b} = t\mathbf{a}$ with either $t > 0$ and $\theta = 0$ or $t < 0$ and $\theta = \pi$. In either case, both sides in Eq. (10) reduce to $t|\mathbf{a}|^2$, so again the conclusion of Theorem 1 follows.

We turn to the general case in which the vector $\mathbf{a} = \overrightarrow{OP}$ and $\mathbf{b} = \overrightarrow{OQ}$ are nonzero and nonparallel. Then

$$\begin{aligned} |\overrightarrow{QP}|^2 &= |\mathbf{a} - \mathbf{b}|^2 = (\mathbf{a} - \mathbf{b}) \cdot (\mathbf{a} - \mathbf{b}) \\ &= \mathbf{a} \cdot \mathbf{a} - \mathbf{a} \cdot \mathbf{b} - \mathbf{b} \cdot \mathbf{a} + \mathbf{b} \cdot \mathbf{b} \\ &= |\mathbf{a}|^2 + |\mathbf{b}|^2 - 2\mathbf{a} \cdot \mathbf{b}. \end{aligned}$$

But $c = |\overrightarrow{QP}|$ is the side of triangle OPQ (Fig. 12.2.12) that is opposite the angle θ included between the sides $a = |\mathbf{a}|$ and $b = |\mathbf{b}|$. Hence the law of cosines (Appendix M) gives

$$\begin{aligned} |\overrightarrow{QP}|^2 &= c^2 = a^2 + b^2 - 2ab\cos\theta \\ &= |\mathbf{a}|^2 + |\mathbf{b}|^2 - 2\,|\mathbf{a}|\,|\mathbf{b}|\cos\theta. \end{aligned}$$

Finally, comparing these two expressions for $|\overrightarrow{QP}|^2$ yields Eq. (10). ◄

This theorem tells us that the angle θ between the nonzero vectors **a** and **b** can be found by using the equation

$$\cos\theta = \frac{\mathbf{a} \cdot \mathbf{b}}{|\mathbf{a}|\,|\mathbf{b}|}. \tag{11}$$

For instance, given the vectors $\mathbf{a} = \langle 8, 5 \rangle$ and $\mathbf{b} = \langle -11, 17 \rangle$ of Example 5 in Section 12.1, we calculate

$$\cos\theta = \frac{\langle 8, 5 \rangle \cdot \langle -11, 17 \rangle}{|\langle 8, 5 \rangle|\,|\langle -11, 17 \rangle|} = \frac{(8)(-11) + (5)(17)}{\sqrt{8^2 + 5^2}\sqrt{(-11)^2 + 17^2}} = \frac{-3}{\sqrt{89}\sqrt{410}}.$$

It follows that $\theta = \arccos(-3/\sqrt{89}\sqrt{410}) \approx 1.5865$ (*radians*) $\approx 90.90^\circ \neq 90^\circ$, so we see again that the vectors **a** and **b** are not perpendicular.

More generally, the two nonzero vectors **a** and **b** are perpendicular if and only if they make a right angle, so that $\theta = \pi/2$. By (11), this in turn is so if and only if $\mathbf{a} \cdot \mathbf{b} = 0$. Hence we have a quick computational check for perpendicularity of vectors.

COROLLARY Test for Perpendicular Vectors

The two nonzero vectors **a** and **b** are perpendicular if and only if $\mathbf{a} \cdot \mathbf{b} = 0$.

EXAMPLE 7 (a) To show that the plane vectors $\mathbf{a} = \langle 8, 5 \rangle$ and $\mathbf{b} = \langle -11, 17 \rangle$ of Example 5 in Section 12.1 were not perpendicular, we need only have calculated their dot product $\mathbf{a} \cdot \mathbf{b} = -88 + 85 = -3$ and observed that its value is not zero. (b) Given the space vectors $\mathbf{a} = \langle 8, 5, -1 \rangle$ and $\mathbf{b} = \langle -11, 17, -3 \rangle$, we find that

$$\mathbf{a} \cdot \mathbf{b} = (8)(-11) + (5)(17) + (-1)(-3) = -88 + 85 + 3 = 0.$$

We may therefore conclude that **a** and **b** *are* perpendicular. ◆

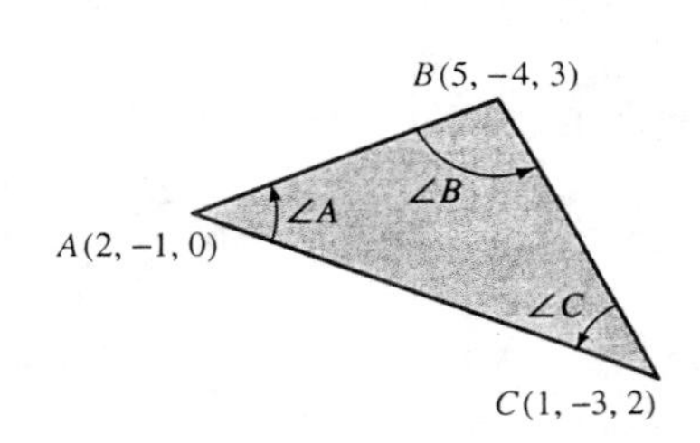

FIGURE 12.2.13 The triangle of Example 8.

EXAMPLE 8 Find the angles shown in the triangle of Fig. 12.2.13 with vertices at $A(2, -1, 0)$, $B(5, -4, 3)$, and $C(1, -3, 2)$.

Solution We apply Eq. (10) with $\theta = \angle A$, $\mathbf{a} = \overrightarrow{AB} = \langle 3, -3, 3 \rangle$, and $\mathbf{b} = \overrightarrow{AC} = \langle -1, -2, 2 \rangle$. This yields

$$\angle A = \cos^{-1}\left(\frac{\overrightarrow{AB} \cdot \overrightarrow{AC}}{|\overrightarrow{AB}|\,|\overrightarrow{AC}|}\right) = \cos^{-1}\left(\frac{\langle 3, -3, 3 \rangle \cdot \langle -1, -2, 2 \rangle}{\sqrt{27}\sqrt{9}}\right)$$

$$= \cos^{-1}\left(\frac{9}{\sqrt{27}\,\sqrt{9}}\right) \approx 0.9553\,(\text{rad}) \approx 54.74^\circ.$$

Similarly,

$$\angle B = \cos^{-1}\left(\frac{\overrightarrow{BA} \cdot \overrightarrow{BC}}{|\overrightarrow{BA}|\,|\overrightarrow{BC}|}\right) = \cos^{-1}\left(\frac{\langle -3, 3, -3 \rangle \cdot \langle -4, 1, -1 \rangle}{\sqrt{27}\sqrt{18}}\right)$$

$$= \cos^{-1}\left(\frac{18}{\sqrt{27}\sqrt{18}}\right) \approx 0.6155\,(\text{rad}) \approx 35.26^\circ.$$

Then $\angle C = 180^\circ - \angle A - \angle B \approx 90^\circ$. As a check, note that

$$\overrightarrow{CA} \cdot \overrightarrow{CB} = \langle 1, 2, -2 \rangle \cdot \langle 4, -1, 1 \rangle = 0.$$

So the angle at C is, indeed, a right angle. ◆

Direction Angles and Projections

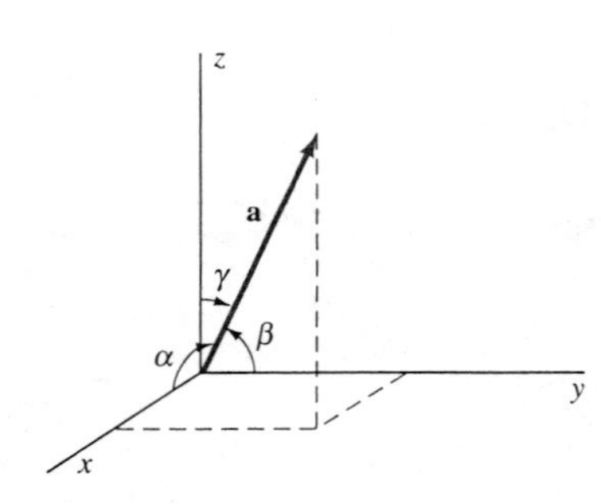

FIGURE 12.2.14 The direction angles of the vector **a**.

The **direction angles** of the nonzero vector $\mathbf{a} = \langle a_1, a_2, a_3 \rangle$ are the angles α, β, and γ that it makes with the vectors **i**, **j**, and **k**, respectively (Fig. 12.2.14). The cosines of these angles, $\cos\alpha$, $\cos\beta$, and $\cos\gamma$, are called the **direction cosines** of the vector **a**. When we replace **b** in Eq. (11) with **i**, **j**, and **k** in turn, we find that

$$\cos\alpha = \frac{\mathbf{a} \cdot \mathbf{i}}{|\mathbf{a}|\,|\mathbf{i}|} = \frac{a_1}{|\mathbf{a}|},$$

$$\cos\beta = \frac{\mathbf{a} \cdot \mathbf{j}}{|\mathbf{a}|\,|\mathbf{j}|} = \frac{a_2}{|\mathbf{a}|}, \quad \text{and} \tag{12}$$

$$\cos\gamma = \frac{\mathbf{a} \cdot \mathbf{k}}{|\mathbf{a}|\,|\mathbf{k}|} = \frac{a_3}{|\mathbf{a}|}.$$

That is, the direction cosines of **a** are the components of the *unit vector* $\mathbf{a}/|\mathbf{a}|$ with the same direction as **a**. Consequently

$$\cos^2\alpha + \cos^2\beta + \cos^2\gamma = 1. \tag{13}$$

EXAMPLE 9 Find the direction angles of the vector $\mathbf{a} = 2\mathbf{i} + 3\mathbf{j} - \mathbf{k}$.

Solution Because $|\mathbf{a}| = \sqrt{14}$, the equations in (12) give

$$\alpha = \cos^{-1}\left(\frac{2}{\sqrt{14}}\right) \approx 57.69^\circ, \qquad \beta = \cos^{-1}\left(\frac{3}{\sqrt{14}}\right) \approx 36.70^\circ,$$

$$\text{and} \qquad \gamma = \cos^{-1}\left(\frac{-1}{\sqrt{14}}\right) \approx 105.50^\circ.$$ ◆

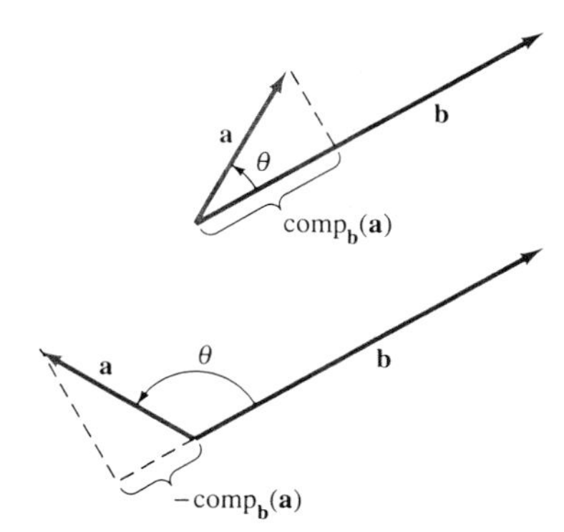

FIGURE 12.2.15 The component of **a** along **b**.

Sometimes we need to find the component of one vector **a** in the direction of another *nonzero* vector **b**. Think of the two vectors located with the same initial point (Fig. 12.2.15). Then the (scalar) **component of a along b**, denoted by $\mathrm{comp}_{\mathbf{b}}\mathbf{a}$, is numerically the length of the perpendicular projection of **a** onto the straight line determined by **b**. The number $\mathrm{comp}_{\mathbf{b}}\mathbf{a}$ is positive if the angle θ between **a** is acute (so **a** and **b** point in the same general direction) and negative if $\theta > \pi/2$. Thus $\mathrm{comp}_{\mathbf{b}}\mathbf{a} = |\mathbf{a}|\cos\theta$ in either case. Equation (10) then gives

$$\mathrm{comp}_{\mathbf{b}}\mathbf{a} = \frac{|\mathbf{a}|\,|\mathbf{b}|\cos\theta}{|\mathbf{b}|} = \frac{\mathbf{a}\cdot\mathbf{b}}{|\mathbf{b}|}. \tag{14}$$

There is no need to memorize this formula, for—in practice—we can always read $\mathrm{comp}_{\mathbf{b}}\mathbf{a} = |\mathbf{a}|\cos\theta$ from the figure and then apply Eq. (10) to eliminate $\cos\theta$. Note that $\mathrm{comp}_{\mathbf{b}}\mathbf{a}$ is a scalar, not a vector.

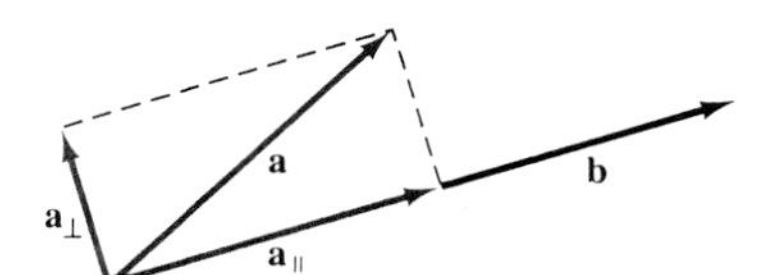

FIGURE 12.2.16 Construction of $\mathbf{a}_\parallel$ and $\mathbf{a}_\perp$.

EXAMPLE 10 Given $\mathbf{a} = \langle 4, -5, 3\rangle$ and $\mathbf{b} = \langle 2, 1, -2\rangle$, express **a** as the sum of a vector $\mathbf{a}_\parallel$ parallel to **b** and a vector $\mathbf{a}_\perp$ perpendicular to **b**.

Solution Our method of solution is motivated by the diagram in Fig. 12.2.16. We take

$$\mathbf{a}_\parallel = (\mathrm{comp}_{\mathbf{b}}\mathbf{a})\frac{\mathbf{b}}{|\mathbf{b}|} = \frac{\mathbf{a}\cdot\mathbf{b}}{|\mathbf{b}|^2}\mathbf{b} = \frac{8-5-6}{9}\mathbf{b}$$

$$= -\frac{1}{3}\langle 2, 1, -2\rangle = \left\langle -\frac{2}{3}, -\frac{1}{3}, \frac{2}{3}\right\rangle,$$

and

$$\mathbf{a}_\perp = \mathbf{a} - \mathbf{a}_\parallel = \langle 4, -5, 3\rangle - \left\langle -\frac{2}{3}, -\frac{1}{3}, \frac{2}{3}\right\rangle = \left\langle \frac{14}{3}, -\frac{14}{3}, \frac{7}{3}\right\rangle.$$

The diagram makes our choice of $\mathbf{a}_\parallel$ plausible, and we have deliberately chosen $\mathbf{a}_\perp$ so that $\mathbf{a} = \mathbf{a}_\parallel + \mathbf{a}_\perp$. To verify that the vector $\mathbf{a}_\parallel$ is indeed parallel to **b**, we simply note that it is a scalar multiple of **b**. To verify that $\mathbf{a}_\perp$ is perpendicular to **b**, we compute the dot product

$$\mathbf{a}_\perp\cdot\mathbf{b} = \tfrac{28}{3} - \tfrac{14}{3} - \tfrac{14}{3} = 0.$$

Thus $\mathbf{a}_\parallel$ and $\mathbf{a}_\perp$ have the required properties. ◆

One important application of vector components is to the definition and computation of *work*. Recall that the work W done by a constant force F exerted along the line of motion in moving a particle a distance d is given by $W = Fd$. But what if the force is a constant vector **F** pointing in some direction other than the line of motion, as when a child pulls a sled against the resistance of friction (Fig. 12.2.17)? Suppose that **F** moves a particle along the line segment from P to Q, and let $\mathbf{D} = \overrightarrow{PQ}$ be the resulting *displacement vector* of the object (Fig. 12.2.18). Then the **work** W done by the force **F** in moving the object along the line from P to Q is, by definition, the product of the component of **F** along **D** and the distance moved:

$$W = (\mathrm{comp}_{\mathbf{D}}\mathbf{F})\,|\mathbf{D}|. \tag{15}$$

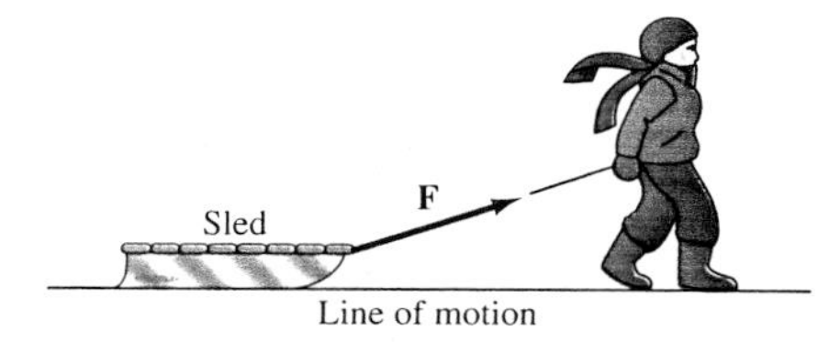

FIGURE 12.2.17 The vector force **F** is constant but acts at an angle to the line of motion (Example 10).

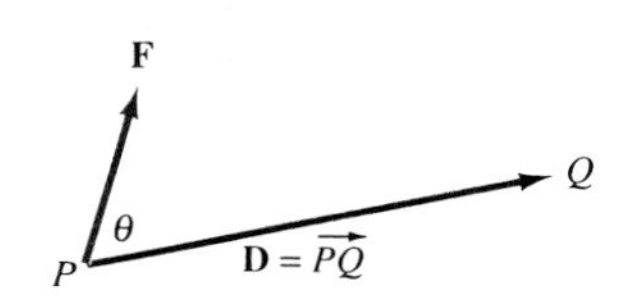

FIGURE 12.2.18 The force vector **F** and displacement vector **D** in Eq. (16).

If we use Eq. (14) and substitute $\text{comp}_{\mathbf{D}}\mathbf{F} = (\mathbf{F} \cdot \mathbf{D})/|\mathbf{D}|$, we get

$$W = \mathbf{F} \cdot \mathbf{D} \tag{16}$$

for the work done by the constant force $\mathbf{F}$ in moving an object along the displacement vector $\mathbf{D} = \overrightarrow{PQ}$. This formula is the vector generalization of the scalar work formula $W = Fd$. Work is measured in foot-pounds (ft·lb) if distance is measured in feet and force in pounds. If metric units of meters (m) for distance and newtons (N) for force are used, then work is measured in joules (J). (One joule is approximately 0.7376 ft·lb.)

EXAMPLE 11 Suppose that the force vector in Fig. 12.2.17 is inclined at an angle of 30° from the ground. If the child exerts a constant force of 20 lb, how much work is done in pulling the sled a distance of one mile?

Solution We are given that $|\mathbf{F}| = 20$ (lb) and $|\mathbf{D}| = 5280$ (ft). Because $\cos 30° = \frac{1}{2}\sqrt{3}$, Eq. (16) yields

$$W = \mathbf{F} \cdot \mathbf{D} = |\mathbf{F}|\,|\mathbf{D}| \cos 30° = (20)(5280)\left(\tfrac{1}{2}\sqrt{3}\right) \approx 91452 \quad \text{(ft·lb)}.$$

This may seem like a lot of work for a child to do. If the 1-mile trip takes an hour, then the child is generating *power* (work per unit time) at the rate of (91452 ft·lb)/(3600 s) ≈ 25.4 ft·lb/s. Because 1 horsepower (hp) is defined to be 550 ft·lb/s, the child's "power rating" is 25.4/550 ≈ $\frac{1}{20}$ hp. By comparison, an adult in excellent physical condition can climb the 2570 steps of the staircase of the CN tower in Toronto in less than 40 minutes. On October 29, 1989, Brendon Keenory of Toronto set the world's record for the fastest stairclimb there with a time of 7 min 52 s. Assuming that he climbed 1672 ft and weighed 160 lb, he generated an average of more than 0.988 hp over this time interval. ◆

12.2 TRUE/FALSE STUDY GUIDE

12.2 CONCEPTS: QUESTIONS AND DISCUSSION

1. Discuss the relation between a 3-dimensional vector and a point in space.
2. How does the dot product of two vectors resemble the ordinary product of two numbers? How do the two products differ?
3. Discuss the analogy between the absolute value of a number and the length of a vector.
4. Give an example of a real-world situation described by a triple of real numbers. In your example, do vector addition and scalar multiplication make any sense?

12.2 PROBLEMS

In Problems 1 through 6, find (a) $2\mathbf{a}+\mathbf{b}$, *(b)* $3\mathbf{a}-4\mathbf{b}$, *(c)* $\mathbf{a}\cdot\mathbf{b}$, *(d)* $|\mathbf{a}-\mathbf{b}|$, *and (e)* $\mathbf{a}/|\mathbf{a}|$.

1. $\mathbf{a} = \langle 2, 5, -4\rangle, \quad \mathbf{b} = \langle 1, -2, -3\rangle$

2. $\mathbf{a} = \langle -1, 0, 2\rangle, \quad \mathbf{b} = \langle 3, 4, -5\rangle$

3. $\mathbf{a} = \mathbf{i}+\mathbf{j}+\mathbf{k}, \quad \mathbf{b} = \mathbf{j}-\mathbf{k}$

4. $\mathbf{a} = 2\mathbf{i}-3\mathbf{j}+5\mathbf{k}, \quad \mathbf{b} = 5\mathbf{i}+3\mathbf{j}-7\mathbf{k}$

5. $\mathbf{a} = 2\mathbf{i}-\mathbf{j}, \quad \mathbf{b} = \mathbf{j}-3\mathbf{k}$

6. $\mathbf{a} = \mathbf{i}-2\mathbf{j}+3\mathbf{k}, \quad \mathbf{b} = \mathbf{i}+3\mathbf{j}-2\mathbf{k}$

7 through **12.** Find, to the nearest degree, the angle between the vectors **a** and **b** in Problems 1 through 6.

13 through **18.** Find $\text{comp}_{\mathbf{a}}\mathbf{b}$ and $\text{comp}_{\mathbf{b}}\mathbf{a}$ for the vectors **a** and **b** given in Problems 1 through 6.

In Problems 19 through 24, write the equation of the indicated sphere.

19. Center (3, 1, 2), radius 5

20. Center (−2, 1, −5), radius $\sqrt{7}$

21. One diameter: the segment joining (3, 5, −3) and (7, 3, 1)

22. Center (4, 5, −2), passing through the point (1, 0, 0)

23. Center (0, 0, 2), tangent to the xy-plane

24. Center (3, −4, 3), tangent to the xz-plane

In Problems 25 through 28, find the center and radius of the sphere with the given equation.

25. $x^2 + y^2 + z^2 + 4x - 6y = 0$

26. $x^2 + y^2 + z^2 - 8x - 9y + 10z + 40 = 0$

27. $3x^2 + 3y^2 + 3z^2 - 18z - 48 = 0$

28. $2x^2 + 2y^2 + 2z^2 = 7x + 9y + 11z$

In Problems 29 through 38, describe the graph of the given equation in geometric terms, using plain, clear language.

29. $z = 0$

30. $x = 0$

31. $z = 10$

32. $xy = 0$

33. $xyz = 0$

34. $x^2 + y^2 + z^2 + 7 = 0$

35. $x^2 + y^2 + z^2 = 0$

36. $x^2 + y^2 + z^2 - 2x + 1 = 0$

37. $x^2 + y^2 + z^2 - 6x + 8y + 25 = 0$

38. $x^2 + y^2 = 0$

Two vectors are ***parallel*** *provided that one is a scalar multiple of the other. Determine whether the vectors* **a** *and* **b** *in Problems 39 through 42 are parallel or perpendicular or neither.*

39. $\mathbf{a} = \langle 4, -2, 6 \rangle$ and $\mathbf{b} = \langle 6, -3, 9 \rangle$

40. $\mathbf{a} = \langle 4, -2, 6 \rangle$ and $\mathbf{b} = \langle 4, 2, 2 \rangle$

41. $\mathbf{a} = 12\mathbf{i} - 20\mathbf{j} + 16\mathbf{k}$ and $\mathbf{b} = -9\mathbf{i} + 15\mathbf{j} - 12\mathbf{k}$

42. $\mathbf{a} = 12\mathbf{i} - 20\mathbf{j} + 17\mathbf{k}$ and $\mathbf{b} = -9\mathbf{i} + 15\mathbf{j} + 24\mathbf{k}$

In Problems 43 and 44, determine whether or not the three given points lie on a single straight line.

43. $P(0, -2, 4)$, $Q(1, -3, 5)$, $R(4, -6, 8)$

44. $P(6, 7, 8)$, $Q(3, 3, 3)$, $R(12, 15, 18)$

In Problems 45 through 48, find (to the nearest degree) the three angles of the triangle with the given vertices.

45. $A(1, 0, 0)$, $B(0, 1, 0)$, $C(0, 0, 1)$

46. $A(1, 0, 0)$, $B(1, 2, 0)$, $C(1, 2, 3)$

47. $A(1, 1, 1)$, $B(3, -2, 3)$, $C(3, 4, 6)$

48. $A(1, 0, 0)$, $B(0, 1, 0)$, $C(-1, -2, -2)$

In Problems 49 through 52, find the direction angles of the vector represented by $\overrightarrow{PQ}$.

49. $P(1, -1, 0)$, $Q(3, 4, 5)$

50. $P(2, -3, 5)$, $Q(1, 0, -1)$

51. $P(-1, -2, -3)$, $Q(5, 6, 7)$

52. $P(0, 0, 0)$, $Q(5, 12, 13)$

In Problems 53 and 54, find the work W done by the force **F** *in moving a particle in a straight line from P to Q.*

53. $\mathbf{F} = \mathbf{i} - \mathbf{k}$; $P(0, 0, 0)$, $Q(3, 1, 0)$

54. $\mathbf{F} = 2\mathbf{i} - 3\mathbf{j} + 5\mathbf{k}$; $P(5, 3, -4)$, $Q(-1, -2, 5)$

55. Suppose that the force vector in Fig. 12.2.17 is inclined at an angle of 40° from the ground. If the child exerts a constant force of 40 N, how much heat energy (in calories) does the child expend in pulling the sled a distance of 1 km along the ground? [*Note:* 1 J of work requires an expenditure of 0.239 calories of energy.]

56. A 1000-lb dog sled has a coefficient of sliding friction of 0.2, so it requires a force with a horizontal component of 200 lb to keep it moving at a constant speed. Suppose that a dog-team harness is attached so that the team's force vector makes an angle of 5° with the horizontal. If the dog team pulls this sled at a speed of 10 mi/h, how much power (in horsepower) are the dogs generating? [*Note:* 1 hp is 550 ft·lb/s.]

57. Suppose that the horizontal and vertical components of the three vectors shown in Fig. 12.2.19 balance (the algebraic sum of the horizontal components is zero, as is the sum of the vertical components). How much work is done by the constant force **F** (parallel to the inclined plane) in pulling the weight mg up the inclined plane a vertical height h?

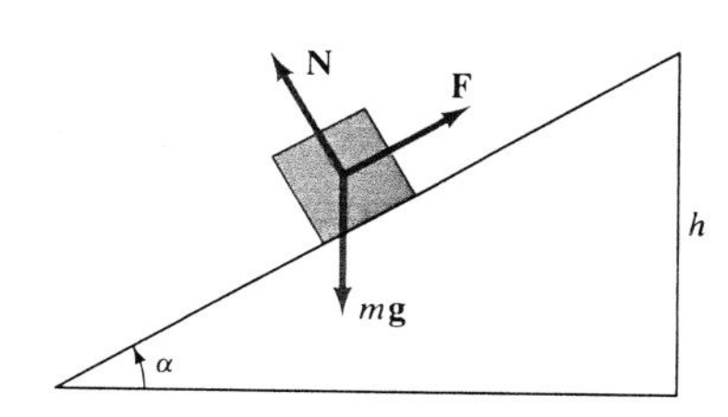

FIGURE 12.2.19 The inclined plane of Problem 57.

58. Prove the **Cauchy-Schwarz inequality:**

$$|\mathbf{a} \cdot \mathbf{b}| \leqq |\mathbf{a}|\,|\mathbf{b}|$$

for all pairs of vectors **a** and **b**.

59. Given two arbitrary vectors **a** and **b**, prove that they satisfy the **triangle inequality,**

$$|\mathbf{a} + \mathbf{b}| \leqq |\mathbf{a}| + |\mathbf{b}|.$$

[*Suggestion:* Square both sides.]

60. Prove that if **a** and **b** are arbitrary vectors, then

$$|\mathbf{a} - \mathbf{b}| \geqq |\mathbf{a}| - |\mathbf{b}|.$$

[*Suggestion:* Write $\mathbf{a} = (\mathbf{a} - \mathbf{b}) + \mathbf{b}$; then apply the triangle inequality of Problem 59.]

61. Use the dot product to construct a nonzero vector $\mathbf{w} = \langle w_1, w_2, w_3 \rangle$ perpendicular to both of the vectors $\mathbf{u} = \langle 1, 2, -3 \rangle$ and $\mathbf{v} = \langle 2, 0, 1 \rangle$.

62. The unit cube in the first octant in space has opposite vertices $O(0, 0, 0)$ and $P(1, 1, 1)$. Find the angle between the edge of the cube on the x-axis and the diagonal OP.

63. Prove that the point M given in Eq. (2) is indeed the midpoint of the segment P_1P_2. [*Note:* You must prove *both* that M is equally distant from P_1 and P_2 *and* that M lies on the segment P_1P_2.]

64. Given vectors **a** and **b**, let $a = |\mathbf{a}|$ and $b = |\mathbf{b}|$. Prove that the vector

$$\mathbf{c} = \frac{(b\mathbf{a} + a\mathbf{b})}{(a + b)}$$

bisects the angle between **a** and **b**.

65. Let **a**, **b**, and **c** be three vectors in the xy-plane with **a** and **b** nonzero and nonparallel. Show that there exist scalars α and β such that $\mathbf{c} = \alpha\mathbf{a} + \beta\mathbf{b}$. [*Suggestion:* Begin by expressing **a**, **b**, and **c** in terms of **i**, **j**, and **k**.]

66. Let $ax + by + c = 0$ be the equation of the line L in the xy-plane with normal vector $\mathbf{n}$. Let $P_0(x_0, y_0)$ be a point on this line and $P_1(x_1, y_1)$ be a point not on L. Prove that the perpendicular distance from P_1 to L is

$$d = \frac{|\mathbf{n} \cdot \overrightarrow{P_0P_1}|}{|\mathbf{n}|} = \frac{|ax_1 + by_1 + c|}{\sqrt{a^2 + b^2}}.$$

67. Given the two points $A(3, -2, 4)$ and $B(5, 7, -1)$, write an equation in x, y, and z that says that the point $P(x, y, z)$ is equally distant from the points A and B. Then simplify this equation and give a geometric description of the set of all such points $P(x, y, z)$.

68. Given the fixed point $A(1, 3, 5)$, the point $P(x, y, z)$, and the vector $\mathbf{n} = \mathbf{i} - \mathbf{j} + 2\mathbf{k}$, use the dot product to help you write an equation in x, y, and z that says this: $\mathbf{n}$ and $\overrightarrow{AP}$ are perpendicular. Then simplify this equation and give a geometric description of all such points $P(x, y, z)$.

69. Prove that the points $(0, 0, 0)$, $(1, 1, 0)$, $(1, 0, 1)$, and $(0, 1, 1)$ are the vertices of a regular tetrahedron by showing that each of the six edges has length $\sqrt{2}$. Then use the dot product to find the angle between any two edges of the tetrahedron.

70. The methane molecule CH_4 is arranged with the four hydrogen atoms at the vertices of a regular tetrahedron and with the carbon atom at its center (Fig. 12.2.20). Suppose that the axes and scale are chosen so that the tetrahedron is that of Problem 69, with its center at $(\frac{1}{2}, \frac{1}{2}, \frac{1}{2})$. Find the *bond angle* α between the lines from the carbon atom to two of the hydrogen atoms.

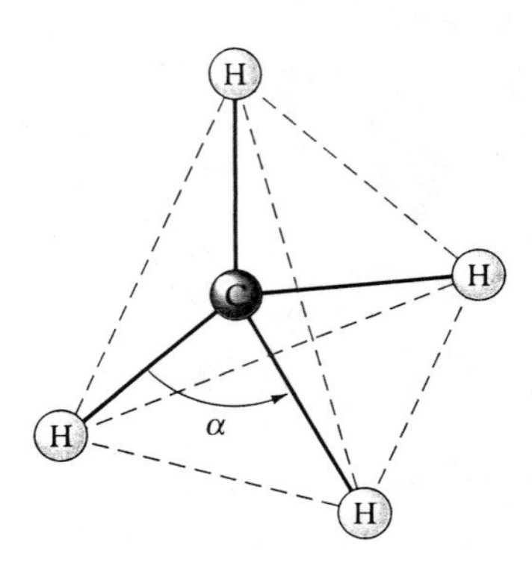

FIGURE 12.2.20 The methane bond angle α of Problem 70.

12.3 THE CROSS PRODUCT OF VECTORS

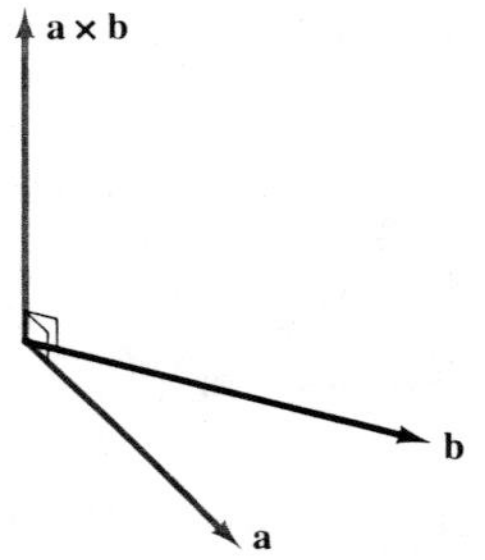

We often need to find a vector that is perpendicular to each of two vectors $\mathbf{a}$ and $\mathbf{b}$ in space. A routine way of doing this is provided by the *cross product* $\mathbf{a} \times \mathbf{b}$ of the vectors $\mathbf{a}$ and $\mathbf{b}$. This vector product is quite unlike the dot product $\mathbf{a} \cdot \mathbf{b}$ in that $\mathbf{a} \cdot \mathbf{b}$ is a *scalar*, whereas $\mathbf{a} \times \mathbf{b}$ is a *vector*. For this reason $\mathbf{a} \times \mathbf{b}$ is sometimes called the *vector product* of the two vectors $\mathbf{a}$ and $\mathbf{b}$.

The **cross product** (or **vector product**) of the vectors $\mathbf{a} = \langle a_1, a_2, a_3 \rangle$ and $\mathbf{b} = \langle b_1, b_2, b_3 \rangle$ is defined algebraically by the formula

$$\mathbf{a} \times \mathbf{b} = \langle a_2b_3 - a_3b_2, a_3b_1 - a_1b_3, a_1b_2 - a_2b_1 \rangle. \tag{1}$$

FIGURE 12.3.1 The cross product $\mathbf{a} \times \mathbf{b}$ is perpendicular to both $\mathbf{a}$ and $\mathbf{b}$.

Although this formula seems unmotivated, it has a redeeming feature: The product $\mathbf{a} \times \mathbf{b}$ is perpendicular both to $\mathbf{a}$ and to $\mathbf{b}$, as suggested in Fig. 12.3.1.

THEOREM 1 Perpendicularity of the Cross Product

The cross product $\mathbf{a} \times \mathbf{b}$ is perpendicular both to $\mathbf{a}$ and to $\mathbf{b}$.

PROOF We show that $\mathbf{a} \times \mathbf{b}$ is perpendicular to $\mathbf{a}$ by showing that the dot product of $\mathbf{a}$ and $\mathbf{a} \times \mathbf{b}$ is zero. With the components as in Eq. (1), we find that

$$\begin{aligned}\mathbf{a} \cdot (\mathbf{a} \times \mathbf{b}) &= a_1(a_2b_3 - a_3b_2) + a_2(a_3b_1 - a_1b_3) + a_3(a_1b_2 - a_2b_1)\\ &= a_1a_2b_3 - a_1a_3b_2 + a_2a_3b_1 - a_2a_1b_3 + a_3a_1b_2 - a_3a_2b_1\\ &= 0.\end{aligned}$$

A similar computation shows that $\mathbf{b} \cdot (\mathbf{a} \times \mathbf{b}) = 0$ as well, so $\mathbf{a} \times \mathbf{b}$ is also perpendicular to the vector $\mathbf{b}$. ◄

You need not memorize Eq. (1), because there is an alternative version involving determinants that is easy both to remember and to use. Recall that a *determinant* of order 2 is defined as follows:

$$\begin{vmatrix} a_1 & a_2 \\ b_1 & b_2 \end{vmatrix} = a_1b_2 - a_2b_1. \tag{2}$$

EXAMPLE 1

$$\begin{vmatrix} 2 & -1 \\ 3 & 4 \end{vmatrix} = 2 \cdot 4 - (-1) \cdot 3 = 11.$$ ◆

A determinant of order 3 can be defined in terms of determinants of order 2:

$$\begin{vmatrix} a_1 & a_2 & a_3 \\ b_1 & b_2 & b_3 \\ c_1 & c_2 & c_3 \end{vmatrix} = +a_1\begin{vmatrix} b_2 & b_3 \\ c_2 & c_3 \end{vmatrix} - a_2\begin{vmatrix} b_1 & b_3 \\ c_1 & c_3 \end{vmatrix} + a_3\begin{vmatrix} b_1 & b_2 \\ c_1 & c_2 \end{vmatrix}. \tag{3}$$

Each element a_i of the first row is multiplied by the 2-by-2 "subdeterminant" obtained by deleting the row *and* column that contain a_i. Note in Eq. (3) that signs are attached to the a_i in accord with the checkerboard pattern

$$\begin{vmatrix} + & - & + \\ - & + & - \\ + & - & + \end{vmatrix}.$$

Equation (3) is an expansion of the 3-by-3 determinant along its first row. It can be expanded along any other row or column as well. For example, its expansion along its second column is

$$\begin{vmatrix} a_1 & a_2 & a_3 \\ b_1 & b_2 & b_3 \\ c_1 & c_2 & c_3 \end{vmatrix} = -a_2\begin{vmatrix} b_1 & b_3 \\ c_1 & c_3 \end{vmatrix} + b_2\begin{vmatrix} a_1 & a_3 \\ c_1 & c_3 \end{vmatrix} - c_2\begin{vmatrix} a_1 & a_3 \\ b_1 & b_3 \end{vmatrix}.$$

In linear algebra it is shown that all such expansions yield the same value for the determinant.

Although we can expand a determinant of order 3 along any row or column, here we will use only expansions along the first row, as in Eq. (3) and Example 2.

EXAMPLE 2

$$\begin{aligned}\begin{vmatrix} 1 & 3 & -2 \\ 2 & -1 & 4 \\ -3 & 7 & 5 \end{vmatrix} &= 1 \cdot \begin{vmatrix} -1 & 4 \\ 7 & 5 \end{vmatrix} - 3 \cdot \begin{vmatrix} 2 & 4 \\ -3 & 5 \end{vmatrix} + (-2) \cdot \begin{vmatrix} 2 & -1 \\ -3 & 7 \end{vmatrix} \\ &= 1 \cdot (-5 - 28) + (-3) \cdot (10 + 12) + (-2) \cdot (14 - 3) \\ &= -33 - 66 - 22 = -121.\end{aligned}$$ ◆

Equation (1) for the cross product of the vectors $\mathbf{a} = a_1\mathbf{i} + a_2\mathbf{j} + a_3\mathbf{k}$ and $\mathbf{b} = b_1\mathbf{i} + b_2\mathbf{j} + b_3\mathbf{k}$ is equivalent to

$$\mathbf{a} \times \mathbf{b} = \begin{vmatrix} a_2 & a_3 \\ b_2 & b_3 \end{vmatrix}\mathbf{i} - \begin{vmatrix} a_1 & a_3 \\ b_1 & b_3 \end{vmatrix}\mathbf{j} + \begin{vmatrix} a_1 & a_2 \\ b_1 & b_2 \end{vmatrix}\mathbf{k}. \tag{4}$$

This is easy to verify by expanding the 2-by-2 determinants on the right-hand side and noting that the three components of the right-hand side of Eq. (1) result. Motivated by Eq. (4), we write

$$\mathbf{a} \times \mathbf{b} = \begin{vmatrix} \mathbf{i} & \mathbf{j} & \mathbf{k} \\ a_1 & a_2 & a_3 \\ b_1 & b_2 & b_3 \end{vmatrix}. \tag{5}$$

The "symbolic determinant" in this equation is to be evaluated by expansion along its first row, just as in Eq. (3) and just as though it were an ordinary determinant with real number entries. The result of this expansion is the right-hand side of Eq. (4). The components of the *first* vector $\mathbf{a}$ in $\mathbf{a} \times \mathbf{b}$ form the *second* row of the 3-by-3 determinant, and the components of the *second* vector $\mathbf{b}$ form the *third* row. The order of the vectors $\mathbf{a}$ and $\mathbf{b}$ is important because, as we soon shall see, $\mathbf{a} \times \mathbf{b}$ is generally *not* equal to $\mathbf{b} \times \mathbf{a}$: The cross product is *not commutative.*

Equation (5) for the cross product is the form most convenient for computational purposes.

EXAMPLE 3 If $\mathbf{a} = 3\mathbf{i} - \mathbf{j} + 2\mathbf{k}$ and $\mathbf{b} = 2\mathbf{i} + 2\mathbf{j} - \mathbf{k}$, then

$$\mathbf{a} \times \mathbf{b} = \begin{vmatrix} \mathbf{i} & \mathbf{j} & \mathbf{k} \\ 3 & -1 & 2 \\ 2 & 2 & -1 \end{vmatrix} = \begin{vmatrix} -1 & 2 \\ 2 & -1 \end{vmatrix} \mathbf{i} - \begin{vmatrix} 3 & 2 \\ 2 & -1 \end{vmatrix} \mathbf{j} + \begin{vmatrix} 3 & -1 \\ 2 & 2 \end{vmatrix} \mathbf{k}$$

$$= (1 - 4)\mathbf{i} - (-3 - 4)\mathbf{j} + (6 - (-2))\mathbf{k}.$$

Thus

$$\mathbf{a} \times \mathbf{b} = -3\mathbf{i} + 7\mathbf{j} + 8\mathbf{k}.$$

You might now pause to verify (by using the dot product) that the vector $-3\mathbf{i} + 7\mathbf{j} + 8\mathbf{k}$ is perpendicular both to $\mathbf{a}$ and to $\mathbf{b}$. ◆

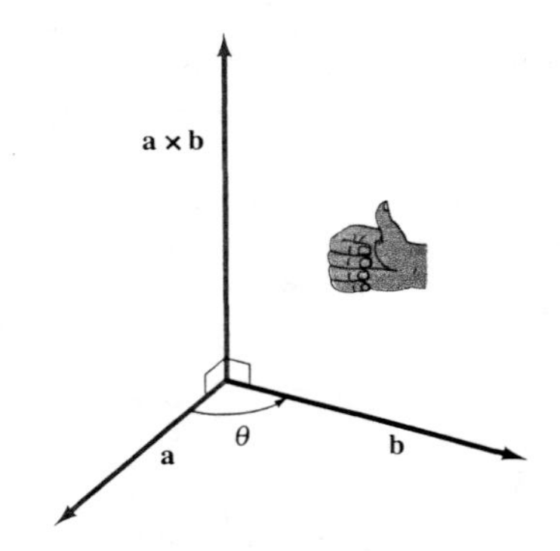

FIGURE 12.3.2 The vectors **a**, **b**, and **a** × **b**—in that order—form a right-handed triple.

If the vectors $\mathbf{a}$ and $\mathbf{b}$ share the same initial point, then Theorem 1 implies that $\mathbf{a} \times \mathbf{b}$ is perpendicular to the plane determined by $\mathbf{a}$ and $\mathbf{b}$ (Fig. 12.3.2). There are still two possible directions for $\mathbf{a} \times \mathbf{b}$, but if $\mathbf{a} \times \mathbf{b} \neq \mathbf{0}$, then the triple $\mathbf{a}, \mathbf{b}, \mathbf{a} \times \mathbf{b}$ is a *right-handed* triple in exactly the same sense as the triple $\mathbf{i}, \mathbf{j}, \mathbf{k}$. Thus if the thumb of your right hand points in the direction of $\mathbf{a} \times \mathbf{b}$, then your fingers curl in the direction of rotation (less than 180°) from $\mathbf{a}$ to $\mathbf{b}$.

Once we have established the direction of $\mathbf{a} \times \mathbf{b}$, we can describe the cross product in completely geometric terms by telling what the length $|\mathbf{a} \times \mathbf{b}|$ of the vector $\mathbf{a} \times \mathbf{b}$ is. This is given by the formula

$$|\mathbf{a} \times \mathbf{b}|^2 = |\mathbf{a}|^2|\mathbf{b}|^2 - (\mathbf{a} \cdot \mathbf{b})^2. \tag{6}$$

We can verify this vector identity routinely (though tediously) by writing $\mathbf{a} = \langle a_1, a_2, a_3 \rangle$ and $\mathbf{b} = \langle b_1, b_2, b_3 \rangle$, computing both sides of Eq. (6), and then noting that the results are equal (Problem 36).

Geometric Significance of the Cross Product

Equation (6) tells us what $|\mathbf{a} \times \mathbf{b}|$ is, but Theorem 2 reveals the geometric significance of the cross product.

THEOREM 2 Length of the Cross Product

Let θ be the angle between the nonzero vectors $\mathbf{a}$ and $\mathbf{b}$ (measured so that $0 \leqq \theta \leqq \pi$). Then

$$|\mathbf{a} \times \mathbf{b}| = |\mathbf{a}|\,|\mathbf{b}| \sin\theta. \tag{7}$$

PROOF We begin with Eq. (6) and use the fact that $\mathbf{a} \cdot \mathbf{b} = |\mathbf{a}|\,|\mathbf{b}| \cos\theta$. Thus

$$|\mathbf{a} \times \mathbf{b}|^2 = |\mathbf{a}|^2|\mathbf{b}|^2 - (\mathbf{a} \cdot \mathbf{b})^2 = |\mathbf{a}|^2|\mathbf{b}|^2 - (|\mathbf{a}|\,|\mathbf{b}| \cos\theta)^2$$

$$= |\mathbf{a}|^2|\mathbf{b}|^2(1 - \cos^2\theta) = |\mathbf{a}|^2|\mathbf{b}|^2 \sin^2\theta.$$

Equation (7) now follows after we take the positive square root of both sides. (This is the correct root on the right-hand side because $\sin\theta \geqq 0$ for $0 \leqq \theta \leqq \pi$.) ◄

COROLLARY Parallel Vectors

Two nonzero vectors **a** and **b** are parallel ($\theta = 0$ or $\theta = \pi$) if and only if $\mathbf{a} \times \mathbf{b} = \mathbf{0}$.

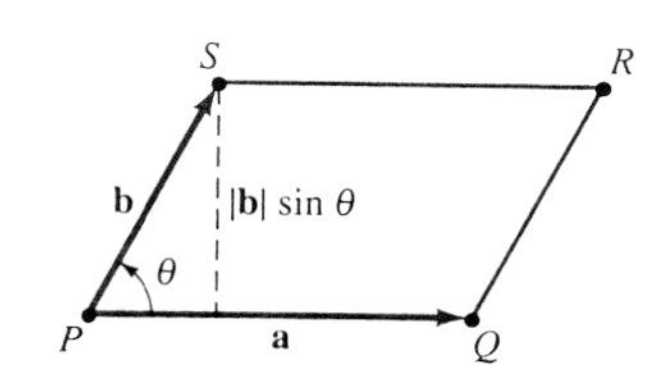

FIGURE 12.3.3 The area of the parallelogram $PQRS$ is $|\mathbf{a} \times \mathbf{b}|$.

In particular, the cross product of any vector with itself is the zero vector. Also, Eq. (1) shows immediately that the cross product of any vector with the zero vector is the zero vector itself. Thus

$$\mathbf{a} \times \mathbf{a} = \mathbf{a} \times \mathbf{0} = \mathbf{0} \times \mathbf{a} = \mathbf{0} \tag{8}$$

for every vector **a**.

Equation (7) has an important geometric interpretation. Suppose that **a** and **b** are represented by adjacent sides of a parallelogram $PQRS$, with $\mathbf{a} = \overrightarrow{PQ}$ and $\mathbf{b} = \overrightarrow{PS}$ (Fig. 12.3.3). The parallelogram then has base of length $|\mathbf{a}|$ and height $|\mathbf{b}| \sin\theta$, so its area is

$$A = |\mathbf{a}|\,|\mathbf{b}| \sin\theta = |\mathbf{a} \times \mathbf{b}|. \tag{9}$$

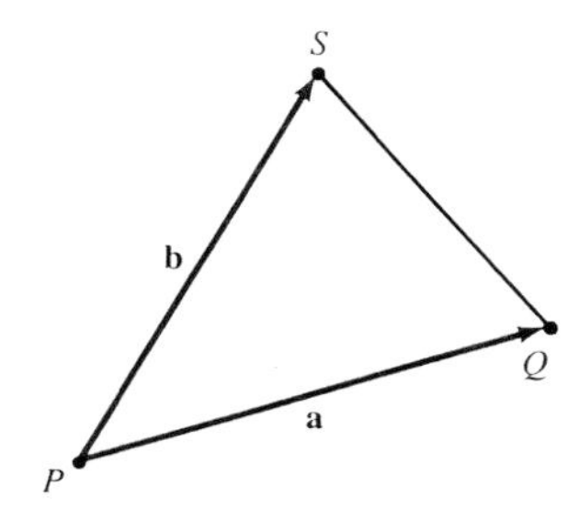

FIGURE 12.3.4 The area of $\triangle PQS$ is $\frac{1}{2}|\mathbf{a} \times \mathbf{b}|$.

Thus *the length of the cross product* $\mathbf{a} \times \mathbf{b}$ *is numerically the same as the area of the parallelogram determined by* **a** *and* **b**. It follows that the area of the triangle PQS in Fig. 12.3.4, whose area is half that of the parallelogram, is

$$\tfrac{1}{2}A = \tfrac{1}{2}|\mathbf{a} \times \mathbf{b}| = \tfrac{1}{2}|\overrightarrow{PQ} \times \overrightarrow{PS}|. \tag{10}$$

Equation (10) gives a quick way to compute the area of a triangle—even one in space—without the need of finding any of its angles.

EXAMPLE 4 Find the area of the triangle with vertices $A(3, 0, -1)$, $B(4, 2, 5)$, and $C(7, -2, 4)$.

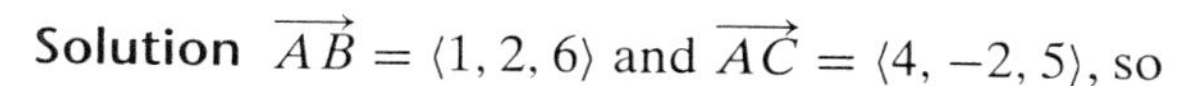

Solution $\overrightarrow{AB} = \langle 1, 2, 6\rangle$ and $\overrightarrow{AC} = \langle 4, -2, 5\rangle$, so

$$\overrightarrow{AB} \times \overrightarrow{AC} = \begin{vmatrix} \mathbf{i} & \mathbf{j} & \mathbf{k} \\ 1 & 2 & 6 \\ 4 & -2 & 5 \end{vmatrix} = 22\mathbf{i} + 19\mathbf{j} - 10\mathbf{k}.$$

Therefore, by Eq. (10), the area of triangle ABC is

$$\tfrac{1}{2}\sqrt{22^2 + 19^2 + (-10)^2} = \tfrac{1}{2}\sqrt{945} \approx 15.37. \qquad \blacklozenge$$

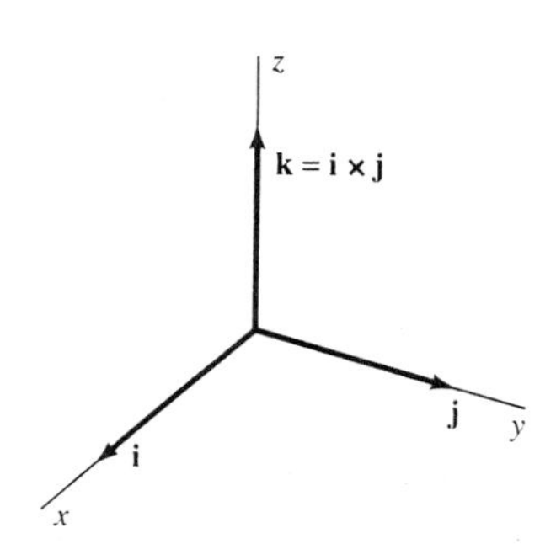

FIGURE 12.3.5 The basic unit vectors in space.

Now let **u**, **v**, **w** be a right-handed triple of mutually perpendicular *unit* vectors. The angle between any two of these is $\theta = \pi/2$, and $|\mathbf{u}| = |\mathbf{v}| = |\mathbf{w}| = 1$. Thus it follows from Eq. (7) that $\mathbf{u} \times \mathbf{v} = \mathbf{w}$. When we apply this observation to the basic unit vectors **i**, **j**, and **k** (Fig. 12.3.5), we see that

$$\mathbf{i} \times \mathbf{j} = \mathbf{k}, \quad \mathbf{j} \times \mathbf{k} = \mathbf{i}, \quad \text{and} \quad \mathbf{k} \times \mathbf{i} = \mathbf{j}. \tag{11a}$$

But

$$\mathbf{j} \times \mathbf{i} = -\mathbf{k}, \quad \mathbf{k} \times \mathbf{j} = -\mathbf{i}, \quad \text{and} \quad \mathbf{i} \times \mathbf{k} = -\mathbf{j}. \tag{11b}$$

These observations, together with the fact that

$$\mathbf{i} \times \mathbf{i} = \mathbf{j} \times \mathbf{j} = \mathbf{k} \times \mathbf{k} = \mathbf{0}, \tag{11c}$$

also follow directly from the original definition of the cross product [in the form in Eq. (5)]. The products in Eq. (11a) are easily remembered in terms of the sequence

$$\mathbf{i}, \quad \mathbf{j}, \quad \mathbf{k}, \quad \mathbf{i}, \quad \mathbf{j}, \quad \mathbf{k}, \quad \ldots.$$

The product of any two consecutive unit vectors, in the order in which they appear in this sequence, is the next one in the sequence.

Note *The cross product is not commutative:* $\mathbf{i} \times \mathbf{j} \neq \mathbf{j} \times \mathbf{i}$. Instead, it is **anticommutative:** For any two vectors **a** and **b**, $\mathbf{a} \times \mathbf{b} = -(\mathbf{b} \times \mathbf{a})$. This is the first part of Theorem 3.

THEOREM 3 Algebraic Properties of the Cross Product

If **a**, **b**, and **c** are vectors and k is a real number, then

1. $\mathbf{a} \times \mathbf{b} = -(\mathbf{b} \times \mathbf{a});$ **(12)**

2. $(k\mathbf{a}) \times \mathbf{b} = \mathbf{a} \times (k\mathbf{b}) = k(\mathbf{a} \times \mathbf{b});$ **(13)**

3. $\mathbf{a} \times (\mathbf{b} + \mathbf{c}) = (\mathbf{a} \times \mathbf{b}) + (\mathbf{a} \times \mathbf{c});$ **(14)**

4. $\mathbf{a} \cdot (\mathbf{b} \times \mathbf{c}) = (\mathbf{a} \times \mathbf{b}) \cdot \mathbf{c};$ **(15)**

5. $\mathbf{a} \times (\mathbf{b} \times \mathbf{c}) = (\mathbf{a} \cdot \mathbf{c})\mathbf{b} - (\mathbf{a} \cdot \mathbf{b})\mathbf{c}.$ **(16)**

The proofs of Eqs. (12) through (15) are straightforward applications of the definition of the cross product in terms of components. See Problem 33 for an outline of the proof of Eq. (16).

We can find cross products of vectors expressed in terms of the basic unit vectors **i**, **j**, and **k** by means of computations that closely resemble those of ordinary algebra. We simply apply the algebraic properties summarized in Theorem 3 together with the relations in Eq. (11) giving the various products of the three unit vectors. We must be careful to preserve the order of factors, because vector multiplication is not commutative—although, of course, we should not hesitate to use Eq. (12).

EXAMPLE 5 $(\mathbf{i} - 2\mathbf{j} + 3\mathbf{k}) \times (3\mathbf{i} + 2\mathbf{j} - 4\mathbf{k})$

$$\begin{aligned}
&= 3(\mathbf{i} \times \mathbf{i}) + 2(\mathbf{i} \times \mathbf{j}) - 4(\mathbf{i} \times \mathbf{k}) - 6(\mathbf{j} \times \mathbf{i}) - 4(\mathbf{j} \times \mathbf{j}) \\
&\quad + 8(\mathbf{j} \times \mathbf{k}) + 9(\mathbf{k} \times \mathbf{i}) + 6(\mathbf{k} \times \mathbf{j}) - 12(\mathbf{k} \times \mathbf{k}) \\
&= 3 \cdot \mathbf{0} + 2\mathbf{k} - 4 \cdot (-\mathbf{j}) - 6 \cdot (-\mathbf{k}) - 4 \cdot \mathbf{0} + 8\mathbf{i} + 9\mathbf{j} + 6 \cdot (-\mathbf{i}) - 12 \cdot \mathbf{0} \\
&= 2\mathbf{i} + 13\mathbf{j} + 8\mathbf{k}.
\end{aligned}$$

◆

Scalar Triple Products

Let us examine the product $\mathbf{a} \cdot (\mathbf{b} \times \mathbf{c})$ that appears in Eq. (15). This expression would not make sense were the parentheses instead around $\mathbf{a} \cdot \mathbf{b}$, because $\mathbf{a} \cdot \mathbf{b}$ is a scalar, and thus we could not form the cross product of $\mathbf{a} \cdot \mathbf{b}$ with the vector **c**. This means that we may omit the parentheses—the expression $\mathbf{a} \cdot \mathbf{b} \times \mathbf{c}$ is not ambiguous—but we keep them for extra clarity. The dot product of the vectors **a** and $\mathbf{b} \times \mathbf{c}$ is a real number, called the **scalar triple product** of the vectors **a**, **b**, and **c**. Equation (15) implies the curious fact that we can interchange the operations $\cdot$ (dot) and $\times$ (cross) without affecting the value of the expression:

$$\mathbf{a} \cdot (\mathbf{b} \times \mathbf{c}) = (\mathbf{a} \times \mathbf{b}) \cdot \mathbf{c}$$

for all vectors **a**, **b**, and **c**.

To compute the scalar triple product in terms of components, write $\mathbf{a} = \langle a_1, a_2, a_3 \rangle$, $\mathbf{b} = \langle b_1, b_2, b_3 \rangle$, and $\mathbf{c} = \langle c_1, c_2, c_3 \rangle$. Then

$$\mathbf{b} \times \mathbf{c} = (b_2c_3 - b_3c_2)\mathbf{i} - (b_1c_3 - b_3c_1)\mathbf{j} + (b_1c_2 - b_2c_1)\mathbf{k},$$

so

$$\mathbf{a} \cdot (\mathbf{b} \times \mathbf{c}) = a_1(b_2c_3 - b_3c_2) - a_2(b_1c_3 - b_3c_1) + a_3(b_1c_2 - b_2c_1).$$

But the expression on the right is the value of the 3-by-3 determinant

$$\mathbf{a} \cdot (\mathbf{b} \times \mathbf{c}) = \begin{vmatrix} a_1 & a_2 & a_3 \\ b_1 & b_2 & b_3 \\ c_1 & c_2 & c_3 \end{vmatrix}. \tag{17}$$

This is the quickest way to compute the scalar triple product.

EXAMPLE 6 If $\mathbf{a} = 2\mathbf{i} - 3\mathbf{k}$, $\mathbf{b} = \mathbf{i} + \mathbf{j} + \mathbf{k}$, and $\mathbf{c} = 4\mathbf{j} - \mathbf{k}$, then

$$\mathbf{a} \cdot (\mathbf{b} \times \mathbf{c}) = \begin{vmatrix} 2 & 0 & -3 \\ 1 & 1 & 1 \\ 0 & 4 & -1 \end{vmatrix}$$

$$= +2 \cdot \begin{vmatrix} 1 & 1 \\ 4 & -1 \end{vmatrix} - 0 \cdot \begin{vmatrix} 1 & 1 \\ 0 & -1 \end{vmatrix} + (-3) \cdot \begin{vmatrix} 1 & 1 \\ 0 & 4 \end{vmatrix}$$

$$= 2 \cdot (-5) + (-3) \cdot 4 = -22.$$ ◆

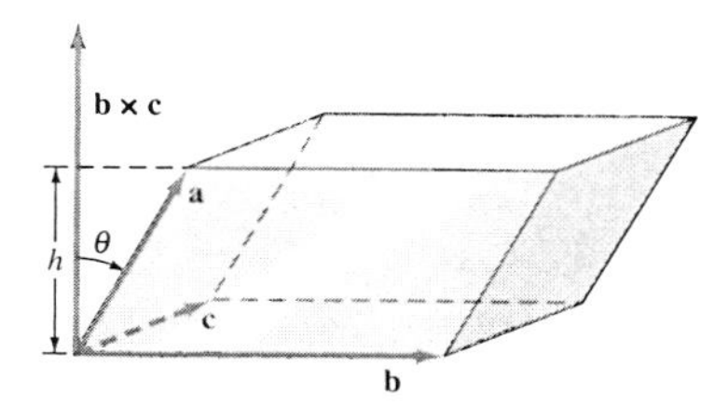

FIGURE 12.3.6 The volume of the parallelepiped is $|\mathbf{a} \cdot (\mathbf{b} \times \mathbf{c})|$.

The importance of the scalar triple product for applications depends on the following geometric interpretation. Let **a**, **b**, and **c** be three vectors with the same initial point. Figure 12.3.6 shows the parallelepiped determined by these vectors—that is, with arrows representing these vectors as adjacent edges. If the vectors **a**, **b**, and **c** are coplanar (lie in a single plane), then the parallelepiped is *degenerate* and its volume is zero. Theorem 4 holds whether or not the three vectors are coplanar, but it is most useful when they are not.

THEOREM 4 Scalar Triple Products and Volume

The volume V of the parallelepiped determined by the vectors **a**, **b**, and **c** is the absolute value of the scalar triple product $\mathbf{a} \cdot (\mathbf{b} \times \mathbf{c})$; that is,

$$V = |\mathbf{a} \cdot (\mathbf{b} \times \mathbf{c})|. \tag{18}$$

PROOF If the three vectors are coplanar, then **a** and $\mathbf{b} \times \mathbf{c}$ are perpendicular, so $V = |\mathbf{a} \cdot (\mathbf{b} \times \mathbf{c})| = 0$. Assume, then, that they are not coplanar. By Eq. (9) the area of the base (determined by **b** and **c**) of the parallelepiped is $A = |\mathbf{b} \times \mathbf{c}|$.

Now let α be the *acute* angle between **a** and the vector $\mathbf{b} \times \mathbf{c}$ that is perpendicular to the base. Then the height of the parallelepiped is $h = |\mathbf{a}| \cos\alpha$. If θ is the angle between the vectors **a** and $\mathbf{b} \times \mathbf{c}$, then either $\theta = \alpha$ or $\theta = \pi - \alpha$. Hence $\cos\alpha = |\cos\theta|$, so

$$V = Ah = |\mathbf{b} \times \mathbf{c}|\,|\mathbf{a}| \cos\alpha = |\mathbf{a}|\,|\mathbf{b} \times \mathbf{c}|\,|\cos\theta| = |\mathbf{a} \cdot (\mathbf{b} \times \mathbf{c})|.$$

Thus we have verified Eq. (18). ◄

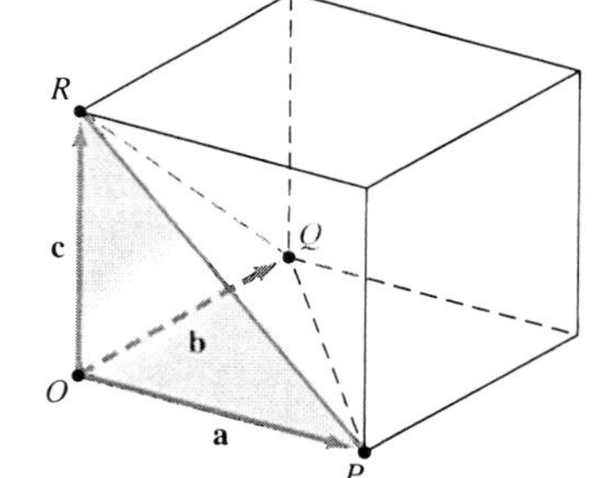

FIGURE 12.3.7 The pyramid (and parallelepiped) of Example 7.

EXAMPLE 7 Figure 12.3.7 shows the pyramid $OPQR$ and the parallelepiped both determined by the vectors

$$\mathbf{a} = \overrightarrow{OP} = \langle 3, 2, -1 \rangle, \quad \mathbf{b} = \overrightarrow{OQ} = \langle -2, 5, 1 \rangle, \quad \text{and} \quad \mathbf{c} = \overrightarrow{OR} = \langle 2, 1, 5 \rangle.$$

The volume of the pyramid is $V = \frac{1}{3}Ah$, where h is its height and the area A of its base OPQ is *half* the area of the corresponding base of the parallelepiped. It therefore follows from Eq. (17) and (18) that V is one-sixth the volume of the parallelepiped:

$$V = \frac{1}{6}|\mathbf{a} \cdot (\mathbf{b} \times \mathbf{c})| = \frac{1}{6}\begin{vmatrix} 3 & 2 & -1 \\ -2 & 5 & 1 \\ 2 & 1 & 5 \end{vmatrix} = \frac{108}{6} = 18.$$ ◆

EXAMPLE 8 Use the scalar triple product to show that the points $A(1, -1, 2)$, $B(2, 0, 1)$, $C(3, 2, 0)$, and $D(5, 4, -2)$ are coplanar.

Solution It's enough to show that the vectors $\overrightarrow{AB} = \langle 1, 1, -1 \rangle$, $\overrightarrow{AC} = \langle 2, 3, -2 \rangle$, and $\overrightarrow{AD} = \langle 4, 5, -4 \rangle$ are coplanar. But their scalar triple product is

$$\begin{vmatrix} 1 & 1 & -1 \\ 2 & 3 & -2 \\ 4 & 5 & -4 \end{vmatrix} = 1 \cdot (-2) - 1 \cdot 0 + (-1) \cdot (-2) = 0,$$

so Theorem 4 guarantees that the parallelepiped determined by these three vectors has volume zero. Hence the four given points are coplanar. ◆

The cross product occurs quite often in scientific applications. For example, suppose that a body in space is free to rotate around the fixed point O. If a force $\mathbf{F}$ acts at a point P of the body, that force causes the body to rotate. This effect is measured by the **torque vector** τ defined by the relation

$$\tau = \mathbf{r} \times \mathbf{F},$$

where $\mathbf{r} = \overrightarrow{OP}$, the straight line through O determined by τ is the axis of rotation, and the length

$$|\tau| = |\mathbf{r}|\,|\mathbf{F}| \sin\theta$$

is the **moment** of the force $\mathbf{F}$ around this axis (Fig. 12.3.8).

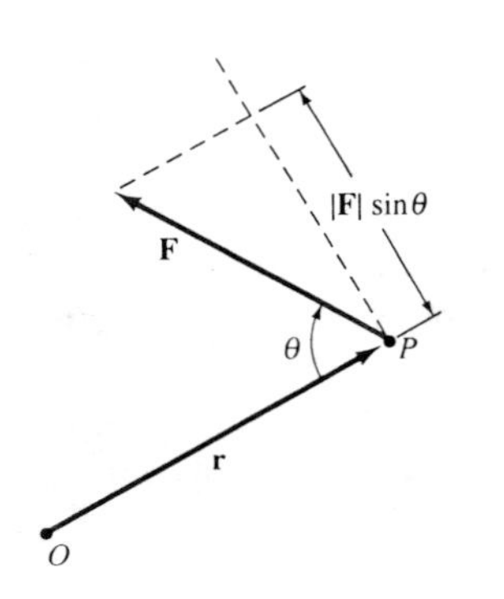

FIGURE 12.3.8 The torque vector τ is normal to both $\mathbf{r}$ and $\mathbf{F}$.

Another application of the cross product involves the force exerted on a moving charged particle by a magnetic field. This force is important in particle accelerators, mass spectrometers, and television picture tubes; controlling the paths of the ions is accomplished through the interplay of electric and magnetic fields. In such circumstances, the force $\mathbf{F}$ on the particle due to a magnetic field depends on three things: the charge q of the particle, its velocity vector $\mathbf{v}$, and the magnetic field vector $\mathbf{B}$ at the instantaneous location of the particle. And it turns out that

$$\mathbf{F} = (q\mathbf{v}) \times \mathbf{B}.$$

12.3 TRUE/FALSE STUDY GUIDE

12.3 CONCEPTS: QUESTIONS AND DISCUSSION

1. How does the cross product of two vectors resemble the ordinary product of two numbers? How do the two products differ?
2. Discuss the differences and the similarities between the dot product and the cross product of two vectors.
3. A surveyor measures a polygonal plot of land by first finding the coordinates of the vertices of its bounding polygon. Outline how the surveyor might then proceed to use cross products to calculate the area of the plot.

12.3 PROBLEMS

Find $\mathbf{a} \times \mathbf{b}$ *in Problems 1 through 4.*

1. $\mathbf{a} = \langle 5, -1, -2 \rangle, \quad \mathbf{b} = \langle -3, 2, 4 \rangle$

2. $\mathbf{a} = \langle 3, -2, 0 \rangle, \quad \mathbf{b} = \langle 0, 3, -2 \rangle$

3. $\mathbf{a} = \mathbf{i} - \mathbf{j} + 3\mathbf{k}, \quad \mathbf{b} = -2\mathbf{i} + 3\mathbf{j} + \mathbf{k}$

4. $\mathbf{a} = 4\mathbf{i} + 2\mathbf{j} - 2\mathbf{k}, \quad \mathbf{b} = 2\mathbf{i} - 5\mathbf{j} + 5\mathbf{k}$

In Problems 5 and 6, find the cross product of the given 2-dimensional vectors $\mathbf{a} = \langle a_1, a_2 \rangle$ *and* $\mathbf{b} = \langle b_1, b_2 \rangle$ *by first "extending" them to 3-dimensional vectors* $\mathbf{a} = \langle a_1, a_2, 0 \rangle$ *and* $\mathbf{b} = \langle b_1, b_2, 0 \rangle$.

5. $\mathbf{a} = \langle 2, -3 \rangle$ and $\mathbf{b} = \langle 4, 5 \rangle$

6. $\mathbf{a} = -5\mathbf{i} + 2\mathbf{j}$ and $\mathbf{b} = 7\mathbf{i} - 11\mathbf{j}$

In Problems 7 and 8, find two different unit vectors $\mathbf{u}$ *and* $\mathbf{v}$ *both of which are perpendicular to both the given vectors* $\mathbf{a}$ *and* $\mathbf{b}$.

7. $\mathbf{a} = \langle 3, 12, 0 \rangle$ and $\mathbf{b} = \langle 0, 4, 3 \rangle$

8. $\mathbf{a} = \mathbf{i} + 2\mathbf{j} + 3\mathbf{k}$ and $\mathbf{b} = 2\mathbf{i} + 3\mathbf{j} + 5\mathbf{k}$

9. Apply Eq. (5) to verify the equations in (11a).

10. Apply Eq. (5) to verify the equations in (11b).

11. Prove that the vector product is not associative by comparing $\mathbf{a} \times (\mathbf{b} \times \mathbf{c})$ with $(\mathbf{a} \times \mathbf{b}) \times \mathbf{c}$ in the case $\mathbf{a} = \mathbf{i}$, $\mathbf{b} = \mathbf{i} + \mathbf{j}$, and $\mathbf{c} = \mathbf{i} + \mathbf{j} + \mathbf{k}$.

12. Find nonzero vectors $\mathbf{a}$, $\mathbf{b}$, and $\mathbf{c}$ such that $\mathbf{a} \times \mathbf{b} = \mathbf{a} \times \mathbf{c}$ but $\mathbf{b} \neq \mathbf{c}$.

13. Suppose that the three vectors $\mathbf{a}$, $\mathbf{b}$, and $\mathbf{c}$ are mutually perpendicular. Prove that $\mathbf{a} \times (\mathbf{b} \times \mathbf{c}) = \mathbf{0}$.

14. Find the area of the triangle with vertices $P(1, 1, 0)$, $Q(1, 0, 1)$, and $R(0, 1, 1)$.

15. Find the area of the triangle with vertices $P(1, 3, -2)$, $Q(2, 4, 5)$, and $R(-3, -2, 2)$.

16. Find the volume of the parallelepiped with adjacent edges $\overrightarrow{OP}$, $\overrightarrow{OQ}$, and $\overrightarrow{OR}$, where P, Q, and R are the points given in Problem 14.

17. (a) Find the volume of the parallelepiped with adjacent edges $\overrightarrow{OP}$, $\overrightarrow{OQ}$, and $\overrightarrow{OR}$, where P, Q, and R are the points given in Problem 15. (b) Find the volume of the pyramid with vertices O, P, Q, and R.

18. Find a unit vector $\mathbf{n}$ perpendicular to the plane through the points P, Q, and R of Problem 15. Then find the distance from the origin to this plane by computing $\mathbf{n} \cdot \overrightarrow{OP}$.

In Problems 19 through 22, determine whether or not the four given points A, B, C, and D are coplanar. If not, find the volume of the pyramid with these four points as its vertices, given that its volume is one-sixth that of the parallelepiped spanned by $\overrightarrow{AB}$, $\overrightarrow{AC}$, *and* $\overrightarrow{AD}$.

19. $A(1, 3, -2)$, $B(3, 4, 1)$, $C(2, 0, -2)$, and $D(4, 8, 4)$

20. $A(13, -25, -37)$, $B(25, -14, -22)$, $C(24, -38, -25)$, and $D(26, 10, -19)$

21. $A(5, 2, -3)$, $B(6, 4, 0)$, $C(7, 5, 1)$, and $D(14, 14, 18)$

22. $A(25, 22, -33)$, $B(36, 34, -20)$, $C(27, 25, -29)$, and $D(34, 34, -12)$

23. Figure 12.3.9 shows a polygonal plot of land, with angles and lengths measured by a surveyor. First find the coordinates of each vertex. Then use the vector product [as in Eq. (10)] to calculate the area of the plot.

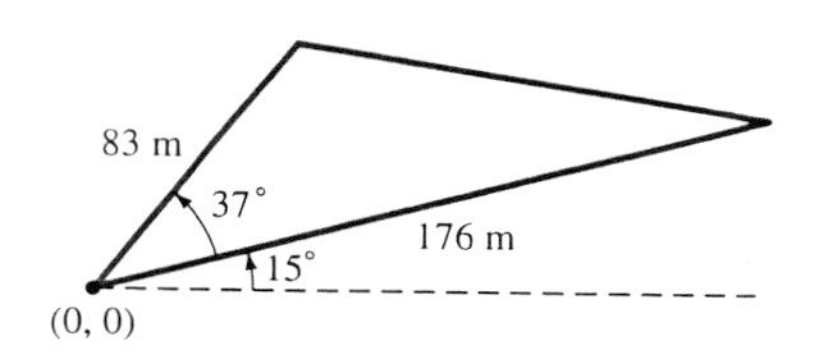

FIGURE 12.3.9 Problem 23.

24. Repeat Problem 23 with the plot shown in Fig. 12.3.10.

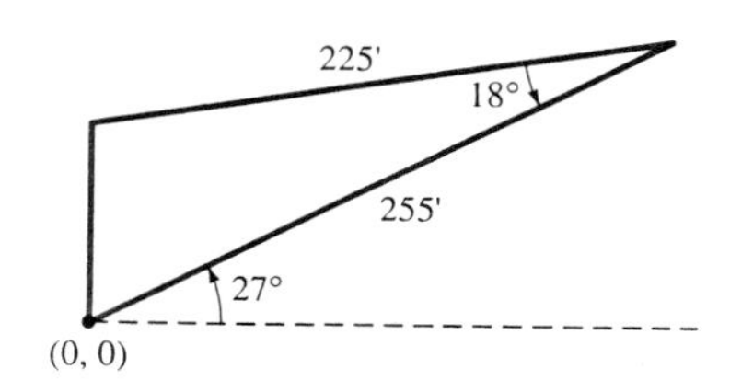

FIGURE 12.3.10 Problem 24.

25. Repeat Problem 23 with the plot shown in Fig. 12.3.11. [*Suggestion:* First divide the plot into two triangles.]

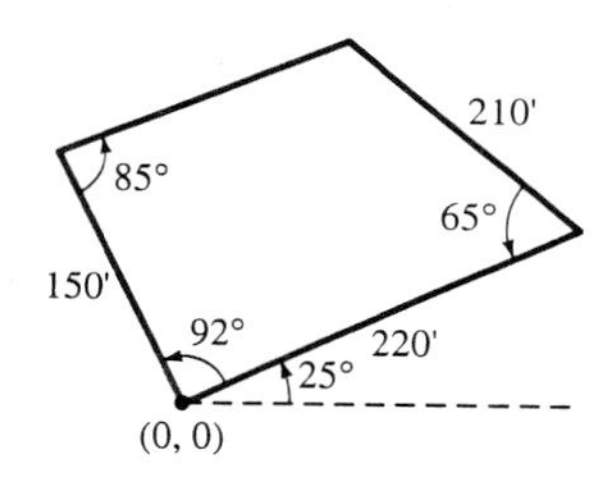

FIGURE 12.3.11 Problem 25.

26. Repeat Problem 23 with the plot shown in Fig. 12.3.12.

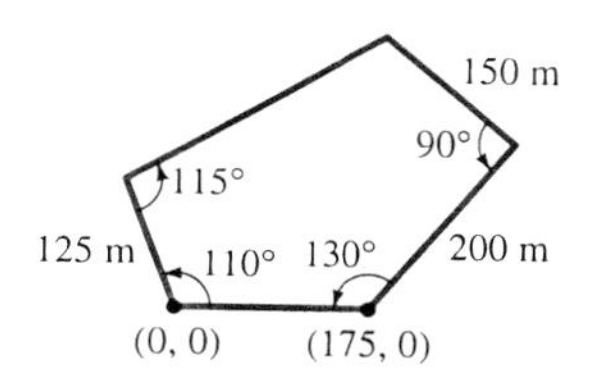

FIGURE 12.3.12 Problem 26.

27. Apply Eq. (5) to verify Eq. (12), the anticommutativity of the vector product.

28. Apply Eq. (17) to verify the identity for scalar triple products stated in Eq. (15).

29. Suppose that P and Q are points on a line L in space. Let A be a point not on L (Fig. 12.3.13). (a) Calculate in two ways the area of the triangle APQ to show that the perpendicular distance from A to the line L is

$$d = \frac{|\overrightarrow{AP} \times \overrightarrow{AQ}|}{|\overrightarrow{PQ}|}.$$

(b) Use this formula to compute the distance from the point $(1, 0, 1)$ to the line through the two points $P(2, 3, 1)$ and $Q(-3, 1, 4)$.

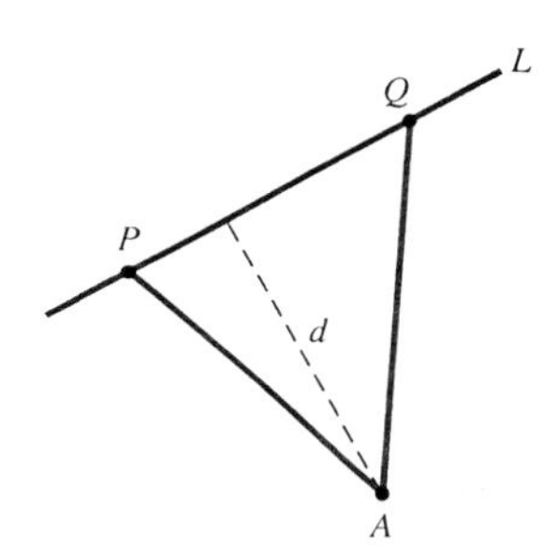

FIGURE 12.3.13 Problem 29.

30. Suppose that A is a point not on the plane determined by the three points P, Q, and R. Calculate in two ways the volume of the pyramid $APQR$ to show that the perpendicular distance from A to this plane is

$$d = \frac{|\overrightarrow{AP} \cdot (\overrightarrow{AQ} \times \overrightarrow{AR})|}{|\overrightarrow{PQ} \times \overrightarrow{PR}|}.$$

Use this formula to compute the distance from the point $(1, 0, 1)$ to the plane through the points $P(2, 3, 1)$, $Q(3, -1, 4)$, and $R(0, 0, 2)$.

31. Suppose that P_1 and Q_1 are two points on the line L_1 and that P_2 and Q_2 are two points on the line L_2. If the lines L_1 and L_2 are not parallel, then the perpendicular distance d between them is the projection of $\overrightarrow{P_1P_2}$ onto a vector $\mathbf{n}$ that is perpendicular both to $\overrightarrow{P_1Q_1}$ and $\overrightarrow{P_2Q_2}$. Prove that

$$d = \frac{|\overrightarrow{P_1P_2} \cdot (\overrightarrow{P_1Q_1} \times \overrightarrow{P_2Q_2})|}{|\overrightarrow{P_1Q_1} \times \overrightarrow{P_2Q_2}|}.$$

32. Use the following method to establish that the **vector triple product** $(\mathbf{a}\times\mathbf{b})\times\mathbf{c}$ is equal to $(\mathbf{a}\cdot\mathbf{c})\mathbf{b}-(\mathbf{b}\cdot\mathbf{c})\mathbf{a}$. (a) Let $\mathbf{I}$ be a unit vector in the direction of $\mathbf{a}$ and let $\mathbf{J}$ be a unit vector perpendicular to $\mathbf{I}$ and parallel to the plane of $\mathbf{a}$ and $\mathbf{b}$. Let $\mathbf{K}=\mathbf{I}\times\mathbf{J}$. Explain why there are scalars a_1, b_1, b_2, c_1, c_2, and c_3 such that

$$\mathbf{a}=a_1\mathbf{I},\quad \mathbf{b}=b_1\mathbf{I}+b_2\mathbf{J},\quad \text{and}\quad \mathbf{c}=c_1\mathbf{I}+c_2\mathbf{J}+c_3\mathbf{K}.$$

(b) Now show that

$$(\mathbf{a}\times\mathbf{b})\times\mathbf{c}=-a_1b_2c_2\mathbf{I}+a_1b_2c_1\mathbf{J}.$$

(c) Finally, substitute for $\mathbf{I}$ and $\mathbf{J}$ in terms of $\mathbf{a}$ and $\mathbf{b}$.

33. By permutation of the vectors $\mathbf{a}$, $\mathbf{b}$, and $\mathbf{c}$, deduce from Problem 32 that

$$\mathbf{a}\times(\mathbf{b}\times\mathbf{c})=(\mathbf{a}\cdot\mathbf{c})\mathbf{b}-(\mathbf{a}\cdot\mathbf{b})\mathbf{c}$$

[this is Eq. (16)].

34. Deduce from the orthogonality properties of the vector product that the vector $(\mathbf{a}\times\mathbf{b})\times(\mathbf{c}\times\mathbf{d})$ can be written in the form $r_1\mathbf{a}+r_2\mathbf{b}$ and in the form $s_1\mathbf{c}+s_2\mathbf{d}$.

35. Consider the triangle in the xy-plane that has vertices $(x_1, y_1, 0)$, $(x_2, y_2, 0)$, and $(x_3, y_3, 0)$. Use the vector product to prove that the area of this triangle is *half* the *absolute value* of the determinant

$$\begin{vmatrix} 1 & x_1 & y_1 \\ 1 & x_2 & y_2 \\ 1 & x_3 & y_3 \end{vmatrix}.$$

36. Given the vectors $\mathbf{a}=\langle a_1, a_2, a_3\rangle$ and $\mathbf{b}=\langle b_1, b_2, b_3\rangle$, verify Eq. (6),

$$|\mathbf{a}\times\mathbf{b}|^2=|\mathbf{a}|^2\,|\mathbf{b}|^2-(\mathbf{a}\cdot\mathbf{b})^2,$$

by computing each side in terms of the components of $\mathbf{a}$ and $\mathbf{b}$.

12.4 LINES AND PLANES IN SPACE

Just as in the plane, a straight line in space is determined by any two points P_0 and P_1 that lie on it. We may write $\mathbf{v}=\overrightarrow{P_0P_1}$—meaning that the directed line segment $\overrightarrow{P_0P_1}$ represents the vector $\mathbf{v}$—to describe the "direction of the line." Thus, alternatively, a line in space can be specified by giving a point P_0 on it *and* a [nonzero] vector $\mathbf{v}$ that determines the direction of the line.

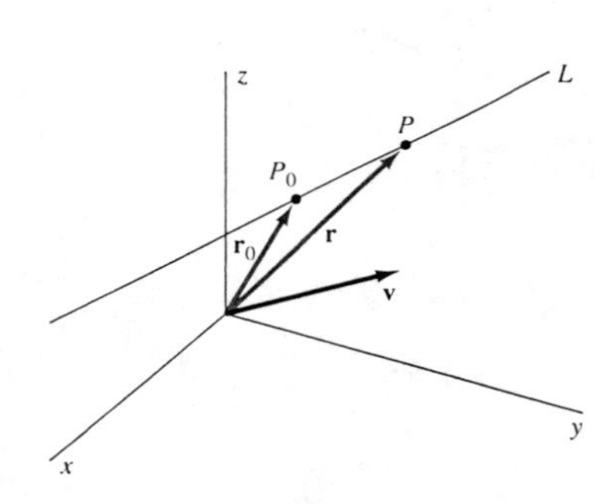

FIGURE 12.4.1 Finding the equation of the line L that passes through the point P_0 and is parallel to the vector $\mathbf{v}$.

To investigate equations that describe lines in space, let us begin with a straight line L that passes through the point $P_0(x_0, y_0, z_0)$ and is parallel to the vector $\mathbf{v}=a\mathbf{i}+b\mathbf{j}+c\mathbf{k}$ (Fig. 12.4.1). Then another point $P(x, y, z)$ lies on the line L if and only if the vectors $\mathbf{v}$ and $\overrightarrow{P_0P}$ are parallel, in which case

$$\overrightarrow{P_0P}=t\mathbf{v} \tag{1}$$

for some real number t. If $\mathbf{r}_0=\overrightarrow{OP_0}$ and $\mathbf{r}=\overrightarrow{OP}$ are the position vectors of the points P_0 and P, respectively, then $\overrightarrow{P_0P}=\mathbf{r}-\mathbf{r}_0$. Hence Eq. (1) gives the **vector equation**

$$\mathbf{r}=\mathbf{r}_0+t\mathbf{v} \tag{2}$$

describing the line L. As indicated in Fig. 12.4.1, $\mathbf{r}$ is the position vector of an *arbitrary* point P on the line L, and Eq. (2) gives $\mathbf{r}$ in terms of the parameter t, the position vector $\mathbf{r}_0$ of a *fixed* point P_0 on L, and the fixed vector $\mathbf{v}$ that determines the direction of L.

The left- and right-hand sides of Eq. (2) are equal, and each side is a vector. So corresponding components are also equal. When we write the resulting equations, we get a scalar description of the line L. Because $\mathbf{r}_0=\langle x_0, y_0, z_0\rangle$ and $\mathbf{r}=\langle x, y, z\rangle$, Eq. (2) thereby yields the three scalar equations

$$x=x_0+at,\qquad y=y_0+bt,\qquad z=z_0+ct. \tag{3}$$

These are **parametric equations** of the line L that passes through the point (x_0, y_0, z_0) and is parallel to the vector $\mathbf{v}=\langle a, b, c\rangle$.

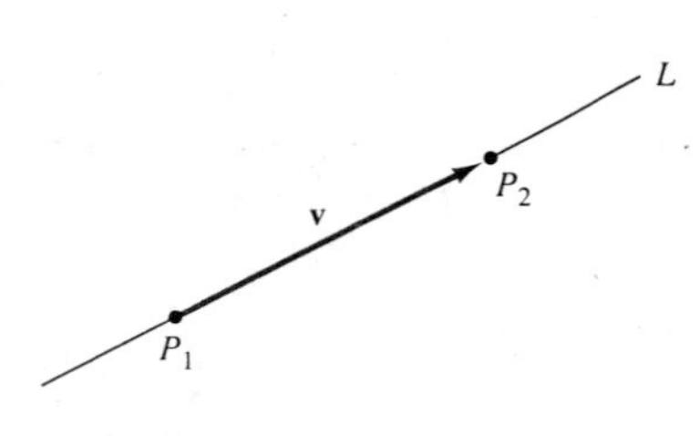

FIGURE 12.4.2 The line L of Example 1.

EXAMPLE 1 Write parametric equations of the line L that passes through the points $P_1(1, 2, 2)$ and $P_2(3, -1, 3)$ of Fig. 12.4.2.

Solution The line L is parallel to the vector

$$\mathbf{v}=\overrightarrow{P_1P_2}=(3\mathbf{i}-\mathbf{j}+3\mathbf{k})-(\mathbf{i}+2\mathbf{j}+2\mathbf{k})=2\mathbf{i}-3\mathbf{j}+\mathbf{k},$$

so we take $a = 2$, $b = -3$, and $c = 1$. With P_1 as the fixed point, the equations in (3) give

$$x = 1 + 2t, \qquad y = 2 - 3t, \qquad z = 2 + t$$

as parametric equations of L. In contrast, with P_2 as the fixed point and with the vector

$$-2\mathbf{v} = -4\mathbf{i} + 6\mathbf{j} - 2\mathbf{k}$$

(parallel to $\mathbf{v}$) as the direction vector, the equations in (3) yield the parametric equations

$$x = 3 - 4t, \qquad y = -1 + 6t, \qquad z = 3 - 2t.$$

Thus the parametric equations of a line are not unique. ◆

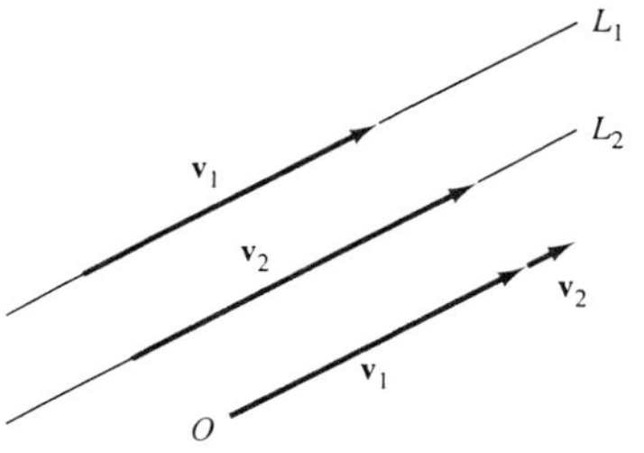

Given two straight lines L_1 and L_2 with parametric equations

$$x = x_1 + a_1 t, \qquad y = y_1 + b_1 t, \qquad z = z_1 + c_1 t \tag{4}$$

and

$$x = x_2 + a_2 s, \qquad y = y_2 + b_2 s, \qquad z = z_2 + c_2 s, \tag{5}$$

respectively, we can see at a glance whether or not L_1 and L_2 are parallel. Because L_1 is parallel to $\mathbf{v}_1 = \langle a_1, b_1, c_1 \rangle$ and L_2 is parallel to $\mathbf{v}_2 = \langle a_2, b_2, c_2 \rangle$, it follows that the lines L_1 and L_2 are parallel if and only if the vectors $\mathbf{v}_1$ and $\mathbf{v}_2$ are scalar multiples of each other (Fig. 12.4.3). If the two lines are not parallel, we can attempt to find a point of intersection by solving the equations

FIGURE 12.4.3 Parallel lines.

$$x_1 + a_1 t = x_2 + a_2 s \quad \text{and} \quad y_1 + b_1 t = y_2 + b_2 s$$

simultaneously for s and t. If these values of s and t also satisfy the equation $z_1 + c_1 t = z_2 + c_2 s$, then we have found a point of intersection. Its rectangular coordinates can be found by substituting the resulting value of t into Eq. (4) [or the resulting value of s into Eq. (5)]. Otherwise, the lines L_1 and L_2 do not intersect. Two nonparallel and nonintersecting lines in space are called **skew lines** (Fig. 12.4.4).

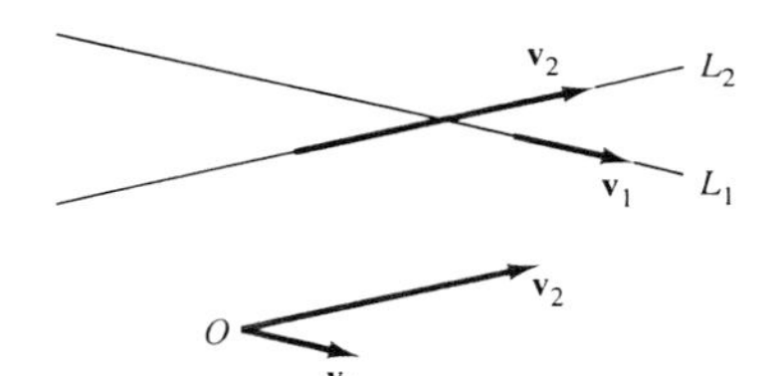

FIGURE 12.4.4 Skew lines.

EXAMPLE 2 The line L_1 with parametric equations

$$x = 1 + 2t, \qquad y = 2 - 3t, \qquad z = 2 + t$$

passes through the point $P_1(1, 2, 2)$ (discovered by substituting $t = 0$) and is parallel to the vector $\mathbf{v}_1 = \langle 2, -3, 1 \rangle$. The line L_2 with parametric equations

$$x = 3 + 4t, \qquad y = 1 - 6t, \qquad z = 5 + 2t$$

passes through the point $P_2(3, 1, 5)$ and is parallel to the vector $\mathbf{v}_2 = \langle 4, -6, 2 \rangle$. Because $\mathbf{v}_2 = 2\mathbf{v}_1$, we see that L_1 and L_2 are parallel.

But are L_1 and L_2 actually different lines, or are we perhaps dealing with two different parametrizations of the same line? To answer this question, we note that $\overrightarrow{P_1P_2} = \langle 2, -1, 3 \rangle$ is not a multiple of, and therefore is not parallel to, $\mathbf{v}_1 = \langle 2, -3, 1 \rangle$. Thus the point P_2 does not lie on the line L_1, and hence the lines L_1 and L_2 are indeed distinct. ◆

If all the coefficients a, b, and c in (3) are nonzero, then we can eliminate the parameter t. Simply solve each equation for t and then set the resulting expressions equal to each other. This gives

$$\frac{x - x_0}{a} = \frac{y - y_0}{b} = \frac{z - z_0}{c}. \tag{6}$$

These are called the **symmetric equations** of the line L. If one or more of a or b or c is zero, this means that L lies in a plane parallel to one of the coordinate planes, and in this case the line does not have symmetric equations. For example, if $c = 0$, then L lies in the horizontal plane $z = z_0$. Of course, it is still possible to write equations for L that don't include the parameter t; if $c = 0$, for instance, but a and b are nonzero, then we could describe the line L as the set of points (x, y, z) satisfying the equations

$$\frac{x - x_0}{a} = \frac{y - y_0}{b}, \qquad z = z_0.$$

EXAMPLE 3 Find both parametric and symmetric equations of the line L through the points $P_0(3, 1, -2)$ and $P_1(4, -1, 1)$. Find also the points at which L intersects the three coordinate planes.

Solution The line L is parallel to the vector $\mathbf{v} = \overrightarrow{P_0P_1} = \langle 1, -2, 3\rangle$, so we take $a = 1$, $b = -2$, and $c = 3$. The equations in (3) then give the parametric equations

$$x = 3 + t, \qquad y = 1 - 2t, \qquad z = -2 + 3t$$

of L, and the equations in (6) give the symmetric equations

$$\frac{x - 3}{1} = \frac{y - 1}{-2} = \frac{z + 2}{3}.$$

To find the point at which L intersects the xy-plane, we set $z = 0$ in the symmetric equations. This gives

$$\frac{x - 3}{1} = \frac{y - 1}{-2} = \frac{2}{3},$$

and so $x = \frac{11}{3}$ and $y = -\frac{1}{3}$. Thus L meets the xy-plane at the point $(\frac{11}{3}, -\frac{1}{3}, 0)$. Similarly, $x = 0$ gives $(0, 7, -11)$ for the point where L meets the yz-plane, and $y = 0$ gives $(\frac{7}{2}, 0, -\frac{1}{2})$ for its intersection with the xz-plane. ◆

Planes in Space

A plane $\mathcal{P}$ in space is determined by a point $P_0(x_0, y_0, z_0)$ through which $\mathcal{P}$ passes and a line through P_0 that is normal to $\mathcal{P}$. Alternatively, we may be given P_0 on $\mathcal{P}$ and a vector $\mathbf{n} = \langle a, b, c\rangle$ normal to the plane $\mathcal{P}$. The point $P(x, y, z)$ lies on the plane $\mathcal{P}$ if and only if the vectors $\mathbf{n}$ and $\overrightarrow{P_0P}$ are perpendicular (Fig. 12.4.5), in which case $\mathbf{n} \cdot \overrightarrow{P_0P} = 0$. We write $\overrightarrow{P_0P} = \mathbf{r} - \mathbf{r}_0$, where $\mathbf{r}$ and $\mathbf{r}_0$ are the position vectors $\mathbf{r} = \overrightarrow{OP}$ and $\mathbf{r}_0 = \overrightarrow{OP_0}$ of the points P and P_0, respectively. Thus we obtain a **vector equation**

$$\mathbf{n} \cdot (\mathbf{r} - \mathbf{r}_0) = 0 \tag{7}$$

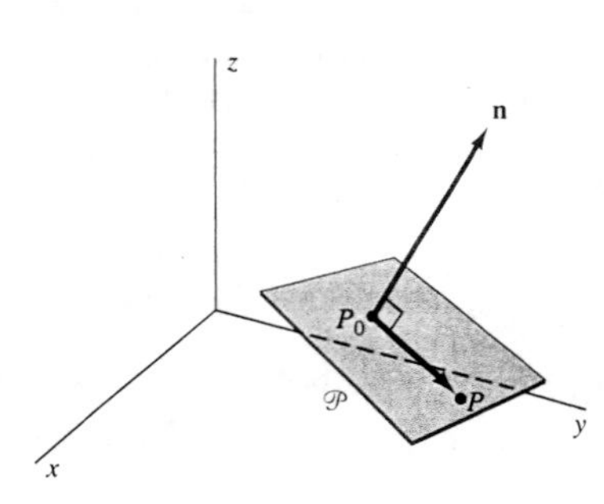

FIGURE 12.4.5 Because **n** is normal to $\mathcal{P}$, it follows that **n** is normal to $\overrightarrow{P_0P}$ for all points P in $\mathcal{P}$.

of the plane $\mathcal{P}$.

If we substitute $\mathbf{n} = \langle a, b, c\rangle$, $\mathbf{r} = \langle x, y, z\rangle$, and $\mathbf{r}_0 = \langle x_0, y_0, z_0\rangle$ into Eq. (7), we thereby obtain a **scalar equation**

$$a(x - x_0) + b(y - y_0) + c(z - z_0) = 0 \tag{8}$$

of the plane through $P_0(x_0, y_0, z_0)$ with **normal vector** $\mathbf{n} = \langle a, b, c\rangle$.

EXAMPLE 4 An equation of the plane through $P_0(-1, 5, 2)$ with normal vector $\mathbf{n} = \langle 1, -3, 2\rangle$ is

$$1 \cdot (x + 1) + (-3) \cdot (y - 5) + 2 \cdot (z - 2) = 0;$$

that is, $x - 3y + 2z = -12$. ◆

IMPORTANT The coefficients of x, y, and z in the last equation are the components of the normal vector. This is always the case, because we can write Eq. (8) in the form

$$ax + by + cz = d, \tag{9}$$

where $d = ax_0 + by_0 + cz_0$. Conversely, every *linear equation* in x, y, and z of the form in Eq. (9) represents a plane in space provided that the coefficients a, b, and c are not all zero. The reason is that if $c \neq 0$ (for instance), then we can choose x_0 and y_0 arbitrarily and solve the equation $ax_0 + by_0 + cz_0 = d$ for z_0. With these values, Eq. (9) takes the form

$$ax + by + cz = ax_0 + by_0 + cz_0;$$

that is,

$$a(x - x_0) + b(y - y_0) + c(z - z_0) = 0,$$

so this equation represents the plane through (x_0, y_0, z_0) with normal vector $\langle a, b, c\rangle$.

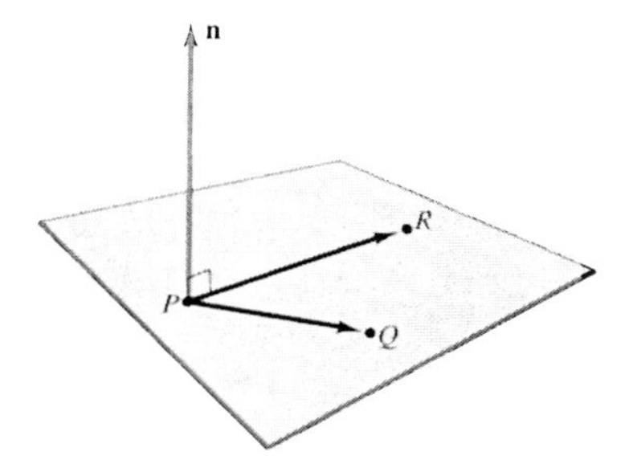

FIGURE 12.4.6 The normal vector **n** as a cross product (Example 5).

EXAMPLE 5 Find an equation for the plane through the three points $P(2, 4, -3)$, $Q(3, 7, -1)$, and $R(4, 3, 0)$.

Solution We want to use Eq. (8), so we first need a vector **n** that is normal to the plane in question. One easy way to obtain such a normal vector is by using the cross product. Let

$$\mathbf{n} = \overrightarrow{PQ} \times \overrightarrow{PR} = \begin{vmatrix} \mathbf{i} & \mathbf{j} & \mathbf{k} \\ 1 & 3 & 2 \\ 2 & -1 & 3 \end{vmatrix} = 11\mathbf{i} + \mathbf{j} - 7\mathbf{k}.$$

Because $\overrightarrow{PQ}$ and $\overrightarrow{PR}$ are in the plane, their cross product **n** is normal to the plane (Fig. 12.4.6). Hence the plane has equation

$$11(x - 2) + (y - 4) - 7(z + 3) = 0.$$

After simplifying, we write the equation as

$$11x + y - 7z = 47.$$

◆

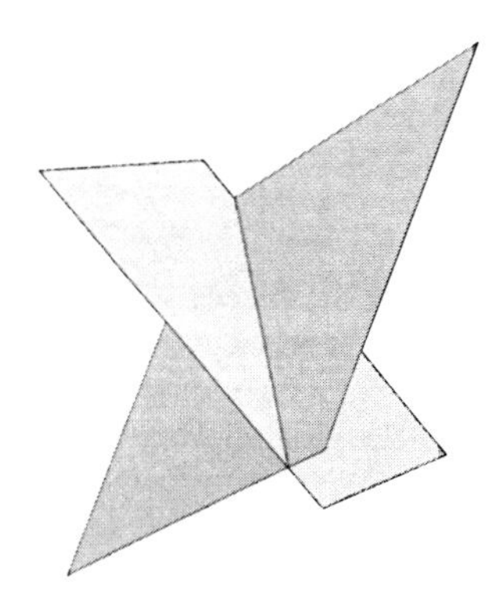

FIGURE 12.4.7 The intersection of two nonparallel planes is a straight line.

Two planes with normal vectors **n** and **m** are said to be **parallel** provided that **n** and **m** are parallel. Otherwise, the two planes meet in a straight line (Fig. 12.4.7), and we can find the angle θ between the normal vectors **n** and **m** (Fig. 12.4.8). We then define the **angle** between the two planes to be either θ or $\pi - \theta$, whichever is an acute angle.

EXAMPLE 6 Find the angle θ between the planes with equations

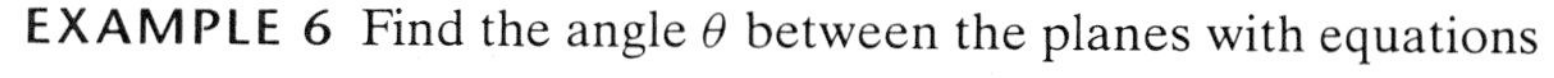

$$2x + 3y - z = -3 \quad \text{and} \quad 4x + 5y + z = 1.$$

Then write symmetric equations of their line of intersection L.

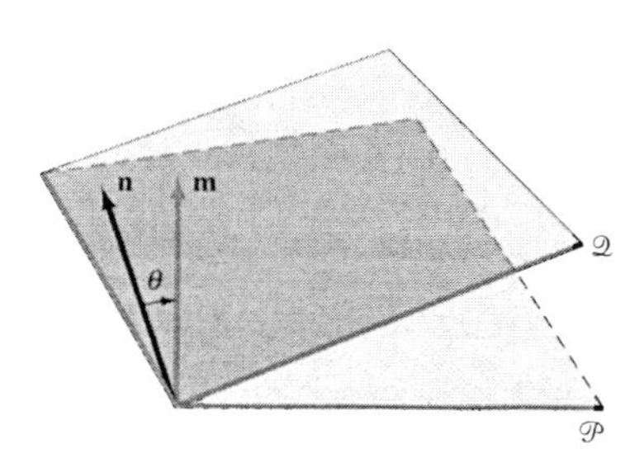

FIGURE 12.4.8 Vectors **m** and **n** normal to the planes $\mathcal{P}$ and $\mathcal{Q}$, respectively.

Solution The vectors $\mathbf{n} = \langle 2, 3, -1\rangle$ and $\mathbf{m} = \langle 4, 5, 1\rangle$ are normal to the two planes, so

$$\cos\theta = \frac{\mathbf{n}\cdot\mathbf{m}}{|\mathbf{n}|\,|\mathbf{m}|} = \frac{22}{\sqrt{14}\sqrt{42}}.$$

Hence $\theta = \cos^{-1}(\frac{11}{21}\sqrt{3}) \approx 24.87°$.

To determine the line of intersection L of the two planes, we need first to find a point P_0 that lies on L. We can do this by substituting an arbitrarily chosen value of x into the equations of the given planes and then solving the resulting equations for y and z. With $x = 1$ we get the equations

$$2 + 3y - z = -3,$$

$$4 + 5y + z = 1.$$

The common solution is $y = -1$, $z = 2$. Thus the point $P_0(1, -1, 2)$ lies on the line L.

Next we need a vector $\mathbf{v}$ parallel to L. The vectors $\mathbf{n}$ and $\mathbf{m}$ normal to the two planes are both perpendicular to L, so their cross product is parallel to L. Alternatively, we can find a second point P_1 on L by substituting a second value of x into the equations of the given planes and solving for y and z, as before. With $x = 5$ we obtain the equations

$$10 + 3y - z = -3,$$
$$20 + 5y + z = 1,$$

with common solution $y = -4$, $z = 1$. Thus we obtain a second point $P_1(5, -4, 1)$ on L and thereby the vector

$$\mathbf{v} = \overrightarrow{P_0P_1} = \langle 4, -3, -1 \rangle$$

parallel to L. From Eq. (6) we now find symmetric equations

$$\frac{x-1}{4} = \frac{y+1}{-3} = \frac{z-2}{-1}$$

of the line of intersection of the two given planes. ◆

Finally, we may note that the symmetric equations of a line L present the line as an intersection of planes: We can rewrite the equations in (6) in the form

$$\begin{aligned} b(x - x_0) - a(y - y_0) &= 0, \\ c(x - x_0) - a(z - z_0) &= 0, \\ c(y - y_0) - b(z - z_0) &= 0. \end{aligned} \tag{10}$$

These are the equations of three planes that intersect in the line L. The first has normal vector $\langle b, -a, 0 \rangle$, a vector parallel to the xy-plane. So the first plane is perpendicular to the xy-plane. Similarly, the second plane is perpendicular to the xz-plane and the third is perpendicular to the yz-plane.

The equations in (10) are symmetric equations of the line that passes through the point $P_0(x_0, y_0, z_0)$ and is parallel to $\mathbf{v} = \langle a, b, c \rangle$. Unlike the equations in (6), these equations are meaningful whether or not all the components a, b, and c are nonzero. They have a special form, though, if one of the three components is zero. If, say, $a = 0$, then the first two equations in (10) take the form $x = x_0$. The line is then the intersection of the two planes $x = x_0$ and $c(y - y_0) = b(z - z_0)$.

EXAMPLE 7 In Example 3 we saw that the line L through the point $P_0(3, 1, -2)$ and $P_1(4, -1, 1)$ has symmetric equations

$$\frac{x-3}{1} = \frac{y-1}{-2} = \frac{z+2}{3}.$$

Proceeding to rewrite these equations as in (10), we obtain first the equations

$$\begin{aligned} -2(x-3) &= y - 1, \\ 3(x-3) &= z + 2, \\ 3(y-1) &= -2(z+2) \end{aligned}$$

and then (upon simplification) the equations

$$\begin{aligned} 2x + y &= 7, \\ 3x - z &= 11, \\ 3y + 2z &= -1 \end{aligned}$$

that represent L as the intersection of three planes, each of them parallel to one of the three coordinate axes in space. Figure 12.4.9 shows a computer plot of these three planes intersecting in the line L. ◆

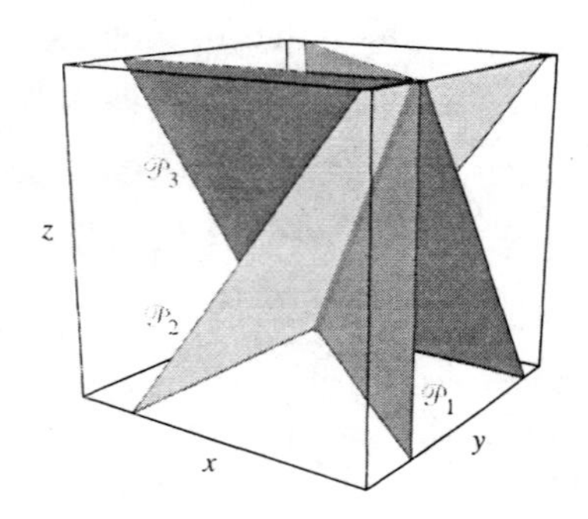

FIGURE 12.4.9 The line L of Example 7 is the intersection of the plane $\mathcal{P}_1$ parallel to the z-axis, the plane $\mathcal{P}_2$ parallel to the y-axis, and the plane $\mathcal{P}_3$ parallel to the x-axis.

12.4 TRUE/FALSE STUDY GUIDE

12.4 CONCEPTS: QUESTIONS AND DISCUSSION

1. Figure 12.4.10 shows the possible configuration of two lines L_1 and L_2 in the xy-plane. We see that the intersection of L_1 and L_2 can consists of either one point, no points, or infinitely many points. Explain why this geometric observation implies that two linear equations $a_1x + b_1y = c_1$ and $a_2x + b_2y = c_2$ in two unknowns x and y can have either a single simultaneous solution (x, y), no solution, or infinitely many different solutions.

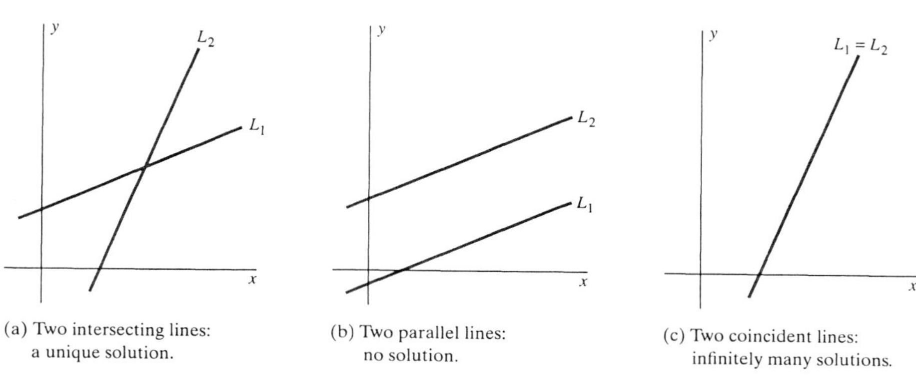

FIGURE 12.4.10 (a) The nonparallel lines L_1 and L_2 intersect in a single point. (b) The distinct parallel lines L_1 and L_2 have no point of intersection. (c) The coincident lines L_1 and L_2 have infinitely many points in common.

In each of the following cases, describe similarly the possible configurations and hence the possible number of points of intersection of the indicated number of lines or planes. Translate your geometric conclusion into a statement about the possible number of solutions of a system of two or three linear equations in two or three unknowns.

2. Three lines in the plane.
3. Two lines in space.
3. Three planes in space.

12.4 PROBLEMS

In Problems 1 through 4, write parametric equations of the straight line that passes through the point P and is parallel to the vector $\mathbf{v}$.

1. $P(0, 0, 0)$, $\mathbf{v} = \mathbf{i} + 2\mathbf{j} + 3\mathbf{k}$
2. $P(3, -4, 5)$, $\mathbf{v} = -2\mathbf{i} + 7\mathbf{j} + 3\mathbf{k}$
3. $P(4, 13, -3)$, $\mathbf{v} = 2\mathbf{i} - 3\mathbf{k}$
4. $P(17, -13, -31)$, $\mathbf{v} = \langle -17, 13, 31\rangle$

In Problems 5 through 8, write parametric equations of the straight line that passes through the points P_1 and P_2.

5. $P_1(0, 0, 0)$, $P_2(-6, 3, 5)$
6. $P_1(3, 5, 7)$, $P_2(6, -8, 10)$
7. $P_1(3, 5, 7)$, $P_2(6, 5, 4)$
8. $P_1(29, -47, 13)$, $P_2(73, 53, -67)$

In Problems 9 through 14, write both parametric and symmetric equations for the indicated straight line.

9. Through $P(2, 3, -4)$ and parallel to $\mathbf{v} = \langle 1, -1, -2\rangle$
10. Through $P(2, 5, -7)$ and $Q(4, 3, 8)$
11. Through $P(1, 1, 1)$ and perpendicular to the xy-plane
12. Through the origin and perpendicular to the plane with equation $x + y + z = 1$
13. Through $P(2, -3, 4)$ and perpendicular to the plane with equation $2x - y + 3z = 4$
14. Through $P(2, -1, 5)$ and parallel to the line with parametric equations $x = 3t$, $y = 2 + t$, $z = 2 - t$

In Problems 15 through 20, determine whether the two lines L_1 and L_2 are parallel, skew, or intersecting. If they intersect, find the point of intersection.

15. L_1: $x - 2 = \frac{1}{2}(y + 1) = \frac{1}{3}(z - 3)$;
    L_2: $\frac{1}{3}(x - 5) = \frac{1}{2}(y - 1) = z - 4$
16. L_1: $\frac{1}{4}(x - 11) = y - 6 = -\frac{1}{2}(z + 5)$;
    L_2: $\frac{1}{6}(x - 13) = -\frac{1}{3}(y - 2) = \frac{1}{8}(z - 5)$
17. L_1: $x = 6 + 2t$, $y = 5 + 2t$, $z = 7 + 3t$;
    L_2: $x = 7 + 3s$, $y = 5 + 3s$, $z = 10 + 5s$
18. L_1: $x = 14 + 3t$, $y = 7 + 2t$, $z = 21 + 5t$;
    L_2: $x = 5 + 3s$, $y = 15 + 5s$, $z = 10 + 7s$

19. L_1: $\frac{1}{6}(x-7) = \frac{1}{4}(y+5) = -\frac{1}{8}(z-9)$;
L_2: $-\frac{1}{9}(x-11) = -\frac{1}{6}(y-7) = \frac{1}{12}(z-13)$

20. L_1: $x = 13 + 12t, \quad y = -7 + 20t, \quad z = 11 - 28t$;
L_2: $x = 22 + 9s, \quad y = 8 + 15s, \quad z = -10 - 21s$

In Problems 21 through 24, write an equation of the plane with normal vector $\mathbf{n}$ *that passes through the point* P.

21. $P(0, 0, 0), \quad \mathbf{n} = \langle 1, 2, 3\rangle$

22. $P(3, -4, 5), \quad \mathbf{n} = \langle -2, 7, 3\rangle$

23. $P(5, 12, 13), \quad \mathbf{n} = \mathbf{i} - \mathbf{k}$

24. $P(5, 12, 13), \quad \mathbf{n} = \mathbf{j}$

In Problems 25 through 32, write an equation of the indicated plane.

25. Through $P(5, 7, -6)$ and parallel to the xz-plane

26. Through $P(1, 0, -1)$ with normal vector $\mathbf{n} = \langle 2, 2, -1\rangle$

27. Through $P(10, 4, -3)$ with normal vector $\mathbf{n} = \langle 7, 11, 0\rangle$

28. Through $P(1, -3, 2)$ with normal vector $\mathbf{n} = \overrightarrow{OP}$

29. Through the origin and parallel to the plane with equation $3x + 4y = z + 10$

30. Through $P(5, 1, 4)$ and parallel to the plane with equation $x + y - 2z = 0$

31. Through the origin and the points $P(1, 1, 1)$ and $Q(1, -1, 3)$

32. Through the points $A(1, 0, -1)$, $B(3, 3, 2)$, and $C(4, 5, -1)$

In Problems 33 and 34, write an equation of the plane that contains both the point P *and the line* L.

33. $P(2, 4, 6)$; $\quad L: x = 7 - 3t, \quad y = 3 + 4t, \quad z = 5 + 2t$

34. $P(13, -7, 29)$; $\quad L: x = 17 - 9t, \quad y = 23 + 14t, \quad z = 35 - 41t$

In Problems 35 through 38, determine whether the line L *and the plane* $\mathcal{P}$ *intersect or are parallel. If they intersect, find the point of intersection.*

35. L: $x = 7 - 4t, \quad y = 3 + 6t, \quad z = 9 + 5t$;
$\mathcal{P}$: $4x + y + 2z = 17$

36. L: $x = 15 + 7t, \quad y = 10 + 12t, \quad z = 5 - 4t$;
$\mathcal{P}$: $12x - 5y + 6z = 50$

37. L: $x = 3 + 2t, \quad y = 6 - 5t, \quad z = 2 + 3t$;
$\mathcal{P}$: $3x + 2y - 4z = 1$

38. L: $x = 15 - 3t, \quad y = 6 - 5t, \quad z = 21 - 14t$;
$\mathcal{P}$: $23x + 29y - 31z = 99$

In Problems 39 through 42, find the angle between the planes with the given equations.

39. $x = 10$ and $x + y + z = 0$

40. $2x - y + z = 5$ and $x + y - z = 1$

41. $x - y - 2z = 1$ and $x - y - 2z = 5$

42. $2x + y + z = 4$ and $3x - y - z = 3$

In Problems 43 through 46, write both parametric and symmetric equations of the line of intersection of the indicated planes.

43. The planes of Problem 39

44. The planes of Problem 40

45. The planes of Problem 41

46. The planes of Problem 42

47. Write symmetric equations for the line through $P(3, 3, 1)$ that is parallel to the line of Problem 46.

48. Find an equation of the plane through $P(3, 3, 1)$ that is perpendicular to the planes $x + y = 2z$ and $2x + z = 10$.

49. Find an equation of the plane through $(1, 1, 1)$ that intersects the xy-plane in the same line as does the plane $3x + 2y - z = 6$.

50. Find an equation for the plane that passes through the point $P(1, 3, -2)$ and contains the line of intersection of the planes $x - y + z = 1$ and $x + y - z = 1$.

51. Find an equation of the plane that passes through the points $P(1, 0, -1)$ and $Q(2, 1, 0)$ and is parallel to the line of intersection of the planes $x + y + z = 5$ and $3x - y = 4$.

52. Prove that the lines $x - 1 = \frac{1}{2}(y + 1) = z - 2$ and $x - 2 = \frac{1}{3}(y - 2) = \frac{1}{2}(z - 4)$ intersect. Find an equation of the [only] plane that contains them both.

53. Prove that the line of intersection of the planes $x + 2y - z = 2$ and $3x + 2y + 2z = 7$ is parallel to the line $x = 1 + 6t$, $y = 3 - 5t$, $z = 2 - 4t$. Find an equation of the plane determined by these two lines.

54. Show that the perpendicular distance D from the point $P_0(x_0, y_0, z_0)$ to the plane $ax + by + cz = d$ is

$$D = \frac{|ax_0 + by_0 + cz_0 - d|}{\sqrt{a^2 + b^2 + c^2}}.$$

[*Suggestion:* The line that passes through P_0 and is perpendicular to the given plane has parametric equations $x = x_0 + at$, $y = y_0 + bt$, $z = z_0 + ct$. Let $P_1(x_1, y_1, z_1)$ be the point of this line, corresponding to $t = t_1$, at which it intersects the given plane. Solve for t_1, and then compute $D = |\overrightarrow{P_0P_1}|$.]

In Problems 55 and 56, use the formula of Problem 54 to find the distance between the given point and the given plane.

55. The origin and the plane $x + y + z = 10$

56. The point $P(5, 12, -13)$ and the plane with equation $3x + 4y + 5z = 12$

57. Prove that any two skew lines lie in parallel planes.

58. Use the formula of Problem 54 to show that the perpendicular distance D between the two parallel planes $ax + by + cz + d_1 = 0$ and $ax + by + cz + d_2 = 0$ is

$$D = \frac{|d_1 - d_2|}{\sqrt{a^2 + b^2 + c^2}}.$$

59. The line L_1 is described by the equations

$$x - 1 = 2y + 2, \qquad z = 4.$$

The line L_2 passes through the points $P(2, 1, -3)$ and $Q(0, 8, 4)$. (a) Show that L_1 and L_2 are skew lines. (b) Use the results of Problems 57 and 58 to find the perpendicular distance between L_1 and L_2.

60. Find the shortest distance between points of the line L_1 with parametric equations

$$x = 7 + 2t, \quad y = 11 - 5t, \quad z = 13 + 6t$$

and the line L_2 of intersection of the planes $3x - 2y + 4z = 10$ and $5x + 3y - 2z = 15$.

12.5 CURVES AND MOTION IN SPACE

In Section 10.4 we discussed parametric curves in the plane. Now think of a point that moves along a curve in three-dimensional space. We can describe this point's changing position by means of *parametric equations*

$$x = f(t), \qquad y = g(t), \qquad z = h(t) \tag{1}$$

that specify its coordinates as functions of time t. A **parametric curve** C in space is (by definition) simply a triple (f, g, h) of such *coordinate functions*. But often it is useful to refer informally to C as the trajectory in space that is traced out by a moving point with these coordinate functions. Space curves exhibit a number of interesting new phenomena that we did not see with plane curves.

FIGURE 12.5.1 A tubular knot whose centerline is the parametric curve of Example 1.

EXAMPLE 1 Figure 12.5.1 shows a common *trefoil knot* in space defined by the parametric equations

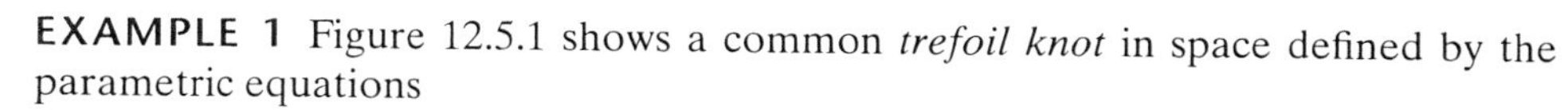

$$x(t) = \left(2 + \cos \tfrac{3}{2}t\right) \cos t, \qquad y(t) = \left(2 + \cos \tfrac{3}{2}t\right) \sin t, \qquad z(t) = \sin \tfrac{3}{2}t.$$

Actually, to enhance the three-dimensional appearance of this curve's shape, we have plotted in the figure a thin tubular surface whose centerline is the knot itself. The viewpoint for the computer plot is so chosen that we are looking down on the curve from a point on the positive z-axis. ◆

EXAMPLE 2 Figure 12.5.2 shows simultaneously the circle

$$x(t) = 4 \cos t, \qquad y(t) = 4 \sin t, \qquad z(t) \equiv 0$$

in the xy-plane, the ellipse

$$x(t) = 5 \cos t, \qquad y(t) \equiv 0, \qquad z(t) = 3 \sin t$$

in the xz-plane, and the ellipse

$$x(t) \equiv 0, \qquad y(t) = 3 \cos t, \qquad z(t) = 5 \sin t$$

in the yz-plane. Here, again, we actually have plotted thin tubular tori having these closed curves as centerlines. Can you see that any two of these curves are unlinked, but that the three together apparently cannot be "pulled apart"? ◆

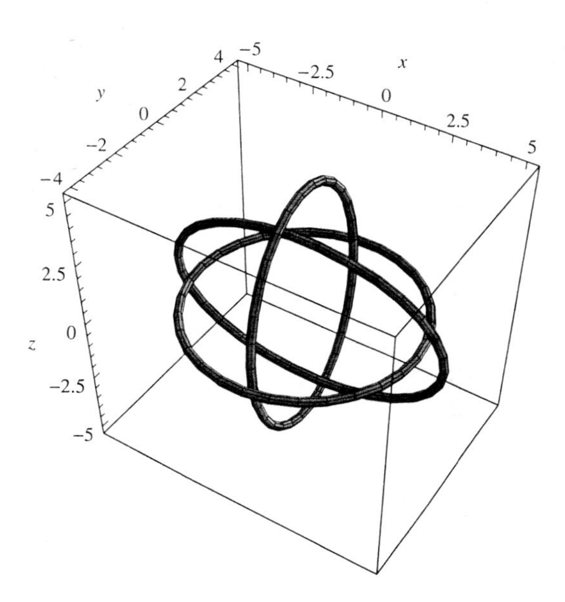

FIGURE 12.5.2 The Borromean rings of Example 2.

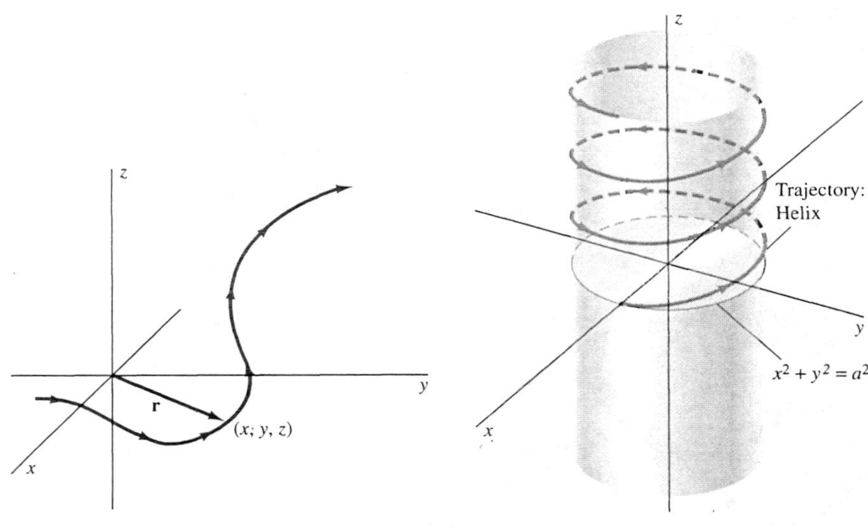

FIGURE 12.5.3 The position vector $\mathbf{r} = \langle x, y, z \rangle$ of a moving particle in space.

FIGURE 12.5.4 The point of Example 3 moves in a helical path.

Vector-Valued Functions

The changing location of a point moving along the parametric curve in (1) can be described by giving its **position vector**

$$\mathbf{r}(t) = x(t)\mathbf{i} + y(t)\mathbf{j} + z(t)\mathbf{k} = \langle x(t), y(t), z(t) \rangle, \tag{2}$$

or simply

$$\mathbf{r} = x\mathbf{i} + y\mathbf{j} + z\mathbf{k} = \langle x, y, z \rangle,$$

whose components are the coordinate functions of the moving point (Fig. 12.5.3). Equation (2) defines a **vector-valued function** that associates with the number t the vector $\mathbf{r}(t)$. In the case of a plane curve described by a two-dimensional position vector, we may suppress the third component in Eq. (2) and write $\mathbf{r}(t) = x(t)\mathbf{i} + y(t)\mathbf{j} = \langle x(t), y(t) \rangle$.

FIGURE 12.5.5 The intertwined helices that model the DNA molecule served as a model for the DNA Tower in Kings Park, Perth, Australia. For a fascinating account of the discovery of the role of the helix as the genetic basis for life itself, see James D. Watson, *The Double Helix* (New York: Atheneum, 1968).

EXAMPLE 3 The position vector

$$\mathbf{r}(t) = \mathbf{i}\cos t + \mathbf{j}\sin t + t\mathbf{k} \tag{3}$$

describes the **helix** of Fig. 12.5.4. Because $x^2 + y^2 = \cos^2 t + \sin^2 t = 1$ for all t, the projection $(x(t), y(t))$ into the xy-plane moves around and around the unit circle. Meanwhile, because $z = t$, the point $(\cos t, \sin t, t)$ steadily moves upward on the vertical cylinder in space that stands above and below the circle $x^2 + y^2 = 1$ in the xy-plane. The familiar corkscrew shape of the helix appears everywhere from the coiled springs of an automobile to the *double helix* model of the DNA molecule that carries the genetic information of living cells (Fig. 12.5.5). ◆

Much of the calculus of (ordinary) real-valued functions applies to vector-valued functions. To begin with, the **limit** of a vector-valued function $\mathbf{r} = \langle f, g, h \rangle$ is *defined* as follows:

$$\begin{aligned} \lim_{t\to a} \mathbf{r}(t) &= \left\langle \lim_{t\to a} f(t), \lim_{t\to a} g(t), \lim_{t\to a} h(t) \right\rangle \\ &= \mathbf{i}\left(\lim_{t\to a} f(t)\right) + \mathbf{j}\left(\lim_{t\to a} g(t)\right) + \mathbf{k}\left(\lim_{t\to a} h(t)\right), \end{aligned} \tag{4}$$

provided that the limits in the last three expressions exist. Thus we take limits of vector-valued functions by taking limits of their component functions.

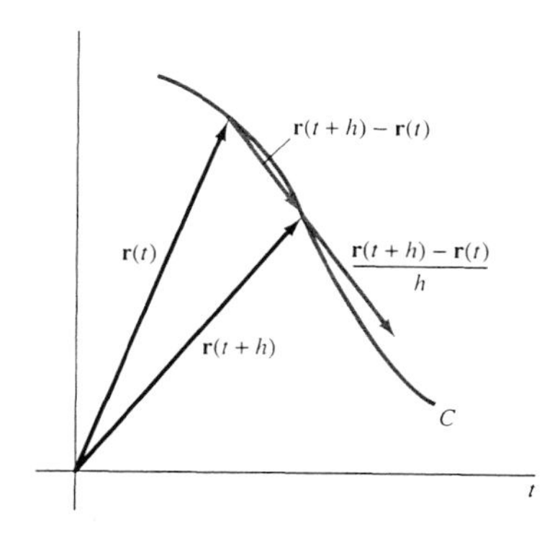

FIGURE 12.5.6 Geometry of the derivative of a vector-valued function.

We say that $\mathbf{r} = \mathbf{r}(t)$ is **continuous** at the number a provided that

$$\lim_{t \to a} \mathbf{r}(t) = \mathbf{r}(a).$$

This amounts to saying that $\mathbf{r}$ is continuous at a if and only if its component functions f, g, and h are continuous at a.

The **derivative** $\mathbf{r}'(t)$ of the vector-valued function $\mathbf{r}(t)$ is defined in almost exactly the same way as the derivative of a real-valued function. Specifically,

$$\mathbf{r}'(t) = \lim_{\Delta t \to 0} \frac{\mathbf{r}(t + \Delta t) - \mathbf{r}(t)}{\Delta t}, \tag{5}$$

provided that this limit exists. Figures 12.5.6 and 12.5.7 correctly suggest that the **derivative vector**

$$\mathbf{r}'(t) = \frac{d\mathbf{r}}{dt} = D_t[\mathbf{r}(t)]$$

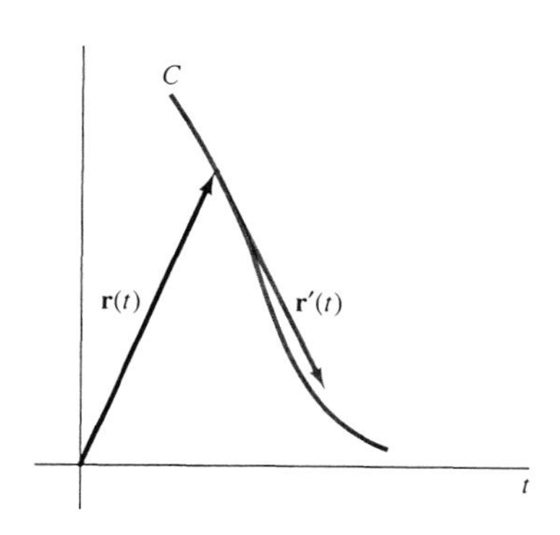

FIGURE 12.5.7 The derivative vector is tangent to the curve at the point of evaluation.

will be tangent to the curve C with position vector $\mathbf{r}(t)$. For this reason, we call $\mathbf{r}'(t)$ a **tangent vector** to the curve C at the corresponding point P provided that $\mathbf{r}'(t)$ exists and is nonzero there. The **tangent line** to C at this point P with position vector $\mathbf{r}(t)$ is then the line through P determined by $\mathbf{r}'(t)$.

Our next result implies the simple *but important* fact that the derivative vector $\mathbf{r}'(t)$ can be calculated by **componentwise differentiation** of $\mathbf{r}(t)$—that is, by differentiating separately the component functions of $\mathbf{r}(t)$.

THEOREM 1 Componentwise Differentiation

Suppose that

$$\mathbf{r}(t) = \langle f(t), g(t), h(t) \rangle = f(t)\mathbf{i} + g(t)\mathbf{j} + h(t)\mathbf{k},$$

where f, g, and h are differentiable functions. Then

$$\mathbf{r}'(t) = \langle f'(t), g'(t), h'(t) \rangle = f'(t)\mathbf{i} + g'(t)\mathbf{j} + h'(t)\mathbf{k}. \tag{6}$$

That is, if $\mathbf{r} = x\mathbf{i} + y\mathbf{j} + z\mathbf{k}$, then

$$\frac{d\mathbf{r}}{dt} = \frac{dx}{dt}\mathbf{i} + \frac{dy}{dt}\mathbf{j} + \frac{dz}{dt}\mathbf{k}.$$

PROOF We take the limit in Eq. (5) simply by taking limits of components. We find that

$$\begin{aligned}
\mathbf{r}'(t) &= \lim_{\Delta t \to 0} \frac{\Delta \mathbf{r}}{\Delta t} = \lim_{\Delta t \to 0} \frac{\mathbf{r}(t + \Delta t) - \mathbf{r}(t)}{\Delta t} \\
&= \lim_{\Delta t \to 0} \frac{f(t + \Delta t)\mathbf{i} + g(t + \Delta t)\mathbf{j} + h(t + \Delta t)\mathbf{k} - f(t)\mathbf{i} - g(t)\mathbf{j} - h(t)\mathbf{k}}{\Delta t} \\
&= \left(\lim_{\Delta t \to 0} \frac{f(t + \Delta t) - f(t)}{\Delta t} \right)\mathbf{i} + \left(\lim_{\Delta t \to 0} \frac{g(t + \Delta t) - g(t)}{\Delta t} \right)\mathbf{j} \\
&\qquad + \left(\lim_{\Delta t \to 0} \frac{h(t + \Delta t) - h(t)}{\Delta t} \right)\mathbf{k} \\
&= f'(t)\mathbf{i} + g'(t)\mathbf{j} + h'(t)\mathbf{k}.
\end{aligned}$$

◄

EXAMPLE 4 Find parametric equations of the line tangent to the helix C of Example 3 at the point $P(-1, 0, \pi)$ where $t = \pi$.

Solution Componentwise differentiation of $\mathbf{r}(t) = \mathbf{i}\cos t + \mathbf{j}\sin t + t\mathbf{k}$ yields

$$\mathbf{r}'(t) = -\mathbf{i}\sin t + \mathbf{j}\cos t + \mathbf{k},$$

so the vector tangent to C at P is $\mathbf{r}'(\pi) = -\mathbf{j} + \mathbf{k} = \langle 0, -1, 1\rangle$. It follows that the parametric equations of the line tangent at P—with its own position vector $\mathbf{r}(\pi) + t\mathbf{r}'(\pi)$—are

$$x = -1, \qquad y = -t, \qquad z = \pi + t.$$

In particular, we see that this tangent line lies in the vertical plane $x = -1$. ◆

Theorem 2 tells us that the formulas for computing derivatives of sums and products of vector-valued functions are formally similar to those for real-valued functions.

THEOREM 2 Differentiation Formulas

Let $\mathbf{u}(t)$ and $\mathbf{v}(t)$ be differentiable vector-valued functions. Let $h(t)$ be a differentiable real-valued function and let c be a (constant) scalar. Then

1. $D_t[\mathbf{u}(t) + \mathbf{v}(t)] = \mathbf{u}'(t) + \mathbf{v}'(t)$,
2. $D_t[c\mathbf{u}(t)] = c\mathbf{u}'(t)$,
3. $D_t[h(t)\mathbf{u}(t)] = h'(t)\mathbf{u}(t) + h(t)\mathbf{u}'(t)$,
4. $D_t[\mathbf{u}(t) \cdot \mathbf{v}(t)] = \mathbf{u}'(t) \cdot \mathbf{v}(t) + \mathbf{u}(t) \cdot \mathbf{v}'(t)$, and
5. $D_t[\mathbf{u}(t) \times \mathbf{v}(t)] = \mathbf{u}'(t) \times \mathbf{v}(t) + \mathbf{u}(t) \times \mathbf{v}'(t)$.

PROOF We'll prove part (4), working with two-dimensional vectors for simplicity, and leave the other parts as exercises. If

$$\mathbf{u}(t) = \langle f_1(t), f_2(t)\rangle \quad \text{and} \quad \mathbf{v}(t) = \langle g_1(t), g_2(t)\rangle,$$

then

$$\mathbf{u}(t) \cdot \mathbf{v}(t) = f_1(t)g_1(t) + f_2(t)g_2(t).$$

Hence the product rule for ordinary real-valued functions gives

$$\begin{aligned} D_t[\mathbf{u}(t) \cdot \mathbf{v}(t)] &= D_t[f_1(t)g_1(t) + f_2(t)g_2(t)] \\ &= [f_1'(t)g_1(t) + f_1(t)g_1'(t)] + [f_2'(t)g_2(t) + f_2(t)g_2'(t)] \\ &= [f_1'(t)g_1(t) + f_2'(t)g_2(t)] + [f_1(t)g_1'(t) + f_2(t)g_2'(t)] \\ &= \mathbf{u}'(t) \cdot \mathbf{v}(t) + \mathbf{u}(t) \cdot \mathbf{v}'(t). \end{aligned}$$ ◄

REMARK The order of the factors in part (5) of Theorem 2 *must* be preserved because the cross product is not commutative.

EXAMPLE 5 The trajectory of the parametric curve $\mathbf{r}(t) = a\mathbf{i}\cos t + a\mathbf{j}\sin t$ is the circle of radius a centered at the origin in the xy-plane. Because $\mathbf{r}(t) \cdot \mathbf{r}(t) = a^2$, a constant, part 4 of Theorem 2 gives

$$0 \equiv \frac{d}{dt}(a^2) = \frac{d}{dt}[\mathbf{r}(t) \cdot \mathbf{r}(t)] = \mathbf{r}'(t) \cdot \mathbf{r}(t) + \mathbf{r}(t) \cdot \mathbf{r}'(t) = 2\mathbf{r}'(t) \cdot \mathbf{r}(t).$$

Because $\mathbf{r}'(t) \cdot \mathbf{r}(t) \equiv 0$, we see that (consistent with elementary geometry) the tangent vector $\mathbf{r}'(t)$ is perpendicular to the position vector $\mathbf{r}(t)$ at every point of the circle (Fig. 12.5.8). ◆

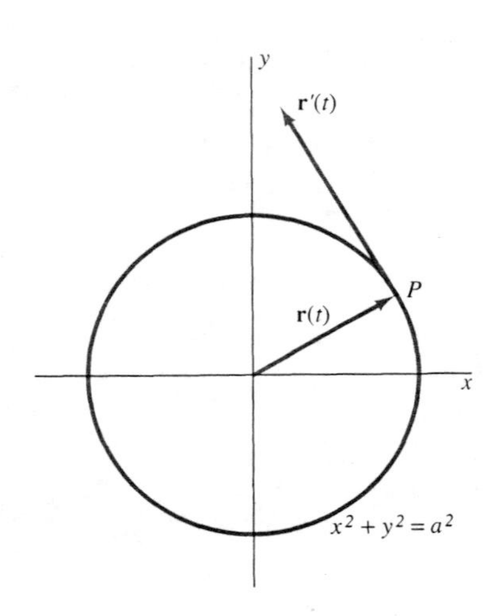

FIGURE 12.5.8 The position and tangent vectors for the circle of Example 5.

Velocity and Acceleration Vectors

Looking at Fig. 12.5.6 and the definition of $\mathbf{r}'(t)$ in Eq. (5), we note that $|\mathbf{r}(t + \Delta t) - \mathbf{r}(t)|$ is the distance from the point with position vector $\mathbf{r}(t)$ to the point with position vector $\mathbf{r}(t + \Delta t)$. It follows that the quotient

$$\frac{|\mathbf{r}(t + \Delta t) - \mathbf{r}(t)|}{\Delta t}$$

is the average speed of a particle that travels from $\mathbf{r}(t)$ to $\mathbf{r}(t + \Delta t)$ in time Δt. Consequently, the limit in Eq. (5) yields both the direction of motion and the instantaneous speed of a particle moving along a curve with position vector $\mathbf{r}(t)$.

We therefore define the **velocity vector** $\mathbf{v}(t)$ at time t of a point moving along a curve with position vector $\mathbf{r}(t)$ as the derivative

$$\mathbf{v}(t) = \mathbf{r}'(t) = f'(t)\mathbf{i} + g'(t)\mathbf{j} + h'(t)\mathbf{k}; \tag{7a}$$

in differential notation,

$$\mathbf{v} = \frac{d\mathbf{r}}{dt} = \frac{dx}{dt}\mathbf{i} + \frac{dy}{dt}\mathbf{j} + \frac{dz}{dt}\mathbf{k}. \tag{7b}$$

Its **acceleration vector** $\mathbf{a}(t)$ is given by

$$\mathbf{a}(t) = \mathbf{v}'(t) = f''(t)\mathbf{i} + g''(t)\mathbf{j} + h''(t)\mathbf{k}; \tag{8a}$$

alternatively,

$$\mathbf{a} = \frac{d\mathbf{v}}{dt} = \frac{d^2x}{dt^2}\mathbf{i} + \frac{d^2y}{dt^2}\mathbf{j} + \frac{d^2z}{dt^2}\mathbf{k}. \tag{8b}$$

Thus, for motion in the plane or in space, just as for motion along a line,

velocity is the **time derivative** of **position;**
acceleration is the **time derivative** of **velocity.**

The **speed** $v(t)$ and **scalar acceleration** $a(t)$ of the moving point are the lengths of its velocity and acceleration vectors, respectively:

$$v(t) = |\mathbf{v}(t)| = \sqrt{\left(\frac{dx}{dt}\right)^2 + \left(\frac{dy}{dt}\right)^2 + \left(\frac{dz}{dt}\right)^2} \tag{9}$$

and

$$a(t) = |\mathbf{a}(t)| = \sqrt{\left(\frac{d^2x}{dt^2}\right)^2 + \left(\frac{d^2y}{dt^2}\right)^2 + \left(\frac{d^2z}{dt^2}\right)^2}. \tag{10}$$

Note The scalar acceleration $a = |d\mathbf{v}/dt|$ is generally *not* equal to the derivative dv/dt of the speed of a moving point. The difference between the two is discussed in Section 12.6.

EXAMPLE 6 A particle moving along the parabola $y = x^2$ in the plane has position vector $\mathbf{r}(t) = t\mathbf{i} + t^2\mathbf{j}$. Find its velocity and acceleration vectors and its speed and scalar acceleration at the instant when $t = 2$.

Solution Because $\mathbf{r}(2) = 2\mathbf{i} + 4\mathbf{j}$, the location of the particle at time $t = 2$ is $(2, 4)$. Its velocity vector and speed are given by

$$\mathbf{v} = \mathbf{i} + 2t\mathbf{j} \quad \text{and} \quad v(t) = |\mathbf{v}(t)| = \sqrt{1 + 4t^2},$$

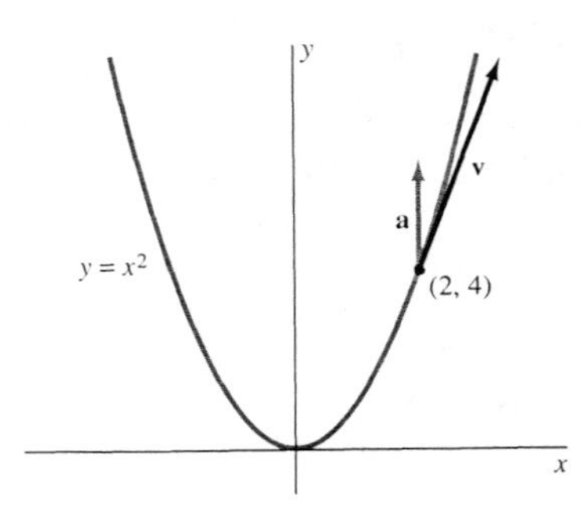

FIGURE 12.5.9 The velocity and acceleration vectors at $t = 2$ (Example 6).

so $\mathbf{v}(2) = \mathbf{i} + 4\mathbf{j}$ (a vector) and $v(2) = \sqrt{17}$ (a scalar). Its acceleration is $\mathbf{a}(t) = \mathbf{v}'(t) = 2\mathbf{j}$ (a constant vector), so $\mathbf{a} = 2\mathbf{j}$ and $a = |\mathbf{a}| = 2$ (scalar acceleration) for all t, including the instant at which $t = 2$. Figure 12.5.9 shows the trajectory of the particle with its velocity and acceleration vectors $\mathbf{v}(2)$ and $\mathbf{a}(2)$ attached at its location $(2, 4)$ when $t = 2$. ◆

EXAMPLE 7 Find the velocity, acceleration, speed, and scalar acceleration of a moving point P whose trajectory is the helix with position vector

$$\mathbf{r}(t) = (a\cos\omega t)\mathbf{i} + (a\sin\omega t)\mathbf{j} + bt\mathbf{k}. \tag{11}$$

Solution Equation (11) is a generalization of the position vector $\mathbf{r}(t) = \mathbf{i}\cos t + \mathbf{j}\sin t + t\mathbf{k}$ of the helix in Example 3. Here $x^2 + y^2 = a^2$, so the xy-projection $(a\cos\omega t, a\sin\omega t)$ of P lies on the circle of radius a centered at the origin. This projection moves around the circle with angular speed ω (radians per unit time). Meanwhile, the point P itself also is moving upward (if $b > 0$) on the vertical cylinder of radius a; the z-component of its velocity is $dz/dt = b$. Except for the radius of the cylinder, the picture looks the same as Fig. 12.5.4.

The derivative of the position vector in (11) is the velocity vector

$$\mathbf{v}(t) = (-a\omega\sin\omega t)\mathbf{i} + (a\omega\cos\omega t)\mathbf{j} + b\mathbf{k}. \tag{12}$$

Another differentiation gives its acceleration vector

$$\begin{aligned}\mathbf{a}(t) &= (-a\omega^2\cos\omega t)\mathbf{i} + (-a\omega^2\sin\omega t)\mathbf{j}\\ &= -a\omega^2(\mathbf{i}\cos\omega t + \mathbf{j}\sin\omega t).\end{aligned} \tag{13}$$

The speed of the moving point is a constant, because

$$v(t) = |\mathbf{v}(t)| = \sqrt{a^2\omega^2 + b^2}.$$

Note that the acceleration vector is a horizontal vector of length $a\omega^2$. Moreover, if we think of $\mathbf{a}(t)$ as attached to the moving point at the time t of evaluation—so that the initial point of $\mathbf{a}(t)$ is the terminal point of $\mathbf{r}(t)$—then $\mathbf{a}(t)$ points directly toward the point $(0, 0, bt)$ on the z-axis. ◆

REMARK The helix of Example 7 is a typical trajectory of a charged particle in a constant magnetic field. Such a particle must satisfy both Newton's law $\mathbf{F} = m\mathbf{a}$ and the magnetic force law $\mathbf{F} = (q\mathbf{v}) \times \mathbf{B}$ mentioned in Section 12.3. Hence its velocity and acceleration vectors must satisfy the equation

$$(q\mathbf{v}) \times \mathbf{B} = m\mathbf{a}. \tag{14}$$

If the constant magnetic field is vertical, $\mathbf{B} = B\mathbf{k}$, then with the velocity vector of Eq. (12) we find that

$$q\mathbf{v} \times \mathbf{B} = q\begin{vmatrix}\mathbf{i} & \mathbf{j} & \mathbf{k}\\ -a\omega\sin\omega t & a\omega\cos\omega t & b\\ 0 & 0 & B\end{vmatrix} = qa\omega B(\mathbf{i}\cos\omega t + \mathbf{j}\sin\omega t).$$

The acceleration vector in Eq. (13) gives

$$m\mathbf{a} = -ma\omega^2(\mathbf{i}\cos\omega t + \mathbf{j}\sin\omega t).$$

When we compare the last two results, we see that the helix of Example 7 satisfies Eq. (14) provided that

$$qa\omega B = -ma\omega^2; \quad \text{that is,} \quad \omega = -\frac{qB}{m}.$$

For example, this equation would determine the angular speed ω for the helical trajectory of electrons ($q < 0$) in a cathode-ray tube placed in a constant magnetic field parallel to the axis of the tube (Fig. 12.5.10).

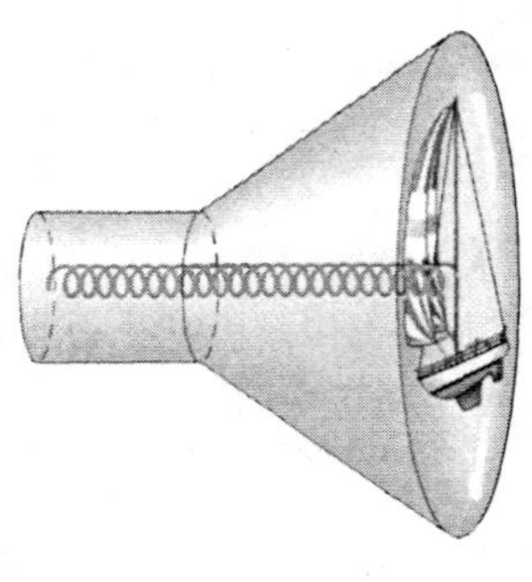

FIGURE 12.5.10 A spiraling electron in a cathode-ray tube.

Integration of Vector-Valued Functions

Integrals of vector-valued functions are defined by analogy with the definition of an integral of a real-valued function:

$$\int_a^b \mathbf{r}(t)\,dt = \lim_{\Delta t \to 0} \sum_{i=1}^{n} \mathbf{r}(t_i^\star)\,\Delta t,$$

where $t_i^\star$ is a point of the ith subinterval of a division of $[a, b]$ into n subintervals, all with the same length $\Delta t = (b-a)/n$.

If $\mathbf{r}(t) = f(t)\mathbf{i} + g(t)\mathbf{j}$ is continuous on $[a, b]$, then—by taking limits componentwise—we get

$$\int_a^b \mathbf{r}(t)\,dt = \lim_{\Delta t \to 0} \sum_{i=1}^{n} \mathbf{r}(t_i^\star)\,\Delta t$$

$$= \mathbf{i}\left(\lim_{\Delta t \to 0} \sum_{i=1}^{n} f(t_i^\star)\,\Delta t\right) + \mathbf{j}\left(\lim_{\Delta t \to 0} \sum_{i=1}^{n} g(t_i^\star)\,\Delta t\right).$$

This gives the result that

$$\int_a^b \mathbf{r}(t)\,dt = \mathbf{i}\left(\int_a^b f(t)\,dt\right) + \mathbf{j}\left(\int_a^b g(t)\,dt\right). \qquad \textbf{(15)}$$

Thus *a vector-valued function may be integrated componentwise.* The three-dimensional version of Eq. (15) is derived in the same way, merely including third components.

Now suppose that $\mathbf{R}(t)$ is an *antiderivative* of $\mathbf{r}(t)$, meaning that $\mathbf{R}'(t) = \mathbf{r}(t)$. That is, if $\mathbf{R}(t) = F(t)\mathbf{i} + G(t)\mathbf{j}$, then

$$\mathbf{R}'(t) = F'(t)\mathbf{i} + G'(t)\mathbf{j} = f(t)\mathbf{i} + g(t)\mathbf{j} = \mathbf{r}(t).$$

Then componentwise integration yields

$$\int_a^b \mathbf{r}(t)\,dt = \mathbf{i}\left(\int_a^b f(t)\,dt\right) + \mathbf{j}\left(\int_a^b g(t)\,dt\right) = \mathbf{i}\Big[F(t)\Big]_a^b + \mathbf{j}\Big[G(t)\Big]_a^b$$

$$= [F(b)\mathbf{i} + G(b)\mathbf{j}] - [F(a)\mathbf{i} + G(a)\mathbf{j}].$$

Thus the *fundamental theorem of calculus* for vector-valued functions takes the form

$$\int_a^b \mathbf{r}(t)\,dt = \Big[\mathbf{R}(t)\Big]_a^b = \mathbf{R}(b) - \mathbf{R}(a), \qquad \textbf{(16)}$$

where $\mathbf{R}'(t) = \mathbf{r}(t)$ is continuous on $[a, b]$.

Indefinite integrals of vector-valued functions may be computed as well. If $\mathbf{R}'(t) = \mathbf{r}(t)$, then every antiderivative of $\mathbf{r}(t)$ is of the form $\mathbf{R}(t) + \mathbf{C}$ for some constant vector $\mathbf{C}$. We therefore write

$$\int \mathbf{r}(t)\,dt = \mathbf{R}(t) + \mathbf{C} \quad \text{if} \quad \mathbf{R}'(t) = \mathbf{r}(t), \qquad \textbf{(17)}$$

on the basis of a componentwise computation similar to the one leading to Eq. (16).

If $\mathbf{r}(t)$, $\mathbf{v}(t)$, and $\mathbf{a}(t)$ are the position, velocity, and acceleration vectors of a point moving in space, then the vector derivatives

$$\frac{d\mathbf{r}}{dt} = \mathbf{v} \quad \text{and} \quad \frac{d\mathbf{v}}{dt} = \mathbf{a}$$

imply the indefinite integrals

$$\mathbf{v}(t) = \int \mathbf{a}(t)\,dt \tag{18}$$

and

$$\mathbf{r}(t) = \int \mathbf{v}(t)\,dt. \tag{19}$$

Both of these integrals involve a *vector* constant of integration.

EXAMPLE 8 Suppose that a moving point has given initial position vector $\mathbf{r}(0) = 2\mathbf{i}$, initial velocity vector $\mathbf{v}(0) = \mathbf{i} - \mathbf{j}$, and acceleration vector $\mathbf{a}(t) = 2\mathbf{i} + 6t\mathbf{j}$. Find its position and velocity at time t.

Solution Equation (18) gives

$$\mathbf{v}(t) = \int \mathbf{a}(t)\,dt = \int (2\mathbf{i} + 6t\mathbf{j})\,dt = 2t\mathbf{i} + 3t^2\mathbf{j} + \mathbf{C}_1.$$

To evaluate the constant vector $\mathbf{C}_1$, we substitute $t = 0$ in this equation and find that $\mathbf{v}(0) = (0)\mathbf{i} + (0)\mathbf{j} + \mathbf{C}_1$, so $\mathbf{C}_1 = \mathbf{v}(0) = \mathbf{i} - \mathbf{j}$. Thus the velocity vector of the moving point at time t is

$$\mathbf{v}(t) = (2t\mathbf{i} + 3t^2\mathbf{j}) + (\mathbf{i} - \mathbf{j}) = (2t + 1)\mathbf{i} + (3t^2 - 1)\mathbf{j}.$$

A second integration, using Eq. (19), gives

$$\begin{aligned}\mathbf{r}(t) &= \int \mathbf{v}(t)\,dt \\ &= \int [(2t + 1)\mathbf{i} + (3t^2 - 1)\mathbf{j}]\,dt = (t^2 + t)\mathbf{i} + (t^3 - t)\mathbf{j} + \mathbf{C}_2.\end{aligned}$$

Again we substitute $t = 0$ and find that $\mathbf{C}_2 = \mathbf{r}(0) = 2\mathbf{i}$. Hence

$$\mathbf{r}(t) = (t^2 + t)\mathbf{i} + (t^3 - t)\mathbf{j} + 2\mathbf{i} = (t^2 + t + 2)\mathbf{i} + (t^3 - t)\mathbf{j}$$

is the position vector of the point at time t. ◆

Vector integration is the basis for at least one method of navigation. If a submarine is cruising beneath the icecap at the North Pole, as in Fig. 12.5.11, and thus can use neither visual nor radio methods to determine its position, there is an alternative. Build a sensitive gyroscope-accelerometer combination and install it in the submarine. The device continuously measures the sub's acceleration vector, beginning at the time $t = 0$ when its position $\mathbf{r}(0)$ and velocity $\mathbf{v}(0)$ are known. Because $\mathbf{v}'(t) = \mathbf{a}(t)$, Eq. (16) gives

$$\int_0^t \mathbf{a}(t)\,dt = \Big[\mathbf{v}(t)\Big]_0^t = \mathbf{v}(t) - \mathbf{v}(0),$$

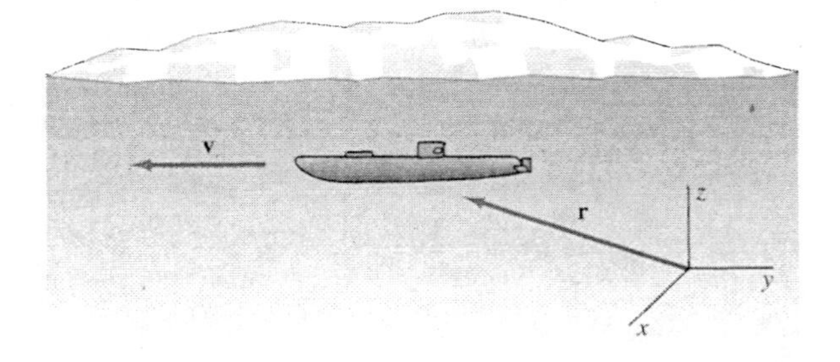

FIGURE 12.5.11 A submarine beneath the polar icecap.

so

$$\mathbf{v}(t) = \mathbf{v}(0) + \int_0^t \mathbf{a}(t)\,dt.$$

Thus the velocity at every time $t \geqq 0$ is known. Similarly, because $\mathbf{r}'(t) = \mathbf{v}(t)$, a second integration gives

$$\mathbf{r}(t) = \mathbf{r}(0) + \int_0^t \mathbf{v}(t)\,dt$$

for the position of the sub at every time $t \geqq 0$. On-board computers can be programmed to carry out these integrations (perhaps by using Simpson's approximation) and continuously provide captain and crew with the submarine's (almost) exact position and velocity.

Motion of Projectiles

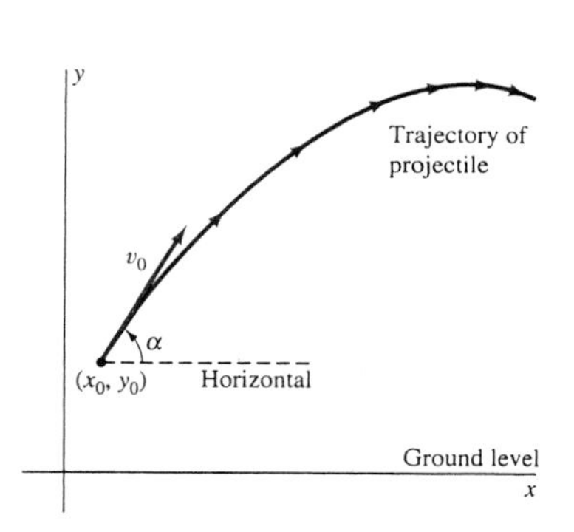

FIGURE 12.5.12 Trajectory of a projectile launched at the angle α.

Suppose that a projectile is launched from the point (x_0, y_0), with y_0 denoting its initial height above the surface of the earth. Let α be the angle of inclination from the horizontal of its initial velocity vector $\mathbf{v}_0$ (Fig. 12.5.12). Then its initial position vector is

$$\mathbf{r}_0 = x_0\mathbf{i} + y_0\mathbf{j}, \tag{20a}$$

and from Fig. 12.5.12 we see that

$$\mathbf{v}_0 = (v_0 \cos\alpha)\mathbf{i} + (v_0 \sin\alpha)\mathbf{j}, \tag{20b}$$

where $v_0 = |\mathbf{v}_0|$ is the initial speed of the projectile.

We suppose that the motion takes place sufficiently close to the surface that we may assume that the earth is flat and that gravity is perfectly uniform. Then, if we also ignore air resistance, the acceleration of the projectile is

$$\mathbf{a} = \frac{d\mathbf{v}}{dt} = -g\mathbf{j},$$

where $g \approx 32\text{ ft/s}^2 \approx 9.8\text{ m/s}^2$. Antidifferentiation gives

$$\mathbf{v}(t) = -gt\mathbf{j} + \mathbf{C}_1.$$

Put $t = 0$ in both sides of this last equation. This shows that $\mathbf{C}_1 = \mathbf{v}_0$ (as expected!) and thus that

$$\mathbf{v}(t) = \frac{d\mathbf{r}}{dt} = -gt\mathbf{j} + \mathbf{v}_0.$$

Another antidifferentiation gives

$$\mathbf{r}(t) = -\tfrac{1}{2}gt^2\mathbf{j} + \mathbf{v}_0 t + \mathbf{C}_2.$$

Now substituting $t = 0$ yields $\mathbf{C}_2 = \mathbf{r}_0$, so the position vector of the projectile at time t is

$$\mathbf{r}(t) = -\tfrac{1}{2}gt^2\mathbf{j} + \mathbf{v}_0 t + \mathbf{r}_0. \tag{21}$$

Equations (20a) and (20b) now give

$$\mathbf{r}(t) = [(v_0\cos\alpha)t + x_0]\mathbf{i} + \left[-\tfrac{1}{2}gt^2 + (v_0\sin\alpha)t + y_0\right]\mathbf{j},$$

so parametric equations of the trajectory of the particle are

$$x(t) = (v_0\cos\alpha)t + x_0, \tag{22}$$

$$y(t) = -\tfrac{1}{2}gt^2 + (v_0\sin\alpha)t + y_0. \tag{23}$$

EXAMPLE 9 An airplane is flying horizontally at an altitude of 1600 ft to pass directly over snowbound cattle on the ground and release hay to land there. The plane's speed is a constant 150 mi/h (220 ft/s). At what angle of sight ϕ (between the horizontal and the direct line to the target) should a bale of hay be released in order to hit the target?

Solution See Fig. 12.5.13. We take $x_0 = 0$ where the bale of hay is released at time $t = 0$. Then $y_0 = 1600$ (ft), $v_0 = 220$ (ft/s), and $\alpha = 0$. Then Eqs. (22) and (23) take the forms

$$x(t) = 220t, \qquad y(t) = -16t^2 + 1600.$$

From the second of these equations we find that $t = 10$ (s) when the bale of hay hits the ground ($y = 0$). It has then traveled a horizontal distance

$$x(10) = 220 \cdot 10 = 2200 \quad \text{(ft)}.$$

Hence the required angle of sight is

$$\phi = \tan^{-1}\left(\frac{1600}{2200}\right) \approx 36°.$$

◆

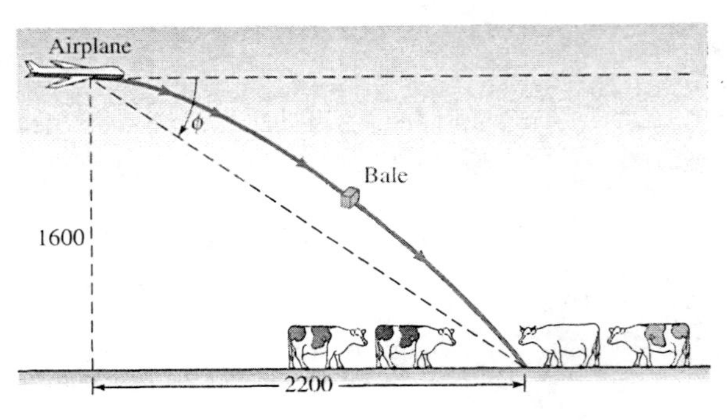

FIGURE 12.5.13 Trajectory of the hay bale of Example 9.

EXAMPLE 10 A ball is thrown northward into the air from the origin in xyz-space (the xy-plane represents the ground and the positive y-axis points north). The initial velocity (vector) of the ball is

$$\mathbf{v}_0 = \mathbf{v}(0) = 80\mathbf{j} + 80\mathbf{k}.$$

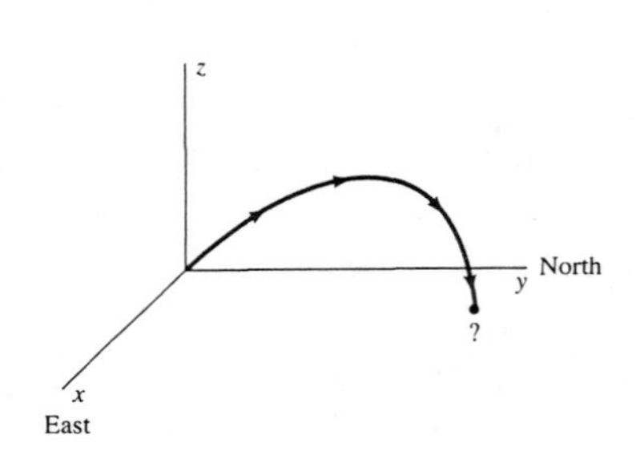

FIGURE 12.5.14 The trajectory of the ball of Example 10.

The spin of the ball causes an eastward acceleration of 2 ft/s^2 in addition to gravitational acceleration. Thus the acceleration vector produced by the combination of gravity and spin is

$$\mathbf{a}(t) = 2\mathbf{i} - 32\mathbf{k}.$$

First find the velocity vector $\mathbf{v}(t)$ of the ball and its position vector $\mathbf{r}(t)$. Then determine where and with what speed the ball hits the ground (Fig. 12.5.14).

Solution When we antidifferentiate $\mathbf{a}(t)$ we get

$$\mathbf{v}(t) = \int \mathbf{a}(t)\,dt = \int (2\mathbf{i} - 32\mathbf{k})\,dt = 2t\mathbf{i} - 32t\mathbf{k} + \mathbf{C}_1.$$

We substitute $t = 0$ to find that $\mathbf{C}_1 = \mathbf{v}_0 = 80\mathbf{j} + 80\mathbf{k}$, so

$$\mathbf{v}(t) = 2t\mathbf{i} + 80\mathbf{j} + (80 - 32t)\mathbf{k}.$$

Another antidifferentiation yields

$$\mathbf{r}(t) = \int \mathbf{v}(t)\,dt = \int [2t\mathbf{i} + 80\mathbf{j} + (80 - 32t)\mathbf{k}]\,dt$$

$$= t^2\mathbf{i} + 80t\mathbf{j} + (80t - 16t^2)\mathbf{k} + \mathbf{C}_2,$$

and substituting $t = 0$ gives $\mathbf{C}_2 = \mathbf{r}(0) = \mathbf{0}$. Hence the position vector of the ball is

$$\mathbf{r}(t) = t^2\mathbf{i} + 80t\mathbf{j} + (80t - 16t^2)\mathbf{k}.$$

The ball hits the ground when $z = 80t - 16t^2 = 0$; that is, when $t = 5$. Its position vector then is

$$\mathbf{r}(5) = 5^2\mathbf{i} + 80 \cdot 5\mathbf{j} = 25\mathbf{i} + 400\mathbf{j},$$

so the ball has traveled 25 ft eastward and 400 ft northward. Its velocity vector at impact is

$$\mathbf{v}(5) = 2 \cdot 5\mathbf{i} + 80\mathbf{j} + (80 - 32 \cdot 5)\mathbf{k} = 10\mathbf{i} + 80\mathbf{j} - 80\mathbf{k},$$

so its speed when it hits the ground is

$$v(5) = |\mathbf{v}(5)| = \sqrt{10^2 + 80^2 + (-80)^2},$$

approximately 113.58 ft/s. Because the ball started with initial speed $v_0 = \sqrt{80^2 + 80^2} \approx 113.14$ ft/s, its eastward acceleration has slightly increased its terminal speed. ◆

12.5 TRUE/FALSE STUDY GUIDE

12.5 CONCEPTS: QUESTIONS AND DISCUSSION

In Questions 1 through 3, let $\mathbf{f} : \boldsymbol{R} \to \boldsymbol{R}^3$ *be a vector-valued function of a real variable t. In each question you are asked for a "coordinate-free" definition. Compare your definition with the corresponding componentwise definition or calculation. Do the two agree?*

1. Give a definition of $\lim_{t \to a} \mathbf{f}(t)$ that does not involve the component functions of $\mathbf{f}$.
2. Give a definition of $\mathbf{f}'(t)$ that does not involve the component functions of $\mathbf{f}$.
3. Give a definition of

$$\int_a^b \mathbf{f}(t)\,dt$$

that does not involve the component functions of $\mathbf{f}$.

12.5 PROBLEMS

In Problems 1 through 4, also match the curves there defined with their three-dimensional plots in Figs. 12.5.15 through 12.5.18.

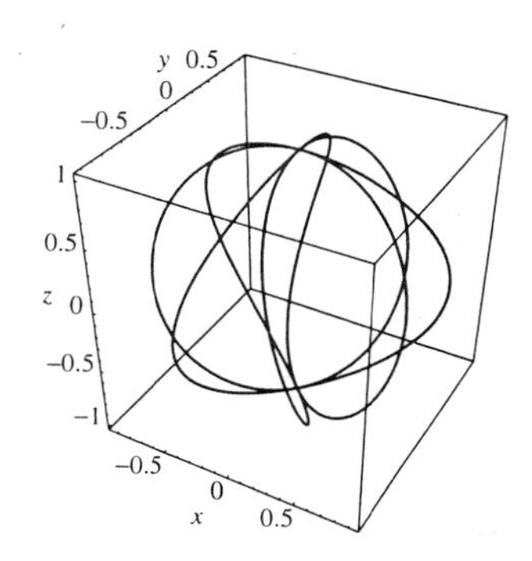

FIGURE 12.5.15

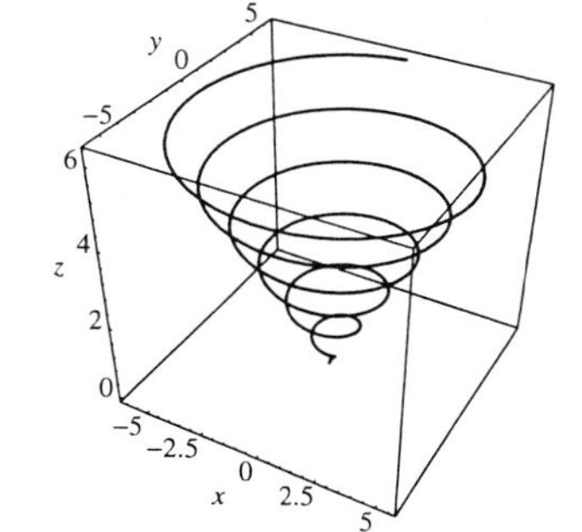

FIGURE 12.5.16

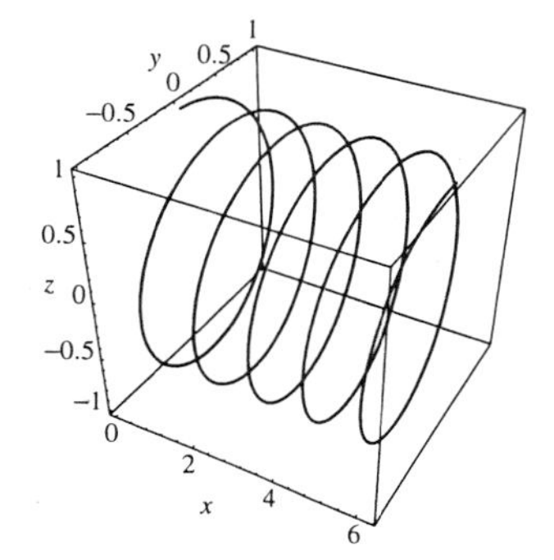

FIGURE 12.5.17

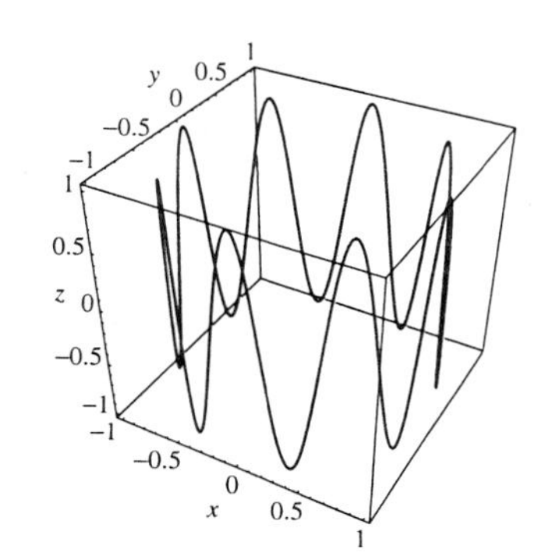

FIGURE 12.5.18

1. Show that the graph of the curve with parametric equations $x = t$, $y = \sin 5t$, $z = \cos 5t$ lies on the circular cylinder $y^2 + z^2 = 1$ centered along the x-axis.

2. Show that the graph of the curve with parametric equations $x = \sin t$, $y = \cos t$, $z = \cos 8t$ lies on the vertical circular cylinder $x^2 + y^2 = 1$.

3. Show that the graph of the curve with parametric equations $x = t \sin 6t, y = t \cos 6t, z = t$ lies on the cone $z = \sqrt{x^2 + y^2}$ with its vertex at the origin and opening upward.

4. Show that the graph of the curve with parametric equations $x = \cos t \sin 4t$, $y = \sin t \sin 4t$, $z = \cos 4t$ lies on the surface of the sphere $x^2 + y^2 + z^2 = 1$.

In Problems 5 through 10, find the values of $\mathbf{r}'(t)$ *and* $\mathbf{r}''(t)$ *for the given values of* t.

5. $\mathbf{r}(t) = 3\mathbf{i} - 2\mathbf{j}; \quad t = 1$

6. $\mathbf{r}(t) = t^2\mathbf{i} - t^3\mathbf{j}; \quad t = 2$

7. $\mathbf{r}(t) = e^{2t}\mathbf{i} + e^{-t}\mathbf{j}; \quad t = 0$

8. $\mathbf{r}(t) = \mathbf{i} \cos t + \mathbf{j} \sin t; \quad t = \pi/4$

9. $\mathbf{r}(t) = 3\mathbf{i} \cos 2\pi t + 3\mathbf{j} \sin 2\pi t; \quad t = 3/4$

10. $\mathbf{r}(t) = 5\mathbf{i} \cos t + 4\mathbf{j} \sin t; \quad t = \pi$

In Problems 11 through 16, the position vector $\mathbf{r}(t)$ *of a particle moving in space is given. Find its velocity and acceleration vectors and its speed at time* t.

11. $\mathbf{r}(t) = t\mathbf{i} + t^2\mathbf{j} + t^3\mathbf{k}$

12. $\mathbf{r}(t) = t^2(3\mathbf{i} + 4\mathbf{j} - 12\mathbf{k})$

13. $\mathbf{r}(t) = t\mathbf{i} + 3e^t\mathbf{j} + 4e^t\mathbf{k}$

14. $\mathbf{r}(t) = e^t\mathbf{i} + e^{2t}\mathbf{j} + e^{3t}\mathbf{k}$

15. $\mathbf{r}(t) = (3 \cos t)\mathbf{i} + (3 \sin t)\mathbf{j} - 4t\mathbf{k}$

16. $\mathbf{r}(t) = 12t\mathbf{i} + (5 \sin 2t)\mathbf{j} - (5 \cos 2t)\mathbf{k}$

Calculate the integrals in Problems 17 through 20.

17. $\displaystyle\int_0^{\pi/4} (\mathbf{i} \sin t + 2\mathbf{j} \cos t)\, dt$

18. $\displaystyle\int_1^e \left(\frac{1}{t}\mathbf{i} - \mathbf{j}\right) dt$

19. $\displaystyle\int_0^2 t^2(1 + t^3)^{3/2}\mathbf{i}\, dt$

20. $\displaystyle\int_0^1 \left(\mathbf{i}e^t - \mathbf{j}te^{-t^2}\right) dt$

In Problems 21 through 24, apply Theorem 2 to compute the derivative $D_t[\mathbf{u}(t) \cdot \mathbf{v}(t)]$.

21. $\mathbf{u}(t) = 3t\mathbf{i} - \mathbf{j}, \quad \mathbf{v}(t) = 2\mathbf{i} - 5t\mathbf{j}$

22. $\mathbf{u}(t) = t\mathbf{i} + t^2\mathbf{j}, \quad \mathbf{v}(t) = t^2\mathbf{i} - t\mathbf{j}$

23. $\mathbf{u}(t) = \langle \cos t, \sin t \rangle, \quad \mathbf{v}(t) = \langle \sin t, -\cos t \rangle$

24. $\mathbf{u} = \langle t, t^2, t^3 \rangle, \quad \mathbf{v} = \langle \cos 2t, \sin 2t, e^{-3t} \rangle$

In Problems 25 through 34, the acceleration vector $\mathbf{a}(t)$*, the initial position* $\mathbf{r}_0 = \mathbf{r}(0)$*, and the initial velocity* $\mathbf{v}_0 = \mathbf{v}(0)$ *of a particle moving in xyz-space are given. Find its position vector* $\mathbf{r}(t)$ *at time* t.

25. $\mathbf{a} = \mathbf{0}; \quad \mathbf{r}_0 = \mathbf{i}; \quad \mathbf{v}_0 = \mathbf{k}$

26. $\mathbf{a} = 2\mathbf{i}; \quad \mathbf{r}_0 = 3\mathbf{j}; \quad \mathbf{v}_0 = 4\mathbf{k}$

27. $\mathbf{a}(t) = 2\mathbf{i} - 4\mathbf{k}; \quad \mathbf{r}_0 = \mathbf{0}; \quad \mathbf{v}_0 = 10\mathbf{j}$

28. $\mathbf{a}(t) = \mathbf{i} - \mathbf{j} + 3\mathbf{k}; \quad \mathbf{r}_0 = 5\mathbf{i}; \quad \mathbf{v}_0 = 7\mathbf{j}$

29. $\mathbf{a}(t) = 2\mathbf{j} - 6t\mathbf{k}; \quad \mathbf{r}_0 = 2\mathbf{i}; \quad \mathbf{v}_0 = 5\mathbf{k}$

30. $\mathbf{a}(t) = 6t\mathbf{i} - 5\mathbf{j} + 12t^2\mathbf{k}; \quad \mathbf{r}_0 = 3\mathbf{i} + 4\mathbf{j}; \quad \mathbf{v}_0 = 4\mathbf{j} - 5\mathbf{k}$

31. $\mathbf{a}(t) = t\mathbf{i} + t^2\mathbf{j} + t^3\mathbf{k}; \quad \mathbf{r}_0 = 10\mathbf{i}; \quad \mathbf{v}_0 = 10\mathbf{j}$

32. $\mathbf{a}(t) = t\mathbf{i} + e^{-t}\mathbf{j}; \quad \mathbf{r}_0 = 3\mathbf{i} + 4\mathbf{j}; \quad \mathbf{v}_0 = 5\mathbf{k}$

33. $\mathbf{a}(t) = \mathbf{i} \cos t + \mathbf{j} \sin t; \quad \mathbf{r}_0 = \mathbf{j}; \quad \mathbf{v}_0 = -\mathbf{i} + 5\mathbf{k}$

34. $\mathbf{a}(t) = 9(\mathbf{i} \sin 3t + \mathbf{j} \cos 3t) + 4\mathbf{k}; \quad \mathbf{r}_0 = 3\mathbf{i} + 4\mathbf{j}; \quad \mathbf{v}_0 = 2\mathbf{i} - 7\mathbf{k}$

35. The parametric equations of a moving point are

$$x(t) = 3 \cos 2t, \qquad y(t) = 3 \sin 2t, \qquad z(t) = 8t.$$

Find its velocity, speed, and acceleration at time $t = 7\pi/8$.

36. Use the equations in Theorem 2 to calculate

$$D_t[\mathbf{u}(t) \cdot \mathbf{v}(t)] \quad \text{and} \quad D_t[\mathbf{u}(t) \times \mathbf{v}(t)]$$

if $\mathbf{u}(t) = \langle t, t^2, t^3 \rangle$ and $\mathbf{v}(t) = \langle e^t, \cos t, \sin t \rangle$.

37. Verify part 5 of Theorem 2 in the special case $\mathbf{u}(t) = \langle 0, 3, 4t \rangle$ and $\mathbf{v}(t) = \langle 5t, 0, -4 \rangle$.

38. Prove part 5 of Theorem 2.

39. A point moves on a sphere centered at the origin. Show that its velocity vector is always tangent to the sphere.

40. A particle moves with constant speed along a curve in space. Show that its velocity and acceleration vectors are always perpendicular.

41. Find the maximum height reached by the ball in Example 10 and also its speed at that height.

42. The **angular momentum** $\mathbf{L}(t)$ and **torque** $\boldsymbol{\tau}(t)$ of a moving particle of mass m with position vector $\mathbf{r}(t)$ are defined to be

$$\mathbf{L}(t) = m\mathbf{r}(t) \times \mathbf{v}(t), \qquad \boldsymbol{\tau}(t) = m\mathbf{r}(t) \times \mathbf{a}(t).$$

Prove that $\mathbf{L}'(t) = \boldsymbol{\tau}(t)$. It follows that $\mathbf{L}(t)$ must be constant if $\boldsymbol{\tau} \equiv \mathbf{0}$; this is the law of conservation of angular momentum.

Problems 43 through 48 deal with a projectile fired from the origin (so $x_0 = y_0 = 0$*) with initial speed* v_0 *and initial angle of inclination* α*. The* ***range*** *of the projectile is the horizontal distance it travels before it returns to the ground.*

43. If $\alpha = 45°$, what value of v_0 gives a range of 1 mi?

44. If $\alpha = 60°$ and the range is $R = 1$ mi, what is the maximum height attained by the projectile?

45. Deduce from Eqs. (22) and (23) the fact that the range is

$$R = \tfrac{1}{16}v_0^2 \sin \alpha \cos \alpha.$$

46. Given the initial speed v_0, find the angle α that maximizes the range. [*Suggestion:* Use the result of Problem 45.]

47. Suppose that $v_0 = 160$ (ft/s). Find the maximum height y_{max} and the range R of the projectile if (a) $\alpha = 30°$; (b) $\alpha = 45°$; (c) $\alpha = 60°$.

48. The projectile of Problem 47 is to be fired at a target 600 ft away, and there is a hill 300 ft high midway between the gun site and this target. At what initial angle of inclination should the projectile be fired?

49. A projectile is to be fired horizontally from the top of a 100-m cliff at a target 1 km from the base of the cliff. What should be the initial velocity of the projectile? (Use $g = 9.8$ m/s^2.)

50. A bomb is dropped (initial speed zero) from a helicopter hovering at a height of 800 m. A projectile is fired from a gun located on the ground 800 m west of the point directly beneath the helicopter. The projectile is supposed to intercept the bomb at a height of exactly 400 m. If the projectile is fired at the same instant that the bomb is dropped, what should be its initial velocity and angle of inclination?

51. Suppose, more realistically, that the projectile of Problem 50 is fired 1 s after the bomb is dropped. What should be its initial velocity and angle of inclination?

52. An artillery gun with a muzzle velocity of 1000 ft/s is located atop a seaside cliff 500 ft high. At what initial inclination angle (or angles) should it fire a projectile in order to hit a ship at sea 20,000 ft from the base of the cliff?

53. Suppose that the vector-valued functions $\mathbf{u}(t)$ and $\mathbf{v}(t)$ both have limits as $t \to a$. Prove:
(a) $\lim_{t\to a}(\mathbf{u}(t) + \mathbf{v}(t)) = \lim_{t\to a}\mathbf{u}(t) + \lim_{t\to a}\mathbf{v}(t)$;
(b) $\lim_{t\to a}(\mathbf{u}(t)\cdot\mathbf{v}(t)) = \left(\lim_{t\to a}\mathbf{u}(t)\right)\cdot\left(\lim_{t\to a}\mathbf{v}(t)\right)$.

54. Suppose that both the vector-valued function $\mathbf{r}(t)$ and the real-valued function $h(t)$ are differentiable. Deduce the chain rule for vector-valued functions,

$$D_t[\mathbf{r}(h(t))] = h'(t)\mathbf{r}'(h(t)),$$

in componentwise fashion from the ordinary chain rule.

55. A point moves with constant speed, so its velocity vector $\mathbf{v}$ satisfies the condition

$$|\mathbf{v}|^2 = \mathbf{v}\cdot\mathbf{v} = C \quad \text{(a constant)}.$$

Prove that the velocity and acceleration vectors of the point are always perpendicular to each other.

56. A point moves on a circle whose center is at the origin. Use the dot product to show that the position and velocity vectors of the moving point are always perpendicular.

57. A point moves on the hyperbola $x^2 - y^2 = 1$ with position vector

$$\mathbf{r}(t) = \mathbf{i}\cosh\omega t + \mathbf{j}\sinh\omega t$$

(the number ω is a constant). Prove that the acceleration vector $\mathbf{a}(t)$ satisfies the equation $\mathbf{a}(t) = c\mathbf{r}(t)$, where c is a positive constant. What sort of external force would produce this kind of motion?

58. Suppose that a point moves on the ellipse

$$\frac{x^2}{a^2} + \frac{y^2}{b^2} = 1$$

with position vector $\mathbf{r}(t) = \mathbf{i}a\cos\omega t + \mathbf{j}b\sin\omega t$ (ω is a constant). Prove that the acceleration vector $\mathbf{a}$ satisfies the equation $\mathbf{a}(t) = c\mathbf{r}(t)$, where c is a negative constant. To what sort of external force $\mathbf{F}(t)$ does this motion correspond?

59. A point moves in the plane with constant acceleration vector $\mathbf{a} = a\mathbf{j}$. Prove that its path is a parabola or a straight line.

60. Suppose that a particle is subject to no force, so its acceleration vector $\mathbf{a}(t)$ is identically zero. Prove that the particle travels along a straight line at constant speed (Newton's first law of motion).

61. *Uniform Circular Motion* Consider a particle that moves counterclockwise around the circle with center $(0, 0)$ and radius r at a constant angular speed of ω radians per second (Fig. 12.5.19). If its initial position is $(r, 0)$, then its position vector is

$$\mathbf{r}(t) = \mathbf{i}\,r\cos\omega t + \mathbf{j}\,r\sin\omega t.$$

(a) Show that the velocity vector of the particle is tangent to the circle and that the speed of the particle is

$$v(t) = |\mathbf{v}(t)| = r\omega.$$

(b) Show that the acceleration vector $\mathbf{a}$ of the particle is directed opposite to $\mathbf{r}$ and that

$$a(t) = |\mathbf{a}(t)| = r\omega^2.$$

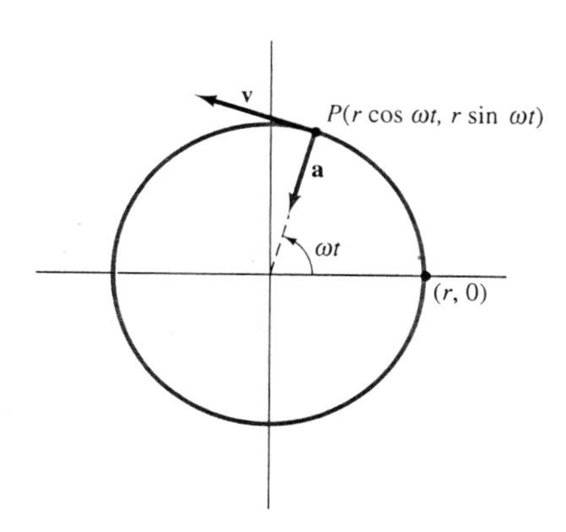

FIGURE 12.5.19 Uniform circular motion (Problem 61).

62. Suppose that a particle is moving under the influence of a *central* force field $\mathbf{R} = k\mathbf{r}$, where k is a scalar function of x, y, and z. Conclude that the trajectory of the particle lies in a *fixed* plane through the origin.

63. A baseball is thrown with an initial velocity of 160 ft/s straight upward from the ground. It experiences a downward gravitational acceleration of 32 ft/s^2. Because of spin, it experiences also a (horizontal) northward acceleration of 0.1 ft/s^2; otherwise, the air has no effect on its motion. How far north of the throwing point will the ball land?

64. A baseball is hit with an initial velocity of 96 ft/s and an initial inclination angle of 15° from ground level straight down a foul line. Because of spin it experiences a horizontal acceleration of 2 ft/s^2 perpendicular to the foul line; otherwise, the air has no effect on its motion. When the ball hits the ground, how far is it from the foul line?

65. A projectile is fired northward (in the positive y-direction) out to sea from the top of a seaside cliff 384 ft high. The projectile's initial velocity vector is $\mathbf{v}_0 = 200\mathbf{j} + 160\mathbf{k}$. In addition to a downward (negative z-direction) gravitational acceleration of 32 ft/s^2, it experiences in flight an eastward (positive x-direction) acceleration of 2 ft/s^2 due to spin.
(a) Find the projectile's velocity and position vectors t seconds after it is fired.
(b) How long is the projectile in the air?
(c) Where does the projectile hit the water ($z = 0$)? Give the answer by telling how far north out to sea and how far east along the coast is its impact position.
(d) What is the maximum height of the projectile above the water?

66. A gun fires a shell with a muzzle velocity of 150 m/s. While the shell is in the air, it experiences a downward (vertical) gravitational acceleration of 9.8 m/s^2 and an eastward (horizontal) Coriolis acceleration of 5 cm/s^2; air resistance may be ignored. The target is 1500 m due north of the gun, and both the gun and target are on level ground. Halfway between them is a hill 600 m high. Tell precisely how to aim the gun—both compass heading and inclination from the horizontal—so that the shell will clear the hill and hit the target.

12.5 Project: Does a Pitched Baseball Really Curve?

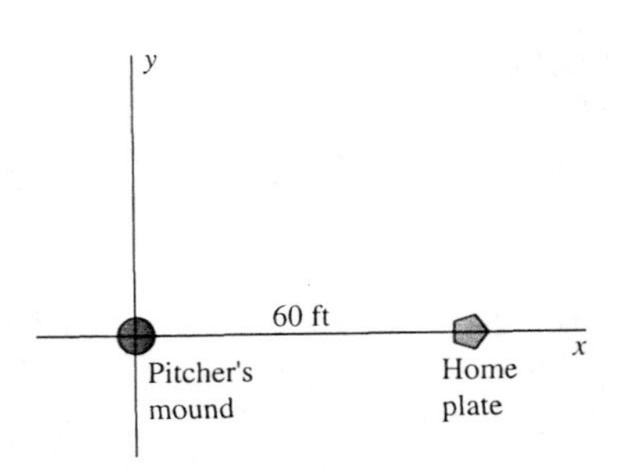

FIGURE 12.5.20 The x-axis points toward home plate.

Have you ever wondered whether a baseball pitch really curves or whether it's some sort of optical illusion? In this project you'll use calculus to illuminate the matter.

Suppose that a pitcher throws a ball toward home plate (60 ft away, as in Fig. 12.5.20) and gives it a spin of S revolutions per second counterclockwise (as viewed from above) about a vertical axis through the center of the ball. This spin is described by the *spin vector* **S** that points along the axis of revolution in the right-handed direction and has length S (Fig. 12.5.21).

We know from studies of aerodynamics that this spin causes a difference in air pressure on the sides of the ball toward and away from this spin. Studies also show that this pressure difference results in a *spin acceleration*

$$\mathbf{a}_S = c\mathbf{S} \times \mathbf{v} \tag{1}$$

of the ball (where c is an empirical constant). The total acceleration of the ball is then

$$\mathbf{a} = (c\mathbf{S} \times \mathbf{v}) - g\mathbf{k}, \tag{2}$$

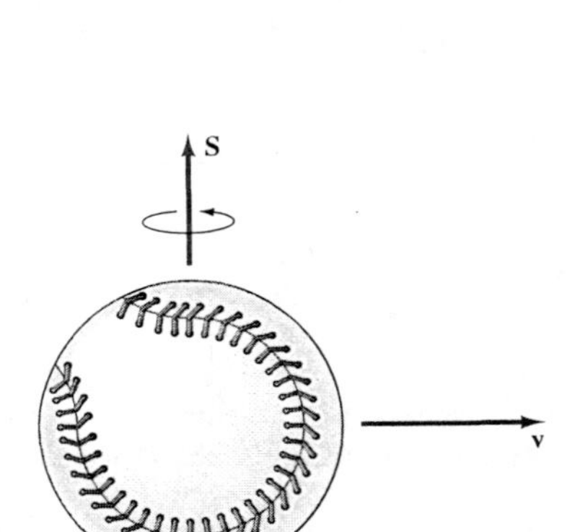

FIGURE 12.5.21 The spin and velocity vectors.

where $g \approx 32$ ft/s^2 is the gravitational acceleration. Here we will ignore any other effects of air resistance.

With the spin vector $\mathbf{S} = S\mathbf{k}$ pointing upward, as in Fig. 12.5.21, show first that

$$\mathbf{S} \times \mathbf{v} = -Sv_y\mathbf{i} + Sv_x\mathbf{j}, \tag{3}$$

where v_x is the component of **v** in the x-direction and v_y is the component of **v** in the y-direction.

For a ball pitched along the x-axis, v_x is much larger than v_y, and so the approximation $\mathbf{S} \times \mathbf{v} = Sv_x\mathbf{j}$ is sufficiently accurate for our purposes. We may then take the acceleration vector of the ball to be

$$\mathbf{a} = cSv_x\mathbf{j} - g\mathbf{k}. \tag{4}$$

Now suppose that the pitcher throws the ball from the initial position $x_0 = y_0 = 0$, $z_0 = 5$ (ft), with initial velocity vector

$$\mathbf{v}_0 = 120\mathbf{i} - 2\mathbf{j} + 4\mathbf{k} \tag{5}$$

(with components in feet per second, so $v_0 \approx 120$ ft/s, about 82 mi/h) and with a spin of $S = \frac{80}{3}$ rev/s. A reasonable value of c is

$$c = 0.005 \text{ ft/s}^2 \quad \text{per ft/s of velocity and rev/s of spin,}$$

although the precise value depends on whether the pitcher has (accidentally, of course) scuffed the ball or administered some foreign substance to it.

Show first that these values of the parameters yield

$$\mathbf{a} = 16\mathbf{j} - 32\mathbf{k}$$

for the ball's acceleration vector. Then integrate twice in succession to find the ball's position vector

$$\mathbf{r}(t) = x(t)\mathbf{i} + y(t)\mathbf{j} + z(t)\mathbf{k}.$$

Use your results to fill in the following table, giving the pitched ball's horizontal deflection y and height z (above the ground) at quarter-second intervals.

t (s)	x (ft)	y (ft)	z (ft)
0.0	0	0	5
0.25	30	?	?
0.50	60	?	?

Suppose that the batter gets a "fix" on the pitch by observing the ball during the first quarter-second and prepares to swing. After 0.25 s does the pitch still appear to be straight on target toward home plate at a height of 5 ft?

What happens to the ball during the final quarter-second of its approach to home plate—*after* the batter has begun to swing the bat? What were the ball's horizontal and vertical deflections during this brief period? What is your conclusion? Does the pitched ball really "curve" or not?

12.6 CURVATURE AND ACCELERATION

The speed of a moving point is closely related to the arc length of its trajectory. The arc-length formula for parametric curves in space (or *space curves*) is a natural generalization of the formula for parametric plane curves [Eq. (8) of Section 10.5]. The **arc length** s along the smooth curve with position vector

$$\mathbf{r}(t) = f(t)\mathbf{i} + g(t)\mathbf{j} + h(t)\mathbf{k} = x\mathbf{i} + y\mathbf{j} + z\mathbf{k} \tag{1}$$

from the point $\mathbf{r}(a)$ to the point $\mathbf{r}(b)$ is, by definition,

$$\begin{aligned} s &= \int_a^b \sqrt{[x'(t)]^2 + [y'(t)]^2 + [z'(t)]^2}\, dt \\ &= \int_a^b \sqrt{\left(\frac{dx}{dt}\right)^2 + \left(\frac{dy}{dt}\right)^2 + \left(\frac{dz}{dt}\right)^2}\, dt. \end{aligned} \tag{2}$$

We see from Eq. (9) in Section 12.5 that the integrand is the speed $v(t) = |\mathbf{r}'(t)|$ of the moving point with position vector $\mathbf{r}(t)$, so

$$s = \int_a^b v(t)\, dt. \tag{3}$$

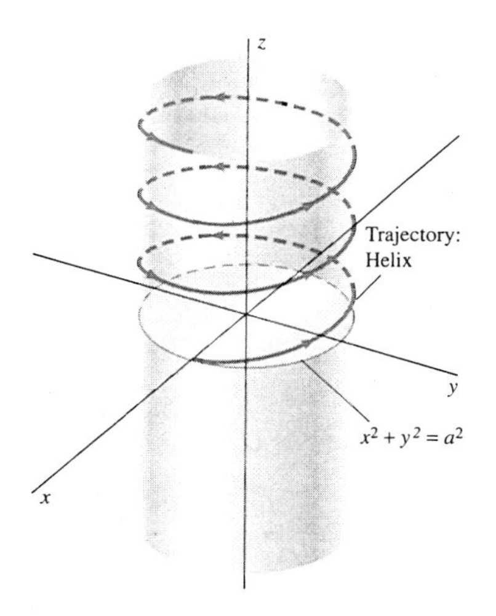

FIGURE 12.6.1 The helix of Example 1.

EXAMPLE 1 Find the arc length of one turn (from $t = 0$ to $t = 2\pi/\omega$) of the helix shown in Fig. 12.6.1. This helix has the parametric equations

$$x(t) = a\cos\omega t, \qquad y(t) = a\sin\omega t, \qquad z(t) = bt.$$

Solution We found in Example 7 of Section 12.5 that

$$v(t) = \sqrt{a^2\omega^2 + b^2}.$$

Hence Eq. (3) gives

$$s = \int_0^{2\pi/\omega} \sqrt{a^2\omega^2 + b^2}\, dt = \frac{2\pi}{\omega}\sqrt{a^2\omega^2 + b^2}.$$

For instance, if $a = b = \omega = 1$, then $s = 2\pi\sqrt{2}$, which is $\sqrt{2}$ times the circumference of the circle in the xy-plane over which the helix lies. ◆

Let $s(t)$ denote the arc length along a smooth curve from its initial point $\mathbf{r}(a)$ to the variable point $\mathbf{r}(t)$. Then, from Eq. (3), we obtain the **arc-length function** $s(t)$

of the curve:

$$s(t) = \int_a^t v(\tau)\,d\tau. \tag{4}$$

The fundamental theorem of calculus then gives

$$\frac{ds}{dt} = v. \tag{5}$$

Thus *the speed of the moving point is the time rate of change of its arc-length function.* If $v(t) > 0$ for all t, then it follows that $s(t)$ is an increasing function of t and therefore has an inverse function $t(s)$. When we replace t with $t(s)$ in the curve's original parametric equations, we obtain the **arc-length parametrization**

$$x = x(s), \qquad y = y(s), \qquad z = z(s).$$

This gives the position of the moving point as a function of arc length measured along the curve from its initial point. (See Fig. 12.6.2.)

FIGURE 12.6.2 A curve parametrized by arc length s.

EXAMPLE 2 If we take $a = 5$, $b = 12$, and $\omega = 1$ for the helix of Example 1, then the velocity formula $v = (a^2\omega^2 + b^2)^{1/2}$ yields

$$v = \sqrt{5^2 \cdot 1^2 + 12^2} = \sqrt{169} = 13.$$

Hence Eq. (5) gives $ds/dt = 13$, so

$$s = 13t,$$

taking $s = 0$ when $t = 0$ and thereby measuring arc length from the natural starting point $(5, 0, 0)$. When we substitute $t = s/13$ and the numerical values of a, b, and ω into the original parametric equations of the helix, we get the arc-length parametrization

$$x(s) = 5\cos\frac{s}{13}, \qquad y(s) = 5\sin\frac{s}{13}, \qquad z(s) = \frac{12s}{13}$$

of the helix. ◆

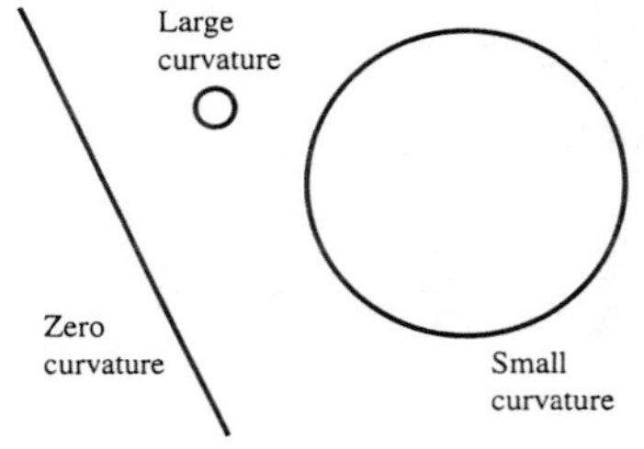
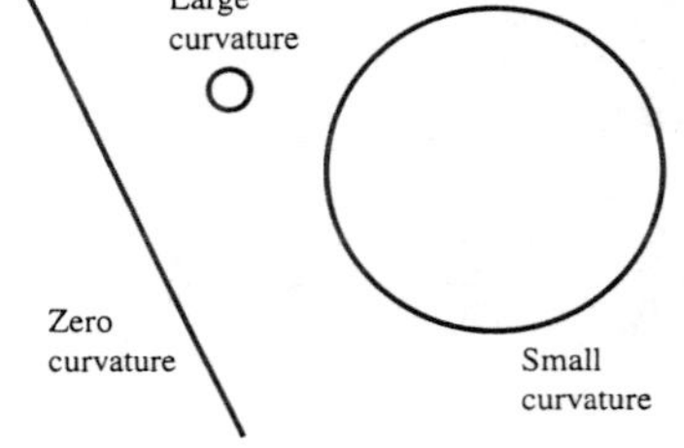

FIGURE 12.6.3 The intuitive idea of curvature.

Curvature of Plane Curves

The word *curvature* has an intuitive meaning that we need to make precise. Most people would agree that a straight line does not curve at all, whereas a circle of small radius is more curved than a circle of large radius (Fig. 12.6.3). This judgment may be based on a conception of curvature as "rate of change of direction." The direction of a curve is determined by its velocity vector, so you would expect the idea of curvature to have something to do with the rate at which the velocity vector is turning.

Let

$$\mathbf{r}(t) = x(t)\mathbf{i} + y(t)\mathbf{j}, \quad a \leqq t \leqq b \tag{6}$$

be the position vector of a differentiable plane curve that is *smooth*—meaning that the velocity vector $\mathbf{v}(t) = \mathbf{r}'(t)$ is *nonzero.* Then the curve's **unit tangent vector** at the point $\mathbf{r}(t)$ is the unit vector

$$\mathbf{T}(t) = \frac{\mathbf{v}(t)}{|\mathbf{v}(t)|} = \frac{\mathbf{v}(t)}{v(t)}, \tag{7}$$

where $v(t) = |\mathbf{v}(t)|$ is the speed. Now denote by ϕ the angle of inclination of T, measured counterclockwise from the positive x-axis (Fig. 12.6.4). Then

$$\mathbf{T} = \mathbf{i}\cos\phi + \mathbf{j}\sin\phi. \tag{8}$$

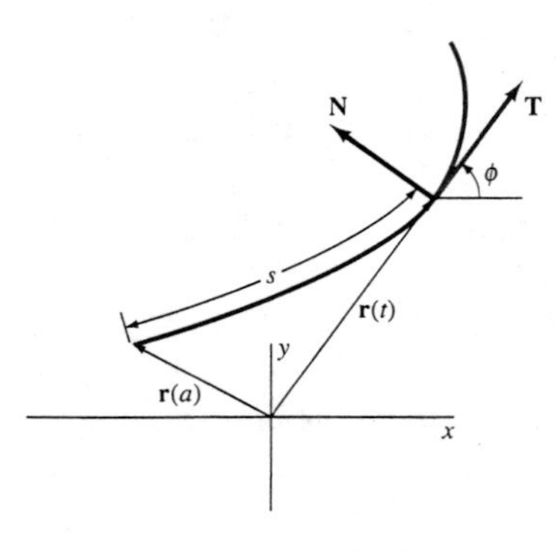

FIGURE 12.6.4 The unit tangent vector **T**.

We can express the unit tangent vector **T** of Eq. (8) as a function of the arc-length parameter s indicated in Fig. 12.6.4. Then the rate at which **T** is turning is measured by the derivative

$$\frac{d\mathbf{T}}{ds} = \frac{d\mathbf{T}}{d\phi} \cdot \frac{d\phi}{ds} = (-\mathbf{i}\sin\phi + \mathbf{j}\cos\phi)\frac{d\phi}{ds}. \tag{9}$$

Note that

$$\left|\frac{d\mathbf{T}}{ds}\right| = \left|\frac{d\phi}{ds}\right| \tag{10}$$

because the vector on the right-hand side of Eq. (9) is a unit vector.

The **curvature** at a point of a plane curve, denoted by κ (lowercase Greek kappa), is therefore defined to be

$$\kappa = \left|\frac{d\phi}{ds}\right|, \tag{11}$$

the absolute value of the rate of change of the angle ϕ with respect to arc length s. We define the curvature κ in terms of $d\phi/ds$ rather than $d\phi/dt$ because the latter depends not only on the shape of the curve, but also on the speed of the moving point $\mathbf{r}(t)$. For a straight line the angle ϕ is a constant, so the curvature given by Eq. (11) is zero. If you imagine a point that is moving with constant speed along a curve, the curvature is greatest at points where ϕ changes the most rapidly, such as the points P and R on the curve of Fig. 12.6.5. The curvature is least at points such as Q and S, where ϕ is changing the least rapidly.

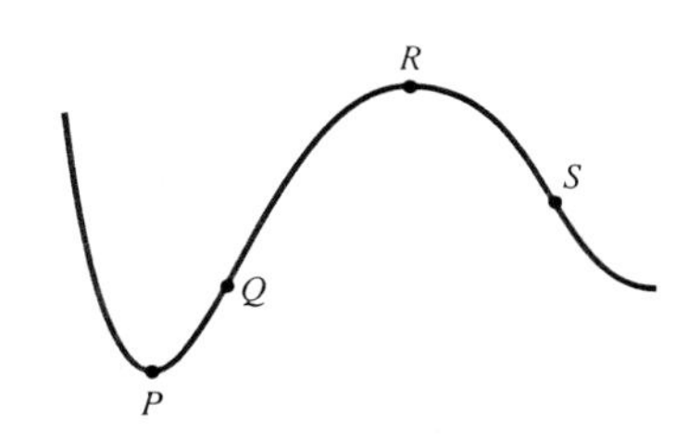

FIGURE 12.6.5 The curvature is large at P and R, small at Q and S.

We need to derive a formula that is effective in computing the curvature of a smooth parametric plane curve $x = x(t)$, $y = y(t)$. First we note that

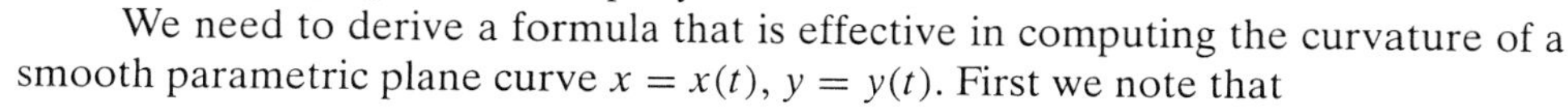

$$\phi = \tan^{-1}\left(\frac{dy}{dx}\right) = \tan^{-1}\left(\frac{y'(t)}{x'(t)}\right)$$

provided $x'(t) \neq 0$. Hence

$$\frac{d\phi}{dt} = \frac{y''x' - y'x''}{(x')^2} \div \left(1 + \left(\frac{y'}{x'}\right)^2\right) = \frac{x'y'' - x''y'}{(x')^2 + (y')^2},$$

where primes denote derivatives with respect to t. Because $v = ds/dt > 0$, Eq. (11) gives

$$\kappa = \left|\frac{d\phi}{ds}\right| = \left|\frac{d\phi}{dt} \cdot \frac{dt}{ds}\right| = \frac{1}{v}\left|\frac{d\phi}{dt}\right|;$$

thus

$$\kappa = \frac{|x'y'' - x''y'|}{[(x')^2 + (y')^2]^{3/2}} = \frac{|x'y'' - x''y'|}{v^3}. \tag{12}$$

At a point where $x'(t) = 0$, we know that $y'(t) \neq 0$, because the curve is smooth. Thus we will obtain the same result if we begin with the equation $\phi = \cot^{-1}(x'/y')$.

An explicitly described curve $y = f(x)$ may be regarded as a parametric curve $x = x$, $y = f(x)$. Then $x' = 1$ and $x'' = 0$, so Eq. (12)—with x in place of t as the parameter—becomes

$$\kappa = \frac{|y''|}{[1 + (y')^2]^{3/2}} = \frac{|d^2y/dx^2|}{[1 + (dy/dx)^2]^{3/2}}. \tag{13}$$

EXAMPLE 3 Show that the curvature at each point of a circle of radius a is $\kappa = 1/a$.

Solution With the familiar parametrization $x = a\cos t$, $y = a\sin t$ of such a circle centered at the origin, we let primes denote derivatives with respect to t and obtain

$$x' = -a\sin t, \qquad y' = a\cos t,$$

$$x'' = -a\cos t, \qquad y'' = -a\sin t.$$

Hence Eq. (12) gives

$$\kappa = \frac{|(-a\sin t)(-a\sin t) - (-a\cos t)(a\cos t)|}{[(-a\sin t)^2 + (a\cos t)^2]^{3/2}} = \frac{a^2}{a^3} = \frac{1}{a}.$$

Alternatively, we could have used Eq. (13). Our point of departure would then be the equation $x^2 + y^2 = a^2$ of the same circle, and we would compute y' and y'' by implicit differentiation. (See Problem 27.) ◆

It follows immediately from Eqs. (8) and (9) that

$$\mathbf{T} \cdot \frac{d\mathbf{T}}{ds} = 0,$$

so the unit tangent vector $\mathbf{T}$ and its derivative vector $d\mathbf{T}/ds$ are perpendicular. If $|d\mathbf{T}/ds| \neq 0$, then the *unit* vector $\mathbf{N}$ that points in the direction of $d\mathbf{T}/ds$ is called the **principal unit normal vector** to the curve. Because $\kappa = |d\phi/ds| = |d\mathbf{T}/ds|$ by Eq. (10), it follows that

$$\frac{d\mathbf{T}}{ds} = \kappa \mathbf{N}. \qquad \textbf{(14)}$$

Intuitively, $\mathbf{N}$ *is the unit normal vector to the curve that points in the direction in which the curve is bending.*

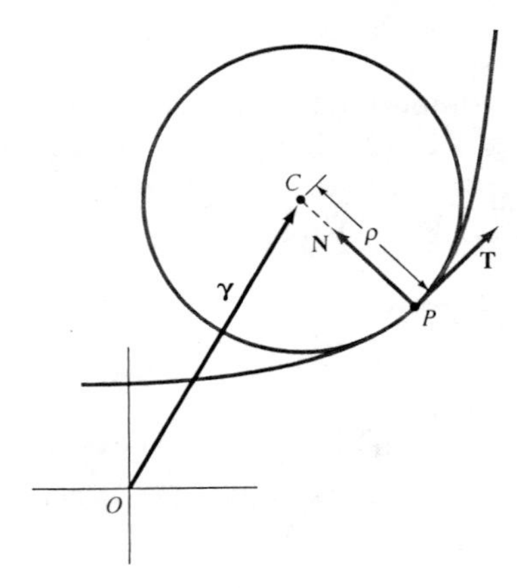

FIGURE 12.6.6 Osculating circle, radius of curvature, and center of curvature.

Suppose that P is a point on a parametrized curve at which $\kappa \neq 0$. Consider the circle that is tangent to the curve at P and has the same curvature there. The center of the circle is to lie on the concave side of the curve—that is, on the side toward which the normal vector $\mathbf{N}$ points. This circle is called the **osculating circle** (or **circle of curvature**) of the curve at the given point because it touches the curve so closely there. (*Osculum* is the Latin word for *kiss.*) Let ρ be the radius of the osculating circle and let $\gamma = \overrightarrow{OC}$ be the position vector of its center C (Fig. 12.6.6). Then ρ is called the **radius of curvature** of the curve at the point P and γ is called the (vector) **center of curvature** of the curve at P.

Example 3 implies that the radius of curvature is

$$\rho = \frac{1}{\kappa}, \qquad \textbf{(15)}$$

and the fact that $|\mathbf{N}| = 1$ implies that the position vector of the center of curvature is

$$\gamma = \mathbf{r} + \rho\mathbf{N} \quad (\mathbf{r} = \overrightarrow{OP}). \qquad \textbf{(16)}$$

EXAMPLE 4 Determine the vectors $\mathbf{T}$ and $\mathbf{N}$, the curvature κ, and the center of curvature of the parabola $y = x^2$ at the point $(1, 1)$.

Solution If the parabola is parametrized by $x = t$, $y = t^2$, then its position vector is $\mathbf{r}(t) = t\mathbf{i} + t^2\mathbf{j}$, so $\mathbf{v}(t) = \mathbf{i} + 2t\mathbf{j}$. The speed is $v(t) = \sqrt{1 + 4t^2}$, so Eq. (7) yields

$$\mathbf{T}(t) = \frac{\mathbf{v}(t)}{v(t)} = \frac{\mathbf{i} + 2t\mathbf{j}}{\sqrt{1 + 4t^2}}.$$

By substituting $t = 1$, we find that the unit tangent vector at $(1, 1)$ is

$$\mathbf{T} = \frac{1}{\sqrt{5}}\mathbf{i} + \frac{2}{\sqrt{5}}\mathbf{j}.$$

Because the parabola is concave upward at $(1, 1)$, the principal unit normal vector is the upward-pointing unit vector

$$\mathbf{N} = -\frac{2}{\sqrt{5}}\mathbf{i} + \frac{1}{\sqrt{5}}\mathbf{j}$$

that is perpendicular to $\mathbf{T}$. (Note that $\mathbf{T}\cdot\mathbf{N} = 0$.) If $y = x^2$, then $dy/dx = 2x$ and $d^2y/dx^2 = 2$, so Eq. (13) yields

$$\kappa = \frac{|y''|}{[1+(y')^2]^{3/2}} = \frac{2}{(1+4x^2)^{3/2}}.$$

So at the point (1, 1) we find the curvature and radius of curvature to be

$$\kappa = \frac{2}{5\sqrt{5}} \quad \text{and} \quad \rho = \frac{5\sqrt{5}}{2},$$

respectively.

Next, Eq. (16) gives the center of curvature as

$$\gamma = \langle 1, 1\rangle + \frac{5\sqrt{5}}{2}\left\langle -\frac{2}{\sqrt{5}}, \frac{1}{\sqrt{5}}\right\rangle = \left\langle -4, \frac{7}{2}\right\rangle.$$

The equation of the osculating circle to the parabola at (1, 1) is, therefore,

$$(x+4)^2 + \left(y - \tfrac{7}{2}\right)^2 = \rho^2 = \tfrac{125}{4}.$$

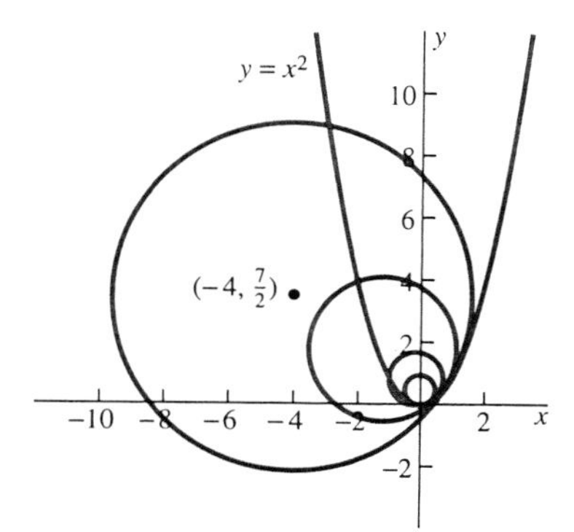

FIGURE 12.6.7 Osculating circles for the parabola of Example 4.

Figure 12.6.7 shows this large osculating circle at the point (1, 1), as well as the smaller osculating circles that are tangent to the parabola at the points (0, 0), $(\frac{1}{3}, \frac{1}{9})$, and $(\frac{2}{3}, \frac{4}{9})$. Is it clear to you which of these osculating circles is which? ◆

Curvature of Space Curves

Consider now a moving particle in space with twice-differentiable position vector $\mathbf{r}(t)$. Suppose also that the velocity vector $\mathbf{v}(t)$ is never zero. The **unit tangent vector** at time t is defined, as before, to be

$$\mathbf{T}(t) = \frac{\mathbf{v}(t)}{|\mathbf{v}(t)|} = \frac{\mathbf{v}(t)}{v(t)}, \tag{17}$$

so

$$\mathbf{v} = v\mathbf{T}. \tag{18}$$

We defined the curvature of a plane curve to be $\kappa = |d\phi/ds|$, where ϕ is the angle of inclination of $\mathbf{T}$ from the positive x-axis. For a space curve, there is no single angle that determines the direction of $\mathbf{T}$, so we adopt the following approach (which leads to the same value for curvature when applied to a space curve that happens to lie in the xy-plane). Differentiating the identity $\mathbf{T}\cdot\mathbf{T} = 1$ with respect to arc length s gives

$$\mathbf{T}\cdot\frac{d\mathbf{T}}{ds} = 0.$$

It follows that the vectors $\mathbf{T}$ and $d\mathbf{T}/ds$ are always perpendicular.

Then we define the **curvature** κ of the curve at the point $\mathbf{r}(t)$ to be

$$\kappa = \left|\frac{d\mathbf{T}}{ds}\right| = \left|\frac{d\mathbf{T}}{dt}\frac{dt}{ds}\right| = \frac{1}{v}\left|\frac{d\mathbf{T}}{dt}\right|. \tag{19}$$

At a point where $\kappa \neq 0$, we define the **principal unit normal vector N** to be

$$\mathbf{N} = \frac{d\mathbf{T}/ds}{|d\mathbf{T}/ds|} = \frac{1}{\kappa}\frac{d\mathbf{T}}{ds}, \tag{20}$$

so

$$\frac{d\mathbf{T}}{ds} = \kappa\mathbf{N}. \tag{21}$$

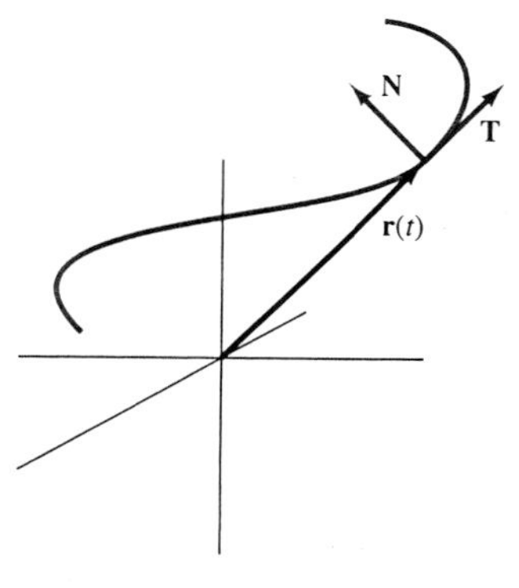

FIGURE 12.6.8 The principal unit normal vector **N** points in the direction in which the curve is turning.

Equation (21) shows that $\mathbf{N}$ has the same direction as $d\mathbf{T}/ds$ (Fig. 12.6.8), and Eq. (20) shows that $\mathbf{N}$ is a unit vector. Because Eq. (21) is the same as Eq. (14), we see that the present definitions of κ and $\mathbf{N}$ agree with those given earlier in the two-dimensional case.

EXAMPLE 5 Compute the curvature κ of the helix of Example 1, the helix with parametric equations

$$x(t) = a\cos\omega t, \qquad y(t) = a\sin\omega t, \qquad z(t) = bt.$$

Solution In Example 7 of Section 12.5, we computed the velocity vector

$$\mathbf{v} = \mathbf{i}(-a\omega\sin\omega t) + \mathbf{j}(a\omega\cos\omega t) + b\mathbf{k}$$

and speed

$$v = |\mathbf{v}| = \sqrt{a^2\omega^2 + b^2}.$$

Hence Eq. (17) gives the unit tangent vector

$$\mathbf{T} = \frac{\mathbf{v}}{v} = \frac{\mathbf{i}(-a\omega\sin\omega t) + \mathbf{j}(a\omega\cos\omega t) + b\mathbf{k}}{\sqrt{a^2\omega^2 + b^2}}.$$

Then

$$\frac{d\mathbf{T}}{dt} = \frac{\mathbf{i}(-a\omega^2\cos\omega t) + \mathbf{j}(-a\omega^2\sin\omega t)}{\sqrt{a^2\omega^2 + b^2}},$$

so Eq. (19) gives

$$\kappa = \frac{1}{v}\left|\frac{d\mathbf{T}}{dt}\right| = \frac{a\omega^2}{a^2\omega^2 + b^2}$$

for the curvature of the helix. Note that the helix has constant curvature. Also note that, if $b = 0$ (so that the helix reduces to a circle of radius a in the xy-plane), our result reduces to $\kappa = 1/a$, in agreement with our computation of the curvature of a circle in Example 3. ◆

Normal and Tangential Components of Acceleration

We may apply Eq. (21) to analyze the meaning of the acceleration vector of a moving particle with velocity vector $\mathbf{v}$ and speed v. Then Eq. (17) gives $\mathbf{v} = v\mathbf{T}$, so the acceleration vector of the particle is

$$\mathbf{a} = \frac{d\mathbf{v}}{dt} = \frac{dv}{dt}\mathbf{T} + v\frac{d\mathbf{T}}{dt} = \frac{dv}{dt}\mathbf{T} + v\frac{d\mathbf{T}}{ds}\frac{ds}{dt}.$$

But $ds/dt = v$, so Eq. (21) gives

$$\mathbf{a} = \frac{dv}{dt}\mathbf{T} + \kappa v^2\mathbf{N}. \tag{22}$$

Because $\mathbf{T}$ and $\mathbf{N}$ are unit vectors tangent and normal to the curve, respectively, Eq. (22) provides a *decomposition of the acceleration vector* into its components tangent to and normal to the trajectory. The **tangential component**

$$a_T = \frac{dv}{dt} \tag{23}$$

is the rate of change of speed of the particle, whereas the **normal component**

$$a_N = \kappa v^2 = \frac{v^2}{\rho} \tag{24}$$

measures the rate of change of its direction of motion. The decomposition

$$\mathbf{a} = a_T\mathbf{T} + a_N\mathbf{N} \tag{25}$$

is illustrated in Fig. 12.6.9.

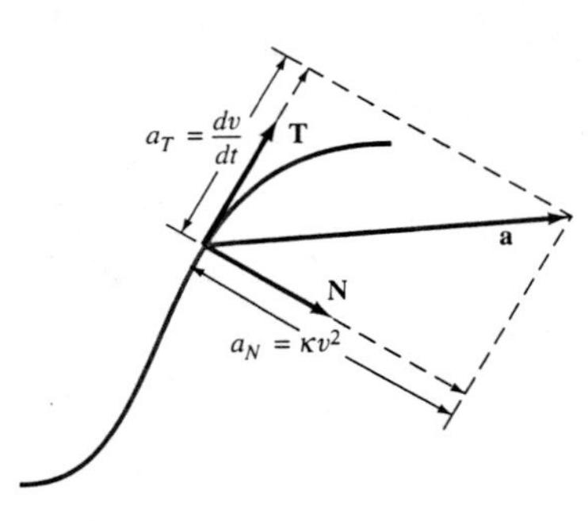

FIGURE 12.6.9 Resolution of the acceleration vector **a** into its tangential and normal components.

As an application of Eq. (22), think of a train moving along a straight track with constant speed v, so that $a_T = 0 = a_N$ (the latter because $\kappa = 0$ for a straight line). Suppose that at time $t = 0$, the train enters a circular curve of radius ρ. At that instant, it will *suddenly* be subjected to a normal acceleration of magnitude v^2/ρ, proportional to the *square* of the speed of the train. A passenger in the train will experience a sudden jerk to the side. If v is large, the stresses may be great enough to damage the track or derail the train. It is for exactly this reason that railroads are built not with curves shaped like arcs of circles but with *approach curves* in which the curvature, and hence the normal acceleration, build up smoothly.

EXAMPLE 6 A particle moves in the xy-plane with parametric equations

$$x(t) = \tfrac{3}{2}t^2, \qquad y(t) = \tfrac{4}{3}t^3.$$

Find the tangential and normal components of its acceleration vector when $t = 1$.

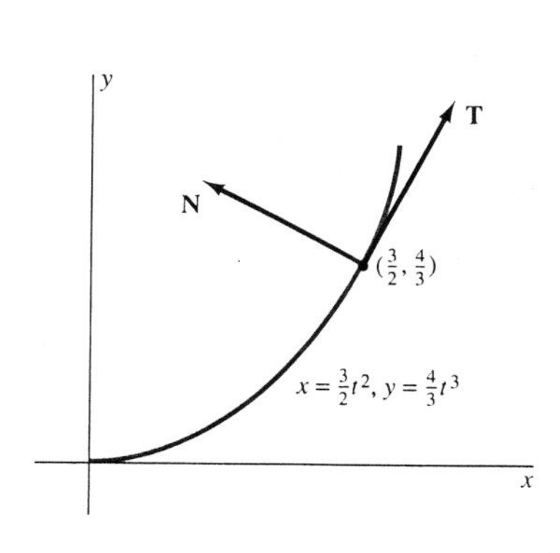

FIGURE 12.6.10 The moving particle of Example 6.

Solution The trajectory and the vectors **N** and **T** appear in Fig. 12.6.10. There **N** and **T** are shown attached at the point of evaluation, at which $t = 1$. The particle has position vector

$$\mathbf{r}(t) = \tfrac{3}{2}t^2\mathbf{i} + \tfrac{4}{3}t^3\mathbf{j}$$

and thus velocity

$$\mathbf{v}(t) = 3t\mathbf{i} + 4t^2\mathbf{j}.$$

Hence its speed is

$$v(t) = \sqrt{9t^2 + 16t^4},$$

from which we calculate

$$a_T = \frac{dv}{dt} = \frac{9t + 32t^3}{\sqrt{9t^2 + 16t^4}}.$$

Thus $v = 5$ and $a_T = \frac{41}{5}$ when $t = 1$.

To use Eq. (12) to compute the curvature at $t = 1$, we compute $dx/dt = 3t$, $dy/dt = 4t^2$, $d^2x/dt^2 = 3$, and $d^2y/dt^2 = 8t$. Thus at $t = 1$ we have

$$\kappa = \frac{|x'y'' - x''y'|}{v^3} = \frac{|3 \cdot 8 - 3 \cdot 4|}{5^3} = \frac{12}{125}.$$

Hence

$$a_N = \kappa v^2 = \tfrac{12}{125} \cdot 5^2 = \tfrac{12}{5}$$

when $t = 1$. As a check (Problem 28), you might compute **T** and **N** when $t = 1$ and verify that

$$\tfrac{41}{5}\mathbf{T} + \tfrac{12}{5}\mathbf{N} = \mathbf{a} = 3\mathbf{i} + 8\mathbf{j}.$$

◆

It remains for us to see how to compute a_T, a_N, and **N** effectively in the case of a space curve. We would prefer to have formulas that explicitly contain only the vectors **r**, **v**, and **a**.

If we compute the dot product of $\mathbf{v} = v\mathbf{T}$ with the acceleration **a** as given in Eq. (22) and use the facts that $\mathbf{T} \cdot \mathbf{T} = 1$ and $\mathbf{T} \cdot \mathbf{N} = 0$, we get

$$\mathbf{v} \cdot \mathbf{a} = v\mathbf{T} \cdot \left(\frac{dv}{dt}\mathbf{T}\right) + (v\mathbf{T}) \cdot (\kappa v^2 \mathbf{N}) = v\frac{dv}{dt}.$$

It follows that

$$a_T = \frac{dv}{dt} = \frac{\mathbf{v} \cdot \mathbf{a}}{v} = \frac{\mathbf{r}'(t) \cdot \mathbf{r}''(t)}{|\mathbf{r}'(t)|}. \tag{26}$$

Similarly, when we compute the cross product of $\mathbf{v} = v\mathbf{T}$ with each side of Eq. (22), we find that

$$\mathbf{v} \times \mathbf{a} = \left(v\mathbf{T} \times \frac{dv}{dt}\mathbf{T} \right) + (v\mathbf{T} \times \kappa v^2 \mathbf{N}) = \kappa v^3 (\mathbf{T} \times \mathbf{N}).$$

Because κ and v are nonnegative and because $\mathbf{T} \times \mathbf{N}$ is a unit vector, we may conclude that

$$\kappa = \frac{|\mathbf{v} \times \mathbf{a}|}{v^3} = \frac{|\mathbf{r}'(t) \times \mathbf{r}''(t)|}{|\mathbf{r}'(t)|^3}. \tag{27}$$

It now follows from Eq. (24) that

$$a_N = \frac{|\mathbf{r}'(t) \times \mathbf{r}''(t)|}{|\mathbf{r}'(t)|}. \tag{28}$$

The curvature of a space curve often is not as easy to compute directly from the definition as we found in the case of the helix of Example 5. It is generally more convenient to use Eq. (27). Once $\mathbf{a}$, $\mathbf{T}$, a_T, and a_N have been computed, we can rewrite Eq. (25) as

$$\mathbf{N} = \frac{\mathbf{a} - a_T \mathbf{T}}{a_N} \tag{29}$$

to find the principal unit normal vector.

EXAMPLE 7 Compute $\mathbf{T}$, $\mathbf{N}$, κ, a_T, and a_N at the point $(1, \frac{1}{2}, \frac{1}{3})$ of the twisted cubic with parametric equations

$$x(t) = t, \qquad y(t) = \tfrac{1}{2}t^2, \qquad z(t) = \tfrac{1}{3}t^3.$$

Solution Differentiating the position vector

$$\mathbf{r}(t) = \left\langle t, \tfrac{1}{2}t^2, \tfrac{1}{3}t^3 \right\rangle$$

gives

$$\mathbf{r}'(t) = \langle 1, t, t^2 \rangle \quad \text{and} \quad \mathbf{r}''(t) = \langle 0, 1, 2t \rangle.$$

When we substitute $t = 1$, we obtain

$$\begin{aligned} \mathbf{v}(1) &= \langle 1, 1, 1 \rangle && \text{(velocity)}, \\ v(1) &= |\mathbf{v}(1)| = \sqrt{3} && \text{(speed)}, \quad \text{and} \\ \mathbf{a}(1) &= \langle 0, 1, 2 \rangle && \text{(acceleration)} \end{aligned}$$

at the point $(1, \frac{1}{2}, \frac{1}{3})$. Then Eq. (26) gives the tangential component of acceleration:

$$a_T = \frac{\mathbf{v} \cdot \mathbf{a}}{v} = \frac{3}{\sqrt{3}} = \sqrt{3}.$$

Because

$$\mathbf{v} \times \mathbf{a} = \begin{vmatrix} \mathbf{i} & \mathbf{j} & \mathbf{k} \\ 1 & 1 & 1 \\ 0 & 1 & 2 \end{vmatrix} = \langle 1, -2, 1 \rangle,$$

Eq. (27) gives the curvature:

$$\kappa = \frac{|\mathbf{v} \times \mathbf{a}|}{v^3} = \frac{\sqrt{6}}{(\sqrt{3})^3} = \frac{\sqrt{2}}{3}.$$

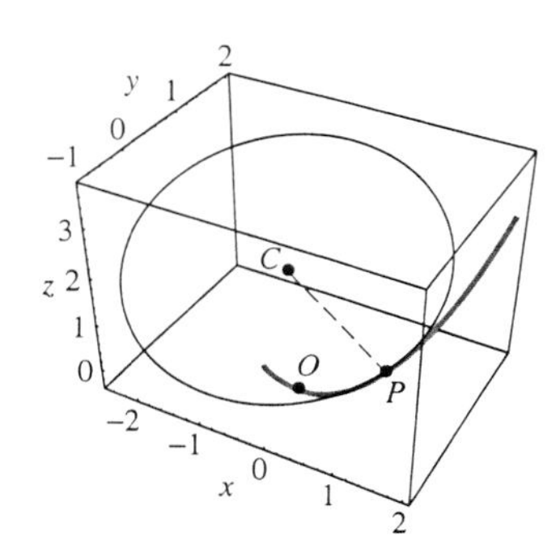

FIGURE 12.6.11 Osculating circle for the twisted cubic of Example 7. It is plotted as the parametric curve with position vector $\mathbf{r}(t) = \overrightarrow{OC} - (a\cos t)\mathbf{N} + (a\sin t)\mathbf{T}$.

The normal component of acceleration is $a_N = \kappa v^2 = \sqrt{2}$. The unit tangent vector is

$$\mathbf{T} = \frac{\mathbf{v}}{v} = \frac{1}{\sqrt{3}}\langle 1, 1, 1\rangle = \frac{\mathbf{i}+\mathbf{j}+\mathbf{k}}{\sqrt{3}}.$$

Finally, Eq. (29) gives

$$\mathbf{N} = \frac{\mathbf{a} - a_T\mathbf{T}}{a_N} = \frac{1}{\sqrt{2}}(\langle 0, 1, 2\rangle - \langle 1, 1, 1\rangle) = \frac{1}{\sqrt{2}}\langle -1, 0, 1\rangle = \frac{-\mathbf{i}+\mathbf{k}}{\sqrt{2}}.$$

Figure 12.6.11 shows the twisted cubic and its osculating circle at the point P. This osculating circle has radius $a = 1/\kappa = \frac{3}{2}\sqrt{2}$ and its center C has position vector $\overrightarrow{OC} = \overrightarrow{OP} + a\mathbf{N} = \langle -\frac{1}{2}, \frac{1}{2}, \frac{11}{6}\rangle$. ◆

Newton, Kepler, and the Solar System

As outlined on the opening page of this chapter, the modern view of our solar system dates back to the formulation by Johannes Kepler (1571–1630) of the following three propositions, now known as **Kepler's laws of planetary motion.**

1. The orbit of each planet is an ellipse with the sun at one focus.
2. The radius vector from the sun to a planet sweeps out area at a constant rate.
3. The *square* of the period of revolution of a planet about the sun is proportional to the *cube* of the major semiaxis of its elliptical orbit.

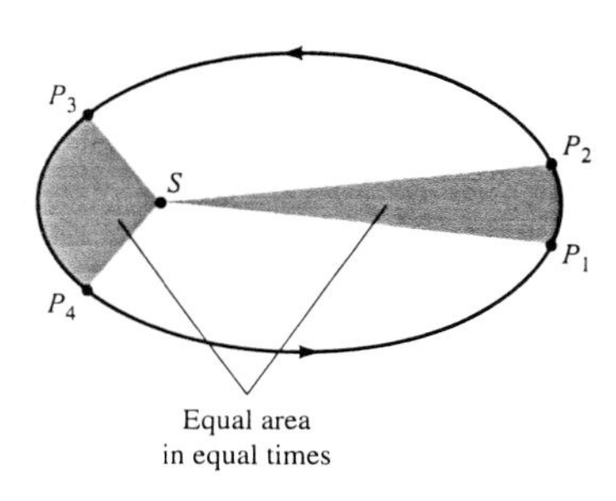

FIGURE 12.6.12 Kepler's law implies that the shaded areas are equal if the planet's times of traversal of the orbital segments P_1P_2 and P_3P_4 are equal.

Figure 12.6.12 illustrates Kepler's second law. If the planet traverses the paths P_1P_2 and P_3P_4 along its orbit in equal times, then the areas of the shaded elliptical sectors SP_1P_2 and SP_3P_4 are equal.

In his *Principia Mathematica* (1687), Newton employed a powerful but now antiquated form of geometrical calculus to show that Kepler's laws follow from the basic principles of mechanics ($F = ma$, and so on) and the inverse-square law of gravitational attraction. In the remainder of this section we apply the modern calculus of vector-valued functions to outline the relation between Newton's laws and Kepler's laws.

Radial and Transverse Components of Acceleration

To begin, we set up a coordinate system in which the sun is located at the origin in the plane of motion of a planet. Let $r = r(t)$ and $\theta = \theta(t)$ be the polar coordinates at time t of the planet as it orbits the sun. We want first to split the planet's position, velocity, and acceleration vectors $\mathbf{r}$, $\mathbf{v}$, and $\mathbf{a}$ into *radial* and *transverse* components. To do so, we introduce at each point (r, θ) of the plane (the origin excepted) the *unit* vectors

$$\mathbf{u}_r = \mathbf{i}\cos\theta + \mathbf{j}\sin\theta, \qquad \mathbf{u}_\theta = -\mathbf{i}\sin\theta + \mathbf{j}\cos\theta. \tag{30}$$

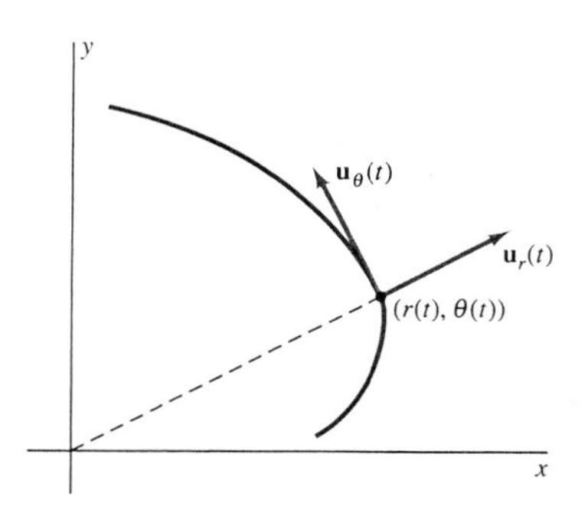

FIGURE 12.6.13 The radial and transverse unit vector $\mathbf{u}_r$ and $\mathbf{u}_\theta$.

If we substitute $\theta = \theta(t)$, then $\mathbf{u}_r$ and $\mathbf{u}_\theta$ become functions of t. The **radial** unit vector $\mathbf{u}_r$ always points directly away from the origin; the **transverse** unit vector $\mathbf{u}_\theta$ is obtained from $\mathbf{u}_r$ by a 90° counterclockwise rotation (Fig. 12.6.13).

In Problem 66 we ask you to verify, by componentwise differentiation of the equations in (30), that

$$\frac{d\mathbf{u}_r}{dt} = \mathbf{u}_\theta\frac{d\theta}{dt} \quad\text{and}\quad \frac{d\mathbf{u}_\theta}{dt} = -\mathbf{u}_r\frac{d\theta}{dt}. \tag{31}$$

The position vector $\mathbf{r}$ points directly away from the origin and has length $|\mathbf{r}| = r$, so

$$\mathbf{r} = r\mathbf{u}_r. \tag{32}$$

Differentiating both sides of Eq. (32) with respect to t gives

$$\mathbf{v} = \frac{d\mathbf{r}}{dt} = \mathbf{u}_r\frac{dr}{dt} + r\frac{d\mathbf{u}_r}{dt}.$$

We use the first equation in (31) and find that the planet's velocity vector is

$$\mathbf{v} = \mathbf{u}_r \frac{dr}{dt} + r \frac{d\theta}{dt} \mathbf{u}_\theta. \tag{33}$$

Thus we have expressed the velocity $\mathbf{v}$ in terms of the radial vector $\mathbf{u}_r$ and the transverse vector $\mathbf{u}_\theta$.

We differentiate both sides of Eq. (33) and thereby find that

$$\mathbf{a} = \frac{d\mathbf{v}}{dt} = \left(\mathbf{u}_r \frac{d^2r}{dt^2} + \frac{dr}{dt}\frac{d\mathbf{u}_r}{dt}\right) + \left(\frac{dr}{dt}\frac{d\theta}{dt}\mathbf{u}_\theta + r\frac{d^2\theta}{dt^2}\mathbf{u}_\theta + r\frac{d\theta}{dt}\frac{d\mathbf{u}_\theta}{dt}\right).$$

Then, by using the equations in (31) and collecting the coefficients of $\mathbf{u}_r$ and $\mathbf{u}_\theta$ (Problem 67), we obtain the decomposition

$$\mathbf{a} = \left[\frac{d^2r}{dt^2} - r\left(\frac{d\theta}{dt}\right)^2\right]\mathbf{u}_r + \left[\frac{1}{r}\frac{d}{dt}\left(r^2\frac{d\theta}{dt}\right)\right]\mathbf{u}_\theta \tag{34}$$

of the acceleration vector into its radial and transverse components.

Planets and Satellites

The key to Newton's analysis was the connection between his law of gravitational attraction and Kepler's *second* law of planetary motion. Suppose that we begin with the inverse-square law of gravitation in its vector form

$$\mathbf{F} = m\mathbf{a} = -\frac{GMm}{r^2}\mathbf{u}_r, \tag{35}$$

where M denotes the mass of the sun and m the mass of the orbiting planet. But the acceleration of the planet is given *also* by

$$\mathbf{a} = -\frac{\mu}{r^2}\mathbf{u}_r, \tag{36}$$

where $\mu = GM$. We equate the transverse components in Eqs. (34) and (36) and thus obtain

$$\frac{1}{r}\cdot\frac{d}{dt}\left(r^2\frac{d\theta}{dt}\right) = 0.$$

We drop the factor $1/r$, then antidifferentiate both sides. We find that

$$r^2\frac{d\theta}{dt} = h \quad (h \text{ a constant}). \tag{37}$$

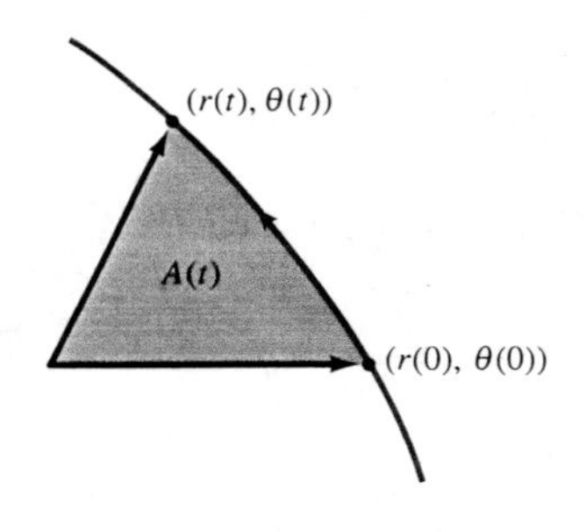

FIGURE 12.6.14 Area swept out by the radius vector.

We know from Section 10.3 that if $A(t)$ denotes the area swept out by the planet's radius vector from time 0 to time t (Fig. 12.6.14), then

$$A(t) = \int_{\theta(0)}^{\theta(t)} \frac{1}{2}r^2\,d\theta = \int_0^t \frac{1}{2}r^2\frac{d\theta}{dt}\,dt.$$

Now we apply the fundamental theorem of calculus, which yields

$$\frac{dA}{dt} = \frac{1}{2}r^2\frac{d\theta}{dt}. \tag{38}$$

When we compare Eqs. (37) and (38), we see that

$$\frac{dA}{dt} = \frac{h}{2}. \tag{39}$$

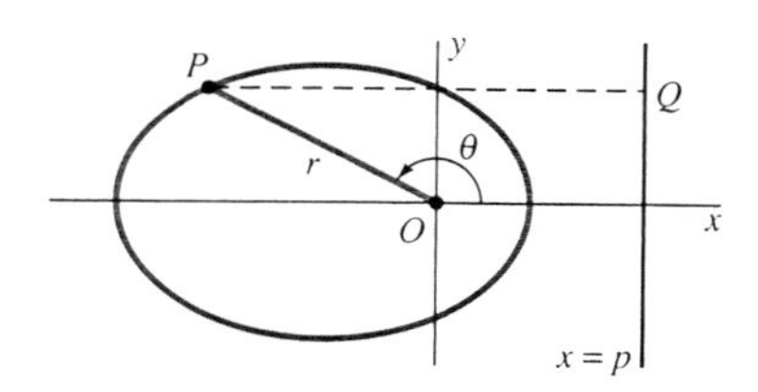

FIGURE 12.6.15 A polar coordinate ellipse with eccentricity $e = |OP|/|PQ|$.

Because $h/2$ is a constant, we have derived Kepler's second law: The radius vector from sun to planet sweeps out area at a constant rate.

Next we outline the derivation of Newton's law of gravitation from Kepler's first and second laws of planetary motion. Figure 12.6.15 shows an ellipse with eccentricity e and focus at the origin. The defining relation $|OP| = e|PQ|$ of this ellipse gives $r = e(p - r\cos\theta)$. Solving this equation then yields the polar-coordinate equation

$$r = \frac{pe}{1 + e\cos\theta} \tag{40}$$

of an ellipse with eccentricity $e < 1$ and directrix $x = p$. In Problem 64 we ask you to show by differentiating twice, using the chain rule and Kepler's second law in the form in Eq. (37), that Eq. (40) implies that

$$\frac{d^2r}{dt^2} = \frac{h^2}{r^2}\left(\frac{1}{r} - \frac{1}{pe}\right). \tag{41}$$

Now if Kepler's second law in the form in Eq. (37) holds, then Eq. (34) gives

$$\mathbf{a} = \left[\frac{d^2r}{dt^2} - r\left(\frac{d\theta}{dt}\right)^2\right]\mathbf{u}_r \tag{42}$$

for the planet's acceleration vector. Finally, upon substituting $d\theta/dt = h/r^2$ from Eq. (37) and the expression in Eq. (41) for d^2r/dt^2, we find (Problem 65) that Eq. (42) can be simplified to the form

$$\mathbf{a} = -\frac{h^2}{per^2}\mathbf{u}_r. \tag{43}$$

This is the inverse-square law of gravitation in the form of Eq. (36) with $\mu = h^2/pe$.

Now suppose that the elliptical orbit of a planet around the sun has major semiaxis a and minor semiaxis b. Then the constant

$$pe = \frac{h^2}{\mu}$$

that appears in Eq. (42) satisfies the equations

$$pe = a(1 - e^2) = a\left(1 - \frac{a^2 - b^2}{a^2}\right) = \frac{b^2}{a}.$$

[See Eq. (29) in Section 10.6.] We equate these two expressions for pe and find that $h^2 = \mu b^2/a$.

Now let T denote the period of revolution of the planet—the time required for it to complete one full revolution in its elliptical orbit around the sun. Then we see from Eq. (39) that the area of the ellipse bounded by this orbit is $A = \frac{1}{2}hT = \pi ab$ and thus that

$$T^2 = \frac{4\pi^2a^2b^2}{h^2} = \frac{4\pi^2a^2b^2}{\mu b^2/a}.$$

Therefore

$$T^2 = \gamma a^3, \tag{44}$$

where the proportionality constant $\gamma = 4\pi^2/\mu = 4\pi^2/GM$ [compare Eqs. (35) and (36)] depends on the gravitational constant G and the sun's mass M. Thus we have derived Kepler's third law of planetary motion from his first two laws and Newton's law of gravitational attraction.

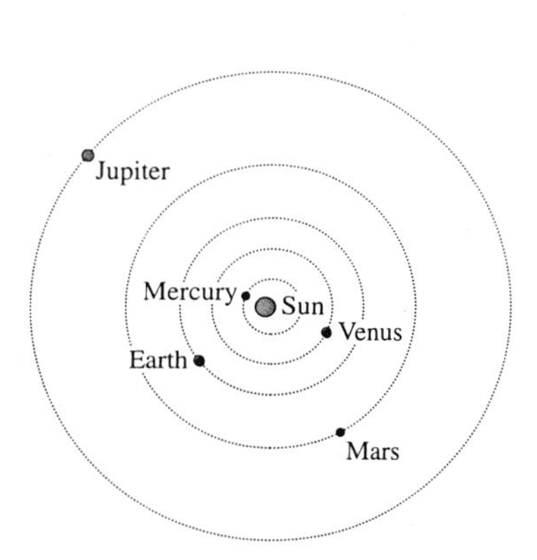

FIGURE 12.6.16 The inner planets of the solar system (Example 8).

EXAMPLE 8 The period of revolution of Mercury in its elliptical orbit around the sun is $T = 87.97$ days, whereas that of the earth is 365.26 days. Compute the major semiaxis (in astronomical units) of the orbit of Mercury. See Fig. 12.6.16.

Solution The major semiaxis of the orbit of the earth is, by definition, 1 AU. So Eq. (44) gives the value of the constant $\gamma = (365.26)^2$ (in day^2/AU^3). Hence the major semiaxis of the orbit of Mercury is

$$a = \left(\frac{T^2}{\gamma}\right)^{1/3} = \left(\frac{(87.97)^2}{(365.26)^2}\right)^{1/3} \approx 0.387 \quad \text{(AU)}.$$ ◆

As yet we have considered only planets in orbits around the sun. But Kepler's laws and the equations of this section apply to bodies in orbit around any common central mass, so long as they move solely under the influence of *its* gravitational attraction. Examples include satellites (artificial or natural) orbiting the earth or the moons of Jupiter.

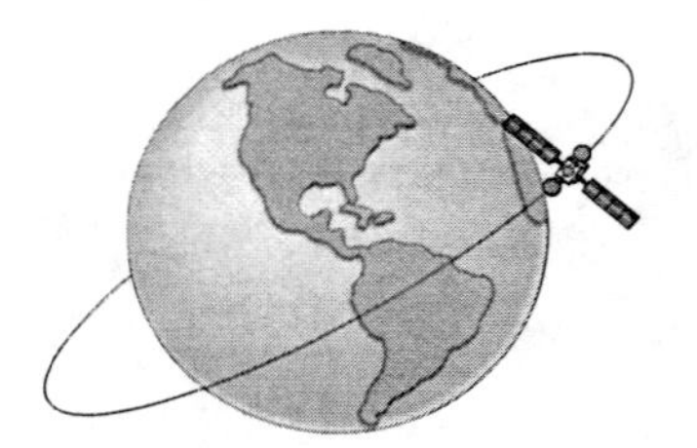

FIGURE 12.6.17 A communications satellite in orbit around the earth (Example 9).

EXAMPLE 9 A communications relay satellite is to be placed in a circular orbit around the earth and is to have a period of revolution of 24 h. This is a *geosynchronous* orbit in which the satellite appears to be stationary in the sky. Assume that the earth's natural moon has a period of 27.32 days in a circular orbit of radius 238,850 mi. What should be the radius of the satellite's orbit? (See Fig. 12.6.17.)

Solution Equation (44), when applied to the moon, yields

$$(27.32)^2 = \gamma(238{,}850)^3.$$

For the stationary satellite that has period $T = 1$ (day), it yields $1^2 = \gamma r^3$, where r is the radius of the geosynchronous orbit. To eliminate γ, we divide the second of these equations by the first and find that

$$r^3 = \frac{(238{,}850)^3}{(27.32)^2}.$$

Thus r is approximately 26,330 mi. The radius of the earth is about 3960 mi, so the satellite will be 22370 mi above the surface. ◆

12.6 TRUE/FALSE STUDY GUIDE

12.6 CONCEPTS: QUESTIONS AND DISCUSSION

1. The curvature of a plane curve is defined in Eq. (11) and the curvature of a space curve is defined in Eq. (21). Do these two definitions agree in the case of a curve that lies in the xy-plane? Explain why.
2. Suppose that two bodies move solely under their mutual gravitational attraction. Then each moves in an elliptical orbit about the other. For instance, in a coordinate system with the earth (rather than the sun) at the origin, the orbit of the sun is an ellipse with the earth at one focus. Which is *really* the center of the solar system? Is this a mathematical or a philosophical question?

12.6 PROBLEMS

Find the arc length of each curve described in Problems 1 through 6.

1. $x = 3\sin 2t,\ y = 3\cos 2t,\ z = 8t;$ from $t = 0$ to $t = \pi$
2. $x = t,\ y = t^2/\sqrt{2},\ z = t^3/3;$ from $t = 0$ to $t = 1$
3. $x = 6e^t\cos t,\ y = 6e^t\sin t,\ z = 17e^t;$ from $t = 0$ to $t = 1$
4. $x = t^2/2,\ y = \ln t,\ z = t\sqrt{2};$ from $t = 1$ to $t = 2$
5. $x = 3t\sin t,\ y = 3t\cos t,\ z = 2t^2;$ from $t = 0$ to $t = 4/5$
6. $x = 2e^t,\ y = e^{-t},\ z = 2t;$ from $t = 0$ to $t = 1$

In Problems 7 through 12, find the curvature of the given plane curve at the indicated point.

7. $y = x^3$ at $(0, 0)$
8. $y = x^3$ at $(-1, -1)$
9. $y = \cos x$ at $(0, 1)$
10. $x = t - 1,\ y = t^2 + 3t + 2$, where $t = 2$
11. $x = 5\cos t,\ y = 4\sin t$, where $t = \pi/4$
12. $x = 5\cosh t,\ y = 3\sinh t$, where $t = 0$

In Problems 13 through 16, find the point or points on the given curve at which the curvature is a maximum.

13. $y = e^x$
14. $y = \ln x$
15. $x = 5\cos t,\ y = 3\sin t$
16. $xy = 1$

For the plane curves in Problems 17 through 21, find the unit tangent and normal vectors at the indicated point.

17. $y = x^3$ at $(-1, -1)$

18. $x = t^3$, $y = t^2$ at $(-1, 1)$

19. $x = 3\sin 2t$, $y = 4\cos 2t$, where $t = \pi/6$

20. $x = t - \sin t$, $y = 1 - \cos t$, where $t = \pi/2$

21. $x = \cos^3 t$, $y = \sin^3 t$, where $t = 3\pi/4$

The position vector of a particle moving in the plane is given in Problems 22 through 26. Find the tangential and normal components of the acceleration vector.

22. $\mathbf{r}(t) = 3\mathbf{i}\sin \pi t + 3\mathbf{j}\cos \pi t$

23. $\mathbf{r}(t) = (2t + 1)\mathbf{i} + (3t^2 - 1)\mathbf{j}$

24. $\mathbf{r}(t) = \mathbf{i}\cosh 3t + \mathbf{j}\sinh 3t$

25. $\mathbf{r}(t) = \mathbf{i}t\cos t + \mathbf{j}t\sin t$

26. $\mathbf{r}(t) = \langle e^t \sin t, e^t \cos t\rangle$

27. Use Eq. (13) to compute the curvature of the circle with equation $x^2 + y^2 = a^2$.

28. Verify the equation $\frac{41}{5}\mathbf{T} + \frac{12}{5}\mathbf{N} = 3\mathbf{i} + 8\mathbf{j}$ given at the end of Example 6.

In Problems 29 through 31, find the equation of the osculating circle for the given plane curve at the indicated point.

29. $y = 1 - x^2$ at $(0, 1)$

30. $y = e^x$ at $(0, 1)$

31. $xy = 1$ at $(1, 1)$

Find the curvature κ of the space curves with position vectors given in Problems 32 through 36.

32. $\mathbf{r}(t) = t\mathbf{i} + (2t - 1)\mathbf{j} + (3t + 5)\mathbf{k}$

33. $\mathbf{r}(t) = t\mathbf{i} + \mathbf{j}\sin t + \mathbf{k}\cos t$

34. $\mathbf{r}(t) = \langle t, t^2, t^3\rangle$

35. $\mathbf{r}(t) = \langle e^t \cos t, e^t \sin t, e^t\rangle$

36. $\mathbf{r}(t) = \mathbf{i}t\sin t + \mathbf{j}t\cos t + \mathbf{k}t$

37 through **41.** Find the tangential and normal components of acceleration a_T and a_N for the curves of Problems 32 through 36, respectively.

*In Problems 42 through 45, find the unit vectors **T** and **N** for the given curve at the indicated point.*

42. The curve of Problem 34 at $(1, 1, 1)$

43. The curve of Problem 33 at $(0, 0, 1)$

44. The curve of Problem 3 at $(6, 0, 17)$

45. The curve of Problem 35 at $(1, 0, 1)$

46. Find $\mathbf{T}$, $\mathbf{N}$, a_T, and a_N as functions of t for the helix of Example 1.

47. Find the arc-length parametrization of the line

$$x(t) = 2 + 4t, \qquad y(t) = 1 - 12t, \qquad z(t) = 3 + 3t$$

in terms of the arc length s measured from the initial point $(2, 1, 3)$.

48. Find the arc-length parametrization of the circle

$$x(t) = 2\cos t, \qquad y(t) = 2\sin t, \qquad z = 0$$

in terms of the arc length s measured counterclockwise from the initial point $(2, 0, 0)$.

49. Find the arc-length parametrization of the helix

$$x(t) = 3\cos t, \qquad y(t) = 3\sin t, \qquad z(t) = 4t$$

in terms of the arc length s measured from the initial point $(3, 0, 0)$.

50. Substitute $x = t$, $y = f(t)$, and $z = 0$ into Eq. (27) to verify that the curvature of the plane curve $y = f(x)$ is

$$\kappa(x) = \frac{|f''(x)|}{[1 + (f'(x))^2]^{3/2}}.$$

51. A particle moves under the influence of a force that is always perpendicular to its direction of motion. Show that the speed of the particle must be constant.

52. Deduce from Eq. (24) that

$$\kappa = \frac{\sqrt{a^2 - (a_T)^2}}{v^2} = \frac{\sqrt{(x''(t))^2 + (y''(t))^2 - (v'(t))^2}}{(x'(t))^2 + (y'(t))^2}.$$

53. Apply the formula of Problem 52 to calculate the curvature of the curve

$$x(t) = \cos t + t\sin t, \qquad y(t) = \sin t - t\cos t.$$

54. The folium of Descartes with equation $x^3 + y^3 = 3xy$ is shown in Fig. 12.6.18. Find the curvature and center of curvature of this folium at the point $(\frac{3}{2}, \frac{3}{2})$. Begin by calculating dy/dx and d^2y/dx^2 by implicit differentiation.

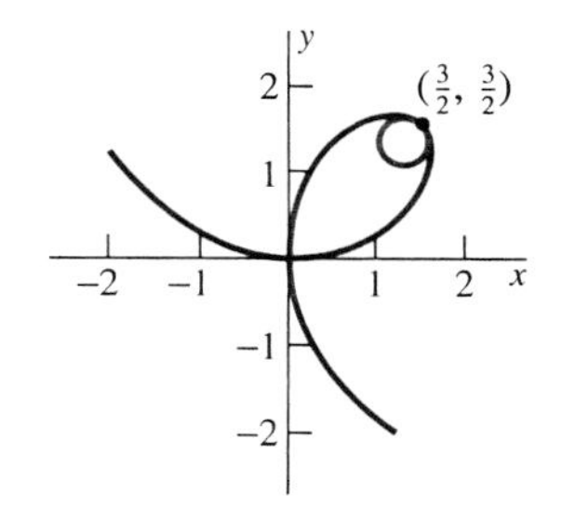

FIGURE 12.6.18 The folium of Descartes (Problem 54).

55. Determine the constants A, B, C, D, E, and F so that the curve

$$y = Ax^5 + Bx^4 + Cx^3 + Dx^2 + Ex + F$$

does, simultaneously, all of the following:

- Joins the two points $(0, 0)$ and $(1, 1)$;
- Has slope 0 at $(0, 0)$ and slope 1 at $(1, 1)$;
- Has curvature 0 at both $(0, 0)$ and $(1, 1)$.

The curve in question is shown in color in Fig. 12.6.19. Why would this be a good curve to join the railroad tracks, shown in black in the figure?

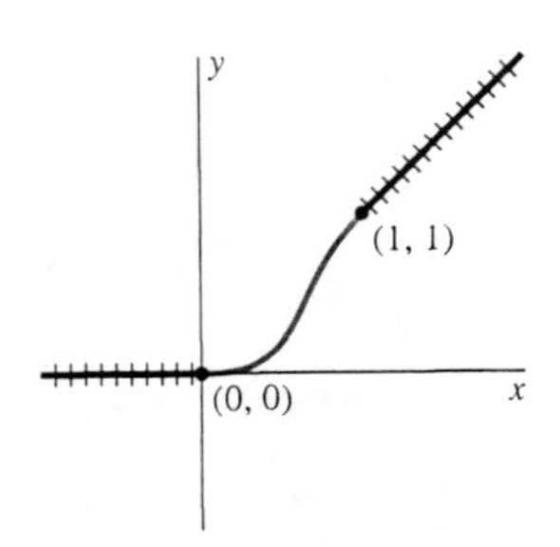

FIGURE 12.6.19 Connecting railroad tracks (Problem 55).

56. Consider a body in an elliptical orbit with major and minor semiaxes a and b and period of revolution T. (a) Deduce from Eq. (33) that $v = r(d\theta/dt)$ when the body is nearest to and farthest from its foci. (b) Then apply Kepler's second law to conclude that $v = 2\pi ab/(rT)$ at the body's nearest and farthest points.

In Problems 57 through 60, apply the equation of part (b) of Problem 56 to compute the speed (in miles per second) of the given body at the nearest and farthest points of its orbit. Convert 1 AU, *the major semiaxis of the Earth's orbit, into* 92,956,000 mi.

57. Mercury: $a = 0.387$ AU, $e = 0.206$, $T = 87.97$ days

58. The earth: $e = 0.0167$, $T = 365.26$ days

59. The earth's moon: $a = 238{,}900$ mi, $e = 0.055$, $T = 27.32$ days

60. An artificial earth satellite: $a = 10{,}000$ mi, $e = 0.5$

61. Assuming the earth to be a sphere with radius 3960 mi, find the altitude above the earth's surface of a satellite in a circular orbit that has a period of revolution of 1 h.

62. Given the fact that Jupiter's period of (almost) circular revolution around the Sun is 11.86 yr, calculate the distance of Jupiter from the Sun.

63. Suppose that an earth satellite in elliptical orbit varies in altitude from 100 to 1000 mi above the earth's surface (assumed spherical). Find this satellite's period of revolution.

64. (a) Beginning with the polar-coordinates equation of an ellipse in Eq. (40), apply the chain rule and Kepler's second law in the form $d\theta/dt = h/r^2$ to differentiate r with respect to t and thereby show that $dr/dt = (h \sin\theta)/p$. (b) Differentiate again to show that $d^2r/dt^2 = (h^2 \cos\theta)/(pr^2)$. (c) Derive Eq. (41) by solving Eq. (40) for $\cos\theta$ and substituting the result in the formula in part (b).

65. Derive Eq. (43) by substituting the expressions for $d\theta/dt$ and d^2r/dt^2 given by Eqs. (37) and (41), respectively, into Eq. (42).

66. Derive both equations in (31) by differentiating the equations in (30).

67. Derive Eq. (34) by differentiating Eq. (33).

12.7 CYLINDERS AND QUADRIC SURFACES

Just as the graph of an equation $f(x, y) = 0$ is generally a curve in the xy-plane, the graph of an equation in three variables is generally a surface in space. A function F of three variables associates a real number $F(x, y, z)$ with each ordered triple (x, y, z) of real numbers. The **graph** of the equation

$$F(x, y, z) = 0 \tag{1}$$

is the set of all points whose coordinates (x, y, z) satisfy this equation. We refer to the graph of such an equation as a **surface.** For instance, the graph of the equation

$$x^2 + y^2 + z^2 - 1 = 0$$

is a familiar surface, the unit sphere centered at the origin. But note that the graph of Eq. (1) does not always agree with our intuitive notion of a surface. For example, the graph of the equation

$$(x^2 + y^2)(y^2 + z^2)(z^2 + x^2) = 0$$

consists of the points lying on the three coordinate axes in space, because

- $x^2 + y^2 = 0$ implies that $x = y = 0$ (the z-axis);
- $y^2 + z^2 = 0$ implies that $y = z = 0$ (the x-axis);
- $z^2 + x^2 = 0$ implies that $z = x = 0$ (the y-axis).

We leave for advanced calculus the precise definition of *surface* as well as the study of conditions sufficient to imply that the graph of Eq. (1) actually is a surface.

Planes and Traces

The simplest example of a surface is a plane with linear equation $Ax + By + Cz + D = 0$. In this section we discuss examples of other simple surfaces that frequently appear in multivariable calculus.

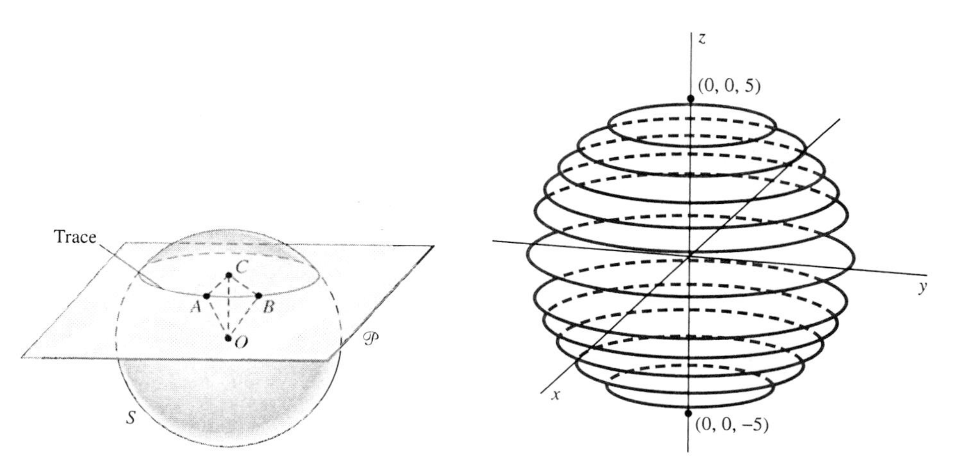

FIGURE 12.7.1 The intersection of the sphere S and the plane $\mathcal{P}$ is a circle.

FIGURE 12.7.2 A sphere as a union of circles (and two points).

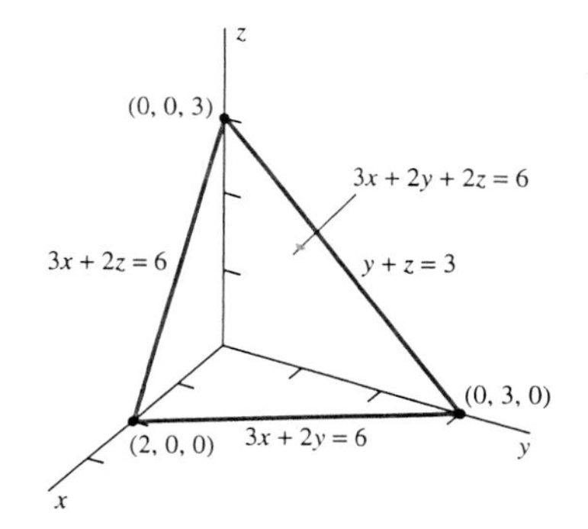

FIGURE 12.7.3 Traces of the plane $3x + 2y + 2z = 6$ in the coordinate planes (Example 1).

In order to sketch a surface S, it is often helpful to examine its intersections with various planes. The **trace** of the surface S in the plane $\mathcal{P}$ is the intersection of $\mathcal{P}$ and S. For example, if S is a sphere, then we can verify by the methods of elementary geometry that the trace of S in the plane $\mathcal{P}$ is a circle (Fig. 12.7.1), provided that $\mathcal{P}$ intersects the sphere but is not merely tangent to it (Problem 49). Figure 12.7.2 illustrates the horizontal trace circles that (together with two "polar points") make up the sphere $x^2 + y^2 + z^2 = 25$.

When we want to visualize a specific surface in space, it often suffices to examine its traces in the coordinate planes and possibly a few planes parallel to them, as in Example 1.

EXAMPLE 1 Consider the plane with equation $3x+2y+2z = 6$. We find its trace in the xy-plane by setting $z = 0$. The equation then reduces to the equation $3x+2y = 6$ of a straight line in the xy-plane. Similarly, when we set $y=0$ we get the line $3x + 2z = 6$ as the trace of the given plane in the xz-plane. To find its trace in the yz-plane, we set $x = 0$, and this yields the line $y + z = 3$. Figure 12.7.3 shows the portions of these three trace lines that lie in the first octant. Together they give us a good picture of how the plane $3x + 2y + 2z = 6$ is situated in space. ◆

Cylinders and Rulings

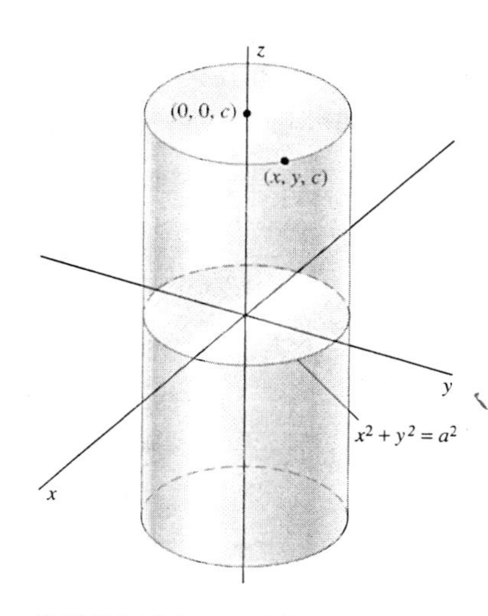

FIGURE 12.7.4 A right circular cylinder.

Let C be a curve in a plane and let L be a line not parallel to that plane. Then the set of points on lines parallel to L that intersect C is called a **cylinder.** These straight lines that make up the cylinder are called **rulings** of the cylinder.

EXAMPLE 2 Figure 12.7.4 shows a vertical cylinder for which C is the circle $x^2+y^2 = a^2$ in the xy-plane. The trace of this cylinder in any horizontal plane $z = c$ is a circle with radius a and center $(0, 0, c)$ on the z-axis. Thus the point (x, y, z) lies on this cylinder if and only if $x^2 + y^2 = a^2$. Hence this cylinder is the graph of the equation $x^2+y^2 = a^2$, an equation in **three** variables—even though the variable z is technically missing.

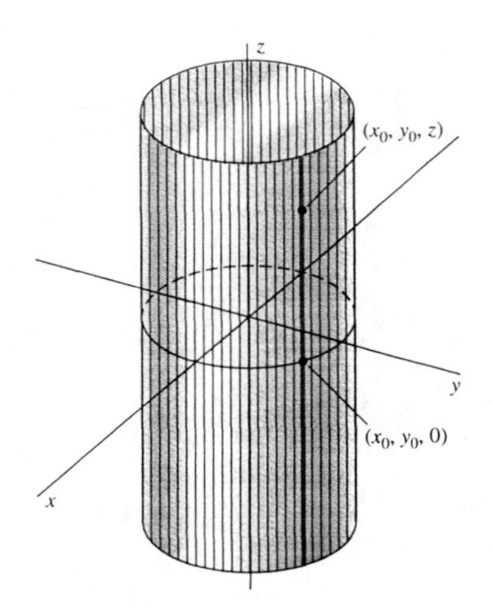

FIGURE 12.7.5 The cylinder $x^2 + y^2 = a^2$; its rulings are parallel to the z-axis.

The fact that the variable z does not appear explicitly in the equation $x^2+y^2 = a^2$ means that given any point $(x_0, y_0, 0)$ on the *circle* $x^2 + y^2 = a^2$ in the xy-plane, the point (x_0, y_0, z) lies on the cylinder for any and all values of z. The set of all such points is the vertical line through the point $(x_0, y_0, 0)$. Thus this vertical line is a ruling of the *cylinder* $x^2 + y^2 = a^2$. Figure 12.7.5 exhibits the cylinder as the union of its rulings. ◆

A cylinder need not be circular—that is, the curve C can be an ellipse, a rectangle, or a quite arbitrary curve.

EXAMPLE 3 Figure 12.7.6 shows both horizontal traces and vertical rulings on a vertical cylinder through the figure-eight curve C in the xy-plane (C has the parametric equations $x = \sin t$, $y = \sin 2t$, $0 \leqq t \leqq 2\pi$). ◆

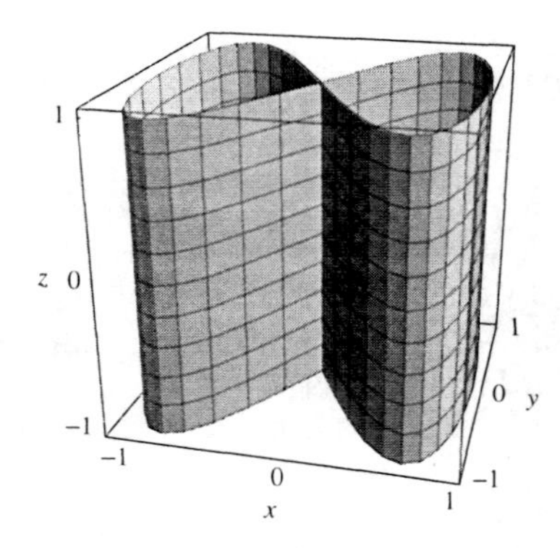

FIGURE 12.7.6 The vertical cylinder through the figure-eight curve $x = \sin t$, $y = \sin 2t$.

If the curve C in the xy-plane has equation

$$f(x, y) = 0, \tag{2}$$

then the cylinder through C with vertical rulings has the same equation in space. This is so because the point $P(x, y, z)$ lies on the cylinder if and only if the point $(x, y, 0)$ lies on the curve C. Similarly, the graph of an equation $g(x, z) = 0$ is a cylinder with rulings parallel to the y-axis, and the graph of an equation $h(y, z) = 0$ is a cylinder with rulings parallel to the x-axis. Thus the graph in space of an equation that includes only two of the three coordinate variables is always a cylinder; its rulings are parallel to the axis corresponding to the *missing* variable.

EXAMPLE 4 The graph of the equation $4y^2 + 9z^2 = 36$ is the **elliptic cylinder** shown in Fig. 12.7.7. Its rulings are parallel to the x-axis, and its trace in every plane perpendicular to the x-axis is an ellipse with semiaxes of lengths 3 and 2 (just like the pictured ellipse $y^2/9 + z^2/4 = 1$ in the yz-plane). ◆

EXAMPLE 5 The graph of the equation $z = 4 - x^2$ is the **parabolic cylinder** shown in Fig. 12.7.8. Its rulings are parallel to the y-axis, and its trace in every plane perpendicular to the y-axis is a parabola that is a parallel translate of the parabola $z = 4-x^2$ in the xz-plane. ◆

Surfaces of Revolution

Another way to use a plane curve C to generate a surface is to revolve the curve in space around a line L in its plane. This gives a **surface of revolution** with **axis** L. For

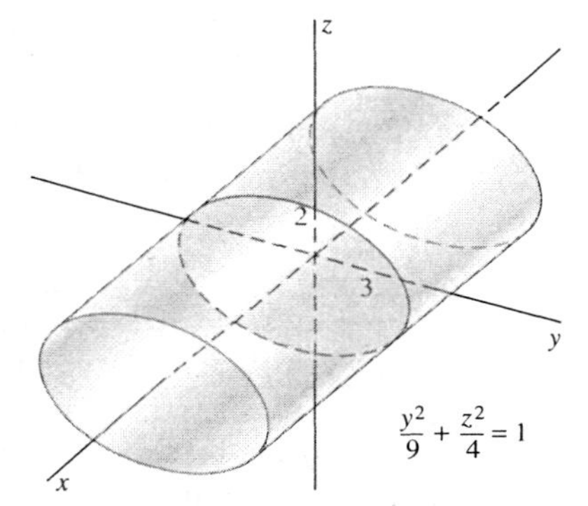

FIGURE 12.7.7 An elliptical cylinder (Example 4).

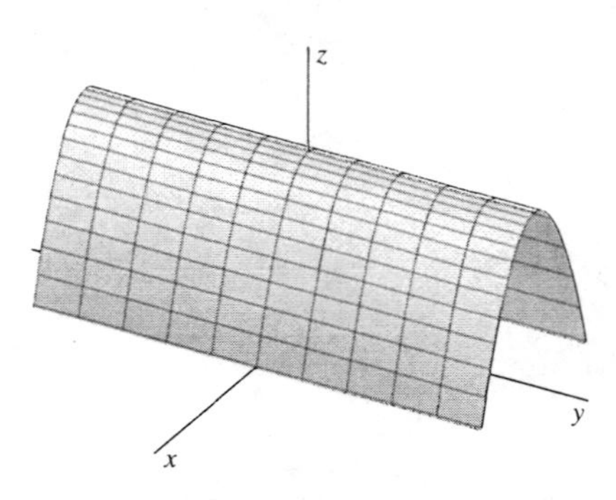

FIGURE 12.7.8 The parabolic cylinder $z = 4 - x^2$ (Example 5).

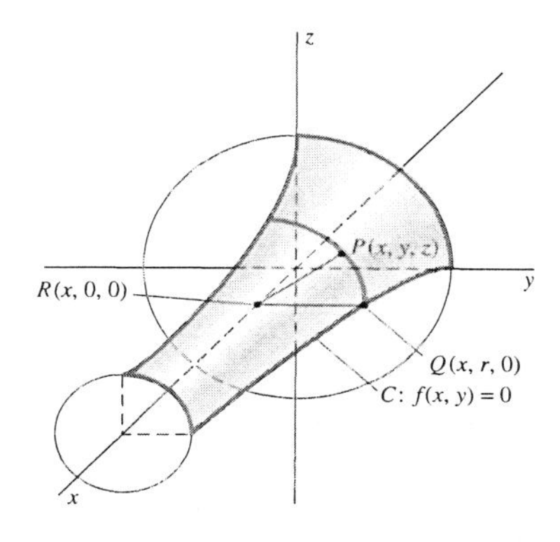

FIGURE 12.7.9 The surface generated by rotating C around the x-axis. (For clarity, only a quarter of the surface is shown.)

example, Fig. 12.7.9 shows the surface generated by revolving the curve $f(x, y) = 0$ in the first quadrant of the xy-plane around the x-axis. The typical point $P(x, y, z)$ lies on this surface of revolution provided that it lies on the vertical circle (parallel to the yz-plane) with center $R(x, 0, 0)$ and radius r such that the point $Q(x, r, 0)$ lies on the given curve C, in which case $f(x, r) = 0$. Because

$$r = |RQ| = |RP| = \sqrt{y^2 + z^2},$$

it is therefore necessary that

$$f(x, \sqrt{y^2 + z^2}) = 0. \tag{3}$$

This, then, is the equation of a **surface of revolution around the x-axis.**

The equations of surfaces of revolution around other coordinate axes are obtained similarly. If the first-quadrant curve $f(x, y) = 0$ is revolved instead around the y-axis, then we replace x with $\sqrt{x^2 + z^2}$ to get the equation $f(\sqrt{x^2 + z^2}, y) = 0$ of the resulting surface of revolution. If the curve $g(y, z) = 0$ in the first quadrant of the yz-plane is revolved around the z-axis, we replace y with $\sqrt{x^2 + y^2}$. Thus the equation of the resulting surface of revolution around the z-axis is $g(\sqrt{x^2 + y^2}, z) = 0$. These assertions are easily verified with the aid of diagrams similar to Fig 12.7.9.

EXAMPLE 6 Write an equation of the **ellipsoid of revolution** obtained by revolving the ellipse $4y^2 + z^2 = 4$ around the z-axis (Fig. 12.7.10).

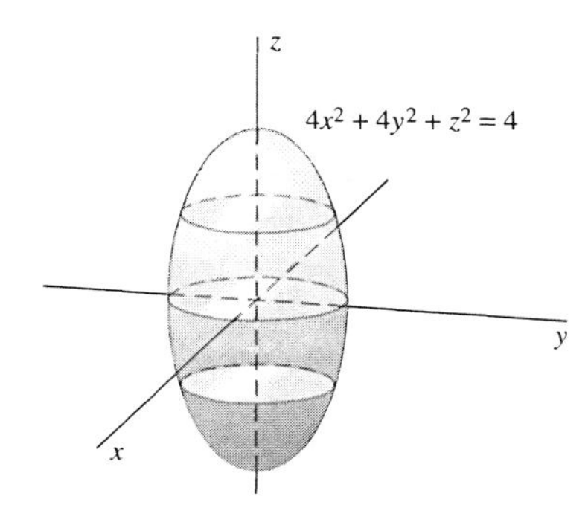

FIGURE 12.7.10 The ellipsoid of revolution of Example 6.

Solution We replace y with $\sqrt{x^2 + y^2}$ in the given equation. This yields $4x^2 + 4y^2 + z^2 = 4$ as an equation of the ellipsoid. ◆

EXAMPLE 7 Determine the graph of the equation $z^2 = x^2 + y^2$.

Solution First we rewrite the given equation in the form $z = \pm\sqrt{x^2 + y^2}$. Thus the surface is symmetric around the xy-plane, and the upper half has equation $z = \sqrt{x^2 + y^2}$. We can obtain this last equation from the simple equation $z = y$ by replacing y with $\sqrt{x^2 + y^2}$. Thus we obtain the upper half of the surface by revolving the line $z = y$ (for $y \geqq 0$) around the z-axis. The graph is the **cone** shown in Fig. 12.7.11. Its upper half has equation $z = \sqrt{x^2 + y^2}$ and its lower half has equation $z = -\sqrt{x^2 + y^2}$. The entire cone $z^2 = x^2 + y^2$ is obtained by revolving the entire line $z = y$ around the z-axis. ◆

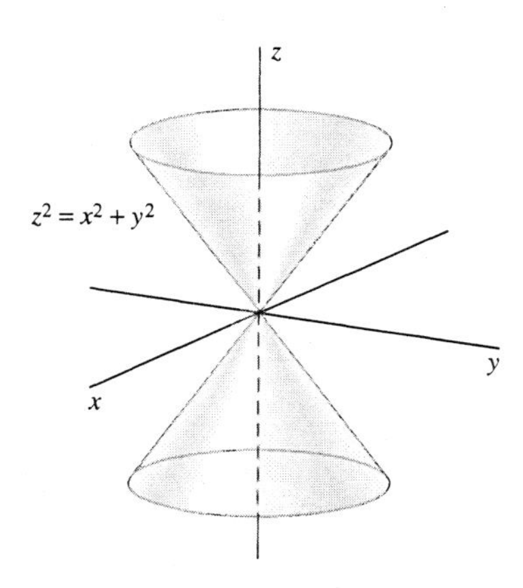

FIGURE 12.7.11 The cone of Example 7.

Quadric Surfaces

Cones, spheres, circular and parabolic cylinders, and ellipsoids of revolution are all surfaces that are graphs of second-degree equations in x, y, and z. The graph of a second-degree equation in three variables is called a **quadric surface.** We discuss here some important special cases of the equation

$$Ax^2 + By^2 + Cz^2 + Dx + Ey + Fz + H = 0. \tag{4}$$

This is a special second-degree equation in that it contains no terms involving the products xy, xz, or yz.

EXAMPLE 8 The **ellipsoid**

$$\frac{x^2}{a^2} + \frac{y^2}{b^2} + \frac{z^2}{c^2} = 1 \tag{5}$$

is symmetric around each of the three coordinate planes and has intercepts $(\pm a, 0, 0)$, $(0, \pm b, 0)$, and $(0, 0, \pm c)$ on the three coordinate axes. (There is no loss of generality in assuming that a, b, and c are positive.) Each trace of this ellipsoid in a plane parallel to one of the coordinate planes is either a single point or an ellipse. For example, if $-c < z_0 < c$, then the trace of the ellipsoid of Eq. (5) in the plane $z = z_0$ has equation

$$\frac{x^2}{a^2} + \frac{y^2}{b^2} = 1 - \frac{z_0^2}{c^2} > 0,$$

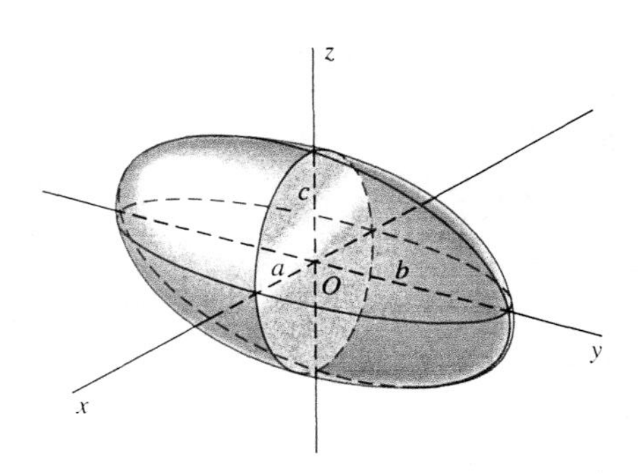

FIGURE 12.7.12 The ellipsoid of Example 8.

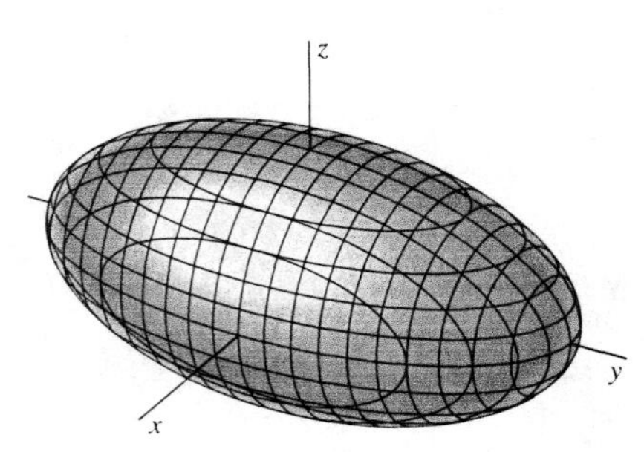

FIGURE 12.7.13 The traces of the ellipsoid $\frac{x^2}{a^2} + \frac{y^2}{b^2} + \frac{z^2}{c^2}$ (Example 8).

which is the equation of an ellipse with semiaxes $(a/c)\sqrt{c^2 - z_0^2}$ and $(b/c)\sqrt{c^2 - z_0^2}$. Figure 12.7.12 shows this ellipsoid with semiaxes a, b, and c labeled. Figure 12.7.13 shows its trace ellipses in planes parallel to the three coordinate planes. ◆

EXAMPLE 9 The **elliptic paraboloid**

$$\frac{x^2}{a^2} + \frac{y^2}{b^2} = \frac{z}{c} \tag{6}$$

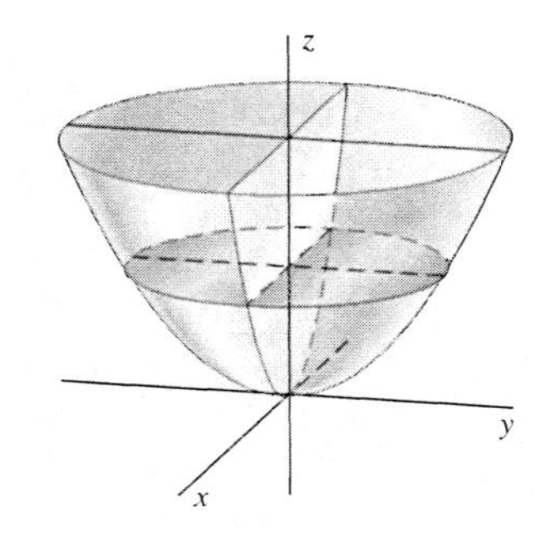

FIGURE 12.7.14 An elliptic paraboloid (Example 9).

is shown in Fig. 12.7.14. Its trace in the horizontal plane $z = z_0 > 0$ is the ellipse $x^2/a^2 + y^2/b^2 = z_0/c$ with semiaxes $a\sqrt{z_0/c}$ and $b\sqrt{z_0/c}$. Its trace in any vertical plane is a parabola. For instance, its trace in the plane $y = y_0$ has equation $x^2/a^2 + y_0^2/b^2 = z/c$, which can be written in the form $z - z_1 = k(x - x_1)^2$ by taking $z_1 = cy_0^2/b^2$ and $x_1 = 0$. The paraboloid opens upward if $c > 0$ and downward if $c < 0$. If $a = b$, then the paraboloid is said to be **circular.** Figure 12.7.15 shows the traces of a circular paraboloid in planes parallel to the xz- and yz-planes. ◆

EXAMPLE 10 The **elliptical cone**

$$\frac{x^2}{a^2} + \frac{y^2}{b^2} = \frac{z^2}{c^2} \tag{7}$$

is shown in Fig. 12.7.16. Its trace in the horizontal plane $z = z_0 \neq 0$ is an ellipse with semiaxes $a|z_0|/c$ and $b|z_0|/c$. ◆

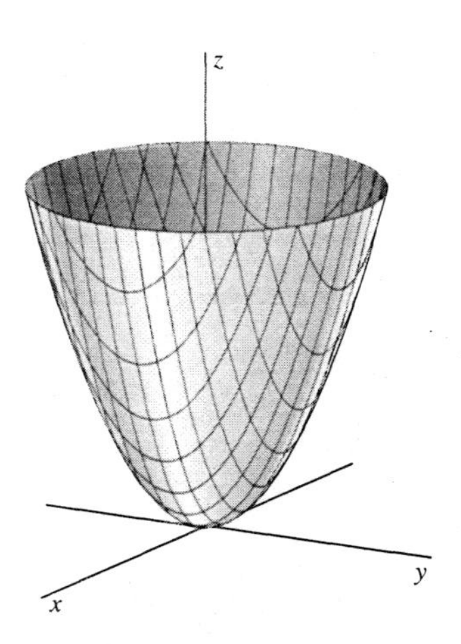

FIGURE 12.7.15 Trace parabolas of a circular paraboloid (Example 9).

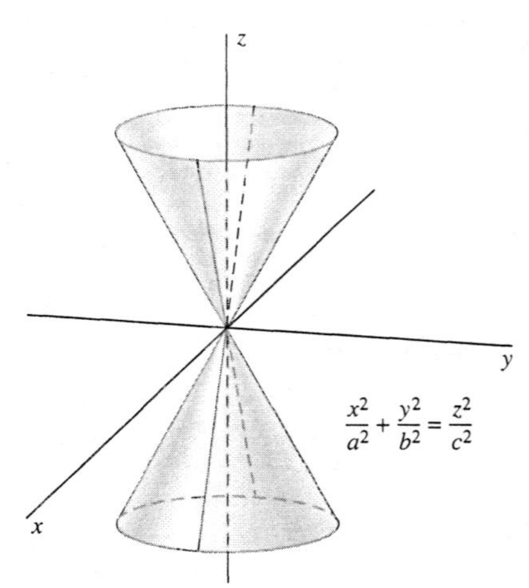

FIGURE 12.7.16 An elliptic cone (Example 10).

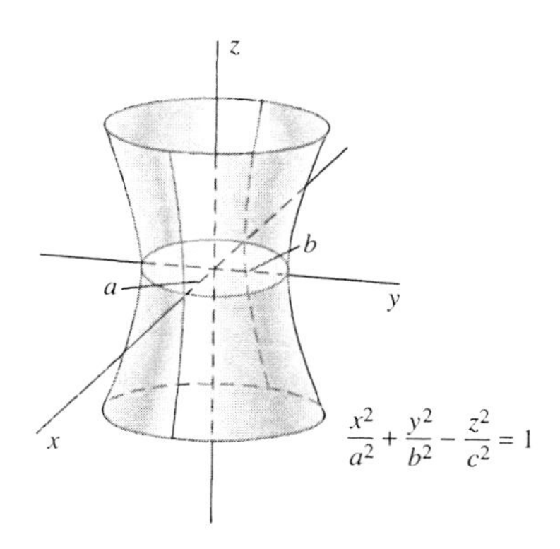

FIGURE 12.7.17 A hyperboloid of one sheet (Example 11).

EXAMPLE 11 The **hyperboloid of one sheet** with equation

$$\frac{x^2}{a^2}+\frac{y^2}{b^2}-\frac{z^2}{c^2}=1 \tag{8}$$

is shown in Fig. 12.7.17. Its trace in the horizontal plane $z = z_0$ is the ellipse $x^2/a^2 + y^2/b^2 = 1 + z_0^2/c^2 > 0$. Its trace in a vertical plane is a hyperbola except when the vertical plane intersects the xy-plane in a line tangent to the ellipse $x^2/a^2 + y^2/b^2 = 1$. In this special case, the trace is a degenerate hyperbola consisting of two intersecting lines. Figure 12.7.18 shows the traces (in planes parallel to the coordinate planes) of a circular ($a = b$) hyperboloid of one sheet.

The graphs of the equations

$$\frac{y^2}{b^2}+\frac{z^2}{c^2}-\frac{z^2}{a^2}=1 \quad \text{and} \quad \frac{x^2}{a^2}+\frac{z^2}{c^2}-\frac{y^2}{c^2}=1$$

are also hyperboloids of one sheet, opening along the x- and y-axes, respectively. ◆

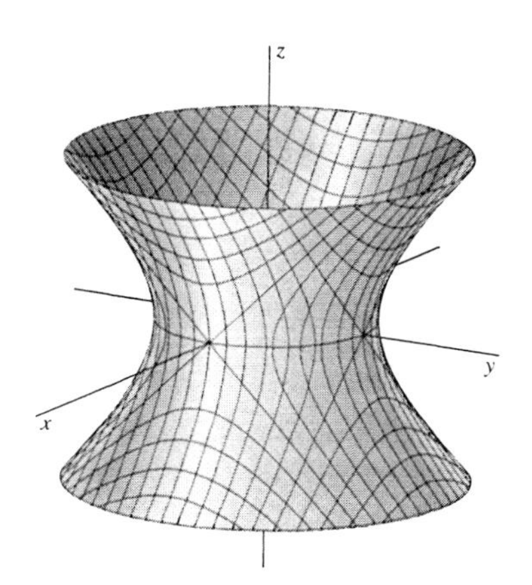

FIGURE 12.7.18 A circular hyperboloid of one sheet (Example 11). Its traces in horizontal planes are circles; its traces in vertical planes are hyperbolas.

EXAMPLE 12 The **hyperboloid of two sheets** with equation

$$\frac{z^2}{c^2}-\frac{x^2}{a^2}-\frac{y^2}{b^2}=1 \tag{9}$$

consists of two separate pieces, or *sheets* (Fig. 12.7.19). The two sheets open along the positive and negative z-axis and intersect it at the points $(0, 0, \pm c)$. The trace of this hyperboloid in a horizontal plane $z = z_0$ with $|z_0| > c$ is the ellipse

$$\frac{x^2}{a^2}+\frac{y^2}{b^2}=\frac{z_0^2}{c^2}-1>0.$$

Its trace in any vertical plane is a nondegenerate hyperbola. Figure 12.7.20 shows traces of a circular hyperboloid of two sheets.

The graphs of the equations

$$\frac{x^2}{a^2}-\frac{y^2}{b^2}-\frac{z^2}{c^2}=1 \quad \text{and} \quad \frac{y^2}{b^2}-\frac{x^2}{a^2}-\frac{z^2}{c^2}=1$$

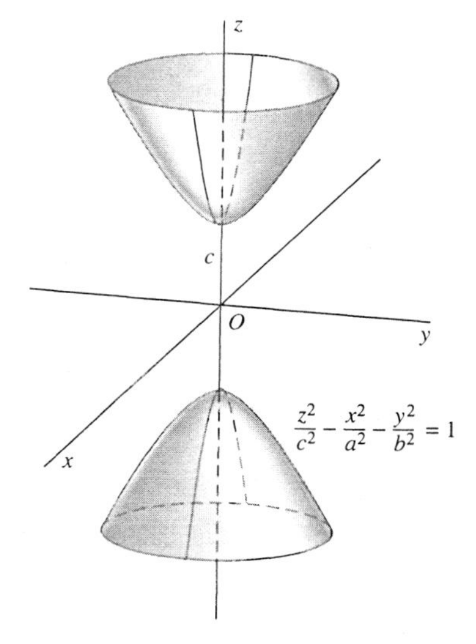

FIGURE 12.7.19 A hyperboloid of two sheets (Example 12).

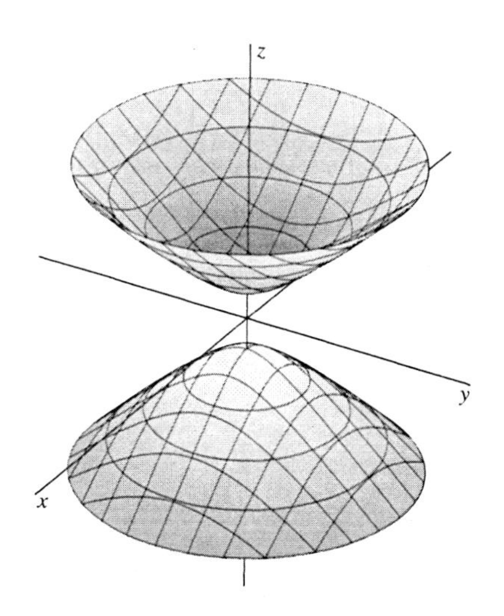

FIGURE 12.7.20 A circular hyperboloid of two sheets (Example 12). Its (nondegenerate) traces in horizontal planes are circles; its traces in vertical planes are hyperbolas.

are also hyperboloids of two sheets, opening along the x-axis and y-axis, respectively. When the equation of a hyperboloid is written in standard form with $+1$ on the right-hand side [as in Eqs. (8) and (9)], then the number of sheets is equal to the number of negative terms on the left-hand side. ◆

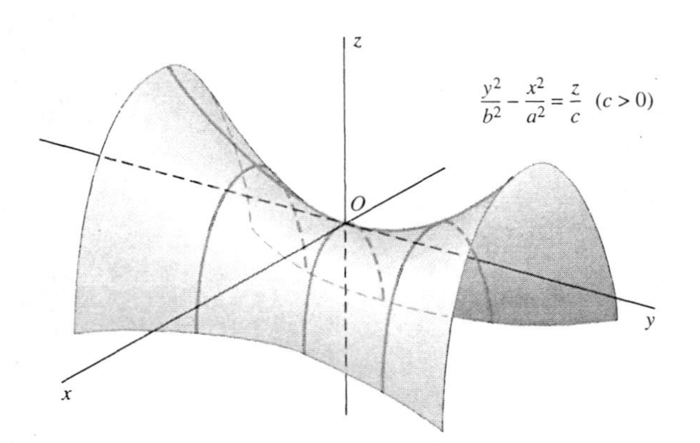

FIGURE 12.7.21 A hyperbolic paraboloid is a saddle-shaped surface (Example 13).

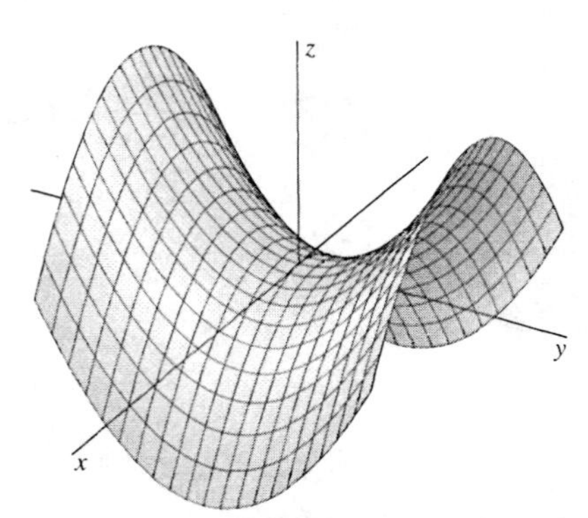

FIGURE 12.7.22 The vertical traces of the hyperbolic paraboloid $z = y^2 - x^2$ (Example 13).

EXAMPLE 13 The **hyperbolic paraboloid**

$$\frac{y^2}{b^2} - \frac{x^2}{a^2} = \frac{z}{c} \quad (c > 0) \tag{10}$$

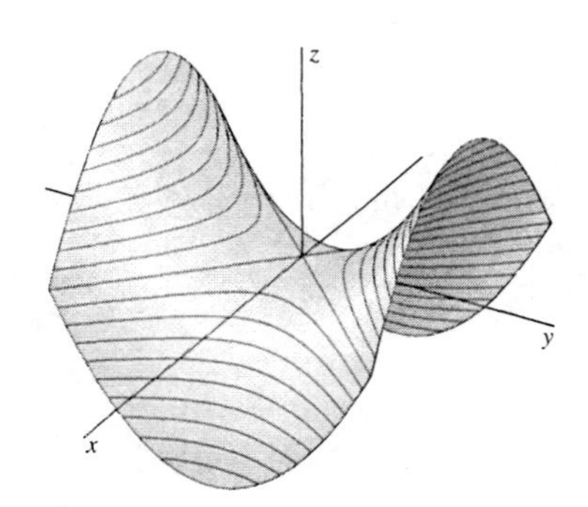

FIGURE 12.7.23 The horizontal traces of the hyperbolic paraboloid $z = y^2 - x^2$ (Example 13).

is saddle shaped, as indicated in Fig. 12.7.21. Its trace in the horizontal plane $z = z_0$ is a hyperbola (or two intersecting lines if $z_0 = 0$). Its trace in a vertical plane parallel to the xz-plane is a parabola that opens downward, whereas its trace in a vertical plane parallel to the yz-plane is a parabola that opens upward. In particular, the trace of the hyperbolic paraboloid in the xz-plane is a parabola opening downward from the origin, whereas its trace in the yz-plane is a parabola opening upward from the origin. Thus the origin looks like a local maximum from one direction but like a local minimum from another. Such a point on a surface is called a **saddle point.**

Figure 12.7.22 shows the parabolic traces in vertical planes of the hyperbolic paraboloid $z = y^2 - x^2$. Figure 12.7.23 shows its hyperbolic traces in horizontal planes. ◆

12.7 TRUE/FALSE STUDY GUIDE

12.7 CONCEPTS: QUESTIONS AND DISCUSSION

The following questions are concerned with possible graphs of the second-degree equation

$$Ax^2 + By^2 + Cz^2 + Dx + Ey + Fz + H = 0. \tag{11}$$

1. Under what conditions on the coefficients A, B, and C is the graph (a) an ellipsoid; (b) a paraboloid; (c) a hyperboloid?
2. Under what conditions on the coefficients is the graph a cone or cylinder?
3. Besides ellipsoids, paraboloids, hyperboloids, cones, and cylinders, what are the other possibilities for the graph of the equation in (11)? Give an example to illustrate each possibility.

12.7 PROBLEMS

Describe and sketch the graphs of the equations given in Problems 1 through 30.

1. $3x + 2y + 10z = 20$

2. $3x + 2y = 30$

3. $x^2 + y^2 = 9$

4. $y^2 = x^2 - 9$

5. $xy = 4$

6. $z = 4x^2 + 4y^2$

7. $z = 4x^2 + y^2$

8. $4x^2 + 9y^2 = 36$

9. $z = 4 - x^2 - y^2$

10. $y^2 + z^2 = 1$

11. $2z = x^2 + y^2$

12. $x = 1 + y^2 + z^2$

13. $z^2 = 4(x^2 + y^2)$

14. $y^2 = 4x$

15. $x^2 = 4z + 8$

16. $x = 9 - z^2$

17. $4x^2 + y^2 = 4$

18. $x^2 + z^2 = 4$

19. $x^2 = 4y^2 + 9z^2$

20. $x^2 - 4y^2 = z$

21. $x^2 + y^2 + 4z = 0$

22. $x = \sin y$

23. $x = 2y^2 - z^2$

24. $x^2 + 4y^2 + 2z^2 = 4$

25. $x^2 + y^2 - 9z^2 = 9$

26. $x^2 - y^2 - 9z^2 = 9$

27. $y = 4x^2 + 9z^2$

28. $y^2 + 4x^2 - 9z^2 = 36$

29. $y^2 - 9x^2 - 4z^2 = 36$

30. $x^2 + 9y^2 + 4z^2 = 36$

Problems 31 through 40 give the equation of a curve in one of the coordinate planes. Write an equation for the surface generated by revolving this curve around the indicated axis. Then sketch the surface.

31. $x = 2z^2$; the x-axis

32. $4x^2 + 9y^2 = 36$; the y-axis

33. $y^2 - z^2 = 1$; the z-axis

34. $z = 4 - x^2$; the z-axis

35. $y^2 = 4x$; the x-axis

36. $yz = 1$; the z-axis

37. $z = \exp(-x^2)$; the z-axis

38. $(y - z)^2 + z^2 = 1$; the z-axis

39. The line $z = 2x$; the z-axis

40. The line $z = 2x$; the x-axis

In Problems 41 through 48, describe the traces of the given surfaces in planes of the indicated type.

41. $x^2 + 4y^2 = 4$; in horizontal planes (those parallel to the xy-plane)

42. $x^2 + 4y^2 + 4z^2 = 4$; in horizontal planes

43. $x^2 + 4y^2 + 4z^2 = 4$; in planes parallel to the yz-plane

44. $z = 4x^2 + 9y^2$; in horizontal planes

45. $z = 4x^2 + 9y^2$; in planes parallel to the yz-plane

46. $z = xy$; in horizontal planes

47. $z = xy$; in vertical planes through the z-axis

48. $x^2 - y^2 + z^2 = 1$; in both horizontal and vertical planes parallel to the coordinate axes

49. Prove that the triangles OAC and OBC in Fig. 12.7.1 are congruent, and thereby conclude that the trace of a sphere in an intersecting plane is a circle.

50. Prove that the projection into the yz-plane of the curve of intersection of the surfaces $x = 1 - y^2$ and $x = y^2 + z^2$ is an ellipse (Fig. 12.7.24).

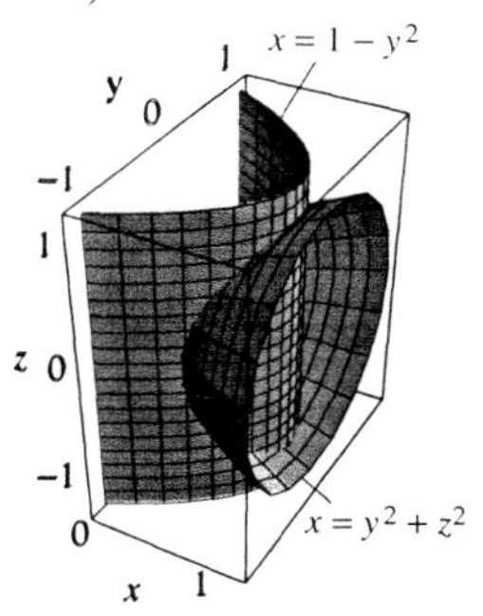

FIGURE 12.7.24 The paraboloid and parabolic cylinder of Problem 50.

51. Show that the projection into the xy-plane of the intersection of the plane $z = y$ and the paraboloid $z = x^2 + y^2$ is a circle (Fig. 12.7.25).

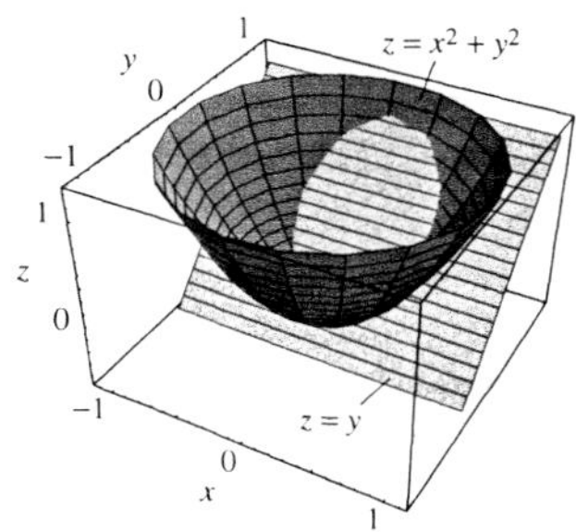

FIGURE 12.7.25 The plane and paraboloid of Problem 51.

52. Prove that the projection into the xz-plane of the intersection of the paraboloids $y = 2x^2 + 3z^2$ and $y = 5 - 3x^2 - 2z^2$ is a circle (Fig. 12.7.26).

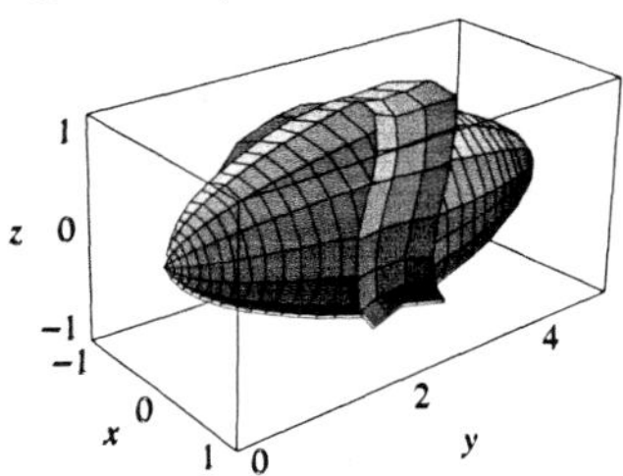

FIGURE 12.7.26 The two paraboloids of Problem 52.

53. Prove that the projection into the xy-plane of the intersection of the plane $x + y + z = 1$ and the ellipsoid $x^2 + 4y^2 + 4z^2 = 4$ is an ellipse.

54. Show that the curve of intersection of the plane $z = ky$ and the cylinder $x^2 + y^2 = 1$ is an ellipse. [*Suggestion:* Introduce uv-coordinates into the plane $z = ky$ as follows: Let the u-axis be the original x-axis and let the v-axis be the line $z = ky$, $x = 0$.]

12.8 CYLINDRICAL AND SPHERICAL COORDINATES

Rectangular coordinates provide only one of several useful ways of describing points, curves, and surfaces in space. Here we discuss two additional coordinate systems in three-dimensional space. Each is a generalization of polar coordinates in the coordinate plane.

Recall from Section 10.2 that the relationship between the rectangular coordinates (x, y) and the polar coordinates (r, θ) of a point in space is given by

$$x = r\cos\theta, \qquad y = r\sin\theta \tag{1}$$

and

$$r^2 = x^2 + y^2, \qquad \tan\theta = \frac{y}{x} \quad \text{if } x \neq 0. \tag{2}$$

Read these relationships directly from the right triangle in Fig. 12.8.1.

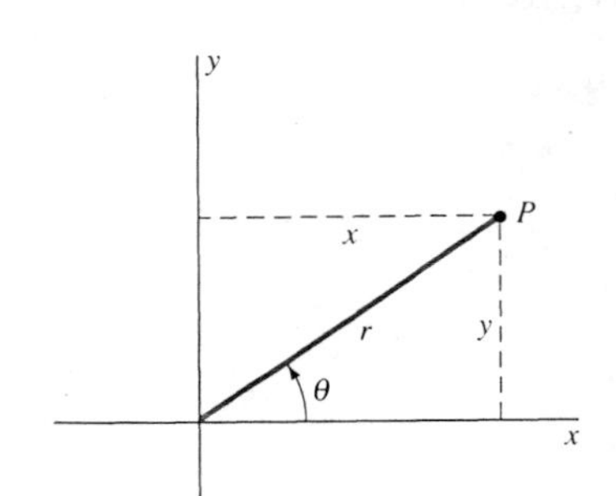

FIGURE 12.8.1 The relation between rectangular and polar coordinates in the xy-plane.

Cylindrical Coordinates

The **cylindrical coordinates** (r, θ, z) of a point P in space are natural hybrids of its polar and rectangular coordinates. We use the polar coordinates (r, θ) of the point in the plane with rectangular coordinates (x, y) and use the same z-coordinate as in rectangular coordinates. (The cylindrical coordinates of a point P in space are illustrated in Fig. 12.8.2.) This means that we can obtain the relations between the rectangular coordinates (x, y, z) of the point P and its cylindrical coordinates (r, θ, z) by simply adjoining the identity $z = z$ to the equations in (1) and (2):

$$x = r\cos\theta, \qquad y = r\sin\theta, \qquad z = z \tag{3}$$

and

$$r^2 = x^2 + y^2, \qquad \tan\theta = \frac{y}{x}, \qquad z = z. \tag{4}$$

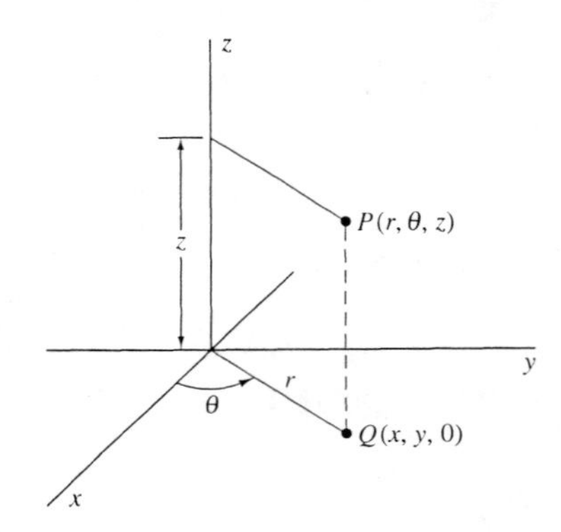

FIGURE 12.8.2 Finding the cylindrical coordinates of the point P.

We can use these equations to convert from rectangular to cylindrical coordinates and vice versa.

EXAMPLE 1 (a) Find the rectangular coordinates of the point P having cylindrical coordinates $(4, \frac{5}{3}\pi, 7)$. (b) Find the cylindrical coordinates of the point Q having rectangular coordinates $(-2, 2, 5)$.

Solution (a) We apply the equations in (3) to write

$$x = 4\cos\left(\tfrac{5}{3}\pi\right) = 4 \cdot \tfrac{1}{2} = 2,$$
$$y = 4\sin\left(\tfrac{5}{3}\pi\right) = 4 \cdot \left(-\tfrac{1}{2}\sqrt{3}\right) = -2\sqrt{3},$$
$$z = 7.$$

Thus the point P has rectangular coordinates $(2, -2\sqrt{3}, 7)$.

(b) Noting first that the point Q is in the second quadrant of the xy-plane, we apply the equations in (4) and write

$$r = \sqrt{(-2)^2 + 2^2} = 2\sqrt{2},$$
$$\tan\theta = \frac{-2}{2} = -1, \quad \text{so} \quad \theta = \frac{3\pi}{4},$$
$$z = 5.$$

Thus the point Q has cylindrical coordinates $(2\sqrt{2}, \frac{3}{4}\pi, 5)$. We can add any even integral multiple of π to θ, so other cylindrical coordinates for Q are $(2\sqrt{2}, \frac{11}{4}\pi, 5)$ and $(2\sqrt{2}, -\frac{5}{4}\pi, 5)$. ◆

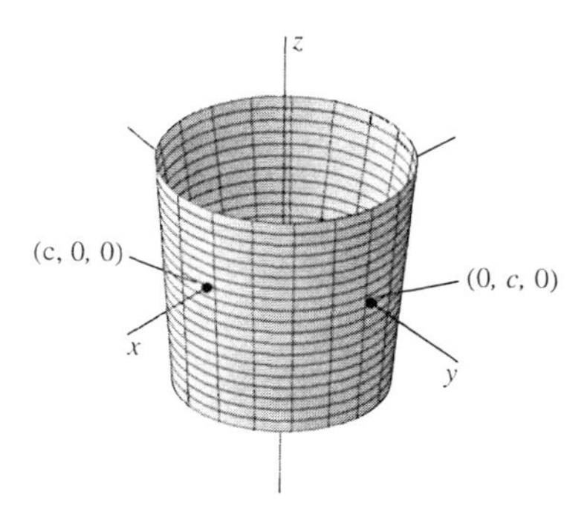

FIGURE 12.8.3 The cylinder $r = c$.

The **graph** of an equation involving r, θ, and z is the set of all points in space having cylindrical coordinates that satisfy the equation. The name *cylindrical coordinates* arises from the fact that the graph in space of the equation $r = c$ (a constant) is a cylinder of radius c symmetric around the z-axis (Fig. 12.8.3). Cylindrical coordinates are useful in describing other surfaces that are symmetric around the z-axis. The rectangular-coordinate equation of such a surface typically involves x and y only in the combination $x^2 + y^2$, for which we can then substitute r^2 to get the cylindrical-coordinate equation.

EXAMPLE 2 (a) The sphere $x^2 + y^2 + z^2 = a^2$ has cylindrical-coordinate equation $r^2 + z^2 = a^2$.

(b) The cone $z^2 = x^2 + y^2$ has cylindrical-coordinate equation $z^2 = r^2$. Taking square roots, we get $z = \pm r$, and the two signs give (for $r \geqq 0$) the two nappes of the cone (Fig. 12.8.4).

(c) The paraboloid $z = x^2 + y^2$ has cylindrical-coordinate equation $z = r^2$ (Fig. 12.8.5).

(d) The ellipsoid $(x/3)^2 + (y/3)^2 + (z/2)^2 = 1$ has cylindrical-coordinate equation $(r/3)^2 + (z/2)^2 = 1$ (Fig. 12.8.6). ◆

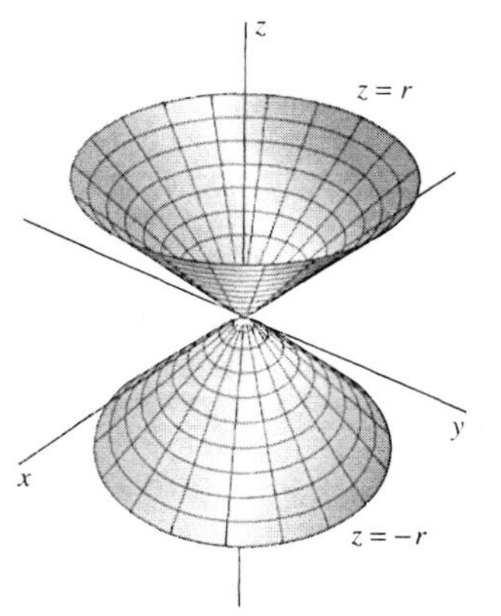

FIGURE 12.8.4 The cone $z^2 = r^2$.

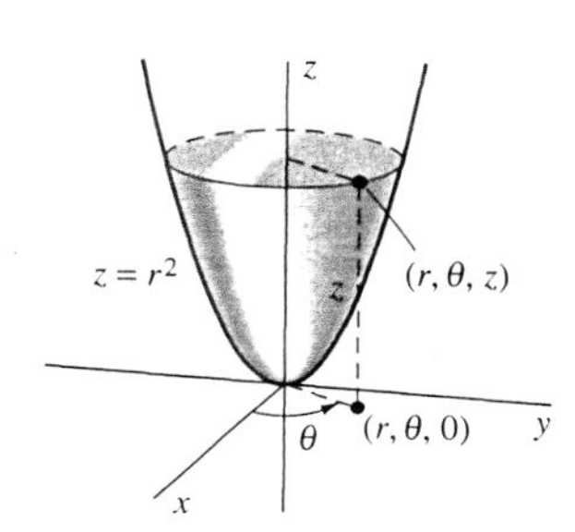

FIGURE 12.8.5 The paraboloid $z = r^2$.

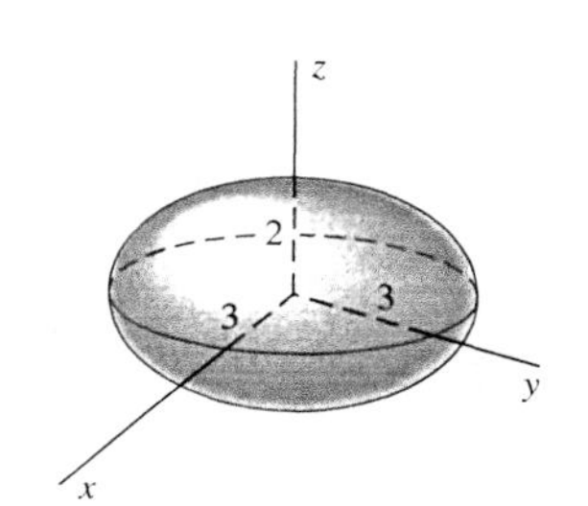

FIGURE 12.8.6 The ellipsoid $\frac{r^2}{9} + \frac{z^2}{4} = 1$.

EXAMPLE 3 Sketch the region that is bounded by the two surfaces with cylindrical-coordinate equations $z = r^2$ and $z = 8 - r^2$.

Solution If we substitute $r^2 = x^2 + y^2$ in the given equations, we get the familiar rectangular equations

$$z = x^2 + y^2 \quad \text{and} \quad z = 8 - x^2 - y^2$$

that describe paraboloids opening upward from (0, 0, 0) and downward from (0, 0, 8), respectively. Figure 12.8.7 shows a computer plot of the region in space that is bounded below by the paraboloid $z = x^2 + y^2$ and above by the paraboloid $z = 8 - x^2 - y^2$. ◆

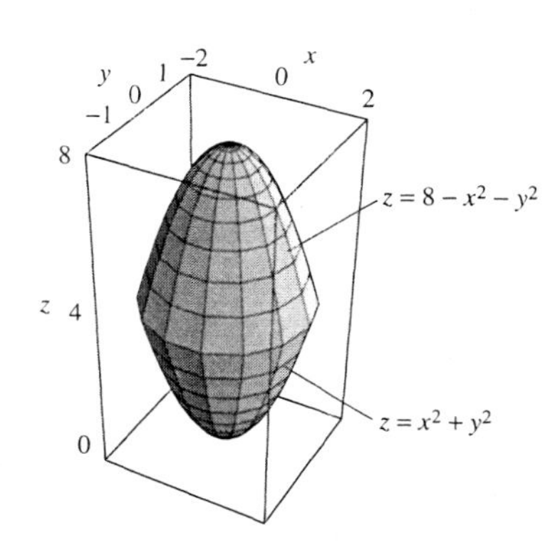

FIGURE 12.8.7 The solid of Example 3.

REMARK The relations $x = r\cos\theta$ and $y = r\sin\theta$ play an important role in the computer plotting of figures symmetric around the z-axis. For instance, the paraboloid $z = 8 - r^2$ of Example 3 can be plotted using computer algebra system syntax like the *Maple* command

```
plot3d( [r*cos(θ), r*sin(θ), 8 - r∧2],
        r=0..2,  θ=0..2*Pi );
```

or the *Mathematica* command

```
ParametricPlot3D[ {r*Cos[θ], r*Sin[θ], 8 - r∧2},
        {r,0,2},  {θ,0,2*Pi} ];
```

In either command the paraboloid is described parametrically by giving x, y, and z in terms of r and θ.

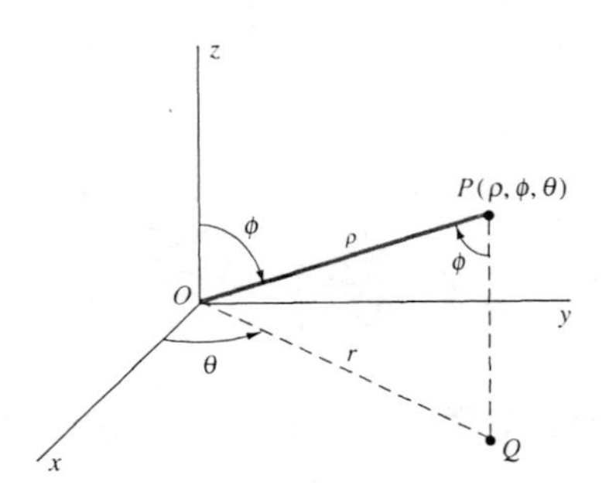

FIGURE 12.8.8 Finding the spherical coordinates of the point P.

Spherical Coordinates

Figure 12.8.8 shows the **spherical coordinates** (ρ, ϕ, θ) of the point P in space. The first spherical coordinate ρ is simply the distance $\rho = |OP|$ from the origin O to P. The second spherical coordinate ϕ is the angle between OP and the positive z-axis. Thus we may always choose ϕ in the interval $[0, \pi]$, although it is not restricted to that domain. Finally, θ is the familiar angle θ of cylindrical coordinates. That is, θ is the angular coordinate of the vertical projection Q of P into the xy-plane. Thus we may always choose θ in the interval $[0, 2\pi]$, although it is not restricted to that domain. Both angles ϕ and θ are always measured in radians.

The name *spherical coordinates* is used because the graph of the equation $\rho = c$ (c is a constant) is a sphere—more precisely, a spherical surface—of radius c centered at the origin. The equation $\phi = c$ (a constant) describes (one nappe of) a cone if $0 < c < \pi/2$ or if $\pi/2 < c < \pi$ (Fig. 12.8.9). The spherical equation of the xy-plane is $\phi = \pi/2$.

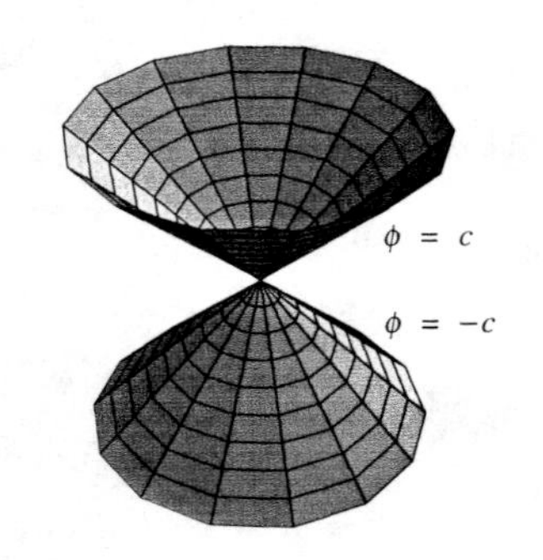

FIGURE 12.8.9 The two nappes of a 45° cone; $\phi = \pi/2$ is the spherical equation of the xy-plane.

From the right triangle OPQ of Fig. 12.8.8, we see that

$$r = \rho \sin\phi \quad \text{and} \quad z = \rho\cos\phi. \tag{5}$$

Indeed, these equations are most easily remembered by visualizing this triangle. Substituting the equations in (5) into those in (3) yields

$$x = \rho\sin\phi\cos\theta, \qquad y = \rho\sin\phi\sin\theta, \qquad z = \rho\cos\phi. \tag{6}$$

These three equations give the relationship between rectangular and spherical coordinates. Also useful is the formula

$$\rho^2 = x^2 + y^2 + z^2, \tag{7}$$

a consequence of the distance formula.

It is important to note the order in which the spherical coordinates (ρ, ϕ, θ) of a point P are written—first the distance ρ of P from the origin, then the angle ϕ down from the positive z-axis, and last the counterclockwise angle θ measured from the positive x-axis. You may find this mnemonic device to be helpful: The consonants in the word "raft" remind us, in order, of *r*ho, *f*ee (for phi), and *t*heta. Warning: In some other physics and mathematics books, a different order, or even different symbols, may be used.

Given the rectangular coordinates (x, y, z) of the point P, one systematic method for finding the spherical coordinates (ρ, ϕ, θ) of P is this. First we find the cylindrical coordinates r and θ of P with the aid of the triangle in Fig. 12.8.10(a). Then we find ρ and ϕ from the triangle in Fig. 12.8.10(b).

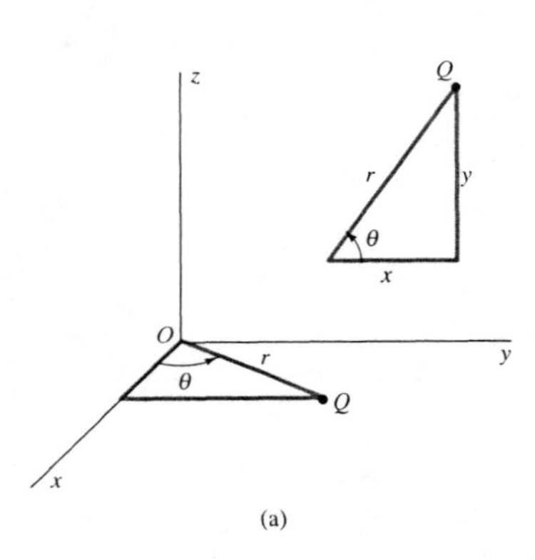

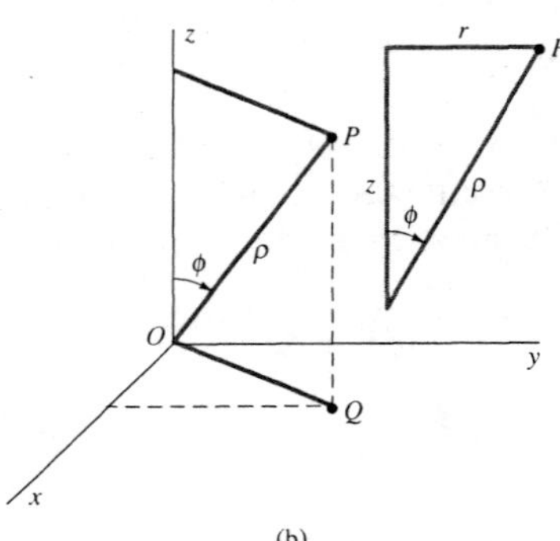

FIGURE 12.8.10 Triangles used in finding spherical coordinates.

EXAMPLE 4 (a) Find the rectangular coordinates of the point P having the spherical coordinates $(8, \frac{5}{6}\pi, \frac{1}{3}\pi)$. (b) Approximate the spherical coordinates of the point Q having rectangular coordinates $(-3, -4, -12)$.

Solution (a) We apply the equations in (6) to write

$$x = 8\sin\left(\tfrac{5}{6}\pi\right)\cos\left(\tfrac{1}{3}\pi\right) = 8\cdot\tfrac{1}{2}\cdot\tfrac{1}{2} = 2,$$

$$y = 8\sin\left(\tfrac{5}{6}\pi\right)\sin\left(\tfrac{1}{3}\pi\right) = 8\cdot\tfrac{1}{2}\cdot\left(\tfrac{1}{2}\sqrt{3}\right) = 2\sqrt{3},$$

$$z = 8\cos\left(\tfrac{5}{6}\pi\right) = 8\cdot\left(-\tfrac{1}{2}\sqrt{3}\right) = -4\sqrt{3}.$$

Thus the point P has rectangular coordinates $(2, 2\sqrt{3}, -4\sqrt{3})$.

(b) First we note that $r = \sqrt{(-3)^2 + (-4)^2} = \sqrt{25} = 5$ and that

$$\rho = \sqrt{(-3)^2 + (-4)^2 + (-12)^2} = \sqrt{169} = 13.$$

Next,

$$\phi = \cos^{-1}\left(\frac{z}{\rho}\right) = \cos^{-1}\left(-\frac{12}{13}\right) \approx 2.7468 \quad \text{(rad)}.$$

Finally, the point $(-3, -4)$ lies in the third quadrant of the xy-plane, so

$$\theta = \pi + \tan^{-1}\left(\frac{3}{4}\right) \approx 3.7851 \quad \text{(rad)}.$$

Thus the approximate spherical coordinates of the point Q are $(13, 2.7468, 3.7851)$. ◆

EXAMPLE 5 Find a spherical-coordinate equation of the paraboloid with rectangular-coordinates equation $z = x^2 + y^2$.

Solution We substitute $z = \rho\cos\phi$ from Eq. (5) and $x^2 + y^2 = r^2 = \rho^2\sin^2\phi$ from Eq. (6). This gives $\rho\cos\phi = \rho^2\sin^2\phi$. Cancelling ρ gives $\cos\phi = \rho\sin^2\phi$; that is,

$$\rho = \csc\phi\cot\phi$$

is the spherical-coordinate equation of the paraboloid. We get the whole paraboloid by using ϕ in the range $0 < \phi \leqq \pi/2$. Note that $\phi = \pi/2$ gives the point $\rho = 0$ that might otherwise have been lost by cancelling ρ. ◆

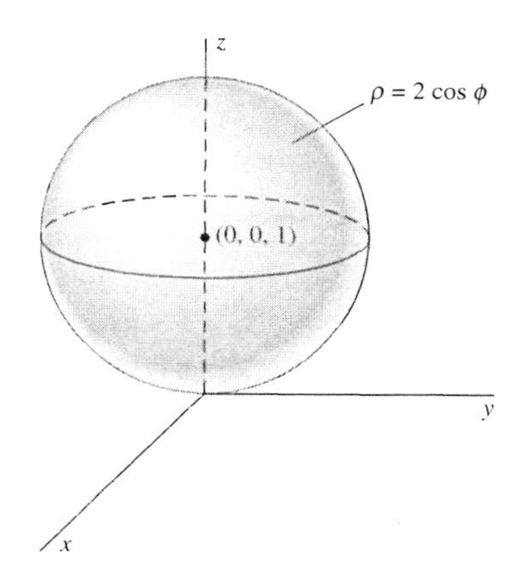

FIGURE 12.8.11 The sphere of Example 6.

EXAMPLE 6 Determine the graph of the spherical-coordinate equation $\rho = 2\cos\phi$.

Solution Multiplying by ρ gives

$$\rho^2 = 2\rho\cos\phi;$$

then substituting $\rho^2 = x^2 + y^2 + z^2$ and $z = \rho\cos\phi$ yields

$$x^2 + y^2 + z^2 = 2z$$

as the rectangular-coordinate equation of the graph. Completing the square in z now gives

$$x^2 + y^2 + (z-1)^2 = 1,$$

so the graph is a sphere with center $(0, 0, 1)$ and radius 1. It is tangent to the xy-plane at the origin (Fig. 12.8.11). ◆

EXAMPLE 7 Determine the graph of the spherical-coordinate equation $\rho = \sin\phi\sin\theta$.

Solution We first multiply each side by ρ and get $\rho^2 = \rho\sin\phi\sin\theta$. We then use Eqs. (6) and (7) and find that $x^2 + y^2 + z^2 = y$. This is a rectangular-coordinate equation of a sphere with center $(0, \frac{1}{2}, 0)$ and radius $\frac{1}{2}$. ◆

REMARK The relations in (6) are used in computer plotting of spherical-coordinate surfaces. For instance, the spherical surface $\rho = 2\cos\phi$ of Example 6 can be plotted using computer algebra system syntax like the *Maple* commands

```
ρ := 2*cos(φ);

plot3d( [ρ*sin(φ)*cos(θ), ρ*sin(φ)*sin(θ), ρ*cos(φ)],

        φ = 0..Pi/2, θ = 0..2*Pi );
```

or the *Mathematica* commands

```
ρ = 2 Cos[φ];

ParametricPlot3D[

    {ρ*Sin[φ]*Cos[θ], ρ*Sin[φ]*Sin[θ], ρ*Cos[φ]},

      {φ, 0, Pi/2}, {θ, 0, 2*Pi} ];
```

In each case the spherical surface is described parametrically by giving x, y, and z in terms of ρ, ϕ, and θ.

Latitude and Longitude

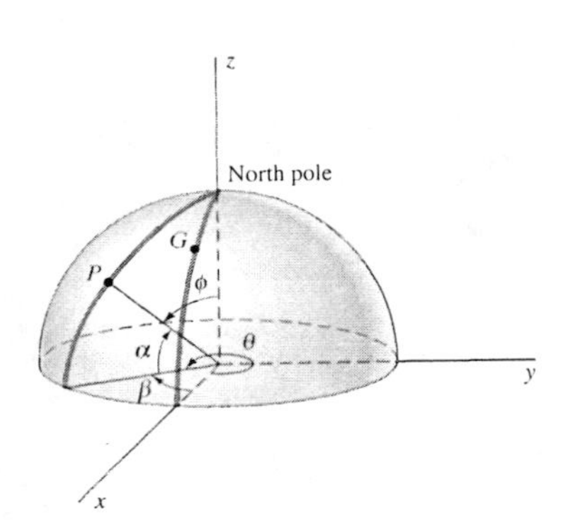

FIGURE 12.8.12 The relations among latitude, longitude, and spherical coordinates.

A **great circle** of a spherical surface is a circle formed by the intersection of the surface with a plane through the center of the sphere. Thus a great circle of a spherical surface is a circle (in the surface) that has the same radius as the sphere. Therefore, a great circle is a circle of maximum possible circumference that lies on the sphere. It's easy to see that any two points on a spherical surface lie on a great circle (uniquely determined unless the two points lie on the ends of a diameter of the sphere). In the calculus of variations, it is shown that the shortest distance between two such points—measured along the curved surface—is the shorter of the two arcs of the great circle that contains them. The surprise may be that the *shortest* distance is found by using the *largest* circle.

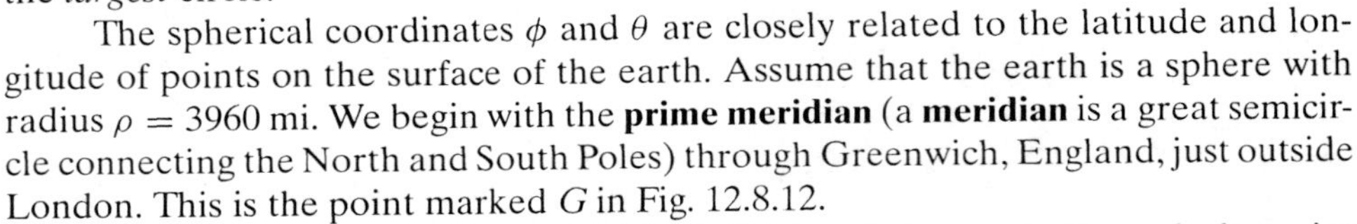

The spherical coordinates ϕ and θ are closely related to the latitude and longitude of points on the surface of the earth. Assume that the earth is a sphere with radius $\rho = 3960$ mi. We begin with the **prime meridian** (a **meridian** is a great semicircle connecting the North and South Poles) through Greenwich, England, just outside London. This is the point marked G in Fig. 12.8.12.

We take the z-axis through the North Pole and the x-axis through the point where the prime meridian intersects the equator. The **latitude** α and (west) **longitude** β of a point P in the Northern Hemisphere are given by the equations

$$\alpha = 90^\circ - \phi^\circ \quad \text{and} \quad \beta = 360^\circ - \theta^\circ, \tag{8}$$

where ϕ° and θ° are the angular spherical coordinates, measured in *degrees*, of P. (That is, ϕ° and θ° denote the degree equivalents of the angles ϕ and θ, respectively, which are measured in radians unless otherwise specified.) Thus the latitude α is measured northward from the equator and the longitude β is measured westward from the prime meridian.

EXAMPLE 8 Find the great-circle distance between New York (latitude 40.75° north, longitude 74° west) and London (latitude 51.5° north, longitude 0°). (See Fig. 12.8.13.)

Solution From the equations in (8) we find that $\phi^\circ = 49.25^\circ$, $\theta^\circ = 286^\circ$ for New York, whereas $\phi^\circ = 38.5^\circ$, $\theta^\circ = 360^\circ$ (or 0°) for London. Hence the angular spherical coordinates of New York are $\phi = (49.25/180)\pi$, $\theta = (286/180)\pi$, and those of London are $\phi = (38.5/180)\pi$, $\theta = 0$. With these values of ϕ and θ and with $\rho = 3960$ (mi),

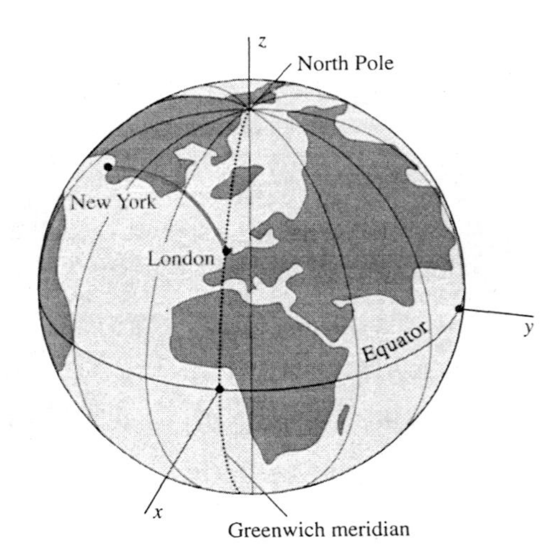

FIGURE 12.8.13 Finding the great-circle distance d from New York to London (Example 8).

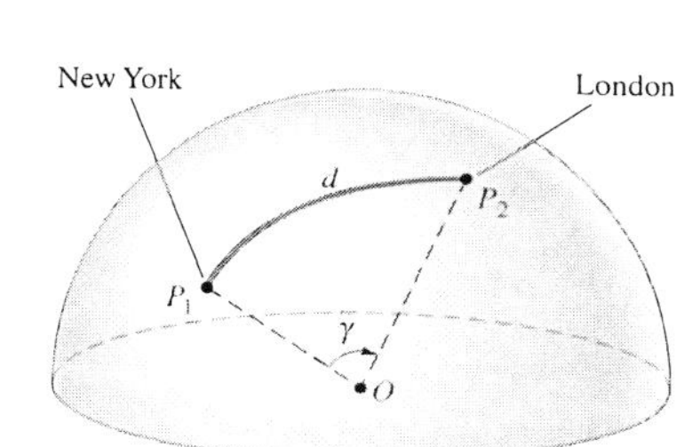

FIGURE 12.8.14 The great-circle arc between New York and London (Example 8).

the equations in (6) give the rectangular coordinates

$$\text{New York:} \quad P_1(826.90, -2883.74, 2584.93)$$

and

$$\text{London:} \quad P_2(2465.16, 0.0, 3099.13).$$

The angle γ between the radius vectors $\mathbf{u} = \overrightarrow{OP_1}$ and $\mathbf{v} = \overrightarrow{OP_2}$ in Fig. 12.8.14 satisfies the equation

$$\begin{aligned}\cos\gamma &= \frac{\mathbf{u}\cdot\mathbf{v}}{|\mathbf{u}|\,|\mathbf{v}|}\\ &= \frac{826.90\cdot 2465.16 - 2883.74\cdot 0 + 2584.93\cdot 3099.13}{(3960)^2} \approx 0.641.\end{aligned}$$

Thus γ is approximately 0.875 (rad). Hence the great-circle distance between New York and London is close to

$$d = 3960\cdot 0.875 \approx 3465 \quad \text{(mi)},$$

about 5576 km. ◆

12.8 TRUE/FALSE STUDY GUIDE

12.8 CONCEPTS: QUESTIONS AND DISCUSSION

1. Give several examples of surfaces that are described more simply in rectangular coordinates than in cylindrical or spherical coordinates.
2. Give several examples of surfaces that are described more simply in cylindrical coordinates than in rectangular or spherical coordinates.
3. Give several examples of surfaces that are described more simply in spherical coordinates than in rectangular or cylindrical coordinates.

12.8 PROBLEMS

In Problems 1 through 6, find the rectangular coordinates of the point with the given cylindrical coordinates.

1. $\left(1, \frac{1}{2}\pi, 2\right)$ **2.** $\left(3, \frac{3}{2}\pi, -1\right)$

3. $\left(2, \frac{3}{4}\pi, 3\right)$ **4.** $\left(3, \frac{7}{6}\pi, -1\right)$

5. $\left(2, \frac{1}{3}\pi, -5\right)$ **6.** $\left(4, \frac{5}{3}\pi, 6\right)$

In Problems 7 through 12, find the rectangular coordinates of the points with the given spherical coordinates (ρ, ϕ, θ).

7. $(2, 0, \pi)$ **8.** $(3, \pi, 0)$

9. $\left(3, \frac{1}{2}\pi, \pi\right)$ **10.** $\left(4, \frac{1}{6}\pi, \frac{2}{3}\pi\right)$

11. $\left(2, \frac{1}{3}\pi, \frac{3}{2}\pi\right)$ **12.** $\left(6, \frac{3}{4}\pi, \frac{4}{3}\pi\right)$

In Problems 13 through 22, find both the cylindrical coordinates and the spherical coordinates of the point P with the given rectangular coordinates.

13. $P(0, 0, 5)$ **14.** $P(0, 0, -3)$

15. $P(1, 1, 0)$ **16.** $P(2, -2, 0)$

17. $P(1, 1, 1)$ **18.** $P(-1, 1, -1)$

19. $P(2, 1, -2)$ **20.** $P(-2, -1, -2)$

21. $P(3, 4, 12)$ **22.** $P(-2, 4, -12)$

In Problems 23 through 38, describe the graph of the given equation. (It is understood that equations including r are in cylindrical coordinates and those including ρ or ϕ are in spherical coordinates.)

23. $r = 5$ **24.** $\theta = 3\pi/4$

25. $\theta = \pi/4$ **26.** $\rho = 5$

27. $\phi = \pi/6$ **28.** $\phi = 5\pi/6$

29. $\phi = \pi/2$ **30.** $\phi = \pi$

31. $z^2 + 2r^2 = 4$ **32.** $z^2 - 2r^2 = 4$

33. $r = 4\cos\theta$ **34.** $\rho = 4\cos\phi$

35. $r^2 - 4r + 3 = 0$ **36.** $\rho^2 - 4\rho + 3 = 0$

37. $z^2 = r^4$ **38.** $\rho^3 + 4\rho = 0$

In Problems 39 through 44, convert the given equation both to cylindrical and to spherical coordinates.

39. $x^2 + y^2 + z^2 = 25$

40. $x^2 + y^2 = 2x$

41. $x + y + z = 1$

42. $x + y = 4$

43. $x^2 + y^2 + z^2 = x + y + z$

44. $z = x^2 - y^2$

In Problems 45 through 52, describe and sketch the surface or solid described by the given equations and/or inequalities.

45. $r = 3, \quad -1 \leqq z \leqq 1$

46. $\rho = 2, \quad 0 \leqq \phi \leqq \pi/2$

47. $\rho = 2, \quad \pi/3 \leqq \phi \leqq 2\pi/3$

48. $0 \leqq r \leqq 3, \quad -2 \leqq z \leqq 2$

49. $1 \leqq r \leqq 3, \quad -2 \leqq z \leqq 2$

50. $0 \leqq \rho \leqq 2, \quad 0 \leqq \phi \leqq \pi/2$

51. $3 \leqq \rho \leqq 5$

52. $0 \leqq \phi \leqq \pi/6, \quad 0 \leqq \rho \leqq 10$

53. The parabola $z = x^2$, $y = 0$ is rotated around the z-axis. Write a cylindrical-coordinate equation for the surface thereby generated.

54. The hyperbola $y^2 - z^2 = 1$, $x = 0$ is rotated around the z-axis. Write a cylindrical-coordinate equation for the surface thereby generated.

55. A sphere of radius 2 is centered at the origin. A hole of radius 1 is drilled through the sphere, with the axis of the hole lying on the z-axis. Describe the solid region that remains (Fig. 12.8.15) in (a) cylindrical coordinates; (b) spherical coordinates.

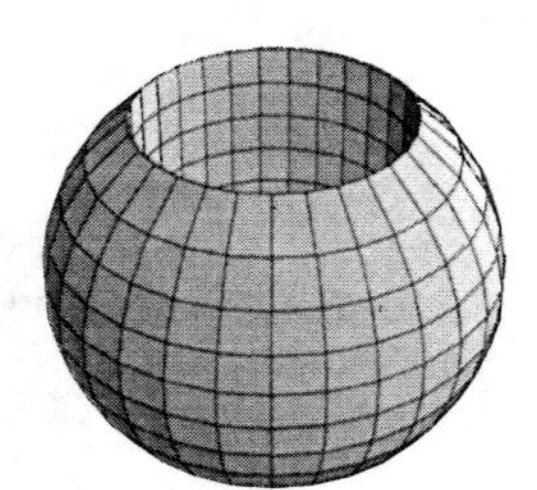

FIGURE 12.8.15 The sphere-with-hole of Problem 55.

56. Find the great-circle distance in miles and in kilometers from Atlanta (latitude 33.75° north, longitude 84.40° west) to San Francisco (latitude 37.78° north, longitude 122.42° west).

57. Find the great-circle distance in miles and in kilometers from Fairbanks (latitude 64.80° north, longitude 147.85° west) to St. Petersburg, Russia (latitude 59.91° north, longitude 30.43° *east* of Greenwich—alternatively, longitude 329.57° west).

58. Because Fairbanks and St. Petersburg, Russia (see Problem 57) are at approximately the same latitude, a plane could fly from one to the other roughly along the 62nd parallel of latitude. Accurately estimate the length of such a trip both in kilometers and in miles.

59. In flying the great-circle route from Fairbanks to St. Petersburg, Russia (see Problem 57), how close in kilometers and in miles to the North Pole would a plane fly?

60. The vertex of a right circular cone of radius R and height H is located at the origin and its axis lies on the nonnegative z-axis. Describe the solid cone in cylindrical coordinates.

61. Describe the cone of Problem 60 in spherical coordinates.

62. In flying the great-circle route from New York to London (Example 8), an airplane initially flies generally east-northeast. Does the plane ever fly at a latitude *higher* than that of London? [*Suggestion:* Express the z-coordinate of the plane's route as a function of x, and then maximize z.]

63. Figure 12.8.16 shows the torus that is obtained by revolving around the z-axis the circle of radius b centered at the point $(a, 0)$ in the yz-plane. Write a radical-free equation describing this torus in (a) rectangular coordinates; (b) cylindrical coordinates; (c) spherical coordinates. (d) Investigate the use of one of these descriptions with a computer algebra system to plot this torus with selected values of a and b.

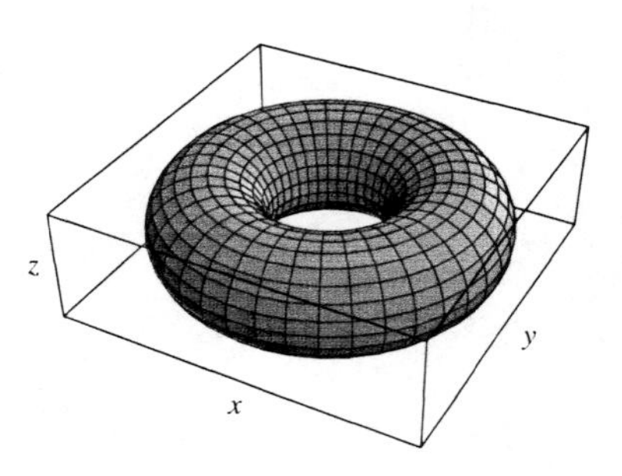

FIGURE 12.8.16 The torus of Problem 63.

64. The bumpy sphere of Fig. 12.8.17 is an exaggerated representation of waves on the surface of a very small planet that is covered by a very deep ocean. Such bumpy or wrinkled spheres may also, perhaps more realistically, be used to model tumors. Use a computer algebra system to plot the spherical-coordinate surface

$$\rho = a + b\cos m\theta \sin n\phi$$

with selected values of the positive numbers a and b and the positive integers m and n. How does the surface depend on the value of each of these four parameters?

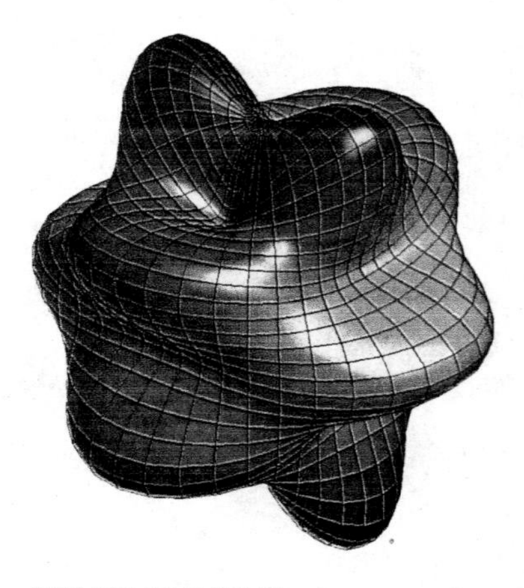

FIGURE 12.8.17 The bumpy sphere of Problem 64.

CHAPTER 12 REVIEW: DEFINITIONS, CONCEPTS, RESULTS

Use the following list as a guide to concepts that you may need to review.

1. Vectors: their definition, length, equality, addition, multiplication by scalars, and dot product.
2. The dot (scalar) product of vectors—definition and geometric interpretation.
3. Use of the dot product to test perpendicularity of vectors and, more generally, to find the angle between two vectors.
4. The cross (vector) product of two vectors—definition and geometric interpretation.
5. The scalar triple product of three vectors—definition and geometric interpretation.
6. The parametric and symmetric equations of the straight line that passes through a given point and is parallel to a given vector.
7. The equation of the plane through a given point normal to a given vector.
8. Vector-valued functions, velocity vectors, and acceleration vectors.
9. Componentwise differentiation and integration of vector-valued functions.
10. The equations of motion of a projectile.
11. The velocity and acceleration vectors of a particle moving along a parametric space curve.
12. Arc length of a parametric space curve.
13. The curvature, unit tangent vector, and principal unit normal vector of a parametric curve in the plane or in space.
14. Tangential and normal components of the acceleration vector of a parametric curve.
15. Kepler's three laws of planetary motion.
16. The radial and transverse unit vectors.
17. Polar decomposition of velocity and acceleration vectors.
18. Outline of the derivation of Kepler's laws from Newton's law of gravitation.
19. Equations of cylinders and surfaces of revolution.
20. The standard types of quadric surfaces.
21. Definitions of the cylindrical-coordinate and spherical-coordinate systems; the equations relating cylindrical and spherical coordinates to rectangular coordinates.

CHAPTER 12 MISCELLANEOUS PROBLEMS

1. Suppose that M is the midpoint of the segment PQ in space and that A is another point. Show that

$$\overrightarrow{AM} = \frac{1}{2}(\overrightarrow{AP} + \overrightarrow{AQ}).$$

2. Let $\mathbf{a}$ and $\mathbf{b}$ be nonzero vectors. Define

$$\mathbf{a}_\parallel = (\text{comp}_\mathbf{b}\mathbf{a})\,\frac{\mathbf{b}}{|\mathbf{b}|} \quad \text{and} \quad \mathbf{a}_\perp = \mathbf{a} - \mathbf{a}_\parallel.$$

Prove that $\mathbf{a}_\perp$ is perpendicular to $\mathbf{b}$.

3. Let P and Q be different points in space. Show that the point R lies on the line through P and Q *if and only if* there exist numbers a and b such that $a + b = 1$ and $\overrightarrow{OR} = a\overrightarrow{OP} + b\overrightarrow{OQ}$. Conclude that

$$\mathbf{r}(t) = t\overrightarrow{OP} + (1-t)\overrightarrow{OQ}$$

is a parametric equation of this line.

4. Conclude from the result of Problem 3 that the points P, Q, and R are collinear if and only if there exist numbers a, b, and c, not all zero, such that $a + b + c = 0$ and $a\overrightarrow{OP} + b\overrightarrow{OQ} + c\overrightarrow{OR} = \mathbf{0}$.

5. Let $P(x_0, y_0)$, $Q(x_1, y_1)$, and $R(x_2, y_2)$ be points in the xy-plane. Use the cross product to show that the area of the triangle PQR is

$$A = \tfrac{1}{2}|(x_1 - x_0)(y_2 - y_0) - (x_2 - x_0)(y_1 - y_0)|.$$

6. Write both symmetric and parametric equations of the line that passes through $P_1(1, -1, 0)$ and is parallel to $\mathbf{v} = \langle 2, -1, 3\rangle$.

7. Write both symmetric and parametric equations of the line that passes through $P_1(1, -1, 2)$ and $P_2(3, 2, -1)$.

8. Write an equation of the plane through $P(3, -5, 1)$ with normal vector $\mathbf{n} = \mathbf{i} + \mathbf{j}$.

9. Show that the lines with symmetric equations

$$x - 1 = 2(y + 1) = 3(z - 2)$$

and

$$x - 3 = 2(y - 1) = 3(z + 1)$$

are parallel. Then write an equation of the plane containing these two lines.

10. Let the lines L_1 and L_2 have symmetric equations

$$\frac{x - x_i}{a_i} = \frac{y - y_i}{b_i} = \frac{z - z_i}{c_i}$$

for $i = 1, 2$. Show that L_1 and L_2 are skew lines if and only if

$$\begin{vmatrix} x_1 - x_2 & y_1 - y_2 & z_1 - z_2 \\ a_1 & b_1 & c_1 \\ a_2 & b_2 & c_2 \end{vmatrix} \neq 0.$$

11. Given the four points $A(2, 3, 2)$, $B(4, 1, 0)$, $C(-1, 2, 0)$, and $D(5, 4, -2)$, find an equation of the plane that passes through A and B and is parallel to the line through C and D.

12. Given the points A, B, C, and D of Problem 11, find points P on the line AB and Q on the line CD such that the line PQ is perpendicular to both AB and CD. What is the perpendicular distance d between the lines AB and CD?

13. Let $P_0(x_0, y_0, z_0)$ be a point of the plane with equation

$$ax + by + cz + d = 0.$$

By projecting $\overrightarrow{OP_0}$ onto the normal vector $\mathbf{n} = \langle a, b, c \rangle$, show that the distance D from the origin to this plane is

$$D = \frac{|d|}{\sqrt{a^2 + b^2 + c^2}}.$$

14. Show that the distance D from the point $P_1(x_1, y_1, z_1)$ to the plane $ax + by + cz + d = 0$ is equal to the distance from the origin to the plane with equation

$$a(x + x_1) + b(y + y_1) + c(z + z_1) + d = 0.$$

Hence conclude from the result of Problem 13 that

$$D = \frac{|ax_1 + by_1 + cz_1 + d|}{\sqrt{a^2 + b^2 + c^2}}.$$

15. Find the perpendicular distance between the parallel planes $2x - y + 2z = 4$ and $2x - y + 2z = 13$.

16. Write an equation of the plane through the point $(1, 1, 1)$ that is normal to the twisted cubic $x = t$, $y = t^2$, $z = t^3$ at this point.

17. Let ABC be an isosceles triangle with $|AB| = |AC|$. Let M be the midpoint of BC. Use the dot product to show that AM and BC are perpendicular.

18. Use the dot product to show that the diagonals of a rhombus (a parallelogram with all four sides of equal length) are perpendicular to each other.

19. The acceleration of a certain particle is

$$\mathbf{a} = \mathbf{i} \sin t - \mathbf{j} \cos t.$$

Assume that the particle begins at time $t = 0$ at the point $(0, 1)$ and has initial velocity $\mathbf{v}_0 = -\mathbf{i}$. Show that its path is a circle.

20. A particle moves in an attracting central force field with force proportional to the distance from the origin. This implies that the particle's acceleration vector is $\mathbf{a} = -\omega^2 \mathbf{r}$, where $\mathbf{r}$ is the position vector of the particle. Assume that the particle's initial position is $\mathbf{r}_0 = p\mathbf{i}$ and that its initial velocity is $\mathbf{v}_0 = q\omega\mathbf{j}$. Show that the trajectory of the particle is the ellipse with equation $x^2/p^2 + y^2/q^2 = 1$. [*Suggestion:* If $x''(t) = -k^2 x(t)$ (where k is constant), then $x(t) = A \cos kt + B \sin kt$ for some constants A and B.]

21. At time $t = 0$, a ground target is 160 ft from a gun and is moving directly away from it with a constant speed of 80 ft/s. If the muzzle velocity of the gun is 320 ft/s, at what angle of elevation α should it be fired in order to strike the moving target?

22. Suppose that a gun with muzzle velocity v_0 is located at the foot of a hill with a 30° slope. At what angle of elevation (from the horizontal) should the gun be fired in order to maximize its range, as measured up the hill?

23. A particle moves in space with parametric equations $x = t$, $y = t^2$, $z = \frac{4}{3}t^{3/2}$. Find the curvature of its trajectory and the tangential and normal components of its acceleration when $t = 1$.

24. The **osculating plane** to a space curve at a point P of that curve is the plane through P that is parallel to the curve's unit tangent and principal unit normal vectors at P. Write an equation of the osculating plane to the curve of Problem 23 at the point $(1, 1, \frac{4}{3})$.

25. Show that the equation of the plane that passes through the point $P_0(x_0, y_0, z_0)$ and is parallel to the vectors $\mathbf{v}_1 = \langle a_1, b_1, c_1 \rangle$ and $\mathbf{v}_2 = \langle a_2, b_2, c_2 \rangle$ can be written in the form

$$\begin{vmatrix} x - x_0 & y - y_0 & z - z_0 \\ a_1 & b_1 & c_1 \\ a_2 & b_2 & c_2 \end{vmatrix} = 0.$$

26. Deduce from Problem 25 that the equation of the osculating plane (Problem 24) to the parametric curve $\mathbf{r}(t)$ at the point $\mathbf{r}(t_0)$ can be written in the form

$$[\mathbf{R} - \mathbf{r}(t_0)] \cdot [\mathbf{r}'(t_0) \times \mathbf{r}''(t_0)] = 0,$$

where $\mathbf{R} = \langle x, y, z \rangle$. Note first that the vectors $\mathbf{T}$ and $\mathbf{N}$ are coplanar with $\mathbf{r}'(t)$ and $\mathbf{r}''(t)$.

27. Use the result of Problem 26 to write an equation of the osculating plane to the twisted cubic $x = t$, $y = t^2$, $z = t^3$ at the point $(1, 1, 1)$.

28. Let a parametric curve in space be described by equations $r = r(t)$, $\theta = \theta(t)$, $z = z(t)$ that give the cylindrical coordinates of a moving point on the curve for $a \leqq t \leqq b$. Use the equations relating rectangular and cylindrical coordinates to show that the arc length of the curve is

$$s = \int_a^b \left[\left(\frac{dr}{dt} \right)^2 + \left(r \frac{d\theta}{dt} \right)^2 + \left(\frac{dz}{dt} \right)^2 \right]^{1/2} dt.$$

29. A point moves on the *unit* sphere $\rho = 1$ with its spherical angular coordinates at time t given by $\phi = \phi(t)$, $\theta = \theta(t)$, $a \leqq t \leqq b$. Use the equations relating rectangular and spherical coordinates to show that the arc length of its path is

$$s = \int_a^b \left[\left(\frac{d\phi}{dt} \right)^2 + (\sin^2 \phi) \left(\frac{d\theta}{dt} \right)^2 \right]^{1/2} dt.$$

30. The vector product $\mathbf{B} = \mathbf{T} \times \mathbf{N}$ of the unit tangent vector and the principal unit normal vector is the **unit binormal vector B** of a curve. (a) Differentiate $\mathbf{B} \cdot \mathbf{T} = 0$ to show that $\mathbf{T}$ is perpendicular to $d\mathbf{B}/ds$. (b) Differentiate $\mathbf{B} \cdot \mathbf{B} = 1$ to show that $\mathbf{B}$ is perpendicular to $d\mathbf{B}/ds$. (c) Conclude from parts (a) and (b) that $d\mathbf{B}/ds = -\tau \mathbf{N}$ for some number τ. Called the **torsion** of the curve, τ measures the amount that the curve twists at each point in space.

31. Show that the torsion of the helix of Example 7 of Section 12.5 is constant by showing that its value is

$$\tau = \frac{b\omega}{a^2\omega^2 + b^2}.$$

32. Deduce from the definition of torsion (Problem 30) that $\tau \equiv 0$ for any curve such that $\mathbf{r}(t)$ lies in a fixed plane.

33. Write an equation in spherical coordinates for the spherical surface with radius 1 and center $x = 0 = y$, $z = 1$.

34. Let C be the circle in the yz-plane with radius 1 and center $y = 1$, $z = 0$. Write equations in both rectangular and cylindrical coordinates of the surface obtained by revolving C around the z-axis.

35. Let C be the curve in the yz-plane with equation $(y^2 + z^2)^2 = 2(z^2 - y^2)$. Write an equation in spherical coordinates of the surface obtained by revolving this curve around the z-axis. Then sketch this surface. [*Suggestion:* Remember that $r^2 = 2 \cos 2\theta$ is the polar equation of a figure-eight curve.]

36. Let A be the area of the parallelogram $PQRS$ in space determined by the vectors $\mathbf{a} = \overrightarrow{PQ}$ and $\mathbf{b} = \overrightarrow{PS}$. Let A' be the area of the perpendicular projection of $PQRS$ into a plane that makes an acute angle γ with the plane of $PQRS$. Assuming that $A' = A\cos\gamma$ in such a situation (this is true), prove that the areas of the perpendicular projections of the parallelogram $PQRS$ into the three coordinate planes are

$$|\mathbf{i} \cdot (\mathbf{a} \times \mathbf{b})|, \qquad |\mathbf{j} \cdot (\mathbf{a} \times \mathbf{b})|, \qquad \text{and} \qquad |\mathbf{k} \cdot (\mathbf{a} \times \mathbf{b})|.$$

Conclude that the square of the area of a parallelogram in space is equal to the sum of the squares of the areas of its perpendicular projections into the three coordinate planes.

37. Take $\mathbf{a} = \langle a_1, a_2, a_3 \rangle$ and $\mathbf{b} = \langle b_1, b_2, b_3 \rangle$ in Problem 36. Show that

$$A^2 = \begin{vmatrix} a_2 & a_3 \\ b_2 & b_3 \end{vmatrix}^2 + \begin{vmatrix} a_3 & a_1 \\ b_3 & b_1 \end{vmatrix}^2 + \begin{vmatrix} a_1 & a_2 \\ b_1 & b_2 \end{vmatrix}^2.$$

38. Let C be a curve in a plane $\mathcal{P}$ that is not parallel to the z-axis. Suppose that the projection of C into the xy-plane is an ellipse. Introduce uv-coordinates into the plane $\mathcal{P}$ to prove that the curve C is itself an ellipse.

39. Conclude from Problem 38 that the intersection of a nonvertical plane and an elliptic cylinder with vertical sides is an ellipse.

40. Use the result of Problem 38 to prove that the intersection of the plane $z = Ax + By$ and the paraboloid $z = a^2x^2 + b^2y^2$ is either empty, a single point, or an ellipse.

41. Use the result of Problem 38 to prove that the intersection of the plane $z = Ax + By$ and the ellipsoid $x^2/a^2 + y^2/b^2 + z^2/c^2 = 1$ is either empty, a single point, or an ellipse.

42. Suppose that $y = f(x)$ is the graph of a function for which f'' is continuous, and suppose also that the graph has an inflection point at $(a, f(a))$. Prove that the curvature of the graph at $x = a$ is zero.

43. Find the points on the curve $y = \sin x$ where the curvature is maximal and those where it is minimal.

44. The right branch of the hyperbola $x^2 - y^2 = 1$ may be parametrized by $x(t) = \cosh t$, $y(t) = \sinh t$. Find the point where its curvature is minimal.

45. Find the vectors $\mathbf{N}$ and $\mathbf{T}$ at the point of the curve $x(t) = t\cos t$, $y(t) = t\sin t$ that corresponds to $t = \pi/2$.

46. Find the points on the ellipse $x^2/a^2 + y^2/b^2 = 1$ (with $a > b > 0$) where the curvature is maximal and those where it is minimal.

47. Suppose that the plane curve $r = f(\theta)$ is given in polar coordinates. Write r' for $f'(\theta)$ and r'' for $f''(\theta)$. Show that its curvature is given by

$$\kappa = \frac{|r^2 + 2(r')^2 - rr'|}{[r^2 + (r')^2]^{3/2}}.$$

48. Use the formula in Problem 47 to calculate the curvature $\kappa(\theta)$ at the point (r, θ) of the spiral of Archimedes with equation $r = \theta$. Then show that $\kappa(\theta) \to 0$ as $\theta \to +\infty$.

49. A railway curve must join two straight tracks, one extending due west from $(-1, -1)$ and the other extending due east from $(1, 1)$. Determine A, B, and C so that the curve $y = Ax + Bx^3 + Cx^5$ joins $(-1, -1)$ and $(1, 1)$ and so that the slope and curvature of this connecting curve are zero at both its endpoints.

50. A plane passing through the origin and not parallel to any coordinate plane has an equation of the form $Ax + By + Cz = 0$ and intersects the spherical surface $x^2 + y^2 + z^2 = R^2$ in a great circle. Find the highest point on this great circle; that is, find the coordinates of the point with the largest z-coordinate.

51. Suppose that a tetrahedron in space has a solid right angle at one vertex (like a corner of a cube). Suppose that A is the area of the side opposite the solid right angle and that B, C, and D are the areas of the other three sides. (a) Prove that

$$A^2 = B^2 + C^2 + D^2.$$

(b) Of what famous theorem is this a three-dimensional version?

PARTIAL DIFFERENTIATION

13

Joseph Louis Lagrange (1736–1813)

Joseph Louis Lagrange is remembered for his great treatises on analytical mechanics and on the theory of functions that summarized much of eighteenth-century pure and applied mathematics. These treatises—*Mécanique analytique* (1788), *Théorie des fonctions analytiques* (1797), and *Leçons sur le calcul des fonctions* (1806)—systematically developed and applied widely the differential and integral calculus of multivariable functions expressed in terms of the rectangular coordinates x, y, z in three-dimensional space. They were written and published in Paris during the last quarter-century of Lagrange's career. But he grew up and spent his first 30 years in Turin, Italy. His father pointed Lagrange toward the law, but by age 17 Lagrange had decided on a career in science and mathematics. Based on his early work in celestial mechanics (the mathematical analysis of the motions of the planets and satellites in our solar system), Lagrange in 1766 succeeded Leonhard Euler as director of the Berlin Academy in Germany.

Lagrange regarded his far-reaching work on maximum-minimum problems as his best work in mathematics. This work, which continued throughout his long career, dated back to a letter to Euler that Lagrange wrote from Turin when he was only 19. This letter outlined a new approach to a certain class of optimization problems that comprise the calculus of variations. A typical example is the *isoperimetric problem,* which asks what curve of a given arc length encloses a plane region with the greatest area. (The answer: a circle.) In the *Mécanique analytique,* Lagrange applied his "method of multipliers" to investigate the motion of a particle in space that is constrained to move on a surface defined by an equation of the form $g(x, y, z) = 0$. Section 13.9 applies the Lagrange multiplier method to the problem of maximizing or minimizing a function $f(x, y, z)$ subject to a "constraint" of the form

$$g(x, y, z) = 0.$$

Today this method has applications that range from minimizing the fuel required for a spacecraft to achieve its desired trajectory to maximizing the productivity of a commercial enterprise limited by the availability of financial, natural, and personnel resources.

Modern scientific visualization often employs computer graphic techniques to present different interpretations of the same data simultaneously in a single figure. The following color graphic shows both a graph of a surface $z = f(x, y)$ and a contour map showing *level curves* that appear to encircle points (x, y) corresponding to *pits and peaks* on the surface. In Section 13.5 we learn how to locate multivariable maximum-minimum points like those visible on this surface.

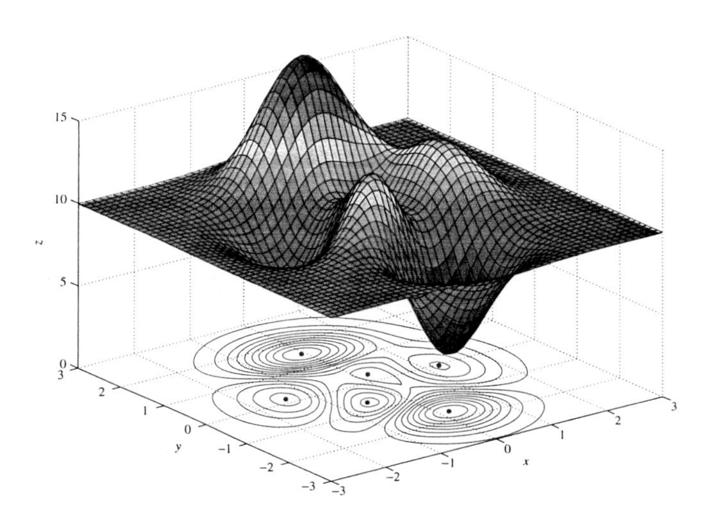

13.1 INTRODUCTION

We turn our attention here and in Chapters 14 and 15 to the calculus of functions of more than one variable. Many real-world functions depend on two or more variables. For example:

- In physical chemistry the ideal gas law $pV = nRT$ (where n and R are constants) is used to express any one of the variables p (pressure), V (volume), and T (temperature) as a function of the other two.
- The altitude above sea level at a particular location on the earth's surface depends on the latitude and longitude of the location.
- A manufacturer's profit depends on sales, overhead costs, the cost of each raw material used, and in many cases, additional variables.
- The amount of usable energy a solar panel can gather depends on its efficiency, its angle of inclination to the sun's rays, the angle of elevation of the sun above the horizon, and other factors.

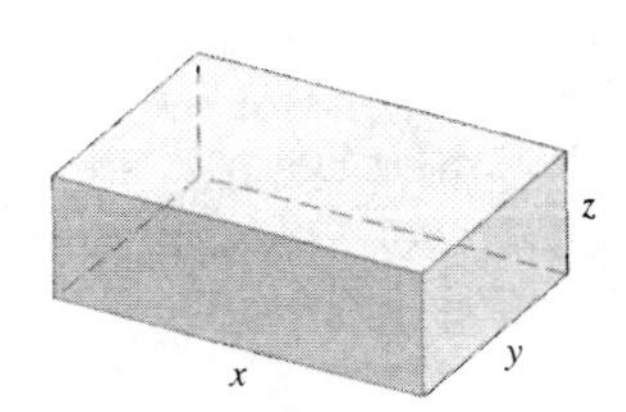

FIGURE 13.1.1 A box whose total cost we want to minimize.

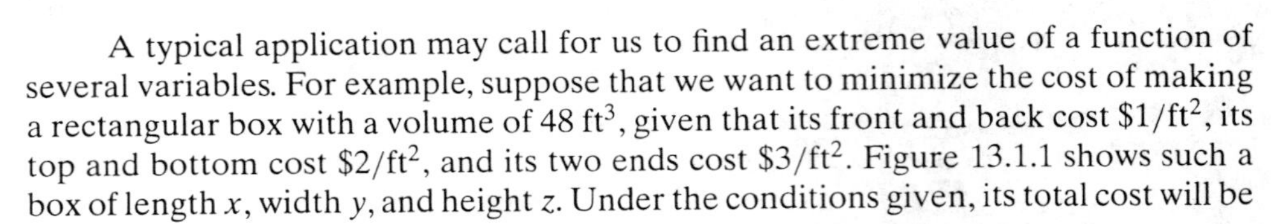

A typical application may call for us to find an extreme value of a function of several variables. For example, suppose that we want to minimize the cost of making a rectangular box with a volume of 48 ft^3, given that its front and back cost $\$1/\text{ft}^2$, its top and bottom cost $\$2/\text{ft}^2$, and its two ends cost $\$3/\text{ft}^2$. Figure 13.1.1 shows such a box of length x, width y, and height z. Under the conditions given, its total cost will be

$$C = 2xz + 4xy + 6yz \quad \text{(dollars)}.$$

But x, y, and z are not independent variables, because the box has fixed volume

$$V = xyz = 48.$$

We eliminate z, for instance, from the first formula by using the second; because $z = 48/(xy)$, the cost we want to minimize is given by

$$C = 4xy + \frac{288}{x} + \frac{96}{y}.$$

Because neither of the variables x or y can be expressed in terms of the other, the single-variable maximum-minimum techniques of Chapter 3 cannot be applied here. We need new optimization techniques applicable to functions of two or more independent variables. In Section 13.5 we shall return to this problem.

The problem of optimization is merely one example. We shall see in this chapter that many of the main ingredients of single-variable differential calculus—limits, derivatives and rates of change, chain rule computations, and maximum-minimum techniques—can be generalized to functions of two or more variables.

13.2 FUNCTIONS OF SEVERAL VARIABLES

Recall from Section 1.1 that a real-valued *function* is a rule or correspondence f that associates a unique real number with each element of a set D. The domain D has always been a subset of the real line for the functions of a single variable that we have studied up to this point. If D is a subset of the plane, then f is a function of *two* variables—for, given a point P of D, we naturally associate with P its rectangular coordinates (x, y).

DEFINITION Functions of Two or Three Variables

A **function of two variables,** defined on the **domain** D in the plane, is a rule f that associates with each point (x, y) in D a unique real number, denoted by $f(x, y)$. A **function of three variables,** defined on the **domain** D in space, is a rule f that associates with each point (x, y, z) in D a unique real number $f(x, y, z)$.

We can typically define a function f of two (or three) variables by giving a formula that specifies $f(x, y)$ in terms of x and y (or $f(x, y, z)$ in terms of x, y, and z). In case the domain D of f is not explicitly specified, we take D to consist of all points for which the given formula is meaningful.

EXAMPLE 1 The domain of the function f with formula

$$f(x, y) = \sqrt{25 - x^2 - y^2}$$

is the set of all (x, y) such that $25 - x^2 - y^2 \geqq 0$—that is, the circular disk $x^2 + y^2 \leqq 25$ of radius 5 centered at the origin. Similarly, the function g defined as

$$g(x, y, z) = \frac{x + y + z}{\sqrt{x^2 + y^2 + z^2}}$$

is defined at all points in space where $x^2 + y^2 + z^2 > 0$. Thus its domain consists of all points in three-dimensional space $\boldsymbol{R}^3$ other than the origin $(0, 0, 0)$. ◆

EXAMPLE 2 Find the domain of definition of the function with formula

$$f(x, y) = \frac{y}{\sqrt{x - y^2}}. \tag{1}$$

Find also the points (x, y) at which $f(x, y) = \pm 1$.

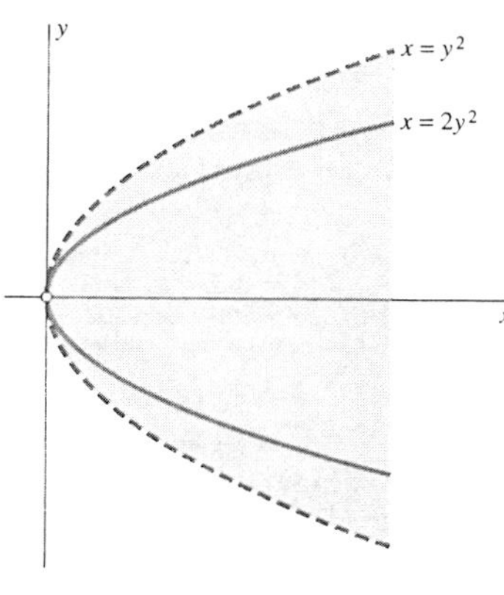

FIGURE 13.2.1 The domain of $f(x, y) = \dfrac{y}{\sqrt{x - y^2}}$ (Example 2).

Solution For $f(x, y)$ to be defined, the *radicand* $x - y^2$ must be positive—that is, $y^2 < x$. Hence the domain of f is the set of points lying strictly to the right of the parabola $x = y^2$. This domain is shaded in Fig. 13.2.1. The parabola in the figure is dashed to indicate that it is not included in the domain of f; any point for which $x = y^2$ would entail division by zero in Eq. (1).

The function $f(x, y)$ has the value ± 1 whenever

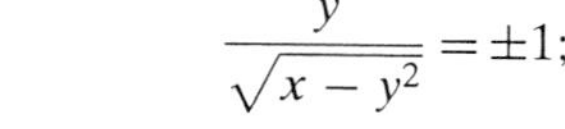

$$\frac{y}{\sqrt{x - y^2}} = \pm 1;$$

that is, when $y^2 = x - y^2$, so $x = 2y^2$. Thus $f(x, y) = \pm 1$ at each point of the parabola $x = 2y^2$ [other than its vertex $(0, 0)$, which is not included in the domain of f]. This parabola is shown as a solid curve in Fig. 13.2.1. ◆

In a geometric, physical, or economic situation, a function typically results from expressing one descriptive variable in terms of others. As we saw in Section 13.1, the cost C of the box discussed there is given by the formula

$$C = 4xy + \frac{288}{x} + \frac{96}{y}$$

in terms of the length x and width y of the box. The value C of this function is a variable that depends on the values of x and y. Hence we call C a **dependent variable,** whereas x and y are **independent variables.** And if the temperature T at the point (x, y, z) in space is given by some formula $T = h(x, y, z)$, then the dependent variable T is a function of the three independent variables x, y, and z.

We can define a function of four or more variables by giving a formula that includes the appropriate number of independent variables. For example, if an amount A of heat is released at the origin in space at time $t = 0$ in a medium with thermal diffusivity k, then—under appropriate conditions—the temperature T at the point (x, y, z) at time $t > 0$ is given by

$$T(x, y, z, t) = \frac{A}{(4\pi kt)^{3/2}} \exp\left(-\frac{x^2 + y^2 + z^2}{4kt}\right).$$

If A and k are constants, then this formula gives the temperature T as a function of the four independent variables x, y, z, and t.

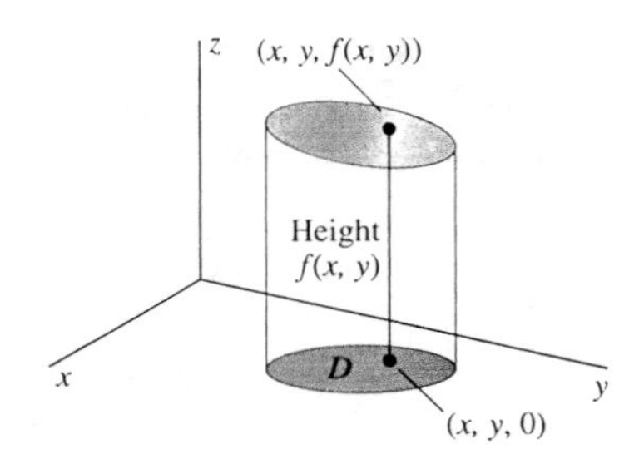

FIGURE 13.2.2 The graph of a function of two variables is typically a surface "over" the domain of the function.

We shall see that the main differences between single-variable and multivariable calculus show up when only two independent variables are involved. Hence many of our results will be stated in terms of functions of two variables. Most of these results readily generalize by analogy to the case of three or more independent variables.

Graphs and Level Curves

We can visualize how a function f of two variables "works" in terms of its graph. The **graph** of f is the graph of the equation $z = f(x, y)$. Thus the graph of f is the set of all points in space with coordinates (x, y, z) that satisfy the equation $z = f(x, y)$. (See Fig. 13.2.2.) The planes and quadric surfaces of Sections 12.4 and 12.7 provide some simple examples of graphs of functions of two variables.

EXAMPLE 3 Sketch the graph of the function $f(x, y) = 2 - \frac{1}{2}x - \frac{1}{3}y$.

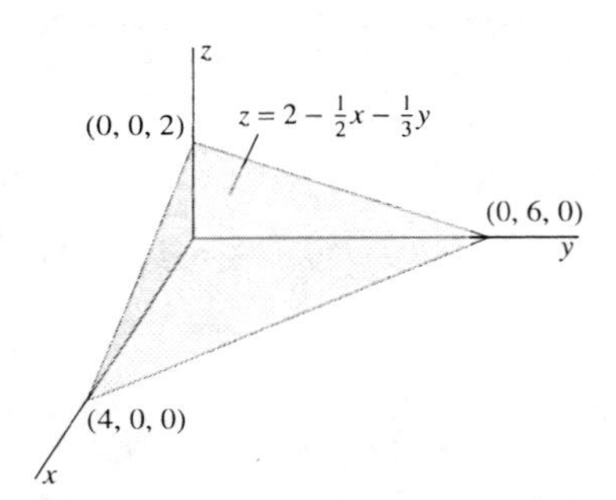

FIGURE 13.2.3 The planar graph of Example 3.

Solution We know from Section 12.4 that the graph of the equation $z = 2 - \frac{1}{2}x - \frac{1}{3}y$ is a plane, and we can visualize it by using its intercepts with the coordinate axes to plot the portion in the first octant of space. Clearly $z = 2$ if $x = y = 0$. Also the equation gives $y = 6$ if $x = z = 0$ and $x = 4$ if $y = z = 0$. Hence the graph looks as pictured in Fig. 13.2.3. ◆

EXAMPLE 4 The graph of the function $f(x, y) = x^2 + y^2$ is the familiar circular paraboloid $z = x^2 + y^2$ (Section 12.7) shown in Fig. 13.2.4. ◆

EXAMPLE 5 Find the domain of definition of the function

$$g(x, y) = \frac{1}{2}\sqrt{4 - 4x^2 - y^2} \tag{2}$$

and sketch its graph.

Solution The function g is defined wherever $4 - 4x^2 - y^2 \geqq 0$—that is, $x^2 + \frac{1}{4}y^2 \leqq 1$—so that Eq. (2) does not involve the square root of a negative number. Thus the domain of g is the set of points in the xy-plane that lie on and within the ellipse $x^2 + \frac{1}{4}y^2 = 1$ (Fig. 13.2.5). If we square both sides of the equation $z = \frac{1}{2}\sqrt{4 - 4x^2 - y^2}$ and simplify the result, we get the equation

$$x^2 + \frac{1}{4}y^2 + z^2 = 1$$

of an ellipsoid with semiaxes $a = 1$, $b = 2$, and $c = 1$ (Section 12.7). But $g(x, y)$ as defined in Eq. (2) is nonnegative wherever it is defined, so the graph of g is the upper half of the ellipsoid (Fig. 13.2.6). ◆

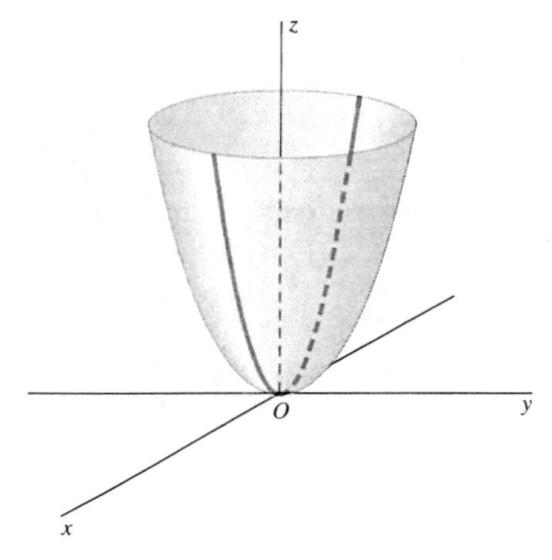

FIGURE 13.2.4 The paraboloid is the graph of the function $f(x, y) = x^2 + y^2$.

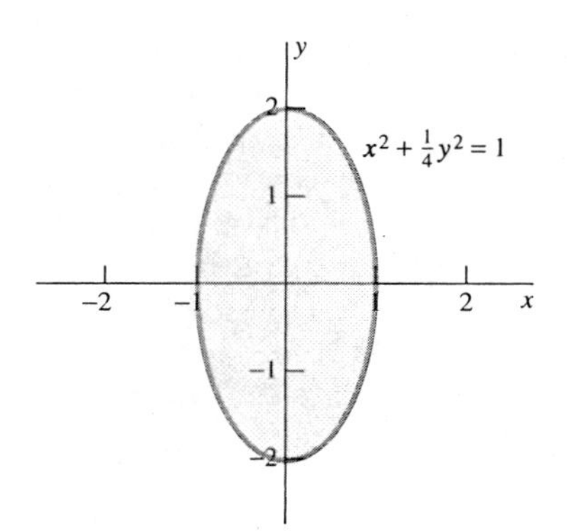

FIGURE 13.2.5 The domain of the function $g(x, y) = \frac{1}{2}\sqrt{4 - 4x^2 - y^2}$.

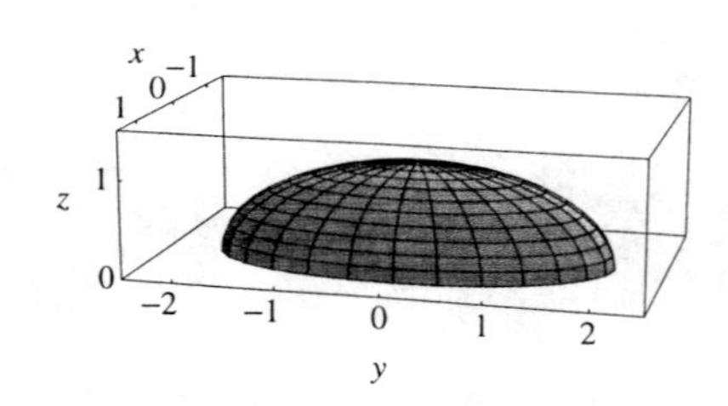

FIGURE 13.2.6 The graph of the function g is the upper half of the ellipsoid.

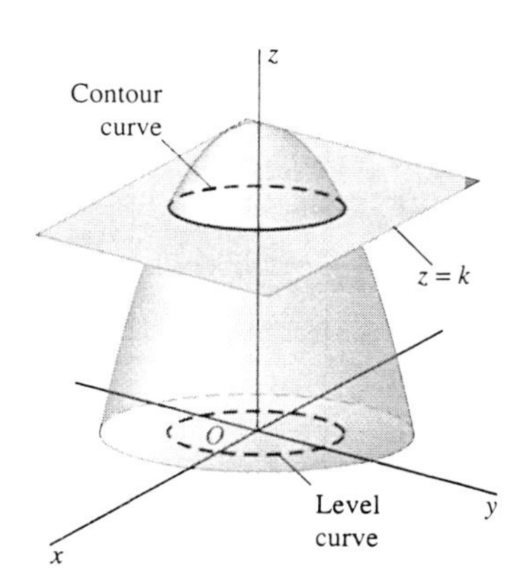

FIGURE 13.2.7 A contour curve and the corresponding level curve.

The intersection of the horizontal plane $z = k$ with the surface $z = f(x, y)$ is called the **contour curve** of **height** k on the surface (Fig. 13.2.7). The vertical projection of this contour curve into the xy-plane is the **level curve** $f(x, y) = k$ of the function f. Thus a level curve of f is simply a set in the xy-plane on which the value $f(x, y)$ is *constant*. On a topographic map, such as the one in Fig. 13.2.8, the level curves are curves of constant height above sea level.

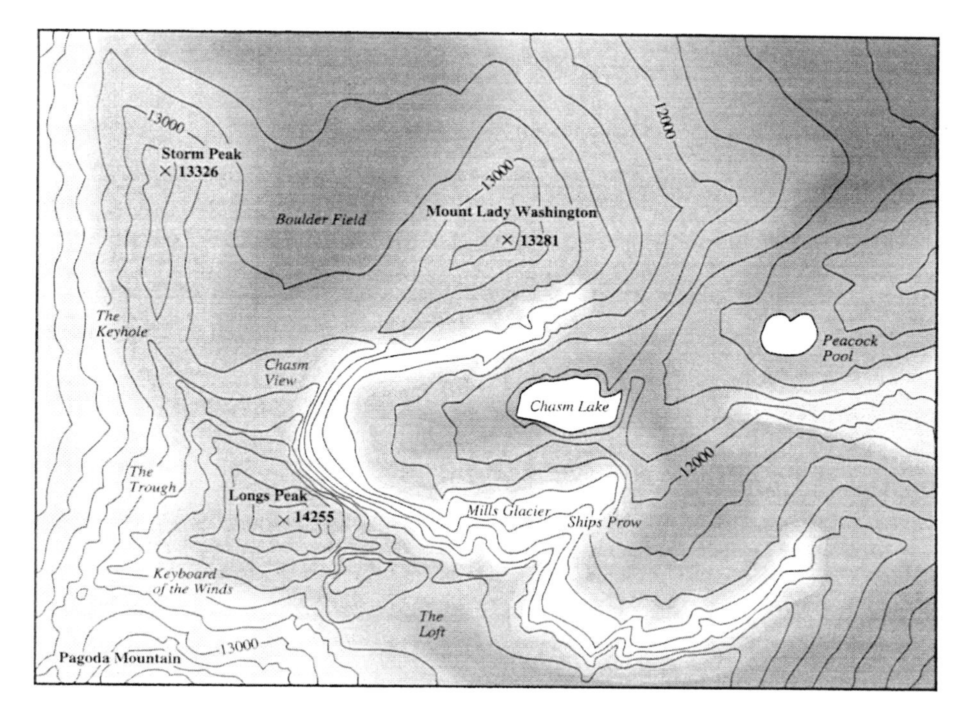

FIGURE 13.2.8 The region near Longs Peak, Rocky Mountain National Park, Colorado, showing contour lines at intervals of 200 feet.

Level curves give a two-dimensional way of representing a three-dimensional surface $z = f(x, y)$, just as the two-dimensional map in Fig. 13.2.8 represents a three-dimensional mountain range. We do this by drawing typical level curves of $z = f(x, y)$ in the xy-plane, labeling each with the corresponding (constant) value of z. Figure 13.2.9 illustrates this process for a simple hill.

EXAMPLE 6 Figure 13.2.10 shows some typical contour curves on the paraboloid $z = 25 - x^2 - y^2$. Figure 13.2.11 shows the corresponding level curves. ◆

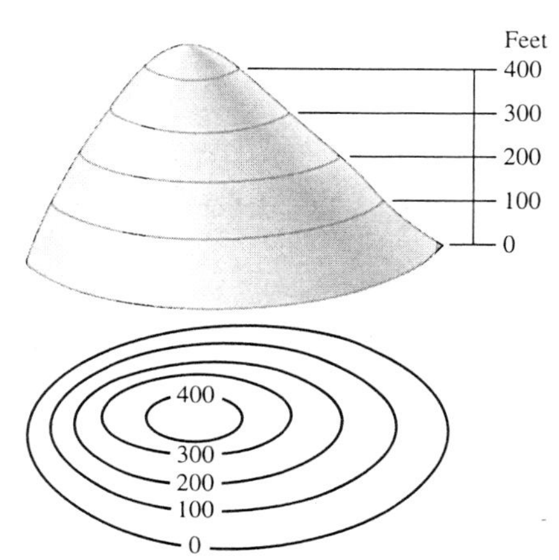

FIGURE 13.2.9 Contour curves and level curves for a hill.

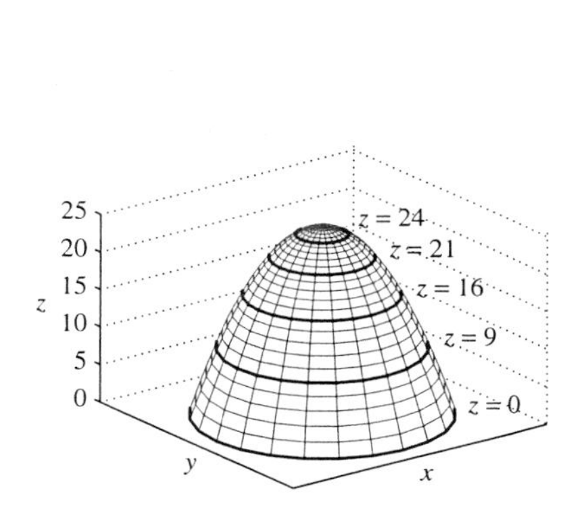

FIGURE 13.2.10 Contour curves on the surface $z = 25 - x^2 - y^2$.

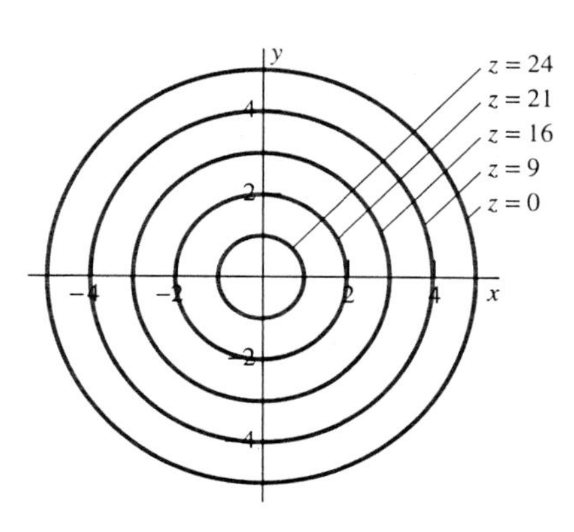

FIGURE 13.2.11 Level curves of the function $f(x, y) = 25 - x^2 - y^2$.

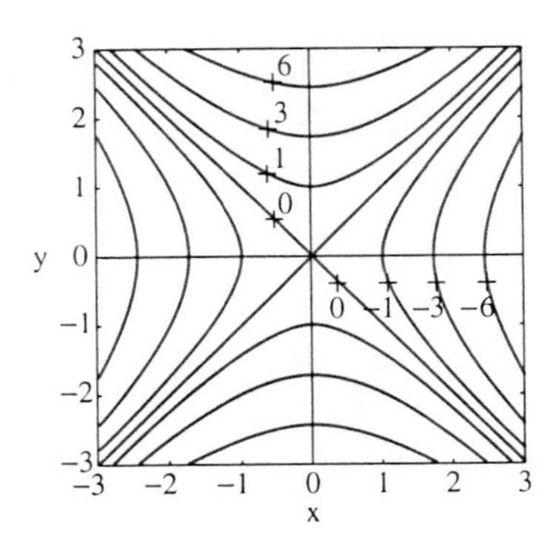

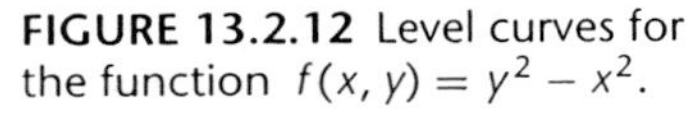

FIGURE 13.2.12 Level curves for the function $f(x, y) = y^2 - x^2$.

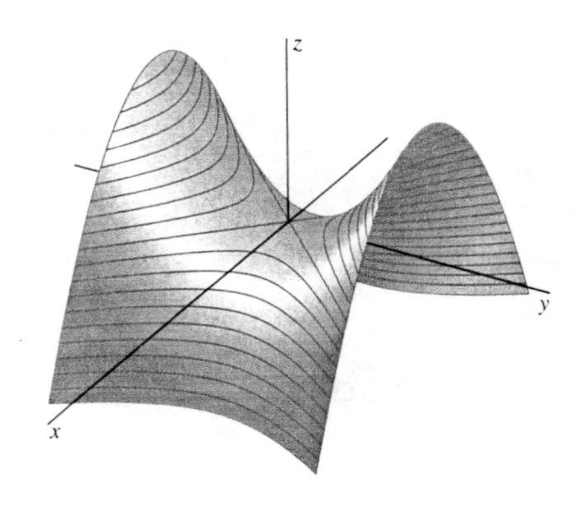

FIGURE 13.2.13 Contour curves on $z = y^2 - x^2$ (Example 7).

EXAMPLE 7 Sketch some typical level curves for the function $f(x, y) = y^2 - x^2$.

Solution If $k \neq 0$ then the curve $y^2 - x^2 = k$ is a hyperbola (Section 10.6). It opens along the y-axis if $k > 0$, along the x-axis if $k < 0$. If $k = 0$ then we have the equation $y^2 - x^2 = 0$, whose graph consists of the two straight lines $y = x$ and $y = -x$. Figure 13.2.12 shows some of the level curves, each labeled with the corresponding constant value of z. Figure 13.2.13 shows contour curves on the hyperbolic paraboloid $z = y^2 - x^2$ (Section 12.7). Note that the saddle point at the origin on the paraboloid corresponds to the intersection point of the two level curves $y = x$ and $y = -x$ in Fig. 13.2.12. ◆

The graph of a function $f(x, y, z)$ of three variables cannot be drawn in three dimensions, but we can readily visualize its **level surfaces** of the form $f(x, y, z) = k$. For example, the level surfaces of the function $f(x, y, z) = x^2 + y^2 + z^2$ are spheres (spherical surfaces) centered at the origin. Thus the level surfaces of f are the sets in space on which the value $f(x, y, z)$ is constant.

If the function f gives the temperature at the location (x, y) or (x, y, z), then its level curves or surfaces are called **isotherms.** A weather map typically includes level curves of the ground-level atmospheric pressure; these are called **isobars.** Even though you may be able to construct the graph of a function of two variables, that graph might be so complicated that information about the function (or the situation it describes) is obscure. Frequently the level curves themselves give more information, as in weather maps. For example, Fig. 13.2.14 shows level curves for the annual

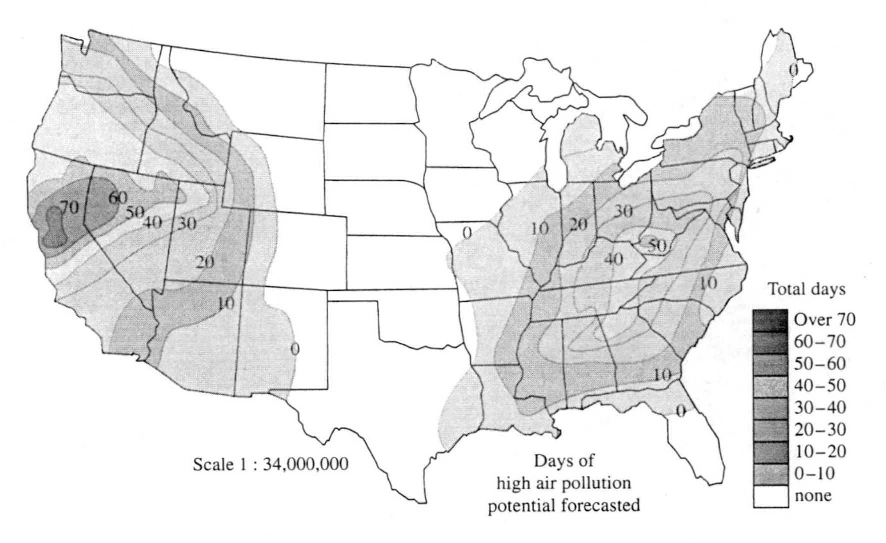

FIGURE 13.2.14 Days of high air pollution forecast in the United States (from National Atlas of the United States, U.S. Department of the Interior, 1970).

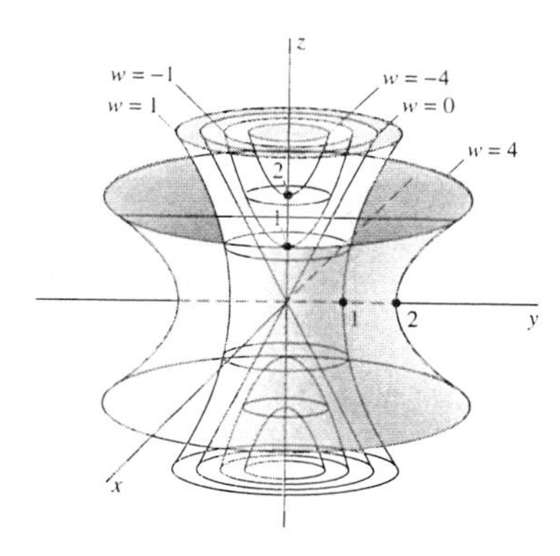

FIGURE 13.2.15 Some level surfaces of the function $w = f(x, y, z) = x^2 + y^2 - z^2$ (Example 8).

numbers of days of *high* air pollution forecast at different localities in the United States. The scale of this figure does not show local variations caused by individual cities. But a glance indicates that western Colorado, south Georgia, and central Illinois all expect the same number (10, in this case) of high-pollution days each year.

EXAMPLE 8 Figure 13.2.15 shows some level surfaces of the function

$$f(x, y, z) = x^2 + y^2 - z^2.$$

If $k > 0$, then the graph of $x^2 + y^2 - z^2 = k$ is a hyperboloid of one sheet, whereas if $k < 0$ it is a hyperboloid of two sheets. The cone $x^2 + y^2 - z^2 = 0$ lies between these two types of hyperboloids. ◆

Computer Plots

Many computer systems have surface and contour plotting routines like the *Maple* commands

```
plot3d(y^2 - x^2, x = -3..3, y = -3..3);

with(plots):  contourplot(y^2 - x^2, x = -3..3, y = -3..3);
```

and the *Mathematica* commands

```
Plot3D[ y^2 - x^2, {x,-3,3}, {y,-3,3} ]

ContourPlot[ y^2 - x^2, {x,-3,3}, {y,-3,3} ]
```

for the function $f(x, y) = y^2 - x^2$ of Example 7.

EXAMPLE 9 Figure 13.2.16 shows both the graph and some projected contour curves of the function

$$f(x, y) = (x^2 - y^2)\exp(-x^2 - y^2).$$

Observe the patterns of nested level curves that indicate "pits" and "peaks" on the surface. In Fig. 13.2.17, the level curves that correspond to surface contours above the xy-plane are shown in red, while those that correspond to contours below the xy-plane are shown in blue. In this way we can distinguish between peaks and pits. It appears likely that the surface has peaks above the points $(\pm 1, 0)$ on the x-axis in the xy-plane, and has pits below the points $(0, \pm 1)$ on the y-axis. Because $f(x, \pm x) \equiv 0$, the two 45° lines $y = \pm x$ in Fig. 13.2.17 are also level curves; they intersect at the point $(0, 0)$ in the plane that corresponds to a saddle point or "pass" (as in *mountain pass*) on the surface. ◆

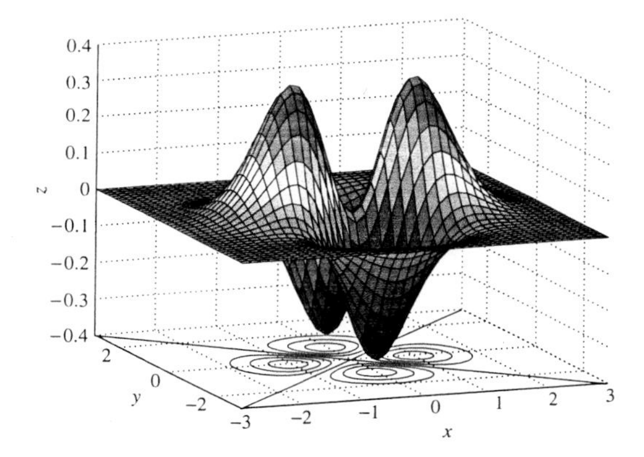

FIGURE 13.2.16 The graph and projected contour curves of the function $f(x, y) = (x^2 - y^2)e^{-x^2-y^2}$.

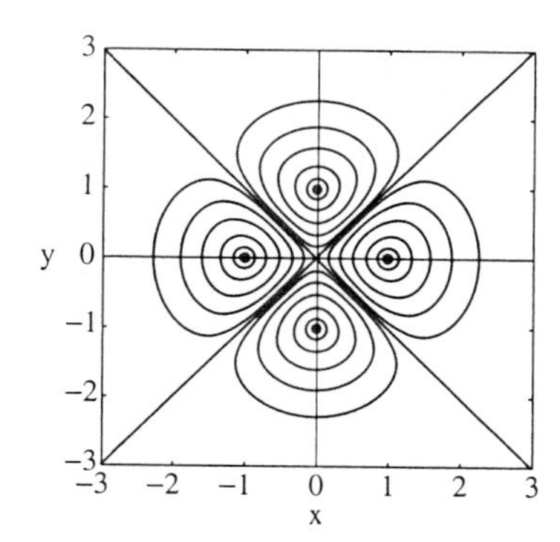

FIGURE 13.2.17 Level curves for the function $f(x, y) = (x^2 - y^2)e^{-x^2-y^2}$.

REMARK In Section 13.5 we will study analytic methods for locating maximum and minimum points of functions of two variables *exactly*. But Example 9 indicates that plots of level curves provide a valuable tool for locating them *approximately*.

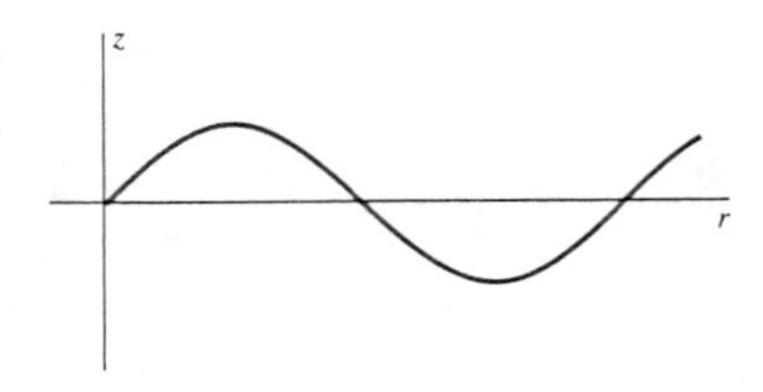

FIGURE 13.2.18 The curve $z = \sin r$ (Example 10).

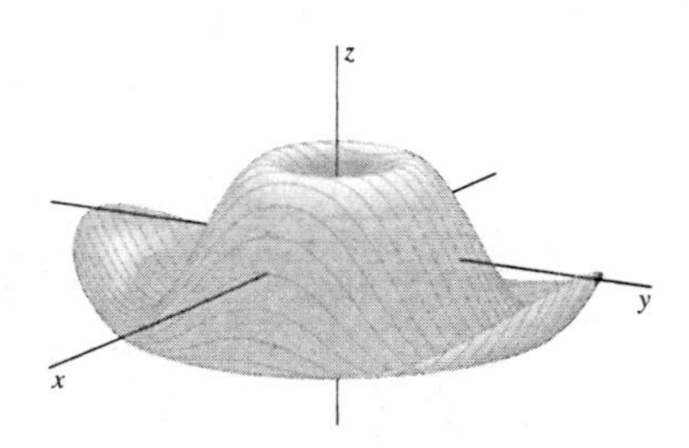

FIGURE 13.2.19 The hat surface $z = \sin\sqrt{x^2+y^2}$ (Example 10).

EXAMPLE 10 The surface

$$z = \sin\sqrt{x^2+y^2} \quad (3)$$

is symmetrical with respect to the z-axis, because Eq. (3) reduces to the equation $z = \sin r$ (Fig. 13.2.18) in terms of the radial coordinate $r = \sqrt{x^2+y^2}$ that measures perpendicular distance from the z-axis. The *surface* $z = \sin r$ is generated by revolving the curve $z = \sin x$ around the z-axis. Hence its level curves are circles centered at the origin in the xy-plane. For instance, $z = 0$ if r is an integral multiple of π, whereas $z = \pm 1$ if r is any odd integral multiple of $\pi/2$. Figure 13.2.19 shows traces of this surface in planes parallel to the yz-plane. The "hat effect" was achieved by plotting (x, y, z) for those points (x, y) that lie within a certain ellipse in the xy-plane. ◆

Given an arbitrary function $f(x, y)$, it can be quite a challenge to construct by hand a picture of the surface $z = f(x, y)$. Example 11 illustrates some special techniques that may be useful. Additional surface-sketching techniques will appear in the remainder of this chapter.

EXAMPLE 11 Investigate the graph of the function

$$f(x, y) = \tfrac{3}{4}y^2 + \tfrac{1}{24}y^3 - \tfrac{1}{32}y^4 - x^2. \quad (4)$$

Solution The key feature in Eq. (4) is that the right-hand side is the *sum* of a function of x and a function of y. If we set $x = 0$, we get the curve

$$z = \tfrac{3}{4}y^2 + \tfrac{1}{24}y^3 - \tfrac{1}{32}y^4 \quad (5)$$

in which the surface $z = f(x, y)$ intersects the yz-plane. But if we set $y = y_0$ in Eq. (4), we get

$$z = \left(\tfrac{3}{4}y_0^2 + \tfrac{1}{24}y_0^3 - \tfrac{1}{32}y_0^4\right) - x^2;$$

that is,

$$z = k - x^2, \quad (6)$$

which is the equation of a parabola in the xz-plane. Hence the trace of $z = f(x, y)$ in each plane $y = y_0$ is a parabola of the form in Eq. (6) (Fig. 13.2.20).

We can use the techniques of Section 4.5 to sketch the curve in Eq. (5). Calculating the derivative of z with respect to y, we get

$$\frac{dz}{dy} = \frac{3}{2}y + \frac{1}{8}y^2 - \frac{1}{8}y^3 = -\frac{1}{8}y(y^2 - y - 12) = -\frac{1}{8}y(y+3)(y-4).$$

Hence the critical points are $y = -3$, $y = 0$, and $y = 4$. The corresponding values of z are

$$f(0, -3) = \tfrac{99}{32} \approx 3.09, \qquad f(0, 0) = 0, \qquad \text{and} \qquad f(0, 4) = \tfrac{20}{3} \approx 6.67.$$

Because $z \to -\infty$ as $y \to \pm\infty$, it follows readily that the graph of Eq. (5) looks like that in Fig. 13.2.21.

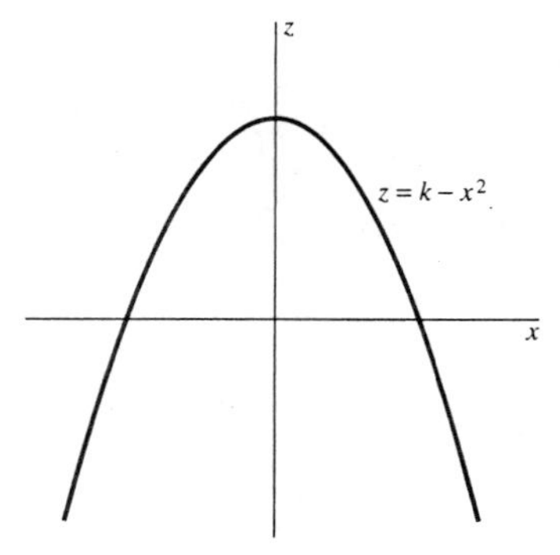

FIGURE 13.2.20 The intersection of $z = f(x, y)$ and the plane $y = y_0$ (Example 11).

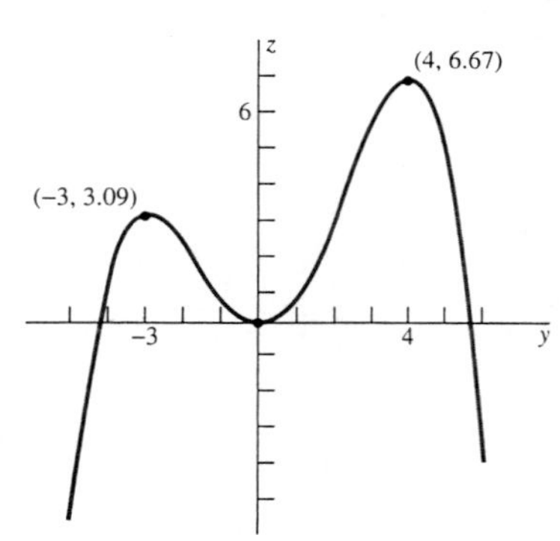

FIGURE 13.2.21 The curve $z = \frac{3}{4}y^2 + \frac{1}{24}y^3 - \frac{1}{32}y^4$ (Example 11).

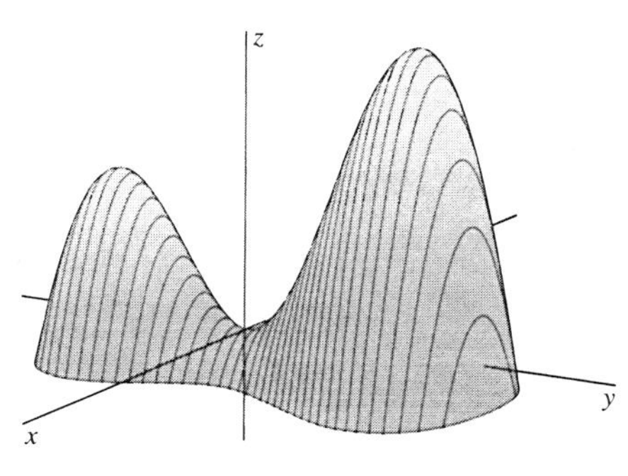

FIGURE 13.2.22 Trace parabolas of $z = f(x, y)$ (Example 11).

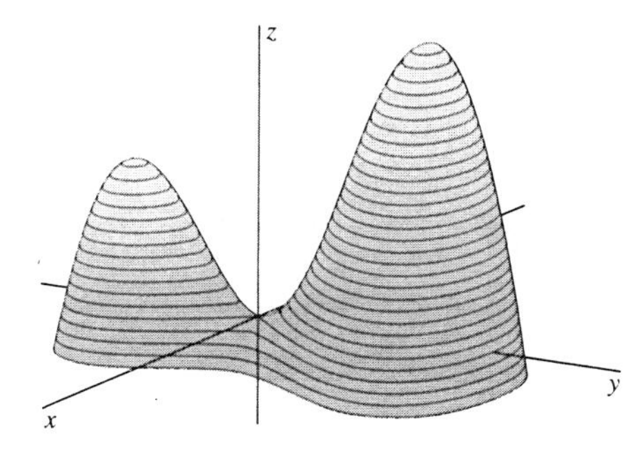

FIGURE 13.2.23 Contour curves on $z = f(x, y)$ (Example 11).

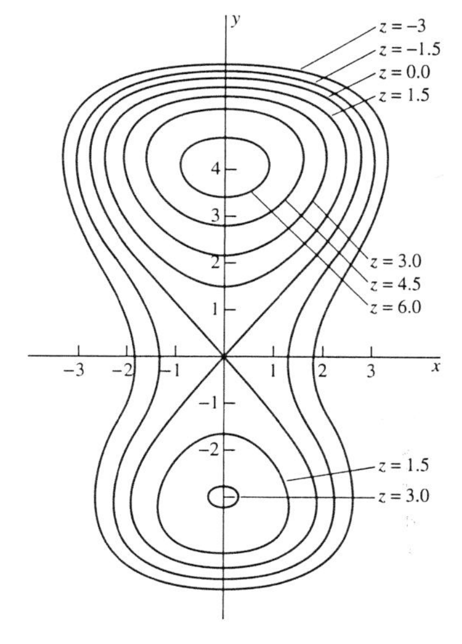

FIGURE 13.2.24 Level curves of the function $f(x, y) = \frac{3}{4}y^2 + \frac{1}{24}y^3 - \frac{1}{32}y^4 - x^2$ (Example 11).

Now we can see what the surface $z = f(x, y)$ looks like. Each vertical plane $y = y_0$ intersects the curve in Eq. (5) at a single point, and this point is the vertex of a parabola that opens downward like that in Eq. (6); this parabola is the intersection of the plane and the surface. Thus the surface $z = f(x, y)$ is generated by translating the vertex of such a parabola along the curve

$$z = \tfrac{3}{4}y^2 + \tfrac{1}{24}y^3 - \tfrac{1}{32}y^4,$$

as indicated in Fig. 13.2.22.

Figure 13.2.23 shows some typical contour curves on this surface. They indicate that the surface resembles two peaks separated by a mountain pass. Figure 13.2.24 shows a computer plot of level curves of the function $f(x, y)$. The nested level curves enclosing the points $(0, -3)$ and $(0, 4)$ correspond to the peaks at the point $(0, -3, \frac{99}{32})$ and $(0, 4, \frac{20}{3})$ on the surface $z = f(x, y)$. The level figure-eight curve through $(0, 0)$ marks the saddle point (or pass) that we see at the origin on the surface in Figs. 13.2.22 and 13.2.23. Extreme values and saddle points of functions of two variables are discussed in Sections 13.5 and 13.10. ◆

13.2 TRUE/FALSE STUDY GUIDE

13.2 CONCEPTS: QUESTIONS AND DISCUSSION

1. Summarize the relationship between the level curves of a function $f(x, y)$ and the pits, peaks, and passes on the surface $z = f(x, y)$. In short, how can you locate likely pits, peaks, and passes by looking at a plot of level curves?
2. Give examples of other types of data for your country that might be presented in the form of a contour (level curve) map like the one shown in Fig. 13.2.14.
3. The function graphed in Example 11 is of the form $z = f(x) + g(y)$, the sum of single-variable functions of the two independent variables x and y. Describe a way of sketching the graph of any such function.

13.2 PROBLEMS

In Problems 1 through 20, state the largest possible domain of definition of the given function f.

1. $f(x, y) = 4 - 3x - 2y$
2. $f(x, y) = \sqrt{x^2 + 2y^2}$
3. $f(x, y) = \dfrac{1}{x^2 + y^2}$
4. $f(x, y) = \dfrac{1}{x - y}$
5. $f(x, y) = \sqrt[3]{y - x^2}$
6. $f(x, y) = \sqrt{2x} + \sqrt[3]{3y}$
7. $f(x, y) = \sin^{-1}(x^2 + y^2)$

8. $f(x, y) = \tan^{-1}\left(\dfrac{y}{x}\right)$

9. $f(x, y) = \exp(-x^2 - y^2)$ (Fig. 13.2.25)

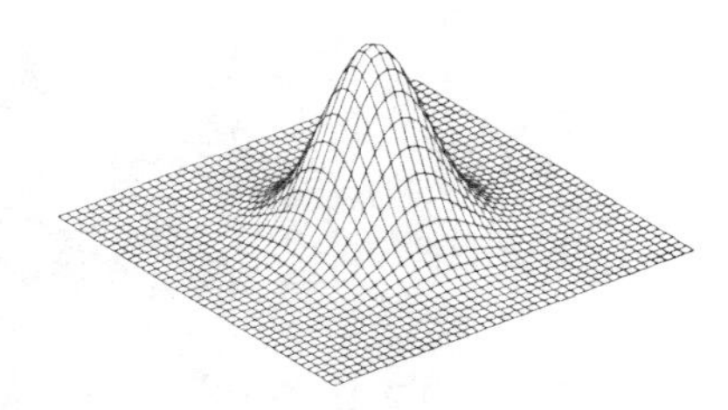

FIGURE 13.2.25 The graph of the function of Problem 9.

10. $f(x, y) = \ln(x^2 - y^2 - 1)$

11. $f(x\ y) = \ln(y - x)$

12. $f(x, y) = \sqrt{4 - x^2 - y^2}$

13. $f(x, y) = \dfrac{1 + \sin xy}{xy}$

14. $f(x, y) = \dfrac{1 + \sin xy}{x^2 + y^2}$ (Fig. 13.2.26)

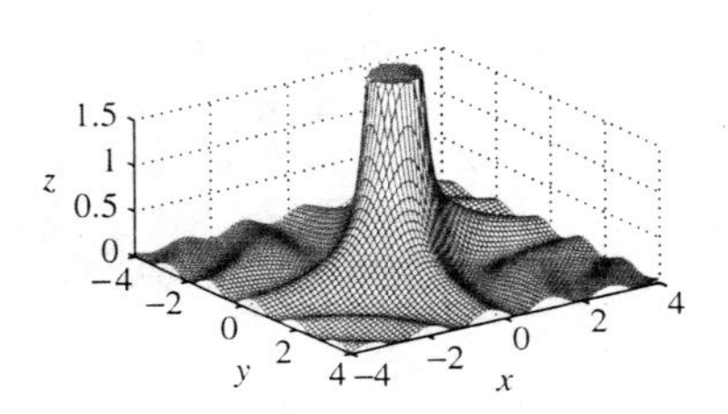

FIGURE 13.2.26 The graph $z = \dfrac{1 + \sin(xy)}{x^2 + y^2}$ of Problem 14.

15. $f(x, y) = \dfrac{xy}{x^2 - y^2}$

16. $f(x, y, z) = \dfrac{1}{\sqrt{z - x^2 - y^2}}$

17. $f(x, y, z) = \exp\left(\dfrac{1}{x^2 + y^2 + z^2}\right)$

18. $f(x, y, z) = \ln(xyz)$

19. $f(x, y, z) = \ln(z - x^2 - y^2)$

20. $f(x, y, z) = \sin^{-1}(3 - x^2 - y^2 - z^2)$

In Problems 21 through 30, describe the graph of the function f.

21. $f(x, y) = 10$

22. $f(x, y) = x$

23. $f(x, y) = x + y$

24. $f(x, y) = \sqrt{x^2 + y^2}$

25. $f(x, y) = x^2 + y^2$

26. $f(x, y) = 4 - x^2 - y^2$

27. $f(x, y) = \sqrt{4 - x^2 - y^2}$

28. $f(x, y) = 16 - y^2$

29. $f(x, y) = 10 - \sqrt{x^2 + y^2}$

30. $f(x, y) = -\sqrt{36 - 4x^2 - 9y^2}$

In Problems 31 through 40, sketch some typical level curves of the function f.

31. $f(x, y) = x - y$

32. $f(x, y) = x^2 - y^2$

33. $f(x, y) = x^2 + 4y^2$

34. $f(x, y) = y - x^2$

35. $f(x, y) = y - x^3$

36. $f(x, y) = y - \cos x$

37. $f(x, y) = x^2 + y^2 - 4x$

38. $f(x, y) = x^2 + y^2 - 6x + 4y + 7$

39. $f(x, y) = \exp(-x^2 - y^2)$

40. $f(x, y) = \dfrac{1}{1 + x^2 + y^2}$

In Problems 41 through 46, describe the level surfaces of the function f.

41. $f(x, y, z) = x^2 + y^2 - z$

42. $f(x, y, z) = z + \sqrt{x^2 + y^2}$

43. $f(x, y, z) = x^2 + y^2 + z^2 - 4x - 2y - 6z$

44. $f(x, y, z) = z^2 - x^2 - y^2$

45. $f(x, y, z) = x^2 + 4y^2 - 4x - 8y + 17$

46. $f(x, y, z) = x^2 + y^2 + 25$

In Problems 47 through 52, the function $f(x, y)$ is the sum of a function of x and a function of y. Hence you can use the method of Example 11 to construct a sketch of the surface $z = f(x, y)$. Match each function with its graph among Figs. 13.2.27 through 13.2.32.

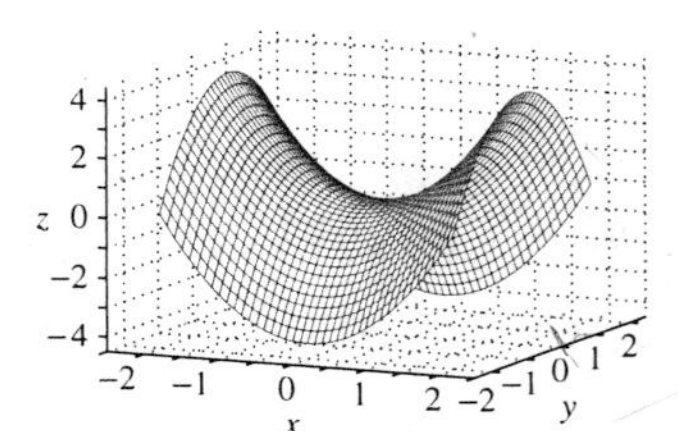

FIGURE 13.2.27

FIGURE 13.2.28

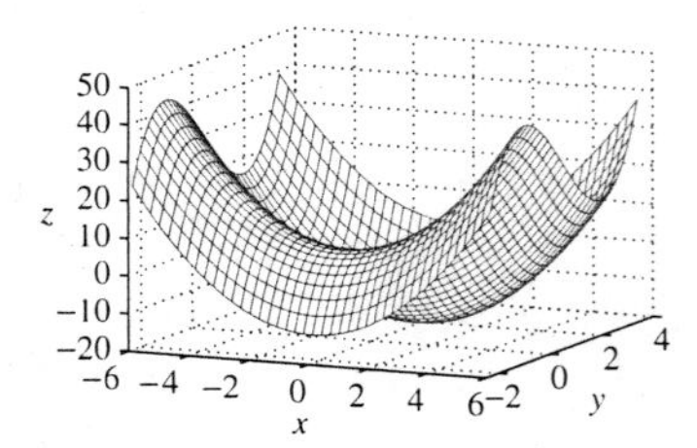

FIGURE 13.2.29

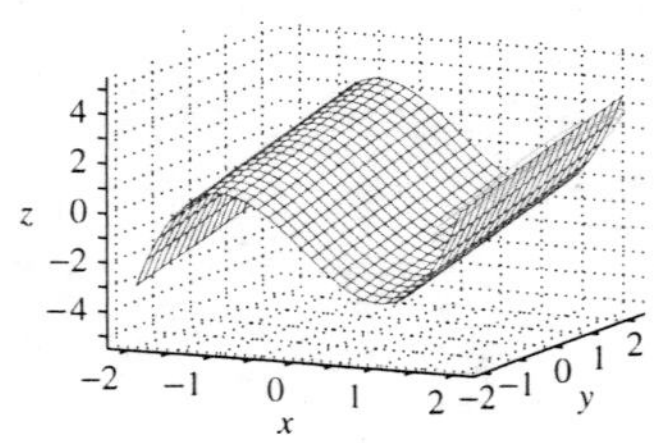

FIGURE 13.2.30

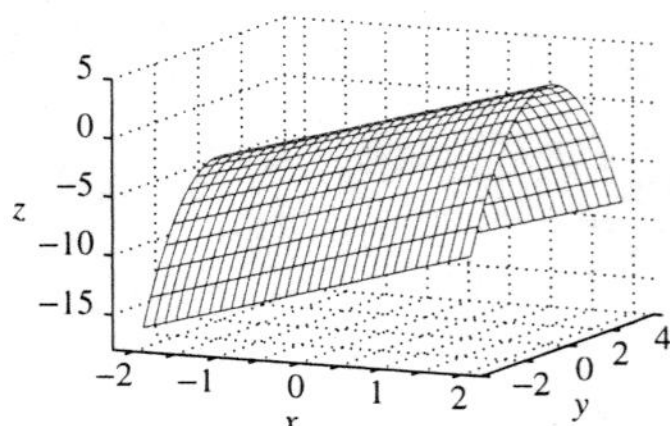

FIGURE 13.2.31

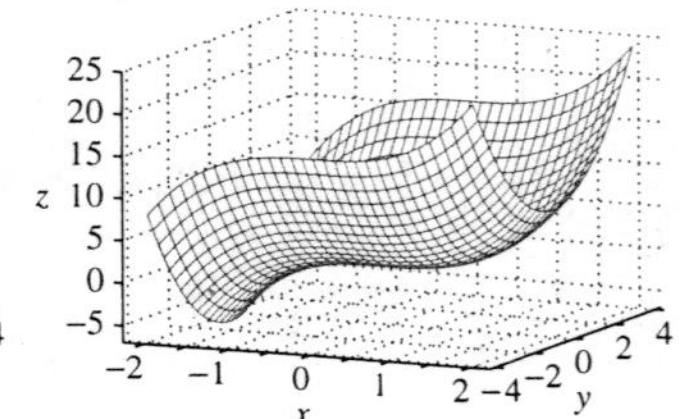

FIGURE 13.2.32

47. $f(x, y) = x^3 + y^2$

48. $f(x, y) = 2x - y^2$

49. $f(x, y) = x^3 - 3x^2 + \frac{1}{2}y^2$

50. $f(x, y) = x^2 - y^2$

51. $f(x, y) = x^2 + y^4 - 4y^2$

52. $f(x, y) = 2y^3 - 3y^2 - 12y + x^2$

Problems 53 through 58 show the graphs of six functions $z = f(x, y)$. Figures 13.2.39 through 13.2.44 show level curve plots for the same functions but in another order; the level curves in each figure correspond to contours at equally spaced heights on the surface $z = f(x, y)$. Match each surface with its level curves.

53. $z = \dfrac{1}{1 + x^2 + y^2}$, $|x| \leqq 2$, $|y| \leqq 2$ (Fig. 13.2.33)

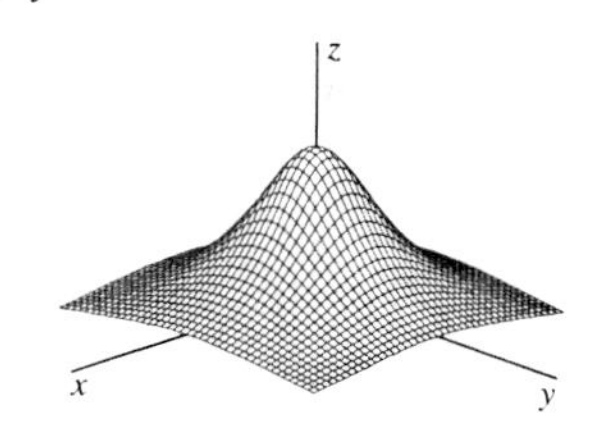

FIGURE 13.2.33 $z = \dfrac{1}{1 + x^2 + y^2}$, $|x| \leqq 2$, $|y| \leqq 2$.

54. $z = r^2 \exp(-r^2) \cos^2\left(\frac{3}{2}\theta\right)$, $|x| \leqq 3$, $|y| \leqq 3$ (Fig. 13.2.34)

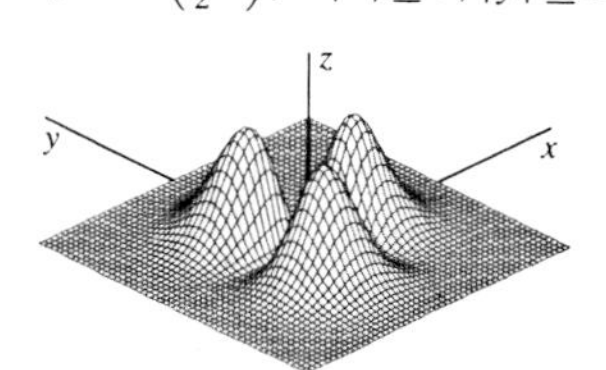

FIGURE 13.2.34 $z = r^2 \exp(-r^2) \cos^2\left(\frac{3}{2}\theta\right)$, $|x| \leqq 3$, $|y| \leqq 3$, $r \geqq 0$.

55. $z = \cos\sqrt{x^2 + y^2}$, $|x| \leqq 10$, $|y| \leqq 10$ (Fig. 13.2.35)

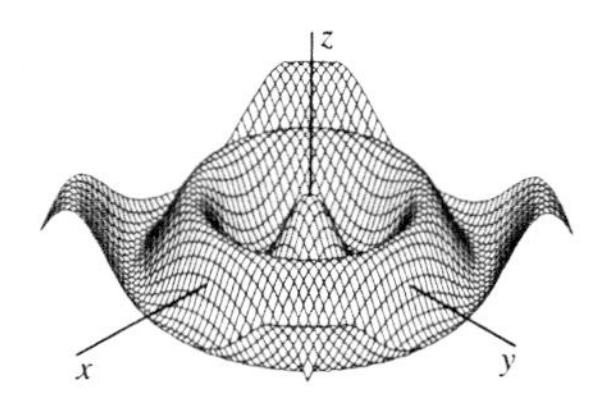

FIGURE 13.2.35 $z = \cos\sqrt{x^2 + y^2}$, $|x| \leqq 10$, $|y| \leqq 10$.

56. $z = x \exp(-x^2 - y^2)$, $|x| \leqq 2$, $|y| \leqq 2$ (Fig. 13.2.36)

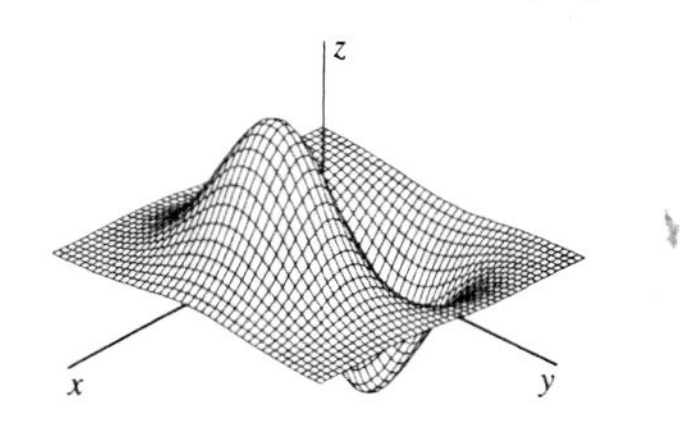

FIGURE 13.2.36 $z = x \exp(-x^2 - y^2)$, $|x| \leqq 2$, $|y| \leqq 2$.

57. $z = 3(x^2 + 3y^2)\exp(-x^2 - y^2)$, $|x| \leqq 2.5$, $|y| \leqq 2.5$ (Fig. 13.2.37)

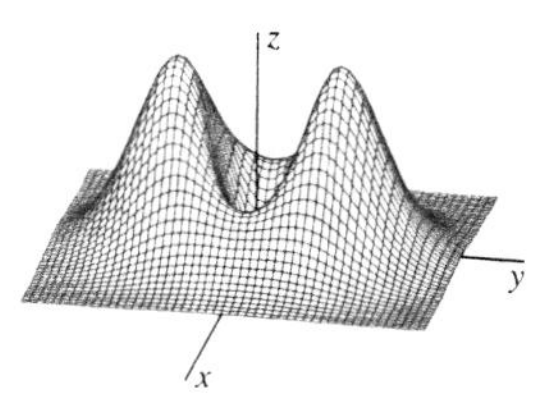

FIGURE 13.2.37 $z = 3(x^2 + 3y^2)\exp(-x^2 - y^2)$, $|x| \leqq 2.5$, $|y| \leqq 2.5$.

58. $z = xy \exp\left(-\frac{1}{2}(x^2 + y^2)\right)$, $|x| \leqq 3.5$, $|y| \leqq 3.5$ (Fig. 13.2.38)

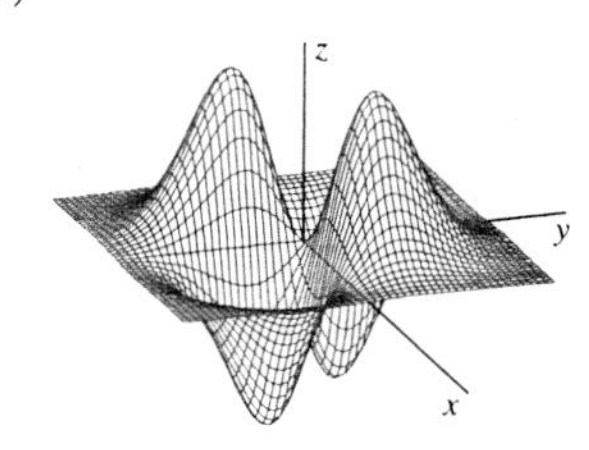

FIGURE 13.2.38 $z = xy \exp\left(-\frac{1}{2}(x^2 + y^2)\right)$, $|x| \leqq 3.5$, $|y| \leqq 3.5$.

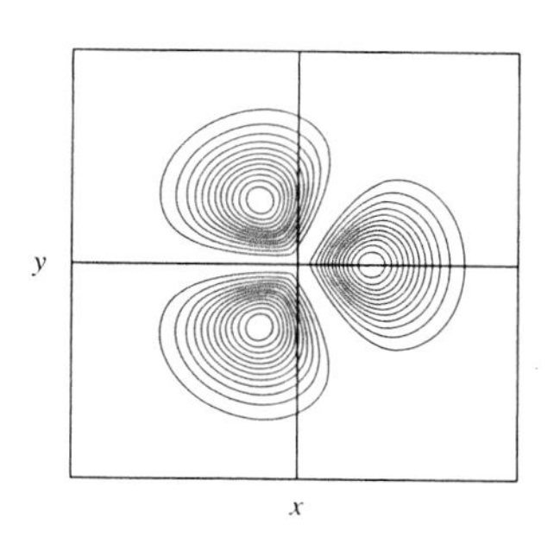

FIGURE 13.2.39

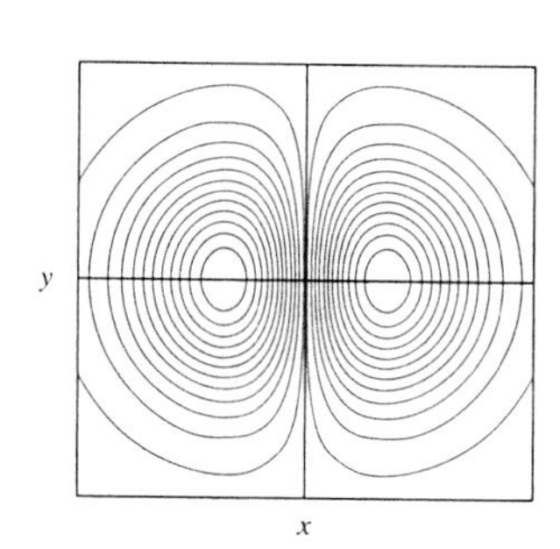

FIGURE 13.2.40

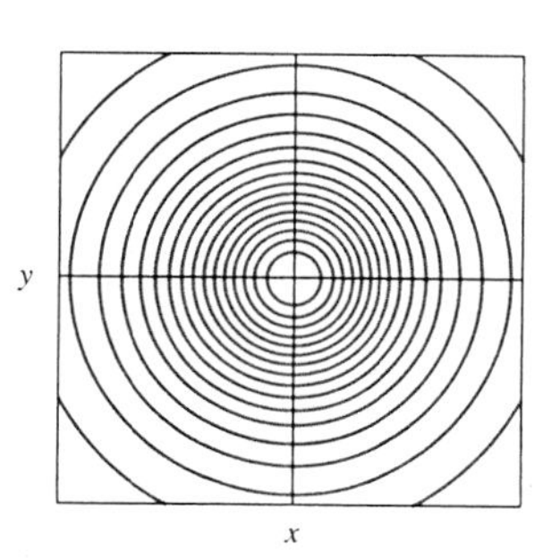

FIGURE 13.2.41

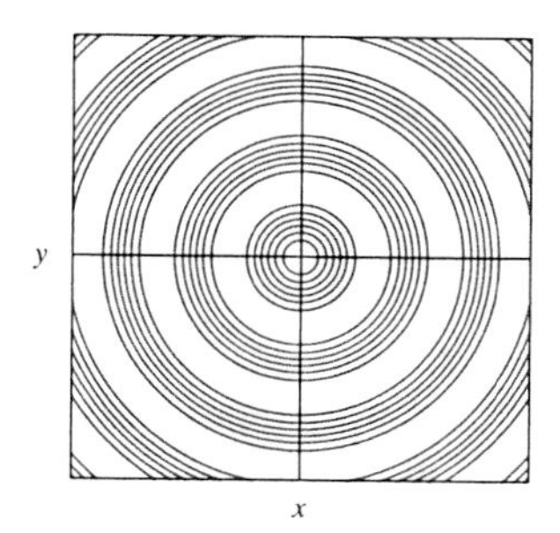

FIGURE 13.2.42

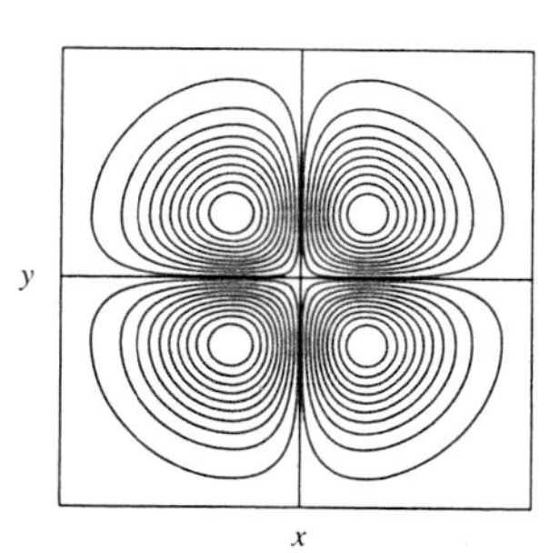

FIGURE 13.2.43

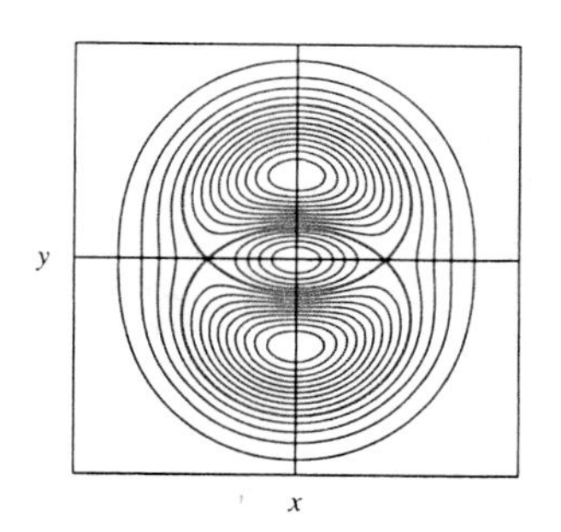

FIGURE 13.2.44

59. Use a computer to investigate surfaces of the form $z = (ax + by)\exp(-x^2 - y^2)$. How do the number and locations of apparent peaks and pits depend on the values of the constants a and b?

60. Use a computer to graph the surface $z = (ax^2 + 2bxy + cy^2)\exp(-x^2 - y^2)$ with different values of the parameters a, b, and c. Describe the different types of surfaces that are obtained in this way. How do the number and locations of apparent peaks and pits depend on the values of the constants a, b, and c?

61. Use a computer to investigate surfaces of the form $z = r^2 \exp(-r^2) \sin n\theta$. How do the number and locations of apparent peaks and pits depend on the value of the integer n?

62. Repeat Problem 61 with surfaces of the form $z = r^2 \exp(-r^2) \cos^2 n\theta$.

13.3 LIMITS AND CONTINUITY

We need limits of functions of several variables for the same reasons that we needed limits of functions of a single variable—so that we can discuss continuity, slopes, and rates of change. Both the definition and the basic properties of limits of functions of several variables are essentially the same as those that we stated in Section 2.2 for functions of a single variable. For simplicity, we shall state them here only for functions of two variables x and y; for a function of three variables, the pair (x, y) should be replaced with the triple (x, y, z).

For a function f of two variables, we ask what number (if any) the values $f(x, y)$ approach as (x, y) approaches the fixed point (a, b) in the coordinate plane. For a function f of three variables, we ask what number (if any) the values $f(x, y, z)$ approach as (x, y, z) approaches the fixed point (a, b, c) in space.

EXAMPLE 1 The numerical data in the table of Fig. 13.3.1 suggest that the values of the function $f(x, y) = xy$ approach 6 as $x \to 2$ and $y \to 3$ simultaneously—that is, as (x, y) approaches the point $(2, 3)$. It therefore is natural to write

$$\lim_{(x,y)\to(2,3)} xy = 6.$$ ◆

x	y	$f(x, y) = xy$ (rounded)
2.2	2.5	5.50000
1.98	3.05	6.03900
2.002	2.995	5.99599
1.9998	3.0005	6.00040
2.00002	2.99995	5.99996
1.999998	3.000005	6.00000
↓	↓	↓
2	3	6

FIGURE 13.3.1 The numerical data of Example 1.

Our intuitive idea of the limit of a function of two variables is this. We say that the number L is the *limit* of the function $f(x, y)$ as (x, y) approaches the point (a, b), and we write

$$\lim_{(x,y)\to(a,b)} f(x, y) = L, \qquad (1)$$

provided that the number $f(x, y)$ can be made as close as we please to L merely by choosing the point (x, y) sufficiently close to—but not equal to—the point (a, b).

To make this intuitive idea precise, we must specify how close to L—within the distance $\epsilon > 0$, say—we want $f(x, y)$ to be, and then how close to (a, b) the

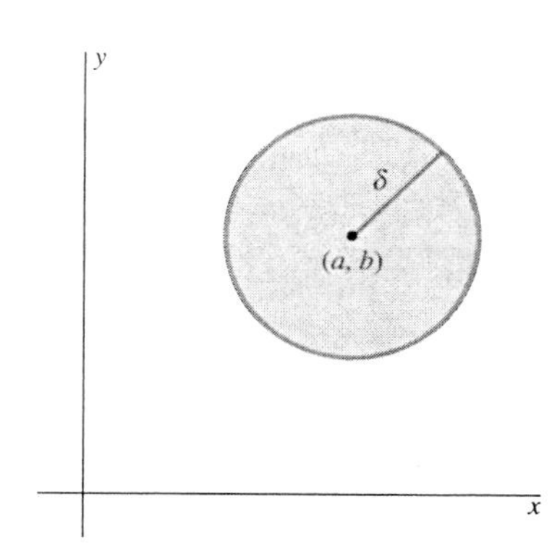

FIGURE 13.3.2 The circular disk with center (a, b) and radius δ.

point (x, y) must be to accomplish this. We think of the point (x, y) as being close to (a, b) provided that it lies within a small circular disk (Fig. 13.3.2) with center (a, b) and radius δ, where δ is a small positive number. The point (x, y) lies within this disk if and only if

$$\sqrt{(x-a)^2+(y-b)^2} < \delta. \tag{2}$$

This observation serves as motivation for the formal definition, with two additional conditions. First, we define the limit of $f(x, y)$ as $(x, y) \to (a, b)$ *only* under the condition that the domain of definition of f contains points $(x, y) \neq (a, b)$ that lie arbitrarily close to (a, b)—that is, within *every* disk of the sort shown in Fig. 13.3.2 and thus within any and every preassigned positive distance of (a, b). Hence we do not speak of the limit of f at an isolated point of its domain D. Finally, we do *not* require that f be defined at the point (a, b) itself. Thus we deliberately exclude the possibility that $(x, y) = (a, b)$.

DEFINITION The Limit of $f(x, y)$

We say that the **limit of** $f(x, y)$ **as** (x, y) **approaches** (a, b) **is** L provided that for every number $\epsilon > 0$, there exists a number $\delta > 0$ with the following property: If (x, y) is a point of the domain of f such that if

$$0 < \sqrt{(x-a)^2+(y-b)^2} < \delta, \tag{2'}$$

then it follows that

$$|f(x, y) - L| < \epsilon. \tag{3}$$

REMARK The "extra" inequality $0 < \sqrt{(x-a)^2+(y-b)^2}$ in Eq. (2′) serves to ensure that $(x, y) \neq (a, b)$.

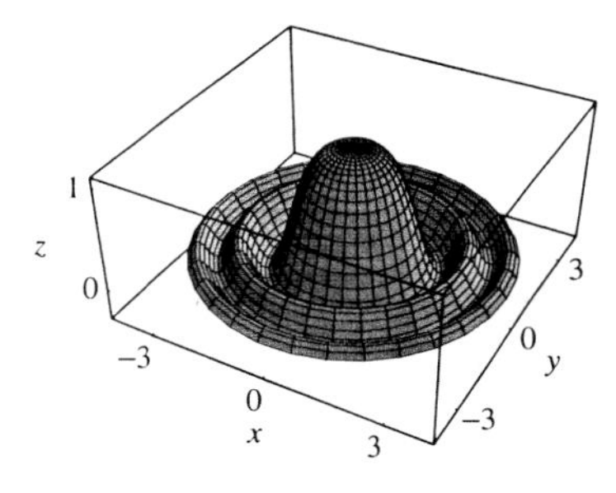

FIGURE 13.3.3 The graph $z = \dfrac{\sin(x^2+y^2)}{x^2+y^2}$ of Example 2.

EXAMPLE 2 The computer-generated graph in Fig. 13.3.3 suggests that

$$\lim_{(x,y)\to(0,0)} \frac{\sin(x^2+y^2)}{x^2+y^2} = 1.$$

Show that this is true.

Solution Here $a = b = 0$ and $L = 1$ in the definition of the limit. Given $\epsilon > 0$, we must find a value $\delta > 0$ such that

$$0 < \sqrt{x^2+y^2} < \delta \quad \text{implies that} \quad \left|\frac{\sin(x^2+y^2)}{x^2+y^2} - 1\right| < \epsilon.$$

But the familiar single-variable limit

$$\lim_{t\to 0} \frac{\sin t}{t} = 1$$

implies the existence of a number δ_1 such that

$$0 < |t| < \delta_1 \quad \text{implies that} \quad \left|\frac{\sin t}{t} - 1\right| < \epsilon.$$

When we substitute $t = x^2 + y^2$, we see that

$$0 < |x^2+y^2| < \delta_1 \quad \text{implies that} \quad \left|\frac{\sin(x^2+y^2)}{x^2+y^2} - 1\right| < \epsilon.$$

Hence we need only choose $\delta = \sqrt{\delta_1}$. Then

$$0 < \sqrt{x^2+y^2} < \delta \quad \text{implies that} \quad 0 < |x^2+y^2| < \delta^2 = \delta_1,$$

$$\text{which (in turn) implies that} \quad \left|\frac{\sin(x^2+y^2)}{x^2+y^2} - 1\right| < \epsilon,$$

as desired. ◆

Continuity and the Limit Laws

We frequently rely on continuity rather than the formal definition of the limit to evaluate limits of functions of several variables. We say that f is **continuous at the point** (a, b) provided that $f(a, b)$ exists and $f(x, y)$ approaches $f(a, b)$ as (x, y) approaches (a, b). That is,

$$\lim_{(x,y)\to(a,b)} f(x, y) = f(a, b).$$

Thus f is continuous at (a, b) if it is defined there and its limit there is equal to its value there, precisely as in the case of a function of a single variable. The function f is said to be **continuous on the set** D if it is continuous at each point of D, again exactly as in the single-variable case.

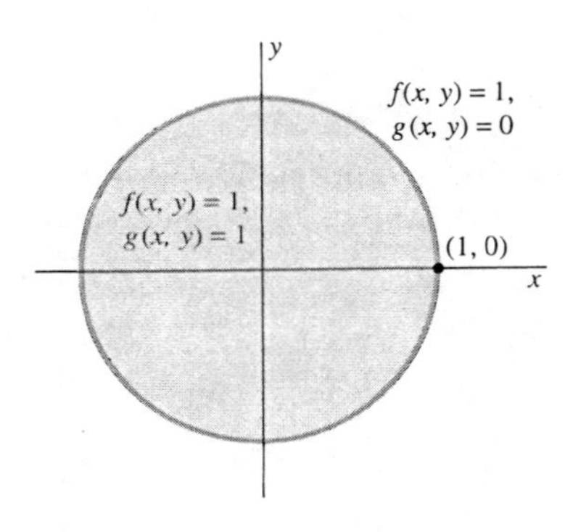

FIGURE 13.3.4 The circular disk of Example 3.

EXAMPLE 3 Let D be the circular disk consisting of the points (x, y) such that $x^2 + y^2 \leqq 1$ and let $f(x, y) = 1$ at each point of D (Fig. 13.3.4). Then the limit of $f(x, y)$ at each point of D is 1, so f is continuous on D. But let the new function $g(x, y)$ be defined on the entire plane $\boldsymbol{R}^2$ as follows:

$$g(x, y) = \begin{cases} f(x, y) & \text{if } (x, y) \text{ is in } D; \\ 0 & \text{otherwise.} \end{cases}$$

Then g is *not* continuous on $\boldsymbol{R}^2$. For instance, the limit of $g(x, y)$ as $(x, y) \to (1, 0)$ does not exist because there exist both points within D arbitrarily close to $(1, 0)$ at which g has the value 1 and points outside of D arbitrarily close to $(1, 0)$ at which g has the value 0. Thus $g(x, y)$ cannot approach any single value as $(x, y) \to (1, 0)$. Because g has no limit at $(1, 0)$, it cannot be continuous there. ◆

The limit laws of Section 2.2 have natural analogues for functions of several variables. If

$$\lim_{(x,y)\to(a,b)} f(x, y) = L \quad \text{and} \quad \lim_{(x,y)\to(a,b)} g(x, y) = M, \tag{4}$$

then the sum, product, and quotient laws for limits are these:

$$\lim_{(x,y)\to(a,b)} [f(x, y) + g(x, y)] = L + M, \tag{5}$$

$$\lim_{(x,y)\to(a,b)} [f(x, y) \cdot g(x, y)] = L \cdot M, \tag{6}$$

and

$$\lim_{(x,y)\to(a,b)} \frac{f(x, y)}{g(x, y)} = \frac{L}{M} \quad \text{if } M \neq 0. \tag{7}$$

EXAMPLE 4 Show that $\displaystyle\lim_{(x,y)\to(a,b)} xy = ab$.

Solution We take $f(x, y) = x$ and $g(x, y) = y$. Then it follows from the definition of limit that

$$\lim_{(x,y)\to(a,b)} f(x, y) = a \quad \text{and} \quad \lim_{(x,y)\to(a,b)} g(x, y) = b.$$

Hence the product law gives

$$\begin{aligned} \lim_{(x,y)\to(a,b)} xy &= \lim_{(x,y)\to(a,b)} [f(x, y)g(x, y)] \\ &= \left[\lim_{(x,y)\to(a,b)} f(x, y)\right]\left[\lim_{(x,y)\to(a,b)} g(x, y)\right] = ab. \end{aligned}$$

◆

More generally, suppose that $P(x, y)$ is a polynomial in the two variables x and y. That is, $P(x, y)$ is a sum of constant multiples of the form $x^i y^j$ where the exponents i and j are nonnegative integers. Thus $P(x, y)$ can be written in the form

$$P(x, y) = \sum c_{ij} x^i y^j.$$

The sum and product laws for limits then imply that

$$\begin{aligned}
\lim_{(x,y)\to(a,b)} P(x, y) &= \lim_{(x,y)\to(a,b)} \sum c_{ij} x^i y^j \\
&= \sum \left(\lim_{(x,y)\to(a,b)} c_{ij} x^i y^j \right) \\
&= \sum c_{ij} \left(\lim_{x\to a} x^i \right) \left(\lim_{y\to b} y^k \right) \\
&= \sum c_{ij} a^i b^j = P(a, b).
\end{aligned}$$

It follows that *every polynomial in two (or more) variables is an everywhere continuous function.*

EXAMPLE 5 The function $f(x, y) = 2x^4y^2 - 7xy + 4x^2y^3 - 5$ is a polynomial, so we can find its limit at any point (a, b) simply by evaluating $f(a, b)$. For instance,

$$\lim_{(x,y)\to(-1,2)} f(x, y) = f(-1, 2) = 2\cdot(-1)^4(2)^2 - 7\cdot(-1)(2) + 4\cdot(-1)^2(2)^3 - 5 = 49.$$

◆

Just as in the single-variable case, any composition of continuous multivariable functions is also a continuous function. For example, suppose that the functions f and g are both continuous at (a, b) and that h is continuous at the point $(f(a, b), g(a, b))$. Then the composite function

$$H(x, y) = h(f(x, y), g(x, y))$$

is also continuous at (a, b). As a consequence, any finite combination involving sums, products, quotients, and compositions of the familiar elementary functions is continuous, except possibly at points where a denominator is zero or where the formula for the function is otherwise meaningless. This general rule suffices for the evaluation of most limits that we shall encounter.

EXAMPLE 6 The function $g(x, y) = \sin(x^2 + y^2)$ is the composition of the continuous function $\sin t$ and the polynomial $x^2 + y^2$, and is therefore continuous everywhere. Hence the function f defined by

$$f(x, y) = \begin{cases} \dfrac{\sin(x^2 + y^2)}{x^2 + y^2} & \text{unless } x = y = 0, \\ 1 & \text{if } x = y = 0 \end{cases}$$

is continuous except possibly at the origin $(0, 0)$, where the denominator is zero. But we saw in Example 2 that

$$\lim_{(x,y)\to(0,0)} f(x, y) = 1 = f(0, 0),$$

so f is continuous at the origin as well. Thus the function f is continuous everywhere.

◆

EXAMPLE 7 If

$$f(x, y) = e^{xy} \sin \frac{\pi y}{4} + xy \ln \sqrt{y - x},$$

then e^{xy} is the composition of continuous functions, thus continuous; $\sin \frac{1}{4}\pi y$ is continuous for the same reason; their product is continuous because each is continuous.

Also $y - x$, a polynomial, is continuous everywhere; $\sqrt{y-x}$ is therefore continuous if $y \geqq x$; $\ln \sqrt{y-x}$ is continuous provided that $y > x$; $xy \ln \sqrt{y-x}$ is the product of functions continuous if $y > x$. And thus the sum

$$f(x, y) = e^{xy} \sin \frac{\pi y}{4} + xy \ln \sqrt{y-x}$$

of functions continuous if $y > x$ is itself continuous if $y > x$. Because $f(x, y)$ is continuous if $y > x$, it follows that

$$\lim_{(x,y)\to(1,2)} \left[e^{xy} \sin \frac{\pi y}{4} + xy \ln \sqrt{y-x} \right] = f(1, 2) = e^2 \cdot 1 + 2 \ln 1 = e^2. \qquad \blacklozenge$$

Examples 8 and 9 illustrate techniques that sometimes are successful in handling cases with denominators that approach zero; in such cases the techniques of Examples 5 through 7 cannot be applied.

EXAMPLE 8 Show that $\displaystyle\lim_{(x,y)\to(0,0)} \frac{xy}{\sqrt{x^2+y^2}} = 0.$

Solution Let (r, θ) be the polar coordinates of the point (x, y). Then $x = r\cos\theta$ and $y = r\sin\theta$, so

$$\frac{xy}{\sqrt{x^2+y^2}} = \frac{(r\cos\theta)(r\sin\theta)}{\sqrt{r^2(\cos^2\theta + \sin^2\theta)}} = r\cos\theta\sin\theta \quad \text{for } r > 0.$$

Because $r = \sqrt{x^2+y^2}$, it is clear that $r \to 0$ as both x and y approach zero. It therefore follows that

$$\lim_{(x,y)\to(0,0)} \frac{xy}{\sqrt{x^2+y^2}} = \lim_{r\to 0} r\cos\theta\sin\theta = 0,$$

because $|\cos\theta\sin\theta| \leqq |\cos\theta|\cdot|\sin\theta| \leqq 1$ for all θ. So if the function f is defined as

$$f(x, y) = \begin{cases} \dfrac{xy}{\sqrt{x^2+y^2}} & \text{if } (x, y) \neq (0, 0), \\ 0 & \text{if } x = y = 0, \end{cases}$$

then it follows that f is continuous at the origin $(0, 0)$. Figure 13.3.5 shows the graph of $z = f(x, y)$. It corroborates the zero limit that we found at $(0, 0)$. Near the origin the graph appears to resemble the saddle point on a hyperbolic paraboloid (Fig. 13.2.13), but this doesn't look like a smooth and comfortable saddle. $\blacklozenge$

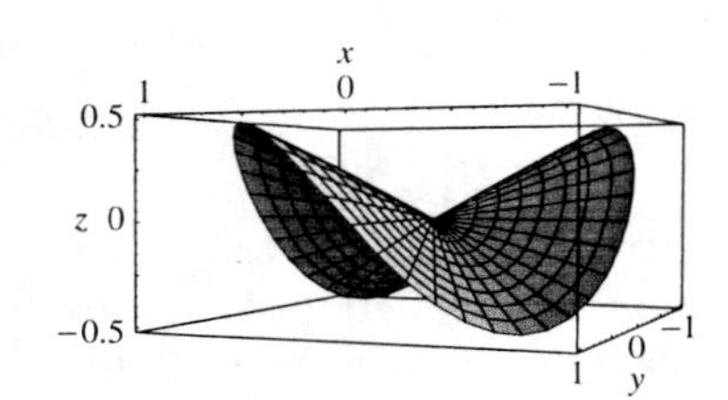

FIGURE 13.3.5 The graph $z = \dfrac{xy}{\sqrt{x^2+y^2}}$ (Example 8).

EXAMPLE 9 Show that

$$\lim_{(x,y)\to(0,0)} \frac{xy}{x^2+y^2}$$

does not exist.

Solution Our plan is to show that $f(x, y) = xy/(x^2+y^2)$ approaches different values as (x, y) approaches $(0, 0)$ from different directions. Suppose that (x, y) approaches $(0, 0)$ along the straight line of slope m through the origin. On this line we have $y = mx$. So, on this line,

$$f(x, y) = f(x, mx) = \frac{x \cdot mx}{x^2 + m^2x^2} = \frac{m}{1+m^2}$$

if $x \neq 0$. If we take $m = 1$, we see that $f(x, y) = \frac{1}{2}$ at every point of the line $y = x$ other than $(0, 0)$. If we take $m = -1$, then $f(x, y) = -\frac{1}{2}$ at every point of the line $y = -x$ other than $(0, 0)$. Thus $f(x, y)$ approaches two different values as (x, y) approaches $(0, 0)$ along these two lines (Fig. 13.3.6). Hence $f(x, y)$ cannot approach any *single* value as (x, y) approaches $(0, 0)$, and this implies that the limit in question cannot exist.

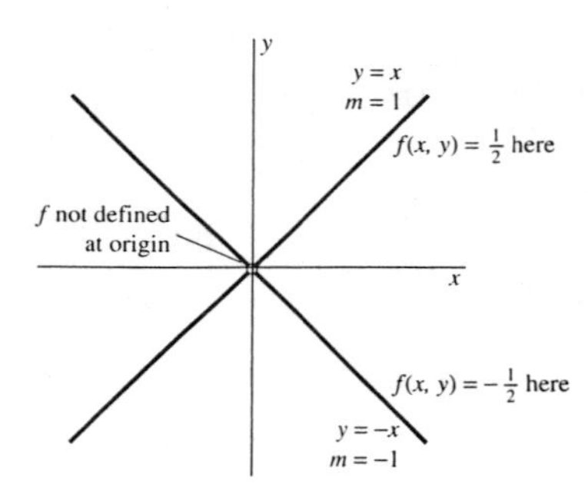

FIGURE 13.3.6 The function f of Example 9 takes on both values $+\frac{1}{2}$ and $-\frac{1}{2}$ at points arbitrarily close to the origin.

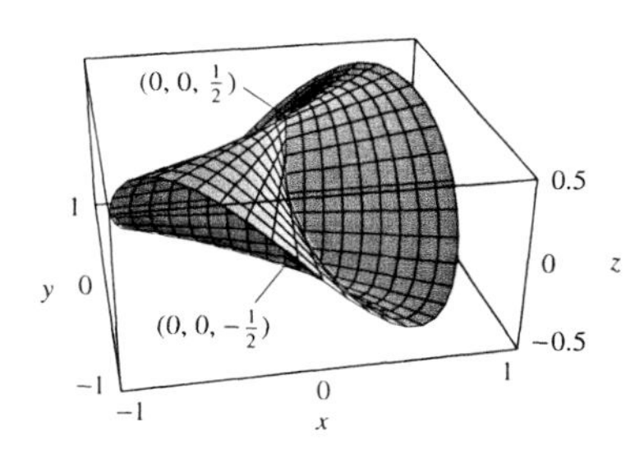

FIGURE 13.3.7 The graph of $f(x, y) = \dfrac{xy}{x^2 + y^2}$ (Example 9).

Figure 13.3.7 shows a computer-generated graph of the function $f(x, y) = xy/(x^2 + y^2)$. It consists of linear rays along each of which the polar angular coordinate θ is constant. For each number z between $-\frac{1}{2}$ and $\frac{1}{2}$ (inclusive), there are rays along which $f(x, y)$ has the constant value z. Hence we can make $f(x, y)$ approach any number we please in $[-\frac{1}{2}, \frac{1}{2}]$ by letting (x, y) approach $(0, 0)$ from the appropriate direction. There are also paths along which (x, y) approaches $(0, 0)$ but the limit of $f(x, y)$ does not exist (Problem 53). ◆

REMARK In order for

$$L = \lim_{(x,y)\to(a,b)} f(x, y)$$

to exist, $f(x, y)$ must approach L for *any and every* mode of approach of (x, y) to (a, b). In Problem 51 we give an example of a function f such that $f(x, y) \to 0$ as $(x, y) \to (0, 0)$ along any straight line through the origin, but $f(x, y) \to 1$ as $(x, y) \to (0, 0)$ along the parabola $y = x^2$. Thus the method of Example 9 cannot be used to show that a limit exists, only that it does not. Fortunately, many important applications, including those we discuss in the remainder of this chapter, involve only functions that exhibit no such exotic behavior as the functions of Problems 51 through 53.

Functions of Three or More Variables

Thus far in this section, we have discussed explicitly only functions of two variables, but the concepts of limits and continuity generalize in a straightforward manner to functions of three or more variables. A **function f of n variables** assigns a single real number $f(x_1, x_2, \ldots, x_n)$ to an n-tuple $(x_1, x_2, \ldots, x_n)$ of real numbers. For instance, the function f might assign to the 4-tuple (x, y, z, t) the temperature $u = f(x, y, z, t)$ at time t at the point (x, y, z) in three-dimensional space.

Just as three-dimensional space $\boldsymbol{R}^3$ is the set of all triples (x_1, x_2, x_3) of real numbers, **n-dimensional space $\boldsymbol{R}^n$** is the set of all n-tuples of real numbers. Thus the temperature function mentioned earlier is defined on four-dimensional space $\boldsymbol{R}^4$. We may therefore write $f : \boldsymbol{R}^4 \to \boldsymbol{R}$, with time t playing the role of the fourth dimension (but without the fanciful implications sometimes enjoyed in science fiction).

It is common practice to identify the n-tuple $(x_1, x_2, \ldots, x_n)$ with the vector $\mathbf{x} = \langle x_1, x_2, \ldots, x_n \rangle$—regarding each notation as simply a way of specifying the same ordered list $x_1, x_2, \ldots, x_n$ of real numbers. Then we may also regard $\boldsymbol{R}^n$ as the set of all n-vectors. This viewpoint enables us to add points in $\boldsymbol{R}^n$ coordinatewise as n-vectors, and similarly to multiply points by scalars. In analogy with lengths of vectors in $\boldsymbol{R}^2$ and $\boldsymbol{R}^3$, we define the **length** $|\mathbf{x}|$ of the vector $\mathbf{x}$ in $\boldsymbol{R}^n$ to be

$$|\mathbf{x}| = \sqrt{x_1^2 + x_2^2 + \cdots + x_n^2}.$$

For instance, the 4-vector $\langle 5, -2, 4, 2 \rangle$ has length $\sqrt{25 + 4 + 16 + 4} = \sqrt{49} = 7$.

The function $f : \boldsymbol{R}^n \to \boldsymbol{R}$ may be regarded either as a function of the n independent real variables $x_1, x_2, \ldots, x_n$ or as a function of the single n-vector $\mathbf{x} = \langle x_1, x_2, \ldots, x_n \rangle$. We may then write either $f(x_1, x_2, \ldots, x_n)$ or $f(\mathbf{x})$, depending on which notation seems most natural in a given situation. For instance, with vector notation the limit concept takes the form of the statement that

$$\lim_{\mathbf{x}\to\mathbf{a}} f(\mathbf{x}) = L \tag{8}$$

provided that, for every number $\epsilon > 0$, there exists a corresponding number $\delta > 0$ such that

$$|f(\mathbf{x}) - L| < \epsilon \quad \text{whenever} \quad 0 < |\mathbf{x} - \mathbf{a}| < \delta. \tag{9}$$

Then the function f is continuous at the point $\mathbf{a} = (a_1, a_2, \ldots, a_n)$ provided that

$$\lim_{\mathbf{x}\to\mathbf{a}} f(\mathbf{x}) = f(\mathbf{a}). \quad \textbf{(10)}$$

An attractive feature of vector notation is that the multidimensional statements in (8), (9), and (10) take precisely the same forms as in the case of functions of a single variable, as do the multidimensional limit laws. (See the discussion questions for this section.)

13.3 TRUE/FALSE STUDY GUIDE

13.3 CONCEPTS: QUESTIONS AND DISCUSSION

1. Give precise statements of the limit laws for real-valued functions of three or more variables. Explain why any polynomial in three or more variables is continuous everywhere.
2. State precisely the general principle of the continuity of compositions of continuous multivariable functions. It should apply, for instance, to a function of four variables each of which is itself a function of three variables.
3. Explain how the reasoning of Examples 2 and 6 applies to the function F defined except at the origin in $\boldsymbol{R}^3$ by

$$F(x, y, z) = \frac{\sin(x^2 + y^2 + z^2)}{x^2 + y^2 + z^2}.$$

What are your conclusions?

4. Give several concrete examples of real-world functions of four or more variables.

13.3 PROBLEMS

Use the limit laws and consequences of continuity to evaluate the limits in Problems 1 through 16.

1. $\lim_{(x,y)\to(0,0)} (7 - x^2 + 5xy)$

2. $\lim_{(x,y)\to(1,-2)} (3x^2 - 4xy + 5y^2)$

3. $\lim_{(x,y)\to(1,-1)} e^{-xy}$

4. $\lim_{(x,y)\to(0,0)} \dfrac{x+y}{1+xy}$

5. $\lim_{(x,y)\to(0,0)} \dfrac{5-x^2}{3+x+y}$

6. $\lim_{(x,y)\to(2,3)} \dfrac{9-x^2}{1+xy}$

7. $\lim_{(x,y)\to(0,0)} \ln\sqrt{1-x^2-y^2}$

8. $\lim_{(x,y)\to(2,-1)} \ln\dfrac{1+x+2y}{3y^2-x}$

9. $\lim_{(x,y)\to(0,0)} e^{x+2y}\cos(3x+4y)$

10. $\lim_{(x,y)\to(0,0)} \dfrac{\cos(x^2+y^2)}{1-x^2-y^2}$

11. $\lim_{(x,y,z)\to(1,1,1)} \dfrac{x^2+y^2+z^2}{1-x-y-z}$

12. $\lim_{(x,y,z)\to(1,1,1)} (x+y+z)\ln xyz$

13. $\lim_{(x,y,z)\to(1,1,0)} \dfrac{xy-z}{\cos xyz}$

14. $\lim_{(x,y,z)\to(2,-1,3)} \dfrac{x+y+z}{x^2+y^2+z^2}$

15. $\lim_{(x,y,z)\to(2,8,1)} \sqrt{xy}\,\tan\dfrac{3\pi z}{4}$

16. $\lim_{(x,y)\to(1,-1)} \arcsin\dfrac{xy}{\sqrt{x^2+y^2}}$

In Problems 17 through 20, evaluate the limits

$$\lim_{h\to 0}\frac{f(x+h, y) - f(x, y)}{h} \quad \text{and} \quad \lim_{k\to 0}\frac{f(x, y+k) - f(x, y)}{k}.$$

17. $f(x, y) = xy$

18. $f(x, y) = x^2 + y^2$

19. $f(x, y) = xy^2 - 2$

20. $f(x, y) = x^2y^3 - 10$

In Problems 21 through 30, find the limit or show that it does not exist.

21. $\lim_{(x,y)\to(1,1)} \dfrac{1-xy}{1+xy}$

22. $\lim_{(x,y)\to(2,-2)} \dfrac{4-xy}{4+xy}$

23. $\lim_{(x,y,z)\to(1,1,1)} \dfrac{xyz}{yz+xz+xy}$

24. $\lim_{(x,y,z)\to(1,-1,1)} \dfrac{yz+xz+xy}{1+xyz}$

25. $\lim_{(x,y)\to(0,0)} \ln(1+x^2+y^2)$

26. $\lim_{(x,y)\to(1,1)} \ln(2-x^2-y^2)$

27. $\lim_{(x,y)\to(0,0)} \dfrac{\cot(x^2+y^2)}{x^2+y^2}$

28. $\lim_{(x,y)\to(0,0)} \sin(\ln(1+x+y))$

29. $\lim_{(x,y)\to(0,0)} \exp\left(-\dfrac{1}{x^2+y^2}\right)$

30. $\lim_{(x,y)\to(0,0)} \arctan\left(-\dfrac{1}{x^2+y^2}\right)$

In Problems 31 through 36, determine the largest set of points in the xy-plane on which the given formula defines a continuous function.

31. $f(x, y) = \sqrt{x+y}$

32. $f(x, y) = \sin^{-1}(x^2+y^2)$

33. $f(x, y) = \ln(x^2+y^2-1)$

34. $f(x, y) = \ln(2x-y)$

35. $f(x, y) = \tan^{-1}\left(\dfrac{1}{x^2+y^2}\right)$

36. $f(x, y) = \tan^{-1}\left(\dfrac{1}{x+y}\right)$

In Problems 37 through 40, evaluate the limit by making the polar coordinates substitution $(x, y) = (r\cos\theta, r\sin\theta)$ *and using the fact that* $r \to 0$ *as* $(x, y) \to (0, 0)$.

37. $\lim_{(x,y)\to(0,0)} \dfrac{x^2-y^2}{\sqrt{x^2+y^2}} = 0$

38. $\lim_{(x,y)\to(0,0)} \dfrac{x^3-y^3}{x^2+y^2} = 0$

39. $\lim_{(x,y)\to(0,0)} \dfrac{x^4+y^4}{(x^2+y^2)^{3/2}} = 0$

40. $\lim_{(x,y)\to(0,0)} \dfrac{\sin\sqrt{x^2+y^2}}{\sqrt{x^2+y^2}}$

41. Determine whether or not

$$\lim_{(x,y,z)\to(0,0,0)} \frac{xyz}{x^2+y^2+z^2}$$

exists; evaluate it if it does exist. [*Suggestion:* Substitute spherical coordinates $x = \rho\sin\phi\cos\theta$, $y = \rho\sin\phi\sin\theta$, $z = \rho\cos\phi$.]

42. Determine whether or not

$$\lim_{(x,y,z)\to(0,0,0)} \arctan\frac{1}{x^2+y^2+z^2}$$

exists; evaluate it if it does exist. [See the *Suggestion* for Problem 41.]

In Problems 43 and 44, investigate the existence of the given limit by making the substitution $y = mx$.

43. $\lim_{(x,y)\to(0,0)} \dfrac{x^2-y^2}{x^2+y^2}$

44. $\lim_{(x,y)\to(0,0)} \dfrac{x^4-y^4}{x^4+x^2y^2+y^4}$

In Problems 45 and 46, show that the given limit does not exist by considering points of the form $(x, 0, 0)$ *or* $(0, y, 0)$ *or* $(0, 0, z)$ *that approach the origin along one of the coordinate axes.*

45. $\lim_{(x,y,z)\to(0,0,0)} \dfrac{x+y+z}{x^2+y^2+z^2}$

46. $\lim_{(x,y,z)\to(0,0,0)} \dfrac{x^2+y^2-z^2}{x^2+y^2+z^2}$

In Problems 47 through 50, use a computer-plotted graph to explain why the given limit does not exist.

47. $\lim_{(x,y)\to(0,0)} \dfrac{x^2-2y^2}{x^2+y^2}$

48. $\lim_{(x,y)\to(0,0)} \dfrac{x^2y^2}{x^4+y^4}$

49. $\lim_{(x,y)\to(0,0)} \dfrac{xy}{2x^2+3y^2}$

50. $\lim_{(x,y)\to(0,0)} \dfrac{x^2+4xy+y^2}{x^2+xy+y^2}$

51. Let

$$f(x, y) = \frac{2x^2y}{x^4+y^2}.$$

(a) Show that $f(x, y) \to 0$ as $(x, y) \to (0, 0)$ along any and every straight line through the origin. (b) Show that $f(x, y) \to 1$ as $(x, y) \to (0, 0)$ along the parabola $y = x^2$. Conclude that the limit of $f(x, y)$ as $(x, y) \to (0, 0)$ does not exist. The graph of f is shown in Fig. 13.3.8.

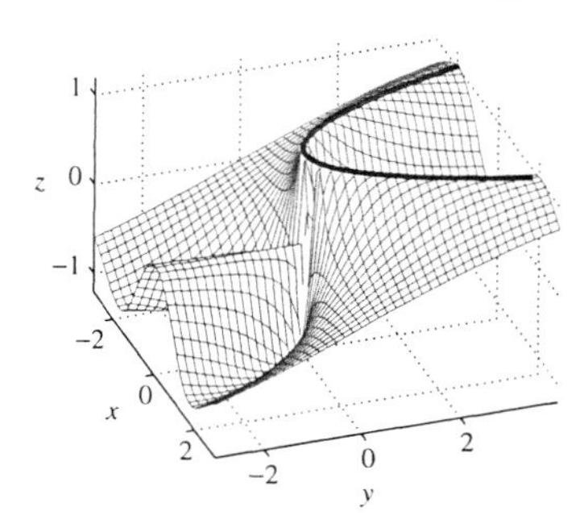

FIGURE 13.3.8 The graph $z = \dfrac{2x^2y}{x^4+y^2}$ of Problem 51; note the curve $y = x^2$, $z = 1$.

52. Suppose that $f(x, y) = (x-y)/(x^3-y)$ except at points of the curve $y = x^3$, where we *define* $f(x, y)$ to be 1. Show that f is not continuous at the point $(1, 1)$. Evaluate the limits of $f(x, y)$ as $(x, y) \to (1, 1)$ along the vertical line $x = 1$ and along the horizontal line $y = 1$. [*Suggestion:* Recall that $a^3 - b^3 = (a-b)(a^2+ab+b^2)$.]

53. Let

$$\lim_{(x,y)\to(0,0)} \frac{xy}{x^2+y^2}$$

be the limit in Example 9. Show that as $(x, y) \to (0, 0)$ along the hyperbolic spiral $r\theta = 1$, the limit of $f(x, y)$ does not exist.

Discuss the continuity of the functions defined in Problems 54 through 56.

54. $f(x, y) = \begin{cases} \dfrac{\sin xy}{xy} & \text{unless } xy = 0, \\ 1 & \text{if } xy = 0. \end{cases}$ (See Fig. 13.3.9.)

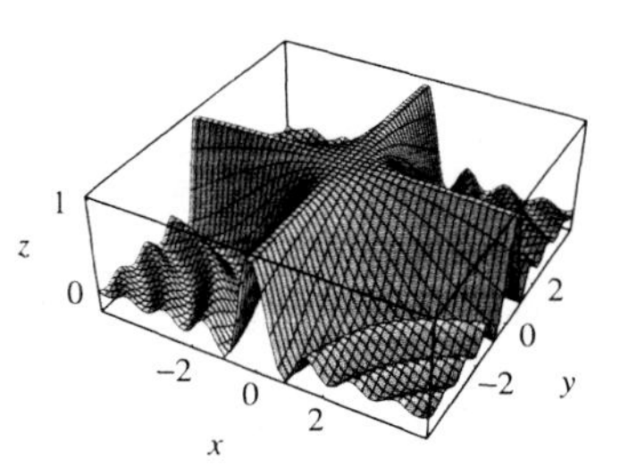

FIGURE 13.3.9 The graph $z = \dfrac{\sin xy}{xy}$ of Problem 54.

55. $g(x, y) = \begin{cases} \dfrac{\sin(x^2-y^2)}{x^2-y^2} & \text{unless } x^2 = y^2, \\ 1 & \text{if } x^2 = y^2. \end{cases}$

56. $h(x, y, z) = \begin{cases} \dfrac{\sin xyz}{xyz} & \text{unless } xyz = 0, \\ 1 & \text{if } xyz = 0. \end{cases}$

13.4 PARTIAL DERIVATIVES

Recall that the derivative of the single-variable function $u = g(x)$ is defined as

$$\frac{du}{dx} = \lim_{\Delta x \to 0} \frac{\Delta u}{\Delta x},$$

where $\Delta u = g(x + h) - g(x)$ is the change in u resulting from the change $h = \Delta x$ in x. This derivative is interpreted as the instantaneous rate of change of u with respect to x. For a function $z = f(x, y)$ of two variables, we need a similar understanding of the rate at which z changes as x and y vary (either singly or simultaneously).

We take a divide-and-conquer approach to this concept. If x is changed by $h = \Delta x$ but y is not changed, then the resulting change in z is

$$\Delta z = f(x + h, y) - f(x, y),$$

and the corresponding instantaneous rate of change of z is

$$\frac{dz}{dx} = \lim_{\Delta x \to 0} \frac{\Delta z}{\Delta x}. \tag{1}$$

On the other hand, if x is not changed but y is changed by the amount $k = \Delta y$, then the resulting change in z is

$$\Delta z = f(x, y + k) - f(x, y),$$

and the corresponding instantaneous rate of change of z is

$$\frac{dz}{dy} = \lim_{\Delta y \to 0} \frac{\Delta z}{\Delta y}. \tag{2}$$

The limits in Eqs. (1) and (2) are the *two* **partial derivatives** of the function $f(x, y)$ with respect to its two independent variables x and y, respectively.

DEFINITION Partial Derivatives

The **partial derivatives** (**with respect to** x **and with respect to** y) of the function $f(x, y)$ are the two functions defined by

$$f_x(x, y) = \lim_{h \to 0} \frac{f(x + h, y) - f(x, y)}{h}, \tag{3}$$

$$f_y(x, y) = \lim_{k \to 0} \frac{f(x, y + k) - f(x, y)}{k} \tag{4}$$

whenever these limits exist.

Note that Eqs. (3) and (4) are simply restatements of Eqs. (1) and (2). Just as with single-variable derivatives, there are several alternative ways of writing partial derivatives.

Notation for Partial Derivatives

If $z = f(x, y)$, then we may express its partial derivatives with respect to x and y, respectively, in these forms:

$$\frac{\partial z}{\partial x} = \frac{\partial f}{\partial x} = f_x(x, y) = \frac{\partial}{\partial x} f(x, y) = D_x[f(x, y)] = D_1[f(x, y)], \tag{5}$$

$$\frac{\partial z}{\partial y} = \frac{\partial f}{\partial y} = f_y(x, y) = \frac{\partial}{\partial y} f(x, y) = D_y[f(x, y)] = D_2[f(x, y)]. \tag{6}$$

Computer algebra systems generally employ variants of the "operator notation" for partial derivatives, such as `diff(f(x,y), x)` and `D[f[x,y], x]` in *Maple* and *Mathematica,* respectively.

Note that if we delete the symbol y throughout Eq. (3), the result is the limit that defines the single-variable derivative $f'(x)$. This means that we can calculate $\partial z/\partial x$ as an "ordinary" derivative with respect to x simply by regarding y as a constant during the process of differentiation. Similarly, we can compute $\partial z/\partial y$ as an ordinary derivative by thinking of y as the *only* variable and treating x as a constant during the computation.

Consequently, we seldom need to evaluate directly the limits in Eqs. (3) and (4) in order to calculate partial derivatives. Ordinarily we simply apply familiar differentiation results to differentiate $f(x, y)$ with respect to either independent variable (x or y) while holding the other variable constant. In short,

- To calculate $\partial f/\partial x$, regard y as a constant and differentiate with respect to x.
- To calculate $\partial f/\partial y$, regard x as a constant and differentiate with respect to y.

EXAMPLE 1 Compute both the partial derivatives $\partial f/\partial x$ and $\partial f/\partial y$ of the function $f(x, y) = x^2 + 2xy^2 - y^3$.

Solution To compute the partial derivative of f with respect to x, we regard y as a constant. Then we differentiate normally and find that

$$\frac{\partial f}{\partial x} = 2x + 2y^2.$$

When we regard x as a constant and differentiate with respect to y, we find that

$$\frac{\partial f}{\partial y} = 4xy - 3y^2.$$ ◆

EXAMPLE 2 Find $\partial z/\partial x$ and $\partial z/\partial y$ if $z = (x^2 + y^2)e^{-xy}$.

Solution Because $\partial z/\partial x$ is calculated as if it were an ordinary derivative with respect to x, with y held constant, we use the product rule. This gives

$$\frac{\partial z}{\partial x} = (2x)(e^{-xy}) + (x^2 + y^2)(-ye^{-xy}) = (2x - x^2y - y^3)e^{-xy}.$$

Because x and y appear symmetrically in the expression for z, we get $\partial z/\partial y$ when we interchange x and y in the expression for $\partial z/\partial x$:

$$\frac{\partial z}{\partial y} = (2y - xy^2 - x^3)e^{-xy}.$$

You should check this result by differentiating with respect to y directly in order to find $\partial z/\partial y$. ◆

Instantaneous Rates of Change

To get an intuitive feel for the meaning of partial derivatives, we can think of $f(x, y)$ as the temperature at the point (x, y) of the plane. Then $f_x(x, y)$ is the instantaneous rate of change of temperature at (x, y) per unit increase in x (with y held constant). Similarly, $f_y(x, y)$ is the instantaneous rate of change of temperature per unit increase in y (with x held constant). For instance, we show in the next example that, with the temperature function $f(x, y) = x^2 + 2xy^2 - y^3$ of Example 1, the rate of change of temperature at the point $(1, -1)$ is $+4^\circ$ per unit distance in the positive x-direction and -7° per unit distance in the positive y-direction.

EXAMPLE 3 Suppose that the xy-plane is somehow heated and that its temperature at the point (x, y) is given by the function $f(x, y) = x^2 + 2xy^2 - y^3$, whose partial derivatives $f_x(x, y) = 2x + 2y^2$ and $f_y(x, y) = 4xy - 3y^2$ were calculated in Example 1. Suppose also that distance is measured in miles and temperature in degrees Celsius (°C). Then at the point $(1, -1)$, one mile east and one mile south of the origin, the rate of change of temperature (in degrees per mile) in the (eastward) positive

x-direction is

$$f_x(1, -1) = 2 \cdot (1) + 2 \cdot (-1)^2 = 4 \quad \text{(deg/mi)},$$

and the rate of change in the (northward) positive y-direction is

$$f_y(1, -1) = 4 \cdot 1 \cdot (-1) - 3 \cdot (-1)^2 = -7 \quad \text{(deg/mi)}.$$

Thus, if we start at the point $(1, -1)$ and walk $\frac{1}{10}$ mi east, we expect to experience a temperature increase of about $4 \cdot (0.1) = 0.4°\text{C}$. If instead we started at $(1, -1)$ and walked 0.2 mi north, we would expect to experience a temperature change of about $(-7) \cdot (0.2) = -1.4°\text{C}$; that is, a temperature decrease of about 1.4°C. ◆

EXAMPLE 4 The volume V (in cubic centimeters) of 1 mole (mol) of an ideal gas is given by

$$V = \frac{(82.06)T}{p},$$

where p is the pressure (in atmospheres) and T is the absolute temperature (in kelvins (K), where $\text{K} = °\text{C} + 273$). Find the rates of change of the volume of 1 mol of an ideal gas with respect to pressure and with respect to temperature when $T = 300$ K and $p = 5$ atm.

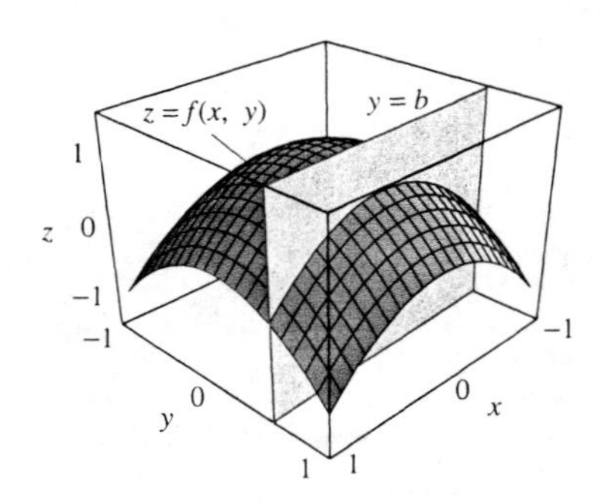

FIGURE 13.4.1 A vertical plane parallel to the xz-plane intersects the surface $z = f(x, y)$ in an x-curve.

Solution The partial derivatives of V with respect to its two variables are

$$\frac{\partial V}{\partial p} = -\frac{(82.06)T}{p^2} \quad \text{and} \quad \frac{\partial V}{\partial T} = \frac{82.06}{p}.$$

With $T = 300$ and $p = 5$, we have the values $\partial V/\partial p = -984.72$ (cm^3/atm) and $\partial V/\partial T = 16.41$ (cm^3/K). These partial derivatives allow us to estimate the effect of a small change in temperature or in pressure on the volume V of the gas, as follows. We are given $T = 300$ and $p = 5$, so the volume of gas with which we are dealing is

$$V = \frac{(82.06)(300)}{5} = 4923.60 \quad (\text{cm}^3).$$

We would expect an increase in pressure of 1 atm (with T held constant) to decrease the volume of gas by appropriately 1 L (1000 cm^3), because $-984.72 \approx -1000$. An increase in temperature of 1 K (or 1°C) would, with p held constant, increase the volume by about 16 cm^3, because $16.41 \approx 16$. ◆

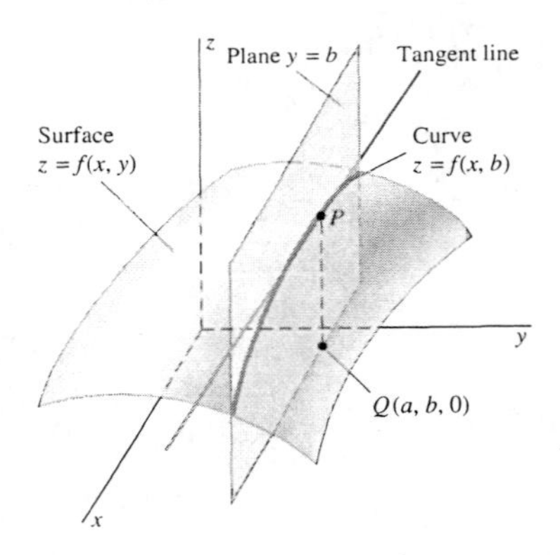

FIGURE 13.4.2 An x-curve and its tangent line at P.

Geometric Interpretation of Partial Derivatives

The partial derivatives f_x and f_y are the slopes of lines tangent to certain curves on the surface $z = f(x, y)$. Figure 13.4.1 illustrates the intersection of this surface with a vertical plane $y = b$ parallel to the xz-coordinate plane. Along the intersection curve, the x-coordinate varies but the y-coordinate is constant: $y = b$ at each point because the curve lies in the vertical plane $y = b$. A curve of intersection of $z = f(x, y)$ with a vertical plane parallel to the xz-plane is therefore called an **x-curve** on the surface.

Figure 13.4.2 shows a point $P(a, b, c)$ in the surface $z = f(x, y)$, the x-curve through P, and the line tangent to this x-curve at P. Figure 13.4.3 shows the parallel projection of the vertical plane $y = b$ onto the xz-plane itself. We can now "ignore" the presence of $y = b$ and regard $z = f(x, b)$ as a function of the *single* variable x. The slope of the line tangent to the original x-curve through P (see Fig. 13.4.2) is equal to the slope of the tangent line in Fig. 13.4.3. But by familiar single-variable calculus, this latter slope is given by

$$\lim_{h \to 0} \frac{f(a + h, b) - f(a, b)}{h} = f_x(a, b).$$

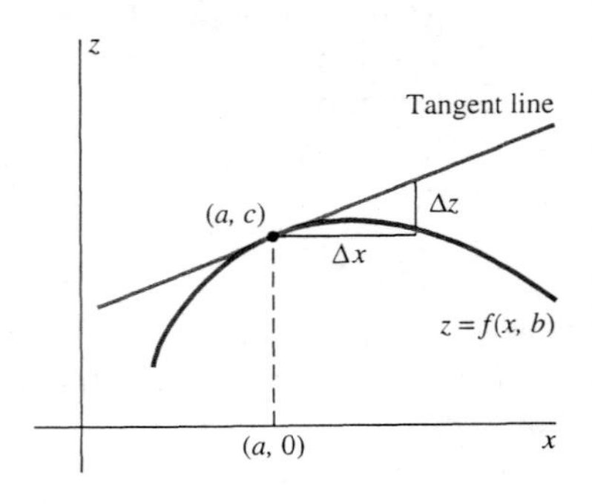

FIGURE 13.4.3 Projection into the xz-plane of the x-curve through $P(a, b, c)$ and its tangent line.

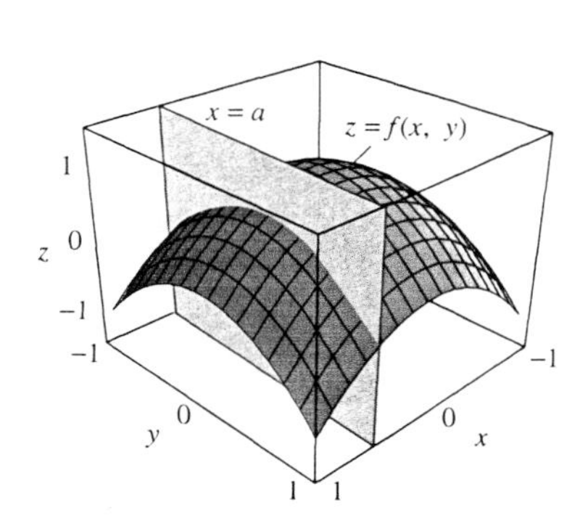

FIGURE 13.4.4 A vertical plane parallel to the yz-plane intersects the surface $z = f(x, y)$ in a y-curve.

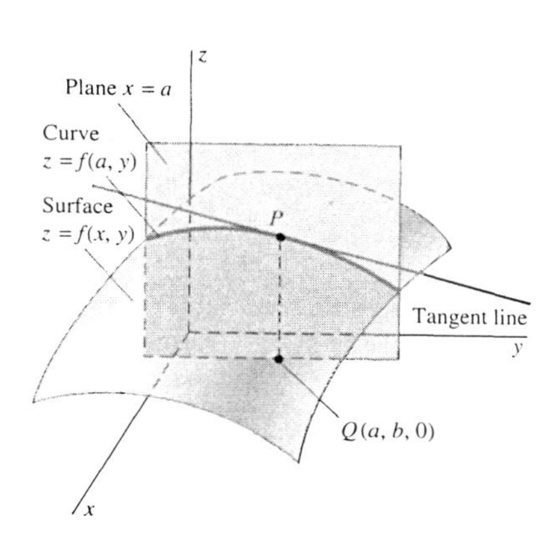

FIGURE 13.4.5 A y-curve and its tangent line at P.

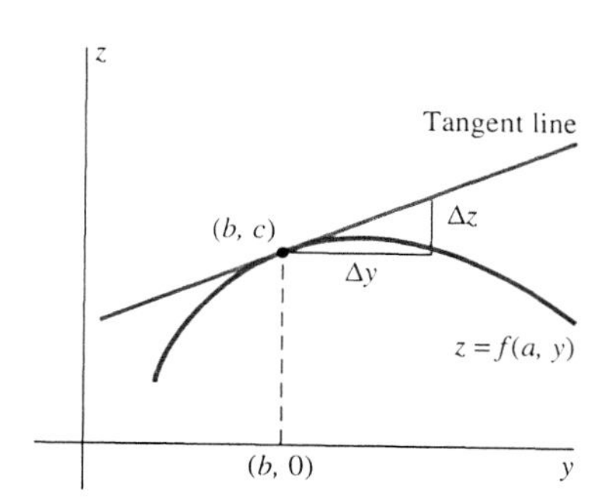

FIGURE 13.4.6 Projection into the yz-plane of the y-curve through $P(a, b, c)$ and its tangent line.

Thus we see that the geometric meaning of f_x is as follows:

Geometric Interpretation of $\partial f/\partial x$

The value $f_x(a, b)$ is the slope of the line tangent at $P(a, b, c)$ to the x-curve through P on the surface $z = f(x, y)$.

We proceed in much the same way to investigate the geometric meaning of partial differentiation with respect to y. Figure 13.4.4 illustrates the intersection with the surface $z = f(x, y)$ of a vertical plane $x = a$ parallel to the yz-coordinate plane. Now the curve of intersection is a y-**curve** along which y varies but $x = a$ is constant. Figure 13.4.5 shows this y-curve $z = f(a, y)$ and its tangent line at P. The projection of the tangent line in the yz-plane (in Fig. 13.4.6) has slope $\partial z/\partial y = f_y(a, b)$. Thus we see that the geometric meaning of f_y is as follows:

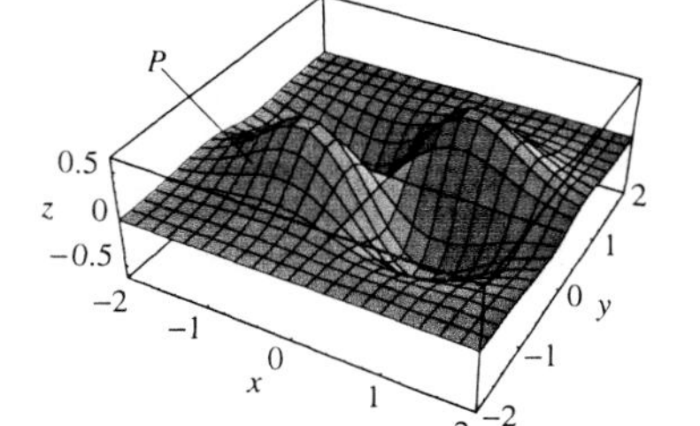

FIGURE 13.4.7 The graph $z = 5xy\exp(-x^2 - 2y^2)$.

Geometric Interpretation of $\partial f/\partial y$

The value $f_y(a, b)$ is the slope of the line tangent at $P(a, b, c)$ to the y-curve through P on the surface $z = f(x, y)$.

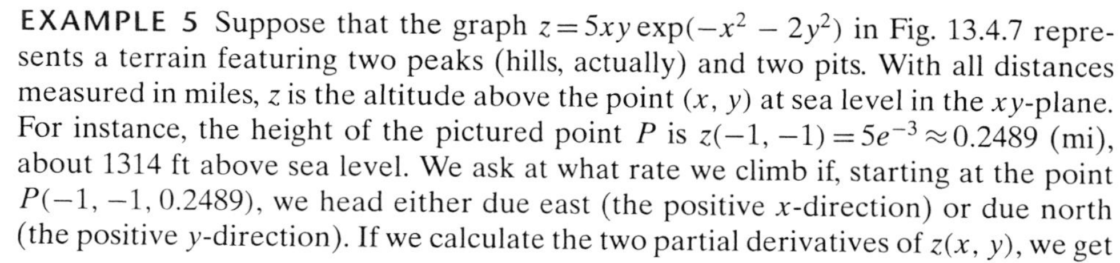

EXAMPLE 5 Suppose that the graph $z = 5xy\exp(-x^2 - 2y^2)$ in Fig. 13.4.7 represents a terrain featuring two peaks (hills, actually) and two pits. With all distances measured in miles, z is the altitude above the point (x, y) at sea level in the xy-plane. For instance, the height of the pictured point P is $z(-1, -1) = 5e^{-3} \approx 0.2489$ (mi), about 1314 ft above sea level. We ask at what rate we climb if, starting at the point $P(-1, -1, 0.2489)$, we head either due east (the positive x-direction) or due north (the positive y-direction). If we calculate the two partial derivatives of $z(x, y)$, we get

$$\frac{\partial z}{\partial x} = 5y(1 - 2x^2)\exp(-x^2 - 2y^2) \quad \text{and} \quad \frac{\partial z}{\partial y} = 5x(1 - 4y^2)\exp(-x^2 - 2y^2).$$

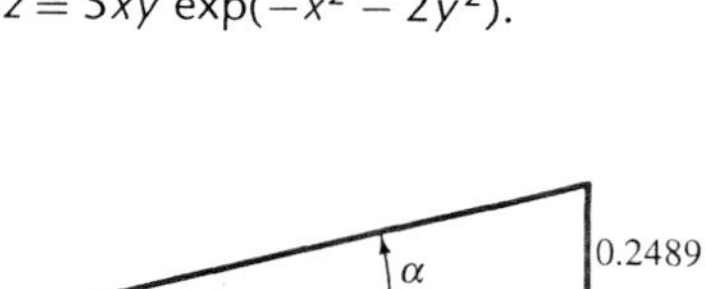

FIGURE 13.4.8 The angle of climb in the x-direction.

(You should check this.) Substituting $x = y = -1$ now gives

$$\left.\frac{\partial z}{\partial x}\right|_{(-1,-1)} = 5e^{-3} \approx 0.2489 \quad \text{and} \quad \left.\frac{\partial z}{\partial y}\right|_{(-1,-1)} = 15e^{-3} \approx 0.7468.$$

The units here are in miles per mile—that is, the ratio of rise to run in vertical miles per horizontal mile. So if we head east, we start climbing at an angle of

$$\alpha = \tan^{-1}(0.2489) \approx 0.2440 \quad \text{(rad)},$$

about 13.98°. (See Fig. 13.4.8.) But if we head north, then we start climbing at an angle of

$$\beta = \tan^{-1}(0.7468) \approx 0.6415 \quad \text{(rad)},$$

approximately 36.75°. (See Fig. 13.4.9.) Do these results appear to be consistent with Fig. 13.4.7? ◆

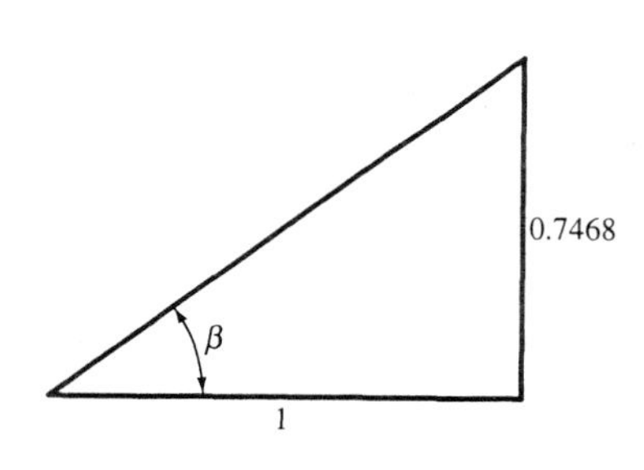

FIGURE 13.4.9 The angle of climb in the y-direction.

Planes Tangent to Surfaces

The two tangent lines illustrated in Figs. 13.4.2 and 13.4.5 determine a unique plane through the point $P(a, b, f(a, b))$. We will see in Section 13.7 that if the partial derivatives f_x and f_y are continuous functions of x and y, then this plane contains the line tangent at P to *every* smooth curve on the surface $z = f(x, y)$ that passes through P. This motivates the following definition of the plane tangent to the surface at P.

DEFINITION Plane Tangent to $z = f(x, y)$

Suppose that the function $f(x, y)$ has continuous partial derivatives on a circular disk centered at the point (a, b). Then the **plane tangent** to the surface $z = f(x, y)$ at the point $P(a, b, f(a, b))$ is the plane through P that contains the lines tangent at P to the two curves

$$z = f(x, b), \quad y = b \quad (x\text{-curve}) \tag{7}$$

and

$$z = f(a, y), \quad x = a \quad (y\text{-curve}). \tag{8}$$

To find an equation of this tangent plane at the point $P(a, b, c)$ where $c = f(a, b)$, recall from Section 12.4 that a typical nonvertical plane in space that passes through the point P has an equation of the form

$$A(x - a) + B(y - b) + C(z - c) = 0 \tag{9}$$

where $C \neq 0$. If we solve for $z - c$ we get the equation

$$z - c = p(x - a) + q(y - b) \tag{10}$$

where $p = -A/C$ and $q = -B/C$. This plane will be tangent to the surface $z = f(x, y)$ at the point $P(a, b, c)$ provided that the line defined in Eq. (10) with $y = b$ is tangent to the x-curve in Eq. (7), and the line defined in (10) with $x = a$ is tangent to the y-curve in Eq. (8). But the substitution $y = b$ reduces Eq. (10) to

$$z - c = p(x - a), \quad \text{so} \quad \frac{\partial z}{\partial x} = p,$$

and the substitution $x = a$ reduces Eq. (10) to

$$z - c = q(y - b), \quad \text{so} \quad \frac{\partial z}{\partial y} = q.$$

Moreover, our discussion of the geometric interpretation of partial derivatives gave

$$\left.\frac{\partial z}{\partial x}\right|_{(a,b)} = f_x(a, b) \quad \text{and} \quad \left.\frac{\partial z}{\partial y}\right|_{(a,b)} = f_y(a, b)$$

for the slopes of the lines through P that are tangent there to the x-curve and y-curve, respectively. Hence we must have $p = f_x(a, b)$ and $q = f_y(a, b)$ in order for the plane in Eq. (10) to be tangent to the surface $z = f(x, y)$ at the point P. Substituting these values in Eq. (10) yields the following result (under the assumption that the partial derivatives are continuous, so that the tangent plane is defined).

The Plane Tangent to a Surface

The plane tangent to the surface $z = f(x, y)$ at the point $P(a, b, f(a, b))$ has equation

$$z - f(a, b) = f_x(a, b)(x - a) + f_y(a, b)(y - b). \tag{11}$$

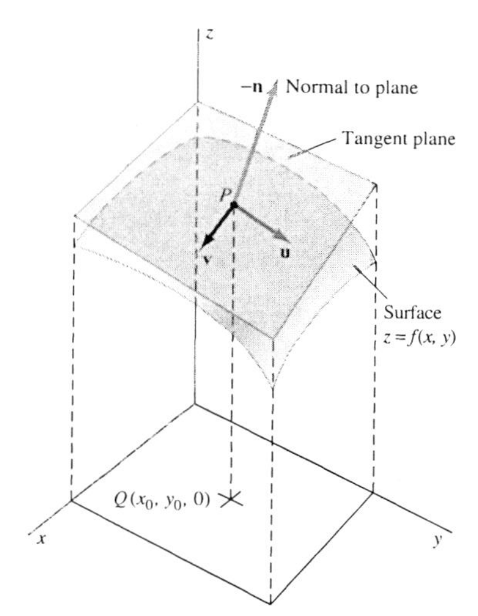

FIGURE 13.4.10 The surface $z = f(x, y)$, its tangent plane at $P(x_0, y_0, z_0)$, and the vector $-\mathbf{n}$ normal to both at P.

If for variety we write (x_0, y_0, z_0) for the coordinates of P, we can rewrite Eq. (11) in the form

$$f_x(x_0, y_0)(x - x_0) + f_y(x_0, y_0)(y - y_0) + (-1)(z - z_0) = 0, \tag{12}$$

from which we see (by consulting Eq. (8) in Section 12.4) that the plane tangent to the surface $z = f(x, y)$ at the point $P(x_0, y_0, z_0)$ has **normal vector**

$$\mathbf{n} = f_x(x_0, y_0)\mathbf{i} + f_y(x_0, y_0)\mathbf{j} - \mathbf{k} = \left\langle \frac{\partial z}{\partial x}, \frac{\partial z}{\partial y}, -1 \right\rangle. \tag{13}$$

Note that $\mathbf{n}$ is a downward-pointing vector (Why?); its negative $-\mathbf{n}$ is the upward-pointing vector shown in Fig. 13.4.10.

EXAMPLE 6 Write an equation of the plane tangent to the paraboloid $z = 5 - 2x^2 - y^2$ at the point $P(1, 1, 2)$.

Solution If $f(x, y) = 5 - 2x^2 - y^2$, then

$$f_x(x, y) = -4x, \qquad f_y(x, y) = -2y;$$
$$f_x(1, 1) = -4, \qquad f_y(1, 1) = -2.$$

Hence Eq. (11) gives

$$z - 2 = -4(x - 1) - 2(y - 1)$$

(when simplified, $z = 8 - 4x - 2y$) as an equation of the plane tangent to the paraboloid at P. The computer plot in Fig. 13.4.11 corroborates this result. ◆

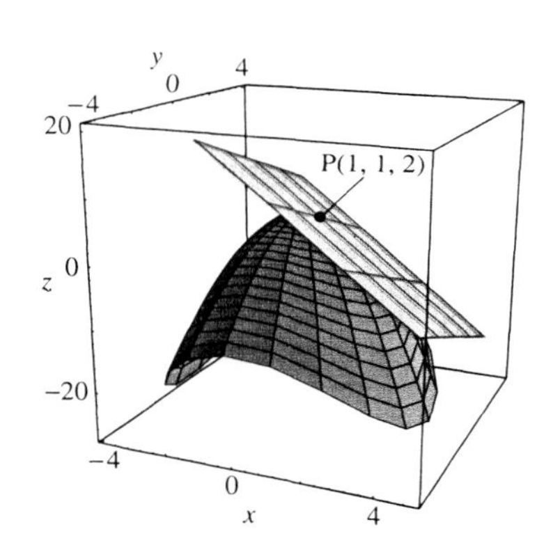

FIGURE 13.4.11 The paraboloid and tangent plane of Example 6.

Functions of Three or More Variables

Just like functions of two variables, a function of three or more variables has partial derivatives with respect to each of its independent variables. The partial derivative with respect to each variable is defined as a limit of a difference quotient involving increments in the selected variable. For instance, a function $f(x, y, z)$ has three partial derivatives, which are defined to be

$$\frac{\partial f}{\partial x} = \lim_{h \to 0} \frac{f(x + h, y, z) - f(x, y, z)}{h},$$
$$\frac{\partial f}{\partial y} = \lim_{h \to 0} \frac{f(x, y + h, z) - f(x, y, z)}{h}, \tag{14}$$
$$\text{and} \quad \frac{\partial f}{\partial z} = \lim_{h \to 0} \frac{f(x, y, z + h) - f(x, y, z)}{h}.$$

Partial derivatives of functions of still more variables are defined in an analogous way. A function $f(x_1, x_2, \dots, x_n)$ of n variables has n partial derivatives, one with respect to each of its independent variables. Limit quotients corresponding to those in (14) can be written more concisely using vector notation. Let us write

$$f(\mathbf{x}) = f(x_1, x_2, \dots, x_n) \quad \text{where} \quad \mathbf{x} = \langle x_1, x_2, \dots, x_n \rangle.$$

If $\mathbf{e}_i = \langle 0, 0, \dots, 1, \dots, 0 \rangle$ is the unit n-vector with ith entry 1, then

$$f(\mathbf{x} + h\mathbf{e}_i) = f(x_1, x_2, \dots, x_{i-1}, x_i + h, x_{i+1}, \dots, x_n).$$

The partial derivative $\partial f/\partial x_i = f_{x_i} = D_i f = D_{x_i} f$ of f with respect to the ith variable x_i is then defined to be

$$\frac{\partial f}{\partial x_i} = \lim_{h \to 0} \frac{f(\mathbf{x} + h\mathbf{e}_i) - f(\mathbf{x})}{h}. \tag{15}$$

The value of $\partial f/\partial x_i$ can be interpreted as the instantaneous rate of change of the function value $f(\mathbf{x})$ per unit change in the ith variable x_i. Just as in the case of two independent variables, each of these partial derivatives is calculated by differentiating with respect to the selected variable, regarding the others as constants.

EXAMPLE 7 The four partial derivatives of the function $g(x, y, u, v) = e^{ux} \sin vy$ are

$$g_x = ue^{ux} \sin vy, \quad g_y = ve^{ux} \cos vy, \quad g_u = xe^{ux} \sin vy, \quad \text{and} \quad g_v = ye^{ux} \cos vy.$$

Observe that we get these partial derivatives by differentiating $e^{ux} \sin vy$ with respect to the variables x, y, u, and v in turn, in each case holding the remaining three variables constant. ◆

Higher-Order Partial Derivatives

The first-order partial derivatives f_x and f_y are themselves functions of x and y, so they may be differentiated with respect to x or to y. The partial derivatives of $f_x(x, y)$ and $f_y(x, y)$ are called the **second-order partial derivatives** of f. There are four of them, because there are four possibilities in the order of differentiation:

$$(f_x)_x = f_{xx} = \frac{\partial f_x}{\partial x} = \frac{\partial}{\partial x}\left(\frac{\partial f}{\partial x}\right) = \frac{\partial^2 f}{\partial x^2},$$

$$(f_x)_y = f_{xy} = \frac{\partial f_x}{\partial y} = \frac{\partial}{\partial y}\left(\frac{\partial f}{\partial x}\right) = \frac{\partial^2 f}{\partial y\,\partial x},$$

$$(f_y)_x = f_{yx} = \frac{\partial f_y}{\partial x} = \frac{\partial}{\partial x}\left(\frac{\partial f}{\partial y}\right) = \frac{\partial^2 f}{\partial x\,\partial y},$$

$$(f_y)_y = f_{yy} = \frac{\partial f_y}{\partial y} = \frac{\partial}{\partial y}\left(\frac{\partial f}{\partial y}\right) = \frac{\partial^2 f}{\partial y^2}.$$

If we write $z = f(x, y)$, then we can replace each occurrence of the symbol f here with z.

Note The function f_{xy} is the second-order partial derivative of f with respect to x first and then to y; f_{yx} is the result of differentiating with respect to y first and x second. Although f_{xy} and f_{yx} are not necessarily equal, it is proved in advanced calculus that these two "mixed" second-order partial derivatives are equal if they are both continuous. More precisely, if f_{xy} and f_{yx} are continuous on a circular disk centered at the point (a, b), then

$$f_{xy}(a, b) = f_{yx}(a, b). \tag{16}$$

But if the mixed second-order derivatives f_{xy} and f_{yx} are merely defined at (a, b) but not necessarily continuous at and near this point, then it is entirely possible at $f_{xy} \neq f_{yx}$ at (a, b). (See Problem 74.)

Because most functions of interest to us have second-order partial derivatives that are continuous everywhere they are defined, we will ordinarily need to deal with only three distinct second-order partial derivatives (f_{xx}, f_{yy}, and $f_{xy} = f_{yx}$) rather than with four. Similarly, if $f(x, y, z)$ is a function of three variables with continuous second-order partial derivatives, then

$$\frac{\partial^2 f}{\partial x\,\partial y} = \frac{\partial^2 f}{\partial y\,\partial x}, \quad \frac{\partial^2 f}{\partial x\,\partial z} = \frac{\partial^2 f}{\partial z\,\partial x}, \quad \text{and} \quad \frac{\partial^2 f}{\partial y\,\partial z} = \frac{\partial^2 f}{\partial z\,\partial y}.$$

Third-order and higher-order partial derivatives are defined similarly, and the order in which the differentiations are performed is unimportant as long as all derivatives involved are continuous. In such a case, for example, the distinct third-order partial derivatives of the function $z = f(x, y)$ are

$$f_{xxx} = \frac{\partial}{\partial x}\left(\frac{\partial^2 f}{\partial x^2}\right) = \frac{\partial^3 f}{\partial x^3},$$

$$f_{xxy} = \frac{\partial}{\partial y}\left(\frac{\partial^2 f}{\partial x^2}\right) = \frac{\partial^3 f}{\partial y\, \partial x^2},$$

$$f_{xyy} = \frac{\partial}{\partial y}\left(\frac{\partial^2 f}{\partial y\, \partial x}\right) = \frac{\partial^3 f}{\partial y^2\, \partial x}, \quad \text{and}$$

$$f_{yyy} = \frac{\partial}{\partial y}\left(\frac{\partial^2 f}{\partial y^2}\right) = \frac{\partial^3 f}{\partial y^3}.$$

EXAMPLE 8 Show that the partial derivatives of third and higher orders of the function $f(x, y) = x^2 + 2xy^2 - y^3$ are constant.

Solution We find that

$$f_x(x, y) = 2x + 2y^2 \quad \text{and} \quad f_y(x, y) = 4xy - 3y^2.$$

So

$$f_{xx}(x, y) = 2, \quad f_{xy}(x, y) = 4y, \quad \text{and} \quad f_{yy}(x, y) = 4x - 6y.$$

Finally,

$$f_{xxx}(x, y) = 0, \quad f_{xxy}(x, y) = 0, \quad f_{xyy}(x, y) = 4, \quad \text{and} \quad f_{yyy}(x, y) = -6.$$

The function f is a polynomial, so all its partial derivatives are polynomials and are, therefore, continuous everywhere. Hence we need not compute any other third-order partial derivatives; each is equal to one of these four. Moreover, because the third-order partial derivatives are all constant, all higher-order partial derivatives of f are zero. ◆

13.4 TRUE/FALSE STUDY GUIDE

13.4 CONCEPTS: QUESTIONS AND DISCUSSION

1. Recall that the absolute value function $f(x) = |x|$ is differentiable except at the single point $x = 0$. Can you define an analogous function of two variables—one that has partial derivatives except at a single point?
2. Suppose that the surface $z = f(x, y)$ has a peak or a pit (that is, either a high point or a low point) at a point (a, b, c) where the surface has a tangent plane. What can you say about this tangent plane? What can you say about the values of the partial derivatives $f_x(a, b)$ and $f_y(a, b)$?
3. Can a surface $z = f(x, y)$ have a pit or a peak at a point where the partial derivatives f_x and f_y do not exist? Supply an example illustrating your answer.

13.4 PROBLEMS

In Problems 1 through 20, compute the first-order partial derivatives of each function.

1. $f(x, y) = x^4 - x^3y + x^2y^2 - xy^3 + y^4$

2. $f(x, y) = x \sin y$

3. $f(x, y) = e^x(\cos y - \sin y)$

4. $f(x, y) = e^2 e^{xy}$

5. $f(x, y) = \dfrac{x + y}{x - y}$

6. $f(x, y) = \dfrac{xy}{x^2 + y^2}$

7. $f(x, y) = \ln(x^2 + y^2)$

8. $f(x, y) = (x - y)^{14}$

9. $f(x, y) = x^y$

10. $f(x, y) = \tan^{-1} xy$

11. $f(x, y, z) = x^2y^3z^4$

12. $f(x, y, z) = x^2 + y^3 + z^4$

13. $f(x, y, z) = e^{xyz}$

14. $f(x, y, z) = x^4 - 16yz$

15. $f(x, y, z) = x^2e^y \ln z$

16. $f(u, v) = (2u^2 + 3v^2)\exp(-u^2 - v^2)$

17. $f(r, s) = \dfrac{r^2 - s^2}{r^2 + s^2}$

18. $f(u, v) = e^{uv}(\cos uv + \sin uv)$

19. $f(u, v, w) = ue^{v} + ve^{w} + we^{u}$

20. $f(r, s, t) = (1 - r^2 - s^2 - t^2)e^{-rst}$

In Problems 21 through 30, verify that $z_{xy} = z_{yx}$.

21. $z = x^2 - 4xy + 3y^2$

22. $z = 2x^3 + 5x^2y - 6y^2 + xy^4$

23. $z = x^2 \exp(-y^2)$

24. $z = xye^{-xy}$

25. $z = \ln(x + y)$

26. $z = (x^3 + y^3)^{10}$

27. $z = e^{-3x} \cos y$

28. $z = (x + y) \sec xy$

29. $z = x^2 \cosh(1/y^2)$

30. $z = \sin xy + \tan^{-1} xy$

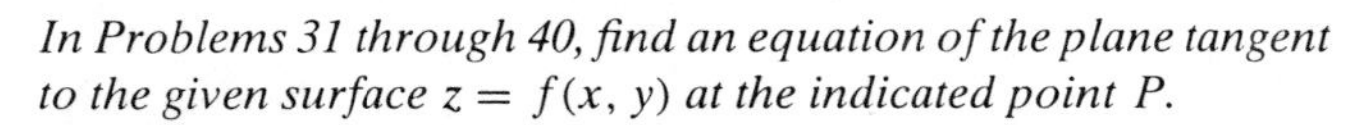

In Problems 31 through 40, find an equation of the plane tangent to the given surface $z = f(x, y)$ *at the indicated point P.*

31. $z = x^2 + y^2$; $P = (3, 4, 25)$

32. $z = \sqrt{50 - x^2 - y^2}$; $P = (4, -3, 5)$

33. $z = \sin \dfrac{\pi xy}{2}$; $P = (3, 5, -1)$

34. $z = \dfrac{4}{\pi} \tan^{-1} xy$; $P = (1, 1, 1)$

35. $z = x^3 - y^3$; $P = (3, 2, 19)$

36. $z = 3x + 4y$; $P = (1, 1, 7)$

37. $z = xy$; $P = (1, -1, -1)$

38. $z = \exp(-x^2 - y^2)$; $P = (0, 0, 1)$

39. $z = x^2 - 4y^2$; $P = (5, 2, 9)$

40. $z = \sqrt{x^2 + y^2}$; $P = (3, -4, 5)$

Recall that $f_{xy} = f_{yx}$ *for a function* $f(x, y)$ *with continuous second-order partial derivatives. In Problems 41 through 44, apply this criterion to determine whether there exists a function* $f(x, y)$ *having the given first-order partial derivatives. If so, try to determine a formula for such a function* $f(x, y)$.

41. $f_x(x, y) = 2xy^3$, $f_y(x, y) = 3x^2y^2$

42. $f_x(x, y) = 5xy + y^2$, $f_y(x, y) = 3x^2 + 2xy$

43. $f_x(x, y) = \cos^2(xy)$, $f_y(x, y) = \sin^2(xy)$

44. $f_x(x, y) = \cos x \sin y$, $f_y(x, y) = \sin x \cos y$

Figures 13.4.12 through 13.4.17 show the graphs of a certain function $f(x, y)$ *and its first- and second-order partial derivatives. In Problems 45 through 50, match that function or partial derivative with its graph.*

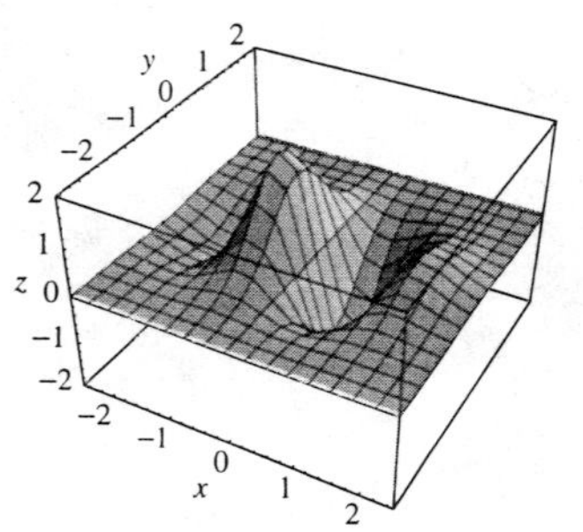

FIGURE 13.4.12

FIGURE 13.4.13

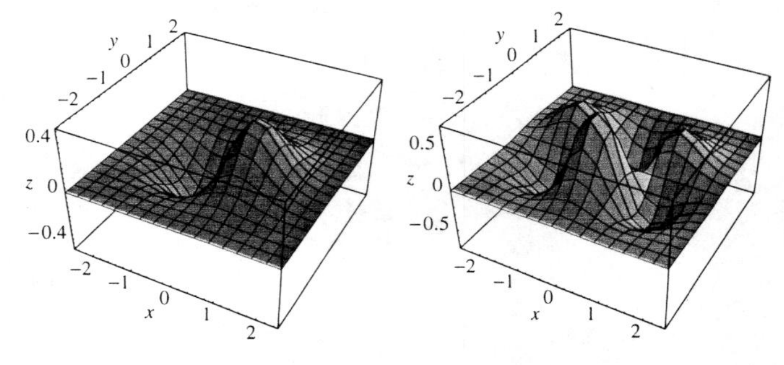

FIGURE 13.4.14 FIGURE 13.4.15

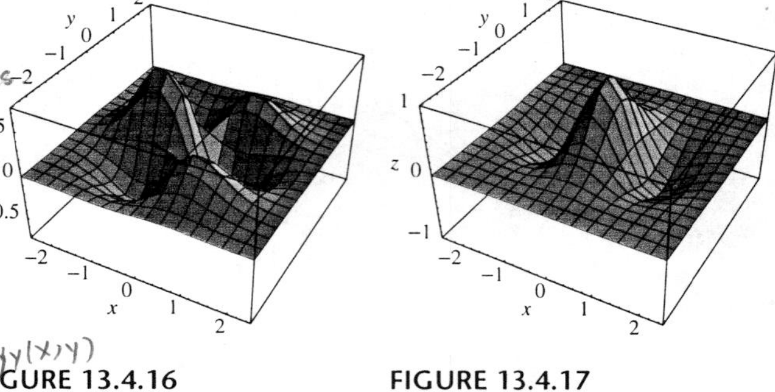

FIGURE 13.4.16 FIGURE 13.4.17

45. $f(x, y)$

46. $f_x(x, y)$

47. $f_y(x, y)$

48. $f_{xx}(x, y)$

49. $f_{xy}(x, y)$

50. $f_{yy}(x, y)$

51. Verify that the mixed second-order partial derivatives f_{xy} and f_{yx} are equal if $f(x, y) = x^m y^n$, where m and n are positive integers.

52. Suppose that $z = e^{x+y}$. Show that e^{x+y} is the result of differentiating z first m times with respect to x, then n times with respect to y.

53. Let $f(x, y, z) = e^{xyz}$. Calculate the distinct second-order partial derivatives of f and the third-order partial derivative f_{xyz}.

54. Suppose that $g(x, y) = \sin xy$. Verify that $g_{xy} = g_{yx}$ and that $g_{xxy} = g_{xyx} = g_{yxx}$.

55. It is shown in physics that the temperature $u(x, t)$ at time t at the point x of a long, insulated rod that lies along the x-axis satisfies the *one-dimensional heat equation*

$$\frac{\partial u}{\partial t} = k\frac{\partial^2 u}{\partial x^2} \qquad (k \text{ is a constant}).$$

Show that the function

$$u = u(x, t) = \exp(-n^2kt) \sin nx$$

satisfies the one-dimensional heat equation for any choice of the constant n.

56. The *two-dimensional heat equation* for an insulated thin plate is

$$\frac{\partial u}{\partial t} = k\left(\frac{\partial^2 u}{\partial x^2} + \frac{\partial^2 u}{\partial y^2}\right).$$

Show that the function

$$u=u(x, y, t)=\exp(-[m^2+n^2]kt)\sin mx\cos ny$$

satisfies this equation for any choice of the constants m and n.

57. A string is stretched along the x-axis, fixed at each end, and then set into vibration. It is shown in physics that the displacement $y=y(x,t)$ of the point of the string at location x at time t satisfies the *one-dimensional wave equation*

$$\frac{\partial^2 y}{\partial t^2}=a^2\frac{\partial^2 y}{\partial x^2},$$

where the constant a depends on the density and tension of the string. Show that the following functions satisfy the one-dimensional wave equation: (a) $y=\sin(x+at)$; (b) $y=\cosh(3[x-at])$; (c) $y=\sin kx\cos kat$ (k is a constant).

58. A steady-state temperature function $u=u(x,y)$ for a thin flat plate satisfies *Laplace's equation*

$$\frac{\partial^2 u}{\partial x^2}+\frac{\partial^2 u}{\partial y^2}=0.$$

Determine which of the following functions satisfy Laplace's equation:

(a) $u=\ln\left(\sqrt{x^2+y^2}\right)$;
(b) $u=\sqrt{x^2+y^2}$;
(c) $u=\arctan(y/x)$;
(d) $u=e^{-x}\sin y$.

59. Suppose that f and g are twice-differentiable functions of a single variable. Show that $y(x,t)=f(x+at)+g(x-at)$ satisfies the one-dimensional wave equation of Problem 57.

60. The electric potential field of a point charge q is defined (in appropriate units) by $\phi(x,y,z)=q/r$ where $r=\sqrt{x^2+y^2+z^2}$. Show that ϕ satisfies the *three-dimensional Laplace equation*

$$\frac{\partial^2\phi}{\partial x^2}+\frac{\partial^2\phi}{\partial y^2}+\frac{\partial^2\phi}{\partial z^2}=0.$$

61. Let $u(x,t)$ denote the underground temperature at depth x and time t at a location where the seasonal variation of surface ($x=0$) temperature is described by

$$u(0,t)=T_0+a_0\cos\omega t,$$

where T_0 is the annual average surface temperature and the constant ω is so chosen that the period of $u(0,t)$ is one year. Show that the function

$$u(x,t)=T_0+a_0\exp\left(-x\sqrt{\omega/2k}\right)\cos\left(\omega t-x\sqrt{\omega/2k}\right)$$

satisfies both the "surface condition" and the one-dimensional heat equation of Problem 55.

62. The aggregate electrical resistance R of three resistances R_1, R_2, and R_3 connected in parallel satisfies the equation

$$\frac{1}{R}=\frac{1}{R_1}+\frac{1}{R_2}+\frac{1}{R_3}.$$

Show that

$$\frac{\partial R}{\partial R_1}+\frac{\partial R}{\partial R_2}+\frac{\partial R}{\partial R_3}=\left(\frac{1}{R_1^2}+\frac{1}{R_2^2}+\frac{1}{R_3^2}\right)\div\left(\frac{1}{R_1}+\frac{1}{R_2}+\frac{1}{R_3}\right)^2.$$

63. The **ideal gas law** $pV=nRT$ (n is the number of moles of the gas, R is a constant) determines each of the three variables p (pressure), V (volume), and T (temperature) as functions of the other two. Show that

$$\frac{\partial p}{\partial V}\cdot\frac{\partial V}{\partial T}\cdot\frac{\partial T}{\partial p}=-1.$$

64. It is geometrically evident that every plane tangent to the cone $z^2=x^2+y^2$ passes through the origin. Show this by methods of calculus.

65. There is only one point at which the plane tangent to the surface

$$z=x^2+2xy+2y^2-6x+8y$$

is horizontal. Find it.

66. Show that the plane tangent to the paraboloid with equation $z=x^2+y^2$ at the point (a,b,c) intersects the xy-plane in the line with equation $2ax+2by=a^2+b^2$. Then show that this line is tangent to the circle with equation $4x^2+4y^2=a^2+b^2$.

67. According to van der Waals' equation, 1 mol of a gas satisfies the equation

$$\left(p+\frac{a}{V^2}\right)(V-b)=(82.06)T$$

where p, V, and T are as in Example 4. For carbon dioxide, $a=3.59\times10^6$ and $b=42.7$, and V is 25,600 cm^3 when p is 1 atm and $T=313$ K. (a) Compute $\partial V/\partial p$ by differentiating van der Waals' equation with T held constant. Then estimate the change in volume that would result from an increase of 0.1 atm of pressure with T held at 313 K. (b) Compute $\partial V/\partial T$ by differentiating van der Waals' equation with p held constant. Then estimate the change in volume that would result from an increase of 1 K in temperature with p held at 1 atm.

68. A *minimal surface* has the least surface area of all surfaces with the same boundary. Figure 13.4.18 shows *Scherk's minimal surface.* It has the equation

$$z=\ln(\cos x)-\ln(\cos y).$$

A minimal surface $z=f(x,y)$ is known to satisfy the partial differential equation

$$\left(1+z_y^2\right)z_{xx}-2z_xz_yz_{xy}+\left(1+z_x^2\right)z_{yy}=0.$$

Verify this in the case of Scherk's minimal surface.

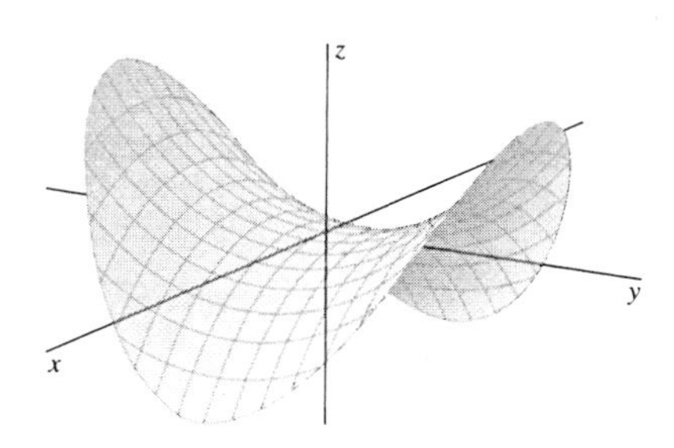

FIGURE 13.4.18 Scherk's minimal surface (Problem 68).

69. We say that the function $z=f(x,y)$ is **harmonic** if it satisfies Laplace's equation $z_{xx}+z_{yy}=0$. (See Problem 58.) Show that each of these four functions is harmonic:

(a) $f_1(x,y)=\sin x\sinh(\pi-y)$;
(b) $f_2(x,y)=\sinh 2x\sin 2y$;
(c) $f_3(x,y)=\sin 3x\sinh 3y$;
(d) $f_4(x,y)=\sinh 4(\pi-x)\sin 4y$.

70. Figure 13.4.19 shows the graph of the sum

$$z(x, y) = \sum_{i=1}^{4} f_i(x, y)$$

of the four functions defined in Problem 69. Explain why $z(x, y)$ is a harmonic function.

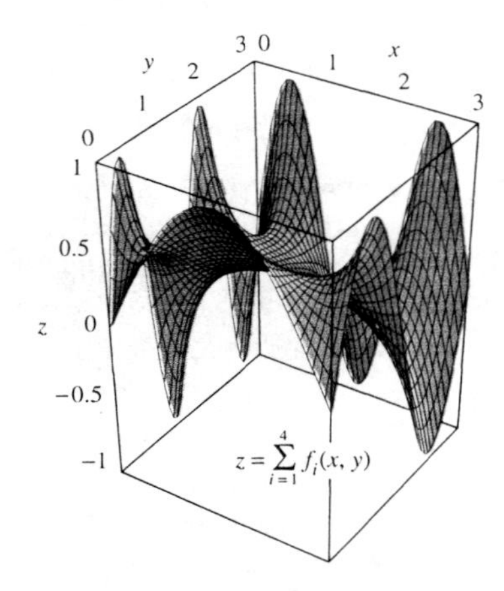

FIGURE 13.4.19 The surface $z = f(x, y)$ of Problem 70.

71. You are standing at the point where $x = y = 100$ (ft) on a hillside whose height (in feet above sea level) is given by

$$z = 100 + \frac{1}{100}(x^2 - 3xy + 2y^2),$$

with the positive x-axis to the east and the positive y-axis to the north. (a) If you head due east, will you initially be ascending or descending? At what angle (in degrees) from the horizontal? (b) If you head due north, will you initially be ascending or descending? At what angle (in degrees) from the horizontal?

72. Answer questions (a) and (b) in Problem 71, except that now you are standing at the point where $x = 150$ and $y = 250$ (ft) on a hillside whose height (in feet above sea level) is given by

$$z = 1000 + \frac{1}{1000}(3x^2 - 5xy + y^2).$$

73. Figure 13.3.7 shows the graph of the function f defined by

$$f(x, y) = \begin{cases} \dfrac{xy}{x^2 + y^2} & \text{unless } x = y = 0, \\ 0 & \text{if } x = y = 0. \end{cases}$$

(a) Show that the first-order partial derivatives f_x and f_y are defined everywhere and are continuous except possibly at the origin. (b) Consider behavior on straight lines to show that neither f_x nor f_y is continuous at the origin. (c) Show that the second-order partial derivatives of f are all defined and continuous except possibly at the origin. (d) Show that the second-order partial derivatives f_{xx} and f_{yy} exist at the origin, but that the mixed partial derivatives f_{xy} and f_{yx} do not.

74. Figure 13.4.20 shows the graph of the function g defined by

$$g(x, y) = \begin{cases} \dfrac{xy(x^2 - y^2)}{x^2 + y^2} & \text{unless } x = y = 0, \\ 0 & \text{if } x = y = 0. \end{cases}$$

(a) Show that the first-order partial derivatives g_x and g_y are defined everywhere and are continuous except possibly at the origin. (b) Use polar coordinates to show that g_x and g_y are continuous at $(0, 0)$ as well. (c) Show that the second-order partial derivatives of g are all defined and continuous except possibly at the origin. (d) Show that all four second-order partial derivatives of g exist at the origin, but that $g_{xy}(0, 0) \neq g_{yx}(0, 0)$. (e) Consider behavior on straight lines to show that none of the four second-order partial derivatives of g is continuous at the origin.

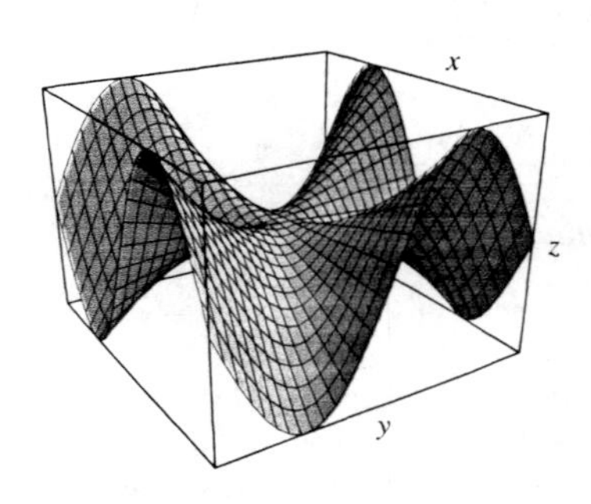

FIGURE 13.4.20 The graph $z = \dfrac{x^3y - xy^3}{x^2 + y^2}$ of Problem 74.

13.5 MULTIVARIABLE OPTIMIZATION PROBLEMS

The single-variable maximum-minimum techniques of Section 3.5 generalize readily to functions of several variables. We consider first a function f of two variables. Suppose that we are interested in the extreme values attained by $f(x, y)$ on a plane region R that consists of the points on and within a simple (nonintersecting) closed curve C (Fig. 13.5.1). We say that the function f attains its **absolute, or global, maximum value** M on R at the point (a, b) of R provided that

$$f(x, y) \leqq M = f(a, b)$$

for all points (x, y) of R. Similarly, f attains its **absolute, or global, minimum value** m at the point (c, d) of R provided that $f(x, y) \geqq m = f(c, d)$ for all points (x, y)

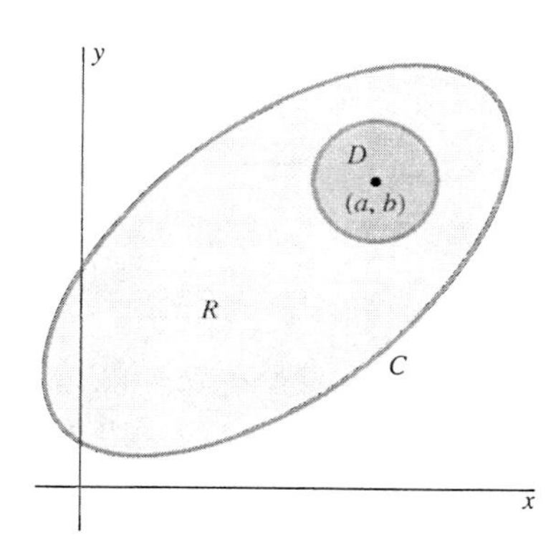

FIGURE 13.5.1 A plane region R bounded by the simple closed curve C and a disk D in R centered at the interior point (a, b) of R.

of R. In plain words, the absolute maximum M and the absolute minimum m are the largest and smallest values (respectively) attained by $f(x, y)$ at points of the domain R of f.

Theorem 1, proved in advanced calculus courses, guarantees the existence of absolute maximum and minimum values in many situations of practical interest.

THEOREM 1 Existence of Extreme Values

Suppose that the function f is continuous on the region R that consists of the points on and within a simple closed curve C in the plane. Then f attains an absolute maximum value at some point (a, b) of R and attains an absolute minimum value at some point (c, d) of R.

We are interested mainly in the case in which the function f attains its absolute maximum (or minimum) value at an interior point of R. The point (a, b) of R is called an **interior point** of R provided that some circular disk centered at (a, b) lies wholly within R (Fig. 13.5.1). The interior points of a region R of the sort described in Theorem 1 are precisely those that do *not* lie on the boundary curve C.

An absolute extreme value attained by the function at an *interior* point of R is necessarily a local extreme value. We say that $f(a, b)$ is a **local maximum value** of $f(x, y)$ provided that it is the absolute maximum value of f on some disk D that is centered at (a, b) and lies wholly within the domain R. Similarly, a **local minimum value** is an absolute minimum value on some such disk. Thus a local maximum (or minimum) value $f(a, b)$ is not necessarily an absolute maximum (or minimum) value, but is the largest (or smallest) value attained by $f(x, y)$ at points near (a, b).

EXAMPLE 1 Figure 13.5.2 shows a computer-generated graph of the function

$$f(x, y) = 3(x-1)^2 e^{-x^2-(y+1)^2} - 10\left(\tfrac{1}{5}x - x^3 - y^5\right)e^{-x^2-y^2} - \tfrac{1}{3}e^{-(x+1)^2-y^2}$$

plotted on the rectangle R for which $-3 \leqq x \leqq 3$ and $-3 \leqq y \leqq 3$. Looking at the labeled extreme values of $f(x, y)$, we see

- A local maximum that is not an absolute maximum,
- A local maximum that is also an absolute maximum,
- A local minimum that is not an absolute minimum, and
- A local minimum that is also an absolute minimum.

We can think of the local maxima on the graph as mountain tops or "peaks" and the local minima as valley bottoms or "pits." ◆

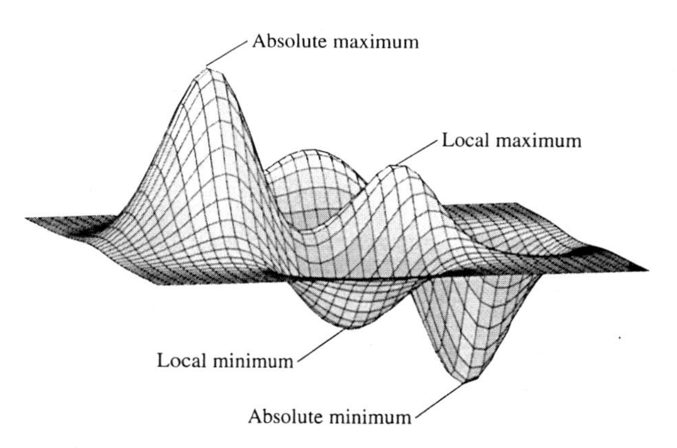

FIGURE 13.5.2 Local extrema contrasted with global extrema.

Finding Local Extrema

We need a criterion that will provide a practical way to find local extrema of functions of two (or more) variables. The desired result—stated in Theorem 2—is analogous to the single-variable criterion of Section 3.5: If $f(c)$ is a local extreme value of the differentiable single-variable function f, then $x = c$ must be a *critical point* where $f'(c) = 0$.

Suppose, for instance, that $f(a, b)$ is a local maximum value of $f(x, y)$ attained at a point (a, b) where both partial derivatives f_x and f_y exist. We consider vertical plane cross-section curves on the graph $z = f(x, y)$, just as when we explored the geometrical interpretation of partial derivatives in Section 13.4. The cross-section curves parallel to the xz- and yz-planes are the graphs (in these planes) of the single-variable functions

$$G(x) = f(x, b) \quad \text{and} \quad H(y) = f(a, y)$$

whose derivatives are the partial derivatives of f:

$$f_x(a, b) = G'(a) \quad \text{and} \quad f_y(a, b) = H'(b). \tag{1}$$

Because $f(a, b)$ is a local maximum value of $f(x, y)$, it follows readily that $G(a)$ and $H(b)$ are local maximum values of $G(x)$ and $H(y)$, respectively. Therefore the single-variable maximum-minimum criterion of Section 3.5 implies that

$$G'(a) = 0 \quad \text{and} \quad H'(b) = 0. \tag{2}$$

Combining (1) and (2), we conclude that

$$f_x(a, b) = 0 \quad \text{and} \quad f_y(a, b) = 0. \tag{3}$$

Essentially the same argument yields the same conclusion if $f(a, b)$ is a local minimum value of $f(x, y)$. This discussion establishes Theorem 2.

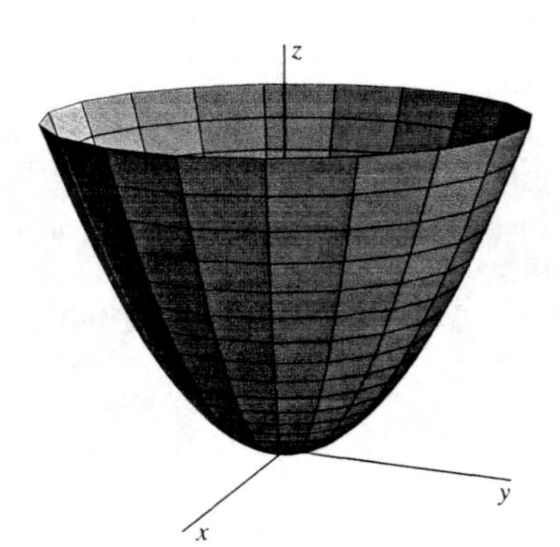

(a) $f(x, y) = x^2 + y^2$, local minimum at (0, 0)

THEOREM 2 Necessary Conditions for Local Extrema
Suppose that $f(x, y)$ attains a local maximum value or a local minimum value at the point (a, b) and that both the partial derivatives $f_x(a, b)$ and $f_y(a, b)$ exist. Then

$$f_x(a, b) = 0 = f_y(a, b). \tag{3}$$

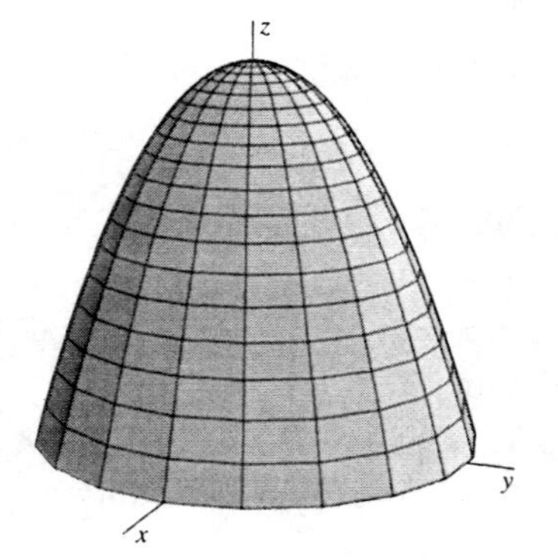

(b) $g(x, y) = 1 - x^2 - y^2$, local maximum at (0, 0)

The equations in (3) imply that the plane tangent to the surface $z = f(x, y)$ must be horizontal at any local maximum or local minimum point $(a, b, f(a, b))$, in perfect analogy to the single-variable case (in which the tangent line is horizontal at any local maximum or minimum point on the graph of a differentiable function).

EXAMPLE 2 Consider the three familiar surfaces

$$z = f(x, y) = x^2 + y^2,$$
$$z = g(x, y) = 1 - x^2 - y^2, \quad \text{and}$$
$$z = h(x, y) = y^2 - x^2$$

shown in Fig. 13.5.3. In each case $\partial z/\partial x = \pm 2x$ and $\partial z/\partial y = \pm 2y$. Thus both partial derivatives are zero at the origin (0, 0) (and only there). It is clear from the figure that $f(x, y) = x^2 + y^2$ has a local minimum at (0, 0). In fact, because a square cannot be negative, $z = x^2 + y^2$ has the global minimum value 0 at (0, 0). Similarly, $g(x, y)$ has a local (indeed, global) maximum value at (0, 0), whereas $h(x, y)$ has neither a local minimum nor a local maximum there—the origin is a *saddle point* of h. This example shows that a point (a, b) where

$$\frac{\partial z}{\partial x} = 0 = \frac{\partial z}{\partial y}$$

may correspond to either a local minimum, a local maximum, or neither. Thus the necessary condition in Eq. (3) is *not* a sufficient condition for a local extremum. ◆

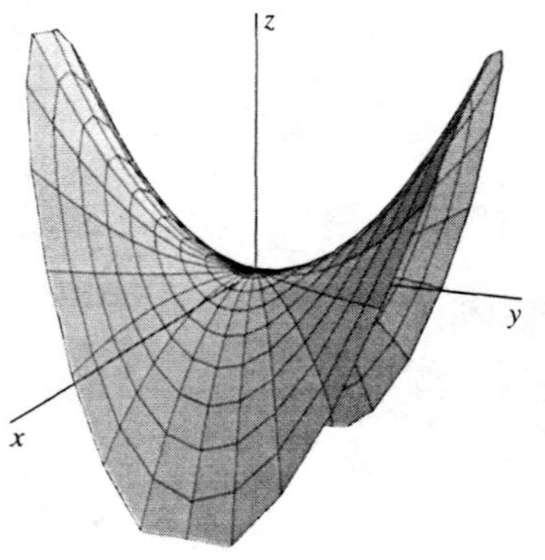

(c) $h(x, y) = y^2 - x^2$, saddle point at (0, 0)

FIGURE 13.5.3 Where both partial derivatives are zero, there may be (a) a minimum, (b) a maximum, or (c) neither.

EXAMPLE 3 Find all points on the surface

$$z = \tfrac{3}{4}y^2 + \tfrac{1}{24}y^3 - \tfrac{1}{32}y^4 - x^2$$

at which the tangent plane is horizontal.

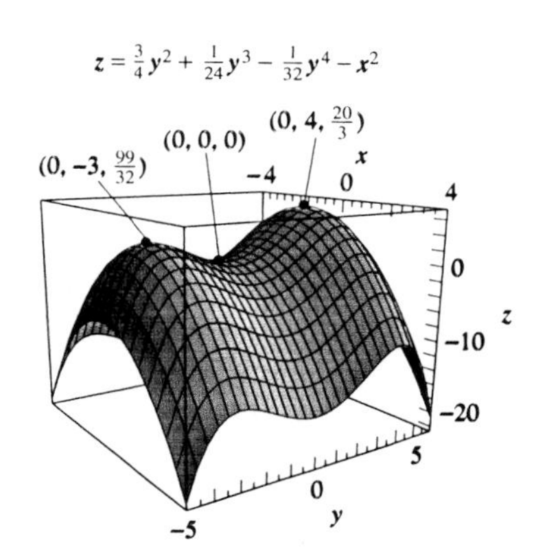

FIGURE 13.5.4 The surface of Example 3.

Solution We first calculate the partial derivatives $\partial z/\partial x$ and $\partial z/\partial y$:

$$\frac{\partial z}{\partial x} = -2x,$$

$$\frac{\partial z}{\partial y} = \tfrac{3}{2}y + \tfrac{1}{8}y^2 - \tfrac{1}{8}y^3$$

$$= -\tfrac{1}{8}y(y^2 - y - 12) = -\tfrac{1}{8}y(y+3)(y-4).$$

We next equate both $\partial z/\partial x$ and $\partial z/\partial y$ to zero. This yields

$$-2x = 0 \quad \text{and} \quad -\tfrac{1}{8}y(y+3)(y-4) = 0.$$

Simultaneous solution of these equations yields exactly three points where both partial derivatives are zero: $(0, -3)$, $(0, 0)$, and $(0, 4)$. The three corresponding points on the surface where the tangent plane is horizontal are $(0, -3, \frac{99}{32})$, $(0, 0, 0)$, and $(0, 4, \frac{20}{3})$. These three points are indicated on the graph in Fig. 13.5.4 of the surface. (Recall that we constructed this surface in Example 11 of Section 13.2.) ◆

Finding Global Extrema

Theorem 2 is a very useful tool for finding the absolute maximum and absolute minimum values attained by a continuous function f on a region R of the type described in Theorem 1. If $f(a, b)$ is the absolute maximum value, for example, then (a, b) is either an interior point of R or a point of the boundary curve C. If (a, b) is an interior point and both the partial derivatives $f_x(a, b)$ and $f_y(a, b)$ exist, then Theorem 2 implies that both these partial derivatives must be zero. Thus we have the following result.

THEOREM 3 Types of Absolute Extrema

Suppose that f is continuous on the plane region R consisting of the points on and within a simple closed curve C. If $f(a, b)$ is either the absolute maximum or the absolute minimum value of $f(x, y)$ on R, then (a, b) is either

1. An interior point of R at which
$$\frac{\partial f}{\partial x} = \frac{\partial f}{\partial y} = 0,$$
2. An interior point of R where not both partial derivatives exist, or
3. A point of the boundary curve C of R.

A point (a, b) where either condition (1) or condition (2) holds is called a **critical point** of the function f. Thus Theorem 3 says that *any extreme value of the continuous function f on the plane region R must occur at an interior critical point or at a boundary point.* Note the analogy with Theorem 3 of Section 3.5, which implies that an extreme value of a single-variable function $f(x)$ on a closed and bounded interval I must occur either at an interior critical point of I or at an endpoint (boundary point) of I.

As a consequence of Theorem 3, we can find the absolute maximum and minimum values of $f(x, y)$ on R as follows:

1. First locate the interior critical points.
2. Next find the possible extreme values of f on the boundary curve C.
3. Finally compare the values of f at the points found in steps 1 and 2.

The technique to be used in the second step will depend on the nature of the boundary curve C, as illustrated in Examples 4 and 5.

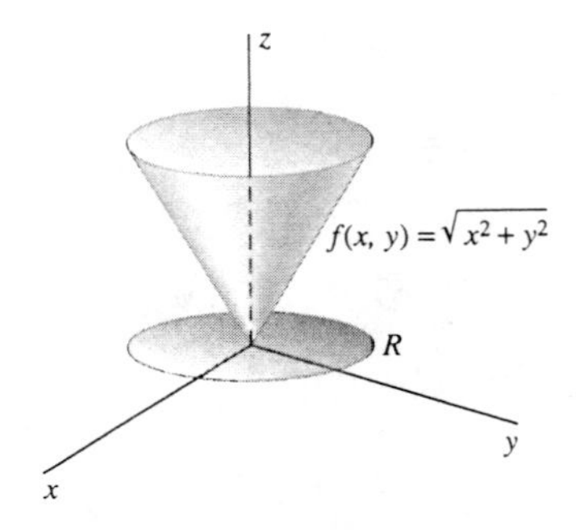

FIGURE 13.5.5 The graph of the function of Example 4.

EXAMPLE 4 Let $f(x, y) = \sqrt{x^2 + y^2}$ on the region R consisting of the points on and within the circle $x^2 + y^2 = 1$ in the xy-plane. The graph of f is shown in Fig. 13.5.5. We see that the minimum value 0 of f occurs at the origin $(0, 0)$, where both the partial derivatives f_x and f_y fail to exist (Why?), whereas the maximum value 1 of f on R occurs at *each and every* point of the boundary circle. ◆

EXAMPLE 5 Find the maximum and minimum values attained by the function

$$f(x, y) = xy - x - y + 3$$

at points of the triangular region R in the xy-plane with vertices at $(0, 0)$, $(2, 0)$, and $(0, 4)$.

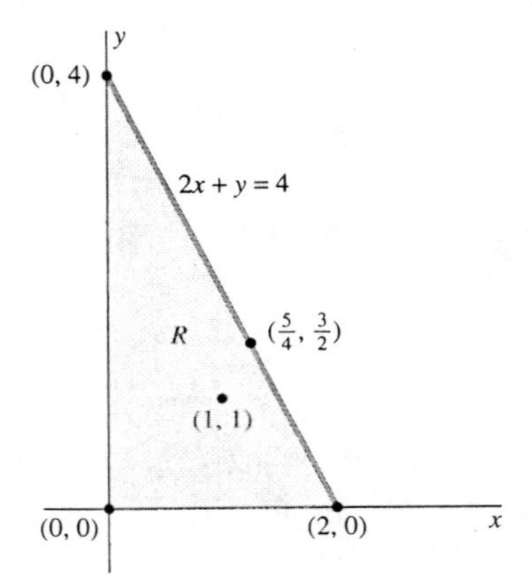

FIGURE 13.5.6 The triangular region of Example 5.

Solution The region R is shown in Fig. 13.5.6. Its boundary "curve" C consists of the segment $0 \leqq x \leqq 2$ on the x-axis, the segment $0 \leqq y \leqq 4$ on the y-axis, and the part of the line $2x + y = 4$ that lies in the first quadrant. Any interior extremum must occur at a point where both

$$\frac{\partial f}{\partial x} = y - 1 \quad \text{and} \quad \frac{\partial f}{\partial y} = x - 1$$

are zero. Hence the only interior critical point is $(1, 1)$.

Along the edge where $y = 0$: The function $f(x, y)$ takes the form

$$\alpha(x) = f(x, 0) = 3 - x, \qquad 0 \leqq x \leqq 2.$$

Because $\alpha(x)$ is a decreasing function, its extrema for $0 \leqq x \leqq 2$ occur at the endpoints $x = 0$ and $x = 2$. This gives the two possibilities $(0, 0)$ and $(2, 0)$ for locations of extrema of $f(x, y)$.

Along the edge where $x = 0$: The function $f(x, y)$ takes the form

$$\beta(y) = f(0, y) = 3 - y, \qquad 0 \leqq y \leqq 4.$$

The endpoints of this interval yield the points $(0, 0)$ and $(0, 4)$ as possibilities for locations of extrema of $f(x, y)$.

On the edge of R *where* $y = 4 - 2x$: We may substitute $4 - 2x$ for y in the formula for $f(x, y)$ and thus express f as a function of a single variable:

$$\begin{aligned}\gamma(x) &= x(4 - 2x) - x - (4 - 2x) + 3 \\ &= -2x^2 + 5x - 1, \qquad 0 \leqq x \leqq 2.\end{aligned}$$

To find the extreme values of $\gamma(x)$, we first calculate

$$\gamma'(x) = -4x + 5;$$

$\gamma'(x) = 0$ where $x = \frac{5}{4}$. Thus each extreme value of $\gamma(x)$ on $[0, 2]$ must occur either at the interior point $x = \frac{5}{4}$ of the interval $[0, 2]$ or at one of the endpoints $x = 0$ and $x = 2$. This gives the possibilities $(0, 4)$, $(\frac{5}{4}, \frac{3}{2})$, and $(2, 0)$ for locations of extrema of $f(x, y)$.

We conclude by evaluating f at each of the points we have found:

$$\begin{aligned} f(0, 0) &= 3, && \longleftarrow \text{maximum} \\ f(\tfrac{5}{4}, \tfrac{3}{2}) &= 2.125 \\ f(1, 1) &= 2, \\ f(2, 0) &= 1, \\ f(0, 4) &= -1. && \longleftarrow \text{minimum} \end{aligned}$$

Thus the maximum value of $f(x, y)$ on the region R is $f(0, 0) = 3$ and the minimum value is $f(0, 4) = -1$. ◆

Note the terminology used in this section. In Example 5, the maximum *value* of f is 3, the maximum *occurs at* the point $(0, 0)$ in the domain of f, and the *highest point* on the graph of f is $(0, 0, 3)$.

Highest and Lowest Points of Surfaces

In applied problems we frequently know in advance that the absolute maximum (or minimum) value of $f(x, y)$ on R occurs at an *interior* point of R where both partial derivatives of f exist. In this important case, Theorem 3 tells us that we can locate every possible point at which the maximum (or minimum) might occur by simultaneously solving the two equations

$$f_x(x, y) = 0 \quad \text{and} \quad f_y(x, y) = 0. \tag{4}$$

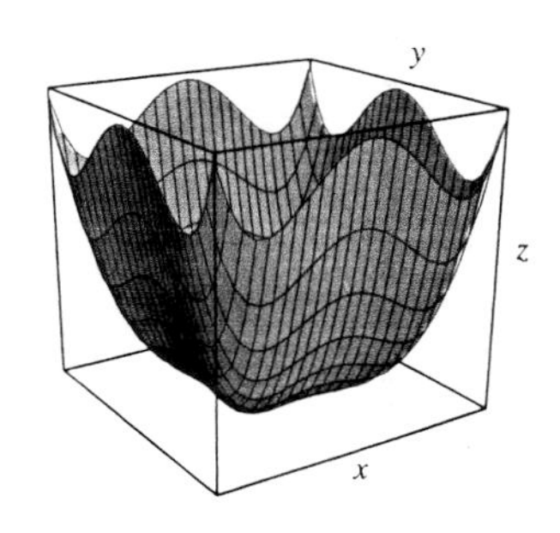

FIGURE 13.5.7 The surface $z = x^4 + y^4 - x^2y^2$ opens upward.

If we are lucky, these equations will have only one simultaneous solution (x, y) interior to R. If so, then *that* solution must be the location of the desired maximum (or minimum). If we find that the equations in (4) have several simultaneous solutions interior to R, then we simply evaluate f at each solution to determine which yields the largest (or smallest) value of $f(x, y)$ and is therefore the desired maximum (or minimum) point.

We can use this method to find the lowest point on a surface $z = f(x, y)$ that opens upward, as in Fig. 13.5.7. If R is a sufficiently large rectangle, then $f(x, y)$ attains large positive values everywhere on the boundary of R but smaller values at interior points. It follows that the minimum value of $f(x, y)$ must be attained at an interior point of R.

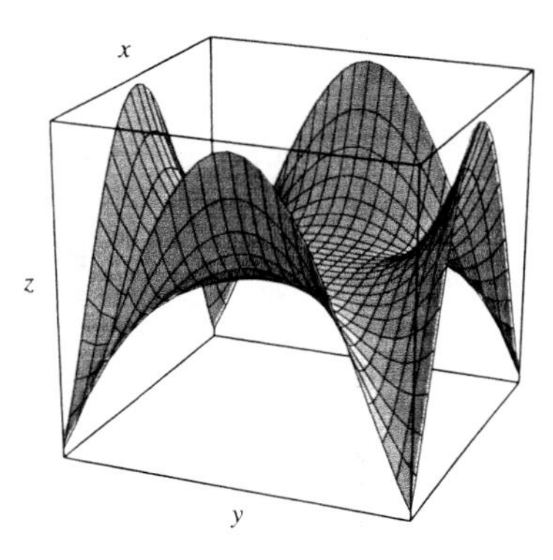

FIGURE 13.5.8 The surface $z = x^4 + y^4 - 3x^2y^2$ opens both upward and downward.

The question of a highest or lowest point is not pertinent for a surface that opens both upward and downward, as in Fig. 13.5.8.

EXAMPLE 6 Find the highest point on the surface

$$z = f(x, y) = \tfrac{8}{3}x^3 + 4y^3 - x^4 - y^4. \tag{5}$$

Solution We observe that the negative fourth-degree terms in $f(x, y)$ clearly predominate when $|x|$ and/or $|y|$ is large, so the surface $z = f(x, y)$ opens downward. (See Fig. 13.5.9.) To verify this observation, we factor out $x^4 + y^4$ and write

$$f(x, y) = (x^4 + y^4)\left[\frac{\frac{8}{3}x^3 + 4y^3}{x^4 + y^4} - 1\right]. \tag{6}$$

Now consider a fixed point (x, y) and let m denote the smaller, and M the larger, of the two numbers $|x|$ and $|y|$. Then

$$\left|\frac{\frac{8}{3}x^3 + 4y^3}{x^4 + y^4}\right| \leqq \frac{4|x|^3 + 4|y|^3}{x^4 + y^4} = \frac{4m^3 + 4M^3}{m^4 + M^4} \leqq \frac{4M^3 + 4M^3}{0^4 + M^4} = \frac{8}{M}.$$

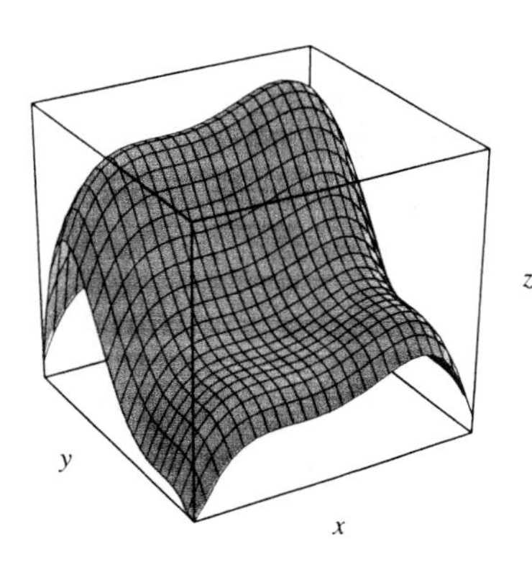

FIGURE 13.5.9 The surface $z = \frac{8}{3}x^3 + 4y^3 - x^4 - y^4$ opens downward.

For instance, if either $|x|$ or $|y|$ is greater than $M = 10$, then the fraction within brackets in Eq. (6) has absolute value less than $\frac{8}{10}$, so it follows that $f(x, y) < 0$.

Thus $f(x, y)$ is negative outside the large square R with vertices $(\pm 10, \pm 10)$ in the xy-plane. But $z = f(x, y)$ certainly attains positive values within R, such as $f(1, 1) = \frac{14}{3}$. Consequently Theorem 1 implies that $f(x, y)$ attains an absolute maximum value at some interior point of R. So let us proceed to find this maximum value.

Because the partial derivatives $\partial z/\partial x$ and $\partial z/\partial y$ exist everywhere, Theorem 3 implies that we need only solve the equations $\partial z/\partial x = 0$ and $\partial z/\partial y = 0$ in Eq. (4)—that is,

$$\frac{\partial z}{\partial x} = 8x^2 - 4x^3 = 4x^2(2 - x) = 0,$$

$$\frac{\partial z}{\partial y} = 12y^2 - 4y^3 = 4y^2(3 - y) = 0.$$

If these two equations are satisfied, then

$$\boxed{\text{Either } x = 0 \text{ or } x = 2} \quad \text{and} \quad \boxed{\text{either } y = 0 \text{ or } y = 3.}$$

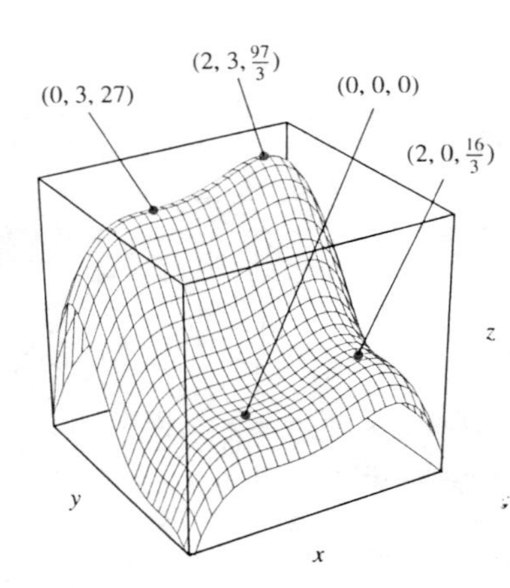

FIGURE 13.5.10 The critical points of Example 6.

It follows that either

$x = 0$ and $y = 0$	or	$x = 0$ and $y = 3$	or	$x = 2$ and $y = 0$	or	$x = 2$ and $y = 3$.

Consequently, we need only inspect the values

$$z(0, 0) = 0,$$
$$z(2, 0) = \tfrac{16}{3} = 5.333\,333\,333\ldots,$$
$$z(0, 3) = 27,$$
$$z(2, 3) = \tfrac{97}{3} = 32.333\,333\,333\ldots. \quad \longleftarrow \quad \text{maximum}$$

Thus the highest point on the surface is the point $\left(2, 3, \frac{97}{3}\right)$. The four critical points on the surface are indicated in Fig. 13.5.10. ◆

Applied Maximum-Minimum Problems

The analysis of a multivariable applied maximum-minimum problem involves the same general steps that we listed at the beginning of Section 3.6. Here, however, we will express the dependent variable—the quantity to be maximized or minimized—as a function $f(x, y)$ of *two* independent variables. Once we have identified the appropriate region in the xy-plane as the domain of f, the methods of this section are applicable. We often find that a preliminary step is required: If the meaningful domain of definition of f is an unbounded region, then we first restrict f to a *bounded* plane region R on which we know the desired extreme value occurs. This procedure is similar to the one we used with open-interval maximum-minimum problems in Section 4.4.

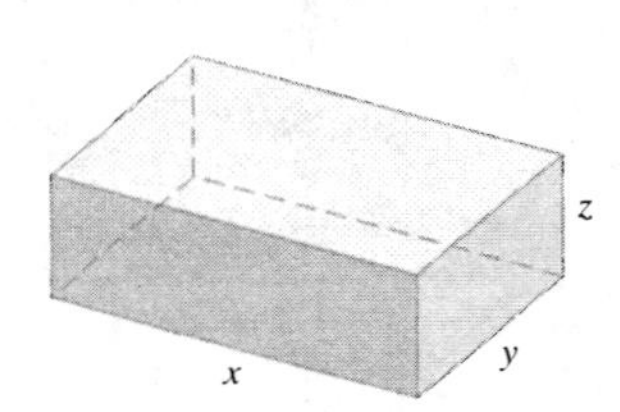

FIGURE 13.5.11 A box whose total cost we want to minimize (Example 7).

EXAMPLE 7 Find the minimum cost of a rectangular box with volume 48 ft^3 if the front and back cost \$1/ft^2, the top and bottom cost \$2/ft^2, and the two ends cost \$3/ft^2. (We first discussed such a box in Section 13.1.) This box is shown in Fig. 13.5.11.

Solution We found in Section 13.1 that the cost C (in dollars) of this box is given by

$$C(x, y) = 4xy + \frac{288}{x} + \frac{96}{y}$$

in terms of its length x and width y. Let R be a square such as the one shown in Fig. 13.5.12. Two sides of R are so close to the coordinate axes that $288/x > 1000$ on the side nearest the y-axis and $96/y > 1000$ on the side nearest the x-axis. Also, the square is so large that $4xy > 1000$ on both of the other two sides. This means that $C(x, y) > 1000$ at every point (x, y) of the first quadrant that lies on or outside the boundary of the square R. Because $C(x, y)$ attains reasonably small values within R (for instance, $C(1, 1) = 388$), it is clear that the absolute minimum of C must occur at an interior point of R. Thus, although the natural domain of the cost function $C(x, y)$ is the entire first quadrant, we have succeeded in restricting its domain to a region R of the sort to which Theorem 3 applies.

We therefore solve the simultaneous equations

$$\frac{\partial C}{\partial x} = 4y - \frac{288}{x^2} = 0,$$
$$\frac{\partial C}{\partial y} = 4x - \frac{96}{y^2} = 0.$$

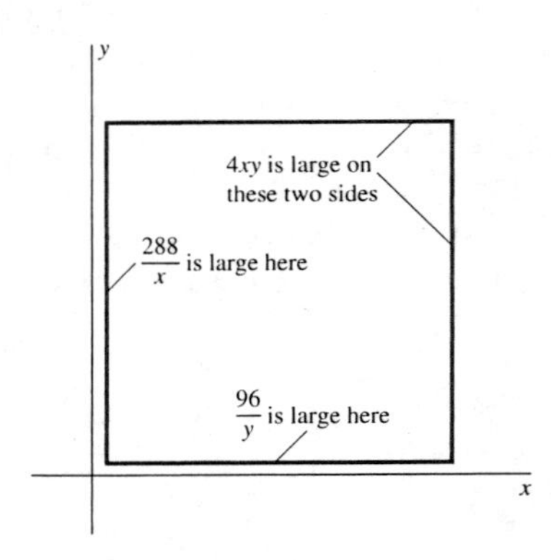

FIGURE 13.5.12 The cost function $C(x, y)$ of Example 7 takes on large positive values on the boundary of the square.

We multiply the first equation by x and the second by y. (*Ad hoc* methods are frequently required in the solution of simultaneous nonlinear equations.) This procedure gives

$$\frac{288}{x} = 4xy = \frac{96}{y},$$

so that $x = 288y/96 = 3y$. We substitute $x = 3y$ into the equation $\partial C/\partial y = 0$ and find that

$$12y - \frac{96}{y^2} = 0, \quad \text{so} \quad 12y^3 = 96.$$

Hence $y = \sqrt[3]{8} = 2$, so $x = 6$. Therefore, the minimum cost of this box is $C(6, 2) = 144$ (dollars). Because the volume of the box is $V = xyz = 48$, its height is $z = 48/(6 \cdot 2) = 4$ when $x = 6$ and $y = 2$. Thus the optimal box is 6 ft wide, 2 ft deep, and 4 ft high. ◆

REMARK As a check, note that the cheapest surfaces (front and back) are the largest, whereas the most expensive surfaces (the sides) are the smallest.

We have seen that if $f_x(a, b) = 0 = f_y(a, b)$, then $f(a, b)$ may be either a maximum value, a minimum value, or neither. In Section 13.10 we discuss conditions that suffice to distinguish between a local maximum, a local minimum, and a saddle point on the surface $z = f(x, y)$. These conditions involve the second-order derivatives of f.

Functions of Three or More Variables

The methods of this section generalize readily to functions of three or more variables. For instance, suppose that the function $f(x, y, z)$ is continuous on a bounded region R in space bounded by a closed surface S. Then (in analogy with Theorem 1), the function f attains an absolute maximum value at some point (a, b, c) of R (and likewise an absolute minimum value at some point of R). If (a, b, c) is an interior point of R at which the partial derivatives of f exist, then (in analogy with Theorem 3) it follows that all three vanish there:

$$f_x(a, b, c) = f_y(a, b, c) = f_z(a, b, c) = 0. \tag{7}$$

We may therefore attempt to find this point by solving the three simultaneous equations

$$f_x(x, y, z) = 0, \quad f_y(x, y, z) = 0, \quad \text{and} \quad f_z(x, y, z) = 0 \tag{8}$$

for the three unknown values $x = a$, $y = b$, and $z = c$. Thus a key step in the method of solution of a three-variable extreme value problem is essentially the same as in the method for a two-variable problem—"set the partial derivatives equal to zero and solve the resulting equations." But see Problems 68 through 70.

Example 8 illustrates a "line-through-the-point" method that we can sometimes use to show that a point (a, b, c) where the conditions in (8) hold is neither a local maximum nor a local minimum point. (The method is also applicable to functions of two or of more than three variables.)

EXAMPLE 8 Determine whether the function $f(x, y, z) = xy + yz - xz$ has any local extrema.

Solution The necessary conditions in Eq. (8) give the equations

$$f_x(x, y, z) = y - z = 0,$$
$$f_y(x, y, z) = x + z = 0,$$
$$f_z(x, y, z) = y - x = 0.$$

We easily find that the simultaneous solution of these equations is $x = y = z = 0$. On the line $x = y = z$ through $(0, 0, 0)$, the function $f(x, y, z)$ reduces to x^2, which is minimal at $x = 0$. But on the line $x = -y = z$, it reduces to $-3x^2$, which is maximal when $x = 0$. Hence f can have neither a local maximum nor a local minimum at $(0, 0, 0)$. Therefore it has no extrema, local *or* global. ◆

13.5 TRUE/FALSE STUDY GUIDE

13.5 CONCEPTS: QUESTIONS AND DISCUSSION

1. Suppose that the function f is continuous on the disk D bounded by the unit circle $x^2 + y^2 = 1$. Is it possible that $f(x, y)$ attains both its maximum and minimum values on D at points of the boundary circle? Illustrate your answer with an example.
2. Give an example of a function that is defined at every point of the unit disk D but attains no maximum value at any point of D.
3. Give an example of a function f defined on the unit disk D that attains its maximum value at an interior point at which the partial derivatives of D do not exist.
4. How would you alter the proof of Theorem 2 to show that the partial derivatives of a function of three variables vanish at an interior local maximum or minimum point? (What would you mean by an *interior point* of a space region?) Does your proof apply to the function $w = f(x, y, z) = \sqrt{x^2 + y^2 + z^2}$?

13.5 PROBLEMS

In Problems 1 through 12, find every point on the given surface $z = f(x, y)$ at which the tangent plane is horizontal.

1. $z = x - 3y + 5$

2. $z = 4 - x^2 - y^2$

3. $z = xy + 5$

4. $z = x^2 + y^2 + 2x$

5. $z = x^2 + y^2 - 6x + 2y + 5$

6. $z = 10 + 8x - 6y - x^2 - y^2$

7. $z = x^2 + 4x + y^3$

8. $z = x^4 + y^3 - 3y$

9. $z = 3x^2 + 12x + 4y^3 - 6y^2 + 5$ (Fig. 13.5.13)

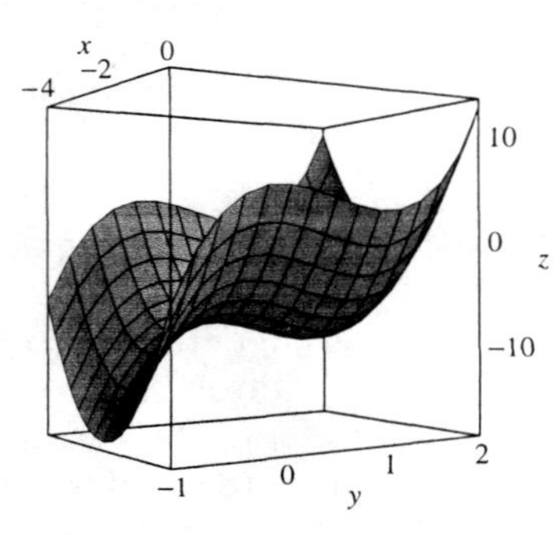

FIGURE 13.5.13 The surface of Problem 9.

10. $z = \dfrac{1}{1 - 2x + 2y + x^2 + y^2}$

11. $z = (2x^2 + 3y^2)\exp(-x^2 - y^2)$ (Fig. 13.5.14)

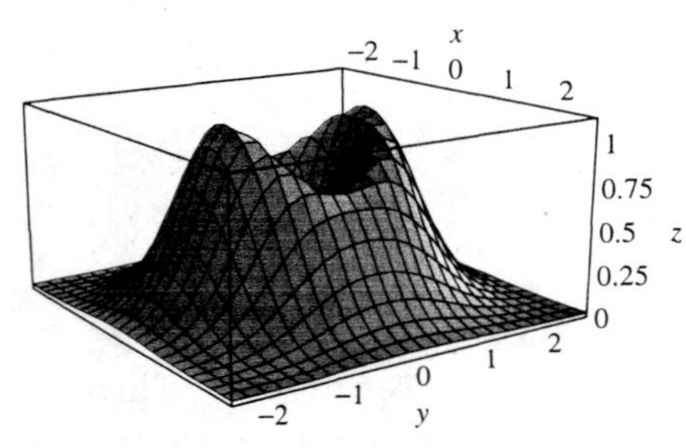

FIGURE 13.5.14 The surface of Problem 11.

12. $z = 2xy\exp\left(-\frac{1}{8}(4x^2 + y^2)\right)$ (Fig. 13.5.15)

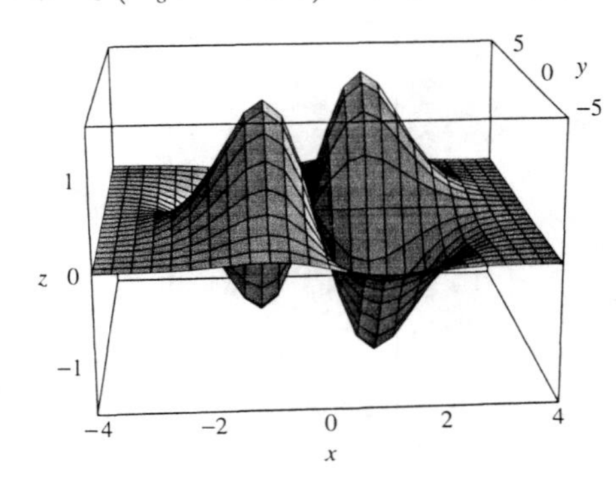

FIGURE 13.5.15 The surface of Problem 12.

Each of the surfaces defined in Problems 13 through 22 either opens downward and has a highest point, or opens upward and has a lowest point. Find this highest or lowest point on the surface $z = f(x, y)$.

13. $z = x^2 - 2x + y^2 - 2y + 3$

14. $z = 6x - 8y - x^2 - y^2$

15. $z = 2x - x^2 + 2y^2 - y^4$

16. $z = 4xy - x^4 - y^4$

17. $z = 3x^4 - 4x^3 - 12x^2 + 2y^2 - 12y$

18. $z = 3x^4 + 4x^3 + 6y^4 - 16y^3 + 12y^2$

19. $z = 2x^2 + 8xy + y^4$

20. $z = \dfrac{1}{10 - 2x - 4y + x^2 + y^4}$

21. $z = \exp(2x - 4y - x^2 - y^2)$

22. $z = (1 + x^2)\exp(-x^2 - y^2)$

In Problems 23 through 28, find the maximum and minimum values attained by the given function $f(x, y)$ on the given plane region R.

23. $f(x, y) = x + 2y$; R is the square with vertices at $(\pm 1, \pm 1)$.

24. $f(x, y) = x^2 + y^2 - x$; R is the square of Problem 23.

25. $f(x, y) = x^2 + y^2 - 2x$; R is the triangular region with vertices at $(0, 0)$, $(2, 0)$, and $(0, 2)$.

26. $f(x, y) = x^2 + y^2 - x - y$; R is the region of Problem 25.

27. $f(x, y) = 2xy$; R is the circular disk $x^2 + y^2 \leqq 1$.

28. $f(x, y) = xy^2$; R is the circular disk $x^2 + y^2 \leqq 3$.

In Problems 29 through 34, the equation of a plane or surface is given. Find the first-octant point $P(x, y, z)$ on the surface closest to the given fixed point $Q(x_0, y_0, z_0)$. [Suggestion: Minimize the squared distance $|PQ|^2$ as a function of x and y.]

29. The plane $12x+4y+3z = 169$ and the fixed point $Q(0, 0, 0)$

30. The plane $2x + 2y + z = 27$ and the fixed point $Q(9, 9, 9)$

31. The plane $2x + 3y + z = 49$ and the fixed point $Q(7, -7, 0)$

32. The surface $xyz = 8$ and the fixed point $Q(0, 0, 0)$

33. The surface $x^2y^2z = 4$ and the fixed point $Q(0, 0, 0)$

34. The surface $x^4y^8z^2 = 8$ and the fixed point $Q(0, 0, 0)$

35. Find the maximum possible product of three positive numbers whose sum is 120.

36. Find the maximum possible volume of a rectangular box if the sum of the lengths of its 12 edges is 6 meters.

37. Find the dimensions of the box with volume 1000 in.^{3} that has minimal total surface area.

38. Find the dimensions of the open-topped box with volume 4000 cm^3 whose bottom and four sides have minimal total surface area.

In Problems 39 through 42, you are to find the dimensions that minimize the total cost of the material needed to construct the rectangular box that is described. It is either closed *(top, bottom, and four sides) or* open-topped *(four sides and a bottom).*

39. The box is to be open-topped with a volume of 600 in.^{3} The material for its bottom costs 6¢/in.^{2} and the material for its four sides costs 5¢/in.^{2}

40. The box is to be closed with a volume of 48 ft^3. The material for its top and bottom costs \$3/ft^2 and the material for its four sides costs \$4/ft^2.

41. The box is to be closed with a volume of 750 in.^{3} The material for its top and bottom costs 3¢/in.^{2}, the material for its front and back costs 6¢/in.^{2}, and the material for its two ends costs 9¢/in.^{2}

42. The box is to be a closed shipping crate with a volume of 12 m^3. The material for its bottom costs *twice* as much (per square meter) as the material for its top and four sides.

43. A rectangular building is to have a volume of 8000 ft^3. Annual heating and cooling costs will amount to \$2/ft^2 for its top, front, and back, and \$4/ft^2 for the two end walls. What dimensions of the building would minimize these annual costs?

44. You want to build a rectangular aquarium with a bottom made of slate costing 28¢/in.^{2} Its sides will be glass, which costs 5¢/in.^{2}, and its top will be stainless steel, which costs 2¢/in.^{2} The volume of this aquarium is to be 24,000 in.^{3} What are the dimensions of the least expensive such aquarium?

45. A rectangular box is inscribed in the first octant with three of its sides in the coordinate planes, their common vertex at the origin, and the opposite vertex on the plane with equation $x+3y+7z=11$. What is the maximum possible volume of such a box?

46. Three sides of a rectangular box lie in the coordinate planes, their common vertex at the origin; the opposite vertex is on the plane with equation

$$\frac{x}{a} + \frac{y}{b} + \frac{z}{c} = 1$$

(a, b, and c are positive constants). In terms of a, b, and c, what is the maximum possible volume of such a box?

47. Find the maximum volume of a rectangular box that a post office will accept for delivery if the sum of its *length* and *girth* cannot exceed 108 in.

48. Repeat Problem 47 for the case of a cylindrical box—one shaped like a hatbox or a fat mailing tube.

49. A rectangular box with its base in the xy-plane is inscribed under the graph of the paraboloid $z=1 - x^2 - y^2$, $z \geqq 0$. Find the maximum possible volume of the box. [*Suggestion:* You may assume that the sides of the box are parallel to the vertical coordinate planes, and it follows that the box is symmetrically placed around these planes.]

50. What is the maximum possible volume of a rectangular box inscribed in a hemisphere of radius R? Assume that one face of the box lies in the planar base of the hemisphere.

51. A buoy is to have the shape of a right circular cylinder capped at each end by identical right circular cones with the same radius as the cylinder. Find the minimum possible surface area of the buoy, given that it has fixed volume V.

52. A pentagonal window is to have the shape of a rectangle surmounted by an isosceles triangle (with horizontal base, so the window is symmetric around its vertical axis), and the perimeter of the window is to be 24 ft. What are the dimensions of such a window that will admit the most light (because its area is the greatest)?

53. Find the point (x, y) in the plane for which the sum of the squares of its distances from $(0, 1)$, $(0, 0)$, and $(2, 0)$ is a minimum.

54. Find the point (x, y) in the plane for which the sum of the squares of its distances from (a_1, b_1), (a_2, b_2), and (a_3, b_3) is a minimum.

55. An A-frame house is to have fixed volume V. Its front and rear walls are in the shape of equal, parallel isosceles triangles with horizontal bases. The roof consists of two rectangles that connect pairs of upper sides of the triangles. To minimize heating and cooling costs, the total area of the A-frame (excluding the floor) is to be minimized. Describe the shape of the A-frame of minimal area.

56. What is the maximum possible volume of a rectangular box whose longest diagonal has fixed length L?

57. A wire 120 cm long is cut into three *or fewer* pieces, and each piece is bent into the shape of a square. How should this be done to minimize the total area of these squares? To maximize it?

58. You must divide a lump of putty of fixed volume V into three or fewer pieces and form the pieces into cubes. How should you do this to maximize the total surface area of the cubes? To minimize it?

59. A very long rectangle of sheet metal has width L and is to be folded to make a rain gutter (Fig. 13.5.16). Maximize its volume by maximizing the cross-sectional area shown in the figure.

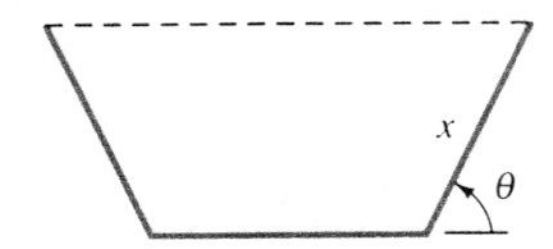

FIGURE 13.5.16 Cross section of the rain gutter of Problem 59.

60. Consider the function $f(x, y) = (y - x^2)(y - 3x^2)$. (a) Show that $f_x(0, 0) = 0 = f_y(0, 0)$. (b) Show that for every straight line $y = mx$ through $(0, 0)$, the function $f(x, mx)$ has a local minimum at $x = 0$. (c) Examine the values of f at points of the parabola $y = 2x^2$ to show that f does *not* have a local minimum at $(0, 0)$. This tells us that we cannot use the line-through-the-point method of Example 8 to show that a point *is* a local extremum.

61. Suppose that Alpha, Inc. and Beta, Ltd. manufacture competitive (but not identical) products, with the weekly sales of each product determined by the selling price of that product *and* the price of its competition. Suppose that Alpha sets a sales price of x dollars per unit for its product, while Beta sets a sales price of y dollars per unit for its product. Market research shows that the weekly profit made by Alpha is then

$$P(x) = -2x^2 + 12x + xy - y - 10$$

and that the weekly profit made by Beta is

$$Q(y) = -3y^2 + 18y + 2xy - 2x - 15$$

(both in thousands of dollars). The peculiar notation arises from the fact that x is the only variable under the control of Alpha and y is the only variable under the control of Beta. (If this disturbs you, feel free to write $P(x, y)$ in place of $P(x)$ and $Q(x, y)$ in place of $Q(y)$.) (a) Assume that both company managers know calculus and that each knows that the *other* knows calculus and has some common sense. What price will each manager set to maximize his company's weekly profit? (b) Now suppose that the two managers enter into an agreement (legal or otherwise) by which they plan to maximize their *total* weekly profit. Now what should be the selling price of each product? (We suppose that they will divide the resulting profit in an equitable way, but the details of this intriguing problem are not the issue.)

62. Three firms—Ajax Products (AP), Behemoth Quicksilver (BQ), and Conglomerate Resources (CR)—produce products in quantities A, B, and C, respectively. The weekly profits that accrue to each, in thousands of dollars, obey the following equations:

$$\text{AP:} \quad P = 1000A - A^2 - 2AB,$$
$$\text{BQ:} \quad Q = 2000B - 2B^2 - 4BC,$$
$$\text{CR:} \quad R = 1500C - 3C^2 - 6AC.$$

(a) If each firm acts independently to maximize its weekly profit, what will those profits be? (b) If firms AP and CR join to maximize their total profit while BQ continues to act alone, what effects will this have? Give a *complete* answer to this problem. Assume that the fact of the merger of AP and CR is known to the management of BQ.

63. A farmer can raise sheep, hogs, and cattle. She has space for 80 sheep or 120 hogs or 60 cattle or any combination using the same amount of space; that is, 8 sheep use as much space as 12 hogs or 6 cattle. The anticipated profits per animal are \$10 per sheep, \$8 per hog, and \$20 for each head of cattle. State law requires that a farmer raise as many hogs as sheep and cattle combined. How does the farmer maximize her profit?

Problems 64 and 65 deal with the quadratic form

$$f(x, y) = ax^2 + 2bxy + cy^2. \tag{9}$$

64. Show that the quadratic form f in (9) has only the single critical point $(0, 0)$ unless $ac - b^2 = 0$, in which case every point on a certain line through the origin is a critical point. Experiment with computer graphs to formulate a conjecture about the shape of the surface $z = f(x, y)$ in the exceptional case $ac - b^2 = 0$. Can you substantiate your conjecture?

65. Use a computer algebra system to graph the quadratic form in (9) for a variety of different values of the coefficients a, b, and c in order to corroborate the following two conclusions. (a) If $ac - b^2 > 0$, then the graph of $z = f(x, y)$ is an elliptic paraboloid and f therefore has either a maximum or a minimum value at $(0, 0)$. (b) If $ac - b^2 < 0$, then the graph of $z = f(x, y)$ is a hyperbolic paraboloid and f therefore has a saddle point at $(0, 0)$.

Figures 13.5.7 and 13.5.8 illustrate (and Problems 66 and 67 deal with) the cases $b = -\frac{1}{2}$ and $b = -\frac{3}{2}$ (respectively) of the special quartic form

$$f(x, y) = x^4 + 2bx^2y^2 + y^4. \tag{10}$$

66. Show that the quartic form f in (10) has only the single critical point $(0, 0)$ unless $b = -1$, in which case every point on a certain pair of lines through the origin is a critical point (Fig. 13.5.17). Experiment with computer graphs to formulate a conjecture about the shape of the graph of f in each of the two cases $b > -1$ and $b < -1$.

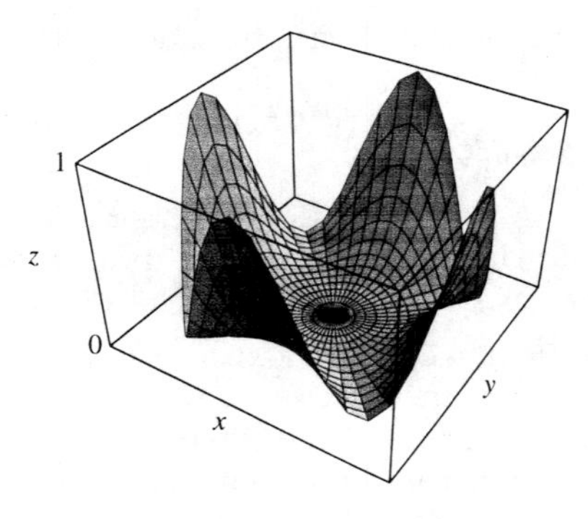

FIGURE 13.5.17 The graph of the function $f(x, y) = x^4 - x^2y^2 + y^4$ having critical points on the lines $y = \pm x$.

67. To show that the quartic form in (10) has a local minimum at the origin if $b > -1$ and a saddle point if $b < -1$, substitute $x = r\cos\theta$, $y = r\sin\theta$ and write $x^4 + 2bx^2y^2 + y^4 = r^4g(\theta)$. Then find the maximum and minimum values of $g(\theta)$ for $0 \leqq \theta \leqq 2\pi$.

68. Find the global maximum and minimum values of
$$f(x, y, z) = x^2 - 6xy + y^2 + 2yz + z^2 + 12.$$
What happens at the point or points at which all three partial derivatives of f are simultaneously zero?

69. Find the global maximum and minimum values of
$$g(x, y, z) = x^4 - 8x^2y^2 + y^4 + z^4 + 12.$$
What happens at the point or points at which all three partial derivatives of g are simultaneously zero?

70. The plane $\mathcal{P}$ with equation $x + y + z = 1$ meets the first octant in the triangle T for which x, y, and z are all nonnegative. Find the maximum value of the expression $E = x - y + z$ on T. You will probably proceed by solving the equation of the plane $\mathcal{P}$ for $z = 1 - x - y$ and substituting for z in the expression E to obtain the quantity $h(x, y) = x - y + (1 - x - y)$ to be maximized. What happens at the point or points at which both partial derivatives of h are simultaneously zero?

13.6 INCREMENTS AND LINEAR APPROXIMATION

In Section 4.2 we used the *differential*
$$df = f'(x)\,\Delta x$$
to approximate the *increment*, or actual change,
$$\Delta f = f(x + \Delta x) - f(x)$$
in the value of a single-variable function that results from the change Δx in the independent variable. Thus $\Delta f \approx df$; that is,
$$f(x + \Delta x) - f(x) \approx f'(x)\,\Delta x. \tag{1}$$

We now describe the use of the partial derivatives $\partial f/\partial x$ and $\partial f/\partial y$ to approximate the **increment**
$$\Delta f = f(x + \Delta x, y + \Delta y) - f(x, y) \tag{2}$$
in the value of a function f (of two variables) that results when its two independent variables are changed simultaneously. If only x were changed and y were held constant, we could temporarily regard $f(x, y)$ as a function of x alone. Then, with $f_x(x, y)$ playing the role of $f'(x)$, the linear approximation in Eq. (1) would give
$$f(x + \Delta x, y) - f(x, y) \approx f_x(x, y)\,\Delta x \tag{3}$$
for the change in f corresponding to the change Δx in x. Similarly, if only y were changed and x were held constant, then—temporarily regarding $f(x, y)$ as a function of y alone—we would get
$$f(x, y + \Delta y) - f(x, y) \approx f_y(x, y)\,\Delta y \tag{4}$$
for the change in f corresponding to the change Δy in y.

If both x and y are changed simultaneously, we might expect the *sum* of the approximations in (3) and (4) to be a good estimate of the resulting increment in the value of f. On this basis we define the **differential**
$$df = f_x(x, y)\,\Delta x + f_y(x, y)\,\Delta y \tag{5}$$
of a function of two independent variables. The approximation $\Delta f \approx df$ then yields the approximation
$$f(x + \Delta x, y + \Delta y) \approx f(x, y) + f_x(x, y)\,\Delta x + f_y(x, y)\,\Delta y. \tag{6}$$

EXAMPLE 1 Find the differential df of the function $f(x, y) = x^2 + 3xy - 2y^2$. Then compare df and the actual increment Δf when (x, y) changes from $P(3, 5)$ to $Q(3.2, 4.9)$.

Solution The differential of f, as given in Eq. (5), is

$$df = \frac{\partial f}{\partial x}\,\Delta x + \frac{\partial f}{\partial y}\,\Delta y = (2x + 3y)\,\Delta x + (3x - 4y)\,\Delta y.$$

At the point $P(3, 5)$ this differential is

$$df = (2\cdot 3 + 3\cdot 5)\,\Delta x + (3\cdot 3 - 4\cdot 5)\,\Delta y = 21\,\Delta x - 11\,\Delta y.$$

With $\Delta x = 0.2$ and $\Delta y = -0.1$, corresponding to the change from $P(3, 5)$ to $Q(3.2, 4.9)$, we get

$$df = 21\cdot(0.2) - 11\cdot(-0.1) = 5.3.$$

The actual change in the value of f from P to Q is the increment

$$\Delta f = f(3.2, 4.9) - f(3, 5) = 9.26 - 4 = 5.26,$$

so in this example the differential seems to be a good approximation to the increment. ◆

At the fixed point $P(a, b)$, the differential

$$df = f_x(a, b)\,\Delta x + f_y(a, b)\,\Delta y \tag{7}$$

is a *linear* function of Δx and Δy; the coefficients $f_x(a, b)$ and $f_y(a, b)$ in this linear function depend on a and b. Thus the differential df is a *linear approximation* to the actual increment Δf. The linear approximation theorem (stated later in this section) implies that if the function f has continuous partial derivatives, then df is a *very good approximation* to Δf when the changes Δx and Δy in x and y are sufficiently small. The **linear approximation**

$$f(a + \Delta x, b + \Delta y) \approx f(a, b) + f_x(a, b)\,\Delta x + f_y(a, b)\,\Delta y \tag{8}$$

may then be used to estimate the value of $f(a + \Delta x, b + \Delta y)$ when Δx and Δy are small and the values $f(a, b)$, $f_x(a, b)$, and $f_y(a, b)$ are all known.

EXAMPLE 2 Use linear approximation to estimate $\sqrt{2\cdot(2.02)^3 + (2.97)^2}$.

Solution We begin by letting $f(x, y) = \sqrt{2x^3 + y^2}$, $a = 2$, and $b = 3$. It is then easy to compute the exact value $f(2, 3) = \sqrt{2\cdot 8 + 9} = \sqrt{25} = 5$. Next,

$$\frac{\partial f}{\partial x} = \frac{3x^2}{\sqrt{2x^3 + y^2}} \quad\text{and}\quad \frac{\partial f}{\partial y} = \frac{y}{\sqrt{2x^3 + y^2}},$$

so

$$f_x(2, 3) = \tfrac{12}{5} \quad\text{and}\quad f_y(2, 3) = \tfrac{3}{5}.$$

Hence Eq. (8) with $\Delta x = 0.02$ and $\Delta y = -0.03$ gives

$$\begin{aligned}\sqrt{2\cdot(2.02)^3 + (2.97)^2} &= f(2.02, 2.97)\\ &\approx f(2, 3) + f_x(2, 3)\cdot(0.02) + f_y(2, 3)\cdot(-0.03)\\ &= 5 + \tfrac{12}{5}\cdot(0.02) - \tfrac{3}{5}\cdot(0.03) = 5.03.\end{aligned}$$

The actual value to four decimal places is 5.0305. ◆

If $z = f(x, y)$, we often write dz in place of df. So the differential of the dependent variable z at the point (a, b) is $dz = f_x(a, b)\,\Delta x + f_y(a, b)\,\Delta y$. At the arbitrary point (x, y) the differential of z takes the form

$$dz = f_x(x, y)\,\Delta x + f_y(x, y)\,\Delta y.$$

More simply, we can write

$$dz = \frac{\partial z}{\partial x}\,\Delta x + \frac{\partial z}{\partial y}\,\Delta y. \tag{9}$$

It is customary to write dx for Δx and dy for Δy in this formula. When this is done, Eq. (9) takes the form

$$dz = \frac{\partial z}{\partial x}\,dx + \frac{\partial z}{\partial y}\,dy. \tag{10}$$

When we use this notation, we must realize that dx and dy have *no* connotation of being "infinitesimal" or even small. The differential dz is still simply a linear function of the ordinary real variables dx and dy, a function that gives a linear approximation to the change in z when x and y are changed by the amounts dx and dy, respectively.

EXAMPLE 3 In Example 4 of Section 13.4, we considered 1 mole of an ideal gas—its volume V in cubic centimeters given in terms of its pressure p in atmospheres and temperature T in kelvins by the formula $V = (82.06)T/p$. Approximate the change in V when p is increased from 5 atm to 5.2 atm and T is increased from 300 K to 310 K.

Solution The differential of $V = V(p, T)$ is

$$dV = \frac{\partial V}{\partial p}\,dp + \frac{\partial V}{\partial T}\,dT = -\frac{82.06 \cdot T}{p^2}\,dp + \frac{82.06}{p}\,dT.$$

With $p = 5$, $T = 300$, $dp = 0.2$, and $dT = 10$, we compute

$$dV = -\frac{82.06 \cdot 300}{5^2} \cdot 0.2 + \frac{82.06}{5} \cdot 10 = -32.8 \quad (\text{cm}^3).$$

This indicates that the gas will decrease in volume by about 33 cm^3. The actual change is

$$\Delta V = \frac{82.06 \cdot 310}{5.2} - \frac{82.06 \cdot 300}{5} = 4892.0 - 4923.6 = -31.6 \quad (\text{cm}^3).$$ ◆

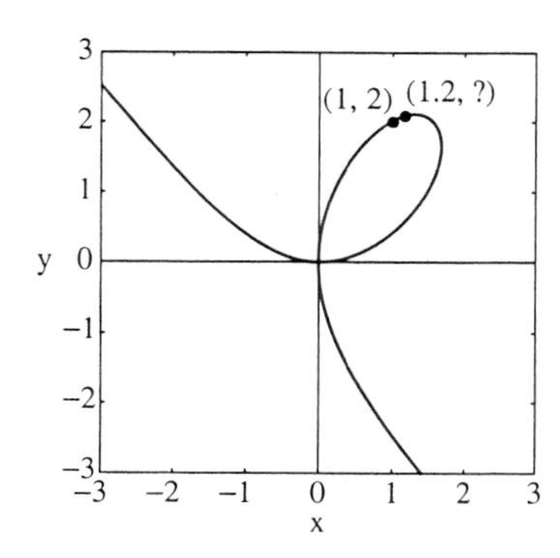

FIGURE 13.6.1 The curve of Example 4.

EXAMPLE 4 The point (1, 2) lies on the curve with equation

$$f(x, y) = 2x^3 + y^3 - 5xy = 0. \tag{11}$$

(See Fig. 13.6.1.) Approximate the y-coordinate of the nearby point (x, y) on this curve for which $x = 1.2$.

Solution The increment between $f(1, 2) = 0$ and $f(x, y) = 0$ on this curve is $\Delta f = 0 \approx df$, so when we compute the differentials in Eq. (11), we get

$$df = \frac{\partial f}{\partial x}\,dx + \frac{\partial f}{\partial y}\,dy = (6x^2 - 5y)\,dx + (3y^2 - 5x)\,dy = 0.$$

Now when we substitute $x = 1$, $y = 2$, and $dx = 0.2$, we obtain the equation $(-4)(0.2) + (7)\,dy = 0$. It then follows that $dy = (0.8)/7 \approx 0.114 \approx 0.1$. This yields (1.2, 2.1) for the approximate coordinates of the nearby point. As a check on the accuracy of this approximation, we can substitute $x = 1.2$ into Eq. (11). This gives the equation

$$2 \cdot (1.2)^3 + y^3 - 5 \cdot (1.2)y = y^3 - 6y + 3.456 = 0.$$

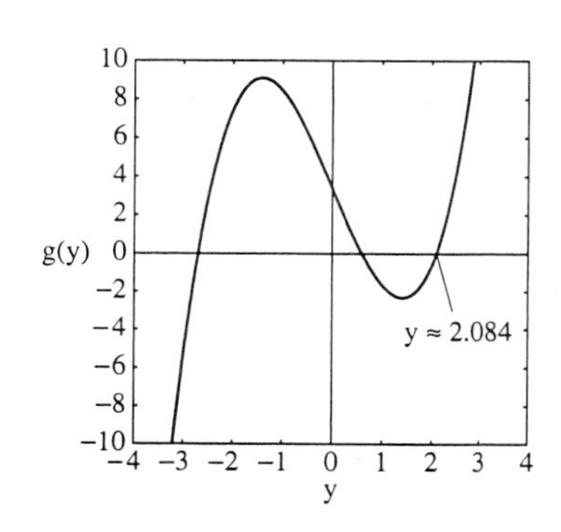

FIGURE 13.6.2 The graph of $g(y) = y^3 - 6y + 3.456$.

The roots of this equation are the x-intercepts of the curve in Fig. 13.6.2. A calculator or computer with an equation solver (or Newton's method) then yields $y \approx 2.084 \approx 2.1$ for the solution near $y = 2$. ◆

Functions of Three or More Variables

Increments and differentials of functions of more than two variables are defined similarly. A function $w = f(x, y, z)$ has **increment**

$$\Delta w = \Delta f = f(x + \Delta x, y + \Delta y, z + \Delta z) - f(x, y, z)$$

and **differential**

$$dw = df = \frac{\partial f}{\partial x}\Delta x + \frac{\partial f}{\partial y}\Delta y + \frac{\partial f}{\partial z}\Delta z;$$

that is,

$$dw = \frac{\partial w}{\partial x}dx + \frac{\partial w}{\partial y}dy + \frac{\partial w}{\partial z}dz$$

if, as in Eq. (10), we write dx for Δx, dy for Δy, and dz for Δz.

EXAMPLE 5 You have constructed a metal cube that is supposed to have edge length 100 mm, but each of its three measured dimensions x, y, and z may be in error by as much as a millimeter. Use differentials to estimate the maximum resulting error in its calculated volume $V = xyz$.

Solution We need to approximate the increment

$$\Delta V = V(100 + dx, 100 + dy, 100 + dz) - V(100, 100, 100)$$

when the errors dx, dy, and dz in x, y, and z are maximal. The differential of $V = xyz$ is

$$dV = yz\ dx + xz\ dy + xy\ dz.$$

When we substitute $x = y = z = 100$ and $dx = \pm 1$, $dy = \pm 1$, and $dz = \pm 1$, we get

$$dV = 100 \cdot 100 \cdot (\pm 1) + 100 \cdot 100 \cdot (\pm 1) + 100 \cdot 100 \cdot (\pm 1) = \pm 30000.$$

It may surprise you to find that an error of only a millimeter in each dimension of a cube can result in an error of 30,000 mm^3 in its volume. (For a cube made of precious metal, an error of 30 cm^3 in its volume could correspond to a difference of hundreds or thousands of dollars in its cost.) ◆

Linear Approximation and Differentiability

Vector notation simplifies the description of differentials and linear approximation for functions of several variables. Let $f(\mathbf{x}) = f(x_1, x_2, \ldots, x_n)$ be a real-valued function of n variables. If

$$\mathbf{x} = \langle x_1, x_2, \ldots, x_n \rangle \quad \text{and} \quad \mathbf{h} = \langle h_1, h_2, \ldots, h_n \rangle,$$

then the linear approximation formula for f takes the form

$$f(\mathbf{x} + \mathbf{h}) \approx f(\mathbf{x}) + \frac{\partial f}{\partial x_1}h_1 + \frac{\partial f}{\partial x_2}h_2 + \cdots + \frac{\partial f}{\partial x_n}h_n \tag{12}$$

with one term for each independent variable. We introduce the **gradient vector**

$$\nabla f(\mathbf{x}) = \langle D_1 f(\mathbf{x}), D_2 f(\mathbf{x}), \ldots, D_n f(\mathbf{x}) \rangle = \left\langle \frac{\partial f}{\partial x_1}, \frac{\partial f}{\partial x_2}, \ldots, \frac{\partial f}{\partial x_n} \right\rangle \tag{13}$$

of the function $f(x_1, x_2, \ldots, x_n)$ of n variables; its elements are the n first-order partial derivatives of f (assuming that they exist). This new vector-valued function is called the *gradient* of f and is denoted by ∇f (pronounced "del f"). In Section 13.8 we explore the meaning of the gradient vector ∇f; here we use it simply as a notational device to simplify the formula in (12).

The dot (or scalar) product of two n-vectors is, exactly as in dimensions 2 and 3, the sum of the products of corresponding elements of the two vectors. That is, if $\mathbf{a} = \langle a_1, a_2, \dots, a_n \rangle$ and $\mathbf{b} = \langle b_1, b_2, \dots, b_n \rangle$, then

$$\mathbf{a} \cdot \mathbf{b} = a_1 b_1 + a_2 b_2 + \cdots + a_n b_n.$$

Consequently, the linear approximation formula in (12) takes the concise form

$$f(\mathbf{x} + \mathbf{h}) \approx f(\mathbf{x}) + \nabla f(\mathbf{x}) \cdot \mathbf{h}, \tag{14}$$

in pleasant analogy with the original single-variable approximation $f(x+h) \approx f(x) + f'(x)h$ (writing h for Δx here). Because $\nabla f(\mathbf{x})$ and $\mathbf{h}$ are both vectors with n components, the dot product on the right-hand side in (14) is defined and gives

$$\nabla f(\mathbf{x}) \cdot \mathbf{h} = D_1 f(\mathbf{x}) h_1 + D_2 f(\mathbf{x}) h_2 + \cdots + D_n f(\mathbf{x}) h_n,$$

thus providing the linear terms on the right-hand side in (12). In analogy with the two-variable case in (5), the sum of these n linear terms is the **differential** $df = \nabla f(\mathbf{x}) \cdot \mathbf{h}$ of the function f of n real variables. With $\mathbf{h} = \mathbf{dx} = \langle dx_1, dx_2, \dots, dx_n \rangle$, this differential takes the form

$$df = \frac{\partial f}{\partial x_1}\,dx_1 + \frac{\partial f}{\partial x_2}\,dx_2 + \cdots + \frac{\partial f}{\partial x_n}\,dx_n$$

that generalizes the two-dimensional differential in Eq. (10).

The gradient vector $\nabla f(\mathbf{x})$ is defined wherever all of the first-order partial derivatives of f exist. In Appendix K we give a proof of the linear approximation theorem stated next. This theorem assures us (in effect) that if the partial derivatives of f are also *continuous*, then the linear approximation in (14) is a *good* approximation when $|\mathbf{h}| = \sqrt{h_1^2 + h_2^2 + \cdots + h_n^2}$ is small.

THEOREM Linear Approximation

Suppose that the function $f(\mathbf{x})$ of n variables has continuous first-order partial derivatives in a region that contains the neighborhood $|\mathbf{x} - \mathbf{a}| < r$ consisting of all points $\mathbf{x}$ at distance less than r from the fixed point $\mathbf{a}$. If $\mathbf{a} + \mathbf{h}$ lies in this neighborhood, then

$$f(\mathbf{a} + \mathbf{h}) = f(\mathbf{a}) + \nabla f(\mathbf{a}) \cdot \mathbf{h} + \boldsymbol{\epsilon}(\mathbf{h}) \cdot \mathbf{h} \tag{15}$$

where $\boldsymbol{\epsilon}(\mathbf{h}) = \langle \epsilon_1(\mathbf{h}), \epsilon_2(\mathbf{h}), \dots, \epsilon_n(\mathbf{h}) \rangle$ is a vector such that each element $\epsilon_i(\mathbf{h})$ approaches zero as $\mathbf{h} \to \mathbf{0}$.

REMARK 1 The multivariable function f is said to be **continuously differentiable** at a point provided that its first-order partial derivatives not only exist but are continuous at the point. Thus the hypothesis of the linear approximation theorem is that the function f is continuously differentiable in the specified neighborhood of the point $\mathbf{a}$.

REMARK 2 The dot product

$$\boldsymbol{\epsilon}(\mathbf{h}) \cdot \mathbf{h} = \epsilon_1(\mathbf{h}) h_1 + \epsilon_2(\mathbf{h}) h_2 + \cdots + \epsilon_n(\mathbf{h}) h_n \tag{16}$$

in (15) is the **error** in the linear approximation—it measures the extent to which the *approximation* $f(\mathbf{a}+\mathbf{h}) \approx f(\mathbf{a}) + \nabla f(\mathbf{a}) \cdot \mathbf{h}$ fails to be an *equality*. We may regard the conclusion of the linear approximation theorem as saying that if $\mathbf{h}$ is "very small," then each element $\epsilon_i(\mathbf{h})$ of $\boldsymbol{\epsilon}(\mathbf{h})$ is also "very small." In this event, each summand in (16) is a product of two very small terms, so we might say that the error $\boldsymbol{\epsilon}(\mathbf{h}) \cdot \mathbf{h}$ is "very very small."

Now let us divide by $|\mathbf{h}|$ in Eq. (16). Then we see that

$$\frac{\boldsymbol{\epsilon}(\mathbf{h}) \cdot \mathbf{h}}{|\mathbf{h}|} = \epsilon_1(\mathbf{h}) \frac{h_1}{|\mathbf{h}|} + \epsilon_2(\mathbf{h}) \frac{h_2}{|\mathbf{h}|} + \cdots + \epsilon_n(\mathbf{h}) \frac{h_n}{|\mathbf{h}|} \to 0 \tag{17}$$

as $\mathbf{h} \to \mathbf{0}$. The reason is that, for each i $(1 \leqq i \leqq n)$,

$$\frac{h_i}{|\mathbf{h}|} \leqq 1 \quad \text{and} \quad \epsilon_i(\mathbf{h}) \to 0$$

as $\mathbf{h} \to \mathbf{0}$. Dividing both sides by $|\mathbf{h}|$ in Eq. (15) therefore gives the limit

$$\lim_{\mathbf{h} \to \mathbf{0}} \frac{f(\mathbf{a}+\mathbf{h}) - f(\mathbf{a}) - \nabla f(\mathbf{a}) \cdot \mathbf{h}}{|\mathbf{h}|} = 0, \tag{18}$$

under the assumption that the function f is continuously differentiable near $\mathbf{a}$.

The condition in Eq. (18) is central to the study of differentiability of multivariable functions. Indeed, the real-valued function $f(\mathbf{x})$ is said to be **differentiable** at the point $\mathbf{a}$ provided that there exists a constant vector $\mathbf{c} = \langle c_1, c_2, \dots, c_n \rangle$ such that

$$\lim_{\mathbf{h} \to \mathbf{0}} \frac{f(\mathbf{a}+\mathbf{h}) - f(\mathbf{a}) - \mathbf{c} \cdot \mathbf{h}}{|\mathbf{h}|} = 0. \tag{19}$$

In effect, this definition means that f is differentiable at $\mathbf{a}$ if there exists a linear function $\mathbf{c} \cdot \mathbf{h} = c_1 h_1 + c_2 h_2 + \cdots + c_n h_n$ (of the components of $\mathbf{h}$) that approximates the increment $f(\mathbf{a}+\mathbf{h}) - f(\mathbf{a})$ so closely that the error is small even in comparison with $|\mathbf{h}|$. Equation (18) implies that if f is continuously differentiable near $\mathbf{a}$, then the gradient vector $\nabla f(\mathbf{a})$ is precisely such a vector $\mathbf{c}$ (and moreover, by Problem 48, is the only such vector).

Thus *a function is differentiable if it is continuously differentiable*. This says little in the case of a single-variable function, which is called differentiable if its derivative merely exists. In contrast, we have as yet said nothing about the existence of partial derivatives of a differentiable multivariable function. The following example treats the case of just $n = 2$ variables.

EXAMPLE 6 Suppose that the function $f(x, y)$ is differentiable at the point (a, b). By Eq. (19), this means that there exists a constant vector $\mathbf{c} = \langle c_1, c_2 \rangle$ such that

$$\lim_{(h_1, h_2) \to (0,0)} \frac{f(a+h_1, b+h_2) - f(a, b) - (c_1 h_1 + c_2 h_2)}{\sqrt{h_1^2 + h_2^2}} = 0. \tag{20}$$

If $h_1 = h$ and $h_2 = 0$, then Eq. (20) implies that

$$\lim_{h \to 0} \frac{f(a+h, b) - f(a, b) - c_1 h}{h} = 0,$$

and hence that

$$\lim_{h \to 0} \frac{f(a+h, b) - f(a, b)}{h} = c_1.$$

Thus the partial derivative $f_x(a, b)$ exists and is equal to the first element c_1 of $\mathbf{c}$. Similarly, if we substitute $h_1 = 0$ and $h_2 = h$ in (20)—do this yourself—we find that the partial derivative $f_y(a, b)$ exists and is equal to the second element c_2 of $\mathbf{c}$. ◆

Example 6 is the case $n = 2$ of the general theorem that *differentiability at a point implies existence of all first-order partial derivatives at that point*. It is also true that differentiability implies continuity (Problem 47). In summary, we have the following implications for a function f of several variables:

- If f is continuously differentiable, then f is differentiable.
- If f is differentiable, then all partial derivatives of f exist.
- If f is differentiable, then f is continuous.

Problems 43 through 45 show that none of these implications can be reversed for a function f of two or more variables. That is, f can have partial derivatives without being differentiable, and can be differentiable without being continuously differentiable. Moreover, f can have partial derivatives without being continuous (and vice versa). Thus the mere existence of partial derivatives—even all of them—appears to imply much less for a function of several variables than it does for a single-variable function. But all these distinctions disappear in the case of polynomials and rational functions of several variables—which have continuous partial derivatives wherever they are defined.

13.6 TRUE/FALSE STUDY GUIDE

13.6 CONCEPTS: QUESTIONS AND DISCUSSION

1. Compare the concept of differentiability for single-variable functions with that for multivariable functions.
2. Compare the roles of the derivative of a single-variable function and the gradient vector of a multivariable function. For instance, what is the value of the gradient vector at a local maximum or minimum point?
3. Does a surface $z = f(x, y)$ always have a tangent plane at a point **a** where f is differentiable? Describe the way this tangent plane approximates the graph near the point $(\mathbf{a}, f(\mathbf{a}))$.

13.6 PROBLEMS

Find the differential dw in Problems 1 through 16.

1. $w = 3x^2 + 4xy - 2y^3$

2. $w = \exp(-x^2 - y^2)$

3. $w = \sqrt{1 + x^2 + y^2}$

4. $w = xye^{x+y}$

5. $w = \arctan\left(\dfrac{x}{y}\right)$

6. $w = xz^2 - yx^2 + zy^2$

7. $w = \ln(x^2 + y^2 + z^2)$

8. $w = \sin xyz$

9. $w = x \tan yz$

10. $w = xye^{uv}$

11. $w = e^{-xyz}$

12. $w = \ln(1 + rs)$

13. $w = u^2 \exp(-v^2)$

14. $w = \dfrac{s + t}{s - t}$

15. $w = \sqrt{x^2 + y^2 + z^2}$

16. $w = pqr \exp(-p^2 - q^2 - r^2)$

In Problems 17 through 24, use the exact value f(P) and the differential df to approximate the value f(Q).

17. $f(x, y) = \sqrt{x^2 + y^2}$; $P(3, 4)$, $Q(2.97, 4.04)$

18. $f(x, y) = \sqrt{x^2 - y^2}$; $P(13, 5)$, $Q(13.2, 4.9)$

19. $f(x, y) = \dfrac{1}{1 + x + y}$; $P(3, 6)$, $Q(3.02, 6.05)$

20. $f(x, y, z) = \sqrt{xyz}$; $P(1, 3, 3)$, $Q(0.9, 2.9, 3.1)$

21. $f(x, y, z) = \sqrt{x^2 + y^2 + z^2}$; $P(3, 4, 12)$, $Q(3.03, 3.96, 12.05)$

22. $f(x, y, z) = \dfrac{xyz}{x + y + z}$; $P(2, 3, 5)$, $Q(1.98, 3.03, 4.97)$

23. $f(x, y, z) = e^{-xyz}$; $P(1, 0, -2)$, $Q(1.02, 0.03, -2.02)$

24. $f(x, y) = (x - y)\cos 2\pi xy$; $P\left(1, \frac{1}{2}\right)$, $Q(1.1, 0.4)$

In Problems 25 through 32, use differentials to approximate the indicated number.

25. $\left(\sqrt{15} + \sqrt{99}\right)^2$

26. $\left(\sqrt{26}\right)\left(\sqrt[3]{28}\right)\left(\sqrt[4]{17}\right)$

27. $e^{0.4} = \exp(1.1^2 - 0.9^2)$

28. $\dfrac{\sqrt[3]{25}}{\sqrt[5]{30}}$

29. $\sqrt{(3.1)^2 + (4.2)^2 + (11.7)^2}$

30. $\sqrt[3]{(5.1)^2 + 2 \cdot (5.2)^2 + 2 \cdot (5.3)^2}$

31. The y-coordinate of the point P near $(1, 2)$ on the curve $2x^3 + 2y^3 = 9xy$, if the x-coordinate of P is 1.1.

32. The x-coordinate of the point P near $(2, 4)$ on the curve $4x^4 + 4y^4 = 17x^2y^2$, if the y-coordinate of P is 3.9.

33. The base and height of a rectangle are measured as 10 cm and 15 cm, respectively, with a possible error of as much as 0.1 cm in each measurement. Use differentials to estimate the maximum resulting error in computing the area of the rectangle.

34. The base radius r and the height h of a right circular cylinder are measured as 3 cm and 9 cm, respectively. There is a possible error of 1 mm in each measurement. Use differentials to estimate the maximum possible error in computing: (a) the volume of the cylinder; (b) the total surface area of the cylinder.

35. The base radius r and height h of a right circular cone are measured as 5 in. and 10 in., respectively. There is a possible error of as much as $\frac{1}{10}$ in. in each measurement. Use differentials to estimate the maximum resulting error that might occur in computing the volume of the cone.

36. The dimensions of a closed rectangular box are found by measurement to be 10 cm by 15 cm by 20 cm, but there is a possible error of 0.1 cm in each. Use differentials to estimate the maximum resulting error in computing the total surface area of the box.

37. A surveyor wants to find the area in acres of a certain field (1 acre is 43,560 ft^2). She measures two different sides, finding them to be $a = 500$ ft and $b = 700$ ft, with a possible error of as much as 1 ft in each measurement. She finds the angle between these two sides to be $\theta = 30°$, with a possible error of as much as $0.25°$. The field is triangular, so its area is given by $A = \frac{1}{2}ab\sin\theta$. Use differentials to estimate the maximum resulting error, in acres, in computing the area of the field by this formula.

38. Use differentials to estimate the change in the volume of the gas of Example 3 if its pressure is decreased from 5 atm to 4.9 atm and its temperature is decreased from 300 K to 280 K.

39. The period of oscillation of a simple pendulum of length L is given (approximately) by the formula $T = 2\pi\sqrt{L/g}$. Estimate the change in the period of a pendulum if its length is increased from 2 ft to 2 ft 1 in. and it is simultaneously moved from a location where g is exactly 32 ft/s^2 to one where $g = 32.2$ ft^2.

40. Given the pendulum of Problem 39, show that the relative error in the determination of T is half the difference of the relative errors in measuring L and g—that is, that

$$\frac{dT}{T} = \frac{1}{2}\left(\frac{dL}{L} - \frac{dg}{g}\right).$$

41. The range of a projectile fired (in a vacuum) with initial velocity v_0 and inclination angle α from the horizontal is $R = \frac{1}{32}v_0^2 \sin 2\alpha$. Use differentials to approximate the change in range if v_0 is increased from 400 to 410 ft/s and α is increased from 30° to 31°.

42. A horizontal beam is supported at both ends and supports a uniform load. The deflection, or sag, at its midpoint is given by

$$S = \frac{k}{wh^3}, \tag{21}$$

where w and h are the width and height, respectively, of the beam and k is a constant that depends on the length and composition of the beam and the amount of the load. Show that

$$dS = -S\left(\frac{1}{w}\,dw + \frac{3}{h}\,dh\right).$$

If $S = 1$ in. when $w = 2$ in. and $h = 4$ in., approximate the sag when $w = 2.1$ in. and $h = 4.1$ in. Compare your approximation with the actual value you compute from Eq. (21).

43. Let the function f be defined on the whole xy-plane by $f(x, y) = 1$ if $x = y \neq 0$, whereas $f(x, y) = 0$ otherwise. (a) Show that f is not continuous at $(0, 0)$. (b) Show that both partial derivatives f_x and f_y exist at $(0, 0)$.

44. Show that the function $f(x, y) = (\sqrt[3]{x} + \sqrt[3]{y})^3$ is continuous and has partial derivatives at the origin $(0, 0)$, but is not differentiable there.

45. Show that the function f defined by $f(x, y) = y^2 + x^3\sin(1/x)$ for $x \neq 0$, and $f(0, y) = y^2$, is differentiable at $(0, 0)$, but is not continuously differentiable there because $f_x(x, y)$ is not continuous at $(0, 0)$.

46. Let $f(x)$ be a function of the single variable x. Show that the ordinary derivative $f'(a)$ exists if and only if f is differentiable in the sense of Eq. (19), meaning that there exists a constant c such that

$$\lim_{h \to 0} \frac{f(a+h) - f(a) - ch}{|h|} = 0,$$

in which case $f'(a) = c$.

47. Deduce from Eq. (19) that the function f is continuous wherever it is differentiable.

48. Deduce from Eq. (19) that if the multivariable function $f(\mathbf{x})$ is differentiable at $\mathbf{a}$, then its first-order partial derivatives at $\mathbf{a}$ exist and are given by $D_i f(\mathbf{a}) = c_i$ for $i = 1, 2, \ldots, n$. Conclude in turn that the vector $\mathbf{c} = \langle c_1, c_2, \ldots, c_n\rangle$ in (19) is unique.

13.7 THE MULTIVARIABLE CHAIN RULE

The single-variable chain rule expresses the derivative of a composite function $f(g(t))$ in terms of the derivatives of f and g:

$$D_t f(g(t)) = f'(g(t) \cdot g'(t). \tag{1}$$

With $w = f(x)$ and $x = g(t)$, the chain rule implies that

$$\frac{dw}{dt} = \frac{dw}{dx}\frac{dx}{dt}. \tag{2}$$

The simplest multivariable chain rule situation involves a function $w = f(x, y)$ where both x and y are functions of the same single variable t: $x = g(t)$ and $y = h(t)$. The composite function $f(g(t), h(t))$ is then a single-variable function of t, and Theorem 1 expresses its derivative in terms of the partial derivatives of f and the ordinary derivatives of g and h. We assume that the stated hypotheses hold on suitable domains such that the composite function is defined.

THEOREM 1 The Chain Rule
Suppose that $w = f(x, y)$ has continuous first-order partial derivatives and that $x = g(t)$ and $y = h(t)$ are differentiable functions. Then w is a differentiable function of t, and

$$\frac{dw}{dt} = \frac{\partial w}{\partial x} \cdot \frac{dx}{dt} + \frac{\partial w}{\partial y} \cdot \frac{dy}{dt}. \tag{3}$$

The variable notation of Eq. (3) ordinarily will be more useful than function notation. Remember, in any case, that the partial derivatives in Eq. (3) are to be evaluated at the point $(g(t), h(t))$, so in function notation Eq. (3) is

$$D_t[f(g(t), h(t))] = f_x(g(t), h(t)) \cdot g'(t) + f_y(g(t), h(t)) \cdot h'(t). \tag{4}$$

A proof of the chain rule is included at the end of this section. In outline, it consists of beginning with the linear approximation

$$\Delta w \approx \frac{\partial w}{\partial x} \Delta x + \frac{\partial w}{\partial y} \Delta y$$

of Section 13.6 and dividing by Δt:

$$\frac{\Delta w}{\Delta t} \approx \frac{\partial w}{\partial x} \frac{\Delta x}{\Delta t} + \frac{\partial w}{\partial y} \frac{\Delta y}{\Delta t}.$$

Then we take the limit as $\Delta t \to 0$ to obtain

$$\frac{dw}{dt} = \frac{\partial w}{\partial x} \cdot \frac{dx}{dt} + \frac{\partial w}{\partial y} \cdot \frac{dy}{dt}.$$

EXAMPLE 1 Suppose that $w = e^{xy}$, $x = t^2$, and $y = t^3$. Then

$$\frac{\partial w}{\partial x} = ye^{xy}, \qquad \frac{\partial w}{\partial y} = xe^{yx}, \qquad \frac{dx}{dt} = 2t, \quad \text{and} \quad \frac{dy}{dt} = 3t^2.$$

So Eq. (3) yields

$$\begin{aligned} \frac{dw}{dt} &= \frac{\partial w}{\partial x} \cdot \frac{dx}{dt} + \frac{\partial w}{\partial y} \cdot \frac{dy}{dt} = (ye^{xy})(2t) + (xe^{xy})(3t^2) \\ &= \left(t^3 e^{t^5}\right)(2t) + \left(t^2 e^{t^5}\right)(3t^2) = 5t^4 e^{t^5}. \end{aligned}$$

◆

REMARK Had our purpose not been to illustrate the multivariable chain rule, we could have obtained the same result $dw/dt = 5t^4 \exp t^5$ more simply by writing

$$w = e^{xy} = e^{(t^2)(t^3)} = e^{t^5}$$

and then differentiating w as a single-variable function of t. But this single-variable approach is available only if the functions $x(t)$ and $y(t)$ are known explicitly. Sometimes, however, we know only the *numerical values* of x and y and/or their rates of change at a given instant. In such cases the multivariable chain rule in (3) can then be used to find the numerical rate of change of w at that instant.

FIGURE 13.7.1 Warm sun melting a cylindrical block of ice (Example 2).

EXAMPLE 2 Figure 13.7.1 shows a melting cylindrical block of ice. Because of the sun's heat beating down from above, its height h is decreasing more rapidly than its radius r. If its height is decreasing at 3 cm/h and its radius is decreasing at 1 cm/h when $r = 15$ cm and $h = 40$ cm, what is the rate of change of the volume V of the block at that instant?

Solution With $V = \pi r^2 h$, the chain rule gives

$$\frac{dV}{dt} = \frac{\partial V}{\partial r}\frac{dr}{dt} + \frac{\partial V}{\partial h}\frac{dh}{dt} = 2\pi r h \frac{dr}{dt} + \pi r^2 \frac{dh}{dt}.$$

Substituting the given numerical values $r = 15, h = 40, dr/dt = -1$, and $dh/dt = -3$, we find that

$$\frac{dV}{dt} = 2\pi(15)(40)(-1) + \pi(15)^2(-3) = -1875\pi \approx -5890.49 \quad (\text{cm}^3/\text{h}).$$

Thus the volume of the cylindrical block is decreasing at slightly less than 6 liters per hour at the given instant. ◆

In the context of Theorem 1, we may refer to w as the **dependent variable,** x and y as **intermediate variables,** and t as the **independent variable.** Then note that the right-hand side of Eq. (3) has two terms, one for each intermediate variable, both terms like the right-hand side of the single-variable chain rule in Eq. (2). If there are more than two intermediate variables, then there is still one term on the right-hand side for each intermediate variable. For example, if $w = f(x, y, z)$ with x, y, and z each a function of t, then the chain rule takes the form

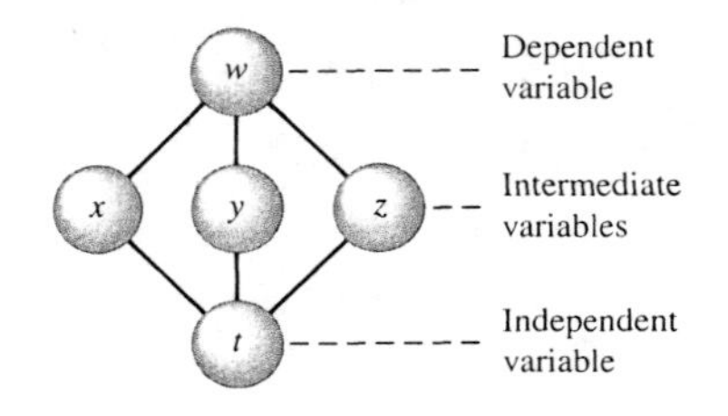

FIGURE 13.7.2 Levels of chain rule variables.

$$\frac{dw}{dt} = \frac{\partial w}{\partial x}\cdot\frac{dx}{dt} + \frac{\partial w}{\partial y}\cdot\frac{dy}{dt} + \frac{\partial w}{\partial z}\cdot\frac{dz}{dt}. \qquad (5)$$

The proof of Eq. (5) is essentially the same as the proof of Eq. (3); it requires the linear approximation theorem for three variables rather than for two variables.

You may find it useful to envision the three types of variables—dependent, intermediate, and independent—as though they were lying at three different levels, as in Fig. 13.7.2, with the dependent variable at the top and the independent variable at the bottom. Each variable then depends (either directly or indirectly) on those that lie below it.

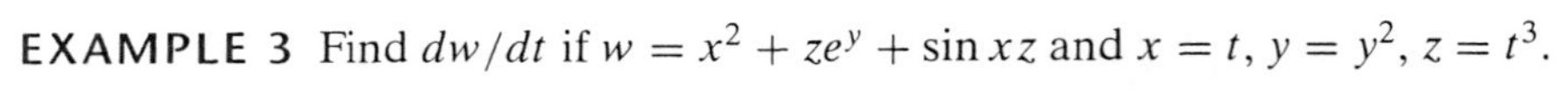

EXAMPLE 3 Find dw/dt if $w = x^2 + ze^y + \sin xz$ and $x = t$, $y = y^2$, $z = t^3$.

Solution Equation (5) gives

$$\begin{aligned}\frac{dw}{dt} &= \frac{\partial w}{\partial x}\cdot\frac{dx}{dt} + \frac{\partial w}{\partial y}\cdot\frac{dy}{dt} + \frac{\partial w}{\partial z}\cdot\frac{dz}{dt}\\ &= (2x + z\cos xz)(1) + (ze^y)(2t) + (e^y + x\cos xz)(3t^2)\\ &= 2t + (3t^2 + 2t^4)e^{t^2} + 4t^3\cos t^4.\end{aligned}$$

◆

In Example 3 we could check the result given by the chain rule by first writing w as an explicit function of t and then computing the ordinary single-variable derivative of w with respect to t.

Several Independent Variables

There may be several independent variables as well as several intermediate variables. For example, if $w = f(x, y, z)$ where $x = g(u, v)$, $y = h(u, v)$, and $z = k(u, v)$, so that

$$w = f(x, y, z) = f(g(u, v), h(u, v), k(u, v)),$$

then we have the three intermediate variables x, y, and z and the two independent variables u and v. In this case we would need to compute the *partial* derivatives $\partial w/\partial u$ and $\partial w/\partial v$ of the composite function. The general chain rule in Theorem 2 implies that each partial derivative of the dependent variable w is given by a chain rule formula such as Eq. (3) or (5). The only difference is that the derivatives with respect to the independent variables are partial derivatives. For instance,

$$\frac{\partial w}{\partial u} = \frac{\partial w}{\partial x}\cdot\frac{\partial x}{\partial u} + \frac{\partial w}{\partial y}\cdot\frac{\partial y}{\partial u} + \frac{\partial w}{\partial z}\cdot\frac{\partial z}{\partial u}.$$

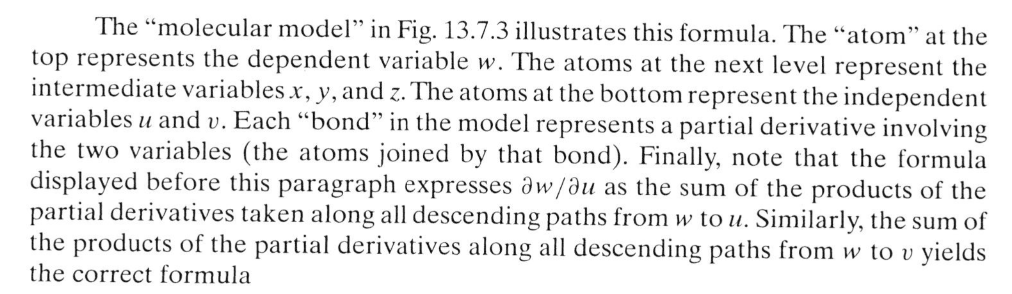

The "molecular model" in Fig. 13.7.3 illustrates this formula. The "atom" at the top represents the dependent variable w. The atoms at the next level represent the intermediate variables x, y, and z. The atoms at the bottom represent the independent variables u and v. Each "bond" in the model represents a partial derivative involving the two variables (the atoms joined by that bond). Finally, note that the formula displayed before this paragraph expresses $\partial w/\partial u$ as the sum of the products of the partial derivatives taken along all descending paths from w to u. Similarly, the sum of the products of the partial derivatives along all descending paths from w to v yields the correct formula

$$\frac{\partial w}{\partial v} = \frac{\partial w}{\partial x}\cdot\frac{\partial x}{\partial v} + \frac{\partial w}{\partial y}\cdot\frac{\partial y}{\partial v} + \frac{\partial w}{\partial z}\cdot\frac{\partial z}{\partial v}.$$

FIGURE 13.7.3 Diagram for $w = w(x, y, z)$, where $x = x(u, v)$, $y = y(u, v)$, and $z = z(u, v)$.

Theorem 2 describes the most general such situation.

THEOREM 2 The General Chain Rule

Suppose that w is a function of the variables $x_1, x_2, \dots, x_m$ and that each of these is a function of the variables $t_1, t_2, \dots, t_n$. If all these functions have continuous first-order partial derivatives, then

$$\frac{\partial w}{\partial t_i} = \frac{\partial w}{\partial x_1}\cdot\frac{\partial x_1}{\partial t_i} + \frac{\partial w}{\partial x_2}\cdot\frac{\partial x_2}{\partial t_i} + \cdots + \frac{\partial w}{\partial x_m}\cdot\frac{\partial x_m}{\partial t_i} \tag{6}$$

for each i, $1 \leqq i \leqq n$.

Thus there is a formula in Eq. (6) for *each* of the independent variables $t_1, t_2, \dots, t_n$, and the right-hand side of each such formula contains one typical chain rule term for each of the intermediate variables $x_1, x_2, \dots, x_m$.

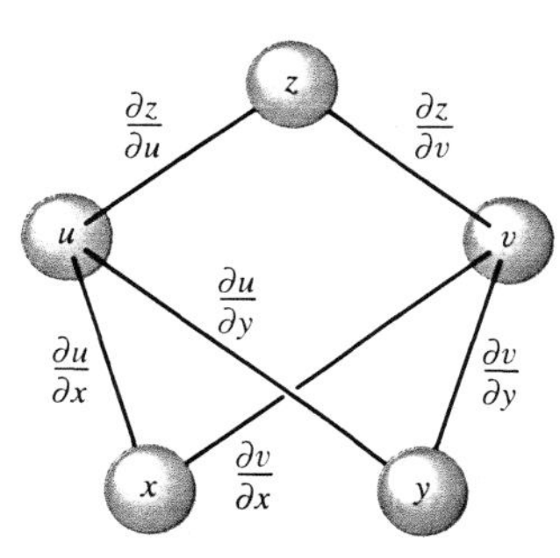

FIGURE 13.7.4 Diagram for $z = z(u, v)$, where $u = u(x, y)$ and $v = v(x, y)$ (Example 4).

EXAMPLE 4 Suppose that

$$z = f(u, v), \qquad u = 2x + y, \qquad v = 3x - 2y.$$

Given the values $\partial z/\partial u = 3$ and $\partial z/\partial v = -2$ at the point $(u, v) = (3, 1)$, find the values $\partial z/\partial x$ and $\partial z/\partial y$ at the corresponding point $(x, y) = (1, 1)$.

Solution The relationships among the variables are shown in Fig. 13.7.4. The chain rule gives

$$\frac{\partial z}{\partial x} = \frac{\partial z}{\partial u}\cdot\frac{\partial u}{\partial x} + \frac{\partial z}{\partial v}\cdot\frac{\partial v}{\partial x} = 3\cdot 2 + (-2)\cdot 3 = 0$$

and

$$\frac{\partial z}{\partial y} = \frac{\partial z}{\partial u}\cdot\frac{\partial u}{\partial y} + \frac{\partial z}{\partial v}\cdot\frac{\partial v}{\partial y} = 3\cdot 1 + (-2)\cdot(-2) = 7$$

at the indicated point $(x, y) = (1, 1)$. ◆

EXAMPLE 5 Let $w = f(x, y)$ where x and y are given in polar coordinates by the equations $x = r\cos\theta$ and $y = r\sin\theta$. Calculate

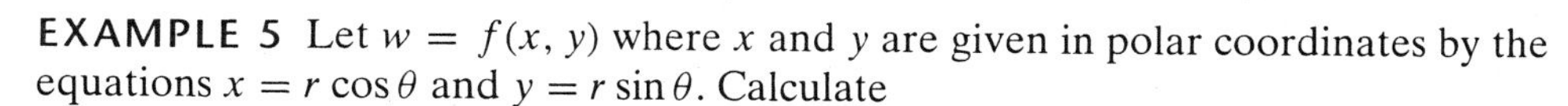

$$\frac{\partial w}{\partial r}, \quad \frac{\partial w}{\partial \theta}, \quad \text{and} \quad \frac{\partial^2 w}{\partial r^2}$$

in terms of r, θ, and the partial derivatives of w with respect to x and y (Fig. 13.7.5).

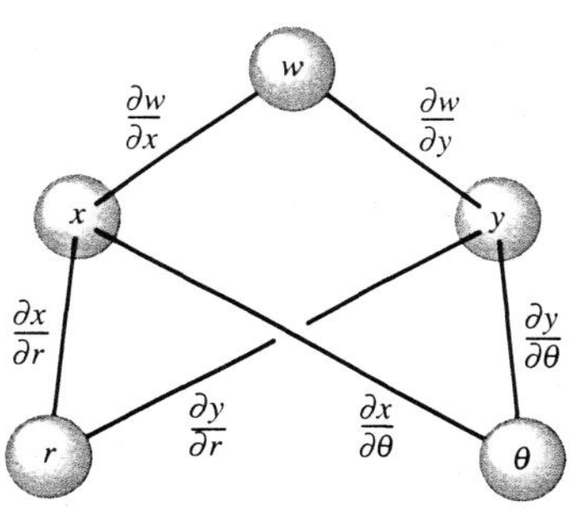

FIGURE 13.7.5 Diagram for $w = w(x, y)$, where $x = x(r, \theta)$ and $y = y(r, \theta)$ (Example 5).

Solution Here x and y are intermediate variables; the independent variables are r and θ. First note that

$$\frac{\partial x}{\partial r} = \cos\theta, \quad \frac{\partial y}{\partial r} = \sin\theta, \quad \frac{\partial x}{\partial \theta} = -r\sin\theta, \quad \text{and} \quad \frac{\partial y}{\partial \theta} = r\cos\theta.$$

Then

$$\frac{\partial w}{\partial r} = \frac{\partial w}{\partial x}\cdot\frac{\partial x}{\partial r} + \frac{\partial w}{\partial y}\cdot\frac{\partial y}{\partial r} = \frac{\partial w}{\partial x}\cos\theta + \frac{\partial w}{\partial y}\sin\theta \tag{7a}$$

and

$$\frac{\partial w}{\partial \theta} = \frac{\partial w}{\partial x} \cdot \frac{\partial x}{\partial \theta} + \frac{\partial w}{\partial y} \cdot \frac{\partial y}{\partial \theta} = -r\frac{\partial w}{\partial x}\sin\theta + r\frac{\partial w}{\partial y}\cos\theta. \tag{7b}$$

Next,

$$\begin{aligned}\frac{\partial^2 w}{\partial r^2} &= \frac{\partial}{\partial r}\left(\frac{\partial w}{\partial r}\right) = \frac{\partial}{\partial r}\left(\frac{\partial w}{\partial x}\cos\theta + \frac{\partial w}{\partial y}\sin\theta\right)\\ &= \frac{\partial w_x}{\partial r}\cos\theta + \frac{\partial w_y}{\partial r}\sin\theta,\end{aligned}$$

where $w_x = \partial w/\partial x$ and $w_y = \partial w/\partial y$. We apply Eq. (7a) to calculate $\partial w_x/\partial r$ and $\partial w_y/\partial r$, and we obtain

$$\begin{aligned}\frac{\partial^2 w}{\partial r^2} &= \left(\frac{\partial w_x}{\partial x} \cdot \frac{\partial x}{\partial r} + \frac{\partial w_x}{\partial y} \cdot \frac{\partial y}{\partial r}\right)\cos\theta + \left(\frac{\partial w_y}{\partial x} \cdot \frac{\partial x}{\partial r} + \frac{\partial w_y}{\partial y} \cdot \frac{\partial y}{\partial r}\right)\sin\theta\\ &= \left(\frac{\partial^2 w}{\partial x^2}\cos\theta + \frac{\partial^2 w}{\partial y \partial x}\sin\theta\right)\cos\theta + \left(\frac{\partial^2 w}{\partial x \partial y}\cos\theta + \frac{\partial^2 w}{\partial y^2}\sin\theta\right)\sin\theta.\end{aligned}$$

Finally, because $w_{yx} = w_{xy}$, we get

$$\frac{\partial^2 w}{\partial r^2} = \frac{\partial^2 w}{\partial x^2}\cos^2\theta + 2\frac{\partial^2 w}{\partial x \partial y}\cos\theta\sin\theta + \frac{\partial^2 w}{\partial y^2}\sin^2\theta. \tag{8}$$

◆

EXAMPLE 6 Suppose that $w = f(u, v, x, y)$, where u and v are functions of x and y. Here x and y play dual roles as intermediate and independent variables. The chain rule yields

$$\begin{aligned}\frac{\partial w}{\partial x} &= \frac{\partial f}{\partial u} \cdot \frac{\partial u}{\partial x} + \frac{\partial f}{\partial v} \cdot \frac{\partial v}{\partial x} + \frac{\partial f}{\partial x} \cdot \frac{\partial x}{\partial x} + \frac{\partial f}{\partial y} \cdot \frac{\partial y}{\partial x}\\ &= \frac{\partial f}{\partial u} \cdot \frac{\partial u}{\partial x} + \frac{\partial f}{\partial v} \cdot \frac{\partial v}{\partial x} + \frac{\partial f}{\partial x},\end{aligned}$$

because $\partial x/\partial x = 1$ and $\partial y/\partial x = 0$. Similarly,

$$\frac{\partial w}{\partial y} = \frac{\partial f}{\partial u} \cdot \frac{\partial u}{\partial y} + \frac{\partial f}{\partial v} \cdot \frac{\partial v}{\partial y} + \frac{\partial f}{\partial y}.$$

These results are consistent with the paths from w to x and from w to y in the molecular model shown in Fig. 13.7.6. ◆

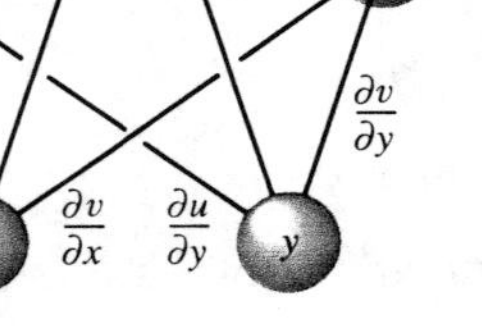

FIGURE 13.7.6 Diagram for $w = f(u, v, x, y)$, where $u = u(x, y)$ and $v = v(x, y)$ (Example 6).

EXAMPLE 7 Consider a parametric curve $x = x(t)$, $y = y(t)$, $z = z(t)$ that lies on the surface $z = f(x, y)$ in space. Recall that if

$$\mathbf{T} = \left\langle \frac{dx}{dt}, \frac{dy}{dt}, \frac{dz}{dt} \right\rangle \quad \text{and} \quad \mathbf{N} = \left\langle \frac{\partial z}{\partial x}, \frac{\partial z}{\partial y}, -1 \right\rangle,$$

then $\mathbf{T}$ is tangent to the curve and $\mathbf{N}$ is normal to the surface. Show that $\mathbf{T}$ and $\mathbf{N}$ are everywhere perpendicular.

Solution The chain rule in Eq. (3) tells us that

$$\frac{dz}{dt} = \frac{\partial z}{\partial x} \cdot \frac{dx}{dt} + \frac{\partial z}{\partial y} \cdot \frac{dy}{dt}.$$

But this equation is equivalent to the vector equation

$$\left\langle \frac{\partial z}{\partial x}, \frac{\partial z}{\partial y}, -1 \right\rangle \cdot \left\langle \frac{dx}{dt}, \frac{dy}{dt}, \frac{dz}{dt} \right\rangle = 0.$$

Thus $\mathbf{N} \cdot \mathbf{T} = 0$, so $\mathbf{N}$ and $\mathbf{T}$ are perpendicular. ◆

Implicit Partial Differentiation

Sometimes we need to investigate a function $z = g(x, y)$ that is not defined explicitly by a formula giving z in terms of x and y, but instead is defined implicitly by an equation of the form $F(x, y, z) = 0$. The following implicit function theorem, proved in advanced calculus, guarantees the existence and differentiability of such implicitly defined functions under certain natural hypotheses.

THEOREM 3 Implicit Function Theorem

Suppose that the function $F(x_1, x_2, \dots, x_n, z)$ is continuously differentiable near the point $(\mathbf{a}, b) = (a_1, a_2, \dots, n, b)$ at which $F(\mathbf{a}, b) = 0$ and $D_z F(\mathbf{a}, b) \neq 0$. Then there exists a continuously differentiable function $z = g(x_1, x_2, \dots, x_n)$ such that $g(\mathbf{a}) = b$ and $F(\mathbf{x}, g(\mathbf{x})) = 0$ for $\mathbf{x}$ near $\mathbf{a}$.

Moreover, the function $g(\mathbf{x})$ is uniquely defined for $\mathbf{x}$ near $\mathbf{a}$. In brief, Theorem 2 implies that the equation $F(x_1, x_2, \dots, x_n, z) = 0$ implicitly defines one and only one continuously differentiable function $z = g(x_1, x_2, \dots, x_n)$ near any point where $\partial F/\partial z \neq 0$. Knowing that the function g exists and is differentiable, we can calculate its partial derivatives by implicit differentiation of the given equation $F(x_1, x_2, \dots, x_n, z) = 0$. Differentiating this equation with respect to x_i yields

$$\frac{\partial F}{\partial x_1} \cdot \frac{\partial x_1}{\partial x_i} + \cdots + \frac{\partial F}{\partial x_i} \cdot \frac{\partial x_i}{\partial x_i} + \cdots + \frac{\partial F}{\partial x_n} \cdot \frac{\partial x_n}{\partial x_i} + \frac{\partial F}{\partial z} \cdot \frac{\partial z}{\partial x_i} = 0. \tag{9}$$

But $\partial x_j/\partial x_i = 0$ unless $j = i$, and $\partial x_i/\partial x_i = 1$, so Eq. (9) reduces to the equation

$$\frac{\partial F}{\partial x_i} + \frac{\partial F}{\partial z} \cdot \frac{\partial z}{\partial x_i} = 0,$$

which (assuming that $\partial F/\partial z \neq 0$) we can solve to obtain the formula

$$\frac{\partial z}{\partial x_i} = -\frac{\partial F/\partial x_i}{\partial F/\partial z} = -\frac{F_{x_i}}{F_z} \tag{10}$$

for the ith partial derivative of $z = g(x_1, x_2, \dots, x_n)$. In a specific example, it usually is just as simple to differentiate the given equation $F(x_1, x_2, \dots, x_n, z) = 0$ as in (9), rather than applying the formula in (10).

EXAMPLE 8 Figure 13.7.7 shows the graph of the equation

$$F(x, y) = x^3 + y^3 - 3xy = 0 \tag{11}$$

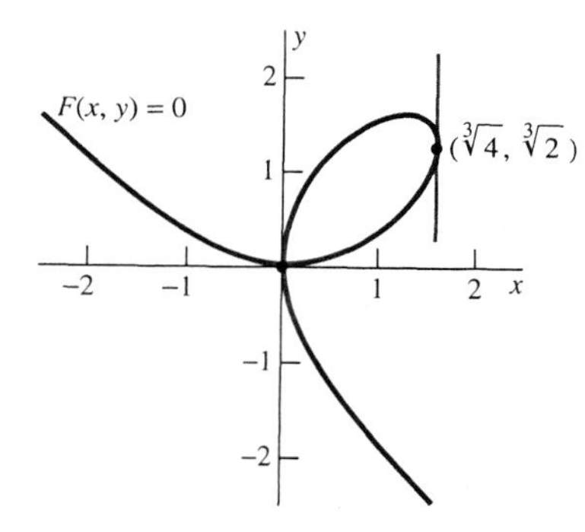

FIGURE 13.7.7 Graph of the equation $F(x, y) = x^3 + y^3 - 3xy = 0$ (Example 8).

(the folium of Descartes that we discussed in Example 3 of Section 4.1). With $n = 1$, x in place of x_1, and y in place of z, the implicit function theorem implies that this equation implicitly defines y as a function of x except possibly where

$$\frac{\partial F}{\partial y} = 3y^2 - 3x = 0.$$

By substituting $y^2 = x$ in Eq. (11), you can show that the only such points on the curve are the origin $(0, 0)$, where two branches of the curve intersect, and the point $(\sqrt[3]{4}, \sqrt[3]{2})$, where the figure shows a vertical tangent line. At any other point on the curve we can differentiate with respect to x in Eq. (11) to obtain

$$\frac{\partial F}{\partial x} \cdot \frac{dx}{dx} + \frac{\partial F}{\partial y} \cdot \frac{dy}{dx} = (3x^2 - 3y) \cdot 1 + (3y^2 - 3x) \cdot \frac{dy}{dx} = 0.$$

We can then solve for the slope

$$\frac{dy}{dx} = -\frac{x^2 - y}{y^2 - x}$$

of the line tangent to the curve at any point where there is not a vertical tangent line. ◆

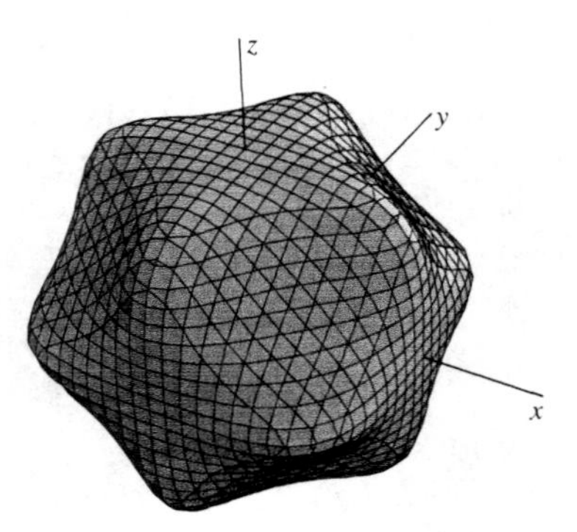

FIGURE 13.7.8 Graph of the equation $F(x, y, z) = x^4 + y^4 + z^4 + 4x^2y^2z^2 - 34 = 0$ (Example 9).

EXAMPLE 9 Figure 13.7.8 shows the graph of the equation

$$F(x, y, z) = x^4 + y^4 + z^4 + 4x^2y^2z^2 - 34 = 0. \tag{12}$$

With $n = 2$ and x and y in place of x_1 and x_2, the implicit function theorem implies that this equation implicitly defines z as a function of x and y except possibly where

$$\frac{\partial F}{\partial z} = 4z^3 + 8x^2y^2z = 4z(z^2 + 2x^2y^2) = 0.$$

The partial derivative is nonzero wherever $z \neq 0$, so it follows that z is defined as a function of x and y except at the points of the curve $x^4 + y^4 = 34$ in which the surface intersects the xy-plane (where $z = 0$). At any other point of the surface we can differentiate with respect to x and y in (12) to obtain

$$\frac{\partial F}{\partial x}\cdot\frac{\partial x}{\partial x} + \frac{\partial F}{\partial y}\cdot\frac{\partial y}{\partial x} + \frac{\partial F}{\partial z}\cdot\frac{\partial z}{\partial x} = (4x^3 + 8xy^2z^2)\cdot 1 + (4z^3 + 8x^2y^2z)\cdot\frac{\partial z}{\partial x} = 0$$

and

$$\frac{\partial F}{\partial x}\cdot\frac{\partial x}{\partial y} + \frac{\partial F}{\partial y}\cdot\frac{\partial y}{\partial y} + \frac{\partial F}{\partial z}\cdot\frac{\partial z}{\partial y} = (4y^3 + 8x^2yz^2)\cdot 1 + (4z^3 + 8x^2y^2z)\cdot\frac{\partial z}{\partial y} = 0.$$

We can then solve for

$$\frac{\partial z}{\partial x} = -\frac{x^3 + 2xy^2z^2}{z^3 + 2x^2y^2z} \quad\text{and}\quad \frac{\partial z}{\partial y} = -\frac{y^3 + 2x^2yz^2}{z^3 + 2x^2y^2z}.$$

For instance, at the point $(2, 1, 1)$ of the surface we find that $\partial z/\partial x = -\frac{4}{3}$ and $\partial z/\partial y = -1$. Hence the plane tangent to the surface at this point has equation

$$z - 1 = -\tfrac{4}{3}(x - 2) + (-1)(y - 1); \quad\text{that is,}\quad 4x + 3y + 3z = 14. \qquad \blacklozenge$$

Matrix Form of the Chain Rule

The case $m = n = 2$ of the chain rule corresponds to the case of two intermediate variables (x and y, say) that are functions of two independent variables (u and v, say),

$$x = f(u, v), \qquad y = g(u, v). \tag{13}$$

These functions describe a **transformation** $T : \mathbf{R}^2_{uv} \to \mathbf{R}^2_{xy}$ from the coordinate plane $\mathbf{R}^2_{uv}$ of (u, v)-pairs to the coordinate plane $\mathbf{R}^2_{xy}$ of (x, y)-pairs. The **image** of the point (u, v) of $\mathbf{R}^2_{uv}$ is the point $T(u, v) = (f(u, v), g(u, v)) = (x, y)$ of $\mathbf{R}^2_{xy}$. The **derivative matrix** of the transformation T at the point (u, v) is then the 2×2 array

$$T'(u, v) = \begin{bmatrix} \dfrac{\partial x}{\partial u} & \dfrac{\partial x}{\partial v} \\ \dfrac{\partial y}{\partial u} & \dfrac{\partial y}{\partial v} \end{bmatrix} \tag{14}$$

of partial derivatives of the component functions in (13) of the transformation T (all evaluated at the point (u, v)).

EXAMPLE 10 The polar coordinate transformation $T : \mathbf{R}^2_{r\theta} \to \mathbf{R}^2_{xy}$ is defined by the familiar equations

$$x = r\cos\theta, \qquad y = r\sin\theta. \tag{15}$$

Its derivative matrix is given by

$$T'(r, \theta) = \begin{bmatrix} \dfrac{\partial x}{\partial r} & \dfrac{\partial x}{\partial \theta} \\ \dfrac{\partial y}{\partial r} & \dfrac{\partial y}{\partial \theta} \end{bmatrix} = \begin{bmatrix} \cos\theta & -r\sin\theta \\ \sin\theta & r\cos\theta \end{bmatrix}. \tag{16}$$

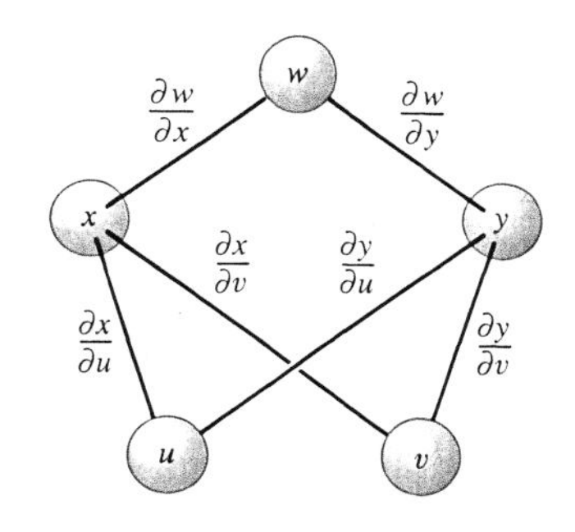

FIGURE 13.7.9 Diagram for $w = F(x, y)$ where $(x, y) = (x(u, v), y(u, v)) = T(u, v)$

Now suppose that the dependent variable w is a function $F(x, y)$ of the intermediate variables x and y, and thereby is given by the composite function

$$G(u, v) = F(T(u, v)) = F(x(u, v), y(u, v)) \tag{17}$$

of the independent variables u and v (Fig. 13.7.9). The derivative matrices

$$F'(x, y) = \begin{bmatrix} \dfrac{\partial w}{\partial x} & \dfrac{\partial w}{\partial y} \end{bmatrix} \quad \text{and} \quad G'(u, v) = \begin{bmatrix} \dfrac{\partial w}{\partial u} & \dfrac{\partial w}{\partial v} \end{bmatrix} \tag{18}$$

of F and G are defined in analogy with (14)—there being a single row in each matrix, corresponding to the single dependent variable w. Those who are familiar with matrix multiplication will recognize that the two chain rule formulas

$$\frac{\partial w}{\partial u} = \frac{\partial w}{\partial x}\frac{\partial x}{\partial u} + \frac{\partial w}{\partial y}\frac{\partial y}{\partial u}, \qquad \frac{\partial w}{\partial v} = \frac{\partial w}{\partial x}\frac{\partial x}{\partial v} + \frac{\partial w}{\partial y}\frac{\partial y}{\partial v}$$

are the "components" of the single matrix equation

$$G'(u, v) = F'(x, y)T'(u, v); \tag{19a}$$

that is,

$$\begin{bmatrix} \dfrac{\partial w}{\partial u} & \dfrac{\partial w}{\partial v} \end{bmatrix} = \begin{bmatrix} \dfrac{\partial w}{\partial x} & \dfrac{\partial w}{\partial y} \end{bmatrix} \begin{bmatrix} \dfrac{\partial x}{\partial u} & \dfrac{\partial x}{\partial v} \\ \dfrac{\partial y}{\partial u} & \dfrac{\partial y}{\partial v} \end{bmatrix}. \tag{19b}$$

Thus the chain rule for the situation indicated in Fig. 13.7.9 implies that *the derivative matrix of the composite function $G = F \circ T$ is the matrix product $G' = F'T'$.* ◆

EXAMPLE 11 With the polar-coordinate derivative matrix $T'(r, \theta)$ in (16), the matrix multiplication in Eq. (19b) yields

$$\begin{aligned} \begin{bmatrix} \dfrac{\partial w}{\partial r} & \dfrac{\partial w}{\partial \theta} \end{bmatrix} &= \begin{bmatrix} \dfrac{\partial w}{\partial x} & \dfrac{\partial w}{\partial y} \end{bmatrix} \begin{bmatrix} \cos\theta & -r\sin\theta \\ \sin\theta & r\cos\theta \end{bmatrix} \\ &= \begin{bmatrix} \dfrac{\partial w}{\partial x}\cos\theta + \dfrac{\partial w}{\partial y}\sin\theta & -r\dfrac{\partial w}{\partial x}\sin\theta + r\dfrac{\partial w}{\partial y}\cos\theta \end{bmatrix}. \end{aligned}$$

The components of this matrix equation are the scalar chain rule formulas

$$\frac{\partial w}{\partial r} = \frac{\partial w}{\partial x}\cos\theta + \frac{\partial w}{\partial y}\sin\theta, \qquad \frac{\partial w}{\partial \theta} = -r\frac{\partial w}{\partial x}\sin\theta + r\frac{\partial w}{\partial y}\cos\theta$$

that we saw previously in Example 5. ◆

We have discussed here the 2×2 case of a general $m \times n$ matrix formulation of the multivariable chain rule. The 3×3 case and its application to spherical coordinates are discussed in Problems 58 through 61.

PROOF OF THE CHAIN RULE Given that $w = f(x, y)$ satisfies the hypotheses of Theorem 1, we choose a point t_0 at which we wish to compute dw/dt and write

$$a = g(t_0), \qquad b = h(t_0).$$

Let

$$\Delta x = g(t_0 + \Delta t) - g(t_0), \qquad \Delta y = h(t_0 + \Delta t) - h(t_0).$$

Then

$$g(t_0 + \Delta t) = a + \Delta x \quad \text{and} \quad h(t_0 + \Delta t) = b + \Delta y.$$

If

$$\begin{aligned} \Delta w &= f(g(t_0 + \Delta t), h(t_0 + \Delta t)) - f(g(t_0), h(t_0)) \\ &= f(a + \Delta x, b + \Delta y) - f(a, b), \end{aligned}$$

then what we need to compute is

$$\frac{dw}{dt} = \lim_{\Delta t \to 0} \frac{\Delta w}{\Delta t}.$$

The linear approximation theorem of Section 13.6 gives

$$\Delta w = f_x(a, b)\,\Delta x + f_y(a, b)\,\Delta y + \epsilon_1\,\Delta x + \epsilon_2\,\Delta y,$$

where ϵ_1 and ϵ_2 approach zero as $\Delta x \to 0$ and $\Delta y \to 0$. We note that both Δx and Δy approach zero as $\Delta t \to 0$, because both the derivatives

$$\frac{dx}{dt} = \lim_{\Delta t \to 0} \frac{\Delta x}{\Delta t} \quad \text{and} \quad \frac{dy}{dt} = \lim_{\Delta t \to 0} \frac{\Delta y}{\Delta t}$$

exist. Therefore,

$$\frac{dw}{dt} = \lim_{\Delta t \to 0} \frac{\Delta w}{\Delta t} = \lim_{\Delta t \to 0} \left[f_x(a, b)\frac{\Delta x}{\Delta t} + f_y(a, b)\frac{\Delta y}{\Delta t} + \epsilon_1 \frac{\Delta x}{\Delta t} + \epsilon_2 \frac{\Delta y}{\Delta t} \right]$$

$$= f_x(a, b)\frac{dx}{dt} + f_y(a, b)\frac{dy}{dt} + 0 \cdot \frac{dx}{dt} + 0 \cdot \frac{dy}{dt}.$$

Hence

$$\frac{dw}{dt} = \frac{\partial w}{\partial x} \cdot \frac{dx}{dt} + \frac{\partial w}{\partial y} \cdot \frac{dy}{dt}.$$

Thus we have established Eq. (3), writing $\partial w/\partial x$ and $\partial w/\partial y$ for the partial derivatives $f_x(a, b)$ and $f_y(a, b)$ in the final step. ◀

13.7 TRUE/FALSE STUDY GUIDE

13.7 CONCEPTS: QUESTIONS AND DISCUSSION

1. Give your own example of a composite function situation illustrating the general chain rule (Theorem 2), but with different numbers of independent, intermediate, and dependent variables than in any of the examples in this section.
2. Let C be a set in the xy-plane $\boldsymbol{R}^2$. We might call C a **smooth curve** provided that every point of C has a neighborhood within which C agrees with the graph of a continuously differentiable function—either $y = f(x)$ or $x = g(y)$. Under what conditions on the function $F(x, y)$ does the implicit function theorem imply that the graph of the equation $F(x, y) = 0$ is a smooth curve? Explain.
3. Let S be a set in xyz-space $\boldsymbol{R}^3$. We might call S a **smooth surface** provided that every point of S has a neighborhood within which S agrees with the graph of a continuously differentiable function—either $z = f(x, y)$ or $x = g(y, z)$ or $y = h(x, z)$. Under what conditions on the function $F(x, y, z)$ does the implicit function theorem imply that the graph of the equation $F(x, y, z) = 0$ is a smooth surface? Explain.

13.7 PROBLEMS

In Problems 1 through 4, find dw/dt both by using the chain rule and by expressing w explicitly as a function of t before differentiating.

1. $w = \exp(-x^2 - y^2)$; $x = t$, $y = \sqrt{t}$

2. $w = \dfrac{1}{u^2 + v^2}$; $u = \cos 2t$, $v = \sin 2t$

3. $w = \sin xyz$; $x = t$, $y = t^2$, $z = t^3$

4. $w = \ln(u + v + z)$; $u = \cos^2 t$, $v = \sin^2 t$, $z = t^2$

In Problems 5 through 8, find $\partial w/\partial s$ and $\partial w/\partial t$.

5. $w = \ln(x^2 + y^2 + z^2)$; $x = s - t$, $y = s + t$, $z = 2\sqrt{st}$

6. $w = pq \sin r$; $p = 2s + t$, $q = s - t$, $r = st$

7. $w = \sqrt{u^2 + v^2 + z^2}$; $u = 3e^t \sin s$, $v = 3e^t \cos s$, $z = 4e^t$

8. $w = yz + zx + xy$; $x = s^2 - t^2$, $y = s^2 + t^2$, $z = s^2 t^2$

In Problems 9 through 12, find $\partial r/\partial x$, $\partial r/\partial y$, and $\partial r/\partial z$.

9. $r = e^{u+v+w}$; $u = yz$, $v = xz$, $w = xy$

10. $r = uvw - u^2 - v^2 - w^2;\quad u = y + z,\ v = x + z,\ w = x + y$

11. $r = \sin(p/q);\quad p = \sqrt{xy^2z^3},\ q = \sqrt{x + 2y + 3z}$

12. $r = \dfrac{p}{q} + \dfrac{q}{s} + \dfrac{s}{p};\quad p = e^{yz},\ q = e^{xz},\ s = e^{xy}$

In Problems 13 through 18, write chain rule formulas giving the partial derivative of the dependent variable p with respect to each independent variable.

13. $p = f(x, y);\quad x = x(u, v, w),\ y = y(u, v, w)$

14. $p = f(x, y, z);\quad x = x(u, v),\ y = y(u, v),\ z = z(u, v)$

15. $p = f(u, v, w);\ u = u(x, y, z),\ v = v(x, y, z),\ w = w(x, y, z)$

16. $p = f(v, w);\quad v = v(x, y, z, t),\ w = w(x, y, z, t)$

17. $p = f(w);\quad w = w(x, y, z, u, v)$

18. $p = f(x, y, u, v);\quad x = x(s, t),\quad y = y(s, t),\quad u = u(s, t),\ v = v(s, t)$

In Problems 19 through 24, find $\partial z/\partial x$ and $\partial z/\partial y$ as functions of x, y, and z, assuming that $z = f(x, y)$ satisfies the given equation.

19. $x^{2/3} + y^{2/3} + z^{2/3} = 1$

20. $x^3 + y^3 + z^3 = xyz$

21. $xe^{xy} + ye^{zx} + ze^{xy} = 3$

22. $x^5 + xy^2 + yz = 5$

23. $\dfrac{x^2}{a^2} + \dfrac{y^2}{b^2} + \dfrac{z^2}{c^2} = 1$

24. $xyz = \sin(x + y + z)$

In Problems 25 through 28, use the method of Example 6 to find $\partial w/\partial x$ and $\partial w/\partial y$ as functions of x and y.

25. $w = u^2 + v^2 + x^2 + y^2;\quad u = x - y,\ v = x + y$

26. $w = \sqrt{uvxy};\quad u = \sqrt{x - y},\ v = \sqrt{x + y}$

27. $w = xy\ln(u + v);\quad u = (x^2 + y^2)^{1/3},\ v = (x^3 + y^3)^{1/2}$

28. $w = uv - xy;\quad u = \dfrac{x}{x^2 + y^2},\ v = \dfrac{y}{x^2 + y^2}$

In Problems 29 through 32, write an equation for the plane tangent at the point P to the surface with the given equation.

29. $x^2 + y^2 + z^2 = 9;\quad P(1, 2, 2)$

30. $x^2 + 2y^2 + 2z^2 = 14;\quad P(2, 1, -2)$

31. $x^3 + y^3 + z^3 = 5xyz;\quad P(2, 1, 1)$

32. $z^3 + (x + y)z^2 + x^2 + y^2 = 13;\quad P(2, 2, 1)$

33. The sun is melting a rectangular block of ice. When the block's height is 1 ft and the edge of its square base is 2 ft, its height is decreasing at 2 in./h and its base edge is decreasing at 3 in./h. What is the block's rate of change of volume V at that instant? $\frac{dh}{dt} = -2\quad \frac{dx}{dt} = -3\quad V = x^2h$

34. A rectangular box has a square base. Find the rate at which its volume and surface area are changing if its base edge is increasing at 2 cm/min and its height is decreasing at 3 cm/min at the instant when each dimension is 1 meter.

35. Falling sand forms a conical sandpile. When the sandpile has a height of 5 ft and its base radius is 2 ft, its height is increasing at 0.4 ft/min and its base radius is increasing at 0.7 ft/min. At what rate is the volume of the sandpile increasing at that moment?

36. A rectangular block has dimensions $x = 3$ m, $y = 2$ m, and $z = 1$ m. If x and y are increasing at 1 cm/min and 2 cm/min, respectively, while z is decreasing at 2 cm/min, are the block's volume and total surface area increasing or decreasing? At what rates?

37. The volume V (in cubic centimeters) and pressure p (in atmospheres) of n moles of an ideal gas satisfy the equation $pV = nRT$, where T is its temperature (in degrees Kelvin) and $R = 82.06$. Suppose that a sample of the gas has a volume of 10 L when the pressure is 2 atm and the temperature is 300°K. If the pressure is increasing at 1 atm/min and the temperature is increasing at 10°K/min, is the volume of the gas sample increasing or decreasing? At what rate?

38. The aggregate resistance R of three variable resistances R_1, R_2, and R_3 connected in parallel satisfies the *harmonic equation*

$$\frac{1}{R} = \frac{1}{R_1} + \frac{1}{R_2} + \frac{1}{R_3}.$$

Suppose that R_1 and R_2 are 100 Ω and are increasing at 1 Ω/s, while R_3 is 200 Ω and is decreasing at 2 Ω/s. Is R increasing or decreasing at that instant? At what rate?

39. Suppose that $x = h(y, z)$ satisfies the equation $F(x, y, z) = 0$ and that $F_x \neq 0$. Show that

$$\frac{\partial x}{\partial y} = -\frac{\partial F/\partial y}{\partial F/\partial x}.$$

40. Suppose that $w = f(x, y)$, $x = r\cos\theta$, and $y = r\sin\theta$. Show that

$$\left(\frac{\partial w}{\partial x}\right)^2 + \left(\frac{\partial w}{\partial y}\right)^2 = \left(\frac{\partial w}{\partial r}\right)^2 + \frac{1}{r^2}\left(\frac{\partial w}{\partial \theta}\right)^2.$$

41. Suppose that $w = f(u)$ and that $u = x + y$. Show that $\partial w/\partial x = \partial w/\partial y$.

42. Suppose that $w = f(u)$ and that $u = x - y$. Show that $\partial w/\partial x = -\partial w/\partial y$ and that

$$\frac{\partial^2 w}{\partial x^2} = \frac{\partial^2 w}{\partial y^2} = -\frac{\partial^2 w}{\partial x\,\partial y}.$$

43. Suppose that $w = f(x, y)$ where $x = u + v$ and $y = u - v$. Show that

$$\frac{\partial^2 w}{\partial x^2} - \frac{\partial^2 w}{\partial y^2} = \frac{\partial^2 w}{\partial u\,\partial v}.$$

44. Assume that $w = f(x, y)$ where $x = 2u + v$ and $y = u - v$. Show that

$$5\frac{\partial^2 w}{\partial x^2} + 2\frac{\partial^2 w}{\partial x\,\partial y} + 2\frac{\partial^2 w}{\partial y^2} = \frac{\partial^2 w}{\partial u^2} + \frac{\partial^2 w}{\partial v^2}.$$

45. Suppose that $w = f(x, y)$, $x = r\cos\theta$, and $y = r\sin\theta$. Show that

$$\frac{\partial^2 w}{\partial x^2} + \frac{\partial^2 w}{\partial y^2} = \frac{\partial^2 w}{\partial r^2} + \frac{1}{r}\frac{\partial w}{\partial r} + \frac{1}{r^2}\frac{\partial^2 w}{\partial \theta^2}.$$

[*Suggestion:* First find $\partial^2 w/\partial\theta^2$ by the method of Example 7. Then combine the result with Eqs. (7) and (8).]

46. Suppose that

$$w = \frac{1}{r} f\left(t - \frac{r}{a}\right)$$

and that $r = \sqrt{x^2 + y^2 + z^2}$. Show that

$$\frac{\partial^2 w}{\partial x^2} + \frac{\partial^2 w}{\partial y^2} + \frac{\partial^2 w}{\partial z^2} = \frac{1}{a^2}\frac{\partial^2 w}{\partial t^2}.$$

47. Suppose that $w = f(r)$ and that $r = \sqrt{x^2 + y^2 + z^2}$. Show that

$$\frac{\partial^2 w}{\partial x^2} + \frac{\partial^2 w}{\partial y^2} + \frac{\partial^2 w}{\partial z^2} = \frac{d^2 w}{dr^2} + \frac{2}{r}\frac{dw}{dr}.$$

48. Suppose that $w = f(u) + g(v)$, that $u = x - at$, and that $v = x + at$. Show that

$$\frac{\partial^2 w}{\partial t^2} = a^2 \frac{\partial^2 w}{\partial x^2}.$$

49. Assume that $w = f(u, v)$ where $u = x + y$ and $v = x - y$. Show that

$$\frac{\partial w}{\partial x}\frac{\partial w}{\partial y} = \left(\frac{\partial w}{\partial u}\right)^2 - \left(\frac{\partial w}{\partial v}\right)^2.$$

50. Given: $w = f(x, y)$, $x = e^u \cos v$, and $y = e^u \sin v$. Show that

$$\left(\frac{\partial w}{\partial x}\right)^2 + \left(\frac{\partial w}{\partial y}\right)^2 = e^{-2u}\left[\left(\frac{\partial w}{\partial u}\right)^2 + \left(\frac{\partial w}{\partial v}\right)^2\right].$$

51. Assume that $w = f(x, y)$ and that there is a constant α such that

$$x = u\cos\alpha - v\sin\alpha \quad \text{and} \quad y = u\sin\alpha + v\cos\alpha.$$

Show that

$$\left(\frac{\partial w}{\partial u}\right)^2 + \left(\frac{\partial w}{\partial v}\right)^2 = \left(\frac{\partial w}{\partial x}\right)^2 + \left(\frac{\partial w}{\partial y}\right)^2.$$

52. Suppose that $w = f(u)$, where

$$u = \frac{x^2 - y^2}{x^2 + y^2}.$$

Show that $xw_x + yw_y = 0$.

Suppose that the equation $F(x, y, z) = 0$ defines implicitly the three functions $z = f(x, y)$, $y = g(x, z)$, and $x = h(y, z)$. To keep track of the various partial derivatives, we use the notation

$$\left(\frac{\partial z}{\partial x}\right)_y = \frac{\partial f}{\partial x}, \quad \left(\frac{\partial z}{\partial y}\right)_x = \frac{\partial f}{\partial y}, \tag{20a}$$

$$\left(\frac{\partial y}{\partial x}\right)_z = \frac{\partial g}{\partial x}, \quad \left(\frac{\partial y}{\partial z}\right)_x = \frac{\partial g}{\partial z}, \tag{20b}$$

$$\left(\frac{\partial x}{\partial y}\right)_z = \frac{\partial h}{\partial y}, \quad \left(\frac{\partial x}{\partial z}\right)_y = \frac{\partial h}{\partial z}, \tag{20c}$$

In short, the general symbol $(\partial w/\partial u)_v$ denotes the derivative of w with respect to u, where w is regarded as a function of the independent variables u and v.

53. Using the notation in the equations in (20), show that

$$\left(\frac{\partial x}{\partial y}\right)_z \left(\frac{\partial y}{\partial z}\right)_x \left(\frac{\partial z}{\partial x}\right)_y = -1.$$

[*Suggestion:* Find the three partial derivatives on the right-hand side in terms of F_x, F_y, and F_z.]

54. Verify the result of Problem 53 for the equation

$$F(x, y, z) = x^2 + y^2 + z^2 - 1 = 0.$$

55. Verify the result of Problem 53 (with p, V, and T in place of x, y, and z) for the equation

$$F(p, V, T) = pV - nRT = 0$$

(n and R are constants), which expresses the ideal gas law.

56. Consider a given quantity of liquid whose pressure p, volume V, and temperature T satisfy a given "state equation" of the form $F(p, V, T) = 0$. The **thermal expansivity** α and **isothermal compressivity** β of the liquid are defined by

$$\alpha = \frac{1}{V}\frac{\partial V}{\partial T} \quad \text{and} \quad \beta = -\frac{1}{V}\frac{\partial V}{\partial p}.$$

Apply Theorem 3 first to calculate $\partial V/\partial p$ and $\partial V/\partial T$, and then to calculate $\partial p/\partial V$ and $\partial p/\partial T$. Deduce from the results that $\partial p/\partial T = \alpha/\beta$.

57. The thermal expansivity and isothermal compressivity of liquid mercury are $\alpha = 1.8 \times 10^{-4}$ and $\beta = 3.9 \times 10^{-6}$, respectively, in L-atm-°C units. Suppose that a thermometer bulb is exactly filled with mercury at 50°C. If the bulb can withstand an internal pressure of no more than 200 atm, can it be heated to 55°C without breaking? *Suggestion:* Apply the result of Problem 56 to calculate the increase in pressure with each increase of one degree in temperature.

58. Suppose that the transformation $T : \mathbf{R}^3_{uvw} \to \mathbf{R}^3_{xyz}$ is defined by the functions $x = x(u, v, w)$, $y = y(u, v, w)$, $z = z(u, v, w)$. Then its derivative matrix is defined by

$$T(u, v, w) = \begin{bmatrix} x_u & x_v & x_w \\ y_u & y_v & y_w \\ z_u & z_v & z_w \end{bmatrix}.$$

Calculate the derivative matrix of the linear transformation defined by $x = a_1u + b_1v + c_1w$, $y = a_2u + b_2v + c_2w$, $z = a_3u + b_3v + c_3w$.

59. Calculate the derivative matrix of the spherical coordinate transformation T defined by $x = \rho\sin\phi\cos\theta$, $y = \rho\sin\phi\sin\theta$, $z = \rho\cos\phi$.

60. Suppose that $q = F(x, y, z)$ with 1×3 derivative matrix $F' = [F_x \ F_y \ F_z]$ and that $(x, y, z) = T(u, v, w)$ as in Problem 58. If $G = F \circ T$, deduce from the chain rule in Theorem 2 that $G' = F'T'$ (matrix product).

61. If $w = F(x, y, z)$, apply the results of Problems 59 and 60 to calculate by matrix multiplication the partial derivatives of w with respect to the spherical coordinates ρ, ϕ, and θ.

13.8 DIRECTIONAL DERIVATIVES AND THE GRADIENT VECTOR

Figure 13.8.1 shows temperatures (in degrees Fahrenheit) recorded at U.S. locations at 2:12 P.M. E.D.T. on Thursday, April 12, 2001. This plot of the U.S. temperature function $T = f(x, y)$ is contoured "by color"—that is, locations with the same temperature are shown in the same color. If we depart from an airport and fly due east (in the positive x-direction), then the rate of change of temperature (in degrees per mile) that we initially observe is given by the partial derivative $\partial T/\partial x = f_x$. If we fly due north, then $\partial T/\partial y = f_y$ gives the initial rate of change of temperature with respect to distance. But we need not fly either due east or due north. The *directional derivative* introduced in this section enables us to calculate the rate of change of a function in any specified direction.

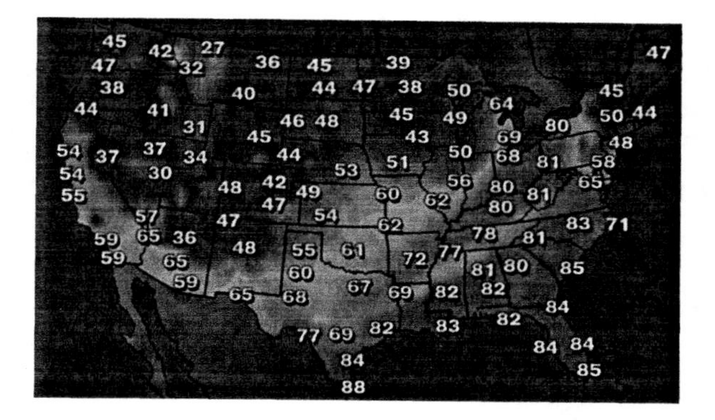

FIGURE 13.8.1 Current temperatures (°F) recorded at 2:12 P.M. on April 12, 2001.

Directional Derivatives

Recall that the first-order partial derivatives of the function $z = f(x, y)$ are defined to be

$$f_x(x, y) = \lim_{h \to 0} \frac{f(x+h, y) - f(x, y)}{h} \quad \text{and} \quad f_y(x, y) = \lim_{h \to 0} \frac{f(x, y+h) - f(x, y)}{h}$$

wherever these limits exist. If we write $\mathbf{x} = \langle x, y \rangle$, then these partial derivatives may be described a bit more concisely in the form

$$f_x(\mathbf{x}) = \lim_{h \to 0} \frac{f(\mathbf{x} + h\mathbf{i}) - f(\mathbf{x})}{h}, \qquad f_y(\mathbf{x}) = \lim_{h \to 0} \frac{f(\mathbf{x} + h\mathbf{j}) - f(\mathbf{x})}{h} \tag{1}$$

where $\mathbf{i} = \langle 1, 0 \rangle$ and $\mathbf{j} = \langle 0, 1 \rangle$ as usual. Thus f_x and f_y represent rates of change of z with respect to distance in the directions of the unit vectors $\mathbf{i}$ and $\mathbf{j}$. We get the definition of the *directional derivative* upon replacing $\mathbf{i}$ or $\mathbf{j}$ in (1) with an arbitrary specified unit vector $\mathbf{u}$.

DEFINITION Directional Derivative

The **directional derivative** of the function f at the point $\mathbf{x}$ in the direction of the unit vector $\mathbf{u}$ is

$$D_{\mathbf{u}} f(\mathbf{x}) = \lim_{h \to 0} \frac{f(\mathbf{x} + h\mathbf{u}) - f(\mathbf{x})}{h} \tag{2}$$

provided that this limit exists.

The function f in Eq. (2) can be a function of two or three or more variables. Comparing Eqs. (1) and (2), we see that the partial derivatives of a function of two variables x and y can be written as

$$f_x(x, y) = D_{\mathbf{i}} f(x, y) \quad \text{and} \quad f_y(x, y) = D_{\mathbf{j}} f(x, y).$$

Thus f_x and f_y are, indeed, the directional derivatives of f in the directions of the unit vectors **i** and **j**. Similarly, if f is a function of the three variables x, y, and z, then its partial derivatives

$$f_x = D_{\mathbf{i}} f, \quad f_y = D_{\mathbf{j}} f, \quad \text{and} \quad f_z = D_{\mathbf{k}} f$$

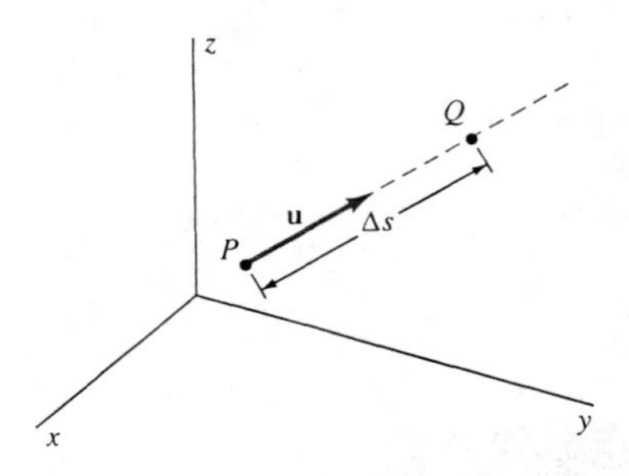

FIGURE 13.8.2 The first step in computing the rate of change of $f(x, y, z)$ in the direction of the unit vector **u**.

are the directional derivatives of f in the directions of the three standard unit vectors $\mathbf{i} = \langle 1, 0, 0 \rangle$, $\mathbf{j} = \langle 0, 1, 0 \rangle$, and $\mathbf{k} = \langle 0, 0, 1 \rangle$ in space.

The limit in Eq. (2) would still make sense if **u** were not a unit vector. But the *meaning* of directional derivatives is easiest to understand when **u** *is* a unit vector, and this is why we define $D_{\mathbf{u}} f(\mathbf{x})$ only when $|\mathbf{u}| = 1$. In Fig. 13.8.2 the unit vector **u** points in the direction from the fixed point P (with position vector **x**) to the point Q (with position vector $\mathbf{x} + h\mathbf{u}$). Then

$$\Delta w = f(Q) - f(P) = f(\mathbf{x} + h\mathbf{u}) - f(\mathbf{x})$$

is the increment in the function value $w = f(x, y, z)$ from the point P to the point Q. If we write $\Delta s = |\overrightarrow{PQ}| = h$ for the distance from P to Q, then the quotient

$$\frac{\Delta w}{\Delta s} = \frac{f(Q) - f(P)}{|\overrightarrow{PQ}|} = \frac{f(\mathbf{x} + h\mathbf{u}) - f(\mathbf{x})}{h}$$

is the *average rate of change* of w with respect to distance from P to Q. It is therefore natural to regard the limit

$$\frac{dw}{ds} = \lim_{\Delta s \to 0} \frac{\Delta w}{\Delta s} = \lim_{h \to 0} \frac{f(\mathbf{x} + h\mathbf{u}) - f(\mathbf{x})}{h} = D_{\mathbf{u}} f(\mathbf{x}) \tag{3}$$

as the **instantaneous rate of change** of w at P with respect to distance in the direction from P to Q. Some science and engineering texts may use the notation

$$\left.\frac{df}{ds}\right|_P = D_{\mathbf{u}} f(P),$$

or simply dw/ds as in Eq. (3), for the instantaneous rate of change of the function $w = f(x, y, z)$ at the point P, with respect to distance s in the direction of the unit vector **u**.

Calculation of Directional Derivatives

Equation (2) *defines* the directional derivative, but how do we actually *calculate* directional derivatives? To answer this question, we recall (from Eq. (18) in Section 13.6) that if the function $f(x_1, x_2, \dots, x_n)$ is differentiable at $\mathbf{x} = \langle x_1, x_2, \dots, x_n \rangle$, then its partial derivatives exist there; moreover,

$$\lim_{\mathbf{h} \to \mathbf{0}} \frac{f(\mathbf{x} + \mathbf{h}) - f(\mathbf{x}) - \nabla f(\mathbf{x}) \cdot \mathbf{h}}{|\mathbf{h}|} = 0 \tag{4}$$

where $\nabla f(\mathbf{x}) = \langle D_1 f(\mathbf{x}), D_2 f(\mathbf{x}), \dots, D_n f(\mathbf{x}) \rangle$ is the gradient vector of f at **x**. If we substitute $\mathbf{h} = h\mathbf{u}$ where **u** is a unit vector and $h > 0$ (so that $|\mathbf{h}| = h$), then Eq. (4) implies that

$$\begin{aligned} &\lim_{h \to 0} \frac{f(\mathbf{x} + \mathbf{h}) - f(\mathbf{x}) - \nabla f(\mathbf{x}) \cdot h\mathbf{u}}{h} \\ &= \lim_{h \to 0} \left(\frac{f(\mathbf{x} + h\mathbf{u}) - f(\mathbf{x})}{h} - \nabla f(\mathbf{x}) \cdot \mathbf{u} \right) = D_{\mathbf{u}} f(\mathbf{x}) - \nabla f(\mathbf{x}) \cdot \mathbf{u} = 0. \end{aligned}$$

In the last step we have used the definition in (2) of the directional derivative $D_{\mathbf{u}} f(\mathbf{x})$ and the fact that **x** and **u** play the role of constants as $h \to 0$. This proves the following theorem.

THEOREM 1 Calculation of Directional Derivatives

If the real-valued function f is differentiable at $\mathbf{x}$ and $\mathbf{u}$ is a unit vector, then the directional derivative $D_{\mathbf{u}} f(\mathbf{x})$ exists and is given by

$$D_{\mathbf{u}} f(\mathbf{x}) = \nabla f(\mathbf{x}) \cdot \mathbf{u}. \tag{5}$$

For instance, if $z = f(x, y)$ is a function of two variables, so that

$$\nabla f(x, y) = \left\langle f_x(x, y), f_y(x, y) \right\rangle \quad \text{and} \quad \mathbf{u} = \langle a, b \rangle,$$

then Eq. (5) gives

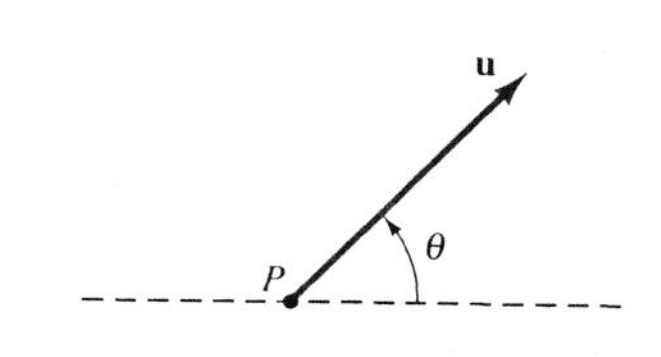

FIGURE 13.8.3 The unit vector $\mathbf{u}$ of Eq. (7).

$$D_{\langle a,b \rangle} f(x, y) = \left\langle f_x(x, y), f_y(x, y) \right\rangle \cdot \langle a, b \rangle = a f_x(x, y) + b f_y(x, y). \tag{6}$$

If the unit vector $\mathbf{u}$ makes the counterclockwise angle θ with the positive x-axis (as in Fig. 13.8.3), then $\mathbf{u} = \langle \cos\theta, \sin\theta \rangle$, so Eq. (6) takes the form

$$D_{\mathbf{u}} f(x, y) = f_x(x, y)\cos\theta + f_y(x, y)\sin\theta = \frac{\partial w}{\partial x}\cos\theta + \frac{\partial w}{\partial y}\sin\theta. \tag{7}$$

If $w = f(x, y, z)$ is a function of three variables and $\mathbf{u} = \langle a, b, c \rangle$ (still a unit vector), then Eq. (5) similarly yields

$$D_{\langle a,b,c \rangle} f(x, y, z) = a f_x(x, y, z) + b f_y(x, y, z) + c f_z(x, y, z). \tag{8}$$

EXAMPLE 1 Suppose that the temperature (in degrees Celsius) at the point (x, y) near an airport is given by

$$f(x, y) = \frac{1}{180}[7400 - 4x - 9y - (0.03)xy]$$

(with distances x and y measured in kilometers). Suppose that your aircraft takes off from this airport at the location $P(200, 200)$ and heads northeast in the direction specified by the vector $\mathbf{v} = \langle 3, 4 \rangle$. What initial rate of change of temperature will you observe?

Solution Because $\mathbf{v}$ is not a unit vector, we must first replace it with the unit vector $\mathbf{u}$ having the same direction:

$$\mathbf{u} = \frac{\mathbf{v}}{|\mathbf{v}|} = \frac{\langle 3, 4 \rangle}{\sqrt{3^2 + 4^2}} = \left\langle \frac{3}{5}, \frac{4}{5} \right\rangle.$$

Now we may use the formula in (6), which yields

$$D_{\mathbf{u}} f(x, y) = \left(\frac{3}{5}\right) \cdot \left(\frac{1}{180}[-4 - (0.03)y]\right) + \left(\frac{4}{5}\right) \cdot \left(\frac{1}{180}[-9 - (0.03)x]\right).$$

When we substitute $x = y = 200$ we find that

$$D_{\mathbf{u}} f(P) = \left(\frac{3}{5}\right) \cdot \left(-\frac{10}{180}\right) + \left(\frac{4}{5}\right) \cdot \left(-\frac{15}{180}\right) = -\frac{18}{180} = -0.1.$$

This instantaneous rate of change -0.1°C/km means that you will observe initially a decrease of 0.1°C in temperature per kilometer traveled. ◆

The Gradient Vector

In Section 13.6 we introduced the gradient vector informally as a notational device for simplifying the expression of certain multivariable formulas. Most of the remainder of this section is devoted to exploration of the meaning and geometric interpretation of gradient vectors, largely in two and three dimensions. We begin with a formal definition.

DEFINITION Gradient Vector

The **gradient** of the differentiable real-valued function $f : \mathbf{R}^n \to \mathbf{R}$ is the vector-valued function $\nabla f : \mathbf{R}^n \to \mathbf{R}^n$ defined by

$$\nabla f(\mathbf{x}) = \langle D_1 f(\mathbf{x}), D_2 f(\mathbf{x}), \dots, D_n f(\mathbf{x})\rangle. \tag{9}$$

In particular, the gradient vectors of functions of two and three variables are given (respectively) by

$$\nabla f(P) = \frac{\partial f}{\partial x}\mathbf{i} + \frac{\partial f}{\partial y}\mathbf{j} \quad \text{and} \quad \nabla f(P) = \frac{\partial f}{\partial x}\mathbf{i} + \frac{\partial f}{\partial y}\mathbf{j} + \frac{\partial f}{\partial z}\mathbf{k}; \tag{10}$$

the partial derivatives in Eq. (10) are to be evaluated at the point P.

EXAMPLE 2 If $f(x, y, z) = yz + \sin xz + e^{xy}$, then the second formula in (10) gives

$$\nabla f(x, y, z) = (z\cos xz + ye^{xy})\mathbf{i} + (z + xe^{xy})\mathbf{j} + (y + x\cos xz)\mathbf{k}.$$

The value of this gradient vector at the point $(0, 7, 3)$ is

$$\nabla f(0, 7, 3) = (3 \cdot 1 + 7 \cdot 1)\mathbf{i} + (3 + 0 \cdot 1)\mathbf{j} + (7 + 0 \cdot 1)\mathbf{k} = 10\mathbf{i} + 3\mathbf{j} + 7\mathbf{k}.$$ ◆

Theorem 1 says that if the function f is differentiable at $\mathbf{x}$ and $\mathbf{u}$ is a unit vector, then the directional derivative of f at $\mathbf{x}$ in the direction $\mathbf{u}$ is given by

$$D_{\mathbf{u}} f(\mathbf{x}) = \nabla f(\mathbf{x}) \cdot \mathbf{u}. \tag{11}$$

The chain rule has a similar gradient vector form. For instance, suppose that the differentiable vector-valued function

$$\mathbf{r}(t) = x(t)\mathbf{i} + y(t)\mathbf{j} + z(y)\mathbf{k}$$

is the position vector of a curve in $\mathbf{R}^3$ and that $f(x, y, z)$ is a differentiable function. Then the composition

$$f(\mathbf{r}(t)) = f(x(t), y(t), z(t))$$

is a differentiable function of t, and its (ordinary) chain-rule derivative with respect to t is

$$D_t[f(\mathbf{r}(t))] = D_t[f(x(t), y(t), z(t))] = \frac{\partial f}{\partial x}\cdot\frac{dx}{dt} + \frac{\partial f}{\partial y}\cdot\frac{dy}{dt} + \frac{\partial f}{\partial z}\cdot\frac{dz}{dt}.$$

We recognize here the dot product

$$D_t[f(\mathbf{r}(t))] = \nabla f(\mathbf{r}(t)) \cdot \mathbf{r}'(t), \tag{12}$$

where

$$\mathbf{r}'(t) = \frac{d\mathbf{r}}{dt} = \frac{dx}{dt}\mathbf{i} + \frac{dy}{dt}\mathbf{j} + \frac{dz}{dt}\mathbf{k}$$

is the velocity vector of the parametric curve $\mathbf{r}(t)$.

If $\mathbf{r}(t)$ is a *smooth* parametric curve with nonzero velocity vector $\mathbf{v}(t) = \mathbf{r}'(t)$, then $\mathbf{v} = v\mathbf{u}$ where $v = |\mathbf{v}|$ is the speed of motion along the curve and $\mathbf{u} = \mathbf{v}/v$ is the unit vector tangent to the curve (Section 12.6). Then Eq. (12) implies that

$$D_t[f(\mathbf{r}(t))] = \nabla f(\mathbf{r}(t)) \cdot \mathbf{r}'(t) = \nabla f(\mathbf{r}(t)) \cdot v\mathbf{u} = v\nabla f(\mathbf{r}(t)) \cdot \mathbf{u},$$

and hence in turn that

$$D_t[f(\mathbf{r}(t))] = v\, D_{\mathbf{u}} f(\mathbf{r}(t)). \tag{13}$$

With $w = f(\mathbf{r}(t))$, we may write $D_{\mathbf{u}} f(\mathbf{r}(t)) = dw/ds$ for the derivative of w with respect to (unit) distance along the parametrized curve, and $v = ds/dt$ for the speed. Then Eq. (13) takes the natural chain rule form

$$\frac{dw}{dt} = \frac{dw}{ds}\cdot\frac{ds}{dt}. \tag{14}$$

EXAMPLE 3 In Example 1 we found that the temperature function

$$w = f(x, y) = \frac{1}{180}[7400 - 4x - 9y - (0.03)xy]$$

(with temperature in degrees Celsius and distance in kilometers) has directional derivative

$$\frac{dw}{ds} = D_{\mathbf{u}} f(P) = -0.1 \ \frac{°\text{C}}{\text{km}}$$

at the point $P(200, 200)$ in the direction of the unit vector $\mathbf{u} = \langle \frac{3}{5}, \frac{4}{5} \rangle$. If a plane departs from an airport at P and flies in the direction $\mathbf{u}$ with speed $v = ds/dt = 5$ km/min, then Eq. (14) gives

$$\frac{dw}{dt} = \frac{dw}{ds} \cdot \frac{ds}{dt} = \left(-0.1 \ \frac{°\text{C}}{\text{km}}\right)\left(5 \ \frac{\text{km}}{\text{min}}\right) = -0.5 \ \frac{°\text{C}}{\text{min}}.$$

Thus an initial rate of decrease of a half-degree of temperature per minute is observed. ◆

EXAMPLE 4 Now suppose that the temperature function of Example 3 is replaced with

$$w = f(x, y, z) = \frac{1}{180}[7400 - 4x - 9y - (0.03)xy] - 2z.$$

The additional term $-2z$ corresponds to a decrease of 2°C in temperature per kilometer of altitude z. Suppose that a hawk hovering at the point $P(200, 200, 5)$ above the airport suddenly dives at a speed of 3 km/min in the direction specified by the vector $\langle 3, 4, -12 \rangle$. What instantaneous rate of change of temperature does the bird experience?

Solution The unit vector in the direction of the given vector $\langle 3, 4, -12 \rangle$ is

$$\mathbf{u} = \frac{3\mathbf{i} + 4\mathbf{j} - 12\mathbf{k}}{\sqrt{3^2 + 4^2 + (-12)^2}} = \frac{3}{13}\mathbf{i} + \frac{4}{13}\mathbf{j} - \frac{12}{13}\mathbf{k}.$$

The temperature gradient vector

$$\nabla f(x, y, z) = -\frac{1}{180}[4 + (0.03)y]\mathbf{i} - \frac{1}{180}[9 + (0.03)x]\mathbf{j} - 2\mathbf{k}$$

has the value

$$\nabla f(P) = -\frac{10}{180}\mathbf{i} - \frac{15}{180}\mathbf{j} - 2\mathbf{k}$$

at the initial position $P(200, 200, 5)$ of the hawk. Therefore the hawk's initial rate of change of temperature with respect to distance is

$$\begin{aligned} \frac{dw}{ds} &= D_{\mathbf{u}} f(P) = \nabla f(P) \cdot \mathbf{u} \\ &= \left(-\frac{10}{180}\right)\left(\frac{3}{13}\right) + \left(-\frac{15}{180}\right)\left(\frac{4}{13}\right) + (-2)\left(-\frac{12}{13}\right) = \frac{47}{26} \approx 1.808 \ \frac{°\text{C}}{\text{km}}. \end{aligned}$$

Its speed is $ds/dt = 3$ km/min, so the time rate of change of temperature experienced by the hawk is

$$\frac{dw}{dt} = \frac{dw}{ds} \cdot \frac{ds}{dt} = \left(1.808 \ \frac{°\text{C}}{\text{km}}\right)\left(3 \ \frac{\text{km}}{\text{min}}\right) = 5.424 \ \frac{°\text{C}}{\text{min}}.$$

Thus the hawk initially gets warmer by almost 5.5 degrees per minute as it dives toward the ground. ◆

Interpretation of the Gradient Vector

The gradient vector ∇f has an important interpretation that involves the maximum possible value of the directional derivative of the differentiable function f at a given point P. If ϕ is the angle between $\nabla f(P)$ and the unit vector $\mathbf{u}$ (Fig. 13.8.4), then Eq. (11) gives

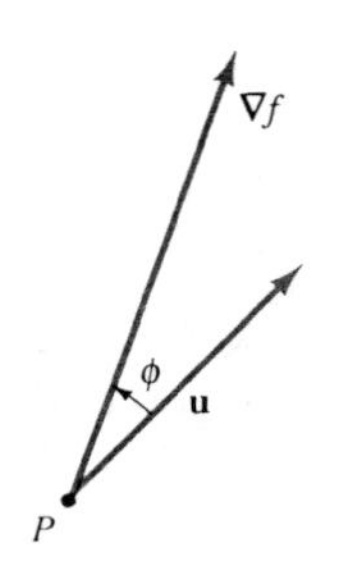

FIGURE 13.8.4 The angle ϕ between ∇f and the unit vector $\mathbf{u}$.

$$D_{\mathbf{u}} f(P) = \nabla f(P) \cdot \mathbf{u} = |\nabla f(P)|\,|\mathbf{u}| \cos\phi = |\nabla f(P)| \cos\phi$$

because $|\mathbf{u}| = 1$. The maximum possible value of $\cos\phi$ is 1, and this occurs when $\phi = 0$. This is so when $\mathbf{u}$ is the particular unit vector $\mathbf{m} = \nabla f(P)/|\nabla f(P)|$ that points in the direction of the gradient vector $\nabla f(P)$ itself. In this case the previous formula yields

$$D_{\mathbf{m}} f(P) = |\nabla f(P)|, \tag{15}$$

so the value of the directional derivative in this direction is equal to the length of the gradient vector. This argument establishes the following result.

THEOREM 2 Significance of the Gradient Vector

The maximum value of the directional derivative $D_{\mathbf{u}} f(P)$ is obtained when $\mathbf{u}$ is the unit vector in the direction of the gradient vector $\nabla f(P)$; that is, when $\mathbf{u} = \nabla f(P)/|\nabla f(P)|$. The value of the maximum directional derivative is $|\nabla f(P)|$, the length of the gradient vector.

Thus *the gradient vector ∇f points in the direction in which the function f increases the most rapidly, and its length is the rate of increase of f (with respect to distance) in that direction.* For instance, if the function f gives the temperature in space, then the gradient vector $\nabla f(P)$ points in the direction in which a hawk at P should initially fly to get warmer the fastest.

EXAMPLE 5 Recall the temperature function

$$w = f(x, y, z) = \frac{1}{180}[7400 - 4x - 9y - (0.03)xy] - 2z$$

of Example 4 (with distance in kilometers and temperature in degrees Celsius). In what direction should a hawk, starting at the point $P(200, 200, 5)$ at an altitude of 5 km, dive in order to get warmer the fastest? How rapidly will its temperature increase as it dives at a speed of 3 km/min? What will be its compass heading and angle of descent as it dives in this particular direction?

Solution In Example 4 we calculated the value

$$\nabla f(P) = -\frac{10}{180}\mathbf{i} - \frac{15}{180}\mathbf{j} - 2\mathbf{k}$$

of the gradient vector of f at the point $P(200, 200, 5)$. By Theorem 2, the maximum value

$$\frac{dw}{ds} = D_{\mathbf{m}} f(P) = |\nabla f(P)| = \sqrt{\left(-\frac{10}{180}\right)^2 + \left(-\frac{15}{180}\right)^2 + (-2)^2} \approx 2.0025$$

(°C/km) of the directional derivative of f at P is attained with the unit vector

$$\mathbf{m} = \frac{\nabla f(P)}{|\nabla f(P)|} \approx \frac{1}{2.0025}\left(-\frac{10}{180}\mathbf{i} - \frac{15}{180}\mathbf{j} - 2\mathbf{k}\right) = \frac{-10\mathbf{i} - 15\mathbf{j} - 360\mathbf{k}}{360.45}.$$

The speed of the hawk is $ds/dt = 3$ km/min, so the time rate of change of temperature experienced by the hawk is

$$\frac{dw}{dt} = \frac{dw}{ds} \cdot \frac{ds}{dt} \approx \left(2.0025\ \frac{°\mathrm{C}}{\mathrm{km}}\right)\left(3\ \frac{\mathrm{km}}{\mathrm{min}}\right) = 6.0075\ \frac{°\mathrm{C}}{\mathrm{min}}.$$

Thus the hawk initially gets warmer by slightly more than 6 °C/min as it dives toward the ground.

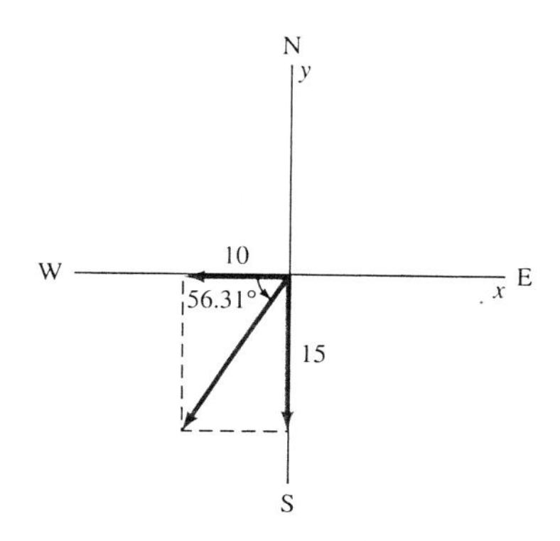

FIGURE 13.8.5 The diving hawk's compass heading.

Figure 13.8.5 shows the third-quadrant vector $-10\mathbf{i} - 15\mathbf{j}$ that represents the hawk's (horizontal) compass heading of $\pi + \tan^{-1}(\frac{15}{10}) \approx 236.31°$ (about 56.31° south of west). The hawk is descending 360 meters vertically for every $\sqrt{10^2 + 15^2} \approx 18.028$ meters it flies horizontally. Hence its angle of descent (measured from the horizontal) is about $\tan^{-1}(360/18.028) \approx 87.13°$. ◆

The Gradient Vector as a Normal Vector

Consider the graph of the equation

$$F(x, y, z) = 0, \tag{16}$$

where the function F is continuously differentiable. The *implicit function theorem* stated in Section 13.7 (Theorem 3 there) implies that, near any point P where the partial derivative $\partial F/\partial z$ is nonzero, Eq. (16) defines z implicitly as a continuously differentiable function f of x and y. Thus the graph $F(x, y, z) = 0$ coincides—near P—with the surface $z = f(x, y)$. Similarly, the graph of Eq. (16) coincides with the surface of the form $x = g(y, z)$ near any point where $\partial F/\partial x$ is nonzero, and with a surface $y = h(x, z)$ near any point where $\partial F/\partial y$ is nonzero. In short, the graph of $F(x, y, z) = 0$ looks like a surface near any point P at which $\nabla F(P) \neq \mathbf{0}$ (so that at least one of the partial derivatives of F is nonzero). The next theorem implies that the gradient vector $\nabla F(P)$ is then normal to the surface $F(x, y, z) = 0$ at the point P.

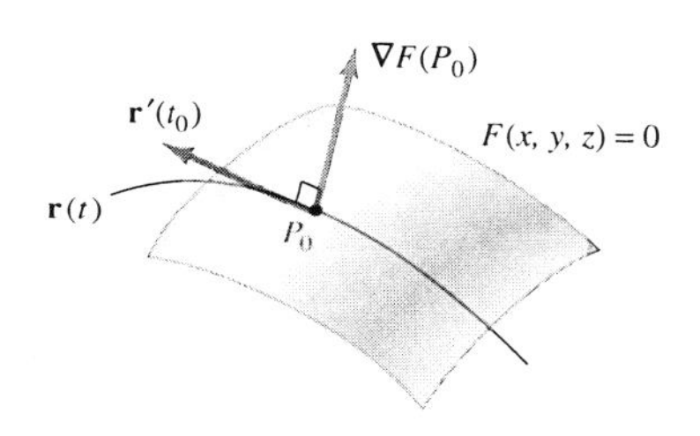

FIGURE 13.8.6 The gradient vector ∇F is normal to every curve in the surface $F(x, y, z) = 0$.

THEOREM 3 Gradient Vector as Normal Vector

Suppose that $F(x, y, z)$ is continuously differentiable and let $P_0(x_0, y_0, z_0)$ be a point of the graph of the equation $F(x, y, z) = 0$ at which $\nabla F(P_0) \neq \mathbf{0}$. If $\mathbf{r}(t)$ is a differentiable curve on this surface with $\mathbf{r}(t_0) = \langle x_0, y_0, z_0 \rangle$ and $\mathbf{r}'(t_0) \neq \mathbf{0}$, then

$$\nabla F(P_0) \cdot \mathbf{r}'(t_0) = 0. \tag{17}$$

Thus $\nabla F(P_0)$ is perpendicular to the tangent vector $\mathbf{r}'(t_0)$, as indicated in Fig. 13.8.6.

PROOF The statement that $\mathbf{r}(t)$ lies on the surface $F(x, y, z) = 0$ implies that $F(\mathbf{r}(t)) = 0$ for all t. Hence

$$0 = D_t F(\mathbf{r}(t_0)) = \nabla F(\mathbf{r}(t_0)) \cdot \mathbf{r}'(t_0) = \nabla F(P_0) \cdot \mathbf{r}'(t_0)$$

by the chain rule in the form in Eq. (12). Therefore the nonzero vectors $\nabla F(P_0)$ and $\mathbf{r}'(t_0)$ are perpendicular. ◄

Because the gradient vector $\nabla F(P_0)$ is perpendicular at P_0 to every curve on the surface through P_0, it is a **normal vector n** to the surface $F(x, y, z) = 0$ at the point P_0:

$$\mathbf{n} = \frac{\partial F}{\partial x}\mathbf{i} + \frac{\partial F}{\partial y}\mathbf{j} + \frac{\partial F}{\partial z}\mathbf{k}. \tag{18}$$

If we write the explicit surface equation $z = f(x, y)$ in the form $F(x, y, z) = f(x, y) - z = 0$, then

$$\frac{\partial F}{\partial x}\mathbf{i} + \frac{\partial F}{\partial y}\mathbf{j} + \frac{\partial F}{\partial z}\mathbf{k} = \frac{\partial f}{\partial x}\mathbf{i} + \frac{\partial f}{\partial y}\mathbf{j} - \mathbf{k}.$$

Thus Eq. (18) agrees with the definition of a normal vector that we gave in Section 13.4 (Eq. (13) there).

If the tangent vector $\mathbf{T}$ to a curve is normal to the vector $\mathbf{n}$ at the point P, then $\mathbf{T}$ lies in the plane through P that is normal to $\mathbf{n}$. If the function F is continuously differentiable, we therefore *define* the **tangent plane** to the surface $F(x, y, z) = 0$ at a point $P(a, b, c)$ at which $\nabla F(P) \neq \mathbf{0}$ to be the plane through P that has the normal vector $\mathbf{n}$ given in Eq. (18). An equation of this tangent plane is then

$$F_x(a, b, c)(x-a) + F_y(a, b, c)(y-b) + F_z(a, b, c)(z-c) = 0. \tag{19}$$

EXAMPLE 6 Write an equation of the plane tangent to the ellipsoid $2x^2 + 4y^2 + z^2 = 45$ at the point $(2, -3, -1)$.

Solution If we write

$$F(x, y, z) = 2x^2 + 4y^2 + z^2 - 45,$$

then $F(x, y, z) = 0$ is an equation of the ellipsoid. Thus, by Theorem 3, a vector normal to the ellipsoidal surface at (x, y, z) is $\nabla F(x, y, z) = \langle 4x, 8y, 2z\rangle$, so

$$\nabla F(2, -3, -1) = 8\mathbf{i} - 24\mathbf{j} - 2\mathbf{k}$$

is normal to the ellipsoid at $(2, -3, -1)$. Equation (19) then gives the answer in the form

$$8(x-2) - 24(y+3) - 2(z+1) = 0;$$

that is, $4x - 12y - z = 45$. ◆

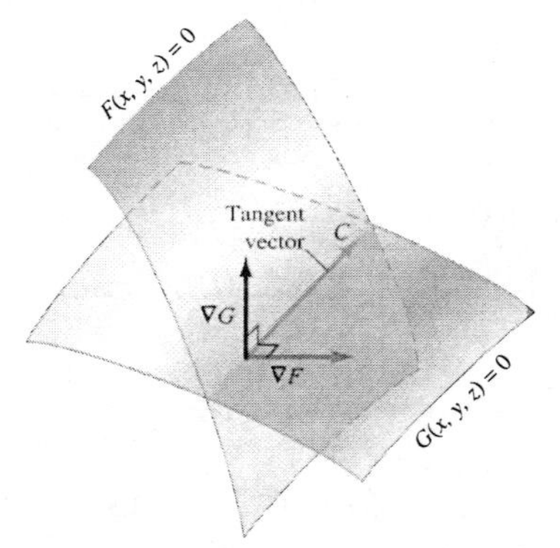

FIGURE 13.8.7 $\nabla F \times \nabla G$ is tangent to the curve C of intersection.

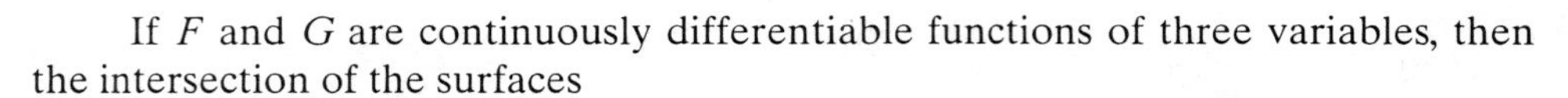

If F and G are continuously differentiable functions of three variables, then the intersection of the surfaces

$$F(x, y, z) = 0 \quad \text{and} \quad G(x, y, z) = 0 \tag{20}$$

will generally be some sort of curve C in space. More precisely, if P is a point of C where the two gradient vectors $\nabla F(P)$ and $\nabla G(P)$ are *not* collinear, then a general multivariable version of the implicit function theorem implies that near P the equations in (20) can be "solved for two of the variables in terms of the third." This means that the two equations implicitly define either x and y as functions of z, or y and z as functions of x, or x and z as functions of y. In any event, C is a smooth curve that passes through P. Because this curve lies on both surfaces, its tangent vector at P is perpendicular to both their normal vectors $\nabla F(P)$ and $\nabla G(P)$. It follows that the vector

$$\mathbf{T} = \nabla F(P) \times \nabla G(P) \tag{21}$$

is tangent at P to the curve C of intersection of the two surfaces $F(x, y, z) = 0$ and $G(x, y, z) = 0$. (See Fig. 13.8.7.)

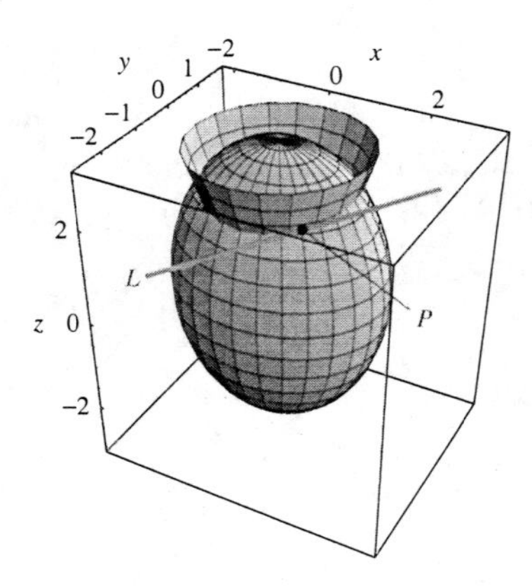

FIGURE 13.8.8 The point $P(1, -1, 2)$ on the curve of intersection of the paraboloid $F(x, y, z) = 0$ and the ellipsoid $G(x, y, z) = 0$ of Example 7, and the tangent line L through P that is parallel to the vector $\mathbf{T} = \nabla F(P) \times \nabla G(P) = \langle -14, -12, -4\rangle$.

EXAMPLE 7 The point $P(1, -1, 2)$ lies on both the paraboloid

$$F(x, y, z) = x^2 + y^2 - z = 0$$

and the ellipsoid

$$G(x, y, z) = 2x^2 + 3y^2 + z^2 - 9 = 0.$$

Write an equation of the plane through P that is normal to the curve of intersection of these two surfaces (Fig. 13.8.8).

Solution First we compute

$$\nabla F = \langle 2x, 2y, -1\rangle \quad \text{and} \quad \nabla G = \langle 4x, 6y, 2z\rangle.$$

At $P(1, -1, 2)$ these two vectors are

$$\nabla F(1, -1, 2) = \langle 2, -2, -1\rangle \quad \text{and} \quad \nabla G(1, -1, 2) = \langle 4, -6, 4\rangle.$$

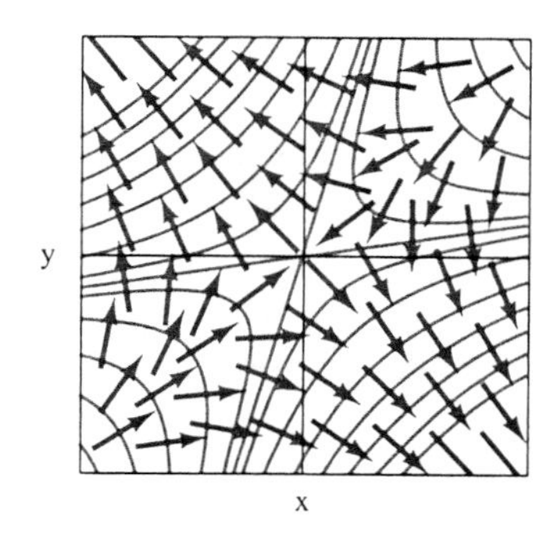

FIGURE 13.8.9 Gradient vectors and level curves for the function $F(x, y) = x^2 - 7xy + 2y^2$.

Hence a vector tangent at P to the curve of intersection of the paraboloid and the ellipsoid is

$$\mathbf{T} = \nabla F \times \nabla G = \begin{vmatrix} \mathbf{i} & \mathbf{j} & \mathbf{k} \\ 2 & -2 & -1 \\ 4 & -6 & 4 \end{vmatrix} = \langle -14, -12, -4 \rangle.$$

A slightly simpler vector parallel to $\mathbf{T}$ is $\mathbf{n} = \langle 7, 6, 2 \rangle$, and $\mathbf{n}$ is also normal to the desired plane through $(1, -1, 2)$. Therefore an equation of the plane is

$$7(x-1) + 6(y+1) + 2(z-2) = 0;$$

that is, $7x + 6y + 2z = 5$. ◆

A result analogous to Theorem 3 holds in two dimensions (and in higher dimensions). If the function F of two variables is continuously differentiable, then the graph of the equation $F(x, y) = 0$ looks like a smooth curve C near each point P at which $\nabla F(P) \neq \mathbf{0}$, and then the gradient vector $\nabla F(P)$ is normal to C at P. Consequently, if we use a computer algebra system to plot both a number of level curves and a "field" of different gradient vectors of the function $F(x, y)$, then (as illustrated in Fig. 13.8.9) the gradient vector at each point is normal to the level curve through that point.

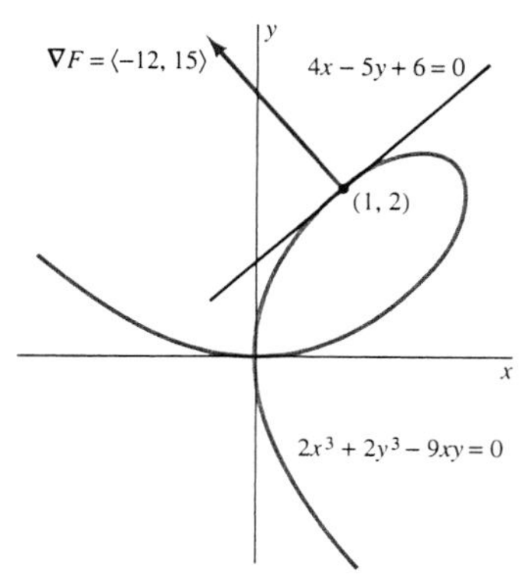

FIGURE 13.8.10 The folium and its tangent (Example 8).

EXAMPLE 8 Write an equation of the line tangent at the point $(1, 2)$ to the folium of Descartes with equation $F(x, y) = 2x^3 + 2y^3 - 9xy = 0$. (See Fig. 13.8.10.)

Solution The gradient of F is

$$\nabla F(x, y) = (6x^2 - 9y)\mathbf{i} + (6y^2 - 9x)\mathbf{j}.$$

So a vector normal to the folium at $(1, 2)$ is $\nabla F(1, 2) = -12\mathbf{i} + 15\mathbf{j}$. Hence the tangent line has equation $-12(x-1) + 15(y-2) = 0$. Simplified, this is $4x - 5y + 6 = 0$. ◆

13.8 TRUE/FALSE STUDY GUIDE

13.8 CONCEPTS: QUESTIONS AND DISCUSSION

1. The partial derivatives $f_x(a, b)$ and $f_y(a, b)$ give the slopes (vertical rise per horizontal run) of the lines tangent to the x-curve $z = f(x, b)$ and the y-curve $z = f(a, y)$ through the point $(a, b, f(a, b))$ on the surface $z = f(x, y)$. What is an analogous interpretation of the directional derivative $D_{\mathbf{u}} f(a, b)$?
2. Suppose that you have a map showing level curves for the function $z = f(x, y)$ describing a mountain you're climbing. How can you use the level curves to sketch a path of "steepest ascent" from your present location on the mountain side to the peak of the mountain? This will be a path that at each point climbs as steeply as possible. Would your compass heading on such a path of steepest ascent always be directly toward the mountain peak?

13.8 PROBLEMS

In Problems 1 through 10, find the gradient vector ∇f at the indicated point P.

1. $f(x, y) = 3x - 7y$; $P(17, 39)$

2. $f(x, y) = 3x^2 - 5y^2$; $P(2, -3)$

3. $f(x, y) = \exp(-x^2 - y^2)$; $P(0, 0)$

4. $f(x, y) = \sin \frac{1}{4}\pi xy$; $P(3, -1)$

5. $f(x, y, z) = y^2 - z^2$; $P(17, 3, 2)$

6. $f(x, y, z) = \sqrt{x^2 + y^2 + z^2}$; $P(12, 3, 4)$

7. $f(x, y, z) = e^x \sin y + e^y \sin z + e^z \sin x$; $P(0, 0, 0)$

8. $f(x, y, z) = x^2 - 3yz + z^3$; $P(2, 1, 0)$

9. $f(x, y, z) = 2\sqrt{xyz}$; $P(3, -4, -3)$

10. $f(x, y, z) = (2x - 3y + 5z)^5$; $P(-5, 1, 3)$

In Problems 11 through 20, find the directional derivative of f at P in the direction of **v**; *that is, find*

$$D_{\mathbf{u}}f(P), \quad \text{where} \quad \mathbf{u} = \frac{\mathbf{v}}{|\mathbf{v}|}.$$

11. $f(x, y) = x^2 + 2xy + 3y^2$; $P(2, 1)$, $\mathbf{v} = \langle 1, 1 \rangle$

12. $f(x, y) = e^x \sin y$; $P(0, \pi/4)$, $\mathbf{v} = \langle 1, -1 \rangle$

13. $f(x, y) = x^3 - x^2y + xy^2 + y^3$; $P(1, -1)$, $\mathbf{v} = 2\mathbf{i} + 3\mathbf{j}$

14. $f(x, y) = \tan^{-1}\left(\dfrac{y}{x}\right)$; $P(-3, 3)$, $\mathbf{v} = 3\mathbf{i} + 4\mathbf{j}$

15. $f(x, y) = \sin x \cos y$; $P(\pi/3, -2\pi/3)$, $\mathbf{v} = \langle 4, -3 \rangle$

16. $f(x, y, z) = xy + yz + zx$; $P(1, -1, 2)$, $\mathbf{v} = \langle 1, 1, 1 \rangle$

17. $f(x, y, z) = \sqrt{xyz}$; $P(2, -1, -2)$, $\mathbf{v} = \mathbf{i} + 2\mathbf{j} - 2\mathbf{k}$

18. $f(x, y, z) = \ln(1+x^2+y^2-z^2)$; $P(1, -1, 1)$, $\mathbf{v} = 2\mathbf{i}-2\mathbf{j}-3\mathbf{k}$

19. $f(x, y, z) = e^{xyz}$; $P(4, 0, -3)$, $\mathbf{v} = \mathbf{j} - \mathbf{k}$

20. $f(x, y, z) = \sqrt{10 - x^2 - y^2 - z^2}$; $P(1, 1, -2)$, $\mathbf{v} = \langle 3, 4, -12 \rangle$

In Problems 21 through 28, find the maximum directional derivative of f at P and the direction in which it occurs.

21. $f(x, y) = 2x^2 + 3xy + 4y^2$; $P(1, 1)$

22. $f(x, y) = \arctan\left(\dfrac{y}{x}\right)$; $P(2, -3)$

23. $f(x, y) = \ln(x^2 + y^2)$; $P(3, 4)$

24. $f(x, y) = \sin(3x - 4y)$; $P(\pi/3, \pi/4)$

25. $f(x, y, z) = 3x^2 + y^2 + 4z^2$; $P(1, 5, -2)$

26. $f(x, y, z) = \exp(x - y - z)$; $P(5, 2, 3)$

27. $f(x, y, z) = \sqrt{xy^2z^3}$; $P(2, 2, 2)$

28. $f(x, y, z) = \sqrt{2x + 4y + 6z}$; $P(7, 5, 5)$

In Problems 29 through 34, use the normal gradient vector to write an equation of the line (or plane) tangent to the given curve (or surface) at the given point P.

29. $\exp(25 - x^2 - y^2) = 1$; $P(3, 4)$

30. $2x^2 + 3y^2 = 35$; $P(2, 3)$

31. $x^4 + xy + y^2 = 19$; $P(2, -3)$

32. $3x^2 + 4y^2 + 5z^2 = 73$; $P(2, 2, 3)$

33. $x^{1/3} + y^{1/3} + z^{1/3} = 1$; $P(1, -1, 1)$

34. $xyz + x^2 - 2y^2 + z^3 = 14$; $P(5, -2, 3)$

The properties of gradient vectors listed in Problems 35 through 38 exhibit the close analogy between the gradient operator ∇ and the single-variable derivative operator D. Verify each, assuming that a and b are constants and that u and v are differentiable functions of x and y.

35. $\nabla(au + bv) = a\nabla u + b\nabla v$.

36. $\nabla(uv) = u\nabla v + v\nabla u$.

37. $\nabla\left(\dfrac{u}{v}\right) = \dfrac{v\nabla u - u\nabla v}{v^2}$ if $v \neq 0$.

38. If n is a positive integer, then $\nabla u^n = nu^{n-1}\nabla u$.

39. Show that the value of a differentiable function f decreases the most rapidly at P in the direction of the vector $-\nabla f(P)$, directly opposite to the gradient vector.

40. Suppose that f is a function of three independent variables x, y, and z. Show that $D_{\mathbf{i}}f = f_x$, $D_{\mathbf{j}}f = f_y$, and $D_{\mathbf{k}}f = f_z$.

41. Show that the equation of the line tangent to the conic section $Ax^2 + Bxy + Cy^2 = D$ at the point (x_0, y_0) is

$$(Ax_0)x + \tfrac{1}{2}B(y_0x + x_0y) + (Cy_0)y = D.$$

42. Show that the equation of the plane tangent to the quadric surface $Ax^2 + By^2 + Cz^2 = D$ at the point (x_0, y_0, z_0) is

$$(Ax_0)x + (By_0)y + (Cz_0)z = D.$$

43. Show that an equation of the plane tangent to the paraboloid $z = Ax^2 + By^2$ at the point (x_0, y_0, z_0) is $z + z_0 = 2Ax_0x + 2By_0y$.

44. Suppose that the temperature at the point (x, y, z) in space, with distance measured in kilometers, is given by

$$w = f(x, y, z) = 10 + xy + xz + yz$$

(in degrees Celsius). Find the rate of change (in degrees Celsius per kilometer) of temperature at the point $P(1, 2, 3)$ in the direction of the vector $\mathbf{v} = \mathbf{i} + 2\mathbf{j} - 2\mathbf{k}$.

45. Suppose that the function

$$w = f(x, y, z) = 10 + xy + xz + yz$$

of Problem 44 gives the temperature at the point (x, y, z) of space. (Units in this problem are in kilometers, degrees Celsius, and minutes.) What time rate of change (in degrees Celsius per minute) will a hawk observe as it flies through $P(1, 2, 3)$ at a speed of 2 km/min, heading directly toward the point $Q(3, 4, 4)$?

46. Suppose that the temperature w (in degrees Celsius) at the point (x, y) is given by

$$w = f(x, y) = 10 + (0.003)x^2 - (0.004)y^2.$$

In what direction **u** should a bumblebee at the point (40, 30) initially fly in order to get warmer the most quickly? Find the directional derivative $D_{\mathbf{u}}f(40, 30)$ in this optimal direction **u**.

47. Suppose that the temperature W (in degrees Celsius) at the point (x, y, z) in space is given by

$$W = 50 + xyz.$$

(a) Find the rate of change of temperature with respect to distance at the point $P(3, 4, 1)$ in the direction of the vector $\mathbf{v} = \langle 1, 2, 2 \rangle$. (The units of distance in space are feet.) (b) Find the maximal directional derivative $D_{\mathbf{u}}W$ at the point $P(3, 4, 1)$ and the direction **u** in which that maximum occurs.

48. Suppose that the temperature (in degrees Celsius) at the point (x, y, z) in space is given by the formula

$$W = 100 - x^2 - y^2 - z^2.$$

The units of distance in space are meters. (a) Find the rate of change of temperature at the point $P(3, -4, 5)$ in the direction of the vector $\mathbf{v} = 3\mathbf{i} - 4\mathbf{j} + 12\mathbf{k}$. (b) In what direction does W increase most rapidly at P? What is the value of the maximal directional derivative at P?

49. Suppose that the altitude z (in miles above sea level) of a certain hill is described by the equation $z = f(x, y)$, where

$$f(x, y) = \frac{1}{10}(x^2 - xy + 2y^2).$$

(a) Write an equation (in the form $z = ax + by + c$) of the plane tangent to the hillside at the point $P(2, 1, 0.4)$. (b) Use $\nabla f(2, 1)$ to approximate the altitude of the hill above the point $(2.2, 0.9)$ in the xy-plane. Compare your result with the actual altitude at this point.

50. Find an equation for the plane tangent to the paraboloid $z = 2x^2 + 3y^2$ and, simultaneously, parallel to the plane $4x - 3y - z = 10$.

51. The cone with equation $z^2 = x^2 + y^2$ and the plane with equation $2x + 3y + 4z + 2 = 0$ intersect in an ellipse. Write an equation of the plane normal to this ellipse at the point $P(3, 4, -5)$ (Fig. 13.8.11).

52. It is apparent from geometry that the highest and lowest points of the ellipse of Problem 51 are those points where its tangent line is horizontal. Find those points.

53. Show that the sphere $x^2 + y^2 + z^2 = r^2$ and the cone $z^2 = a^2x^2 + b^2y^2$ are orthogonal (that is, have perpendicular tangent planes) at every point of their intersection (Fig. 13.8.12).

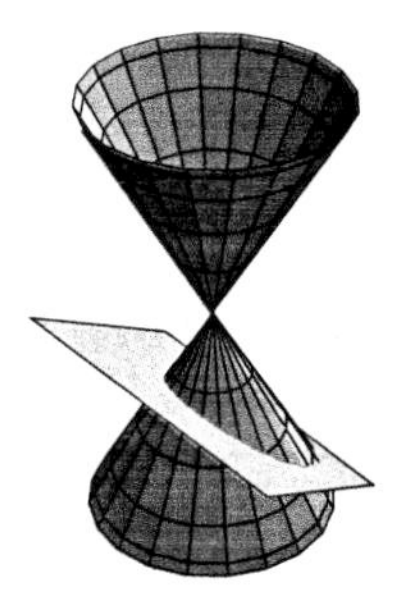

FIGURE 13.8.11 The cone and plane of Problems 51 and 52.

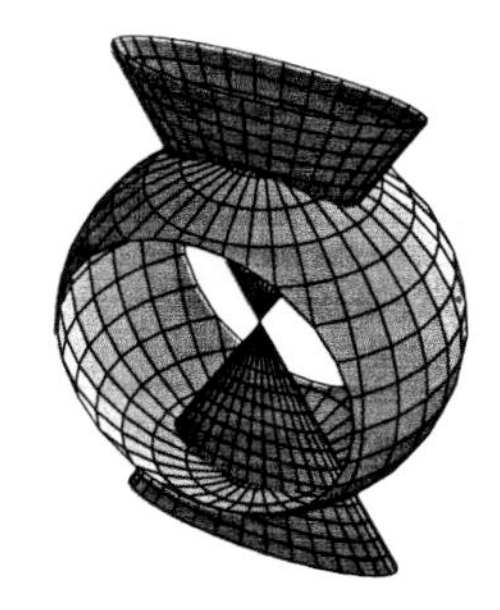

FIGURE 13.8.12 A cut-away view of the cone and sphere of Problem 53.

54. Suppose that $\mathcal{P}_1$ and $\mathcal{P}_2$ are planes tangent to the circular ellipsoid $x^2 + y^2 + 2z^2 = 2$ at the two points P_1 and P_2 having the same z-coordinate. Show that $\mathcal{P}_1$ and $\mathcal{P}_2$ intersect the z-axis at the same point.

55. A plane tangent to the surface $xyz = 1$ at a point in the first octant cuts off a pyramid from the first octant. Show that any two such pyramids have the same volume.

In Problems 56 through 61, the function $z = f(x, y)$ describes the shape of a hill; $f(P)$ is the altitude of the hill above the point $P(x, y)$ in the xy-plane. If you start at the point $(P, f(P))$ of this hill, then $D_{\mathbf{u}} f(P)$ is your rate of climb (rise per unit of horizontal distance) as you proceed in the horizontal *direction* $\mathbf{u} = a\mathbf{i} + b\mathbf{j}$. *And the angle at which you climb while you walk in this direction is $\gamma = \tan^{-1}(D_{\mathbf{u}} f(P))$, as shown in Fig. 13.8.13.*

56. You are standing at the point $(-100, -100, 430)$ on a hill that has the shape of the graph of

$$z = 500 - (0.003)x^2 - (0.004)y^2,$$

with x, y, and z given in feet. (a) What will be your rate of climb (*rise* over *run*) if you head northwest? At what angle from the horizontal will you be climbing? (b) Repeat part (a), except now you head northeast.

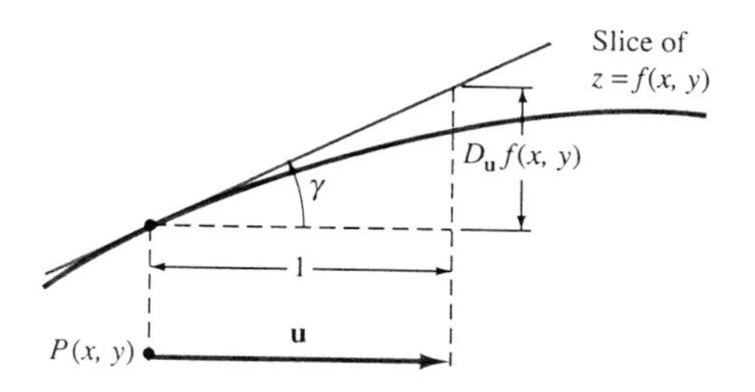

FIGURE 13.8.13 The cross section of the part of the graph above $\mathbf{u}$ (Problems 56 through 61).

57. You are standing at the point $(-100, -100, 430)$ on the hill of Problem 56. In what direction (that is, with what compass heading) should you proceed in order to climb the most steeply? At what angle from the horizontal will you initially be climbing?

58. Repeat Problem 56, but now you are standing at the point $P(100, 100, 500)$ on the hill described by

$$z = \frac{1000}{1 + (0.00003)x^2 + (0.00007)y^2}.$$

59. Repeat Problem 57, except begin at the point $P(100, 100, 500)$ of the hill of Problem 58.

60. You are standing at the point $(30, 20, 5)$ on a hill with the shape of the surface

$$z = 100 \exp\left(-\frac{x^2 + 3y^2}{701}\right).$$

(a) In what direction (with what compass heading) should you proceed in order to climb the most steeply? At what angle from the horizontal will you initially be climbing? (b) If, instead of climbing as in part (a), you head directly west (the negative x-direction), then at what angle will you be climbing initially?

61. (a) You are standing at the point where $x = y = 100$ (ft) on the side of a mountain whose height (in feet above sea level) is given by

$$z = \frac{1}{1000}(3x^2 - 5xy + y^2),$$

with the x-axis pointing east and the y-axis pointing north. If you head northeast, will you be ascending or descending? At what angle (in degrees) from the horizontal? (b) If you head 30° north of east, will you be ascending or descending? At what angle (in degrees) from the horizontal?

62. Suppose that the two surfaces $f(x, y, z) = 0$ and $g(x, y, z) = 0$ both pass through the point P where both gradient vectors $\nabla f(P)$ and $\nabla g(P)$ exist. (a) Show that the two surfaces are tangent at P if and only if $\nabla f(P) \times \nabla g(P) = \mathbf{0}$. (b) Show that the two surfaces are orthogonal at P if and only if $\nabla f(P) \cdot \nabla g(P) = 0$.

63. Suppose that the plane vectors $\mathbf{u}$ and $\mathbf{v}$ are not collinear and that the function $f(x, y)$ is differentiable at P. Show that the values of the directional derivatives $D_{\mathbf{u}} f(P)$ and $D_{\mathbf{v}} f(P)$ determine the value of the directional derivative of f at P in every other direction.

64. Show that the function $f(x, y) = (\sqrt[3]{x} + \sqrt[3]{y})^3$ is continuous at the origin and has directional derivatives in all directions there, but is not differentiable at the origin.

13.9 LAGRANGE MULTIPLIERS AND CONSTRAINED OPTIMIZATION

In Section 13.5 we discussed the problem of finding the maximum and minimum values attained by a function $f(x, y)$ at points of the plane region R, in the simple case in which R consists of the points on and within the simple closed curve C. We saw that any local maximum or minimum in the *interior* of R occurs at a point where $f_x(x, y) = 0 = f_y(x, y)$ or at a point where f is not differentiable (the latter usually signaled by the failure of f_x or f_y to exist). Here we discuss the very different matter of finding the maximum and minimum values attained by f at points of the *boundary* curve C.

If the curve C is the graph of the equation $g(x, y) = 0$, then our task is to maximize or minimize the function $f(x, y)$ subject to the **constraint, or side condition,**

$$g(x, y) = 0. \tag{1}$$

We could in principle try to solve this constraint equation for $y = \phi(x)$ and then maximize or minimize the single-variable function $f(x, \phi(x))$ by the standard method of finding its critical points. But what if it is impractical or impossible to solve Eq. (1) explicitly for y in terms of x? An alternative approach that does not require that we first solve this equation is the **method of Lagrange multipliers.** It is named for its discoverer, the Italian-born French mathematician Joseph Louis Lagrange (1736–1813). The method is based on Theorem 1.

THEOREM 1 Lagrange Multiplier (with one constraint)

Let $f(x, y)$ and $g(x, y)$ be continuously differentiable functions. If the maximum (or minimum) value of $f(x, y)$ subject to the constraint

$$g(x, y) = 0 \tag{1}$$

occurs at a point P where $\nabla g(P) \neq \mathbf{0}$, then

$$\nabla f(P) = \lambda \nabla g(P) \tag{2}$$

for some constant λ.

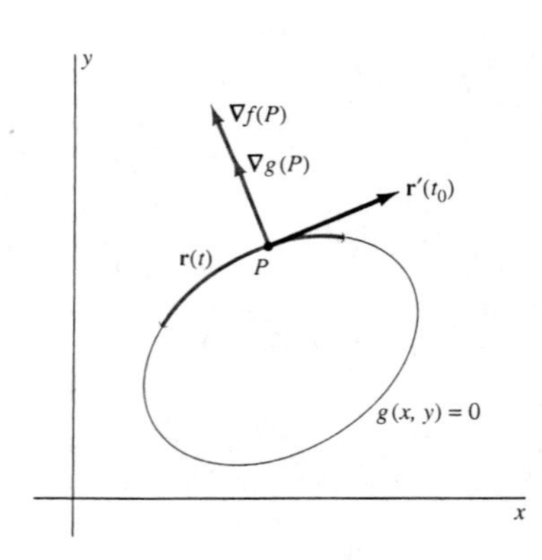

FIGURE 13.9.1 The conclusion of Theorem 1 illustrated.

PROOF Because $\nabla g(P) \neq \mathbf{0}$, the implicit function theorem implies that the graph C of the constraint equation $g(x, y) = 0$ agrees near $P(x_0, y_0)$ with the graph of a continuously differentiable single-variable function—either $y = \alpha(x)$ or $x = \beta(y)$. Either case provides a smooth parametric curve $\mathbf{r}(t)$ whose image agrees near P with C. For instance, in the case $y = \alpha(x)$ we define $\mathbf{r}(t) = \langle t, \alpha(t) \rangle$. If $\mathbf{r}(t_0) = \langle x_0, y_0 \rangle$, then $\mathbf{r}'(t_0) = \langle 1, \alpha'(t_0) \rangle \neq \mathbf{0}$ as indicated in Fig. 13.9.1. If $f(x, y)$ attains its maximum (or minimum) value on C at $P(x_0, y_0)$, then the composite function $F(t) = f(\mathbf{r}(t))$ attains its maximum (or minimum) value at $t = t_0$, so that $F'(t_0) = 0$. Therefore

$$F'(t_0) = \nabla f(\mathbf{r}(t_0)) \cdot \mathbf{r}'(t_0) = \nabla f(P) \cdot \mathbf{r}'(t_0) = 0 \tag{3}$$

by the gradient vector form of the chain rule of Eq. (12) in Section 13.8.

Because $\mathbf{r}(t)$ lies on the curve $g(x, y) = 0$, the composite function $G(t) = g(\mathbf{r}(t))$ is constant-valued—$G(t) \equiv 0$—so $G'(t) \equiv 0$. Therefore

$$G'(t_0) = \nabla g(\mathbf{r}(t_0)) \cdot \mathbf{r}'(t_0) = \nabla g(P) \cdot \mathbf{r}'(t_0) = 0. \tag{4}$$

Equations (3) and (4), when taken together, imply that the two-dimensional plane vectors $\nabla f(P)$ and $\nabla g(P)$ are both perpendicular to the nonzero vector $\mathbf{r}'(t_0)$, and are therefore collinear. Because $\nabla g(P) \neq \mathbf{0}$, it now follows that $\nabla f(P)$ must be a scalar multiple of $\nabla g(P)$, just as claimed in Eq. (2). ◂

The Method

Let's see what steps we should follow to solve a problem by using Theorem 1—the method of Lagrange multipliers. First we need to identify a quantity $z = f(x, y)$ to be maximized or minimized, subject to the constraint $g(x, y) = 0$. Then Eq. (1) and the two scalar components of Eq. (2) yield three equations:

$$g(x, y) = 0, \tag{1}$$

$$f_x(x, y) = \lambda g_x(x, y), \quad \text{and} \tag{2a}$$

$$f_y(x, y) = \lambda g_y(x, y). \tag{2b}$$

Thus we have three equations that we can attempt to solve for the three unknowns x, y, and λ. The points (x, y) that we find (assuming that our efforts are successful) are the only possible locations for the extrema of f subject to the constraint $g(x, y) = 0$. The associated values of λ, called **Lagrange multipliers,** may be revealed as well but often are not of much interest. Finally, we calculate the value $f(x, y)$ at each of the solution points (x, y) in order to identify its maximum and minimum values.

We must bear in mind the additional possibility that the maximum or minimum (or both) of f may occur at a point where $g_x(x, y) = 0 = g_y(x, y)$. The Lagrange multiplier method may fail to locate these exceptional points, but they can usually be recognized as points where the graph of $g(x, y) = 0$ fails to be a smooth curve.

EXAMPLE 1 Find the points of the rectangular hyperbola $xy = 1$ that are closest to the origin $(0, 0)$.

Solution We need to minimize the distance $d = \sqrt{x^2 + y^2}$ from the origin of a point $P(x, y)$ on the curve $xy = 1$. But the algebra is simpler if instead we minimize the square

$$f(x, y) = x^2 + y^2$$

of this distance subject to the constraint

$$g(x, y) = xy - 1 = 0$$

that the point P lies on the hyperbola. Because

$$\frac{\partial f}{\partial x} = 2x, \quad \frac{\partial f}{\partial y} = 2y, \quad \text{and} \quad \frac{\partial g}{\partial x} = y, \quad \frac{\partial g}{\partial y} = x,$$

the Lagrange multiplier equations in (2a) and (2b) take the form

$$2x = \lambda y, \qquad 2y = \lambda x.$$

If we multiply the first of these equations by x and the second by y, we can conclude that

$$2x^2 = \lambda xy = 2y^2$$

at $P(x, y)$. But the fact that $xy = 1 > 0$ implies that x and y have the same sign. Hence the fact that $x^2 = y^2$ implies that $x = y$. Substituting in $xy = 1$ then gives $x^2 = 1$, so it follows finally that either $x = y = 1$ or $x = y = -1$. The two resulting possibilities $(1, 1)$ and $(-1, -1)$ are indicated in Fig. 13.9.2. ◆

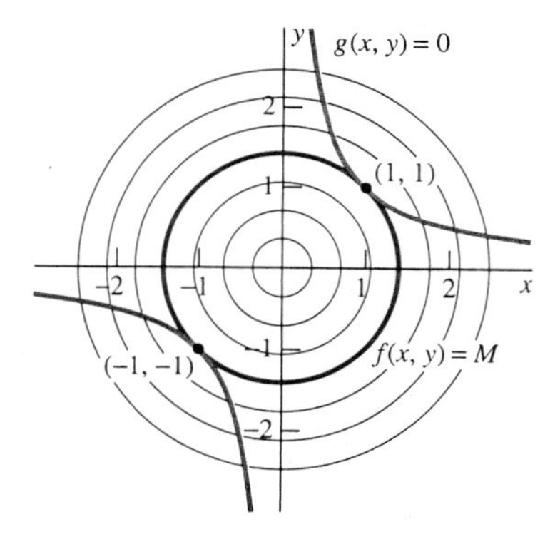

FIGURE 13.9.2 The level curve $f(x, y) = M$ and the constraint curve $g(x, y) = 0$ are tangent at a point P where the maximum or minimum value M is attained.

REMARK Example 1 illustrates an interesting geometric interpretation of Theorem 1. We see in Fig. 13.9.2 the constraint curve $g(x, y) = 0$ together with typical level curves of the function $f(x, y)$. Because the gradient vectors ∇f and ∇g are normal to the level curves of the functions f and g, respectively, it follows that the curves $f(x, y) = M$ and $g(x, y) = 0$ are tangent to one another at the point P where the two gradient vectors are collinear and f attains its maximum (or minimum) value M. In effect, the Lagrange multiplier criterion serves to select, from among the level curves of f, the one that is tangent to the constraint curve at P. Thus we see in Fig. 13.9.2 that the circle $x^2 + y^2 = 2$ and the hyperbola $xy = 1$ are, indeed, tangent at the two points $(1, 1)$ and $(-1, -1)$ where the squared distance $f(x, y) = x^2 + y^2$ is minimal subject to the constraint $g(x, y) = xy - 1$.

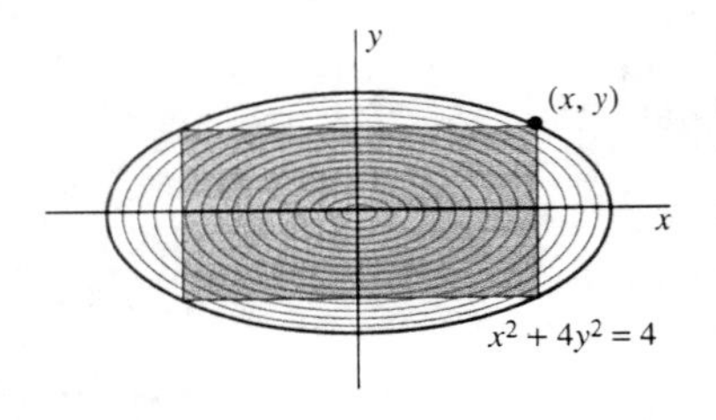

FIGURE 13.9.3 Cutting a rectangular beam from an elliptical log (Example 2).

EXAMPLE 2 In the sawmill problem of Example 5 in Section 3.6, we maximized the cross-sectional area of a rectangular beam cut from a circular log. Now we consider the elliptical log of Fig. 13.9.3, with semiaxes of lengths $a = 2$ ft and $b = 1$ ft. What is the maximal cross-sectional area of a rectangular beam cut as indicated from this elliptical log?

Solution The log is bounded by the ellipse $(x/2)^2 + y^2 = 1$; that is, $x^2 + 4y^2 = 4$. So with the coordinate system indicated in Fig. 13.9.3, we want to maximize the cross-sectional area

$$A = f(x, y) = 4xy \tag{5}$$

of the beam subject to the constraint

$$g(x, y) = x^2 + 4y^2 - 4 = 0. \tag{6}$$

Because

$$\frac{\partial f}{\partial x} = 4y, \quad \frac{\partial f}{\partial y} = 4x \quad \text{and} \quad \frac{\partial g}{\partial x} = 2x, \quad \frac{\partial g}{\partial y} = 8y,$$

Eqs. (2a) and (2b) give

$$4y = 2\lambda x, \qquad 4x = 8\lambda y.$$

It is clear that neither $x = 0$ nor $y = 0$ gives the maximum area, so we can solve these two multiplier equations for

$$\frac{2y}{x} = \lambda = \frac{x}{2y}.$$

Thus $x^2 = 4y^2$ at the desired maximum. Because $x^2 + 4y^2 = 4$, it follows that $x^2 = 4y^2 = 2$. Because we seek (as in Fig. 13.9.3) a first-quadrant solution point (x, y), we conclude that $x = \sqrt{2}$, $y = 1/\sqrt{2}$ gives the maximum possible cross-sectional area $A_{\text{max}} = 4(\sqrt{2})(1/\sqrt{2}) = 4$ ft² of a rectangular beam cut from the elliptical log. Note that this maximum area of 4 ft² is about 64% of the total cross-sectional area $A = \pi ab = 2\pi$ ft² of the original log. ◆

REMARK If we consider all four quadrants, then the condition $x^2 = 4y^2 = 2$ yields the *four* points $(\sqrt{2}, 1/\sqrt{2})$, $(-\sqrt{2}, 1/\sqrt{2})$, $(-\sqrt{2}, -1/\sqrt{2})$, and $(\sqrt{2}, -1/\sqrt{2})$. The function $f(x, y) = 4xy$ in Eq. (5) attains its maximum value $+4$ on the ellipse $x^2 + 4y^2 = 4$ at the first and third of these points and its minimum value -4 at the second and fourth points. The Lagrange multiplier method thus locates all of the global extrema of $f(x, y)$ on the ellipse.

In the applied maximum-minimum problems of Section 3.6, we typically began with a *formula* such as Eq. (5) of this section, expressing the quantity to be maximized in terms of *two* variables x and y, for example. We then used some available *relation* such as Eq. (6) between the variables x and y to eliminate one of them, such as y. Thus we finally obtained a single-variable *function* by substituting for y in terms of x in the original formula. As in Example 2, the Lagrange multiplier method frees us from the necessity of formulating the problem in terms of a single-variable function, and frequently leads to a solution process that is algebraically simpler and easier.

Lagrange Multipliers in Three Dimensions

Now suppose that $f(x, y, z)$ and $g(x, y, z)$ are continuously differentiable functions and that we want to find the points on the *surface*

$$g(x, y, z) = 0 \tag{7}$$

at which the function $f(x, y, z)$ attains its maximum and minimum values. Theorem 1 holds precisely as we have stated it, except with three independent variables rather than two. We leave the details to Problem 45, but an argument similar to the proof

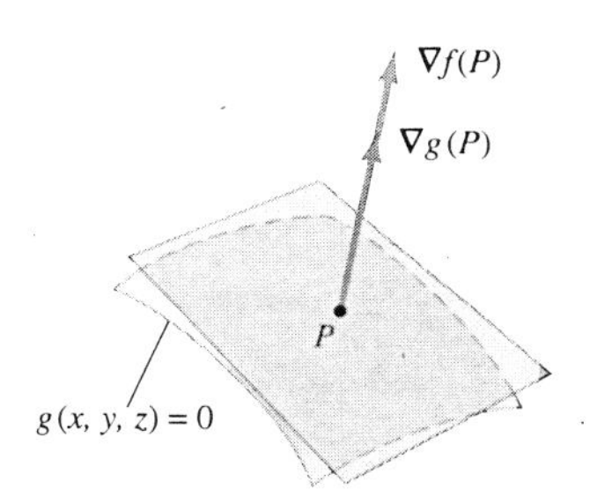

FIGURE 13.9.4 The natural generalization of Theorem 1 holds for functions of three variables.

of Theorem 1 shows that—at a maximum or minimum point P of $f(x, y, z)$ on the surface $g(x, y, z) = 0$—the two gradient vectors $\nabla f(P)$ and $\nabla g(P)$ are both perpendicular to every smooth curve on the surface through P. Hence they are both normal to the surface at P, and are therefore collinear. (See Fig. 13.9.4.) Because $\nabla g(P) \neq \mathbf{0}$, it follows that

$$\nabla f(P) = \lambda \nabla g(P) \tag{8}$$

for some scalar λ. This vector equation corresponds to three scalar equations. To find the possible locations of the extrema of f subject to the constraint g, we can attempt to solve simultaneously the four equations

$$g(x, y, z) = 0, \tag{7}$$

$$f_x(x, y, z) = \lambda g_x(x, y, z), \tag{8a}$$

$$f_y(x, y, z) = \lambda g_y(x, y, z), \tag{8b}$$

$$f_z(x, y, z) = \lambda g_z(x, y, z) \tag{8c}$$

for the four unknowns x, y, z, and λ. If successful, we then evaluate $f(x, y, z)$ at each of the solution points (x, y, z) to see at which it attains its maximum and minimum values. In analogy to the two-dimensional case, we also check points at which the surface $g(x, y, z) = 0$ fails to be smooth. Thus the Lagrange multiplier method with one constraint is essentially the same in dimension three as in dimension two.

EXAMPLE 3 Find the maximum volume of a rectangular box inscribed in the ellipsoid $x^2/a^2 + y^2/b^2 + z^2/c^2 = 1$ with its faces parallel to the coordinate planes (Fig. 13.9.5).

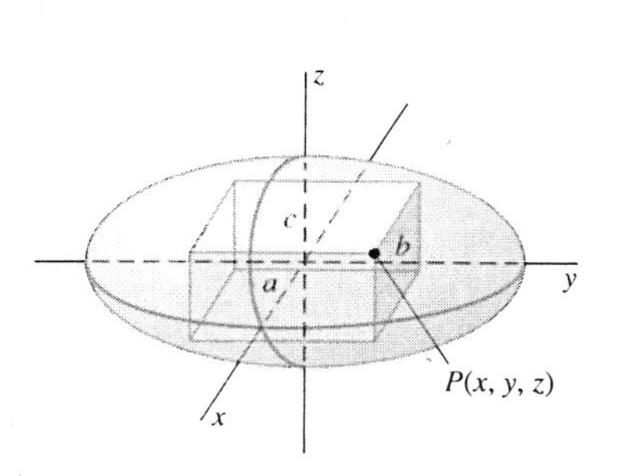

FIGURE 13.9.5 A rectangular $2x \times 2y \times 2z$ box inscribed in an ellipsoid with semiaxes a, b, and c. The whole box is determined by its first-octant vertex $P(x, y, z)$.

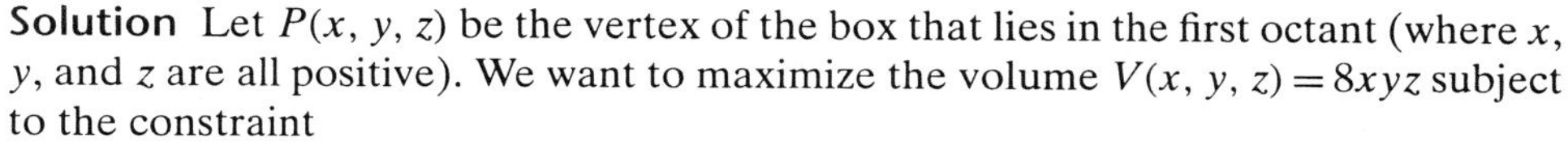

Solution Let $P(x, y, z)$ be the vertex of the box that lies in the first octant (where x, y, and z are all positive). We want to maximize the volume $V(x, y, z) = 8xyz$ subject to the constraint

$$g(x, y, z) = \frac{x^2}{a^2} + \frac{y^2}{b^2} + \frac{z^2}{c^2} - 1 = 0.$$

Equations (8a), (8b), and (8c) give

$$8yz = \frac{2\lambda x}{a^2}, \quad 8xz = \frac{2\lambda y}{b^2}, \quad 8xy = \frac{2\lambda z}{c^2}.$$

Part of the art of mathematics lies in pausing for a moment to find an elegant way to solve a problem rather than rushing in headlong with brute force methods. Here, if we multiply the first equation by x, the second by y, and the third by z, we find that

$$2\lambda\frac{x^2}{a^2} = 2\lambda\frac{y^2}{b^2} = 2\lambda\frac{z^2}{c^2} = 8xyz.$$

Now $\lambda \neq 0$ because (at maximum volume) x, y, and z are nonzero. We conclude that

$$\frac{x^2}{a^2} = \frac{y^2}{b^2} = \frac{z^2}{c^2}.$$

The sum of the last three expressions is 1, because that is precisely the constraint condition in this problem. Thus each of these three expressions is equal to $\frac{1}{3}$. All three of x, y, and z are positive, and therefore

$$x = \frac{a}{\sqrt{3}}, \quad y = \frac{b}{\sqrt{3}}, \quad \text{and} \quad z = \frac{c}{\sqrt{3}}.$$

Therefore, the box of maximum volume has volume

$$V = V_{\max} = \frac{8}{3\sqrt{3}}abc.$$

Note that this answer is dimensionally correct—the product of the three *lengths* a, b, and c yields a *volume*. But because the volume of the ellipsoid is $V = \frac{4}{3}\pi abc$, and $[8/(3\sqrt{3})]/(4\pi/3) = 2/(\pi\sqrt{3}) \approx 0.37$, it follows that the maximal box occupies only about 37% of the volume of the circumscribed ellipsoid. Considering the 64% result in Example 2, would you consider this result plausible, or surprising? ◆

Problems that have Two Constraints

Suppose that we want to find the maximum and minimum values of the function $f(x, y, z)$ at points of the curve of intersection of the two surfaces

$$g(x, y, z) = 0 \quad \text{and} \quad h(x, y, z) = 0. \tag{9}$$

This is a maximum-minimum problem with *two* constraints. The Lagrange multiplier method for such situations is based on Theorem 2.

THEOREM 2 Lagrange Multipliers (with two constraints)

Suppose that $f(x, y, z)$, $g(x, y, z)$, and $h(x, y, z)$ are continuously differentiable functions. If the maximum (or minimum) value of $f(x, y, z)$ subject to the two constraints

$$g(x, y, z) = 0 \quad \text{and} \quad h(x, y, z) = 0 \tag{9}$$

occurs at a point P where the vectors $\nabla g(P)$ and $\nabla h(P)$ are nonzero and nonparallel, then

$$\nabla f(P) = \lambda_1 \nabla g(P) + \lambda_2 \nabla h(P) \tag{10}$$

for some two constants λ_1 and λ_2.

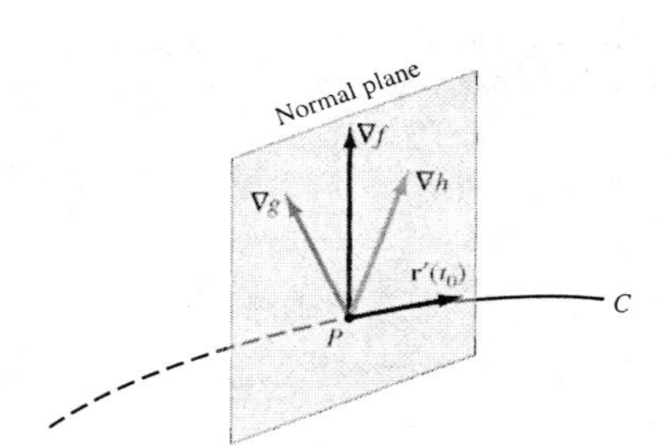

FIGURE 13.9.6 The relation between the gradient vectors in the proof of Theorem 2.

OUTLINE OF PROOF By an appropriate version of the implicit function theorem, the curve C of intersection of the two surfaces (Fig. 13.9.6) may be represented near P by a parametric curve $\mathbf{r}(t)$ with nonzero tangent vector $\mathbf{r}'(t)$. Let t_0 be the value of t such that $\mathbf{r}(t_0) = \overrightarrow{OP}$. We compute the derivatives at t_0 of the composite functions $f(\mathbf{r}(t))$, $g(\mathbf{r}(t))$, and $h(\mathbf{r}(t))$. We find—exactly as in the proof of Theorem 1—that

$$\nabla f(P) \cdot \mathbf{r}'(t_0) = 0, \quad \nabla g(P) \cdot \mathbf{r}'(t_0) = 0, \quad \text{and} \quad \nabla h(P) \cdot \mathbf{r}'(t_0) = 0.$$

These three equations imply that all three gradient vectors are perpendicular to the curve C at P and thus that they all lie in a single plane, the plane normal to the curve C at the point P.

Now $\nabla g(P)$ and $\nabla h(P)$ are nonzero and nonparallel, so $\nabla f(P)$ is the sum of its projections onto $\nabla g(P)$ and $\nabla h(P)$. (See Problem 65 of Section 12.2.) As illustrated in Fig. 13.9.7, this fact implies Eq. (10). ◀

In examples we prefer to avoid subscripts by writing λ and μ for the Lagrange multipliers λ_1 and λ_2 in the statement of Theorem 2. The equations in (9) and the three scalar components of the vector equation in (10) then give rise to the five simultaneous equations

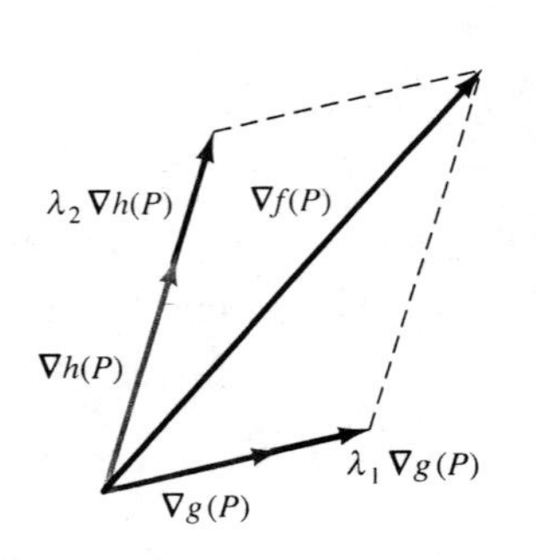

FIGURE 13.9.7 Geometry of the equation $\nabla f(P) = \lambda_1 \nabla g(P) + \lambda_2 \nabla h(P)$.

$$g(x, y, z) = 0, \tag{9a}$$

$$h(x, y, z) = 0, \tag{9b}$$

$$f_x(x, y, z) = \lambda g_x(x, y, z) + \mu h_x(x, y, z), \tag{10a}$$

$$f_y(x, y, z) = \lambda g_y(x, y, z) + \mu h_y(x, y, z), \tag{10b}$$

$$f_z(x, y, z) = \lambda g_z(x, y, z) + \mu h_z(x, y, z) \tag{10c}$$

in the five unknowns x, y, z, λ, and μ.

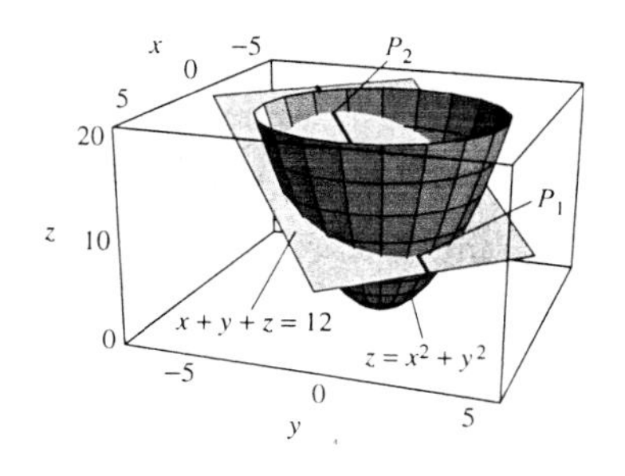

FIGURE 13.9.8 The plane and paraboloid intersecting in the ellipse of Example 4.

EXAMPLE 4 The plane $x + y + z = 12$ intersects the paraboloid $z = x^2 + y^2$ in an ellipse (Fig. 13.9.8). Find the highest and lowest points on this ellipse.

Solution The height of the point (x, y, z) is z, so we want to find the maximum and minimum values of

$$f(x, y, z) = z \tag{11}$$

subject to the two conditions

$$g(x, y, z) = x + y + z - 12 = 0 \tag{12}$$

and

$$h(x, y, z) = x^2 + y^2 - z = 0. \tag{13}$$

The conditions in (10a) through (10c) yield

$$0 = \lambda + 2\mu x, \tag{14a}$$

$$0 = \lambda + 2\mu y, \tag{14b}$$

and

$$1 = \lambda - \mu. \tag{14c}$$

If μ were zero, then Eq. (14a) would imply that $\lambda = 0$, which contradicts Eq. (14c). Hence $\mu \neq 0$, and therefore the equations

$$2\mu x = -\lambda = 2\mu y$$

imply that $x = y$. Substituting $x = y$ into Eq. (13) gives $z = 2x^2$, and then Eq. (12) yields

$$2x^2 + 2x - 12 = 0;$$

$$2(x + 3)(x - 2) = 0.$$

Thus we obtain the two solutions $x = -3$ and $x = 2$. Because $y = x$ and $z = 2x^2$, the corresponding points of the ellipse are $P_1(2, 2, 8)$ and $P_2(-3, -3, 18)$. It's clear which is the lowest and which is the highest. ◆

More Variables, More Constraints

Many practical constrained optimization problems have more than three variables and/or more than two constraints. For instance, Problem 48 is a concrete plane geometry problem with four independent variables.

There is a general form of the Lagrange multiplier condition that applies to any such problem, whatever the numbers of variables and constraints. We need only adjoin an additional term to the right-hand side in Eq. (10) for each additional constraint. The resulting condition for maximizing or minimizing the value $f(x_1, x_2, \ldots, x_n)$ of a function of n variables subject to the k constraints

$$\begin{aligned} g_1(x_1, x_2, \ldots, x_n) &= 0, \\ g_2(x_1, x_2, \ldots, x_n) &= 0, \\ &\vdots \\ g_k(x_1, x_2, \ldots, x_n) &= 0 \end{aligned} \tag{15}$$

is

$$\nabla f(P) = \lambda_1 \nabla g_1(P) + \lambda_2 \nabla g_2(P) + \cdots + \lambda_k \nabla g_k(P), \tag{16}$$

where we write $P = (x_1, x_2, \ldots, x_n)$. This condition holds under the assumptions that the functions $f, g_1, g_2, \ldots,$ and g_k are continuously differentiable near the optimal point P, and that—in the language of linear algebra—the gradient vectors

$\nabla g_1(P), \nabla g_2(P), \dots, \nabla g_k(P)$ are linearly independent in $\mathbf{R}^n$. The latter hypothesis means that no one of these k vectors can be expressed as a linear combination of the other $k-1$. The corresponding theorem is stated and proved in Chapter II of Edwards: *Advanced Calculus of Several Variables* (New York: Dover Publications, 1994).

Each of the gradient vectors in Eq. (16) has n components. When the resulting n "scalar component equations" are combined (Problem 61) with the k scalar equations in (15), we obtain the $k+n$ scalar equations

$$\begin{aligned}
g_1(x_1, x_2, \dots, x_n) = 0, \dots, g_k(x_1, x_2, \dots, x_n) &= 0,\\
D_1 f(x_1, x_2, \dots, x_n) = \lambda_1 D_1 g_1(x_1, x_2, \dots, x_n) &+ \cdots + \lambda_k D_1 g_k(x_1, x_2, \dots, x_n),\\
D_2 f(x_1, x_2, \dots, x_n) = \lambda_1 D_2 g_1(x_1, x_2, \dots, x_n) &+ \cdots + \lambda_k D_2 g_k(x_1, x_2, \dots, x_n),\\
&\vdots\\
D_n f(x_1, x_2, \dots, x_n) = \lambda_1 D_n g_1(x_1, x_2, \dots, x_n) &+ \cdots + \lambda_k D_n g_k(x_1, x_2, \dots, x_n)
\end{aligned} \tag{17}$$

to solve for the $k+n$ unknowns $\lambda_1, \lambda_2, \dots, \lambda_k, x_1, x_2, \dots, x_n$.

For instance, suppose that we ask for the minimal distance between points $P(x, y, z)$ and $Q(u, v, w)$ on two different space curves, each of which is presented as the intersection of two surfaces. We have the six coordinates x, y, z, u, v, and w of the two points and the four constraint equations of the four given surfaces. Then the system in (17) becomes a system of ten equations in the ten unknowns $x, y, z, u, v, w, \lambda_1, \lambda_2, \lambda_3$, and λ_4. See Problem 65, where the two curves are skew lines in space. This is a comparatively simple case, but you surely will want to use a computer algebra system to solve the problem. (See the CD-ROM project material for this section.)

13.9 TRUE/FALSE STUDY GUIDE

13.9 CONCEPTS: QUESTIONS AND DISCUSSION

Give examples of continuously differentiable functions $f, g: \mathbf{R}^2 \to \mathbf{R}$ satisfying the conditions in Questions 1 through 3.

1. $f(x, y)$ attains a minimum value but no maximum value subject to the constraint $g(x, y) = 0$.
2. $f(x, y)$ attains neither a maximum value nor a minimum value subject to the constraint $g(x, y) = 0$.
3. $f(x, y)$ attains its maximum value subject to the constraint $g(x, y) = 0$ at a point P where $\nabla f(P) \neq \lambda \nabla g(P)$ for any λ. (In view of Theorem 1, how is this possible?)

13.9 PROBLEMS

In Problems 1 through 18, find the maximum and minimum values—if any—of the given function f subject to the given constraint or constraints.

1. $f(x, y) = 2x + y; \quad x^2 + y^2 = 1$

2. $f(x, y) = x + y; \quad x^2 + 4y^2 = 1$

3. $f(x, y) = x^2 - y^2; \quad x^2 + y^2 = 4$

4. $f(x, y) = x^2 + y^2; \quad 2x + 3y = 6$

5. $f(x, y) = xy; \quad 4x^2 + 9y^2 = 36$

6. $f(x, y) = 4x^2 + 9y^2; \quad x^2 + y^2 = 1$

7. $f(x, y, z) = x^2 + y^2 + z^2; \quad 3x + 2y + z = 6$

8. $f(x, y, z) = 3x + 2y + z; \quad x^2 + y^2 + z^2 = 1$

9. $f(x, y, z) = x + y + z; \quad x^2 + 4y^2 + 9z^2 = 36$

10. $f(x, y, z) = xyz; \quad x^2 + y^2 + z^2 = 1$

11. $f(x, y, z) = xy + 2z; \quad x^2 + y^2 + z^2 = 36$

12. $f(x, y, z) = x - y + z; \quad z = x^2 - 6xy + y^2$

13. $f(x, y, z) = x^2y^2z^2; \quad x^2 + 4y^2 + 9z^2 = 27$

14. $f(x, y, z) = x^2 + y^2 + z^2; \quad x^4 + y^4 + z^4 = 3$

15. $f(x, y, z) = x^2 + y^2 + z^2$; $x + y + z = 1$ and $x + 2y + 3z = 6$

16. $f(x, y, z) = z$; $x^2 + y^2 = 1$ and $2x + 2y + z = 5$

17. $f(x, y, z) = z; \quad x + y + z = 1$ and $x^2 + y^2 = 1$

18. $f(x, y, z) = x; \quad x + y + z = 12$ and $4y^2 + 9z^2 = 36$

19. Find the point on the line $3x + 4y = 100$ that is closest to the origin. Use Lagrange multipliers to minimize the *square* of the distance.

20. A rectangular open-topped box is to have volume 700 in.^{3} The material for its bottom costs 7¢/in.^{2} and the material for its four vertical sides costs 5¢/in.^{2} Use the method of Lagrange multipliers to find what dimensions will minimize the cost of the material used in constructing this box.

In Problems 21 through 34, use the method of Lagrange multipliers to solve the indicated problem from Section 13.5.

21. Problem 29
22. Problem 30
23. Problem 31
24. Problem 32
25. Problem 33
26. Problem 34
27. Problem 35
28. Problem 36
29. Problem 37
30. Problem 38
31. Problem 39
32. Problem 40
33. Problem 41
34. Problem 42

35. Find the point or points of the surface $z = xy + 5$ closest to the origin. [*Suggestion:* Minimize the *square* of the distance.]

36. A triangle with sides x, y, and z has fixed perimeter $2s = x + y + z$. Its area A is given by *Heron's formula:*

$$A = \sqrt{s(s-a)(s-b)(s-c)}.$$

Use the method of Lagrange multipliers to show that, among all triangles with the given perimeter, the one of largest area is equilateral. [*Suggestion:* Consider maximizing A^2 rather than A.]

37. Use the method of Lagrange multipliers to show that, of all triangles inscribed in the unit circle, the one of greatest area is equilateral. [*Suggestion:* Use Fig. 13.9.9 and the fact that the area of a triangle with sides a and b and included angle θ is given by the formula $A = \frac{1}{2}ab\sin\theta$.]

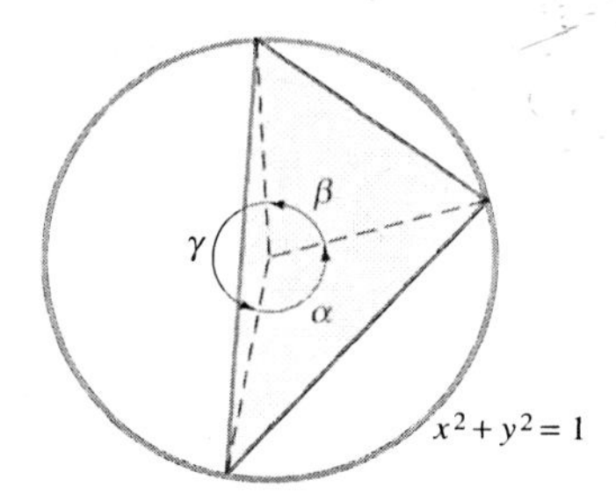

FIGURE 13.9.9 A triangle inscribed in a circle (Problem 37).

38. Find the points on the rotated ellipse $x^2 + xy + y^2 = 3$ that are closest to and farthest from the origin. [*Suggestion:* Write the Lagrange multiplier equations in the form

$$ax + by = 0,$$

$$cx + dy = 0.$$

These equations have a nontrivial solution *only if* $ad - bc = 0$. Use this fact to solve first for λ.]

39. Use the method of Problem 38 to find the points of the rotated hyperbola $x^2 + 12xy + 6y^2 = 130$ that are closest to the origin.

40. Find the points of the ellipse $4x^2 + 9y^2 = 36$ that are closest to the point $(1, 1)$ as well as the point or points farthest from it.

41. Find the highest and lowest points on the ellipse formed by the intersection of the cylinder $x^2 + y^2 = 1$ and the plane $2x + y - z = 4$.

42. Apply the method of Example 4 to find the highest and lowest points on the ellipse formed by the intersection of the cone $z^2 = x^2 + y^2$ and the plane $x + 2y + 3z = 3$.

43. Find the points on the ellipse of Problem 42 that are nearest the origin and those that are farthest from it.

44. The ice tray shown in Fig. 13.9.10 is to be made from material that costs 1¢/in.^{2} Minimize the cost function $f(x, y, z) = xy + 3xz + 7yz$ subject to the constraints that each of the 12 compartments is to have a square horizontal cross section and that the total volume (ignoring the partitions) is to be 12 in.^{3}

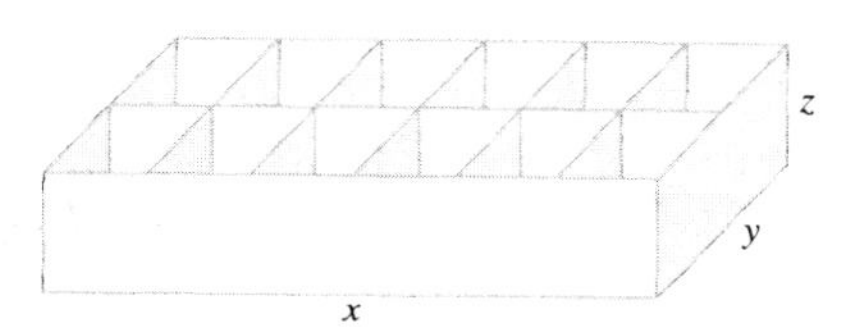

FIGURE 13.9.10 The ice tray of Problem 44.

45. Prove Theorem 1 for functions of three variables by showing that both of the vectors $\nabla f(P)$ and $\nabla g(P)$ are perpendicular at P to every curve on the surface $g(x, y, z) = 0$.

46. Find the lengths of the semiaxes of the ellipse of Example 4.

47. Figure 13.9.11 shows a right triangle with sides x, y, and z and fixed perimeter P. Maximize its area $A = \frac{1}{2}xy$ subject to the constraints $x + y + z = P$ and $x^2 + y^2 = z^2$. In particular, show that the optimal such triangle is isosceles (by showing that $x = y$).

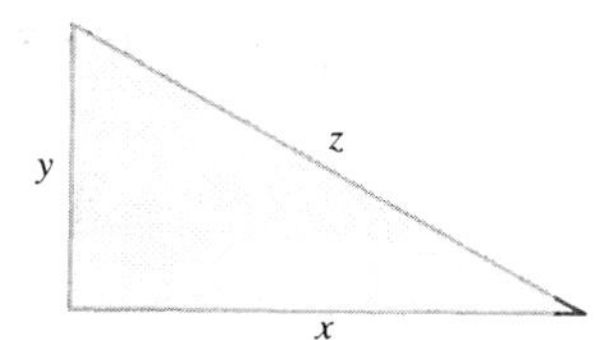

FIGURE 13.9.11 A right triangle with fixed perimeter P (Problem 47).

FIGURE 13.9.12 A general triangle with fixed perimeter P (Problem 48).

48. Figure 13.9.12 shows a general triangle with sides x, y, and z and fixed perimeter P. Maximize its area

$$A = f(x, y, z, \alpha) = \tfrac{1}{2}xy\sin\alpha$$

subject to the constraints $x + y + z = P$ and

$$z^2 = x^2 + y^2 - 2xy\cos\alpha$$

(the law of cosines). In particular, show that the optimal such triangle is equilateral (by showing that $x = y = z$).

49. Figure 13.9.13 shows a hexagon with vertices $(0, \pm 1)$ and $(\pm x, \pm y)$ inscribed in the unit circle $x^2 + y^2 = 1$. Show that its area is maximal when it is a *regular* hexagon with equal sides and angles.

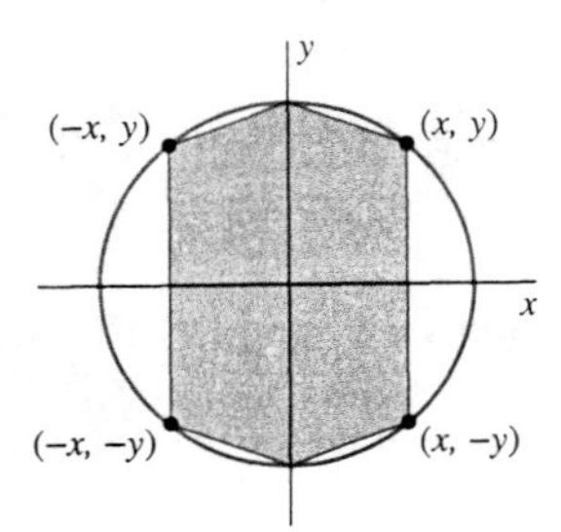

FIGURE 13.9.13 The inscribed hexagon of Problem 49.

50. When the hexagon of Fig. 13.9.13 is rotated around the y-axis, it generates a solid of revolution consisting of a cylinder and two cones (Fig. 13.9.14). What radius and cylinder height maximize the volume of this solid?

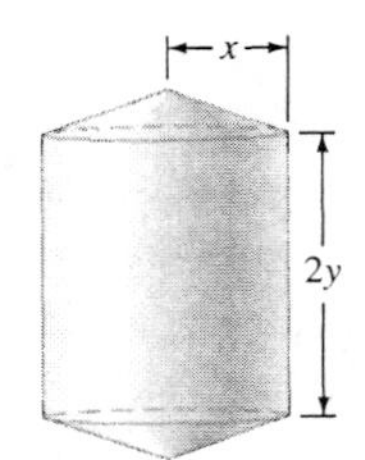

FIGURE 13.9.14 The solid of Problem 50.

In Problems 51 through 58, consider the square of the distance to be maximized or minimized. Use the numerical solution command in a computer algebra system as needed to solve the appropriate Lagrange multiplier equations.

51. Find the points of the parabola $y = (x - 1)^2$ that are closest to the origin.

52. Find the points of the ellipse $4x^2 + 9y^2 = 36$ that are closest to and farthest from the point $(3, 2)$.

53. Find the first-quadrant point of the curve $xy = 24$ that is closest to the point $(1, 4)$.

54. Find the point of the surface $xyz = 1$ that is closest to the point $(1, 2, 3)$.

55. Find the points on the sphere with center $(1, 2, 3)$ and radius 6 that are closest to and farthest from the origin.

56. Find the points of the ellipsoid $4x^2 + 9y^2 + z^2 = 36$ that are closest to and farthest from the origin.

57. Find the points of the ellipse $4x^2 + 9y^2 = 36$ that are closest to and farthest from the straight line $x + y = 10$.

58. Find the points on the ellipsoid $4x^2 + 9y^2 + z^2 = 36$ that are closest to and farthest from the plane $2x + 3y + z = 10$.

59. Find the maximum possible volume of a rectangular box that has its base in the xy-plane and its upper vertices on the elliptic paraboloid $z = 9 - x^2 - 2y^2$.

60. The plane $4x + 9y + z = 0$ intersects the elliptic paraboloid $z = 2x^2 + 3y^2$ in an ellipse. Find the highest and lowest points on this ellipse.

61. Explain carefully how the equations in (17) result from those in (15) and (16). If you wish, consider only a nontrivial special case, such as the case $n = 4$ and $k = 3$.

62. (a) Suppose that $x_1, x_2, \dots,$ and x_n are positive. Show that the minimum value of $f(\mathbf{x}) = x_1 + x_2 + \cdots + x_n$ subject to the constraint $x_1 x_2 \cdots x_n = 1$ is n. (b) Given n positive numbers $a_1, a_2, \dots, a_n$, let

$$x_i = \frac{a_i}{(a_1 a_2 \cdots a_n)^{1/n}}$$

for $1 \leqq i \leqq n$ and apply the result in part (a) to deduce the **arithmetic-geometric mean inequality**

$$\sqrt[n]{a_1 a_2 \cdots a_n} \leqq \frac{a_1 + a_2 + \cdots + a_n}{n}.$$

63. Figure 13.9.15 shows a moat of width $a = 10$ ft, filled with alligators, and bounded on each side by a wall of height $b = 6$ ft. Soldiers plan to bridge this moat by scaling a ladder placed across the nearer wall as indicated, anchored at the ground with a handy boulder, and with the upper end directly above the far wall on the opposite side of the moat. They naturally wonder what is the minimal length L of a ladder that will suffice for this purpose. This is a particular case of the problem of minimizing the length of a line segment in the uv-plane that joins the points $P(x, 0)$ and $Q(0, y)$ on the two coordinate axes and passes through the given first-quadrant point (a, b). Show that $L_{\min} = (a^{2/3} + b^{2/3})^{3/2}$ by minimizing the squared length $f(x, y) = x^2 + y^2$ subject to the constraint that $u = a$ and $v = b$ satisfy the uv-equation $u/x + v/y = 1$ of the line through P and Q.

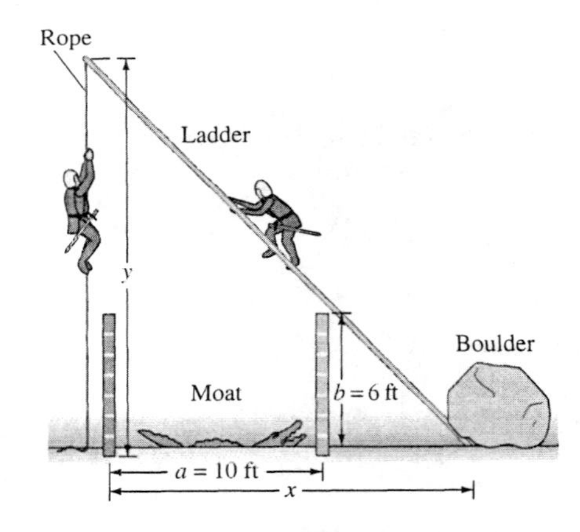

FIGURE 13.9.15 The alligator-filled moat of Problem 63.

64. A three-dimensional analog of the two-dimensional problem in Problem 63 asks for the minimal area A of the triangle in uvw-space with vertices $P(x, 0, 0)$, $Q(0, y, 0)$, and $R(0, 0, z)$ on the three coordinate axes and passing through the given first-octant point (a, b, c). (a) First deduce from Miscellaneous Problem 51 of Chapter 12 that $A^2 = \frac{1}{4}(x^2y^2 + x^2z^2 + y^2z^2)$. (b) If $a = b = c = 1$ then, by symmetry, $x = y = z$. Show in this case that $x = y = z = 3$, and thus that $A = \frac{9}{2}\sqrt{3}$. (c) Set up the Lagrange multiplier equations for minimizing the squared area A^2 subject to the constraint that the given coordinates (a, b, c) satisfy the uvw-equation $u/x + v/y + w/z = 1$ of the plane through the points P, Q, and R. In general, these equations have no known closed-form solution. Nevertheless, you can use a computer algebra system (as in the CD-ROM project material for this section) to approximate numerically the minimum value of A with given numerical values of a, b, and c. Show first that with $a = b = c = 1$ you get an accurate approximation to the exact value in part (b). Then repeat the process with your own selection of values of a, b, and c. [*Note:* This three-dimensional problem was motivated by the investigation of the n-dimensional version in David Spring's article "Solution of a Calculus Problem on Minimal Volume" in *The American Mathematical Monthly* (March 2001, pp. 217–221), where a Lagrange system of $n+1$ equations is reduced to a single nonlinear equation in a single unknown.]

65. Suppose that L_1 is the line of intersection of the planes $2x + y + 2z = 15$ and $x + 2y + 3z = 30$, and that L_2 is the line of intersection of the planes $x - y - 2z = 15$ and $3x - 2y - 3z = 20$. Find the closest points P_1 and P_2 on these two skew lines. Use a computer to solve the corresponding Lagrange multiplier system of 10 linear equations in 10 unknowns.

13.9 Project: Numerical Solution of Lagrange Multiplier Systems

The Lagrange multiplier problems in Examples 1 through 4 of this section are somewhat atypical in that the equations in these examples can be solved exactly and without great effort. Frequently a Lagrange multiplier problem leads to a system of equations that can be solved only numerically and approximately. The CD-ROM material for this section supplies typical computer algebra system commands for the numerical solution of such systems, plus a two-ladder moat problem that leads to a system of 12 nonlinear equations in seven coordinate variables and five Lagrange multipliers.

13.10 CRITICAL POINTS OF FUNCTIONS OF TWO VARIABLES

We saw in Section 13.5 that in order for the differentiable function $f(x, y)$ to have either a local minimum or a local maximum at an interior critical point $P(a, b)$ of its domain, it is a *necessary* condition that P be a *critical point* of f—that is, that

$$f_x(a, b) = 0 = f_y(a, b).$$

Here we give conditions *sufficient* to ensure that f has a local extremum at a critical point. The criterion stated in Theorem 1 involves the second-order partial derivatives of f at (a, b) and plays the role of the single-variable second derivative test (Section 4.6) for functions of two variables. To simplify the statement of this result, we use the following abbreviations:

$$A = f_{xx}(a, b), \qquad B = f_{xy}(a, b), \qquad C = f_{yy}(a, b), \tag{1}$$

and

$$\Delta = AC - B^2 = f_{xx}(a, b)\, f_{yy}(a, b) - [f_{xy}(a, b)]^2. \tag{2}$$

We outline a proof of Theorem 1 at the end of this section.

THEOREM 1 Two-Variable Second Derivative Test

Suppose that the function $f(x, y)$ has continuous second-order partial derivatives in a neighborhood of the critical point (a, b) at which its first-order partial derivatives all vanish. Let A, B, C, and Δ be defined as in Eqs. (1) and (2). Then:

- $f(a, b)$ is a local minimum value of f if $A > 0$ and $\Delta > 0$;
- $f(a, b)$ is a local maximum value of f if $A < 0$ and $\Delta > 0$;
- $f(a, b)$ is neither a local minimum nor a local maximum if $\Delta < 0$.

Thus f has *either* a local maximum *or* a local minimum at the critical point (a, b) provided that the **discriminant** $\Delta = AC - B^2$ is *positive*. In this case, $A = f_{xx}(a, b)$ plays the role of the second derivative of a single-variable function: There is a local minimum at (a, b) if $A > 0$ and a local maximum if $A < 0$.

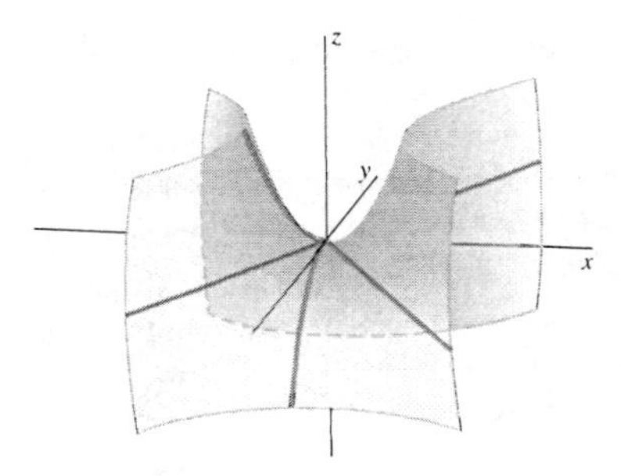

FIGURE 13.10.1 The origin is a saddle point of the surface with equation $z = x^2 - y^2$.

If $\Delta < 0$, then f has *neither* a local maximum *nor* a local minimum at (a, b). In this case we call (a, b) a **saddle point** of f, thinking of the appearance of the hyperbolic paraboloid $f(x, y) = x^2 - y^2$ (Fig. 13.10.1), a typical example of this case.

Theorem 1 does not answer the question of what happens when $\Delta = 0$. In this case, the two-variable second derivative test fails—it gives no information. Moreover, at such a point (a, b), *anything* can happen, ranging from the local (indeed global) minimum of $f(x, y) = x^4 + y^4$ at $(0, 0)$ to the "monkey saddle" of Example 2.

In the case of a function $f(x, y)$ with several critical points, we must compute the quantities A, B, C, and Δ separately at each critical point in order to apply the test.

EXAMPLE 1 Locate and classify the critical points of

$$f(x, y) = 3x - x^3 - 3xy^2.$$

Solution This function is a polynomial, so all its partial derivatives exist and are continuous everywhere. When we equate its first partial derivatives to zero (to locate the critical points of f), we get

$$f_x(x, y) = 3 - 3x^2 - 3y^2 = 0 \quad \text{and} \quad f_y(x, y) = -6xy = 0.$$

The second of these equations implies that x or y must be zero; then the first implies that the other must be ± 1. Thus there are four critical points: $(1, 0)$, $(-1, 0)$, $(0, 1)$, and $(0, -1)$.

Critical Point	A	B	C	Δ	Type of Extremum
$(1, 0)$	-6	0	-6	36	Local maximum
$(-1, 0)$	6	0	6	36	Local minimum
$(0, 1)$	0	-6	0	-36	Saddle point
$(0, -1)$	0	6	0	-36	Saddle point

FIGURE 13.10.2 Critical-point analysis for the function of Example 1.

The second-order partial derivatives of f are

$$A = f_{xx}(x, y) = -6x, \quad B = f_{xy}(x, y) = -6y, \quad C = f_{yy}(x, y) = -6x.$$

Hence $\Delta = 36(x^2 - y^2)$ at each of the critical points. The table in Fig. 13.10.2 summarizes the situation at each of the four critical points, which are labeled in the contour plot in Fig. 13.10.3. Near the points $(\pm 1, 0)$ we see the nested "ellipse-like" contours that signal local extrema (Fig. 13.10.4), and near the points $(0, \pm 1)$ we see "hyperbola-like" contours that signal saddle points (Fig. 13.10.5). Figure 13.10.6 shows the critical points on the graph of $z = f(x, y)$. ◆

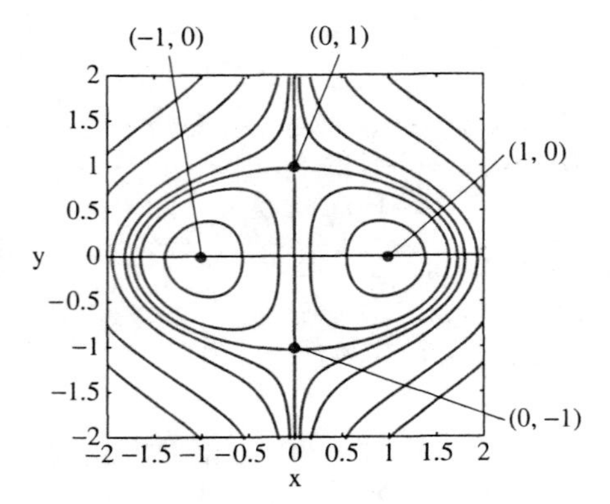

FIGURE 13.10.3 Level curves for the function of Example 1.

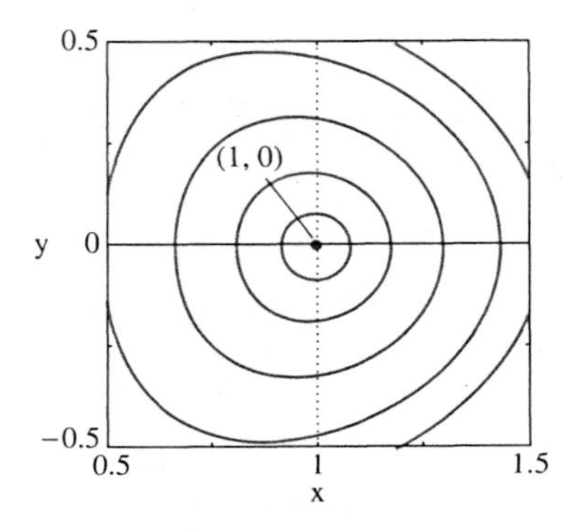

FIGURE 13.10.4 Level curves near the critical point (1, 0).

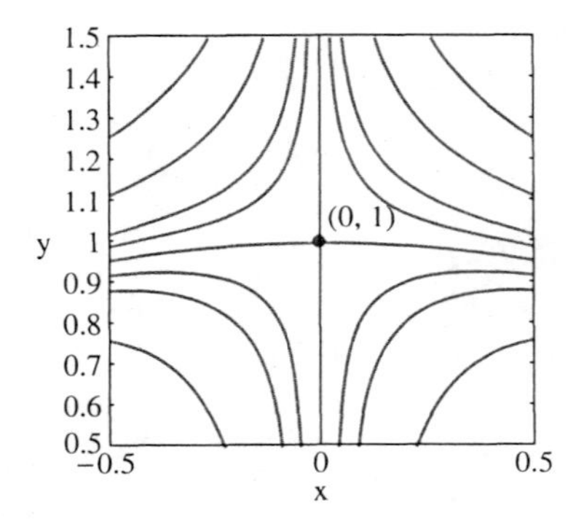

FIGURE 13.10.5 Level curves near the critical point (0, 1).

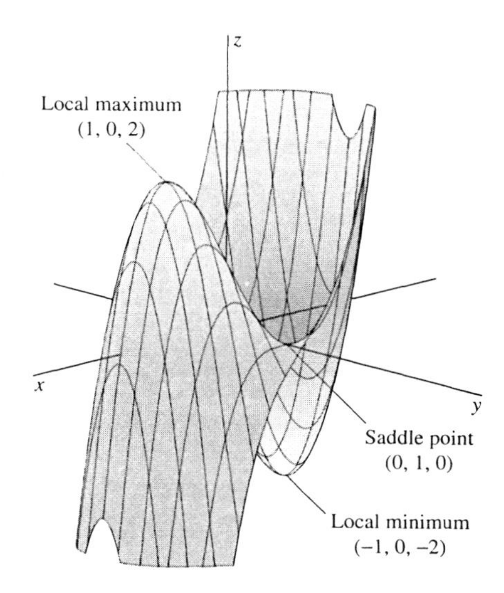

FIGURE 13.10.6 Graph of the function of Example 1.

EXAMPLE 2 Find and classify the critical points of the function

$$f(x, y) = 6xy^2 - 2x^3 - 3y^4.$$

Solution When we equate the first-order partial derivatives to zero, we get the equations

$$f_x(x, y) = 6y^2 - 6x^2 = 0 \quad \text{and} \quad f_y(x, y) = 12xy - 12y^3 = 0.$$

It follows that

$$x^2 = y^2 \quad \text{and} \quad y(x - y^2) = 0.$$

The first of these equations gives $x = \pm y$. If $x = y$, the second equation implies that $y = 0$ or $y = 1$. If $x = -y$, the second equation implies that $y = 0$ or $y = -1$. Hence there are three critical points: $(0, 0)$, $(1, 1)$, and $(1, -1)$.

The second-order partial derivatives of f are

$$A = f_{xx}(x, y) = -12x, \quad B = f_{xy}(x, y) = 12y, \quad C = f_{yy}(x, y) = 12x - 36y^2.$$

These expressions give the data shown in the table in Fig. 13.10.7. The critical point test fails at $(0, 0)$, so we must find another way to test this point.

Critical Point	A	B	C	Δ	Type of Extremum
$(0, 0)$	0	0	0	0	Test fails
$(1, 1)$	-12	12	-24	144	Local maximum
$(1, -1)$	-12	-12	-24	144	Local maximum

FIGURE 13.10.7 Critical-point analysis for the function of Example 2.

We observe that $f(x, 0) = -2x^3$ and that $f(0, y) = -3y^4$. Hence, as we move away from the origin in the

Positive x-direction:	f decreases;
Negative x-direction:	f increases;
Positive y-direction:	f decreases;
Negative y-direction:	f decreases.

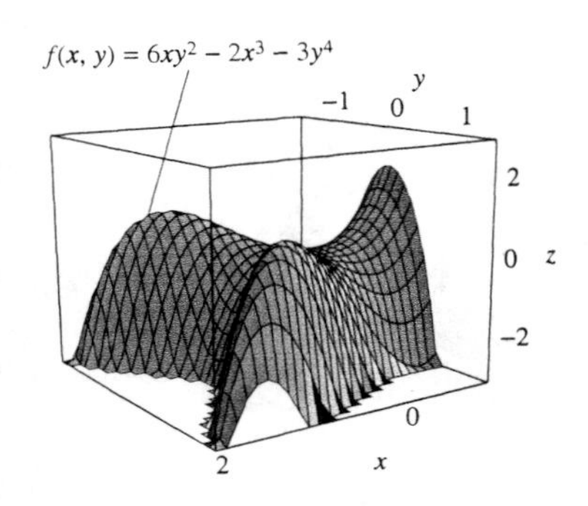

FIGURE 13.10.8 The monkey saddle of Example 2.

Consequently, f has neither a local maximum nor a local minimum at the origin. The graph of f is shown in Fig. 13.10.8. If a monkey were to sit with its rump at the origin and face the negative x-direction, then the directions in which $f(x, y)$ decreases would provide places for both its tail and its two legs to hang. That's why this particular surface is called a *monkey saddle* (Fig. 13.10.9). ◆

EXAMPLE 3 Find and classify the critical points of the function

$$f(x, y) = \tfrac{1}{3}x^4 + \tfrac{1}{2}y^4 - 4xy^2 + 2x^2 + 2y^2 + 3.$$

Solution When we equate to zero the first-order partial derivatives of f, we obtain the equations

$$f_x(x, y) = \tfrac{4}{3}x^3 - 4y^2 + 4x = 0, \tag{3}$$

$$f_y(x, y) = 2y^3 - 8xy + 4y = 0, \tag{4}$$

which are not as easy to solve as the corresponding equations in Examples 1 and 2. But if we write Eq. (4) in the form

FIGURE 13.10.9 The monkey in its saddle (Example 2).

$$2y(y^2 - 4x + 2) = 0,$$

we see that either $y = 0$ or

$$y^2 = 4x - 2. \tag{5}$$

If $y = 0$, then Eq. (3) reduces to the equation

$$\tfrac{4}{3}x^3 + 4x = \tfrac{4}{3}x(x^2 + 3) = 0,$$

whose only solution is $x = 0$. Thus one critical point of f is $(0, 0)$.

If $y \neq 0$, we substitute $y^2 = 4x - 2$ into Eq. (3) to obtain

$$\tfrac{4}{3}x^3 - 4(4x - 2) + 4x = 0;$$

that is,

$$\tfrac{4}{3}x^3 - 12x + 8 = 0.$$

Thus we need to solve the cubic equation

$$\phi(x) = x^3 - 9x + 6 = 0. \tag{6}$$

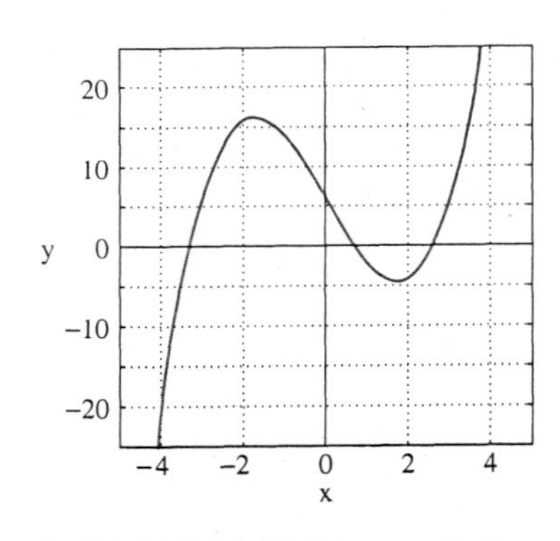

FIGURE 13.10.10 The graph of $\phi(x) = x^3 - 9x + 6$ (Example 3).

The graph of $\phi(x)$ in Fig. 13.10.10 shows that this equation has three real solutions with approximate values $x \approx -3$, $x \approx 1$, and $x \approx 3$. Using either graphical techniques or Newton's method (Section 3.8), you can obtain the values

$$x \approx -3.2899, \quad x \approx 0.7057, \quad x \approx 2.5842, \tag{7}$$

accurate to four decimal places. The corresponding values of y are given from Eq. (10) by

$$y = \pm\sqrt{4x - 2}, \tag{8}$$

but the first value of x in (7) yields *no* real value at all for y. Thus the two positive values of x in (7) add *four* critical points of $f(x, y)$ to the one critical point $(0, 0)$ already found.

These five critical points are listed in the table in Fig. 13.10.11, together with the corresponding values of

$$A = f_{xx}(x, y) = 4x^2 + 4, \qquad B = f_{xy}(x, y) = -8y,$$

$$C = f_{yy}(x, y) = 6y^2 - 8x + 4, \quad \Delta = AC - B^2$$

Critical Point	1	2	3	4	5
x	0.0000	0.7057	0.7057	2.5842	2.5842
y	0.0000	0.9071	−0.9071	2.8874	−2.8874
z	3.0000	3.7402	3.7402	−3.5293	−3.5293
A	4.00	5.99	5.99	30.71	30.71
B	0.00	−7.26	7.26	−23.10	23.10
C	4.00	3.29	3.29	33.35	33.35
Δ	16.00	−32.94	−32.94	490.62	490.62
Type	Local minimum	Saddle point	Saddle point	Local minimum	Local minimum

FIGURE 13.10.11 Classification of the critical points in Example 3.

(rounded to two decimal places) at each of these critical points. We see that $\Delta > 0$ and $A > 0$ at $(0, 0)$ and at $(2.5482, \pm 2.8874)$, so these points are local minimum points. But $\Delta < 0$ at $(0.7057, \pm 0.9071)$, so these two are saddle points. The level curve diagram in Fig. 13.10.12 shows how these five critical points fit together.

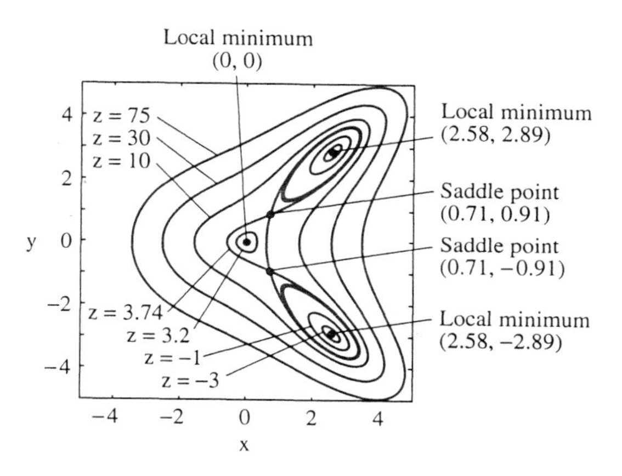

FIGURE 13.10.12 Level curves for the function of Example 3.

Finally, we observe that the behavior of $f(x, y)$ is approximately that of $\frac{1}{3}x^4 + \frac{1}{2}y^4$ when $|x|$ or $|y|$ is large, so the surface $z = f(x, y)$ must open upward and, therefore, have a global low point (but no global high point). Examining the values

$$f(0, 0) = 3 \quad \text{and} \quad f(2.5842, \pm 2.8874) \approx -3.5293,$$

we see that the global minimum value of $f(x, y)$ is approximately -3.5293. ◆

Proof of Theorem 1

It happens that the behavior of the function $f(x, y)$ near its critical point (a, b) is determined by the behavior near the origin $(0, 0)$ of the **quadratic form**

$$q(h, k) = Ah^2 + 2Bhk + Ck^2 \tag{9}$$

in h and k (A, B, and C are computed as in Eq. (1)). If $A \neq 0$, then you can verify readily that

$$q(h, k) = \frac{1}{A}[(Ah + Bk)^2 + \Delta k^2], \tag{10}$$

either by expanding the right-hand side in (10) or by completing the square in Eq. (9). The three parts of the following proposition correspond to the three cases in the conclusion of Theorem 1.

PROPOSITION Behavior of Quadratic Forms

1. If $\Delta > 0$ and $A > 0$, then $q(h, k) > 0$ unless h and k are both zero.
2. If $\Delta > 0$ and $A < 0$, then $q(h, k) < 0$ unless h and k are both zero.
3. If $\Delta < 0$, then every neighborhood of $(0, 0)$ contains points at which $q(h, k) > 0$ and points at which $q(h, k) < 0$.

The three parts of this proposition can be visualized by thinking of the graph of q as an upward-opening elliptic paraboloid in part 1, as a downward-opening paraboloid in part 2, and as a hyperbolic paraboloid with a saddle point in part 3.

PROOF Parts 1 and 2 of the proposition follow immediately by consideration of signs in Eq. (10), because the quantity within the brackets is positive if $\Delta > 0$ and h and k are not both zero, in which case the sign of $q(x, y)$ is the same as the sign of A.

Part 3 leads to several cases depending on the possible values of A, B, and C. If $A = C = 0$ and $\Delta = -B^2 < 0$, then $q(h, k) = 2Bhk$, so the conclusion in part 3 follows at once.

If $B = 0$ and $\Delta = AC < 0$, then A and C have different signs and $Q(h, k) = Ah^2 + Ck^2$, so again the conclusion in part 3 follows at once.

If $B \neq 0$ and $A \neq 0$, then the values $q(h, 0) = Ah^2$ and $q(h, -Ah/B) = \Delta k^2/A$ have different signs if $\Delta < 0$, so again the conclusion of part 3 follows. The analysis of the remaining case, in which $B \neq 0$ and $C \neq 0$, is similar. ◀

Now let us consider the critical point (a, b) of the function $f(x, y)$ of Theorem 1. Draw a circular disk centered at (a, b) as in Fig. 13.10.13. Because the second-order partial derivatives of f are continuous, we can make the radius of this disk so small that the quantity $f_{xx}(x, y) f_{yy}(x, y) - [f_{xy}(x, y)]^2$ has the same sign as the constant $\Delta = f_{xx}(a, b) f_{yy}(a, b) - [f_{xy}(a, b)]^2$ at every point (x, y) of the disk.

Now consider the single-variable function g defined by

$$g(t) = f(a + th, b + tk)$$

for $0 \leqq t \leqq 1$. Application of Taylor's formula (Section 11.4) to $g(t)$ gives

$$g(1) = g(0) + g'(0) + \tfrac{1}{2}g''(\bar{t}) \tag{11}$$

for some number $\bar{t}$ between 0 and 1. But the chain rule gives first

$$g'(t) = \frac{\partial f}{\partial x}\frac{dx}{dt} + \frac{\partial f}{\partial y}\frac{dy}{dt} = hf_x + kf_y,$$

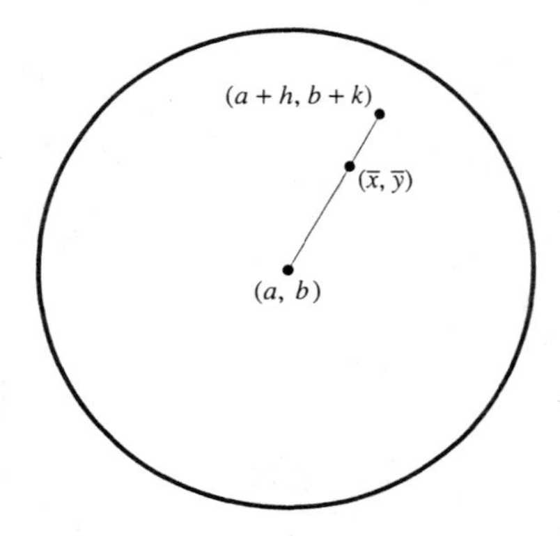

FIGURE 13.10.13 The circular disk centered at the point (a, b).

and then

$$g''(t) = \frac{\partial}{\partial x}(hf_x + kf_y)\frac{dx}{dt} + \frac{\partial}{\partial y}(hf_x + kf_y)\frac{dy}{dt}$$
$$= h^2 f_{xx} + 2hkf_{xy} + k^2 f_{yy},$$

where the indicated partial derivatives of f are to be evaluated at the point $(x, y) = (a + th, b + tk)$. Consequently $g'(0) = 0$ because $f_x(a, b) = 0 = f_y(a, b)$, and

$$g''(\bar{t}) = \overline{A}h^2 + 2\overline{B}hk + \overline{C}k^2 \qquad \textbf{(12)}$$

where the coefficients $\overline{A}$, $\overline{B}$, and $\overline{C}$ in this quadratic form denote the values of the second derivatives f_{xx}, f_{xy}, and f_{yy} (respectively) at the point $(\overline{x}, \overline{y}) = (a+\bar{t}h, b+\bar{t}k)$.

Because $g(0) = f(a, b)$ and $g(1) = f(a + h, b + k)$, Eqs. (11) and (12) imply that

$$f(a + h, b + k) = f(a, b) + \tfrac{1}{2}(\overline{A}h^2 + 2\overline{B}hk + \overline{C}k^2). \qquad \textbf{(13)}$$

Now $\overline{\Delta} = \overline{A}\,\overline{C} - \overline{B}^2$ has the same sign as $\Delta = AC - B^2$. And if $A \neq 0$, then we may assume that the circular disk in Fig. 13.10.13 is so small that $\overline{A}$ has the same sign as A. Then the quadratic form

$$\overline{q}(h, k) = \overline{A}h^2 + 2\overline{B}hk + \overline{C}k^2$$

that appears in Eq. (13) exhibits the same behavior as the quadratic form $q(h, k)$ of Eq. (9). Theorem 1 now follows from the proposition on the behavior of quadratic forms. For instance, if Δ and A are both positive, then the values $q(h, k)$ and hence $\overline{q}(h, k)$ are positive unless h and k are both zero. Therefore Eq. (13) gives

$$f(a + h, b + k) = f(a, b) + \tfrac{1}{2}\overline{q}(h, k) > f(a, b)$$

at each point $(a + h, b + k)$—other than (a, b) itself—of the circular disk of Fig. 13.10.13. Thus $f(a, b)$ is a local minimum value in this first case of Theorem 1. The other two cases follow from similar arguments.

13.10 TRUE/FALSE STUDY GUIDE

13.10 CONCEPTS: QUESTIONS AND DISCUSSION

Give simple examples of your own, different from any that appear in this section, that illustrate the following situations.

1. The three cases in Theorem 1.
2. The fact that either a maximum or a minimum, or neither, can occur at a critical point at which $\Delta = AC - B^2 = 0$.

13.10 PROBLEMS

Find and classify the critical points of the functions in Problems 1 through 22. If a computer algebra system is available, check your results by means of contour plots like those in Figs. 13.10.14–13.10.17.

1. $f(x, y) = 2x^2 + y^2 + 4x - 4y + 5$
2. $f(x, y) = 10 + 12x - 12y - 3x^2 - 2y^2$
3. $f(x, y) = 2x^2 - 3y^2 + 2x - 3y + 7$
4. $f(x, y) = xy + 3x - 2y + 4$
5. $f(x, y) = 2x^2 + 2xy + y^2 + 4x - 2y + 1$
6. $f(x, y) = x^2 + 4xy + 2y^2 + 4x - 8y + 3$
7. $f(x, y) = x^3 + y^3 + 3xy + 3$ (Fig. 13.10.14)
8. $f(x, y) = x^2 - 2xy + y^3 - y$

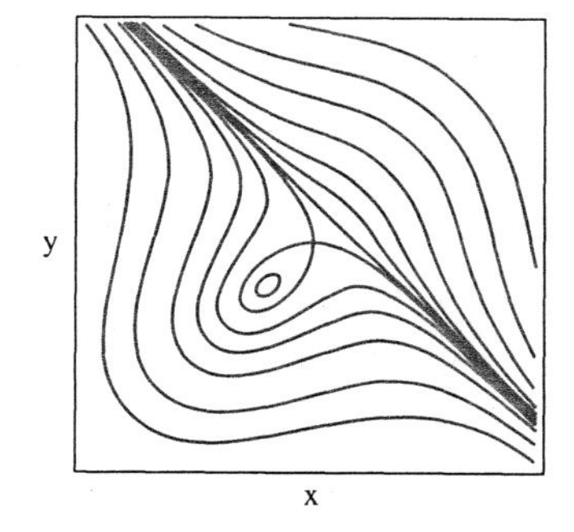

FIGURE 13.10.14 Contour plot for Problem 7.

9. $f(x, y) = 6x - x^3 - y^3$
10. $f(x, y) = 3xy - x^3 - y^3$

11. $f(x, y) = x^4 + y^4 - 4xy$

12. $f(x, y) = x^3 + 6xy + 3y^2$

13. $f(x, y) = x^3 + 6xy + 3y^2 - 9x$ (Fig. 13.10.15)

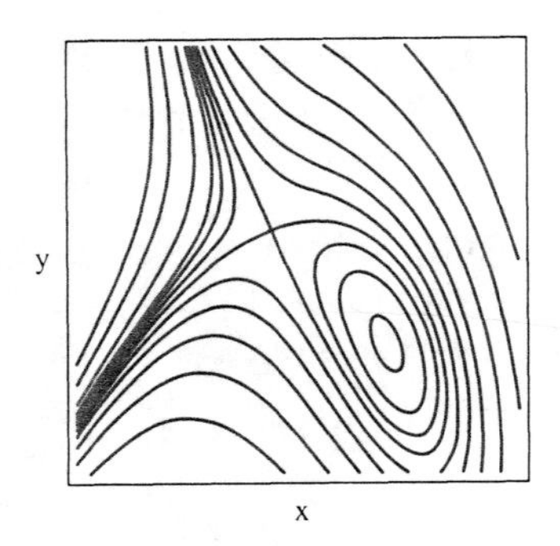

FIGURE 13.10.15 Contour plot for Problem 13.

14. $f(x, y) = x^3 + 6xy + 3y^2 + 6x$

15. $f(x, y) = 3x^2 + 6xy + 2y^3 + 12x - 24y$

16. $f(x, y) = 3x^2 + 12xy + 2y^3 - 6x + 6y$

17. $f(x, y) = 4xy - 2x^4 - y^2$ (Fig. 13.10.16)

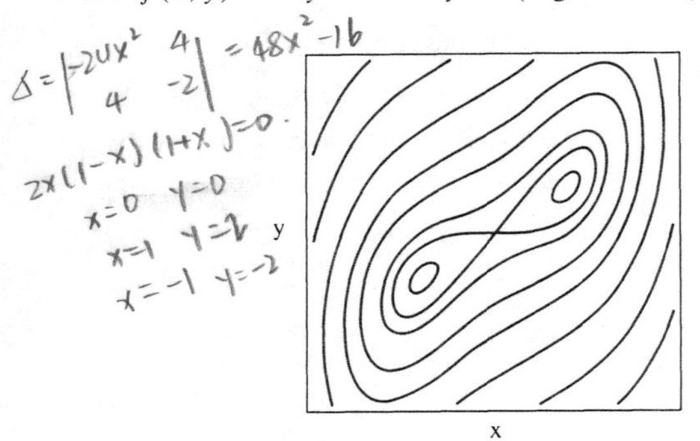

FIGURE 13.10.16 Contour plot for Problem 17.

18. $f(x, y) = 8xy - 2x^2 - y^4$

19. $f(x, y) = 2x^3 - 3x^2 + y^2 - 12x + 10$

20. $f(x, y) = 2x^3 + y^3 - 3x^2 - 12x - 3y$ (Fig. 13.10.17)

21. $f(x, y) = xy\exp(-x^2 - y^2)$

22. $f(x, y) = (x^2 + y^2)\exp(x^2 - y^2)$

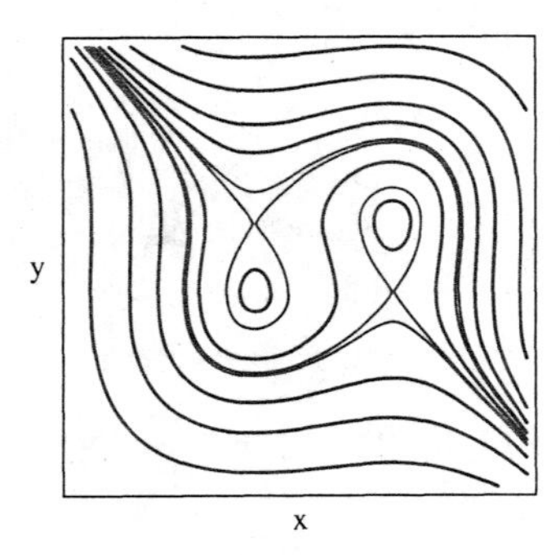

FIGURE 13.10.17 Contour plot for Problem 20.

In Problems 23 through 25, first show that $\Delta = f_{xx}f_{yy} - (f_{xy})^2$ is zero at the origin. Then classify this critical point by visualizing the surface $z = f(x, y)$.

23. $f(x, y) = x^4 + y^4$

24. $f(x, y) = x^3 + y^3$

25. $f(x, y) = \exp(-x^4 - y^4)$

26. Let $f(s, t)$ denote the *square* of the distance between a typical point of the line $x = t$, $y = t + 1$, $z = 2t$ and a typical point of the line $x = 2s$, $y = s - 1$, $z = s + 1$. Show that the single critical point of f is a local minimum. Hence find the closest points on these two skew lines.

27. Let $f(x, y)$ denote the square of the distance from $(0, 0, 2)$ to a typical point of the surface $z = xy$. Find and classify the critical points of f.

28. Show that the surface

$$z = (x^2 + 2y^2)\exp(1 - x^2 - y^2)$$

looks like two mountain peaks joined by two ridges with a pit between them.

29. A wire 120 cm long is cut into three pieces of lengths x, y, and $120 - x - y$, and each piece is bent into the shape of a square. Let $f(x, y)$ denote the sum of the areas of these squares. Show that the single critical point of f is a local minimum. But surely it is possible to *maximize* the sum of the areas. Explain.

30. Show that the graph of the function

$$f(x, y) = xy\exp\left(\tfrac{1}{8}[x^2 + 4y^2]\right)$$

has a saddle point but no local extrema.

31. Find and classify the critical points of the function

$$f(x, y) = \sin\frac{\pi x}{2}\sin\frac{\pi y}{2}.$$

32. Let $f(x, y) = x^3 - 3xy^2$. (a) Show that its only critical point is $(0, 0)$ and that $\Delta = 0$ there. (b) By examining the behavior of $x^3 - 3xy^2$ on straight lines through the origin, show that the surface $z = x^3 - 3xy^2$ qualifies as a monkey saddle (Fig. 13.10.18).

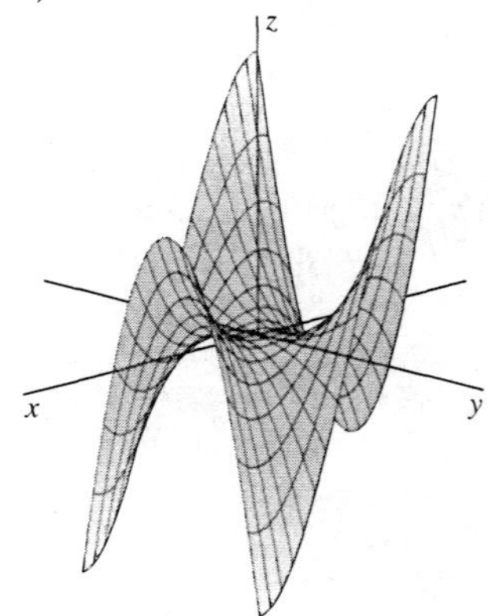

FIGURE 13.10.18 The monkey saddle of Problem 32.

33. Repeat Problem 32 with $f(x, y) = 4xy(x^2 - y^2)$. Show that near the critical point $(0, 0)$ the surface $z = f(x, y)$

qualifies as a "dog saddle" for a dog with a very short tail (Fig. 13.10.19).

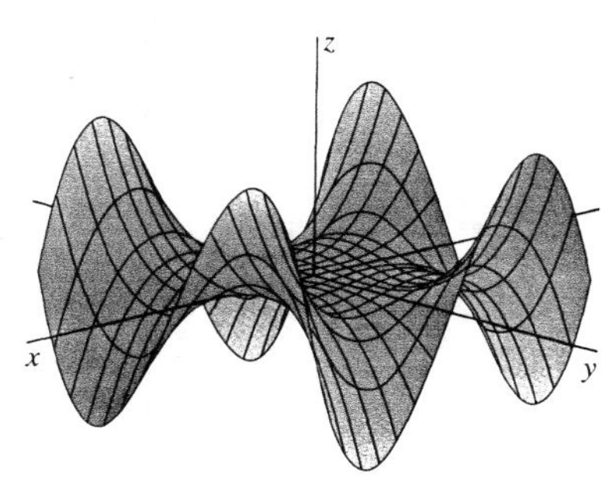

FIGURE 13.10.19 The dog saddle of Problem 33.

34. Let

$$f(x, y) = \frac{xy(x^2 - y^2)}{x^2 + y^2}.$$

Classify the behavior of f near the critical point $(0, 0)$.

In Problems 35 through 39, use a computer algebra program (as illustrated in the project material for this section) to approximate numerically and classify the critical point of the given function.

35. $f(x, y) = 2x^4 - 12x^2 + y^2 + 8x$

36. $f(x, y) = x^4 + 4x^2 - y^2 - 16x$

37. $f(x, y) = x^4 + 12xy + 6y^2 + 4x + 10$

38. $f(x, y) = x^4 + 8xy - 4y^2 - 16x + 10$

39. $f(x, y) = x^4 + 2y^4 - 12xy^2 - 20y^2$

13.10 Project: Critical Point Investigations

In the CD-ROM material for this project, the function

$$f(x, y) = 10 \exp(-x^2 - \tfrac{1}{2}xy - \tfrac{1}{2}y^2) \sin x \sin y \tag{1}$$

is used to illustrate computer algebra system techniques for the location and classification of critical points for functions of two variables, as follows:

- First, a surface graph shows the "big picture" that we want to investigate in detail. In Fig. 13.10.20 we see two peaks and two pits, as well as an apparent saddle point.
- Next, a contour graph as in Fig. 13.10.21 reveals the approximate location of each of these critical points.
- Then we set up the equations $f_x(x, y) = 0$ and $f_y(x, y) = 0$; we use a computer algebra **`solve`** command to approximate the critical points accurately—with the known approximate location of each critical point providing an initial guess for its calculation.
- Finally, we compute that information about the second-order partial derivatives needed to apply Theorem 1 to classify each critical point. And a contour plot in a small neighborhood of a critical point (as in Figs. 13.10.3 through 13.10.5) can provide satisfying visual corroboration of our results.

You can follow this agenda to investigate a function such as

$$f(x, y) = (ax^2 + 2bxy + cy^2) \exp(-x^2 - y^2) \tag{2}$$

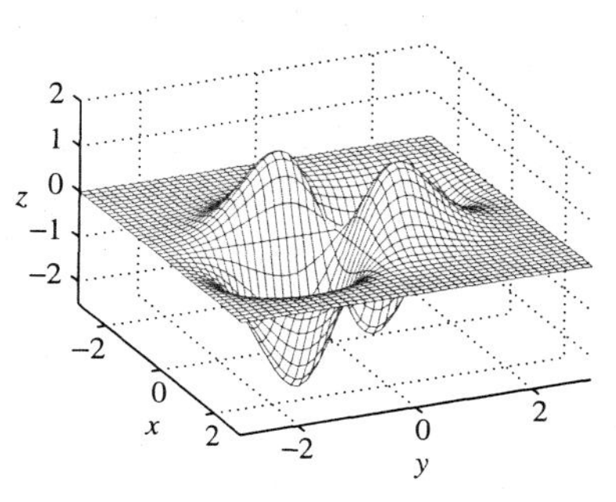

FIGURE 13.10.20 Graph of the function in Eq. (1).

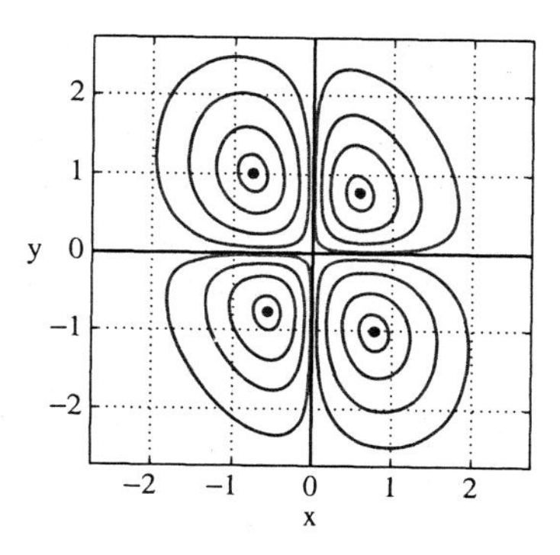

FIGURE 13.10.21 Contour plot for the function in Eq. (1).

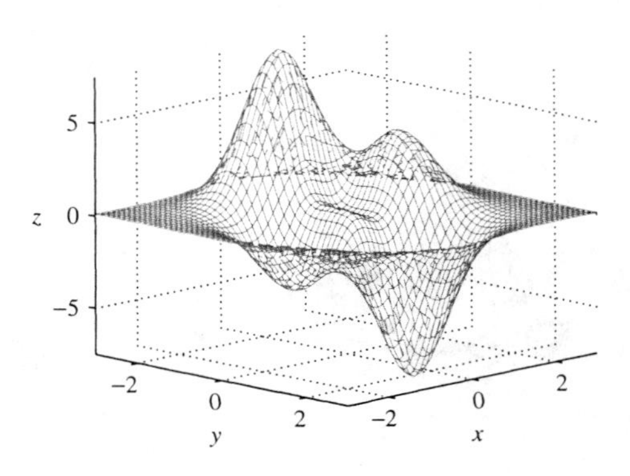

FIGURE 13.10.22 Graph of the function in Eq. (3), with $p = 5$ and the minus sign.

where a, b, and c are selected integers, or the more exotic function

$$f(x, y) = 10\left(x^3 + y^5 \pm \frac{x}{p}\right)\exp(-x^2 - y^2) + \frac{1}{3}\exp(-(x-1)^2 - y^2) \qquad \textbf{(3)}$$

where p is a small positive integer. With the plus sign in Eq. (3) you are likely to see a half-dozen critical points, but with the minus sign you can expect to see more (as in Fig. 13.10.22, in which $p = 5$ and there appears to be some "action" near the origin, in addition to the pairs of pits, peaks, and passes that are most evident).

CHAPTER 13 REVIEW: DEFINITIONS, CONCEPTS, RESULTS

Use the following list as a guide to concepts that you may need to review.

1. Graphs and level curves of functions of two variables.
2. Limits and continuity of functions of two or three variables.
3. Partial derivatives—definition and computation.
4. Geometric interpretation of partial derivatives and the plane tangent to the surface $z = f(x, y)$.
5. Absolute and local maxima and minima.
6. Necessary conditions for a local extremum.
7. Increments and differentials of functions of two or three variables.
8. The linear approximation theorem.
9. The chain rule for functions of several variables.
10. Directional derivatives—definition and computation.
11. The gradient vector and the chain rule.
12. Significance of the length and direction of the gradient vector.
13. The gradient vector as a normal vector; tangent plane to a surface $F(x, y, z) = 0$.
14. Constrained maximum-minimum problems and the Lagrange multiplier method.
15. Sufficient conditions for a local extremum of a function of two variables.

CHAPTER 13 MISCELLANEOUS PROBLEMS

1. Use polar coordinates to show that

$$\lim_{(x,y)\to(0,0)} \frac{x^2y^2}{x^2+y^2} = 0.$$

2. Use spherical coordinates to show that

$$\lim_{(x,y,z)\to(0,0,0)} \frac{x^3+y^3-z^3}{x^2+y^2+z^2} = 0.$$

3. Suppose that

$$g(x, y) = \frac{xy}{x^2+y^2}$$

if $(x, y) \neq (0, 0)$; we *define* $g(0, 0)$ to be zero. Show that g is not continuous at $(0, 0)$.

4. Compute $g_x(0, 0)$ and $g_y(0, 0)$ for the function g of Problem 3.

5. Find a function $f(x, y)$ such that

$$f_x(x, y) = 2xy^3 + e^x \sin y$$

and

$$f_y(x, y) = 3x^2y^2 + e^x \cos y + 1.$$

6. Prove that there is *no* function f with continuous second-order partial derivatives such that $f_x(x, y) = 6xy^2$ and $f_y(x, y) = 8x^2y$.

7. Find the point or points on the paraboloid $z = x^2 + y^2$ at which the normal line passes through the point $(0, 0, 1)$.

8. Write an equation of the plane tangent to the surface

$$\sin xy + \sin yz + \sin xz = 1$$

at the point $(1, \pi/2, 0)$.

9. Prove that every line normal to the cone with equation $z = \sqrt{x^2 + y^2}$ intersects the z-axis.

10. Show that the function

$$u(x, t) = \frac{1}{\sqrt{4\pi kt}} \exp\left(-\frac{x^2}{4kt}\right)$$

satisfies the one-dimensional heat equation

$$\frac{\partial u}{\partial t} = k\frac{\partial^2 u}{\partial x^2}.$$

11. Show that the function

$$u(x, y, t) = \frac{1}{4\pi kt} \exp\left(-\frac{x^2 + y^2}{4kt}\right)$$

satisfies the two-dimensional heat equation

$$\frac{\partial u}{\partial t} = k\left(\frac{\partial^2 u}{\partial x^2} + \frac{\partial^2 u}{\partial y^2}\right).$$

12. Suppose that $f(x, y, z) = \sqrt[5]{xyz}$. (a) Show that the partial derivatives f_x, f_y, and f_z all exist at the origin. (b) Show that the directional derivative $D_{\mathbf{u}} f(0, 0, 0)$ exists if and only if the unit vector $\mathbf{u}$ is a linear combination of some two of the standard unit vectors $\mathbf{i}$, $\mathbf{j}$, and $\mathbf{k}$.

13. Define the partial derivatives $\mathbf{r}_x$ and $\mathbf{r}_y$ of the vector-valued function $\mathbf{r}(x, y) = \mathbf{i}x + \mathbf{j}y + \mathbf{k}f(x, y)$ by componentwise partial differentiation. Then show that the vector $\mathbf{r}_x \times \mathbf{r}_y$ is normal to the surface $z = f(x, y)$.

14. An open-topped rectangular box is to have total surface area 300 cm^2. Find the dimensions that maximize its volume.

15. You must build a rectangular shipping crate with volume 60 ft^3. Its sides cost \$1/ft^2, its top costs \$2/ft^2, and its bottom costs \$3/ft^2. What dimensions would minimize the total cost of the box?

16. A pyramid is bounded by the three coordinate planes and by the plane tangent to the surface $xyz = 1$ at a point in the first octant. Find the volume of this pyramid (it is independent of the point of tangency).

17. Two resistors have resistances R_1 and R_2, respectively. When they are connected in parallel, the total resistance R of the resulting circuit satisfies the equation

$$\frac{1}{R} = \frac{1}{R_1} + \frac{1}{R_2}.$$

Suppose that R_1 and R_2 are measured to be 300 and 600 Ω (ohms) respectively, with a maximum error of 1% in each measurement. Use differentials to estimate the maximum error (in ohms) in the calculated value of R.

18. Consider a gas that satisfies van der Waals' equation. (See Problem 67 of Section 13.4.) Use differentials to approximate the change in its volume if p is increased from 1 atm to 1.1 atm and T is decreased from 313 K to 303 K.

19. Each of the semiaxes a, b, and c of an ellipsoid with volume $V = \frac{4}{3}\pi abc$ is measured with a maximum percentage error of 1%. Use differentials to estimate the maximum percentage error in the calculated value of V.

20. Two spheres have radii a and b, and the distance between their centers is $c < a + b$. Thus the spheres meet in a common circle. Let P be a point on this circle, and let $\mathcal{P}_1$ and $\mathcal{P}_2$ be the planes tangent at P to the two spheres. Find the angle between $\mathcal{P}_1$ and $\mathcal{P}_2$ in terms of a, b, and c. [*Suggestion:* Recall that the angle between two planes is, by definition, the angle between their normal vectors.]

21. Find every point on the surface of the ellipsoid $x^2 + 4y^2 + 9z^2 = 16$ at which the normal line at the point passes through the center $(0, 0, 0)$ of the ellipsoid.

22. Suppose that

$$F(x) = \int_{g(x)}^{h(x)} f(t)\,dt.$$

Show that

$$F'(x) = f(h(x))h'(x) - f(g(x))g'(x).$$

[*Suggestion:* Write $w = \int_u^v f(t)\,dt$ where $u = g(x)$ and $v = h(x)$.]

23. Suppose that $\mathbf{a}$, $\mathbf{b}$, and $\mathbf{c}$ are mutually perpendicular unit vectors in space and that f is a function of the three independent variables x, y, and z. Show that

$$\nabla f = \mathbf{a}(D_{\mathbf{a}} f) + \mathbf{b}(D_{\mathbf{b}} f) + \mathbf{c}(D_{\mathbf{c}} f).$$

24. Let $\mathbf{R} = \langle \cos\theta, \sin\theta, 0\rangle$ and $\Theta = \langle -\sin\theta, \cos\theta, 0\rangle$ be the polar-coordinates unit vectors. Given $f(x, y, z) = w(r, \theta, z)$, show that

$$D_{\mathbf{R}} f = \frac{\partial w}{\partial r} \quad \text{and} \quad D_{\Theta} f = \frac{1}{r}\frac{\partial w}{\partial \theta}.$$

Then conclude from Problem 23 that the gradient vector is given in cylindrical coordinates by

$$\nabla f = \frac{\partial w}{\partial r}\mathbf{R} + \frac{1}{r}\frac{\partial w}{\partial \theta}\Theta + \frac{\partial w}{\partial z}\mathbf{k}.$$

25. Suppose that you are standing at the point with coordinates $(-100, -100, 430)$ on a hill that has the shape of the graph of

$$z = 500 - (0.003)x^2 - (0.004)y^2$$

(in units of meters). In what (horizontal) direction should you move in order to maintain a constant altitude—that is, to neither climb nor descend the hill?

26. Suppose that the blood concentration in the ocean at the point (x, y) is given by

$$f(x, y) = A\exp(-k[x^2 + 2y^2]),$$

where A and k are positive constants. A shark always swims in the direction of ∇f. Show that its path is a parabola $y = cx^2$. [*Suggestion:* Show that the condition that $\langle dx/dt, dy/dt\rangle$ is a multiple of ∇f implies that

$$\frac{1}{x}\frac{dx}{dt} = \frac{1}{2y}\frac{dy}{dt}.$$

Then antidifferentiate this equation.]

27. Consider a plane tangent to the surface with equation $x^{2/3} + y^{2/3} + z^{2/3} = 1$. Find the sum of the squares of the x-, y-, and z-intercepts of this plane.

28. Find the points on the ellipse $x^2/a^2+y^2/b^2=1$ (with $a \neq b$) where the normal line passes through the origin.

29. Let

$$f(x, y)=\frac{x^2 y^2}{x^2+y^2}$$

if $(x, y) \neq (0, 0)$ and define $f(0, 0)$ to be 0. First show that f is differentiable at the origin. Then classify the origin as a critical point of f.

30. Find the point of the surface $z=xy+1$ that is closest to the origin.

31. Use the method of Problem 38 in Section 13.9 to find the semiaxes of the rotated ellipse

$$73x^2+72xy+52y^2=100.$$

32. Use the Lagrange multiplier method to show that the longest chord of the sphere $x^2+y^2+z^2=1$ has length 2. [*Suggestion:* There is no loss of generality in assuming that $(1, 0, 0)$ is one endpoint of the chord.]

33. Use the method of Lagrange multipliers, the law of cosines, and Fig. 13.9.9 to find the triangle of minimum perimeter inscribed in the unit circle.

34. When a current I enters two resistors, with resistances R_1 and R_2, that are connected in parallel, it splits into two currents I_1 and I_2 (with $I=I_1+I_2$) in such a way to minimize the total power loss $R_1 I_1^2+R_2 I_2^2$. Express I_1 and I_2 in terms of R_1, R_2, and I. Then derive the formula in Problem 17.

35. Use the method of Lagrange multipliers to find the points of the ellipse $x^2+2y^2=1$ that are closest to and farthest from the line $x+y=2$. [*Suggestion:* Let $f(x, y, u, v)$ denote the square of the distance between the point (x, y) of the ellipse and the point (u, v) of the line.]

36. (a) Show that the maximum of

$$f(x, y, z)=x+y+z$$

at points of the sphere $x^2+y^2+z^2=a^2$ is $a\sqrt{3}$. (b) Conclude from the result in part (a) that

$$(x+y+z)^2 \leqq 3(x^2+y^2+z^2)$$

for any three numbers x, y, and z.

37. Generalize the method of Problem 36 to show that, for any n arbitrary real numbers $x_1, x_2, \ldots,$ and x_n,

$$\frac{x_1+x_2+\cdots+x_n}{n} \leqq \sqrt{\frac{x_1^2+x_2^2+\cdots+x_n^2}{n}}.$$

Thus the *arithmetic mean* of the real numbers $x_1, x_2, \ldots, x_n$ is no greater than their *root-square mean.*

38. Find the maximum and minimum values of $f(x, y)=xy-x-y$ at points on and within the plane triangle with vertices $(0, 0)$, $(0, 1)$, and $(3, 0)$.

39. Find the maximum and minimum values of $f(x, y, z)=x^2-yz$ at points of the sphere $x^2+y^2+z^2=1$.

40. Find the maximum and minimum values of $f(x, y)=x^2y^2$ at points of the ellipse $x^2+4y^2=24$.

Locate and classify the critical points (local maxima, local minima, saddle points, and other points at which the tangent plane is horizontal) of the functions in Problems 41 through 50.

41. $f(x, y)=x^3y-3xy+y^2$

42. $f(x, y)=x^2+xy+y^2-6x+2$

43. $f(x, y)=x^3-6xy+y^3$

44. $f(x, y)=x^2y+xy^2+x+y$

45. $f(x, y)=x^3y^2(1-x-y)$

46. $f(x, y)=x^4-2x^2+y^2+4y+3$

47. $f(x, y)=e^{xy}-2xy$

48. $f(x, y)=x^3-y^3+x^2+y^2$

49. $f(x, y)=(x-y)(xy-1)$

50. $f(x, y)=(2x^2+y^2)\exp(-x^2-y^2)$

51. Given the data points (x_i, y_i) for $i=1, 2, \ldots, n$, the **least-squares straight line** $y=mx+b$ is the line that best fits these data in the following sense. Let $d_i=y_i-(mx_i+b)$ be the *deviation* of the predicted value mx_i+b from the true value y_i. Let

$$f(m, b)=d_1^2+d_2^2+\cdots+d_n^2=\sum_{i=1}^{n}[y_i-(mx_i+b)]^2$$

be the sum of the squares of the deviations. The least-squares straight line is the one that minimizes this sum (Fig. 13.MP.1). Show how to choose m and b by minimizing f. [*Note:* The only variables in this computation are m and b.]

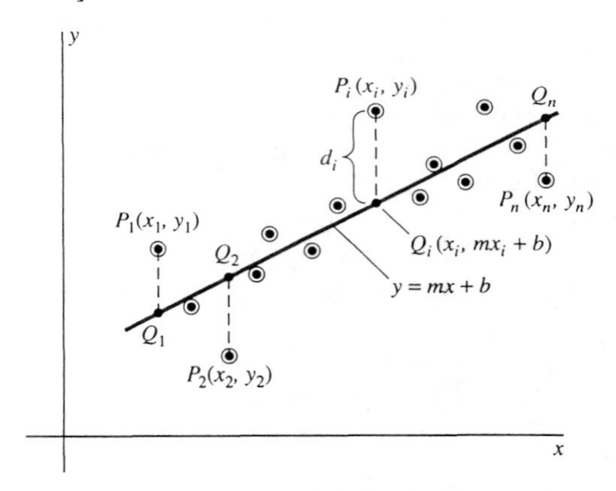

FIGURE 13.MP.1 Fitting the best straight line to the data points (x_i, y_i), $1 \leqq i \leqq n$ (Problem 51).

52. Let $f: \boldsymbol{R}^{2n} \to \boldsymbol{R}$ be defined for $(\mathbf{x}, \mathbf{y})$ in $\boldsymbol{R}^{2n}$ by

$$f(\mathbf{x}, \mathbf{y})=\mathbf{x}^T\mathbf{y}=\sum_{i=1}^{n} x_i y_i.$$

Use Lagrange multipliers to show that the maximum value of $f(\mathbf{x}, \mathbf{y})$ subject to the constraints $|\mathbf{x}|=1$ and $|\mathbf{y}|=1$ is 1. Given any two vectors $\mathbf{a}$ and $\mathbf{b}$ in $\boldsymbol{R}^2$, write $\mathbf{x}=\mathbf{a}/|\mathbf{a}|$ and $\mathbf{y}=\mathbf{b}/|\mathbf{b}|$ to conclude that

$$\mathbf{a}^T\mathbf{b} \leqq |\mathbf{a}|\,|\mathbf{b}|$$

(the **Cauchy-Schwarz inequality**).

MULTIPLE INTEGRALS

14

Henri Lebesgue (1875–1941)

Geometric problems of *measure*—dealing with concepts of length, area, and volume—can be traced back 40 centuries to the rise of civilizations in the fertile river valleys of Africa and Asia, when such issues as areas of fields and volumes of granaries became important. These problems led ultimately to the *integral,* which is used to calculate (among other things) areas and volumes of curvilinear figures. But only in the early twentieth century were certain long-standing difficulties with measure and integration finally resolved, largely as a consequence of the work of the French mathematician Henri Lebesgue.

In his 1902 thesis presented at the Sorbonne in Paris, Lebesgue introduced a new definition of the integral, generalizing Riemann's definition. In essence, to define the integral of the function f from $x = a$ to $x = b$, Lebesgue replaced Riemann's subdivision of the interval $[a, b]$ into nonoverlapping subintervals with a partition of $[a, b]$ into disjoint measurable sets $\{E_i\}$. The Riemann sum $\sum f(x_i^\star)\,\Delta x$ was thereby replaced with a sum of the form $\sum f(x_i^\star)\,m_i$, where m_i is the measure of the ith set E_i and $x_i^\star$ is a number in E_i. To see the advantage of the "Lebesgue integral," consider the fact that there exist differentiable functions whose derivatives are not integrable in the sense of Riemann. For such a function, the fundamental theorem of calculus in the form

$$\int_a^b f'(x)\,dx = f(b) - f(a)$$

fails to hold. But with his new definition of the integral, Lebesgue showed that a derivative function f' is integrable and that the fundamental theorem holds. Similarly, the equality of double and iterated integrals (Section 14.1) holds only under rather drastic restrictions if the Riemann definition of multiple integrals is used, but the Lebesgue integral resolves the difficulty.

For such reasons, the Lebesgue theory of measure and integration predominates in modern mathematical research, both pure and applied. For instance, the Lebesgue integral is basic to such diverse realms as applied probability and mathematical biology, the quantum theory of atoms and nuclei, and the information theory and electric signals processing of modern computer technology.

The Section 14.5 Project illustrates the application of multiple integrals to such concrete problems as the optimal design of race-car wheels.

We could use multiple integrals to determine the best design for the wheels of these soapbox derby cars.

14.1 DOUBLE INTEGRALS

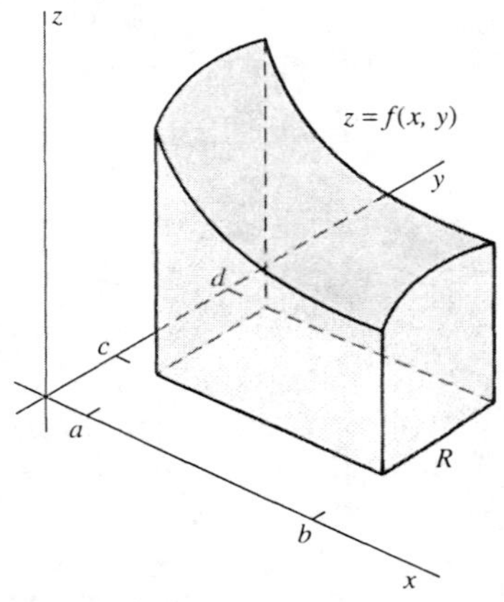

This chapter is devoted to integrals of functions of two or three variables. Such integrals are called **multiple integrals.** The applications of multiple integrals include computation of area, volume, mass, and surface area in a wider variety of situations than can be handled with the single integral of Chapters 5 and 6.

The simplest sort of multiple integral is the *double integral*

$$\iint_R f(x, y)\, dA$$

of a continuous function $f(x, y)$ over the *rectangle*

$$R = [a, b] \times [c, d] = \{(x, y) \mid a \leqq x \leqq b, c \leqq y \leqq d\}$$

in the xy-plane. (We will see that dA represents here a differential element of area A.) Just as the definition of the single integral is motivated by the problem of computing areas, the definition of the double integral is motivated by the problem of computing the volume V of the solid of Fig. 14.1.1—a solid bounded above by the graph $z = f(x, y)$ of the nonnegative function f over the rectangle R in the xy-plane.

FIGURE 14.1.1 We will use a double integral to compute the volume V.

To define the *value*

$$V = \iint_R f(x, y)\, dA$$

of such a double integral, we begin with an approximation to V. To obtain this approximation, the first step is to construct a **partition** $\mathcal{P}$ of R into subrectangles $R_1, R_2, \ldots, R_k$ determined by the points

$$a = x_0 < x_1 < x_2 < \cdots < x_m = b$$

of $[a, b]$ and

$$c = y_0 < y_1 < y_2 < \cdots < y_n = d$$

of $[c, d]$. Such a partition of R into $k = mn$ rectangles is shown in Fig. 14.1.2. The order in which these rectangles are labeled makes no difference.

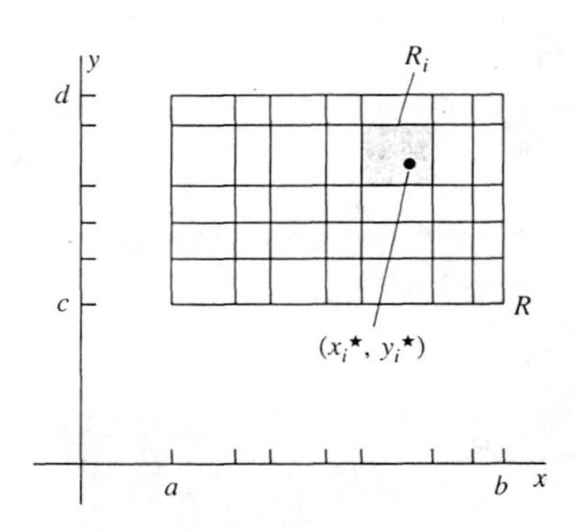

FIGURE 14.1.2 A partition $\mathcal{P}$ of the rectangle R.

Next we choose an arbitrary point $(x_i^\star, y_i^\star)$ of the ith rectangle R_i for each i (where $1 \leqq i \leqq k$). The collection of points $S = \{(x_i^\star, y_i^\star) \mid 1 \leqq i \leqq k\}$ is called a **selection** for the partition $\mathcal{P} = \{R_i \mid 1 \leqq i \leqq k\}$. As a measure of the size of the rectangles of the partition $\mathcal{P}$, we define its **norm** $|\mathcal{P}|$ to be the maximum of the lengths of the diagonals of the rectangles $\{R_i\}$.

Now consider a rectangular column that rises straight up from the xy-plane. Its base is the rectangle R_i and its height is the value $f(x_i^\star, y_i^\star)$ of f at the selected point $(x_i^\star, y_i^\star)$ of R_i. One such column is shown in Fig. 14.1.3. If ΔA_i denotes the area of R_i, then the volume of the ith column is $f(x_i^\star, y_i^\star)\, \Delta A_i$. The sum of the volumes of all such columns (Fig. 14.1.4) is the **Riemann sum**

$$\sum_{i=1}^{k} f(x_i^\star, y_i^\star)\, \Delta A_i, \tag{1}$$

an approximation to the volume V of the solid region that lies above the rectangle R and under the graph $z = f(x, y)$.

We would expect to determine the exact volume V by taking the limit of the Riemann sum in Eq. (1) as the norm $|\mathcal{P}|$ of the partition $\mathcal{P}$ approaches zero. We therefore define the **(double) integral** of the function f over the rectangle R to be

$$\iint_R f(x, y)\, dA = \lim_{|\mathcal{P}| \to 0} \sum_{i=1}^{k} f(x_i^\star, y_i^\star)\, \Delta A_i, \tag{2}$$

provided that this limit exists. (We will make the concept of the existence of such a

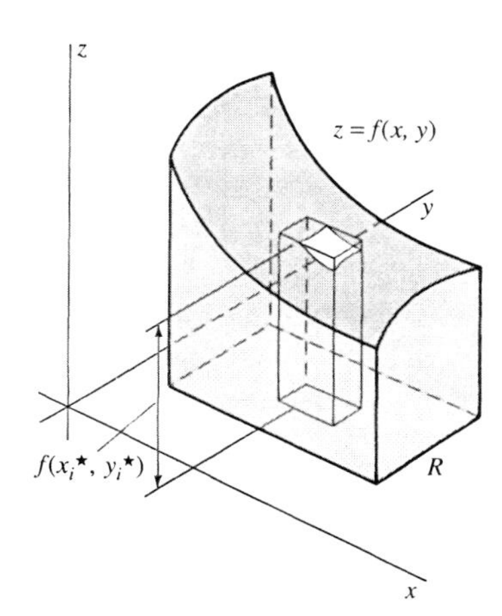

FIGURE 14.1.3 Approximating the volume under the surface by summing volumes of columns with rectangular bases.

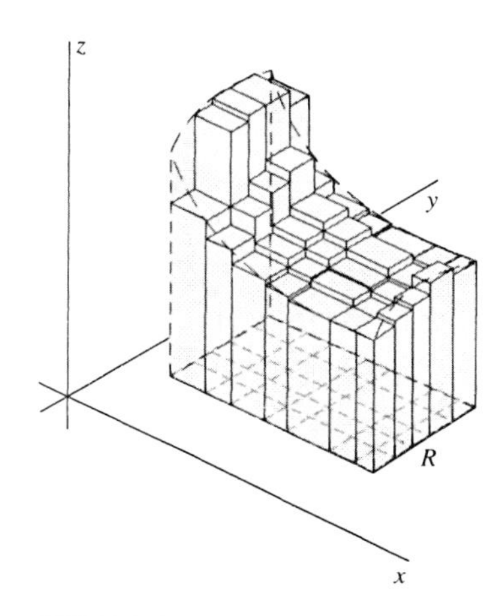

FIGURE 14.1.4 Columns corresponding to a partition of the rectangle R.

limit more precise in Section 14.2.) It is proved in advanced calculus that the limit in Eq. (2) *does* exist if f is continuous on R. To motivate the introduction of the Riemann sum in Eq. (1), we assumed that f was nonnegative on R, but Eq. (2) serves to define the double integral over a rectangle whether or not f is nonnegative.

EXAMPLE 1 Approximate the value of the integral

$$\iint_R (4x^3 + 6xy^2)\, dA$$

over the rectangle $R = [1, 3] \times [-2, 1]$, by calculating the Riemann sum in (1) for the partition illustrated in Fig. 14.1.5, with the ith point $(x_i^\star, y_i^\star)$ selected as the center of the ith rectangle R_i (for each i, $1 \leqq i \leqq 6$).

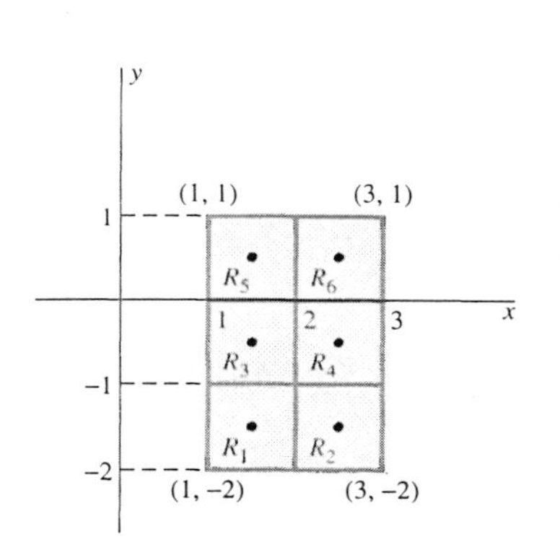

FIGURE 14.1.5 The partition in Example 1.

Solution Each of the six partition rectangles shown in Fig. 14.1.5 is a unit square with area $\Delta A_i = 1$. With $f(x, y) = 4x^3 + 6xy^2$, the desired Riemann sum is therefore

$$\begin{aligned}
\sum_{i=1}^{6} f(x_i^\star, y_i^\star)\,\Delta A_i &= f(x_1^\star, y_1^\star)\,\Delta A_1 + f(x_2^\star, y_2^\star)\,\Delta A_2 + f(x_3^\star, y_3^\star)\,\Delta A_3 \\
&\quad + f(x_4^\star, y_4^\star)\,\Delta A_4 + f(x_5^\star, y_5^\star)\,\Delta A_5 + f(x_6^\star, y_6^\star)\,\Delta A_6 \\
&= f\left(\tfrac{3}{2}, -\tfrac{3}{2}\right)(1) + f\left(\tfrac{5}{2}, -\tfrac{3}{2}\right)(1) + f\left(\tfrac{3}{2}, -\tfrac{1}{2}\right)(1) \\
&\quad + f\left(\tfrac{5}{2}, -\tfrac{1}{2}\right)(1) + f\left(\tfrac{3}{2}, \tfrac{1}{2}\right)(1) + f\left(\tfrac{5}{2}, \tfrac{1}{2}\right)(1) \\
&= \tfrac{135}{4} \cdot 1 + \tfrac{385}{4} \cdot 1 + \tfrac{63}{4} \cdot 1 + \tfrac{265}{4} \cdot 1 + \tfrac{63}{4} \cdot 1 + \tfrac{265}{4} \cdot 1 = 294.
\end{aligned}$$

Thus we find that

$$\iint_R (4x^3 + 6xy^2)\, dA \approx 294,$$

but our calculation provides no information about the accuracy of this approximation. ◆

REMARK 1 The single-variable approximation methods of Section 5.9 all have analogs for double integrals. In Example 1 we calculated the **midpoint approximation** to the given double integral

$$\iint_R f(x, y)\, dA$$

Number of Subrectangles	Midpoint Approximation
6	294.00
24	307.50
96	310.88
384	311.72
1536	311.93
6144	311.98

FIGURE 14.1.6 Midpoint approximations to the integral in Example 1.

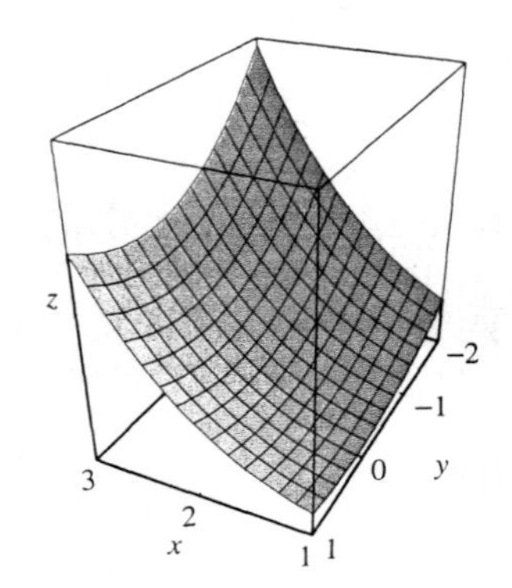

FIGURE 14.1.7 The surface $z = 4x^3 + 6xy^2$ over the rectangle R.

using a partition of the rectangle R into six subrectangles. The Riemann sum we calculated is the sum of the volumes of six rectangular columns or blocks. Each of these columns has a base consisting of one of the subrectangles in Fig. 14.1.5 and has height equal to the value $f(x_i^\star, y_i^\star)$ of the function at the *midpoint* of that subrectangle.

REMARK 2 If we subdivide each rectangle in Fig. 14.1.5 into four equal smaller rectangles, we get a partition of R into 24 subrectangles, and the corresponding Riemann sum is the sum of the volume of 24 rectangular columns with bases these 24 subrectangles. Suppose that we continue in this way, quadrupling the number of subrectangles (and of rectangular columns) at each step, and use a computer to calculate each time the Riemann sum defined by selecting the center of each subrectangle to calculate the height of the corresponding rectangular column. Then we get the midpoint approximations listed in Fig. 14.1.6 to the actual volume V that lies over the rectangle R and under the surface $z = f(x, y)$. (See Fig. 14.1.7.) Figure 14.1.8 shows the "rectangular block approximations" to V that correspond to partitions of R into 24, 96, and 384 subrectangles. In Example 2 we will see (much more easily) that the exact value of V is given by

$$V = \iint_R (4x^3 + 6xy^2)\, dA = 312.$$

Iterated Integrals

The direct evaluation of the limit in Eq. (2) is generally even less practical than the direct evaluation of the limit we used in Section 5.4 to define the single-variable integral. In practice, we shall calculate double integrals over rectangles by means of the **iterated integrals** that appear in Theorem 1.

THEOREM 1 Double Integrals as Iterated Single Integrals

Suppose that $f(x, y)$ is continuous on the rectangle $R = [a, b] \times [c, d]$. Then

$$\iint_R f(x, y)\, dA = \int_a^b \left(\int_c^d f(x, y)\, dy \right) dx = \int_c^d \left(\int_a^b f(x, y)\, dx \right) dy. \qquad \textbf{(3)}$$

Theorem 1 tells us how to compute a double integral by means of two successive (or *iterated*) single-variable integrations, each of which we can compute by using the fundamental theorem of calculus (if the function f is sufficiently well-behaved on R).

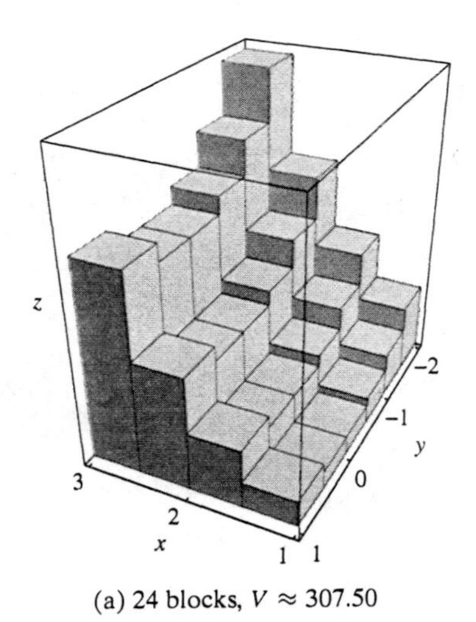

(a) 24 blocks, $V \approx 307.50$

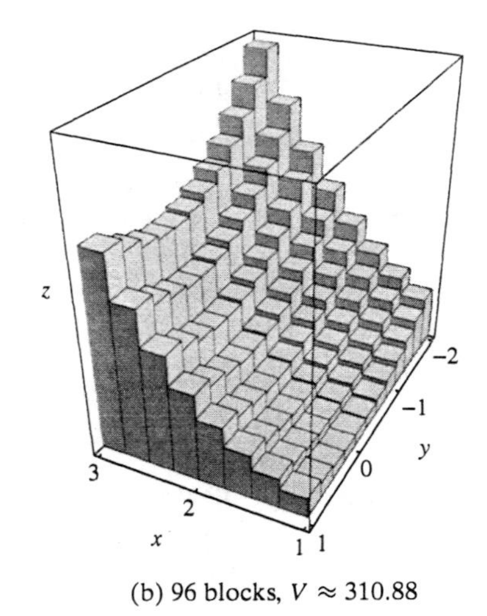

(b) 96 blocks, $V \approx 310.88$

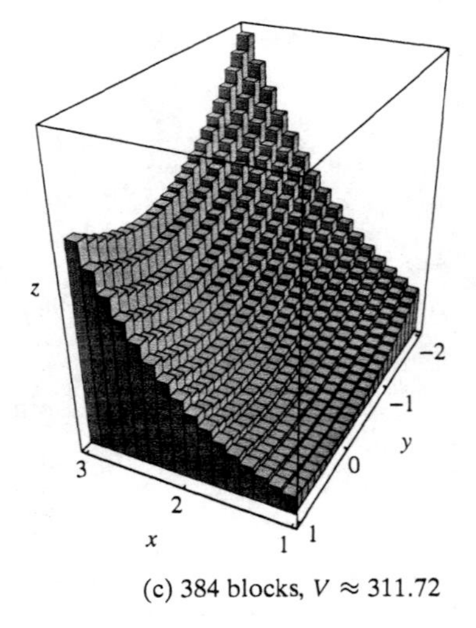

(c) 384 blocks, $V \approx 311.72$

FIGURE 14.1.8 Midpoint sum approximations to the volume V under the surface $z = 4x^3 + 6xy^2$ with 24, 96, and 384 subrectangles.

Let us explain what we mean by the parentheses in the iterated integral

$$\int_a^b \int_c^d f(x, y)\, dy\, dx = \int_a^b \left(\int_c^d f(x, y)\, dy \right) dx. \tag{4}$$

First we hold x constant and integrate with respect to y, from $y = c$ to $y = d$. The result of this first integration is the **partial integral of f with respect to** y, denoted by

$$\int_c^d f(x, y)\, dy,$$

and it is a function of x alone. Then we integrate this latter function with respect to x, from $x = a$ to $x = b$.

Similarly, we calculate the iterated integral

$$\int_c^d \int_a^b f(x, y)\, dx\, dy = \int_c^d \left(\int_a^b f(x, y)\, dx \right) dy \tag{5}$$

by first integrating from a to b with respect to x (while holding y fixed) and then integrating the result from c to d with respect to y. The order of integration (either first with respect to x and then with respect to y, or the reverse) is determined by the order in which the differentials dx and dy appear in the iterated integrals in Eqs. (4) and (5). We almost always work "from the inside out." Theorem 1 guarantees that the value obtained is independent of the order of integration provided that f is continuous on R.

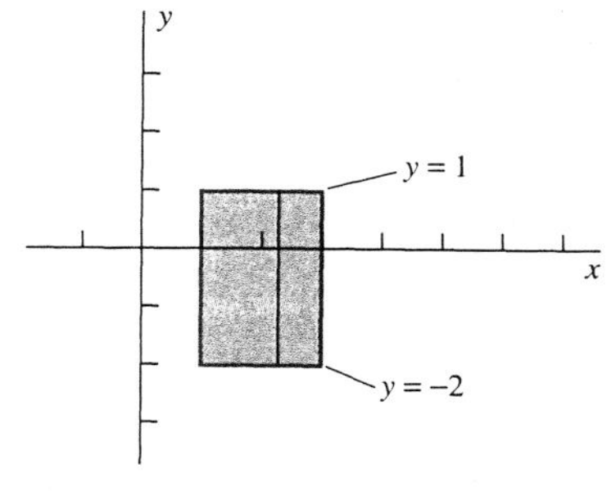

FIGURE 14.1.9 The inner limits of the first iterated integral (Example 2).

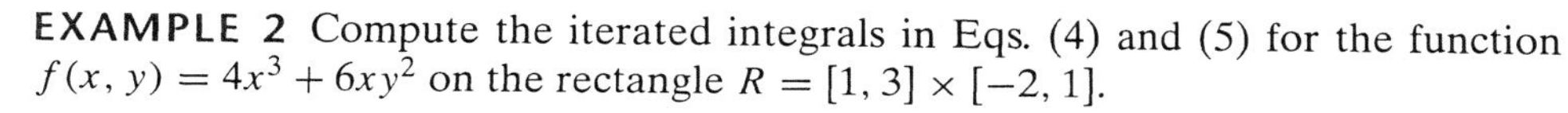

EXAMPLE 2 Compute the iterated integrals in Eqs. (4) and (5) for the function $f(x, y) = 4x^3 + 6xy^2$ on the rectangle $R = [1, 3] \times [-2, 1]$.

Solution The rectangle R is shown in Fig. 14.1.9, where the vertical segment (on which x is constant) corresponds to the inner integral in Eq. (4). Its endpoints lie at heights $y = -2$ and $y = 1$, which are, therefore, the limits on the inner integral. So Eq. (4) yields

$$\begin{aligned} \int_1^3 \left(\int_{-2}^1 (4x^3 + 6xy^2)\, dy \right) dx &= \int_1^3 \left[4x^3y + 2xy^3 \right]_{y=-2}^1 dx \\ &= \int_1^3 [(4x^3 + 2x) - (-8x^3 - 16x)]\, dx \\ &= \int_1^3 (12x^3 + 18x)\, dx \\ &= \left[3x^4 + 9x^2 \right]_1^3 = 312. \end{aligned}$$

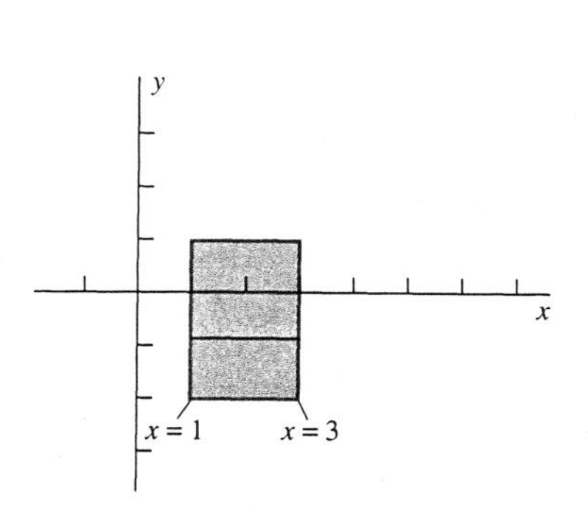

FIGURE 14.1.10 The inner limits of the second iterated integral (Example 2).

The horizontal segment (on which y is constant) in Fig. 14.1.10 corresponds to the inner integral in Eq. (5). Its endpoints lie at $x = 1$ and $x = 3$ (the limits of integration for x), so Eq. (5) gives

$$\begin{aligned} \int_{-2}^1 \left(\int_1^3 (4x^3 + 6xy^2)\, dx \right) dy &= \int_{-2}^1 \left[x^4 + 3x^2y^2 \right]_{x=1}^3 dy \\ &= \int_{-2}^1 [(81 + 27y^2) - (1 + 3y^2)]\, dy \\ &= \int_{-2}^1 (80 + 24y^2)\, dy \\ &= \left[80y + 8y^3 \right]_{-2}^1 = 312. \end{aligned}$$

◆

When we note that iterated double integrals are almost always evaluated from the inside out, it becomes clear that the parentheses appearing on the right-hand

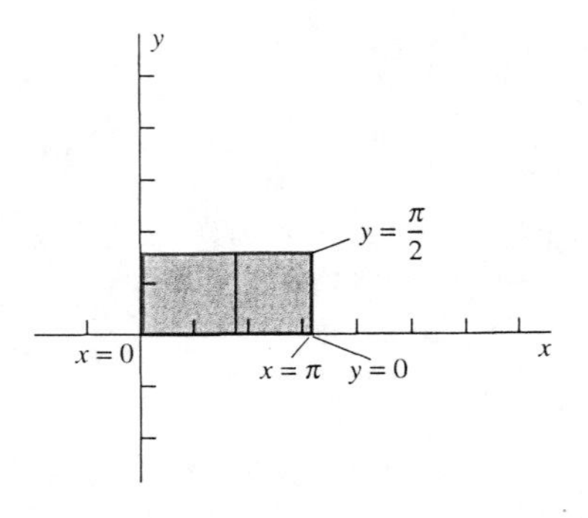

FIGURE 14.1.11 Example 3.

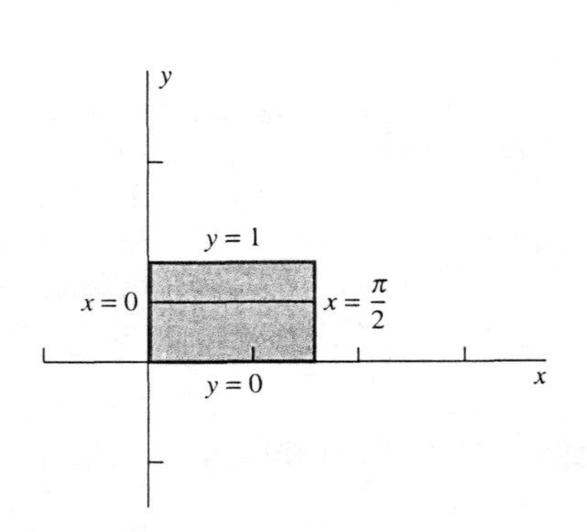

FIGURE 14.1.12 Example 4.

sides in Eqs. (4) and (5) are unnecessary. They are, therefore, generally omitted, as in Examples 3 and 4. When $dy\,dx$ appears in the integrand, we integrate first with respect to y, whereas the appearance of $dx\,dy$ tells us to integrate first with respect to x.

EXAMPLE 3 See Fig. 14.1.11.

$$\int_0^{\pi}\int_0^{\pi/2} \cos x \cos y \, dy\, dx = \int_0^{\pi} \Big[\cos x \sin y\Big]_{y=0}^{\pi/2} dx$$
$$= \int_0^{\pi} \cos x\, dx = \Big[\sin x\Big]_0^{\pi} = 0. \qquad \blacklozenge$$

EXAMPLE 4 See Fig. 14.1.12.

$$\int_0^1\int_0^{\pi/2} (e^y + \sin x)\, dx\, dy = \int_0^1 \Big[xe^y - \cos x\Big]_{x=0}^{\pi/2} dy$$
$$= \int_0^1 \left(\frac{1}{2}\pi e^y + 1\right) dy$$
$$= \left[\frac{1}{2}\pi e^y + y\right]_0^1 = \frac{1}{2}\pi(e-1) + 1. \qquad \blacklozenge$$

Iterated Integrals and Cross Sections

An outline of the proof of Theorem 1 illuminates the relationship between iterated integrals and the method of cross sections (for computing volumes) discussed in Section 6.2. First we partition $[a, b]$ into n equal subintervals, each of length $\Delta x = (b-a)/n$, and we also partition $[c, d]$ into n equal subintervals, each of length $\Delta y = (d-c)/n$. This gives n^2 rectangles, each of which has area $\Delta A = \Delta x\, \Delta y$. Choose a point $x_i^\star$ in $[x_{i-1}, x_i]$ for each i, $1 \leqq i \leqq n$. Then the average value theorem for single integrals (Section 5.6) gives a point $y_{ij}^\star$ in $[y_{j-1}, y_j]$ such that

$$\int_{y_{j-1}}^{y_j} f(x_i^\star, y)\, dy = f(x_i^\star, y_{ij}^\star)\, \Delta y.$$

This gives us the selected point $(x_i^\star, y_{ij}^\star)$ in the rectangle $[x_{i-1}, x_i] \times [y_{j-1}, y_j]$. Then

$$\iint_R f(x, y)\, dA \approx \sum_{i,j=1}^{n} f(x_i^\star, y_{ij}^\star)\, \Delta A = \sum_{i=1}^{n}\sum_{j=1}^{n} f(x_i^\star, y_{ij}^\star)\, \Delta y\, \Delta x$$
$$= \sum_{i=1}^{n}\left(\sum_{j=1}^{n}\int_{y_{j-1}}^{y_j} f(x_i^\star, y)\, dy\right)\Delta x$$
$$= \sum_{i=1}^{n}\left(\int_c^d f(x_i^\star, y)\, dy\right)\Delta x$$
$$= \sum_{i=1}^{n} A(x_i^\star)\, \Delta x,$$

where

$$A(x) = \int_c^d f(x, y)\, dy.$$

Moreover, the last sum is a Riemann sum for the integral

$$\int_a^b A(x)\, dx,$$

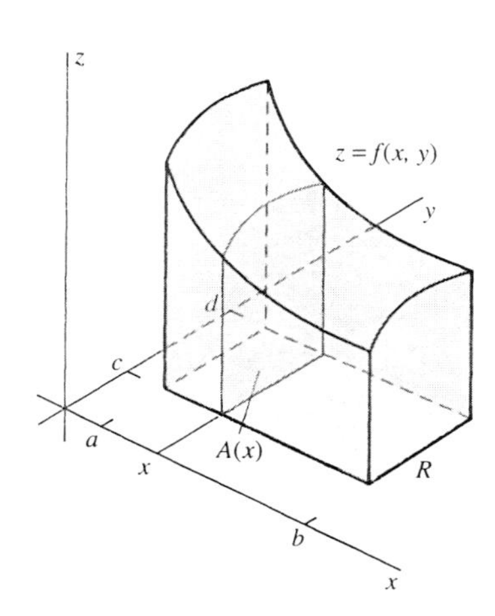

FIGURE 14.1.13 The area of the cross section at x is
$$A(x) = \int_c^d f(x, y)\,dy.$$

so the result of our computation is

$$\iint_R f(x, y)\,dA \approx \sum_{i=1}^{n} A(x_i^\star)\,\Delta x$$
$$\approx \int_a^b A(x)\,dx = \int_a^b \left(\int_c^d f(x, y)\,dy\right)dx.$$

We can convert this outline into a complete proof of Theorem 1 by showing that the preceding approximations become equalities when we take limits as $n \to +\infty$.

In case the function f is nonnegative on R, the function $A(x)$ introduced here gives the area of the vertical cross section of R perpendicular to the x-axis (Fig. 14.1.13). Thus the iterated integral in Eq. (4) expresses the volume V as the integral from $x = a$ to $x = b$ of the cross-sectional area function $A(x)$. Similarly, the iterated integral in Eq. (5) expresses V as the integral from $y = c$ to $y = d$ of the function

$$A(y) = \int_a^b f(x, y)\,dx,$$

which gives the area of a vertical cross section in a plane perpendicular to the y-axis. [Although it seems appropriate to use the notation $A(y)$ here, note that $A(x)$ and $A(y)$ are by no means the same function!]

14.1 TRUE/FALSE STUDY GUIDE

14.1 CONCEPTS: QUESTIONS AND DISCUSSION

1. Describe as completely as possible the analogy between
   - a single-variable integral $\int_I f(x)\,dx$ over an interval $I = [a, b]$

   and

   - a double integral $\iint_R f(x, y)\,dA$ over a rectangle $R = [a, b] \times [c, d]$.

   Discuss both the similarities and the differences.
2. Write the "double" Riemann sum
$$\sum_{i=1}^{m}\sum_{j=1}^{n}\cdots$$
corresponding to subdivision of $[a, b]$ and $[c, d]$ into m subintervals each of length Δx and into n subintervals each of length Δy (respectively), together with selections $\{x_i^\star\}_{i=1}^{m}$ and $\{y_j^\star\}_{j=1}^{n}$ of points in these subintervals. What choices of these selections might correspond to left-hand, right-hand, and midpoint sums for single-variable integrals?
3. Can you describe a way of generalizing the idea of trapezoidal approximations for single-variable integrals to double-sum approximations for double integrals? Think of using trapezoidal approximations for the cross-sectional area integrals discussed at the end of this section.

14.1 PROBLEMS

1. Approximate the integral
$$\iint_R (4x^3 + 6xy^2)\,dA$$
of Example 1 using the partition shown in Fig. 14.1.5, but selecting each $(x_i^\star, y_i^\star)$ as (a) the lower left corner of the rectangle R_i; (b) the upper right corner of the rectangle R_i.

2. Approximate the integral
$$\iint_R (4x^3 + 6xy^2)\,dA$$
as in Problem 1, but selecting each $(x_i^\star, y_i^\star)$ as (a) the upper left corner of the rectangle R_i; (b) the lower right corner of the rectangle R_i.

In Problems 3 through 8, calculate the Riemann sum for

$$\iint_R f(x, y)\, dA$$

using the given partition and selection of points $(x_i^\star, y_i^\star)$ for the rectangle R.

3. $f(x, y) = x + y$; $R = [0, 2] \times [0, 2]$; the partition $\mathcal{P}$ consists of four unit squares; each $(x_i^\star, y_i^\star)$ is the center point of the ith rectangle R_i.

4. $f(x, y) = xy$; $R = [0, 2] \times [0, 2]$; the partition $\mathcal{P}$ consists of four unit squares; each $(x_i^\star, y_i^\star)$ is the center point of the ith rectangle R_i.

5. $f(x, y) = x^2 - 2y$; $R = [2, 6] \times [-1, 1]$; the partition $\mathcal{P}$ consists of four equal rectangles of width $\Delta x = 2$ and height $\Delta y = 1$; each $(x_i^\star, y_i^\star)$ is the lower left corner of the ith rectangle R_i.

6. $f(x, y) = x^2 + y^2$; $R = [0, 2] \times [0, 3]$; the partition $\mathcal{P}$ consists of six unit squares; each $(x_i^\star, y_i^\star)$ is the upper right corner of the ith rectangle R_i.

7. $f(x, y) = \sin x \sin y$; $R = [0, \pi] \times [0, \pi]$; the partition $\mathcal{P}$ consists of four equal squares; each $(x_i^\star, y_i^\star)$ is the center point of the ith rectangle R_i.

8. $f(x, y) = \sin 4xy$; $R = [0, 1] \times [0, \pi]$; the partition $\mathcal{P}$ consists of six equal rectangles of width $\Delta x = \frac{1}{2}$ and height $\Delta y = \frac{1}{3}\pi$; each $(x_i^\star, y_i^\star)$ is the center point of the ith rectangle R_i.

In Problems 9 and 10, let L, M, and U denote the Riemann sums calculated for the given function f and the indicated partition $\mathcal{P}$ by selecting the lower left corners, midpoints, and upper right corners (respectively) of the rectangles in $\mathcal{P}$. Without actually calculating any of these Riemann sums, arrange them in increasing order of size.

9. $f(x, y) = x^2y^2$; $R = [1, 3] \times [2, 5]$; the partition $\mathcal{P}$ consists of six unit squares.

10. $f(x, y) = \sqrt{100 - x^2 - y^2}$; $R = [1, 4] \times [2, 5]$; the partition $\mathcal{P}$ consists of nine unit squares.

Evaluate the iterated integrals in Problems 11 through 30.

11. $\int_0^2 \int_0^4 (3x + 4y)\, dx\, dy$

12. $\int_0^3 \int_0^2 x^2 y\, dx\, dy$

13. $\int_{-1}^2 \int_1^3 (2x - 7y)\, dy\, dx$

14. $\int_{-2}^1 \int_2^4 x^2 y^3\, dy\, dx$

15. $\int_0^3 \int_0^3 (xy + 7x + y)\, dx\, dy$

16. $\int_0^2 \int_2^4 (x^2y^2 - 17)\, dx\, dy$

17. $\int_{-1}^2 \int_{-1}^2 (2xy^2 - 3x^2y)\, dy\, dx$

18. $\int_1^3 \int_{-3}^{-1} (x^3y - xy^3)\, dy\, dx$

19. $\int_0^{\pi/2} \int_0^{\pi/2} (\sin x \cos y)\, dx\, dy$

20. $\int_0^{\pi/2} \int_0^{\pi/2} (\cos x \sin y)\, dy\, dx$

21. $\int_0^1 \int_0^1 xe^y\, dy\, dx$

22. $\int_0^1 \int_{-2}^2 x^2 e^y\, dx\, dy$

23. $\int_0^1 \int_0^\pi e^x \sin y\, dy\, dx$

24. $\int_0^1 \int_0^1 e^{x+y}\, dx\, dy$

25. $\int_0^\pi \int_0^\pi (xy + \sin x)\, dx\, dy$

26. $\int_0^{\pi/2} \int_0^{\pi/2} (y - 1) \cos x\, dx\, dy$

27. $\int_0^{\pi/2} \int_1^e \frac{\sin y}{x}\, dx\, dy$

28. $\int_1^e \int_1^e \frac{1}{xy}\, dy\, dx$

29. $\int_0^1 \int_0^1 \left(\frac{1}{x+1} + \frac{1}{y+1}\right) dx\, dy$

30. $\int_1^2 \int_1^3 \left(\frac{x}{y} + \frac{y}{x}\right) dy\, dx$

In Problems 31 through 34, verify that the values of

$$\iint_R f(x, y)\, dA$$

given by the iterated integrals in Eqs. (4) and (5) are indeed equal.

31. $f(x, y) = 2xy - 3y^2$; $R = [-1, 1] \times [-2, 2]$

32. $f(x, y) = \sin x \cos y$; $R = [0, \pi] \times [-\pi/2, \pi/2]$

33. $f(x, y) = \sqrt{x + y}$; $R = [0, 1] \times [1, 2]$

34. $f(x, y) = e^{x+y}$; $R = [0, \ln 2] \times [0, \ln 3]$

35. Prove that

$$\lim_{n\to\infty} \int_0^1 \int_0^1 x^n y^n\, dx\, dy = 0.$$

36. Suppose that $f(x, y) = k$ is a constant-valued function and $R = [a, b] \times [c, d]$. Use Riemann sums to prove that

$$\iint_R k\, dA = k(b - a)(d - c).$$

37. Use Riemann sums to show, without calculating the value of the integral, that

$$0 \leqq \int_0^{\pi} \int_0^{\pi} \sin \sqrt{xy}\, dx\, dy \leqq \pi^2.$$

Problems 38 through 40 list properties of double integrals that are analogous to familiar properties of single integrals. In each case state the corresponding relation between Riemann sums associated with a given partition and selection for the rectangle R.

38. $\displaystyle\iint_R cf(x, y)\, dA = c \iint_R f(x, y)\, dA$ (c is a constant).

39. $\displaystyle\iint_R [f(x, y) + g(x, y)]\, dA$

$$= \iint_R f(x, y)\, dA + \iint_R g(x, y)\, dA.$$

40. If $f(x, y) \leqq g(x, y)$ at each point of R, then

$$\iint_R f(x, y)\, dA \leqq \iint_R g(x, y)\, dA.$$

14.1 Project: Midpoint Sums Approximating Double Integrals

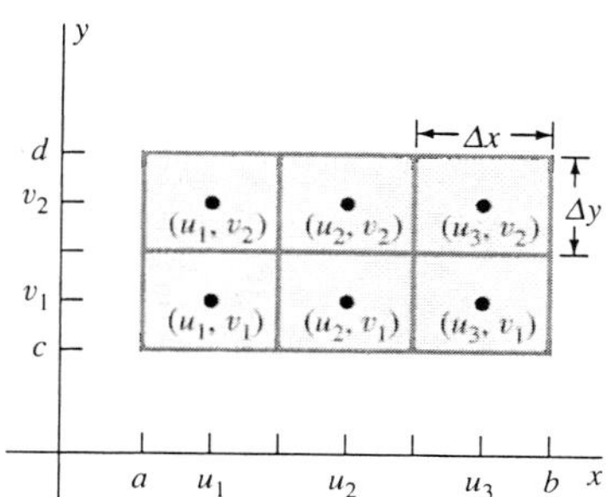

FIGURE 14.1.14 The points used in the midpoint approximation.

Suppose that we divide the intervals $[a, b]$ and $[c, d]$ into m subintervals of length Δx and into n subintervals of length Δy (respectively). If u_i and v_j denote the midpoints of the ith subinterval of $[a, b]$ and the jth subinterval of $[c, d]$ (respectively), then (u_i, v_j) is the midpoint of the ijth subrectangle $[x_{i-1}, x_i] \times [y_{j-1}, y_j]$. We thereby obtain the midpoint sum approximation

$$S_{mn} = \sum_{i=1}^{m} \sum_{j-1}^{n} f(u_i, v_j)\, \Delta x\, \Delta y \approx \iint_R f(x, y)\, dA$$

to the double integral of the function f over the rectangle $R = [a, b] \times [c, d]$. Figure 14.1.14 illustrates the case in which $m = 3$ and $n = 2$. In the CD-ROM material for this project we illustrate the use of computer algebra systems to calculate midpoint sum approximations rapidly and efficiently.

14.2 DOUBLE INTEGRALS OVER MORE GENERAL REGIONS

Now we want to define and compute double integrals over regions more general than rectangles. Let the function f be defined on the plane region R, and suppose that R is **bounded**—that is, that R lies within some rectangle S. To define the (double) integral of f over R, we begin with a partition $\mathcal{Q}$ of the rectangle S into subrectangles. Some of the rectangles of $\mathcal{Q}$ will lie wholly within R, some will be outside R, and some will lie partly within and partly outside R. We consider the collection $\mathcal{P} = \{R_1, R_2, \dots, R_k\}$ of all those rectangles in $\mathcal{Q}$ that lie *completely within* the region R. This collection $\mathcal{P}$ is called the **inner partition** of the region R determined by the partition $\mathcal{Q}$ of the rectangle S (Fig. 14.2.1). By the **norm** $|\mathcal{P}|$ of the inner partition $\mathcal{P}$ we mean the norm of the partition $\mathcal{Q}$ that determines $\mathcal{P}$. Note that $|\mathcal{P}|$ depends not only on $\mathcal{P}$ but on $\mathcal{Q}$ as well.

Using the inner partition $\mathcal{P}$ of the region R, we can proceed in much the same way as in Section 14.1. By choosing an arbitrary point $(x_i^\star, y_i^\star)$ in the ith rectangle R_i of $\mathcal{P}$ for $i = 1, 2, 3, \dots, k$, we obtain a **selection** for the inner partition $\mathcal{P}$. Let us denote by ΔA_i the area of R_i. Then this selection gives the **Riemann sum**

$$\sum_{i=1}^{k} f(x_i^\star, y_i^\star)\, \Delta A_i$$

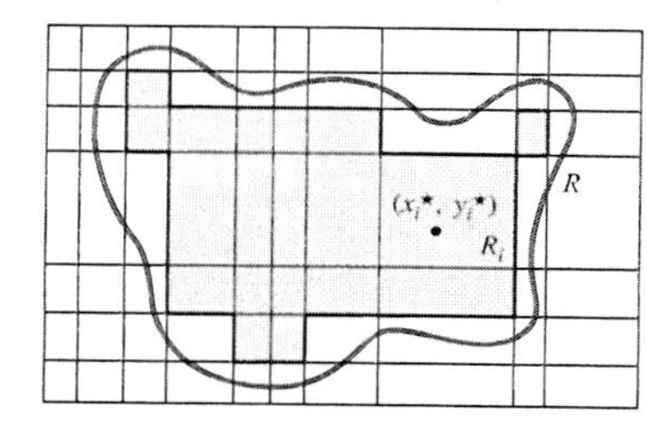

FIGURE 14.2.1 The rectangular partition of S produces an associated inner partition (shown shaded) of the region R.

associated with the inner partition $\mathcal{P}$. In case f is nonnegative on R, this Riemann sum approximates the volume of the three-dimensional region that lies under the surface $z = f(x, y)$ and above the region R in the xy-plane. We therefore define the double integral of f over the region R by taking the limit of this Riemann sum as the norm $|\mathcal{P}|$ approaches zero. Thus

$$\iint_R f(x, y)\,dA = \lim_{|\mathcal{P}|\to 0} \sum_{i=1}^{k} f(x_i^\star, y_i^\star)\,\Delta A_i, \tag{1}$$

provided that this limit exists in the sense of the following definition.

DEFINITION The Double Integral

The **double integral** of the bounded function f over the plane region R is the number

$$I = \iint_R f(x, y)\,dA$$

provided that, for every $\epsilon > 0$, there exists a number $\delta > 0$ such that

$$\left| \sum_{i=1}^{k} f(x_i^\star, y_i^\star)\,\Delta A_i - I \right| < \epsilon$$

for every inner partition $\mathcal{P} = \{R_1, R_2, \dots, R_k\}$ of R that has norm $|\mathcal{P}| < \delta$ and every selection of points $(x_i^\star, y_i^\star)$ in R_i $(i = 1, 2, \dots, k)$.

Thus the meaning of the limit in Eq. (1) is that the Riemann sum can be made arbitrarily close to the number

$$I = \iint_R f(x, y)\,dA$$

merely by choosing the norm of the inner partition $\mathcal{P}$ sufficiently small. In this case we say that the function f is **integrable** on the region R.

Note If R is a rectangle and we choose $S = R$ (so that an inner partition of R is simply a partition of R), then the preceding definition reduces to our earlier definition of a double integral over a rectangle. In advanced calculus the double integral of the function f over the bounded plane region R is shown to exist provided that f is continuous on R and the *boundary* of R is reasonably well-behaved. In particular, it suffices for the boundary of R to consist of a finite number of piecewise smooth simple closed curves (that is, each boundary curve consists of a finite number of smooth arcs).

EXAMPLE 1 Approximate the value of the integral

$$\iint_R (x + y)\,dA$$

where R is the region in the first quadrant bounded by the unit circle and the coordinate axes. Do so by calculating the sum in Eq. (1) for the inner partition and midpoint selection indicated in Fig. 14.2.2(a).

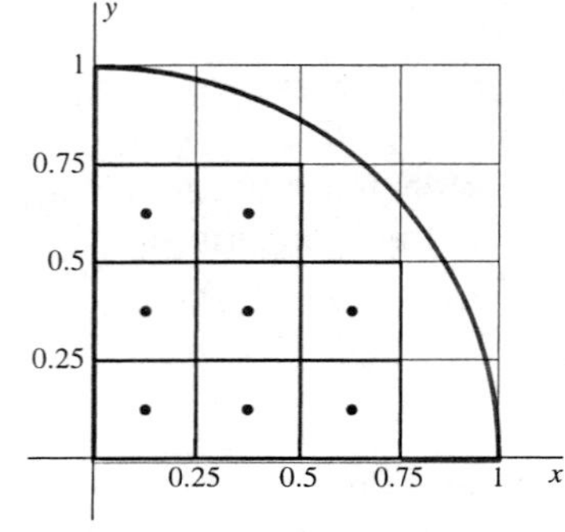

(a) 8 interior squares with $\Delta x = \Delta y = \frac{1}{4}$

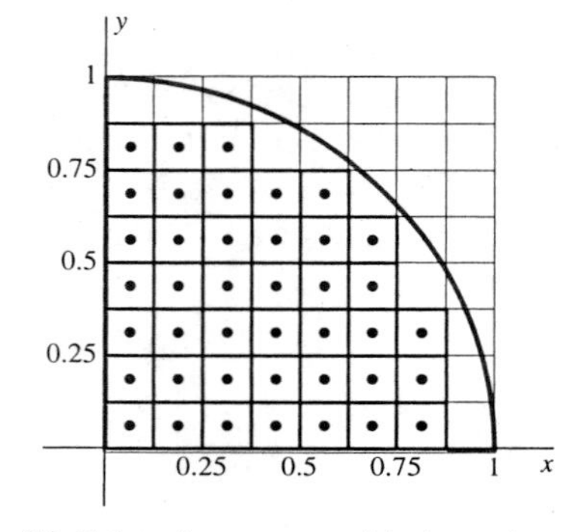

(b) 41 interior squares with $\Delta x = \Delta y = \frac{1}{8}$

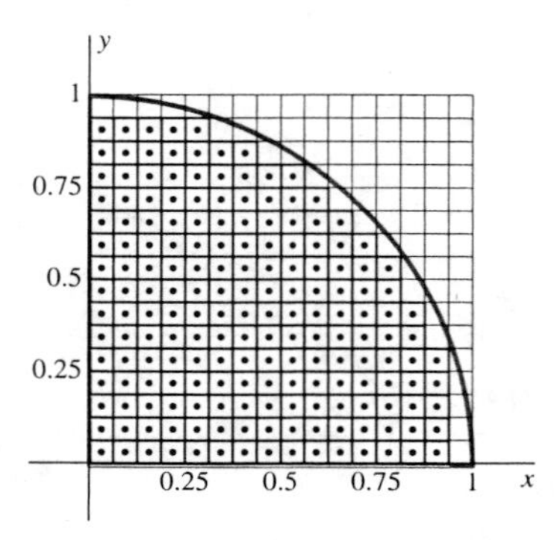

(c) 183 interior squares with $\Delta x = \Delta y = \frac{1}{16}$

FIGURE 14.2.2 Inner partitions of the quarter-circle R with $\Delta x = \Delta y = \frac{1}{4}$, $\Delta x = \Delta y = \frac{1}{8}$, and $\Delta x = \Delta y = \frac{1}{16}$.

Solution The figure shows a partition of the unit square into 16 smaller squares each with side length $\Delta x = \Delta y = \frac{1}{4}$. The inner partition we use consists of the 8 small squares that are contained wholly within the quarter-circular region R. The midpoints of these squares are the 8 points $(\frac{1}{8}, \frac{1}{8})$, $(\frac{3}{8}, \frac{1}{8})$, $(\frac{5}{8}, \frac{1}{8})$, $(\frac{1}{8}, \frac{3}{8})$, $(\frac{3}{8}, \frac{3}{8})$, $(\frac{5}{8}, \frac{3}{8})$, $(\frac{1}{8}, \frac{5}{8})$, and $(\frac{3}{8}, \frac{5}{8})$. The corresponding Riemann sum is

$$\begin{aligned} S = \Big[f\left(\tfrac{1}{8}, \tfrac{1}{8}\right) &+ f\left(\tfrac{3}{8}, \tfrac{1}{8}\right) + f\left(\tfrac{5}{8}, \tfrac{1}{8}\right) + f\left(\tfrac{1}{8}, \tfrac{3}{8}\right) \\ &+ f\left(\tfrac{3}{8}, \tfrac{3}{8}\right) + f\left(\tfrac{5}{8}, \tfrac{3}{8}\right) + f\left(\tfrac{1}{8}, \tfrac{5}{8}\right) + f\left(\tfrac{3}{8}, \tfrac{5}{8}\right)\Big]\,\Delta x\,\Delta y \\ = \Big[\left(\tfrac{1}{8} + \tfrac{1}{8}\right) &+ \left(\tfrac{3}{8} + \tfrac{1}{8}\right) + \left(\tfrac{5}{8} + \tfrac{1}{8}\right) + \left(\tfrac{1}{8} + \tfrac{3}{8}\right) \\ &+ \left(\tfrac{3}{8} + \tfrac{3}{8}\right) + \left(\tfrac{5}{8} + \tfrac{3}{8}\right) + \left(\tfrac{1}{8} + \tfrac{5}{8}\right) + \left(\tfrac{3}{8} + \tfrac{5}{8}\right)\Big] \cdot \tfrac{1}{4} \cdot \tfrac{1}{4}, \end{aligned}$$

and thus

$$S = \frac{11}{32} = 0.34375 \approx 0.344.$$

◆

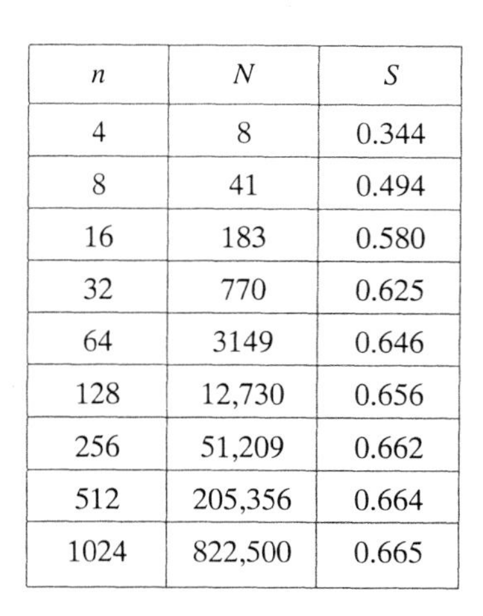

n	N	S
4	8	0.344
8	41	0.494
16	183	0.580
32	770	0.625
64	3149	0.646
128	12,730	0.656
256	51,209	0.662
512	205,356	0.664
1024	822,500	0.665

FIGURE 14.2.3 The number n of subintervals in each direction, the number N of small squares in the inner partition, and the corresponding approximate Riemann sum S.

REMARK In Fig. 14.2.2(a) we began by dividing the unit intervals on the x- and y-axes into $n = 4$ subintervals each. Figures 14.2.2(b) and 14.2.2(c) show the inner partitions that result when we begin with $n = 8$ and $n = 16$ subintervals (respectively) in each direction. Suppose that we continue in this way, doubling the number n of subintervals in each direction at each step, and use a computer to calculate each time the midpoint Riemann sum corresponding to the resulting inner partition of the quarter-circular region R. Figure 14.2.3 shows the resulting approximations to the integral

$$\iint_R (x + y)\,dA;$$

we also show the total number N of interior squares used at each step. In Problem 51 we ask you to show (using a comparatively simple computation with iterated integrals) that the exact value of this integral is $\frac{2}{3}$. (Thus the approximation in Example 1 is not every impressive.)

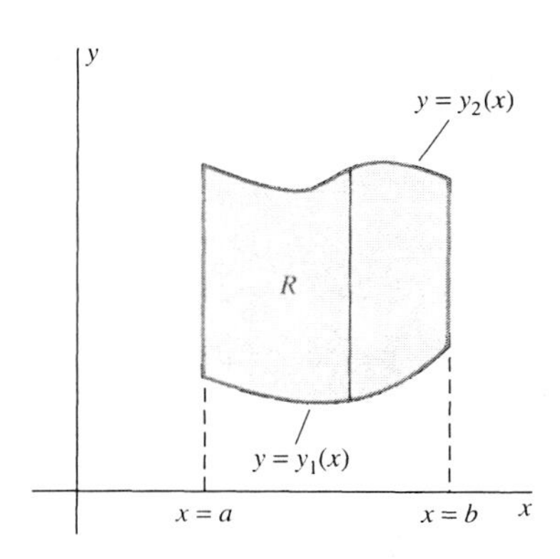

FIGURE 14.2.4 A vertically simple region R.

Evaluation of Double Integrals

The explicit evaluation of Riemann sums as in Example 1 is cumbersome and inefficient. But for certain common types of regions, we can evaluate double integrals by using iterated integrals in much the same way as we do when the region is a rectangle. The plane region R is called **vertically simple** if it can be described by means of the inequalities

$$a \leqq x \leqq b, \qquad y_1(x) \leqq y \leqq y_2(x), \tag{2}$$

where $y_1(x)$ and $y_2(x)$ are continuous functions of x on $[a, b]$. Such a region appears in Fig. 14.2.4. The region R is called **horizontally simple** if it can be described by the inequalities

$$c \leqq y \leqq d, \qquad x_1(y) \leqq x \leqq x_2(y), \tag{3}$$

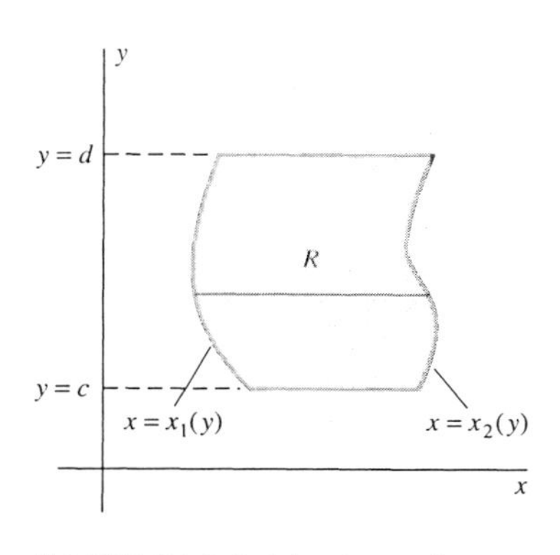

FIGURE 14.2.5 A horizontally simple region R.

where $x_1(y)$ and $x_2(y)$ are continuous functions of y on $[c, d]$. The region in Fig. 14.2.5 is horizontally simple.

Theorem 1 tells us how to compute by iterated integration a double integral over a region R that is either vertically simple or horizontally simple.

THEOREM 1 Evaluation of Double Integrals

Suppose that $f(x, y)$ is continuous on the region R. If R is the vertically simple region given in (2), then

$$\iint_R f(x, y)\,dA = \int_a^b \int_{y_1(x)}^{y_2(x)} f(x, y)\,dy\,dx. \tag{4}$$

If R is the horizontally simple region given in (3), then

$$\iint_R f(x, y)\,dA = \int_c^d \int_{x_1(y)}^{x_2(y)} f(x, y)\,dx\,dy. \tag{5}$$

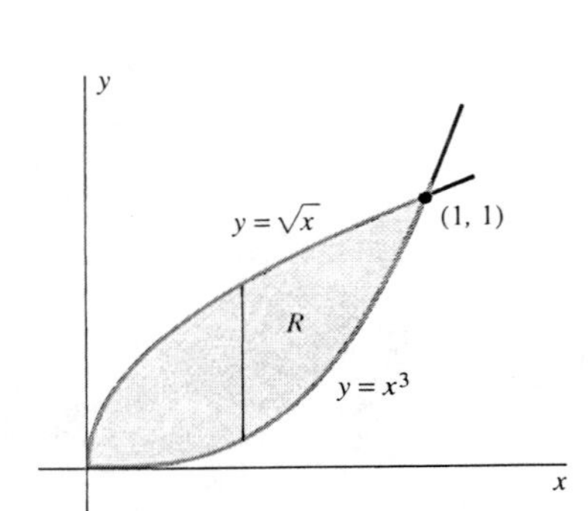

FIGURE 14.2.6 The vertically simple region of Example 2.

Theorem 1 here includes Theorem 1 of Section 14.1 as a special case (when R is a rectangle), and it can be proved by a generalization of the argument we outlined there.

EXAMPLE 2 Compute in two different ways the integral

$$\iint_R xy^2\,dA,$$

where R is the first-quadrant region bounded by the two curves $y = \sqrt{x}$ and $y = x^3$.

Solution *Always sketch the region R of integration before attempting to evaluate a double integral.* As indicated in Figs. 14.2.6 and 14.2.7, the given region R is both vertically and horizontally simple. The vertical segment in Fig. 14.2.6 with endpoints on the curves $y = x^3$ and $y = \sqrt{x}$ corresponds to integrating first with respect to y:

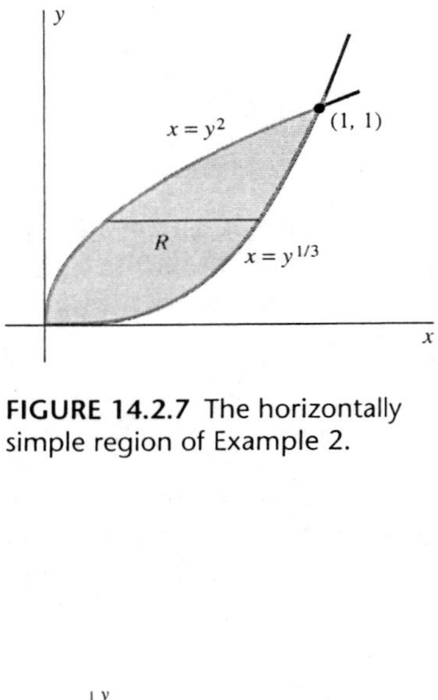

$$\iint_R xy^2\,dA = \int_0^1 \int_{x^3}^{\sqrt{x}} xy^2\,dy\,dx = \int_0^1 \left[\frac{1}{3}xy^3\right]_{y=x^3}^{\sqrt{x}} dx$$

$$= \int_0^1 \left(\frac{1}{3}x^{5/2} - \frac{1}{3}x^{10}\right) dx = \frac{2}{21} - \frac{1}{33} = \frac{5}{77}.$$

FIGURE 14.2.7 The horizontally simple region of Example 2.

We obtain $x = y^2$ and $x = y^{1/3}$ when we solve the equations $y = \sqrt{x}$ and $y = x^3$ for x in terms of y. The horizontal segment in Fig. 14.2.7 corresponds to integrating first with respect to x:

$$\iint_R xy^2\,dA = \int_0^1 \int_{y^2}^{y^{1/3}} xy^2\,dx\,dy = \int_0^1 \left[\frac{1}{2}x^2y^2\right]_{x=y^2}^{y^{1/3}} dy$$

$$= \int_0^1 \left(\frac{1}{2}y^{8/3} - \frac{1}{2}y^6\right) dy = \frac{3}{22} - \frac{1}{14} = \frac{5}{77}.$$ ◆

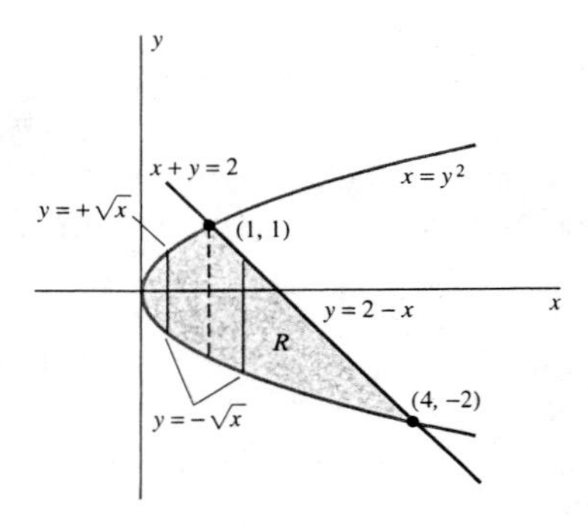

FIGURE 14.2.8 The vertically simple region of Example 3.

EXAMPLE 3 Evaluate

$$\iint_R (6x + 2y^2)\,dA,$$

where R is the region bounded by the parabola $x = y^2$ and the straight line $x + y = 2$.

Solution The region R appears in Fig. 14.2.8. It is both horizontally and vertically simple. If we wished to integrate first with respect to y and then with respect to x, we would need to evaluate two integrals:

$$\iint_R f(x, y)\,dA = \int_0^1 \int_{-\sqrt{x}}^{\sqrt{x}} (6x + 2y^2)\,dy\,dx + \int_1^4 \int_{-\sqrt{x}}^{2-x} (6x + 2y^2)\,dy\,dx.$$

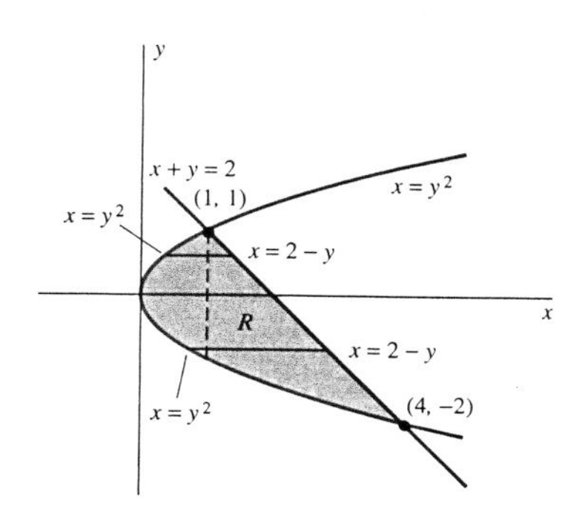

FIGURE 14.2.9 The horizontally simple region of Example 3.

The reason is that the formula of the function $y = y_2(x)$ describing the "top boundary curve" of R changes at the point $(1, 1)$, from $y = \sqrt{x}$ on the left to $y = 2 - x$ on the right. But as we see in Fig. 14.2.9, every *horizontal* segment in R extends from $x = y^2$ on the left to $x = 2 - y$ on the right. Therefore, integrating first with respect to x requires us to evaluate only *one* iterated integral:

$$\iint_R f(x, y)\, dA = \int_{-2}^{1} \int_{y^2}^{2-y} (6x + 2y^2)\, dx\, dy$$

$$= \int_{-2}^{1} \Big[3x^2 + 2xy^2\Big]_{x=y^2}^{2-y} dy$$

$$= \int_{-2}^{1} [3(2-y)^2 + 2(2-y)y^2 - 3(y^2)^2 - 2y^4]\, dy$$

$$= \int_{-2}^{1} (12 - 12y + 7y^2 - 2y^3 - 5y^4)\, dy$$

$$= \left[12y - 6y^2 + \frac{7}{3}y^3 - \frac{1}{2}y^4 - y^5\right]_{-2}^{1} = \frac{99}{2}.$$ ◆

Example 3 shows that even when the region R is both vertically and horizontally simple, it may be easier to integrate in one order rather than the other because of the shape of R. We naturally prefer the easier route. The choice of the preferable order of integration may be influenced also by the nature of the function $f(x, y)$. It may be difficult—or even impossible—to compute a given iterated integral but easy to do so *after we reverse the order of integration.* Example 4 shows that the key to reversing the order of integration is this:

> Find and sketch the region R over which the integration is to be performed.

EXAMPLE 4 Evaluate

$$\int_0^2 \int_{y/2}^{1} y e^{x^3}\, dx\, dy.$$

Solution We cannot integrate first with respect to x, as indicated, because $\exp(x^3)$ is known to have no elementary antiderivative. So we try to evaluate the integral by first reversing the order of integration. To do so, we sketch the region of integration specified by the limits in the given iterated integral.

The region R is determined by the inequalities

$$\tfrac{1}{2}y \leqq x \leqq 1 \quad \text{and} \quad 0 \leqq y \leqq 2.$$

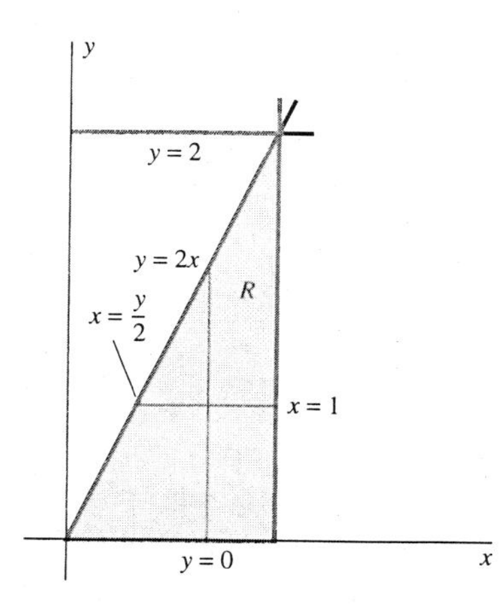

FIGURE 14.2.10 The region of Example 4.

Thus all points (x, y) of R lie between the horizontal lines $y = 0$ and $y = 2$ and between the two lines $x = y/2$ and $x = 1$. We draw the four lines $y = 0$, $y = 2$, $x = y/2$, and $x = 1$ and find that the region of integration is the shaded triangle that appears in Fig. 14.2.10.

Integrating first with respect to y, from $y_1(x) \equiv 0$ to $y_2(x) = 2x$, we obtain

$$\int_0^2 \int_{y/2}^{1} y e^{x^3}\, dx\, dy = \int_0^1 \int_0^{2x} y e^{x^3}\, dy\, dx = \int_0^1 \left[\frac{1}{2}y^2\right]_{y=0}^{2x} e^{x^3}\, dx$$

$$= \int_0^1 2x^2 e^{x^3}\, dx = \left[\frac{2}{3}e^{x^3}\right]_{x=0}^{1} = \frac{2}{3}(e - 1).$$ ◆

Properties of Double Integrals

We conclude this section by listing some formal properties of double integrals. Let c be a constant and f and g be continuous functions on a region R on which $f(x, y)$ attains a minimum value m and a maximum value M. Let $a(R)$ denote the area of the region R. If all the indicated integrals exist, then:

$$\iint_R cf(x, y)\,dA = c\iint_R f(x, y)\,dA, \tag{6}$$

$$\iint_R [f(x, y) + g(x, y)]\,dA = \iint_R f(x, y)\,dA + \iint_R g(x, y)\,dA, \tag{7}$$

$$m \cdot a(R) \leqq \iint_R f(x, y)\,dA \leqq M \cdot a(R), \tag{8}$$

$$\iint_R f(x, y)\,dA = \iint_{R_1} f(x, y)\,dA + \iint_{R_2} f(x, y)\,dA. \tag{9}$$

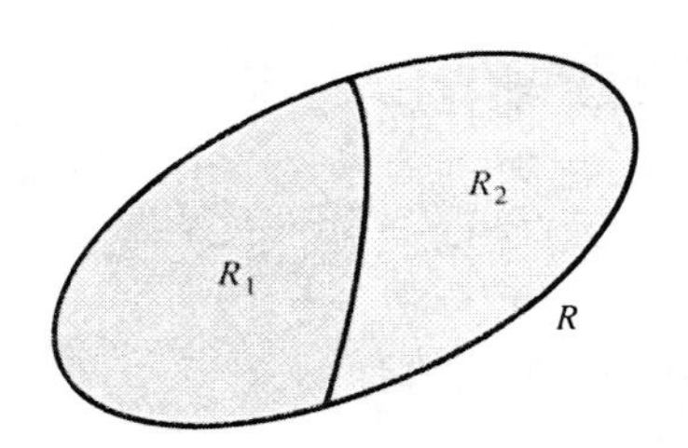

FIGURE 14.2.11 The regions of Eq. (9).

In Eq. (9), R_1 and R_2 are simply two nonoverlapping regions (with disjoint interiors) with union R (Fig. 14.2.11). We indicate in Problems 45 through 48 proofs of the properties in (6) through (9) for the special case in which R is a rectangle.

The property in Eq. (9) enables us to evaluate double integrals over a region R that is neither vertically nor horizontally simple. All that is necessary is to divide R into a finite number of simple regions $R_1, R_2, \dots, R_n$. Then we integrate over each (converting each double integral into an iterated integral, as in the examples of this section) and add the results.

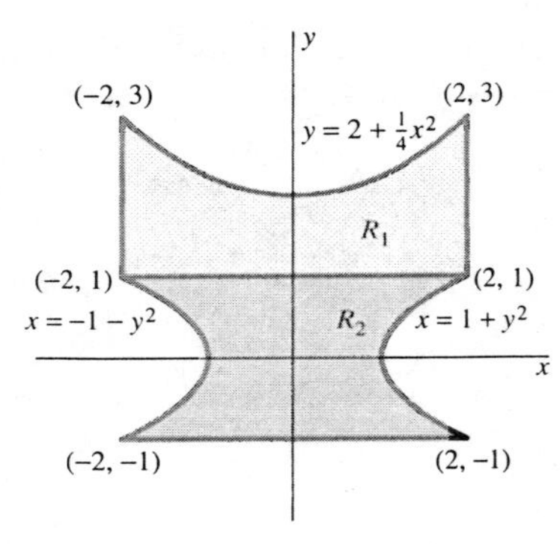

FIGURE 14.2.12 The nonsimple region R is the union of the nonoverlapping simple regions R_1 and R_2.

EXAMPLE 5 Let f be a function that is integrable on the region R of Fig. 14.2.12. Note that R is not simple, but is the union of the vertically simple region R_1 and the horizontally simple region R_2. Using the boundary curves labeled in the figure and the appropriate order of integration for each region, we see that

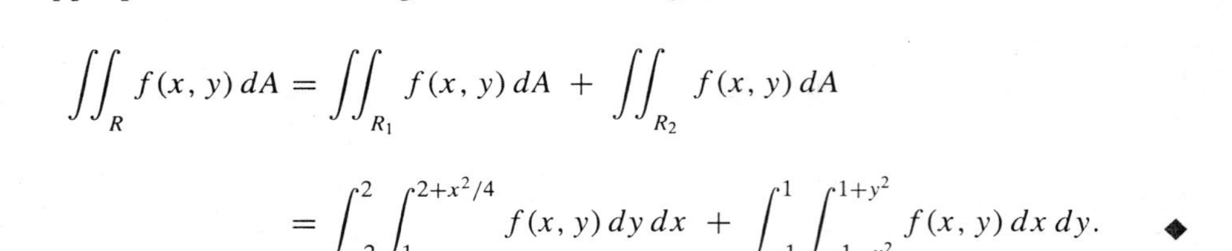

$$\iint_R f(x, y)\,dA = \iint_{R_1} f(x, y)\,dA + \iint_{R_2} f(x, y)\,dA$$

$$= \int_{-2}^{2}\int_{1}^{2+x^2/4} f(x, y)\,dy\,dx + \int_{-1}^{1}\int_{-1-y^2}^{1+y^2} f(x, y)\,dx\,dy. \qquad \blacklozenge$$

14.2 TRUE/FALSE STUDY GUIDE

14.2 CONCEPTS: QUESTIONS AND DISCUSSION

1. Sketch a plane region that is (a) both horizontally simple and vertically simple; (b) horizontally simple but not vertically simple; (c) vertically simple but not horizontally simple; (d) neither horizontally nor vertically simple.
2. Sketch several different regions that are neither horizontally nor vertically simple but can be subdivided into different numbers of nonoverlapping regions, each of which is either horizontally simple or vertically simple. What about an annular region bounded by two concentric circles?
3. Construct several examples of double integrals that are readily evaluated by integrating in one order but not in the reverse order.

14.2 PROBLEMS

Evaluate the iterated integrals in Problems 1 through 14.

1. $\displaystyle\int_0^1 \int_0^x (1+x)\,dy\,dx$

2. $\displaystyle\int_0^2 \int_0^{2x} (1+y)\,dy\,dx$

3. $\displaystyle\int_0^1 \int_y^1 (x+y)\,dx\,dy$ (Fig. 14.2.13)

4. $\displaystyle\int_0^2 \int_{y/2}^1 (x+y)\,dx\,dy$ (Fig. 14.2.14)

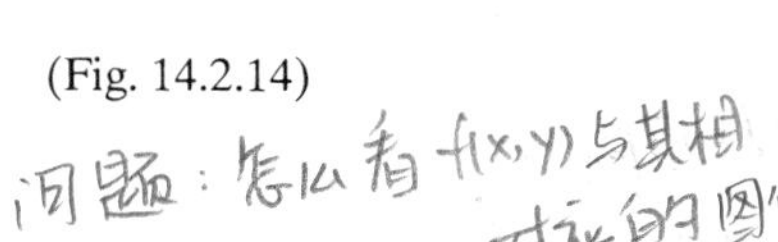

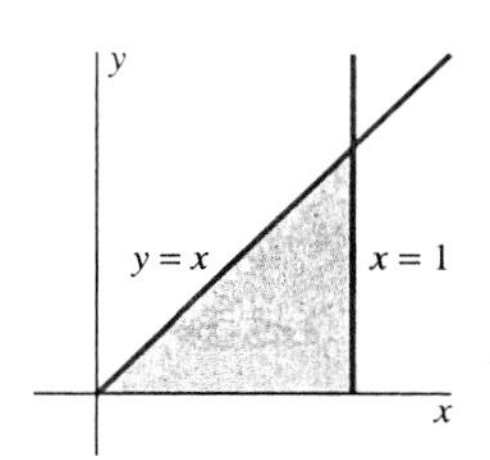

FIGURE 14.2.13
Problem 3.

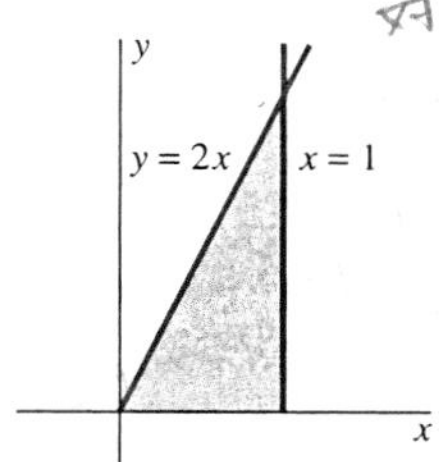

FIGURE 14.2.14
Problem 4.

5. $\displaystyle\int_0^1 \int_0^{x^2} xy\,dy\,dx$

6. $\displaystyle\int_0^1 \int_y^{\sqrt{y}} (x+y)\,dx\,dy$

7. $\displaystyle\int_0^1 \int_x^{\sqrt{x}} (2x-y)\,dy\,dx$ (Fig. 14.2.15)

8. $\displaystyle\int_0^2 \int_{-\sqrt{2y}}^{\sqrt{2y}} (3x+2y)\,dx\,dy$ (Fig. 14.2.16)

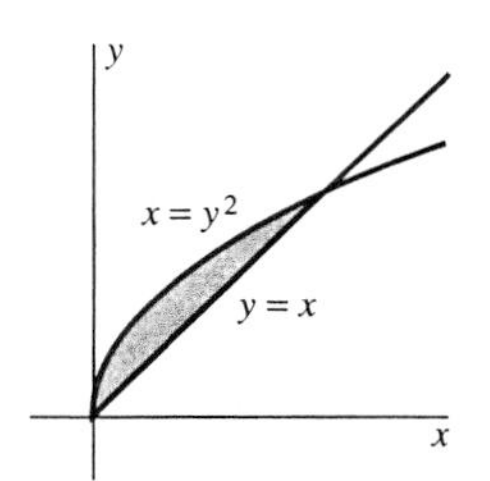

FIGURE 14.2.15
Problem 7.

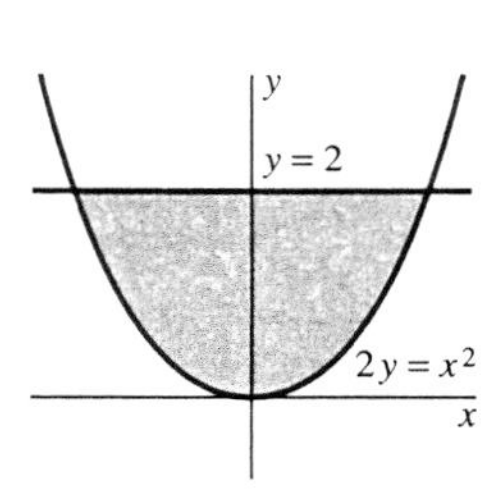

FIGURE 14.2.16
Problem 8.

9. $\displaystyle\int_0^1 \int_{x^4}^x (y-x)\,dy\,dx$

10. $\displaystyle\int_{-1}^2 \int_{-y}^{y+2} (x+2y^2)\,dx\,dy$ (Fig. 14.2.17)

11. $\displaystyle\int_0^1 \int_0^{x^3} e^{y/x}\,dy\,dx$

12. $\displaystyle\int_0^\pi \int_0^{\sin x} y\,dy\,dx$ (Fig. 14.2.18)

13. $\displaystyle\int_0^3 \int_0^y \sqrt{y^2+16}\,dx\,dy$

14. $\displaystyle\int_1^{e^2} \int_0^{1/y} e^{xy}\,dx\,dy$

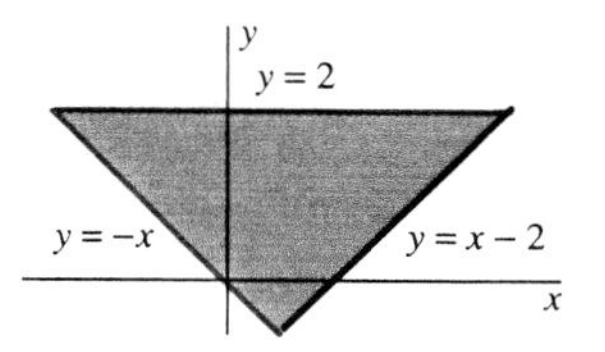

FIGURE 14.2.17
Problem 10.

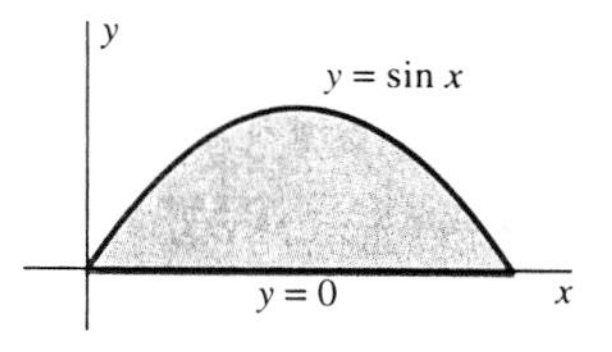

FIGURE 14.2.18
Problem 12.

In Problems 15 through 24, evaluate the integral of the given function $f(x, y)$ over the plane region R that is described.

15. $f(x, y) = xy$; R is bounded by the parabola $y = x^2$ and the line $y = 4$.

16. $f(x, y) = x^2$; R is bounded by the parabola $y = 2 - x^2$ and the line $y = -4$.

17. $f(x, y) = x$; R is bounded by the parabolas $y = x^2$ and $y = 8 - x^2$.

18. $f(x, y) = y$; R is bounded by the parabolas $x = 1 - y^2$ and $x = y^2 - 1$.

19. $f(x, y) = x$; R is bounded by the x-axis and the curve $y = \sin x$, $0 \leqq x \leqq \pi$.

20. $f(x, y) = \sin x$; R is bounded by the x-axis and the curve $y = \cos x$, $-\pi/2 \leqq x \leqq \pi/2$.

21. $f(x, y) = 1/y$; R is the triangle bounded by the lines $y = 1$, $x = e$, and $y = x$.

22. $f(x, y) = xy$; R is the first-quadrant quarter circle bounded by $x^2 + y^2 = 1$ and the coordinate axes.

23. $f(x, y) = 1 - x$; R is the triangle with vertices $(0, 0)$, $(1, 1)$, and $(-2, 1)$.

24. $f(x, y) = 9 - y$; R is the triangle with vertices $(0, 0)$, $(0, 9)$, and $(3, 6)$.

In Problems 25 through 34, first sketch the region of integration, reverse the order of integration as in Examples 3 and 4, and finally evaluate the resulting integral.

25. $\displaystyle\int_{-2}^2 \int_{x^2}^4 x^2 y\,dy\,dx$

26. $\displaystyle\int_0^1 \int_{x^4}^x (x-1)\,dy\,dx$

27. $\displaystyle\int_{-1}^3 \int_{x^2}^{2x+3} x\,dy\,dx$

28. $\displaystyle\int_{-2}^2 \int_{y^2-4}^{4-y^2} y\,dx\,dy$

29. $\displaystyle\int_0^2 \int_{2x}^{4x-x^2} 1\,dy\,dx$

30. $\displaystyle\int_0^1 \int_y^1 e^{-x^2}\,dx\,dy$

31. $\displaystyle\int_0^\pi \int_x^\pi \frac{\sin y}{y}\,dy\,dx$

32. $\displaystyle\int_0^{\sqrt{\pi}} \int_y^{\sqrt{\pi}} \sin x^2\,dx\,dy$

33. $\displaystyle\int_0^1 \int_y^1 \frac{1}{1+x^4}\,dx\,dy$

34. $\displaystyle\int_0^1 \int_{\tan^{-1} y}^{\pi/4} \sec x\,dx\,dy$

In Problems 35 through 40, find the approximate value of

$$\iint_R x\,dA,$$

where R is the region bounded by the two given curves. Before integrating, use a calculator or computer to approximate (graphically or otherwise) the coordinates of the points of intersection of the given curves.

35. $y = x^3 + 1, \quad y = 3x^2$

36. $y = x^4, \quad y = x + 4$

37. $y = x^2 - 1, \quad y = \dfrac{1}{1 + x^2}$

38. $y = x^4 - 16, \quad y = 2x - x^2$

39. $y = x^2, \quad y = \cos x$

40. $y = x^2 - 2x, \quad y = \sin x$

In Problems 41 through 44, the region R is the square with vertices $(\pm 1, 0)$ and $(0, \pm 1)$. Use the symmetry of this region around the coordinate axes to reduce the labor of evaluating the given integrals.

41. $\displaystyle\iint_R x\, dA$

42. $\displaystyle\iint_R x^2\, dA$

43. $\displaystyle\iint_R xy\, dA$

44. $\displaystyle\iint_R (x^2 + y^2)\, dA$

45. Use Riemann sums to prove Eq. (6) for the case in which R is a rectangle with sides parallel to the coordinate axes.

46. Use iterated integrals and familiar properties of single integrals to prove Eq. (7) for the case in which R is a rectangle with sides parallel to the coordinate axes.

47. Use Riemann sums to prove the inequalities in (8) for the case in which R is a rectangle with sides parallel to the coordinate axes.

48. Use iterated integrals and familiar properties of single integrals to prove Eq. (9) if R_1 and R_2 are rectangles with sides parallel to the coordinate axes and the right-hand edge of R_1 is the left-hand edge of R_2.

49. Use Riemann sums to prove that

$$\iint_R f(x, y)\, dA \leqq \iint_R g(x, y)\, dA$$

if $f(x, y) \leqq g(x, y)$ at each point of the region R, a rectangle with sides parallel to the coordinate axes.

50. Suppose that the continuous function f is integrable on the plane region R and that f attains a minimum value m and a maximum value M on R. Assume that R is *connected* in the following sense: For any two points (x_0, y_0) and (x_1, y_1) of R, there is a continuous parametric curve $\mathbf{r}(t)$ in R for which $\mathbf{r}(0) = \langle x_0, y_0 \rangle$ and $\mathbf{r}(1) = \langle x_1, y_1 \rangle$. Let $a(R)$ denote the area of R. Then deduce from (8) the *average value property* of double integrals:

$$\iint_R f(x, y)\, dA = f(\overline{x}, \overline{y}) \cdot a(R)$$

for some point $(\overline{x}, \overline{y})$ of R. [*Suggestion:* If $m = f(x_0, y_0)$ and $M = f(x_1, y_1)$, then you may apply the intermediate value property of the continuous function $f(\mathbf{r}(t))$.]

51. Show by iterated integration that the exact value of the integral in Example 1 is $\frac{2}{3}$.

In Problems 52 and 53, first approximate (as in Example 1) the integral

$$\iint_R f(x, y)\, dA$$

of the given function over the region R bounded by the unit circle and the coordinate axes in the first quadrant, except—unlike Example 1—use an inner partition resulting from the use of $n = 5$ subintervals in each direction. Then use iterated integrals to calculate the exact value of the double integral.

52. $f(x, y) = xy$

53. $f(x, y) = xy \exp(y^2)$

14.3 AREA AND VOLUME BY DOUBLE INTEGRATION

Our definition of the double integral $\iint_R f(x, y)\, dA$ was *motivated* in Section 14.2 by the problem of computing the volume of the solid

$$T = \{(x, y, z) \mid (x, y) \in R \text{ and } 0 \leqq z \leqq f(x, y)\}$$

that lies below the surface $z = f(x, y)$ and above the region R in the xy-plane. Such a solid appears in Fig. 14.3.1. Despite this geometric motivation, the actual definition of the double integral as a limit of Riemann sums does not depend on the concept of volume. We may, therefore, turn matters around and use the double integral to *define* volume.

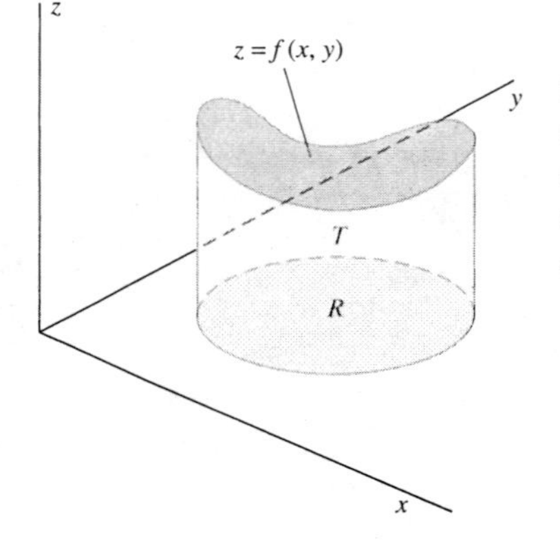

FIGURE 14.3.1 A solid region T with vertical sides and base R in the xy-plane.

DEFINITION Volume below $z = f(x, y)$

Suppose that the function f is continuous and nonnegative on the bounded plane region R. Then the **volume** V of the solid that lies below the surface $z = f(x, y)$ and above the region R is defined to be

$$V = \iint_R f(x, y)\, dA, \tag{1}$$

provided that this integral exists.

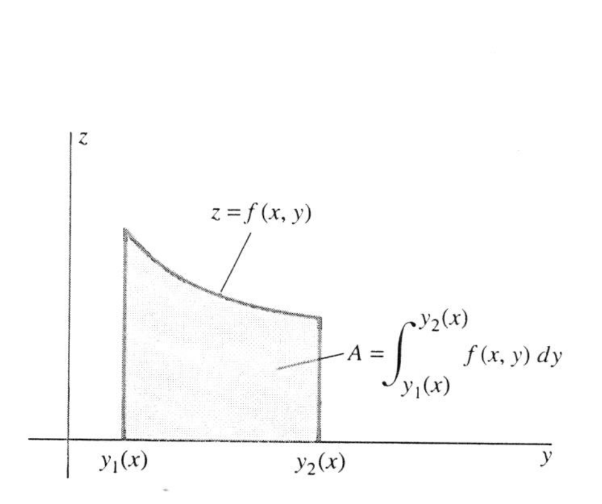

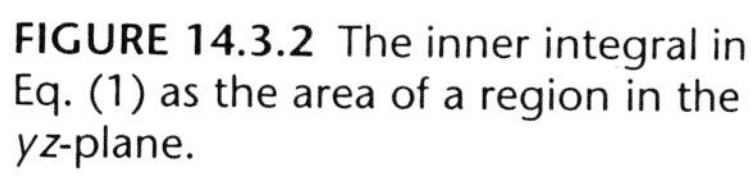

FIGURE 14.3.2 The inner integral in Eq. (1) as the area of a region in the yz-plane.

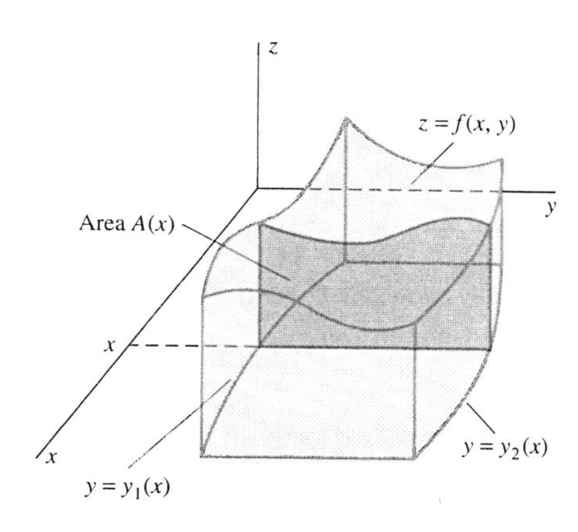

FIGURE 14.3.3 The cross-sectional area is $A = \int_{y_1(x)}^{y_2(x)} f(x, y)\, dy.$

It is of interest to note the connection between this definition and the cross-sectional approach to volume that we discussed in Section 6.2. If, for example, the region R is vertically simple, then the volume integral in Eq. (1) takes the form

$$V = \iint_R z\, dA = \int_a^b \int_{y_1(x)}^{y_2(x)} f(x, y)\, dy\, dx$$

in terms of iterated integrals. The inner integral

$$A(x) = \int_{y_1(x)}^{y_2(x)} f(x, y)\, dy$$

is equal to the area of the region in the yz-plane that lies below the curve

$$z = f(x, y) \quad (x \text{ fixed})$$

and above the interval $y_1(x) \leqq y \leqq y_2(x)$ (Fig. 14.3.2). But this is the projection of the cross section shown in Fig. 14.3.3. Hence the value of the inner integral is simply the area of the cross section of the solid region T in a plane perpendicular to the x-axis. Thus

$$V = \int_a^b A(x)\, dx,$$

and so in this case Eq. (1) reduces to "volume is the integral of cross-sectional area."

Volume by Iterated Integrals

A three-dimensional region T is typically described in terms of the surfaces that bound it. The first step in applying Eq. (1) to compute the volume V of such a region is to determine the region R in the xy-plane over which T lies. The second step is to determine the appropriate order of integration. This may be done in the following way:

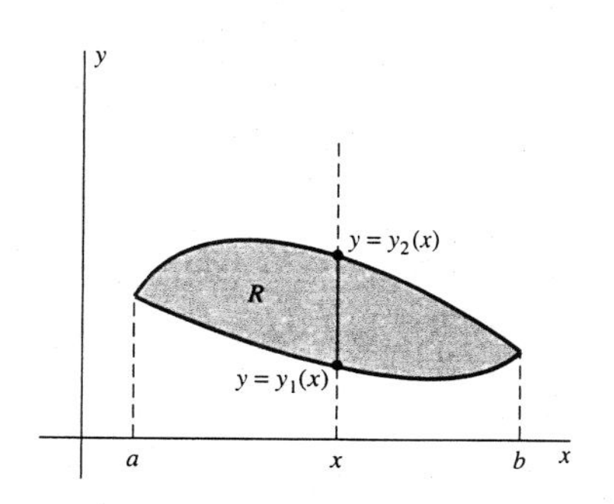

FIGURE 14.3.4 A vertically simple region.

If each vertical line in the xy-plane meets R in a *single* line segment, then R is vertically simple, and you may integrate first with respect to y. The limits on y will be the y-coordinates $y_1(x)$ and $y_2(x)$ of the endpoints of this line segment. (See Fig. 14.3.4.) The limits on x will be the endpoints a and b of the interval on the x-axis onto which R projects. Theorem 2 of Section 14.2 then gives

$$V = \iint_R f(x, y)\, dA = \int_a^b \int_{y_1(x)}^{y_2(x)} f(x, y)\, dy\, dx. \qquad \textbf{(2)}$$

Alternatively,

> If each horizontal line in the xy-plane meets R in a *single* line segment, then R is horizontally simple, and you may integrate first with respect to x. In this case
>
> $$V = \iint_R f(x, y)\,dA = \int_c^d \int_{x_1(y)}^{x_2(y)} f(x, y)\,dx\,dy. \tag{3}$$
>
> As indicated in Fig. 14.3.5, $x_1(y)$ and $x_2(y)$ are the x-coordinates of the endpoints of this horizontal line segment, and c and d are the endpoints of the corresponding interval on the y-axis.

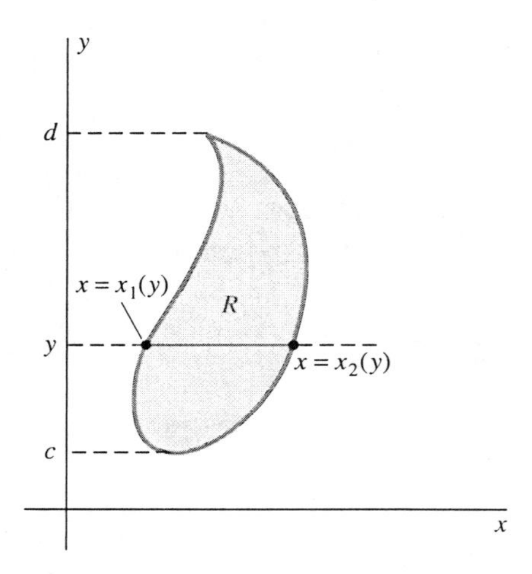

FIGURE 14.3.5 A horizontally simple region.

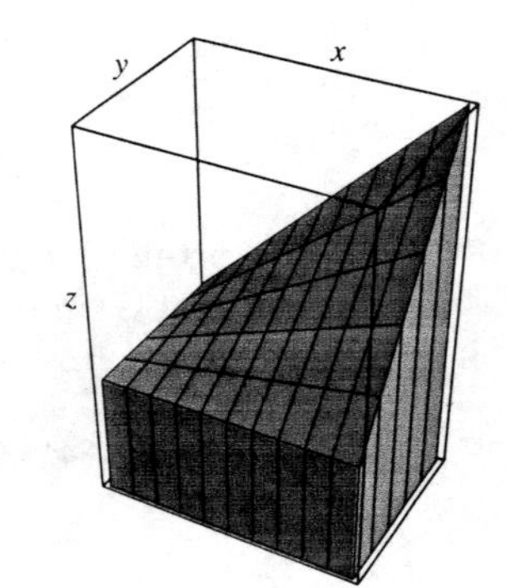

FIGURE 14.3.6 The solid of Example 1.

EXAMPLE 1 The rectangle R in the xy-plane consists of those points (x, y) for which $0 \leqq x \leqq 2$ and $0 \leqq y \leqq 1$. Find the volume V of the solid that lies below the surface $z = 1 + xy$ and above R (Fig. 14.3.6).

Solution Here $f(x, y) = 1 + xy$, so Eq. (1) yields

$$V = \iint_R z\,dA = \int_0^2 \int_0^1 (1 + xy)\,dy\,dx$$

$$= \int_0^2 \left[y + \frac{1}{2}xy^2 \right]_{y=0}^{1} dx = \int_0^2 \left(1 + \frac{1}{2}x \right) dx = \left[x + \frac{1}{4}x^2 \right]_0^2 = 3. \quad \blacklozenge$$

The special case $f(x, y) \equiv 1$ in Eq. (1) gives the area

$$A = a(R) = \iint_R 1\,dA = \iint_R dA \tag{4}$$

of the plane region R. In this case the solid region T resembles a desert mesa (Fig. 14.3.7)—a solid cylinder with base R of area A and height 1. The volume of any such cylinder—not necessarily circular—is the product of its height and the area of its base. In this case, the iterated integrals in Eqs. (2) and (3) reduce to

$$A = \int_a^b \int_{y_{\text{bot}}}^{y_{\text{top}}} 1\,dy\,dx \quad \text{and} \quad A = \int_c^d \int_{x_{\text{left}}}^{x_{\text{right}}} 1\,dx\,dy,$$

respectively.

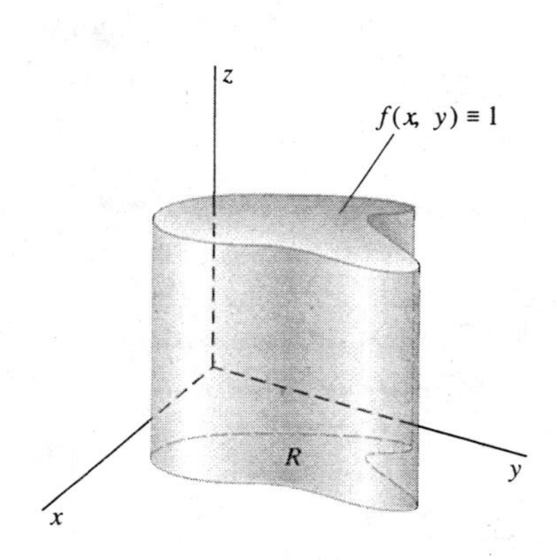

FIGURE 14.3.7 The mesa.

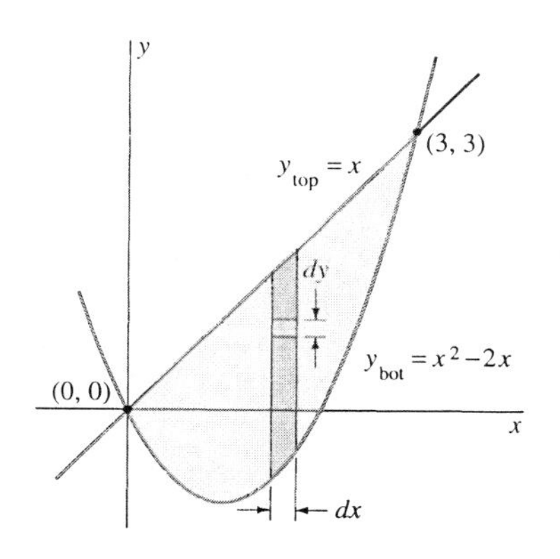

FIGURE 14.3.8 The region R of Example 2.

EXAMPLE 2 Compute by double integration the area A of the region R in the xy-plane that is bounded by the parabola $y = x^2 - 2x$ and the line $y = x$.

Solution As indicated in Fig. 14.3.8, the line $y_{top} = x$ and the parabola $y_{bot} = x^2 - 2x$ intersect at the points (0, 0) and (3, 3). (These coordinates are easy to find by solving the equation $y_{top} = y_{bot}$.) Therefore,

$$A = \int_a^b \int_{y_{bot}}^{y_{top}} 1\,dy\,dx = \int_0^3 \int_{x^2-2x}^{x} 1\,dy\,dx$$

$$= \int_0^3 \Big[y\Big]_{y=x^2-2x}^{x} dx = \int_0^3 (3x - x^2)\,dx = \left[\frac{3}{2}x^2 - \frac{1}{3}x^3\right]_0^3 = \frac{9}{2}.$$ ◆

EXAMPLE 3 Find the volume of the wedge-shaped solid T that lies above the xy-plane, below the plane $z = x$, and within the cylinder $x^2 + y^2 = 4$. This wedge is shown in Fig. 14.3.9.

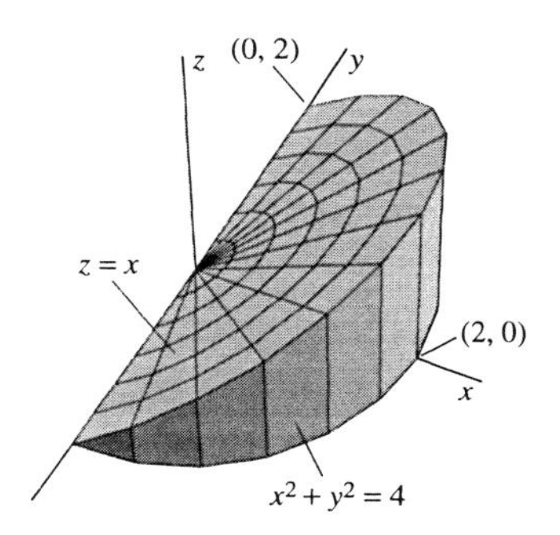

FIGURE 14.3.9 The wedge of Example 3.

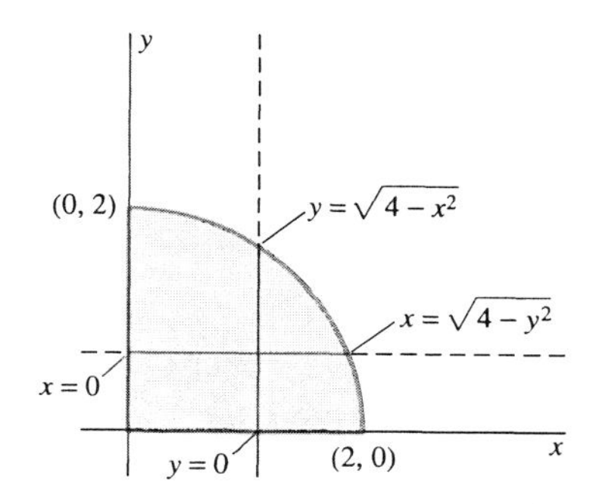

FIGURE 14.3.10 *Half* of the base R of the wedge (Example 3).

Solution The base region R is a semicircle of radius 2, but by symmetry we may integrate over the first-quadrant quarter circle S alone and then double the result. A sketch of the quarter circle (Fig. 14.3.10) helps establish the limits of integration. We could integrate in either order, but integrating with respect to x first gives a slightly simpler computation of the volume V:

$$V = \iint_S z\,dA = 2\int_0^2 \int_0^{\sqrt{4-y^2}} x\,dx\,dy = 2\int_0^2 \left[\frac{1}{2}x^2\right]_{x=0}^{\sqrt{4-y^2}} dy$$

$$= \int_0^2 (4 - y^2)\,dy = \left[4y - \frac{1}{3}y^3\right]_0^2 = \frac{16}{3}.$$

As an exercise, you should integrate in the other order and verify that the result is the same. ◆

Volume Between Two Surfaces

Suppose now that the solid region T lies above the plane region R, as before, but *between* the surfaces $z = z_1(x, y)$ and $z = z_2(x, y)$, where $z_1(x, y) \leqq z_2(x, y)$ for all (x, y) in R (Fig. 14.3.11). Then we get the volume V of T by subtracting the volume below $z = z_1(x, y)$ from the volume below $z = z_2(x, y)$, so

$$V = \iint_R [z_2(x, y) - z_1(x, y)]\,dA. \tag{5}$$

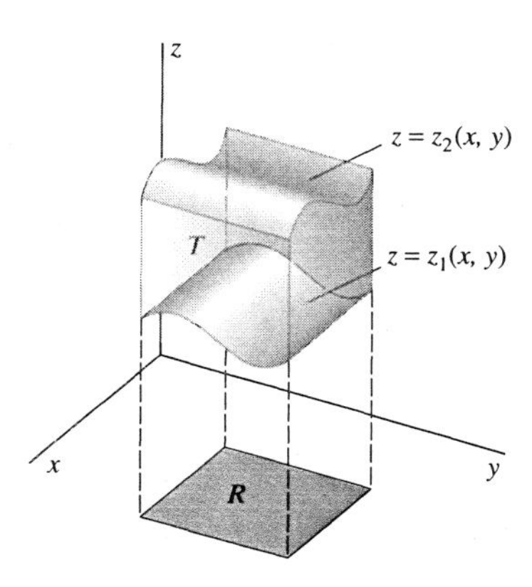

FIGURE 14.3.11 The solid T has vertical sides and is bounded above and below by surfaces.

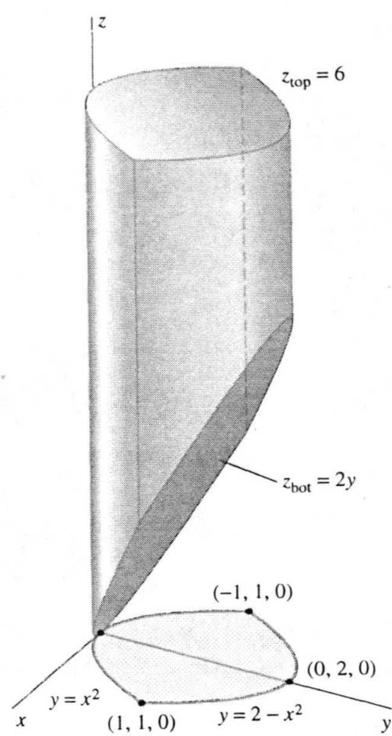

FIGURE 14.3.12 The solid T of Example 4.

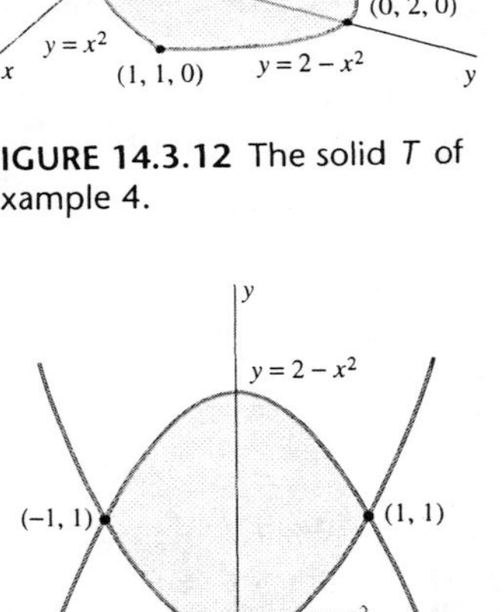
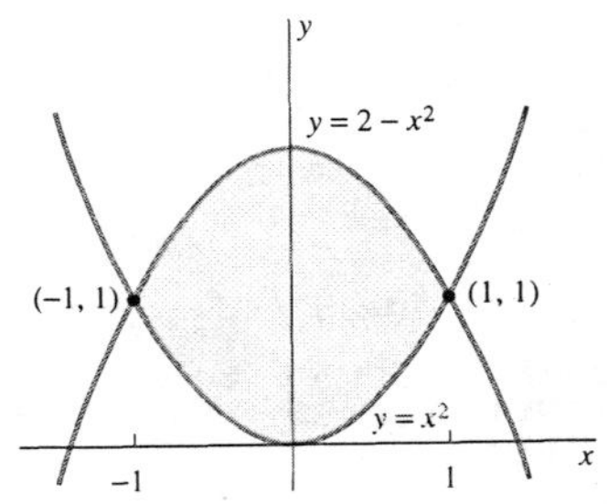

FIGURE 14.3.13 The region R of Example 4.

More briefly,

$$V = \iint_R (z_{\text{top}} - z_{\text{bot}})\, dA$$

where $z_{\text{top}} = z_2(x, y)$ describes the top surface and $z_{\text{bot}} = z_1(x, y)$ the bottom surface of T. This is a natural generalization of the formula for the area of the plane region between the curves $y = z_1(x)$ and $y = z_2(x)$ over the interval $[a, b]$. Moreover, like that formula, Eq. (5) is valid even if $z_1(x, y)$, or both $z_1(x, y)$ and $z_2(x, y)$, are negative over part or all of the region R.

EXAMPLE 4 Find the volume V of the solid T bounded by the planes $z = 6$ and $z = 2y$ and by the parabolic cylinders $y = x^2$ and $y = 2 - x^2$. This solid is sketched in Fig. 14.3.12.

Solution Because the given parabolic cylinders are perpendicular to the xy-plane, the solid T has vertical sides. Thus we may think of T as lying between the planes $z_{\text{top}} = 6$ and $z_{\text{bot}} = 2y$ and above the xy-plane region R that is bounded by the parabolas $y = x^2$ and $y = 2 - x^2$. As indicated in Fig. 14.3.13, these parabolas intersect at the points $(-1, 1)$ and $(1, 1)$.

Integrating first with respect to y (for otherwise we would need two integrals), we get

$$\begin{aligned} V = \iint_R (z_{\text{top}} - z_{\text{bot}})\, dA &= \int_{-1}^{1} \int_{x^2}^{2-x^2} (6 - 2y)\, dy\, dx \\ &= 2 \int_0^1 \Big[6y - y^2 \Big]_{y=x^2}^{2-x^2} dx \qquad \text{(by symmetry)} \\ &= 2 \int_0^1 \left([6 \cdot (2 - x^2) - (2 - x^2)^2] - [6x^2 - x^4] \right) dx \\ &= 2 \int_0^1 (8 - 8x^2)\, dx = 16 \Big[x - \tfrac{1}{3}x^3 \Big]_0^1 = \frac{32}{3}. \end{aligned}$$

◆

14.3 TRUE/FALSE STUDY GUIDE

14.3 CONCEPTS: QUESTIONS AND DISCUSSION

These questions involve "trick integrals." In each case the region R of integration is the unit disk $x^2 + y^2 \leqq 1$ in the xy-plane, and the evaluation of the double integral by means of iterated single integrals might be tedious. But you should be able to evaluate the integral mentally *either by visualizing the volume represented by the integral or by exploiting symmetry (or both). Do so.*

1. $\displaystyle\iint_R \sqrt{1 - x^2 - y^2}\, dA$

2. $\displaystyle\iint_R (10 - x + y)\, dA$

3. $\displaystyle\iint_R \left(1 - \sqrt{x^2 + y^2}\right) dA$

4. $\displaystyle\iint_R \sqrt{x^2 + y^2}\, dA$

5. $\displaystyle\iint_R (5 - x^2 \sin x + y^3 \cos y)\, dA$

14.3 PROBLEMS

In Problems 1 through 10, use double integration to find the area of the region in the xy-plane bounded by the given curves.

1. $y = x, \quad y^2 = x$

2. $y = x, \quad y = x^4$

3. $y = x^2, \quad y = 2x + 3$ (Fig. 14.3.14)

4. $y = 2x + 3, \quad y = 6x - x^2$ (Fig. 14.3.15)

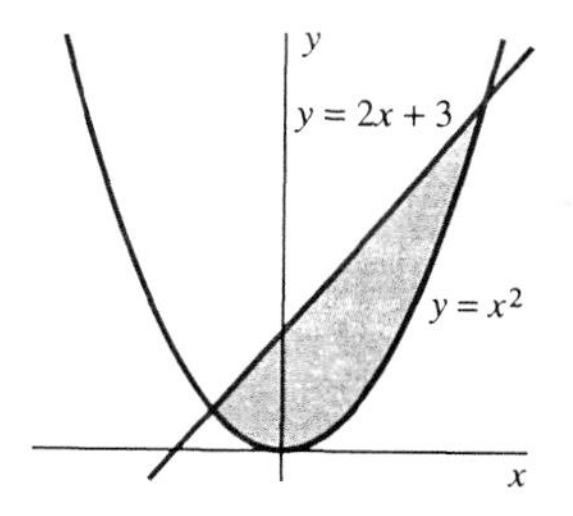

FIGURE 14.3.14 Problem 3.

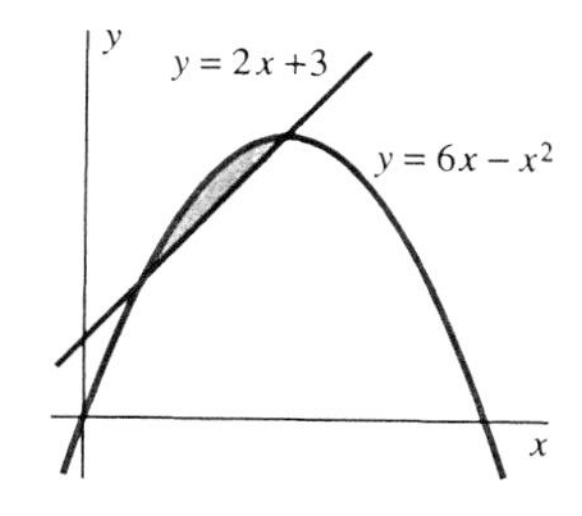

FIGURE 14.3.15 Problem 4.

5. $y = x^2, \quad x + y = 2, \quad y = 0$

6. $y = (x - 1)^2, \quad y = (x + 1)^2, \quad y = 0$

7. $y = x^2 + 1, \quad y = 2x^2 - 3$ (Fig. 14.3.16)

8. $y = x^2 + 1, \quad y = 9 - x^2$ (Fig. 14.3.17)

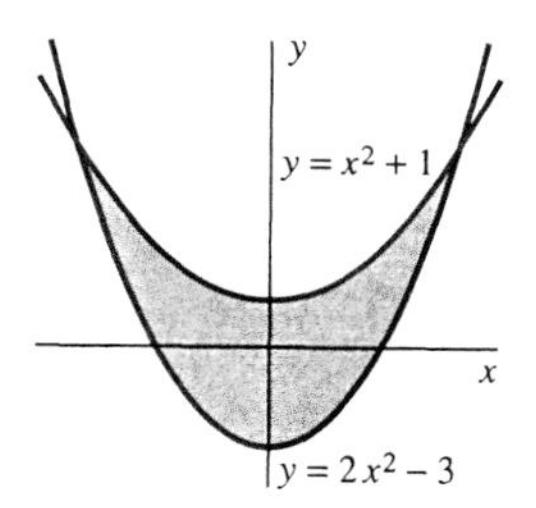

FIGURE 14.3.16 Problem 7.

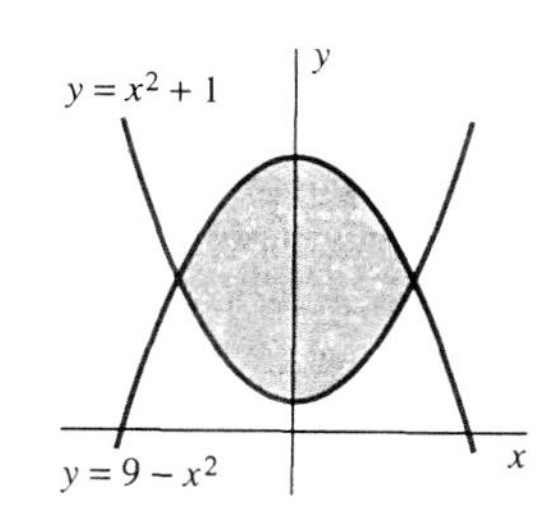

FIGURE 14.3.17 Problem 8.

9. $y = x, \quad y = 2x, \quad xy = 2$

10. $y = x^2, \quad y = \dfrac{2}{1 + x^2}$

In Problems 11 through 26, find the volume of the solid that lies below the surface $z = f(x, y)$ and above the region in the xy-plane bounded by the given curves.

11. $z = 1 + x + y; \quad x = 0, \ x = 1, \ y = 0, \ y = 1$

12. $z = 2x + 3y; \quad x = 0, \ x = 3, \ y = 0, \ y = 2$

13. $z = y + e^x; \quad x = 0, \ x = 1, \ y = 0, \ y = 2$

14. $z = 3 + \cos x + \cos y; \quad x = 0, \ x = \pi, \ y = 0, \ y = \pi$ (Fig. 14.3.18)

15. $z = x + y; \quad x = 0, \ y = 0, \ x + y = 1$

16. $z = 3x + 2y; \quad x = 0, \ y = 0, \ x + 2y = 4$

17. $z = 1 + x + y; \quad x = 1, \ y = 0, \ y = x^2$

18. $z = 2x + y; \quad x = 0, \ y = 1, \ x = \sqrt{y}$

19. $z = x^2; \quad y = x^2, \ y = 1$

20. $z = y^2; \quad x = y^2, \ x = 4$

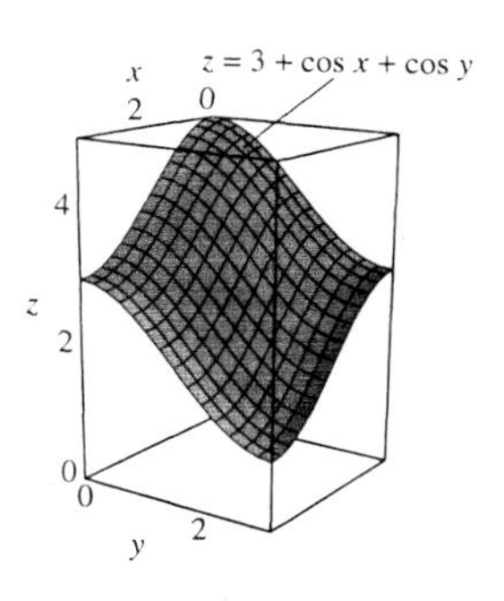

FIGURE 14.3.18 The surface of Problem 14.

21. $z = x^2 + y^2; \quad x = 0, \ x = 1, \ y = 0, \ y = 2$

22. $z = 1 + x^2 + y^2; \quad y = x, \ y = 2 - x^2$

23. $z = 9 - x - y; \quad y = 0, \ x = 3, \ y = \frac{2}{3}x$

24. $z = 10 + y - x^2; \quad y = x^2, \ x = y^2$

25. $z = 4x^2 + y^2; \quad x = 0, \ y = 0, \ 2x + y = 2$

26. $z = 2x + 3y; \quad y = x^2, \ y = x^3$

In Problems 27 through 30, find the volume of the given solid.

27. The solid is bounded by the planes $x = 0$, $y = 0$, $z = 0$, and $3x + 2y + z = 6$.

28. The solid is bounded by the planes $y = 0$, $z = 0$, $y = 2x$, and $4x + 2y + z = 8$.

29. The solid lies under the hyperboloid $z = xy$ and above the triangle in the xy-plane with vertices (1, 2), (1, 4), and (5, 2).

30. The solid lies under the paraboloid $z = 25 - x^2 - y^2$ and above the triangle in the xy-plane with vertices (−3, −4), (−3, 4), and (5, 0).

In Problems 31 through 34, first set up an iterated integral that gives the volume of the given solid. Then use a computer algebra system (if available) to evaluate this integral.

31. The solid lies inside the cylinder $x^2 + y^2 = 1$, above the xy-plane, and below the plane $z = x + 1$ (Fig. 14.3.19).

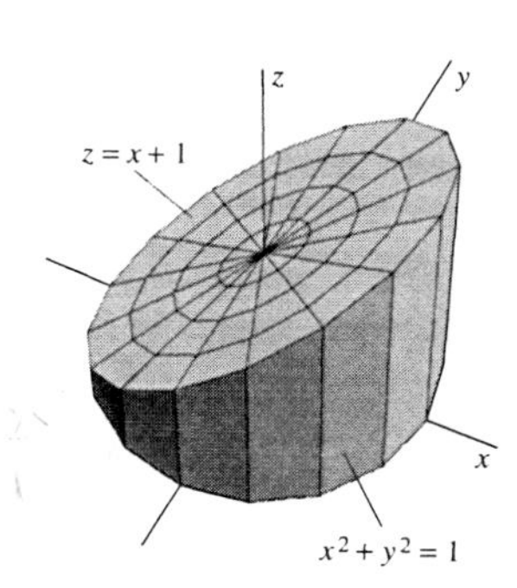

FIGURE 14.3.19 The solid of Problem 31.

32. The solid lies above the xy-plane and below the paraboloid $z = 9 - x^2 - y^2$.

33. The solid lies inside both the cylinder $x^2 + y^2 = 1$ and the sphere $x^2 + y^2 + z^2 = 4$.

34. The solid lies inside the sphere $x^2 + y^2 + z^2 = 2$ and above the paraboloid $z = x^2 + y^2$.

35. Use double integration to find the volume of the tetrahedron in the first octant that is bounded by the coordinate planes and the plane with equation

$$\frac{x}{a} + \frac{y}{b} + \frac{z}{c} = 1$$

(Fig. 14.3.20). The numbers a, b, and c are positive constants.

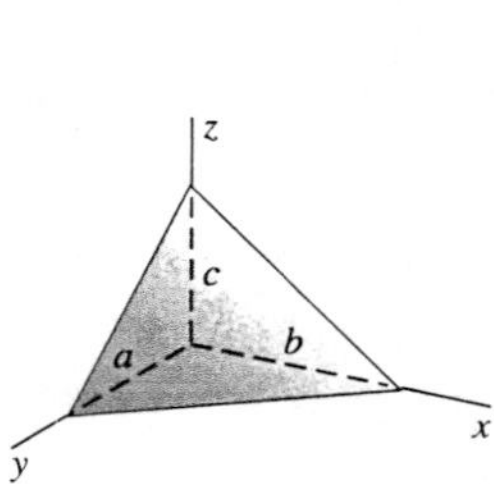

FIGURE 14.3.20 The tetrahedron of Problem 35.

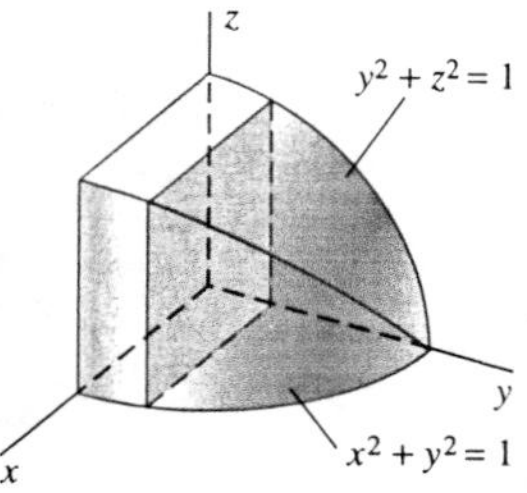

FIGURE 14.3.21 The solid of Problem 37.

36. Suppose that $h > a > 0$. Show that the volume of the solid bounded by the cylinder $x^2 + y^2 = a^2$, the plane $z = 0$, and the plane $z = x + h$ is $V = \pi a^2 h$.

37. Find the volume of the first octant part of the solid bounded by the cylinders $x^2 + y^2 = 1$ and $y^2 + z^2 = 1$ (Fig. 14.3.21). [*Suggestion:* One order of integration is considerably easier than the other.]

38. Find by double integration the volume of the solid bounded by the surfaces $y = \sin x$, $y = -\sin x$, $z = \sin x$, and $z = -\sin x$ for $0 \leqq x \leqq \pi$.

In Problems 39 through 45, you may consult Chapter 7 or the integral table inside the back cover of this book to find antiderivatives of such expressions as $(a^2 - x^2)^{3/2}$.

39. Find the volume of a sphere of radius a by double integration.

40. Use double integration to find the formula $V = V(a, b, c)$ for the volume of an ellipsoid with semiaxes of lengths a, b, and c.

41. Find the volume of the solid bounded below by the xy-plane and above by the paraboloid $z = 25 - x^2 - y^2$ by evaluating a double integral (Fig. 14.3.22).

42. Find the volume of the solid bounded by the two paraboloids $z = x^2 + 2y^2$ and $z = 12 - 2x^2 - y^2$ (Fig. 14.3.23).

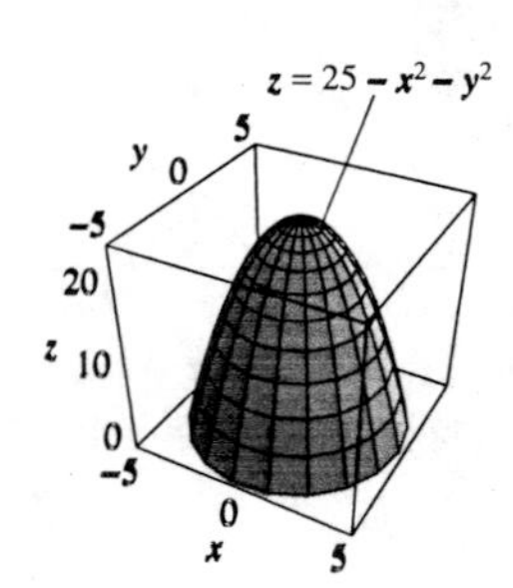

FIGURE 14.3.22 The solid paraboloid of Problem 41.

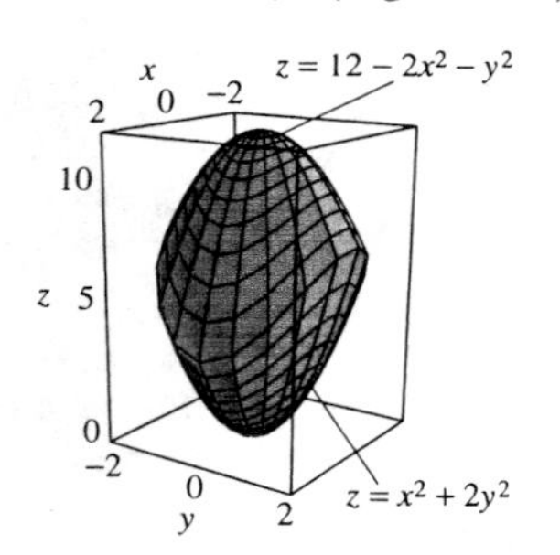

FIGURE 14.3.23 The solid of Problem 42.

43. Find the volume removed when a vertical square hole of edge length R is cut directly through the center of a long horizontal solid cylinder of radius R.

44. Find the volume of the solid bounded by the two surfaces $z = x^2 + 3y^2$ and $z = 4 - y^2$ (Fig. 14.3.24).

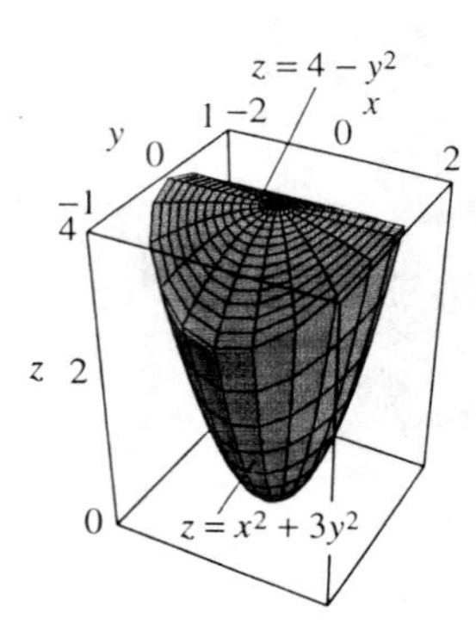

FIGURE 14.3.24 The solid of Problem 44.

45. Find the volume V of the solid T bounded by the parabolic cylinders $z = x^2$, $z = 2x^2$, $y = x^2$, and $y = 8 - x^2$.

In Problems 46 and 47, use a computer algebra system to find (either approximately or exactly) the volume of the solid that lies under the surface $z = f(x, y)$ *and above the region in the* xy*-plane that is bounded by* $y = \cos x$ *and* $y = -\cos x$ *for* $-\pi/2 \leqq x \leqq \pi/2$.

46. $f(x, y) = 4 - x^2 - y^2$ **47.** $f(x, y) = \cos y$

48. Repeat Problem 47, but with $f(x, y) = |\sin x| \cos x$. Also try to exploit symmetry to evaluate the volume integral manually.

In Problems 49 through 51, the equations of a plane and a paraboloid are given. Use a computer algebra system to evaluate the double integral that gives the volume of the solid bounded by the two surfaces.

49. $z = 2x + 3$ and $z = x^2 + y^2$

50. $z = 4x + 4y$ and $z = x^2 + y^2 - 1$

51. $16x + 18y + z = 0$ and $z = 11 - 4x^2 - 9y^2$

52. Suppose that a square hole with sides of length 2 is cut symmetrically through the center of a sphere of radius 2. Use a computer algebra system to compute the volume thereby removed. Show that your result is (exactly or approximately) equal to the exact value

$$V = \frac{4}{3}\left(19\pi + 2\sqrt{2} - 54\tan^{-1}\sqrt{2}\right).$$

53. Suppose that a square hole with sides of length 2 is cut off-center through a sphere of radius 4. Let S be the square cross section of the hole in an equatorial plane of the sphere. The midpoint C of S is at distance 2 from the center of the sphere, and the radius of the sphere that passes through C is perpendicular to two sides of S. Use a computer algebra system to show that about 10% of the whole volume of the sphere is removed when the hole is cut.

14.4 DOUBLE INTEGRALS IN POLAR COORDINATES

A double integral may be easier to evaluate after it has been transformed from rectangular xy-coordinates into polar $r\theta$-coordinates. This is likely to be the case when the region R of integration is a *polar rectangle.* A **polar rectangle** is a region described in polar coordinates by the inequalities

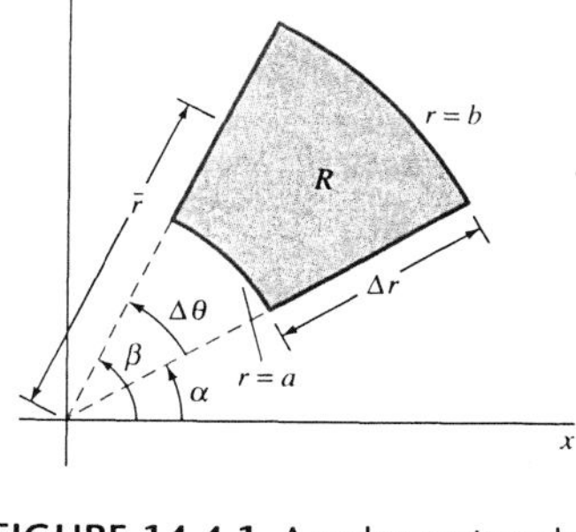

FIGURE 14.4.1 A polar rectangle.

$$a \leqq r \leqq b, \qquad \alpha \leqq \theta \leqq \beta. \tag{1}$$

This polar rectangle is shown in Fig. 14.4.1. If $a = 0$, it is a sector of a circular disk of radius b. If $0 < a < b$, $\alpha = 0$, and $\beta = 2\pi$, it is an annular ring of inner radius a and outer radius b. Because the area of a circular sector with radius r and central angle θ is $\frac{1}{2}r^2\theta$, the area of the polar rectangle in (1) is

$$\begin{aligned} A &= \tfrac{1}{2}b^2(\beta - \alpha) - \tfrac{1}{2}a^2(\beta - \alpha) \\ &= \tfrac{1}{2}(a + b)(a - b)(\beta - \alpha) = \bar{r}\,\Delta r\,\Delta\theta, \end{aligned} \tag{2}$$

where $\Delta r = b - a$, $\Delta\theta = \beta - \alpha$, and $\bar{r} = \frac{1}{2}(a + b)$ is the *average radius* of the polar rectangle.

Suppose that we want to compute the value of the double integral

$$\iint_R f(x, y)\,dA,$$

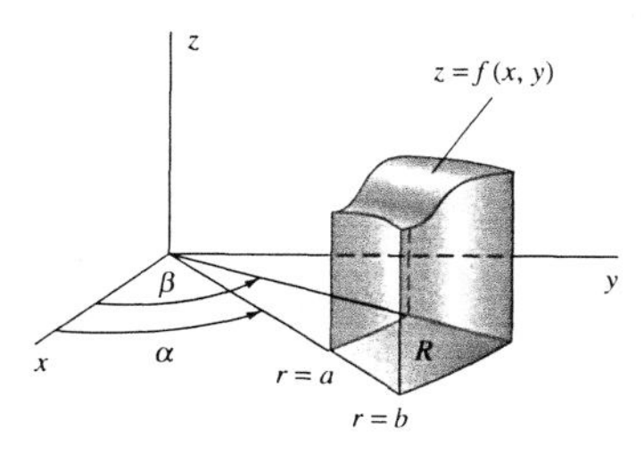

FIGURE 14.4.2 A solid region whose base is the polar rectangle R.

where R is the polar rectangle in (1). Thus we want the volume of the solid with base R that lies below the surface $z = f(x, y)$ (Fig. 14.4.2). We defined in Section 14.1 the double integral as a limit of Riemann sums associated with partitions consisting of ordinary rectangles. We can define the double integral in terms of *polar partitions* as well, made up of polar rectangles. We begin with a partition

$$a = r_0 < r_1 < r_2 < \cdots < r_m = b$$

of $[a, b]$ into m subintervals all having the same length $\Delta r = (b - a)/m$ and a partition

$$\alpha = \theta_0 < \theta_1 < \theta_2 < \cdots < \theta_n = \beta$$

of $[\alpha, \beta]$ into n subintervals all having the same length $\Delta\theta = (\beta - \alpha)/n$. This gives the **polar partition** $\mathcal{P}$ of R into the $k = mn$ polar rectangles $R_1, R_2, \ldots, R_k$ indicated in Fig. 14.4.3. The **norm** $|\mathcal{P}|$ of this polar partition is the maximum of the lengths of the diagonals of its polar subrectangles.

Let the center point of R_i have polar coordinates $(r_i^\star, \theta_i^\star)$, where $r_i^\star$ is the average radius of R_i. Then the rectangular coordinates of this point are $x_i^\star = r_i^\star \cos\theta_i^\star$ and

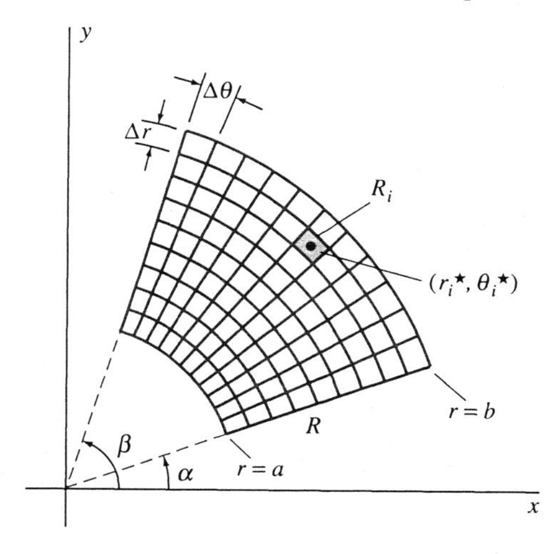

FIGURE 14.4.3 A polar partition of the polar rectangle R.

$y_i^\star = r_i^\star \sin\theta_i^\star$. Therefore the Riemann sum for the function $f(x, y)$ associated with the polar partition $\mathcal{P}$ is

$$\sum_{i=1}^{k} f(x_i^\star, y_i^\star)\,\Delta A_i,$$

where $\Delta A_i = r_i^\star \Delta r\,\Delta\theta$ is the area of the polar rectangle R_i [in part a consequence of Eq. (2)]. When we express this Riemann sum in polar coordinates, we obtain

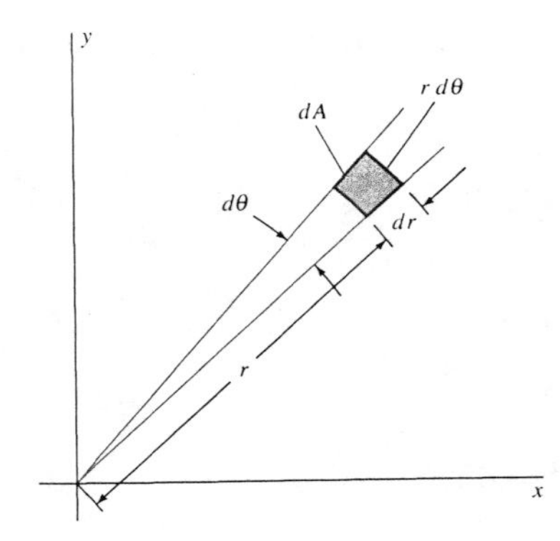

FIGURE 14.4.4 The dimensions of the small polar rectangle suggest that its area is $dA = r\,dr\,d\theta$.

$$\sum_{i=1}^{k} f(x_i^\star, y_i^\star)\,\Delta A_i = \sum_{i=1}^{k} f(r_i^\star\cos\theta_i^\star, r_i^\star\sin\theta_i^\star)\,r_i^\star\,\Delta r\,\Delta\theta$$
$$= \sum_{i=1}^{k} g(r_i^\star, \theta_i^\star)\,\Delta r\,\Delta\theta,$$

where $g(r, \theta) = r \cdot f(r\cos\theta, r\sin\theta)$. This last sum is simply a Riemann sum for the double integral

$$\int_\alpha^\beta \int_a^b g(r, \theta)\,dr\,d\theta = \int_\alpha^\beta \int_a^b f(r\cos\theta, r\sin\theta)\,r\,dr\,d\theta,$$

so it finally follows that

$$\iint_R f(x, y)\,dA = \lim_{|\mathcal{P}|\to 0} \sum_{i=1}^{k} f(x_i^\star, y_i^\star)\,\Delta A_i$$
$$= \lim_{\Delta r, \Delta\theta \to 0} \sum_{i=1}^{k} g(r_i^\star, \theta_i^\star)\,\Delta r\,\Delta\theta = \int_\alpha^\beta \int_a^b g(r, \theta)\,dr\,d\theta.$$

That is,

$$\iint_R f(x, y)\,dA = \int_\alpha^\beta \int_a^b f(r\cos\theta, r\sin\theta)\,r\,dr\,d\theta. \tag{3}$$

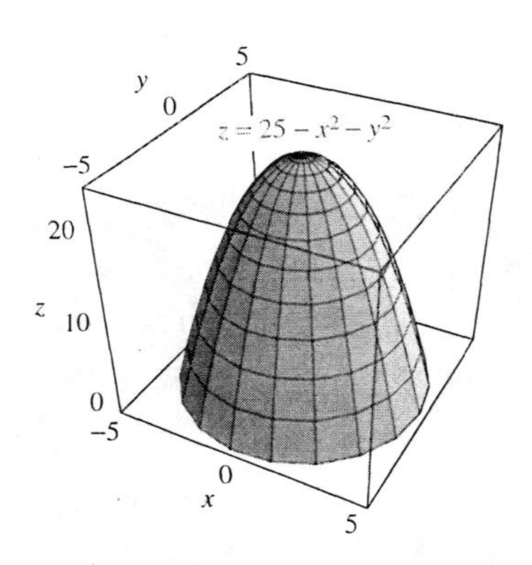

FIGURE 14.4.5 The paraboloid of Example 1.

Thus we formally transform into polar coordinates a double integral over a polar rectangle of the form in (1) by substituting

$$x = r\cos\theta, \qquad y = r\sin\theta, \qquad dA = r\,dr\,d\theta \tag{4}$$

and inserting the appropriate limits of integration on r and θ. In particular, *note the "extra" r on the right-hand side* of Eq. (3). You may remember it by visualizing the "infinitesimal polar rectangle" of Fig. 14.4.4, with "area" $dA = r\,dr\,d\theta$.

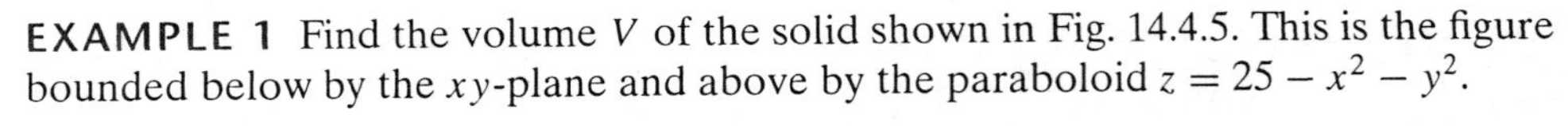

EXAMPLE 1 Find the volume V of the solid shown in Fig. 14.4.5. This is the figure bounded below by the xy-plane and above by the paraboloid $z = 25 - x^2 - y^2$.

Solution The paraboloid intersects the xy-plane in the circle $x^2 + y^2 = 25$. We can compute the volume of the solid by integrating over the quarter of that circle that lies in the first quadrant (Fig. 14.4.6) and then multiplying the result by 4. Thus

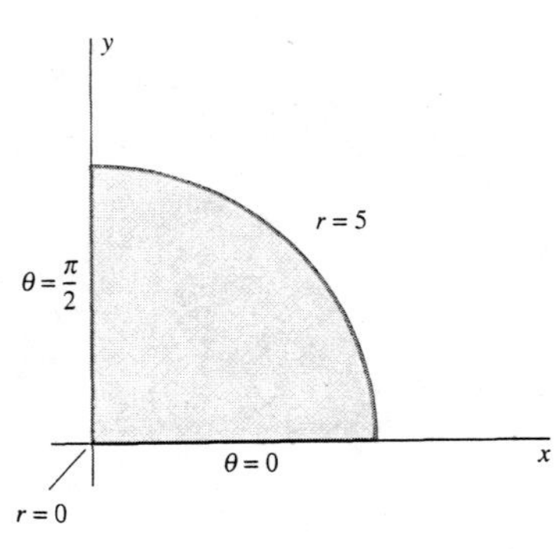

FIGURE 14.4.6 One-fourth of the domain of the integral of Example 1.

$$V = 4\int_0^5 \int_0^{\sqrt{25-x^2}} (25 - x^2 - y^2)\,dy\,dx.$$

There is no difficulty in performing the integration with respect to y, but then we are confronted with the integrals

$$\int \sqrt{25 - x^2}\,dx, \qquad \int x^2\sqrt{25 - x^2}\,dx, \qquad \text{and} \qquad \int (25 - x^2)^{3/2}\,dx.$$

Let us instead transform the original integral into polar coordinates. Because $25 - x^2 - y^2 = 25 - r^2$ and because the quarter of the circular disk in the first quadrant is described by

$$0 \leqq r \leqq 5, \qquad 0 \leqq \theta \leqq \pi/2,$$

Eq. (3) yields the volume

$$\begin{aligned} V &= 4\int_0^{\pi/2}\int_0^5 (25 - r^2)\, r\, dr\, d\theta \\ &= 4\int_0^{\pi/2}\left[\frac{25}{2}r^2 - \frac{1}{4}r^4\right]_{r=0}^5 d\theta = 4\cdot\frac{625}{4}\cdot\frac{\pi}{2} = \frac{625\pi}{2}. \end{aligned}$$ ◆

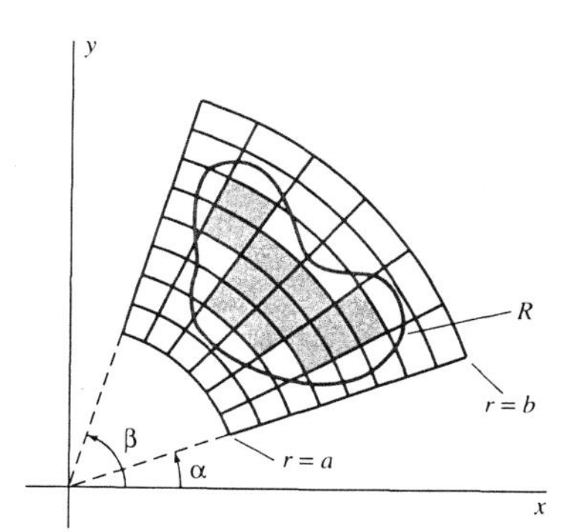

FIGURE 14.4.7 A polar inner partition of the region R.

More General Polar-Coordinate Regions

If R is a more general region, then we can transform into polar coordinates the double integral

$$\iint_R f(x, y)\, dA$$

by expressing it as a limit of Riemann sums associated with "polar inner partitions" of the sort indicated in Fig. 14.4.7. Instead of giving the detailed derivation—a generalization of the preceding derivation of Eq. (3)—we shall simply give the results in one special case of practical importance.

Figure 14.4.8 shows a *radially simple* region R consisting of those points with polar coordinates that satisfy the inequalities

$$\alpha \leqq \theta \leqq \beta, \qquad r_1(\theta) \leqq r \leqq r_2(\theta).$$

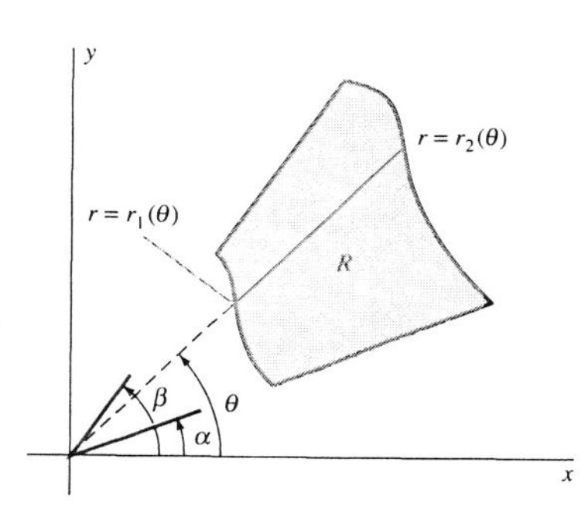

FIGURE 14.4.8 A radially simple region R.

In this case, the formula

$$\iint_R f(x, y)\, dA = \int_\alpha^\beta \int_{r_1(\theta)}^{r_2(\theta)} f(r\cos\theta, r\sin\theta)\, r\, dr\, d\theta \tag{5}$$

gives the evaluation in polar coordinates of a double integral over R (under the usual assumption that the indicated integrals exist). Note that we integrate first with respect to r, with the limits $r_1(\theta)$ and $r_2(\theta)$ being the r-coordinates of a typical radial segment in R (Fig. 14.4.8).

Figure 14.4.9 shows how we can set up the iterated integral on the right-hand side of Eq. (5) in a formal way. First, a typical area element $dA = r\, dr\, d\theta$ is swept radially from $r = r_1(\theta)$ to $r = r_2(\theta)$. Second, the resulting strip is rotated from $\theta = \alpha$ to $\theta = \beta$ to sweep out the region R. Equation (5) yields the volume formula

$$V = \int_\alpha^\beta \int_{r_{\text{inner}}}^{r_{\text{outer}}} z\, r\, dr\, d\theta \tag{6}$$

for the volume V of the solid that lies above the region R of Fig. 14.4.8 and below the surface $z = f(x, y) = f(r\cos\theta, r\sin\theta)$.

Observe that Eqs. (3) and (5) for the evaluation of a double integral in polar coordinates take the form

$$\iint_R f(x, y)\, dA = \iint_S f(r\cos\theta, r\sin\theta)\, r\, dr\, d\theta. \tag{7}$$

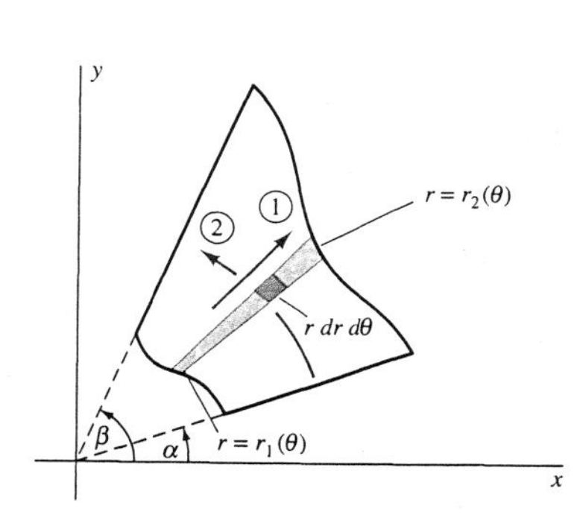

FIGURE 14.4.9 Integrating first with respect to r and then with respect to θ.

The symbol S on the right-hand side represents the appropriate limits on r and θ such that the region R is swept out in the manner indicated in Fig. 14.4.9.

With $f(x, y) \equiv 1$, Eq. (7) reduces to the formula

$$A = a(R) = \iint_S r\, dr\, d\theta \tag{8}$$

for computing the area $a(R)$ of R by double integration in polar coordinates. Note again that the symbol S refers not to a new region in the xy-plane, but to a new description—in terms of polar coordinates—of the original region R.

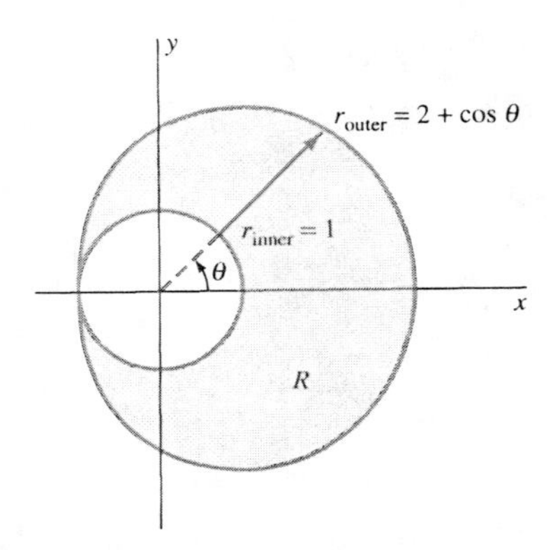

FIGURE 14.4.10 The region R of Example 2.

EXAMPLE 2 Figure 14.4.10 shows the region R bounded on the inside by the circle $r = 1$ and on the outside by the limaçon $r = 2 + \cos\theta$. By following a typical radial line outward from the origin, we see that $r_{\text{inner}} = 1$ and $r_{\text{outer}} = 2 + \cos\theta$. Hence the area of R is

$$
\begin{aligned}
A &= \int_{\alpha}^{\beta}\int_{r_{\text{inner}}}^{r_{\text{outer}}} r\,dr\,d\theta \\
&= 2\int_0^{\pi}\int_1^{2+\cos\theta} r\,dr\,d\theta \qquad \text{(symmetry)} \\
&= 2\int_0^{\pi} \frac{1}{2}[(2+\cos\theta)^2 - 1^2]\,d\theta \\
&= \int_0^{\pi} (3 + 4\cos\theta + \cos^2\theta)\,d\theta \\
&= \int_0^{\pi} \left(3 + 4\cos\theta + \frac{1}{2} + \frac{1}{2}\cos 2\theta\right)d\theta \\
&= \int_0^{\pi} \left(3 + \frac{1}{2}\right)d\theta = \frac{7}{2}\pi.
\end{aligned}
$$

The cosine terms in the next-to-last integral contribute nothing, because upon integration they yield sine terms that are zero at both limits of integration. ◆

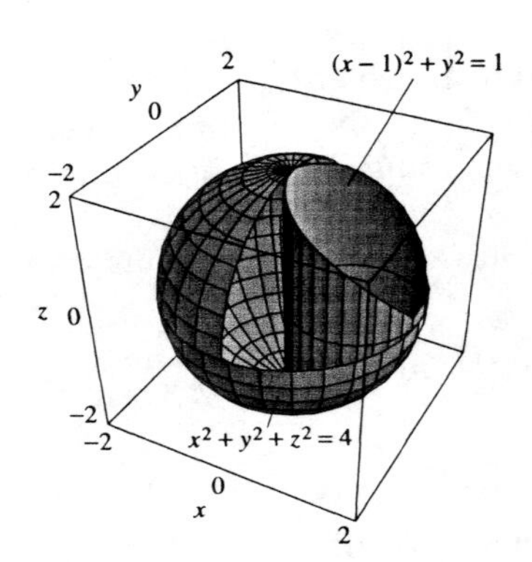

FIGURE 14.4.11 The sphere with off-center hole (Example 3).

EXAMPLE 3 Find the volume of the solid region that is interior to both the sphere $x^2 + y^2 + z^2 = 4$ of radius 2 and the cylinder $(x-1)^2 + y^2 = 1$. This is the volume of material removed when an off-center hole of radius 1 is bored just tangent to a diameter all the way through a sphere of radius 2 (Fig. 14.4.11).

Solution We need to integrate the function $f(x, y) = \sqrt{4 - x^2 - y^2}$ over the disk R that is bounded by the circle with center (1, 0) and radius 1 (Fig. 14.4.12). The desired volume V is twice that of the part above the xy-plane, so

$$
V = 2\iint_R \sqrt{4 - x^2 - y^2}\,dA.
$$

But this integral would be awkward to evaluate in rectangular coordinates, so we change to polar coordinates.

The circle of radius 1 in Fig. 14.4.12 is familiar from Section 10.2; its polar equation is $r = 2\cos\theta$. Therefore the region R is described by the inequalities

$$
0 \leqq r \leqq 2\cos\theta, \qquad -\pi/2 \leqq \theta \leqq \pi/2.
$$

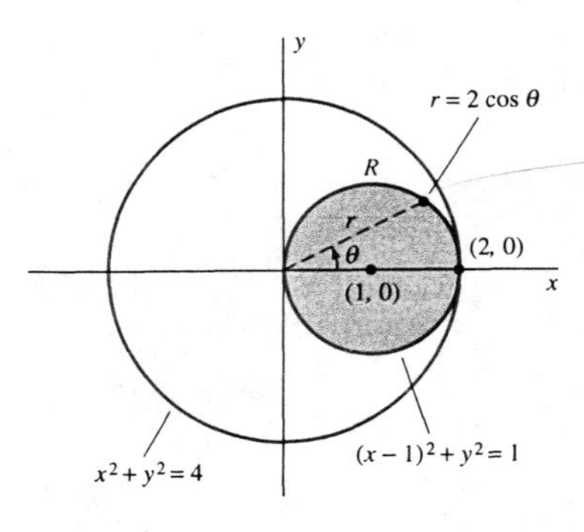

FIGURE 14.4.12 The small circle is the domain R of the integral of Example 3.

We shall integrate only over the upper half of R, taking advantage of the symmetry of the sphere-with-hole. This involves doubling, for a second time, the integral we write. So—using Eq. (5)—we find that

$$
\begin{aligned}
V &= 4\int_0^{\pi/2}\int_0^{2\cos\theta} \sqrt{4 - r^2}\,r\,dr\,d\theta \\
&= 4\int_0^{\pi/2} \left[-\frac{1}{3}(4 - r^2)^{3/2}\right]_{r=0}^{2\cos\theta} d\theta = \frac{32}{3}\int_0^{\pi/2} (1 - \sin^3\theta)\,d\theta.
\end{aligned}
$$

Now we see from Formula (113) inside the back cover that

$$
\int_0^{\pi/2} \sin^3\theta\,d\theta = \frac{2}{3},
$$

and therefore

$$V = \frac{32}{3} \cdot \left(\frac{\pi}{2} - \frac{2}{3} \right) = \frac{16}{3}\pi - \frac{64}{9} \approx 9.64405. \quad \blacklozenge$$

In Example 4 we use a polar-coordinates version of the familiar volume formula

$$V = \iint_R (z_{\text{top}} - z_{\text{bot}})\, dA.$$

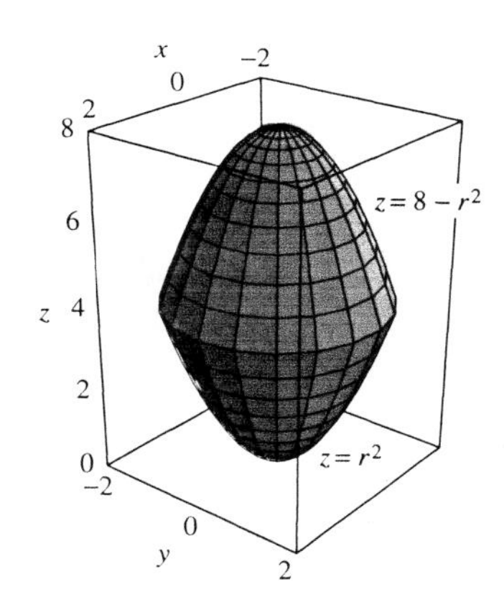

FIGURE 14.4.13 The solid of Example 4.

EXAMPLE 4 Find the volume of the solid that is bounded above by the paraboloid $z = 8 - r^2$ and below by the paraboloid $z = r^2$ (Fig. 14.4.13).

Solution The curve of intersection of the two paraboloids is found by simultaneous solution of the equations of the two surfaces. We eliminate z to obtain

$$r^2 = 8 - r^2; \quad \text{that is,} \quad r^2 = 4.$$

Hence the solid lies above the plane circular disk D with polar description $0 \leqq r \leqq 2$, and so the volume of the solid is

$$V = \iint_D (z_{\text{top}} - z_{\text{bot}})\, dA = \int_0^{2\pi} \int_0^2 [(8 - r^2) - r^2]\, r\, dr\, d\theta$$

$$= \int_0^{2\pi} \int_0^2 (8r - 2r^3)\, dr\, d\theta = 2\pi \left[4r^2 - \frac{1}{2} r^4 \right]_0^2 = 16\pi. \quad \blacklozenge$$

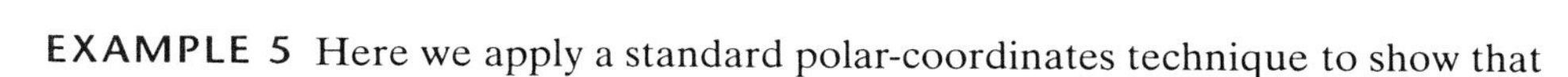

EXAMPLE 5 Here we apply a standard polar-coordinates technique to show that

$$\int_0^{\infty} e^{-x^2}\, dx = \frac{\sqrt{\pi}}{2}. \tag{9}$$

REMARK This important improper integral converges because

$$\int_1^b e^{-x^2}\, dx \leqq \int_1^b e^{-x}\, dx \leqq \int_1^{\infty} e^{-x}\, dx = \frac{1}{e}.$$

(The first inequality is valid because $e^{-x^2} \leqq e^{-x}$ for $x \geqq 1$.) It follows that

$$\int_1^b e^{-x^2}\, dx$$

is a bounded and increasing function of b.

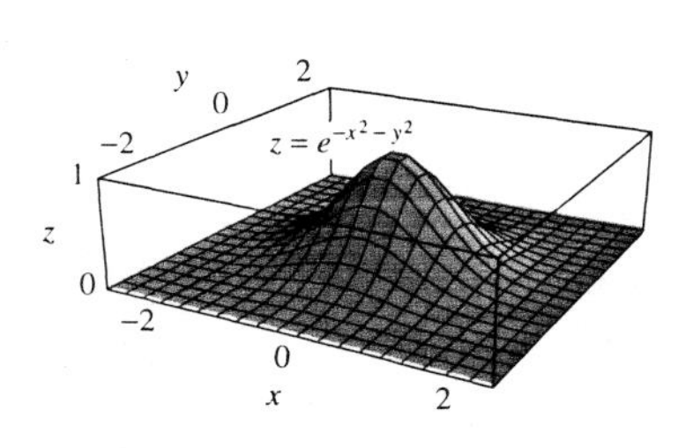

FIGURE 14.4.14 The surface $z = e^{-x^2-y^2}$ (Example 5).

Solution Let V_b denote the volume of the region that lies below the surface $z = e^{-x^2-y^2}$ and above the square with vertices $(\pm b, \pm b)$ in the xy-plane (Fig. 14.4.14). Then

$$V_b = \int_{-b}^{b} \int_{-b}^{b} e^{-x^2-y^2}\, dx\, dy = \int_{-b}^{b} e^{-y^2} \left(\int_{-b}^{b} e^{-x^2}\, dx \right) dy$$

$$= \left(\int_{-b}^{b} e^{-x^2}\, dx \right) \left(\int_{-b}^{b} e^{-y^2}\, dy \right) = \left(\int_{-b}^{b} e^{-x^2}\, dx \right)^2 = 4 \left(\int_0^b e^{-x^2}\, dx \right)^2.$$

It follows that the volume below $z = e^{-x^2-y^2}$ and above the entire xy-plane is

$$V = \lim_{b \to \infty} V_b = \lim_{b \to \infty} 4 \left(\int_0^b e^{-x^2}\, dx \right)^2 = 4 \left(\int_0^{\infty} e^{-x^2}\, dx \right)^2 = 4I^2,$$

where I denotes the value of the improper integral in (9).

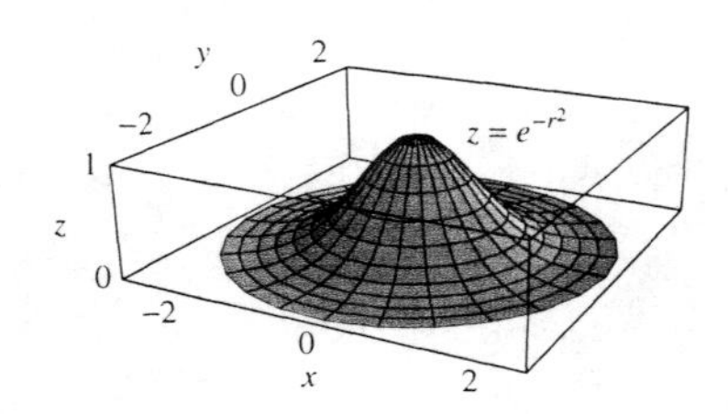

FIGURE 14.4.15 The surface $z = e^{-r^2}$ (Example 5).

Now we compute V by another method—by using polar coordinates. We take the limit, as $b \to +\infty$, of the volume below $z = e^{-x^2-y^2} = e^{-r^2}$ and above the circular disk with center $(0, 0)$ and radius b (Fig. 14.4.15). This disk is described by $0 \leqq r \leqq b, 0 \leqq \theta \leqq 2\pi$, so we obtain

$$V = \lim_{b\to\infty} \int_0^{2\pi} \int_0^b re^{-r^2}\,dr\,d\theta = \lim_{b\to\infty} \int_0^{2\pi} \left[-\frac{1}{2}e^{-r^2}\right]_{r=0}^{b} d\theta$$

$$= \lim_{b\to\infty} \int_0^{2\pi} \frac{1}{2}\left[1 - e^{-b^2}\right] d\theta = \lim_{b\to\infty} \pi\left(1 - e^{-b^2}\right) = \pi.$$

We equate these two values of V, and it follows that $4I^2 = \pi$. Therefore, $I = \frac{1}{2}\sqrt{\pi}$, as desired. ◆

14.4 TRUE/FALSE STUDY GUIDE

14.4 CONCEPTS: QUESTIONS AND DISCUSSION

1. Describe a plane region R such that evaluation of $\iint_R f\,dA$ by iterated integration without subdividing the region R would require the use of rectangular coordinates, and another region such that this would require the use of polar coordinates.
2. Can you describe an integral $\iint_R f\,dA$ such that R is the unit square $0 \leqq x \leqq 1, 0 \leqq y \leqq 1$, but the integral is more easily evaluated in polar coordinates than in rectangular coordinates?
3. Can you describe an integral $\iint_R f\,dA$ such that R is the unit disk $0 \leqq r \leqq 1$ but the integral is more easily evaluated in rectangular coordinates than in polar coordinates?

14.4 PROBLEMS

In Problems 1 through 7, find the indicated area by double integration in polar coordinates.

1. The area bounded by the circle $r = 1$
2. The area bounded by the circle $r = 3\sin\theta$
3. The area bounded by the cardioid $r = 1 + \cos\theta$ (Fig. 14.4.16)
4. The area bounded by one loop of $r = 2\cos 2\theta$ (Fig. 14.4.17)
5. The area inside both the circles $r = 1$ and $r = 2\sin\theta$
6. The area inside $r = 2\cos\theta$ and outside the circle $r = 2$

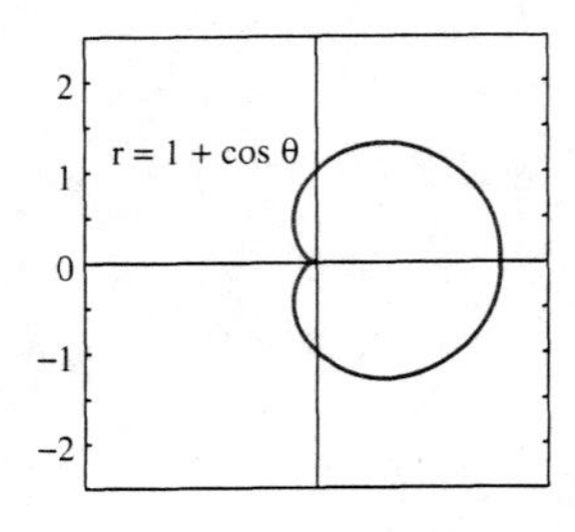

FIGURE 14.4.16 The cardioid of Problem 3.

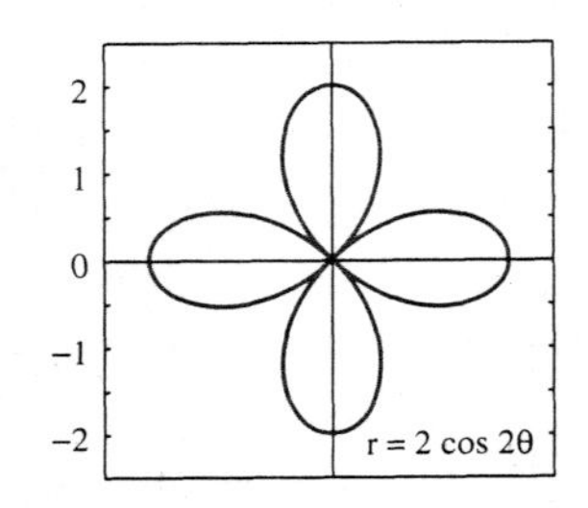

FIGURE 14.4.17 The rose of Problem 4.

7. The area inside the smaller loop of $r = 1 - 2\sin\theta$ (Fig. 14.4.18)

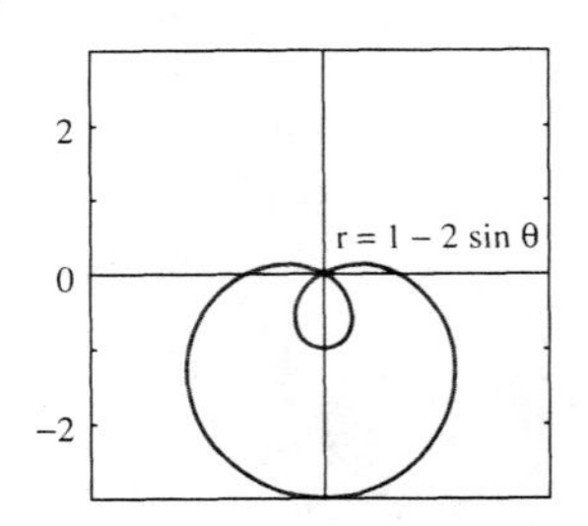

FIGURE 14.4.18 The limaçon of Problem 7.

In Problems 8 through 12, use double integration in polar coordinates to find the volume of the solid that lies below the given surface and above the plane region R bounded by the given curve.

8. $z = x^2 + y^2$; $r = 3$
9. $z = \sqrt{x^2 + y^2}$; $r = 2$
10. $z = x^2 + y^2$; $r = 2\cos\theta$
11. $z = 10 + 2x + 3y$; $r = \sin\theta$
12. $z = a^2 - x^2 - y^2$; $r = a$

In Problems 13 through 18, evaluate the given integral by first converting to polar coordinates.

13. $\displaystyle\int_0^1 \int_0^{\sqrt{1-y^2}} \frac{1}{1+x^2+y^2}\,dx\,dy$ (Fig. 14.4.19)

14. $\displaystyle\int_0^1 \int_0^{\sqrt{1-x^2}} \frac{1}{\sqrt{4-x^2-y^2}}\,dy\,dx$ (Fig. 14.4.19)

15. $\displaystyle\int_0^2 \int_0^{\sqrt{4-x^2}} (x^2+y^2)^{3/2}\,dy\,dx$

16. $\displaystyle\int_0^1 \int_x^1 x^2\,dy\,dx$

17. $\displaystyle\int_0^1 \int_0^{\sqrt{1-y^2}} \sin(x^2+y^2)\,dx\,dy$

18. $\displaystyle\int_1^2 \int_0^{\sqrt{2x-x^2}} \frac{1}{\sqrt{x^2+y^2}}\,dy\,dx$ (Fig. 14.4.20)

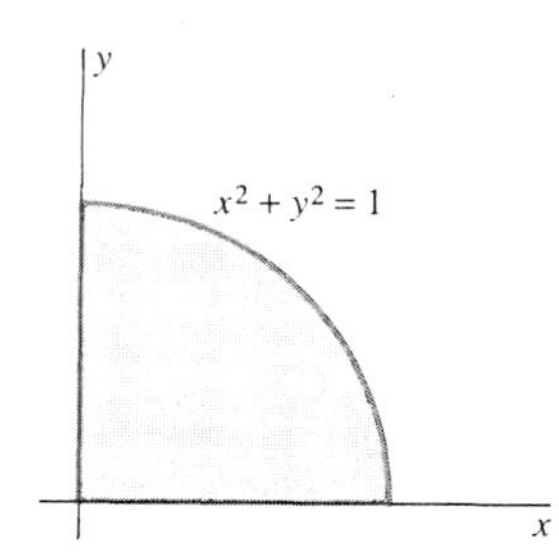

FIGURE 14.4.19 The quarter-circle of Problems 13 and 14.

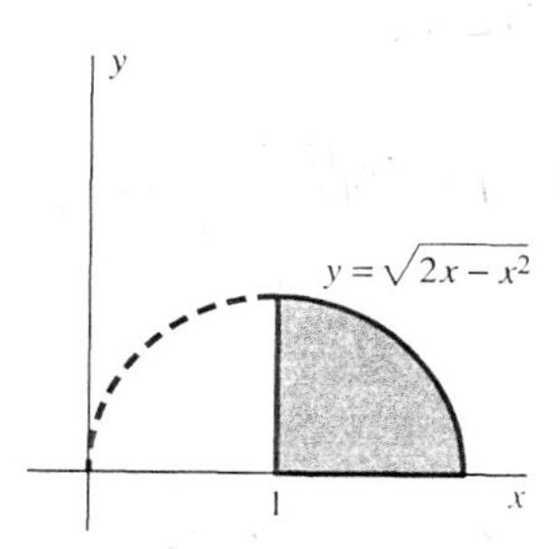

FIGURE 14.4.20 The quarter-circle of Problem 18.

In Problems 19 through 22, find the volume of the solid that is bounded above and below by the given surfaces $z = z_1(x, y)$ and $z = z_2(x, y)$ and lies above the plane region R bounded by the given curve $r = g(\theta)$.

19. $z = 1,\ z = 3 + x + y;\quad r = 1$

20. $z = 2 + x,\ z = 4 + 2x;\quad r = 2$

21. $z = 0,\ z = 3 + x + y;\quad r = 2\sin\theta$

22. $z = 0,\ z = 1 + x;\quad r = 1 + \cos\theta$

Solve Problems 23 through 32 by double integration in polar coordinates.

23. Find the volume of a sphere of radius a by double integration.

24. Find the volume of the solid bounded by the paraboloids $z = 12 - 2x^2 - y^2$ and $z = x^2 + 2y^2$.

25. Suppose that $h > a > 0$. Show that the volume of the solid bounded by the cylinder $x^2 + y^2 = a^2$, the plane $z = 0$, and the plane $z = x + h$ is $V = \pi a^2 h$.

26. Find the volume of the wedge-shaped solid described in Example 3 of Section 14.3 (Fig. 14.4.21).

27. Find the volume bounded by the paraboloids $z = x^2 + y^2$ and $z = 4 - 3x^2 - 3y^2$.

28. Find the volume bounded by the paraboloids $z = x^2 + y^2$ and $z = 2x^2 + 2y^2 - 1$.

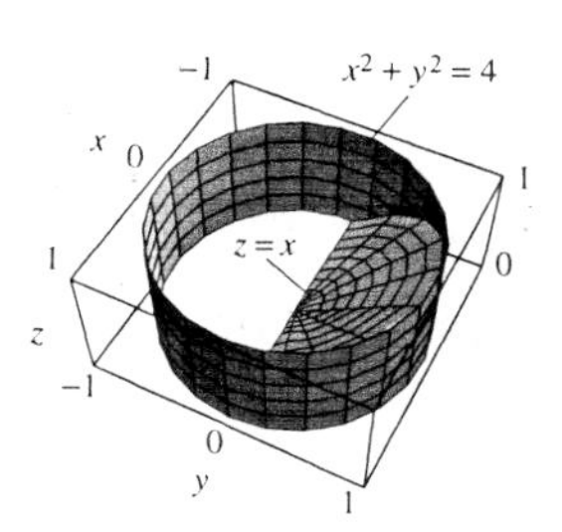

FIGURE 14.4.21 The wedge of Problem 26.

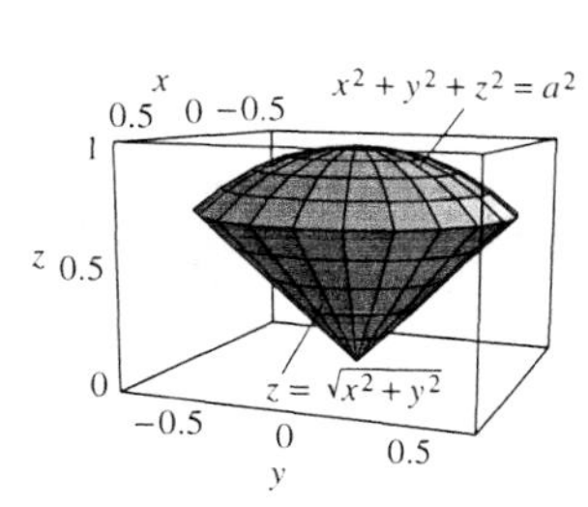

FIGURE 14.4.22 The fat ice-cream cone of Problem 29.

29. Find the volume of the "ice-cream cone" bounded by the sphere $x^2 + y^2 + z^2 = a^2$ and the cone $z = \sqrt{x^2 + y^2}$. When $a = 1$ this solid is the one shown in Fig. 14.4.22.

30. Find the volume bounded by the paraboloid $z = r^2$, the cylinder $r = 2a\sin\theta$, and the plane $z = 0$.

31. Find the volume that lies below the paraboloid $z = r^2$ and above one loop of the lemniscate with equation $r^2 = 2\sin\theta$.

32. Find the volume that lies inside both the cylinder $x^2+y^2 = 4$ and the ellipsoid $2x^2 + 2y^2 + z^2 = 18$.

33. If $0 < h < a$, then the plane $z = a-h$ cuts off a spherical segment of height h and radius b from the sphere $x^2+y^2+z^2 = a^2$ (Fig. 14.4.23). (a) Show that $b^2 = 2ah - h^2$. (b) Show that the volume of the spherical segment is $V = \frac{1}{6}\pi h(3b^2 + h^2)$.

34. Show by the method of Example 5 that

$$\int_0^\infty \int_0^\infty \frac{1}{(1+x^2+y^2)^2}\,dx\,dy = \frac{\pi}{4}.$$

35. Find the volume of the solid torus obtained by revolving the disk $r \leqq a$ around the line $x = b > a$ (Fig. 14.4.24). [*Suggestion:* If the area element $dA = r\,dr\,d\theta$ is revolved around the line, the volume generated is $dV = 2\pi(b-x)\,dA$. Express everything in polar coordinates.]

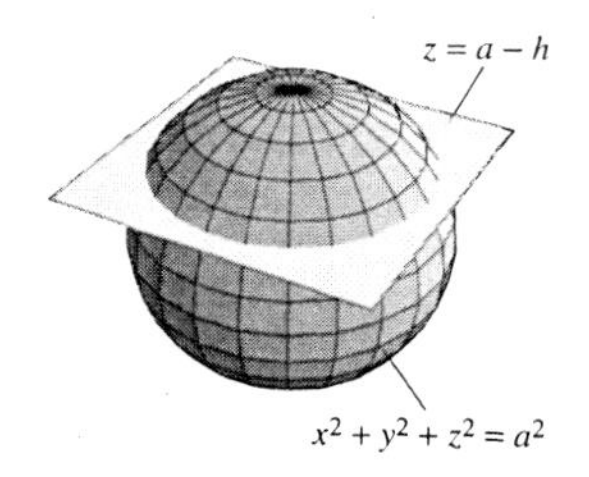

FIGURE 14.4.23 The spherical segment of Problem 33.

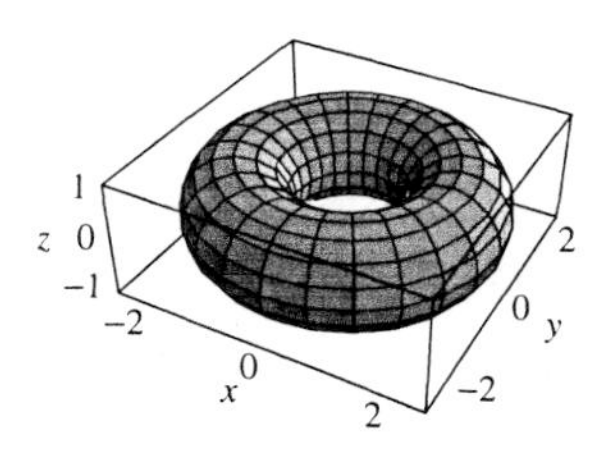

FIGURE 14.4.24 The torus of Problem 35 (the case $a = 1$, $b = 2$ is shown).

In Problems 36 through 40, use double integrals in polar coordinates to find the volumes of the indicated solids.

36. The solid lies above the plane $z = -3$ and below the paraboloid $z = 15 - 2x^2 - 2y^2$.

37. The solid is bounded above by the plane $z = y + 4$ and below by the paraboloid $z = x^2 + y^2 + y$.

38. The solid lies inside the cylinder $x^2 + y^2 = 4$, above the xy-plane, and below the plane $z = x + y + 3$.

39. The solid is bounded by the elliptical paraboloids $z = x^2 + 2y^2$ and $z = 12 - 2x^2 - y^2$.

40. The solid lies inside the ellipsoid $4x^2 + 4y^2 + z^2 = 80$ and above the paraboloid $z = 2x^2 + 2y^2$.

41. Find the volume removed when a circular hole of radius $a < b$ is bored symmetrically through the center of a sphere of radius b. [*Check:* It's about 35% of the volume of the sphere when $a = 1$ and $b = 2$.]

42. Suppose that a circular hole with radius 1 is cut off-center through a sphere of radius 4. The axis of the hole is at distance 2 from the center of the sphere. Use a computer algebra system to show that the volume of material removed is about 8% of the volume of the sphere.

43. Suppose that a hexagonal hole is cut symmetrically through the center of a sphere of radius 2. The cross section of the hole is a unit regular hexagon—a six-sided equiangular polygon with each side and "radius" of length 1. Use a computer algebra system to show that the volume of material removed is about 29% of the volume of the sphere. [To give your computer algebra system a more vigorous workout, you could try a pentagonal (five-sided) or heptagonal (seven-sided) hole, each with "radius" 1. With a unit 17-sided polygon, the volume of the material removed is over 34% of that of the sphere, close to the 35% figure cited in Problem 41.]

14.5 APPLICATIONS OF DOUBLE INTEGRALS

In Section 6.6 we discussed the *mass* m and *centroid* $(\overline{x}, \overline{y})$ of a plane region that corresponds to a thin plate or *lamina* of uniform (constant) density. This special case is amenable to calculation using single-variable integrals. Nevertheless, the double integral provides the proper setting for the general case of a lamina with variable density that occupies a bounded region R in the xy-plane. We suppose that the density of the lamina (in units of mass per unit area) at the point (x, y) is given by the continuous function $\delta(x, y)$.

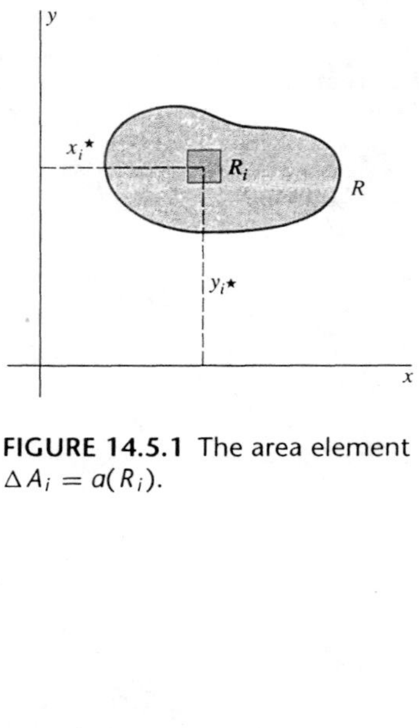

Let $\mathcal{P} = \{R_1, R_2, \dots, R_n\}$ be an inner partition of R, and choose a point $(x_i^\star, y_i^\star)$ in each subrectangle R_i (Fig. 14.5.1). Then the mass of the part of the lamina occupying R_i is approximately $\delta(x_i^\star, y_i^\star)\,\Delta A_i$, where ΔA_i denotes the area $a(R_i)$ of R_i. Hence the mass of the entire lamina is given approximately by

$$m \approx \sum_{i=1}^{n} \delta(x_i^\star, y_i^\star)\,\Delta A_i.$$

FIGURE 14.5.1 The area element $\Delta A_i = a(R_i)$.

As the norm $|\mathcal{P}|$ of the inner partition $\mathcal{P}$ approaches zero, this Riemann sum approaches the corresponding double integral over R. We therefore *define* the **mass** m of the lamina by means of the formula

$$m = \iint_R \delta(x, y)\,dA. \tag{1}$$

In brief,

$$m = \iint_R \delta\,dA = \iint_R dm$$

in terms of the density δ and the mass element

$$dm = \delta\,dA.$$

The coordinates $(\overline{x}, \overline{y})$ of the **centroid,** or *center of mass,* of the lamina are defined to be

$$\overline{x} = \frac{1}{m}\iint_R x\delta(x, y)\,dA, \tag{2}$$

$$\overline{y} = \frac{1}{m}\iint_R y\delta(x, y)\,dA. \tag{3}$$

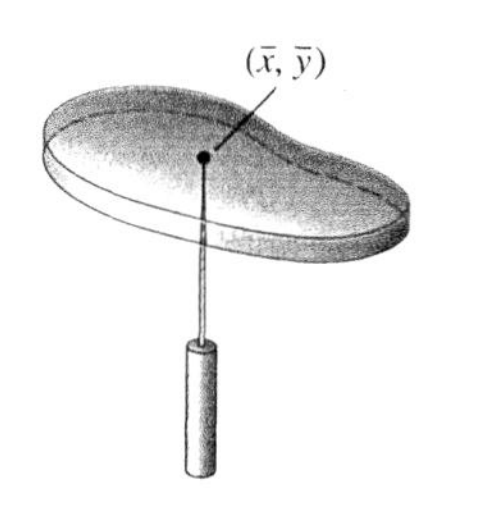

FIGURE 14.5.2 A lamina balanced on its centroid.

You may prefer to remember these formulas in the form

$$\bar{x} = \frac{1}{m} \iint_R x\,dm, \qquad \bar{y} = \frac{1}{m} \iint_R y\,dm.$$

Thus $\bar{x}$ and $\bar{y}$ are the *average values* of x and y *with respect to mass* in the region R. The centroid $(\bar{x}, \bar{y})$ is the point of the lamina where it would balance horizontally if placed on the point of an ice pick (Fig. 14.5.2).

If the density function δ has the *constant* value $k > 0$, then the coordinates of $\bar{x}$ and $\bar{y}$ are independent of the specific value of k. (Why?) In such a case we will generally take $\delta \equiv 1$ in our computations. Moreover, in this case m will have the same numerical value as the area A of R, and $(\bar{x}, \bar{y})$ is then called the **centroid of the plane region** R.

Generally, we must calculate all three integrals in Eqs. (1) through (3) in order to find the centroid of a lamina. But sometimes we can take advantage of the following *symmetry principle:* If the plane region R (considered to be a lamina of constant density) is symmetric with respect to the line L—that is, if R is carried onto itself when the plane is rotated through an angle of 180° around the line L—then the centroid of R lies on L (Fig. 14.5.3). For example, the centroid of a rectangle (Fig. 14.5.4) is the point where the perpendicular bisectors of its sides meet, because these bisectors are also lines of symmetry.

In the case of a nonconstant density function δ, we require (for symmetry) that δ—as well as the region itself—be symmetric about the geometric line L of symmetry. That is, we require that $\delta(P) = \delta(Q)$ if (as in Fig. 14.5.3) the points P and Q are symmetrically located with respect to L. Then the centroid of the lamina R will lie on the line L of symmetry.

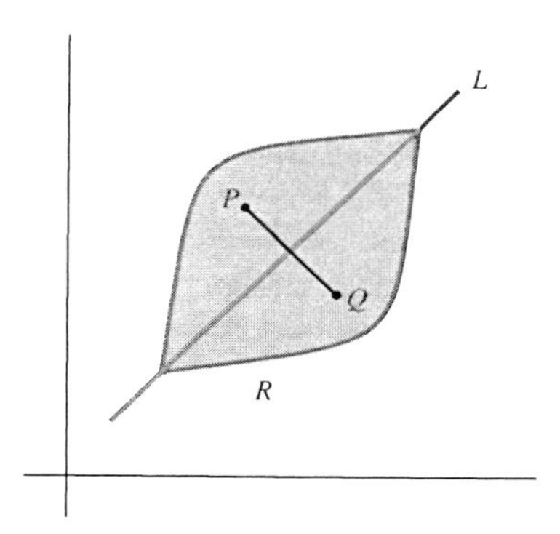

FIGURE 14.5.3 A line of symmetry.

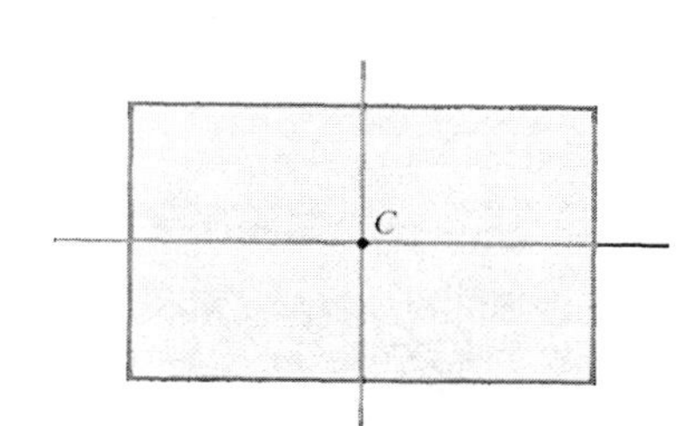

FIGURE 14.5.4 The centroid of a rectangle.

EXAMPLE 1 Consider the semicircular disk of radius a shown in Fig. 14.5.5. If it has constant density $\delta \equiv 1$, then its mass is $m = \frac{1}{2}\pi a^2$ (numerically equal to its area), and by symmetry its centroid $C(\bar{x}, \bar{y})$ lies on the y-axis. Hence $\bar{x} = 0$, and we need only compute

$$\begin{aligned}
\bar{y} &= \frac{1}{m} \iint_R y\,dm \\
&= \frac{2}{\pi a^2} \int_0^{\pi} \int_0^a (r \sin\theta)\, r\,dr\,d\theta \qquad \text{(polar coordinates)} \\
&= \frac{2}{\pi a^2} \Big[-\cos\theta\Big]_0^{\pi} \left[\frac{1}{3}r^3\right]_0^a = \frac{2}{\pi a^2} \cdot 2 \cdot \frac{a^3}{3} = \frac{4a}{3\pi}.
\end{aligned}$$

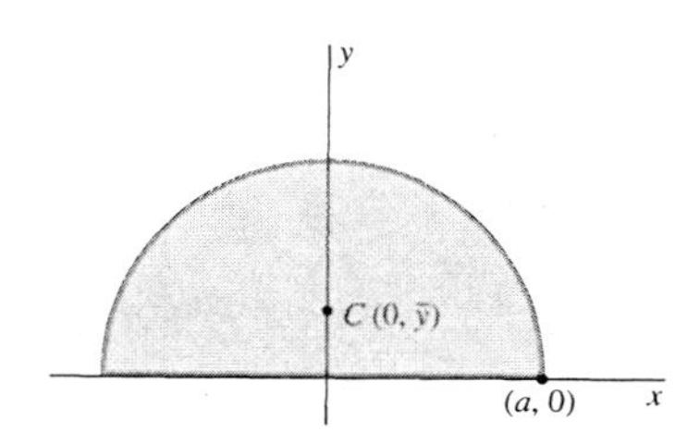

FIGURE 14.5.5 The centroid of a semicircular disk (Example 1).

Thus the centroid of the semicircular lamina is located at the point $(0, 4a/3\pi)$. Note that the computed value for $\bar{y}$ has the dimensions of length (because a is a length), as it should. Any answer that has other dimensions would be suspect. ◆

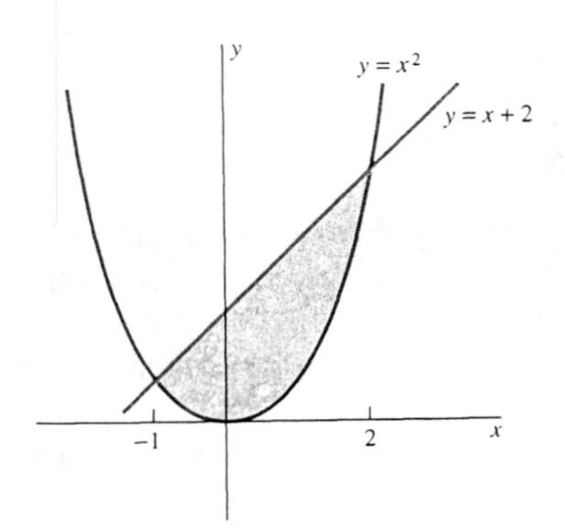

FIGURE 14.5.6 The lamina of Example 2.

EXAMPLE 2 A lamina occupies the region bounded by the line $y = x + 2$ and the parabola $y = x^2$ (Fig. 14.5.6). The density of the lamina at the point $P(x, y)$ is proportional to the square of the distance of P from the y-axis—thus $\delta(x, y) = kx^2$ (where k is a positive constant). Find the mass and centroid of the lamina.

Solution The line and the parabola intersect in the two points $(-1, 1)$ and $(2, 4)$, so Eq. (1) gives mass

$$m = \int_{-1}^{2}\int_{x^2}^{x+2} kx^2\,dy\,dx = k\int_{-1}^{2}\Big[x^2y\Big]_{y=x^2}^{x+2}dx$$
$$= k\int_{-1}^{2}(x^3 + 2x^2 - x^4)\,dx = \frac{63}{20}k.$$

Then Eqs. (2) and (3) give

$$\overline{x} = \frac{20}{63k}\int_{-1}^{2}\int_{x^2}^{x+2} kx^3\,dy\,dx = \frac{20}{63}\int_{-1}^{2}\Big[x^3y\Big]_{y=x^2}^{x+2}dx$$
$$= \frac{20}{63}\int_{-1}^{2}(x^4 + 2x^3 - x^5)\,dx = \frac{20}{63}\cdot\frac{18}{5} = \frac{8}{7};$$

$$\overline{y} = \frac{20}{63k}\int_{-1}^{2}\int_{x^2}^{x+2} kx^2y\,dy\,dx = \frac{20}{63}\int_{-1}^{2}\left[\frac{1}{2}x^2y^2\right]_{y=x^2}^{x+2}$$
$$= \frac{10}{63}\int_{-1}^{2}(x^4 + 4x^3 + 4x^2 - x^6)\,dx = \frac{10}{63}\cdot\frac{531}{35} = \frac{118}{49}.$$

Thus the lamina of this example has mass $63k/20$, and its centroid is located at the point $\left(\frac{8}{7}, \frac{118}{49}\right)$. ◆

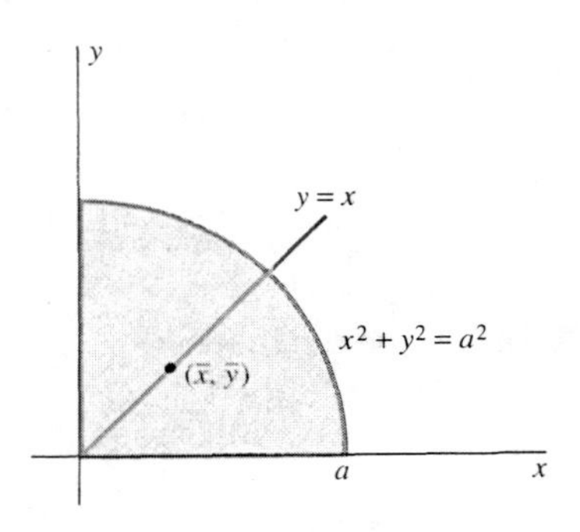

FIGURE 14.5.7 Finding mass and centroid (Example 3).

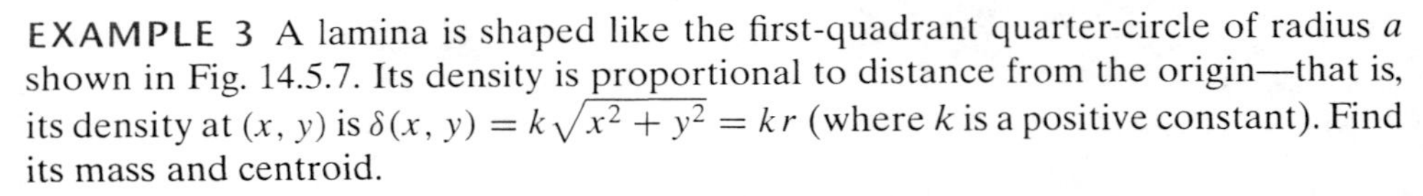

EXAMPLE 3 A lamina is shaped like the first-quadrant quarter-circle of radius a shown in Fig. 14.5.7. Its density is proportional to distance from the origin—that is, its density at (x, y) is $\delta(x, y) = k\sqrt{x^2 + y^2} = kr$ (where k is a positive constant). Find its mass and centroid.

Solution First we change to polar coordinates, because both the shape of the boundary of the lamina and the formula for its density suggest that this will make the computations much simpler. Equation (1) then yields the mass to be

$$m = \iint_R \delta\,dA = \int_0^{\pi/2}\int_0^a kr^2\,dr\,d\theta$$
$$= k\int_0^{\pi/2}\left[\frac{1}{3}r^3\right]_{r=0}^{a} = k\int_0^{\pi/2}\frac{1}{3}a^3\,d\theta = \frac{k\pi a^3}{6}.$$

By symmetry of the lamina and its density function, the centroid lies on the line $y = x$. So Eq. (3) gives

$$\overline{x} = \overline{y} = \frac{1}{m}\iint_R y\delta\,dA = \frac{6}{k\pi a^3}\int_0^{\pi/2}\int_0^a kr^3\sin\theta\,dr\,d\theta$$
$$= \frac{6}{\pi a^3}\int_0^{\pi/2}\left[\frac{1}{4}r^4\sin\theta\right]_{r=0}^{a}d\theta = \frac{6}{\pi a^3}\cdot\frac{a^4}{4}\int_0^{\pi/2}\sin\theta\,d\theta = \frac{3a}{2\pi}.$$

Thus the given lamina has mass $\frac{1}{6}k\pi a^3$; its centroid is located at the point $(3a/2\pi, 3a/2\pi)$. ◆

Volume and the First Theorem of Pappus

Now we can give a more general proof of the first theorem of Pappus, which was discussed from a single-variable viewpoint in Section 6.6.

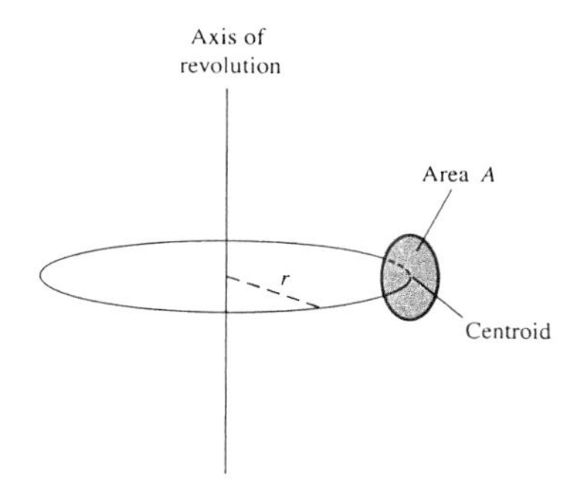

FIGURE 14.5.8 A solid of volume $V = A \cdot d$ is generated by the area A as its centroid travels the distance $d = 2\pi r$ around a circle of radius r.

FIRST THEOREM OF PAPPUS Volume of Revolution
Suppose that a plane region R is revolved around an axis in its plane (Fig. 14.5.8), generating a solid of revolution with volume V. Assume that the axis does not intersect the interior of R. Then the volume

$$V = A \cdot d$$

of the solid is the product of the area A of R and the distance d traveled by the centroid of R.

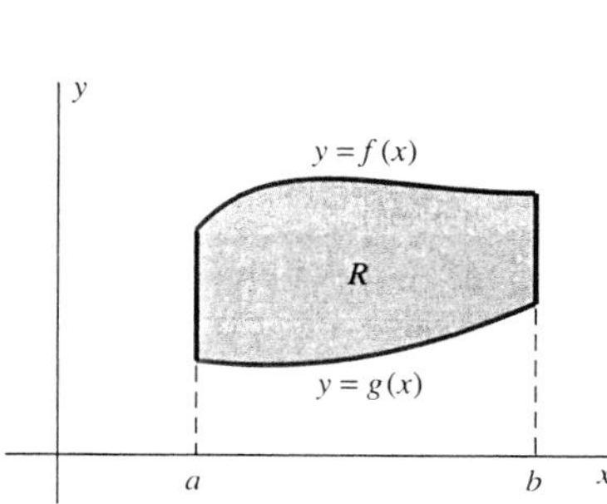

FIGURE 14.5.9 A region R between the graphs of two functions.

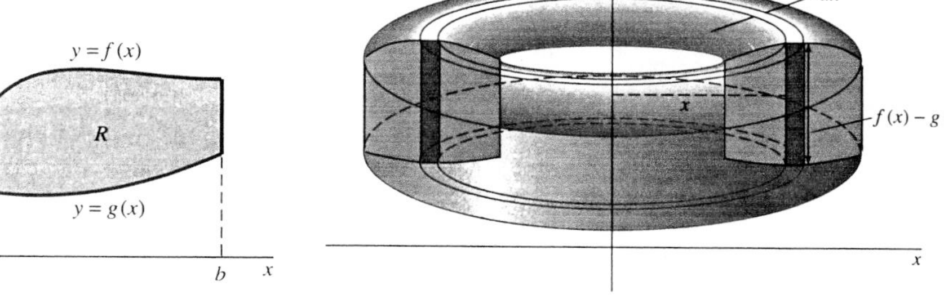

FIGURE 14.5.10 A solid of revolution consisting of cylindrical shells.

PROOF In Section 6.6 we treated the special case of a vertically simple region of the form illustrated in Fig. 14.5.9 and the corresponding volume of revolution illustrated in Fig. 14.5.10. More generally, let $\mathcal{P} = \{R_1, R_2, \dots, R_n\}$ be an inner partition of R, let $(x_i^\star, y_i^\star)$ be the center of the rectangle R_i, and let ΔA_i denote the area of R_i. Then, by the formula for the volume of a cylindrical shell (Eq. (1) in Section 6.3), the volume obtained by revolving the rectangle R_i in a circle of radius $x_i^\star$ around the y-axis (for instance) is $\Delta V_i = 2\pi x_i^\star \, \Delta A_i$. Hence the volume of the entire solid of revolution is given approximately by

$$V \approx \sum_{i=1}^{n} \Delta V_i = \sum_{i=1}^{n} 2\pi x_i^\star \, \Delta A_i.$$

We see here a Riemann sum approximating the integral

$$V = \iint_R 2\pi x \, dA = 2\pi A \cdot \frac{1}{A} \iint_R x \, dA = 2\pi A \cdot \overline{x}$$

(using Eq. (2) with $\delta = 1$). But $d = 2\pi \overline{x}$ is the distance traveled by the centroid, so we conclude that $V = A \cdot d$, as desired. ◄

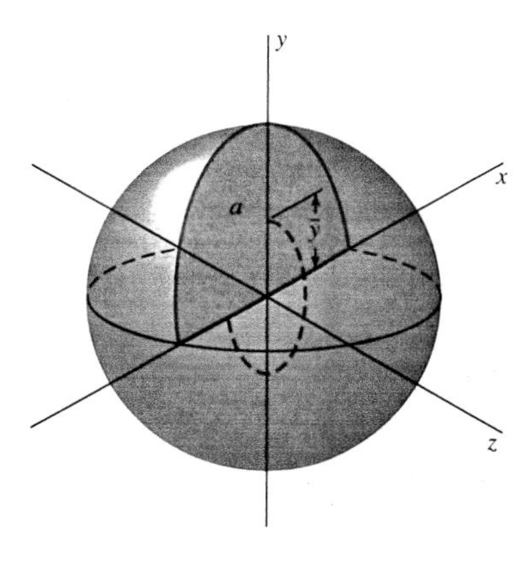

FIGURE 14.5.11 A sphere of radius a generated by revolving a semicircular region of area $A = \frac{1}{2}\pi a^2$ around its diameter on the x-axis (Example 4). The centroid of the semicircle travels along a circle of circumference $d = 2\pi \overline{y}$.

EXAMPLE 4 Find the volume V of the sphere of radius a generated by revolving around the x-axis the semicircular region D of Example 1. See Fig. 14.5.11.

Solution The area of D is $A = \frac{1}{2}\pi a^2$, and we found in Example 1 that $\overline{y} = 4a/3\pi$. Hence Pappus's theorem gives

$$V = 2\pi \overline{y} A = 2\pi \cdot \frac{4a}{3\pi} \cdot \frac{\pi a^2}{2} = \frac{4}{3}\pi a^3.$$ ◆

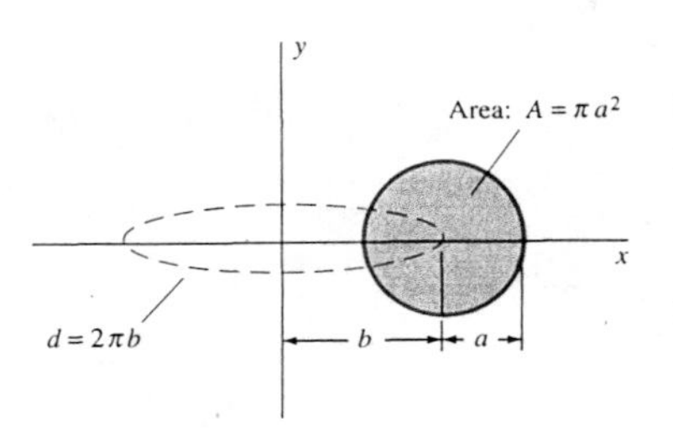

FIGURE 14.5.12 Rotating the circular disk around the y-axis to generated a torus (Example 5).

EXAMPLE 5 Consider the circular disk of Fig. 14.5.12, with radius a and center at the point $(b, 0)$ with $0 < a < b$. Find the volume V of the solid torus generated by revolving this disk around the y-axis. Such a torus is shown in Fig. 14.4.24.

Solution The centroid of the circle is at its center $(b, 0)$, so $\overline{x} = b$. Hence the centroid is revolved through the distance $d = 2\pi b$. Consequently,

$$V = d \cdot A = 2\pi b \cdot \pi a^2 = 2\pi^2 a^2 b.$$

Note that this result is dimensionally correct. ◆

Surface Area and the Second Theorem of Pappus

Centroids of plane *curves* are defined in analogy with centroids of plane regions, so we shall present this topic in less detail. It will suffice for us to treat only the case of constant density $\delta \equiv 1$ (such as a thin wire with unit mass per unit length). Then the centroid $(\overline{x}, \overline{y})$ of the plane curve C is defined by the formulas

$$\overline{x} = \frac{1}{s}\int_C x\,ds, \qquad \overline{y} = \frac{1}{s}\int_C y\,ds \tag{4}$$

where s is the arc length of C.

The meaning of the integrals in Eq. (4) is that of the notation of Section 6.4. That is, ds is a symbol to be replaced (before the integral is evaluated) with either

$$ds = \sqrt{1 + \left(\frac{dy}{dx}\right)^2}\,dx \quad \text{or} \quad ds = \sqrt{1 + \left(\frac{dx}{dy}\right)^2}\,dy,$$

depending on whether C is a smooth arc of the form $y = f(x)$ or one of the form $x = g(y)$. Alternatively, we may have

$$ds = \sqrt{(dx)^2 + (dy)^2} = \sqrt{\left(\frac{dx}{dt}\right)^2 + \left(\frac{dy}{dt}\right)^2}\,dt$$

if C is presented in parametric form, as in Section 10.5.

EXAMPLE 6 Let J denote the upper half of the *circle* (not the disk) of radius a and center $(0, 0)$, represented parametrically by

$$x = a\cos t, \quad y = a\sin t, \qquad 0 \leqq t \leqq \pi.$$

The arc J is shown in Fig. 14.5.13. Find its centroid.

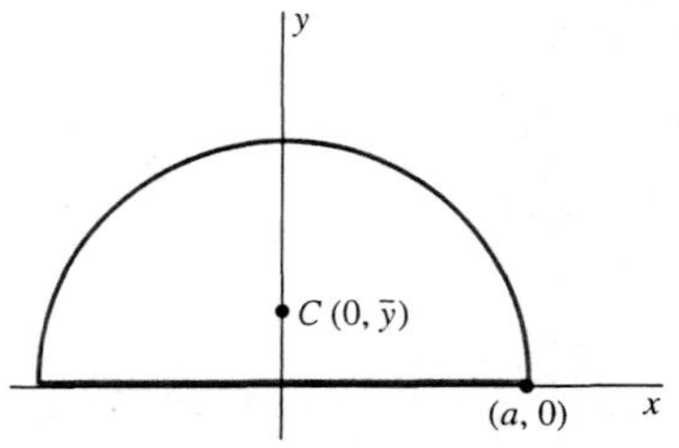

FIGURE 14.5.13 The semicircular arc of Example 6.

Solution Note first that $\overline{x} = 0$ by symmetry. The arc length of J is $s = \pi a$; the arc-length element is

$$ds = \sqrt{(-a\sin t)^2 + (a\cos t)^2}\,dt = a\,dt.$$

Hence the second formula in (4) yields

$$\overline{y} = \frac{1}{\pi a}\int_0^{\pi} (a\sin t)a\,dt = \frac{a}{\pi}\Big[-\cos t\Big]_0^{\pi} = \frac{2a}{\pi}.$$

Thus the centroid of the semicircular arc is located at the point $(0, 2a/\pi)$ on the y-axis. Note that the answer is both plausible and dimensionally correct. ◆

The first theorem of Pappus has an analogue for surface area of revolution.

SECOND THEOREM OF PAPPUS Surface Area of Revolution
Let the plane curve C be revolved around an axis in its plane that does not intersect the curve. Then the area

$$A = s \cdot d$$

of the surface of revolution generated is equal to the product of the length s of C and the distance d traveled by the centroid of C.

PROOF Let C be a smooth arc parametrized by $x = f(t)$, $y = g(t)$, $a \leqq t \leqq b$. If C is revolved around the y-axis (for instance), then by Eqs. (4) and (8) in Section 10.5 the resulting surface area of revolution is given by

$$A = \int_{t=a}^{b} 2\pi x \, ds = 2\pi s \cdot \frac{1}{s} \int_{t=a}^{b} x \, ds$$

$$= 2\pi s \cdot \overline{x} \qquad \left(\text{where} \quad ds = \sqrt{[f'(t)]^2 + [g'(t)]^2} \, dt\right)$$

(using Eq. (4)). But $d = 2\pi \overline{x}$ is the distance traveled by the centroid, so we see that $A = s \cdot d$, and this concludes the proof. ◀

EXAMPLE 7 Find the surface area A of the sphere of radius a generated by revolving around the x-axis the semicircular arc of Example 6.

Solution Because we found that $\overline{y} = 2a/\pi$ and we know that $s = \pi a$, the second theorem of Pappus gives

$$A = 2\pi \overline{y} s = 2\pi \cdot \frac{2a}{\pi} \cdot \pi a = 4\pi a^2.$$ ◆

EXAMPLE 8 Find the surface area A of the torus of Example 5.

Solution Now we think of revolving around the y-axis the circle (*not* the disk) of radius a centered at the point $(b, 0)$. Of course, the centroid of the circle is located at its center $(b, 0)$; this follows from the symmetry principle or can be verified by using computations such as those in Example 6. Hence the distance traveled by the centroid is $d = 2\pi b$. Because the circumference of the circle is $s = 2\pi a$, the second theorem of Pappus gives

$$A = 2\pi b \cdot 2\pi a = 4\pi^2 ab.$$ ◆

Moments of Inertia

Let R be a plane lamina and L a straight line that may or may not lie in the xy-plane. Then the **moment of inertia** I of R around the axis L is defined to be

$$I = \iint_R p^2 \, dm, \tag{5}$$

where $p = p(x, y)$ denotes the perpendicular distance to L from the point (x, y) of R.

The most important case is that in which the axis of revolution is the z-axis, so $p = r = \sqrt{x^2 + y^2}$ (Fig. 14.5.14). In this case we call $I = I_0$ the **polar moment of inertia** of the lamina R. Thus the polar moment of inertia of R is defined to be

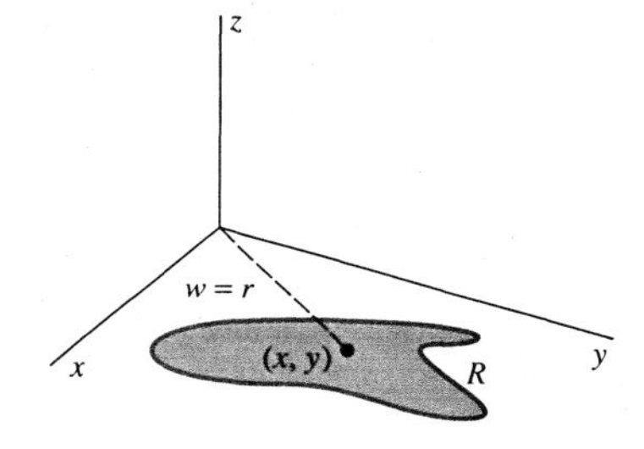

FIGURE 14.5.14 A lamina in the xy-plane in space.

$$I_0 = \iint_R r^2 \delta(x, y) \, dA = \iint_R r^2 \, dm = \iint_R (x^2 + y^2) \, dm. \tag{6}$$

It follows that

$$I_0 = I_x + I_y,$$

where

$$I_x = \iint_R y^2\,dm = \iint_R y^2\delta\,dA \tag{7}$$

and

$$I_y = \iint_R x^2\,dm = \iint_R x^2\delta\,dA. \tag{8}$$

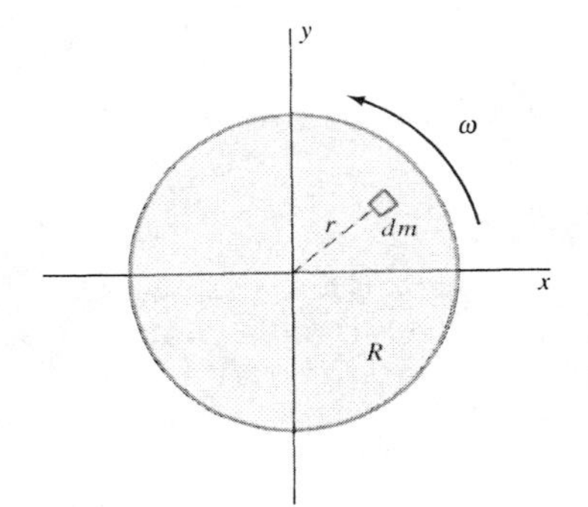

FIGURE 14.5.15 The rotating disk.

Here I_x is the moment of inertia of the lamina around the x-axis and I_y is its moment of inertia around the y-axis.

An important application of moments of inertia involves *kinetic energy of rotation*. Consider a circular disk that is revolving around its center (the origin) with angular speed ω radians per second. A mass element dm at distance r from the origin is moving with (linear) velocity $v = r\omega$ (Fig. 14.5.15). Thus the kinetic energy of the mass element is

$$\tfrac{1}{2}(dm)v^2 = \tfrac{1}{2}\omega^2 r^2\,dm.$$

Summing by integration over the whole disk, we find that its kinetic energy due to rotation at angular speed ω is

$$\text{KE}_{\text{rot}} = \iint_R \frac{1}{2}\omega^2 r^2\,dm = \frac{1}{2}\omega^2 \iint_R r^2\,dm;$$

that is,

$$\text{KE}_{\text{rot}} = \tfrac{1}{2}I_0\omega^2. \tag{9}$$

Because linear kinetic energy has the formula $\text{KE} = \frac{1}{2}mv^2$, Eq. (9) suggests (correctly) that moment of inertia is the rotational analogue of mass.

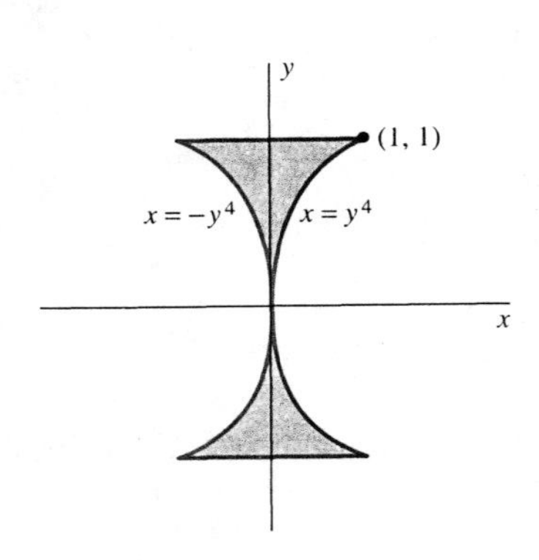

FIGURE 14.5.16 The lamina of Example 9.

EXAMPLE 9 Compute I_x for a lamina of constant density $\delta \equiv 1$ that occupies the region bounded by the curves $x = \pm y^4$, $-1 \leqq y \leqq 1$ (Fig. 14.5.16).

Solution Equation (7) gives

$$I_x = \int_{-1}^{1}\int_{-y^4}^{y^4} y^2\,dx\,dy = \int_{-1}^{1}\Big[xy^2\Big]_{x=-y^4}^{y^4}\,dy = \int_{-1}^{1} 2y^6\,dy = \frac{4}{7}. \quad ◆$$

The region of Example 9 resembles the cross section of an I beam. It is known that the stiffness, or resistance to bending, of a horizontal beam is proportional to the moment of inertia of its cross section with respect to a horizontal axis through the centroid of the cross section of the beam. Let us compare our I beam with a rectangular beam of equal height 2 and equal area

$$A = \int_{-1}^{1}\int_{-y^4}^{y^4} 1\,dx\,dy = \frac{4}{5}.$$

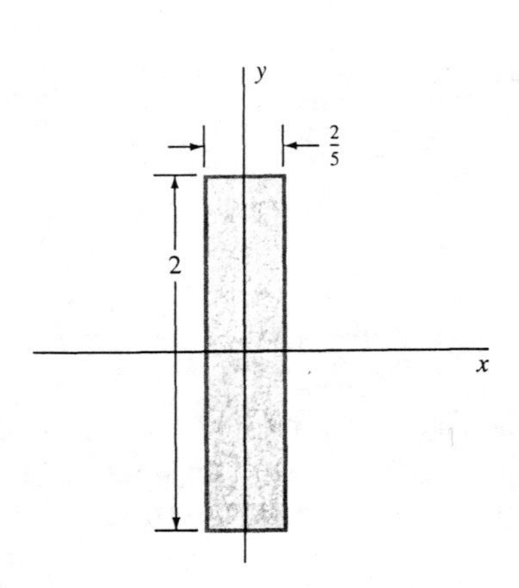

FIGURE 14.5.17 A rectangular beam for comparison with the I beam of Example 9.

The cross section of such a rectangular beam is shown in Fig. 14.5.17. Its width is $\frac{2}{5}$ and the moment of inertia of its cross section is

$$I_x = \int_{-1}^{1}\int_{-1/5}^{1/5} y^2\,dx\,dy = \frac{4}{15}.$$

Because the ratio of $\frac{4}{7}$ to $\frac{4}{15}$ is $\frac{15}{7}$, we see that the I beam is more than twice as strong as a rectangular beam of the same cross-sectional area. This strength is why I beams are commonly used in construction.

EXAMPLE 10 Find the polar moment of inertia of a circular lamina R of radius a and constant density δ centered at the origin.

Solution In Cartesian coordinates, the lamina R occupies the plane region $x^2 + y^2 \leqq a^2$; in polar coordinates, this region has the much simpler description $0 \leqq r \leqq a$, $0 \leqq \theta \leqq 2\pi$. Equation (6) then gives

$$I_0 = \iint_R r^2 \delta \, dA = \int_0^{2\pi} \int_0^a \delta r^3 \, dr \, d\theta = \frac{\delta \pi a^4}{2} = \frac{1}{2} m a^2,$$

where $m = \delta \pi a^2$ is the mass of the circular lamina. ◆

Finally, the **radius of gyration** $\hat{r}$ of a lamina of mass m around an axis is defined to be

$$\hat{r} = \sqrt{\frac{I}{m}}, \tag{10}$$

where I is the moment of inertia of the lamina around that axis. For example, the radii of gyration $\hat{x}$ and $\hat{y}$ around the y-axis and x-axis, respectively, are given by

$$\hat{x} = \sqrt{\frac{I_y}{m}} \quad \text{and} \quad \hat{y} = \sqrt{\frac{I_x}{m}}. \tag{11}$$

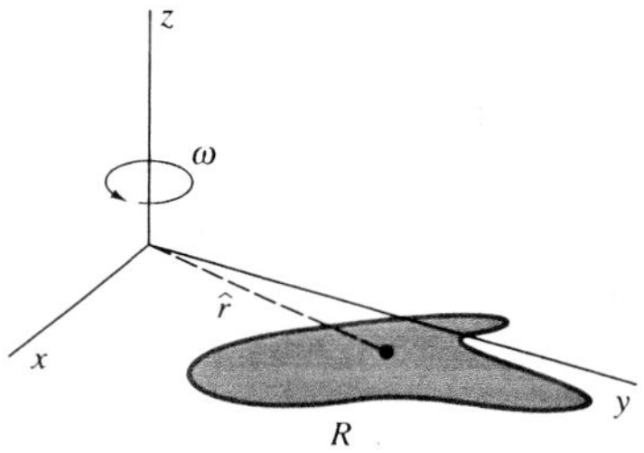

Now suppose that this lamina lies in the right half-plane $x > 0$ and is symmetric around the x-axis. If it represents the face of a tennis racquet whose handle (considered of negligible weight) extends along the x-axis from the origin to the face, then the point $(\hat{x}, 0)$ is a plausible candidate for the racquet's "sweet spot" that delivers the maximum impact and control. (See Problem 56.)

FIGURE 14.5.18 A plane lamina rotating around the z-axis.

The definition in Eq. (10) is motivated by considerating a plane lamina R rotating with angular speed ω around the z-axis (Fig. 14.5.18). Then Eq. (10) yields

$$I_0 = m\hat{r}^2,$$

so it follows from Eq. (9) that the kinetic energy of the lamina is

$$\text{KE} = \tfrac{1}{2} m (\hat{r}\omega)^2.$$

Thus the kinetic energy of the rotating lamina equals that of a single particle of mass m revolving at the distance $\hat{r}$ from the axis of revolution.

14.5 TRUE/FALSE STUDY GUIDE

14.5 CONCEPTS: QUESTIONS AND DISCUSSION

1. Suppose that a plane lamina has a line of symmetry. Must the centroid of the lamina lie on this line?
2. Must the centroid of a plane curve lie on the curve? Must the centroid of a plane region lie within the region? If not, provide counterexamples.

14.5 PROBLEMS

In Problems 1 through 10, find the centroid of the plane region bounded by the given curves. Assume that the density is $\delta \equiv 1$ for each region.

1. $x = 0, \quad x = 4, \quad y = 0, \quad y = 6$
2. $x = 1, \quad x = 3, \quad y = 2, \quad y = 4$
3. $x = -1, \quad x = 3, \quad y = -2, \quad y = 4$
4. $x = 0, \quad y = 0, \quad x + y = 3$
5. $x = 0, \quad y = 0, \quad x + 2y = 4$
6. $y = 0, \quad y = x, \quad x + y = 2$

7. $y = 0, \quad y = x^2, \quad x = 2$
8. $y = x^2, \quad y = 9$
9. $y = 0, \quad y = x^2 - 4$
10. $x = -2, \quad x = 2, \quad y = 0, \quad y = x^2 + 1$

In Problems 11 through 30, find the mass and centroid of the plane lamina with the indicated shape and density.

11. The triangular region bounded by $x = 0$, $y = 0$, and $x + y = 1$, with $\delta(x, y) = xy$

12. The triangular region of Problem 11, with $\delta(x, y) = x^2$

13. The region bounded by $y = 0$ and $y = 4 - x^2$, with $\delta(x, y) = y$

14. The region bounded by $x = 0$ and $x = 9 - y^2$, with $\delta(x, y) = x^2$

15. The region bounded by the parabolas $y = x^2$ and $x = y^2$, with $\delta(x, y) = xy$

16. The region of Problem 15, with $\delta(x, y) = x^2 + y^2$

17. The region bounded by the parabolas $y = x^2$ and $y = 2 - x^2$, with $\delta(x, y) = y$

18. The region bounded by $x = 0$, $x = e$, $y = 0$, and $y = \ln x$ for $1 \leqq x \leqq e$, with $\delta(x, y) \equiv 1$

19. The region bounded by $y = 0$ and $y = \sin x$ for $0 \leqq x \leqq \pi$, with $\delta(x, y) \equiv 1$

20. The region bounded by $y = 0$, $x = -1$, $x = 1$, and $y = \exp(-x^2)$, with $\delta(x, y) = |xy|$

21. The square with vertices $(0, 0)$, $(0, a)$, (a, a), and $(a, 0)$, with $\delta(x, y) = x + y$

22. The triangular region bounded by the coordinate axes and the line $x + y = a$ $(a > 0)$, with $\delta(x, y) = x^2 + y^2$

23. The region bounded by $y = x^2$ and $y = 4$; $\delta(x, y) = y$

24. The region bounded by $y = x^2$ and $y = 2x + 3$; $\delta(x, y) = x^2$

25. The region of Problem 19; $\delta(x, y) = x$

26. The semicircular region $x^2 + y^2 \leqq a^2$, $y \geqq 0$; $\delta(x, y) = y$

27. The region of Problem 26; $\delta(x, y) = r$ (the radial polar coordinate)

28. The region bounded by the cardioid with polar equation $r = 1 + \cos\theta$; $\delta(r, \theta) = r$ (Fig. 14.5.19)

29. The region inside the circle $r = 2\sin\theta$ and outside the circle $r = 1$; $\delta(x, y) = y$

30. The region inside the limaçon $r = 1 + 2\cos\theta$ and outside the circle $r = 2$; $\delta(r, \theta) = r$ (Fig. 14.5.20)

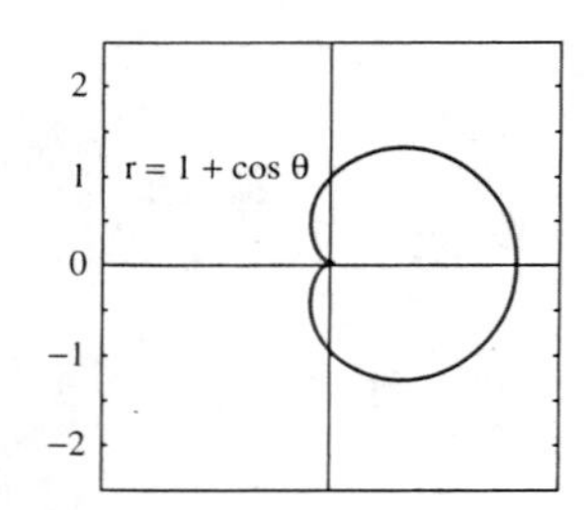

FIGURE 14.5.19 The cardiod of Problem 28.

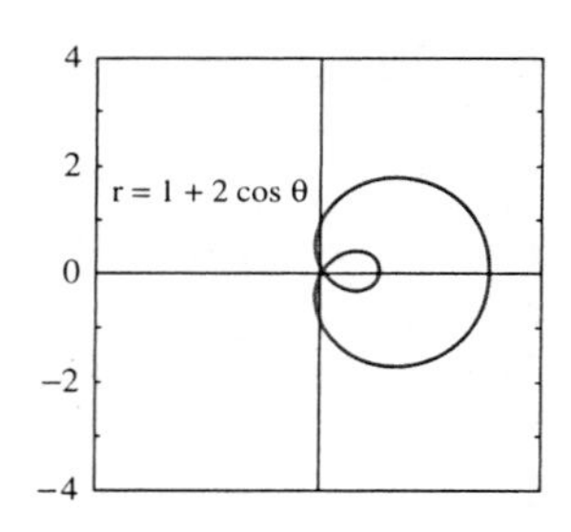

FIGURE 14.5.20 The limaçon of Problem 30.

In Problems 31 through 35, find the polar moment of inertia I_0 of the indicated lamina.

31. The region bounded by the circle $r = a$; $\delta(x, y) = r^n$, where n is a fixed positive integer

32. The lamina of Problem 26

33. The disk bounded by $r = 2\cos\theta$; $\delta(x, y) = k$ (a positive constant)

34. The lamina of Problem 29

35. The region bounded by the right-hand loop of the lemniscate $r^2 = \cos 2\theta$; $\delta(x, y) = r^2$ (Fig. 14.5.21)

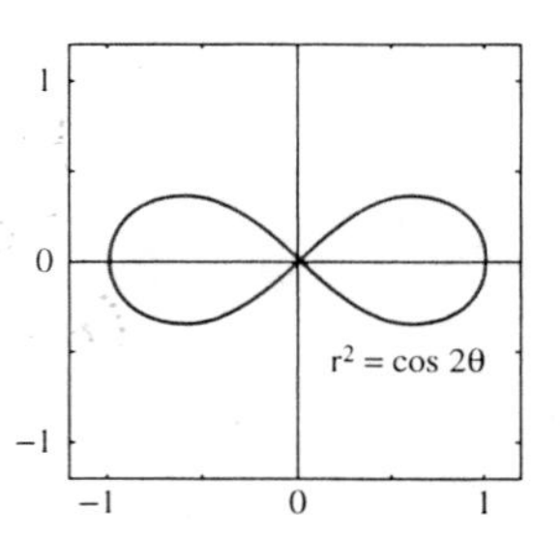

FIGURE 14.5.21 The leminscate of Problem 35.

In Problems 36 through 40, find the radii of gyration $\hat{x}$ and $\hat{y}$ of the indicated lamina around the coordinate axes.

36. The lamina of Problem 21

37. The lamina of Problem 23

38. The lamina of Problem 24

39. The lamina of Problem 27

40. The lamina of Problem 33

41. Find the centroid of the first quadrant of the circular disk $x^2 + y^2 \leqq r^2$ by direct computation, as in Example 1.

42. Apply the first theorem of Pappus to find the centroid of the first quadrant of the circular disk $x^2 + y^2 \leqq r^2$. Use the facts that $\overline{x} = \overline{y}$ (by symmetry) and that revolution of this quarter-disk around either coordinate axis gives a solid hemisphere with volume $V = \frac{2}{3}\pi r^3$.

43. Find the centroid of the arc that consists of the first-quadrant portion of the circle $x^2 + y^2 = r^2$ by direct computation, as in Example 6.

44. Apply the second theorem of Pappus to find the centroid of the quarter-circular arc of Problem 43. Note that $\overline{x} = \overline{y}$ (by symmetry) and that rotation of this arc around either coordinate axis gives a hemisphere with surface area $A = 2\pi r^2$.

45. Show by direct computation that the centroid of the triangle with vertices $(0, 0)$, $(r, 0)$, and $(0, h)$ is the point $(r/3, h/3)$. Verify that this point lies on the line from the vertex $(0, 0)$ to the midpoint of the opposite side of the triangle and two-thirds of the way from the vertex to the midpoint.

46. Apply the first theorem of Pappus and the result of Problem 45 to verify the formula $V = \frac{1}{3}\pi r^2 h$ for the volume of the cone obtained by revolving the triangle around the y-axis.

47. Apply the second theorem of Pappus to show that the lateral surface area of the cone of Problem 46 is $A = \pi r L$, where $L = \sqrt{r^2 + h^2}$ is the slant height of the cone.

48. (a) Find the centroid of the trapezoid shown in Fig. 14.5.22. (b) Apply the first theorem of Pappus and the result of part (a) to show that the volume of the conical frustum generated by revolving the trapezoid around the y-axis is

$$V = \frac{\pi h}{3}\left(r_1^2 + r_1 r_2 + r_2^2\right).$$

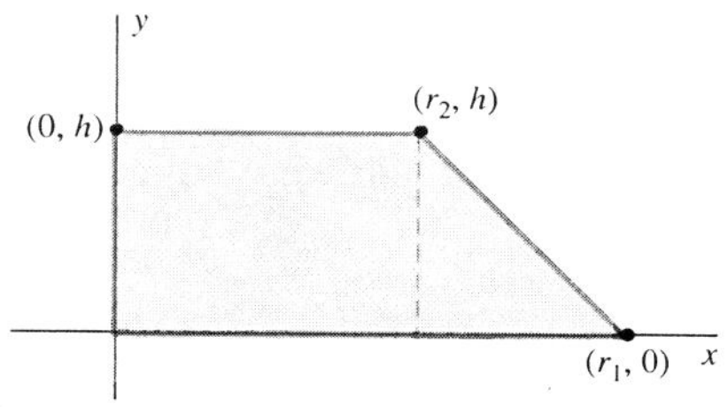

FIGURE 14.5.22 The trapezoid of Problem 48.

49. Apply the second theorem of Pappus to show that the lateral surface area of the conical frustum of Problem 48 is $a = \pi(r_1 + r_2)L$, where

$$L = \sqrt{(r_1 - r_2)^2 + h^2}$$

is its slant height.

50. (a) Apply the second theorem of Pappus to verify that the curved surface area of a right circular cylinder of height h and base radius r is $A = 2\pi rh$. (b) Explain how this follows also from the result of Problem 49.

51. (a) Find the centroid of the plane region shown in Fig. 14.5.23, which consists of a semicircular region of radius a sitting atop a rectangular region of width $2a$ and height b whose base is on the x-axis. (b) Then apply the first theorem of Pappus to find the volume generated by rotating this region around the x-axis.

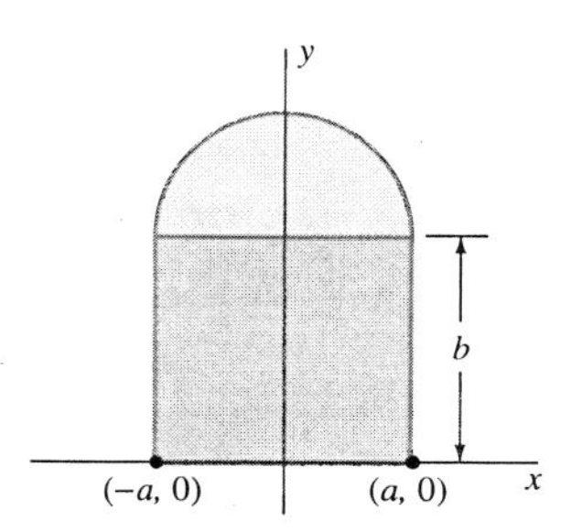

FIGURE 14.5.23 The plane region of Problem 51(a).

52. (a) Consider the plane region of Fig. 14.5.24, bounded by $x^2 = 2py$, $x = 0$, and $y = h = r^2/2p$ $(p > 0)$. Show that its area is $A = \frac{2}{3}rh$ and that the x-coordinate of its centroid is $\overline{x} = \frac{3}{8}r$. (b) Use Pappus's theorem and the result of part (a) to show that the volume of a paraboloid of revolution with radius r and height h is $V = \frac{1}{2}\pi r^2 h$.

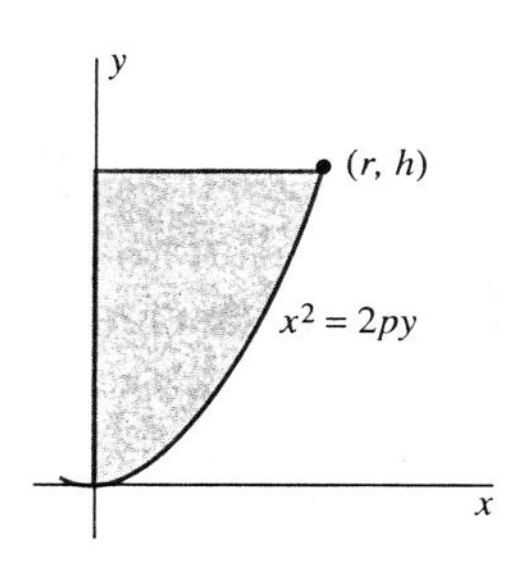

FIGURE 14.5.24 The region of Problem 52.

53. A uniform rectangular plate with base length a, height b, and mass m is centered at the origin. Show that its polar moment of inertia is $I_0 = \frac{1}{12}m(a^2 + b^2)$.

54. The centroid of a uniform plane region is at $(0, 0)$, and the region has total mass m. Show that its moment of inertia about an axis perpendicular to the xy-plane at the point (x_0, y_0) is

$$I = I_0 + m(x_0^2 + y_0^2).$$

55. Suppose that a plane lamina consists of two nonoverlapping laminae. Show that its polar moment of inertia is the sum of theirs. Use this fact together with the results of Problems 53 and 54 to find the polar moment of inertia of the T-shaped lamina of constant density $\delta = k > 0$ shown in Fig. 14.5.25.

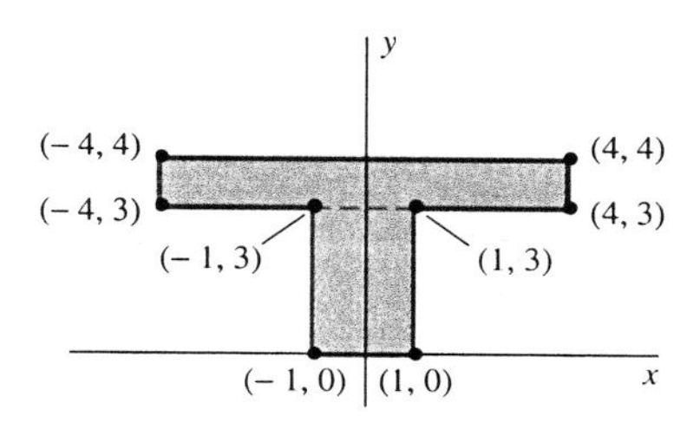

FIGURE 14.5.25 One lamina made of two simpler ones (Problem 55).

56. A racquet consists of a uniform lamina that occupies the region inside the right-hand loop of $r^2 = \cos 2\theta$ on the end of a handle (assumed to be of negligible mass) corresponding to the interval $-1 \leqq x \leqq 0$ (Fig. 14.5.26). Find the radius of gyration of the racquet around the line $x = -1$. Where is its sweet spot?

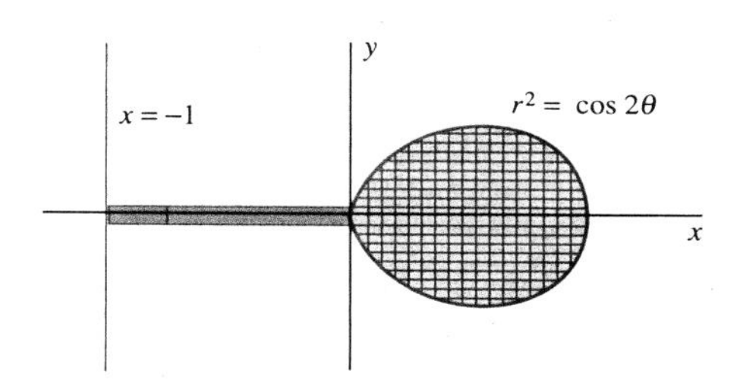

FIGURE 14.5.26 The racquet of Problem 56.

In Problems 57 through 60, find the mass m and centroid $(\overline{x}, \overline{y})$ of the indicated plane lamina R. You may use either a computer algebra system or the sine-cosine integrals of Formula (113) inside the back cover.

57. R is bounded by the circle with polar equation $r = 2\sin\theta$ and has density function $\delta(x, y) = y$

58. R is bounded by the circle with polar equation $r = 2\sin\theta$ and has density function $\delta(x, y) = y\sqrt{x^2 + y^2}$

59. R is the semicircular disk bounded by the x-axis and the upper half of the circle with polar equation $r = 2\cos\theta$ and has density function $\delta(x, y) = x$

60. R is the semicircular disk bounded by the x-axis and the upper half of the circle with polar equation $r = 2\cos\theta$ and has density function $\delta(x, y) = x^2y^2$

14.5 Project: Optimal Design of Downhill Race-Car Wheels

To see moments of inertia in action, suppose that your club is designing an unpowered race car for the annual downhill derby. You have a choice of solid wheels, bicycle wheels with thin spokes, or even solid spherical wheels (like giant ball bearings). Which wheels will make the race car go the fastest?

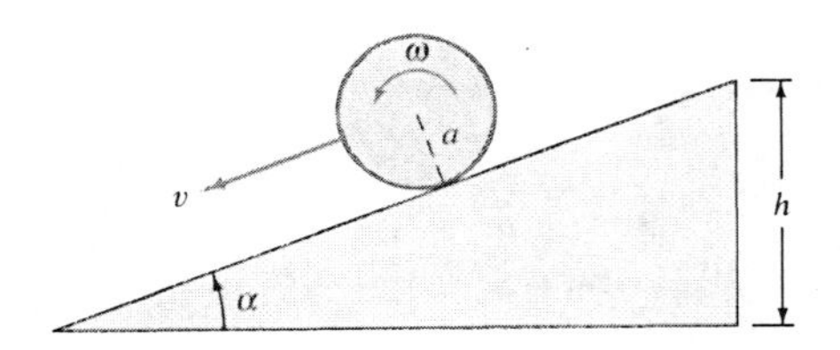

FIGURE 14.5.27 A circular object rolling down an incline.

Imagine an experiment in which you roll various types of wheels down an incline to see which reaches the bottom the fastest (Fig. 14.5.27). Suppose that a wheel of radius a and mass M starts from rest at the top with potential energy $\text{PE} = Mgh$ and reaches the bottom with angular speed ω and (linear) velocity $v = a\omega$. Then (by conservation of energy) the wheel's initial potential energy has been transformed into a sum $\text{KE}_{\text{tr}} + \text{KE}_{\text{rot}}$ of translation kinetic energy $\text{KE}_{\text{tr}} = \frac{1}{2}Mv^2$ and rotational kinetic energy

$$\text{KE}_{\text{rot}} = \frac{1}{2}I_0\omega^2 = \frac{I_0 v^2}{2a^2}, \tag{1}$$

a consequence of Eq. (9) of this section. Thus

$$Mgh = \frac{1}{2}Mv^2 + \frac{I_0 v^2}{2a^2}. \tag{2}$$

Problems 1 through 8 explore the implications of this formula.

1. Suppose that the wheel's (polar) moment of inertia is given by

$$I_0 = kMa^2 \tag{3}$$

for some constant k. (For instance, Example 10 gives $k = \frac{1}{2}$ for a wheel in the shape of a uniform solid disk.) Then deduce from Eq. (2) that

$$v = \sqrt{\frac{2gh}{1+k}}. \tag{4}$$

Thus the smaller k is (and hence the smaller the wheel's moment of inertia), the faster the wheel will roll down the incline.

In Problems 2 through 8, take $g = 32$ ft/s^2 *and assume that the vertical height of the incline is* $h = 100$ ft.

2. Why does it follow from Eq. (4) that, whatever the wheel's design, the maximum velocity a circular wheel can attain on this incline is 80 ft/s (just under 55 mi/h)?

3. If the wheel is a uniform solid disk (like a medieval wooden wagon wheel) with $I_0 = \frac{1}{2}Ma^2$, what is its speed v at the bottom of the incline?

4. Answer Problem 3 if the wheel is shaped like a narrow bicycle tire, with its entire mass, in effect, concentrated at the distance a from its center. In this case, $I_0 = Ma^2$. (Why?)

5. Answer Problem 3 if the wheel is shaped like an annular ring (or washer) with outer radius a and inner radius b.

Example 3 and Problems 41 and 42 in Section 14.7 provide the moments of inertia needed in Problems 6 through 8. In each of these problems, find the velocity of the wheel when it reaches the bottom of the incline.

6. The wheel is a uniform solid sphere of radius a.
7. The wheel is a very thin, spherical shell whose entire mass is, in effect, concentrated at the distance a from its center.
8. The wheel is a spherical shell with outer radius a and inner radius $b = \frac{1}{2}a$.

Finally, what is your conclusion? What is the shape of the wheels that will yield the fastest downhill race car?

14.6 TRIPLE INTEGRALS

The definition of the triple integral is the three-dimensional version of the definition of the double integral of Section 14.2. Let $f(x, y, z)$ be continuous on the bounded space region T and suppose that T lies inside the rectangular block R determined by the inequalities

$$a \leqq x \leqq b, \quad c \leqq y \leqq d, \quad \text{and} \quad p \leqq z \leqq q.$$

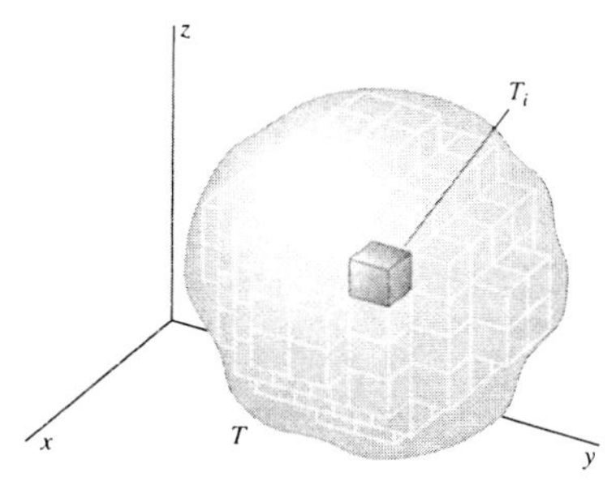

FIGURE 14.6.1 One small block in an inner partition of the bounded space region T.

We divide $[a, b]$ into subintervals of equal length Δx, $[c, d]$ into subintervals of equal length Δy, and $[p, q]$ into subintervals of equal length Δz. This generates a partition of R into smaller rectangular blocks (as in Fig. 14.6.1), each of volume $\Delta V = \Delta x\, \Delta y\, \Delta z$. Let $\mathcal{P} = \{T_1, T_2, \dots, T_n\}$ be the collection of these smaller blocks that lie wholly within T. Then $\mathcal{P}$ is called an **inner partition** of the region T. The **norm** $|\mathcal{P}|$ of $\mathcal{P}$ is the length of a longest diagonal of any of the blocks T_i. If $(x_i^\star, y_i^\star, z_i^\star)$ is an arbitrarily selected point of T_i (for each $i = 1, 2, \dots, n$), then the **Riemann sum**

$$\sum_{i=1}^{n} f(x_i^\star, y_i^\star, z_i^\star)\, \Delta V$$

is an approximation to the triple integral of f over the region T.

For example, if T is a solid body with density function f, then such a Riemann sum approximates its total mass. We define the **triple integral of f over T** by means of the equation

$$\iiint_T f(x, y, z)\, dV = \lim_{|\mathcal{P}| \to 0} \sum_{i=1}^{n} f(x_i^\star, y_i^\star, z_i^\star)\, \Delta V. \tag{1}$$

It is proved in advanced calculus that this limit of Riemann sums exists as the norm $|\mathcal{P}|$ approaches zero provided that f is continuous on T and that the boundary of the region T is reasonably well-behaved. (For instance, it suffices for the boundary of T to consist of a finite number of smooth surfaces.)

Just as with double integrals, we ordinarily compute triple integrals by means of iterated integrals. If the region of integration is a rectangular block, as in Example 1, then we can integrate in any order we wish.

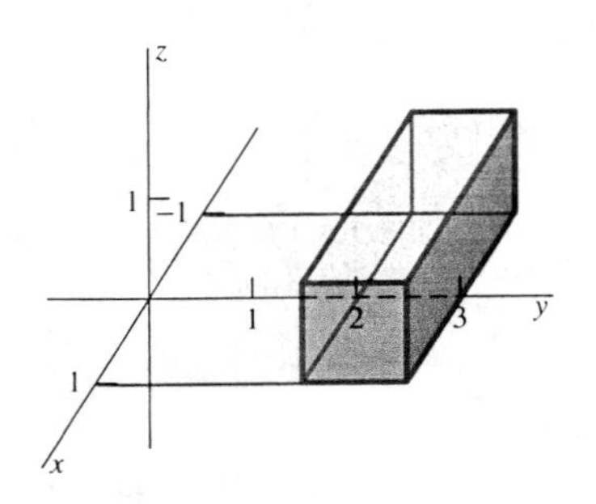

FIGURE 14.6.2 The rectangular block T of Example 1, for which $-1 \leqq x \leqq 1$, $2 \leqq y \leqq 3$, and $0 \leqq z \leqq 1$.

EXAMPLE 1 If $f(x, y, z) = xy + yz$ and T consists of those points (x, y, z) in space that satisfy the inequalities

$$-1 \leqq x \leqq 1, \quad 2 \leqq y \leqq 3, \quad \text{and} \quad 0 \leqq z \leqq 1$$

(Fig. 14.6.2), then

$$\begin{aligned}
\iiint_T f(x, y, z)\,dV &= \int_{-1}^{1}\int_{2}^{3}\int_{0}^{1} (xy + yz)\,dz\,dy\,dx \\
&= \int_{-1}^{1}\int_{2}^{3} \left[xyz + \frac{1}{2}yz^2 \right]_{z=0}^{1} dy\,dx \\
&= \int_{-1}^{1}\int_{2}^{3} \left(xy + \frac{1}{2}y \right) dy\,dx \\
&= \int_{-1}^{1} \left[\frac{1}{2}xy^2 + \frac{1}{4}y^2 \right]_{y=2}^{3} dx \\
&= \int_{-1}^{1} \left(\frac{5}{2}x + \frac{5}{4} \right) dx = \left[\frac{5}{4}x^2 + \frac{5}{4}x \right]_{-1}^{1} = \frac{5}{2}.
\end{aligned}$$

◆

The applications of double integrals that we saw in earlier sections generalize immediately to triple integrals. If T is a solid body with the density function $\delta(x, y, z)$, then its **mass** m is given by

$$m = \iiint_T \delta\,dV. \tag{2}$$

The case $\delta \equiv 1$ gives the **volume**

$$V = \iiint_T dV \tag{3}$$

of T. The coordinates of its **centroid** are

$$\overline{x} = \frac{1}{m}\iiint_T x\delta\,dV, \tag{4a}$$

$$\overline{y} = \frac{1}{m}\iiint_T y\delta\,dV, \quad \text{and} \tag{4b}$$

$$\overline{z} = \frac{1}{m}\iiint_T z\delta\,dV. \tag{4c}$$

The **moments of inertia** of T around the three coordinate axes are

$$I_x = \iiint_T (y^2 + z^2)\delta\,dV, \tag{5a}$$

$$I_y = \iiint_T (x^2 + z^2)\delta\,dV, \quad \text{and} \tag{5b}$$

$$I_z = \iiint_T (x^2 + y^2)\delta\,dV. \tag{5c}$$

Iterated Triple Integrals

As indicated previously, we almost always evaluate triple integrals by iterated single integration. Suppose that the region T with piecewise smooth boundary is **z-simple:** Each line parallel to the z-axis intersects T (if at all) in a single line segment. In effect, this means that T can be described by the inequalities

$$z_1(x, y) \leqq z \leqq z_2(x, y), \quad (x, y) \text{ in } R,$$

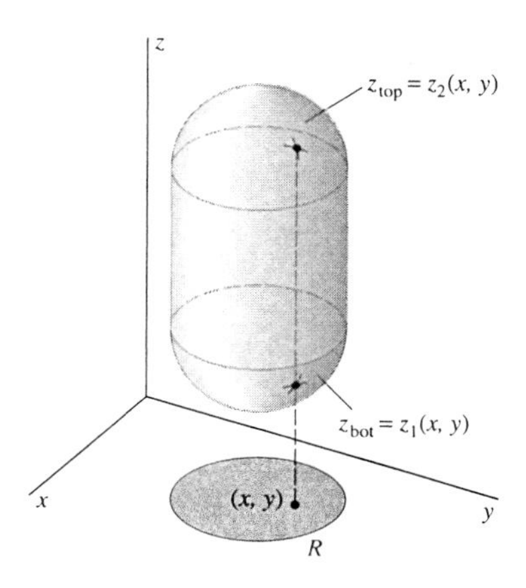

FIGURE 14.6.3 Obtaining the limits of integration for z.

where R is the vertical projection of T into the xy-plane. Then

$$\iiint_T f(x, y, z)\, dV = \iint_R \left(\int_{z_1(x,y)}^{z_2(x,y)} f(x, y, z)\, dz \right) dA. \tag{6}$$

In Eq. (6), we take $dA = dx\, dy$ or $dA = dy\, dx$, depending on the preferred order of integration over the set R. The limits $z_1(x, y)$ and $z_2(x, y)$ are the z-coordinates of the endpoints of the line segment in which the vertical line at (x, y) meets T (Fig. 14.6.3).

If the region R has the description

$$y_1(x) \leqq y \leqq y_2(x), \quad a \leqq x \leqq b,$$

then (integrating last with respect to x),

$$\iiint_T f(x, y, z)\, dV = \int_a^b \int_{y_1(x)}^{y_2(x)} \int_{z_1(x,y)}^{z_2(x,y)} f(x, y, z)\, dz\, dy\, dx.$$

Thus the triple integral reduces in this case to three iterated single integrals. These can (in principle) be evaluated by using the fundamental theorem of calculus.

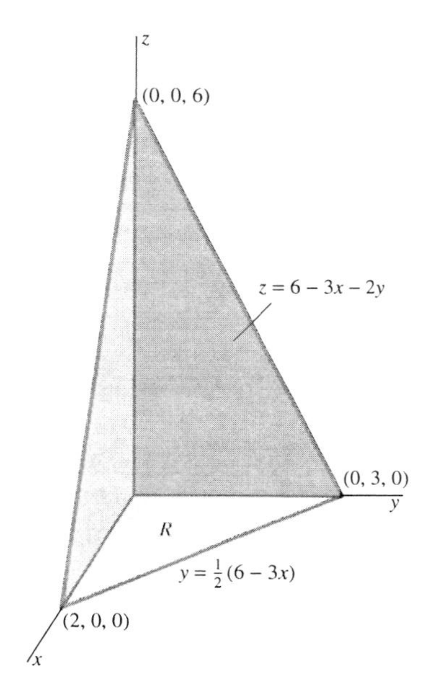

FIGURE 14.6.4 The pyramid T of Example 2; its base is the triangle R in the xy-plane.

EXAMPLE 2 Find the mass m of the pyramid T of Fig. 14.6.4 if its density function is given by $\delta(x, y, z) = z$.

Solution The region T is bounded below by the xy-plane $z = 0$ and above by the plane $z = 6 - 3x - 2y$. Its base is the plane region R bounded by the x- and y-axes and the line $y = \frac{1}{2}(6 - 3x)$. Hence Eqs. (2) and (6) yield

$$\begin{aligned} m &= \int_0^2 \int_0^{(6-3x)/2} \int_0^{6-3x-2y} z\, dz\, dy\, dx = \int_0^2 \int_0^{(6-3x)/2} \left[\frac{1}{2} z^2 \right]_{z=0}^{6-3x-2y} dy\, dx \\ &= \frac{1}{2} \int_0^2 \int_0^{(6-3x)/2} (6 - 3x - 2y)^2\, dy\, dx = \frac{1}{2} \int_0^2 \left[-\frac{1}{6}(6 - 3x - 2y)^3 \right]_{y=0}^{(6-3x)/2} dx \\ &= \frac{1}{12} \int_0^2 (6 - 3x)^3\, dx = \frac{1}{12} \left[-\frac{1}{12}(6 - 3x)^4 \right]_{x=0}^{2} = \frac{6^4}{12^2} = 9. \end{aligned}$$

We leave as an exercise (Problem 45) to show that the coordinates of the centroid $(\overline{x}, \overline{y}, \overline{z})$ of the pyramid are given by

$$\overline{x} = \frac{1}{9} \int_0^2 \int_0^{(6-3x)/2} \int_0^{6-3x-2y} xz\, dz\, dy\, dx = \frac{2}{5},$$

$$\overline{y} = \frac{1}{9} \int_0^2 \int_0^{(6-3x)/2} \int_0^{6-3x-2y} yz\, dz\, dy\, dx = \frac{3}{5},$$

$$\overline{z} = \frac{1}{9} \int_0^2 \int_0^{(6-3x)/2} \int_0^{6-3x-2y} z^2\, dz\, dy\, dx = \frac{12}{5}.$$

◆

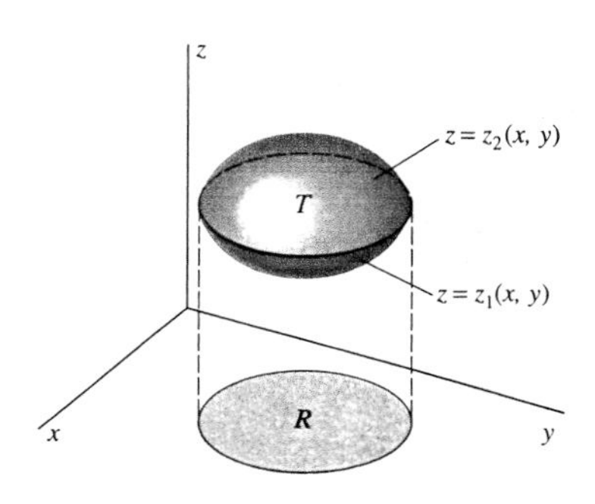

FIGURE 14.6.5 To find the boundary of R, solve the equation $z_1(x, y) = z_2(x, y)$.

If the solid T is bounded by the *two* surfaces $z = z_1(x, y)$ and $z = z_2(x, y)$ (as in Fig. 14.6.5), then we can find the "base region" R in Eq. (6) as follows. Note that the equation $z_1(x, y) = z_2(x, y)$ determines a vertical cylinder (not necessarily circular) that passes through the curve of intersection of the two surfaces. (Why?) This cylinder intersects the xy-plane in the boundary curve C of the plane region R. In essence, we obtain the equation of the curve C by equating the height functions of the surfaces that form the top and bottom of the space region T.

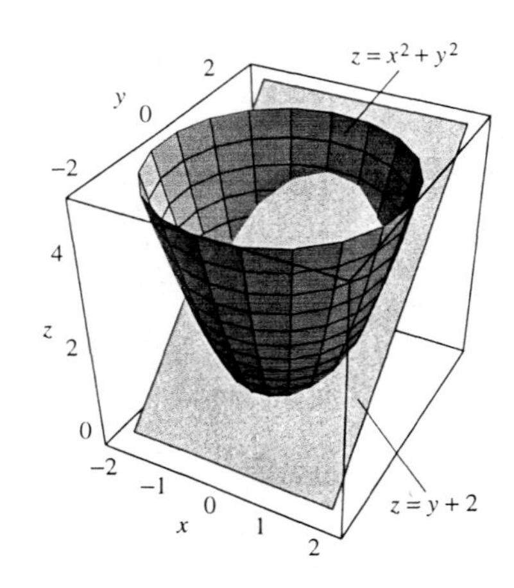

FIGURE 14.6.6 The solid T of Example 3.

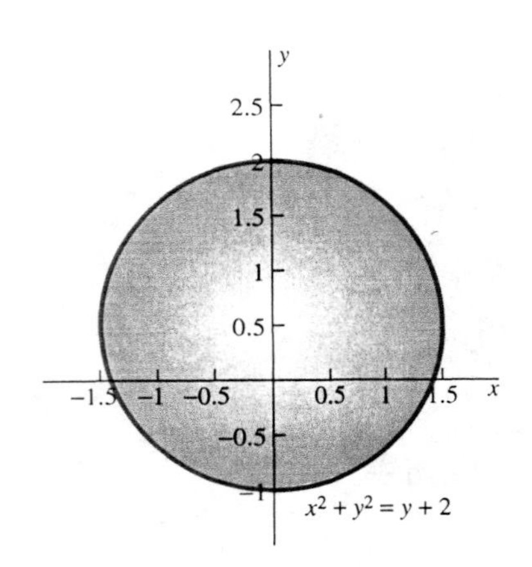

FIGURE 14.6.7 The circular disk R of Example 3.

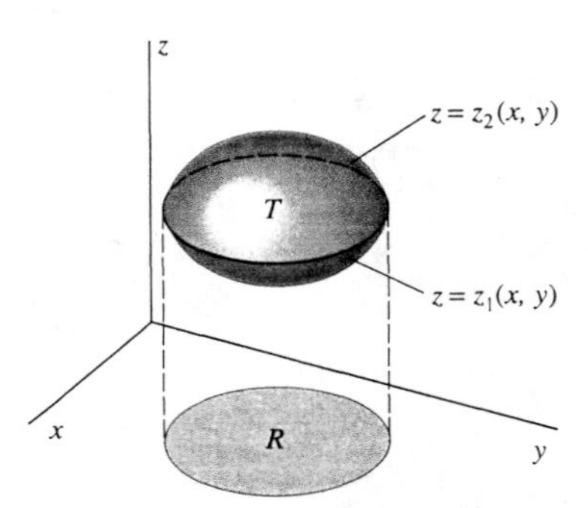

(a) T is z-simple

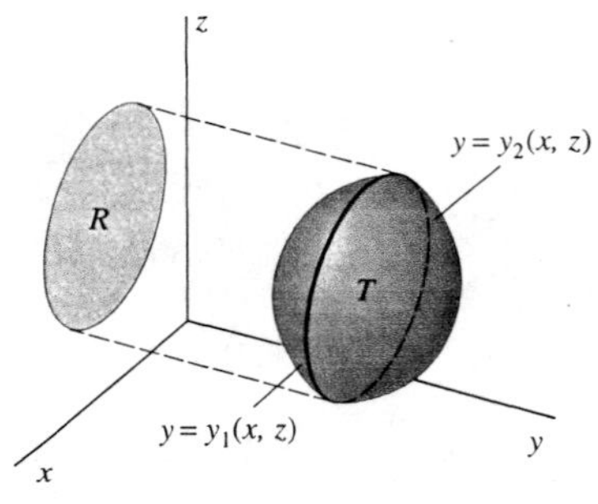

(b) T is y-simple

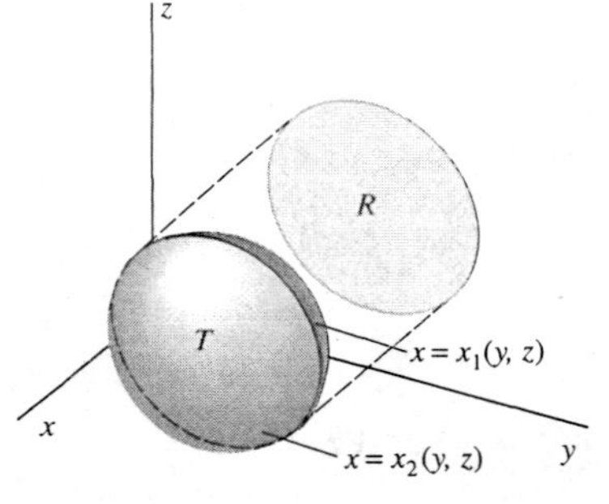

(c) T is x-simple

FIGURE 14.6.8 Solids that are (a) z-simple, (b) y-simple, and (c) x-simple.

EXAMPLE 3 Figure 14.6.6 shows the solid T bounded above by the plane $z = y + 2$ and below by the paraboloid $z = x^2 + y^2$. The equation

$$x^2 + y^2 = y + 2; \quad \text{that is,} \quad x^2 + \left(y - \tfrac{1}{2}\right)^2 = \tfrac{9}{4}$$

describes the boundary circle of the disk R of radius $\frac{3}{2}$ and with center $(0, \frac{1}{2})$ in the xy-plane (Fig. 14.6.7). Because this disk is not centered at the origin, the volume integral

$$V = \iint_R \left(\int_{z=x^2+y^2}^{y+2} dz \right) dA$$

is awkward to evaluate directly. In Example 5 we calculate V by integrating in a different order. ◆

We may integrate first with respect to either x or y if the space region T is either **x-simple** or **y-simple.** Such situations, as well as a z-simple solid, appear in Fig. 14.6.8. For example, suppose that T is y-simple, so that it has a description of the form

$$y_1(x, z) \leqq y \leqq y_2(x, z), \quad (x, z) \text{ in } R,$$

where R is the projection of T into the xz-plane. Then

$$\iiint_T f(x, y, z)\, dV = \iint_R \left(\int_{y_1(x,z)}^{y_2(x,z)} f(x, y, z)\, dy \right) dA, \tag{7}$$

where $dA = dx\,dz$ or $dA = dz\,dx$ and the limits $y_1(x, z)$ and $y_2(x, z)$ are the y-coordinates of the endpoints of the line segment in which a typical line parallel to the y-axis intersects T. If T is x-simple, we have

$$\iiint_T f(x, y, z)\, dA = \iint_R \left(\int_{x_1(y,z)}^{x_2(y,z)} f(x, y, z)\, dx \right) dA, \tag{8}$$

where $dA = dy\,dz$ or $dA = dz\,dy$ and R is the projection of T into the yz-plane.

EXAMPLE 4 Compute by triple integration the volume of the region T that is bounded by the parabolic cylinder $x = y^2$ and the planes $z = 0$ and $x + z = 1$. Also find the centroid of T given that it has constant density $\delta \equiv 1$.

COMMENT The three segments in Fig. 14.6.9 parallel to the coordinate axes indicate that the region T is simultaneously x-simple, y-simple, and z-simple. We may therefore integrate in any order we choose, so there are six ways to evaluate the integral. Here are three computations of the volume V of T.

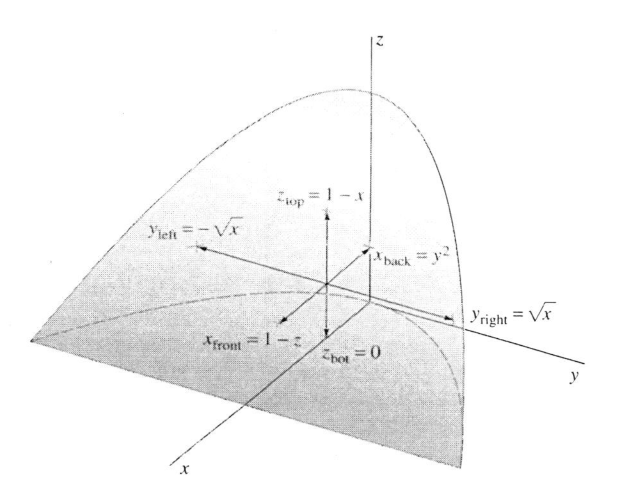

FIGURE 14.6.9 The region T of Example 4 is x-simple, y-simple, and z-simple.

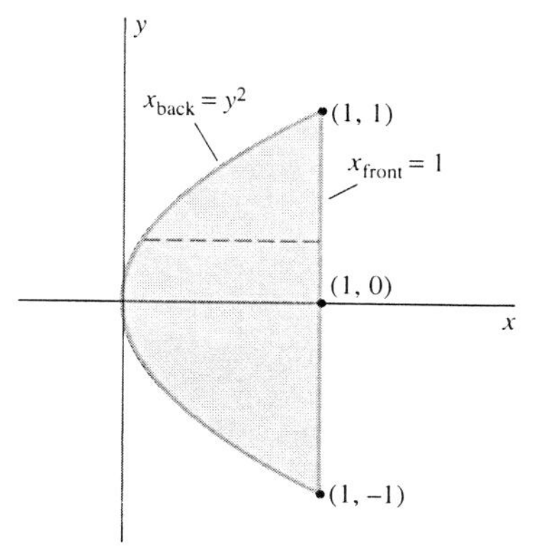

FIGURE 14.6.10 The vertical projection of the solid region T into the xy-plane (Example 4, Solution 1).

Solution 1 The projection of T into the xy-plane is the region shown in Fig. 14.6.10, bounded by $x = y^2$ and $x = 1$. So Eq. (6) gives

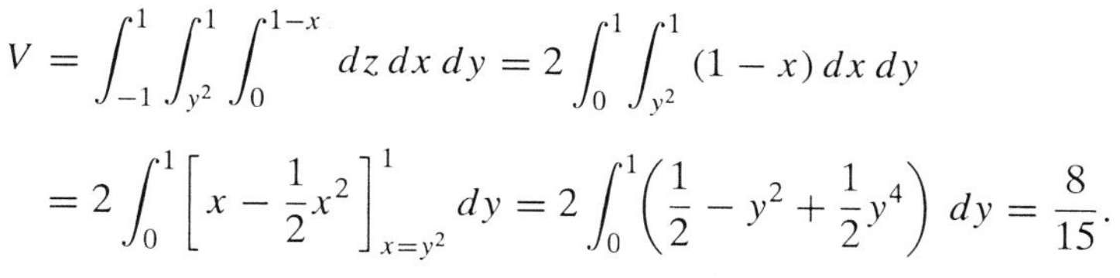

$$V = \int_{-1}^{1}\int_{y^2}^{1}\int_{0}^{1-x} dz\,dx\,dy = 2\int_{0}^{1}\int_{y^2}^{1}(1-x)\,dx\,dy$$
$$= 2\int_{0}^{1}\left[x - \frac{1}{2}x^2\right]_{x=y^2}^{1} dy = 2\int_{0}^{1}\left(\frac{1}{2} - y^2 + \frac{1}{2}y^4\right) dy = \frac{8}{15}.$$

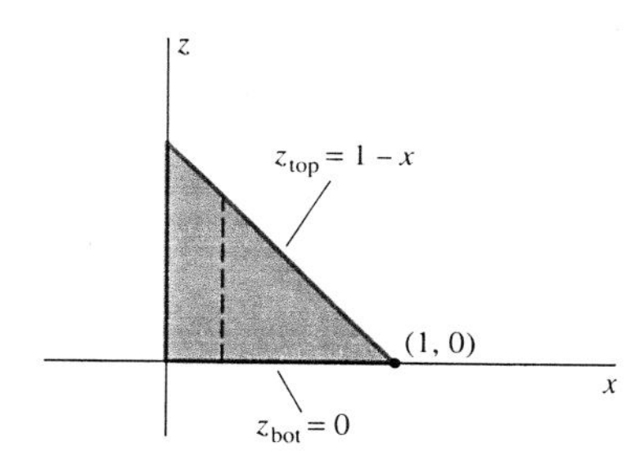

FIGURE 14.6.11 The vertical projection of the solid region T into the xz-plane (Example 4, Solution 2).

Solution 2 The projection of T into the xz-plane is the triangle bounded by the coordinate axes and the line $x + z = 1$ (Fig. 14.6.11), so Eq. (7) gives

$$V = \int_{0}^{1}\int_{0}^{1-x}\int_{-\sqrt{x}}^{\sqrt{x}} dy\,dz\,dx = 2\int_{0}^{1}\int_{0}^{1-x}\sqrt{x}\,dz\,dx$$
$$= 2\int_{0}^{1}(x^{1/2} - x^{3/2})\,dx = \frac{8}{15}.$$

Solution 3 The projection of T into the yz-plane is the triangle bounded by the y-axis and the parabola $z = 1 - y^2$ (Fig. 14.6.12), so Eq. (8) yields

$$V = \int_{-1}^{1}\int_{0}^{1-y^2}\int_{y^2}^{1-z} dx\,dz\,dy,$$

and evaluation of this integral again gives $V = \frac{8}{15}$.

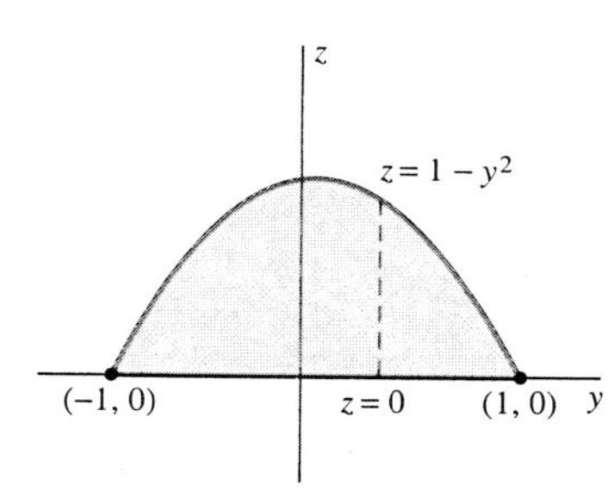

FIGURE 14.6.12 The vertical projection of the solid region T into the yz-plane (Example 4, Solution 3).

Now for the centroid of T. Because the region T is symmetric with respect to the xz-plane, its centroid lies in this plane, and so $\bar{y} = 0$. We compute $\bar{x}$ and $\bar{z}$ by integrating first with respect to y:

$$\bar{x} = \frac{1}{V}\iiint_T x\,dV = \frac{15}{8}\int_{0}^{1}\int_{0}^{1-x}\int_{-\sqrt{x}}^{\sqrt{x}} x\,dy\,dz\,dx$$
$$= \frac{15}{4}\int_{0}^{1}\int_{0}^{1-x} x^{3/2}\,dz\,dx = \frac{15}{4}\int_{0}^{1}(x^{3/2} - x^{5/2})\,dx = \frac{3}{7};$$

similarly,

$$\bar{z} = \frac{1}{V} \iiint_T z\,dV = \frac{15}{8} \int_0^1 \int_0^{1-x} \int_{-\sqrt{x}}^{\sqrt{x}} z\,dy\,dz\,dx = \frac{2}{7}.$$

Thus the centroid of T is located at the point $\left(\frac{3}{7}, 0, \frac{2}{7}\right)$. ◆

EXAMPLE 5 Find the volume of the *oblique segment of a paraboloid* bounded by the paraboloid $z = x^2 + y^2$ and the plane $z = y + 2$ (Fig. 14.6.13).

Solution The given region T is z-simple, but its projection into the xy-plane is bounded by the graph of the equation $x^2 + y^2 = y + 2$, which is a translated circle. It would be possible to integrate first with respect to z, but perhaps another choice will yield a simpler integral.

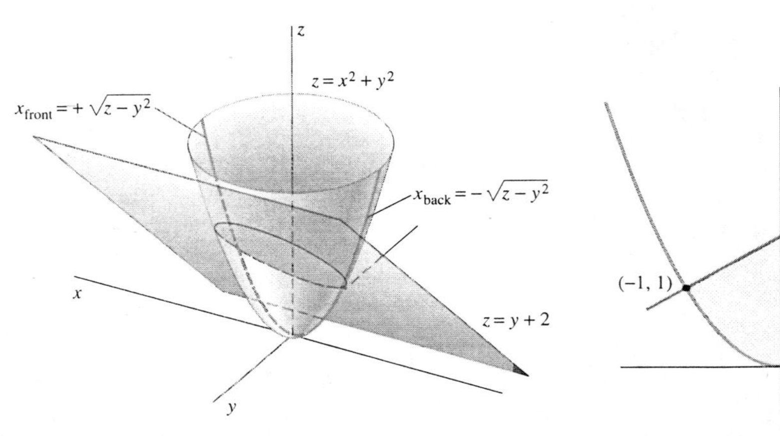

FIGURE 14.6.13 An oblique segment of a paraboloid (Example 5).

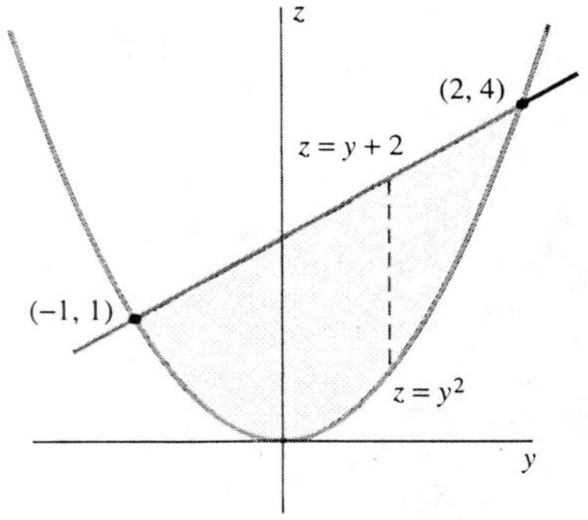

FIGURE 14.6.14 Projection of the segment of the paraboloid into the *yz*-plane (Example 5).

The region T is also x-simple, so we may integrate first with respect to x. The projection of T into the yz-plane is bounded by the line $z = y + 2$ and the parabola $z = y^2$, which intersect at the points $(-1, 1)$ and $(2, 4)$ (Fig. 14.6.14). The endpoints of a line segment in T parallel to the x-axis have x-coordinates $x = \pm\sqrt{z - y^2}$. Because T is symmetric with respect to the yz-plane, we can integrate from $x = 0$ to $x = \sqrt{z - y^2}$ and double the result. Hence T has volume

$$\begin{aligned}
V &= 2 \int_{-1}^{2} \int_{y^2}^{y+2} \int_0^{\sqrt{z-y^2}} dx\,dz\,dy = 2 \int_{-1}^{2} \int_{y^2}^{y+2} \sqrt{z - y^2}\,dz\,dy \\
&= 2 \int_{-1}^{2} \left[\frac{2}{3} \left(z - y^2 \right)^{3/2} \right]_{z=y^2}^{y+2} dy = \frac{4}{3} \int_{-1}^{2} (2 + y - y^2)^{3/2}\,dy \\
&= \frac{4}{3} \int_{-3/2}^{3/2} \left(\frac{9}{4} - u^2 \right)^{3/2} du \qquad \left(\text{completing the square;} \quad u = y - \frac{1}{2} \right) \\
&= \frac{27}{4} \int_{-\pi/2}^{\pi/2} \cos^4\theta\,d\theta \qquad \left(u = \frac{3}{2} \sin\theta \right) \\
&= \frac{27}{4} \cdot 2 \cdot \frac{1}{2} \cdot \frac{3}{4} \cdot \frac{\pi}{2} = \frac{81\pi}{32}.
\end{aligned}$$

In the final evaluation, we used symmetry—integrating from $\theta = 0$ to $\theta = \pi/2$ and doubling—and then Formula (113) (inside the back cover). ◆

14.6 TRUE/FALSE STUDY GUIDE

14.6 CONCEPTS: QUESTIONS AND DISCUSSION

1. Describe a region T in space such that you can calculate its volume by iterated integration in at least three different orders—integrating with respect to x first in one order, with respect to y first in another order, and finally with respect to z first. Then find its volume in each of these three ways.
2. (a) Give an example of a space region whose volume is most easily calculated by integrating first with respect to x. (b) Repeat, but with respect to y first. (c) Repeat, but with respect to z first.

14.6 PROBLEMS

In Problems 1 through 10, compute the value of the triple integral

$$\iiint_T f(x, y, z)\, dV.$$

1. $f(x, y, z) = x + y + z$; T is the rectangular box $0 \leqq x \leqq 2$, $0 \leqq y \leqq 3$, $0 \leqq z \leqq 1$.
2. $f(x, y, z) = xy \sin z$; T is the cube $0 \leqq x \leqq \pi$, $0 \leqq y \leqq \pi$, $0 \leqq z \leqq \pi$.
3. $f(x, y, z) = xyz$; T is the rectangular block $-1 \leqq x \leqq 3$, $0 \leqq y \leqq 2$, $-2 \leqq z \leqq 6$.
4. $f(x, y, z) = x + y + z$; T is the rectangular block of Problem 3.
5. $f(x, y, z) = x^2$; T is the tetrahedron bounded by the coordinate planes and the first octant part of the plane with equation $x + y + z = 1$.
6. $f(x, y, z) = 2x + 3y$; T is a first-octant tetrahedron as in Problem 5, except that the plane has equation $2x + 3y + z = 6$.
7. $f(x, y, z) = xyz$; T lies below the surface $z = 1 - x^2$ and above the rectangle $-1 \leqq x \leqq 0$, $0 \leqq y \leqq 2$ in the xy-plane.
8. $f(x, y, z) = 2y + z$; T lies below the surface with equation $z = 4 - y^2$ and above the rectangle $-1 \leqq x \leqq 1$, $-2 \leqq y \leqq 2$ in the xy-plane.
9. $f(x, y, z) = x + y$; T is the region between the surfaces $z = 2 - x^2$ and $z = x^2$ for $0 \leqq y \leqq 3$ (Fig. 14.6.15).
10. $f(x, y, z) = z$; T is the region between the surfaces $z = y^2$ and $z = 8 - y^2$ for $-1 \leqq x \leqq 1$.

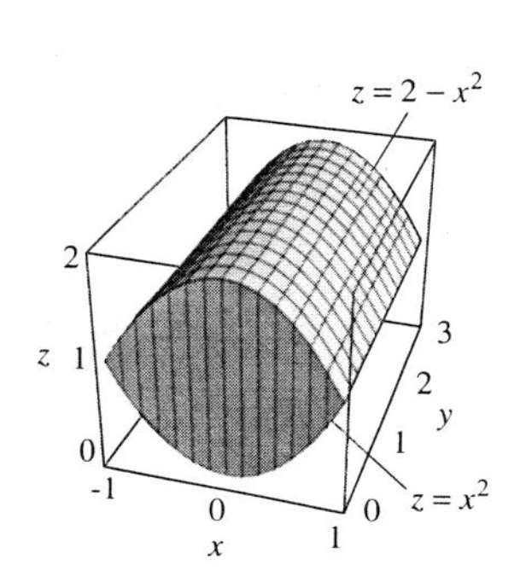

FIGURE 14.6.15 The solid of Problem 9.

In Problems 11 through 20, sketch the solid bounded by the graphs of the given equations. Then find its volume by triple integration.

11. $2x + 3y + z = 6$, $x = 0$, $y = 0$, $z = 0$
12. $z = y$, $y = x^2$, $y = 4$, $z = 0$ (Fig. 14.6.16)

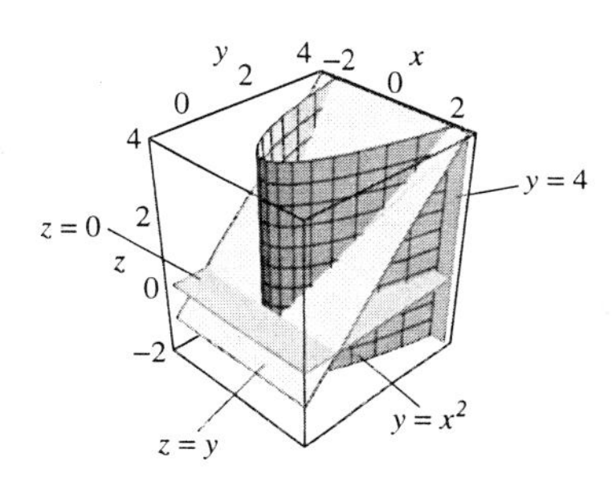

FIGURE 14.6.16 The surfaces of Problem 12.

13. $y + z = 4$, $y = 4 - x^2$, $y = 0$, $z = 0$
14. $z = x^2 + y^2$, $z = 0$, $x = 0$, $y = 0$, $x + y = 1$
15. $z = 10 - x^2 - y^2$, $y = x^2$, $x = y^2$, $z = 0$
16. $x = z^2$, $x = 8 - z^2$, $y = -1$, $y = -3$
17. $z = x^2$, $y + z = 4$, $y = 0$, $z = 0$
18. $z = 1 - y^2$, $z = y^2 - 1$, $x + z = 1$, $x = 0$ (Fig. 14.6.17)
19. $y = z^2$, $z = y^2$, $x + y + z = 2$, $x = 0$
20. $y = 4 - x^2 - z^2$, $x = 0$, $y = 0$, $z = 0$, $x + z = 2$

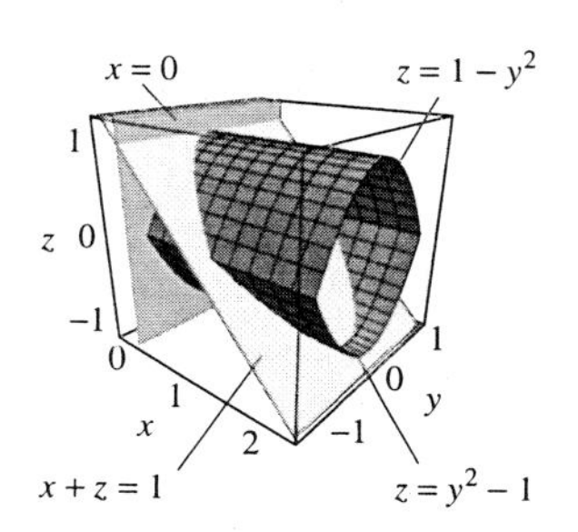

FIGURE 14.6.17 The surfaces of Problem 18.

In Problems 21 through 32, assume that the indicated solid has constant density $\delta \equiv 1$.

21. Find the centroid of the solid of Problem 12.

22. Find the centroid of the hemisphere

$$x^2 + y^2 + z^2 \leqq R^2, \quad z \geqq 0.$$

23. Find the centroid of the solid of Problem 17.

24. Find the centroid of the solid bounded by $z = 1 - x^2$, $z = 0$, $y = -1$, and $y = 1$.

25. Find the centroid of the solid bounded by $z = \cos x$, $x = -\pi/2$, $x = \pi/2$, $y = 0$, $z = 0$, and $y + z = 1$.

26. Find the moment of inertia around the z-axis of the solid of Problem 12.

27. Find the moment of inertia around the y-axis of the solid of Problem 24.

28. Find the moment of inertia around the z-axis of the solid cylinder $x^2 + y^2 \leqq R^2$, $0 \leqq z \leqq H$.

29. Find the moment of inertia around the z-axis of the solid bounded by $x + y + z = 1$, $x = 0$, $y = 0$, and $z = 0$.

30. Find the moment of inertia around the z-axis of the cube with vertices $(\pm\frac{1}{2}, 3, \pm\frac{1}{2})$ and $(\pm\frac{1}{2}, 4, \pm\frac{1}{2})$.

31. Consider the solid paraboloid bounded by $z = x^2 + y^2$ and the plane $z = h > 0$. Show that its centroid lies on its axis of symmetry, two-thirds of the way from its "vertex" $(0, 0, 0)$ to its base.

32. Show that the centroid of a right circular cone lies on the axis of the cone and three-fourths of the way from the vertex to the base.

In Problems 33 through 40, the indicated solid has uniform density $\delta \equiv 1$ unless otherwise indicated.

33. For a cube with edge length a, find the moment of inertia around one of its edges.

34. The density at $P(x, y, z)$ of the first-octant cube with edge length a, faces parallel to the coordinate planes, and opposite vertices $(0, 0, 0)$ and (a, a, a) is proportional to the square of the distance from P to the origin. Find the coordinates of the centroid of this cube.

35. Find the moment of inertia around the z-axis of the cube of Problem 34.

36. The cube bounded by the coordinate planes and the planes $x = 1$, $y = 1$, and $z = 1$ has density $\delta = kz$ at the point $P(x, y, z)$ (k is a positive constant). Find its centroid.

37. Find the moment of inertia around the z-axis of the cube of Problem 36.

38. Find the moment of inertia around a diameter of a solid sphere of radius a.

39. Find the centroid of the first-octant region that is interior to the two cylinders $x^2 + z^2 = 1$ and $y^2 + z^2 = 1$ (Figs. 14.6.18 and 14.6.19).

40. Find the moment of inertia around the z-axis of the solid of Problem 39.

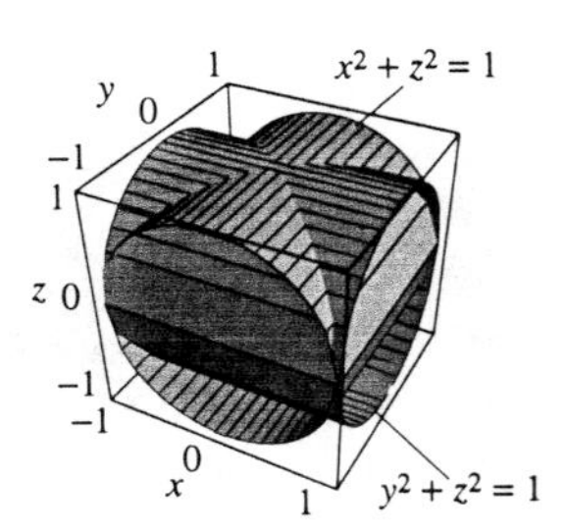

FIGURE 14.6.18 The intersecting cylinders of Problem 39.

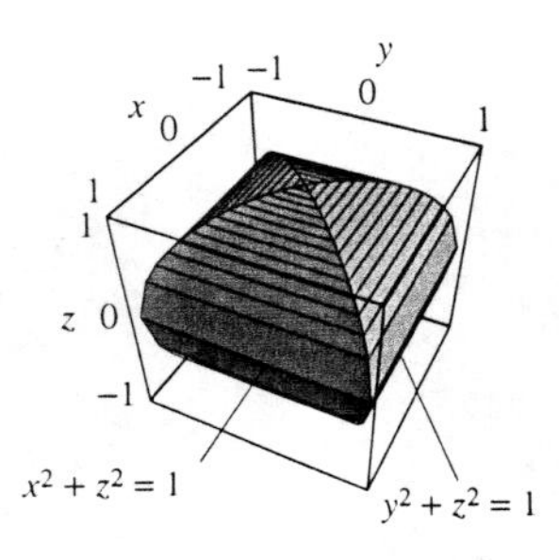

FIGURE 14.6.19 The solid of intersection in Problem 39.

41. Find the volume bounded by the elliptic paraboloids $z = 2x^2 + y^2$ and $z = 12 - x^2 - 2y^2$. Note that this solid projects onto a circular disk in the xy-plane.

42. Find the volume bounded by the elliptic paraboloid $y = x^2 + 4z^2$ and the plane $y = 2x + 3$.

43. Find the volume of the elliptical cone bounded by $z = \sqrt{x^2 + 4y^2}$ and the plane $z = 1$. [*Suggestion:* Integrate first with respect to x.]

44. Find the volume of the region bounded by the paraboloid $x = y^2 + 2z^2$ and the parabolic cylinder $x = 2 - y^2$ (Fig. 14.6.20).

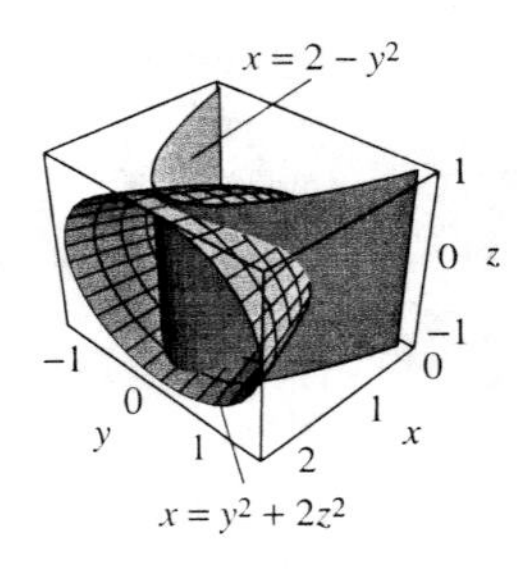

FIGURE 14.6.20 The surfaces of Problem 44.

45. Find the centroid of the pyramid in Example 2 with density $\delta(x, y, z) = z$.

46. Find the centroid of the parabolic segment (with density $\delta \equiv 1$) in Example 5.

For Problems 47 through 52, the ***average value*** $\overline{f}$ *of the function $f(x, y, z)$ at points of the space region T is defined to be*

$$\overline{f} = \frac{1}{V}\iiint_T f(x, y, z)\,dV$$

where V is the volume of T. For instance, if T is a solid with density $\delta \equiv 1$, then the coordinates $\overline{x}$, $\overline{y}$, and $\overline{z}$ of its centroid are the average values of the "coordinate functions" x, y, and z at points of T.

47. Find the average value of the density function $\delta(x, y, z) = z$ at points of the pyramid T of Example 2.

48. Suppose that T is the unit cube in the first octant with diagonally opposite vertices $(0, 0, 0)$ and $(1, 1, 1)$. Find the average of the "squared distance" $f(x, y, z) = x^2 + y^2 + z^2$ of points of T from the origin.

49. Let T be the cube of Problem 48. Find the average squared distance of points of T from its centroid.

50. Let T be the cube of Problem 48, but with density function $\delta(x, y, z) = x + y + z$ that varies linearly from 0 at the origin to 3 at the opposite vertex of T. Find the average value $\overline{\delta}$ of the density of T. Can you guess the value of $\overline{\delta}$ before evaluating the triple integral?

51. Find the average squared distance from the origin of points of the pyramid of Example 2.

52. Suppose that T is the pyramid of Example 2, but with density function $\delta \equiv 1$. Find the average squared distance of points of T from its centroid.

53. Use a computer algebra system to find the average distance of points of the cube T of Problem 48 from the origin. [*Answer:*

$$\frac{1}{24}\left[6\sqrt{3} - \pi + 8\ln\left(\sqrt{3} + \frac{1}{2}\sqrt{2}\right) - 8\ln 2 + 16\ln\left(1 + \sqrt{3}\right)\right] \approx 0.960591956455.]$$

14.6 Project: Archimedes' Floating Paraboloid

Archimedes was interested in floating bodies and studied the possible position (see Fig. 14.6.21) of a floating right circular paraboloid of uniform density. For a paraboloid that floats in an "inclined position," he discovered how to determine its angle of inclination in terms of the volume and centroid of the "oblique segment" of the paraboloid that lies beneath the water line. The principles he introduced for this investigation (over 22 centuries ago) are still important in modern naval architecture.

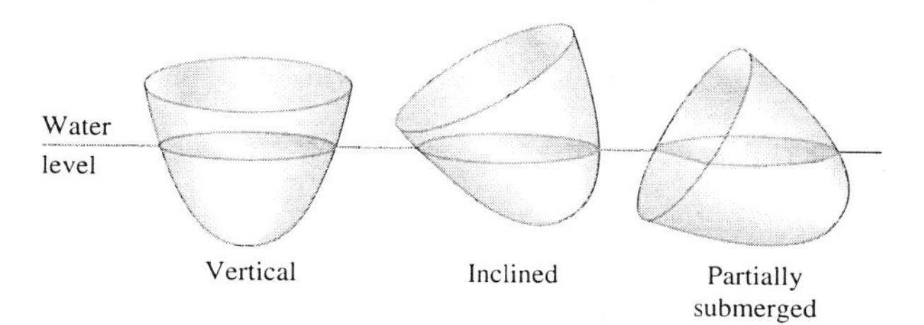

FIGURE 14.6.21 How a uniform solid paraboloid might float.

For your own personal paraboloid to investigate, let T be the three-dimensional solid region bounded below by the paraboloid $z = x^2 + y^2$ and above by the plane $z = (b - a)y + ab$, where a and b are the smallest and largest nonzero digits (respectively) of your student I.D. number. (If $a = 1$ and $b = 2$ then T is the solid of Example 5.) In the following problems you can evaluate the triple integrals either by hand—consulting an integral table if you wish—or by using a computer algebra system.

1. Find the volume V of the solid oblique paraboloid T. Sketch a picture of T similar to Fig. 14.6.13. Can you see that T is symmetric with respect to the yz-plane? Describe the region R in the yz-plane that is the vertical projection of T. This plane region will determine the z-limits and the y-limits of your triple integral (as in Example 5).

2. Find the coordinates $(\overline{x}, \overline{y}, \overline{z})$ of the centroid C of T (assume that T has density $\delta \equiv 1$).

3. Find the coordinates of the point P at which a plane parallel to the original top plane $z = (b - a)y + ab$ is tangent to the paraboloid. Also find the coordinates of the point Q in which a vertical line through P intersects the top plane. According to Archimedes, the centroid C of Problem 2 should lie on the line PQ two-thirds of the way from P to Q. Is this so, according to your computations? (Compare with Problem 31 of this section.)

14.7 INTEGRATION IN CYLINDRICAL AND SPHERICAL COORDINATES

Suppose that $f(x, y, z)$ is a continuous function defined on the z-simple region T, which—because it is z-simple—can be described by

$$z_1(x, y) \leqq z \leqq z_2(x, y) \quad \text{for } (x, y) \text{ in } R$$

(where R is the projection of T into the xy-plane, as usual). We saw in Section 14.6 that

$$\iiint_T f(x, y, z)\, dV = \iint_R \left(\int_{z_1(x,y)}^{z_2(x,y)} f(x, y, z)\, dz \right) dA. \tag{1}$$

If we can describe the region R more naturally in polar coordinates than in rectangular coordinates, then it is likely that the integration over the plane region R will be simpler if it is carried out in polar coordinates.

We first express the inner partial integral of Eq. (1) in terms of r and θ by writing

$$\int_{z_1(x,y)}^{z_2(x,y)} f(x, y, z)\, dz = \int_{Z_1(r,\theta)}^{Z_2(r,\theta)} F(r, \theta, z)\, dz, \tag{2}$$

where

$$F(r, \theta, z) = f(r \cos\theta, r \sin\theta, z) \tag{3a}$$

and

$$Z_i(r, \theta) = z_i(r \cos\theta, r \sin\theta) \tag{3b}$$

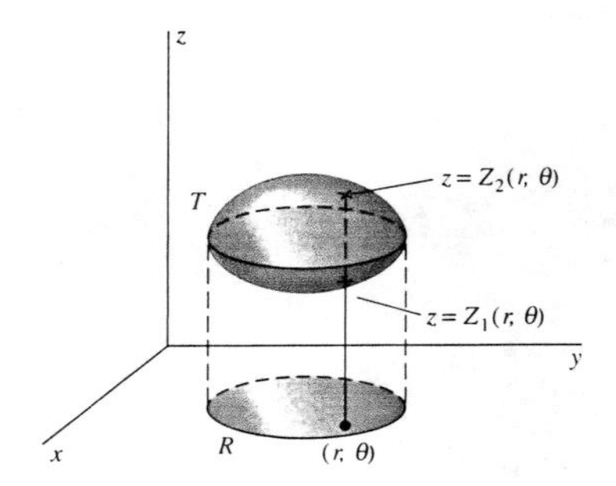

FIGURE 14.7.1 The limits on z in a triple integral in cylindrical coordinates are determined by the lower and upper surfaces.

for $i = 1, 2$. Substituting Eq. (2) into Eq. (1) with $dA = r\, dr\, d\theta$ (**important**) gives

$$\iiint_T f(x, y, z)\, dV = \iint_S \left(\int_{Z_1(r,\theta)}^{Z_2(r,\theta)} F(r, \theta, z)\, dz \right) r\, dr\, d\theta, \tag{4}$$

where F, Z_1, and Z_2 are the functions given in (3) and S represents the appropriate limits on r and θ needed to describe the plane region R in polar coordinates (as discussed in Section 14.4). The limits on z are simply the z-coordinates (in terms of r and θ) of a typical line segment joining the lower and upper boundary surfaces of T, as indicated in Fig. 14.7.1.

Thus the general formula for **triple integration in cylindrical coordinates** is

$$\iiint_T f(x, y, z)\, dV = \iiint_U f(r \cos\theta, r \sin\theta, z)\, r\, dz\, dr\, d\theta, \tag{5}$$

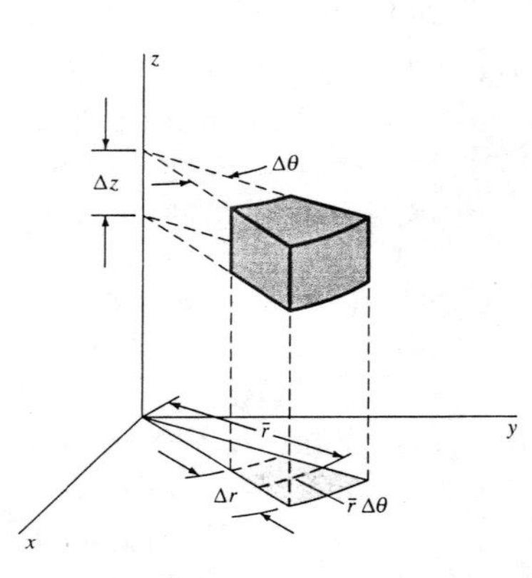

FIGURE 14.7.2 The volume of the cylindrical block is $\Delta V = \bar{r}\, \Delta z\, \Delta r\, \Delta\theta$.

where U is not a region in xyz-space, but—as in Section 14.4—a representation of limits on z, r, and θ appropriate to describe the space region T in cylindrical coordinates. Before we integrate, we must replace the variables x and y with $r \cos\theta$ and $r \sin\theta$, respectively, but z is left unchanged. The cylindrical-coordinates volume element

$$dV = r\, dz\, dr\, d\theta$$

may be regarded informally as the product of dz and the polar-coordinates area element $dA = r\, dr\, d\theta$. It is a consequence of the formula $\Delta V = \bar{r}\, \Delta z\, \Delta r\, \Delta\theta$ for the volume of the *cylindrical block* shown in Fig. 14.7.2.

Integration in cylindrical coordinates is particularly useful for computations associated with solids of revolution. So that the limits of integration will be the simplest, the solid should usually be placed so that the axis of revolution is the z-axis.

EXAMPLE 1 Find the centroid of the first-octant portion T of the solid ball bounded by the sphere $r^2 + z^2 = a^2$. The solid T appears in Fig. 14.7.3.

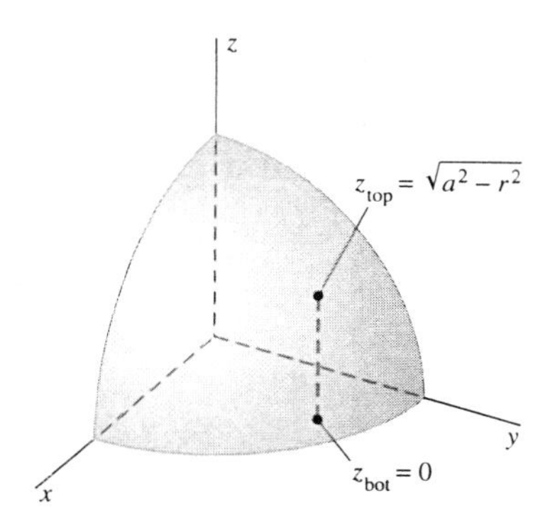

FIGURE 14.7.3 The first octant of the sphere (Example 1).

Solution The volume of the first octant of the solid ball is $V = \frac{1}{8} \cdot \frac{4}{3}\pi a^3 = \frac{1}{6}\pi a^3$. Because $\bar{x} = \bar{y} = \bar{z}$ by symmetry, we need calculate only

$$\begin{aligned}
\bar{z} &= \frac{1}{V}\iiint_T z\,dV = \frac{6}{\pi a^3}\int_0^{\pi/2}\int_0^a\int_0^{\sqrt{a^2-r^2}} zr\,dz\,dr\,d\theta \\
&= \frac{6}{\pi a^3}\int_0^{\pi/2}\int_0^a \frac{1}{2}r(a^2 - r^2)\,dr\,d\theta \\
&= \frac{3}{\pi a^3}\int_0^{\pi/2}\left[\frac{1}{2}a^2r^2 - \frac{1}{4}r^4\right]_{r=0}^a d\theta = \frac{3}{\pi a^3}\cdot\frac{\pi}{2}\cdot\frac{a^4}{4} = \frac{3a}{8}.
\end{aligned}$$

Thus the centroid is located at the point $(\frac{3}{8}a, \frac{3}{8}a, \frac{3}{8}a)$. Observe that the answer is both plausible and dimensionally correct. ◆

EXAMPLE 2 Find the volume and centroid of the solid T that is bounded by the paraboloid $z = b(x^2 + y^2)$ $(b > 0)$ and the plane $z = h$ $(h > 0)$.

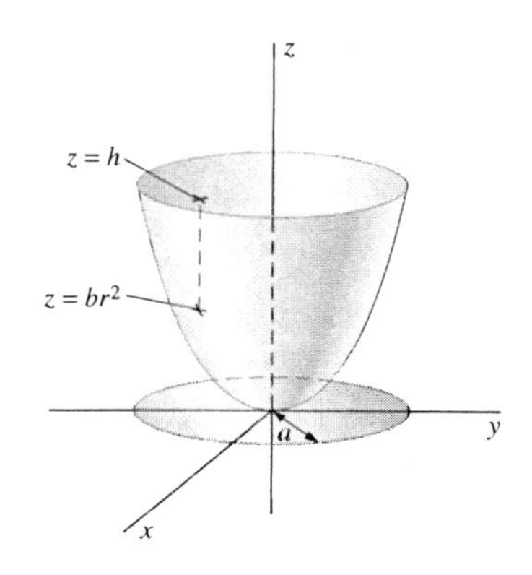

FIGURE 14.7.4 The paraboloid of Example 2.

Solution Figure 14.7.4 makes it clear that we get the radius of the circular top of T by equating $z = b(x^2 + y^2) = br^2$ and $z = h$. This gives $a = \sqrt{h/b}$ for the radius of the circle over which the solid lies. Hence Eq. (4), with $f(x, y, z) \equiv 1$, gives the volume:

$$\begin{aligned}
V &= \iiint_T dV = \int_0^{2\pi}\int_0^a\int_{br^2}^h r\,dz\,dr\,d\theta = \int_0^{2\pi}\int_0^a (hr - br^3)\,dr\,d\theta \\
&= 2\pi\left(\frac{1}{2}ha^2 - \frac{1}{4}ba^4\right) = \frac{\pi h^2}{2b} = \frac{1}{2}\pi a^2 h
\end{aligned}$$

(because $a^2 = h/b$).

By symmetry, the centroid of T lies on the z-axis, so all that remains is to compute $\bar{z}$:

$$\begin{aligned}
\bar{z} &= \frac{1}{V}\iiint_T z\,dV = \frac{2}{\pi a^2 h}\int_0^{2\pi}\int_0^a\int_{br^2}^h rz\,dz\,dr\,d\theta \\
&= \frac{2}{\pi a^2 h}\int_0^{2\pi}\int_0^a\left(\frac{1}{2}h^2 r - \frac{1}{2}b^2 r^5\right)dr\,d\theta \\
&= \frac{4}{a^2 h}\left(\frac{1}{4}h^2 a^2 - \frac{1}{12}b^2 a^6\right) = \frac{2}{3}h,
\end{aligned}$$

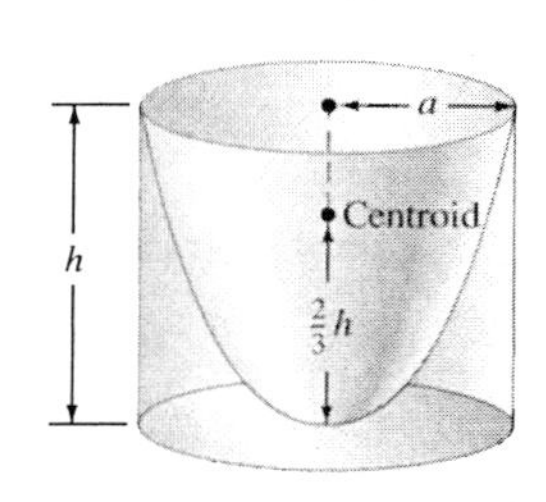

FIGURE 14.7.5 Volume and centroid of a right circular paraboloid in terms of the circumscribed cylinder.

again using the fact that $a^2 = h/b$. Therefore the centroid of T is located at the point $(0, 0, \frac{2}{3}h)$. Again, this answer is both plausible and dimensionally correct. ◆

We can summarize the results of Example 2 as follows: The volume of a right circular paraboloid is *half* that of the circumscribed cylinder (Fig. 14.7.5), and its centroid lies on its axis of symmetry *two-thirds* of the way from the "vertex" at $(0, 0, 0)$ to its circular "base" at the top.

Spherical Coordinate Integrals

When the boundary surfaces of the region T of integration are spheres, cones, or other surfaces with simple descriptions in spherical coordinates, it is generally advantageous to transform a triple integral over T into spherical coordinates. Recall from Section 12.8 that the relationship between spherical coordinates (ρ, ϕ, θ) (shown in

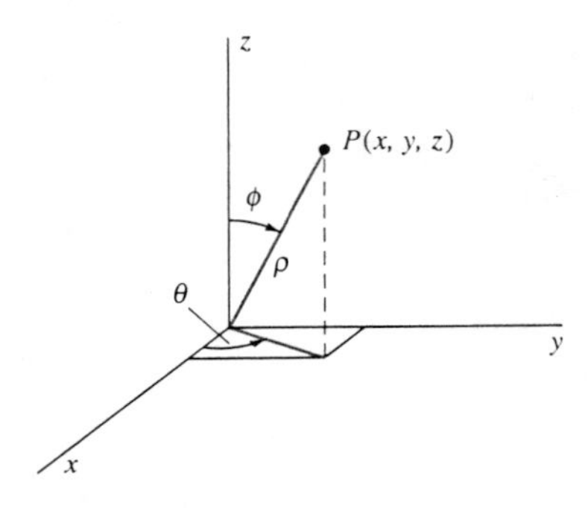

FIGURE 14.7.6 The spherical coordinates (ρ, ϕ, θ) of the point P.

Fig. 14.7.6) and rectangular coordinates (x, y, z) is given by

$$x = \rho \sin\phi \cos\theta, \qquad y = \rho \sin\phi \sin\theta, \qquad z = \rho \cos\phi. \tag{6}$$

Suppose, for example, that T is the **spherical block** determined by the simple inequalities

$$\begin{aligned} \rho_1 &\leqq \rho \leqq \rho_2 = \rho_1 + \Delta\rho, \\ \phi_1 &\leqq \phi \leqq \phi_2 = \phi_1 + \Delta\phi, \\ \theta_1 &\leqq \theta \leqq \theta_2 = \theta_1 + \Delta\theta. \end{aligned} \tag{7}$$

As indicated by the dimensions labeled in Fig. 14.7.7, this spherical block is (if $\Delta\rho$, $\Delta\phi$, and $\Delta\theta$ are small) *approximately* a rectangular block with dimensions $\Delta\rho$, $\rho_1\,\Delta\phi$, and $\rho_1 \sin\phi_2\,\Delta\theta$. Thus its volume is approximately $\rho_1^2 \sin\phi_2\,\Delta\rho\,\Delta\phi\,\Delta\theta$. It can be shown (see Problem 19 of Section 14.8) that the *exact* volume of the spherical block described in (7) is

$$\Delta V = \hat{\rho}^2 \sin\hat{\phi}\,\Delta\rho\,\Delta\phi\,\Delta\theta \tag{8}$$

for certain numbers $\hat{\rho}$ and $\hat{\phi}$ such that $\rho_1 < \hat{\rho} < \rho_2$ and $\phi_1 < \hat{\phi} < \phi_2$.

Now suppose that we partition each of the intervals $[\rho_1, \rho_2]$, $[\phi_1, \phi_2]$, and $[\theta_1, \theta_2]$ into n subintervals of lengths

$$\Delta\rho = \frac{\rho_2 - \rho_1}{n}, \qquad \Delta\phi = \frac{\phi_2 - \phi_1}{n}, \qquad \text{and} \qquad \Delta\theta = \frac{\theta_2 - \theta_1}{n},$$

respectively. This produces a **spherical partition** $\mathcal{P}$ of the spherical block T into $k = n^3$ smaller spherical blocks $T_1, T_2, \dots, T_k$; see Fig. 14.7.8. By Eq. (8), there exists a point $(\hat{\rho}_i, \hat{\phi}_i, \hat{\theta}_i)$ of the spherical block T_i such that its volume is $\Delta V_i = \hat{\rho}_i^2 \sin\hat{\phi}_i\,\Delta\rho\,\Delta\phi\,\Delta\theta$. The **norm** $|\mathcal{P}|$ of $\mathcal{P}$ is the length of the longest diagonal of any of the small spherical blocks $T_1, T_2, \dots, T_k$.

If $(x_i^\star, y_i^\star, z_i^\star)$ are the rectangular coordinates of the point with spherical coordinates $(\hat{\rho}_i, \hat{\phi}_i, \hat{\theta}_i)$, then the definition of the triple integral as a limit of Riemann sums as the norm $|\mathcal{P}|$ approaches zero gives

$$\begin{aligned} \iiint_T f(x, y, z)\,dV &= \lim_{|\mathcal{P}|\to 0} \sum_{i=1}^{k} f(x_i^\star, y_i^\star, z_i^\star)\,\Delta V_i \\ &= \lim_{|\mathcal{P}|\to 0} \sum_{i=1}^{k} F(\hat{\rho}_i, \hat{\phi}_i, \hat{\theta}_i)\hat{\rho}_i^2 \sin\hat{\phi}_i\,\Delta\rho\,\Delta\phi\,\Delta\theta, \end{aligned} \tag{9}$$

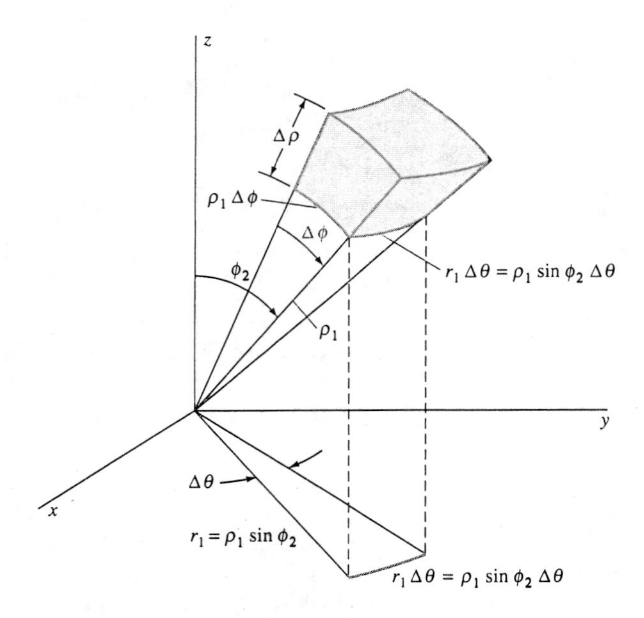

FIGURE 14.7.7 The volume of the spherical block is approximately $\rho_1^2 \sin\phi_2\,\Delta\rho\,\Delta\phi\,\Delta\theta$.

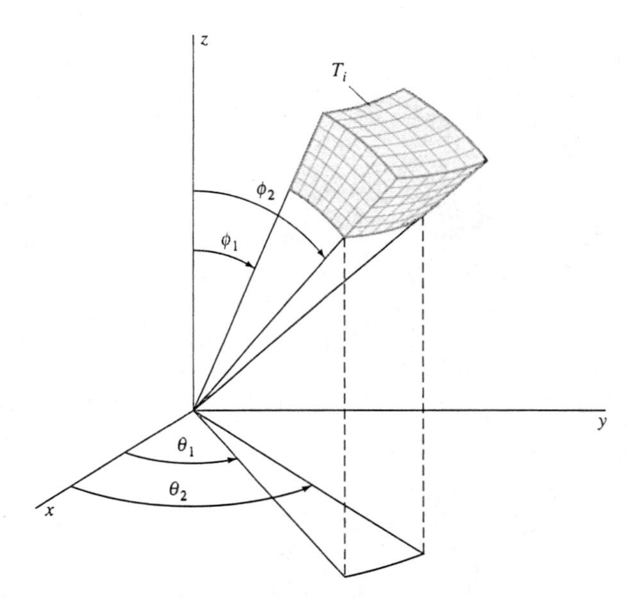

FIGURE 14.7.8 The spherical block T divided into k smaller spherical blocks.

where

$$F(\rho, \phi, \theta) = f(\rho \sin\phi \cos\theta, \rho \sin\phi \sin\theta, \rho \cos\phi) \tag{10}$$

is the result of substituting Eq. (6) into $f(x, y, z)$. But the right-hand sum in Eq. (9) is simply a Riemann sum for the triple integral

$$\int_{\theta_1}^{\theta_2} \int_{\phi_1}^{\phi_2} \int_{\rho_1}^{\rho_2} F(\rho, \phi, \theta)\, \rho^2 \sin\phi \; d\rho \, d\phi \, d\theta.$$

It therefore follows that

$$\iiint_T f(x, y, z)\, dV = \int_{\theta_1}^{\theta_2} \int_{\phi_1}^{\phi_2} \int_{\rho_1}^{\rho_2} F(\rho, \phi, \theta)\, \rho^2 \sin\phi \; d\rho \, d\phi \, d\theta. \tag{11}$$

Thus we transform the integral

$$\iiint_T f(x, y, z)\, dV$$

into spherical coordinates by replacing the rectangular-coordinate variables x, y, and z with their expressions in Eq. (6) in terms of the spherical-coordinate variables ρ, ϕ, and θ. In addition, we write

$$dV = \rho^2 \sin\phi \; d\rho \, d\phi \, d\theta$$

for the volume element in spherical coordinates.

More generally, we can transform the triple integral

$$\iiint_T f(x, y, z)\, dV$$

into spherical coordinates whenever the region T is **centrally simple**—that is, whenever it has a spherical-coordinates description of the form

$$\rho_1(\phi, \theta) \leqq \phi \leqq \rho_2(\phi, \theta), \qquad \phi_1 \leqq \phi \leqq \phi_2, \qquad \theta_1 \leqq \theta \leqq \theta_2. \tag{12}$$

If so, then

$$\iiint_T f(x, y, z)\, dV = \int_{\theta_1}^{\theta_2} \int_{\phi_1}^{\phi_2} \int_{\rho_1(\phi,\theta)}^{\rho_2(\phi,\theta)} F(\rho, \phi, \theta)\, \rho^2 \sin\phi \; d\rho \, d\phi \, d\theta. \tag{13}$$

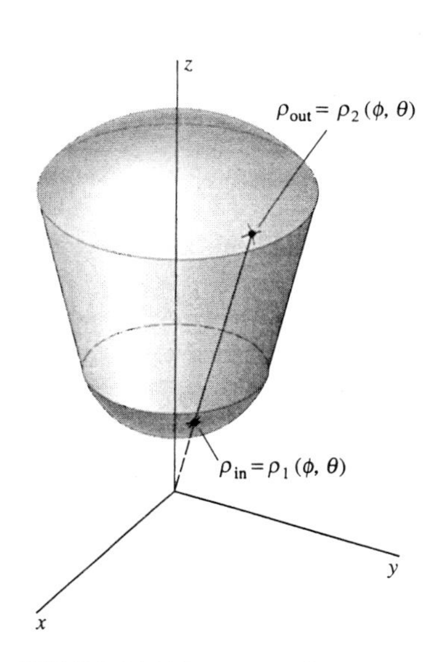

FIGURE 14.7.9 A centrally simple region.

The limits on ρ in Eq. (13) are simply the ρ-coordinates (in terms of ϕ and θ) of the endpoints of a typical radial segment that joins the "inner" and "outer" parts of the boundary of T (Fig. 14.7.9). Thus the general formula for **triple integration in spherical coordinates** is

$$\iiint_T f(x, y, z)\, dV = \iiint_U f(\rho \sin\phi \cos\theta, \rho \sin\phi \sin\theta, \rho \cos\phi)\, \rho^2 \sin\phi \; d\rho \, d\phi \, d\theta, \tag{14}$$

where, as before, U does not denote a region in xyz-space but rather indicates limits on ρ, ϕ, and θ appropriate to describe the region T in spherical coordinates.

EXAMPLE 3 A solid ball T with constant density δ is bounded by the spherical surface with equation $\rho = a$. Use spherical coordinates to compute its volume V and its moment of inertia I_z around the z-axis.

Solution The points of the ball T are described by the inequalities

$$0 \leqq \rho \leqq a, \qquad 0 \leqq \phi \leqq \pi, \qquad 0 \leqq \theta \leqq 2\pi.$$

We take $f = F \equiv 1$ in Eq. (11) and thereby obtain

$$\begin{aligned} V &= \iiint_T dV = \int_0^{2\pi}\int_0^{\pi}\int_0^{a} \rho^2 \sin\phi \, d\rho \, d\phi \, d\theta \\ &= \frac{1}{3}a^3 \int_0^{2\pi}\int_0^{\pi} \sin\phi \, d\phi \, d\theta \\ &= \frac{1}{3}a^3 \int_0^{2\pi} \Big[-\cos\phi\Big]_{\phi=0}^{\pi} d\theta = \frac{2}{3}a^3 \int_0^{2\pi} d\theta = \frac{4}{3}\pi a^3. \end{aligned}$$

The distance from the typical point (ρ, ϕ, θ) of the sphere to the z-axis is $r = \rho \sin\phi$, so the moment of inertia of the sphere around that axis is

$$\begin{aligned} I_z &= \iiint_T r^2 \delta \, dV = \int_0^{2\pi}\int_0^{\pi}\int_0^{a} \delta\rho^4 \sin^3\phi \, d\rho \, d\phi \, d\theta \\ &= \frac{1}{5}\delta a^5 \int_0^{2\pi}\int_0^{\pi} \sin^3\phi \, d\phi \, d\theta \\ &= \frac{2}{5}\pi\delta a^5 \int_0^{\pi} \sin^3\phi \, d\phi = \frac{2}{5}\pi\delta a^5 \cdot 2 \cdot \frac{2}{3} = \frac{2}{5}ma^2, \end{aligned}$$

where $m = \frac{4}{3}\pi a^3 \delta$ is the mass of the ball. (In evaluating the final integral, we used symmetry and Formula (113) inside the back cover.) The answer is dimensionally correct because it is the product of mass and the square of a distance. The answer is plausible because it implies that, for purposes of rotational inertia, the sphere acts as if its mass were concentrated about 63% of the way from the axis to the equator. ◆

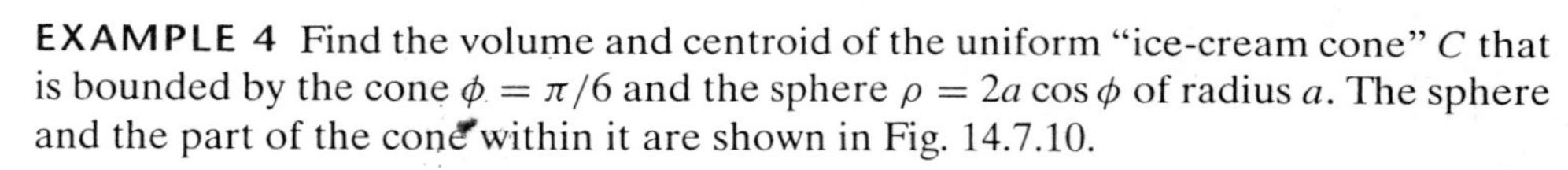

EXAMPLE 4 Find the volume and centroid of the uniform "ice-cream cone" C that is bounded by the cone $\phi = \pi/6$ and the sphere $\rho = 2a\cos\phi$ of radius a. The sphere and the part of the cone within it are shown in Fig. 14.7.10.

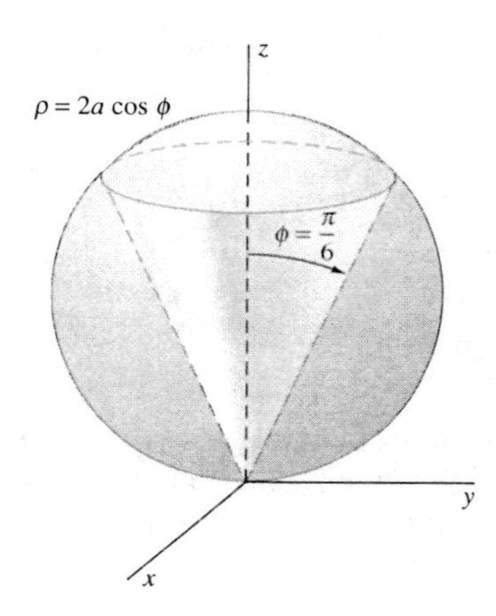

FIGURE 14.7.10 The ice-cream cone of Example 4 is the part of the cone that lies within the sphere.

Solution The ice-cream cone is described by the inequalities

$$0 \leqq \theta \leqq 2\pi, \qquad 0 \leqq \phi \leqq \frac{\pi}{6}, \qquad 0 \leqq \rho \leqq 2a\cos\phi.$$

Using Eq. (13) to compute its volume, we get

$$\begin{aligned} V &= \int_0^{2\pi}\int_0^{\pi/6}\int_0^{2a\cos\phi} \rho^2 \sin\phi \, d\rho \, d\phi \, d\theta \\ &= \frac{8}{3}a^3 \int_0^{2\pi}\int_0^{\pi/6} \cos^3\phi \sin\phi \, d\phi \, d\theta \\ &= \frac{16}{3}\pi a^3 \left[-\frac{1}{4}\cos^4\phi\right]_0^{\pi/6} = \frac{7}{12}\pi a^3. \end{aligned}$$

Now for the centroid. It is clear by symmetry that $\overline{x} = \overline{y} = 0$. We may also assume that C has density $\delta \equiv 1$, so that the mass of C is numerically the same as its volume. Because $z = \rho\cos\phi$, the z-coordinate of the centroid of C is

$$\begin{aligned} \overline{z} &= \frac{1}{V}\iiint_C z \, dV = \frac{12}{7\pi a^3}\int_0^{2\pi}\int_0^{\pi/6}\int_0^{2a\cos\phi} \rho^3 \cos\phi \sin\phi \, d\rho \, d\phi \, d\theta \\ &= \frac{48a}{7\pi}\int_0^{2\pi}\int_0^{\pi/6} \cos^5\phi \sin\phi \, d\phi \, d\theta = \frac{96a}{7}\left[-\frac{1}{6}\cos^6\phi\right]_0^{\pi/6} = \frac{37a}{28}. \end{aligned}$$

Hence the centroid of the ice-cream cone is located at the point $(0, 0, \frac{37}{28}a)$. ◆

14.7 TRUE/FALSE STUDY GUIDE

14.7 CONCEPTS: QUESTIONS AND DISCUSSION

1. Give examples of triple integrals that are most easily evaluated using (a) cylindrical rather than rectangular or spherical coordinates; (b) spherical rather than rectangular or cylindrical coordinates.
2. Describe a triple integral that you can most easily evaluate by using cylindrical coordinates and integrating first with respect to θ. Then evaluate it.
3. Describe a triple integral that you can most easily evaluate by using spherical coordinates and integrating first with respect to ϕ. Then evaluate it.

14.7 PROBLEMS

Solve Problems 1 through 20 by triple integration in cylindrical coordinates. Assume throughout that each solid has unit density unless another density function is specified.

1. Find the volume of the solid bounded above by the plane $z = 4$ and below by the paraboloid $z = r^2$.
2. Find the centroid of the solid of Problem 1.
3. Derive the formula for the volume of a sphere of radius a.
4. Find the moment of inertia around the z-axis of the solid sphere of Problem 3 given that the z-axis passes through its center.
5. Find the volume of the region that lies inside both the sphere $x^2 + y^2 + z^2 = 4$ and the cylinder $x^2 + y^2 = 1$.
6. Find the centroid of the half of the region of Problem 5 that lies on or above the xy-plane.
7. Find the mass of the cylinder $0 \leqq r \leqq a, 0 \leqq z \leqq h$ if its density at (x, y, z) is z.
8. Find the centroid of the cylinder of Problem 7.
9. Find the moment of inertia around the z-axis of the cylinder of Problem 7.
10. Find the volume of the region that lies inside both the sphere $x^2 + y^2 + z^2 = 4$ and the cylinder $x^2 + y^2 - 2x = 0$ (Fig. 14.7.11).

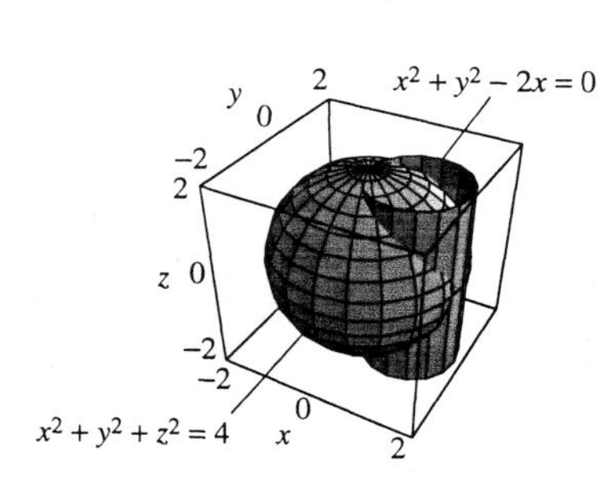

FIGURE 14.7.11 The sphere and cylinder of Problem 10.

11. Find the volume and centroid of the region bounded by the plane $z = 0$ and the paraboloid $z = 9 - x^2 - y^2$.
12. Find the volume and centroid of the region bounded by the paraboloids $z = x^2 + y^2$ and $z = 12 - 2x^2 - 2x^2$.
13. Find the volume of the region bounded by the paraboloids $z = 2x^2 + y^2$ and $z = 12 - x^2 - 2y^2$.
14. Find the volume of the region bounded below by the paraboloid $z = x^2 + y^2$ and above by the plane $z = 2x$ (Fig. 14.7.12).

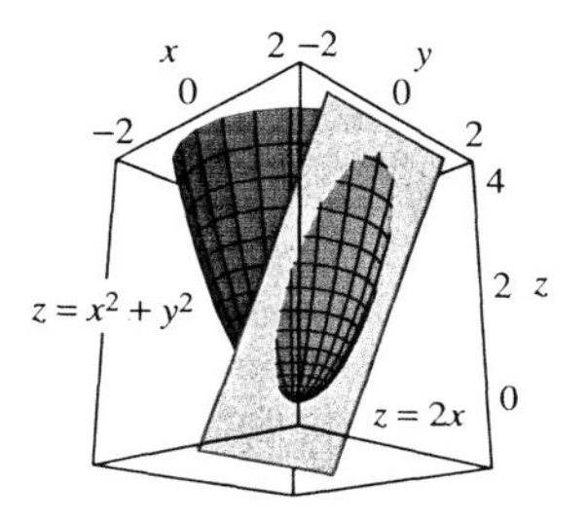

FIGURE 14.7.12 The plane and paraboloid of Problem 14.

15. Find the volume of the region bounded above by the spherical surface $x^2 + y^2 + z^2 = 2$ and below by the paraboloid $z = x^2 + y^2$ (Fig. 14.7.13).

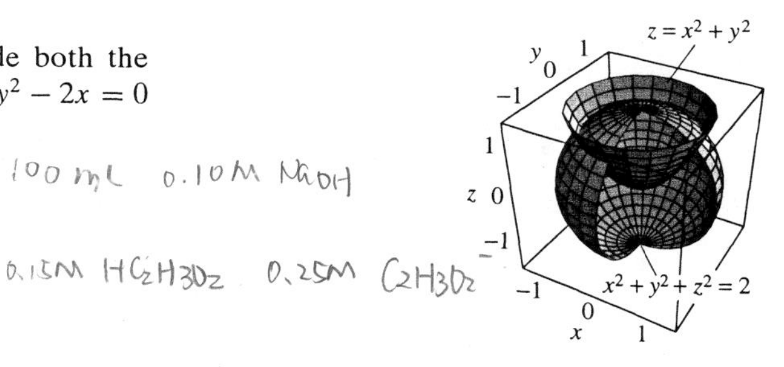

FIGURE 14.7.13 The sphere and paraboloid of Problem 15.

16. A homogeneous solid cylinder has mass m and radius a. Show that its moment of inertia around its axis of symmetry is $\frac{1}{2}ma^2$.
17. Find the moment of inertia I of a homogeneous solid right circular cylinder around a diameter of its base. Express I in terms of the radius a, the height h, and the (constant) density δ of the cylinder.
18. Find the centroid of a homogeneous solid right circular cylinder of radius a and height h.

19. Find the volume of the region bounded by the plane $z = 1$ and the cone $z = r$.

20. Show that the centroid of a homogeneous solid right circular cone lies on its axis three-quarters of the way from its vertex to its base.

Solve Problems 21 through 30 by triple integration in spherical coordinates.

21. Find the centroid of a homogeneous solid hemisphere of radius a.

22. Find the mass and centroid of the solid hemisphere $x^2 + y^2 + z^2 \leqq a^2$, $z \geqq 0$ if its density δ is proportional to distance z from its base—so $\delta = kz$ (where k is a positive constant).

23. Solve Problem 19 by triple integration in spherical coordinates.

24. Solve Problem 20 by triple integration in spherical coordinates.

25. Find the volume and centroid of the uniform solid that lies inside the sphere $\rho = a$ and above the cone $r = z$.

26. Find the moment of inertia I_z of the solid of Problem 25.

27. Find the moment of inertia around a tangent line of a solid homogeneous sphere of radius a and total mass m.

28. A spherical shell of mass m is bounded by the spheres $\rho = a$ and $\rho = 2a$, and its density function is $\delta = \rho^2$. Find its moment of inertia around a diameter.

29. Describe the surface $\rho = 2a \sin\phi$ and compute the volume of the region it bounds.

30. Describe the surface $\rho = 1 + \cos\phi$ and compute the volume of the region it bounds. Figure 14.7.14 may be useful.

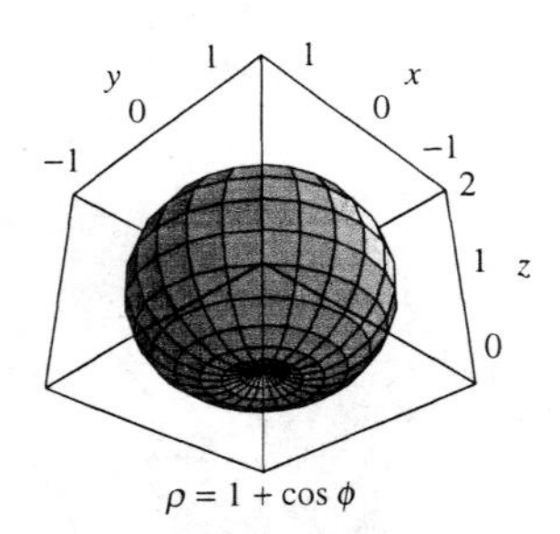

FIGURE 14.7.14 The surface of Problem 30.

31. Find the moment of inertia around the x-axis of the region that lies inside both the cylinder $r = a$ and the sphere $\rho = 2a$.

32. Find the moment of inertia around the z-axis of the ice-cream cone of Example 4.

33. Find the mass and centroid of the ice-cream cone of Example 4 if its density at (x, y, z) is $\delta(x, y, z) = z$.

34. Find the moment of inertia of the ice-cream cone of Problem 33 around the z-axis.

35. Suppose that a gaseous spherical star of radius a has density function $\delta = k(1 - \rho^2/a^2)$, so its density varies from $\delta = k$ at its center to $\delta = 0$ at its boundary $\rho = a$. Show that its mass is $\frac{2}{5}$ that of a similar star with uniform density k.

36. Find the moment of inertia around a diameter of the gaseous spherical star of Problem 35.

37. (a) Use spherical coordinates to evaluate the integral

$$\iiint_B \exp(-\rho^3)\, dV$$

where B is the solid ball of radius a centered at the origin. (b) Let $a \to \infty$ in the result of part (a) to show that

$$\int_{-\infty}^{\infty}\int_{-\infty}^{\infty}\int_{-\infty}^{\infty} \exp(-(x^2 + y^2 + z^2)^{3/2})\, dx\, dy\, dz = \tfrac{4}{3}\pi.$$

38. Use the method of Problem 37 to show that

$$\int_{-\infty}^{\infty}\int_{-\infty}^{\infty}\int_{-\infty}^{\infty} (x^2 + y^2 + z^2)^{1/2} \times \exp(-x^2 - y^2 - z^2)\, dx\, dy\, dz = 2\pi.$$

39. Find the average distance of points of a solid ball of radius a from the center of the ball. (The definition of the average value of a function precedes Problem 47 in Section 14.6.)

40. Find the average distance of the points of a solid ball of radius a from a fixed boundary point of the ball.

Problems 41 and 42 provide results that are needed in the Section 14.5 project.

41. A spherical shell of radius a and negligible thickness has area density δ, so its mass is $m = 4\pi\delta a^2$. Show that its moment of inertia about an axis of symmetry is $I_0 = \frac{2}{3}ma^2$.

42. A spherical shell has inner radius a, outer radius b, and uniform density δ. Show that its moment of inertia about an axis of symmetry is $I_0 = \frac{2}{5}mc^2$, where m is the mass of the shell and

$$c^2 = \frac{b^5 - a^5}{b^3 - a^3}.$$

43. A hole of radius $a < b$ is bored symmetrically through the center of a solid sphere of radius b and uniform density δ, leaving a "ring" of mass m. Show that the moment of inertia of this ring about its axis of symmetry is $I_0 = \frac{1}{5}m(3a^2 + 2b^2)$.

44. The three cylinders $x^2 + y^2 = 1$, $x^2 + z^2 = 1$, and $y^2 + z^2 = 1$ intersect as illustrated in Fig. 14.7.15(a); Fig. 14.7.15(b) shows a view directly from above, looking downward along the z-axis. Find the volume of the region that lies within all three cylinders.

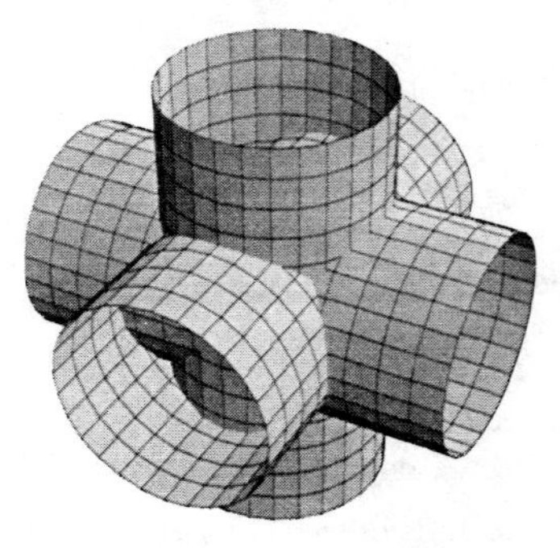

FIGURE 14.7.15(a) The three intersecting cylinders of Problem 44.

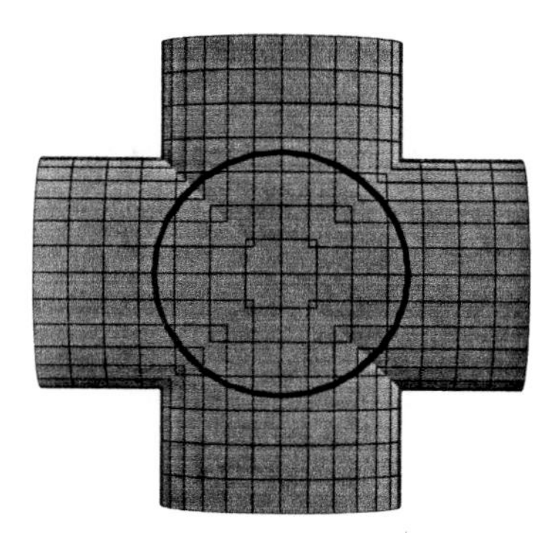

FIGURE 14.7.15(b) The view looking down from a point high on the z-axis.

45. Figure 14.7.16 shows the bumpy sphere with spherical-coordinates equation $\rho = 6 + 3\cos 3\theta \sin 5\phi$. Use a computer algebra system to find the volume of the region enclosed by this bumpy sphere.

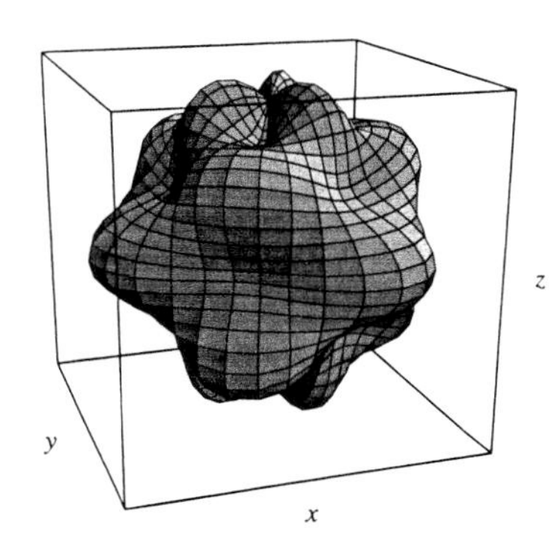

FIGURE 14.7.16 The bumpy sphere of Problem 45.

46. The bumpy sphere of Fig. 14.7.16 looks somewhat symmetrical. Is its centroid actually at the origin?

A crucial discovery of Newton (proved in his Principia Mathematica*) was the fact that the gravitational attraction of a uniform solid sphere (such as an idealized planet) is the same as though all of the mass of the planet were concentrated at its center. Problems 47 and 48 deal with this and a related fact.*

47. Consider a homogeneous spherical ball of radius a centered at the origin, with density δ and mass $M = \frac{4}{3}\pi a^3 \delta$. Show that the gravitational force $\mathbf{F}$ exerted by this ball on a point mass m located at the point $(0, 0, c)$, where $c > a$ (Fig. 14.7.17), is the same as though all the mass of the ball were concentrated at its center $(0, 0, 0)$. That is, show that $|\mathbf{F}| = GMm/c^2$. [*Suggestion:* By symmetry you may assume that the force is vertical, so that $\mathbf{F} = F_z\mathbf{k}$. Set up the integral

$$F_z = -\int_0^{2\pi}\int_0^{a}\int_0^{\pi} \frac{Gm\,\delta\cos\alpha}{w^2}\,\rho^2\sin\phi\;d\phi\,d\rho\,d\theta.$$

Change the first variable of integration from ϕ to w by using the law of cosines:

$$w^2 = \rho^2 + c^2 - 2\rho c\cos\phi.$$

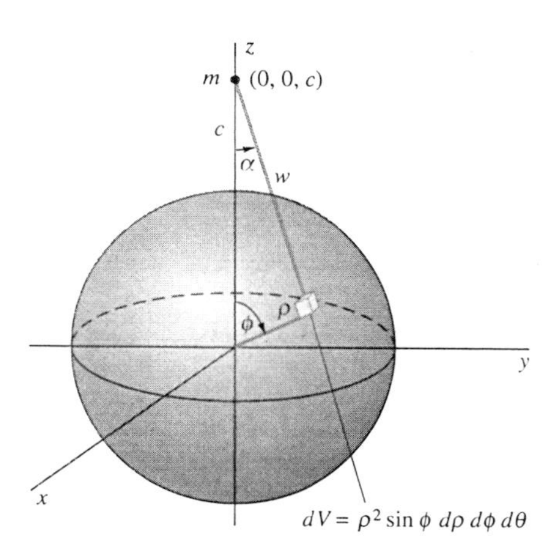

FIGURE 14.7.17 The system of Problem 47.

Then $2w\,dw = 2\rho c\sin\phi\,d\phi$ and $w\cos\alpha + \rho\cos\phi = c$. (Why?)]

48. Consider now the spherical shell $a \leqq r \leqq b$ with uniform density δ. Show that this shell exerts *no* net force on a point mass m located at the point $(0, 0, c)$ *inside* it—that is, with $|c| < a$. The computation will be the same as in Problem 47 except for the limits of integration on ρ and w.

49. If the earth were perfectly spherical with radius $R = 6370$ km, *uniform* density δ, and mass $M = \frac{4}{3}\pi\delta R^3$, then (according to Example 3) its moment of inertia about its polar axis would be $I = \frac{2}{5}MR^2$. In actuality, however, measurements from satellites indicate that

$$I = kMR^2 \qquad \textbf{(15)}$$

where $k \approx 0.371 < \frac{2}{5}$. The reason is that, instead of having a uniform interior, a more realistic model of the earth has a dense core covered with a lighter mantle a few thousand kilometers thick (Fig. 14.7.18). The density of the core is $\delta_1 \approx 11 \times 10^3$ kg/m^3 and that of the mantle is $\delta_2 \approx 5 \times 10^3$ kg/m^3. (a) With this core-mantle model, calculate the mass M of the earth and its polar moment of inertia I (using Problem 42) in terms of the unknown radius x of the spherical core. (b) Substitute your calculated values of M and I in Eq. (15) and use a computer algebra system to solve the resulting equation for x.

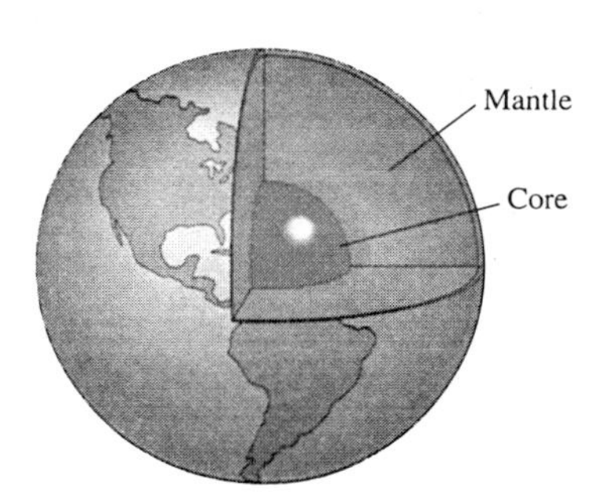

FIGURE 14.7.18 The core and mantle of the earth.

14.8 SURFACE AREA

Until now our concept of a surface has been the graph $z = f(x, y)$ of a function of two variables. Occasionally we have seen such a surface defined implicitly by an equation of the form $F(x, y, z) = 0$. Now we want to introduce the more precise concept of a *parametric surface*—the two-dimensional analogue of a parametric curve.

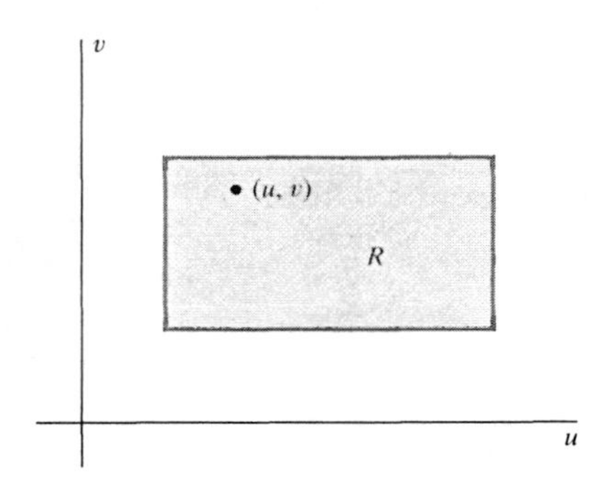

FIGURE 14.8.1 The *uv*-region *R* on which the transformation **r** is defined.

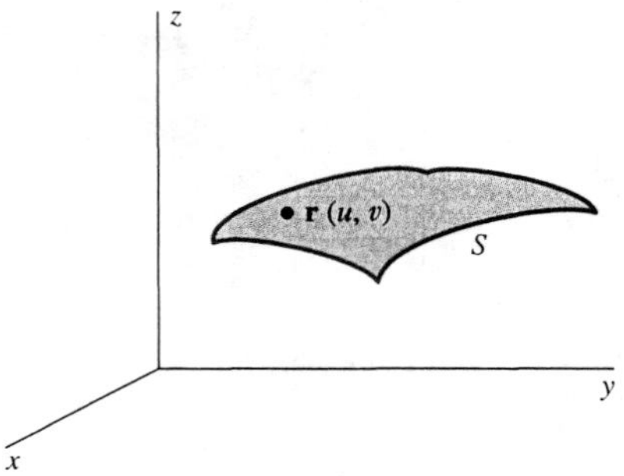

FIGURE 14.8.2 The parametric surface *S* in *xyz*-space.

A **parametric surface** S is the *image* of a function or transformation $\mathbf{r}$ that is defined on a region R in the uv-plane (Fig. 14.8.1) and has values in xyz-space (Fig. 14.8.2). The **image** under $\mathbf{r}$ of each point (u, v) in R is the point in xyz-space with position vector

$$\mathbf{r}(u, v) = \langle x(u, v), y(u, v), z(u, v) \rangle. \tag{1}$$

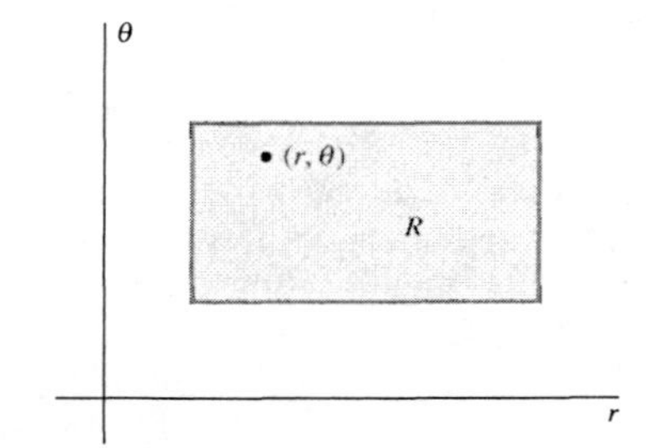

FIGURE 14.8.3 A rectangle in the $r\theta$-plane; the domain of the function $z = g(r, \theta)$ of Example 1.

The parametric surface S is called **smooth** provided that the component functions of $\mathbf{r}$ have continuous partial derivatives with respect to u and v and, moreover, the vectors

$$\mathbf{r}_u = \frac{\partial \mathbf{r}}{\partial u} = \langle x_u, y_u, z_u \rangle = \frac{\partial x}{\partial u}\mathbf{i} + \frac{\partial y}{\partial u}\mathbf{j} + \frac{\partial z}{\partial u}\mathbf{k} \tag{2}$$

and

$$\mathbf{r}_v = \frac{\partial \mathbf{r}}{\partial v} = \langle x_v, y_v, z_v \rangle = \frac{\partial x}{\partial v}\mathbf{i} + \frac{\partial y}{\partial v}\mathbf{j} + \frac{\partial z}{\partial v}\mathbf{k} \tag{3}$$

are nonzero and nonparallel at each interior point of R. (Compare this with the definition of *smooth* parametric curve $\mathbf{r}(t)$ in Section 10.4.) We call the variables u and v the *parameters* for the surface S, in analogy with the single parameter t for a parametric curve.

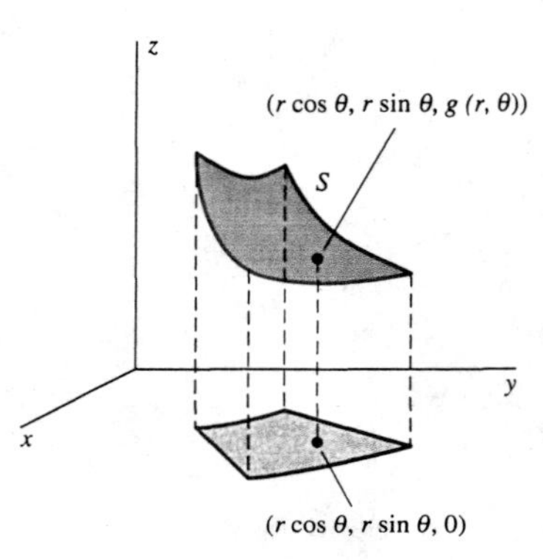

FIGURE 14.8.4 A cylindrical-coordinates surface in *xyz*-space (Example 1).

EXAMPLE 1 (a) We may regard the graph $z = f(x, y)$ of a function as a parametric surface with parameters x and y. In this case the transformation $\mathbf{r}$ from the xy-plane to xyz-space has the component functions

$$x = x, \qquad y = y, \qquad z = f(x, y). \tag{4}$$

(b) Similarly, we may regard a surface given in cylindrical coordinates by the graph of $z = g(r, \theta)$ as a parametric surface with parameters r and θ. The transformation $\mathbf{r}$ from the $r\theta$-plane (Fig. 14.8.3) to xyz-space (Fig. 14.8.4) is then given by

$$x = r\cos\theta, \qquad y = r\sin\theta, \qquad z = g(r, \theta). \tag{5}$$

(c) We may regard a surface given in spherical coordinates by $\rho = h(\phi, \theta)$ as a parametric surface with parameters ϕ and θ, and the corresponding transformation from the $\phi\theta$-plane to xyz-space is then given by

$$x = h(\phi, \theta)\sin\phi\cos\theta, \qquad y = h(\phi, \theta)\sin\phi\sin\theta, \qquad z = h(\phi, \theta)\cos\phi. \tag{6}$$

The concept of a parametric surface lets us treat all these special cases, and many others, with the same techniques. ◆

Surface Area of Parametric Surfaces

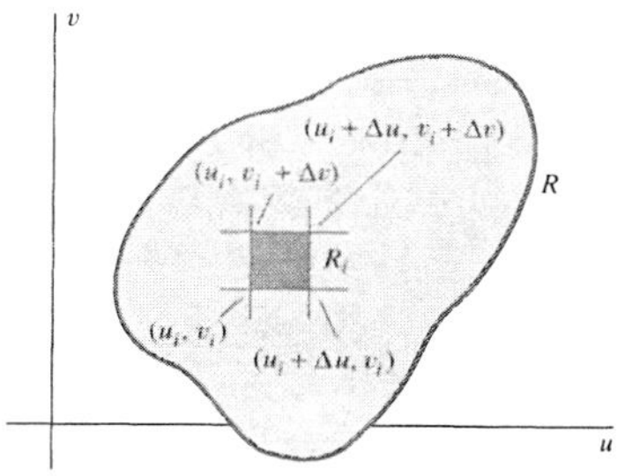

Now we want to define the *surface area* of the general smooth parametric surface given in Eq. (1). We begin with an inner partition of the region R—the domain of $\mathbf{r}$ in the uv-plane—into rectangles $R_1, R_2, \ldots, R_n$, each with dimensions Δu and Δv. Let (u_i, v_i) be the lower left-hand corner of R_i (as in Fig. 14.8.5). The image S_i of R_i under $\mathbf{r}$ will not generally be a rectangle in xyz-space; it will look more like a *curvilinear figure* on the image surface S, with $\mathbf{r}(u_i, v_i)$ as one "vertex" (Fig. 14.8.6). Let ΔS_i denote the area of this curvilinear figure S_i.

FIGURE 14.8.5 The rectangle R_i in the uv-plane.

The parametric curves $\mathbf{r}(u, v_i)$ and $\mathbf{r}(u_i, v)$—with parameters u and v, respectively—lie on the surface S and meet at the point $\mathbf{r}(u_i, v_i)$. At this point of intersection, these two curves have the tangent vectors $\mathbf{r}_u(u_i, v_i)$ and $\mathbf{r}_v(u_i, v_i)$ shown in Fig. 14.8.7. Hence their vector product

$$\mathbf{N}(u_i, v_i) = \mathbf{r}_u(u_i, v_i) \times \mathbf{r}_v(u_i, v_i) \tag{7}$$

is a vector normal to S at the point $\mathbf{r}(u_i, v_i)$.

Now suppose that both Δu and Δv are small. Then the area ΔS_i of the curvilinear figure S_i should be approximately equal to the area ΔP_i of the parallelogram with adjacent sides $\mathbf{r}_u(u_i, v_i)\,\Delta u$ and $\mathbf{r}_v(u_i, v_i)\,\Delta v$ (Fig. 14.8.8). But the area of this parallelogram is

$$\Delta P_i = |\mathbf{r}_u(u_i, v_i)\,\Delta u \times \mathbf{r}_v(u_i, v_i)\,\Delta v| = |\mathbf{N}(u_i, v_i)|\,\Delta u\,\Delta v.$$

This means that the area $a(S)$ of the surface S should be given approximately by

$$a(S) = \sum_{i=1}^{n} \Delta S_i \approx \sum_{i=1}^{n} \Delta P_i,$$

so

$$a(S) \approx \sum_{i=1}^{n} |\mathbf{N}(u_i, v_i)|\,\Delta u\,\Delta v.$$

But this last sum is a Riemann sum for the double integral

$$\iint_R |\mathbf{N}(u, v)|\,du\,dv.$$

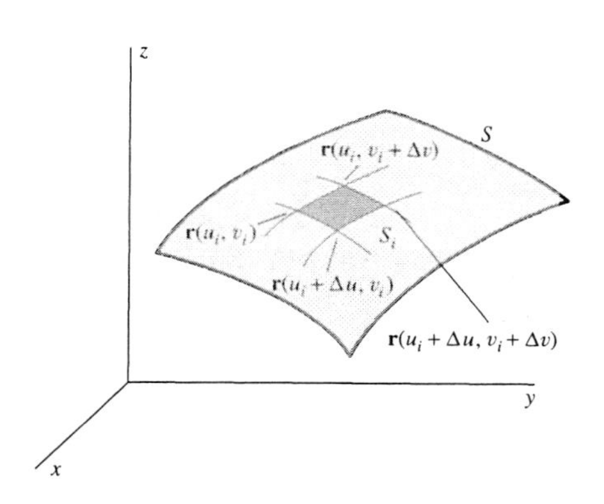

FIGURE 14.8.6 The image of R_i is a curvilinear figure.

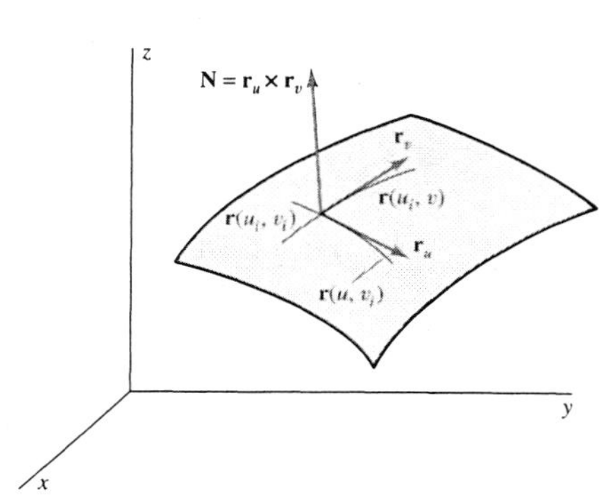

FIGURE 14.8.7 The vector **N** normal to the surface at $\mathbf{r}(u_i, v_i)$.

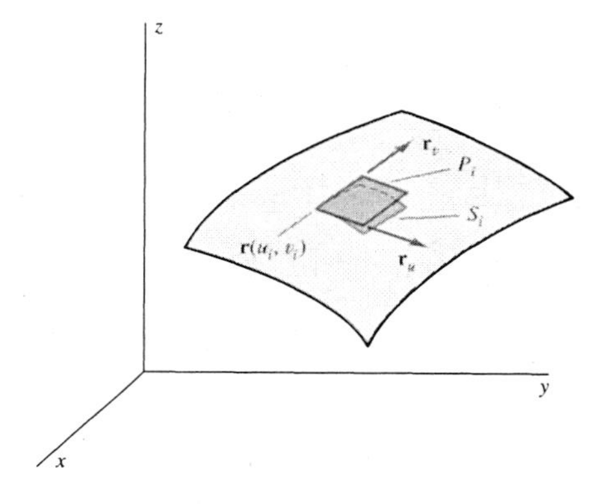

FIGURE 14.8.8 The area of the parallelogram P_i is an approximation to the area of the curvilinear figure S_i.

We are therefore motivated to *define* the **surface area** A of the smooth parametric surface S by

$$A = a(S) = \iint_R |\mathbf{N}(u, v)|\, du\, dv = \iint_R \left| \frac{\partial \mathbf{r}}{\partial u} \times \frac{\partial \mathbf{r}}{\partial v} \right| du\, dv. \tag{8}$$

Surface Area in Rectangular Coordinates

In the case of the surface $z = f(x, y)$ for (x, y) in the region R in the xy-plane, the component functions of $\mathbf{r}$ are given by the equations in (4) with parameters x and y (in place of u and v). Then

$$\mathbf{N} = \frac{\partial \mathbf{r}}{\partial x} \times \frac{\partial \mathbf{r}}{\partial y} = \begin{vmatrix} \mathbf{i} & \mathbf{j} & \mathbf{k} \\ 1 & 0 & \dfrac{\partial f}{\partial x} \\ 0 & 1 & \dfrac{\partial f}{\partial y} \end{vmatrix} = -\frac{\partial f}{\partial x}\mathbf{i} - \frac{\partial f}{\partial y}\mathbf{j} + \mathbf{k},$$

so Eq. (8) takes the special form

$$A = a(S) = \iint_R \sqrt{1 + \left(\frac{\partial f}{\partial x}\right)^2 + \left(\frac{\partial f}{\partial y}\right)^2}\, dx\, dy = \iint_R \sqrt{1 + z_x^2 + z_y^2}\, dx\, dy. \tag{9}$$

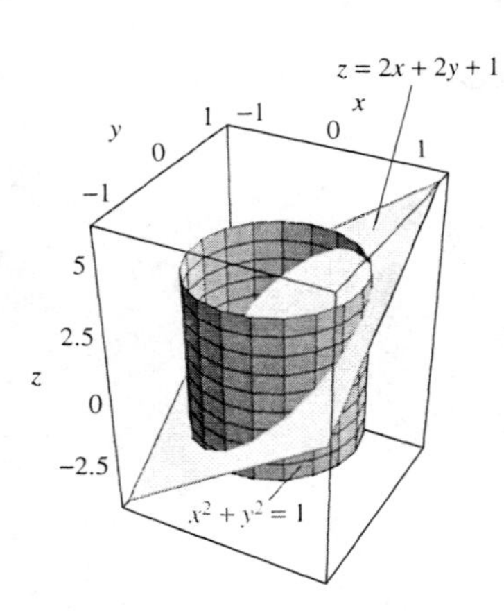

FIGURE 14.8.9 The cylinder and plane of Example 2.

EXAMPLE 2 Find the area of the ellipse cut from the plane $z = 2x + 2y + 1$ by the cylinder $x^2 + y^2 = 1$ (Fig. 14.8.9).

Solution Here, R is the unit circle in the xy-plane with area

$$\iint_R 1\, dx\, dy = \pi,$$

so Eq. (9) gives the area of the ellipse to be

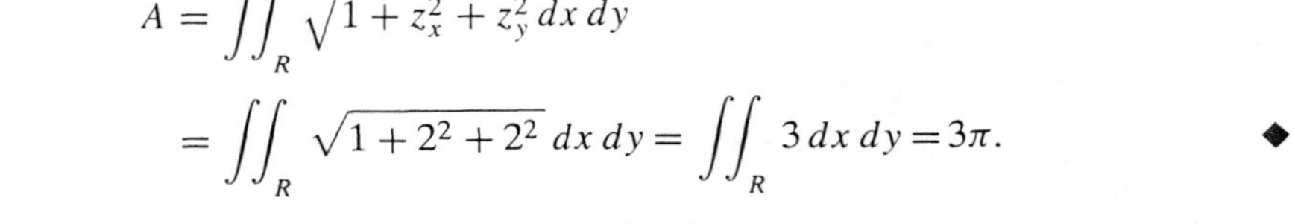

$$A = \iint_R \sqrt{1 + z_x^2 + z_y^2}\, dx\, dy$$

$$= \iint_R \sqrt{1 + 2^2 + 2^2}\, dx\, dy = \iint_R 3\, dx\, dy = 3\pi.$$

◆

REMARK Computer-generated figures such as Fig. 14.8.9 could not be constructed easily without using parametric surfaces. For example, the vertical cylinder in Fig. 14.8.9 was generated by instructing the computer to plot the parametric surface defined on the $r\theta$-rectangle $-5 \leqq z \leqq 5$, $0 \leqq \theta \leqq 2\pi$ by

$$\mathbf{r}(z, \theta) = \langle \cos\theta, \sin\theta, z \rangle.$$

Is it clear that the image of this transformation is the cylinder $x^2 + y^2 = 1$, $-5 \leqq z \leqq 5$?

Surface Area in Cylindrical Coordinates

Now consider a cylindrical-coordinates surface $z = g(r, \theta)$ parametrized by the equations in (5) for (r, θ) in a region R of the $r\theta$-plane. Then the normal vector is

$$\mathbf{N} = \frac{\partial \mathbf{r}}{\partial r} \times \frac{\partial \mathbf{r}}{\partial \theta} = \begin{vmatrix} \mathbf{i} & \mathbf{j} & \mathbf{k} \\ \cos\theta & \sin\theta & \dfrac{\partial z}{\partial r} \\ -r\sin\theta & r\cos\theta & \dfrac{\partial z}{\partial \theta} \end{vmatrix}$$

$$= \mathbf{i}\left(\frac{\partial z}{\partial \theta}\sin\theta - r\frac{\partial z}{\partial r}\cos\theta\right) - \mathbf{j}\left(\frac{\partial z}{\partial \theta}\cos\theta + r\frac{\partial z}{\partial r}\sin\theta\right) + r\mathbf{k}.$$

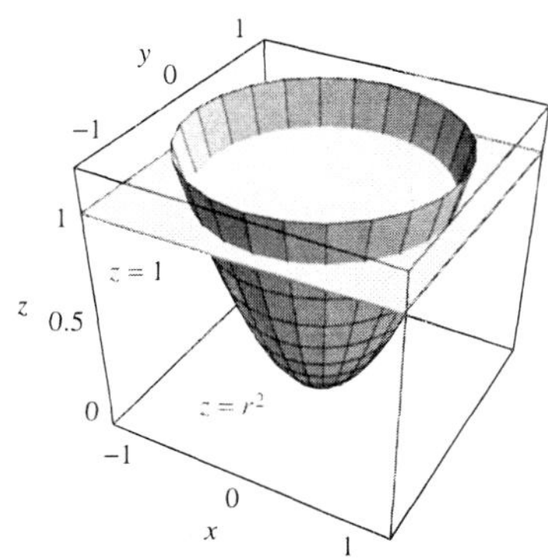
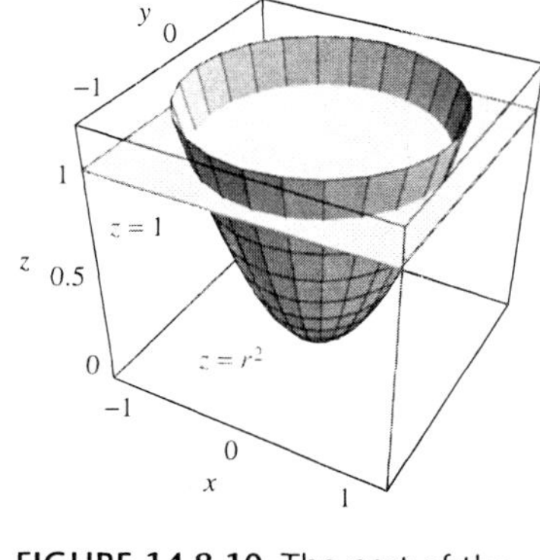

FIGURE 14.8.10 The part of the paraboloid $z = r^2$ inside the cylinder $r = 1$ (Example 3) is the same as the part beneath the plane $z = 1$. (Why?).

After some simplifications, we find that

$$|\mathbf{N}| = \sqrt{r^2 + r^2\left(\frac{\partial z}{\partial r}\right)^2 + \left(\frac{\partial z}{\partial \theta}\right)^2}.$$

Then Eq. (8) yields the formula

$$A = \iint_R \sqrt{r^2 + (rz_r)^2 + (z_\theta)^2}\, dr\, d\theta \tag{10}$$

for surface area in cylindrical coordinates.

EXAMPLE 3 Find the surface area cut from the paraboloid $z = r^2$ by the cylinder $r = 1$ (Fig. 14.8.10).

Solution Equation (10) gives area

$$A = \int_0^{2\pi}\int_0^1 \sqrt{r^2 + r^2\cdot(2r)^2}\, dr\, d\theta = 2\pi\int_0^1 r\sqrt{1+4r^2}\, dr$$

$$= 2\pi\left[\frac{2}{3}\cdot\frac{1}{8}(1+4r^2)^{3/2}\right]_0^1 = \frac{\pi}{6}(5\sqrt{5}-1) \approx 5.3304.$$

◆

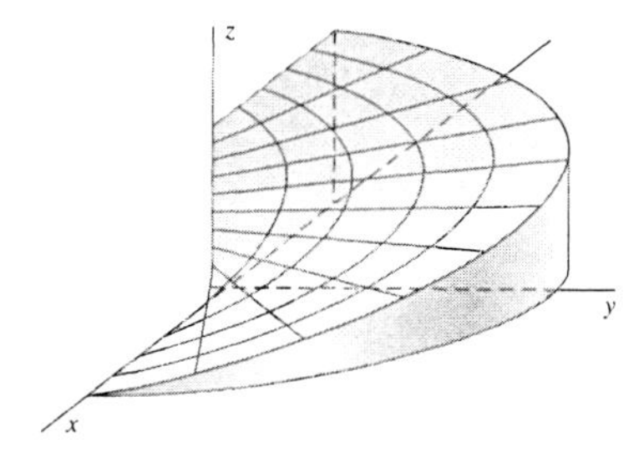

FIGURE 14.8.11 The spiral ramp of Example 4.

In Example 3, you would get the same result if you first wrote $z = x^2 + y^2$, used Eq. (9), which gives

$$A = \iint_R \sqrt{1 + 4x^2 + 4y^2}\, dx\, dy,$$

and then changed to polar coordinates. In Example 4 it would be less convenient to begin with rectangular coordinates.

EXAMPLE 4 Find the area of the *spiral ramp* $z = \theta$, $0 \leqq r \leqq 1$, $0 \leqq \theta \leqq \pi$. This is the upper surface of the solid shown in Fig. 14.8.11.

Solution Equation (10) gives area

$$A = \int_0^{\pi}\int_0^1 \sqrt{r^2+1}\, dr\, d\theta = \frac{\pi}{2}\left[\sqrt{2} + \ln\left(1+\sqrt{2}\right)\right] \approx 3.6059.$$

We avoided a trigonometric substitution by using the table of integrals inside the back cover.

◆

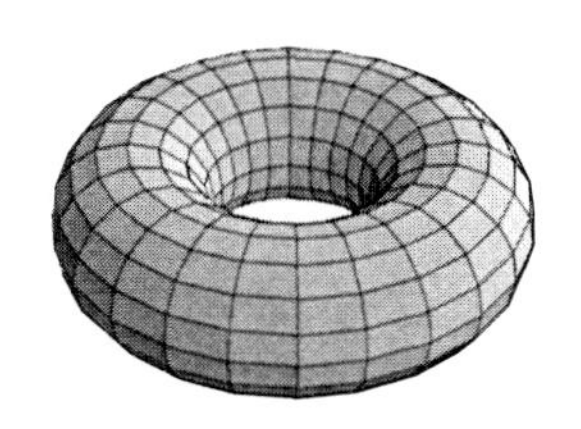

FIGURE 14.8.12 The torus of Example 5.

EXAMPLE 5 Find the surface area of the torus generated by revolving the circle

$$(x-b)^2 + z^2 = a^2 \quad (0 < a < b)$$

in the xz-plane around the z-axis (Fig. 14.8.12).

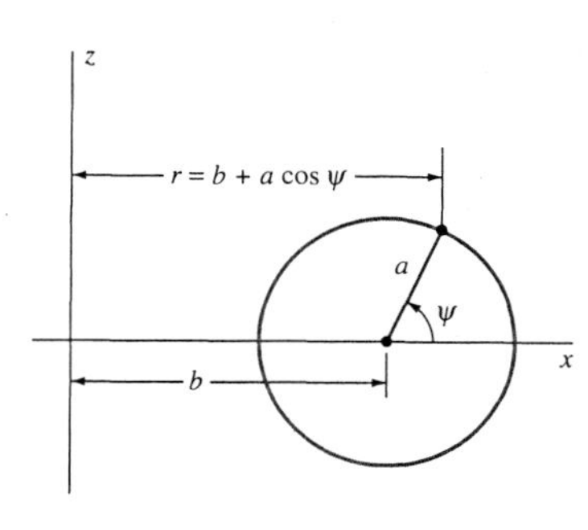

FIGURE 14.8.13 The circle that generates the torus of Example 5.

Solution With the ordinary polar coordinate θ and the angle ψ of Fig. 14.8.13, the torus is described for $0 \leqq \theta \leqq 2\pi$ and $0 \leqq \psi \leqq 2\pi$ by the parametric equations

$$x = r\cos\theta = (b + a\cos\psi)\cos\theta,$$

$$y = r\sin\theta = (b + a\cos\psi)\sin\theta,$$

$$z = a\sin\psi.$$

When we compute $\mathbf{N} = \mathbf{r}_\theta \times \mathbf{r}_\psi$ and simplify, we find that

$$|\mathbf{N}| = a(b + a\cos\psi).$$

Hence the general surface-area formula, Eq. (8), gives area

$$A = \int_0^{2\pi}\int_0^{2\pi} a(b + a\cos\psi)\,d\theta\,d\psi = 2\pi a\Big[b\psi + a\sin\psi\Big]_0^{2\pi} = 4\pi^2 ab.$$

We obtained the same result in Section 14.5 with the aid of Pappus's first theorem. ◆

14.8 TRUE/FALSE STUDY GUIDE

14.8 CONCEPTS: QUESTIONS AND DISCUSSION

1. Compare the calculations of the surface area of a sphere using (a) rectangular coordinates; (b) cylindrical coordinates; (c) spherical coordinates. (See Problem 18 of this section.)
2. We know that the volume of a solid ball of radius r is $V(r) = \frac{4}{3}\pi r^3$ and that the area of its spherical surface is $S(r) = 4\pi r^2$. Is it a coincidence that $V'(r) = S(r)$? Think about a thin spherical shell as "volume as product of thickness and area."

14.8 PROBLEMS

1. Find the area of the portion of the plane $z = x + 3y$ that lies inside the elliptical cylinder with equation

$$\frac{x^2}{4} + \frac{y^2}{9} = 1.$$

2. Find the area of the region in the plane $z = 1 + 2x + 2y$ that lies directly above the region in the xy-plane bounded by the parabolas $y = x^2$ and $x = y^2$.

3. Find the area of the part of the paraboloid $z = 9 - x^2 - y^2$ that lies above the plane $z = 5$.

4. Find the area of the part of the surface $2z = x^2$ that lies directly above the triangle in the xy-plane with vertices at $(0, 0)$, $(1, 0)$, and $(1, 1)$.

5. Find the area of the surface that is the graph of $z = x + y^2$ for $0 \leqq x \leqq 1$, $0 \leqq y \leqq 2$.

6. Find the area of that part of the surface of Problem 5 that lies above the triangle in the xy-plane with vertices at $(0, 0)$, $(0, 1)$, and $(1, 1)$.

7. Find by integration the area of the part of the plane $2x + 3y + z = 6$ that lies in the first octant.

8. Find the area of the ellipse that is cut from the plane $2x + 3y + z = 6$ by the cylinder $x^2 + y^2 = 2$.

9. Find the area that is cut from the saddle-shaped surface $z = xy$ by the cylinder $x^2 + y^2 = 1$.

10. Find the area that is cut from the surface $z = x^2 - y^2$ by the cylinder $x^2 + y^2 = 4$.

11. Find the surface area of the part of the paraboloid $z = 16 - x^2 - y^2$ that lies above the xy-plane.

12. Show by integration that the surface area of the conical surface $z = br$ between the planes $z = 0$ and $z = h = ab$ is given by $A = \pi a L$, where L is the slant height $\sqrt{a^2 + h^2}$ and a is the radius of the base of the cone.

13. Let the part of the cylinder $x^2 + y^2 = a^2$ between the planes $z = 0$ and $z = h$ be parametrized by $x = a\cos\theta$, $y = a\sin\theta$, $z = z$. Apply Eq. (8) to show that the area of this zone is $A = 2\pi a h$.

14. Consider the meridional zone of height $h = c - b$ that lies on the sphere $r^2 + z^2 = a^2$ between the planes $z = b$ and $z = c$, where $0 \leqq b < c \leqq a$. Apply Eq. (10) to show that the area of this zone is $A = 2\pi a h$.

15. Find the area of the part of the cylinder $x^2 + z^2 = a^2$ that lies within the cylinder $r^2 = x^2 + y^2 = a^2$.

16. Find the area of the part of the sphere $r^2 + z^2 = a^2$ that lies within the cylinder $r = a\sin\theta$.

17. (a) Apply Eq. (8) to show that the surface area of the surface $y = f(x, z)$, for (x, z) in the region R of the xz-plane, is given by

$$A = \iint_R \sqrt{1 + \left(\frac{\partial f}{\partial x}\right)^2 + \left(\frac{\partial f}{\partial z}\right)^2}\,dx\,dz.$$

(b) State and derive a similar formula for the area of the surface $x = f(y, z)$ for (y, z) in the region R of the yz-plane.

18. Suppose that R is a region in the $\phi\theta$-plane. Consider the part of the sphere $\rho = a$ that corresponds to (ϕ, θ) in R, parametrized by the equations in (6) with $h(\phi, \theta) = a$. Apply Eq. (8) to show that the surface area of this part of the sphere is

$$A = \iint_R a^2 \sin\phi\,d\phi\,d\theta.$$

19. (a) Consider the "spherical rectangle" defined by

$$\rho = a, \quad \phi_1 \leqq \phi \leqq \phi_2 = \phi_1 + \Delta\phi, \quad \theta_1 \leqq \theta \leqq \theta_2 = \theta_1 + \Delta\theta.$$

Apply the formula of Problem 18 and the average value property (see Problem 50 in Section 14.2) to show that the area of this spherical rectangle is $A = a^2 \sin\hat{\phi}\,\Delta\phi\,\Delta\theta$ for some $\hat{\phi}$ in (ϕ_1, ϕ_2). (b) Conclude from the result in part (a) that the volume of the spherical block defined by

$$\rho_1 \leqq \rho \leqq \rho_2 = \rho_1 + \Delta\rho, \quad \phi_1 \leqq \phi \leqq \phi_2, \quad \theta_1 \leqq \theta \leqq \theta_2$$

is

$$\Delta V = \tfrac{1}{3}(\rho_2^3 - \rho_1^3)\sin\hat{\phi}\,\Delta\phi\,\Delta\theta.$$

Finally, derive Eq. (8) of Section 14.7 by applying the mean value theorem to the function $f(\rho) = \rho^3$ on the interval $[\rho_1, \rho_2]$.

20. Describe the surface $\rho = 2a\sin\phi$. Why is it called a *pinched torus*? It is parametrized as in Eq. (6) with $h(\phi, \theta) = 2a\sin\phi$. Show that its surface area is $A = 4\pi^2 a^2$. Figure 14.8.14 may be helpful.

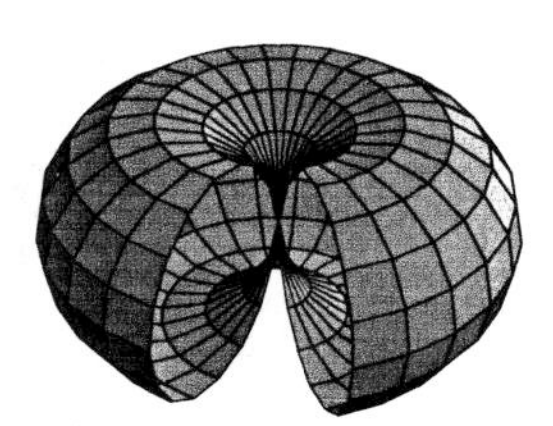

FIGURE 14.8.14 Cutaway view of the pinched torus of Problem 20.

21. The surface of revolution obtained when we revolve the curve $x = f(z)$, $a \leqq z \leqq b$, around the z-axis is parametrized in terms of θ $(0 \leqq \theta \leqq 2\pi)$ and z $(a \leqq z \leqq b)$ by $x = f(z)\cos\theta$, $y = f(z)\sin\theta$, $z = z$. From Eq. (8) derive the surface-area formula

$$A = \int_0^{2\pi}\int_a^b f(z)\sqrt{1 + [f'(z)]^2}\,dz\,d\theta.$$

This formula agrees with the area of a surface of revolution as defined in Section 6.4.

22. Apply the formula of Problem 18 in both parts of this problem. (a) Verify the formula $A = 4\pi a^2$ for the surface area of a sphere of radius a. (b) Find the area of that part of a sphere of radius a and center $(0, 0, 0)$ that lies inside the cone $\phi = \pi/6$.

23. Apply the result of Problem 21 to verify the formula $A = 2\pi rh$ for the lateral surface area of a right circular cylinder of radius r and height h.

24. Apply Eq. (9) to verify the formula $A = 2\pi rh$ for the lateral surface area of the cylinder $x^2 + z^2 = r^2$, $0 \leqq y \leqq h$ of radius r and height h.

In Problems 25 through 28, use a computer algebra system first to plot and then to approximate (with four-place accuracy) the area of the part of the given surface S that lies above the square in the xy-plane defined by (a) $-1 \leqq x \leqq 1$, $-1 \leqq y \leqq 1$; (b) $|x| + |y| \leqq 1$.

25. S is the paraboloid $z = x^2 + y^2$.

26. S is the cone $z = \sqrt{x^2 + y^2}$.

27. S is the hyperboloid $z = 1 + xy$.

28. S is the sphere $x^2 + y^2 + z^2 = 4$.

In Problems 29 through 32, a parametrization of a quadric surface is given. Use identities such as $\cos^2 t + \sin^2 t = 1$ and $\cosh^2 t - \sinh^2 t = 1$ to identify these surfaces by means of the quadric surface equations listed in Section 12.7. For visual corroboration you can use the parametric plot command in a computer algebra system to plot each surface (with selected numerical values of the coefficients a, b, and c).

29. $x = au\cos v$, $y = bu\sin v$, $z = cu^2$; $0 \leqq u \leqq 1$, $0 \leqq v \leqq 2\pi$

30. $x = a\sin u\cos v$, $y = b\sin u\sin v$, $z = c\cos u$; $0 \leqq u \leqq \pi$, $0 \leqq v \leqq 2\pi$

31. $x = a\sinh u\cos v$, $y = b\sinh u\sin v$, $z = c\cosh u$; $0 \leqq u \leqq 1$, $0 \leqq v \leqq 2\pi$

32. $x = a\cosh u\cos v$, $y = b\cosh u\sin v$, $z = c\sinh u$; $-1 \leqq u \leqq 1$, $0 \leqq v \leqq 2\pi$

33. An ellipsoid with semiaxes a, b, and c is defined by the parametrization

$$x = a\sin\phi\cos\theta, \qquad y = b\sin\phi\sin\theta, \qquad z = c\cos\phi$$

$(0 \leqq \phi \leqq \pi, 0 \leqq \theta \leqq 2\pi)$ in terms of the angular spherical coordinates ϕ and θ. Use a computer algebra system to approximate (to four-place accuracy) the area of the ellipsoid with $a = 4$, $b = 3$, and $c = 2$.

34. (a) Generalize Example 5 to derive the parametric equations

$$x = (b + a\cos\psi)\cos\theta, \quad y = (b + a\cos\psi)\sin\theta, \quad z = c\sin\psi$$

$(0 \leqq \psi \leqq 2\pi, 0 \leqq \theta \leqq 2\pi)$ of the "elliptical torus" obtained by revolving around the z-axis the ellipse $(x-b)^2/a^2 + z^2/c^2 = 1$ (where $0 < a < b$) in the xz-plane. (b) Use a computer algebra system to approximate (to four-place accuracy) the area of the elliptical torus obtained as in part (a) with $a = 2$, $b = 3$, and $c = 1$. (c) Also approximate the perimeter of the ellipse of part (a). Are your results consistent with Pappus's theorem for the area of a surface of revolution?

14.9 CHANGE OF VARIABLES IN MULTIPLE INTEGRALS

We have seen in preceding sections that we can evaluate certain multiple integrals by transforming them from rectangular coordinates into polar or spherical coordinates. The technique of changing coordinate systems to evaluate a multiple integral is the multivariable analogue of substitution in a single integral. Recall from Section 5.7 that if $x = g(u)$, then

$$\int_a^b f(x)\,dx = \int_c^d f(g(u))\,g'(u)\,du, \tag{1}$$

where $a = g(c)$ and $b = g(d)$. The method of substitution involves a "change of variables" that is tailored to the evaluation of a given integral.

Suppose that we want to evaluate the double integral

$$\iint_R F(x, y)\,dx\,dy.$$

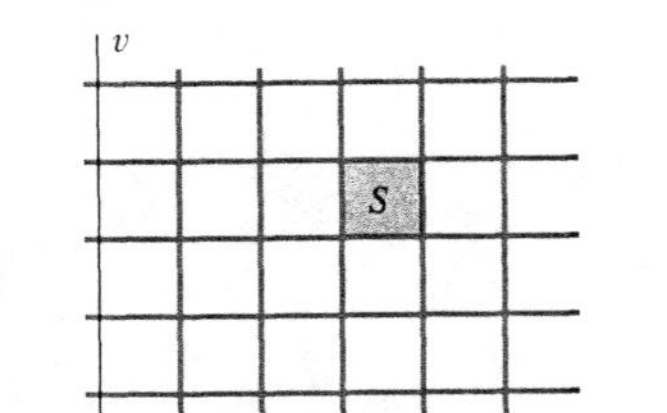

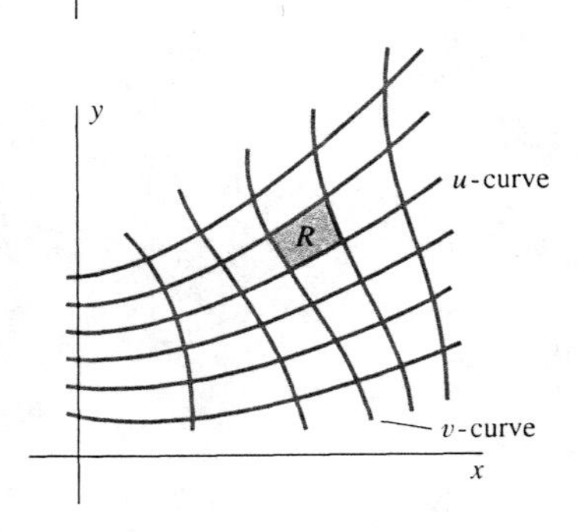

FIGURE 14.9.1 The transformation T turns the rectangle S into the curvilinear figure R.

A *change of variables* for this integral is determined by a continuously differentiable **transformation** T from the uv-plane to the xy-plane—that is, a function T that associates with the point (u, v) a point $(x, y) = T(u, v)$ given by equations of the form

$$x = f(u, v), \qquad y = g(u, v). \tag{2}$$

The point (x, y) is called the **image** of the point (u, v) under the transformation T. If no two different points in the uv-plane have the same image point in the xy-plane, then the transformation T is said to be **one-to-one.** In this case it may be possible to solve the equations in (2) for u and v in terms of x and y and thus obtain the equations

$$u = h(x, y), \qquad v = k(x, y) \tag{3}$$

of the **inverse transformation** T^{-1} from the xy-plane to the uv-plane.

Often it is convenient to visualize the transformation T geometrically in terms of its *u-curves* and *v-curves.* The **u-curves** of T are the images in the xy-plane of the *horizontal* lines in the uv-plane—on each such curve the value of u varies but v is constant. The **v-curves** of T are the images of the *vertical* lines in the uv-plane—on each of these, the value of v varies but u is constant. Note that the image under T of a rectangle bounded by horizontal and vertical lines in the uv-plane is a *curvilinear figure* bounded by u-curves and v-curves in the xy-plane (Fig. 14.9.1). If we know the equations in (3) of the inverse transformation, then we can find the u-curves and the v-curves quite simply by writing the equations

$$k(x, y) = C_1 \qquad (u\text{-curve on which } v = C_1 \text{ is constant})$$

and

$$h(x, y) = C_2 \qquad (v\text{-curve on which } u = C_2 \text{ is constant}).$$

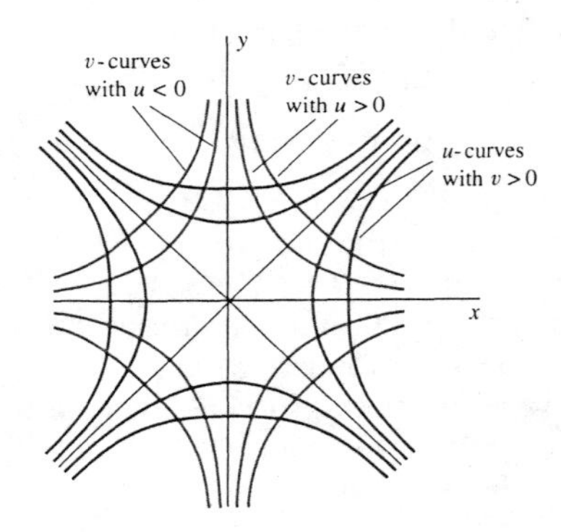

FIGURE 14.9.2 The u-curves and v-curves of Example 1.

EXAMPLE 1 Determine the u-curves and the v-curves of the transformation T whose inverse T^{-1} is specified by the equations $u = xy$, $v = x^2 - y^2$.

Solution The u-curves are the hyperbolas

$$x^2 - y^2 = v = C_1 \quad \text{(constant)},$$

and the v-curves are the rectangular hyperbolas

$$xy = u = C_2 \quad \text{(constant)}.$$

These two familiar families of hyperbolas are shown in Fig. 14.9.2. ◆

Change of Variables in Double Integrals

Now we shall describe the change of variables in a double integral that corresponds to the transformation T specified by the equations in (2). Let the region R in the xy-plane be the image under T of the region S in the uv-plane. Let $F(x, y)$ be continuous on R and let $\{S_1, S_2, \dots, S_n\}$ be an inner partition of S into rectangles each with dimensions Δu by Δv. Each rectangle S_i is transformed by T into a curvilinear figure R_i in the xy-plane (Fig. 14.9.3). The images $\{R_1, R_2, \dots, R_n\}$ under T of the rectangles S_i then constitute an inner partition of the region R (though into curvilinear figures rather than rectangles).

Let $(u_i^\star, v_i^\star)$ be the lower left-hand corner point of S_i, and write

$$(x_i^\star, y_i^\star) = (f(u_i^\star, v_i^\star), g(u_i^\star, v_i^\star))$$

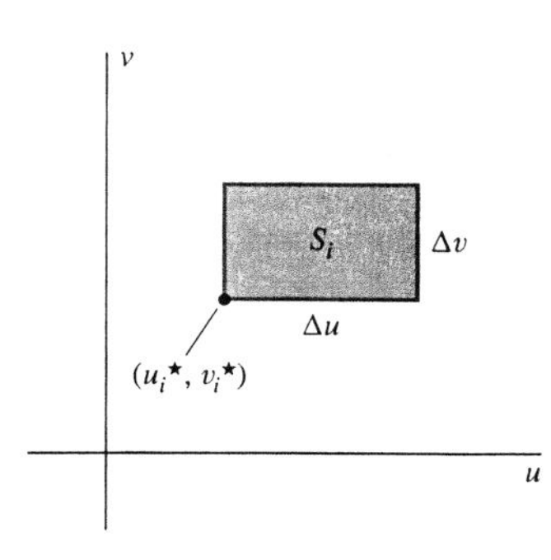

for its image under T. The u-curve through $(x_i^\star, y_i^\star)$ has velocity vector

$$\mathbf{t}_u = \mathbf{i} f_u(u_i^\star, v_i^\star) + \mathbf{j} g_u(u_i^\star, v_i^\star) = \frac{\partial x}{\partial u}\mathbf{i} + \frac{\partial y}{\partial u}\mathbf{j},$$

and the v-curve through $(x_i^\star, y_i^\star)$ has velocity vector

$$\mathbf{t}_v = \mathbf{i} f_v(u_i^\star, v_i^\star) + \mathbf{j} g_v(u_i^\star, v_i^\star) = \frac{\partial x}{\partial v}\mathbf{i} + \frac{\partial y}{\partial v}\mathbf{j}.$$

Thus we can approximate the curvilinear figure R_i by a parallelogram P_i with edges that are "copies" of the vectors $\mathbf{t}_u\,\Delta u$ and $\mathbf{t}_v\,\Delta v$. These edges and the approximating parallelogram appear in Fig. 14.9.3.

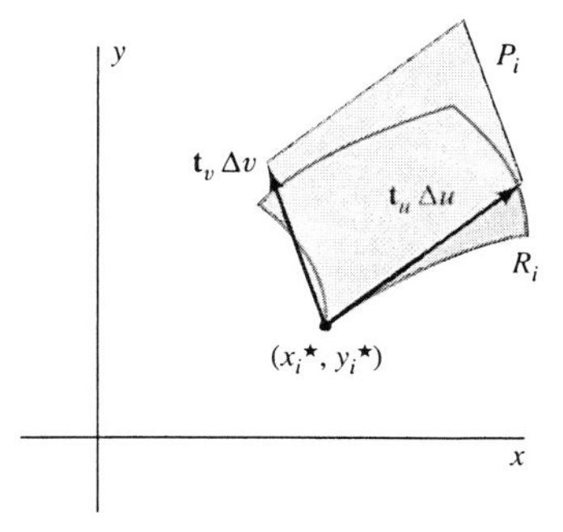

FIGURE 14.9.3 The effect of the transformation T. We estimate the area of $R_i = T(S_i)$ by computing the area of the parallelogram P_i.

Our strategy is to approximate the area ΔA_i of the curvilinear figure R_i with the area $a(P_i)$ of the parallelogram P_i. To calculate this approximating area, we recall from Section 12.3 that the area of a parallelogram spanned by two vectors **a** and **b** is the length $|\mathbf{a} \times \mathbf{b}|$ of their cross product. Therefore

$$\Delta A_i \approx a(P_i) = |(\mathbf{t}_u\,\Delta u) \times (\mathbf{t}_v\,\Delta v)| = |\mathbf{t}_u \times \mathbf{t}_v|\,\Delta u\,\Delta v. \tag{4}$$

But

$$\mathbf{t}_u \times \mathbf{t}_v = \begin{vmatrix} \mathbf{i} & \mathbf{j} & \mathbf{k} \\ \dfrac{\partial x}{\partial u} & \dfrac{\partial y}{\partial u} & 0 \\ \dfrac{\partial x}{\partial v} & \dfrac{\partial y}{\partial v} & 0 \end{vmatrix} = \begin{vmatrix} \dfrac{\partial x}{\partial u} & \dfrac{\partial x}{\partial v} \\ \dfrac{\partial y}{\partial u} & \dfrac{\partial y}{\partial v} \end{vmatrix}\mathbf{k}. \tag{5}$$

The 2×2 determinant on the right in Eq. (5) is called the *Jacobian* of the transformation $T : \boldsymbol{R}^2_{uv} \to \boldsymbol{R}^2_{xy}$, after the German mathematician Carl Jacobi (1804–1851), who first investigated general changes of variables in double integrals.

DEFINITION The Jacobian

The **Jacobian** of the continuously differentiable transformation $T : \boldsymbol{R}^2_{uv} \to \boldsymbol{R}^2_{xy}$ is the (real-valued) function $J_T : \boldsymbol{R}^2_{uv} \to \boldsymbol{R}$ defined by

$$J_T(u, v) = \begin{vmatrix} x_u(u, v) & x_v(u, v) \\ y_u(u, v) & y_v(u, v) \end{vmatrix}. \tag{6}$$

Another common and particularly suggestive notation for the Jacobian is

$$J_T = \frac{\partial(x, y)}{\partial(u, v)},$$

where the 2×2 pattern reminds us that both the dependent variables x and y are differentiated with respect to both the independent variables u and v.

Recall that we began with an inner partition $\{S_1, S_2, \dots, S_n\}$ of the region S in the uv-plane, with the images of these rectangles forming a curvilinear partition $\{R_1, R_2, \dots, R_n\}$ of the region $R = T(S)$ in the xy-plane. Now Eqs. (4) and (5) imply that the area ΔA_i of R_i is given approximately by

$$\Delta A_i \approx |J_T(u_i^\star, v_i^\star)|\,\Delta u\,\Delta v$$

in terms of the *absolute value* of the Jacobian determinant and the area $a(S_i) = \Delta u\,\Delta v$. Therefore, when we set up Riemann sums for approximating double integrals, we

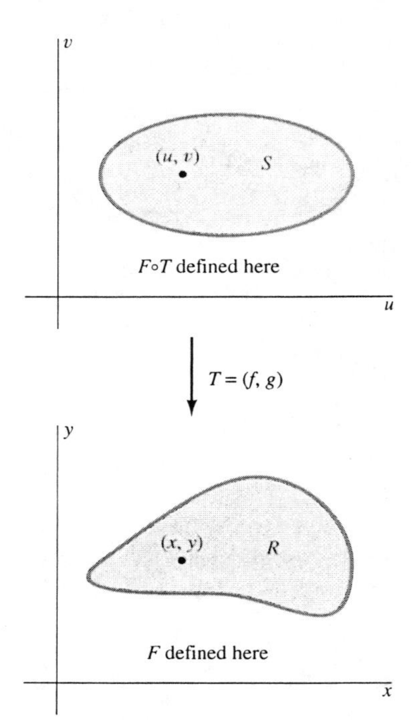

FIGURE 14.9.4 The domains of $F(x, y)$ and $F(T(u, v)) = F(f(x, y), g(x, y))$.

find that

$$\begin{aligned}\iint_R F(x, y)\,dx\,dy &\approx \sum_{i=1}^{n} F(x_i^\star, y_i^\star)\,\Delta A_i \\ &\approx \sum_{i=1}^{m} F(f(u_i^\star, v_i^\star), g(u_i^\star, v_i^\star))|J_T(u_i^\star, v_i^\star)|\,\Delta u\,\Delta v \\ &\approx \iint_S F(f(u, v), g(u, v))\,|J_T(u, v)|\,du\,dv.\end{aligned}$$

This discussion is, in fact, an outline of a proof of the following general **change-of-variables theorem.** To ensure the existence of the indicated double integrals, we assume that the boundaries of both regions R and S consist of a finite number of piecewise smooth curves. (See Fig. 14.9.4.)

THEOREM 1 Change of Variables

Suppose that the continuously differentiable transformation $T : \mathbf{R}^2_{uv} \to \mathbf{R}^2_{xy}$ takes the bounded region S in the uv-plane onto the bounded region R in the xy-plane, and is one-to-one from the interior of S to the interior of R. If $F(x, y)$ is continuous on R, then

$$\iint_R F(x, y)\,dx\,dy = \iint_S F(T(u, v))\,|J_T(u, v)|\,du\,dv. \tag{7}$$

If we write $G(u, v) = F(T(u, v))$ for the result of substituting $x(u, v)$ and $y(u, v)$ for x and y in the original integrand $F(x, y)$, then the change-of-variables formula in (7) takes the form

$$\iint_R F(x, y)\,dx\,dy = \iint_S G(u, v)\left|\frac{\partial(x, y)}{\partial(u, v)}\right|du\,dv. \tag{8}$$

Thus we formally transform the integral $\iint_R F(x, y)\,dA$ by replacing the original variables x and y with $x(u, v)$ and $y(u, v)$, respectively, and writing

$$dA = \left|\frac{\partial(x, y)}{\partial(u, v)}\right|du\,dv$$

for the area element in terms of u and v.

Note the analogy between Eq. (8) and the single-variable formula in Eq. (1). In fact, if $g'(x) \neq 0$ on $[c, d]$ and we denote by α the smaller, and by β the larger, of the two limits c and d in Eq. (1), then Eq. (1) takes the form

$$\int_a^b f(x)\,dx = \int_\alpha^\beta f(g(u))\,|g'(u)|\,du. \tag{1a}$$

Thus the Jacobian in Eq. (8) plays the role of the derivative $g'(u)$ in Eq. (1).

EXAMPLE 2 Suppose that the transformation T from the $r\theta$-plane to the xy-plane is determined by the polar equations

$$x = f(r, \theta) = r\cos\theta, \qquad y = g(r, \theta) = r\sin\theta.$$

The Jacobian of T is

$$\frac{\partial(x, y)}{\partial(r, \theta)} = \begin{vmatrix}\cos\theta & -r\sin\theta \\ \sin\theta & r\cos\theta\end{vmatrix} = r > 0,$$

so Eq. (8) reduces to the familiar formula

$$\iint_R F(x, y)\,dx\,dy = \iint_S F(r\cos\theta,\ r\sin\theta)\,r\,dr\,d\theta. \qquad \blacklozenge$$

Given a particular double integral $\iint_R f(x, y)\,dx\,dy$, how do we find a *productive* change of variables? One standard approach is to choose a transformation T such that the boundary of R consists of u-curves and v-curves. In case it is more convenient to express u and v in terms of x and y, we can first compute $\partial(u, v)/\partial(x, y)$ explicitly and then find the needed Jacobian $\partial(x, y)/\partial(u, v)$ from the formula

$$\frac{\partial(x, y)}{\partial(u, v)} \cdot \frac{\partial(u, v)}{\partial(x, y)} = 1. \tag{9}$$

Equation (9) is a consequence of the chain rule. (See Problem 18.)

EXAMPLE 3 Suppose that R is the plane region of unit density that is bounded by the hyperbolas

$$xy = 1, \quad xy = 3 \qquad \text{and} \qquad x^2 - y^2 = 1, \quad x^2 - y^2 = 4.$$

Find the polar moment of inertia

$$I_0 = \iint_R (x^2 + y^2)\,dx\,dy$$

of this region.

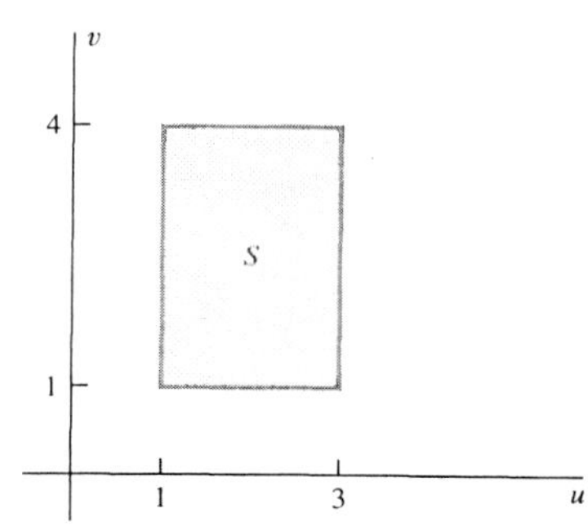

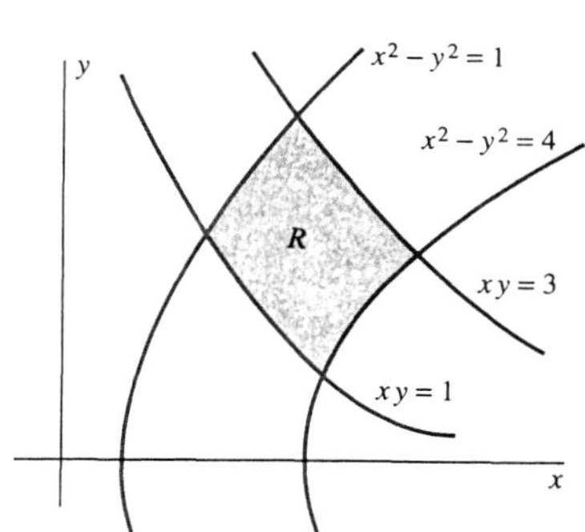

FIGURE 14.9.5 The transformation T and the new region S constructed in Example 3.

Solution The hyperbolas bounding R are u-curves and v-curves if $u = xy$ and $v = x^2 - y^2$, as in Example 1. We can most easily write the integrand $x^2 + y^2$ in terms of u and v by first noting that

$$4u^2 + v^2 = 4x^2y^2 + (x^2 - y^2)^2 = (x^2 + y^2)^2,$$

so $x^2 + y^2 = \sqrt{4u^2 + v^2}$. Now

$$\frac{\partial(u, v)}{\partial(x, y)} = \begin{vmatrix} y & x \\ 2x & -2y \end{vmatrix} = -2(x^2 + y^2).$$

Hence Eq. (9) gives

$$\frac{\partial(x, y)}{\partial(u, v)} = -\frac{1}{2(x^2 + y^2)} = -\frac{1}{2\sqrt{4u^2 + v^2}}.$$

We are now ready to apply the change-of-variables theorem, with the regions S and R as shown in Fig. 14.9.5. With $F(x, y) = x^2 + y^2$, Eq. (8) gives

$$I_0 = \iint_R (x^2 + y^2)\,dx\,dy = \int_1^4 \int_1^3 \sqrt{4u^2 + v^2}\,\frac{1}{2\sqrt{4u^2 + v^2}}\,du\,dv$$

$$= \int_1^4 \int_1^3 \frac{1}{2}\,du\,dv = 3. \qquad \blacklozenge$$

Example 4 is motivated by an important application. Consider an engine with an operating cycle that consists of alternate expansion and compression of gas in a piston. During one cycle the point (p, V), which gives the pressure and volume of this gas, traces a closed curve in the pV-plane. The work done by the engine—ignoring friction and related losses—is then equal (in appropriate units) to the area *enclosed by this curve,* called the *indicator diagram* of the engine. The indicator diagram for an ideal *Carnot engine* consists of two *isotherms* $xy = a$, $xy = b$ and two *adiabatics* $xy^\gamma = c$, $xy^\gamma = d$, where γ is the heat capacity ratio of the working gas in the piston. A typical value is $\gamma = 1.4$.

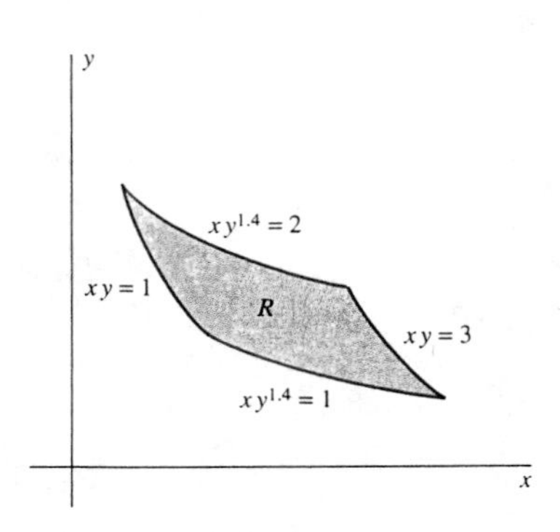

FIGURE 14.9.6 Finding the area of the region R (Example 4).

EXAMPLE 4 Find the area of the region R bounded by the curves $xy = 1$, $xy = 3$ and $xy^{1.4} = 1$, $xy^{1.4} = 2$ (Fig. 14.9.6).

Solution To force the given curves to be u-curves and v-curves, we define our change-of-variables transformation by $u = xy$ and $v = xy^{1.4}$. Then

$$\frac{\partial(u, v)}{\partial(x, y)} = \begin{vmatrix} y & x \\ y^{1.4} & (1.4)xy^{0.4} \end{vmatrix} = (0.4)xy^{1.4} = (0.4)v.$$

So

$$\frac{\partial(x, y)}{\partial(u, v)} = \frac{1}{\partial(u, v)/\partial(x, y)} = \frac{2.5}{v}.$$

Consequently, the change-of-variables theorem gives the formula

$$A = \iint_R 1\,dx\,dy = \int_1^2 \int_1^3 \frac{2.5}{v}\,du\,dv = 5 \ln 2.$$

◆

Change of Variables in Triple Integrals

The change-of-variables formula for triple integrals is similar to Eq. (8). Suppose that S and $R = T(S)$ are regions that correspond under the continuously differentiable transformation $T : \mathbf{R}^3_{uvw} \to \mathbf{R}^3_{xyz}$. Then the Jacobian of T is the determinant

$$J_T(u, v, w) = \frac{\partial(x, y, z)}{\partial(u, v, w)} = \begin{vmatrix} \dfrac{\partial x}{\partial u} & \dfrac{\partial x}{\partial v} & \dfrac{\partial x}{\partial w} \\ \dfrac{\partial y}{\partial u} & \dfrac{\partial y}{\partial v} & \dfrac{\partial y}{\partial w} \\ \dfrac{\partial z}{\partial u} & \dfrac{\partial z}{\partial v} & \dfrac{\partial z}{\partial w} \end{vmatrix}. \tag{10}$$

Then (under assumptions equivalent to those stated in Theorem 1) the change-of-variables formula for triple integrals is

$$\iiint_R F(x, y, z)\,dx\,dy\,dz = \iint_S F(T(u, v, w))\,|J_T(u, v, w)|\,du\,dv\,dw, \tag{11}$$

in direct analogy to Eq. (7) for double integrals. That is,

$$\iiint_R F(x, y, z)\,dx\,dy\,dz = \iiint_S G(u, v, w) \left| \frac{\partial(x, y, z)}{\partial(u, v, w)} \right| du\,dv\,dw, \tag{12}$$

where $G(u, v, w) = F(T(u, v, w)) = F(x(u, v, w), y(u, v, w), z(u, v, w))$ is the function obtained from $F(x, y, z)$ upon expressing the original variables x, y, and z in terms of the new variables u, v, and w.

EXAMPLE 5 If T is the spherical-coordinates transformation given by

$$x = \rho \sin\phi \cos\theta, \qquad y = \rho \sin\phi \sin\theta, \qquad z = \rho \cos\phi,$$

then the Jacobian of T is

$$\frac{\partial(x, y, z)}{\partial(\rho, \phi, \theta)} = \begin{vmatrix} \sin\phi\cos\theta & \rho\cos\phi\cos\theta & -\rho\sin\phi\sin\theta \\ \sin\phi\sin\theta & \rho\cos\phi\sin\theta & \rho\sin\phi\cos\theta \\ \cos\phi & -\rho\sin\phi & 0 \end{vmatrix} = \rho^2 \sin\phi.$$

Thus Eq. (11) reduces to the familiar formula

$$\iiint_R F(x, y, z)\,dx\,dy\,dz = \iiint_S G(\rho, \phi, \theta)\,\rho^2 \sin\phi\,d\rho\,d\phi\,d\theta.$$

The sign is correct because $\rho^2 \sin\phi \geqq 0$ for ϕ in $[0, \pi]$.

◆

EXAMPLE 6 Find the volume of the solid torus R obtained by revolving around the z-axis the circular disk

$$(x-b)^2+z^2 \leqq a^2, \quad 0<a<b \tag{13}$$

in the xz-plane.

Solution This is the torus of Example 5 of Section 14.8. Let us write u for the ordinary polar coordinate angle θ, v for the angle ψ of Fig. 14.8.13, and w for the distance from the center of the circular disk described by the inequality in (13). We then define the transformation T by means of the equations

$$x=(b+w\cos v)\cos u, \qquad y=(b+w\cos v)\sin u, \qquad z=w\sin v.$$

Then the solid torus R is the image under T of the region in uvw-space described by the inequalities

$$0 \leqq u \leqq 2\pi, \qquad 0 \leqq v \leqq 2\pi, \qquad 0 \leqq w \leqq a.$$

By a routine computation, we find that the Jacobian of T is

$$\frac{\partial(x, y, z)}{\partial(u, v, w)}=w(b+w\cos v).$$

Hence Eq. (11) with $F(x, y, z) \equiv 1$ yields volume

$$\begin{aligned} V &= \iiint_T 1\,dx\,dy\,dz=\int_0^{2\pi}\int_0^{2\pi}\int_0^a (bw+w^2\cos v)\,dw\,du\,dv \\ &= 2\pi\int_0^{2\pi}\left(\frac{1}{2}a^2 b+\frac{1}{3}a^3\cos v\right)dv=2\pi^2 a^2 b, \end{aligned}$$

which agrees with the value $V=2\pi b\cdot\pi a^2$ given by Pappus's first theorem (Section 14.5). ◆

14.9 TRUE/FALSE STUDY GUIDE

14.9 CONCEPTS: QUESTIONS AND DISCUSSION

1. Explain why the change-of-variables formula involves the *absolute value* of the Jacobian, rather than the Jacobian itself.
2. Suppose that R is a given parallelogram in the xy-plane. Explain how to transform an integral $\iint_R F(x, y)\,dA$ into an integral over a rectangle in the uv-plane.
3. Suppose that your pocket computer contains a routine for the numerical evaluation of double integrals, but requires that the domain of integration be a rectangle. Given an integral $\iint_R F(x, y)\,dA$ where R is a region of the form $a \leqq x \leqq b$, $f(x) \leqq y \leqq g(x)$, describe a transformation that converts this integral into one that your pocket computer can evaluate.
4. Describe a strategy for evaluating an integral over the region R in the xy-plane that is bounded by a given rotated ellipse $ax^2+bxy+cy^2=1$.

14.9 PROBLEMS

In Problems 1 through 6, solve for x and y in terms of u and v. Then compute the Jacobian $\partial(x, y)/\partial(u, v)$.

1. $u=x+y, \quad v=x-y$

2. $u=x-2y, \quad v=3x+y$

3. $u=xy, \quad v=y/x$

4. $u=2(x^2+y^2), \quad v=2(x^2-y^2)$

5. $u=x+2y^2, \quad v=x-2y^2$

6. $u=\dfrac{2x}{x^2+y^2}, \quad v=-\dfrac{2y}{x^2+y^2}$

7. Let R be the parallelogram bounded by the lines $x+y=1$, $x+y=2$ and $2x-3y=2$, $2x-3y=5$. Substitute $u=x+y$, $v=2x-3y$ to find its area

$$A=\iint_R 1\,dx\,dy.$$

8. Substitute $u = xy$, $v = y/x$ to find the area of the first-quadrant region bounded by the lines $y = x$, $y = 2x$ and the hyperbolas $xy = 1$, $xy = 2$ (Fig. 14.9.7).

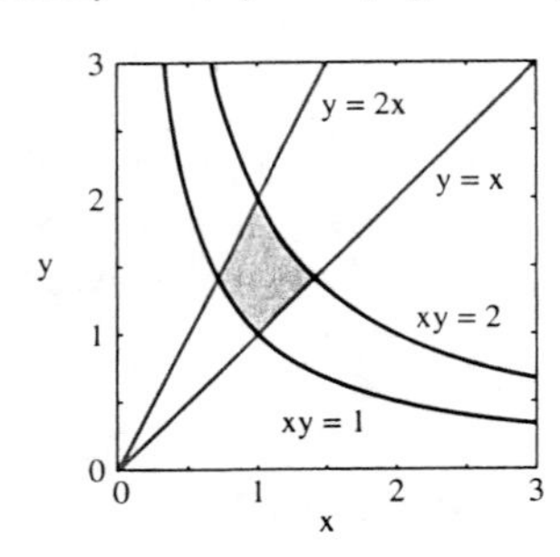

FIGURE 14.9.7 The region of Problem 8.

9. Substitute $u = xy$, $v = xy^3$ to find the area of the first-quadrant region bounded by the curves $xy = 2$, $xy = 4$ and $xy^3 = 3$, $xy^3 = 6$ (Fig. 14.9.8).

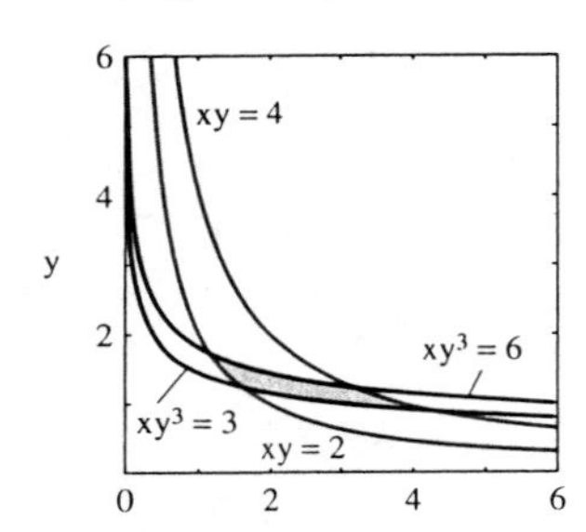

FIGURE 14.9.8 The region of Problem 9.

10. Find the area of the first-quadrant region bounded by the curves $y = x^2$, $y = 2x^2$ and $x = y^2$, $x = 4y^2$ (Fig. 14.9.9). [*Suggestion:* Let $y = ux^2$ and $x = vy^2$.]

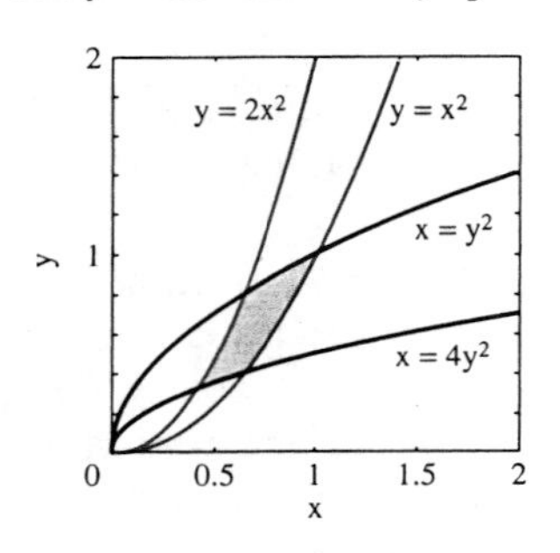

FIGURE 14.9.9 The region of Problem 10.

11. Use the method of Problem 10 to find the area of the first-quadrant region bounded by the curves $y = x^3$, $y = 2x^3$ and $x = y^3$, $x = 4y^3$.

12. Let R be the first-quadrant region bounded by the circles $x^2 + y^2 = 2x$, $x^2 + y^2 = 6x$ and the circles $x^2 + y^2 = 2y$, $x^2 + y^2 = 8y$. Use the transformation

$$u = \frac{2x}{x^2 + y^2}, \qquad v = \frac{2y}{x^2 + y^2}$$

to evaluate the integral

$$\iint_R \frac{1}{(x^2 + y^2)^2}\,dx\,dy.$$

13. Use elliptical coordinates $x = 3r\cos\theta$, $y = 2r\sin\theta$ to find the volume of the region bounded by the xy-plane, the paraboloid $z = x^2 + y^2$, and the elliptic cylinder

$$\frac{x^2}{9} + \frac{y^2}{4} = 1.$$

14. Let R be the solid ellipsoid with outer boundary surface

$$\frac{x^2}{a^2} + \frac{y^2}{b^2} + \frac{z^2}{c^2} = 1.$$

Use the transformation $x = au$, $y = bv$, $z = cw$ to show that the volume of this ellipsoid is

$$V = \iiint_R 1\,dx\,dy\,dz = \frac{4}{3}\pi abc.$$

15. Find the volume of the region in the first octant that is bounded by the hyperbolic cylinders $xy = 1$, $xy = 4$; $xz = 1$, $xz = 9$; and $yz = 4$, $yz = 9$. [*Suggestion:* Let $u = xy$, $v = xz$, $w = yz$, and note that $uvw = x^2y^2z^2$.]

16. Use the transformation

$$x = \frac{r}{t}\cos\theta, \qquad y = \frac{r}{t}\sin\theta, \qquad z = r^2$$

to find the volume of the region R that lies between the paraboloids $z = x^2 + y^2$, $z = 4(x^2 + y^2)$ and the planes $z = 1$, $z = 4$.

17. Let R be the rotated elliptical region bounded by the graph of $x^2 + xy + y^2 = 3$. Let $x = u + v$ and $y = u - v$. Show that

$$\iint_R \exp(-x^2 - xy - y^2)\,dx\,dy$$
$$= 2\iint_S \exp(-3u^2 - v^2)\,du\,dv.$$

Then substitute $u = r\cos\theta$, $v = \sqrt{3}\,(r\sin\theta)$ to evaluate the latter integral.

18. From the chain rule and from the following property of determinants, derive the relation in Eq. (9) between the Jacobians of a transformation and its inverse.

$$\begin{vmatrix} a_1 & b_1 \\ c_1 & d_1 \end{vmatrix} \cdot \begin{vmatrix} a_2 & b_2 \\ c_2 & d_2 \end{vmatrix} = \begin{vmatrix} a_1a_2 + b_1c_2 & a_1b_2 + b_1d_2 \\ a_2c_1 + c_2d_1 & b_2c_1 + d_1d_2 \end{vmatrix}.$$

19. Change to spherical coordinates to show that, for $k > 0$,

$$\int_{-\infty}^{+\infty}\int_{-\infty}^{+\infty}\int_{-\infty}^{+\infty} \sqrt{x^2 + y^2 + z^2}$$
$$\times \exp(-k(x^2 + y^2 + z^2))\,dx\,dy\,dz = \frac{2\pi}{k^2}.$$

20. Let R be the solid ellipsoid with constant density δ and boundary surface

$$\frac{x^2}{a^2} + \frac{y^2}{b^2} + \frac{z^2}{c^2} = 1.$$

Use ellipsoidal coordinates $x = a\rho\sin\phi\cos\theta$, $y = b\rho\sin\phi\sin\theta$, $z = c\rho\cos\phi$ to show that the mass of R is $M = \frac{4}{3}\pi\delta abc$.

21. Show that the moment of inertia of the ellipsoid of Problem 20 with respect to the z-axis is $I_z = \frac{1}{5}M(a^2 + b^2)$.

In Problems 22 through 26, use a computer algebra system (if necessary) to find the indicated centroids and moments of inertia.

22. The centroid of the plane region of Problem 8 (Fig. 14.9.7)

23. The centroid of the plane region of Problem 9 (Fig. 14.9.8)

24. The centroid of the plane region of Problem 10 (Fig. 14.9.9)

25. The moment of inertia around each coordinate axis of the solid ellipsoid of Problem 20

26. The centroid of the solid of Problem 16 and its moments of inertia around the coordinate axes

27. Write the triple integral that gives the average distance of points of the solid ellipsoid of Problem 20 from the origin. Then approximate that integral in the case $a = 4$, $b = 3$, and $c = 2$.

Problems 28 and 29 outline the use of double integrals to evaluate the famous infinite series

$$\zeta(2) = \sum_{n=1}^{\infty} \frac{1}{n^2} = 1 + \frac{1}{2^2} + \frac{1}{3^2} + \cdots$$

mentioned earlier in the Section 11.5 project. These problems are based on a calculation presented by Dirk Huylebrouck in his article "Similarity in Irrationality Proofs for π, ln 2, $\zeta(2)$, *and* $\zeta(3)$," The American Mathematical Monthly *(March 2001), 222–231.*

28. Substitute the geometric series for $(1 - xy)^{-1}$ to show that

$$\int_0^1 \int_0^1 \frac{1}{1 - xy}\, dx\, dy = \zeta(2),$$

assuming the validity of termwise integration of the resulting series in powers of xy.

29. (a) First find a common denominator in the integrand, then make the substitution $u = x^2$, $v = y^2$ to show that

$$\int_0^1 \int_0^1 \left(\frac{1}{1 - xy} - \frac{1}{1 + xy} \right) dx\, dy = \frac{1}{2}\zeta(2).$$

(b) Add the equation in part (a) and the identity

$$\int_0^1 \int_0^1 \left(\frac{1}{1 - xy} + \frac{1}{1 + xy} \right) dx\, dy = 2 \int_0^1 \int_0^1 \frac{1}{1 - x^2 y^2}\, dx\, dy$$

to show that

$$\zeta(2) = \frac{4}{3} \int_0^1 \int_0^1 \frac{1}{1 - x^2 y^2}\, dx\, dy.$$

(c) Finally, use the transformation $T : \mathbf{R}^2_{uv} \to \mathbf{R}^2_{xy}$ defined by $x = (\sin u)/(\cos v)$, $y = (\sin v)/(\cos u)$ to evaluate the final integral in part (b) and thereby obtain Euler's result that $\zeta(2) = \pi^2/6$. As indicated in Fig. 14.9.10, the transformation T carries the interior of the triangle $0 \leqq u \leqq (\pi/2) - v$, $0 \leqq v \leqq \pi/2$ in the uv-plane one-to-one to the interior of the unit square in the xy-plane.

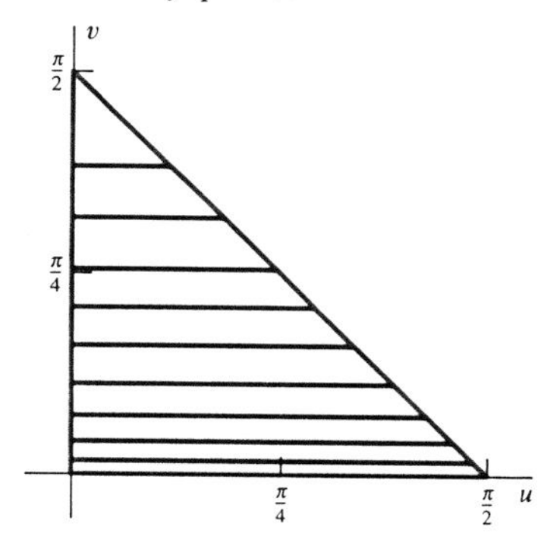

FIGURE 14.9.10(a) Horizontal u-lines in the domain of the transformation T.

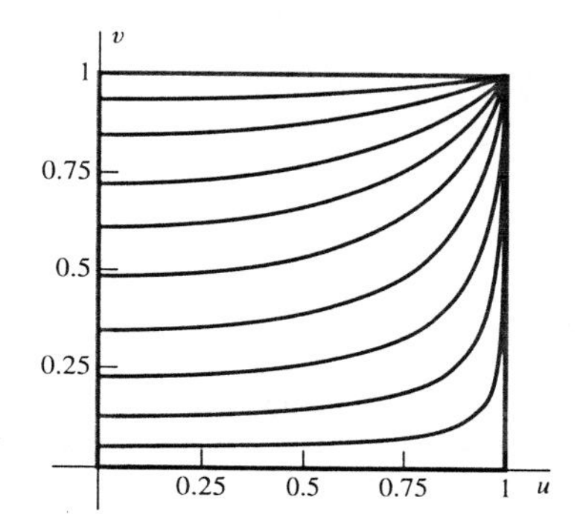

FIGURE 14.9.10(b) Their image u-curves in the range of the transformation T.

CHAPTER 14 REVIEW: DEFINITIONS, CONCEPTS, RESULTS

Use the following list as a guide to concepts that you may need to review.

1. Definition of the double integral as a limit of Riemann sums
2. Evaluation of double integrals by iterated single integration
3. Use of the double integral to find the volume between two surfaces above a given plane region
4. Transformation of the double integral $\iint_R f(x, y)\, dA$ into polar coordinates
5. Application of double integrals to find mass, centroids, and moments of inertia of plane laminae
6. The two theorems of Pappus
7. Definition of the triple integral as a limit of Riemann sums
8. Evaluation of triple integrals by iterated single integration
9. Applications of triple integrals to find volume, mass, centroids, and moments of inertia
10. Transformation of the triple integral $\iiint_T f(x, y, z)\, dV$ into cylindrical and spherical coordinates
11. The surface area of a parametric surface
12. The area of a surface $z = f(x, y)$ for (x, y) in the plane region R
13. The Jacobian of a transformation of coordinates
14. The transformation of a double or triple integral corresponding to a given change of variables

CHAPTER 14 MISCELLANEOUS PROBLEMS

In Problems 1 through 5, evaluate the given integral by first reversing the order of integration.

1. $\displaystyle\int_0^1\int_{y^{1/3}}^1 \frac{1}{\sqrt{1+x^2}}\,dx\,dy$

2. $\displaystyle\int_0^1\int_y^1 \frac{\sin x}{x}\,dx\,dy$

3. $\displaystyle\int_0^1\int_x^1 \exp(-y^2)\,dy\,dx$

4. $\displaystyle\int_0^8\int_{x^{2/3}}^4 x\cos y^4\,dy\,dx$

5. $\displaystyle\int_0^4\int_{\sqrt{y}}^2 \frac{y\exp(x^2)}{x^3}\,dx\,dy$

6. The double integral

$$\int_0^\infty\int_x^\infty \frac{e^{-y}}{y}\,dy\,dx$$

is an improper integral over the unbounded region in the first quadrant between the lines $y = x$ and $x = 0$. Assuming that it is valid (it is) to reverse the order of integration, evaluate this integral by integrating first with respect to x.

7. Find the volume of the solid T that lies below the paraboloid $z = x^2 + y^2$ and above the triangle R in the xy-plane that has vertices at $(0, 0, 0)$, $(1, 1, 0)$, and $(2, 0, 0)$.

8. Find by integration in cylindrical coordinates the volume bounded by the paraboloids $z = 2x^2 + 2y^2$ and $z = 48 - x^2 - y^2$.

9. Use integration in spherical coordinates to find the volume and centroid of the solid region that is inside the sphere $\rho = 3$, below the cone $\phi = \pi/3$, and above the xy-plane $\phi = \pi/2$.

10. Find the volume of the solid bounded by the elliptic paraboloids $z = x^2 + 3y^2$ and $z = 8 - x^2 - 5y^2$.

11. Find the volume bounded by the paraboloid $y = x^2 + 3z^2$ and the parabolic cylinder $y = 4 - z^2$.

12. Find the volume of the region bounded by the parabolic cylinders $z = x^2$, $z = 2 - x^2$ and the planes $y = 0$, $y + z = 4$.

13. Find the volume of the region bounded by the elliptical cylinder $y^2 + 4z^2 = 4$ and the planes $x = 0$, $x = y + 2$.

14. Show that the volume of the solid bounded by the elliptical cylinder

$$\frac{x^2}{a^2} + \frac{y^2}{b^2} = 1$$

and the planes $z = 0$, $z = h + x$ (where $h > a > 0$) is $V = \pi abh$.

15. Let R be the first-quadrant region bounded by the curve $x^4 + x^2y^2 = y^2$ and the line $y = x$. Use polar coordinates to evaluate

$$\iint_R \frac{1}{(1 + x^2 + y^2)^2}\,dA.$$

In Problems 16 through 20, find the mass and centroid of a plane lamina with the given shape and density δ.

16. The region bounded by $y = x^2$ and $x = y^2$; $\delta(x, y) = x^2 + y^2$

17. The region bounded by $x = 2y^2$ and $y^2 = x - 4$; $\delta(x, y) = y^2$

18. The region between $y = \ln x$ and the x-axis over the interval $1 \leqq x \leqq 2$; $\delta(x, y) = 1/x$

19. The circle bounded by $r = 2\cos\theta$; $\delta(r, \theta) = k$ (a constant)

20. The region of Problem 19; $\delta(r, \theta) = r$

21. Use the first theorem of Pappus to find the y-coordinate of the centroid of the upper half of the ellipse

$$\frac{x^2}{a^2} + \frac{y^2}{b^2} = 1.$$

Employ the facts that the area of this semiellipse is $A = \pi ab/2$ and the volume of the ellipsoid it generates when rotated around the x-axis is $V = \frac{4}{3}\pi ab^2$.

22. (a) Use the first theorem of Pappus to find the centroid of the first-quadrant portion of the annular ring with boundary circles $x^2 + y^2 = a^2$ and $x^2 + y^2 = b^2$ (where $0 < a < b$). (b) Show that the limiting position of this centroid as $b \to a$ is the centroid of a quarter-circular arc, as we found in Problem 44 of Section 14.5.

23. Find the centroid of the region in the xy-plane bounded by the x-axis and the parabola $y = 4 - x^2$.

24. Find the volume of the solid that lies below the parabolic cylinder $z = x^2$ and above the triangle in the xy-plane bounded by the coordinate axes and the line $x + y = 1$.

25. Use cylindrical coordinates to find the volume of the ice-cream cone bounded above by the sphere $x^2 + y^2 + z^2 = 5$ and below by the cone $z = 2\sqrt{x^2 + y^2}$.

26. Find the volume and centroid of the ice-cream cone bounded above by the sphere $\rho = a$ and below by the cone $\phi = \pi/3$.

27. A homogeneous solid circular cone has mass M and base radius a. Find its moment of inertia around its axis of symmetry.

28. Find the mass of the first octant of the ball $\rho \leqq a$ if its density at (x, y, z) is $\delta(x, y, z) = xyz$.

29. Find the moment of inertia around the x-axis of the homogeneous solid ellipsoid with unit density and boundary surface

$$\frac{x^2}{a^2} + \frac{y^2}{b^2} + \frac{z^2}{c^2} = 1.$$

30. Find the volume of the region in the first octant that is bounded by the sphere $\rho = a$, the cylinder $r = a$, the plane $z = a$, the xz-plane, and the yz-plane.

31. Find the moment of inertia around the z-axis of the homogeneous region of unit density that lies inside both the sphere $\rho = 2$ and the cylinder $r = 2\cos\theta$.

In Problems 32 through 34, a volume is generated by revolving a plane region R around an axis. To find the volume, set up a double integral over R by revolving an area element dA around the indicated axis to generate a volume element dV.

32. Find the volume of the solid obtained by revolving around the y-axis the region inside the circle $r = 2a\cos\theta$.

33. Find the volume of the solid obtained by revolving around the x-axis the region enclosed by the cardioid $r = 1 + \cos\theta$.

34. Find the volume of the solid torus obtained by revolving the disk $0 \leqq r \leqq a$ around the line $x = -b$, $|b| \geqq a$.

35. Assume that the torus of Problem 34 has uniform density δ. Find its moment of inertia around its natural axis of symmetry.

Problems 36 through 42 deal with average distance. *The **average distance** $\bar{d}$ of the point (x_0, y_0) from the points of the plane region R with area A is defined to be*

$$\bar{d} = \frac{1}{A}\iint_R \sqrt{(x - x_0)^2 + (y - y_0)^2}\, dA.$$

The average distance of a point (x_0, y_0, z_0) from the points of a space region is defined analogously.

36. Show that the average distance of the points of a disk of radius a from its center is $2a/3$.

37. Show that the average distance of the points of a disk of radius a from a fixed point on its boundary is $32a/9\pi$.

38. A circle of radius 1 is interior to and tangent to a circle of radius 2. Find the average distance of the point of tangency from the points that lie between the two circles.

39. Show that the average distance of the points of a spherical ball of radius a from its center is $3a/4$.

40. Show that the average distance of the points of a spherical ball of radius a from a fixed point on its surface is $6a/5$.

41. A sphere of radius 1 is interior to and tangent to a sphere of radius 2. Find the average distance of the point of tangency from the set of all points between the two spheres.

42. A right circular cone has radius R and height H. Find the average distance of points of the cone from its vertex.

43. Find the surface area of the part of the paraboloid $z = 10 - r^2$ that lies between the two planes $z = 1$ and $z = 6$.

44. Find the surface area of the part of the surface $z = y^2 - x^2$ that is inside the cylinder $x^2 + y^2 = 4$.

45. Let A be the surface area of the zone on the sphere $\rho = a$ between the planes $z = z_1$ and $z = z_2$ (where $-a \leqq z_1 < z_2 \leqq a$). Use the formula of Problem 18 in Section 14.8 to show that $A = 2\pi a h$, where $h = z_2 - z_1$.

46. Find the surface area of the part of the sphere $\rho = 2$ that is inside the cylinder $x^2 + y^2 = 2x$.

47. A square hole with side length 2 is cut through a cone of height 2 and base radius 2; the centerline of the hole is the axis of symmetry of the cone. Find the area of the surface removed from the cone.

48. Numerically approximate the surface area of the part of the parabolic cylinder $2z = x^2$ that lies inside the cylinder $x^2 + y^2 = 1$.

49. A "fence" of variable height $h(t)$ stands above the plane curve $(x(t), y(t))$. Thus the fence has the parametrization $x = x(t)$, $y = y(t)$, $z = z$ for $a \leqq t \leqq b$, $0 \leqq z \leqq h(t)$. Apply Eq. (8) of Section 14.8 to show that the area of the fence is

$$A = \int_a^b \int_0^{h(t)} \left[\left(\frac{dx}{dt}\right)^2 + \left(\frac{dy}{dt}\right)^2\right]^{1/2} dz\, dt.$$

50. Apply the formula of Problem 49 to compute the area of the part of the cylinder $r = a\sin\theta$ that lies inside the sphere $r^2 + z^2 = a^2$.

51. Find the polar moment of inertia of the first-quadrant region of constant density δ that is bounded by the hyperbolas $xy = 1$, $xy = 3$ and $x^2 - y^2 = 1$, $x^2 - y^2 = 4$.

52. Substitute $u = x - y$ and $v = x + y$ to evaluate

$$\iint_R \exp\left(\frac{x - y}{x + y}\right) dx\, dy,$$

where R is bounded by the coordinate axes and the line $x + y = 1$.

53. Use ellipsoidal coordinates $x = a\rho\sin\phi\cos\theta$, $y = b\rho\sin\phi\sin\theta$, $z = c\rho\cos\phi$ to find the mass of the solid ellipsoid

$$\frac{x^2}{a^2} + \frac{y^2}{b^2} + \frac{z^2}{c^2} \leqq 1$$

if its density at the point (x, y, z) is given by

$$\delta(x, y, z) = 1 - \frac{x^2}{a^2} - \frac{y^2}{b^2} - \frac{z^2}{c^2}.$$

54. Let R be the first-quadrant region bounded by the lemniscates $r^2 = 3\cos 2\theta$, $r^2 = 4\cos 2\theta$ and $r^2 = 3\sin 2\theta$, $r^2 = 4\sin 2\theta$ (Fig. 14.MP.1). Show that its area is

$$A = \frac{2\sqrt{17} - 5\sqrt{2}}{4}.$$

[*Suggestion:* Define the transformation T from the uv-plane to the $r\theta$-plane by $r^2 = u^{1/2}\cos 2\theta$, $r^2 = v^{1/2}\sin 2\theta$. Show first that

$$r^2 = \frac{uv}{u + v}, \qquad \theta = \frac{1}{2}\arctan\frac{u^{1/2}}{v^{1/2}}.$$

Then show that

$$\frac{\partial(r, \theta)}{\partial(u, v)} = -\frac{1}{16r(u + v)^{3/2}}.]$$

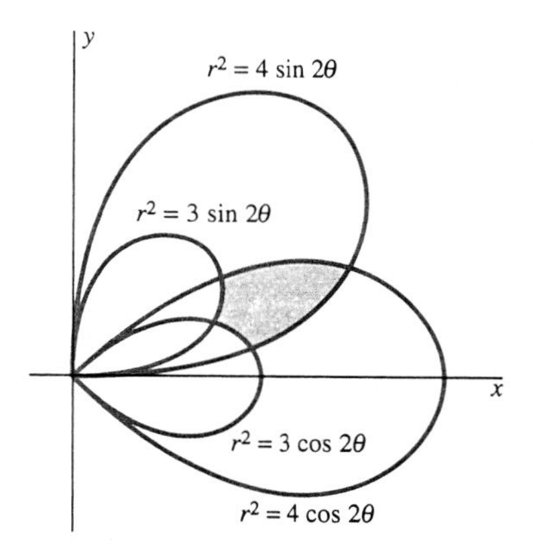

FIGURE 14.MP.1 The region R of Problem 54.

55. A 2-by-2 square hole is cut symmetrically through a sphere of radius $\sqrt{3}$. (See Fig. 14.MP.2.) (a) Show that the total surface area of the two pieces cut from the sphere is

$$A = \int_0^1 8\sqrt{3} \arcsin\left(\frac{1}{\sqrt{3 - x^2}}\right) dx.$$

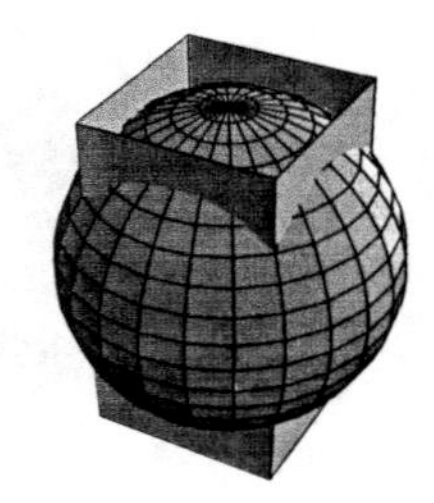

FIGURE 14.MP.2 Cutting a square hole through the sphere of Problem 55.

Then use Simpson's rule to approximate this integral. (b) (Difficult!) Show that the exact value of the integral in part (a) is $A = 4\pi(\sqrt{3} - 1)$. [*Suggestion:* First integrate by parts, then substitute $x = \sqrt{2} \sin\theta$.]

56. Show that the volume enclosed by the surface

$$x^{2/3} + y^{2/3} + z^{2/3} = a^{2/3}$$

is $V = \frac{4}{35}\pi a^3$. [*Suggestion:* Substitute $y = b \sin^3 \theta$.]

57. Show that the volume enclosed by the surface

$$x^{1/3} + y^{1/3} + z^{1/3} = a^{1/3}$$

is $V = \frac{1}{210}a^3$. [*Suggestion:* Substitute $y = b \sin^6 \theta$.]

58. Find the average of the *square* of the distance of points of the solid ellipsoid $(x/a)^2 + (y/b)^2 + (z/c)^2 \leqq 1$ from the origin.

59. A cube C of edge length 1 is rotated around a line passing through two opposite vertices, thereby sweeping out a solid S of revolution. Find the volume of S. (*Answer:* $\pi/\sqrt{3} \approx 1.8138$.)

15 VECTOR CALCULUS

C. F. Gauss (1777–1855)

It is customary to list Archimedes, Newton, and Carl Friedrich Gauss as history's three preeminent mathematicians. Gauss was a precocious infant in a poor and uneducated family. He learned to calculate before he could talk and taught himself to read before beginning school in his native Brunswick, Germany. At age 14 he was already familiar with elementary geometry, algebra, and analysis. By age 18, when he entered the University of Göttingen, he had discovered empirically the *prime number theorem,* which implies that the number of primes p between 1 and n is about $n/(\ln n)$. This theorem was not proved rigorously until a century later.

During his first year at university, Gauss discovered conditions for the ruler-and-compass construction of regular polygons and demonstrated the constructability of the regular 17-gon (the first advance in this area since the similar construction of the regular pentagon in Euclid's *Elements* 2000 years earlier). In 1801 Gauss published his great treatise *Disquisitiones arithmeticae,* which summarized number theory to that time and set the pattern for nineteenth-century research in that area. This book established Gauss as a mathematician of uncommon stature, but another event thrust him into the public eye. On January 1, 1801, the new asteroid Ceres was observed, but it disappeared behind the sun a month later. In the following weeks, astronomers searched the skies in vain for Ceres' reappearance. It was Gauss who developed the method of least-squares approximations to predict the asteroid's future orbit on the basis of a handful of observations. When Gauss's three-month long computation was finished, Ceres was soon spotted in the precise location he had predicted. All this made Gauss famous as a mathematician and astronomer at the age of 25.

In 1807 Gauss became director of the astronomical observatory in Göttingen, where he remained until his death. His published work thereafter dealt mainly with physical science, although his unpublished papers show that he continued to work on theoretical mathematics ranging from infinite series and special functions to non-Euclidean geometry. His work on the shape of the earth's surface established the new subject of differential geometry, and his studies of the earth's magnetic and gravitational fields involved results such as the divergence theorem (Section 15.6).

The concept of curved space-time in Albert Einstein's general relativity theory traces back to the discovery of non-Euclidean geometry and Gauss's early investigations of differential geometry. A current application of relativity theory is the study of black holes. Space is itself thought to be severely warped in the vicinity of a black hole, with its immense gravitational attraction, and the mathematics required to analyze such a situation begins with the vector calculus of Chapter 15.

Schematic of mass swirling into a supermassive black hole at the center of the Milky Way galaxy.

15.1 VECTOR FIELDS

This chapter is devoted to topics in the calculus of vector fields of importance in science and engineering. A **vector field** defined on a region T in space is a vector-valued function $\mathbf{F}$ that associates with each point (x, y, z) of T a vector

$$\mathbf{F}(x, y, z) = \mathbf{i}P(x, y, z) + \mathbf{j}Q(x, y, z) + \mathbf{k}R(x, y, z). \tag{1}$$

We may more briefly describe the vector field $\mathbf{F}$ in terms of its *component functions* P, Q, and R by writing

$$\mathbf{F} = P\mathbf{i} + Q\mathbf{j} + R\mathbf{k} \quad \text{or} \quad \mathbf{F} = \langle P, Q, R\rangle.$$

Note that the components P, Q, and R of a vector function are *scalar* (real-valued) functions.

A **vector field** in the plane is similar except that neither z-components nor z-coordinates are involved. Thus a vector field on the plane region R is a vector-valued function $\mathbf{F}$ that associates with each point (x, y) of R a vector

$$\mathbf{F}(x, y) = \mathbf{i}P(x, y) + \mathbf{j}Q(x, y) \tag{2}$$

or, briefly, $\mathbf{F} = P\mathbf{i} + Q\mathbf{j}$ or $\mathbf{F} = \langle P, Q\rangle$.

It is useful to be able to visualize a given vector field $\mathbf{F}$. One common way is to sketch a collection of typical vectors $\mathbf{F}(x, y)$, each represented by an arrow of length $|\mathbf{F}(x, y)|$ and located with (x, y) as its initial point. This procedure is illustrated in Example 1.

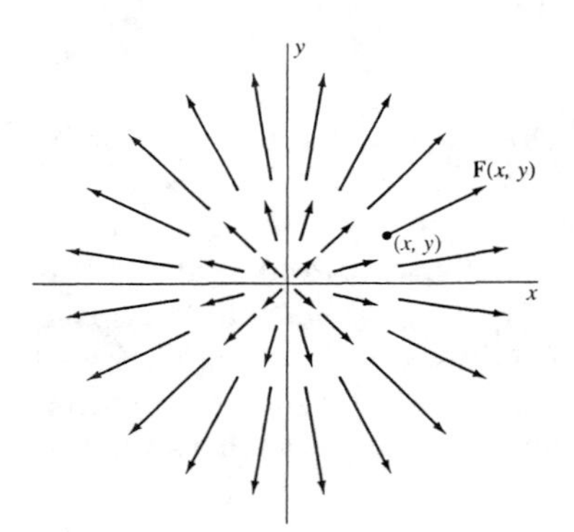

FIGURE 15.1.1 The vector field $\mathbf{F}(x, y) = x\mathbf{i} + y\mathbf{j}$.

EXAMPLE 1 Describe the vector field $\mathbf{F}(x, y) = x\mathbf{i} + y\mathbf{j}$.

Solution For each point (x, y) in the coordinate plane, $\mathbf{F}(x, y)$ is simply its position vector. It points directly away from the origin and has length

$$|\mathbf{F}(x, y)| = |x\mathbf{i} + y\mathbf{j}| = \sqrt{x^2 + y^2} = r,$$

equal to the distance from the origin to (x, y). Figure 15.1.1 shows some typical vectors representing this vector field. ◆

Among the most important vector fields in applications are velocity vector fields. Imagine the steady flow of a fluid, such as the water in a river or the solar wind. By a *steady flow* we mean that the velocity vector $\mathbf{v}(x, y, z)$ of the fluid flowing through each point (x, y, z) is independent of time (although not necessarily independent of x, y, and z), so the pattern of the flow remains constant. Then $\mathbf{v}(x, y, z)$ is the **velocity vector field** of the fluid flow.

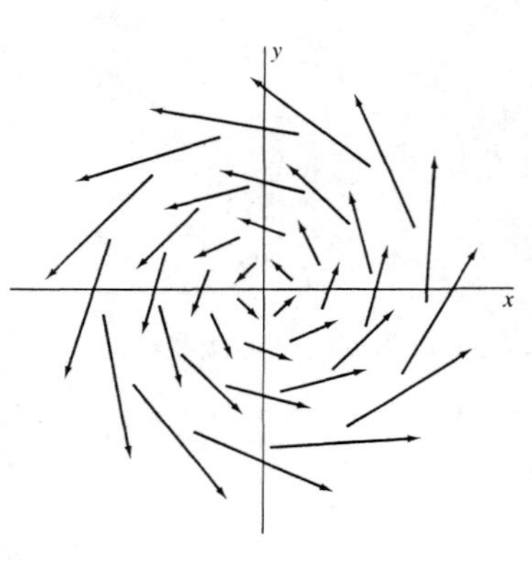

FIGURE 15.1.2 The velocity vector field $\mathbf{v}(x, y) = \omega(-y\mathbf{i} + x\mathbf{j})$, drawn for $\omega = 1$ (Example 2).

EXAMPLE 2 Suppose that the horizontal xy-plane is covered with a thin sheet of water that is revolving (rather like a whirlpool) around the origin with constant angular speed ω radians per second in the counterclockwise direction. Describe the associated velocity vector field.

Solution In this case we have a two-dimensional vector field $\mathbf{v}(x, y)$. At each point (x, y) the water is moving with speed $v = r\omega$ and tangential to the circle of radius $r = \sqrt{x^2 + y^2}$. The vector field

$$\mathbf{v}(x, y) = \omega(-y\mathbf{i} + x\mathbf{j}) \tag{3}$$

has length $r\omega$ and points in a generally counterclockwise direction, and

$$\mathbf{v} \cdot \mathbf{r} = \omega(-y\mathbf{i} + x\mathbf{j}) \cdot (x\mathbf{i} + y\mathbf{j}) = 0,$$

so $\mathbf{v}$ is tangent to the circle just mentioned. The velocity field determined by Eq. (3) is illustrated in Fig. 15.1.2. ◆

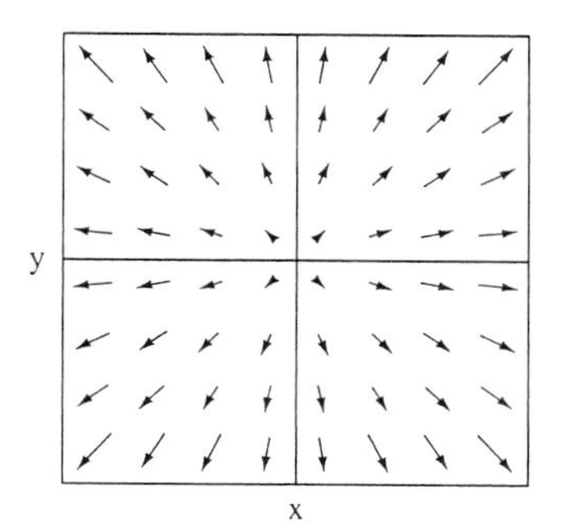

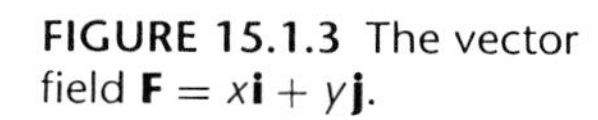
FIGURE 15.1.3 The vector field $\mathbf{F} = x\mathbf{i} + y\mathbf{j}$.

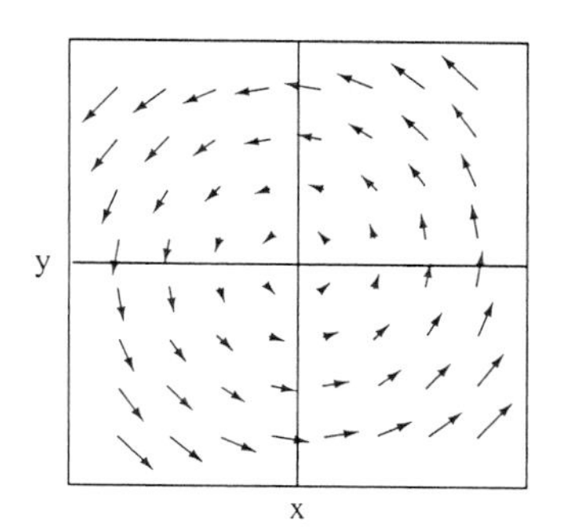

FIGURE 15.1.4 The vector field $\mathbf{F} = -y\mathbf{i} + x\mathbf{j}$.

REMARK Most computer algebra systems have the facility to plot vector fields. For instance, either the *Maple* command `fieldplot( [x,y], x=-2..2, y=-2..2 )` or the *Mathematica* command `PlotVectorField[ {x,y}, {x,-2,2}, {y,-2,2} ]` generates a computer plot like Fig. 15.1.3 of the vector field $\mathbf{F} = x\mathbf{i} + y\mathbf{j}$ of Example 1. The computer has scaled the vectors to a fixed maximum length so that the length of each vector as plotted is proportional to its actual length. Figure 15.1.4 shows a similar computer plot of the vector field $\mathbf{F} = -y\mathbf{i} + x\mathbf{j}$ of Example 2.

Equally important in physical applications are *force fields*. Suppose that some circumstance (perhaps gravitational or electrical in character) causes a force $\mathbf{F}(x, y, z)$ to act on a particle when it is placed at the point (x, y, z). Then we have a force field $\mathbf{F}$. Example 3 deals with what is perhaps the most common force field perceived by human beings.

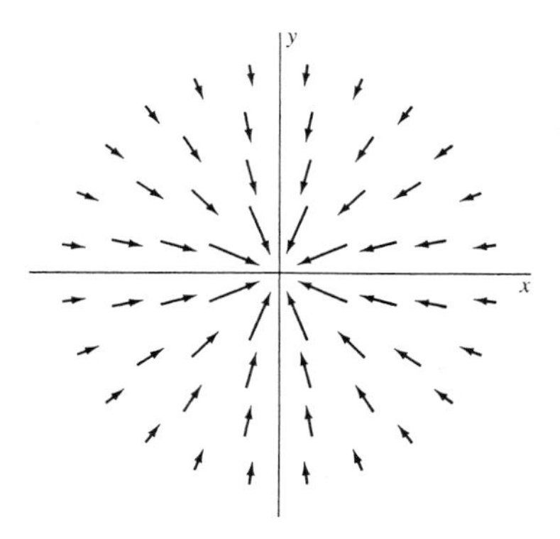

FIGURE 15.1.5 An inverse-square force field (Example 3).

EXAMPLE 3 Suppose that a mass M is fixed at the origin in space. When a particle of unit mass is placed at the point (x, y, z) other than the origin, it is subjected to a force $\mathbf{F}(x, y, z)$ of gravitational attraction directed toward the mass M at the origin. By Newton's inverse-square law of gravitation, the magnitude of $\mathbf{F}$ is $F = GM/r^2$, where $r = \sqrt{x^2 + y^2 + z^2}$ is the length of the position vector $\mathbf{r} = x\mathbf{i} + y\mathbf{j} + z\mathbf{k}$. It follows immediately that

$$\mathbf{F}(x, y, z) = -\frac{k\mathbf{r}}{r^3}, \tag{4}$$

where $k = GM$, because this vector has both the correct magnitude and the correct direction (toward the origin, for $\mathbf{F}$ is a multiple of $-\mathbf{r}$). A force field of the form in Eq. (4) is called an *inverse-square* force field. Note that $\mathbf{F}(x, y, z)$ is not defined at the origin and that $|\mathbf{F}| \to +\infty$ as $r \to 0^+$. Figure 15.1.5 illustrates an inverse-square force field. ◆

The Gradient Vector Field

In Section 13.8 we introduced the gradient vector of the differentiable real-valued function $f(x, y, z)$. It is the vector ∇f defined as follows:

$$\nabla f = \mathbf{i}\frac{\partial f}{\partial x} + \mathbf{j}\frac{\partial f}{\partial y} + \mathbf{k}\frac{\partial f}{\partial z}. \tag{5}$$

The partial derivatives on the right-hand side of Eq. (5) are evaluated at the point (x, y, z). Thus $\nabla f(x, y, z)$ is a vector field: It is the **gradient vector field** of the function f and is sometimes denoted by grad f. According to Theorem 1 of Section 13.8, the vector $\nabla f(x, y, z)$ points in the direction in which the maximal directional derivative of f at (x, y, z) is obtained. For example, if $f(x, y, z)$ is the temperature at the point (x, y, z) in space, then you should move in the direction $\nabla f(x, y, z)$ in order to warm up the most quickly.

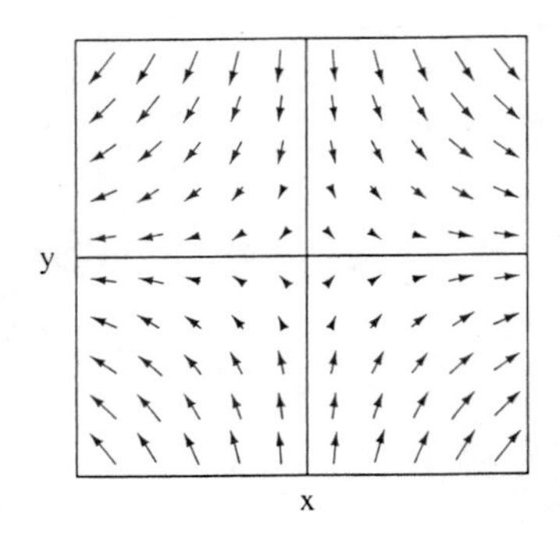

FIGURE 15.1.6 The gradient vector field $\nabla f = 2x\mathbf{i} - 2y\mathbf{j}$ of Example 4.

In the case of a two-variable scalar function $f(x, y)$ we suppress the third component in Eq. (5), so $\nabla f = \langle f_x, f_y \rangle = f_x\mathbf{i} + f_y\mathbf{j}$ defines a plane vector field.

EXAMPLE 4 With $f(x, y) = x^2 - y^2$, the gradient vector field $\nabla f = 2x\mathbf{i} - 2y\mathbf{j}$ plotted in Fig. 15.1.6 should remind you of a contour plot near a saddle point. ◆

The notation in Eq. (5) suggests the formal expression

$$\nabla = \mathbf{i}\frac{\partial}{\partial x} + \mathbf{j}\frac{\partial}{\partial y} + \mathbf{k}\frac{\partial}{\partial z}. \tag{6}$$

It is fruitful to think of ∇ as a *vector differential operator.* That is, ∇ is the operation that, when applied to the scalar function f, yields its gradient vector field ∇f. This operation behaves in several familiar and important ways like the operation D_x of single-variable differentiation. For a familiar example of this phenomenon, recall that in Chapter 13 we found the critical points of a function f of several variables to be those points at which $\nabla f(x, y, z) = \mathbf{0}$ and those at which $\nabla f(x, y, z)$ does not exist. As a computationally useful instance, suppose that f and g are functions and that a and b are constants. It then follows readily from Eq. (5) and from the linearity of partial differentiation that

$$\nabla(af + bg) = a\,\nabla f + b\,\nabla g. \tag{7}$$

Thus the gradient operator is *linear.* It also satisfies the product rule, as demonstrated in Example 5.

EXAMPLE 5 Given the differentiable functions $f(x, y, z)$ and $g(x, y, z)$, show that

$$\nabla(fg) = f\,\nabla g + g\,\nabla f. \tag{8}$$

Solution We apply the definition in Eq. (5) and the product rule for partial differentiation. Thus

$$\begin{aligned}\nabla(fg) &= \mathbf{i}\frac{\partial(fg)}{\partial x} + \mathbf{j}\frac{\partial(fg)}{\partial y} + \mathbf{k}\frac{\partial(fg)}{\partial z}\\ &= \mathbf{i}(fg_x + gf_x) + \mathbf{j}(fg_y + gf_y) + \mathbf{k}(fg_z + gf_z)\\ &= f\cdot(\mathbf{i}g_x + \mathbf{j}g_y + \mathbf{k}g_z) + g\cdot(\mathbf{i}f_x + \mathbf{j}f_y + \mathbf{k}f_z) = f\,\nabla g + g\,\nabla f,\end{aligned}$$

as desired. ◆

The Divergence of a Vector Field

Suppose that we are given the vector-valued function

$$\mathbf{F}(x, y, z) = \mathbf{i}P(x, y, z) + \mathbf{j}Q(x, y, z) + \mathbf{k}R(x, y, z)$$

with differentiable component functions P, Q, and R. Then the **divergence** of $\mathbf{F}$ is the scalar function div $\mathbf{F}$ defined as follows:

$$\operatorname{div}\mathbf{F} = \nabla\cdot\mathbf{F} = \frac{\partial P}{\partial x} + \frac{\partial Q}{\partial y} + \frac{\partial R}{\partial z}. \tag{9}$$

Here *div* is an abbreviation for "divergence," and the alternative notation $\nabla\cdot\mathbf{F}$ is consistent with the formal expression for ∇ in Eq. (6). That is,

$$\nabla\cdot\mathbf{F} = \left\langle \frac{\partial}{\partial x}, \frac{\partial}{\partial y}, \frac{\partial}{\partial z} \right\rangle \cdot \langle P, Q, R \rangle = \frac{\partial P}{\partial x} + \frac{\partial Q}{\partial y} + \frac{\partial R}{\partial z}.$$

We will see in Section 15.6 that if $\mathbf{v}$ is the velocity vector field of a steady fluid flow, then the value of div $\mathbf{v}$ at the point (x, y, z) is essentially the net rate per unit volume at which fluid mass is flowing away (or "diverging") from the point (x, y, z).

EXAMPLE 6 If the vector field $\mathbf{F}$ is given by

$$\mathbf{F}(x, y, z) = (xe^y)\mathbf{i} + (z \sin y)\mathbf{j} + (xy \ln z)\mathbf{k},$$

then $P(x, y, z) = xe^y$, $Q(x, y, z) = z \sin y$, and $R(x, y, z) = xy \ln z$. Hence Eq. (9) yields

$$\operatorname{div} \mathbf{F} = \frac{\partial}{\partial x}(xe^y) + \frac{\partial}{\partial y}(z \sin y) + \frac{\partial}{\partial z}(xy \ln z) = e^y + z \cos y + \frac{xy}{z}.$$

For instance, the value of div $\mathbf{F}$ at the point $(-3, 0, 2)$ is

$$\nabla \cdot \mathbf{F}(-3, 0, 2) = e^0 + 2 \cos 0 + 0 = 3.$$

◆

The analogues of Eqs. (7) and (8) for divergence are the formulas

$$\nabla \cdot (a\mathbf{F} + b\mathbf{G}) = a\, \nabla \cdot \mathbf{F} + b\, \nabla \cdot \mathbf{G} \tag{10}$$

and

$$\nabla \cdot (f\mathbf{G}) = (f)(\nabla \cdot \mathbf{G}) + (\nabla f) \cdot \mathbf{G}. \tag{11}$$

We ask you to verify these formulas in the problems. Note that Eq. (11)—in which f is a scalar function and $\mathbf{G}$ is a vector field—is consistent in that f and $\nabla \cdot \mathbf{G}$ are scalar functions, whereas ∇f and $\mathbf{G}$ are vector fields, so the sum on the right-hand side makes sense (and is a scalar function).

The Curl of a Vector Field

The **curl** of the differentiable vector field $\mathbf{F} = P\mathbf{i} + Q\mathbf{j} + R\mathbf{k}$ is the following vector field, abbreviated as curl $\mathbf{F}$:

$$\operatorname{curl} \mathbf{F} = \nabla \times \mathbf{F} = \begin{vmatrix} \mathbf{i} & \mathbf{j} & \mathbf{k} \\ \dfrac{\partial}{\partial x} & \dfrac{\partial}{\partial y} & \dfrac{\partial}{\partial z} \\ P & Q & R \end{vmatrix}. \tag{12}$$

When we evaluate the formal determinant in Eq. (12), we obtain

$$\operatorname{curl} \mathbf{F} = \mathbf{i}\left(\frac{\partial R}{\partial y} - \frac{\partial Q}{\partial z}\right) + \mathbf{j}\left(\frac{\partial P}{\partial z} - \frac{\partial R}{\partial x}\right) + \mathbf{k}\left(\frac{\partial Q}{\partial x} - \frac{\partial P}{\partial y}\right). \tag{13}$$

Although you may wish to memorize this formula, we recommend—because you will generally find it simpler—that in practice you set up and evaluate directly the formal determinant in Eq. (12). Example 7 shows how easy this is.

EXAMPLE 7 For the vector field $\mathbf{F}$ of Example 6, Eq. (12) yields

$$\begin{aligned} \operatorname{curl} \mathbf{F} &= \begin{vmatrix} \mathbf{i} & \mathbf{j} & \mathbf{k} \\ \dfrac{\partial}{\partial x} & \dfrac{\partial}{\partial y} & \dfrac{\partial}{\partial z} \\ xe^y & z \sin y & xy \ln z \end{vmatrix} \\ &= \mathbf{i}(x \ln z - \sin y) + \mathbf{j}(-y \ln z) + \mathbf{k}(-xe^y). \end{aligned}$$

In particular, the value of curl $\mathbf{F}$ at the point $(3, \pi/2, e)$ is

$$\nabla \times \mathbf{F}(3, \pi/2, e) = 2\mathbf{i} - \tfrac{1}{2}\pi\mathbf{j} - 3e^{\pi/2}\mathbf{k}.$$

◆

We will see in Section 15.7 that if $\mathbf{v}$ is the velocity vector of a fluid flow, then the value of the vector curl $\mathbf{v}$ at the point (x, y, z) (where that vector is nonzero) determines the axis through (x, y, z) about which the fluid is rotating (or whirling or "curling") as well as the angular velocity of the rotation.

The analogues of Eqs. (10) and (11) for curl are the formulas

$$\nabla \times (a\mathbf{F} + b\mathbf{G}) = a(\nabla \times \mathbf{F}) + b(\nabla \times \mathbf{G}) \tag{14}$$

and

$$\nabla \times (f\mathbf{G}) = (f)(\nabla \times \mathbf{G}) + (\nabla f) \times \mathbf{G} \tag{15}$$

that we ask you to verify in the problems.

EXAMPLE 8 If the function $f(x, y, z)$ has continuous second-order partial derivatives, show that

$$\operatorname{curl}(\operatorname{grad} f) = \mathbf{0}.$$

Solution Direct computation yields

$$\nabla \times \nabla f = \begin{vmatrix} \mathbf{i} & \mathbf{j} & \mathbf{k} \\ \dfrac{\partial}{\partial x} & \dfrac{\partial}{\partial y} & \dfrac{\partial}{\partial z} \\ \dfrac{\partial f}{\partial x} & \dfrac{\partial f}{\partial y} & \dfrac{\partial f}{\partial z} \end{vmatrix}$$

$$= \mathbf{i}\left(\frac{\partial^2 f}{\partial y\,\partial z} - \frac{\partial^2 f}{\partial z\,\partial y}\right) + \mathbf{j}\left(\frac{\partial^2 f}{\partial z\,\partial x} - \frac{\partial^2 f}{\partial x\,\partial z}\right) + \mathbf{k}\left(\frac{\partial^2 f}{\partial x\,\partial y} - \frac{\partial^2 f}{\partial y\,\partial x}\right).$$

Therefore,

$$\nabla \times \nabla f = \mathbf{0}$$

because of the equality of continuous mixed second-order partial derivatives. ◆

15.1 TRUE/FALSE STUDY GUIDE

15.1 CONCEPTS: QUESTIONS AND DISCUSSION

1. Discuss the analogy between the differential operators

$$D = \frac{d}{dx} \quad \text{and} \quad \nabla = \mathbf{i}\frac{\partial}{\partial x} + \mathbf{j}\frac{\partial}{\partial y}$$

for functions of one and two variables (respectively).

2. What are the similarities and differences between the divergence and the curl of a vector field?

15.1 PROBLEMS

In Problems 1 through 10, illustrate the given vector field **F** *by sketching several typical vectors in the field.*

1. $\mathbf{F}(x, y) = \mathbf{i} + \mathbf{j}$
2. $\mathbf{F}(x, y) = 3\mathbf{i} - 2\mathbf{j}$
3. $\mathbf{F}(x, y) = x\mathbf{i} - y\mathbf{j}$
4. $\mathbf{F}(x, y) = 2\mathbf{i} + x\mathbf{j}$
5. $\mathbf{F}(x, y) = (x^2 + y^2)^{1/2}(x\mathbf{i} + y\mathbf{j})$
6. $\mathbf{F}(x, y) = (x^2 + y^2)^{-1/2}(x\mathbf{i} + y\mathbf{j})$
7. $\mathbf{F}(x, y, z) = \mathbf{j} + \mathbf{k}$
8. $\mathbf{F}(x, y, z) = \mathbf{i} + \mathbf{j} - \mathbf{k}$
9. $\mathbf{F}(x, y, z) = -x\mathbf{i} - y\mathbf{j}$
10. $\mathbf{F}(x, y, z) = x\mathbf{i} + y\mathbf{j} + z\mathbf{k}$

Match the gradient vector fields of the functions in Problems 11 through 14 with the computer-generated plots in Figs. 15.1.7 through 15.1.10.

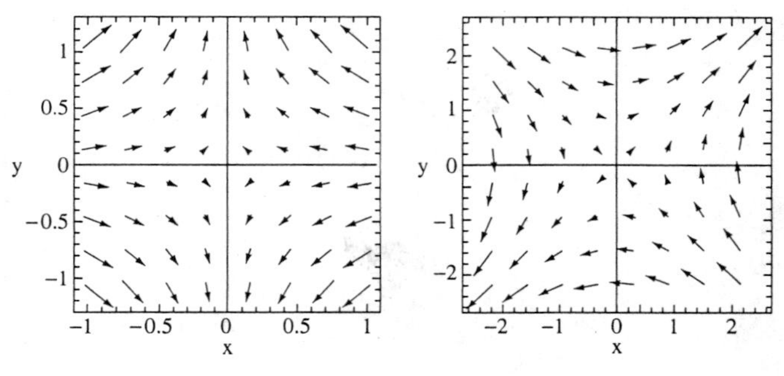

FIGURE 15.1.7 FIGURE 15.1.8

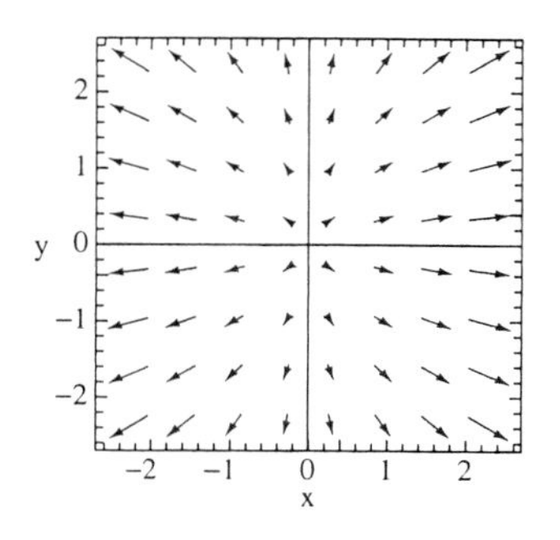

FIGURE 15.1.9

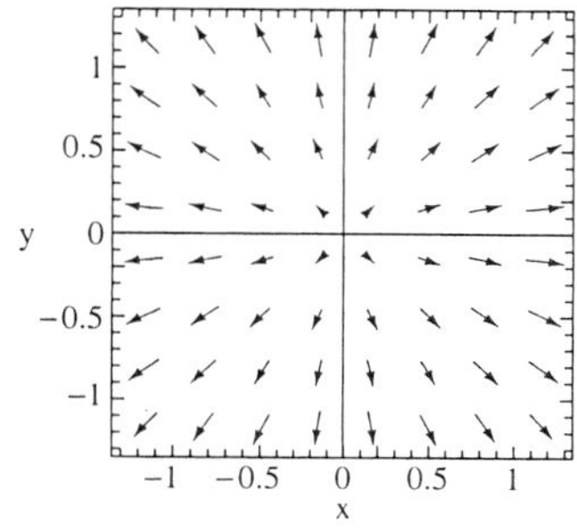

FIGURE 15.1.10

11. $f(x, y) = xy$

12. $f(x, y) = 2x^2 + y^2$

13. $f(x, y) = \sin \frac{1}{2}(x^2 + y^2)$

14. $f(x, y) = \sin \frac{1}{2}(y^2 - x^2)$

In Problems 15 through 24, calculate the divergence and curl of the given vector field **F**.

15. $\mathbf{F}(x, y, z) = x\mathbf{i} + y\mathbf{j} + z\mathbf{k}$

16. $\mathbf{F}(x, y, z) = 3x\mathbf{i} - 2y\mathbf{j} - 4z\mathbf{k}$

17. $\mathbf{F}(x, y, z) = yz\mathbf{i} + xz\mathbf{j} + xy\mathbf{k}$

18. $\mathbf{F}(x, y, z) = x^2\mathbf{i} + y^2\mathbf{j} + z^2\mathbf{k}$

19. $\mathbf{F}(x, y, z) = xy^2\mathbf{i} + yz^2\mathbf{j} + zx^2\mathbf{k}$

20. $\mathbf{F}(x, y, z) = (2x - y)\mathbf{i} + (3y - 2z)\mathbf{j} + (7z - 3x)\mathbf{k}$

21. $\mathbf{F}(x, y, z) = (y^2 + z^2)\mathbf{i} + (x^2 + z^2)\mathbf{j} + (x^2 + y^2)\mathbf{k}$

22. $\mathbf{F}(x, y, z) = (e^{xz} \sin y)\mathbf{j} + (e^{xy} \cos z)\mathbf{k}$

23. $\mathbf{F}(x, y, z) = (x + \sin yz)\mathbf{i} + (y + \sin xz)\mathbf{j} + (z + \sin xy)\mathbf{k}$

24. $\mathbf{F}(x, y, z) = (x^2 e^{-z})\mathbf{i} + (y^3 \ln x)\mathbf{j} + (z \cosh y)\mathbf{k}$

Apply the definitions of gradient, divergence, and curl to establish the identities in Problems 25 through 31, in which a and b denote constants, f and g denote differentiable scalar functions, and **F** *and* **G** *denote differentiable vector fields.*

25. $\nabla(af + bg) = a\,\nabla f + b\,\nabla g$

26. $\nabla \cdot (a\mathbf{F} + b\mathbf{G}) = a\,\nabla \cdot \mathbf{F} + b\,\nabla \cdot \mathbf{G}$

27. $\nabla \times (a\mathbf{F} + b\mathbf{G}) = a(\nabla \times \mathbf{F}) + b(\nabla \times \mathbf{G})$

28. $\nabla \cdot (f\mathbf{G}) = (f)(\nabla \cdot \mathbf{G}) + (\nabla f) \cdot \mathbf{G}$

29. $\nabla \times (f\mathbf{G}) = (f)(\nabla \times \mathbf{G}) + (\nabla f) \times \mathbf{G}$

30. $\nabla\left(\dfrac{f}{g}\right) = \dfrac{g\,\nabla f - f\,\nabla g}{g^2}$

31. $\nabla \cdot (\mathbf{F} \times \mathbf{G}) = \mathbf{G} \cdot (\nabla \times \mathbf{F}) - \mathbf{F} \cdot (\nabla \times \mathbf{G})$

Establish the identities in Problems 32 through 34 under the assumption that the scalar functions f and g and the vector field **F** *are twice continuously differentiable.*

32. $\operatorname{div}(\operatorname{curl} \mathbf{F}) = 0$

33. $\operatorname{div}(\nabla fg) = f \operatorname{div}(\nabla g) + g \operatorname{div}(\nabla f) + 2(\nabla f) \cdot (\nabla g)$

34. $\operatorname{div}(\nabla f \times \nabla g) = 0$

Verify the identities in Problems 35 through 44, in which **a** *is a constant vector,* $\mathbf{r} = x\mathbf{i} + y\mathbf{j} + z\mathbf{k}$, *and* $r = |\mathbf{r}|$. *Problems 37 and 38 imply that both the divergence and the curl of an inverse-square vector field vanish identically.*

35. $\nabla \cdot \mathbf{r} = 3$ and $\nabla \times \mathbf{r} = \mathbf{0}$

36. $\nabla \cdot (\mathbf{a} \times \mathbf{r}) = 0$ and $\nabla \times (\mathbf{a} \times \mathbf{r}) = 2\mathbf{a}$

37. $\nabla \cdot \dfrac{\mathbf{r}}{r^3} = 0$

38. $\nabla \times \dfrac{\mathbf{r}}{r^3} = \mathbf{0}$

39. $\nabla r = \dfrac{\mathbf{r}}{r}$

40. $\nabla\left(\dfrac{1}{r}\right) = -\dfrac{\mathbf{r}}{r^3}$

41. $\nabla \cdot (r\mathbf{r}) = 4r$

42. $\nabla \cdot (\nabla r) = \dfrac{2}{r}$

43. $\nabla(\ln r) = \dfrac{\mathbf{r}}{r^2}$

44. $\nabla(r^{10}) = 10r^8\mathbf{r}$

15.2 LINE INTEGRALS

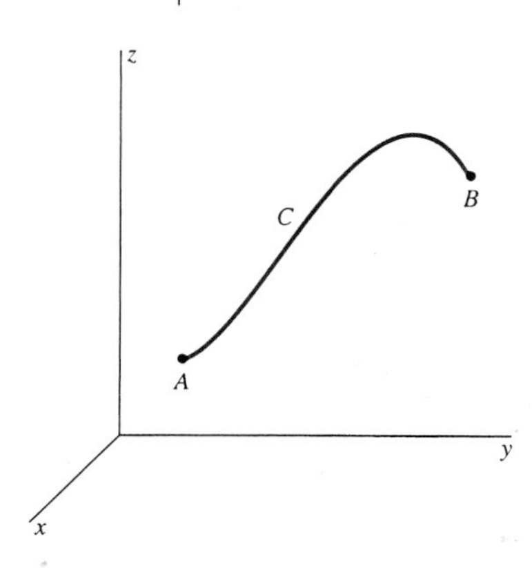

FIGURE 15.2.1 A wire of variable density in the shape of the smooth curve C from A (where $t = a$) to B (where $t = b$).

The single integral $\int_a^b f(x)\,dx$ might be described as an integral along the x-axis. We now define integrals along curves in space (or in the plane). Such integrals are called *line integrals* (although the phrase "curve integrals" might be more appropriate).

To motivate the definition of the line integral of the function f along the smooth space curve C with parametrization

$$x = x(t), \qquad y = y(t), \qquad z = z(t) \tag{1}$$

for $a \leqq t \leqq b$, we imagine a thin wire shaped like C (Fig. 15.2.1). Suppose that $f(x, y, z)$ denotes the density of the wire at the point (x, y, z), measured in units of mass per unit length—for example, grams per centimeter. Then we expect to compute the total mass m of the curved wire as some kind of integral of the function f. To *approximate* m, we begin with a partition

$$a = t_0 < t_1 < t_2 < \cdots < t_{n-1} < t_n = b$$

of $[a, b]$ into n subintervals, all with the same length $\Delta t = (b - a)/n$. These subdivision points of $[a, b]$ produce, via our parametrization, a physical subdivision of the wire

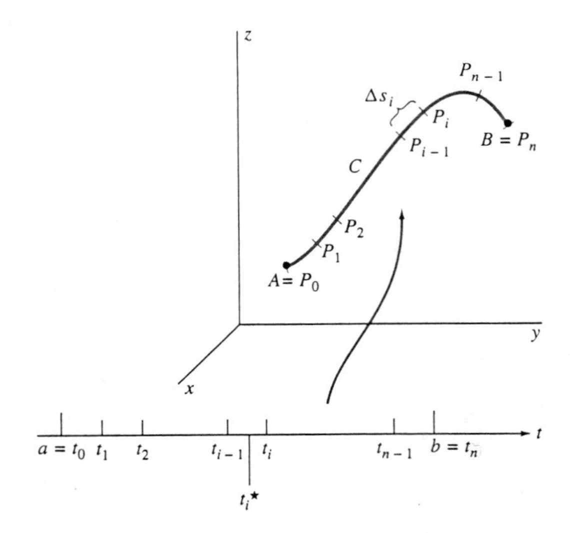

FIGURE 15.2.2 The partition of the interval $[a, b]$ determines a related partition of the curve C into short arcs.

into short curve segments (Fig. 15.2.2). We let P_i denote the point $(x(t_i), y(t_i), z(t_i))$ for $i = 0, 1, 2, \ldots, n$. Then the points $P_0, P_1, \ldots, P_n$ are the subdivision points of C.

From our study of arc length in Sections 10.4 and 12.6, we know that the arc length Δs_i of the segment of C from P_{i-1} to P_i is

$$\Delta s_i = \int_{t_{i-1}}^{t_i} \sqrt{[x'(t)]^2 + [y'(t)]^2 + [z'(t)]^2}\, dt$$

$$= \sqrt{[x'(t_i^\star)]^2 + [y'(t_i^\star)]^2 + [z'(t_i^\star)]^2}\, \Delta t \tag{2}$$

for some number $t_i^\star$ in the interval $[t_{i-1}, t_i]$. This is a consequence of the average value theorem for integrals of Section 5.5.

Denote $x(t_i^\star)$ by $x_i^\star$ and similarly define $y_i^\star$ and $z_i^\star$. If we multiply the density of the wire at the point $(x_i^\star, y_i^\star, z_i^\star)$ by the length Δs_i of the segment of C containing that point, we obtain an estimate of the mass of that segment of C. So, after we sum over all the segments, we have an estimate of the total mass m of the wire:

$$m \approx \sum_{i=1}^{n} f(x(t_i^\star), y(t_i^\star), z(t_i^\star))\, \Delta s_i.$$

The limit of this sum as $\Delta t \to 0$ should be the actual mass m. This is our motivation for the definition of the line integral of the function f along the curve C, denoted by

$$\int_C f(x, y, z)\, ds.$$

DEFINITION Line Integral of a Function along a Curve

Suppose that the function $f(x, y, z)$ is defined at each point of the smooth curve C parametrized as in (1). Then the **line integral of f along C** is defined by

$$\int_C f(x, y, z)\, ds = \lim_{\Delta t \to 0} \sum_{i=1}^{n} f(x(t_i^\star), y(t_i^\star), z(t_i^\star))\, \Delta s_i, \tag{3}$$

provided that this limit exists.

REMARK It can be shown that the limit in (3) always exists if the function f is continuous at each point of C. Recall from Section 10.4 that the curve C is *smooth* provided that the component functions in its parametrization have continuous derivatives that are never simultaneously zero.

When we substitute Eq. (2) into Eq. (3), we recognize the result as the limit of a Riemann sum. Therefore

$$\int_C f(x, y, z)\, ds = \int_a^b f(x(t), y(t), z(t))\sqrt{[x'(t)]^2 + [y'(t)]^2 + [z'(t)]^2}\, dt. \qquad \textbf{(4)}$$

Thus we may evaluate the line integral $\int_C f(x, y, z)\, ds$ by expressing everything in terms of the parameter t, including the symbolic arc-length element

$$ds = \sqrt{\left(\frac{dx}{dt}\right)^2 + \left(\frac{dy}{dt}\right)^2 + \left(\frac{dz}{dt}\right)^2}\, dt.$$

As a consequence, the right-hand side in Eq. (4) is evaluated as *an ordinary single integral with respect to the real variable t*.

Line Integral with Respect to Arc Length

Because of the appearance of the arc-length element ds in Eq. (4), the line integral $\int_C f(x, y, z)\, ds$ is sometimes called the **line integral of the function f with respect to arc length along the curve C.**

A curve C that lies in the xy-plane may be regarded as a space curve for which z [and $z'(t)$] are zero. In this case we simply suppress the variable z in Eq. (4) and write

$$\int_C f(x, y)\, ds = \int_a^b f(x(t), y(t))\sqrt{[x'(t)]^2 + [y'(t)]^2}\, dt. \qquad \textbf{(5)}$$

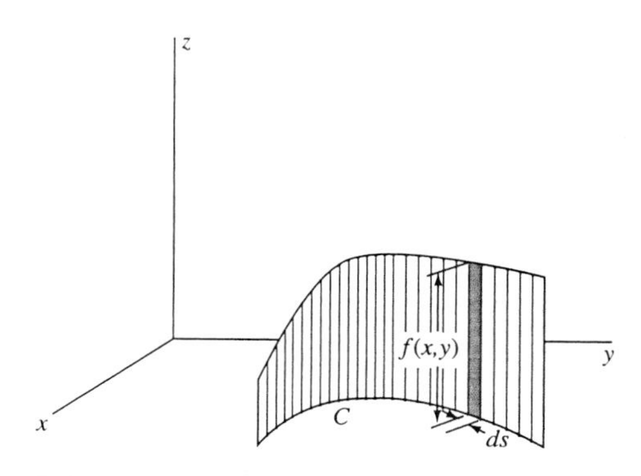

FIGURE 15.2.3 The vertical strip with base ds and height $f(x, y)$ has area $dA = f(x, y)\, ds$, so the whole fence with base curve C has area $A = \int dA = \int_C f(x, y)\, ds$.

In the case that f is positive-valued, Fig. 15.2.3 illustrates an interpretation of the line integral in Eq. (5) as the area of a "fence" whose base is the curve C in the xy-plane, with the height of the fence above the point (x, y) given by $f(x, y)$.

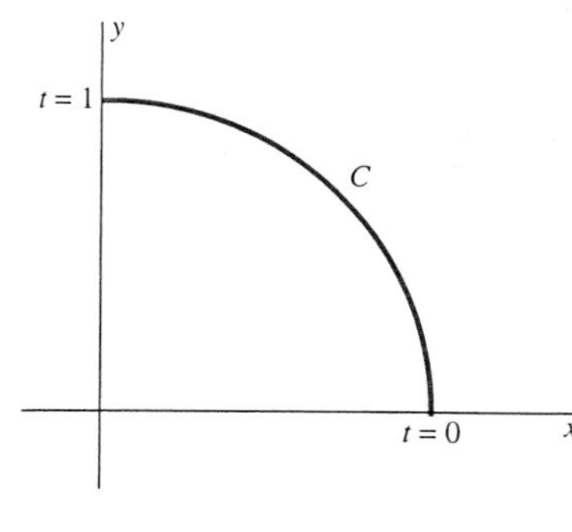

FIGURE 15.2.4 The quarter-circle of Example 1.

EXAMPLE 1 Evaluate the line integral

$$\int_C xy\, ds,$$

where C is the first-quadrant quarter-circle of radius 1 parametrized by $x = \cos t$, $y = \sin t$, $0 \leqq t \leqq \pi/2$ (Fig. 15.2.4).

Solution Here

$$ds = \sqrt{(-\sin t)^2 + (\cos t)^2}\, dt = dt,$$

so Eq. (5) yields

$$\int_C xy\, ds = \int_{t=0}^{\pi/2} \cos t \sin t\, dt = \left[\frac{1}{2}\sin^2 t\right]_0^{\pi/2} = \frac{1}{2}.$$ ◆

Let us now return to the physical wire and denote its density function by $\delta(x, y, z)$. The mass of a small piece of length Δs is $\Delta m = \delta\, \Delta s$, so we write

$$dm = \delta(x, y, z)\, ds$$

for its (symbolic) element of mass. Then the **mass** m of the wire and its **centroid** $(\overline{x}, \overline{y}, \overline{z})$ are defined as follows:

$$m = \int_C dm = \int_C \delta\, ds, \qquad \overline{x} = \frac{1}{m}\int_C x\, dm,$$
$$\overline{y} = \frac{1}{m}\int_C y\, dm, \qquad \overline{z} = \frac{1}{m}\int_C z\, dm. \tag{6}$$

Note the analogy with Eqs. (2) and (4) of Section 14.6. The **moment of inertia** of the wire around a given axis is

$$I = \int_C p^2\, dm, \tag{7}$$

where $p = p(x, y, z)$ denotes the perpendicular distance from the point (x, y, z) of the wire to the axis in question.

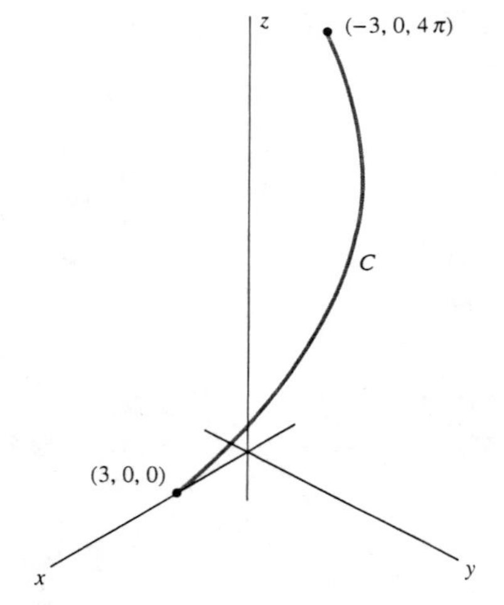

FIGURE 15.2.5 The helical wire of Example 2. Does the centroid (−1.22, 1.91, 8.38) lie on the wire?

EXAMPLE 2 Find the centroid of a wire that has density $\delta = kz$ and the shape of the helix C (Fig. 15.2.5) with parametrization

$$x = 3\cos t, \quad y = 3\sin t, \quad z = 4t, \quad 0 \leqq t \leqq \pi.$$

Solution The mass element of the wire is

$$dm = \delta\, ds = kz\, ds = 4kt\sqrt{(-3\sin t)^2 + (3\cos t)^2 + 4^2}\, dt = 20kt\, dt.$$

Hence the formulas in (6) yield

$$m = \int_C \delta\, ds = \int_0^{\pi} 20kt\, dt = 10k\pi^2;$$

$$\overline{x} = \frac{1}{m}\int_C \delta x\, ds = \frac{1}{10k\pi^2}\int_0^{\pi} 60kt\cos t\, dt$$
$$= \frac{6}{\pi^2}\Big[\cos t + t\sin t\Big]_0^{\pi} = -\frac{12}{\pi^2} \approx -1.22;$$

$$\overline{y} = \frac{1}{m}\int_C \delta y\, ds = \frac{1}{10k\pi^2}\int_0^{\pi} 60kt\sin t\, dt$$
$$= \frac{6}{\pi^2}\Big[\sin t - t\cos t\Big]_0^{\pi} = \frac{6}{\pi} \approx 1.91;$$

$$\overline{z} = \frac{1}{m}\int_C \delta z\, ds = \frac{1}{10k\pi^2}\int_0^{\pi} 80kt^2\, dt$$
$$= \frac{8}{\pi^2}\left[\frac{1}{3}t^3\right]_0^{\pi} = \frac{8\pi}{3} \approx 8.38.$$

So the centroid of the wire is located close to the point $(-1.22, 1.91, 8.38)$. ◆

Line Integrals with Respect to Coordinate Variables

We obtain a different kind of line integral by replacing Δs_i in Eq. (3) with

$$\Delta x_i = x(t_i) - x(t_{i-1}) = x'(t_i^\star)\,\Delta t.$$

The **line integral of f along C with respect to x** is defined to be

$$\int_C f(x, y, z)\,dx = \lim_{\Delta t \to 0} \sum_{i=1}^{n} f(x(t_i^\star), y(t_i^\star), z(t_i^\star))\,\Delta x_i.$$

Thus

$$\int_C f(x, y, z)\,dx = \int_a^b f(x(t), y(t), z(t))x'(t)\,dt. \qquad \textbf{(8a)}$$

Similarly, the **line integrals of f along C with respect to y** and **with respect to z** are given by

$$\int_C f(x, y, z)\,dy = \int_a^b f(x(t), y(t), z(t))y'(t)\,dt \qquad \textbf{(8b)}$$

and

$$\int_C f(x, y, z)\,dz = \int_a^b f(x(t), y(t), z(t))z'(t)\,dt. \qquad \textbf{(8c)}$$

The three integrals in (8) typically occur together. If P, Q, and R are continuous functions of the variables x, y, and z, then we write (indeed, *define*)

$$\int_C P\,dx + Q\,dy + R\,dz = \int_C P\,dx + \int_C Q\,dy + \int_C R\,dz. \qquad \textbf{(9)}$$

REMARK Although it would be natural enough to write $\int_C (P\,dx + Q\,dy + R\,dz)$ on the right-hand side in Eq. (9), the parentheses are customarily omitted. Indeed, in more advanced vector calculus the *differential form* $P\,dx + Q\,dy + R\,dz$ is regarded as a single object that "hangs together" all by itself. For instance, one may see the abbreviations $\omega = P\,dx + Q\,dy + R\,dz$ for the differential form and $\int_C \omega$ for the line integral.

The line integrals in Eqs. (8) and (9) are evaluated by expressing x, y, z, dx, dy, and dz in terms of t as determined by a suitable parametrization of the curve C. The result is an ordinary single-variable integral. For instance, if C is a parametric plane curve parametrized over the interval $[a, b]$ by $\mathbf{r}(t) = \langle x(t), y(t)\rangle$, then

$$\int_C P\,dx + Q\,dy = \int_a^b [P(x(t), y(t)) \cdot x'(t) + Q(x(t), y(t)) \cdot y'(t)]\,dt.$$

EXAMPLE 3 Evaluate the line integral

$$\int_C y\,dx + z\,dy + x\,dz,$$

where C is the parametric curve $x = t$, $y = t^2$, $z = t^3$, $0 \leqq t \leqq 1$.

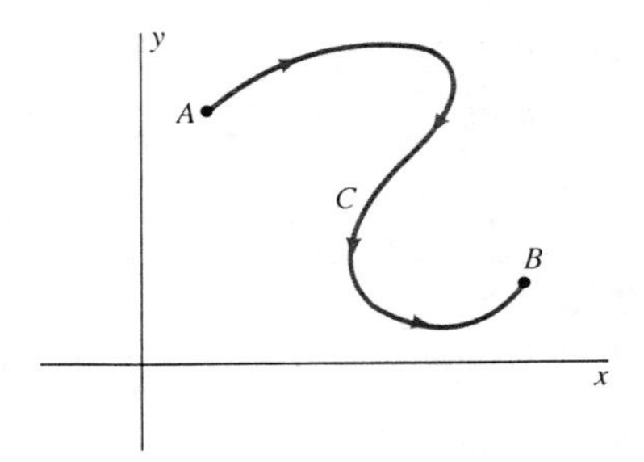

Solution Because $dx = dt$, $dy = 2t\,dt$, and $dz = 3t^2\,dt$, substitution in terms of t yields

$$\int_C y\,dx + z\,dy + x\,dz = \int_0^1 t^2\,dt + t^3(2t\,dt) + t(3t^2\,dt)$$
$$= \int_0^1 (t^2 + 3t^3 + 2t^4)\,dt = \left[\frac{1}{3}t^3 + \frac{3}{4}t^4 + \frac{2}{5}t^5\right]_0^1 = \frac{89}{60}.$$ ◆

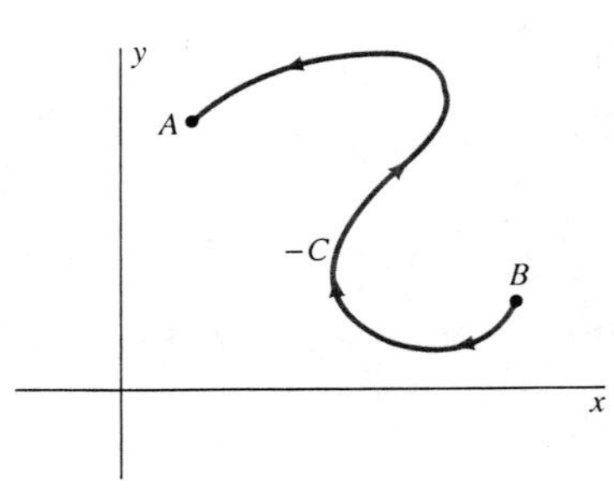

FIGURE 15.2.6 $\int_{-C} f\,ds = \int_C f\,ds$ but $\int_{-C} P\,dx + Q\,dy = -\int_C P\,dx + Q\,dy$.

The given parametrization of a smooth curve C determines an **orientation** or "positive direction" along the curve. As the parameter t increases from $t = a$ to $t = b$, the point $(x(t), y(t))$ moves along the curve from its initial point A to its terminal point B. Now think of a curve $-C$ with the *opposite orientation.* This new curve consists of the same points as C, but the parametrization of $-C$ traces these points in the opposite direction, from initial point B to terminal point A (Fig. 15.2.6). Because the arc-length differential $ds = \sqrt{[x'(t)]^2 + [y'(t)]^2 + [z'(t)]^2}\,dt$ is always positive (the square root is positive), the value of the line integral with respect to arc length is not affected by the reversal of orientation. That is,

$$\int_{-C} f(x, y, z)\,ds = \int_C f(x, y, z)\,ds. \qquad \textbf{(10)}$$

In contrast, the signs of the derivatives $x'(t)$, $y'(t)$, and $z'(t)$ in Eqs. (8a), (8b), and (8c) are changed when the direction of the parametrization is reversed, so it follows that

$$\int_{-C} P\,dx + Q\,dy + R\,dz = -\int_C P\,dx + Q\,dy + R\,dz. \qquad \textbf{(11)}$$

Thus changing the orientation of the curve changes the *sign* of a line integral with respect to coordinate variables, but does not affect the value of a line integral with respect to arc length. It is proved in advanced calculus that, for either type of line integral, two one-to-one parametrizations of the same smooth curve give the same value if they agree in orientation.

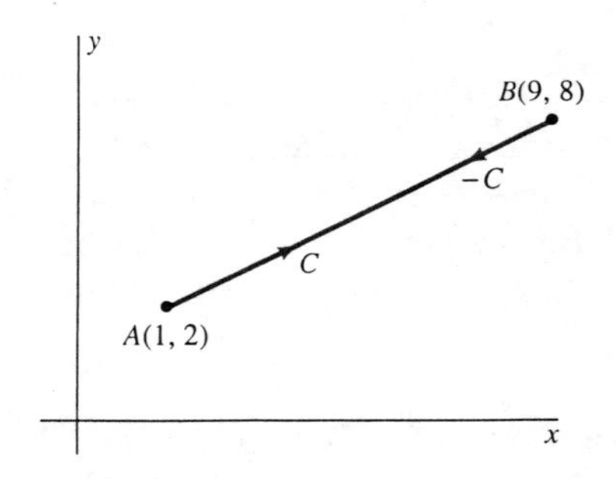

FIGURE 15.2.7 The line segment of Example 4.

EXAMPLE 4 The parametrization $x = 1 + 8t$, $y = 2 + 6t$ $(0 \leqq t \leqq 1)$ of the line segment C from $A(1, 2)$ to $B(9, 8)$ in Fig. 15.2.7 gives $dx = 8\,dt$, $dy = 6\,dt$, and $ds = 10\,dt$. Hence we easily verify that

$$\int_C xy\,ds = \int_0^1 (1 + 8t)(2 + 6t) \cdot 10\,dt = 290$$

and

$$\int_C y\,dx + x\,dy = \int_0^1 [(2 + 6t) \cdot 8 + (1 + 8t) \cdot 6]\,dt = 70.$$

The parametrization $x = 9 - 4t$, $y = 8 - 3t$ $(0 \leqq t \leqq 2)$ of the oppositely oriented segment $-C$ from $B(9, 8)$ to $A(1, 2)$ gives $dx = -4\,dt$, $dy = -3\,dt$, and $ds = 5\,dt$, and we easily verify that

$$\int_{-C} xy\,ds = \int_0^2 (9 - 4t)(8 - 3t) \cdot 5\,dt = 290,$$

whereas

$$\int_{-C} y\,dx + x\,dy = \int_0^2 [(8 - 3t) \cdot (-4) + (9 - 4t) \cdot (-3)]\,dt = -70.$$ ◆

If the curve C consists of a finite number of smooth curves joined at consecutive corner points, then we say that C is **piecewise smooth.** In such a case the value of a line integral along C is defined to be the sum of its values along the smooth segments

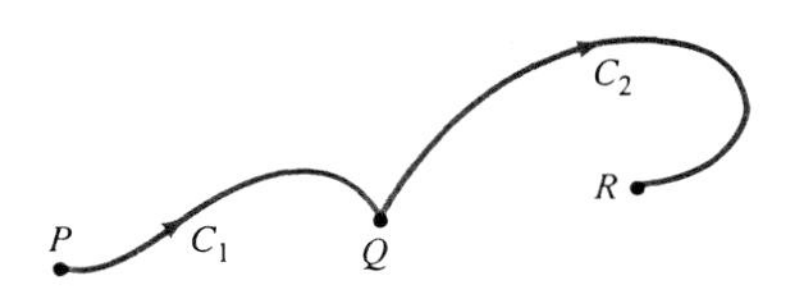

FIGURE 15.2.8 The curve $C = C_1 + C_2$ from P to R.

of C. For instance, with the piecewise smooth curve $C = C_1 + C_2$ of Fig. 15.2.8, we have

$$\int_C f(x, y, z)\,ds = \int_{C_1+C_2} f(x, y, z)\,ds = \int_{C_1} f(x, y, z)\,ds + \int_{C_2} f(x, y, z)\,ds.$$

EXAMPLE 5 Evaluate the line integral

$$\int_C y\,dx + 2x\,dy$$

for each of these three curves C (Fig. 15.2.9):

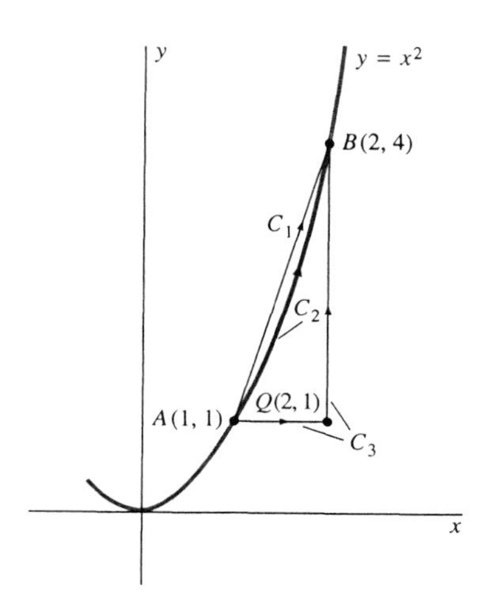

FIGURE 15.2.9 The three arcs of Example 5.

C_1 The straight line segment in the plane from $A(1, 1)$ to $B(2, 4)$;

C_2 The plane path from $A(1, 1)$ to $B(2, 4)$ along the graph of the parabola $y = x^2$; and

C_3 The straight line in the plane from $A(1, 1)$ to $Q(2, 1)$ followed by the straight line from $Q(2, 1)$ to $B(2, 4)$.

Solution The straight line segment C_1 from A to B can be parametrized by $x = 1+t$, $y = 1 + 3t$, $0 \leqq t \leqq 1$. Hence

$$\int_{C_1} y\,dx + 2x\,dy = \int_0^1 (1 + 3t)\,dt + 2(1 + t)(3\,dt)$$
$$= \int_0^1 (7 + 9t)\,dt = \frac{23}{2}.$$

Next, the arc C_2 of the parabola $y = x^2$ from A to B is "self-parametrizing": It has the parametrization $x = x$, $y = x^2$, $1 \leqq x \leqq 2$. So

$$\int_{C_2} y\,dx + 2x\,dy = \int_1^2 (x^2)(dx) + 2(x)(2x\,dx) = \int_1^2 5x^2\,dx = \frac{35}{3}.$$

Finally, along the straight line segment from $(1, 1)$ to $(2, 1)$ we have $y \equiv 1$ and (because y is a constant) $dy = 0$. Along the vertical segment from $(2, 1)$ to $(2, 4)$ we have $x \equiv 2$ and $dx = 0$. Therefore

$$\int_{C_3} y\,dx + 2x\,dy = \int_{x=1}^2 [(1)(dx) + (2x)(0)] + \int_{y=1}^4 [(y)(0) + (4)(dy)]$$
$$= \int_{x=1}^2 1\,dx + \int_{y=1}^4 4\,dy = 13.$$ ◆

Example 5 shows that we may well obtain different values for the line integral from A to B if we evaluate it along different curves from A to B. Thus this line integral is *path-dependent.* We shall give in Section 15.3 a sufficient condition for the line integral

$$\int_C P\,dx + Q\,dy + R\,dz$$

to have the same value for *all* smooth or piecewise smooth curves C from A to B, and thus for the integral to be *independent of path.*

Line Integrals and Vector Fields

Suppose now that $\mathbf{F} = P\mathbf{i} + Q\mathbf{j} + R\mathbf{k}$ is a force field defined on a region that contains the curve C from the point A to the point B. Suppose also that C has a parametrization

$$\mathbf{r}(t) = \mathbf{i}x(t) + \mathbf{j}y(t) + \mathbf{k}z(t), \quad t \text{ in } [a, b],$$

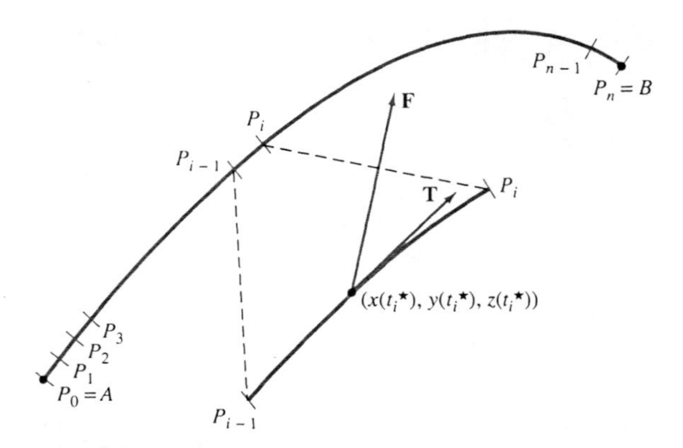

FIGURE 15.2.10 The component of **F** along C from P_{i-1} to P_i is $\mathbf{F} \cdot \mathbf{T}$.

with a *nonzero* velocity vector

$$\mathbf{v} = \mathbf{i}\frac{dx}{dt} + \mathbf{j}\frac{dy}{dt} + \mathbf{k}\frac{dz}{dt}.$$

The speed associated with this velocity vector is

$$v = |\mathbf{v}| = \sqrt{\left(\frac{dx}{dt}\right)^2 + \left(\frac{dy}{dt}\right)^2 + \left(\frac{dz}{dt}\right)^2}.$$

Recall from Section 12.6 that the *unit tangent vector* to the curve C is

$$\mathbf{T} = \frac{\mathbf{v}}{v} = \frac{1}{v}\left(\frac{dx}{dt}\mathbf{i} + \frac{dy}{dt}\mathbf{j} + \frac{dz}{dt}\mathbf{k}\right).$$

We want to approximate the work W done by the force field **F** in moving a particle along the curve C from A to B. Subdivide C as indicated in Fig. 15.2.10. Think of **F** moving the particle from P_{i-1} to P_i, two consecutive division points of C. The work ΔW_i done is approximately the product of the distance Δs_i from P_{i-1} to P_i (measured along C) and the tangential component $\mathbf{F} \cdot \mathbf{T}$ of the force **F** at a typical point $(x(t_i^\star), y(t_i^\star), z(t_i^\star))$ between P_{i-1} and P_i. Thus

$$\Delta W_i \approx \mathbf{F}(x(t_i^\star), y(t_i^\star), z(t_i^\star)) \cdot \mathbf{T}(t_i^\star)\, \Delta s_i,$$

so the total work W is given approximately by

$$W \approx \sum_{i=1}^{n} \mathbf{F}(x(t_i^\star), y(t_i^\star), z(t_i^\star)) \cdot \mathbf{T}(t_i^\star)\, \Delta s_i.$$

This approximation suggests that we *define* the **work** W as

$$W = \int_C \mathbf{F} \cdot \mathbf{T}\, ds. \qquad \textbf{(12)}$$

Thus *work is the integral with respect to arc length of the tangential component of the force.* Intuitively, we may regard $dW = \mathbf{F} \cdot \mathbf{T}\, ds$ as the infinitesimal element of work done by the tangential component $\mathbf{F} \cdot \mathbf{T}$ of the force in moving the particle along the arc-length element ds. The line integral in Eq. (12) is then the "sum" of all these infinitesimal elements of work.

It is customary to write

$$\mathbf{r} = x\mathbf{i} + y\mathbf{j} + z\mathbf{k} \quad \text{and} \quad d\mathbf{r} = \mathbf{i}\,dx + \mathbf{j}\,dy + \mathbf{k}\,dz.$$

Then

$$\mathbf{T}\,ds = \frac{\mathbf{v}}{v}\cdot v\,dt = \mathbf{v}\,dt$$
$$= \left(\frac{dx}{dt}\mathbf{i} + \frac{dy}{dt}\mathbf{j} + \frac{dz}{dt}\mathbf{k}\right)dt = \mathbf{i}\,dx + \mathbf{j}\,dy + \mathbf{k}\,dz,$$

so

$$\mathbf{T}\,ds = d\mathbf{r}.$$

With this notation, Eq. (12) takes the form

$$W = \int_C \mathbf{F}\cdot d\mathbf{r} \tag{13}$$

that is common in engineering and physics texts.

To evaluate the line integral in Eq. (12) or (13), we express its integrand and limit of integration in terms of the parameter t, as usual. Thus

$$W = \int_C \mathbf{F}\cdot\mathbf{T}\,ds$$
$$= \int_a^b (P\mathbf{i} + Q\mathbf{j} + R\mathbf{k})\cdot\frac{1}{v}\left(\frac{dx}{dt}\mathbf{i} + \frac{dy}{dt}\mathbf{j} + \frac{dz}{dt}\mathbf{k}\right)v\,dt$$
$$= \int_a^b \left(P\frac{dx}{dt} + Q\frac{dy}{dt} + R\frac{dz}{dt}\right)dt.$$

Therefore,

$$W = \int_C P\,dx + Q\,dy + R\,dz. \tag{14}$$

This computation reveals an important relation between the two types of line integrals we have defined here.

THEOREM 1 Equivalent Line Integrals

Suppose that the vector field $\mathbf{F} = P\mathbf{i} + Q\mathbf{j} + R\mathbf{k}$ has continuous component functions and that $\mathbf{T}$ is the unit tangent vector to the smooth curve C. Then

$$\int_C \mathbf{F}\cdot\mathbf{T}\,ds = \int_C P\,dx + Q\,dy + R\,dz. \tag{15}$$

REMARK If the orientation of the curve C is reversed, then the sign of the right-hand integral in Eq. (15) is changed according to Eq. (11), whereas the sign of the left-hand integral is changed because $\mathbf{T}$ is replaced with $-\mathbf{T}$.

EXAMPLE 6 The work done by the force field $\mathbf{F} = y\mathbf{i} + z\mathbf{j} + x\mathbf{k}$ in moving a particle from (0, 0, 0) to (1, 1, 1) along the twisted cubic $x = t$, $y = t^2$, $z = t^3$ is given by the line integral

$$W = \int_C \mathbf{F}\cdot d\mathbf{r} = \int_C \mathbf{F}\cdot\mathbf{T}\,ds = \int_C y\,dx + z\,dy + x\,dz,$$

and we computed the value of this integral in Example 3. Hence $W = \frac{89}{60}$. ◆

EXAMPLE 7 Find the work done by the inverse-square force field

$$\mathbf{F}(x, y, z) = \frac{k\mathbf{r}}{r^3} = \frac{k(x\mathbf{i} + y\mathbf{j} + z\mathbf{k})}{(x^2 + y^2 + z^2)^{3/2}}$$

in moving a particle along the straight line segment C from (0, 4, 0) to (0, 4, 3).

Solution Along C we have $x = 0$, $y = 4$, and z varying from 0 to 3. Thus we choose z as the parameter:

$$x \equiv 0, \quad y \equiv 4, \quad \text{and} \quad z = z, \qquad 0 \leqq z \leqq 3.$$

Because $dx = 0 = dy$, Eq. (14) gives

$$W = \int_C \frac{k(x\,dx + y\,dy + z\,dz)}{(x^2 + y^2 + z^2)^{3/2}}$$

$$= \int_0^3 \frac{kz}{(16 + z^2)^{3/2}}\,dz = \left[\frac{-k}{\sqrt{16 + z^2}}\right]_0^3 = \frac{k}{20}. \qquad \blacklozenge$$

15.2 TRUE/FALSE STUDY GUIDE

15.2 CONCEPTS: QUESTIONS AND DISCUSSION

1. Contrast the definitions of $\int_C f\,ds$ and $\int_C f\,dx$ where C is a smooth curve in space. Describe physical measurements appropriate to both types of line integrals. Explain why one depends on the orientation or direction of integration along the curve C and the other does not.
2. Explain the relation between the line integrals

$$\int_C \mathbf{F}\cdot\mathbf{T}\,ds \quad \text{and} \quad \int_C P\,dx + Q\,dy + R\,dz.$$

In what way does each depend on the orientation of the curve C?

15.2 PROBLEMS

In Problems 1 through 5, evaluate the line integrals

$$\int_C f(x, y)\,ds, \quad \int_C f(x, y)\,dx, \quad \text{and} \quad \int_C f(x, y)\,dy$$

along the indicated parametric curve.

1. $f(x, y) = x^2 + y^2$; $x = 4t - 1$, $y = 3t + 1$, $-1 \leqq t \leqq 1$
2. $f(x, y) = x$; $x = t$, $y = t^2$, $0 \leqq t \leqq 1$
3. $f(x, y) = x + y$; $x = e^t + 1$, $y = e^t - 1$, $0 \leqq t \leqq \ln 2$
4. $f(x, y) = 2x - y$; $x = \sin t$, $y = \cos t$, $0 \leqq t \leqq \pi/2$
5. $f(x, y) = xy$; $x = 3t$, $y = t^4$, $0 \leqq t \leqq 1$

In Problems 6 through 10, evaluate

$$\int_C P(x, y)\,dx + Q(x, y)\,dy.$$

6. $P(x, y) = xy$, $Q(x, y) = x + y$; C is the part of the graph of $y = x^2$ from $(-1, 1)$ to $(2, 4)$
7. $P(x, y) = y^2$, $Q(x, y) = x$; C is the part of the graph of $x = y^3$ from $(-1, -1)$ to $(1, 1)$
8. $P(x, y) = y\sqrt{x}$, $Q(x, y) = x\sqrt{x}$; C is the part of the graph of $y^2 = x^3$ from $(1, 1)$ to $(4, 8)$
9. $P(x, y) = x^2y$, $Q(x, y) = xy^3$; C consists of the line segments from $(-1, 1)$ to $(2, 1)$ and from $(2, 1)$ to $(2, 5)$
10. $P(x, y) = x + 2y$, $Q(x, y) = 2x - y$; C consists of the line segments from $(3, 2)$ to $(3, -1)$ and from $(3, -1)$ to $(-2, -1)$

In Problems 11 through 15, evaluate the line integral

$$\int_C \mathbf{F}\cdot\mathbf{T}\,ds$$

along the indicated path C.

11. $\mathbf{F}(x, y, z) = z\mathbf{i} + x\mathbf{j} - y\mathbf{k}$; C is parametrized by $x = t$, $y = t^2$, $z = t^3$, $0 \leqq t \leqq 1$.
12. $\mathbf{F}(x, y, z) = yz\mathbf{i} + xz\mathbf{j} + xy\mathbf{k}$; C is the straight line segment from $(2, -1, 3)$ to $(4, 2, -1)$.
13. $\mathbf{F}(x, y, z) = y\mathbf{i} - x\mathbf{j} + z\mathbf{k}$; $x = \sin t$, $y = \cos t$, $z = 2t$, $0 \leqq y \leqq \pi$.
14. $\mathbf{F}(x, y, z) = (2x + 3y)\mathbf{i} + (3x + 2y)\mathbf{j} + 3z^2\mathbf{k}$; C is the path from $(0, 0, 0)$ to $(4, 2, 3)$ that consists of three line segments parallel to the x-axis, the y-axis, and the z-axis, in that order.
15. $\mathbf{F}(x, y, z) = yz^2\mathbf{i} + xz^2\mathbf{j} + 2xyz\,\mathbf{k}$; C is the path from $(-1, 2, -2)$ to $(1, 5, 2)$ that consists of three line segments parallel to the z-axis, the x-axis, and the y-axis, in that order.

In Problems 16 through 18, evaluate

$$\int_C f(x, y, z)\,ds$$

for the given function $f(x, y, z)$ and the given path C.

16. $f(x, y, z) = xyz$; C is the straight line segment from $(1, -1, 2)$ to $(3, 2, 5)$.
17. $f(x, y, z) = 2x + 9xy$; C is the curve $x = t$, $y = t^2$, $z = t^3$, $0 \leqq t \leqq 1$.

18. $f(x, y, z) = xy$; C is the elliptical helix $x = 4\cos t$, $y = 9\sin t$, $z = 7t$, $0 \leqq t \leqq 5\pi/2$.

19. Find the centroid of a uniform thin wire shaped like the semicircle $x^2 + y^2 = a^2$, $a > 0$, $y \geqq 0$.

20. Find the moments of inertia around the x- and y-axes of the wire of Problem 19.

21. Find the mass and centroid of a wire that has constant density $\delta = k$ and is shaped like the helix $x = 3\cos t$, $y = 3\sin t$, $z = 4t$, $0 \leqq t \leqq 2\pi$.

22. Find the moment of inertia $I_z = \int_C (x^2 + y^2)\,dm$ around the z-axis of the helical wire of Problem 21.

23. A wire shaped like the first-quadrant portion of the circle $x^2 + y^2 = a^2$ has density $\delta = kxy$ at the point (x, y). Find its mass, centroid, and moment of inertia around each coordinate axis.

24. A wire is shaped like the arch $x = t - \sin t$, $y = 1 - \cos t$ $(0 \leqq t \leqq 2\pi)$ of a cycloid C and has constant density $\delta(x, y) \equiv k$. Find its mass, centroid, and moment of inertia $I_x = \int_C y^2\,dm$ around the x-axis.

25. A wire is shaped like the astroid $x = \cos^3 t$, $y = \sin^3 t$ $(0 \leqq t \leqq 2\pi)$ and has constant density $\delta(x, y) \equiv k$. Find its moment of inertia $I_0 = \int_C (x^2 + y^2)dm$ around the origin.

The ***average distance*** $\overline{D}$ *from the fixed point P to points of the parametrized curve C is defined by*

$$\overline{D} = \frac{1}{s}\int_C D(x, y)\,ds$$

where s is the length of C and D(x, y) denotes the distance from P to the variable point (x, y) of C. In Problems 26 through 31, compute the average distance, exactly if possible, else by using a computer algebra system to find it (either symbolically or, if necessary, numerically).

26. Use the standard trigonometric parametrization of a circle C of radius a to verify that the average distance of points of C from its center is $\overline{D} = a$.

27. Find (exactly) the average distance from the point $(a, 0)$ to points of the circle of radius a centered at the origin. [*Suggestion:* Use the law of cosines to find $D(x, y)$.]

28. Find the average distance from the origin to points of the cycloidal arch of Problem 24.

29. Find the average distance from the origin to points of the astroid of Problem 25.

30. Find the average distance from the origin to points of the helix of Problem 21.

31. The spiral parametrized by $x = e^{-t}\cos t$, $y = e^{-t}\sin t$ starts at $(1, 0)$ when $t = 0$ and closes in on the origin as $t \to \infty$. Use improper integrals to calculate the average distance from the origin to points of this spiral.

32. Find the work done by the inverse-square force field of Example 7 in moving a particle from $(1, 0, 0)$ to $(0, 3, 4)$. Integrate first along the line segment from $(1, 0, 0)$ to $(5, 0, 0)$ and then along a path on the sphere with equation $x^2 + y^2 + z^2 = 25$. The second integral is automatically zero. (Why?)

33. Imagine an infinitely long and uniformly charged wire that coincides with the z-axis. The electric force that it exerts on a unit charge at the point $(x, y) \neq (0, 0)$ in the xy-plane is

$$\mathbf{F}(x, y) = \frac{k(x\mathbf{i} + y\mathbf{j})}{x^2 + y^2}.$$

Find the work done by $\mathbf{F}$ in moving a unit charge along the straight line segment from (a) $(1, 0)$ to $(1, 1)$; (b) $(1, 1)$ to $(0, 1)$.

34. Show that if $\mathbf{F}$ is a *constant* force field, then it does zero work on a particle that moves once uniformly counterclockwise around the unit circle in the xy-plane.

35. Show that if $\mathbf{F} = k\mathbf{r} = k(x\mathbf{i} + y\mathbf{j})$, then $\mathbf{F}$ does zero work on a particle that moves once uniformly counterclockwise around the unit circle in the xy-plane.

36. Find the work done by the force field $\mathbf{F} = -y\mathbf{i} + x\mathbf{j}$ in moving a particle counterclockwise once around the unit circle in the xy-plane.

37. Let C be a curve on the unit sphere $x^2 + y^2 + z^2 = 1$. Explain why the inverse-square force field of Example 7 does zero work in moving a particle along C.

In Problems 38 through 40, the given curve C joins the points P and Q in the xy-plane. The point P represents the top of a ten-story building, and Q is a point on the ground 100 ft *from the base of the building. A* 150-lb *person slides down a frictionless slide shaped like the curve C from P to Q under the influence of the gravitational force* $\mathbf{F} = -150\mathbf{j}$. *In each problem show that* $\mathbf{F}$ *does the same amount of work on the person,* $W = 15000$ ft·lb, *as if he or she had dropped straight down to the ground.*

38. C is the straight line segment $y = x$ from $P(100, 100)$ to $Q(0, 0)$.

39. C is the circular arc $x = 100\sin t$, $y = 100\cos t$ from $P(0, 100)$ to $Q(100, 0)$.

40. C is the parabolic arc $y = x^2/100$ from $P(100, 100)$ to $Q(0, 0)$.

41. Now suppose that the 100-ft ten-story building of Problems 38 through 40 is a circular tower with a radius of 25 ft, and the fire-escape slide is a spiral (helical) ramp that encircles the tower once every two floors. Use a line integral to compute the work done by the gravitational force field $\mathbf{F} = -200\mathbf{k}$ on a 200-lb person who slides down this (frictionless) ramp from the top of the building to the ground.

42. An electric current I in a long straight wire generates a magnetic field $\mathbf{B}$ in the space surrounding the wire. The vector $\mathbf{B}$ is tangent to any circle C that is centered on the wire and lies in a plane perpendicular to the wire. *Ampere's law* implies that

$$\int_C \mathbf{B}\cdot d\mathbf{r} = \mu I$$

where μ is a certain electromagnetic constant. Deduce from this fact that the magnitude $B = |\mathbf{B}|$ of the magnetic field is proportional to the current I and inversely proportional to the distance r from the wire.

15.3 THE FUNDAMENTAL THEOREM AND INDEPENDENCE OF PATH

The fundamental theorem of calculus says, in effect, that differentiation and integration are inverse processes for single-variable functions. Specifically, part 2 of the fundamental theorem in Section 5.6 implies that

$$\int_a^b G'(t)\,dt = G(b) - G(a) \tag{1}$$

if the derivative G' is continuous on $[a, b]$. Theorem 1 here can be interpreted as saying that "gradient vector differentiation" and "line integration" are, similarly, inverse processes for multivariable functions.

THEOREM 1 The Fundamental Theorem for Line Integrals
Let f be a function of two or three variables and let C be a smooth curve parametrized by the vector-valued function $\mathbf{r}(t)$ for $a \leqq t \leqq b$. If f is continuously differentiable at each point of C, then

$$\int_C \nabla f \cdot d\mathbf{r} = f(\mathbf{r}(b)) - f(\mathbf{r}(a)). \tag{2}$$

PROOF We consider the three-dimensional case $f(x, y, z)$ illustrated in Fig. 15.3.1. Then $\nabla f = \langle \partial f/\partial x, \partial f/\partial y, \partial f/\partial z \rangle$, so Theorem 1 in Section 15.2 yields

$$\int_C \nabla f \cdot d\mathbf{r} = \int_C \frac{\partial f}{\partial x}\,dx + \frac{\partial f}{\partial y}\,dy + \frac{\partial f}{\partial z}\,dz$$
$$= \int_a^b \left(\frac{\partial f}{\partial x}\frac{dx}{dt} + \frac{\partial f}{\partial y}\frac{dy}{dt} + \frac{\partial f}{\partial z}\frac{dz}{dt} \right) dt.$$

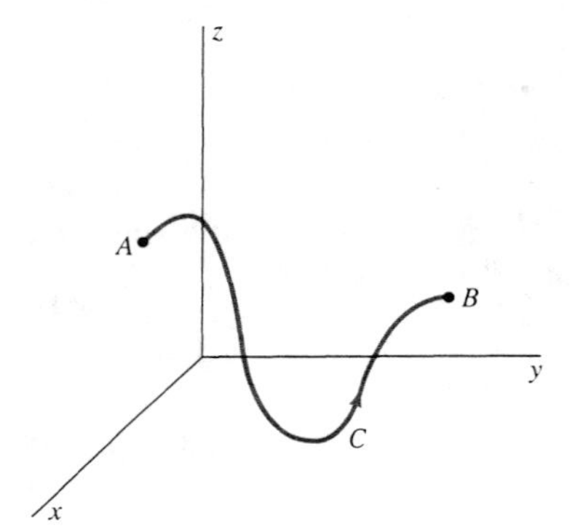

FIGURE 15.3.1 The path C of Theorem 1.

By the multivariable chain rule (Section 13.7), the integrand here is the derivative $G'(t)$ of the composite function $G(t) = f(\mathbf{r}(t)) = f(x(t), y(t), z(t))$. Therefore it follows that

$$\int_C \nabla f \cdot d\mathbf{r} = \int_a^b G'(t)\,dt = G(b) - G(a) \quad \text{(by Eq. (1))}$$
$$= f(\mathbf{r}(b)) - f(\mathbf{r}(a)),$$

and so we have established Eq. (2), as desired. ◂

REMARK If we write A and B for the endpoints $\mathbf{r}(a)$ and $\mathbf{r}(b)$ of C, then Eq. (2) takes the form

$$\int_C \nabla f \cdot d\mathbf{r} = f(B) - f(A), \tag{3}$$

which is quite similar to Eq. (1).

EXAMPLE 1 If

$$f(x, y, z) = -\frac{k}{r} = -\frac{k}{\sqrt{x^2 + y^2 + z^2}},$$

then a brief computation shows that $\nabla f = \mathbf{F}$ is the inverse-square force field

$$\mathbf{F}(x, y, z) = \frac{k(x\mathbf{i} + y\mathbf{j} + z\mathbf{k})}{(x^2 + y^2 + z^2)^{3/2}}$$

of Example 7 in Section 15.2, where we calculated directly the work $W = k/20$ done by the force field $\mathbf{F}$ in moving a particle along a straight line segment from the point $A(0, 4, 0)$ to the point $B(0, 4, 3)$. Indeed, using Theorem 1 we find that the work done by **F** in moving a particle along *any* smooth path from A to B (that does not pass through the origin) is given by

$$\begin{aligned} W &= \int_C \mathbf{F} \cdot d\mathbf{r} = \int_C \nabla f \cdot d\mathbf{r} \\ &= f(0, 4, 3) - f(0, 4, 0) \quad \text{(by Eq. (3))} \\ &= \left(-\frac{k}{5}\right) - \left(-\frac{k}{4}\right) = \frac{k}{20}. \end{aligned}$$

◆

Independence of Path

We next apply the fundamental theorem for line integrals to discuss the question whether the integral

$$\int_C \mathbf{F} \cdot \mathbf{T}\, ds = \int_C \mathbf{F} \cdot d\mathbf{r} = \int_C P\, dx + Q\, dt + R\, dz \tag{4}$$

(where $\mathbf{F} = \langle P, Q, R \rangle$) has the *same value* for *any* two curves with the same initial and terminal points.

DEFINITION Independence of Path

The line integral in Eq. (4) is said to be **independent of path in the region** D provided that, given any two points A and B of D, the integral has the same value along every piecewise smooth curve, or **path,** in D from A to B. In this case we may write

$$\int_C \mathbf{F} \cdot \mathbf{T}\, ds = \int_A^B \mathbf{F} \cdot \mathbf{T}\, ds \tag{5}$$

because the value of the integral depends only on the points A and B, not on the particular choice of the path C joining them.

For a tangible interpretation of independence of path, let us think of walking along the curve C from point A to point B in the plane where a wind with velocity vector $\mathbf{w}(x, y)$ is blowing. Suppose that when we are at (x, y), the wind exerts a force $\mathbf{F} = k\mathbf{w}(x, y)$ on us, k being a constant that depends on our size and shape (and perhaps other factors as well). Then, by Eq. (12) of Section 15.2, the amount of work the wind does on us as we walk along C is given by

$$W = \int_C \mathbf{F} \cdot \mathbf{T}\, ds = k \int_C \mathbf{w} \cdot \mathbf{T}\, ds. \tag{6}$$

This is the wind's contribution to our trip from A to B. In this context, the question of independence of path is whether or not the wind's work W depends on *which* path we choose from point A to point B.

EXAMPLE 2 Suppose that a steady wind blows toward the northeast with velocity vector $\mathbf{w} = 10\mathbf{i} + 10\mathbf{j}$ in fps units; its speed is $|\mathbf{w}| = 10\sqrt{2} \approx 14$ ft/s—about 10 mi/h. Assume that $k = 0.5$, so the wind exerts 0.5 lb of force for each foot per second of its velocity. Then $\mathbf{F} = 5\mathbf{i} + 5\mathbf{j}$, so Eq. (6) yields

$$W = \int_C \langle 5, 5 \rangle \cdot \mathbf{T}\, ds = \int_C 5\, dx + 5\, dy \tag{7}$$

for the work done on us by the wind as we walk along C.

For instance, if C is the straight path $x = 10t$, $y = 10t$ $(0 \leqq x \leqq 1)$ from $(0, 0)$ to $(10, 10)$, then Eq. (7) gives

$$W = \int_0^1 5 \cdot 10\,dt + 5 \cdot 10\,dt = 100 \int_0^1 1\,dt = 100$$

ft·lb of work. Or, if C is the parabolic path $y = \frac{1}{10}x^2$, $0 \leqq x \leqq 10$ from the same initial point $(0, 0)$ to the same terminal point $(10, 10)$, then Eq. (7) yields

$$W = \int_0^{10} 5\,dx + 5 \cdot \frac{1}{5}x\,dx = \int_0^{10} (5 + x)\,dx$$

$$= \left[5x + \frac{1}{2}x^2\right]_0^{10} = 100$$

ft·lb of work, the same as before. We shall see that it follows from Theorem 2 of this section that the line integral in Eq. (7) is independent of path, so the wind does 100 ft·lb of work along *any* path from $(0, 0)$ to $(10, 10)$. ◆

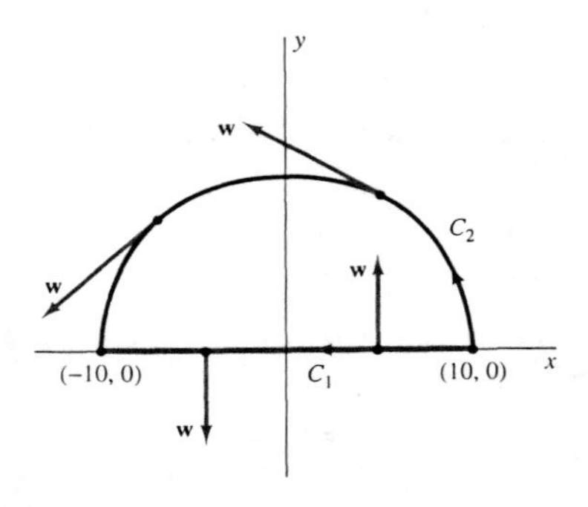

FIGURE 15.3.2 Around and through the eye of the hurricane (Example 2).

EXAMPLE 3 Suppose that $\mathbf{w} = -2y\mathbf{i} + 2x\mathbf{j}$. This wind is blowing counterclockwise around the origin, as in a hurricane with its eye at $(0, 0)$. With $k = 0.5$ as before, $\mathbf{F} = -y\mathbf{i} + x\mathbf{j}$, so the work integral is

$$W = \int_C \mathbf{F} \cdot \mathbf{T}\,ds = \int_C -y\,dx + x\,dy. \tag{8}$$

If we walk from $(10, 0)$ to $(-10, 0)$ along the straight path C_1 through the eye of the hurricane, then the wind is always perpendicular to our unit tangent vector $\mathbf{T}$ (Fig. 15.3.2). Hence $\mathbf{F} \cdot \mathbf{T} = 0$, and therefore

$$W = \int_{C_1} \mathbf{F} \cdot \mathbf{T}\,ds = \int_{C_1} -y\,dx + x\,dy = 0.$$

But if we walk along the semicircular path C_2 shown in Fig. 15.3.2, then $\mathbf{w}$ remains tangent to our path, so $\mathbf{F} \cdot \mathbf{T} = |\mathbf{F}| = 10$ at each point. In this case,

$$W = \int_{C_2} -y\,dx + x\,dy = \int_{C_2} \mathbf{F} \cdot \mathbf{T}\,ds = 10 \cdot 10\pi = 100\pi.$$

The fact that we get different values along different paths from $(10, 0)$ to $(-10, 0)$ shows that the line integral in Eq. (8) is *not* independent of path. ◆

Theorem 2 tells us when a given line integral is independent of path and when it is not.

THEOREM 2 Independence of Path

The line integral $\int_C \mathbf{F} \cdot \mathbf{T}\,ds$ of the continuous vector field $\mathbf{F}$ is independent of path in the plane or space region D if and only if $\mathbf{F} = \nabla f$ for some function f defined on D.

PROOF Suppose that $\mathbf{F} = \nabla f = \langle \partial f/\partial x, \partial f/\partial y, \partial f/\partial z \rangle$ and that C is a piecewise smooth curve from A to B in D parametrized as usual with parameter t in $[a, b]$. Then the fundamental theorem in the form in Eq. (3) gives

$$\int_C \mathbf{F} \cdot \mathbf{T}\,ds = \int_C \nabla f \cdot \mathbf{T}\,ds = \int_C \nabla f \cdot d\mathbf{r} = f(B) - f(A).$$

This result shows that the value of the line integral depends only on the points A and B and is therefore independent of the choice of the particular path C. This proves the *if* part of Theorem 2.

To prove the *only if* part of Theorem 2, we suppose that the line integral is independent of path in D. Choose a *fixed* point $A_0 = A_0(x_0, y_0, z_0)$ in D, and let

$B = B(x, y, z)$ be an arbitrary point in D. Given any path C from A_0 to B in D, we *define* the function f by means of the equation

$$f(x, y, z) = \int_C \mathbf{F} \cdot \mathbf{T} \, ds = \int_{(x_0, y_0, z_0)}^{(x, y, z)} \mathbf{F} \cdot \mathbf{T} \, ds. \tag{9}$$

Because of the hypothesis of independence of path, the resulting value of $f(x, y, z)$ depends only on (x, y, z) and not on the particular path C used.

To verify that $\nabla f = \mathbf{F}$, suppose that $\mathbf{F} = P\mathbf{i} + Q\mathbf{j} + R\mathbf{k}$. To show that $\partial f/\partial x = P$, we write

$$\begin{aligned} f(x+h, y, z) - f(x, y, z) &= \int_{(x_0, y_0, z_0)}^{(x+h, y, z)} \mathbf{F} \cdot \mathbf{T} \, ds - \int_{(x_0, y_0, z_0)}^{(x, y, z)} \mathbf{F} \cdot \mathbf{T} \, ds \\ &= \int_{(x, y, z)}^{(x+h, y, z)} \mathbf{F} \cdot \mathbf{T} \, ds. \end{aligned}$$

In the last integral, we may take the path of integration to be the parametrized line segment L from (x, y, z) to $(x + h, y, z)$ defined by $\phi(t) = (x + th, y, z)$. Then $\phi'(t) = (h, 0, 0) = h\mathbf{i}$, so the unit tangent vector is $\mathbf{T} = \mathbf{i}$. Also, $dy = dz = 0$, so $ds = dx = h\,dt$ along this path. Therefore

$$\begin{aligned} f(x+h, y, z) - f(x, y, z) &= \int_L \mathbf{F} \cdot \mathbf{T} \, ds = \int_0^1 P(\phi(t))h \, dt \\ &= hP(\phi(\bar{t})) = hP(x + \bar{t}h, y, z) \end{aligned}$$

for some $\bar{t}$ between 0 and 1 (by the average value theorem for single-variable integrals in Section 5.6). It follows that

$$\frac{\partial f}{\partial x} = \lim_{h \to 0} \frac{f(x+h, y, z) - f(x, y, z)}{h} = \lim_{h \to 0} P(x + \bar{t}h, y, z) = P(x, y, z)$$

by the continuity of P. It follows similarly that $\partial f/\partial y = Q$ and $\partial f/\partial z = R$, so $\nabla f = \mathbf{F}$ as desired. ◂

For a first application of Theorem 2, consider the force field $\mathbf{F} = -y\mathbf{i} + x\mathbf{j}$ that corresponds to the counterclockwise wind discussed in Example 3. Because the line integral $\int_C \mathbf{F} \cdot \mathbf{T} \, ds$ is *not* independent of path in any plane region that either includes or encloses the origin, it follows that $\mathbf{F}$ is not the gradient of any scalar function f.

By contrast, the force field $\mathbf{F} = 5\mathbf{i} + 5\mathbf{j}$, which corresponds to a constant wind blowing northeast, is obviously the gradient of the function $f(x, y) = 5x + 5y$. Therefore Theorem 2 implies that the line integral $\int_C \mathbf{F} \cdot \mathbf{T} \, ds$ *is* independent of path in the entire plane $\mathbf{R}^2$.

Similarly, the inverse-square force field $\mathbf{F}(x, y, z) = k\mathbf{r}/r^3$ of Example 1 is the gradient of the function $f(x, y, z) = -k/r$ (where $r = \sqrt{x^2 + y^2 + z^2}$). Hence Theorem 2 implies that the line integral $\int_C \mathbf{F} \cdot \mathbf{T} \, ds$ is independent of path in $\mathbf{R}^3$ minus the origin.

Conservative Vector Fields

DEFINITION Conservative Fields and Potential Functions

The vector field $\mathbf{F}$ defined on a region D is **conservative** provided that there exists a scalar function f defined on D such that

$$\mathbf{F} = \nabla f \tag{10}$$

at each point of D. In this case f is called a **potential function** for the vector field $\mathbf{F}$.

For instance, Example 1 implies that $f(x, y, z) = -k/r$ is a potential function for the inverse-square force field $\mathbf{F}(x, y, z) = k\mathbf{r}/r^3$ (where $r = \sqrt{x^2 + y^2 + z^2}$).

COMMENT In some physical applications the scalar function f is called a *potential function* for the vector field $\mathbf{F}$ provided that $\mathbf{F} = -\nabla f$.

If the line integral $\int_C \mathbf{F} \cdot \mathbf{T}\, ds$ is known to be independent of path, then Theorem 2 guarantees that the vector field $\mathbf{F}$ is conservative and that Eq. (9) yields a potential function f for $\mathbf{F}$. In this case—because the value of the integral does not depend on the specific curve C from A to B—we may well write Eq. (3) in the form

$$\int_A^B \mathbf{F} \cdot \mathbf{T}\, ds = \int_A^B \nabla f \cdot d\mathbf{r} = f(B) - f(A) \tag{11}$$

that is still more reminiscent of the ordinary fundamental theorem.

EXAMPLE 4 Find a potential function for the conservative vector field

$$\mathbf{F}(x, y) = (6xy - y^3)\mathbf{i} + (4y + 3x^2 - 3xy^2)\mathbf{j}. \tag{12}$$

Solution Because we are given the information that $\mathbf{F}$ is a conservative field, the line integral $\int \mathbf{F} \cdot \mathbf{T}\, ds$ is independent of path by Theorem 2. Therefore we may apply Eq. (9) to find a scalar potential function f. Let C be the straight-line path from $A(0, 0)$ to $B(x_1, y_1)$ parametrized by $x = x_1 t$, $y = y_1 t$, $0 \leqq t \leqq 1$. Then Eq. (9) yields

$$\begin{aligned}
f(x_1, y_1) &= \int_A^B \mathbf{F} \cdot \mathbf{T}\, ds \\
&= \int_A^B (6xy - y^3)\, dx + (4y + 3x^2 - 3xy^2)\, dy \\
&= \int_0^1 \left(6x_1 y_1 t^2 - y_1^3 t^3\right)(x_1\, dt) + \left(4y_1 t + 3x_1^2 t^2 - 3x_1 y_1^2 t^3\right)(y_1\, dt) \\
&= \int_0^1 \left(4y_1^2 t + 9x_1^2 y_1 t^2 - 4x_1 y_1^3 t^3\right) dt \\
&= \Big[2y_1^2 t^2 + 3x_1^2 y_1 t^3 - x_1 y_1^3 t^4\Big]_0^1 = 2y_1^2 + 3x_1^2 y_1 - x_1 y_1^3.
\end{aligned}$$

At this point we delete the subscripts, because (x_1, y_1) is an arbitrary point of the plane. Thus we obtain the potential function

$$f(x, y) = 2y^2 + 3x^2 y - xy^3$$

for the vector field $\mathbf{F}$ in Eq. (12). As a check, we can differentiate f to obtain

$$\frac{\partial f}{\partial x} = 6xy - y^3, \qquad \frac{\partial f}{\partial y} = 4y + 3x^2 - 3xy^2. \qquad ◆$$

But how did we know in advance that the vector field $\mathbf{F}$ in Example 4 was conservative? The answer is provided by Theorem 3; a proof of this theorem based on Green's theorem is outlined in the next section.

THEOREM 3 Conservative Fields and Potential Functions

Suppose that the vector field $\mathbf{F} = P\mathbf{i} + Q\mathbf{j}$ is continuously differentiable in an open rectangle R in the xy-plane. Then $\mathbf{F}$ is conservative in R—and hence has a potential function $f(x, y)$ defined on R—if and only if, at each point of R,

$$\frac{\partial P}{\partial y} = \frac{\partial Q}{\partial x}. \tag{13}$$

Observe that the vector field $\mathbf{F}$ in Eq. (12), where $P(x, y) = 6xy - y^3$ and $Q(x, y) = 4y + 3x^2 - 3xy^2$, satisfies the criterion in Eq. (13) because

$$\frac{\partial P}{\partial y} = 6x - 3y^2 = \frac{\partial Q}{\partial x}.$$

When this sufficient condition for the existence of a potential function is satisfied, the method illustrated in Example 5 is usually an easier way to find a potential function than the evaluation of the line integral in Eq. (9)—the method used in Example 4.

EXAMPLE 5 Given

$$P(x, y) = 6xy - y^3 \quad \text{and} \quad Q(x, y) = 4y + 3x^2 - 3xy^2,$$

note that P and Q satisfy the condition $\partial P/\partial y = \partial Q/\partial x$. Find a potential function $f(x, y)$ such that

$$\frac{\partial f}{\partial x} = 6xy - y^3 \quad \text{and} \quad \frac{\partial f}{\partial y} = 4y + 3x^2 - 3xy^2. \tag{14}$$

Solution Upon integrating the first of these equations with respect to x, we get

$$f(x, y) = 3x^2 y - xy^3 + \xi(y), \tag{15}$$

where $\xi(y)$ is an "arbitrary function" of y alone; it acts as a "constant of integration" with respect to x, because its derivative with respect to x is zero. We next determine $\xi(y)$ by imposing the second condition in (14):

$$\frac{\partial f}{\partial y} = 3x^2 - 3xy^2 + \xi'(y) = 4y + 3x^2 - 3xy^2.$$

It follows that $\xi'(y) = 4y$, so $\xi(y) = 2y^2 + C$. When we set $C = 0$ and substitute the result into Eq. (15), we get the same potential function

$$f(x, y) = 3x^2 y - xy^3 + 2y^2$$

that we found by entirely different methods in Example 4. ◆

Conservative Force Fields and Conservation of Energy

Given a conservative force field $\mathbf{F}$, it is customary in physics to introduce a minus sign and write $\mathbf{F} = -\nabla V$. Then $V(x, y, z)$ is called the **potential energy** at the point (x, y, z). With $f = -V$ in Eq. (11), we have

$$W = \int_A^B \mathbf{F} \cdot \mathbf{T}\, ds = V(A) - V(B), \tag{16}$$

and this means that the work W done by $\mathbf{F}$ in moving a particle from A to B is equal to the *decrease* in potential energy.

Here is the reason why the expression *conservative field* is used. Suppose that a particle of mass m moves from A to B under the influence of the conservative force $\mathbf{F}$, with position vector $\mathbf{r}(t)$, $a \leqq t \leqq b$. Then Newton's law $\mathbf{F}(\mathbf{r}(t)) = m\mathbf{r}''(t) = m\mathbf{v}'(t)$ with $d\mathbf{r} = \mathbf{r}'(t)\, dt = \mathbf{v}(t)\, dt$ gives

$$\begin{aligned} \int_A^B \mathbf{F} \cdot \mathbf{T}\, ds &= \int_a^b m\mathbf{v}'(t) \cdot \mathbf{v}(t)\, dt \\ &= \int_a^b m D_t \left[\frac{1}{2}\mathbf{v}(t) \cdot \mathbf{v}(t)\right] dt = \left[\frac{1}{2}m[v(t)]^2\right]_a^b. \end{aligned}$$

Thus with the abbreviations v_A for $v(a)$ and v_B for $v(b)$, we see that

$$\int_A^B \mathbf{F} \cdot d\mathbf{r} = \frac{1}{2}m(v_B)^2 - \frac{1}{2}m(v_A)^2. \tag{17}$$

By equating the right-hand sides of Eqs. (16) and (17), we get the formula

$$\tfrac{1}{2}m(v_A)^2 + V(A) = \tfrac{1}{2}m(v_B)^2 + V(B). \tag{18}$$

This is the law of **conservation of mechanical energy** for a particle moving under the influence of a *conservative* force field: Its **total energy**—the sum of its kinetic energy $\frac{1}{2}mv^2$ and its potential energy V—remains *constant.*

EXAMPLE 6 If we take $k = GMm$ in the calculation in Example 1, we see that the inverse-square gravitational force

$$\mathbf{F}(x, y, z) = -\frac{GMm\mathbf{r}}{r^3}$$

(exerted on a mass particle m by a mass M fixed at the origin) is the *negative* of the gradient of the potential energy function

$$V(x, y, z) = -\frac{GMm}{r}.$$

Hence Eq. (18) implies that the total energy of the mass particle moving with velocity v at distance r from the origin is the constant

$$E = \frac{1}{2}mv^2 - \frac{GMm}{r}.$$

It follows (for instance) that if—for whatever reason—the particle approaches the origin ($r \to 0$), then its velocity must increase without bound ($v \to +\infty$). ◆

15.3 TRUE/FALSE STUDY GUIDE

15.3 CONCEPTS: QUESTIONS AND DISCUSSION

1. Give several examples of line integrals that are *not* independent of the path.
2. Give several examples of a vector field $\mathbf{F}$ and a closed path C such that

$$\int_C \mathbf{F}\cdot\mathbf{T}\,ds \neq 0.$$

3. Give several examples of vector fields that are not conservative.

15.3 PROBLEMS

Determine whether the vector fields in Problems 1 through 16 are conservative. Find potential functions for those that are conservative (either by inspection or by using the method of Example 5).

1. $\mathbf{F}(x, y) = (2x + 3y)\mathbf{i} + (3x + 2y)\mathbf{j}$

2. $\mathbf{F}(x, y) = (4x - y)\mathbf{i} + (6y - x)\mathbf{j}$

3. $\mathbf{F}(x, y) = (3x^2 + 2y^2)\mathbf{i} + (4xy + 6y^2)\mathbf{j}$

4. $\mathbf{F}(x, y) = (2xy^2 + 3x^2)\mathbf{i} + (2x^2y + 4y^3)\mathbf{j}$

5. $\mathbf{F}(x, y) = (2y + \sin 2x)\mathbf{i} + (3x + \cos 3y)\mathbf{j}$

6. $\mathbf{F}(x, y) = (4x^2y - 5y^4)\mathbf{i} + (x^3 - 20xy^3)\mathbf{j}$

7. $\mathbf{F}(x, y) = \left(x^3 + \frac{y}{x}\right)\mathbf{i} + (y^2 + \ln x)\mathbf{j}$

8. $\mathbf{F}(x, y) = (1 + ye^{xy})\mathbf{i} + (2y + xe^{xy})\mathbf{j}$

9. $\mathbf{F}(x, y) = (\cos x + \ln y)\mathbf{i} + \left(\frac{x}{y} + e^y\right)\mathbf{j}$

10. $\mathbf{F}(x, y) = (x + \arctan y)\mathbf{i} + \dfrac{x + y}{1 + y^2}\mathbf{j}$

11. $\mathbf{F}(x, y) = (x \cos y + \sin y)\mathbf{i} + (y \cos x + \sin x)\mathbf{j}$

12. $\mathbf{F}(x, y) = e^{x-y}\left[(xy + y)\mathbf{i} + (xy + x)\mathbf{j}\right]$

13. $\mathbf{F}(x, y) = (3x^2y^3 + y^4)\mathbf{i} + (3x^3y^2 + y^4 + 4xy^3)\mathbf{j}$

14. $\mathbf{F}(x, y) = (e^x \sin y + \tan y)\mathbf{i} + (e^x \cos y + x \sec^2 y)\mathbf{j}$

15. $\mathbf{F}(x, y) = \left(\dfrac{2x}{y} - \dfrac{3y^2}{x^4}\right)\mathbf{i} + \left(\dfrac{2y}{x^3} - \dfrac{x^2}{y^2} + \dfrac{1}{\sqrt{y}}\right)\mathbf{j}$

16. $\mathbf{F}(x, y) = \dfrac{2x^{5/2} - 3y^{5/3}}{2x^{5/2}y^{2/3}}\mathbf{i} + \dfrac{3y^{5/3} - 2x^{5/2}}{3x^{3/2}y^{5/3}}\mathbf{j}$

In Problems 17 through 20, apply the method of Example 4 to find a potential function for the indicated vector field.

17. The vector field of Problem 3

18. The vector field of Problem 4

19. The vector field of Problem 13

20. The vector field of Problem 8

In Problems 21 through 26, show that the given line integral is independent of path in the entire xy-plane, then calculate the value of the line integral.

21. $\int_{(0,0)}^{(1,2)} (y^2 + 2xy)\,dx + (x^2 + 2xy)\,dy$

22. $\int_{(0,0)}^{(1,1)} (2x - 3y)\,dx + (2y - 3x)\,dy$

23. $\int_{(0,0)}^{(1,-1)} 2xe^y\,dx + x^2e^y\,dy$

24. $\int_{(0,0)}^{(2,\pi)} \cos y\,dx - x\sin y\,dy$

25. $\int_{(\pi/2,\pi/2)}^{(\pi,\pi)} (\sin y + y\cos x)\,dx + (\sin x + x\cos y)\,dy$

26. $\int_{(0,0)}^{(1,-1)} (e^y + ye^x)\,dx + (e^x + xe^y)\,dy$

Find a potential function for each of the conservative vector fields in Problems 27 through 29.

27. $\mathbf{F}(x, y, z) = yz\,\mathbf{i} + xz\,\mathbf{j} + xy\,\mathbf{k}$

28. $\mathbf{F}(x, y, z) = (2x - y - z)\mathbf{i} + (2y - x)\mathbf{j} + (2z - x)\mathbf{k}$

29. $\mathbf{F}(x, y, z) = (y\cos z - yze^x)\mathbf{i} + (x\cos z - ze^x)\mathbf{j} - (xy\sin z + ye^x)\mathbf{k}$

30. Let $\mathbf{F}(x, y) = (-y\mathbf{i} + x\mathbf{j})/(x^2 + y^2)$ for x and y not both zero. Calculate the values of

$$\int_C \mathbf{F}\cdot\mathbf{T}\,ds$$

along both the upper and the lower halves of the circle $x^2 + y^2 = 1$ from $(1, 0)$ to $(-1, 0)$. Is there a function $f = f(x, y)$ defined for x and y not both zero such that $\nabla f = \mathbf{F}$? Why?

31. Show that if the force field $\mathbf{F} = P\,\mathbf{i} + Q\,\mathbf{j}$ is conservative, then $\partial P/\partial y = \partial Q/\partial x$. Show that the force field of Problem 30 satisfies the condition $\partial P/\partial y = \partial Q/\partial x$ but nevertheless is *not* conservative.

32. Suppose that the force field $\mathbf{F} = P\,\mathbf{i} + Q\,\mathbf{j} + R\,\mathbf{k}$ is conservative. Show that

$$\frac{\partial P}{\partial y} = \frac{\partial Q}{\partial x}, \quad \frac{\partial P}{\partial z} = \frac{\partial R}{\partial x}, \quad \text{and} \quad \frac{\partial Q}{\partial z} = \frac{\partial R}{\partial y}.$$

33. Apply Theorem 2 and the result of Problem 32 to show that

$$\int_C 2xy\,dx + x^2\,dy + y^2\,dz$$

is not independent of path.

34. Let $\mathbf{F}(x, y, z) = yz\,\mathbf{i} + (xz + y)\mathbf{j} + (xy + 1)\mathbf{k}$. Define the function f by

$$f(x, y, z) = \int_C \mathbf{F}\cdot\mathbf{T}\,ds,$$

where C is the straight line segment from $(0, 0, 0)$ to (x, y, z). Determine f by evaluating this line integral, then show that $\nabla f = \mathbf{F}$.

35. Let $f(x, y) = \tan^{-1}(y/x)$, which if $x > 0$ equals the polar angle θ for the point (x, y). (a) Show that

$$\mathbf{F} = \nabla f = \frac{-y\mathbf{i} + x\mathbf{j}}{x^2 + y^2}.$$

(b) Suppose that $A(x_1, y_1) = (r_1, \theta_1)$ and $B(x_2, y_2) = (r_2, \theta_2)$ are two points in the right half-plane $x > 0$ and that C is a smooth curve from A to B. Explain why it follows from the fundamental theorem for line integrals that $\int_C \mathbf{F}\cdot\mathbf{T}\,ds = \theta_2 - \theta_1$. (c) Suppose that C_1 is the upper half of the unit circle from $(1, 0)$ to $(-1, 0)$ and that C_2 is the lower half, oriented also from $(1, 0)$ to $(-1, 0)$. Show that

$$\int_{C_1} \mathbf{F}\cdot\mathbf{T}\,ds = \pi \quad \text{whereas} \quad \int_{C_2} \mathbf{F}\cdot\mathbf{T}\,ds = -\pi.$$

Why does this not contradict the fundamental theorem?

36. Let $\mathbf{F} = k\mathbf{r}/r^3$ be the inverse-square force field of Example 7 in Section 15.2. Show that the work done by $\mathbf{F}$ in moving a particle from a point at distance r_1 from the origin to a point at distance r_2 from the origin is given by

$$W = k\left(\frac{1}{r_1} - \frac{1}{r_2}\right).$$

37. Suppose that an earth satellite with mass $m = 10000$ kg travels in an elliptical orbit whose apogee (farthest point) and perigee (closest point) are, respectively, 11000 km and 9000 km from the center of the earth. Calculate the work done against the earth's gravitational force field $\mathbf{F} = -GMm\mathbf{r}/r^3$ in lifting the satellite from perigee to apogee. Use the values $M = 5.97 \times 10^{24}$ kg for the mass of the earth and $G = 6.67 \times 10^{-11}$ N·m²/kg² for the universal gravitational constant.

38. Calculate the work that must be done against the sun's gravitational force field in transporting the satellite of Problem 37 from the earth to Mars. Use the values $M = 1.99 \times 10^{30}$ kg for the mass of the sun, $r_E = 1.50 \times 10^8$ km for the distance from the sun to earth, and $r_M = 2.29 \times 10^8$ km for the distance from the sun to Mars.

15.4 GREEN'S THEOREM

Green's theorem relates a line integral around a simple closed plane curve C to an ordinary double integral over the plane region R with boundary C. [The curve C parametrized by $\mathbf{r}: [a, b] \to \mathbf{R}^2$ is called **closed** if $\mathbf{r}(a) = \mathbf{r}(b)$ and **simple** if it has no other "self-intersections."] Suppose that the curve C is piecewise smooth—it consists of a finite number of parametric arcs with continuous nonzero velocity vectors. Then C has a unit tangent vector $\mathbf{T}$ everywhere except possibly at a finite number of *corner points.* The **positive,** or **counterclockwise,** direction along C is the

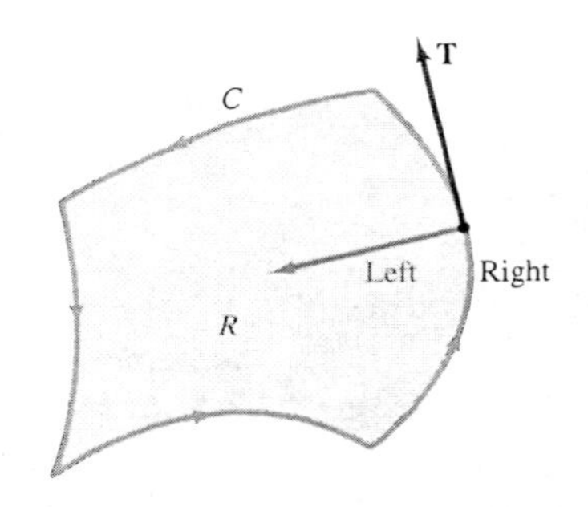

FIGURE 15.4.1 Positive orientation of the curve C: The region R within C is to the *left* of the unit tangent vector **T**.

direction determined by a parametrization $\mathbf{r}(t)$ of C such that the region R remains on the *left* as the point $\mathbf{r}(t)$ traces the boundary curve C. That is, the vector obtained from the unit tangent vector $\mathbf{T}$ by a counterclockwise rotation through 90° always points *into* the region R (Fig. 15.4.1). The boundary curve C is said to be **positively oriented** if it is equipped with such a parametrization. The symbol

$$\oint_C P\,dx + Q\,dy$$

then denotes a line integral around C in the positive direction—that is, using a parametrization consistent with the positive orientation of the curve.

The following result first appeared (in an equivalent form) in a booklet on the applications of mathematics to electricity and magnetism, published privately in 1828 by the self-taught English mathematical physicist George Green (1793–1841).

GREEN'S THEOREM

Let C be a positively oriented piecewise-smooth simple closed curve that bounds the region R in the plane. Suppose that the functions $P(x, y)$ and $Q(x, y)$ have continuous first-order partial derivatives on R. Then

$$\oint_C P\,dx + Q\,dy = \iint_R \left(\frac{\partial Q}{\partial x} - \frac{\partial P}{\partial y}\right) dA. \tag{1}$$

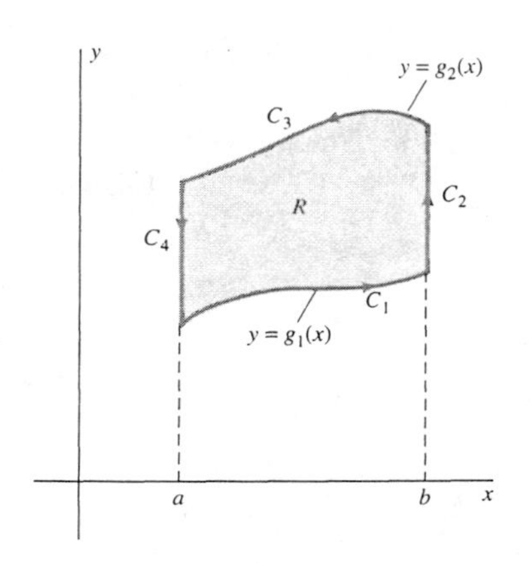

FIGURE 15.4.2 The boundary curve C is the union of the four arcs C_1, C_2, C_3, and C_4.

PROOF First we give a proof for the case in which the region R is both horizontally simple and vertically simple. Then we indicate how to extend the result to more general regions.

Recall from Section 14.2 that if R is vertically simple, then it has a description of the form $g_1(x) \leqq y \leqq g_2(y)$, $a \leqq x \leqq b$. The boundary curve C is then the union of the four arcs C_1, C_2, C_3, and C_4 of Fig. 15.4.2, positively oriented as indicated there. Hence

$$\oint_C P\,dx = \int_{C_1} P\,dx + \int_{C_2} P\,dx + \int_{C_3} P\,dx + \int_{C_4} P\,dx.$$

The integrals along both C_2 and C_4 are zero, because on those two curves $x(t)$ is constant, so that $dx = x'(t)\,dt = 0$. Thus we need compute only the integrals along C_1 and C_3.

The point $(x, g_1(x))$ traces C_1 as x increases from a to b, whereas the point $(x, g_2(x))$ traces C_3 as x *decreases* from b to a. Hence

$$\oint_C P\,dx = \int_a^b P(x, g_1(x))\,dx + \int_b^a P(x, g_2(x))\,dx$$

$$= -\int_a^b [P(x, g_2(x)) - P(x, g_1(x))]\,dx = -\int_a^b \int_{g_1(x)}^{g_2(x)} \frac{\partial P}{\partial y}\,dy\,dx$$

by the fundamental theorem of calculus. Thus

$$\oint_C P\,dx = -\iint_R \frac{\partial P}{\partial y}\,dA. \tag{2}$$

In Problem 36 we ask you to show in a similar way that

$$\oint_C Q\,dy = +\iint_R \frac{\partial Q}{\partial x}\,dA \tag{3}$$

if the region R is horizontally simple. We then obtain Eq. (1), the conclusion of Green's theorem, by adding Eqs. (2) and (3). ◄

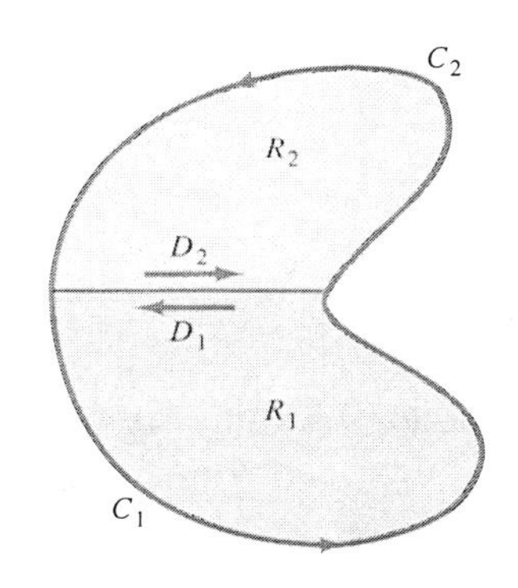

FIGURE 15.4.3 Decomposing the region R into two horizontally and vertically simple regions by using a crosscut.

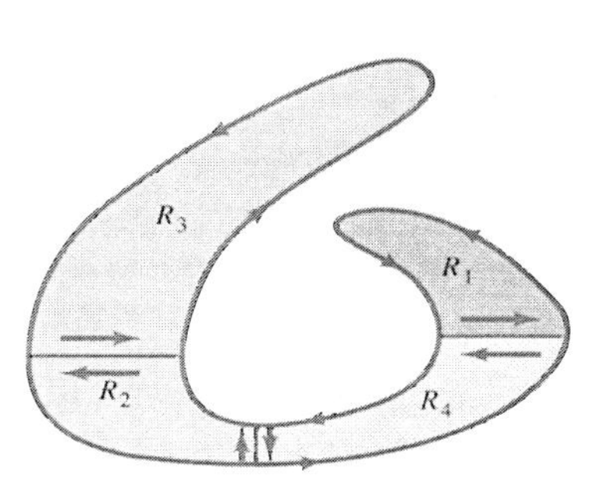

FIGURE 15.4.4 Many important regions can be decomposed into simple regions by using one or more crosscuts.

The complete proof of Green's theorem for more general regions is beyond the scope of an elementary text. But the typical region R that appears in practice can be divided into smaller regions $R_1, R_2, \dots, R_k$ that are both vertically and horizontally simple. Green's theorem for the region R then follows from the fact that it holds for each of the regions $R_1, R_2, \dots, R_k$. (See Problem 37.)

For example, we can divide the horseshoe-shaped region R of Fig. 15.4.3 into the two regions R_1 and R_2, both of which are horizontally simple and vertically simple. We also subdivide the boundary C of R accordingly and write $C_1 \cup D_1$ for the boundary of R_1 and $C_2 \cup D_2$ for the boundary of R_2 (Fig. 15.4.3). Applying Green's theorem separately to the regions R_1 and R_2, we get

$$\oint_{C_1 \cup D_1} P\,dx + Q\,dy = \iint_{R_1} \left(\frac{\partial Q}{\partial x} - \frac{\partial P}{\partial y}\right) dA$$

and

$$\oint_{C_2 \cup D_2} P\,dx + Q\,dy = \iint_{R_2} \left(\frac{\partial Q}{\partial x} - \frac{\partial P}{\partial y}\right) dA.$$

When we add these two equations, the result is Eq. (1), Green's theorem for the region R, because the two line integrals along D_1 and D_2 cancel. This occurs because D_1 and D_2 represent the same curve with opposite orientations, so

$$\int_{D_2} P\,dx + Q\,dy = -\int_{D_1} P\,dx + Q\,dy$$

by Eq. (11) of Section 15.2. It therefore follows that

$$\int_{C_1 \cup D_1 \cup C_2 \cup D_2} P\,dx + Q\,dy = \oint_{C_1 \cup C_2} P\,dx + Q\,dy = \oint_C P\,dx + Q\,dy.$$

Similarly, we could establish Green's theorem for the region shown in Fig. 15.4.4 by dividing it into the four simple regions indicated there.

EXAMPLE 1 Use Green's theorem to evaluate the line integral

$$\oint_C \left(2y + \sqrt{9 + x^3}\right) dx + (5x + e^{\arctan y})\,dy,$$

where C is the positively oriented circle $x^2 + y^2 = 4$.

Solution With $P(x, y) = 2y + \sqrt{9 + x^3}$ and $Q(x, y) = 5x + e^{\arctan y}$, we see that

$$\frac{\partial Q}{\partial x} - \frac{\partial P}{\partial y} = 5 - 2 = 3.$$

Because C bounds R, a circular disk with area 4π, Green's theorem therefore implies that the given line integral is equal to

$$\iint_R 3\,dA = 3 \cdot 4\pi = 12\pi.$$

◆

REMARK Suppose that the force field $\mathbf{F}$ is defined by

$$\mathbf{F}(x, y) = \left(2y + \sqrt{9 + x^3}\right)\mathbf{i} + (5x + e^{\arctan y})\mathbf{j} = P(x, y)\,\mathbf{i} + Q(x, y)\,\mathbf{j},$$

using the notation in Example 1. Then (as in Section 15.2) the work W done by the force field $\mathbf{F}$ in moving a particle counterclockwise once around the circle C of radius 2 is given by

$$W = \oint_C \mathbf{F} \cdot \mathbf{T}\,ds = \oint_C P\,dx + Q\,dy = \iint_R \left(\frac{\partial Q}{\partial x} - \frac{\partial P}{\partial y}\right) dA = \iint_R 3\,dA = 12\pi$$

as in Example 1.

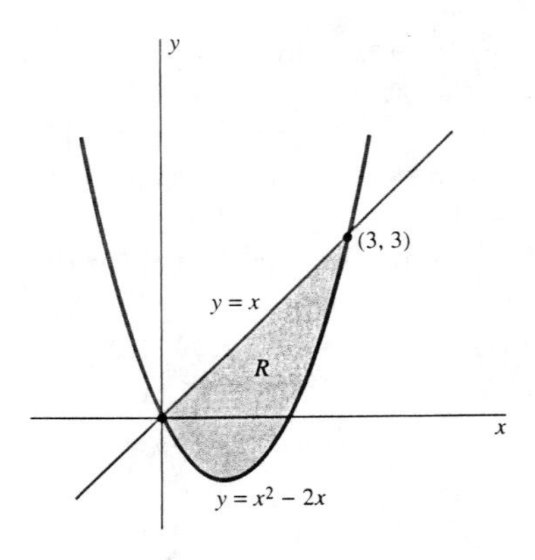

FIGURE 15.4.5 The region of Example 2.

EXAMPLE 2 Evaluate the line integral

$$\oint_C 3xy\,dx + 2x^2\,dy,$$

where C is the positively oriented boundary of the region R shown in Fig. 15.4.5. It is bounded above by the line $y = x$ and below by the parabola $y = x^2 - 2x$.

Solution To evaluate the line integral directly, we would need to parametrize separately the line and the parabola. Instead, we apply Green's theorem with $P(x, y) = 3xy$ and $Q(x, y) = 2x^2$, so

$$\frac{\partial Q}{\partial x} - \frac{\partial P}{\partial y} = 4x - 3x = x.$$

Then

$$\begin{aligned}\oint_C 3xy\,dx + 2x^2\,dy &= \iint_R x\,dA \\ &= \int_0^3 \int_{x^2-2x}^{x} x\,dy\,dx = \int_0^3 \Big[xy\Big]_{y=x^2-2x}^{x} dx \\ &= \int_0^3 (3x^2 - x^3)\,dx = \left[x^3 - \frac{1}{4}x^4\right]_0^3 = \frac{27}{4}.\end{aligned}$$

◆

In Examples 1 and 2 we found the double integral easier to evaluate directly than the line integral. Sometimes the situation is the reverse. The following consequence of Green's theorem illustrates the technique of evaluating a double integral $\iint_R f(x, y)\,dA$ by converting it into a line integral

$$\oint_C P\,dx + Q\,dy.$$

To do this we must be able to find functions $P(x, y)$ and $Q(x, y)$ such that

$$\frac{\partial Q}{\partial x} - \frac{\partial P}{\partial y} = f(x, y).$$

As in the proof of the following result, this is sometimes easy.

COROLLARY TO GREEN'S THEOREM

The area A of the region R bounded by the positively oriented piecewise-smooth simple closed curve C is given by

$$A = \frac{1}{2}\oint_C -y\,dx + x\,dy = -\oint_C y\,dx = \oint_C x\,dy. \tag{4}$$

PROOF With $P(x, y) = -y$ and $Q(x, y) \equiv 0$, Green's theorem gives

$$-\oint_C y\,dx = \iint_R 1\,dA = A.$$

Similarly, with $P(x, y) \equiv 0$ and $Q(x, y) = x$, we obtain

$$\oint_C x\,dy = \iint_R 1\,dA = A.$$

The third result may be obtained by averaging the left- and right-hand sides in the last two equations. Alternatively, with $P(x, y) = -y/2$ and $Q(x, y) = x/2$, Green's theorem gives

$$\frac{1}{2}\oint_C -y\,dx + x\,dy = \iint_R \left(\frac{1}{2} + \frac{1}{2}\right) dA = A.$$

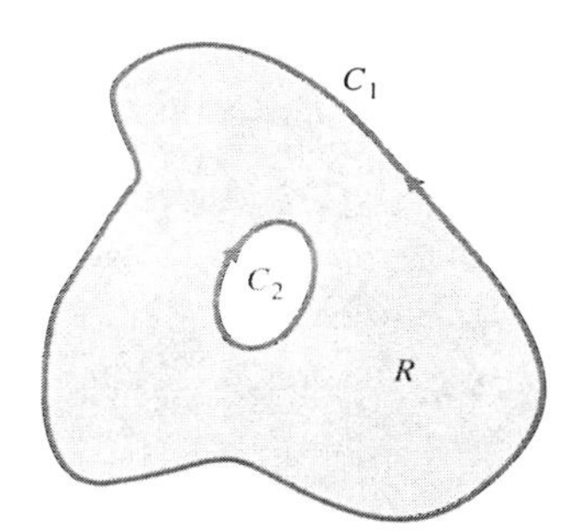

EXAMPLE 3 Apply the corollary to Green's theorem to find the area A of the region R bounded by the ellipse $x^2/a^2 + y^2/b^2 = 1$.

Solution With the parametrization $x = a\cos t$, $y = b\sin t$, $0 \leqq t \leqq 2\pi$, Eq. (4) gives

$$A = \oint_R x\,dy = \int_0^{2\pi} (a\cos t)(b\cos t\,dt)$$

$$= \frac{1}{2}ab\int_0^{2\pi} (1 + \cos 2t)\,dt = \pi ab.$$

FIGURE 15.4.6 An annular region—the boundary consists of two simple closed curves, one within the other.

By using the technique of subdividing a region into simpler ones, we can extend Green's theorem to regions with boundaries that consist of two or more simple closed curves. For example, consider the annular region R of Fig. 15.4.6, with boundary C consisting of the two simple closed curves C_1 and C_2. The positive direction along C—the direction for which the region R always lies on the left—is counterclockwise on the outer curve C_1 but clockwise on the inner curve C_2.

We divide R into two regions R_1 and R_2 by using two crosscuts, as shown in Fig. 15.4.7. Applying Green's theorem to each of these subregions, and noting cancellation of line integrals in opposite directions along the crosscuts, we get

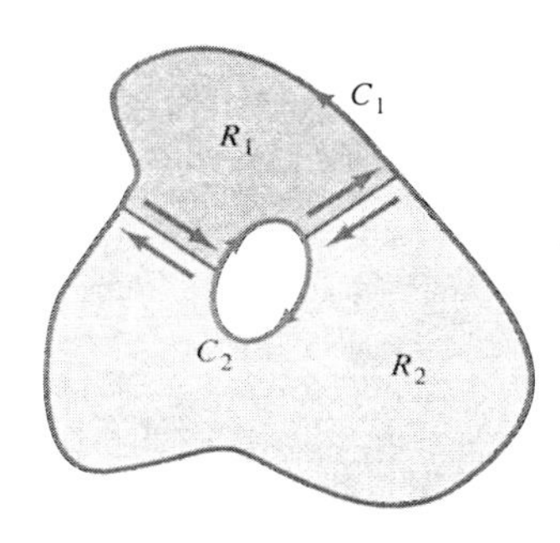

$$\iint_R \left(\frac{\partial Q}{\partial x} - \frac{\partial P}{\partial y}\right) dA = \iint_{R_1} \left(\frac{\partial Q}{\partial x} - \frac{\partial P}{\partial y}\right) dA + \iint_{R_2} \left(\frac{\partial Q}{\partial x} - \frac{\partial P}{\partial y}\right) dA$$

$$= \oint_{C_1} (P\,dx + Q\,dy) + \oint_{C_2} (P\,dx + Q\,dy)$$

$$= \oint_C P\,dx + Q\,dy.$$

FIGURE 15.4.7 Two crosscuts convert the annular region into the union of two ordinary regions R_1 and R_2, each bounded by a single closed curve.

Thus we obtain Green's theorem for the given region R. What makes this proof work is that the opposite line integrals along the two crosscuts cancel each other. You may, of course, use any finite number of crosscuts.

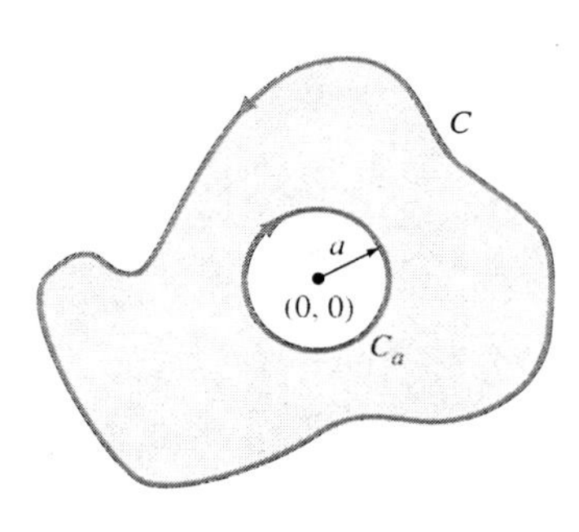

EXAMPLE 4 Suppose that C is a positively oriented piecewise-smooth simple closed curve that encloses the origin (0, 0). Show that

$$\oint_C \frac{-y\,dx + x\,dy}{x^2 + y^2} = 2\pi,$$

and also show that this integral is zero if C does *not* enclose the origin.

Solution With $P(x, y) = -y/(x^2 + y^2)$ and $Q(x, y) = x/(x^2 + y^2)$, a brief computation gives $\partial Q/\partial x - \partial P/\partial y \equiv 0$ when x and y are not both zero. If the region R bounded by C does not contain the origin, then P and Q and their derivatives are continuous on R. Hence Green's theorem implies that the integral in question is zero.

FIGURE 15.4.8 Use the small circle C_a if C encloses the origin (Example 4).

If C does enclose the origin, then we enclose the origin in a small circle C_a of radius a so small that C_a lies wholly within C (Fig. 15.4.8). We parametrize this circle by $x = a\cos t$, $y = a\sin t$, $0 \leqq t \leqq 2\pi$. Then Green's theorem, applied to the region R between C and C_a, gives

$$\oint_C \frac{-y\,dx + x\,dy}{x^2 + y^2} - \oint_{C_a} \frac{-y\,dx + x\,dy}{x^2 + y^2} = \iint_R 0\,dA = 0.$$

IMPORTANT The reason for the minus sign on the left-hand side is the fact that the positive orientation of C_a with respect to the region R is clockwise (as indicated in Fig. 15.4.8), whereas the selected parametrization determines a counterclockwise orientation of C_a. Therefore,

$$\oint_C \frac{-y\,dx + x\,dy}{x^2 + y^2} = \oint_{C_a} \frac{-y\,dx + x\,dy}{x^2 + y^2}$$

$$= \int_0^{2\pi} \frac{(-a \sin t)(-a \sin t\,dt) + (a \cos t)(a \cos t\,dt)}{(a \cos t)^2 + (a \sin t)^2}$$

$$= \int_0^{2\pi} 1\,dt = 2\pi.$$

◆

REMARK The result of Example 4 can be interpreted in terms of the polar-coordinate angle $\theta = \arctan(y/x)$. Because

$$d\theta = \frac{-y\,dx + x\,dy}{x^2 + y^2},$$

the line integral of Example 4 measures the net change in θ as we go around the curve C once in a counterclockwise direction. This net change is 2π if C encloses the origin and is zero otherwise.

PROOF OF THEOREM 3 IN SECTION 15.3 We now sketch a proof that if the vector field $\mathbf{F} = P\mathbf{i} + Q\mathbf{j}$ is continuously differentiable in an open rectangle R in the xy-plane, then $\mathbf{F}$ is conservative in R if and only if $\partial P/\partial y = \partial Q/\partial x$ at each point of R.

First suppose that $\mathbf{F}$ is conservative in R. Then there exists a function $f(x, y)$ defined on R such that

$$\nabla f = \frac{\partial f}{\partial x}\mathbf{i} + \frac{\partial f}{\partial y}\mathbf{j} = P\mathbf{i} + Q\mathbf{j} = \mathbf{F}$$

at each point of R. Then $P = \partial f/\partial x$ and $Q = \partial f/\partial y$, so it follows that

$$\frac{\partial P}{\partial y} = \frac{\partial}{\partial y}\left(\frac{\partial f}{\partial x}\right) = \frac{\partial^2 f}{\partial y\,\partial x} = \frac{\partial^2 f}{\partial x\,\partial y} = \frac{\partial}{\partial x}\left(\frac{\partial f}{\partial y}\right) = \frac{\partial Q}{\partial x},$$

as desired. The equality in the middle—where the order of partial differentiation is reversed—follows from the equality of mixed second-order partial derivatives of a function whose second-order partial derivatives are continuous.

Next, for the converse, suppose that $\partial P/\partial y = \partial Q/\partial x$ at each point of R. If we can show that line integrals of $\mathbf{F}$ are independent of the path in R, then it will follow from Theorem 2 in Section 15.3 that $\mathbf{F}$ is conservative in R, as desired. So let C_1 and C_2 be two smooth paths in R with the same initial point A and the same terminal point B, and let $C = C_1 \cup (-C_2)$ be the closed path that first follows C_1 from A to B and then follows the reversed path $-C_2$ from B back to A. If C is a *simple* closed path (one with no crossing points) and D is the region bounded by C, then Green's theorem gives

$$\oint_C \mathbf{F}\cdot\mathbf{T}\,ds = \oint_C P\,dx + Q\,dy = \iint_D \left(\frac{\partial Q}{\partial x} - \frac{\partial P}{\partial y}\right) dA = 0$$

because of the hypothesis that $\partial P/\partial y = \partial Q/\partial x$. But then

$$\oint_C \mathbf{F}\cdot\mathbf{T}\,ds = \int_{C_1} \mathbf{F}\cdot\mathbf{T}\,ds + \int_{-C_2} \mathbf{F}\cdot\mathbf{T}\,ds = \int_{C_1} \mathbf{F}\cdot T\,ds - \int_{C_2} \mathbf{F}\cdot\mathbf{T}\,ds = 0.$$

Thus we may conclude that

$$\int_{C_1} \mathbf{F}\cdot\mathbf{T}\,ds = \int_{C_2} \mathbf{F}\cdot\mathbf{T}\,ds$$

for two such smooth paths C_1 and C_2 that together form a simple closed path. The more general case in which C has finitely many self-intersections can be treated by decomposing C into finitely many simple closed paths. Then the line integral around each of these simple closed paths is zero, so it follows by addition that

$$\oint_C \mathbf{F} \cdot \mathbf{T}\, ds = 0,$$

as needed to conclude that line integrals of **F** are independent of the path, and hence that **F** is conservative. A complete proof would require discussion of the possibility that C has infinitely many self-intersections. ◄

The Divergence and Flux of a Vector Field

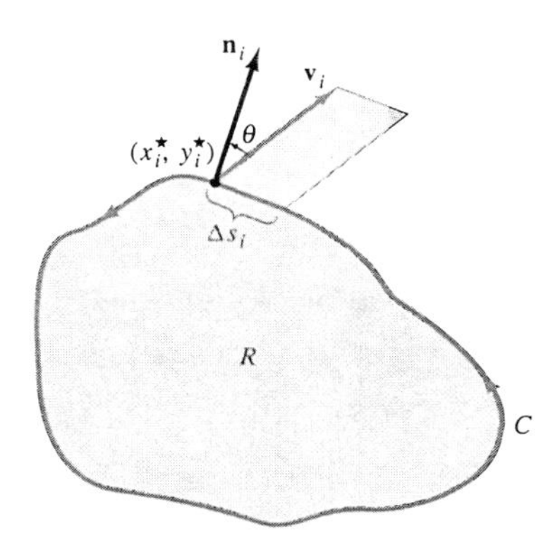

FIGURE 15.4.9 The area of the parallelogram approximates the fluid flow across Δs_i in unit time.

Now let us consider the steady flow of a thin layer of fluid in the plane (perhaps like a sheet of water spreading across a floor). Let $\mathbf{v}(x, y)$ be its velocity vector field and $\delta(x, y)$ be the density of the fluid at the point (x, y). The term *steady flow* means that $\mathbf{v}$ and δ depend only on x and y and *not* on time t. We want to compute the rate at which the fluid flows out of the region R bounded by a positively oriented simple closed curve C (Fig. 15.4.9). We seek the net rate of outflow—the actual outflow minus the inflow.

Let Δs_i be a short segment of the curve C, and let $(x_i^\star, y_i^\star)$ be an endpoint of Δs_i. Then the area of the portion of the fluid that flows out of R across Δs_i per unit time is approximately the area of the parallelogram in Fig. 15.4.9. This is the parallelogram spanned by the segment Δs_i and the vector $\mathbf{v}_i = \mathbf{v}(x_i^\star, y_i^\star)$. Suppose that $\mathbf{n}_i$ is the unit normal vector to C at the point $(x_i^\star, y_i^\star)$, the normal that points *out* of R. Then the area of this parallelogram is

$$(|\mathbf{v}_i| \cos\theta)\ \Delta s_i = \mathbf{v}_i \cdot \mathbf{n}_i\ \Delta s_i,$$

where θ is the angle between $\mathbf{n}_i$ and $\mathbf{v}_i$.

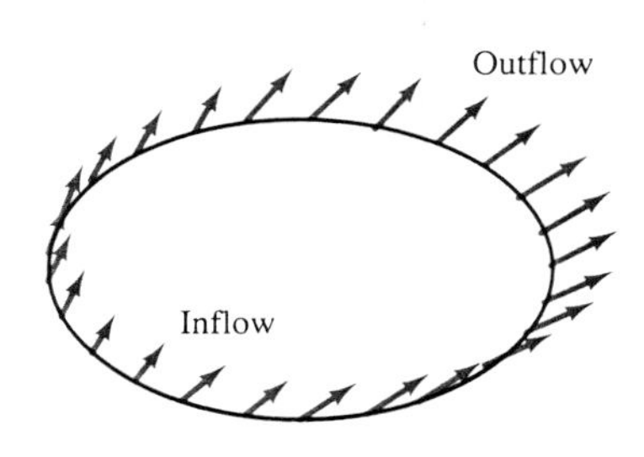

FIGURE 15.4.10 The flux Φ of the vector field **F** across the curve C is the net outflow minus the net inflow.

We multiply this area by the density $\delta_i = \delta(x_i^\star, y_i^\star)$ and then add these terms over those values of i that correspond to a subdivision of the entire curve C. This gives the (net) total mass of fluid leaving R per unit of time; it is approximately

$$\sum_{i=1}^{n} \delta_i \mathbf{v}_i \cdot \mathbf{n}_i\ \Delta s_i = \sum_{i=1}^{n} \mathbf{F}_i \cdot \mathbf{n}_i\ \Delta s_i,$$

where $\mathbf{F} = \delta\mathbf{v}$. The line integral around C that this sum approximates is called the **flux of the vector field F across the curve** C. Thus the flux Φ of **F** across C is given by

$$\Phi = \oint_C \mathbf{F} \cdot \mathbf{n}\, ds, \tag{5}$$

where **n** is the *outer* unit normal vector to C (Fig. 15.4.10).

In the present case of fluid flow with velocity vector **v**, the flux Φ of $\mathbf{F} = \delta\mathbf{v}$ is the rate at which the fluid is flowing out of R across the boundary curve C, in units of mass per unit of time. But the same terminology is used for an arbitrary vector field $\mathbf{F} = M\mathbf{i} + N\mathbf{j}$. For example, we may speak of the flux of an electric or gravitational field across a curve C.

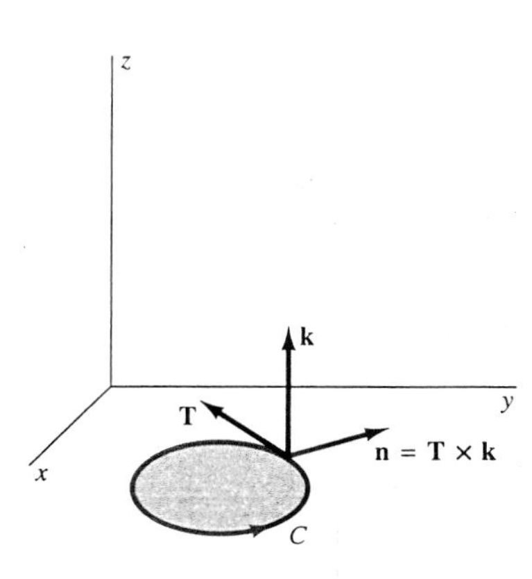

FIGURE 15.4.11 Computing the outer unit normal vector **n** from the unit tangent vector **T**.

From Fig. 15.4.11 we see that the outer unit normal vector **n** is equal to $\mathbf{T} \times \mathbf{k}$. The unit tangent vector **T** to the curve C is

$$\mathbf{T} = \frac{1}{v}\left(\mathbf{i}\frac{dx}{dt} + \mathbf{j}\frac{dy}{dt}\right) = \mathbf{i}\frac{dx}{ds} + \mathbf{j}\frac{dy}{ds}$$

because $v = ds/dt$. Hence

$$\mathbf{n} = \mathbf{T} \times \mathbf{k} = \left(\mathbf{i}\frac{dx}{ds} + \mathbf{j}\frac{dy}{ds}\right) \times \mathbf{k}.$$

But $\mathbf{i} \times \mathbf{k} = -\mathbf{j}$ and $\mathbf{j} \times \mathbf{k} = \mathbf{i}$. Thus we find that

$$\mathbf{n} = \mathbf{i}\frac{dy}{ds} - \mathbf{j}\frac{dx}{ds}. \tag{6}$$

Substituting the expression in Eq. (6) into the flux integral of Eq. (5) gives

$$\oint_C \mathbf{F} \cdot \mathbf{n}\, ds = \oint_C (M\mathbf{i} + N\mathbf{j}) \cdot \left(\mathbf{i}\frac{dy}{ds} - \mathbf{j}\frac{dx}{ds}\right) ds = \oint_C -N\, dx + M\, dy.$$

Applying Green's theorem to the last line integral with $P = -N$ and $Q = M$, we get

$$\oint_C \mathbf{F} \cdot \mathbf{n}\, ds = \iint_R \left(\frac{\partial M}{\partial x} + \frac{\partial N}{\partial y}\right) dA \tag{7}$$

for the flux of $\mathbf{F} = M\mathbf{i} + N\mathbf{j}$ across C.

The scalar function $\partial M/\partial x + \partial N/\partial y$ that appears in Eq. (7) is the **divergence** of the two-dimensional vector field $\mathbf{F} = M\mathbf{i} + N\mathbf{j}$ as defined in Section 15.1 and denoted by

$$\operatorname{div} \mathbf{F} = \nabla \cdot \mathbf{F} = \frac{\partial M}{\partial x} + \frac{\partial N}{\partial y}. \tag{8}$$

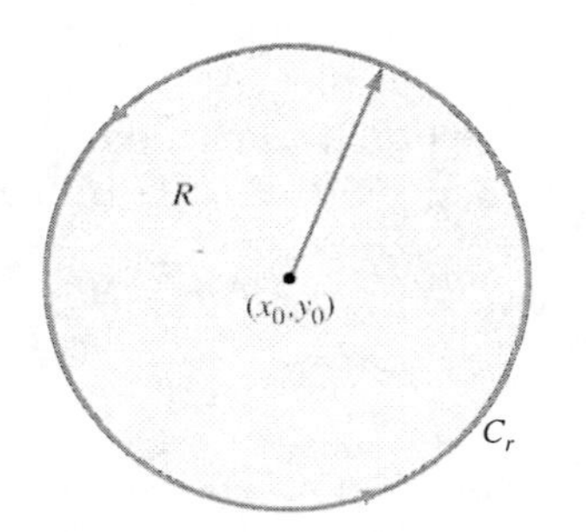

FIGURE 15.4.12 The circular disk R of radius r centered at (x_0, y_0).

When we substitute Eq. (8) into Eq. (7), we obtain a **vector form of Green's theorem:**

$$\oint_C \mathbf{F} \cdot \mathbf{n}\, ds = \iint_R \nabla \cdot \mathbf{F}\, dA, \tag{9}$$

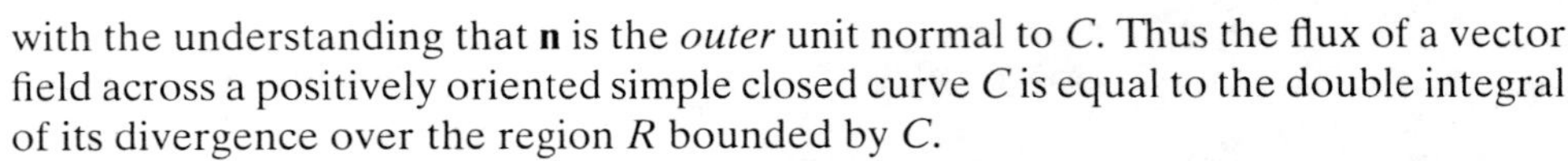
with the understanding that $\mathbf{n}$ is the *outer* unit normal to C. Thus the flux of a vector field across a positively oriented simple closed curve C is equal to the double integral of its divergence over the region R bounded by C.

If the disk R with area $a(R)$ is bounded by a positively oriented circle C with radius r and center (x_0, y_0) (Fig. 15.4.12), then the average value property of double integrals (see Problem 50 in Section 14.2) gives

$$\oint_{C_r} \mathbf{F} \cdot \mathbf{n}\, ds = \iint_R \nabla \cdot \mathbf{F}\, dA = [\nabla \cdot \mathbf{F}(\bar{x}, \bar{y})] \cdot a(R)$$

for some point $(\bar{x}, \bar{y})$ in R. We assume that $\mathbf{F}$ is continuously differentiable, so it follows that

$$\nabla \cdot \mathbf{F}(\bar{x}, \bar{y}) \to \nabla \cdot \mathbf{F}(x_0, y_0) \quad \text{as} \quad (\bar{x}, \bar{y}) \to (x_0, y_0).$$

If we first divide both sides by $a(R) = \pi r^2$ and then take the limit as $r \to 0$, we see that

$$\nabla \cdot \mathbf{F}(x_0, y_0) = \lim_{r \to 0} \frac{1}{\pi r^2} \oint_{C_r} \mathbf{F} \cdot \mathbf{n}\, ds \tag{10}$$

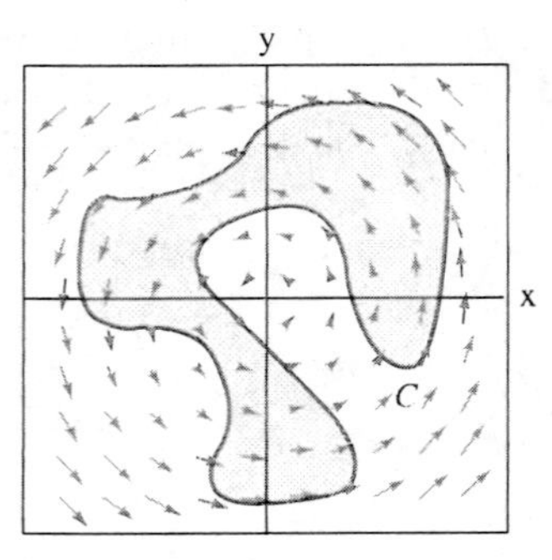

FIGURE 15.4.13 The flux $\oint \mathbf{F} \cdot \mathbf{n}\, ds$ of the vector field $\mathbf{F} = -y\mathbf{i} + x\mathbf{j}$ across the curve C is zero.

because $(\bar{x}, \bar{y}) \to (x_0, y_0)$ as $r \to 0$.

In the case of our original fluid flow, with $\mathbf{F} = \delta\mathbf{v}$, Eq. (10) implies that the value of $\nabla \cdot \mathbf{F}$ at (x_0, y_0) is a measure of the rate at which the fluid is "diverging away" from the point (x_0, y_0).

EXAMPLE 5 The vector field $\mathbf{F} = -y\mathbf{i} + x\mathbf{j}$ is the velocity field of a steady-state counterclockwise rotation around the origin. Show that the flux of $\mathbf{F}$ across any simple closed curve C is zero (Fig. 15.4.13).

Solution This follows immediately from Eq. (9) because

$$\nabla \cdot \mathbf{F} = \frac{\partial}{\partial x}(-y) + \frac{\partial}{\partial y}(x) = 0.$$

◆

15.4 TRUE/FALSE STUDY GUIDE

15.4 CONCEPTS: QUESTIONS AND DISCUSSION

1. Sketch several different examples of bounded plane regions with multiple boundary curves. In each case indicate how to introduce crosscuts to divide the given region R into a union of simpler regions $R_1, R_2, \ldots, R_k$, each of which is bounded by a single simple closed curve, so that

$$\iint_R f(x, y)\, dA = \sum_{i=1}^{k} \left(\iint_{R_i} f(x, y)\, dA \right).$$

2. Sketch several examples of smooth closed paths in the plane that have self-intersections. In each case indicate how to decompose the given self-intersecting path C into simple closed paths $C_1, C_2, \ldots, C_k$ so that

$$\int_C \mathbf{F} \cdot \mathbf{T}\, ds = \sum_{i=1}^{k} \left(\int_{C_i} \mathbf{F} \cdot \mathbf{T}\, ds \right).$$

15.4 PROBLEMS

In Problems 1 through 12, apply Green's theorem to evaluate the integral

$$\oint_C P\, dx + Q\, dy$$

around the specified positively oriented closed curve C.

1. $P(x, y) = x + y^2$, $Q(x, y) = y + x^2$; C is the square with vertices $(\pm 1, \pm 1)$.
2. $P(x, y) = x^2 + y^2$, $Q(x, y) = -2xy$; C is the boundary of the triangle bounded by the lines $x = 0$, $y = 0$, and $x + y = 1$.
3. $P(x, y) = y + e^x$, $Q(x, y) = 2x^2 + \cos y$; C is the boundary of the triangle with vertices $(0, 0)$, $(1, 1)$, and $(2, 0)$.
4. $P(x, y) = x^2 - y^2$, $Q(x, y) = xy$; C is the boundary of the region bounded by the line $y = x$ and the parabola $y = x^2$.
5. $P(x, y) = -y^2 + \exp(e^x)$, $Q(x, y) = \arctan y$; C is the boundary of the region between the parabolas $y = x^2$ and $x = y^2$.
6. $P(x, y) = y^2$, $Q(x, y) = 2x - 3y$; C is the circle $x^2 + y^2 = 9$.
7. $P(x, y) = x - y$, $Q(x, y) = y$; C is the boundary of the region between the x-axis and the graph of $y = \sin x$ for $0 \leqq x \leqq \pi$.
8. $P(x, y) = e^x \sin y$, $Q(x, y) = e^x \cos y$; C is the right-hand loop of the graph of the polar equation $r^2 = 4\cos\theta$.
9. $P(x, y) = y^2$, $Q(x, y) = xy$; C is the ellipse with equation $x^2/9 + y^2/4 = 1$.
10. $P(x, y) = y/(1 + x^2)$, $Q(x, y) = \arctan x$; C is the oval with equation $x^4 + y^4 = 1$.
11. $P(x, y) = xy$, $Q(x, y) = x^2$; C is the first-quadrant loop of the graph of the polar equation $r = \sin 2\theta$.
12. $P(x, y) = x^2$, $Q(x, y) = -y^2$; C is the cardioid with polar equation $r = 1 + \cos\theta$.

In Problems 13 through 16, use the corollary to Green's theorem to find the area of the indicated region.

13. The circle bounded by $x = a\cos t$, $y = a\sin t$, $0 \leqq t \leqq 2\pi$
14. The region between the x-axis and one arch of the cycloid with parametric equations $x = a(t - \sin t)$, $y = a(1 - \cos t)$
15. The region bounded by the astroid with parametric equations $x = \cos^3 t$, $y = \sin^3 t$, $0 \leqq t \leqq 2\pi$
16. The region between the graphs of $y = x^2$ and $y = x^3$

In Problems 17 through 20, use Green's theorem to calculate the work

$$W = \oint_C \mathbf{F} \cdot \mathbf{T}\, ds$$

done by the given force field **F** *in moving a particle counterclockwise once around the indicated curve C.*

17. $\mathbf{F} = -2y\mathbf{i} + 3x\mathbf{j}$ and C is the ellipse $x^2/9 + y^2/4 = 1$.
18. $\mathbf{F} = (y^2 - x^2)\mathbf{i} + 2xy\mathbf{j}$ and C is the circle $x^2 + y^2 = 9$.
19. $\mathbf{F} = 5x^2y^3\mathbf{i} + 7x^3y^2\mathbf{j}$ and C is the triangle with vertices $(0, 0)$, $(3, 0)$, and $(0, 6)$.
20. $\mathbf{F} = xy^2\mathbf{i} + 3x^2y\mathbf{j}$ and C is the boundary of the semicircular disk bounded by the x-axis and the circular arc $y = \sqrt{4 - x^2} \geqq 0$.

In Problems 21 through 24, use Green's theorem in the vector form in Eq. (9) to calculate the outward flux

$$\Phi = \oint_C \mathbf{F} \cdot \mathbf{n}\, ds$$

of the given vector field across the indicated positively oriented closed curve C.

21. $\mathbf{F} = 2x\mathbf{i} + 3y\mathbf{j}$ and C is the ellipse of Problem 17.
22. $\mathbf{F} = x^3\mathbf{i} + y^3\mathbf{j}$ and C is the circle of Problem 18.

23. $\mathbf{F} = (3x + \sqrt{1+y^2})\mathbf{i} + (2y - \sqrt[3]{1+x^4})\mathbf{j}$ and C is the triangle of Problem 19.

24. $\mathbf{F} = (3xy^2 + 4x)\mathbf{i} + (3x^2y - 4y)\mathbf{j}$ and C is the closed curve of Problem 20.

25. Suppose that f is a twice differentiable scalar function of x and y. Show that

$$\nabla^2 f = \operatorname{div}(\nabla f) = \frac{\partial^2 f}{\partial x^2} + \frac{\partial^2 f}{\partial y^2}.$$

26. Show that $f(x, y) = \ln(x^2 + y^2)$ satisfies **Laplace's equation** $\nabla^2 f = 0$ except at the point $(0, 0)$.

27. Suppose that f and g are twice-differentiable functions. Show that

$$\nabla^2(fg) = f\,\nabla^2 g + g\,\nabla^2 f + 2\,\nabla f \cdot \nabla g.$$

28. Suppose that the function $f(x, y)$ is twice continuously differentiable in the region R bounded by the positively oriented piecewise-smooth curve C. Prove that

$$\oint_C \frac{\partial f}{\partial x}\,dy - \frac{\partial f}{\partial y}\,dx = \iint_R \nabla^2 f\,dx\,dy.$$

29. Let R be the plane region with area A enclosed by the positively oriented piecewise-smooth simple closed curve C. Use Green's theorem to show that the coordinates of the centroid of R are

$$\bar{x} = \frac{1}{2A}\oint_C x^2\,dy, \qquad \bar{y} = -\frac{1}{2A}\oint_C y^2\,dx.$$

30. Use the result of Problem 29 to find the centroid of (a) a semicircular region of radius a; (b) a quarter-circular region of radius a.

31. Suppose that a lamina shaped like the region of Problem 29 has constant density δ. Show that its moments of inertia around the coordinate axes are

$$I_x = -\frac{\delta}{3}\oint_C y^3\,dx, \qquad I_y = \frac{\delta}{3}\oint_C x^3\,dy.$$

32. Use the result of Problem 31 to show that the polar moment of inertia $I_0 = I_x + I_y$ of a circular lamina of radius a, centered at the origin and of constant density δ, is $\frac{1}{2}Ma^2$, where M is the mass of the lamina.

33. The loop of the folium of Descartes (with equation $x^3 + y^3 = 3xy$) appears in Fig. 15.4.14. Apply the corollary to Green's theorem (Eq. (4)) to find the area of this loop. [*Suggestion:* Set $y = tx$ to discover a parametrization of the loop. To obtain the area of the loop, use values of t that lie in the interval $[0, 1]$. This gives the half of the loop that lies below the line $y = x$.]

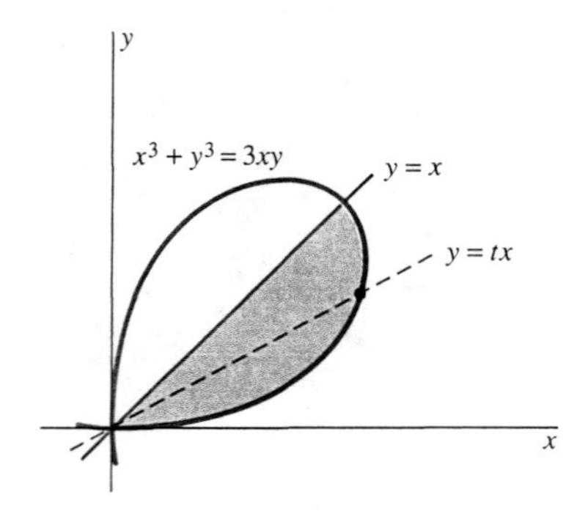

FIGURE 15.4.14 The loop of Problem 33.

34. Find the area bounded by one loop of the curve $x = \sin 2t$, $y = \sin t$.

35. Let f and g be functions with continuous second-order partial derivatives in the region R bounded by the positively oriented piecewise-smooth simple closed curve C. Apply Green's theorem in vector form to show that

$$\oint_C f\,\nabla g \cdot \mathbf{n}\,ds = \iint_R [f(\nabla \cdot \nabla g) + (\nabla f) \cdot (\nabla g)]\,dA.$$

It was this formula rather than Green's theorem itself that appeared in Green's book of 1828.

36. Complete the proof of the simple case of Green's theorem by showing directly that

$$\oint_C Q\,dy = \iint_R \frac{\partial Q}{\partial x}\,dA$$

if the region R is horizontally simple.

37. Suppose that the bounded plane region R is divided into the nonoverlapping subregions $R_1, R_2, \ldots, R_k$. If Green's theorem, Eq. (1), holds for each of these subregions, explain why it follows that Green's theorem holds for R. State carefully any assumptions that you need to make.

38. (a) If C is the line segment from (x_1, y_1) to (x_2, y_2), show by direct evaluation of the line integral that

$$\int_C x\,dy - y\,dx = x_1 y_2 - x_2 y_1.$$

(b) Let $(0, 0)$, (x_1, y_1), and (x_2, y_2) be the vertices of a triangle taken in counterclockwise order. Deduce from part (a) and Green's theorem that the area of this triangle is $A = \frac{1}{2}(x_1 y_2 - x_2 y_1)$.

39. Use the result of Problem 38 to find the area of (a) the equilateral triangle with vertices $(1, 0)$, $(\cos\frac{2}{3}\pi, \sin\frac{2}{3}\pi)$, and $(\cos\frac{4}{3}\pi, \sin\frac{4}{3}\pi)$; (b) the regular pentagon with vertices $(1, 0)$, $(\cos\frac{2}{5}\pi, \sin\frac{2}{5}\pi)$, $(\cos\frac{4}{5}\pi, \sin\frac{4}{5}\pi)$, $(\cos\frac{6}{5}\pi, \sin\frac{6}{5}\pi)$, and $(\cos\frac{8}{5}\pi, \sin\frac{8}{5}\pi)$.

40. Let T be a one-to-one transformation from the region S (with boundary curve J) in the uv-plane to the region R (with boundary curve C) in the xy-plane. Then the change-of-variables formula in Section 14.9 implies that the area A of the region R is given by

$$\iint_R dx\,dy = \iint_S \left|\frac{\partial(x, y)}{\partial(u, v)}\right| du\,dv. \qquad \textbf{(11)}$$

Establish this formula by carrying out the following steps. (a) Use Eq. (4) to convert the left-hand side in Eq. (11) to a line integral around C. (b) Use the coordinate functions $x(u, v)$ and $y(u, v)$ of the transformation T to convert the line integral in part (a) to a line integral around J. (c) Apply Green's theorem to the line integral in part (b).

41. Figure 15.4.15 shows the first-quadrant loop of the generalized folium of Descartes defined implicitly by the equation

$$x^{2n+1} + y^{2n+1} = (2n+1)x^n y^n \tag{12}$$

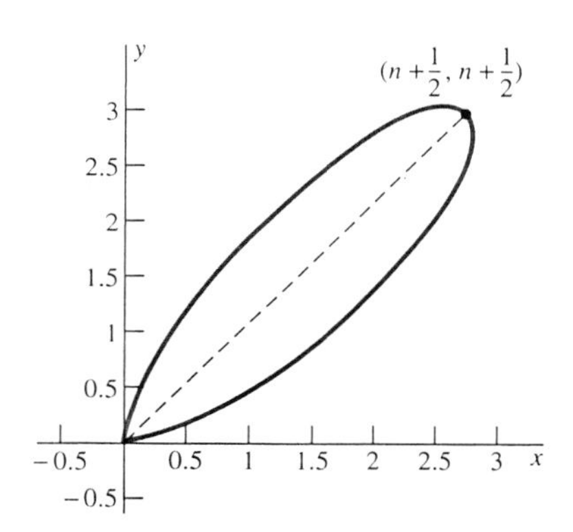

FIGURE 15.4.15 The first-quadrant loop of the generalized folium.

(where n is a positive integer). Your task here is to calculate the area A_n of the region bounded by this loop. Begin by substituting $y = tx$ to discover the parametrization

$$x = \frac{(2n+1)t^n}{t^{2n+1}+1}, \quad y = \frac{(2n+1)t^{n+1}}{t^{2n+1}+1} \quad (0 \leqq t < \infty) \tag{13}$$

of the loop. A computer algebra system may be useful in showing that with this parametrization the area formula in Eq. (4) of this section yields

$$A_n = (2n+1)^2 \int_0^{\infty} \left[\frac{(2n+1)t^{2n}}{(t^{2n+1}+1)^3} - \frac{nt^{2n}}{(t^{2n+1}+1)^2} \right] dt. \tag{14}$$

You can now calculate A_n for $n = 1, 2, 3, \dots$; you should find that $A_n = n + \frac{1}{2}$. (Do you need a computer algebra system for this?) But the improper integral in Eq. (14) should give you pause. Check your result by calculating (and then doubling) the area of the lower half of the loop indicated in Fig. 15.4.15—this involves only the integral from $t = 0$ to $t = 1$. (Why?)

15.5 SURFACE INTEGRALS

A *surface integral* is to surfaces in space what a line (or "curve") integral is to curves in the plane. Consider a curved, thin metal sheet shaped like the surface S. Suppose that this sheet has variable density, given at the point (x, y, z) by the known continuous function $f(x, y, z)$ in units such as grams per square centimeter of surface. We want to define the surface integral

$$\iint_S f(x, y, z)\, dS$$

in such a way that—upon evaluation—it gives the total mass of the thin metal sheet. In case $f(x, y, z) \equiv 1$, the numerical value of the integral should also equal the surface area of S.

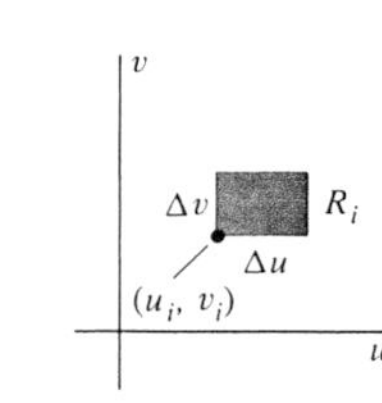

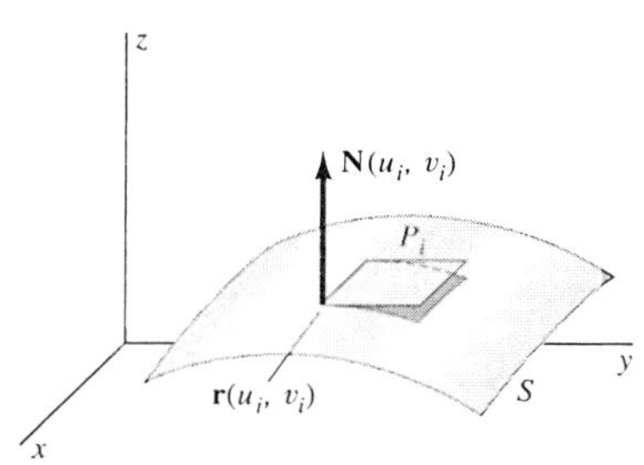

FIGURE 15.5.1 Approximating surface area with parallelograms.

As in Section 14.8, we assume that S is a smooth parametric surface described by the function or transformation

$$\mathbf{r}(u, v) = \langle x(u, v), y(u, v), z(u, v) \rangle = x\mathbf{i} + y\mathbf{j} + z\mathbf{k}$$

for (u, v) in a region D in the uv-plane. We suppose throughout that the component functions of $\mathbf{r}$ have continuous partial derivatives and also that the vectors $\mathbf{r}_u = \partial\mathbf{r}/\partial u$ and $\mathbf{r}_v = \partial\mathbf{r}/\partial v$ are nonzero and nonparallel at each interior point of D.

Recall how we computed the surface area A of S in Section 14.8. We began with an inner partition of D consisting of n rectangles $R_1, R_2, \dots, R_n$, each Δu by Δv in size. The images under $\mathbf{r}$ of the rectangles are curvilinear figures filling most or all of the surface S, and these pieces of S are themselves approximated by parallelograms P_i of the sort shown in Fig. 15.5.1. This gave us the approximation

$$A \approx \sum_{i=1}^{n} \Delta S_i = \sum_{i=1}^{n} |\mathbf{N}(u_i, v_i)|\, \Delta u\, \Delta v, \tag{1}$$

where the vector

$$\mathbf{N} = \frac{\partial \mathbf{r}}{\partial u} \times \frac{\partial \mathbf{r}}{\partial v} = \begin{vmatrix} \mathbf{i} & \mathbf{j} & \mathbf{k} \\ \dfrac{\partial x}{\partial u} & \dfrac{\partial y}{\partial u} & \dfrac{\partial z}{\partial u} \\ \dfrac{\partial x}{\partial v} & \dfrac{\partial y}{\partial v} & \dfrac{\partial z}{\partial v} \end{vmatrix} \tag{2}$$

is normal to S at the point $\mathbf{r}(u, v)$ and $\Delta S_i = |\mathbf{N}(u_i, v_i)|\, \Delta u\, \Delta v$ is the area of the parallelogram P_i that is tangent to the surface S at the point $\mathbf{r}(u_i, v_i)$.

If the surface S also has a density function $f(x, y, z)$, then we can approximate the total mass m of the surface by first multiplying each parallelogram area ΔS_i in Eq. (1) by the density $f(\mathbf{r}(u_i, v_i))$ at $\mathbf{r}(u_i, v_i)$ and then summing these estimates over all such parallelograms. Thus we obtain the approximation

$$m \approx \sum_{i=1}^{n} f(\mathbf{r}(u_i, v_i))\, \Delta S_i = \sum_{i=1}^{n} f(\mathbf{r}(u_i, v_i))\, |\mathbf{N}(u_i, v_i)|\, \Delta u\, \Delta v. \tag{3}$$

This approximation is a Riemann sum for the **surface integral of the function f over the surface S**, denoted by

$$\begin{aligned} \iint_S f(x, y, z)\, dS &= \iint_D f(\mathbf{r}(u, v))\, |\mathbf{N}(u, v)|\, du\, dv \\ &= \iint_D f(\mathbf{r}(u, v)) \left| \frac{\partial \mathbf{r}}{\partial u} \times \frac{\partial \mathbf{r}}{\partial v} \right| du\, dv. \end{aligned} \tag{4}$$

To evaluate the surface integral $\iint_S f(x, y, z)\, dS$, we simply use the parametrization $\mathbf{r}$ to express the variables x, y, and z in terms of u and v and formally replace the **surface area element dS** with

$$dS = |\mathbf{N}(u, v)|\, du\, dv = \left| \frac{\partial \mathbf{r}}{\partial u} \times \frac{\partial \mathbf{r}}{\partial v} \right| du\, dv. \tag{5}$$

Expanding the cross product determinant in Eq. (2) gives

$$\mathbf{N} = \frac{\partial \mathbf{r}}{\partial u} \times \frac{\partial \mathbf{r}}{\partial v} = \frac{\partial(y, z)}{\partial(u, v)}\mathbf{i} + \frac{\partial(z, x)}{\partial(u, v)}\mathbf{j} + \frac{\partial(x, y)}{\partial(u, v)}\mathbf{k} \tag{6}$$

in the Jacobian notation of Section 14.9, so the surface integral in Eq. (4) takes the form

$$\begin{aligned} &\iint_S f(x, y, z)\, dS \\ &= \iint_D f(x(u, v), y(u, v), z(u, v)) \sqrt{\left[\frac{\partial(y, z)}{\partial(u, v)}\right]^2 + \left[\frac{\partial(z, x)}{\partial(u, v)}\right]^2 + \left[\frac{\partial(x, y)}{\partial(u, v)}\right]^2}\, du\, dv. \end{aligned} \tag{7}$$

This formula converts the surface integral into an *ordinary double integral* over the region D in the uv-plane, and is analogous to the formula (Eq. (4) of Section 15.2)

$$\int_C f(x, y, z)\, ds = \int_a^b f(x(t), y(t), z(t)) \sqrt{\left(\frac{dx}{dt}\right)^2 + \left(\frac{dy}{dt}\right)^2 + \left(\frac{dz}{dt}\right)^2}\, dt$$

that converts a line integral into an ordinary single integral.

In the important special case of a surface S described as a graph $z = h(x, y)$ of a function h defined on a region D in the xy-plane, we may use x and y (rather than u and v) as the parameters. The surface area element then takes the form

$$dS = \sqrt{1 + \left(\frac{\partial h}{\partial x}\right)^2 + \left(\frac{\partial h}{\partial y}\right)^2}\, dx\, dy \tag{8}$$

(as in Eq. (9) of Section 14.8). The surface integral of f over S is then given by

$$\iint_S f(x, y, z)\, dS = \iint_D f(x, y, h(x, y)) \sqrt{1 + \left(\frac{\partial h}{\partial x}\right)^2 + \left(\frac{\partial h}{\partial y}\right)^2}\, dx\, dy. \tag{9}$$

Centroids and moments of inertia for surfaces are computed in much the same way as for curves (see Section 15.2), using surface integrals in place of line integrals. For example, suppose that the surface S has density $\delta(x, y, z)$ at the point (x, y, z) and total mass m. Then the z-component $\bar{z}$ of its centroid and its moment of inertia I_z around the z-axis are given by

$$\bar{z} = \frac{1}{m} \iint_S z\,\delta(x, y, z)\, dS \quad \text{and} \quad I_z = \iint_S (x^2 + y^2)\,\delta(x, y, z)\, dS.$$

EXAMPLE 1 Find the centroid of the unit-density hemispherical surface

$$z = \sqrt{a^2 - x^2 - y^2}, \quad x^2 + y^2 \leqq a^2.$$

Solution By symmetry, $\bar{x} = 0 = \bar{y}$. A simple computation gives $\partial z/\partial x = -x/z$ and $\partial z/\partial y = -y/z$, so Eq. (8) takes the form

$$\begin{aligned} dS &= \sqrt{1 + \left(\frac{\partial z}{\partial x}\right)^2 + \left(\frac{\partial z}{\partial y}\right)^2}\, dx\, dy = \sqrt{1 + \left(\frac{x}{z}\right)^2 + \left(\frac{y}{z}\right)^2}\, dx\, dy \\ &= \frac{1}{z}\sqrt{x^2 + y^2 + z^2}\, dx\, dy = \frac{a}{z}\, dx\, dy. \end{aligned}$$

Hence

$$\bar{z} = \frac{1}{2\pi a^2} \iint_D z \cdot \frac{a}{z}\, dx\, dy = \frac{1}{2\pi a} \iint_D 1\, dx\, dy = \frac{a}{2}.$$

Note in the final step that D is a circular disk of radius a in the xy-plane. This simplifies the computation of the last integral. ◆

EXAMPLE 2 Find the moment of inertia around the z-axis of the spherical surface $x^2 + y^2 + z^2 = a^2$, assuming that it has constant density $\delta = k$.

Solution The spherical surface of radius a is most easily parametrized in spherical coordinates:

$$x = a \sin\phi \cos\theta, \qquad y = a \sin\phi \sin\theta, \qquad z = a \cos\phi$$

for $0 \leqq \phi \leqq \pi$ and $0 \leqq \theta \leqq 2\pi$. Hence the sphere S is defined parametrically by

$$\mathbf{r}(\phi, \theta) = \mathbf{i}\, a \sin\phi \cos\theta + \mathbf{j}\, a \sin\phi \sin\theta + \mathbf{k}\, a \cos\phi.$$

As in Problem 18 of Section 14.8, the surface area element is then

$$dS = \left| \frac{\partial \mathbf{r}}{\partial \phi} \times \frac{\partial \mathbf{r}}{\partial \theta} \right| = a^2 \sin\phi\, d\phi\, d\theta.$$

Because

$$x^2 + y^2 = a^2 \sin^2\phi \cos^2\phi + a^2 \sin^2\phi \sin^2\phi = a^2 \sin^2\phi,$$

it follows that

$$\begin{aligned} I_z &= \iint_S (x^2 + y^2)\,\delta\, dS = \int_0^{2\pi} \int_0^{\pi} k(a^2 \sin^2\phi)\, a^2 \sin\phi\, d\phi\, d\theta \\ &= 2\pi \cdot ka^4 \cdot 2 \int_0^{\pi/2} \sin^3\phi\, d\phi = 4\pi k a^4 \cdot \frac{2}{3} \quad \text{(by integral formula 113)} \\ &= \frac{2}{3} \cdot 4\pi k a^2 \cdot a^2 = \frac{2}{3} m a^2, \end{aligned}$$

using in the final step the fact that the mass of the spherical surface with density k is $m = 4\pi k a^2$. Is this result both plausible and dimensionally correct? ◆

Surface Integrals With Respect to Coordinate Elements

The surface integral $\iint_S f(x, y, z)\, dS$ is an integral **respect to surface area,** and thus is analogous to the line integral $\int_C f(x, y)\, ds$ with respect to arc length. A second type of surface integral of the form

$$\iint_S P\, dy\, dz + Q\, dz\, dx + R\, dx\, dy$$

is analogous to the line integral $\int_C P\, dx + Q\, dy$ with respect to coordinate variables.

The definition of the integral $\iint_S R\, dx\, dy$, for instance—with $R(x, y, z)$ a scalar function (instead of f) and $dx\, dy$ an area element in the xy-plane (instead of the area element dS on the surface S)—is motivated by replacing the area element $\Delta S_i = |\mathbf{N}(u_i, v_i)|\, \Delta u\, \Delta v$ in the Riemann sum in Eq. (3) with the area $\Delta S_i \cos\gamma$ of its projection into the xy-plane (Fig. 15.5.2). The result is the Riemann sum

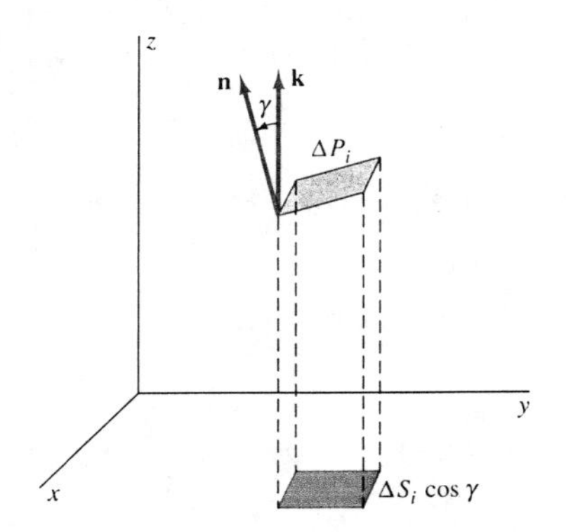

FIGURE 15.5.2 Finding the area of the projected parallelogram.

$$\sum_{i=1}^{n} R(\mathbf{r}(u_i, v_i)) \cos\gamma\, |\mathbf{N}(u_i, v_i)|\, \Delta u\, \Delta v \approx \iint_D R(\mathbf{r}(u, v)) \cos\gamma\, |\mathbf{N}(u, v)|\, du\, dv. \quad \textbf{(10)}$$

To calculate the factor $\cos\gamma$ in Eq. (10), we consider the unit normal vector

$$\mathbf{n} = \frac{\mathbf{N}}{|\mathbf{N}|} = \mathbf{i}\cos\alpha + \mathbf{j}\cos\beta + \mathbf{k}\cos\gamma \quad \textbf{(11)}$$

with direction cosines $\cos\alpha$, $\cos\beta$, and $\cos\gamma$. Using Eq. (6) we find that

$$\cos\alpha = \mathbf{n}\cdot\mathbf{i} = \frac{\mathbf{N}\cdot\mathbf{i}}{|\mathbf{N}|} = \frac{1}{|\mathbf{N}|}\frac{\partial(y, z)}{\partial(u, v)} \quad \text{and, similarly,}$$

$$\cos\beta = \frac{1}{|\mathbf{N}|}\frac{\partial(z, x)}{\partial(u, v)}, \qquad \cos\gamma = \frac{1}{|\mathbf{N}|}\frac{\partial(x, y)}{\partial(u, v)}. \quad \textbf{(12)}$$

Substitution for $\cos\gamma$ in (10) now yields the *definition*

$$\iint_S R(x, y, z)\, dx\, dy = \iint_S R(x, y, z) \cos\gamma\, dS = \iint_D R(\mathbf{r}(u, v)) \frac{\partial(x, y)}{\partial(u, v)}\, du\, dv. \quad \textbf{(13)}$$

Similarly, we *define*

$$\iint_S P(x, y, z)\, dy\, dz = \iint_S P(x, y, z) \cos\alpha\, dS = \iint_D P(\mathbf{r}(u, v)) \frac{\partial(y, z)}{\partial(u, v)}\, du\, dv \quad \textbf{(14)}$$

and

$$\iint_S Q(x, y, z)\, dz\, dx = \iint_S Q(x, y, z) \cos\beta\, dS = \iint_D Q(\mathbf{r}(u, v)) \frac{\partial(z, x)}{\partial(u, v)}\, du\, dv. \quad \textbf{(15)}$$

Note The symbols z and x appear in the reverse of alphabetical order in Eq. (15). It is important to write them in the correct order because

$$\frac{\partial(x, z)}{\partial(u, v)} = \begin{vmatrix} x_u & x_v \\ z_u & z_v \end{vmatrix} = -\begin{vmatrix} z_u & z_v \\ x_u & x_v \end{vmatrix} = -\frac{\partial(z, x)}{\partial(u, v)}.$$

This implies that

$$\iint_S f(x, y, z)\,dx\,dz = -\iint_S f(x, y, z)\,dz\,dx.$$

In an ordinary *double integral,* the order in which the differentials are written simply indicates the order of integration. But in a *surface integral,* it instead indicates the order of appearance of the corresponding variables in the Jacobians in Eqs. (13) through (15).

The three integrals in Eqs. (13) through (15) typically occur together, and the general **surface integral with respect to coordinate area elements** is the sum

$$\iint_S P\,dy\,dz + Q\,dz\,dx + R\,dx\,dy = \iint_S (P\cos\alpha + Q\cos\beta + R\cos\gamma)\,dS; \qquad \textbf{(16)}$$

that is,

$$\iint_S P\,dy\,dz + Q\,dz\,dx + R\,dx\,dy = \iint_D \left(P\frac{\partial(y, z)}{\partial(u, v)} + Q\frac{\partial(z, x)}{\partial(u, v)} + R\frac{\partial(x, y)}{\partial(u, v)}\right) du\,dv. \qquad \textbf{(17)}$$

Equation (17) gives the evaluation procedure for the surface integral in Eq. (16): Substitute for x, y, z, and their derivatives in terms of u and v, then integrate over the appropriate region D in the uv-plane.

The relation between surface integrals with respect to surface area and with respect to coordinate areas is somewhat analogous to the formula

$$\int_C \mathbf{F}\cdot\mathbf{T}\,ds = \int_C P\,dx + Q\,dy + R\,dz$$

relating line integrals with respect to arc length and with respect to coordinates. Given the vector field $\mathbf{F} = P\mathbf{i} + Q\mathbf{j} + R\mathbf{k}$, Eq. (11) implies that

$$\mathbf{F}\cdot\mathbf{n} = P\cos\alpha + Q\cos\beta + R\cos\gamma, \qquad \textbf{(18)}$$

so the equations in (12) yield

$$\iint_S \mathbf{F}\cdot\mathbf{n}\,dS = \iint_S P\,dy\,dz + Q\,dz\,dx + R\,dx\,dy. \qquad \textbf{(19)}$$

Only the *sign* of the right-hand surface integral in Eq. (19) depends on the parametrization of S. The unit normal vector on the left-hand side is the vector provided by the parametrization of S via the equations in (12). In the case of a surface given by $z = h(x, y)$, with x and y used as the parameters u and v, this will be the *upper* normal, as you will see in Example 3.

EXAMPLE 3 Suppose that S is the surface $z = h(x, y)$, (x, y) in D. Then show that

$$\iint_S P\,dy\,dz + Q\,dz\,dx + R\,dx\,dy = \iint_D \left(-P\frac{\partial z}{\partial x} - Q\frac{\partial z}{\partial y} + R\right) dx\,dy, \qquad \textbf{(20)}$$

where P, Q, and R in the second integral are evaluated at $(x, y, h(x, y))$.

Solution This is simply a matter of computing the three Jacobians in Eq. (17) with the parameters x and y. We note first that $\partial x/\partial x = 1 = \partial y/\partial y$ and that $\partial x/\partial y = 0 = \partial y/\partial x$. Hence

$$\frac{\partial(y, z)}{\partial(x, y)} = \begin{vmatrix} y_x & y_y \\ z_x & z_y \end{vmatrix} = -\frac{\partial z}{\partial x}, \qquad \frac{\partial(z, x)}{\partial(x, y)} = \begin{vmatrix} z_x & z_y \\ x_x & x_y \end{vmatrix} = -\frac{\partial z}{\partial y},$$

and

$$\frac{\partial(x, y)}{\partial(x, y)} = \begin{vmatrix} x_x & x_y \\ y_x & y_y \end{vmatrix} = 1.$$

Equation (20) is an immediate consequence. ◆

The Flux of a Vector Field

One of the most important applications of surface integrals involves the computation of the flux of a vector field. To define the flux of the vector field **F** across the surface S, we assume that S has a *unit* normal vector field **n** that varies *continuously* from point to point of S. This condition excludes from our consideration one-sided (*nonorientable*) surfaces, such as the Möbius strip of Fig. 15.5.3. If S is a two-sided (*orientable*) surface, then there are two possible choices for **n**. For example, if S is a closed surface (such as a torus or sphere) that separates space, then we may choose for **n** either the outer normal vector (at each point of S) or the inner normal vector (Fig. 15.5.4). The unit normal vector defined in Eq. (11) may be either the outer normal or the inner normal; which of the two it is depends on how S has been parametrized.

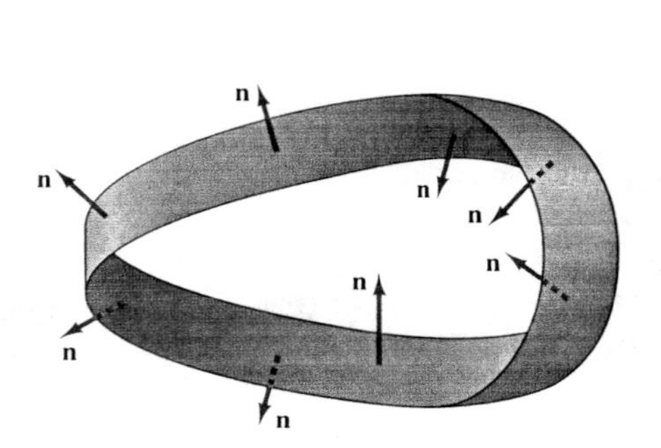

FIGURE 15.5.3 The Möbius strip is an example of a one-sided surface.

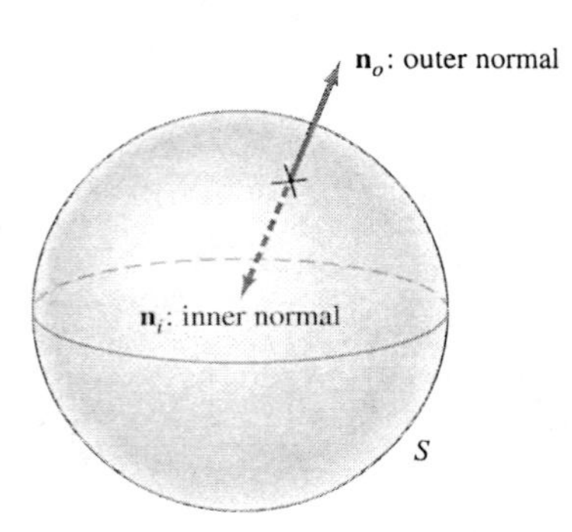

FIGURE 15.5.4 Inner and outer normal vectors to a two-sided closed surface.

To define the concept of flux, suppose that we are given the vector field **F**, the orientable surface S, and a continuous unit normal vector field **n** on S. Then, in analogy with Eq. (5) in Section 15.4, we define the **flux Φ across S in the direction of n** by

$$\Phi = \iint_S \mathbf{F} \cdot \mathbf{n} \, dS. \tag{21}$$

For example, if $\mathbf{F} = \delta\mathbf{v}$, where **v** is the velocity vector field corresponding to the steady flow in space of a fluid of density δ and **n** is the *outer* unit normal vector for a closed surface S that bounds the space region T, then the flux determined by Eq. (21) is the net rate of flow of the fluid *out of* T across its boundary surface S in units such as grams per second.

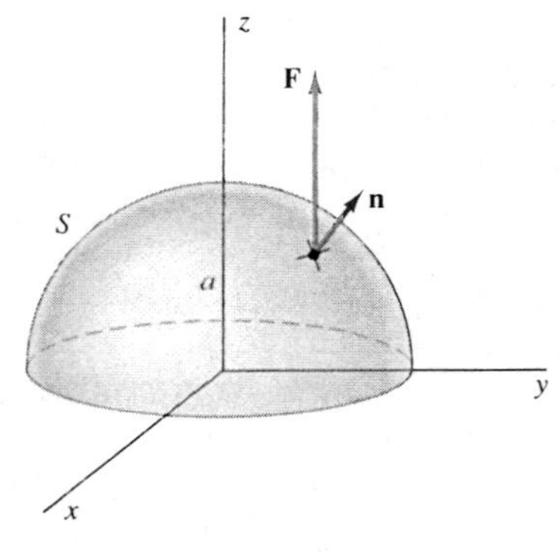

FIGURE 15.5.5 The hemisphere S of Example 4.

EXAMPLE 4 Calculate the flux $\iint_S \mathbf{F} \cdot \mathbf{n}\, dS$, where $\mathbf{F} = v_0\mathbf{k}$ and S is the hemispherical surface of radius a with equation $z = \sqrt{a^2 - x^2 - y^2}$ and with outer unit normal vector **n**. (See Fig. 15.5.5.)

Solution If we think of $\mathbf{F} = v_0\mathbf{k}$ as the velocity vector field of a fluid that is flowing upward with constant speed v_0, then we can interpret the flux in question as the rate of flow (in cubic centimeters per second, for example) of the fluid across S. To calculate this flux, we note that

$$\mathbf{n} = \frac{x\mathbf{i} + y\mathbf{j} + z\mathbf{k}}{\sqrt{x^2 + y^2 + z^2}} = \frac{1}{a}(x\mathbf{i} + y\mathbf{j} + z\mathbf{k}).$$

Hence

$$\mathbf{F}\cdot\mathbf{n} = v_0\mathbf{k}\cdot\frac{1}{a}(x\mathbf{i}+y\mathbf{j}+z\mathbf{k}) = \frac{v_0}{a}z,$$

so

$$\iint_S \mathbf{F}\cdot\mathbf{n}\,dS = \iint_S \frac{v_0}{a}z\,dS.$$

If we introduce spherical coordinates $z = a\cos\phi$, $dS = a^2\sin\phi\,d\phi\,d\theta$ on the hemispherical surface, we get

$$\begin{aligned}\iint_S \mathbf{F}\cdot\mathbf{n}\,dS &= \frac{v_0}{a}\int_0^{2\pi}\int_0^{\pi/2}(a\cos\phi)(a^2\sin\phi)\,d\phi\,d\theta\\ &= 2\pi a^2 v_0\int_0^{\pi/2}\cos\phi\sin\phi\,d\phi = 2\pi a^2 v_0\left[\frac{1}{2}\sin^2\phi\right]_0^{\pi/2};\end{aligned}$$

thus

$$\iint_S \mathbf{F}\cdot\mathbf{n}\,dS = \pi a^2 v_0.$$

This last quantity is equal to the flux of $\mathbf{F} = v_0\mathbf{k}$ across the disk $x^2+y^2 \leqq a^2$ of area πa^2. If we think of the hemispherical region T bounded by the hemisphere S and the circular disk D that forms its base, it should be no surprise that the rate of inflow of an incompressible fluid across the disk D is equal to its rate of outflow across the hemisphere S. ◆

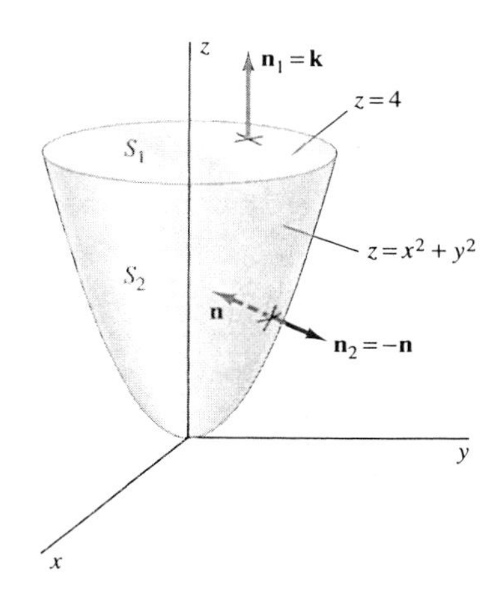

FIGURE 15.5.6 The surface of Example 5.

EXAMPLE 5 Find the flux of the vector field $\mathbf{F} = x\mathbf{i} + y\mathbf{j} + 3\mathbf{k}$ out of the region T bounded by the paraboloid $z = x^2 + y^2$ and the plane $z = 4$ (Fig. 15.5.6).

Solution Let S_1 denote the circular top, which has outer unit normal vector $\mathbf{n}_1 = \mathbf{k}$. Let S_2 be the parabolic part of this surface, with outer unit normal vector $\mathbf{n}_2$. The flux across S_1 is

$$\iint_{S_1}\mathbf{F}\cdot\mathbf{n}_1\,dS = \iint_{S_1} 3\,dS = 12\pi$$

because S_1 is a circular disk of radius 2.

Next, the computation in Example 3 gives

$$\mathbf{N} = \left\langle -\frac{\partial z}{\partial x}, -\frac{\partial z}{\partial y}, 1\right\rangle = \langle -2x, -2y, 1\rangle$$

for a vector normal to the paraboloid $z = x^2 + y^2$. Then $\mathbf{n} = \mathbf{N}/|\mathbf{N}|$ is an upper—and thus an *inner*—unit normal vector to the surface S_2. The unit *outer* normal vector is, therefore, $\mathbf{n}_2 = -\mathbf{n} = -\mathbf{N}/|\mathbf{N}|$, opposite to the direction of $\mathbf{N} = \langle -2x, -2y, 1\rangle$. With parameters (x, y) in the circular disk $x^2 + y^2 \leqq 4$ in the xy-plane, the surface-area element is $dS = |\mathbf{N}|\,dx\,dy$. Therefore, the outward flux across S_2 is

$$\begin{aligned}\iint_{S_2}\mathbf{F}\cdot\mathbf{n}_2\,dS &= -\iint_{S_2}\mathbf{F}\cdot\mathbf{n}\,dS = -\iint_D \mathbf{F}\cdot\frac{\mathbf{N}}{|\mathbf{N}|}|\mathbf{N}|\,dx\,dy\\ &= -\iint_D \mathbf{F}\cdot\mathbf{N}\,dx\,dy = -\iint_D \langle x, y, 3\rangle\cdot\langle -2x, -2y, 1\rangle\,dx\,dy\\ &= -\iint_D (3 - 2x^2 - 2y^2)\,dx\,dy.\end{aligned}$$

We change to polar coordinates in the disk D of radius 2—so that $3 - 2x^2 - 2y^2 = 3 - 2r^2$ and $dx\,dy = r\,dr\,d\theta$—and find that

$$\iint_{S_2} \mathbf{F} \cdot \mathbf{n}_2\, dS = \int_0^{2\pi} \int_0^2 (2r^2 - 3)\, r\, dr\, d\theta = 2\pi \left[\frac{1}{2}r^4 - \frac{3}{2}r^2\right]_0^2 = 4\pi.$$

Hence the total flux of $\mathbf{F}$ out of T is $16\pi \approx 50.27$. ◆

Another physical application of flux is to the flow of heat, which is mathematically quite similar to the flow of a fluid. Suppose that a body has temperature $u = u(x, y, z)$ at the point (x, y, z). Experiments indicate that the flow of heat in the body is described by the heat-flow vector

$$\mathbf{q} = -K\, \nabla u. \tag{22}$$

The number K—normally, but not always, a constant—is the *heat conductivity* of the body. The vector $\mathbf{q}$ points in the direction of heat flow, and its length is the rate of flow of heat across a unit area normal to $\mathbf{q}$. This flow rate is measured in units such as calories per second per square centimeter. If S is a closed surface within the body bounding the solid region T and $\mathbf{n}$ denotes the outer unit normal vector for S, then

$$\iint_S \mathbf{q} \cdot \mathbf{n}\, dS = -\iint_S K\, \nabla u \cdot \mathbf{n}\, dS \tag{23}$$

is the net rate of heat flow (in calories per second, for example) out of the region T across its boundary surface S.

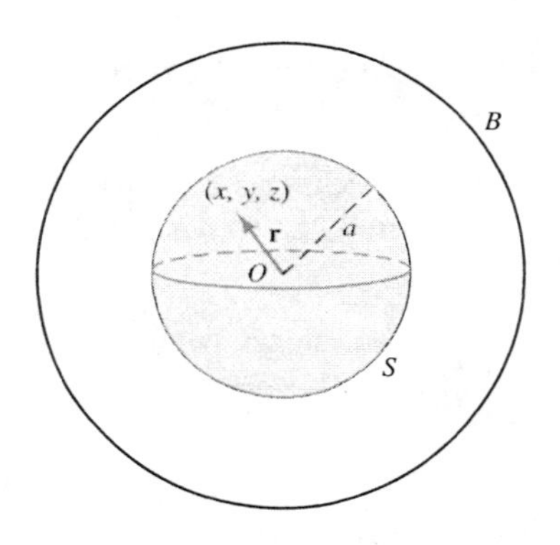

FIGURE 15.5.7 The solid ball B of Example 6.

EXAMPLE 6 Suppose that a uniform solid ball B of radius R is centered at the origin (Fig. 15.5.7) and that the temperature u within it is given by

$$u(x, y, z) = c(R^2 - x^2 - y^2 - z^2).$$

Thus the temperature of B is maximal at its center and is zero on its boundary. Find the rate of flow of heat across a sphere S of radius $a < R$ centered at the origin.

Solution Writing $\mathbf{r} = x\mathbf{i} + y\mathbf{j} + z\mathbf{k}$ for the position vector of a point (x, y, z) of B, we find that the heat flow vector $\mathbf{q}$ in Eq. (22) is

$$\mathbf{q} = -K\, \nabla u = -K \cdot c(-2x\mathbf{i} - 2y\mathbf{j} - 2z\mathbf{k}) = 2Kc\mathbf{r}.$$

Obviously the outer unit normal vector $\mathbf{n}$ at a point (x, y, z) of the sphere S is $\mathbf{n} = \mathbf{r}/a$, so

$$\mathbf{q} \cdot \mathbf{n} = 2Kc\mathbf{r} \cdot \frac{\mathbf{r}}{a} = 2Kca$$

because $\mathbf{r} \cdot \mathbf{r} = a^2$ at points of S. Therefore the heat flow across the sphere S (with area $A(S) = 4\pi a^2$) is

$$\iint_S \mathbf{q} \cdot \mathbf{n}\, dS = \iint_S 2Kca\, dS = 2Kca \cdot 4\pi a^2 = 8Kc\pi a^3.$$ ◆

Still other applications of flux involve force fields rather than flow fields. For example, suppose that $\mathbf{F}$ is the gravitational field of a collection of fixed masses in space, so $\mathbf{F}(\mathbf{r})$ is the net force exerted on a unit mass located at $\mathbf{r}$. Then **Gauss's law** for inverse-square gravitational fields says that the (outward) flux of $\mathbf{F}$ across the closed surface S is

$$\Phi = \iint_S \mathbf{F} \cdot \mathbf{n}\, dS = -4\pi GM \tag{24}$$

where M is the total mass enclosed by S and G is the universal gravitational constant.

Gauss's law also applies to inverse-square electric fields. The electric field at $\mathbf{r}$ of a charge q located at the origin is $\mathbf{E} = q\mathbf{r}/(4\pi\epsilon_0|\mathbf{r}|^3)$, where $\epsilon_0 \approx 8.85 \times 10^{-12}$ in mks units (charge in coulombs). Then **Gauss's law** for electric fields says that the (outward) flux of $\mathbf{E}$ across the closed surface S is

$$\Phi = \iint_S \mathbf{E}\cdot\mathbf{n}\,dS = \frac{Q}{\epsilon_0} \tag{25}$$

where Q is the net (positive minus negative) charge enclosed by S.

15.5 TRUE/FALSE STUDY GUIDE

15.5 CONCEPTS: QUESTIONS AND DISCUSSION

1. Explain why an integral of the form $\iint_S f\,dS$ is defined even if S is a surface such as the Möbius strip in Fig. 15.5.3—which has no continuous unit normal vector field defined on the entire surface. Can the Möbius strip be subdivided into parametrized surfaces that *do* have continuous unit normal vector fields?
2. Explain why the value of a surface integral with respect to surface area does not depend on the orientation—that is, the direction of the normal vector $\mathbf{N} = \mathbf{r}_u \times \mathbf{r}_v$ provided by the surface parametrization—whereas the value of a surface integral with respect to coordinate area elements *does* depend on the orientation of the surface.

15.5 PROBLEMS

In Problems 1 through 6, evaluate the surface integral $\iint_S f(x, y, z)\,dS$.

1. $f(x, y, z) = x + y$; S is the first-octant part of the plane $x + y + z = 1$.
2. $f(x, y, z) = xyz$; S is the triangle with vertices $(3, 0, 0)$, $(0, 2, 0)$, and $(0, 0, 6)$.
3. $f(x, y, z) = y + z + 3$; S is the part of the plane $z = 2x + 3y$ that lies inside the cylinder $x^2 + y^2 = 9$.
4. $f(x, y, z) = z^2$; S is the part of the cone $z = \sqrt{x^2 + y^2}$ that lies inside the cylinder $x^2 + y^2 = 4$.
5. $f(x, y, z) = xy + 1$; S is the part of the paraboloid $z = x^2 + y^2$ that lies inside the cylinder $x^2 + y^2 = 4$.
6. $f(x, y, z) = (x^2 + y^2)z$; S is the hemisphere $z = \sqrt{1 - x^2 - y^2}$.

In Problems 7 through 12, find the moment of inertia $\iint_S (x^2 + y^2)\,dS$ *of the given surface S with respect to the z-axis. Assume that S has constant density* $\delta \equiv 1$.

7. S is the part of the plane $z = x + y$ that lies inside the cylinder $x^2 + y^2 = 9$.
8. S is the part of the surface $z = xy$ that lies inside the cylinder $x^2 + y^2 = 25$.
9. S is the part of the cylinder $x^2 + z^2 = 1$ that lies between the planes $y = -1$ and $y = 1$. As parameters on the cylinder use y and the polar angular coordinate in the xz-plane.
10. S is the part of the cone $z = \sqrt{x^2 + y^2}$ that lies between the planes $z = 2$ and $z = 5$.
11. S is the part of the sphere $x^2 + y^2 + z^2 = 25$ that lies above the plane $z = 3$.
12. S is the part of the sphere $x^2 + y^2 + z^2 = 25$ that lies outside the cylinder $x^2 + y^2 = 9$.

In Problems 13 through 18, evaluate the surface integral $\iint_S \mathbf{F}\cdot\mathbf{n}\,dS$, *where* $\mathbf{n}$ *is the upward-pointing unit normal vector to the given surface S.*

13. $\mathbf{F} = x\mathbf{i} + y\mathbf{j}$; S is the hemisphere $z = \sqrt{9 - x^2 - y^2}$.
14. $\mathbf{F} = x\mathbf{i} + y\mathbf{j} + z\mathbf{k}$; S is the first-octant part of the plane $2x + 2y + z = 3$.
15. $\mathbf{F} = 2y\mathbf{j} + 3z\mathbf{k}$; S is the part of the plane $z = 3x + 2$ that lies within the cylinder $x^2 + y^2 = 4$.
16. $\mathbf{F} = z\mathbf{k}$; S is the upper half of the spherical surface $\rho = 2$. [*Suggestion:* Use spherical coordinates.]
17. $\mathbf{F} = y\mathbf{i} - x\mathbf{j}$; S is the part of the cone $z = r$ that lies within the cylinder $r = 3$.
18. $\mathbf{F} = 2x\mathbf{i} + 2y\mathbf{j} + 3\mathbf{k}$; S is the part of the paraboloid $z = 4 - x^2 - y^2$ that lies above the xy-plane.

In Problems 19 through 24, calculate the outward flux of the vector field $\mathbf{F}$ *across the given closed surface S.*

19. $\mathbf{F} = x\mathbf{i} + 2y\mathbf{j} + 3z\mathbf{k}$; S is the boundary of the first-octant unit cube with opposite vertices $(0, 0, 0)$ and $(1, 1, 1)$.
20. $\mathbf{F} = 2x\mathbf{i} - 3y\mathbf{j} + z\mathbf{k}$; S is the boundary of the solid hemisphere $0 \leqq z \leqq \sqrt{4 - x^2 - y^2}$.
21. $\mathbf{F} = x\mathbf{i} - y\mathbf{j}$; S is the boundary of the solid first-octant pyramid bounded by the coordinate planes and the plane $3x + 4y + z = 12$.
22. $\mathbf{F} = 2x\mathbf{i} + 2y\mathbf{j} + 3\mathbf{k}$; S is the boundary of the solid paraboloid bounded by the xy-plane and $z = 4 - x^2 - y^2$.

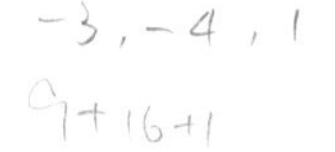

23. $\mathbf{F} = z^2\mathbf{k}$; S is the boundary of the solid bounded by the paraboloids $z = x^2 + y^2$ and $z = 18 - x^2 - y^2$.

24. $\mathbf{F} = x^2\mathbf{i} + 2y^2\mathbf{j} + 3z^2\mathbf{k}$; S is the boundary of the solid bounded by the cone $z = \sqrt{x^2 + y^2}$ and the plane $z = 3$.

25. The first-octant part of the spherical surface $\rho = a$ has unit density. Find its centroid.

26. The conical surface $z = r$, $r \leqq a$, has constant density $\delta = k$. Find its centroid and its moment of inertia around the z-axis.

27. The paraboloid $z = r^2$, $r \leqq a$, has constant density δ. Find its centroid and moment of inertia around the z-axis.

28. Find the centroid of the part of the spherical surface $\rho = a$ that lies within the cone $r = z$.

29. Find the centroid of the part of the spherical surface $x^2 + y^2 + z^2 = 4$ that lies both inside the cylinder $x^2 + y^2 = 2x$ and above the xy-plane.

30. Suppose that the toroidal surface of Example 5 of Section 14.8 has uniform density and total mass M. Show that its moment of inertia around the z-axis is $\frac{1}{2}M(3a^2 + 2b^2)$.

In Problems 31 and 32, use a table of integrals or a computer algebra system (if necessary) to find the moment of inertia around the z-axis of the given surface S. Assume that S has constant density $\delta \equiv 1$.

31. S is the part of the parabolic cylinder $z = 4 - y^2$ that lies inside the rectangular cylinder $-1 \leqq x \leqq 1$, $-2 \leqq y \leqq 2$.

32. S is the part of the paraboloid $z = 4 - x^2 - y^2$ that lies inside the square cylinder $-1 \leqq x \leqq 1$, $-1 \leqq y \leqq 1$.

33. Let S denote the surface $z = h(x, y)$ for (x, y) in the region D in the xy-plane, and let γ be the angle between $\mathbf{k}$ and the upper normal vector $\mathbf{N}$ to S. Prove that

$$\iint_S f(x, y, z)\, dS = \iint_S f(x, y, h(x, y)) \sec\gamma\, dx\, dy.$$

34. Find a formula for

$$\iint_S P\, dy\, dz + Q\, dz\, dx + R\, dx\, dy$$

analogous to Eq. (20), but for the case of a surface S described explicitly by $x = h(y, z)$.

35. A uniform solid ball has radius 5 and its temperature u is proportional to the square of the distance from its center, with $u = 100$ at the boundary of the ball. If the heat conductivity of the ball is $K = 2$, find the rate of flow of heat across a concentric sphere of radius 3.

36. A uniform solid cylinder has radius 5 and height 10, and its temperature u is proportional to the square of the distance from its vertical axis, with $u = 100$ at the outer curved boundary of the cylinder. If the heat conductivity of the cylinder is $K = 2$, find the rate of flow of heat across a concentric cylinder of radius 3 and height 10.

In Problems 37 through 39, set up integrals giving the area and moment of inertia around the z-axis of the given surface S (assuming that S has constant density $\delta \equiv 1$*). Use a computer algebra system—as illustrated in the project material for this section—to evaluate these integrals, symbolically if possible, numerically if necessary (with the numerical values* $a = 4$, $b = 3$, *and* $c = 2$ *of the given parameters).*

37. S is the elliptic paraboloid $z = (x/a)^2 + (y/b)^2$ with parametrization $x = au\cos v$, $y = bu\sin v$, $z = u^2$, $0 \leqq u \leqq c$, $0 \leqq v \leqq 2\pi$.

38. S is the ellipsoid $(x/a)^2 + (y/b)^2 + (z/c)^2 = 1$ with parametrization $x = a\sin u\cos v$, $y = b\sin u\sin v$, $z = c\cos u$, $0 \leqq u \leqq \pi$, $0 \leqq v \leqq 2\pi$.

39. S is the hyperboloid $(x/a)^2 + (y/b)^2 - z^2 = 1$ with parametrization $x = a\cosh u\cos v$, $y = b\cosh u\sin v$, $z = \sinh u$, $-c \leqq u \leqq c$, $0 \leqq v \leqq 2\pi$. See Fig. 15.5.8, where the u-curves are hyperbolas and the v-curves are ellipses.

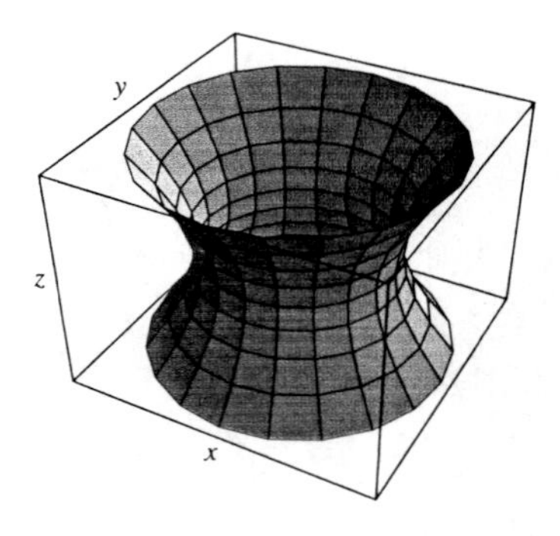

FIGURE 15.5.8 The hyperboloid of Problem 39.

40. The Möbius strip in Fig. 15.5.9 was generated by plotting the points

$$x = \left(4 + t\cos\tfrac{1}{2}\theta\right)\cos\theta, \qquad y = \left(4 + t\cos\tfrac{1}{2}\theta\right)\sin\theta,$$

$$z = t\sin\tfrac{1}{2}\theta$$

for $-1 \leqq t \leqq 1$, $0 \leqq \theta \leqq 2\pi$. This Möbius strip has width 2 and a circular centerline of radius 4. Set up integrals giving its area and moment of inertia (assume constant density $\delta \equiv 1$) around the z-axis, and use a computer algebra system to evaluate them numerically.

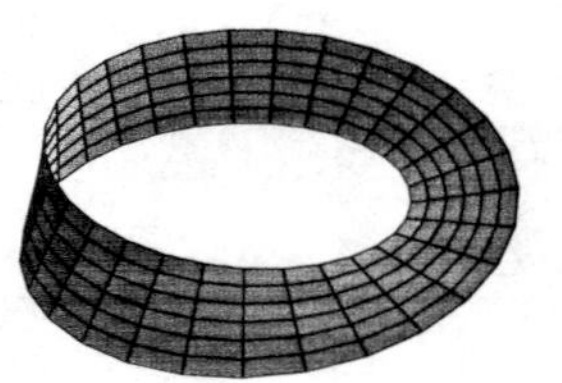

FIGURE 15.5.9 The Möbius strip of Problem 40.

41. Consider a homogeneous thin spherical shell S of radius a centered at the origin, with density δ and total mass $M = 4\pi a^2\delta$. A particle of mass m is located at the point $(0, 0, c)$ with $c > a$. Use the method and notation of Problem 41 of Section 14.7 to show that the gravitational force of attraction between the particle and the spherical shell is

$$F = \iint_S \frac{Gm\delta}{w^2}\, dS = \frac{GMm}{c^2}.$$

15.5 Project: Surface Integrals and Rocket Nose Cones

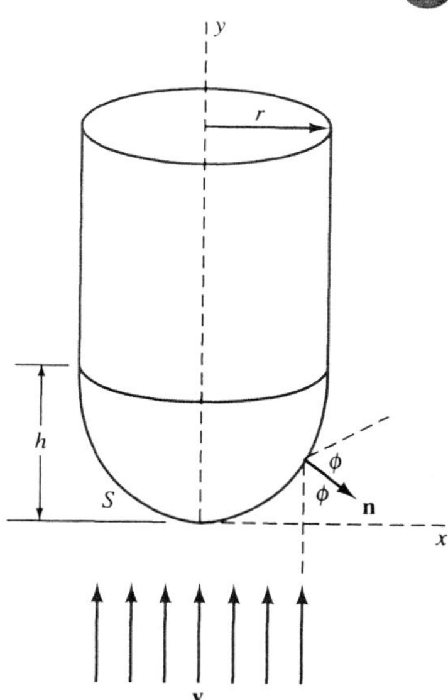

FIGURE 15.5.10 The nose cone S of height h and radius r.

Figure 15.5.10 shows a (curved) nose cone S of height $h = 1$ attached to a cylindrical rocket of radius $r = 1$ that is moving downward with velocity v through air of density δ (or, equivalently, the rocket is stationary and the air is streaming upward). In the *Principia Mathematica* Newton showed that (under plausible assumptions) the force of air resistance the rocket experiences is given by $F = 2\pi R\,\delta v^2$, and thus is proportional both to the density of the air and to the square of its velocity. The *drag coefficient* R is given by the surface integral

$$R = \frac{1}{\pi}\iint_S \cos^3\phi\, dS,$$

where ϕ is the angle between the unit normal **n** and the direction of motion.

The integrals involved in Problems 1 through 5 of this project reduce to single-variable integrals that you should be able to evaluate by hand. The CD-ROM project material illustrates the use of a computer algebra system for the evaluation both of "nose cone integrals" and of more general surface integrals.

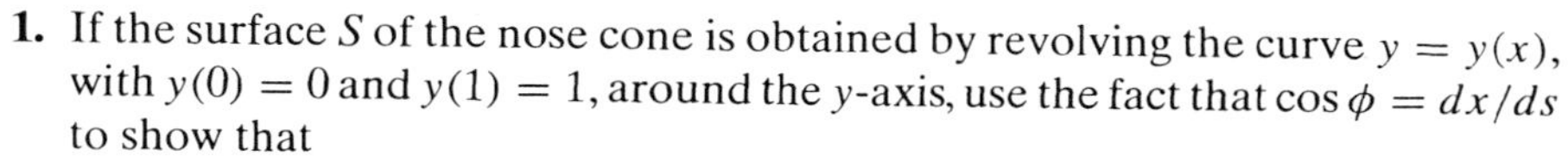

1. If the surface S of the nose cone is obtained by revolving the curve $y = y(x)$, with $y(0) = 0$ and $y(1) = 1$, around the y-axis, use the fact that $\cos\phi = dx/ds$ to show that

$$R = \int_0^1 \frac{2x}{1 + [y'(x)]^2}\, dx.$$

Use this integral to calculate the numerical value of R in the particular cases that follow.

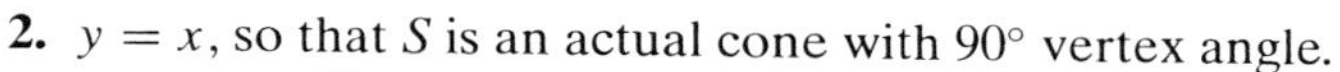

2. $y = x$, so that S is an actual cone with 90° vertex angle.
3. $y = 1 - \sqrt{1 - x^2}$, so S is a hemisphere.
4. $y = x^2$, so S is a paraboloid.
5. For the flat-tipped conical frustum nose cone illustrated in Fig. 15.5.11 (still with $r = h = 1$), show that

$$R = \cos^2\alpha - 2\cos\alpha\sin\alpha + 2\sin^2\alpha$$

where α is the indicated angle. Then show that this drag coefficient is minimal when $\tan 2\alpha = 2$.

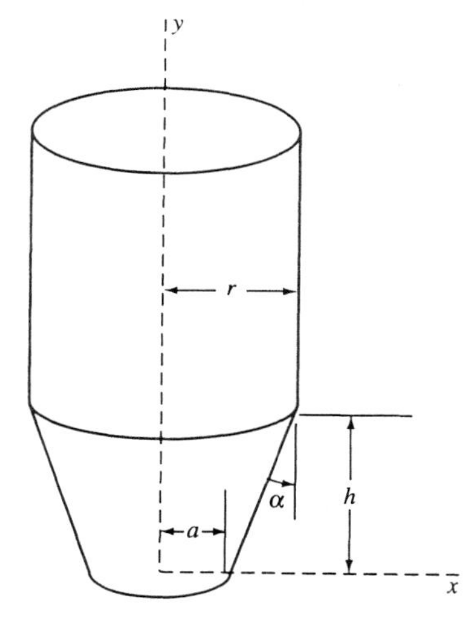

FIGURE 15.5.11 The flat-tipped nose cone.

If you compare your numerical results, you should find that

- the cone and hemisphere offer the same resistance;
- the paraboloid offers less resistance than either; and
- the optimal flat-tipped conical frustum offers still less!

In an extraordinary *tour de force*, Newton determined the nose cone with minimum possible air resistance, allowing both a circular flat tip and a curved surface of resolution connecting the tip to the cylindrical body of the rocket—see C. Henry Edwards, "Newton's Nose-Cone Problem," *The Mathematica Journal* **7** (Winter 1997), pp. 75–82.

15.6 THE DIVERGENCE THEOREM

The *divergence theorem* is to surface integrals what Green's theorem is to line integrals. It lets us convert a surface integral over a closed surface into a triple integral over the enclosed region, or vice versa. The divergence theorem is known also as *Gauss's theorem,* and in some eastern European countries it is called *Ostrogradski's theorem.* The German "prince of mathematics" Carl Friedrich Gauss (1777–1855) used it to study inverse-square force fields; the Russian Michel Ostrogradski (1801–1861) used it to study heat flow. Both did their work in the 1830s.

The surface S is called **piecewise smooth** if it consists of a finite number of smooth parametric surfaces. It is called **closed** if it is the boundary of a bounded region in space. For example, the boundary of a cube is a closed piecewise smooth surface, as are the boundary of a pyramid and the boundary of a solid cylinder.

THE DIVERGENCE THEOREM

Suppose that S is a closed piecewise smooth surface that bounds the space region T and let the *outer* unit normal vector field $\mathbf{n}$ be continuous on each smooth piece of S. If the vector field $\mathbf{F}$ is continuously differentiable on T, then

$$\iint_S \mathbf{F}\cdot\mathbf{n}\,dS = \iiint_T \nabla\cdot\mathbf{F}\,dV. \tag{1}$$

Equation (1) is a three-dimensional analogue of the vector form of Green's theorem that we saw in Eq. (9) of Section 15.4:

$$\oint_C \mathbf{F}\cdot\mathbf{n}\,ds = \iint_R \nabla\cdot\mathbf{F}\,dA,$$

where $\mathbf{F}$ is a vector field in the plane, C is a piecewise smooth curve that bounds the plane region R, and $\mathbf{n}$ is the outer unit normal vector to C. The left-hand side of Eq. (1) is the flux of $\mathbf{F}$ across S in the direction of the outer unit normal vector $\mathbf{n}$.

Recall from Section 15.1 that the *divergence* $\nabla\cdot\mathbf{F}$ of the vector field $\mathbf{F} = P\mathbf{i} + Q\mathbf{j} + R\mathbf{k}$ is given in the three-dimensional case by

$$\operatorname{div}\mathbf{F} = \nabla\cdot\mathbf{F} = \frac{\partial P}{\partial x} + \frac{\partial Q}{\partial y} + \frac{\partial R}{\partial z}. \tag{2}$$

If $\mathbf{n}$ is given in terms of its direction cosines, as $\mathbf{n} = \langle\cos\alpha, \cos\beta, \cos\gamma\rangle$, then (by Eq. (18) in Section 15.5) we can write the divergence theorem in scalar form:

$$\iint_S (P\cos\alpha + Q\cos\beta + R\cos\gamma)\,dS = \iiint_T \left(\frac{\partial P}{\partial x} + \frac{\partial Q}{\partial y} + \frac{\partial R}{\partial z}\right)dV. \tag{3}$$

It is best to parametrize S so that the normal vector given by the parametrization is the outer normal. Then we can write Eq. (3) entirely in Cartesian form:

$$\iint_S P\,dy\,dz + Q\,dz\,dx + R\,dx\,dy = \iiint_T \left(\frac{\partial P}{\partial x} + \frac{\partial Q}{\partial y} + \frac{\partial R}{\partial z}\right)dV. \tag{4}$$

PARTIAL PROOF OF THE DIVERGENCE THEOREM We shall prove the divergence theorem only for the case in which the region T is simultaneously x-simple, y-simple, and z-simple. This guarantees that every straight line parallel to a coordinate axis intersects T, if at all, in a single point or a single line segment. It suffices for us to derive separately the equations

$$\iint_S P\,dy\,dz = \iiint_T \frac{\partial P}{\partial x}\,dV,$$

$$\iint_S Q\,dz\,dx = \iiint_T \frac{\partial Q}{\partial y}\,dV, \quad\text{and} \tag{5}$$

$$\iint_S R\,dx\,dy = \iiint_T \frac{\partial R}{\partial z}\,dV.$$

Then the sum of the equations in (5) is Eq. (4).

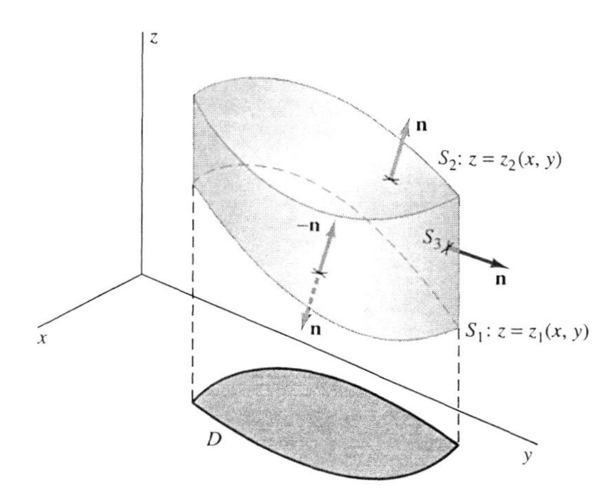

FIGURE 15.6.1 A z-simple space region bounded by the surfaces S_1, S_2, and S_3.

Because T is z-simple, it has the description

$$z_1(x, y) \leqq z \leqq z_2(x, y)$$

for (x, y) in D, the projection of T into the xy-plane. As in Fig. 15.6.1, we denote the lower surface $z = z_1(x, y)$ of T by S_1, the upper surface $z = z_2(x, y)$ by S_2, and the lateral surface between S_1 and S_2 by S_3. In the case of some simple surfaces, such as a spherical surface, there may be no surface S_3 to consider. But even if there is,

$$\iint_{S_3} R\,dx\,dy = \iint_{S_3} R\cos\gamma\,dS = 0, \tag{6}$$

because $\gamma = 90^\circ$ at each point of the vertical cylinder S_3.

On the upper surface S_2, the unit upper normal vector corresponding to the parametrization $z = z_2(x, y)$ is the given outer unit normal vector $\mathbf{n}$, so Eq. (20) of Section 15.5 yields

$$\iint_{S_2} R\,dx\,dy = \iint_{D} R\,(x, y, z_2(x, y))\,dx\,dy. \tag{7}$$

But on the lower surface S_1, the unit upper normal vector corresponding to the parametrization $z = z_1(x, y)$ is the inner normal vector $-\mathbf{n}$, so we must reverse the sign. Thus

$$\iint_{S_1} R\,dx\,dy = -\iint_{D} R\,(x, y, z_1(x, y))\,dx\,dy. \tag{8}$$

We add Eqs. (6), (7), and (8). The result is that

$$\iint_S R\,dx\,dy = \iint_D [R(x, y, z_2(x, y)) - R(x, y, z_1(x, y))]\,dx\,dy$$

$$= \iint_D \left(\int_{z_1(x,y)}^{z_2(x,y)} \frac{\partial R}{\partial z}\,dz\right) dx\,dy.$$

Therefore

$$\iint_S R\,dx\,dy = \iiint_T \frac{\partial R}{\partial z}\,dV.$$

This is the third equation in (5), and we can derive the other two in the same way. ◀

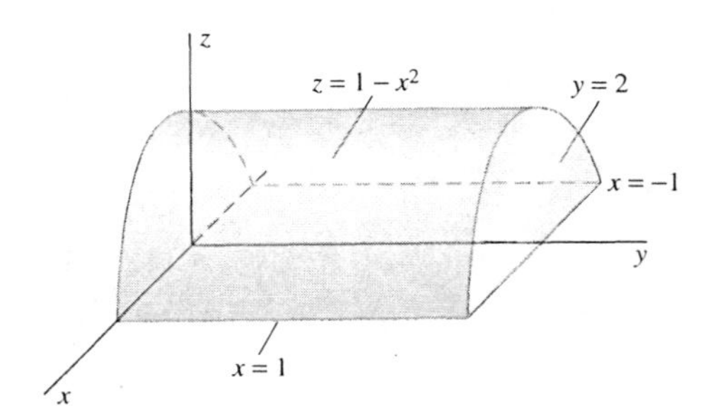

FIGURE 15.6.2 The region of Example 1.

EXAMPLE 1 Let S be the surface (with outer unit normal vector $\mathbf{n}$) of the region T bounded by the planes $z = 0$, $y = 0$, $y = 2$, and the parabolic cylinder $z = 1 - x^2$ (Fig. 15.6.2). Apply the divergence theorem to compute

$$\iint_S \mathbf{F} \cdot \mathbf{n}\, dS$$

given $\mathbf{F} = (x + \cos y)\mathbf{i} + (y + \sin z)\mathbf{j} + (z + e^x)\mathbf{k}$.

Solution To evaluate the surface integral directly would be a lengthy project. But $\operatorname{div} \mathbf{F} = 1 + 1 + 1 = 3$, so we can apply the divergence theorem easily:

$$\iint_S \mathbf{F} \cdot \mathbf{n}\, dS = \iiint_T \operatorname{div} \mathbf{F}\, dV = \iiint_T 3\, dV.$$

We examine Fig. 15.6.2 to find the limits for the volume integral and thus obtain

$$\iint_S \mathbf{F} \cdot \mathbf{n}\, dS = \int_{-1}^{1} \int_0^2 \int_0^{1-x^2} 3\, dz\, dy\, dx = 12 \int_0^1 (1 - x^2)\, dx = 8.$$ ◆

EXAMPLE 2 Let S be the surface of the solid cylinder T bounded by the planes $z = 0$ and $z = 3$ and the cylinder $x^2 + y^2 = 4$. Calculate the outward flux

$$\iint_S \mathbf{F} \cdot \mathbf{n}\, dS$$

given $\mathbf{F} = (x^2 + y^2 + z^2)(x\mathbf{i} + y\mathbf{j} + z\,\mathbf{k})$.

Solution If we denote by P, Q, and R the component functions of the vector field $\mathbf{F}$, we find that

$$\frac{\partial P}{\partial x} = 2x \cdot x + (x^2 + y^2 + z^2) \cdot 1 = 3x^2 + y^2 + z^2.$$

Similarly,

$$\frac{\partial Q}{\partial y} = 3y^2 + z^2 + x^2 \quad \text{and} \quad \frac{\partial R}{\partial z} = 3z^2 + x^2 + y^2,$$

so

$$\operatorname{div} \mathbf{F} = 5(x^2 + y^2 + z^2).$$

Therefore the divergence theorem yields

$$\iint_S \mathbf{F} \cdot \mathbf{n}\, dS = \iiint_T 5(x^2 + y^2 + z^2)\, dV.$$

Using cylindrical coordinates to evaluate the triple integral, we get

$$\begin{aligned}
\iint_S \mathbf{F} \cdot \mathbf{n}\, dS &= \int_0^{2\pi} \int_0^2 \int_0^3 5(r^2 + z^2)\, r\, dz\, dr\, d\theta \\
&= 10\pi \int_0^2 \left[r^3 z + \frac{1}{3} r z^3 \right]_{z=0}^{3} dr \\
&= 10\pi \int_0^2 (3r^3 + 9r)\, dr = 10\pi \left[\frac{3}{4} r^4 + \frac{9}{2} r^2 \right]_0^2 = 300\pi.
\end{aligned}$$ ◆

EXAMPLE 3 Suppose that the space region T is bounded by the smooth closed surface S with a parametrization that gives the outer unit normal vector to S. Show that the volume V of T is given by

$$V = \frac{1}{3}\iint_S x\,dy\,dz + y\,dz\,dx + z\,dx\,dy. \tag{9}$$

Solution Equation (9) follows immediately from Eq. (4) if we take $P(x, y, z) = x$, $Q(x, y, z) = y$, and $R(x, y, z) = z$. For example, if S is the spherical surface $x^2 + y^2 + z^2 = a^2$ with volume V, surface area A, and outer unit normal vector

$$\mathbf{n} = \langle \cos\alpha, \cos\beta, \cos\gamma \rangle = \left\langle \frac{x}{a}, \frac{y}{b}, \frac{z}{c} \right\rangle,$$

then Eq. (9) yields

$$\begin{aligned} V &= \frac{1}{3}\iint_S x\,dy\,dz + y\,dz\,dx + z\,dx\,dy \\ &= \frac{1}{3}\iint_S (x\cos\alpha + y\cos\beta + z\cos\gamma)\,dS \\ &= \frac{1}{3}\iint_S \frac{x^2 + y^2 + z^2}{a}\,dS = \frac{1}{3}a\iint_S 1\,dS = \frac{1}{3}aA. \end{aligned}$$

You should confirm that this result is consistent with the familiar formulas $V = \frac{4}{3}\pi a^3$ and $A = 4\pi a^2$. ◆

EXAMPLE 4 Show that the divergence of the vector field $\mathbf{F}$ at the point P is given by

$$\nabla \cdot \mathbf{F}(P) = \lim_{r \to 0} \frac{1}{v(B_r)} \iint_{S_r} \mathbf{F} \cdot \mathbf{n}\,dS, \tag{10}$$

where S_r is the sphere of radius r centered at P and $v(B_r) = \frac{4}{3}\pi r^3$ is the volume of the ball B_r that the sphere bounds.

Solution The divergence theorem gives

$$\iint_{S_r} \mathbf{F} \cdot \mathbf{n}\,dS = \iiint_{B_r} \nabla \cdot \mathbf{F}\,dV.$$

Then we apply the average value property of triple integrals, a result analogous to the double integral result of Problem 50 in Section 14.2. This yields

$$\iiint_{B_r} \nabla \cdot \mathbf{F}\,dV = [\nabla \cdot \mathbf{F}(\overline{P})] \cdot v(B_r)$$

for some point $\overline{P}$ of B_r. We assume that $\mathbf{F}$ is continuously differentiable, so it follows that

$$\nabla \cdot \mathbf{F}(\overline{P}) \to \nabla \cdot \mathbf{F}(P) \quad \text{as} \quad \overline{P} \to P.$$

Equation (10) follows after we divide both sides by $v(B_r)$ and then take the limit as $r \to 0$. ◆

For instance, suppose that $\mathbf{F} = \delta\mathbf{v}$ is the vector field of a fluid flow. We can interpret Eq. (10) as saying that $\nabla \cdot \mathbf{F}(P)$ is the net rate per unit volume that fluid mass is flowing away (or "diverging") from the point P. For this reason the point P is called a **source** if $\nabla \cdot \mathbf{F}(P) > 0$ but a **sink** if $\nabla \cdot \mathbf{F}(P) < 0$.

Heat in a conducting body can be treated mathematically as though it were a fluid flowing through the body. Miscellaneous Problems 25 through 27 at the end of this chapter ask you to apply the divergence theorem to show that if $u = u(x, y, z, t)$ is the temperature at the point (x, y, z) at the time t in a body through which heat is flowing, then the function u must satisfy the equation

$$\frac{\partial^2 u}{\partial x^2} + \frac{\partial^2 u}{\partial y^2} + \frac{\partial^2 u}{\partial z^2} = \frac{1}{k} \cdot \frac{\partial u}{\partial t}, \tag{11}$$

where k is a constant (the *thermal diffusivity* of the body). This is a *partial differential equation* called the **heat equation.** If both the initial temperature $u(x, y, z, 0)$ and the temperature on the boundary of the body are given, then its interior temperatures at future times are determined by the heat equation. A large part of advanced applied mathematics consists of techniques for solving such partial differential equations.

More General Regions and Gauss's Law

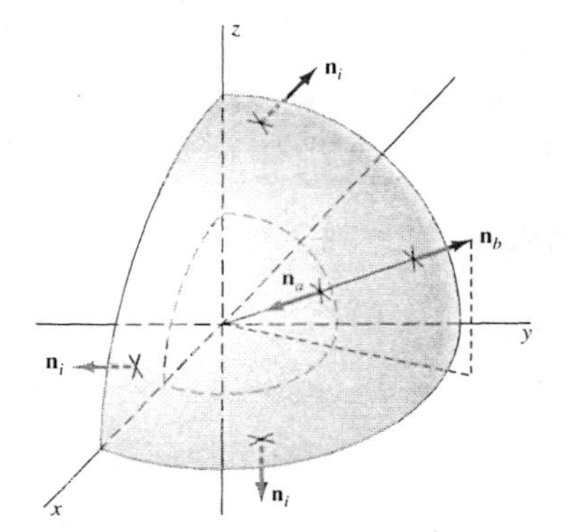

FIGURE 15.6.3 One octant of the shell between S_a and S_b.

We can establish the divergence theorem for more general regions by the device of subdividing T into simpler regions, regions for which the preceding proof holds. For example, suppose that T is the shell between the concentric spherical surfaces S_a and S_b of radii a and b, with $0 < a < b$. The coordinate planes separate T into eight regions $T_1, T_2, \dots, T_8$, each shaped as in Fig. 15.6.3. Let Σ_i denote the boundary of T_i and let $\mathbf{n}_i$ be the outer unit normal vector to Σ_i. We apply the divergence theorem to each of these eight regions and obtain

$$\begin{aligned} \iiint_T \nabla \cdot \mathbf{F}\, dV &= \sum_{i=1}^{8} \iiint_{T_i} \nabla \cdot \mathbf{F}\, dV \\ &= \sum_{i=1}^{8} \iint_{\Sigma_i} \mathbf{F} \cdot \mathbf{n}_i \, dS \qquad \text{(divergence theorem)} \\ &= \iint_{S_a} \mathbf{F} \cdot \mathbf{n}_a \, dS + \iint_{S_b} \mathbf{F} \cdot \mathbf{n}_b \, dS. \end{aligned}$$

Here we write $\mathbf{n}_a$ for the inner normal vector on S_a and $\mathbf{n}_b$ for the outer normal vector on S_b. The last equality holds because the surface integrals over the internal boundary surfaces (the surfaces in the coordinate planes) cancel in pairs—the normals are oppositely oriented there. As the boundary S of T is the union of the spherical surfaces S_a and S_b, it now follows that

$$\iiint_T \nabla \cdot \mathbf{F}\, dV = \iint_S \mathbf{F} \cdot \mathbf{n}\, dS.$$

This is the divergence theorem for the spherical shell T.

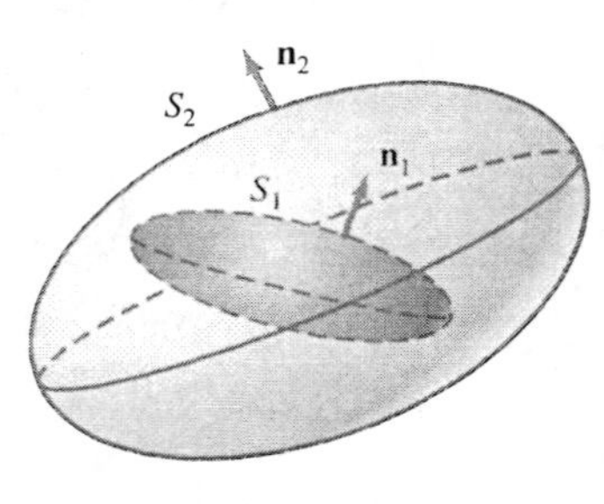

FIGURE 15.6.4 Nested closed surfaces S_1 and S_2.

In a similar manner, the divergence theorem can be established for a region T that is bounded by two smooth closed surfaces S_1 and S_2 with S_1 interior to S_2, as in Fig. 15.6.4, where $\mathbf{n}_1$ and $\mathbf{n}_2$ denote the outward-pointing unit normal vectors to the two surfaces. Then the boundary S of T is the union of S_1 and S_2, and the outer unit normal vector field $\mathbf{n}$ on S consists of $-\mathbf{n}_1$ on the inner surface S_1 and $\mathbf{n}_2$ on the outer surface S_2 (both pointing out of T). Hence the divergence theorem takes the form

$$\iiint_T \nabla \cdot \mathbf{F}\, dV = \iint_S \mathbf{F} \cdot \mathbf{n}\, dS = \iint_{S_2} \mathbf{F} \cdot \mathbf{n}_2 \, dS - \iint_{S_1} \mathbf{F} \cdot \mathbf{n}_1 \, dS. \tag{12}$$

For example, suppose that $\mathbf{F} = -GM\mathbf{r}/|\mathbf{r}|^3$ is the inverse-square gravitational force field of a mass M located at the origin. According to Problem 22, $\nabla \cdot \mathbf{F} = 0$ except at the origin. If S is a smooth surface enclosing M and S_a is a small sphere of radius a within S enclosing M, it then follows from Eq. (12) that (with $\mathbf{n}$ denoting the outer unit normal on each surface)

$$\begin{aligned}\iint_S \mathbf{F}\cdot\mathbf{n}\,dS &= \iint_{S_a} \mathbf{F}\cdot\mathbf{n}\,dS \\ &= \iint_{S_a} -\frac{GM\mathbf{r}}{|\mathbf{r}|^3}\cdot\frac{\mathbf{r}}{|\mathbf{r}|}\,dS = -\frac{GM}{a^2}\iint_{S_a} 1\,dS = -4\pi GM. \qquad (13)\end{aligned}$$

Thus we have established Gauss's law (Eq. (24) in Section 15.5) for the special case of a single point mass. The more general case of a collection of point masses within S can be established by enclosing each in its own small sphere. If we replace the constant GM in Eq. (13) with $Q/4\pi\epsilon_0$, we obtain similarly Gauss's law

$$\iint_S \mathbf{E}\cdot\mathbf{n}\,dS = \frac{Q}{\epsilon_0} \qquad (14)$$

for the inverse-square electric field $\mathbf{E} = Q\mathbf{r}/(4\pi\epsilon_0|\mathbf{r}|^3)$ of a charge Q lying within S.

Another impressive consequence of the divergence theorem is Archimedes' law of buoyancy; see Problem 21 here and Problem 22 of Section 15.7.

15.6 TRUE/FALSE STUDY GUIDE

15.6 CONCEPTS: QUESTIONS AND DISCUSSION

1. The third equation in (5) is established in the "proof of the divergence theorem" included in this section. What changes in the derivation would be required to establish the first and second equations in (5)?
2. Let S_1 and S_2 be piecewise smooth surfaces in space with S_1 lying interior to S_2. Under what conditions on the vector field $\mathbf{F}$ can we conclude that

$$\iint_{S_1} \mathbf{F}\cdot\mathbf{n}\,dS = \iint_{S_2} \mathbf{F}\cdot\mathbf{n}\,dS?$$

If $\mathbf{F} = -k\mathbf{r}/|\mathbf{r}|^3$ is an inverse-square force field directed toward the origin, under what conditions on S_1 and S_2 can we conclude that the two flux integrals are equal?

15.6 PROBLEMS

In Problems 1 through 5, verify the divergence theorem by direct computation of both the surface integral and the triple integral of Eq. (1).

1. $\mathbf{F} = x\mathbf{i} + y\mathbf{j} + z\,\mathbf{k}$; S is the spherical surface with equation $x^2 + y^2 + z^2 = 1$.

2. $\mathbf{F} = |\mathbf{r}|\mathbf{r}$, where $\mathbf{r} = x\mathbf{i} + y\mathbf{j} + z\,\mathbf{k}$; S is the spherical surface with equation $x^2 + y^2 + z^2 = 9$.

3. $\mathbf{F} = x\mathbf{i} + y\mathbf{j} + z\,\mathbf{k}$; S is the surface of the cube bounded by the three coordinates planes and the three planes $x = 2$, $y = 2$, and $z = 2$.

4. $\mathbf{F} = xy\mathbf{i} + yz\mathbf{j} + xz\,\mathbf{k}$; S is the surface of Problem 3.

5. $\mathbf{F} = (x + y)\mathbf{i} + (y + z)\mathbf{j} + (z + x)\mathbf{k}$; S is the surface of the tetrahedron bounded by the three coordinates planes and the plane $x + y + z = 1$.

In Problems 6 through 14, use the divergence theorem to evaluate $\iint_S \mathbf{F}\cdot\mathbf{n}\,dS$, where $\mathbf{n}$ is the outer unit normal vector to the surface S.

6. $\mathbf{F} = x^2\mathbf{i} + y^2\mathbf{j} + z^2\mathbf{k}$; S is the surface of Problem 3.

7. $\mathbf{F} = x^3\mathbf{i} + y^3\mathbf{j} + z^3\mathbf{k}$; S is the surface of the cylinder bounded by $x^2 + y^2 = 9$, $z = -1$, and $z = 4$.

8. $\mathbf{F} = (x^2 + y^2)(x\mathbf{i} + y\mathbf{j})$; S is the surface of the region bounded by the plane $z = 0$ and the paraboloid $z = 25 - x^2 - y^2$.

9. $\mathbf{F} = (x^2 + e^{-yz})\mathbf{i} + (y + \sin xz)\mathbf{j} + (\cos xy)\mathbf{k}$; S is the surface of Problem 5.

10. $\mathbf{F} = (xy^2 + e^{-y}\sin z)\mathbf{i} + (x^2y + e^{-x}\cos z)\mathbf{j} + (\tan^{-1} xy)\mathbf{k}$; S is the surface of the region bounded by the paraboloid $z = x^2 + y^2$ and the plane $z = 9$.

11. $\mathbf{F} = (x^2+y^2+z^2)(x\mathbf{i}+y\mathbf{j}+z\mathbf{k})$; S is the surface of Problem 8.

12. $\mathbf{F} = \mathbf{r}/|\mathbf{r}|$, where $\mathbf{r} = x\mathbf{i} + y\mathbf{j} + z\mathbf{k}$; S is the sphere $\rho = 2$ of radius 2 centered at the origin.

13. $\mathbf{F} = x\mathbf{i} + y\mathbf{j} + 3\mathbf{k}$; S is the boundary of the region bounded by the paraboloid $z = x^2 + y^2$ and the plane $z = 4$.

14. $\mathbf{F} = (x^3 + e^z)\mathbf{i} + x^2y\mathbf{j} + (\sin xy)\mathbf{k}$; S is the boundary of the region bounded by the parabolic cylinder $z = 4 - x^2$ and the planes $y = 0$, $z = 0$, and $y + z = 5$.

15. The **Laplacian** of the twice-differentiable scalar function f is defined to be $\nabla^2 f = \operatorname{div}(\operatorname{grad} f) = \nabla \cdot \nabla f$. Show that

$$\nabla^2 f = \frac{\partial^2 f}{\partial x^2} + \frac{\partial^2 f}{\partial y^2} + \frac{\partial^2 f}{\partial z^2}.$$

16. Let $\partial f/\partial n = \nabla f \cdot \mathbf{n}$ denote the directional derivative of the scalar function f in the direction of the outer unit normal vector $\mathbf{n}$ to the surface S that bounds the region T. Show that

$$\iint_S \frac{\partial f}{\partial n}\, dS = \iiint_T \nabla^2 f\, dV.$$

Use the notation of Problems 15 and 16 in Problems 17 through 19.

17. Suppose that $\nabla^2 f \equiv 0$ in the region T with boundary surface S. Show that

$$\iint_S f\frac{\partial f}{\partial n}\, dV = \iiint_T |\nabla f|^2\, dV.$$

18. Apply the divergence theorem to $\mathbf{F} = f\,\nabla g$ to establish **Green's first identity,**

$$\iint_S f\frac{\partial g}{\partial n}\, dS = \iiint_T (f\,\nabla^2 g + \nabla f \cdot \nabla g)\, dV.$$

19. Interchange f and g in Green's first identity (Problem 18) to establish **Green's second identity,**

$$\iint_S \left(f\frac{\partial g}{\partial n} - g\frac{\partial f}{\partial n}\right) dS = \iiint_T (f\,\nabla^2 g - g\,\nabla^2 f)\, dV.$$

20. Suppose that f is a differentiable scalar function defined on the region T of space and that S is the boundary of T. Prove that

$$\iint_S f\mathbf{n}\, dS = \iiint_T \nabla f\, dV.$$

[*Suggestion:* Apply the divergence theorem to $\mathbf{F} = f\mathbf{a}$, where $\mathbf{a}$ is an arbitrary constant vector. *Note:* Integrals of vector-valued functions are defined by componentwise integration.]

21. *Archimedes' Law of Buoyancy* Let S be the surface of a body T submerged in a fluid of constant density δ. Set up coordinates so that positive values of z are measured *downward* from the surface. Then the pressure at depth z is $p = \delta g z$. The buoyant force exerted on the body by the fluid is

$$\mathbf{B} = -\iint_S p\mathbf{n}\, dS.$$

(Why?) Apply the result of Problem 20 to show that $\mathbf{B} = -W\mathbf{k}$, where W is the weight of the fluid displaced by the body. Because z is measured downward, the vector $\mathbf{B}$ is directed upward.

22. Let $\mathbf{r} = \langle x, y, z\rangle$, let $\mathbf{r}_0 = \langle x_0, y_0, z_0\rangle$ be a fixed point, and suppose that

$$\mathbf{F}(x, y, z) = \frac{\mathbf{r} - \mathbf{r}_0}{|\mathbf{r} - \mathbf{r}_0|^3}.$$

Show that $\operatorname{div} \mathbf{F} = 0$ except at the point $\mathbf{r}_0$.

23. Apply the divergence theorem to compute the outward flux

$$\iint_S \mathbf{F} \cdot \mathbf{n}\, dS,$$

where $\mathbf{F} = |\mathbf{r}|\mathbf{r}$, $\mathbf{r} = x\mathbf{i} + y\mathbf{j} + z\mathbf{k}$, and S is the surface of Problem 8. [*Suggestion:* Integrate in cylindrical coordinates, first with respect to r and then with respect to z. For the latter integration, make a trigonometric substitution and then consult Eq. (9) of Section 8.3 for the antiderivative of $\sec^5 \theta$.]

24. Assume that Gauss's law in (13) holds for a uniform solid ball of mass M centered at the origin. Also assume by symmetry that the force $\mathbf{F}$ exerted by this mass on an exterior particle of unit mass is directed toward the origin. Apply Gauss's law with S being a sphere of radius r to show that $|\mathbf{F}| = GM/r^2$ at each point of S. Thus it follows that the solid ball acts (gravitationally) like a single point-mass M concentrated at its center.

25. Let $\mathbf{F}$ be the gravitational force field due to a uniform distribution of mass in the shell bounded by the concentric spherical surfaces $\rho = a$ and $\rho = b > a$. Apply Gauss's law in (13), with S being the sphere $\rho = r < a$, to show that $\mathbf{F}$ is zero at all points interior to this spherical shell.

26. Consider a solid spherical ball of radius a and constant density δ. Apply Gauss's law to show that the gravitational force on a particle of unit mass located at a distance $r < a$ from the center of the ball is given by $F = GM_r/r^2$, where $M_r = \frac{4}{3}\pi\delta r^3$ is the mass enclosed by a sphere of radius r. Thus the mass at a greater distance from the center of the ball exerts no net gravitational force on the particle.

27. Consider an infinitely long vertical straight wire with a uniform positive charge of q coulombs per meter. Assume by symmetry that the electric field vector $\mathbf{E}$ is at each point of space a horizontal radial vector directed away from the wire. Apply Gauss's law in (14), with S being a cylinder with the wire as its axis, to show that $|\mathbf{E}| = q/(2\pi\epsilon_0 r)$. Thus the electric field intensity is inversely proportional to distance r from the wire.

15.7 STOKES' THEOREM

In Section 15.4 we gave Green's theorem,

$$\oint_C P\,dx + Q\,dy = \iint_R \left(\frac{\partial Q}{\partial x} - \frac{\partial P}{\partial y} \right) dA, \tag{1}$$

in a vector form that is equivalent to a two-dimensional version of the divergence theorem. Another vector form of Green's theorem involves the curl of a vector field. Recall from Section 15.1 that if $\mathbf{F} = P\mathbf{i} + Q\mathbf{j} + R\mathbf{k}$ is a vector field, then curl $\mathbf{F}$ is the vector field given by

$$\text{curl}\,\mathbf{F} = \nabla \times \mathbf{F} = \begin{vmatrix} \mathbf{i} & \mathbf{j} & \mathbf{k} \\ \dfrac{\partial}{\partial x} & \dfrac{\partial}{\partial y} & \dfrac{\partial}{\partial z} \\ P & Q & R \end{vmatrix}$$

$$= \left(\frac{\partial R}{\partial y} - \frac{\partial Q}{\partial z} \right)\mathbf{i} + \left(\frac{\partial P}{\partial z} - \frac{\partial R}{\partial x} \right)\mathbf{j} + \left(\frac{\partial Q}{\partial x} - \frac{\partial P}{\partial y} \right)\mathbf{k}. \tag{2}$$

The $\mathbf{k}$-component of $\nabla \times \mathbf{F}$ is the integrand of the double integral in Eq. (1). We know from Section 15.2 that we can write the line integral in Eq. (1) as

$$\oint_C \mathbf{F} \cdot \mathbf{T}\,ds,$$

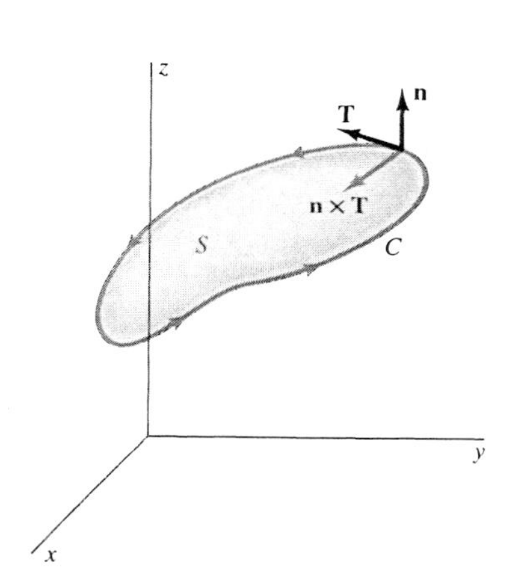

FIGURE 15.7.1 Vectors, surface, and boundary curve mentioned in the statement of Stokes' theorem.

where $\mathbf{T}$ is the positive-directed unit tangent vector to C. Consequently, we can rewrite Green's theorem in the form

$$\oint_C \mathbf{F} \cdot \mathbf{T}\,ds = \iint_R (\text{curl}\,\mathbf{F}) \cdot \mathbf{k}\,dA. \tag{3}$$

Stokes' theorem is the generalization of Eq. (3) that we get by replacing the plane region R with a floppy two-dimensional version: an oriented bounded surface S in three-dimensional space with boundary C that consists of one or more simple closed curves in space.

An **oriented surface** is a piecewise smooth surface S together with a chosen *unit* normal vector field $\mathbf{n}$ that is continuous (that is, continuously turning) on each smooth piece of S. The positive orientation of the boundary C of an oriented surface S corresponds to the unit tangent vector $\mathbf{T}$ such that $\mathbf{n} \times \mathbf{T}$ always points *into* S (Fig. 15.7.1). Check that for a plane region with unit normal vector $\mathbf{k}$, the positive orientation of its outer boundary is counterclockwise.

STOKES' THEOREM

Let S be an oriented, bounded, and piecewise smooth surface in space with positively oriented boundary C and unit normal vector field $\mathbf{n}$. Suppose that $\mathbf{T}$ is a positively oriented unit vector field tangent to C. If the vector field $\mathbf{F}$ is continuously differentiable in a space region containing S, then

$$\oint_C \mathbf{F} \cdot \mathbf{T}\,ds = \iint_S (\text{curl}\,\mathbf{F}) \cdot \mathbf{n}\,dS. \tag{4}$$

Thus Stokes' theorem means that *the line integral around the boundary curve of the tangential component of* $\mathbf{F}$ *equals the surface integral of the normal component of* curl $\mathbf{F}$. Compare Eqs. (3) and (4).

This result first appeared publicly as a problem posed by George Stokes (1819–1903) on a prize examination for Cambridge University students in 1854. It had been stated in an 1850 letter to Stokes from the physicist William Thomson (Lord Kelvin, 1824–1907).

In terms of the components of $\mathbf{F} = P\mathbf{i} + Q\mathbf{j} + R\mathbf{k}$ and those of curl $\mathbf{F}$, we can recast Stokes' theorem—with the aid of Eq. (19) of Section 15.5—in its scalar form

$$\oint_C P\,dx + Q\,dy + R\,dz = \iint_S \left(\frac{\partial R}{\partial y} - \frac{\partial Q}{\partial z}\right) dy\,dz + \left(\frac{\partial P}{\partial z} - \frac{\partial R}{\partial x}\right) dz\,dx + \left(\frac{\partial Q}{\partial x} - \frac{\partial P}{\partial y}\right) dx\,dy. \quad \textbf{(5)}$$

Here, as usual, the parametrization of S must correspond to the given unit normal vector $\mathbf{n}$.

To prove Stokes' theorem, we need only establish the equation

$$\oint_C P\,dx = \iint_S \left(\frac{\partial P}{\partial z}\,dz\,dx - \frac{\partial P}{\partial y}\,dx\,dy\right) \quad \textbf{(6)}$$

and the corresponding two equations that are the Q and R "components" of Eq. (5). Equation (5) itself then follows by adding the three results.

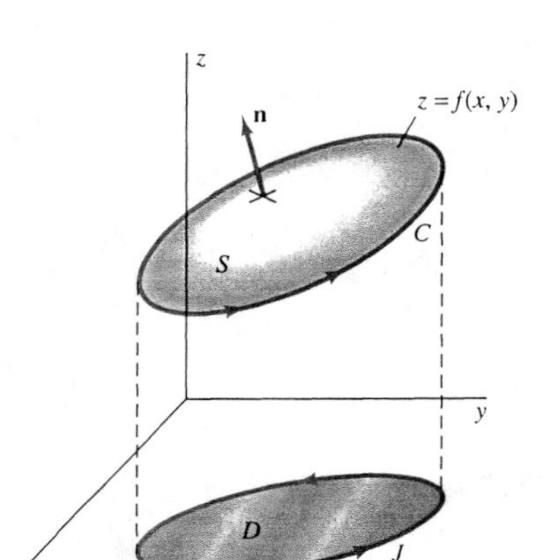

FIGURE 15.7.2 The surface S.

PARTIAL PROOF Suppose first that S is the graph of a function $z = f(x, y)$, (x, y) in D, where S has an upper unit normal vector and D is a region in the xy-plane bounded by the simple closed curve J (Fig. 15.7.2). Then

$$\begin{aligned}\oint_C P\,dx &= \oint_J P(x, y, f(x, y))\,dx \\ &= \oint_J p(x, y)\,dx \quad [\text{where } p(x, y) \equiv P(x, y, f(x, y))] \\ &= -\iint_D \frac{\partial p}{\partial y}\,dx\,dy \quad (\text{by Green's theorem}).\end{aligned}$$

We now use the chain rule to compute $\partial p/\partial y$ and find that

$$\oint_C P\,dx = -\iint_D \left(\frac{\partial P}{\partial y} + \frac{\partial P}{\partial z}\frac{\partial z}{\partial y}\right) dx\,dy. \quad \textbf{(7)}$$

Next, we use Eq. (20) of Section 15.5:

$$\iint_S P\,dy\,dz + Q\,dz\,dx + R\,dx\,dy = \iint_D \left(-P\frac{\partial z}{\partial x} - Q\frac{\partial z}{\partial y} + R\right) dx\,dy.$$

In this equation we replace P with 0, Q with $\partial P/\partial z$, and R with $-\partial P/\partial y$. This gives

$$\iint_S \left(\frac{\partial P}{\partial z}\,dz\,dx - \frac{\partial P}{\partial y}\,dx\,dy\right) = \iint_D \left(-\frac{\partial P}{\partial z}\frac{\partial z}{\partial y} - \frac{\partial P}{\partial y}\right) dx\,dy. \quad \textbf{(8)}$$

Finally, we compare Eqs. (7) and (8) and see that we have established Eq. (6). If we can write the surface S in the forms $y = g(x, z)$ and $x = h(y, z)$, then we can derive the Q and R "components" of Eq. (5) in much the same way. This proves Stokes' theorem for the special case of a surface S that can be represented as a graph in all three coordinate directions. Stokes' theorem may then be extended to a more general oriented surface by the now-familiar method of subdividing it into simpler surfaces, to each of which the preceding proof is applicable. ◀

EXAMPLE 1 Apply Stokes' theorem to evaluate

$$\oint_C \mathbf{F}\cdot\mathbf{T}\,ds,$$

where C is the ellipse in which the plane $z = y + 3$ intersects the cylinder $x^2 + y^2 = 1$. Orient the ellipse counterclockwise as viewed from above and take $\mathbf{F}(x, y, z) = 3z\mathbf{i} + 5x\mathbf{j} - 2y\mathbf{k}$.

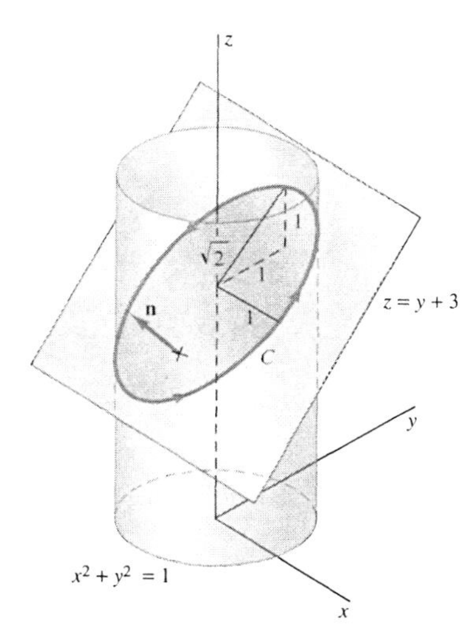

FIGURE 15.7.3 The ellipse of Example 1.

Solution The plane, cylinder, and ellipse appear in Fig. 15.7.3. The given orientation of C corresponds to the upward unit normal vector $\mathbf{n} = (-\mathbf{j}+\mathbf{k})/\sqrt{2}$ to the elliptical region S in the plane $z = y+3$ bounded by C. Now

$$\operatorname{curl}\mathbf{F} = \begin{vmatrix} \mathbf{i} & \mathbf{j} & \mathbf{k} \\ \dfrac{\partial}{\partial x} & \dfrac{\partial}{\partial y} & \dfrac{\partial}{\partial z} \\ 3z & 5x & -2y \end{vmatrix} = -2\mathbf{i}+3\mathbf{j}+5\mathbf{k},$$

so

$$(\operatorname{curl}\mathbf{F})\cdot\mathbf{n} = (-2\mathbf{i}+3\mathbf{j}+5\mathbf{k})\cdot\frac{1}{\sqrt{2}}(-\mathbf{j}+\mathbf{k}) = \frac{-3+5}{\sqrt{2}} = \sqrt{2}.$$

Hence by Stokes' theorem,

$$\oint_C \mathbf{F}\cdot\mathbf{T}\,ds = \iint_S (\operatorname{curl}\mathbf{F})\cdot\mathbf{n}\,dS = \iint_S \sqrt{2}\,dS = \sqrt{2}\cdot\operatorname{area}(S) = 2\pi,$$

because we can see from Fig. 15.7.3 that S is an ellipse with semiaxes 1 and $\sqrt{2}$. Thus its area is $\pi\sqrt{2}$. ◆

EXAMPLE 2 Apply Stokes' theorem to evaluate

$$\iint_S (\nabla\times\mathbf{F})\cdot\mathbf{n}\,dS,$$

where $\mathbf{F} = 3z\mathbf{i}+5x\mathbf{j}-2y\mathbf{k}$ and S is the part of the parabolic surface $z = x^2+y^2$ that lies below the plane $z = 4$ and whose orientation is given by the upper unit normal vector (Fig. 15.7.4).

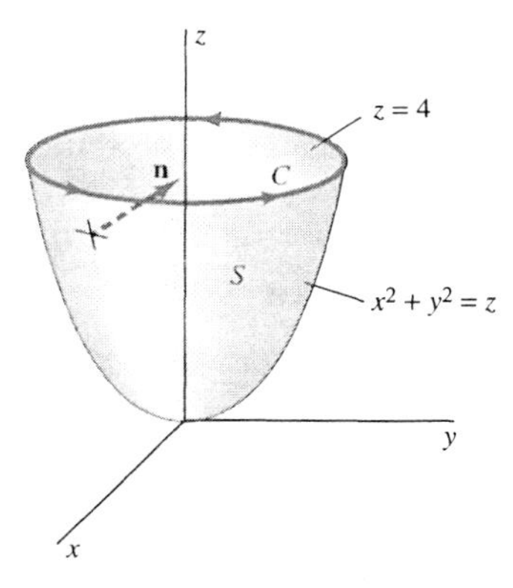

FIGURE 15.7.4 The parabolic surface of Example 2.

Solution We parametrize the boundary circle C of S by $x = 2\cos t$, $y = 2\sin t$, $z = 4$ for $0 \leqq t \leqq 2\pi$. Then $dx = -2\sin t\,dt$, $dy = 2\cos t\,dt$, and $dz = 0$. So Stokes' theorem yields

$$\iint_S (\nabla\times\mathbf{F})\cdot\mathbf{n}\,dS = \oint_C \mathbf{F}\cdot\mathbf{T}\,ds = \oint_C 3z\,dx+5x\,dy-2y\,dz$$

$$= \int_0^{2\pi} 3\cdot4\cdot(-2\sin t\,dt)+5\cdot(2\cos t)(2\cos t\,dt)+2\cdot(2\sin t)\cdot 0$$

$$= \int_0^{2\pi}(-24\sin t+20\cos^2 t)\,dt = \int_0^{2\pi}(-24\sin t+10+10\cos 2t)\,dt$$

$$= \Big[24\cos t+10t+5\sin 2t\Big]_0^{2\pi} = 20\pi.$$ ◆

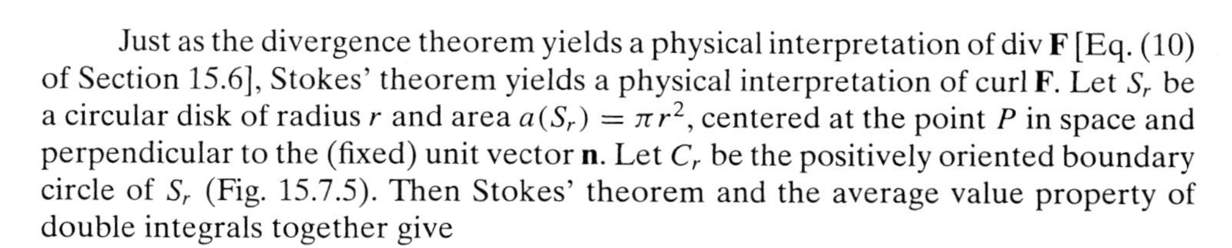

Just as the divergence theorem yields a physical interpretation of div $\mathbf{F}$ [Eq. (10) of Section 15.6], Stokes' theorem yields a physical interpretation of curl $\mathbf{F}$. Let S_r be a circular disk of radius r and area $a(S_r) = \pi r^2$, centered at the point P in space and perpendicular to the (fixed) unit vector $\mathbf{n}$. Let C_r be the positively oriented boundary circle of S_r (Fig. 15.7.5). Then Stokes' theorem and the average value property of double integrals together give

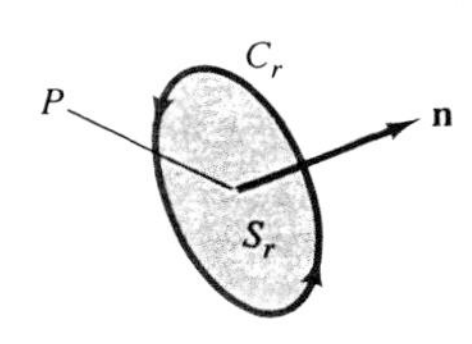

FIGURE 15.7.5 A physical interpretation of the curl of a vector field.

$$\oint_{C_r}\mathbf{F}\cdot\mathbf{T}\,ds = \iint_{S_r}(\nabla\times\mathbf{F})\cdot\mathbf{n}\,dS = a(S_r)[\nabla\times\mathbf{F}(\overline{P})]\cdot\mathbf{n}$$

for some point $\overline{P}$ of S_r. We assume that $\mathbf{F}$ is continuously differentiable at P, so it follows that

$$\nabla \times \mathbf{F}(\overline{P}) \to \nabla \times \mathbf{F}(P) \quad \text{as} \quad \overline{P} \to P.$$

If we first divide both sides by $a(S_r)$ and then take the limit as $r \to 0$, we get

$$[\nabla \times \mathbf{F}(P)] \cdot \mathbf{n} = \lim_{r \to 0} \frac{1}{\pi r^2} \oint_{C_r} \mathbf{F} \cdot \mathbf{T}\, ds. \tag{9}$$

Equation (9) has a natural physical meaning. Suppose that $\mathbf{F} = \delta \mathbf{v}$, where $\mathbf{v}$ is the velocity vector field of the steady-state flow of a fluid with constant density δ. Then the value of the integral

$$\Gamma(C) = \oint_C \mathbf{F} \cdot \mathbf{T}\, ds \tag{10}$$

measures the rate of flow of fluid mass *around* the curve C and is therefore called the **circulation** of $\mathbf{F}$ around C. We see from Eq. (9) that

$$[\nabla \times \mathbf{F}(P)] \cdot \mathbf{n} \approx \frac{\Gamma(C_r)}{\pi r^2}$$

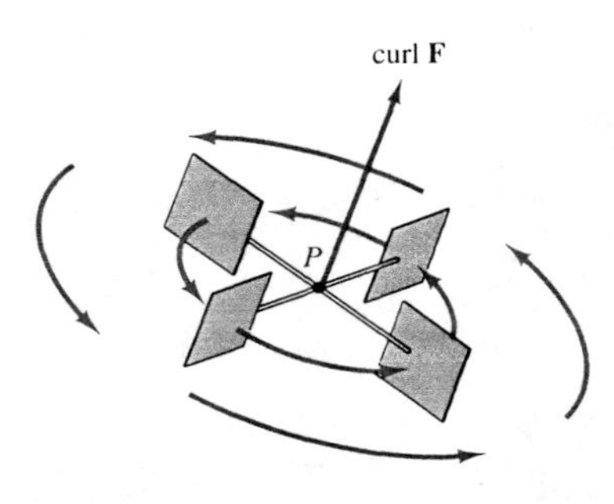

FIGURE 15.7.6 The paddle-wheel interpretation of curl **F**.

if C_r is a circle of very small radius r centered at P and perpendicular to $\mathbf{n}$. If it should happen that $\nabla \times \mathbf{F}(P) \neq \mathbf{0}$, it follows that $\Gamma(C_r)$ is greatest (for r fixed and small) when the unit vector $\mathbf{n}$ points in the direction of $\nabla \times \mathbf{F}(P)$. Hence the line through P determined by $\nabla \times \mathbf{F}(P)$ is the axis about which the fluid near P is revolving the most rapidly. A tiny paddle wheel placed in the fluid at P (see Fig. 15.7.6) would rotate the fastest if its axis lay along this line. It follows from Miscellaneous Problem 32 at the end of this chapter that $|\text{curl}\,\mathbf{F}| = 2\delta\omega$ in the case of a fluid revolving steadily around a fixed axis with constant angular speed ω (in radians per second). Thus $\nabla \times \mathbf{F}(P)$ indicates both the direction *and* rate of rotation of the fluid near P. Because of this interpretation, some older books use the notation "rot **F**" for the curl, an abbreviation that has disappeared from general use.

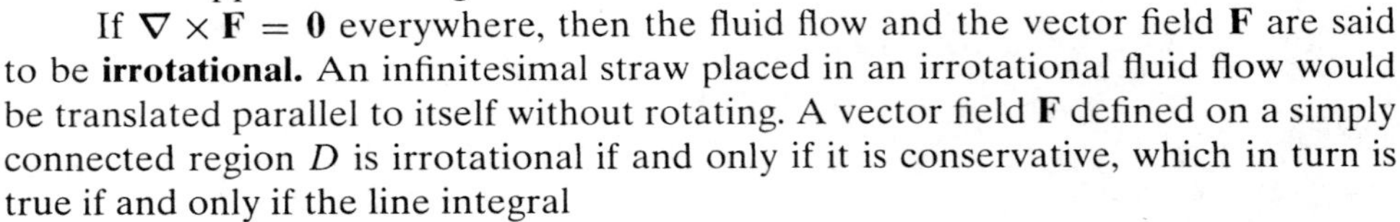

If $\nabla \times \mathbf{F} = \mathbf{0}$ everywhere, then the fluid flow and the vector field $\mathbf{F}$ are said to be **irrotational.** An infinitesimal straw placed in an irrotational fluid flow would be translated parallel to itself without rotating. A vector field $\mathbf{F}$ defined on a simply connected region D is irrotational if and only if it is conservative, which in turn is true if and only if the line integral

$$\int_C \mathbf{F} \cdot \mathbf{T}\, ds$$

is independent of the path in D. (The region D is said to be **simply connected** if every simple closed curve in D can be continuously shrunk to a point while staying inside D. The interior of a torus is an example of a space region that is *not* simply connected. It is true, though not obvious, that any piecewise smooth simple closed curve in a simply connected region D is the boundary of a piecewise smooth oriented surface in D.)

THEOREM 1 Conservative and Irrotational Fields

Let $\mathbf{F}$ be a vector field with continuous first-order partial derivatives in a simply connected region D in space. Then the vector field $\mathbf{F}$ is irrotational if and only if it is conservative; that is, $\nabla \times \mathbf{F} \equiv \mathbf{0}$ in D if and only if $\mathbf{F} = \nabla\phi$ for some scalar function ϕ defined on D.

OUTLINE OF PARTIAL PROOF A complete proof of the *if* part of Theorem 1 is easy; by Example 8 of Section 15.1, $\nabla \times (\nabla\phi) = \mathbf{0}$ for any twice-differentiable scalar function ϕ.

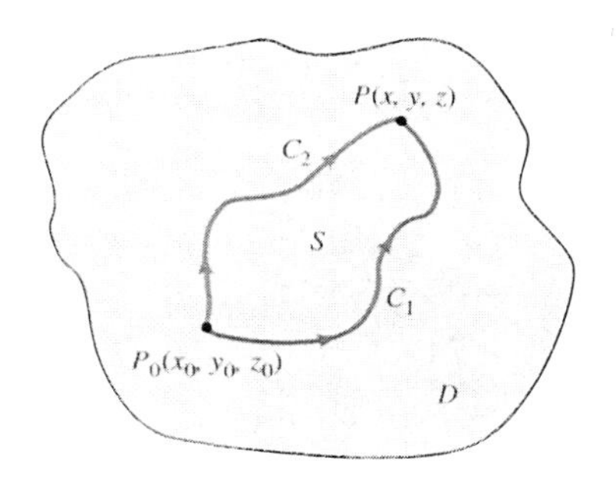

FIGURE 15.7.7 Two paths from P_0 to P in the simply connected space region D.

Here is a description of how we might show the *only if* part of the proof of Theorem 1. Assume that $\mathbf{F}$ is irrotational. Let $P_0(x_0, y_0, z_0)$ be a fixed point of D. Given an arbitrary point $P(x, y, z)$ of D, we would like to define

$$\phi(x, y, z) = \int_{C_1} \mathbf{F} \cdot \mathbf{T}\, ds, \tag{11}$$

where C_1 is a path in D from P_0 to P. But we must show that any *other* path C_2 from P_0 to P would give the *same* value for $\phi(x, y, z)$.

We assume, as suggested by Fig. 15.7.7, that C_1 and C_2 intersect only at their endpoints. Let $C = C_1 \cup (-C_2)$ be the closed path that first follows C_1 from P_0 to P and then follows the reversed path $-C_2$ from P back to P_0. Then it can be shown that the oriented closed curve C bounds an oriented surface S contained in D, and Stokes' theorem then gives

$$\int_{C_1} \mathbf{F} \cdot \mathbf{T}\, ds - \int_{C_2} \mathbf{F} \cdot \mathbf{T}\, ds = \oint_C \mathbf{F} \cdot \mathbf{T}\, ds = \iint_S (\nabla \times \mathbf{F}) \cdot \mathbf{n}\, dS = 0$$

because of the hypothesis that $\nabla \times \mathbf{F} \equiv \mathbf{0}$. This shows that the line integral $\int_C \mathbf{F} \cdot \mathbf{T}\, ds$ is *independent of path*, just as desired. In Problem 21 we ask you to complete this proof by showing that the function ϕ of Eq. (11) is the one whose existence is claimed in Theorem 1. That is, $\mathbf{F} = \nabla \phi$. ◂

EXAMPLE 3 Show that the vector field $\mathbf{F} = 3x^2\mathbf{i} + 5z^2\mathbf{j} + 10yz\,\mathbf{k}$ is irrotational. Then find a potential function $\phi(x, y, z)$ such that $\nabla \phi = \mathbf{F}$.

Solution To show that $\mathbf{F}$ is irrotational, we calculate

$$\nabla \times \mathbf{F} = \begin{vmatrix} \mathbf{i} & \mathbf{j} & \mathbf{k} \\ \dfrac{\partial}{\partial x} & \dfrac{\partial}{\partial y} & \dfrac{\partial}{\partial z} \\ 3x^2 & 5z^2 & 10yz \end{vmatrix} = (10z - 10z)\mathbf{i} = \mathbf{0}.$$

Hence Theorem 1 implies that $\mathbf{F}$ has a potential function ϕ. We can apply Eq. (11) to find ϕ explicitly. If C_1 is the straight line segment from $(0, 0, 0)$ to (u, v, w) that is parametrized by $x = ut$, $y = vt$, $z = wt$ for $0 \leqq t \leqq 1$, then Eq. (11) yields

$$\begin{aligned} \phi(u, v, w) &= \int_{C_1} \mathbf{F} \cdot \mathbf{T}\, ds = \int_{(0,0,0)}^{(u,v,w)} 3x^2\, dx + 5z^2\, dy + 10yz\, dz \\ &= \int_{t=0}^{1} (3u^2t^2)(u\, dt) + (5w^2t^2)(v\, dt) + (10vtwt)(w\, dt) \\ &= \int_{t=0}^{1} (3u^3t^2 + 15vw^2t^2)\, dt = \Big[u^3t^3 + 5vwt^3\Big]_{t=0}^{1}, \end{aligned}$$

and thus

$$\phi(u, v, w) = u^3 + 5vw^2.$$

But because (u, v, w) is an arbitrary point of space, we have found that $\phi(x, y, z) = x^3 + 5yz^2$ is a scalar potential for $\mathbf{F}$. As a check, we note that $\phi_x = 3x^2$, $\phi_y = 5z^2$, and $\phi_z = 10yz$, so $\nabla \phi = \mathbf{F}$, as desired. ◆

Application Suppose that $\mathbf{v}$ is the velocity field of a steady fluid flow that is both irrotational and incompressible—the density δ of the fluid is constant. Suppose that S is any closed surface that bounds a region T. Then, because of conservation of mass, the flux of $\mathbf{v}$ across S must be zero; the mass of fluid within S remains constant.

Hence the divergence theorem gives

$$\iiint_T \operatorname{div} \mathbf{v}\, dV = \iint_S \mathbf{v} \cdot \mathbf{n}\, dS = 0.$$

Because this holds for *any* region T, it follows from the usual average value property argument that $\operatorname{div} \mathbf{v} = 0$ everywhere. The scalar function ϕ provided by Theorem 1, for which $\mathbf{v} = \nabla\phi$, is called the **velocity potential** of the fluid flow. We substitute $\mathbf{v} = \nabla\phi$ into the equation $\operatorname{div} \mathbf{v} = 0$ and thereby obtain

$$\operatorname{div}(\nabla\phi) = \frac{\partial^2 \phi}{\partial x^2} + \frac{\partial^2 \phi}{\partial y^2} + \frac{\partial^2 \phi}{\partial z^2} = 0. \tag{12}$$

Thus the velocity potential ϕ of an irrotational and incompressible fluid flow satisfies *Laplace's equation.*

Laplace's equation appears in numerous other applications. For example, consider a heated body whose temperature function $u = u(x, y, z)$ is independent of time t. Then $\partial u/\partial t \equiv 0$ in the heat equation, Eq. (11) of Section 15.6, shows that the "steady-state temperature" function $u(x, y, z)$ satisfies Laplace's equation

$$\frac{\partial^2 u}{\partial x^2} + \frac{\partial^2 u}{\partial y^2} + \frac{\partial^2 u}{\partial z^2} = 0. \tag{13}$$

These brief remarks should indicate how the mathematics of this chapter forms the starting point for investigations in a number of areas, including acoustics, aerodynamics, electromagnetism, meteorology, and oceanography. Indeed, the entire subject of vector calculus stems historically from its mathematical applications rather than from abstract mathematical considerations. The modern form of the subject is due primarily to J. Willard Gibbs (1839–1903), the first great American physicist, and the English electrical engineer Oliver Heaviside (1850–1925).

15.7 TRUE/FALSE STUDY GUIDE

15.7 CONCEPTS: QUESTIONS AND DISCUSSION

1. The P "component" of Eq. (5) was established in the partial proof of Stokes' theorem included in this section. What changes in the derivation would be required to establish the Q and R "components" of Eq. (5)?
2. Suppose that C_1 and C_2 are piecewise smooth closed curves in space, each being the boundary of a parametrized piecewise smooth disk. (Do C_1 and C_2 then necessarily form the boundary of some curvilinear annular ring?) Under what conditions on the vector field $\mathbf{F}$ can we conclude that

$$\oint_{C_1} \mathbf{F} \cdot \mathbf{T}\, ds = \oint_{C_2} \mathbf{F} \cdot \mathbf{T}\, ds?$$

15.7 PROBLEMS

In Problems 1 through 5, use Stokes' theorem for the evaluation of

$$\iint_S (\operatorname{curl} \mathbf{F}) \cdot \mathbf{n}\, dS.$$

1. $\mathbf{F} = 3y\mathbf{i} - 2x\mathbf{j} + xyz\mathbf{k}$; S is the hemispherical surface $z = \sqrt{4 - x^2 - y^2}$ with upper unit normal vector.

2. $\mathbf{F} = 2y\mathbf{i} + 3x\mathbf{j} + e^z\mathbf{k}$; S is the part of the paraboloid $z = x^2 + y^2$ below the plane $z = 4$ and with upper unit normal vector.

3. $\mathbf{F} = \langle xy, -2, \arctan x^2 \rangle$; S is the part of the paraboloid $z = 9 - x^2 - y^2$ above the xy-plane and with upper unit normal vector.

4. $\mathbf{F} = yz\mathbf{i} + xz\mathbf{j} + xy\mathbf{k}$; S is the part of the cylinder $x^2 + y^2 = 1$ between the two planes $z = 1$ and $z = 3$ and with outer unit normal vector.

5. $\mathbf{F} = \langle yz, -xz, z^3 \rangle$; S is the part of the cone $z = \sqrt{x^2 + y^2}$ between the two planes $z = 1$ and $z = 3$ and with upper unit normal vector.

In Problems 6 through 10, use Stokes' theorem to evaluate

$$\oint_C \mathbf{F} \cdot \mathbf{T}\, ds.$$

6. $\mathbf{F} = 3y\mathbf{i} - 2x\mathbf{j} + 3y\mathbf{k}$; C is the circle $x^2 + y^2 = 9$, $z = 4$, oriented counterclockwise as viewed from above.

7. $\mathbf{F} = 2z\,\mathbf{i} + x\mathbf{j} + 3y\mathbf{k}$; C is the ellipse in which the plane $z = x$ meets the cylinder $x^2 + y^2 = 4$, oriented counterclockwise as viewed from above.

8. $\mathbf{F} = y\mathbf{i} + z\mathbf{j} + x\mathbf{k}$; C is the boundary of the triangle with vertices $(0, 0, 0)$, $(2, 0, 0)$, and $(0, 2, 2)$, oriented counterclockwise as viewed from above.

9. $\mathbf{F} = \langle y - x, x - z, x - y \rangle$; C is the boundary of the part of the plane $x + 2y + z = 2$ that lies in the first octant, oriented counterclockwise as viewed from above.

10. $\mathbf{F} = y^2\mathbf{i} + z^2\mathbf{j} + x^2\mathbf{k}$; C is the intersection of the plane $z = y$ and the cylinder $x^2 + y^2 = 2y$, oriented counterclockwise as viewed from above.

In Problems 11 through 14, first show that the given vector field $\mathbf{F}$ *is irrotational; then apply the method of Example 3 to find a potential function* $\phi = \phi(x, y, z)$ *for* $\mathbf{F}$.

11. $\mathbf{F} = (3y - 2z)\mathbf{i} + (3x + z)\mathbf{j} + (y - 2x)\mathbf{k}$

12. $\mathbf{F} = (3y^3 - 10xz^2)\mathbf{i} + 9xy^2\mathbf{j} - 10x^2z\mathbf{k}$

13. $\mathbf{F} = (3e^z - 5y\sin x)\mathbf{i} + (5\cos x)\mathbf{j} + (17 + 3xe^z)\mathbf{k}$

14. $\mathbf{F} = r^3\mathbf{r}$, where $\mathbf{r} = x\mathbf{i} + y\mathbf{j} + z\mathbf{k}$ and $r = |\mathbf{r}|$

15. Suppose that $\mathbf{r} = x\mathbf{i} + y\mathbf{j} + z\,\mathbf{k}$ and that $\mathbf{a}$ is a constant vector. Show that
(a) $\nabla \cdot (\mathbf{a} \times \mathbf{r}) = 0$;
(b) $\nabla \times (\mathbf{a} \times \mathbf{r}) = 2\mathbf{a}$;
(c) $\nabla \cdot [(\mathbf{r} \cdot \mathbf{r})\mathbf{a}] = 2\mathbf{r} \cdot \mathbf{a}$;
(d) $\nabla \times [(\mathbf{r} \cdot \mathbf{r})\mathbf{a}] = 2(\mathbf{r} \times \mathbf{a})$.

16. Prove that

$$\iint_S (\text{curl}\,\mathbf{F}) \cdot \mathbf{n}\, dS$$

has the same value for all oriented surfaces S that have the same oriented boundary curve C.

17. Suppose that S is a closed surface. Prove in two different ways that

$$\iint_S (\text{curl}\,\mathbf{F}) \cdot \mathbf{n}\, dS = 0:$$

(a) by using the divergence theorem, with T the region bounded outside by S, and (b) by using Stokes' theorem, with the aid of a simple closed curve C on S.

Line integrals, surface integrals, and triple integrals of vector-valued functions are defined by componentwise integration. Such integrals appear in Problems 18 through 20.

18. Suppose that C and S are as described in the statement of Stokes' theorem and that ϕ is a scalar function. Prove that

$$\oint_C \phi\mathbf{T}\, ds = \iint_S \mathbf{n} \times \nabla\phi\, dS.$$

[*Suggestion:* Apply Stokes' theorem with $\mathbf{F} = \phi\mathbf{a}$, where $\mathbf{a}$ is an arbitrary constant vector.]

19. Suppose that $\mathbf{a}$ and $\mathbf{r}$ are as in Problem 15. Prove that

$$\oint_C (\mathbf{a} \times \mathbf{r}) \cdot \mathbf{T}\, ds = 2\mathbf{a} \cdot \iint_S \mathbf{n}\, dS.$$

20. Suppose that S is a closed surface that bounds the region T. Prove that

$$\iint_S \mathbf{n} \times \mathbf{F}\, dS = \iiint_T \nabla \times \mathbf{F}\, dV.$$

[*Suggestion:* Apply the divergence theorem to $\mathbf{F} \times \mathbf{a}$, where $\mathbf{a}$ is an arbitrary constant vector.]

REMARK The formulas in Problem 20, the divergence theorem, and Problem 20 of Section 15.6 all fit the pattern

$$\iint_S \mathbf{n} \odot (\quad)\, dS = \iiint_T \nabla \odot (\quad)\, dV,$$

where $\odot$ denotes either ordinary multiplication, the dot product, or the vector product, and either a scalar function or a vector-valued function is placed within the parentheses, as appropriate.

21. Suppose that the line integral $\int_C \mathbf{F} \cdot \mathbf{T}\, ds$ is independent of path. If

$$\phi(x, y, z) = \int_{P_0}^{P_1} \mathbf{F} \cdot \mathbf{T}\, ds$$

as in Eq. (11), show that $\nabla\phi = \mathbf{F}$. [*Suggestion:* If L is the line segment from (x, y, z) to $(x + \Delta x, y, z)$, then

$$\phi(x + \Delta x, y, z) - \phi(x, y, z) = \int_L \mathbf{F} \cdot \mathbf{T}\, ds = \int_x^{x+\Delta x} P\, dx.]$$

22. Let T be the submerged body of Problem 21 in Section 15.6, with centroid

$$\mathbf{r}_0 = \frac{1}{V} \iiint_T \mathbf{r}\, dV.$$

The torque about $\mathbf{r}_0$ of Archimedes' buoyant force $\mathbf{B} = -W\mathbf{k}$ is given by

$$\mathbf{L} = \iint_S (\mathbf{r} - \mathbf{r}_0) \times (-\delta g z\,\mathbf{n})\, dS.$$

(Why?) Apply the result of Problem 20 of this section to prove that $\mathbf{L} = \mathbf{0}$. It follows that $\mathbf{B}$ acts along the vertical line through the centroid $\mathbf{r}_0$ of the submerged body. (Why?)

CHAPTER 15 REVIEW: DEFINITIONS, CONCEPTS, RESULTS

Use the following list as a guide to concepts that you may need to review.

1. Definition and evaluation of the line integral
$$\int_C f(x, y, z)\, ds$$
2. Definition and evaluation of the line integral
$$\int_C P\,dx + Q\,dy + R\,dz$$
3. Relationship between the two types of line integrals; the line integral of the tangential component of a vector field
4. Line integrals and independence of path
5. Green's theorem
6. Flux and the vector form of Green's theorem
7. The divergence of a vector field
8. Definition and evaluation of the surface integral
$$\iint_S f(x, y, z)\, dS$$
9. Definition and evaluation of the surface integral
$$\iint_S P\,dy\,dz + Q\,dz\,dx + R\,dx\,dy$$
10. Relationship between the two types of surface integrals; the flux of a vector field across a surface
11. The divergence theorem in vector and in scalar notation
12. The curl of a vector field
13. Stokes' theorem in vector and in scalar notation
14. The circulation of a vector field around a simple closed curve
15. Physical interpretation of the divergence and the curl of a vector field

CHAPTER 15 MISCELLANEOUS PROBLEMS

1. Evaluate the line integral
$$\int_C (x^2 + y^2)\, ds,$$
where C is the straight line segment from $(0, 0)$ to $(3, 4)$.
2. Evaluate the line integral
$$\int_C y^2\, dx + x^2\, dy,$$
where C is the part of the graph of $y = x^2$ from $(-1, 1)$ to $(1, 1)$.
3. Evaluate the line integral
$$\int_C \mathbf{F}\cdot\mathbf{T}\, ds,$$
where $\mathbf{F} = x\mathbf{i} + y\mathbf{j} + z\mathbf{k}$ and C is the curve $x = e^{2t}$, $y = e^t$, $z = e^{-t}$, $0 \leqq t \leqq \ln 2$.
4. Evaluate the line integral
$$\int_C xyz\, ds,$$
where C is the path from $(1, 1, 2)$ to $(2, 3, 6)$ consisting of three straight line segments, the first parallel to the x-axis, the second parallel to the y-axis, and the third parallel to the z-axis.
5. Evaluate the line integral
$$\int_C \sqrt{z}\, dx + \sqrt{x}\, dy + y^2\, dz,$$
where C is the curve $x = t$, $y = t^{3/2}$, $z = t^2$, $0 \leqq t \leqq 4$.
6. Apply Theorem 2 of Section 15.3 to show that the line integral
$$\int_C y^2\, dx + 2xy\, dy + z\, dz$$
from the fixed point A to the fixed point B is independent of the path C from A to B.
7. Apply Theorem 2 of Section 15.3 to show that the line integral
$$\int_C x^2y\, dx + xy^2\, dy$$
is not independent of the path C from $(0, 0)$ to $(1, 1)$.
8. A wire shaped like the circle $x^2 + y^2 = a^2$, $z = 0$ has constant density and total mass M. Find its moment of inertia around (a) the z-axis; (b) the x-axis.
9. A wire shaped like the parabola $y = \frac{1}{2}x^2$, $0 \leqq x \leqq 2$, has density function $\delta = x$. Find its mass and its moment of inertia around the y-axis.
10. Find the work done by the force field $\mathbf{F} = z\mathbf{i} - x\mathbf{j} + y\mathbf{k}$ in moving a particle from $(1, 1, 1)$ to $(2, 4, 8)$ along the curve $y = x^2$, $z = x^3$.
11. Apply Green's theorem to evaluate the line integral
$$\oint_C x^2y\, dx + xy^2\, dy,$$
where C is the boundary of the region between the two curves $y = x^2$ and $y = 8 - x^2$.
12. Evaluate the line integral
$$\oint_C x^2\, dy,$$

where C is the cardioid with polar equation $r = 1+\cos\theta$, by first applying Green's theorem and then changing to polar coordinates.

13. Let C_1 be the circle $x^2 + y^2 = 1$ and C_2 the circle $(x-1)^2 + y^2 = 9$. Show that if $\mathbf{F} = x^2y\mathbf{i} - xy^2\mathbf{j}$, then

$$\oint_{C_1} \mathbf{F}\cdot\mathbf{n}\,ds = \oint_{C_2} \mathbf{F}\cdot\mathbf{n}\,ds$$

where $\mathbf{n}$ is an outer unit normal in the plane.

14. (a) Let C be the straight line segment from (x_1, y_1) to (x_2, y_2). Show that

$$\int_C -y\,dx + x\,dy = x_1y_2 - x_2y_1.$$

(b) Suppose that the vertices of a polygon are (x_1, y_1), $(x_2, y_2), \dots, (x_n, y_n)$, named in counterclockwise order around the polygon. Apply the result in part (a) to show that the area of the polygon is

$$A = \frac{1}{2}\sum_{i=1}^{n}(x_iy_{i+1} - x_{i+1}y_i),$$

where $x_{n+1} = x_1$ and $y_{n+1} = y_1$.

15. Suppose that the line integral $\int_C P\,dx + Q\,dy$ is independent of the path in the plane region D. Prove that

$$\oint_C P\,dx + Q\,dy = 0$$

for every piecewise smooth simple closed curve C in D.

16. Use Green's theorem to prove that

$$\oint_C P\,dx + Q\,dy = 0$$

for every piecewise smooth simple closed curve C in the plane region D if and only if $\partial P/\partial y = \partial Q/\partial x$ at each point of D.

17. Evaluate the surface integral

$$\iint_S (x^2 + y^2 + 2z)\,dS,$$

where S is the part of the paraboloid $z = 2 - x^2 - y^2$ that lies above the xy-plane.

18. Suppose that $\mathbf{F} = (x^2 + y^2 + z^2)(x\mathbf{i} + y\mathbf{j} + z\,\mathbf{k})$ and that S is the spherical surface $x^2 + y^2 + z^2 = a^2$. Evaluate

$$\iint_S \mathbf{F}\cdot\mathbf{n}\,dS$$

without actually performing an antidifferentiation.

19. Let T be the solid bounded by the paraboloids

$$z = x^2 + 2y^2 \quad\text{and}\quad z = 12 - 2x^2 - y^2,$$

and suppose that $\mathbf{F} = x\mathbf{i} + y\mathbf{j} + z\mathbf{k}$. Find by evaluation of surface integrals the outward flux of $\mathbf{F}$ across the boundary of T.

20. Give a reasonable definition—in terms of a surface integral—of the average distance of the point P from points of the surface S. Then show that the average distance of a fixed point of a spherical surface of radius a from all points of the surface is $\frac{4}{3}a$.

21. Suppose that the surface S is the graph of the equation $x = g(y, z)$ for (y, z) in the region D of the yz-plane. Prove that

$$\iint_S P\,dy\,dz + Q\,dz\,dx + R\,dx\,dy = \iint_D \left(P - Q\frac{\partial x}{\partial y} - R\frac{\partial x}{\partial z}\right)dy\,dz.$$

22. Suppose that the surface S is the graph of the equation $y = g(x, z)$ for (x, z) in the region D of the xz-plane. Prove that

$$\iint_S f(x, y, z)\,dS = \iint_D f(x, g(x, z), z)\sec\beta\,dx\,dz,$$

where $\sec\beta = \sqrt{1 + (\partial y/\partial x)^2 + (\partial y/\partial z)^2}$.

23. Let T be a region in space with volume V, boundary surface S, and centroid $(\bar{x}, \bar{y}, \bar{z})$. Use the divergence theorem to show that

$$\bar{z} = \frac{1}{2V}\iint_S z^2\,dx\,dy.$$

24. Apply the result of Problem 23 to find the centroid of the solid hemisphere

$$x^2 + y^2 + z^2 \leqq a^2, \qquad z \geqq 0.$$

Problems 25 through 27 outline the derivation of the heat equation for a body with temperature $u = u(x, y, z, t)$ at the point (x, y, z) at time t. Denote by K its heat conductivity and by c its heat capacity, both assumed constant, and let $k = K/c$. Let B be a small solid ball within the body, and let S denote the boundary sphere of B.

25. Deduce from the divergence theorem and Eq. (23) of Section 15.5 that the rate of heat flow across S into B is

$$R = \iiint_B K\,\nabla^2 u\,dV.$$

26. The meaning of heat capacity is that, if Δu is small, then $(c\,\Delta u)\,\Delta V$ calories of heat are required to raise the temperature of the volume ΔV by Δu degrees. It follows that the rate at which the volume ΔV is absorbing heat is $c(\partial u/\partial t)\,\Delta V$. (Why?) Conclude that the rate of heat flow into B is

$$R = \iiint_B c\frac{\partial u}{\partial t}\,dV.$$

27. Equate the results of Problem 25 and 26, apply the average value property of triple integrals, and then take the limit as the radius of the ball B approaches zero. You should thereby obtain the heat equation

$$\frac{\partial u}{\partial t} = k\,\nabla^2 u.$$

28. For a *steady-state* temperature function (one that is independent of time t), the heat equation reduces to Laplace's equation,

$$\nabla^2 u = \frac{\partial^2 u}{\partial x^2} + \frac{\partial^2 u}{\partial y^2} + \frac{\partial^2 u}{\partial z^2} = 0.$$

(a) Suppose that u_1 and u_2 are two solutions of Laplace's equation in the region T and that u_1 and u_2 agree on its boundary surface S. Apply Problem 17 of Section 15.6 to the function $f = u_1 - u_2$ to conclude that $\nabla f = \mathbf{0}$ at each point of T. (b) From the facts that $\nabla f = \mathbf{0}$ in T and $f \equiv 0$ on S, conclude that $f \equiv 0$, so $u_1 \equiv u_2$. Thus the steady-state temperatures within a region are *determined* by the boundary-value temperatures.

29. Suppose that $\mathbf{r} = x\mathbf{i} + y\mathbf{j} + z\mathbf{k}$ and that $\phi(r)$ is a scalar function of $r = |\mathbf{r}|$. Compute
(a) $\nabla\phi(r)$;
(b) $\operatorname{div}[\phi(r)\mathbf{r}]$;
(c) $\operatorname{curl}[\phi(r)\mathbf{r}]$.

30. Let S be the upper half of the torus obtained by revolving around the z-axis the circle $(y-a)^2+z^2 = b^2$ $(a > b > 0)$ in the yz-plane, with upper unit normal vector. Describe how to subdivide S to establish Stokes' theorem for it. How are the two boundary circles oriented?

31. Explain why the method of subdivision is not sufficient to establish Stokes' theorem for the Möbius strip of Fig. 15.5.3.

32. (a) Suppose that a fluid or a rigid body is rotating with angular speed ω radians per second around the line through the origin determined by the unit vector $\mathbf{u}$. Show that the velocity of the point with position vector $\mathbf{r}$ is $\mathbf{v} = \boldsymbol{\omega} \times \mathbf{r}$, where $\boldsymbol{\omega} = \omega\mathbf{u}$ is the angular velocity vector. Note that $|\mathbf{v}| = \omega|\mathbf{r}|\sin\theta$, where θ is the angle between $\mathbf{r}$ and $\boldsymbol{\omega}$. (b) Use the fact that $\mathbf{v} = \boldsymbol{\omega} \times \mathbf{r}$, established in part(a), to show that $\operatorname{curl}\mathbf{v} = 2\boldsymbol{\omega}$.

33. Consider an incompressible fluid flowing in space (no sources or sinks) with variable density $\delta(x, y, z, t)$ and velocity field $\mathbf{v}(x, y, z, t)$. Let B be a small ball with radius r and spherical surface S centered at the point (x_0, y_0, z_0). Then the amount of fluid within S at time t is

$$Q(t) = \iiint_B \delta\, dV,$$

and differentiation under the integral sign yields

$$Q'(t) = \iiint_B \frac{\partial \delta}{\partial t}\, dV.$$

(a) Consider fluid flow across S to get

$$Q'(t) = -\iint_S \delta\mathbf{v}\cdot\mathbf{n}\, dS,$$

where $\mathbf{n}$ is the outer unit normal vector to S. Now apply the divergence theorem to convert this into a volume integral. (b) Equate your two volume integrals for $Q'(t)$, apply the mean value theorem for integrals, and finally take limits as $r \to 0$ to obtain the **continuity equation**

$$\frac{\partial \delta}{\partial t} + \nabla\cdot(\delta\mathbf{v}) = 0.$$

APPENDICES

APPENDIX A: REAL NUMBERS AND INEQUALITIES

The **real numbers** are already familiar to you. They are just those numbers ordinarily used in most measurements. The mass, velocity, temperature, and charge of a body are measured with real numbers. Real numbers can be represented by **terminating** or **nonterminating** decimal expansions; in fact, every real number has a nonterminating decimal expansion because a terminating expansion can be padded with infinitely many zeros:

$$\frac{3}{8} = 0.375 = 0.375000000\ldots.$$

Any **repeating** decimal, such as

$$\frac{7}{22} = 0.31818181818\ldots,$$

represents a **rational** number, one that is the ratio of two integers. Conversely, every rational number is represented by a repeating decimal like the two displayed above. But the decimal expansion of an **irrational** number (a real number that is not rational), such as

$$\sqrt{2} = 1.414213562\ldots \quad \text{or} \quad \pi = 3.14159265358979\ldots,$$

is both nonterminating and nonrepeating.

The geometric interpretation of real numbers as points on the **real line** (or *real number line*) $\boldsymbol{R}$ should also be familiar to you. Each real number is represented by precisely one point of $\boldsymbol{R}$, and each point of $\boldsymbol{R}$ represents precisely one real number. By convention, the positive numbers lie to the right of zero and the negative numbers to the left, as in Fig. A.1.

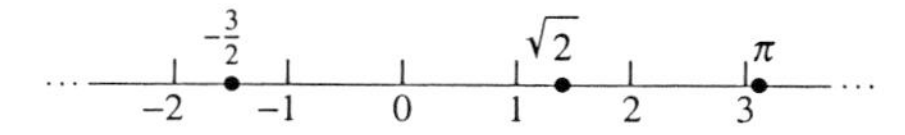

FIGURE A.1 The real line $\boldsymbol{R}$.

The following properties of inequalities of real numbers are fundamental and often used:

$$\begin{aligned}
&\text{If } a < b \text{ and } b < c, \text{ then } a < c.\\
&\text{If } a < b, \text{ then } a + c < b + c.\\
&\text{If } a < b \text{ and } c > 0, \text{ then } ac < bc.\\
&\text{If } a < b \text{ and } c < 0, \text{ then } ac > bc.
\end{aligned} \tag{1}$$

The last two statements mean that an inequality is preserved when its members are multiplied by a *positive* number but is *reversed* when they are multiplied by a *negative* number.

ABSOLUTE VALUE

The (nonnegative) distance along the real line between zero and the real number a is the **absolute value** of a, written $|a|$. Equivalently,

$$|a| = \begin{cases} a & \text{if } a \geqq 0; \\ -a & \text{if } a < 0. \end{cases} \quad \textbf{(2)}$$

The notation $a \geqq 0$ means that a is *either* greater than zero *or* equal to zero. Equation (2) implies that $|a| \geqq 0$ for every real number a and that $|a| = 0$ if and only if $a = 0$.

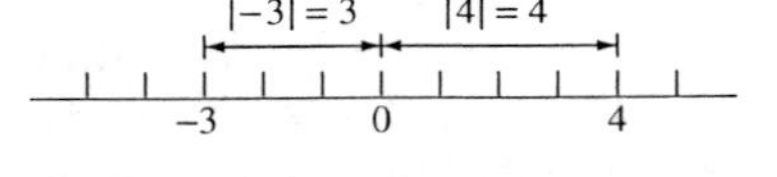

FIGURE A.2 The absolute value of a real number is simply its distance from zero (Example 1).

EXAMPLE 1 As Fig. A.2 shows,

$$|4| = 4 \quad \text{and} \quad |-3| = 3.$$

Moreover, $|0| = 0$ and $|\sqrt{2} - 2| = 2 - \sqrt{2}$, the latter being true because $2 > \sqrt{2}$. Thus $\sqrt{2} - 2 < 0$, and hence

$$|\sqrt{2} - 2| = -(\sqrt{2} - 2) = 2 - \sqrt{2}.$$

◆

The following properties of absolute values are frequently used:

$$\begin{aligned} |a| &= |-a| = \sqrt{a^2} \geqq 0, \\ |ab| &= |a|\,|b|, \\ -|a| &\leqq a \leqq |a|, \\ \text{and} \quad |a| < b \quad &\text{if and only if} \quad -b < a < b. \end{aligned} \quad \textbf{(3)}$$

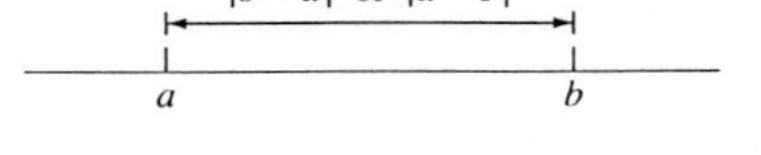

FIGURE A.3 The distance between a and b.

The **distance** between the real numbers a and b is defined to be $|a - b|$ (or $|b - a|$; there's no difference). This distance is simply the length of the line segment of the real line $\boldsymbol{R}$ with endpoints a and b (Fig. A.3).

The properties of inequalities and of absolute values in Eqs. (1) through (3) imply the following important theorem.

THEOREM 1 Triangle Inequality
For all real numbers a and b,

$$|a + b| \leqq |a| + |b|. \quad \textbf{(4)}$$

PROOF There are several cases to consider, depending upon whether the two numbers a and b are positive or negative and which has the larger absolute value. If both are positive, then so is $a + b$; in this case,

$$|a + b| = a + b = |a| + |b|. \quad \textbf{(5)}$$

If $a > 0$ but $b < 0$ and $|b| < |a|$, then

$$0 < a + b < a,$$

so

$$|a + b| = a + b < a = |a| < |a| + |b|, \quad \textbf{(6)}$$

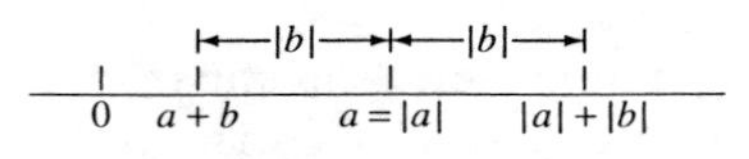

FIGURE A.4 The triangle inequality with $a > 0$, $b < 0$, and $|b| < |a|$.

as illustrated in Fig. A.4. The other cases are similar. In particular, we see that the triangle inequality is actually an equality [as in Eq. (5)] unless a and b have different signs, in which case it is a strict inequality [as in Eq. (6)]. ◄

INTERVALS

Suppose that S is a set (collection) of real numbers. It is common to describe S by the notation

$$S = \{x : \text{condition}\},$$

where the "condition" is true for those numbers x in S and false for those numbers x not in S. The most important sets of real numbers in calculus are *intervals.* If $a < b$, then the **open interval** (a, b) is defined to be the set

$$(a, b) = \{x : a < x < b\}$$

of real numbers, and the **closed interval** $[a, b]$ is

$$[a, b] = \{x : a \leqq x \leqq b\}.$$

Thus a closed interval contains its endpoints, whereas an open interval does not. We also use the **half-open intervals**

$$[a, b) = \{x : a \leqq x < b\} \quad \text{and} \quad (a, b] = \{x : a < x \leqq b\}.$$

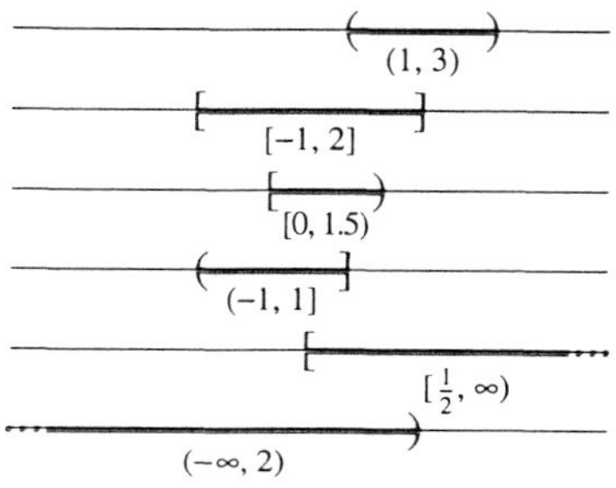

Thus the open interval $(1, 3)$ is the set of those real numbers x such that $1 < x < 3$, the closed interval $[-1, 2]$ is the set of those real numbers x such that $-1 \leqq x \leqq 2$, and the half-open interval $(-1, 2]$ is the set of those real numbers x such that $-1 < x \leqq 2$. In Fig. A.5 we show examples of such intervals as well as some **unbounded** intervals, which have forms such as

$$[a, +\infty) = \{x : x \geqq a\},$$

$$(-\infty, a] = \{x : x \leqq a\},$$

$$(a, +\infty) = \{x : x > a\},$$

$$\text{and} \quad (-\infty, a) = \{x : x < a\}.$$

$(1, 3)$

$[-1, 2]$

$[0, 1.5)$

$(-1, 1]$

$[\frac{1}{2}, \infty)$

$(-\infty, 2)$

FIGURE A.5 Some examples of intervals of real numbers.

The symbols $+\infty$ and $-\infty$, denoting "plus infinity" and "minus infinity," are merely notational conveniences and do *not* represent real numbers—the real line $\boldsymbol{R}$ does *not* have "endpoints at infinity." The use of these symbols is motivated by the brief and natural descriptions $[\pi, +\infty)$ and $(-\infty, 2)$ for the sets

$$\{x : x \geqq \pi\} \quad \text{and} \quad \{x : x < 2\}$$

of all real numbers x such that $x \geqq \pi$ and $x < 2$, respectively.

INEQUALITIES

The set of solutions of an inequality involving a variable x is often an interval or a union of intervals, as in the next examples. The **solution set** of such an inequality is simply the set of all those real numbers x that satisfy the inequality.

EXAMPLE 2 Solve the inequality $2x - 1 < 4x + 5$.

Solution Using the properties of inequalities listed in (1), we proceed much as if we were solving an equation for x: We isolate x on one side of the inequality. Here we begin with

$$2x - 1 < 4x + 5$$

and it follows that

$$-1 < 2x + 5;$$

$$-6 < 2x;$$

$$-3 < x.$$

Hence the solution set is the unbounded interval $(-3, +\infty)$. ◆

EXAMPLE 3 Solve the inequality $-13 < 1 - 4x \leqq 7$.

Solution We simplify the given inequality as follows:

$$-13 < 1 - 4x \leqq 7;$$
$$-7 \leqq 4x - 1 < 13;$$
$$-6 \leqq 4x < 14;$$
$$-\tfrac{3}{2} \leqq x < \tfrac{7}{2}.$$

Thus the solution set of the given inequality is the half-open interval $[-\frac{3}{2}, \frac{7}{2})$. ◆

EXAMPLE 4 Solve the inequality $|3 - 5x| < 2$.

Solution From the fourth property of absolute values in (3), we see that the given inequality is equivalent to

$$-2 < 3 - 5x < 2.$$

We now simplify as in the previous two examples:

$$-5 < -5x < -1;$$
$$\tfrac{1}{5} < x < 1.$$

Thus the solution set is the open interval $\left(\frac{1}{5}, 1\right)$. ◆

EXAMPLE 5 Solve the inequality

$$\frac{5}{|2x - 3|} < 1.$$

Solution It is usually best to begin by eliminating a denominator containing the unknown. Here we multiply each term by the *positive* quantity $|2x - 3|$ to obtain the equivalent inequality

$$|2x - 3| > 5.$$

It follows from the last property in (3) that this is so if and only if either

$$2x - 3 < -5 \quad \text{or} \quad 2x - 3 > 5.$$

The solutions of these *two* inequalities are the open intervals $(-\infty, -1)$ and $(4, +\infty)$, respectively. Hence the solution set of the original inequality consists of all those numbers x that lie in *either* of these two open intervals. ◆

The **union** of the two sets S and T is the set $S \cup T$ given by

$$S \cup T = \{x : \text{either } x \in S \text{ or } x \in T \text{ or both}\}.$$

Thus the solution set in Example 5 can be written in the form $(-\infty, -1) \cup (4, +\infty)$.

EXAMPLE 6 In accord with Boyle's law, the pressure p (in pounds per square inch) and volume V (in cubic inches) of a certain gas satisfy the condition $pV = 100$. Suppose that $50 \leqq V \leqq 150$. What is the range of possible values of the pressure p?

Solution If we substitute $V = 100/p$ in the given inequality $50 \leqq V \leqq 150$, we get

$$50 \leqq \frac{100}{p} \leqq 150.$$

It follows that *both*

$$50 \leqq \frac{100}{p} \quad \textit{and} \quad \frac{100}{p} \leqq 150;$$

that is, that both

$$p \leqq 2 \quad \text{and} \quad p \geqq \tfrac{2}{3}.$$

Thus the pressure p must lie in the closed interval $[\frac{2}{3}, 2]$. ◆

The **intersection** of the two sets S and T is the set $S \cap T$ defined as follows:

$$S \cap T = \{x : \text{both } x \in S \text{ and } x \in T\}.$$

Thus the solution set in Example 6 is the set $(-\infty, 2] \cap [\frac{2}{3}, +\infty) = [\frac{2}{3}, 2]$.

APPENDIX A PROBLEMS

Simplify the expressions in Problems 1 through 12 by writing each without using absolute value symbols.

1. $|3 - 17|$

2. $|-3| + |17|$

3. $\left|-0.25 - \frac{1}{4}\right|$

4. $|5| - |-7|$

5. $|(-5)(4 - 9)|$

6. $\dfrac{|-6|}{|4| + |-2|}$

7. $|(-3)^3|$

8. $\left|3 - \sqrt{3}\right|$

9. $\left|\pi - \frac{22}{7}\right|$

10. $-|7 - 4|$

11. $|x - 3|$, given $x < 3$

12. $|x - 5| + |x - 10|$, given $|x - 7| < 1$

Solve the inequalities in Problems 13 through 31. Write each solution set in interval notation.

13. $2x - 7 < -3$

14. $1 - 4x > 2$

15. $3x - 4 \geqq 17$

16. $2x + 5 \leqq 9$

17. $2 - 3x < 7$

18. $6 - 5x > -9$

19. $-3 < 2x + 5 < 7$

20. $4 \leqq 3x - 5 \leqq 10$

21. $-6 \leqq 5 - 2x < 2$

22. $3 < 1 - 5x < 7$

23. $|3 - 2x| < 5$

24. $|5x + 3| \leqq 4$

25. $|1 - 3x| > 2$

26. $1 < |7x - 1| < 3$

27. $2 \leqq |4 - 5x| \leqq 4$

28. $\dfrac{1}{2x + 1} > 3$

29. $\dfrac{2}{7 - 3x} \leqq -5$

30. $\dfrac{2}{|3x - 4|} < 1$

31. $\dfrac{1}{|1 - 5x|} \geqq -\dfrac{1}{3}$

32. Solve the inequality $x^2 - x - 6 > 0$. [*Suggestion:* Conclude from the factorization $x^2 - x - 6 = (x - 3)(x + 2)$ that the quantities $x - 3$ and $x + 2$ are either both positive or both negative. Consider the two cases separately to deduce that the solution set is $(-\infty, -2) \cup (3, \infty)$.]

Use the method of Problem 32 to solve the inequalities in Problems 33 through 36.

33. $x^2 - 2x - 8 > 0$

34. $x^2 - 3x + 2 < 0$

35. $4x^2 - 8x + 3 \geqq 0$

36. $2x \geqq 15 - x^2$

37. In accord with Boyle's law, the pressure p (in pounds per square inch) and volume V (in cubic inches) of a certain gas satisfy the condition $pV = 800$. What is the range of possible values of the pressure, given $100 \leqq V \leqq 200$?

38. The relationship between the Fahrenheit temperature F and the Celsius temperature C is given by $F = 32 + \frac{9}{5}C$. If the temperature on a certain day ranged from a low of 70°F to a high of 90°F, what was the range of the temperature in degrees Celsius?

39. An electrical circuit contains a battery supplying E volts in series with a resistance of R ohms, as shown in Fig. A.6. Then the current of I amperes that flows in the circuit satisfies Ohm's law, $E = IR$. If $E = 100$ and $25 < R < 50$, what is the range of possible values of I?

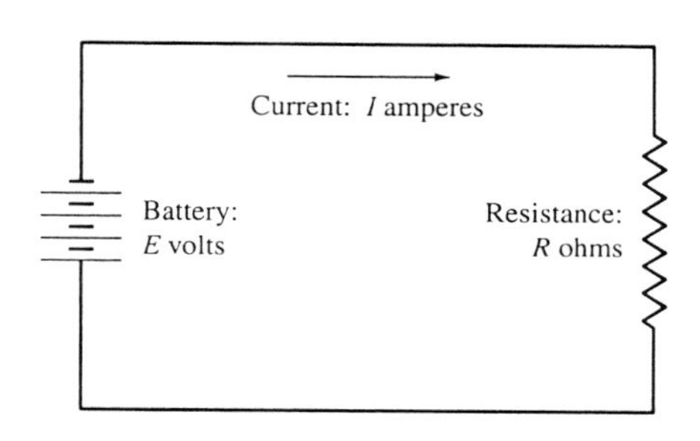

FIGURE A.6 A simple electric circuit.

40. The period T (in seconds) of a simple pendulum of length L (in feet) is given by $T = 2\pi\sqrt{L/32}$. If $3 < L < 4$, when is the range of possible values of T?

41. Use the properties of inequalities in (1) to show that the sum of two positive numbers is positive.

42. Use the properties of inequalities in (1) to show that the product of two positive numbers is positive.

43. Prove that the product of two negative numbers is positive and that the product of a positive number and a negative number is negative.

44. Suppose that $a < b$ and that a and b are either both positive or both negative. Prove that $1/a > 1/b$.

45. Apply the triangle inequality twice to show that

$$|a + b + c| \leqq |a| + |b| + |c|$$

for arbitrary real numbers a, b, and c.

46. Write $a = (a - b) + b$ to deduce from the triangle inequality that

$$|a| - |b| \leqq |a - b|$$

for arbitrary real numbers a and b.

47. Deduce from the definition in (2) that $|a| < b$ if and only if $-b < a < b$.

APPENDIX B: THE COORDINATE PLANE AND STRAIGHT LINES

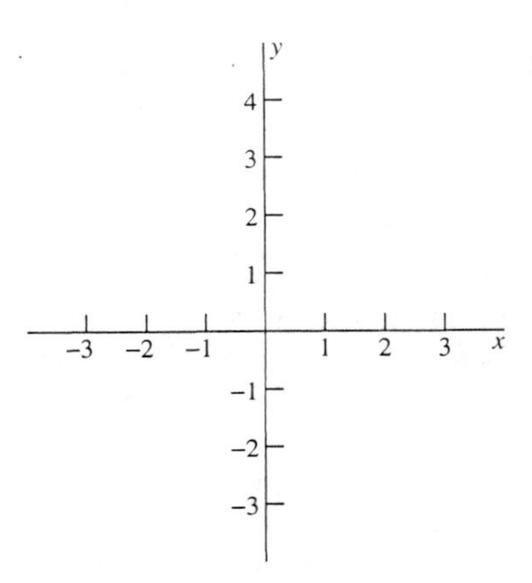

FIGURE B.1 The coordinate plane.

Imagine the flat, featureless, two-dimensional plane of Euclid's geometry. Install a copy of the real number line $\boldsymbol{R}$, with the line horizontal and the positive numbers to the right. Add another copy of $\boldsymbol{R}$ perpendicular to the first, with the two lines crossing where the number zero is located on each. The vertical line should have the positive numbers above the horizontal line, as in Fig. B.1; the negative numbers thus will be below it. The horizontal line is called the x-**axis** and the vertical line is called the y-**axis.**

With these added features, we call the plane the **coordinate plane,** because it's now possible to locate any point there by a pair of numbers, called the *coordinates of the point*. Here's how: If P is a point in the plane, draw perpendiculars from P to the coordinate axes, as shown in Fig. B.2. One perpendicular meets the x-axis at the x-**coordinate** (or **abscissa**) of P, labeled x_1 in Fig. B.2. The other meets the y-axis in the y-**coordinate** (or **ordinate**) y_1 of P. The pair of numbers (x_1, y_1), in that order, is called the **coordinate pair** for P, or simply the **coordinates** of P. To be concise, we speak of "the point $P(x_1, y_1)$."

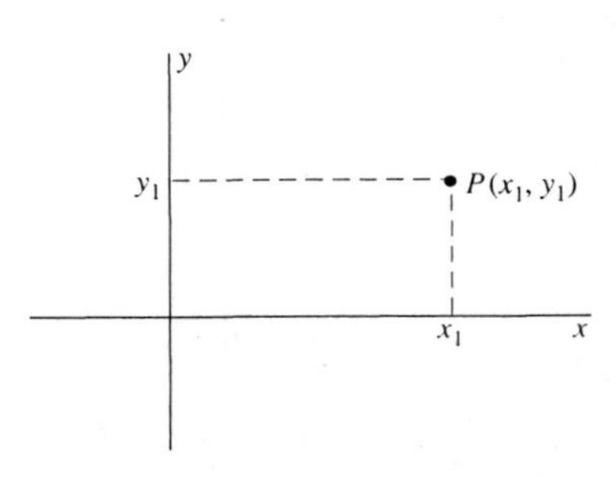

FIGURE B.2 The point P has rectangular coordinates (x_1, y_1).

This coordinate system is called the **rectangular coordinate system,** or the **Cartesian coordinate system** (because its use was popularized, beginning in the 1630s, by the French mathematician and philosopher René Descartes [1596–1650]). The plane, thus coordinatized, is denoted by $\boldsymbol{R}^2$ because two copies of $\boldsymbol{R}$ are used; it is known also as the **Cartesian plane.**

Rectangular coordinates are easy to use, because $P(x_1, y_1)$ and $Q(x_2, y_2)$ denote the same point if and only if $x_1 = x_2$ and $y_1 = y_2$. Thus when you know that P and Q are two different points, you may conclude that P and Q have different abscissas, different ordinates, or both.

The point of symmetry $(0, 0)$ where the coordinate axes meet is called the **origin.** All points on the x-axis have coordinates of the form $(x, 0)$. Although the *real number* x is not the same as the geometric point $(x, 0)$, there are situations in which it is useful to think of the two as the same. Similar remarks apply to points $(0, y)$ on the y-axis.

The concept of distance in the coordinate plane is based on the **Pythagorean theorem:** If ABC is a right triangle with its right angle at the point C, with hypotenuse of length c and the other two sides of lengths a and b (as in Fig. B.3), then

$$c^2 = a^2 + b^2. \tag{1}$$

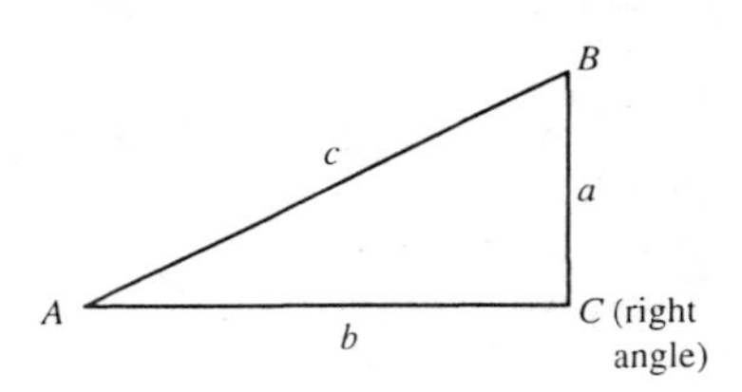

FIGURE B.3 The Pythagorean theorem.

The converse of the Pythagorean theorem is also true: If the three sides of a given triangle satisfy the Pythagorean relation in Eq. (1), then the angle opposite side c must be a right angle.

The *distance* $d(P_1, P_2)$ between the points P_1 and P_2 is, by definition, the length of the straight-line segment joining P_1 and P_2. The following formula gives $d(P_1, P_2)$ in terms of the coordinates of the two points.

Distance Formula

The **distance** between the two points $P_1(x_1, y_1)$ and $P_2(x_2, y_2)$ is

$$d(P_1, P_2) = \sqrt{(x_2 - x_1)^2 + (y_2 - y_1)^2}. \tag{2}$$

PROOF If $x_1 \neq x_2$ and $y_1 \neq y_2$, then Eq. (2) follows from the Pythagorean theorem. Use the right triangle with vertices P_1, P_2, and $P_3(x_2, y_1)$ shown in Fig. B.4.

If $x_1 = x_2$, then P_1 and P_2 lie in a vertical line. In this case

$$d(P_1, P_2) = |y_1 - y_2| = \sqrt{(y_1 - y_2)^2}.$$

This agrees with Eq. (2) because $x_1 = x_2$. The remaining case $(y_1 = y_2)$ is similar. ◄

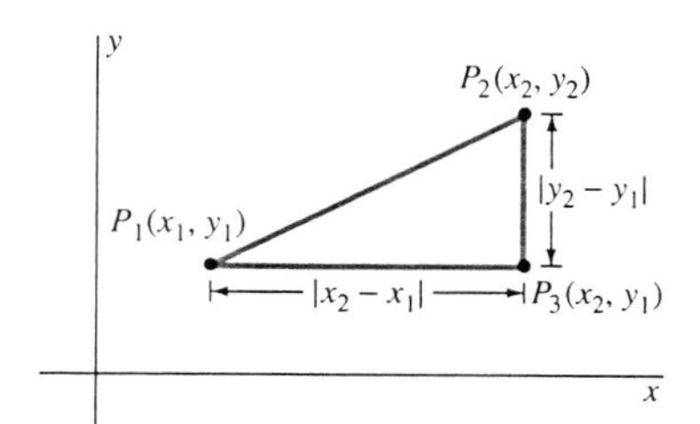

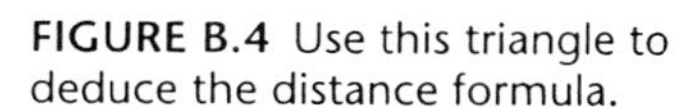

FIGURE B.4 Use this triangle to deduce the distance formula.

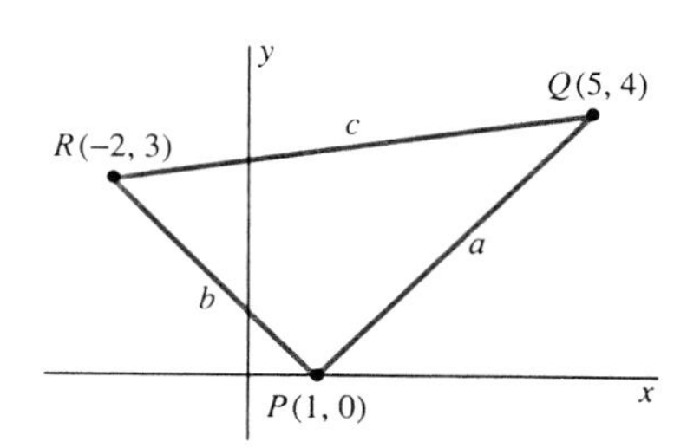

FIGURE B.5 Is this a right triangle (Example 1)?

EXAMPLE 1 Show that the triangle PQR with vertices $P(1, 0)$, $Q(5, 4)$, and $R(-2, 3)$ is a right triangle (Fig. B.5).

Solution The distance formula gives

$$\begin{aligned} a^2 &= [d(P, Q)]^2 = (5-1)^2 + (4-0)^2 = 32, \\ b^2 &= [d(P, R)]^2 = (-2-1)^2 + (3-0)^2 = 18, \quad \text{and} \\ c^2 &= [d(Q, R)]^2 = (-2-5)^2 + (3-4)^2 = 50. \end{aligned}$$

Because $a^2 + b^2 = c^2$, the *converse* of the Pythagorean theorem implies that RPQ is a right angle. (The right angle is at P because P is the vertex opposite the longest side, QR.) ◆

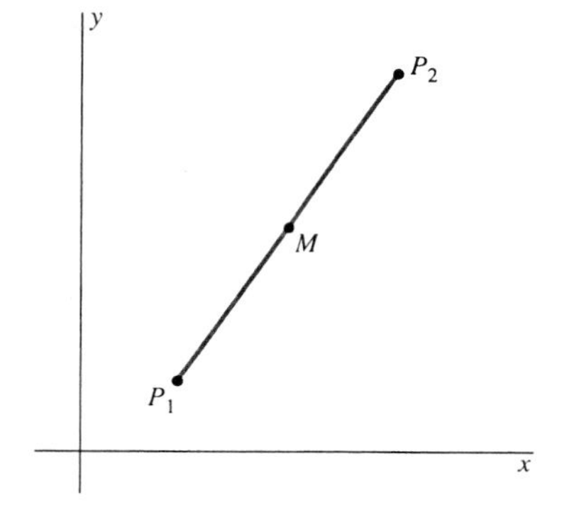

FIGURE B.6 The midpoint M.

Another application of the distance formula is an expression for the coordinates of the midpoint M of the line segment P_1P_2 with endpoints P_1 and P_2 (Fig. B.6). Recall from geometry that M is the one (and only) point of the line segment P_1P_2 that is equally distant from P_1 and P_2. The following formula tells us that the coordinates of M are the *averages* of the corresponding coordinates of P_1 and P_2.

Midpoint Formula

The **midpoint** of the line segment with endpoints $P_1(x_1, y_1)$ and $P_2(x_2, y_2)$ is the point $M(\overline{x}, \overline{y})$ with coordinates

$$\overline{x} = \tfrac{1}{2}(x_1 + x_2) \quad \text{and} \quad \overline{y} = \tfrac{1}{2}(y_1 + y_2). \tag{3}$$

PROOF If you substitute the coordinates of P_1, M, and P_2 in the distance formula, you find that $d(P_1, M) = d(P_2, M)$. All that remains is to show that M lies on the line segment P_1P_2. We ask you to do this, and thus complete the proof, in Problem 31. ◀

STRAIGHT LINES AND SLOPE

We want to define the *slope* of a straight line, a measure of its rate of rise or fall from left to right. Given a nonvertical straight line L in the coordinate plane, choose two points $P_1(x_1, y_1)$ and $P_2(x_2, y_2)$ on L. Consider the **increments** Δx and Δy (read "delta x" and "delta y") in the x- and y-coordinates from P_1 to P_2. These are defined as follows:

$$\Delta x = x_2 - x_1 \quad \text{and} \quad \Delta y = y_2 - y_1. \tag{4}$$

FIGURE B.7 The slope of a straight line.

Engineers (and others) call Δx the **run** from P_1 to P_2 and Δy the **rise** from P_1 to P_2, as in Fig. B.7. The **slope** m of the nonvertical line L is then defined to be the ratio of the rise to the run:

$$m = \frac{\Delta y}{\Delta x} = \frac{y_2 - y_1}{x_2 - x_1}. \tag{5}$$

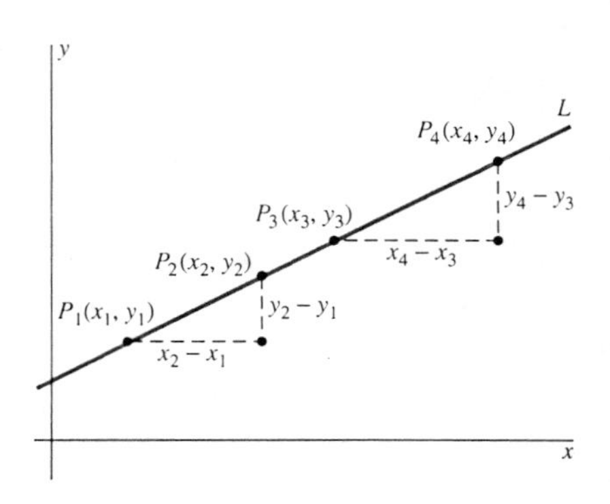

FIGURE B.8 The result of the slope computation does not depend on which two points of L are used.

This is also the definition of a line's slope in civil engineering (and elsewhere). In a surveying text you are likely to find the memory aid

$$\text{"slope} = \frac{\text{rise}}{\text{run}}.\text{"}$$

Recall that corresponding sides of similar (that is, equal-angled) triangles have equal ratios. Hence, if $P_3(x_3, y_3)$ and $P_4(x_4, y_4)$ are two other points of L, then the similarity of the triangles in Fig. B.8 implies that

$$\frac{y_4 - y_3}{x_4 - x_3} = \frac{y_2 - y_1}{x_2 - x_1}.$$

Therefore, the slope m as defined in Eq. (5) does *not* depend on the particular choice of P_1 and P_2.

If the line L is horizontal, then $\Delta y = 0$. In this case Eq. (5) gives $m = 0$. If L is vertical, then $\Delta x = 0$, so the slope of L is *not defined*. Thus we have the following statements:

- Horizontal lines have slope zero.
- Vertical lines have no defined slope.

EXAMPLE 2

(a) The slope of the line through the points $(3, -2)$ and $(-1, 4)$ is

$$m = \frac{4 - (-2)}{(-1) - 3} = \frac{6}{-4} = -\frac{3}{2}.$$

(b) The points $(3, -2)$ and $(7, -2)$ have the same y-coordinate. Therefore, the line through them is horizontal and thus has slope zero.

(c) The points $(3, -2)$ and $(3, 4)$ have the same x-coordinate. Thus the line through them is vertical, and so its slope is undefined. ◆

EQUATIONS OF STRAIGHT LINES

Our immediate goal is to be able to write equations of given straight lines. That is, if L is a straight line in the coordinate plane, we wish to construct a mathematical sentence—an equation—about points (x, y) in the plane. We want this equation to be *true* when (x, y) is a point on L and *false* when (x, y) is not a point on L. Clearly this equation will involve x and y and some numerical constants determined by L itself. For us to write this equation, the concept of the slope of L is essential.

Suppose, then, that $P(x_0, y_0)$ is a fixed point on the nonvertical line L of slope m. Let $P(x, y)$ be any *other* point on L. We apply Eq. (5) with P and P_0 in place of P_1 and P_2 to find that

$$m = \frac{y - y_0}{x - x_0};$$

that is,

$$y - y_0 = m(x - x_0). \tag{6}$$

Because the point (x_0, y_0) satisfies Eq. (6), as does every other point of L, and because no other points of the plane can do so, Eq. (6) is indeed an equation for the given line L. In summary, we have the following result.

The Point-Slope Equation

The point $P(x, y)$ lies on the line with slope m through the fixed point (x_0, y_0) if and only if its coordinates satisfy the equation

$$y - y_0 = m(x - x_0). \tag{6}$$

Equation (6) is called the **point-slope** equation of L, partly because the coordinates of the point (x_0, y_0) and the slope m of L may be read directly from this equation.

EXAMPLE 3 Write an equation for the straight line L through the points $P_1(1, -1)$ and $P_2(3, 5)$.

Solution The slope m of L may be obtained from the two given points:

$$m = \frac{5 - (-1)}{3 - 1} = 3.$$

Either P_1 or P_2 will do for the fixed point. We use $P_1(1, -1)$. Then, with the aid of Eq. (6), the point-slope equation of L is

$$y + 1 = 3(x - 1).$$

If simplification is appropriate, we may write $3x - y = 4$ or $y = 3x - 4$. ◆

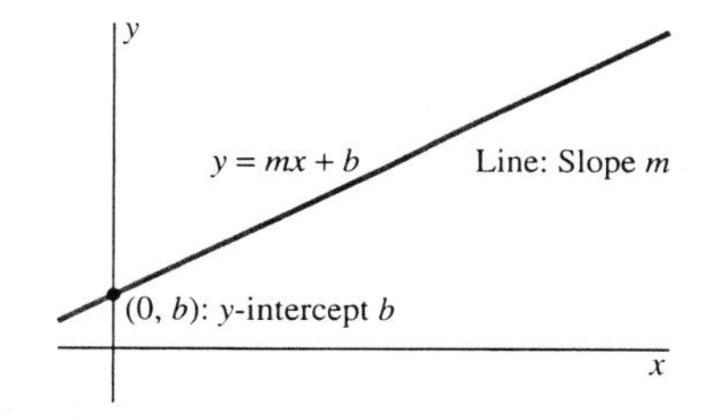

FIGURE B.9 The straight line with equation $y = mx + b$ has a slope m and y-intercept b.

Equation (6) can be written in the form

$$y = mx + b \tag{7}$$

where $b = y_0 - mx_0$ is a constant. Because $y = b$ when $x = 0$, the **y-intercept** of L is the point $(0, b)$ shown in Fig. B.9. Equations (6) and (7) are different forms of the equation of a straight line.

The Slope-Intercept Equation

The point $P(x, y)$ lies on the line with slope m and y-intercept b if and only if the coordinates of P satisfy the equation

$$y = mx + b. \tag{7}$$

Perhaps you noticed that both Eq. (6) and Eq. (7) can be written in the form of the general linear equation

$$Ax + By = C, \tag{8}$$

where A, B, and C are constants. Conversely, if $B \neq 0$, then Eq. (8) can be written in the form of Eq. (7) if we divide each term by B. Therefore Eq. (8) represents a straight line with its slope being the coefficient of x *after* solution of the equation for y. If $B = 0$, then Eq. (8) reduces to the equation of a vertical line: $x = K$ (where K is a constant). If $A = 0$ and $B \neq 0$, then Eq. (8) reduces to the equation of a horizontal line: $y = H$ (where H is a constant). Thus we see that Eq. (8) is always an equation of a straight line unless $A = B = 0$. Conversely, every straight line in the coordinate plane—even a vertical one—has an equation of the form in (8).

PARALLEL LINES AND PERPENDICULAR LINES

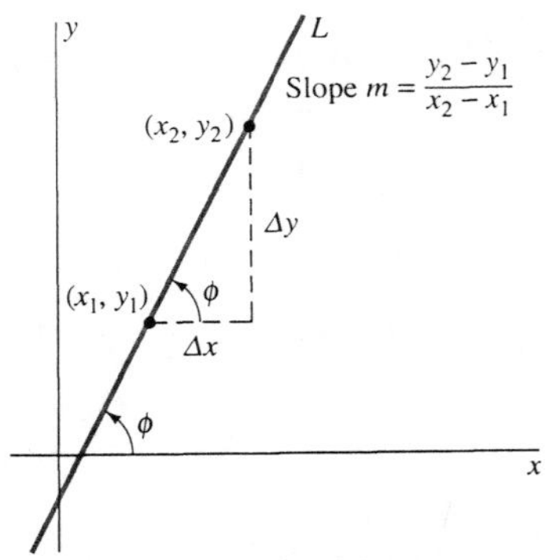

FIGURE B.10 How is the angle of inclination ϕ related to the slope m?

If the line L is not horizontal, then it must cross the x-axis. Then its **angle of inclination** is the angle ϕ measured counterclockwise from the positive x-axis to L. It follows that $0^\circ < \phi < 180^\circ$ if ϕ is measured in degrees. Figure B.10 makes it clear that this angle ϕ and the slope m of a nonvertical line are related by the equation

$$m = \frac{\Delta y}{\Delta x} = \tan\phi. \tag{9}$$

This is true because if ϕ is an acute angle in a right triangle, then $\tan\phi$ is the ratio of the leg opposite ϕ to the leg adjacent to ϕ.

Your intuition correctly assures you that two lines are parallel if and only if they have the same angle of inclination. So it follows from Eq. (9) that two parallel nonvertical lines have the same slope and that two lines with the same slope must be parallel. This completes the proof of Theorem 1.

THEOREM 1 Slopes of Parallel Lines
Two nonvertical lines are parallel if and only if they have the same slope.

Theorem 1 can also be proved without the use of the tangent function. The two lines shown in Fig. B.11 are parallel if and only if the two right triangles are similar, which is equivalent to the slopes of the lines being equal.

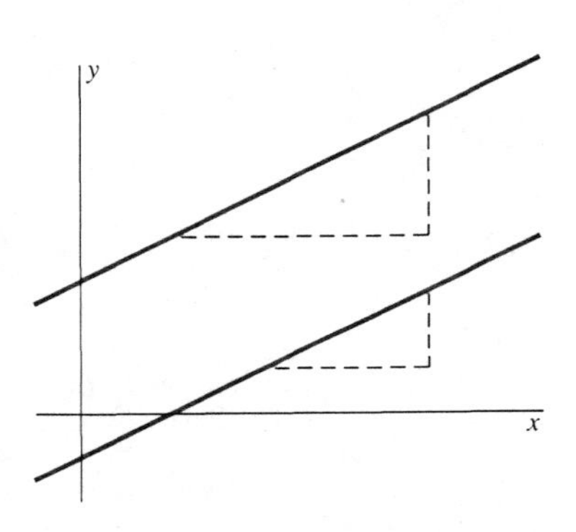

FIGURE B.11 Two parallel lines.

EXAMPLE 4 Write an equation of the line L that passes through the point $P(3, -2)$ and is parallel to the line L' with the equation $x + 2y = 6$.

Solution When we solve the equation of L' for y, we get $y = -\frac{1}{2}x + 3$. So L' has slope $m = -\frac{1}{2}$. Because L has the same slope, its point-slope equation is then

$$y + 2 = -\tfrac{1}{2}(x - 3);$$

if you prefer, $x + 2y = -1$. ◆

THEOREM 2 Slopes of Perpendicular Lines
Two lines L_1 and L_2 with slopes m_1 and m_2, respectively, are perpendicular if and only if

$$m_1 m_2 = -1. \tag{10}$$

That is, the slope of each is the *negative reciprocal* of the slope of the other.

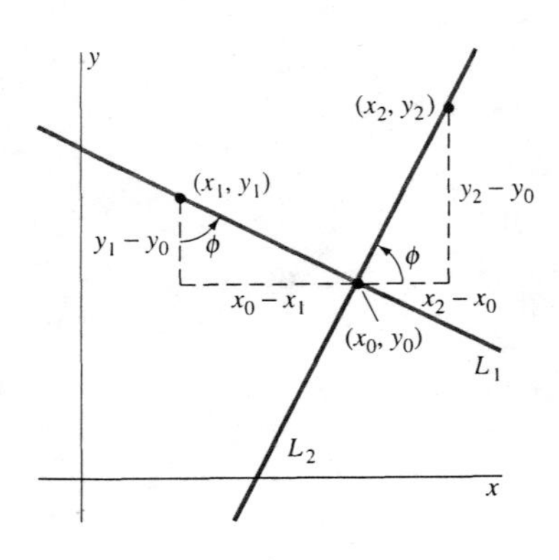

FIGURE B.12 Illustration of the proof of Theorem 2.

PROOF If the two lines L_1 and L_2 are perpendicular and the slope of each exists, then neither is horizontal or vertical. Thus the situation resembles the one shown in Fig. B.12, in which the two lines meet at the point (x_0, y_0). It is easy to see that the two right triangles of the figure are similar, so equality of ratios of corresponding sides yields

$$m_2 = \frac{y_2 - y_0}{x_2 - x_0} = \frac{x_0 - x_1}{y_1 - y_0} = -\frac{x_1 - x_0}{y_1 - y_0} = -\frac{1}{m_1}.$$

Thus Eq. (10) holds if the two lines are perpendicular. This argument can be reversed to prove the converse—that the lines are perpendicular if $m_1 m_2 = -1$. ◄

EXAMPLE 5 Write an equation of the line L through the point $P(3, -2)$ that is perpendicular to the line L' with equation $x + 2y = 6$.

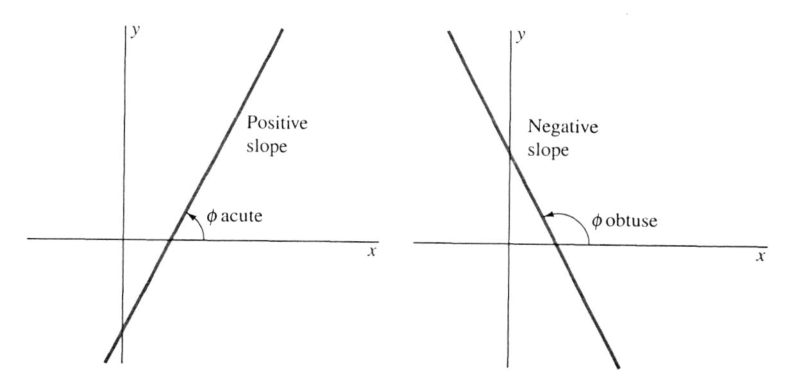

FIGURE B.13 Positive and negative slope; effect on ϕ.

Solution As we saw in Example 4, the slope of L' is $m' = -\frac{1}{2}$. By Theorem 2, the slope of L is $m = -1/m' = 2$. Thus L has the point-slope equation

$$y + 2 = 2(x - 3);$$

equivalently, $2x - y = 8$. ◆

You will find it helpful to remember that the *sign* of the slope m of the line L indicates whether L runs upward or downward as your eyes move from left to right. If $m > 0$, then the angle of inclination ϕ of L must be an acute angle, because $m = \tan\phi$. In this case, L "runs upward" to the right. If $m < 0$, then ϕ is obtuse, so L "runs downward." Figure B.13 shows the geometry behind these observations.

GRAPHICAL INVESTIGATION

Many mathematical problems require the simultaneous solution of a pair of linear equations of the form

$$a_1x + b_1y = c_1,$$
$$a_2x + b_2y = c_2. \tag{11}$$

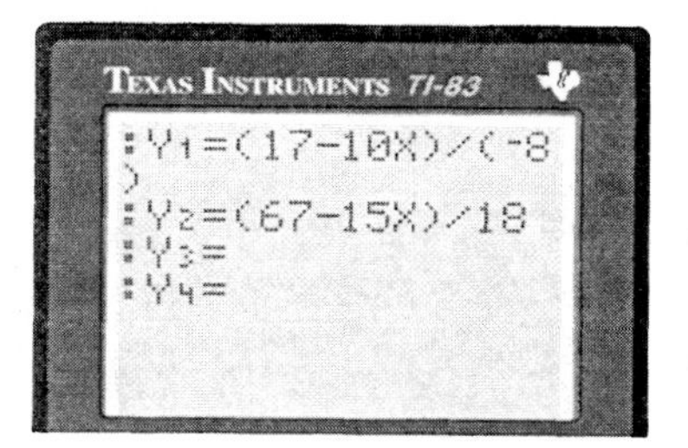

FIGURE B.14 A calculator prepared to graph the lines in Eq. (12) (Example 6).

The graph of these two equations consists of a pair of straight lines in the xy-plane. If these two lines are not parallel, then they must intersect at a single point whose coordinates (x_0, y_0) constitute the solution of (11). That is, $x = x_0$ and $y = y_0$ are the (only) values of x and y for which both equations in (11) are true.

In elementary algebra you studied various elimination and substitution methods for solving linear systems such as the one in (11). Example 6 illustrates an alternative *graphical method* that is sometimes useful when a graphing utility—a graphics calculator or a computer with a graphing program—is available.

EXAMPLE 6 We want to investigate the simultaneous solution of the linear equations

$$10x - 8y = 17,$$
$$15x + 18y = 67. \tag{12}$$

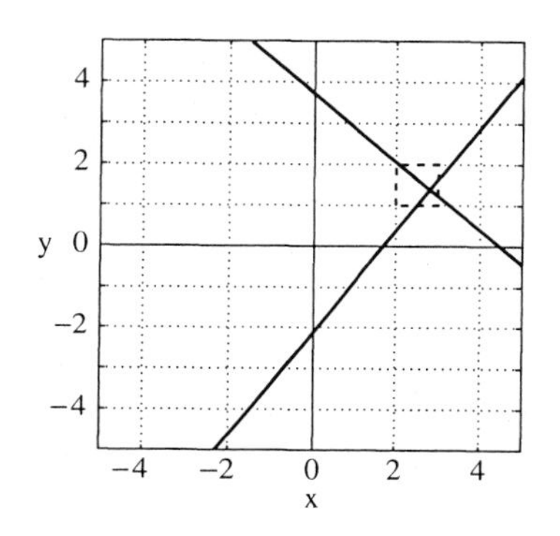

FIGURE B.15 $-5 \leqq x \leqq 5$, $-5 \leqq y \leqq 5$ (Example 6).

With many graphics calculators, it is necessary first to solve each equation for y:

$$y = (17 - 10x)/(-8),$$
$$y = (67 - 15x)/18. \tag{13}$$

Figure B.14 shows a calculator prepared to graph the two lines represented by the equations in (12), and Fig. B.15 shows the result in the *viewing window* $-5 \leqq x \leqq 5$, $-5 \leqq y \leqq 5$.

Before proceeding, note that in Fig. B.15 the two lines *appear* to be perpendicular. But their slopes, $(-10)/(-8) = \frac{5}{4}$ and $(-15)/18 = -\frac{5}{6}$, are *not* negative

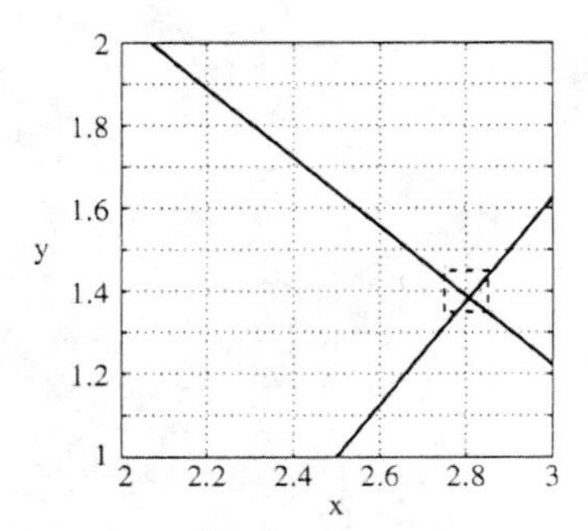

FIGURE B.16 $2 \leq x \leq 3$, $1 \leq y \leq 2$ (Example 6).

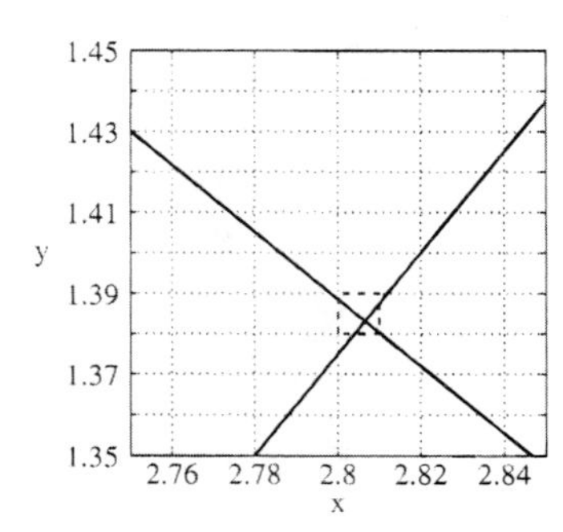

FIGURE B.17 $2.75 \leq x \leq 2.85$, $1.35 \leq y \leq 1.45$ (Example 6).

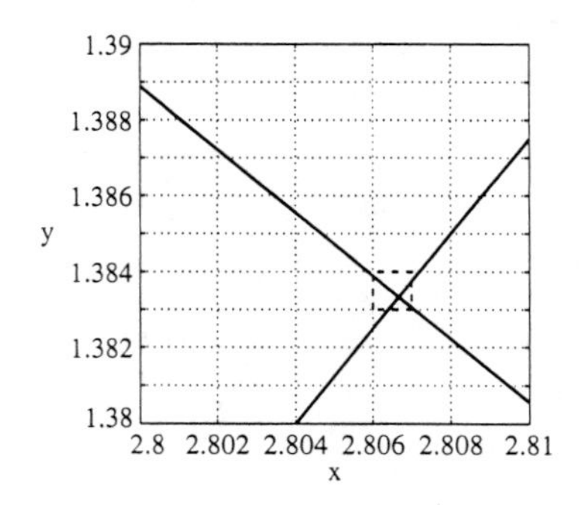

FIGURE B.18 $2.80 \leq x \leq 2.81$, $1.38 \leq y \leq 1.39$ (Example 6).

reciprocals of one another. It follows from Theorem 2 that the two linear are *not* perpendicular.

Figures B.16, B.17, and B.18 show successive magnifications produced by "zooming in" on the point of intersection of the two lines. The dashed-line box in each figure is the viewing window for the next figure. Looking at Fig. B.18, we see that the intersection point is given by the approximations

$$x \approx 2.807, \qquad y \approx 1.383, \tag{14}$$

rounded to three decimal places.

The result in (14) can be checked by equating the right-hand sides in (13) and solving for x. This gives $x = 421/150 \approx 2.8067$. Substituting the exact value of x into either equation in (13) then yields $y = 83/60 \approx 1.3833$.

The graphical method illustrated by Example 6 typically produces approximate solutions that are sufficiently accurate for practical purposes. But the method is especially useful for *nonlinear* equations, for which exact algebraic techniques of solution may not be available. ◆

APPENDIX B PROBLEMS

Three points A, B, and C lie on a single straight line if and only if the slope of AB is equal to the slope of BC. In Problems 1 through 4, plot the three given points and then determine whether or not they lie on a single line.

1. $A(-1, -2)$, $B(2, 1)$, $C(4, 3)$

2. $A(-2, 5)$, $B(2, 3)$, $C(8, 0)$

3. $A(-1, 6)$, $B(1, 2)$, $C(4, -2)$

4. $A(-3, 2)$, $B(1, 6)$, $C(8, 14)$

In Problems 5 and 6, use the concept of slope to show that the four points given are the vertices of a parallelogram.

5. $A(-1, 3)$, $B(5, 0)$, $C(7, 4)$, $D(1, 7)$

6. $A(7, -1)$, $B(-2, 2)$, $C(1, 4)$, $D(10, 1)$

In Problems 7 and 8, show that the three given points are the vertices of a right triangle.

7. $A(-2, -1)$, $B(2, 7)$, $C(4, -4)$

8. $A(6, -1)$, $B(2, 3)$, $C(-3, -2)$

In Problems 9 through 13, find the slope m and y-intercept b of the line with the given equation. Then sketch the line.

9. $2x = 3y$

10. $x + y = 1$

11. $2x - y + 3 = 0$

12. $3x + 4y = 6$

13. $2x = 3 - 5y$

In Problems 14 through 23, write an equation of the straight line L described.

14. L is vertical and has x-intercept 7.

15. L is horizontal and passes through $(3, -5)$.

16. L has x-intercept 2 and y-intercept -3.

17. L passes through $(2, -3)$ and $(5, 3)$.

18. L passes through $(-1, -4)$ and has slope $\frac{1}{2}$.

19. L passes through $(4, 2)$ and has angle of inclination $135°$.

20. L has slope 6 and y-intercept 7.

21. L passes through $(1, 5)$ and is parallel to the line with equation $2x + y = 10$.

22. L passes through $(-2, 4)$ and is perpendicular to the line with equation $x + 2y = 17$.

23. L is the perpendicular bisector of the line segment that has endpoints $(-1, 2)$ and $(3, 10)$.

24. Find the perpendicular distance from the point $(2, 1)$ to the line with equation $y = x + 1$.

25. Find the perpendicular distance between the parallel lines $y = 5x + 1$ and $y = 5x + 9$.

26. The points $A(-1, 6)$, $B(0, 0)$, and $C(3, 1)$ are three consecutive vertices of a parallelogram. What are the coordinates of the fourth vertex? (What happens if the word *consecutive* is omitted?)

27. Prove that the diagonals of the parallelogram of Problem 26 bisect each other.

28. Show that the points $A(-1, 2)$, $B(3, -1)$, $C(6, 3)$, and $D(2, 6)$ are the vertices of a *rhombus*—a parallelogram with all four sides having the same length. Then prove that the diagonals of this rhombus are perpendicular to each other.

29. The points $A(2, 1)$, $B(3, 5)$, and $C(7, 3)$ are the vertices of a triangle. Prove that the line joining the midpoints of AB and BC is parallel to AC.

30. A **median** of a triangle is a line joining a vertex to the midpoint of the opposite side. Prove that the medians of the triangle of Problem 29 intersect in a single point.

31. Complete the proof of the midpoint formula in Eq. (3). It is necessary to show that the point M lies on the segment P_1P_2. One way to do this is to show that the slope of MP_1 is equal to the slope of MP_2.

32. Let $P(x_0, y_0)$ be a point of the circle with center $C(0, 0)$ and radius r. Recall that the line tangent to the circle at the point P is perpendicular to the radius CP. Prove that the equation of this tangent line is $x_0x + y_0y = r^2$.

33. The Fahrenheit temperature F and the absolute temperature K satisfy a linear equation. Moreover, $K = 273.16$ when $F = 32$, and $K = 373.16$ when $F = 212$. Express K in terms of F. What is the value of F when $K = 0$?

34. The length L (in centimeters) of a copper rod is a linear function of its Celsius temperature C. If $L = 124.942$ when $C = 20$ and $L = 125.134$ when $C = 110$, express L in terms of C.

35. The owner of a grocery store finds that she can sell 980 gal of milk each week at \$1.69/gal and 1220 gal of milk each week at \$1.49/gal. Assume a linear relationship between price and sales. How many gallons would she then expect to sell each week at \$1.56/gal?

36. Figure B.19 shows the graphs of the equations

$$17x - 10y = 57,$$
$$25x - 15y = 17.$$

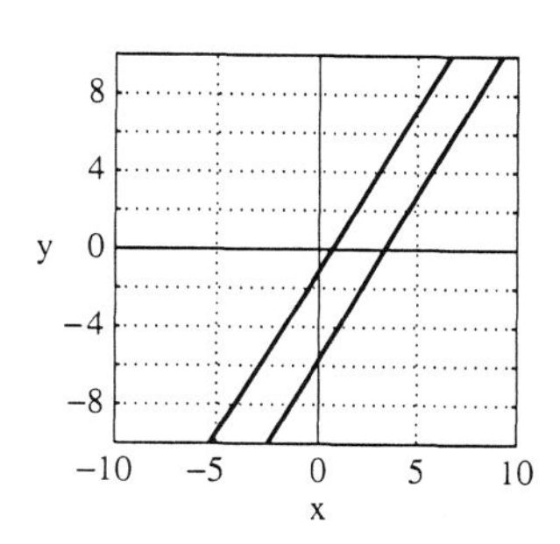

FIGURE B.19 The lines of Problem 36.

Are these two lines parallel? If not, find their point of intersection. If you have a graphing utility, find the solution by graphical approximation as well as by exact algebraic methods.

In Problems 37 through 46, use a graphics calculator or computer to approximate graphically (with three digits to the right of the decimal correct or correctly rounded) the solution of the given linear equation. Then check your approximate solution by solving the system by an exact algebraic method.

37. $2x + 3y = 5$; $2x + 5y = 12$

38. $6x + 4y = 5$; $8x - 6y = 13$

39. $3x + 3y = 17$; $3x + 5y = 16$

40. $2x + 3y = 17$; $2x + 5y = 20$

41. $4x + 3y = 17$; $5x + 5y = 21$

42. $4x + 3y = 15$; $5x + 5y = 29$

43. $5x + 6y = 16$; $7x + 10y = 29$

44. $5x + 11y = 21$; $4x + 10y = 19$

45. $6x + 6y = 31$; $9x + 11y = 37$

46. $7x + 6y = 31$; $11x + 11y = 47$

47. Justify the phrase "no other point of the plane can do so" that follows the first appearance of Eq. (6).

48. The discussion of the linear equation $Ax+By = C$ in Eq. (8) does not include a description of the graph of this equation if $A = B = 0$. What is the graph in this case?

APPENDIX C: REVIEW OF TRIGONOMETRY

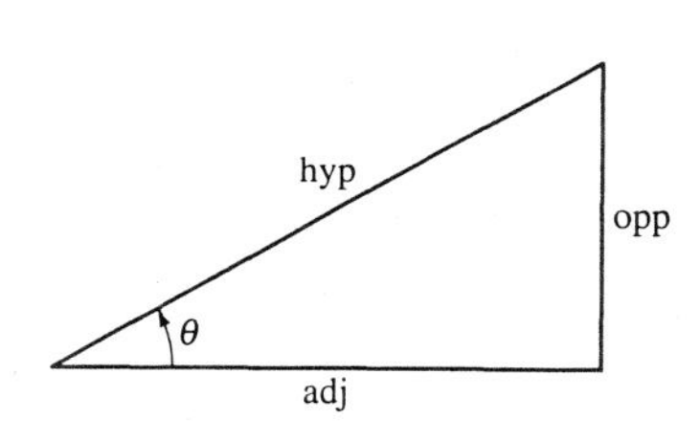

FIGURE C.1 The sides and angle θ of a right triangle.

In elementary trigonometry, the six basic trigonometric functions of an acute angle θ in a right triangle are defined as ratios between pairs of sides of the triangle. As in Fig. C.1, where "adj" stands for "adjacent," "opp" for "opposite," and "hyp" for "hypotenuse,"

$$\cos\theta = \frac{\text{adj}}{\text{hyp}}, \quad \sin\theta = \frac{\text{opp}}{\text{hyp}}, \quad \tan\theta = \frac{\text{opp}}{\text{adj}},$$
$$\sec\theta = \frac{\text{hyp}}{\text{adj}}, \quad \csc\theta = \frac{\text{hyp}}{\text{opp}}, \quad \cot\theta = \frac{\text{adj}}{\text{opp}}. \tag{1}$$

We generalize these definitions to *directed* angles of arbitrary size in the following way. Suppose that the initial side of the angle θ is the positive x-axis, so its vertex is

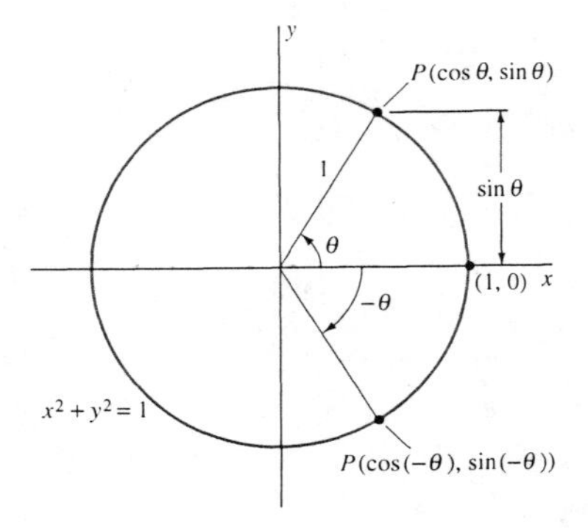

FIGURE C.2 Using the unit circle to define the trigonometric functions.

at the origin. The angle is **directed** if a direction of rotation from its initial side to its terminal side is specified. We call θ a **positive angle** if this rotation is counterclockwise and a **negative angle** if it is clockwise.

Let $P(x, y)$ be the point at which the terminal side of θ intersects the *unit* circle $x^2 + y^2 = 1$. Then we define

$$\cos\theta = x, \qquad \sin\theta = y, \qquad \tan\theta = \frac{y}{x},$$
$$\sec\theta = \frac{1}{x}, \qquad \csc\theta = \frac{1}{y}, \qquad \cot\theta = \frac{x}{y}. \tag{2}$$

We assume that $x \neq 0$ in the case of $\tan\theta$ and $\sec\theta$ and that $y \neq 0$ in the case of $\cot\theta$ and $\csc\theta$. If the angle θ is positive and acute, then it is clear from Fig. C.2 that the definitions in (2) agree with the right triangle definitions in (1) in terms of the coordinates of P. A glance at the figure also shows which of the functions are positive for angles in each of the four quadrants. Figure C.3 summarizes this information.

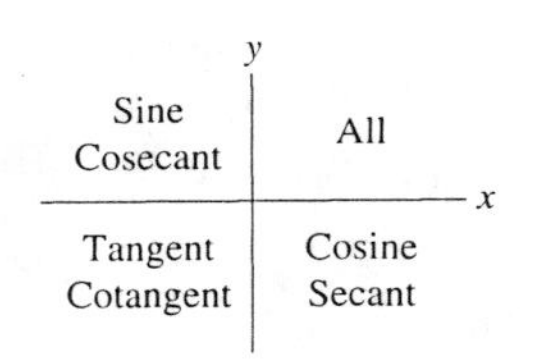

Positive in quadrants shown

FIGURE C.3 The signs of the trigonometric functions.

Here we discuss primarily the two most basic trigonometric functions, the sine and the cosine. From (2) we see immediately that the other four trigonometric functions are defined in terms of $\sin\theta$ and $\cos\theta$ by

$$\tan\theta = \frac{\sin\theta}{\cos\theta}, \qquad \sec\theta = \frac{1}{\cos\theta},$$
$$\cot\theta = \frac{\cos\theta}{\sin\theta}, \qquad \csc\theta = \frac{1}{\sin\theta}. \tag{3}$$

Next, we compare the angles θ and $-\theta$ in Fig. C.4. We see that

$$\cos(-\theta) = \cos\theta \quad \text{and} \quad \sin(-\theta) = -\sin\theta. \tag{4}$$

Because $x = \cos\theta$ and $y = \sin\theta$ in (2), the equation $x^2 + y^2 = 1$ of the unit circle translates immediately into the **fundamental identity of trigonometry,**

$$\cos^2\theta + \sin^2\theta = 1. \tag{5}$$

Dividing each term of this fundamental identity by $\cos^2\theta$ gives the identity

$$1 + \tan^2\theta = \sec^2\theta. \tag{5$'$}$$

Similarly, dividing each term in Eq. (5) by $\sin^2\theta$ yields the identity

$$1 + \cot^2\theta = \csc^2\theta. \tag{5$''$}$$

(See Problem 9 of this appendix.)

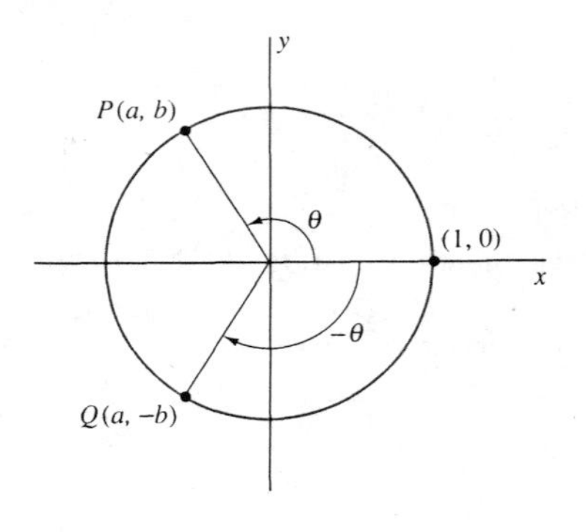

FIGURE C.4 The effect of replacing θ with $-\theta$ in the sine and cosine functions.

In Problems 41 and 42 we outline derivations of the **addition formulas**

$$\sin(\alpha + \beta) = \sin\alpha\cos\beta + \cos\alpha\sin\beta, \tag{6}$$
$$\cos(\alpha + \beta) = \cos\alpha\cos\beta - \sin\alpha\sin\beta. \tag{7}$$

With $\alpha = \theta = \beta$ in Eqs. (6) and (7), we get the **double-angle formulas**

$$\sin 2\theta = 2\sin\theta\cos\theta, \tag{8}$$
$$\cos 2\theta = \cos^2\theta - \sin^2\theta \tag{9}$$
$$= 2\cos^2\theta - 1 \tag{9a}$$
$$= 1 - 2\sin^2\theta, \tag{9b}$$

where Eqs. (9a) and (9b) are obtained from Eq. (9) by use of the fundamental identity in Eq. (5).

If we solve Eq. (9a) for $\cos^2\theta$ and Eq. (9b) for $\sin^2\theta$, we get the **half-angle formulas**

$$\cos^2\theta = \tfrac{1}{2}(1 + \cos 2\theta), \tag{10}$$
$$\sin^2\theta = \tfrac{1}{2}(1 - \cos 2\theta). \tag{11}$$

Equations (10) and (11) are especially important in integral calculus.

RADIAN MEASURE

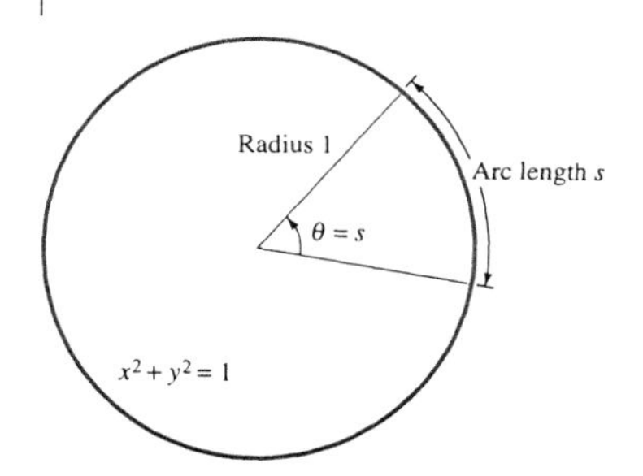

FIGURE C.5 The radian measure of an angle.

In elementary mathematics, angles frequently are measured in *degrees,* with 360° in one complete revolution. In calculus it is more convenient—and often essential—to measure angles in *radians.* The **radian measure** of an angle is the length of the arc it subtends in (that is, the arc it cuts out of) the unit circle when the vertex of the angle is at the center of the circle (Fig. C.5).

Recall that the area A and circumference C of a circle of radius r are given by the formulas

$$A = \pi r^2 \quad \text{and} \quad C = 2\pi r,$$

where the irrational number π is approximately 3.14159. Because the circumference of the unit circle is 2π and its central angle is 360°, it follows that

$$2\pi \text{ rad} = 360^\circ; \qquad 180^\circ = \pi \text{ rad} \approx 3.14159 \text{ rad}. \tag{12}$$

Using Eq. (12) we can easily convert back and forth between radians and degrees:

$$1 \text{ rad} = \frac{180^\circ}{\pi} \approx 57^\circ\, 17'\, 44.8'', \tag{12a}$$

$$1^\circ = \frac{\pi}{180} \text{ rad} \approx 0.01745 \text{ rad}. \tag{12b}$$

Radians	Degrees
0	0
$\pi/6$	30
$\pi/4$	45
$\pi/3$	60
$\pi/2$	90
$2\pi/3$	120
$3\pi/4$	135
$5\pi/6$	150
π	180
$3\pi/2$	270
2π	360
4π	720

FIGURE C.6 Some radian-degree conversions.

Figure C.6 shows radian-degree conversions for some common angles.

Now consider an angle of θ radians at the center of a circle of radius r (Fig. C.7). Denote by s the length of the arc subtended by θ; denote by A the area of the sector of the circle bounded by this angle. Then the proportions

$$\frac{s}{2\pi r} = \frac{A}{\pi r^2} = \frac{\theta}{2\pi}$$

give the formulas

$$s = r\theta \qquad (\theta \text{ in radians}) \tag{13}$$

and

$$A = \tfrac{1}{2}r^2\theta \qquad (\theta \text{ in radians}). \tag{14}$$

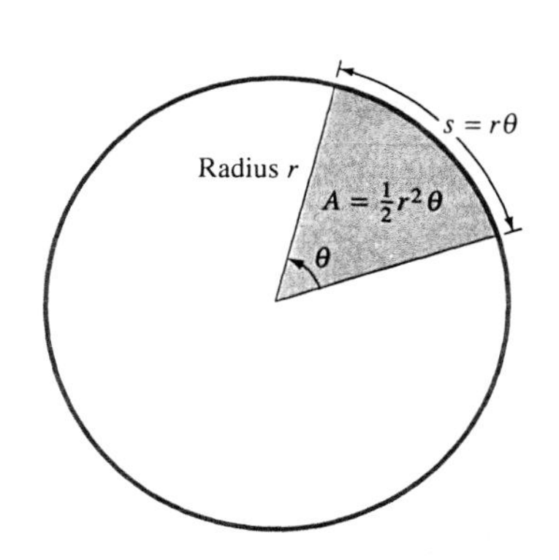

FIGURE C.7 The area of a sector and arc length of a circular arc.

The definitions in (2) refer to trigonometric functions of *angles* rather than trigonometric functions of *numbers.* Suppose that t is a real number. Then the number $\sin t$ is, *by definition,* the sine of an angle of t radians—recall that a positive angle is directed counterclockwise from the positive x-axis, whereas a negative angle is directed clockwise. Briefly, $\sin t$ is the sine of an angle of t *radians*. The other trigonometric functions of the number t have similar definitions. Hence, when we write $\sin t$, $\cos t$, and so on, with t a real number, it is *always* in reference to an angle to t *radians.*

When we need to refer to the sine of an angle of t *degrees,* we will henceforth write $\sin t^\circ$. The point is that $\sin t$ and $\sin t^\circ$ are quite different functions of the variable t. For example, you would get

$$\sin 1^\circ \approx 0.0175 \quad \text{and} \quad \sin 30^\circ = 0.5$$

on a calculator set in degree mode. But in radian mode, a calculator would give

$$\sin 1 \approx 0.8415 \quad \text{and} \quad \sin 30 \approx -0.9880.$$

The relationship between the functions $\sin t$ and $\sin t^\circ$ is

$$\sin t^\circ = \sin\left(\frac{\pi t}{180}\right). \tag{15}$$

The distinction extends even to programming languages. In FORTRAN, the function `SIN` is the radian sine function, and you must write $\sin t^\circ$ in the form `SIND(T)`. In BASIC you must write `SIN(Pi*T/180)` to get the correct value of the sine of an angle of t degrees.

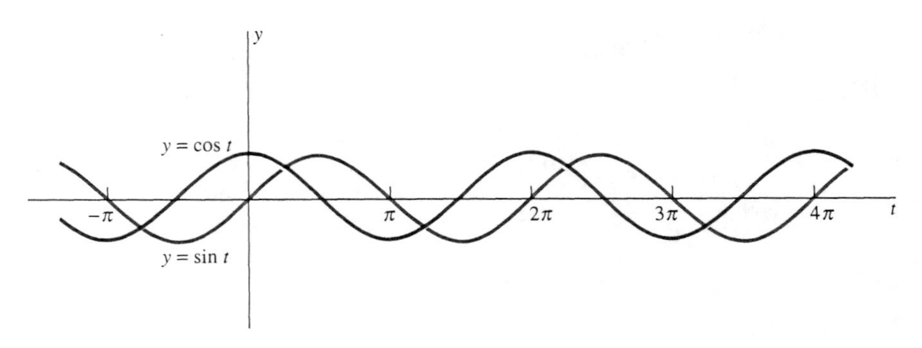

FIGURE C.8 Periodicity of the sine and cosine functions.

An angle of 2π rad corresponds to one revolution around the unit circle. This implies that the sine and cosine functions have **period** 2π, meaning that

$$\begin{aligned} \sin(t + 2\pi) &= \sin t, \\ \cos(t + 2\pi) &= \cos t. \end{aligned} \tag{16}$$

It follows from the equations in (16) that

$$\sin(t + 2n\pi) = \sin t \quad \text{and} \quad \cos(t + 2n\pi) = \cos t \tag{17}$$

for every integer n. This periodicity of the sine and cosine functions is evident in their graphs (Fig. C.8). From the equations in (3), the other four trigonometric functions also must be periodic, as their graphs in Figs. C.9 and C.10 show.

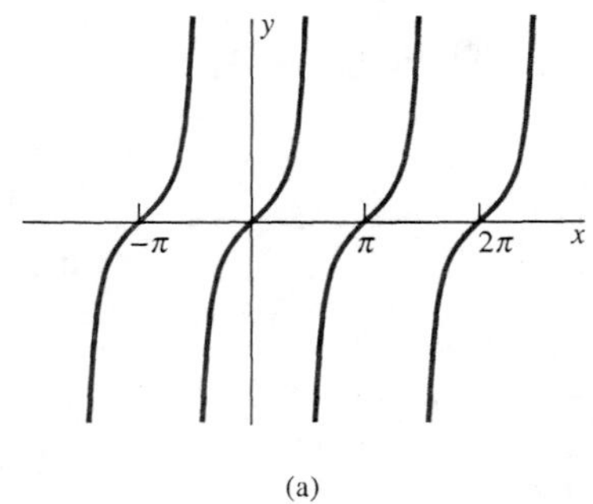

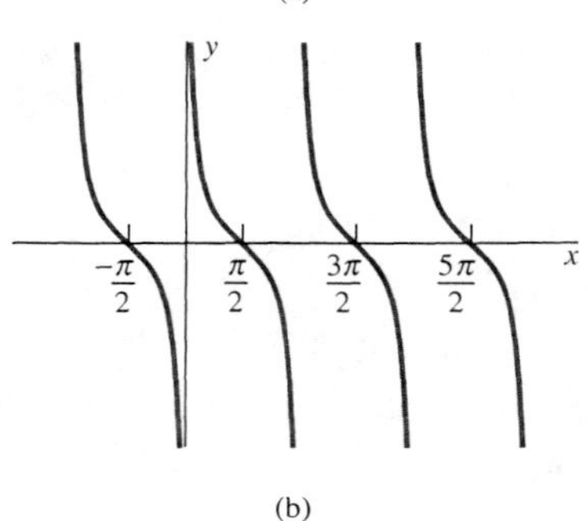

FIGURE C.9 The graphs of (a) the tangent function and (b) the cotangent function.

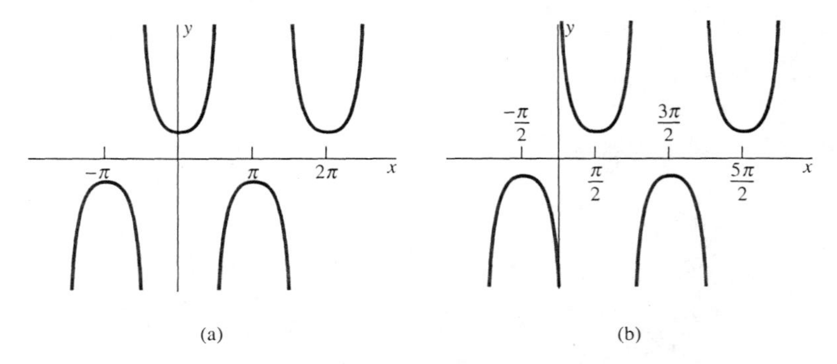

FIGURE C.10 The graphs of (a) the secant function and (b) the cosecant function.

We see from the equations in (2) that

$$\begin{aligned} &\sin 0 = 0, \qquad \sin\frac{\pi}{2} = 1, \qquad \sin\pi = 0, \\ &\cos 0 = 1, \qquad \cos\frac{\pi}{2} = 0, \qquad \cos\pi = -1. \end{aligned} \tag{18}$$

The trigonometric functions of $\pi/6$, $\pi/4$, and $\pi/3$ (the radian equivalents of 30°, 45°, and 60°, respectively) are easy to read from the well-known triangles of Fig. C.11. For instance,

$$\begin{aligned} &\sin\frac{\pi}{6} = \cos\frac{\pi}{3} = \frac{1}{2} = \frac{\sqrt{1}}{2}, \\ &\sin\frac{\pi}{4} = \cos\frac{\pi}{4} = \frac{1}{\sqrt{2}} = \frac{\sqrt{2}}{2}, \quad \text{and} \\ &\sin\frac{\pi}{3} = \cos\frac{\pi}{6} = \frac{\sqrt{3}}{2}. \end{aligned} \tag{19}$$

2 1 π/6 √3

√2 1 π/4 1

2 √3 π/3 1

FIGURE C.11 Familiar right triangles.

To find the values of trigonometric functions of angles larger than $\pi/2$, we can use their periodicity and the identities

$$\begin{aligned}\sin(\pi \pm \theta) &= \mp\sin\theta,\\ \cos(\pi \pm \theta) &= -\cos\theta \quad \text{and}\\ \tan(\pi \pm \theta) &= \pm\tan\theta\end{aligned} \tag{20}$$

(Problems 38, 39, and 40) as well as similar identities for the cosecant, secant, and cotangent functions.

EXAMPLE 1

$$\sin\frac{5\pi}{4} = \sin\left(\pi + \frac{\pi}{4}\right) = -\sin\frac{\pi}{4} = -\frac{\sqrt{2}}{2};$$

$$\cos\frac{2\pi}{3} = \cos\left(\pi - \frac{\pi}{3}\right) = -\cos\frac{\pi}{3} = -\frac{1}{2};$$

$$\tan\frac{2\pi}{4} = \tan\left(\pi - \frac{\pi}{4}\right) = -\tan\frac{\pi}{4} = -1;$$

$$\sin\frac{7\pi}{6} = \sin\left(\pi + \frac{\pi}{6}\right) - \sin\frac{\pi}{6} = -\frac{1}{2};$$

$$\cos\frac{5\pi}{3} = \cos\left(2\pi - \frac{\pi}{3}\right) = \cos\left(-\frac{\pi}{3}\right) = \cos\frac{\pi}{3} = \frac{1}{2};$$

$$\begin{aligned}\sin\frac{17\pi}{6} &= \sin\left(2\pi + \frac{5\pi}{6}\right) = \sin\frac{5\pi}{6}\\ &= \sin\left(\pi - \frac{\pi}{6}\right) = \sin\frac{\pi}{6} = \frac{1}{2}.\end{aligned}$$

◆

EXAMPLE 2 Find the solutions (if any) of the equation

$$\sin^2 x - 3\cos^2 x + 2 = 0$$

that lie in the interval $[0, \pi]$.

Solution Using the fundamental identity in Eq. (5), we substitute $\cos^2 x = 1 - \sin^2 x$ into the given equation to obtain

$$\begin{aligned}\sin^2 x - 3(1 - \sin^2 x) + 2 &= 0;\\ 4\sin^2 x - 1 &= 0;\\ \sin x &= \pm\tfrac{1}{2}.\end{aligned}$$

Because $\sin x \geqq 0$ for x in $[0, \pi]$, $\sin x = -\frac{1}{2}$ is impossible. But $\sin x = \frac{1}{2}$ for $x = \pi/6$ and for $x = \pi - \pi/6 = 5\pi/6$. These are the solutions of the given equation that lie in $[0, \pi]$. ◆

APPENDIX C PROBLEMS

Express in radian measure the angles in Problems 1 through 5.

1. 40°

2. −270°

3. 315°

4. 210°

5. −150°

In Problems 6 through 10, express in degrees the angles given in radian measure.

6. $\frac{\pi}{10}$

7. $\frac{2\pi}{5}$

8. 3π

9. $\frac{15\pi}{4}$

10. $\frac{23\pi}{60}$

In Problems 11 through 14, evaluate the six trigonometric functions of x at the given values.

11. $x = -\frac{\pi}{3}$

12. $x = \frac{3\pi}{4}$

13. $x = \dfrac{7\pi}{6}$

14. $x = \dfrac{5\pi}{3}$

Find all solutions x of each equation in Problems 15 through 23.

15. $\sin x = 0$

16. $\sin x = 1$

17. $\sin x = -1$

18. $\cos x = 0$

19. $\cos x = 1$

20. $\cos x = -1$

21. $\tan x = 0$

22. $\tan x = 1$

23. $\tan x = -1$

24. Suppose that $\tan x = \frac{3}{4}$ and that $\sin x < 0$. Find the values of the other five trigonometric functions of x.

25. Suppose that $\csc x = -\frac{5}{3}$ and that $\cos x > 0$. Find the values of the other five trigonometric functions of x.

Deduce the identities in Problems 26 and 27 from the fundamental identity

$$\cos^2\theta + \sin^2\theta = 1$$

and from the definitions of the other four trigonometric functions.

26. $1 + \tan^2\theta = \sec^2\theta$

27. $1 + \cot^2\theta = \csc^2\theta$

28. Deduce from the addition formulas for the sine and cosine the addition formula for the tangent:

$$\tan(x + y) = \frac{\tan x + \tan y}{1 - \tan x \tan y}.$$

In Problems 29 through 36, use the method of Example 1 to find the indicated values.

29. $\sin \dfrac{5\pi}{6}$

30. $\cos \dfrac{7\pi}{6}$

31. $\sin \dfrac{11\pi}{6}$

32. $\cos \dfrac{19\pi}{6}$

33. $\sin \dfrac{2\pi}{3}$

34. $\cos \dfrac{4\pi}{3}$

35. $\sin \dfrac{5\pi}{3}$

36. $\cos \dfrac{10\pi}{3}$

37. Apply the addition formula for the sine, cosine, and tangent functions (the latter from Problem 28) to show that if $0 < \theta < \pi/2$, then

(a) $\cos\left(\dfrac{\pi}{2} - \theta\right) = \sin\theta$;

(b) $\sin\left(\dfrac{\pi}{2} - \theta\right) = \cos\theta$;

(c) $\cot\left(\dfrac{\pi}{2} - \theta\right) = \tan\theta$.

The prefix *co-* is an abbreviation for the adjective *complementary,* which describes two angles whose sum is $\pi/2$. For example, $\pi/6$ and $\pi/3$ are complementary angles, so (a) implies that $\cos\pi/6 = \sin\pi/3$.

Suppose that $0 < \theta < \pi/2$. *Derive the identities in Problems 38 through 40.*

38. $\sin(\pi \pm \theta) = \mp\sin\theta$

39. $\cos(\pi \pm \theta) = -\cos\theta$

40. $\tan(\pi \pm \theta) = \pm\tan\theta$

41. The points $A(\cos\theta, -\sin\theta)$, $B(1, 0)$, $C(\cos\phi, \sin\phi)$, and $D(\cos(\theta + \phi), \sin(\theta + \phi))$ are shown in Fig. C.12; all are points on the unit circle. Deduce from the fact that the line segments AC and BD have the same length (because they are subtended by the same central angle $\theta + \phi$) that

$$\cos(\theta + \phi) = \cos\theta\,\cos\phi - \sin\theta\,\sin\phi.$$

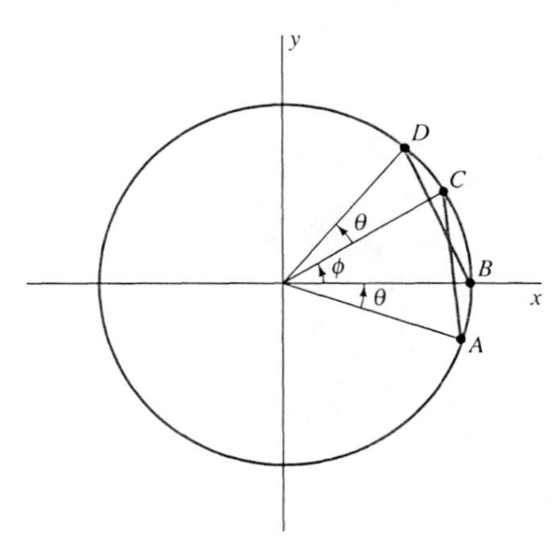

FIGURE C.12 Deriving the cosine addition formula (Problem 41).

42. (a) Use the triangles shown in Fig. C.13 to deduce that

$$\sin\left(\theta + \frac{\pi}{2}\right) = \cos\theta \quad\text{and}\quad \cos\left(\theta + \frac{\pi}{2}\right) = -\sin\theta.$$

(b) Use the results of Problem 41 and part (a) to derive the addition formula for the sine function.

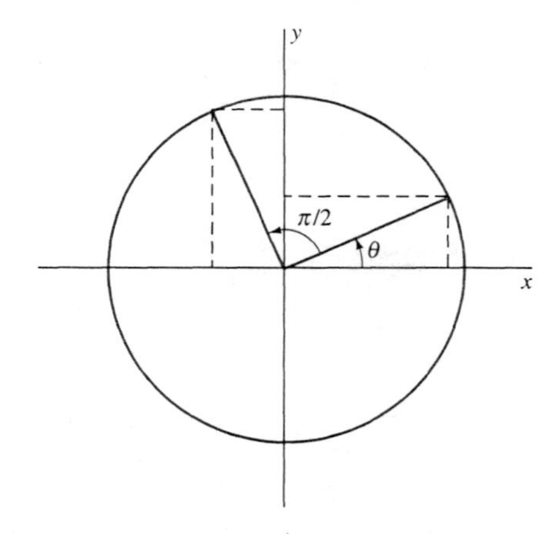

FIGURE C.13 Deriving the identities of Problem 42.

In Problems 43 through 48, find all solutions of the given equation that lie in the interval $[0, \pi]$.

43. $3\sin^2 x - \cos^2 x = 2$

44. $\sin^2 x = \cos^2 x$

45. $2\cos^2 x + 3\sin^2 x = 3$

46. $2\sin^2 x + \cos x = 2$

47. $8\sin^2 x\cos^2 x = 1$

48. $\cos 2\theta - 3\cos\theta = -2$

APPENDIX D: PROOFS OF THE LIMIT LAWS

Recall the definition of the limit:

$$\lim_{x \to a} F(x) = L$$

provided that, given $\epsilon > 0$, there exists a number $\delta > 0$ such that

$$0 < |x - a| < \delta \quad \text{implies that} \quad |F(x) - L| < \epsilon. \tag{1}$$

Note that the number ϵ comes *first*. *Then* a value of $\delta > 0$ must be found so that the implication in (1) holds. To prove that $F(x) \to L$ as $x \to a$, you must, in effect, be able to stop the next person you see and ask him or her to pick a positive number ϵ at random. Then you must *always* be ready to respond with a positive number δ. This number δ must have the property that the implication in (1) holds for your number δ and the given number ϵ. The **only** restriction on x is that

$$0 < |x - a| < \delta,$$

as given in (1).

To do all this, you will ordinarily need to give an explicit method—a recipe or formula—for producing a value of δ that works for each value of ϵ. As Examples 1 through 3 show, the method will depend on the particular function F under study as well as the values of a and L.

EXAMPLE 1 Prove that $\lim_{x \to 3} (2x - 1) = 5$.

Solution Given $\epsilon > 0$, we must find $\delta > 0$ such that

$$|(2x - 1) - 5)| < \epsilon \quad \text{if} \quad 0 < |x - 3| < \delta.$$

Now

$$|(2x - 1) - 5| = |2x - 6| = 2|x - 3|,$$

so

$$0 < |x - 3| < \frac{\epsilon}{2} \quad \text{implies that} \quad |(2x - 1) - 5| < 2 \cdot \frac{\epsilon}{2} = \epsilon.$$

Hence, given $\epsilon > 0$, it suffices to choose $\delta = \epsilon/2$. This illustrates the observation that the required number δ is generally a function of the given number ϵ. ◆

EXAMPLE 2 Prove that $\lim_{x \to 2} (3x^2 + 5) = 17$.

Solution Given $\epsilon > 0$, we must find $\delta > 0$ such that

$$0 < |x - 2| < \delta \quad \text{implies that} \quad |(3x^2 + 5) - 17| < \epsilon.$$

Now

$$|(3x^2 + 5) - 17| = |3x^2 - 12| = 3 \cdot |x + 2| \cdot |x - 2|.$$

Our problem, therefore, is to show that $|x + 2| \cdot |x - 2|$ can be made as small as we please by choosing $x - 2$ sufficiently small. The idea is that $|x + 2|$ cannot be too large if $|x - 2|$ is fairly small. For example, if $|x - 2| < 1$, then

$$|x + 2| = |(x - 2) + 4| \leqq |x - 2| + 4 < 5.$$

Therefore,

$$0 < |x - 2| < 1 \quad \text{implies that} \quad |(3x^2 + 5) - 17| < 15 \cdot |x - 2|.$$

Consequently, let us choose δ to be the minimum of the two numbers 1 and $\epsilon/15$. Then

$$0 < |x-2| < \delta \quad \text{implies that} \quad |(3x^2+5)-17| < 15 \cdot \frac{\epsilon}{15} = \epsilon,$$

as desired. ◆

EXAMPLE 3 Prove that

$$\lim_{x \to a} \frac{1}{x} = \frac{1}{a} \quad \text{if} \quad a \neq 0.$$

Solution For simplicity, we will consider only the case in which $a > 0$ (the case $a < 0$ is similar).

Suppose that $\epsilon > 0$ is given. We must find a number δ such that

$$0 < |x-a| < \delta \quad \text{implies that} \quad \left| \frac{1}{x} - \frac{1}{a} \right| < \epsilon.$$

Now

$$\left| \frac{1}{x} - \frac{1}{a} \right| = \left| \frac{a-x}{ax} \right| = \frac{|x-a|}{a|x|}.$$

The idea is that $1/|x|$ cannot be too large if $|x-a|$ is fairly small. For example, if $|x-a| < a/2$, then $a/2 < x < 3a/2$. Therefore,

$$|x| > \frac{a}{2}, \quad \text{so} \quad \frac{1}{|x|} < \frac{2}{a}.$$

In this case it would follow that

$$\left| \frac{1}{x} - \frac{1}{a} \right| < \frac{2}{a^2} \cdot |x-a|$$

if $|x-a| < a/2$. Thus, if we choose δ to be the minimum of the two numbers $a/2$ and $a^2\epsilon/2$, then

$$0 < |x-a| < \delta \quad \text{implies that} \quad \left| \frac{1}{x} - \frac{1}{a} \right| < \frac{2}{a^2} \cdot \frac{a^2\epsilon}{2} = \epsilon.$$

Therefore

$$\lim_{x \to a} \frac{1}{x} = \frac{1}{a} \quad \text{if} \quad a \neq 0,$$

as desired. ◆

We are now ready to give proofs of the limit laws stated in Section 2.2.

Constant Law

If $f(x) \equiv C$, a constant, then

$$\lim_{x \to a} f(x) = \lim_{x \to a} C = C.$$

PROOF Because $|C - C| = 0$, we merely choose $\delta = 1$, regardless of the previously given value of $\epsilon > 0$. Then, if $0 < |x-a| < \delta$, it is automatic that $|C - C| < \epsilon$. ◀

Addition Law

If $\lim_{x \to a} F(x) = L$ and $\lim_{x \to a} G(x) = M$, then

$$\lim_{x \to a} [F(x) + G(x)] = L + M.$$

PROOF Let $\epsilon > 0$ be given. Because L is the limit of $F(x)$ as $x \to a$, there exists a number $\delta_1 > 0$ such that

$$0 < |x - a| < \delta_1 \quad \text{implies that} \quad |F(x) - L| < \frac{\epsilon}{2}.$$

Because M is the limit of $G(x)$ as $x \to a$, there exists a number $\delta_2 > 0$ such that

$$0 < |x - a| < \delta_2 \quad \text{implies that} \quad |G(x) - M| < \frac{\epsilon}{2}.$$

Let $\delta = \min\{\delta_1, \delta_2\}$. Then $0 < |x - a| < \delta$ implies that

$$|(F(x) + G(x)) - (L + M)| \leqq |F(x) - L| + |G(x) - M| < \frac{\epsilon}{2} + \frac{\epsilon}{2} = \epsilon.$$

Therefore

$$\lim_{x \to a} [F(x) + G(x)] = L + M,$$

as desired. ◄

Product Law

If $\lim_{x \to a} F(x) = L$ and $\lim_{x \to a} G(x) = M$, then

$$\lim_{x \to a} [F(x) \cdot G(x)] = L \cdot M.$$

PROOF Given $\epsilon > 0$, we must find a number $\delta > 0$ such that

$$0 < |x - a| < \delta \quad \text{implies that} \quad |F(x) \cdot G(x) - L \cdot M| < \epsilon.$$

But first, the triangle inequality gives the result

$$\begin{aligned} |F(x) \cdot G(x) - L \cdot M| &= |F(x) \cdot G(x) - L \cdot G(x) + L \cdot G(x) - L \cdot M| \\ &\leqq |G(x)| \cdot |F(x) - L| + |L| \cdot |G(x) - M|. \end{aligned} \tag{2}$$

Because $\lim_{x \to a} F(x) = L$, there exists $\delta_1 > 0$ such that

$$0 < |x - a| < \delta_1 \quad \text{implies that} \quad |F(x) - L| < \frac{\epsilon}{2(|M| + 1)}. \tag{3}$$

And because $\lim_{x \to a} G(x) = M$, there exists $\delta_2 > 0$ such that

$$0 < |x - a| < \delta_2 \quad \text{implies that} \quad |G(x) - M| < \frac{\epsilon}{2(|L| + 1)}. \tag{4}$$

Moreover, there is a *third* number $\delta_3 > 0$ such that

$$0 < |x - a| < \delta_3 \quad \text{implies that} \quad |G(x) - M| < 1,$$

which in turn implies that

$$|G(x)| < |M| + 1. \tag{5}$$

We now choose $\delta = \min\{\delta_1, \delta_2, \delta_3\}$. Then we substitute (3), (4), and (5) into (2) and, finally, see that $0 < |x - a| < \delta$ implies that

$$\begin{aligned} |F(x) \cdot G(x) - L \cdot M| &< (|M| + 1) \cdot \frac{\epsilon}{2(|M| + 1)} + |L| \cdot \frac{\epsilon}{2(|L| + 1)} \\ &< \frac{\epsilon}{2} + \frac{\epsilon}{2} = \epsilon, \end{aligned}$$

as desired. The use of $|M| + 1$ and $|L| + 1$ in the denominators avoids the technical difficulty that arises should either L or M be zero. ◄

Substitution Law

If $\lim_{x\to a} g(x) = L$ and $\lim_{x\to L} f(x) = f(L)$, then

$$\lim_{x\to a} f(g(x)) = f(L).$$

PROOF Let $\epsilon > 0$ be given. We must find a number $\delta > 0$ such that

$$0 < |x - a| < \delta \quad \text{implies that} \quad |f(g(x)) - f(L)| < \epsilon.$$

Because $f(y) \to f(L)$ as $y \to L$, there exists $\delta_1 > 0$ such that

$$0 < |y - L| < \delta_1 \quad \text{implies that} \quad |f(y) - f(L)| < \epsilon. \tag{6}$$

Also, because $g(x) \to L$ as $x \to a$, we can find $\delta > 0$ such that

$$0 < |x - a| < \delta \quad \text{implies that} \quad |g(x) - L| < \delta_1;$$

that is, such that

$$|y - L| < \delta_1,$$

where $y = g(x)$. From (6) we see that $0 < |x - a| < \delta$ implies that

$$|f(g(x)) - f(L)| = |f(y) - f(L)| < \epsilon,$$

as desired. ◄

Reciprocal Law

If $\lim_{x\to a} g(x) = L$ and $L \neq 0$, then

$$\lim_{x\to a} \frac{1}{g(x)} = \frac{1}{L}.$$

PROOF Let $f(x) = 1/x$. Then, as we saw in Example 3,

$$\lim_{x\to a} f(x) = \lim_{x\to a} \frac{1}{x} = \frac{1}{L} = f(L).$$

Hence the substitution law gives the result

$$\lim_{x\to a} \frac{1}{g(x)} = \lim_{x\to a} f(g(x)) = f(L) = \frac{1}{L},$$

as desired. ◄

Quotient Law

If $\lim_{x\to a} F(x) = L$ and $\lim_{x\to a} G(x) = M \neq 0$, then

$$\lim_{x\to a} \frac{F(x)}{G(x)} = \frac{L}{M}.$$

PROOF It follows immediately from the product and reciprocal laws that

$$\lim_{x\to a} \frac{F(x)}{G(x)} = \lim_{x\to a} F(x) \cdot \frac{1}{G(x)} = \left(\lim_{x\to a} F(x)\right)\left(\lim_{x\to a} \frac{1}{G(x)}\right) = L \cdot \frac{1}{M} = \frac{L}{M},$$

as desired. ◄

Squeeze Law

Suppose that $f(x) \leqq g(x) \leqq h(x)$ in some deleted neighborhood of a and that

$$\lim_{x\to a} f(x) = L = \lim_{x\to a} h(x).$$

Then

$$\lim_{x\to a} g(x) = L.$$

PROOF Given $\epsilon > 0$, we choose $\delta_1 > 0$ and $\delta_2 > 0$ such that

$$0 < |x - a| < \delta_1 \quad \text{implies that} \quad |f(x) - L| < \epsilon$$

and

$$0 < |x - a| < \delta_2 \quad \text{implies that} \quad |h(x) - L| < \epsilon.$$

Let $\delta = \min\{\delta_1, \delta_2\}$. Then $\delta > 0$. Moreover, if $0 < |x - a| < \delta$, then both $f(x)$ and $h(x)$ are points of the open interval $(L - \epsilon, L + \epsilon)$. So

$$L - \epsilon < f(x) \leqq g(x) \leqq h(x) < L + \epsilon.$$

Thus

$$0 < |x - a| < \delta \quad \text{implies that} \quad |g(x) - L| < \epsilon,$$

as desired. ◄

APPENDIX D PROBLEMS

In Problems 1 through 10, apply the definition of the limit to establish the given equality.

1. $\lim_{x \to a} x = a$

2. $\lim_{x \to 2} 3x = 6$

3. $\lim_{x \to 2} (x + 3) = 5$

4. $\lim_{x \to -3} (2x + 1) = -5$

5. $\lim_{x \to 1} x^2 = 1$

6. $\lim_{x \to a} x^2 = a^2$

7. $\lim_{x \to -1} (2x^2 - 1) = 1$

8. $\lim_{x \to a} \dfrac{1}{x^2} = \dfrac{1}{a^2}$

9. $\lim_{x \to a} \dfrac{1}{x^2 + 1} = \dfrac{1}{a^2 + 1}$

10. $\lim_{x \to a} \dfrac{1}{\sqrt{x}} = \dfrac{1}{\sqrt{a}}$ if $a > 0$

11. Suppose that

$$\lim_{x \to a} f(x) = L \quad \text{and} \quad \lim_{x \to a} f(x) = M.$$

Apply the definition of the limit to prove that $L = M$. Thus the limit of the function f at $x = a$ is unique if it exists.

12. Suppose that C is a constant and that $f(x) \to L$ as $x \to a$. Apply the definition of the limit to prove that

$$\lim_{x \to a} C \cdot f(x) = C \cdot L.$$

13. Suppose that $L \neq 0$ and that $f(x) \to L$ as $x \to a$. Use the method of Example 3 and the definition of the limit to show directly that

$$\lim_{x \to a} \frac{1}{f(x)} = \frac{1}{L}.$$

14. Use the algebraic identity

$$x^n - a^n = (x - a)(x^{n-1} + x^{n-2}a + x^{n-3}a^2 + \cdots + xa^{n-2} + a^{n-1})$$

to show directly from the definition of the limit that $\lim_{x \to a} x^n = a^n$ if n is a positive integer.

15. Apply the identity

$$\left|\sqrt{x} - \sqrt{a}\right| = \frac{|x - a|}{\sqrt{x} + \sqrt{a}}$$

to show directly from the definition of the limit that $\lim_{x \to a} \sqrt{x} = \sqrt{a}$ if $a > 0$.

16. Suppose that $f(x) \to f(a) > 0$ as $x \to a$. Prove that there exists a neighborhood of a on which $f(x) > 0$; that is, prove that there exists $\delta > 0$ such that

$$|x - a| < \delta \quad \text{implies that} \quad f(x) > 0.$$

APPENDIX E: THE COMPLETENESS OF THE REAL NUMBER SYSTEM

Here we present a self-contained treatment of those consequences of the completeness of the real number system that are relevant to this text. Our principal objective is to prove the intermediate value theorem and the maximum value theorem. We begin with the least upper bound property of the real numbers, which we take to be an axiom.

DEFINITION Upper Bound and Lower Bound

The set S of real numbers is said to be **bounded above** if there is a number b such that $x \leqq b$ for every number x in S, and the number b is then called an **upper bound** for S. Similarly, if there is a number a such that $x \geqq a$ for every number x in S, then S is said to be **bounded below,** and a is called a **lower bound** for S.

DEFINITION Least Upper Bound and Greatest Lower Bound
The number λ is said to be a **least upper bound** for the set S of real numbers provided that

1. λ is an upper bound for S, and
2. If b is an upper bound for S, then $\lambda \leqq b$.

Similarly, the number γ is said to be a **greatest lower bound** for S if γ is a lower bound for S and $\gamma \geqq a$ for every lower bound a of S.

EXERCISE Prove that if the set S has a least upper bound λ, then it is unique. That is, prove that if λ and μ are least upper bounds for S, then $\lambda = \mu$.

It is easy to show that the greatest lower bound γ of a set S, if any, is also unique. At this point you should construct examples to illustrate that a set with a least upper bound λ may or may not contain λ and that a similar statement is true of the set's greatest lower bound.

We now state the *completeness axiom* of the real number system.

Least Upper Bound Axiom
If the nonempty set S of real numbers has an upper bound, then it has a least upper bound.

By working with the set T consisting of the numbers $-x$, where x is in S, it is not difficult to show the following consequence of the least upper bound axiom: If the nonempty set S of real numbers is bounded below, then S has a greatest lower bound. Because of this symmetry, we need only one axiom, not two; results for least upper bounds also hold for greatest lower bounds, provided that some attention is paid to the directions of the inequalities.

The restriction that S be nonempty is annoying but necessary. If S is the "empty" set of real numbers, then 15 is an upper bound for S, but S has no least upper bound because $14, 13, 12, \ldots, 0, -1, -2, \ldots$ are also upper bounds for S.

DEFINITION Increasing, Decreasing, and Monotonic Sequences
The infinite sequence $x_1, x_2, x_3, \ldots, x_k, \ldots$ is said to be **nondecreasing** if $x_n \leqq x_{n+1}$ for every $n \geqq 1$. This sequence is said to be **nonincreasing** if $x_n \geqq x_{n+1}$ for every $n \geqq 1$. If the sequence $\{x_n\}$ is either nonincreasing or nondecreasing, then it is said to be **monotonic.**

Theorem 1 gives the **bounded monotonic sequence property** of the set of real numbers. (Recall that a set S of real numbers is said to be **bounded** if it is contained in an interval of the form $[a, b]$.)

THEOREM 1 Bounded Monotonic Sequences
Every bounded monotonic sequence of real numbers converges.

PROOF Suppose that the sequence

$$S = \{x_n\} = \{x_1, x_2, x_3, \ldots, x_k, \ldots\}$$

is bounded and nondecreasing. By the least upper bound axiom, S has a least upper bound λ. We claim that λ is the limit of the sequence $\{x_n\}$. Consider an open interval centered at λ—that is, an interval of the form $I = (\lambda - \epsilon, \lambda + \epsilon)$, where $\epsilon > 0$. Some terms of the sequence must lie within I, else $\lambda - \epsilon$ would be an upper bound for

S that is less than its least upper bound λ. But if x_N is in I, then—because we are dealing with a nondecreasing sequence—$x_N \leqq x_k \leqq \lambda$ for all $k \geqq N$. That is, x_k is in I for all $k \geqq N$. Because ϵ is an arbitrary positive number, λ is—almost by definition—the limit of the sequence $\{x_n\}$. Thus we have shown that a bounded nonincreasing sequence converges. A similar proof can be constructed for nonincreasing sequences by working with the greatest lower bound. ◄

Therefore, the least upper bound axiom implies the bounded monotonic sequence property of the real numbers. With just a little effort, you can prove that the two are logically equivalent. That is, if you take the bounded monotonic sequence property as an axiom, then the least upper bound property follows as a theorem. The *nested interval property* of Theorem 2 is also equivalent to the least upper bound property, but we shall prove only that it follows from the least upper bound property, because we have chosen the latter as the fundamental completeness axiom for the real number system.

THEOREM 2 Nested Interval Property of the Real Numbers

Suppose that $I_1, I_2, I_3, \dots, I_n, \dots$ is a sequence of closed intervals (so I_n is of the form $[a_n, b_n]$ for each positive integer n) such that

1. I_n contains I_{n+1} for each $n \geqq 1$, and
2. $\lim_{n\to\infty} (b_n - a_n) = 0$.

Then there exists exactly one real number c such that c belongs to I_n for all n. Thus

$$\{c\} = I_1 \cap I_2 \cap I_3 \cap \cdots.$$

PROOF It is clear from hypothesis (2) of Theorem 2 that there is at most one such number c. The sequence $\{a_n\}$ of the left-hand endpoints of the intervals is a bounded (by b_1) nondecreasing sequence and thus has a limit a by the bounded monotonic sequence property. Similarly, the sequence $\{b_n\}$ has a limit b. Because $a_n \leqq b_n$ for all n, it follows easily that $a \leqq b$. It is clear that $a_n \leqq a \leqq b \leqq b_n$ for all $n \geqq 1$, so a and b belong to every interval I_n. But then hypothesis (2) of Theorem 2 implies that $a = b$, and clearly this common value—call it c—is the number satisfying the conclusion of Theorem 2. ◄

We can now use these results to prove several important theorems used in the text.

THEOREM 3 Intermediate Value Property of Continuous Functions

If the function f is continuous on the interval $[a, b]$ and $f(a) < K < f(b)$, then $K = f(c)$ for some number c in (a, b).

PROOF Let $I_1 = [a, b]$. Suppose that I_n has been defined for $n \geqq 1$. We describe (inductively) how to define I_{n+1}, and this shows in particular how to define I_2, I_3, and so forth. Let a_n be the left-hand endpoint of I_n, b_n be its right-hand endpoint, and m_n be its midpoint. There are now three cases to consider: $f(m_n) > K$, $f(m_n) < K$, and $f(m_n) = K$.

If $f(m_n) > K$, then $f(a_n) < K < f(m_n)$; in this case, let $a_{n+1} = a_n$, $b_{n+1} = m_n$, and $I_{n+1} = [a_{n+1}, b_{n+1}]$.

If $f(m_n) < K$, then let $a_{n+1} = m_n$, $b_{n+1} = b_n$, and $I_{n+1} = [a_{n+1}, b_{n+1}]$.

If $f(m_n) = K$, then we simply let $c = m_n$ and the proof is complete. Otherwise, at each stage we bisect I_n and let I_{n+1} be the half of I_n on which f takes on values both above and below K.

It is easy to show that the sequence $\{I_n\}$ of intervals satisfies the hypotheses of Theorem 2. Let c be the (unique) real number common to all the intervals I_n. We will show that $f(c) = K$, and this will conclude the proof.

The sequence $\{b_n\}$ has limit c, so by the continuity of f, the sequence $\{f(b_n)\}$ has limit $f(c)$. But $f(b_n) > K$ for all n, so the limit of $\{f(b_n)\}$ can be no less than K; that is, $f(c) \geqq K$. By considering the sequence $\{a_n\}$, it follows that $f(c) \leqq K$ as well. Therefore, $f(c) = K$. ◄

LEMMA 1

If f is continuous on the closed interval $[a, b]$, then f is bounded there.

PROOF Suppose by way of contradiction that f is not bounded on $I_1 = [a, b]$. Bisect I_1 and let I_2 be either half of I_1 on which f is unbounded. (If f is unbounded on both halves, let $I_2 = I_1$.) In general, let I_{n+1} be a half of I_n on which f is unbounded.

Again it is easy to show that the sequence $\{I_n\}$ of closed intervals satisfies the hypotheses of Theorem 2. Let c be the number common to them all. Because f is continuous, there exists a number $\epsilon > 0$ such that f is bounded on the interval $(c - \epsilon, c + \epsilon)$. But for sufficiently large values of n, I_n is a subset of $(c - \epsilon, c + \epsilon)$. This contradiction shows that f must be bounded on $[a, b]$. ◄

THEOREM 4 Maximum Value Property of Continuous Functions

If the function f is continuous on the closed and bounded interval $[a, b]$, then there exists a number c in $[a, b]$ such that $f(x) \leqq f(c)$ for all x in $[a, b]$.

PROOF Consider the set $S = \{f(x) \mid a \leqq x \leqq b\}$. By Lemma 1, this set is bounded, and it is certainly nonempty. Let λ be the least upper bound of S. Our goal is to show that λ is a value $f(x)$ of f.

With $I_1 = [a, b]$, bisect I_1 as before. Note that λ is the least upper bound of the values of f on at least one of the two halves of I_1; let I_2 be that half. Having defined I_n, let I_{n+1} be the half of I_n on which λ is the least upper bound of the values of f. Let c be the number common to all these intervals. It then follows from the continuity of f, much as in the proof of Theorem 3, that $f(c) = \lambda$. And it is clear that $f(x) \leqq \lambda$ for all x in $[a, b]$. ◄

The technique we are using in these proof is called the *method of bisection.* We now use it once again to establish the *Bolzano–Weierstrass property* of the real number system.

DEFINITION Limit Point

Let S be a set of real numbers. The number p is said to be a **limit point** of S if every open interval containing p also contains points of S other than p.

BOLZANO–WEIERSTRASS THEOREM

Every bounded infinite set of real numbers has a limit point.

PROOF Let I_0 be a closed interval containing the bounded infinite set S of real numbers. Bisect I_0. Let I_1 be one of the resulting closed half-intervals of I_0 that contains infinitely many points of S. If I_n has been chosen, let I_{n+1} be one of the closed half-intervals of I_n containing infinitely many points of S. An application of Theorem 2 yields a number p common to all the intervals I_n. If J is an open interval containing p, then J contains I_n for some sufficiently large value of n and thus contains infinitely many points of S. Therefore p is a limit point of S. ◄

Our final goal is in sight: We can now prove that a sequence of real numbers converges if and only if it is a *Cauchy sequence*.

DEFINITION Cauchy Sequence

The sequence $\{a_n\}_1^\infty$ is said to be a **Cauchy sequence** if, for every $\epsilon > 0$, there exists an integer N such that

$$|a_m - a_n| < \epsilon$$

for all $m, n \geqq N$.

LEMMA 2 Convergent Subsequences

Every bounded sequence of real numbers has a convergent subsequence.

PROOF If $\{a_n\}$ has only a finite number of values, then the conclusion of Lemma 2 follows easily. We therefore focus our attention on the case in which $\{a_n\}$ is an infinite set. It is easy to show that this set is also bounded, and thus we may apply the Bolzano–Weierstrass theorem to obtain a limit point p of $\{a_n\}$.

For each integer $k \geqq 1$, let $a_{n(k)}$ be a term of the sequence $\{a_n\}$ such that

1. $n(k+1) > n(k)$ for all $k \geqq 1$, and
2. $|a_{n(k)} - p| < \dfrac{1}{k}$.

It is then easy to show that $\{a_{n(k)}\}$ is a convergent (to p) subsequence of $\{a_n\}$. ◄

THEOREM 6 Convergence of Cauchy Sequences

A sequence of real numbers converges if and only if it is a Cauchy sequence.

PROOF It follows immediately from the triangle inequality that every convergent sequence is a Cauchy sequence. Thus suppose that the sequence $\{a_n\}$ is a Cauchy sequence.

Choose N such that

$$|a_m - a_n| < 1$$

if $m, n \geqq N$. It follows that if $n \geqq N$, then a_n lies in the closed interval $[a_N - 1, a_N + 1]$. This implies that the sequence $\{a_n\}$ is bounded, and thus by Lemma 2 it has a convergent subsequence $\{a_{n(k)}\}$. Let p be the limit of this subsequence.

We claim that $\{a_n\}$ itself converges to p. Given $\epsilon > 0$, choose M such that

$$|a_m - a_n| < \frac{\epsilon}{2}$$

if $m, n \geqq M$. Next choose K such that $n(K) \geqq M$ and

$$|a_{n(K)} - p| < \frac{\epsilon}{2}.$$

Then if $n \geqq M$,

$$|a_n - p| \leqq |a_n - a_{n(K)}| + |a_{n(K)} - p| < \epsilon.$$

Therefore, $\{a_n\}$ converges to p by definition. ◄

APPENDIX F: EXISTENCE OF THE INTEGRAL

When the basic computational algorithms of the calculus were discovered by Newton and Leibniz in the latter half of the seventeenth century, the logical rigor that had been a feature of the Greek method of exhaustion was largely abandoned. When computing the area A under the curve $y = f(x)$, for example, Newton took it as intuitively obvious that the area function existed, and he proceeded to compute it as the antiderivative of the height function $f(x)$. Leibniz regarded A as an infinite sum of infinitesimal area elements, each of the form $dA = f(x)\,dx$, but in practice computed the area

$$A = \int_a^b f(x)\,dx$$

by antidifferentiation just as Newton did—that is, by computing

$$A = \Big[D^{-1} f(x)\Big]_a^b.$$

The question of the *existence* of the area function—one of the conditions that a function f must satisfy in order for its integral to exist—did not at first seem to be of much importance. Eighteenth-century mathematicians were mainly occupied (and satisfied) with the impressive applications of calculus to the solution of real-world problems and did not concentrate on the logical foundations of the subject.

The first attempt at a precise definition of the integral and a proof of its existence for continuous functions was that of the French mathematician Augustin Louis Cauchy (1789–1857). Curiously enough, Cauchy was trained as an engineer, and much of his research in mathematics was in fields that we today regard as applications-oriented: hydrodynamics, waves in elastic media, vibrations of elastic membranes, polarization of light, and the like. But he was a prolific researcher, and his writings cover the entire spectrum of mathematics, with occasional essays into almost unrelated fields.

Around 1824, Cauchy defined the integral of a continuous function in a way that is familiar to us, as a limit of left-endpoint approximations:

$$\int_a^b f(x)\,dx = \lim_{\Delta x \to 0} \sum_{i=1}^{n} f(x_{i-1})\,\Delta x.$$

This is a much more complicated sort of limit than the ones we discussed in Chapter 2. Cauchy was not entirely clear about the nature of the limit process involved in this equation, nor was he clear about the precise role that the hypothesis of the continuity of f played in proving that the limit exists.

A complete definition of the integral, as we gave in Section 5.4, was finally produced in the 1850s by the German mathematician Georg Bernhard Riemann. Riemann was a student of Gauss; he met Gauss upon his arrival at Göttingen, Germany, for the purpose of studying theology, when he was about 20 years old and Gauss was about 70. Riemann soon decided to study mathematics and became known as one of the truly great mathematicians of the nineteenth century. Like Cauchy, he was particularly interested in applications of mathematics to the real world; his research emphasized electricity, heat, light, acoustics, fluid dynamics, and—as you might infer from the fact that Wilhelm Weber was a major influence on Riemann's education—magnetism. Riemann also made significant contributions to mathematics itself, particularly in the field of complex analysis. A major conjecture of his, involving the zeta function

$$\zeta(s) = \sum_{n=1}^{\infty} \frac{1}{n^s}, \qquad \textbf{(1)}$$

remains unsolved to this day. This conjecture has important consequences in the theory of the distribution of prime numbers because

$$\zeta(k) = \prod \left(1 - \frac{1}{p^k}\right)^{-1},$$

where the product $\prod$ is taken over all primes p. [The zeta function is defined in Eq. (1) for complex numbers s to the right of the vertical line at $x = 1$ and is extended to other complex numbers by the requirement that it be differentiable.] Riemann died of tuberculosis shortly before his fortieth birthday.

Here we give a proof of the existence of the integral of a continuous function. We will follow Riemann's approach. Specifically, suppose that the function f is continuous on the closed and bounded interval $[a, b]$. We will prove that the definite integral

$$\int_a^b f(x)\,dx$$

exists. That is, we will demonstrate the existence of a number I that satisfies the following condition: For every $\epsilon > 0$ there exists $\delta > 0$ such that, for *every* Riemann sum R associated with *any* partition P with $|P| < \delta$,

$$|I - R| < \epsilon.$$

(Recall that the norm $|P|$ of the partition P is the length of the longest subinterval in the partition.) In other words, every Riemann sum associated with every sufficiently "fine" partition is close to the number I. If this happens, then the definite integral

$$\int_a^b f(x)\,dx$$

is said to **exist,** and I is its **value.**

Now we begin the proof. Suppose throughout that f is a function continuous on the closed interval $[a, b]$. Given $\epsilon > 0$, we need to show the existence of a number $\delta > 0$ such that

$$\left| I - \sum_{i=1}^{n} f(x_i^\star)\,\Delta x_i \right| < \epsilon \qquad \textbf{(2)}$$

for every Riemann sum associated with any partition P of $[a, b]$ with $|P| < \delta$.

Given a partition P of $[a, b]$ into n subintervals that are *not necessarily of equal length,* let p_i be a point in the subinterval $[x_{i-1}, x_i]$ at which f attains its minimum value $f(p_i)$. Similarly, let $f(q_i)$ be its maximum value there. These numbers exist for $i = 1, 2, 3, \dots, n$ because of the maximum value property of continuous functions (Theorem 4 of Appendix E).

In what follows we will denote the resulting lower and upper Riemann sums associated with P by

$$L(P) = \sum_{i=1}^{n} f(p_i)\,\Delta x_i \qquad \textbf{(3a)}$$

and

$$U(P) = \sum_{i=1}^{n} f(q_i)\,\Delta x_i, \qquad \textbf{(3b)}$$

respectively. Then Lemma 1 is obvious.

LEMMA 1

For any partition P of $[a, b]$, $L(P) \leqq U(P)$.

Now we need a definition. The partition P' is called a **refinement** of the partition P if each subinterval of P' is contained in some subinterval of P. That is, P' is obtained from P by adding more points of subdivision to P.

LEMMA 2

Suppose that P' is a refinement of P. Then

$$L(P) \leqq L(P') \leqq U(P') \leqq U(P). \qquad \textbf{(4)}$$

PROOF The inequality $L(P') \leqq U(P')$ is a consequence of Lemma 1. We will show that $L(P) \leqq L(P')$; the proof that $U(P') \leqq U(P)$ is similar.

The refinement P' is obtained from P by adding one or more points of subdivision to P. So all we need show is that the Riemann sum $L(P)$ cannot be decreased by adding a single point of subdivision. Thus we will suppose that the partition P' is obtained from P by dividing the kth subinterval $[x_{k-1}, x_k]$ of P into two subintervals $[x_{k-1}, z]$ and $[z, x_k]$ by means of the new subdivision point z.

The only resulting effect on the corresponding Riemann sum is to replace the term

$$f(p_k) \cdot (x_k - x_{k-1})$$

in $L(P)$ with the two-term sum

$$f(u) \cdot (z - x_{k-1}) + f(v) \cdot (x_k - z),$$

where $f(u)$ is the minimum of f on $[x_{k-1}, z]$ and $f(v)$ is the minimum of f on $[z, x_k]$. But

$$f(p_k) \leqq f(u) \quad \text{and} \quad f(p_k) \leqq f(v).$$

Hence

$$\begin{aligned} f(u) \cdot (z - x_{k-1}) + f(v) \cdot (x_k - z) &\geqq f(p_k) \cdot (z - x_{k-1}) + f(p_k) \cdot (x_k - z) \\ &= f(p_k) \cdot (z - x_{k-1} + x_k - z) \\ &= f(p_k) \cdot (x_k - x_{k-1}). \end{aligned}$$

So the replacement of $f(p_k) \cdot (x_k - x_{k-1})$ cannot decrease the sum $L(P)$ in question, and therefore $L(P) \leqq L(P')$. Because this is all we needed to show, we have completed the proof of Lemma 2. ◄

To prove that all the Riemann sums for sufficiently fine partitions are close to some number I, we must first give a construction of I. This is accomplished through Lemma 3.

LEMMA 3

Let P_n denote the regular partition of $[a, b]$ into 2^n subintervals of equal length. Then the (sequential) limit

$$I = \lim_{n \to \infty} L(P_n) \qquad \textbf{(5)}$$

exists.

PROOF We begin with the observation that each partition P_{n+1} is a refinement of P_n, so (by Lemma 2)

$$L(P_1) \leqq L(P_2) \leqq \cdots \leqq L(P_n) \leqq \cdots.$$

Therefore $\{L(P_n)\}$ is a nondecreasing sequence of real numbers. Moreover,

$$L(P_n) = \sum_{i=1}^{2^n} f(p_i)\Delta x_i \leqq M \sum_{i-1}^{2^n} \Delta x_i = M(b-a),$$

where M is the maximum value of f on $[a, b]$.

Theorem 1 of Appendix E guarantees that a bounded monotonic sequence of real numbers must converge. Thus the number

$$I = \lim_{n \to \infty} L(P_n)$$

exists. This establishes Eq. (5), and the proof of Lemma 3 is complete. ◄

It is proved in advanced calculus that if f is continuous on $[a, b]$, then—for every number $\epsilon > 0$—there exists a number $\delta > 0$ such that

$$|f(u) - f(v)| < \epsilon$$

for every two points u and v of $[a, b]$ such that

$$|u - v| < \delta.$$

This property of a function is called **uniform continuity** of f on the interval $[a, b]$. Thus the theorem from advanced calculus that we need to use states that every continuous function on a closed and bounded interval is uniformly continuous there.

Note The fact that f is continuous on $[a, b]$ means that for each number u in the interval and each $\epsilon > 0$, there exists $\delta > 0$ such that if v is a number in the interval with $|u - v| < \delta$, then $|f(u) - f(v)| < \epsilon$. But *uniform* continuity is a more stringent condition. It means that given $\epsilon > 0$, you can find not only a value δ_1 that "works" for u_1, a value δ_2 that works for u_2, and so on, but more: You can find a universal value of $\delta > 0$ that works for *all* values of u in the interval. This should not be obvious when you notice the possibility that $\delta_1 = 1$, $\delta_2 = \frac{1}{2}$, $\delta_3 = \frac{1}{3}$, and so on. In any case, it is clear that uniform continuity of f on an interval implies its continuity there.

Remember that throughout we have a continuous function f defined on the closed interval $[a, b]$.

LEMMA 4

Suppose that $\epsilon > 0$ is given. Then there exists a number $\delta > 0$ such that if P is a partition of $[a, b]$ with $|P| < \delta$ and P' is a refinement of P, then

$$|R(P) - R(P')| < \frac{\epsilon}{3} \tag{6}$$

for any two Riemann sums $R(P)$ associated with P and $R(P')$ associated with P'.

PROOF Because f must be uniformly continuous on $[a, b]$, there exists a number $\delta > 0$ such that if

$$|u - v| < \delta, \quad \text{then} \quad |f(u) - f(v)| < \frac{\epsilon}{3(b-a)}.$$

Suppose now that P is a partition of $[a, b]$ with $|P| < \delta$. Then

$$|U(P) - L(P)| = \sum_{i=1}^{n} |f(q_i) - f(p_i)|\Delta x_i < \frac{\epsilon}{3(b-a)} \sum_{i=1}^{n} \Delta x_i = \frac{\epsilon}{3}.$$

This is valid because $|p_i - q_i| < \delta$, for both p_i and q_i belong to the same subinterval $[x_{i-1}, x_i]$ of P, and $|P| < \delta$.

Now, as shown in Fig. F.1, we know that $L(P)$ and $U(P)$ differ by less than $\epsilon/3$. We know also that

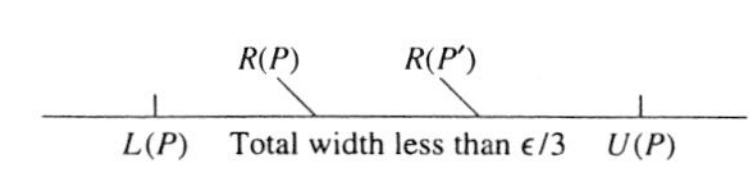

FIGURE F.1 Part of the proof of Lemma 4.

$$L(P) \leqq R(P) \leqq U(P)$$

for every Riemann sum $R(P)$ associated with P. But

$$L(P) \leqq L(P') \leqq U(P') \leqq U(P)$$

by Lemma 2, because P' is a refinement of P; moreover,

$$L(P') \leqq R(P') \leqq U(P')$$

for every Riemann sum $R(P')$ associated with P'.

As Fig. F.1 shows, both the numbers $R(P)$ and $R(P')$ lie in the interval $[L(P), U(P)]$ of length less than $\epsilon/3$, so Eq. (6) follows. This concludes the proof of Lemma 4. ◀

THEOREM 1 Existence of the Integral
If f is continuous on the closed and bounded interval $[a, b]$, then the integral

$$\int_a^b f(x)\,dx$$

exists.

PROOF Suppose that $\epsilon > 0$ is given. We must show the existence of a number $\delta > 0$ such that, for every partition P of $[a, b]$ with $|P| < \delta$, we have

$$|I - R(P)| < \epsilon,$$

where I is the number given in Lemma 3 and $R(P)$ is an arbitrary Riemann sum for f associated with P.

We choose the number δ provided by Lemma 4 such that

$$|R(P) - R(P')| < \frac{\epsilon}{3}$$

if $|P| < \delta$ and P' is a refinement of P.

By Lemma 3, we can choose an integer N so large that

$$|P_N| < \delta \quad \text{and} \quad |L(P_N) - I| < \frac{\epsilon}{3}. \tag{7}$$

Given an arbitrary partition P such that $|P| < \delta$, let P' be a common refinement of both P and P_N. You can obtain such a partition P', for example, by using all the points of subdivision of both P and P_N to form the subintervals of $[a, b]$ that constitute P'.

Because P' is a refinement of both P and P_N and both the latter partitions have mesh less than δ, Lemma 4 implies that

$$|R(P) - R(P')| < \frac{\epsilon}{3} \quad \text{and} \quad |L(P_N) - R(P')| < \frac{\epsilon}{3}. \tag{8}$$

Here $R(P)$ and $R(P')$ are (arbitrary) Riemann sums associated with P and P', respectively.

Given an arbitrary Riemann sum $R(P)$ associated with the partition P with mesh less than δ, we see that

$$\begin{aligned}|I - R(P)| &= |I - L(P_N) + L(P_N) - R(P') + R(P') - R(P)| \\ &\leqq |I - L(P_N)| + |L(P_N) - R(P')| + |R(P') - R(P)|.\end{aligned}$$

In the last sum, both of the last two terms are less than $\epsilon/3$ by virtue of the inequalities in (8). We also know, by (7), that the first term is less than $\epsilon/3$. Consequently,

$$|I - R(P)| < \epsilon.$$

This establishes Theorem 1. ◀

We close with an example that shows that some hypothesis of continuity is required for integrability.

EXAMPLE 1 Suppose that f is defined for $0 \leqq x \leqq 1$ as follows:

$$f(x) = \begin{cases} 1 & \text{if } x \text{ is irrational;} \\ 0 & \text{if } x \text{ is rational.} \end{cases}$$

Then f is not continuous anywhere. (Why?) Given a partition P of $[0, 1]$, let p_i be a rational point and q_i an irrational point of the ith subinterval of P for each i, $1 \leqq i \leqq n$. As before, f attains its minimum value 0 at each p_i and its maximum value 1 at each q_i. Also

$$L(P) = \sum_{i=1}^{n} f(p_i)\Delta x_i = 0, \quad \text{whereas} \quad U(P) = \sum_{i=1}^{n} f(q_i)\Delta x_i = 1.$$

Thus if we choose $\epsilon = \frac{1}{2}$, then there is *no* number I that can lie within ϵ of both $L(P)$ and $U(P)$, no matter how small the mesh of P. It follows that f is *not* integrable on $[0, 1]$. ◆

REMARK This is not the end of the story of the integral. Integrals of highly discontinuous functions are important in many applications of physics, and near the beginning of the twentieth century a number of mathematicians, most notably Henri Lebesgue (1875–1941), developed more powerful integrals. The Lebesgue integral, in particular, always exists when the Riemann integral does, and gives the same value; but the Lebesgue integral is sufficiently powerful to integrate even functions that are continuous nowhere. It reports that

$$\int_0^1 f(x)\,dx = 1$$

for the function f of Example 1. Other mathematicians have developed integrals with domains far more general than sets of real numbers or subsets of the plane or space.

APPENDIX G: APPROXIMATIONS AND RIEMANN SUMS

Several times in Chapter 6 our attempt to compute some quantity Q led to the following situation. Beginning with a regular partition of an appropriate interval $[a, b]$ into n subintervals, each of length Δx, we found an approximation A_n to Q of the form

$$A_n = \sum_{i=1}^{n} g(u_i)h(v_i)\,\Delta x, \tag{1}$$

where u_i and v_i are two (generally different) points of the ith subinterval $[x_{i-1}, x_i]$. For example, in our discussion of surface area of revolution that precedes Eq. (8) of Section 6.4, we found the approximation

$$\sum_{i=1}^{n} 2\pi f(u_i)\sqrt{1 + [f'(v_i)]^2}\,\Delta x \tag{2}$$

to the area of the surface generated by revolving the curve $y = f(x)$, $a \leqq x \leqq b$, around the x-axis. (In Section 6.4 we wrote $x_i^{\star\star}$ for u_i and $x_i^{\star}$ for v_i.) Note that the expression in (2) is the same as the right-hand side in Eq. (1); take $g(x) = 2\pi f(x)$ and $h(x) = \sqrt{1 + [f'(x)]^2}$.

In such a situation we observe that if u_i and v_i were the *same* point $x_i^{\star}$ of $[x_{i-1}, x_i]$ for each i ($i = 1, 2, 3, \ldots, n$), then the approximation in Eq. (1) would be a Riemann sum for the function $g(x)h(x)$ on $[a, b]$. This leads us to suspect that

$$\lim_{\Delta x \to 0} \sum_{i=1}^{n} g(u_i)h(v_i)\,\Delta x = \int_a^b g(x)h(x)\,dx. \tag{3}$$

In Section 6.4, we assumed the validity of Eq. (3) and concluded from the approximation in (2) that the surface area of revolution ought to be defined to be

$$A = \lim_{\Delta x \to 0} \sum_{i=1}^{n} 2\pi f(u_i)\sqrt{1 + [f'(v_i)]^2}\,\Delta x = \int_a^b 2\pi f(x)\sqrt{1 + [f'(x)]^2}\,dx.$$

Theorem 1 guarantees that Eq. (3) holds under mild restrictions on the functions g and h.

THEOREM 1 A Generalization of Riemann Sums
Suppose that h and g' are continuous on $[a, b]$. Then

$$\lim_{\Delta x \to 0} \sum_{i=1}^{n} g(u_i)h(v_i)\,\Delta x = \int_a^b g(x)h(x)\,dx, \tag{3}$$

where u_i and v_i are arbitrary points of the ith subinterval of a regular partition of $[a, b]$ into n subintervals, each of length Δx.

PROOF Let M_1 and M_2 denote the maximum values on $[a, b]$ of $|g'(x)|$ and $|h(x)|$, respectively. Note that

$$\sum_{i=1}^{n} g(u_i)h(v_i)\,\Delta x = R_n + S_n, \quad \text{where} \quad R_n = \sum_{i=1}^{n} g(v_i)h(v_i)\,\Delta x$$

is a Riemann sum approaching $\int_a^b g(x)h(x)dx$ as $\Delta x \to 0$, and

$$S_n = \sum_{i=1}^{n} [g(u_i) - g(v_i)]\,h(v_i)\,\Delta x.$$

To prove Eq. (3) it is sufficient to show that $S_n \to 0$ as $\Delta x \to 0$. The mean value theorem gives

$$\begin{aligned} |g(u_i) - g(v_i)| &= |g'(\overline{x}_i)| \cdot |u_i - v_i| \qquad [\overline{x}_i \text{ in } (u_i, v_i)] \\ &\leqq M_1\,\Delta x, \end{aligned}$$

because both u_i and v_i are points of the interval $[x_{i-1}, x_i]$ of length Δx. Then

$$\begin{aligned} |S_n| &\leqq \sum_{i=1}^{n} |g(u_i) - g(v_i)| \cdot |h(v_i)|\,\Delta x \leqq \sum_{i=1}^{n} (M_1\,\Delta x) \cdot (M_2\,\Delta x) \\ &= (M_1 M_2\,\Delta x) \sum_{i=1}^{n} \Delta x = M_1 M_2 (b - a)\,\Delta x, \end{aligned}$$

from which it follows that $S_n \to 0$ as $\Delta x \to 0$, as desired. ◀

As an application of Theorem 1, let us give a rigorous derivation of Eq. (2) of Section 6.3,

$$V = \int_a^b 2\pi x f(x)\,dx, \tag{4}$$

for the volume of the solid generated by revolving around the y-axis the region between the graph of $y = f(x)$ and the x-axis for $a \leqq x \leqq b$. Beginning with the usual regular partition of $[a, b]$, let $f(x_i^\flat)$ and $f(x_i^\sharp)$ denote the minimum and maximum values of f on the ith subinterval $[x_{i-1}, x_i]$. Denote by $x_i^\star$ the midpoint of this subinterval. From Fig. G.1, we see that the part of the solid generated by revolving the region below $y = f(x)$, $x_{i-1} \leqq x \leqq x_i$, contains a cylindrical shell with average radius $x_i^\star$, thickness Δx, and height $f(x_i^\flat)$ and is contained in another cylindrical shell with the same average radius and thickness but with height $f(x_i^\sharp)$. Hence the volume ΔV_i of this part of the solid satisfies the inequalities

$$2\pi x_i^\star f(x_i^\flat)\,\Delta x \leqq \Delta V_i \leqq 2\pi x_i^\star f(x_i^\sharp)\,\Delta x.$$

We add these inequalities for $i = 1, 2, 3, \ldots, n$ and find that

$$\sum_{i=1}^{n} 2\pi x_i^\star f(x_i^\flat)\,\Delta x \leqq V \leqq \sum_{i=1}^{n} 2\pi x_i^\star f(x_i^\sharp)\,\Delta x.$$

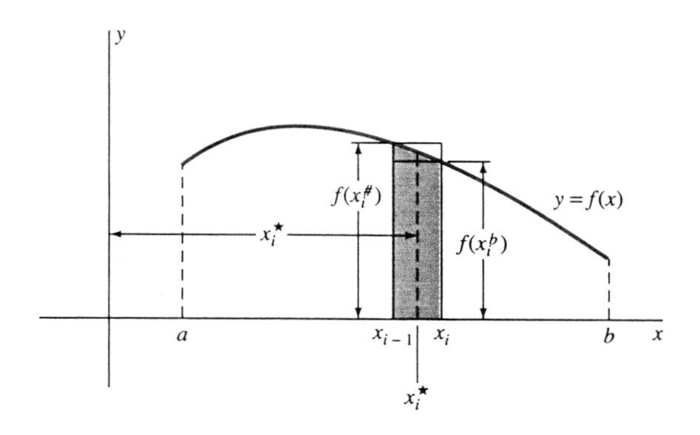

FIGURE G.1 A careful estimate of the volume of a solid of revolution around the y-axis.

Because Theorem 1 implies that both of the last two sums approach $\int_a^b 2\pi f(x)dx$, the squeeze law of limits now implies Eq. (4).

We will occasionally need a generalization of Theorem 1 that involves the notion of a continuous function $F(x, y)$ of two variables. We say that F is *continuous* at the point (x_0, y_0) provided that the value $F(x, y)$ can be made arbitrarily close to $F(x_0, y_0)$ merely by choosing the point (x, y) sufficiently close to (x_0, y_0). We discuss continuity of functions of two variables in Chapter 13. Here it will suffice to accept the following facts: If $g(x)$ and $h(y)$ are continuous functions of the single variables x and y, respectively, then simple combinations such as

$$g(x) \pm h(y), \quad g(x)h(y), \quad \text{and} \quad \sqrt{[g(x)]^2 + [h(y)]^2}$$

are continuous functions of the two variables x and y.

Now consider a regular partition of $[a, b]$ into n subintervals, each of length Δx, and let u_i and v_i denote arbitrary points of the ith subinterval $[x_{i-1}, x_i]$. Theorem 2—we omit the proof—tells us how to find the limit as $\Delta x \to 0$ of a sum such as

$$\sum_{i=1}^{n} F(u_i, v_i)\,\Delta x.$$

THEOREM 2 A Further Generalization

Let $F(x, y)$ be continuous for x and y both in the interval $[a, b]$. Then, in the notation of the preceding paragraph,

$$\lim_{\Delta x \to 0} \sum_{i=1}^{n} F(u_i, v_i)\,\Delta x = \int_a^b F(x, x)\,dx. \qquad \textbf{(5)}$$

Theorem 1 is the special case $F(x, y) = g(x)h(y)$ of Theorem 2. Moreover, the integrand $F(x, x)$ on the right in Eq. (5) is merely an ordinary function of the single variable x. As a formal matter, the integral corresponding to the sum in Eq. (5) is obtained by replacing the summation symbol with an integral sign, changing both u_i and v_i to x, replacing Δx with dx, and inserting the correct limits of integration. For example, if the interval $[a, b]$ is $[0, 4]$, then

$$\begin{aligned}
\lim_{\Delta x \to 0} \sum_{i=1}^{n} \sqrt{9u_i^2 + v_i^4}\,\Delta x &= \int_0^4 \sqrt{9x^2 + x^4}\,dx \\
&= \int_0^4 x(9 + x^2)^{1/2}\,dx = \left[\frac{1}{3}(9 + x^2)^{3/2}\right]_0^4 \\
&= \frac{1}{3}\left[(25)^{3/2} - (9)^{3/2}\right] = \frac{98}{3}.
\end{aligned}$$

APPENDIX G PROBLEMS

In Problems 1 through 7, u_i and v_i are arbitrary points of the ith subinterval of a regular partition of $[a, b]$ into n subintervals, each of length Δx. Express the given limit as an integral from a to b, then compute the value of this integral.

1. $\lim_{\Delta x \to 0} \sum_{i=1}^{n} u_i v_i \, \Delta x; \quad a = 0, \ b = 1$

2. $\lim_{\Delta x \to 0} \sum_{j=1}^{n} (3u_j + 5v_j) \, \Delta x; \quad a = -1, \ b = 3$

3. $\lim_{\Delta x \to 0} \sum_{i=1}^{n} u_i \sqrt{4 - v_i^2} \, \Delta x; \quad a = 0, \ b = 2$

4. $\lim_{\Delta x \to 0} \sum_{i=1}^{n} \frac{u_i}{\sqrt{16 + v_i^2}} \Delta x; \quad a = 0, \ b = 3$

5. $\lim_{\Delta x \to 0} \sum_{i=1}^{n} \sin u_i \cos v_i \, \Delta x; \quad a = 0, \ b = \pi/2$

6. $\lim_{\Delta x \to 0} \sum_{i=1}^{n} \sqrt{\sin^2 u_i + \cos^2 v_i} \, \Delta x; \quad a = 0, \ b = \pi$

7. $\lim_{\Delta x \to 0} \sum_{k=1}^{n} \sqrt{u_k^4 + v_k^7} \, \Delta x; \quad a = 0, \ b = 2$

8. Explain how Theorem 1 applies to show that Eq. (8) of Section 6.4 follows from the discussion that precedes it in that section.

9. Use Theorem 1 to derive Eq. (10) of Section 6.4.

APPENDIX H: L'HÔPITAL'S RULE AND CAUCHY'S MEAN VALUE THEOREM

Here we give a proof of l'Hôpital's rule,

$$\lim_{x \to a} \frac{f(x)}{g(x)} = \lim_{x \to a} \frac{f'(x)}{g'(x)}, \tag{1}$$

under the hypotheses of Theorem 1 in Section 7.2. The proof is based on a generalization of the mean value theorem due to the French mathematician Augustin Louis Cauchy. Cauchy used this generalization in the early nineteenth century to give rigorous proofs of several calculus results not previously established firmly.

CAUCHY'S MEAN VALUE THEOREM

Suppose that the functions f and g are continuous on the closed and bounded interval $[a, b]$ and differentiable on (a, b). Then there exists a number c in (a, b) such that

$$[f(b) - f(a)]\, g'(c) = [g(b) - g(a)]\, f'(c). \tag{2}$$

REMARK 1 To see that this theorem is indeed a generalization of the (ordinary) mean value theorem, we take $g(x) = x$. Then $g'(x) \equiv 1$, and the conclusion in Eq. (2) reduces to the fact that

$$f(b) - f(a) = (b - a)\, f'(c)$$

for some number c in (a, b).

REMARK 2 Equation (2) has a geometric interpretation like that of the ordinary mean value theorem. Let us think of the equations $x = g(t)$, $y = f(t)$ as describing the motion of a point $P(x, y)$ moving along a curve C in the xy-plane as t increases from a to b (Fig. H.1). That is, $P(x, y) = P(g(t), f(t))$ is the location of the point P at time t. Under the assumption that $g(b) \neq g(a)$, the slope of the line L connecting the endpoints of the curve C is

$$m = \frac{f(b) - f(a)}{g(b) - g(a)}. \tag{3}$$

But if $g'(c) \neq 0$, then the chain rule gives

$$\frac{dy}{dx} = \frac{dy/dt}{dx/dt} = \frac{f'(c)}{g'(c)} \tag{4}$$

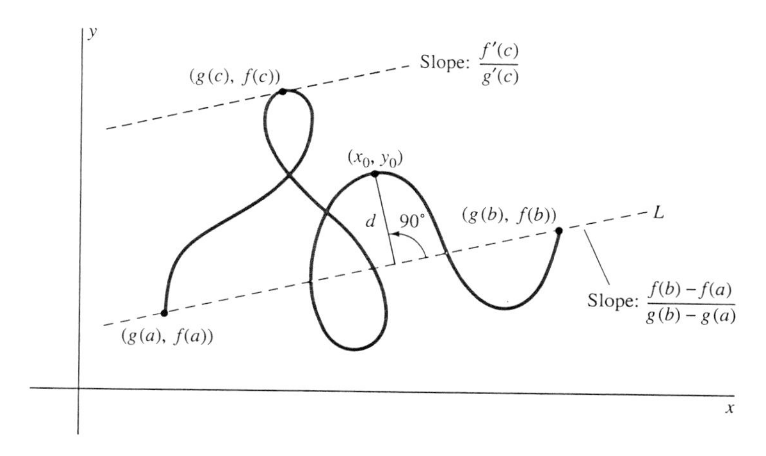

FIGURE H.1 The idea of Cauchy's mean value theorem.

for the slope of the line tangent to the curve C at the point $(g(c), f(c))$. But if $g(b) \neq g(a)$ and $g'(c) \neq 0$, then Eq. (2) may be written in the form

$$\frac{f(b) - f(a)}{g(b) - g(a)} = \frac{f'(c)}{g'(c)}, \tag{5}$$

so the two slopes in Eqs. (3) and (4) are equal. Thus Cauchy's mean value theorem implies that (under our assumptions) there is a point on the curve C where the tangent line is *parallel* to the line joining the endpoints of C. This is exactly what the (ordinary) mean value theorem says for an explicitly defined curve $y = f(x)$. This geometric interpretation motivates the following proof of Cauchy's mean value theorem.

PROOF The line L through the endpoints in Fig. H.1 has point-slope equation

$$y - f(a) = \frac{f(b) - f(a)}{g(b) - g(a)}[x - g(a)],$$

which can be rewritten in the form $Ax + By + C = 0$ with

$$A = g(b) - f(a), \quad B = -[g(b) - g(a)], \quad \text{and}$$
$$C = f(a)[g(b) - g(a)] - g(a)[f(b) - f(a)]. \tag{6}$$

According to Miscellaneous Problem 93 of Chapter 3, the (perpendicular) distance from the point (x_0, y_0) to the line L is

$$d = \frac{|Ax_0 + By_0 + C|}{\sqrt{A^2 + B^2}}.$$

Figure H.1 suggests that the point $(g(c), f(c))$ will maximize this distance d for points on the curve C.

We are motivated, therefore, to define the auxiliary function

$$\phi(t) = Ag(t) + Bf(t) + C, \tag{7}$$

with the constants A, B, and C as defined in (6). Thus $\phi(t)$ is essentially a constant multiple of the distance from $(g(t), f(t))$ to the line L in Fig. H.1.

Now $\phi(a) = 0 = \phi(b)$ (why?), so Rolle's theorem (Section 4.3) implies the existence of a number c in (a, b) such that

$$\phi'(c) = Ag'(c) + Bf'(c) = 0. \tag{8}$$

We substitute the values of A and B from Eq. (6) into (8) and obtain the equation

$$[f(b) - f(a)]\, g'(c) - [g(b) - g(a)]\, f'(c) = 0.$$

This is the same as Eq. (2) in the conclusion of Cauchy's mean value theorem, and the proof is complete. ◀

Note Although the assumptions that $g(b) \neq g(a)$ and $g'(c) \neq 0$ were needed for our geometric interpretation of the theorem, they were not used in its proof—only in the motivation for the method of proof.

PROOF OF L'HÔPITAL'S RULE Suppose that $f(x)/g(x)$ has the indeterminate form $0/0$ at $x = a$. We may invoke continuity of f and g to allow the assumption that $f(a) = 0 = f(b)$. That is, we simply define $f(a)$ and $g(a)$ to be zero in case their values at $x = a$ are not originally given.

Now we restrict our attention to values of $x \neq a$ in a fixed neighborhood of a on which both f and g are differentiable. Choose one such value of x and hold it temporarily constant. Then apply Cauchy's mean value theorem on the interval $[a, x]$. (If $x < a$, use the interval $[x, a]$.) We find that there is a number z between a and x that behaves as c does in Eq. (2). Hence, by virtue of Eq. (2), we obtain the equation

$$\frac{f(x)}{g(x)} = \frac{f(x) - f(a)}{g(x) - g(a)} = \frac{f'(z)}{g'(z)}.$$

Now z depends on x, but z is trapped between x and a, so z is forced to approach a as $x \to a$. We conclude that

$$\lim_{x \to a} \frac{f(x)}{g(x)} = \lim_{z \to a} \frac{f'(z)}{g'(z)} = \lim_{x \to a} \frac{f'(x)}{g'(x)},$$

under the assumption that the right-hand limit exists. Thus we have verified l'Hôpital's rule in the form of Eq. (1). ◂

APPENDIX I: PROOF OF TAYLOR'S FORMULA

Several different proofs of Taylor's formula (Theorem 2 of Section 11.4) are known, but none of them seems very well motivated—each requires some "trick" to begin the proof. The trick we employ here (suggested by C. R. MacCluer) is to begin by introducing an auxiliary function $F(x)$, defined as follows:

$$F(x) = f(b) - f(x) - f'(x)(b - x) - \frac{f''(x)}{2!}(b - x)^2 - \cdots - \frac{f^{(n)}(x)}{n!}(b - x)^n - K(b - x)^{n+1}, \quad \textbf{(1)}$$

where the *constant* K is chosen so that $F(a) = 0$. To see that there *is* such a value of k, we could substitute $x = a$ on the right and $F(x) = F(a) = 0$ on the left in Eq. (1) and then solve routinely for K, but we have no need to do this explicitly.

Equation (1) makes it quite obvious that $F(b) = 0$ as well. Therefore, Rolle's theorem (Section 4.3) implies that

$$F'(z) = 0 \quad \textbf{(2)}$$

for some point z of the open interval (a, b) (under the assumption that $a < b$). To see what Eq. (2) means, we differentiate both sides of Eq. (1) and find that

$$\begin{aligned} F'(x) = {} & -f'(x) + [f'(x) - f''(x)(b - x)] \\ & + \left[f''(x)(b - x) - \frac{1}{2!}f^{(3)}(x)(b - x)^2\right] \\ & + \left[\frac{1}{2!}f^{(3)}(x)(b - x)^2 - \frac{1}{3!}f^{(4)}(x)(b - x)^3\right] \\ & + \cdots + \left[\frac{1}{(n-1)!}f^{(n)}(x)(b - x)^{n-1} - \frac{1}{n!}f^{(n+1)}(x)(b - x)^n\right] \\ & + (n + 1)K(b - x)^n. \end{aligned}$$

Upon careful inspection of this result, we see that all terms except the final two cancel in pairs. Thus the sum "telescopes" to give

$$F'(x) = (n+1)K(b-x)^n - \frac{f^{(n+1)}(x)}{n!}(b-x)^n. \quad \textbf{(3)}$$

Hence Eq. (2) means that

$$(n+1)K(b-z)^n - \frac{f^{(n+1)}(z)}{n!}(b-z)^n = 0.$$

Consequently we can cancel $(b-z)^n$ and solve for

$$K = \frac{f^{(n+1)}(z)}{(n+1)!}. \quad \textbf{(4)}$$

Finally, we return to Eq. (1) and substitute $x = a$, $f(x) = 0$, and the value of K given in Eq. (4). The result is the equation

$$0 = f(b) - f(a) - f'(a)(b-a) - \frac{f''(a)}{2!}(b-a)^2 - \cdots - \frac{f^{(n)}(a)}{n!}(b-a)^n - \frac{f^{(n+1)}(z)}{(n+1)!}(b-a)^{n+1},$$

which is equivalent to the desired Taylor's formula, Eq. (11) of Section 10.4. ◄

APPENDIX J: CONIC SECTIONS AS SECTIONS OF A CONE

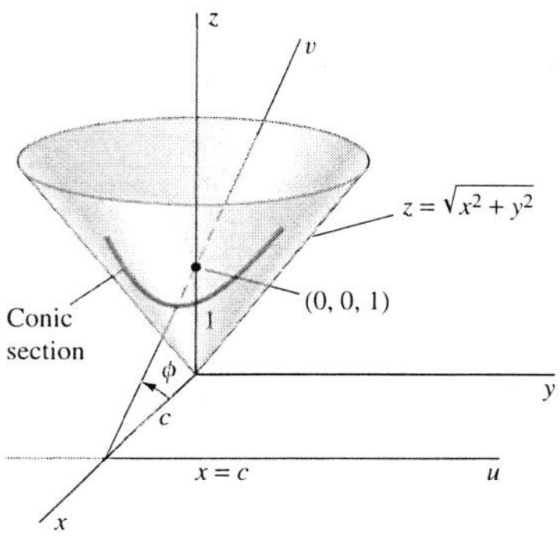

The parabola, hyperbola, and ellipse that we studied in Chapter 10 were originally introduced by the ancient Greek mathematicians as plane sections (traces) of a right circular cone. Here we show that the intersection of a plane and a cone is, indeed, one of the three conic sections as defined in Chapter 10.

Figure J.1 shows the cone with equation $z = \sqrt{x^2 + y^2}$ and its intersection with a plane $\mathcal{P}$ that passes through the point (0, 0, 1) and the line $x = c > 0$ in the xy-plane. An equation of $\mathcal{P}$ is

$$z = 1 - \frac{x}{c}. \quad \textbf{(1)}$$

FIGURE J.1 Finding an equation for a conic section.

The angle between $\mathcal{P}$ and the xy-plane is $\phi = \tan^{-1}(1/c)$. We want to show that the conic section obtained by intersecting the cone and the plane is

A parabola if $\phi = 45°$ $(c = 1)$,

An ellipse if $\phi < 45°$ $(c > 1)$,

A hyperbola if $\phi > 45°$ $(c < 1)$.

We begin by introducing uv-coordinates in the plane $\mathcal{P}$ as follows. The u-coordinate of the point (x, y, z) of $\mathcal{P}$ is $u = y$. The v-coordinate of the same point is its perpendicular distance from the line $x = c$. This explains the u- and v-axes indicated in Fig. J.1. Figure J.2 shows the cross section in the plane $y = 0$ exhibiting the relation between v, x, and z. We see that

$$z = v \sin\phi = \frac{v}{\sqrt{1+c^2}}. \quad \textbf{(2)}$$

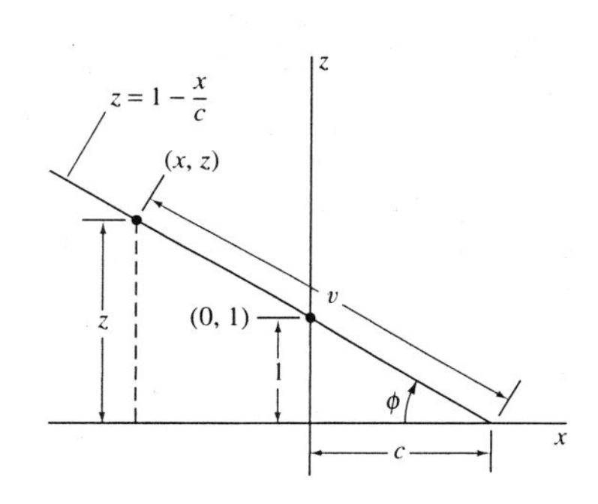

FIGURE J.2 Computing coordinates in the uv-plane.

Equations (1) and (2) give

$$x = c(1-z) = c\left(1 - \frac{v}{\sqrt{1+c^2}}\right). \quad \textbf{(3)}$$

We had $z^2 = x^2 + y^2$ for the equation of the cone. We make the following substitutions in this equation: Replace y with u, and replace z and x with the expressions on the right-hand sides of Eqs. (2) and (3), respectively. These replacements yield

$$\frac{v^2}{1+c^2} = c^2\left(1 - \frac{v}{\sqrt{1+c^2}}\right)^2 + u^2.$$

After we simplify, this last equation takes the form

$$u^2 + \frac{c^2-1}{c^2+1}v^2 - \frac{2c^2}{\sqrt{1+c^2}}v + c^2 = 0. \tag{4}$$

This is the equation of the curve in the uv-plane. We examine the three cases for the angle ϕ.

Suppose first that $\phi = 45°$. Then $c = 1$, so Eq. (4) contains a term that includes u^2, another term that includes v, and a constant term. So the curve is a parabola; see Eq. (6) of Section 9.6.

Suppose next that $\phi < 45°$. Then $c > 1$, and both the coefficients of u^2 and v^2 in Eq.(4) are positive. Thus the curve is an ellipse; see Eq. (17) of Section 9.6.

Finally, if $\phi > 45°$, then $c < 1$, and the coefficients of u^2 and v^2 in Eq. (4) have opposite signs. So the curve is a hyperbola; see Eq. (26) of Section 9.6.

APPENDIX K: PROOF OF THE LINEAR APPROXIMATION THEOREM

Under the hypothesis of continuous differentiability of the linear approximation theorem stated in Section 13.6, we want to prove that the increment

$$\Delta f = f(\mathbf{a}+\mathbf{h}) - f(\mathbf{a})$$

is given by

$$\Delta f = \nabla f(\mathbf{a}) \cdot \mathbf{h} + \epsilon(\mathbf{h}) \cdot \mathbf{h} \tag{1}$$

where $\epsilon(\mathbf{h}) = \langle \epsilon_1(\mathbf{h}), \epsilon_2(\mathbf{h}), \dots, \epsilon_n(\mathbf{h})\rangle$ is a vector such that each element $\epsilon_i(\mathbf{h})$ approaches zero as $\mathbf{h} \to \mathbf{0}$. [Note the symbol Δ for "increment" and the inverted ∇ for "gradient" on the right-hand side in Eq. (1).]

To analyze the increment Δf, we split the jump from $\mathbf{a}$ to $\mathbf{a}+\mathbf{h}$ into n separate steps, in each of which only a single coordinate is changed. Let $\mathbf{e}_i$ denote the unit n-vector with 1 in the ith position, and write

$$\mathbf{a}_0 = \mathbf{a} \quad \text{and} \quad \mathbf{a}_i = \mathbf{a}_{i-1} + h_i\mathbf{e}_i \tag{2}$$

for $i = 1, 2, \dots, n$, so that $\mathbf{a}_n = \mathbf{a}+\mathbf{h}$. Then

$$\begin{aligned}\Delta f &= f(\mathbf{a}_n) - f(\mathbf{a}_0)\\ &= [f(\mathbf{a}_n) - f(\mathbf{a}_{n-1})] + [f(\mathbf{a}_{n-1}) - f(\mathbf{a}_{n-2})] + \cdots\\ &\quad + [f(\mathbf{a}_2) - f(\mathbf{a}_1)] + [f(\mathbf{a}_1) - f(\mathbf{a}_0)];\end{aligned}$$

that is,

$$\Delta f = \sum_{i=1}^{n} [f(\mathbf{a}_i) - f(\mathbf{a}_{i-1})]. \tag{3}$$

The ith term in this sum is given by

$$\begin{aligned}f(\mathbf{a}_i) - f(\mathbf{a}_{i-1}) &= f(a_1+h_1, \dots, a_{i-1}+h_{i-1}, a_i+h_i, a_{i+1}, \dots, a_n)\\ &\quad - f(a_1+h_1, \dots, a_{i-1}+h_{i-1}, a_i, a_{i+1}, \dots, a_n)\\ &= g_i(1) - g_i(0),\end{aligned}$$

where the differentiable function g_i is defined by

$$g_i(t) = f(a_1+h_1, \dots, a_{i-1}+h_{i-1}, a_i + th_i, a_{i+1}, \dots, a_n).$$

The mean value theorem then yields

$$\begin{aligned} f(\mathbf{a}_i) - f(\mathbf{a}_{i-1}) = g_i(1) - g_i(0) &= g_i'(\overline{t}_i)(1-0) \\ &= D_i f(a_1 + h_1, \ldots, a_{i-1} + h_{i-1}, a_i + \overline{t}_i h_i, a_{i+1}, \ldots, a_n) \cdot h_i \\ &= D_i f(\mathbf{a}_{i-1} + \overline{t}_i h_i \mathbf{e}_i) \cdot h_i \end{aligned}$$

for some $\overline{t}_i$ between 0 and 1. Substitution in (2) then gives

$$\begin{aligned} \Delta f &= \sum_{i=1}^{n} D_i f(\mathbf{a}_{i-1} + \overline{t}_i h_i \mathbf{e}_i) \cdot h_i \\ &= \sum_{i=1}^{n} [D_i f(\mathbf{a}) + D_i f(\mathbf{a}_{i-1} + \overline{t}_i h_i \mathbf{e}_i) - D_i f(\mathbf{a})] \cdot h_i. \end{aligned}$$

Thus

$$\begin{aligned} \Delta f &= \sum_{i=1}^{n} [D_i f(\mathbf{a}) + \epsilon_i(\mathbf{h})] \cdot h_i \\ &= \nabla f(\mathbf{a}) \cdot \mathbf{h} + \langle \epsilon_i(\mathbf{h}), \epsilon_i(\mathbf{h}), \ldots, \epsilon_i(\mathbf{h}) \rangle \cdot \mathbf{h} \end{aligned}$$

where

$$\epsilon_i(\mathbf{h}) = D_i f(\mathbf{a}_{i-1} + \overline{t}_i h_i \mathbf{e}_i) - D_i f(\mathbf{a}) \to 0$$

(by continuity of $D_i f$ at $\mathbf{a}$) as $\mathbf{h} \to \mathbf{0}$ (and hence $\mathbf{a}_{i-1} \to \mathbf{a}$ by (2)). We have therefore established (1) and hence completed the proof. ◄

APPENDIX L: UNITS OF MEASUREMENT AND CONVERSION FACTORS

MKS SCIENTIFIC UNITS

- *Length* in meters (m); *mass* in kilograms (kg), *time* in seconds (s)
- *Force* in newtons (N); a force of 1 N imparts an acceleration of 1 m/s^2 to a mass of 1 kg.
- *Work* in joules (J); 1 J is the work done by a force of 1 N acting through a distance of 1 m.
- *Power* in watts (W); 1 W is 1 J/s.

BRITISH ENGINEERING UNITS (FPS)

- *Length* in feet (ft), *force* in pounds (lb), *time* in seconds (s)
- *Mass* in slugs; 1 lb of force imparts an acceleration of 1 ft/s^2 to a mass of 1 slug. A mass of m slugs at the surface of the earth has a *weight* of $w = mg$ pounds (lb), where $g \approx 32.17$ ft/s^2.
- *Work* in ft · lb, *power* in ft · lb/s.

CONVERSION FACTORS

$$1 \text{ in.} = 2.54 \text{ cm} = 0.0254\text{m}, \quad 1 \text{ m} \approx 3.2808 \text{ ft}$$

$$1 \text{ mi} = 5280 \text{ ft}; \quad 60 \text{ mi/h} = 88 \text{ ft/s}$$

$$1 \text{ lb} \approx 4.4482 \text{ N}; \quad 1 \text{ slug} \approx 14.594 \text{ kg}$$

$$1 \text{ hp} = 550 \text{ ft} \cdot \text{lb/s} \approx 745.7 \text{ W}$$

- *Gravitational acceleration:* $g \approx 32.17 \text{ ft/s}^2 \approx 9.807 \text{ m/s}^2$
- *Atmospheric pressure:* 1 atm is the pressure exerted by a column of mercury 76 cm high; 1 atm $\approx 14.70 \text{ lb/in.}^2 \approx 1.013 \times 10^5 \text{ N/m}^2$
- *Heat energy:* 1 Btu $\approx$ 778 ft · lb $\approx$ 252 cal, 1 cal $\approx$ 4.184 J

APPENDIX M: FORMULAS FROM ALGEBRA, GEOMETRY, AND TRIGONOMETRY

LAWS OF EXPONENTS

$$a^m a^n = a^{m+n}, \qquad (a^m)^n = a^{mn}, \qquad (ab)^n = a^n b^n, \qquad a^{m/n} = \sqrt[n]{a^m};$$

in particular,

$$a^{1/2} = \sqrt{a}.$$

If $a \neq 0$, then

$$a^{m-n} = \frac{a^m}{a^n}, \qquad a^{-n} = \frac{1}{a^n}, \qquad \text{and} \quad a^0 = 1.$$

QUADRATIC FORMULA

The quadratic equation

$$ax^2 + bx + c = 0 \quad (a \neq 0)$$

has solutions

$$x = \frac{-b \pm \sqrt{b^2 - 4ac}}{2a}.$$

FACTORING

$$\begin{aligned}
a^2 - b^2 &= (a - b)(a + b) \\
a^3 - b^3 &= (a - b)(a^2 + ab + b^2) \\
a^4 - b^4 &= (a - b)(a^3 + a^2 b + ab^2 + b^3) \\
&= (a - b)(a + b)(a^2 + b^2) \\
a^5 - b^5 &= (a - b)(a^4 + a^3 b + a^2 b^2 + ab^3 + b^4)
\end{aligned}$$

(The pattern continues.)

$$\begin{aligned}
a^3 + b^3 &= (a + b)(a^2 - ab + b^2) \\
a^5 + b^5 &= (a + b)(a^4 - a^3 b + a^2 b^2 - ab^3 + b^4) \\
a^7 + b^7 &= (a + b)(a^6 - a^5 b + a^4 b^2 - a^3 b^3 + a^2 b^4 - ab^5 + b^6)
\end{aligned}$$

(The pattern continues for odd exponents.)

BINOMIAL FORMULA

$$(a + b)^n = a^n + na^{n-1}b + \frac{n(n-1)}{1 \cdot 2} a^{n-2} b^2 + \frac{n(n-1)(n-2)}{1 \cdot 2 \cdot 3} a^{n-3} b^3 + \cdots + nab^{n-1} + b^n$$

if n is a positive integer.

AREA AND VOLUME

In Fig. M.1, the symbols have the following meanings.

A: area	b: length of base	r: radius
B: area of base	C: circumference	V: volume
h: height	ℓ: length	w: width

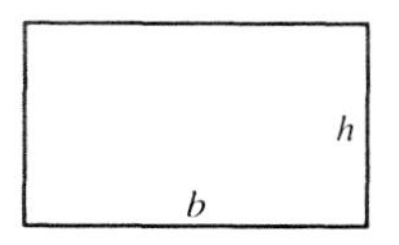

Rectangle: $A = bh$

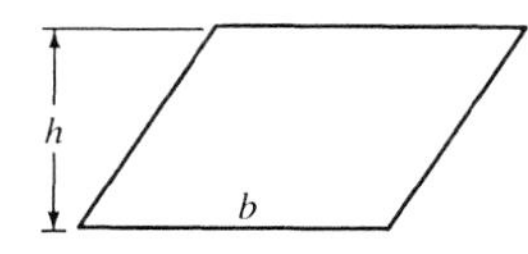

Parallelogram: $A = bh$

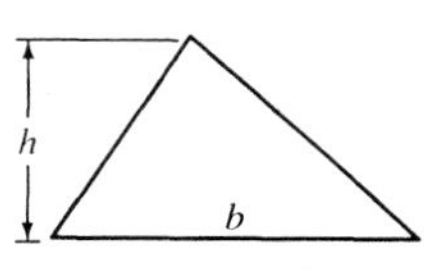

Triangle: $A = \frac{1}{2}bh$

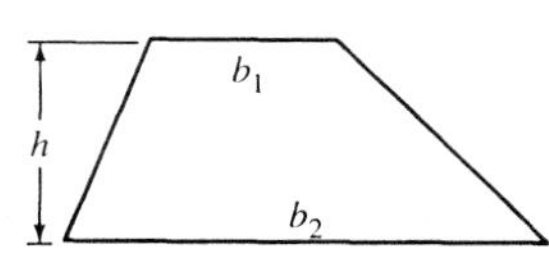

Trapezoid: $A = \frac{1}{2}(b_1 + b_2)h$

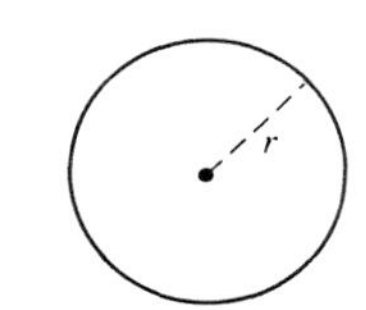

Circle: $C = 2\pi r$ and $A = \pi r^2$

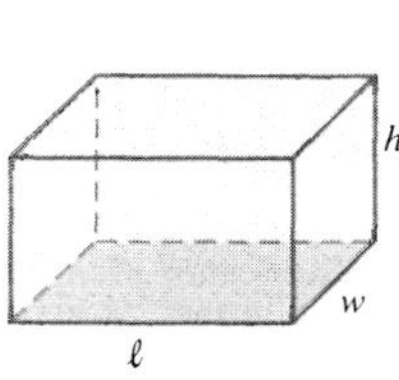

Rectangular parallelepiped: $V = \ell wh$

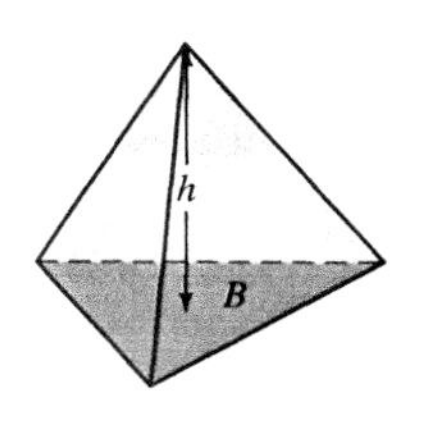

Pyramid: $V = \frac{1}{3}Bh$

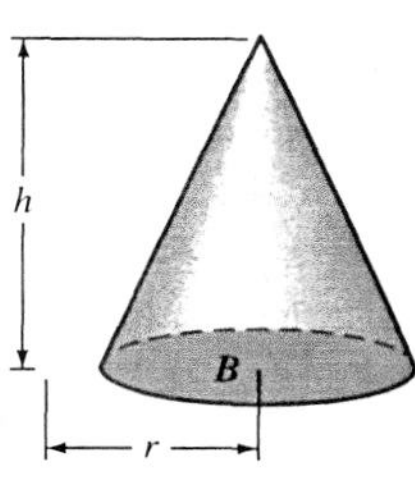

Right circular cone: $V = \frac{1}{3}\pi r^2 h = \frac{1}{3}Bh$

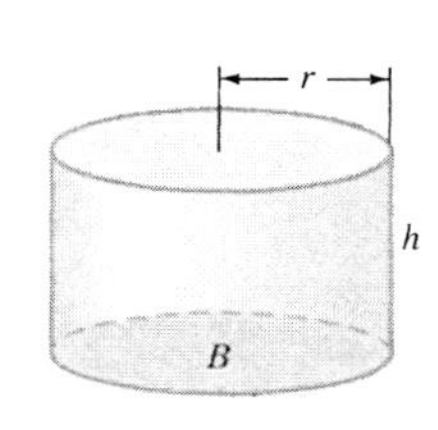

Right circular cylinder: $V = \pi r^2 h = Bh$

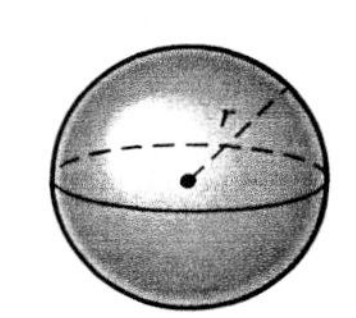

Sphere: $V = \frac{4}{3}\pi r^3$ and $A = 4\pi r^2$

FIGURE M.1 The basic geometric shapes.

PYTHAGOREAN THEOREM

In a right triangle with legs a and b and hypotenuse c,

$$a^2 + b^2 = c^2.$$

FORMULAS FROM TRIGONOMETRY

$$\sin(-\theta) = -\sin\theta$$

$$\cos(-\theta) = \cos\theta$$

$$\sin^2\theta + \cos^2\theta = 1$$

$$\sin 2\theta = 2\sin\theta\cos\theta$$

$$\cos 2\theta = \cos^2\theta - \sin^2\theta$$

$$\sin(\alpha+\beta) = \sin\alpha\cos\beta + \cos\alpha\sin\beta$$

$$\cos(\alpha+\beta) = \cos\alpha\cos\beta - \sin\alpha\sin\beta$$

$$\tan(\alpha+\beta) = \frac{\tan\alpha + \tan\beta}{1 - \tan\alpha\tan\beta}$$

$$\sin^2\frac{\theta}{2} = \frac{1-\cos\theta}{2}$$

$$\cos^2\frac{\theta}{2} = \frac{1+\cos\theta}{2}$$

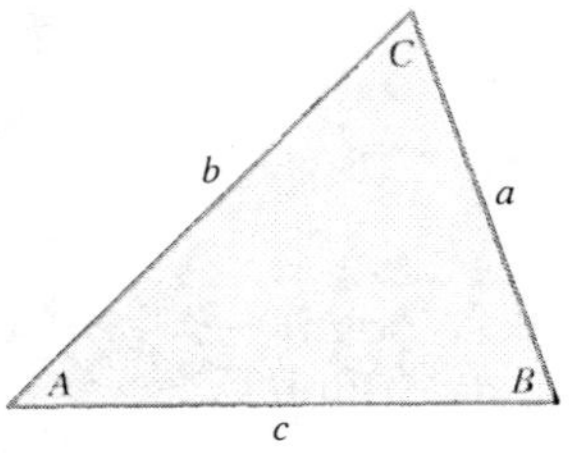

For an arbitrary triangle (Fig. M.2):

Law of cosines: $c^2 = a^2 + b^2 - 2ab \cos C.$

Law of sines: $$\frac{\sin A}{a} = \frac{\sin B}{b} = \frac{\sin C}{c}.$$

FIGURE M.2 An arbitrary triangle.

APPENDIX N: THE GREEK ALPHABET

Α	α	alpha	Ι	ι	iota	Ρ	ρ	rho
Β	β	beta	Κ	κ	kappa	Σ	σ	sigma
Γ	γ	gamma	Λ	λ	lambda	Τ	τ	tau
Δ	δ	delta	Μ	μ	mu	Υ	υ	upsilon
Ε	ϵ	epsilon	Ν	ν	nu	Φ	ϕ	phi
Ζ	ζ	zeta	Ξ	ξ	xi	Χ	χ	chi
Η	η	eta	Ο	o	omicron	Ψ	ψ	psi
Θ	θ	theta	Π	π	pi	Ω	ω	omega

ANSWERS TO ODD-NUMBERED PROBLEMS

SECTION 10.1 (PAGE 628)

1. $x + 2y + 3 = 0$

3. $3x - 4y = 25$

5. $x + y = 1$

7. Center $(-1, 0)$, radius $\sqrt{5}$

9. Center $(2, -3)$, radius 4

11. Center $(\frac{1}{2}, 0)$, radius 1

13. Center $(\frac{1}{2}, -\frac{3}{2})$, radius 3

15. Center $(-\frac{1}{3}, \frac{4}{3})$, radius 2

17. The single point $(3, 2)$

19. There are no points on the graph.

21. $(x+1)^2 + (y+2)^2 = 34$

23. $(x-6)^2 + (y-6)^2 = \frac{4}{5}$

25. The locus is the perpendicular bisector of the segment joining the two given points; it has equation $2x + y = 13$.

27. The circle with center $(6, 11)$ and radius $3\sqrt{2}$

29. The locus has equation $9x^2 + 25y^2 = 225$; it is an ellipse with center $(0, 0)$, horizontal major axis of length 10, vertical minor axis of length 6, and intercepts $(\pm 5, 0)$ and $(0, \pm 3)$.

31. There are two such lines, with equations $y - 1 = \left(4 \pm 2\sqrt{3}\right) \cdot (x - 2)$.

33. There are two such lines, with equations $y - 1 = 4(x - 4)$ and $y + 1 = 4(x + 4)$.

SECTION 10.2 (PAGE 635)

1. **a.** $\left(\frac{1}{2}\sqrt{2}, \frac{1}{2}\sqrt{2}\right)$; **b.** $\left(1, -\sqrt{3}\right)$; **c.** $\left(\frac{1}{2}, -\frac{1}{2}\sqrt{3}\right)$; **d.** $(0, -3)$; **e.** $\left(\sqrt{2}, -\sqrt{2}\right)$; **f.** $\left(\sqrt{3}, -1\right)$; **g.** $\left(-\sqrt{3}, 1\right)$

3. $r = 4\sec\theta$

5. $\theta = \tan^{-1}\left(\dfrac{1}{3}\right)$

7. $r^2 = \sec\theta \csc\theta$

9. $r = \sec\theta \tan\theta$

11. $x^2 + y^2 = 9$

13. $x^2 + y^2 + 5x = 0$

15. $(x^2 + y^2)^3 = 4y^4$

17. $x = 3$

19. $r = 2\sec\theta$; $x = 2$

21. $r = \dfrac{1}{\cos\theta + \sin\theta}$; $x + y = 1$

23. $r = \dfrac{2}{\sin\theta - \cos\theta}$; $y = x + 2$

25. $r + 8\sin\theta = 0$; $x^2 + y^2 + 8y = 0$

27. $r = 2(\cos\theta + \sin\theta)$; $x^2 + y^2 = 2x + 2y$

29. Matches Fig. 10.2.23.

31. Matches Fig. 10.2.24.

33. Matches Fig. 10.2.25.

35. Matches Fig. 10.2.26.

37. Circle, center $(\frac{1}{2}a, \frac{1}{2}b)$, radius $\frac{1}{2}\sqrt{a^2 + b^2}$

39. Circle, center $(1, 0)$, radius 1, symmetric around the x-axis

41. Cardioid, cusp at the origin (where $\theta = \pi$), symmetric around the x-axis

43. Limaçon, symmetric around the y-axis

45. Lemniscate lying in the first and third quadrants, symmetric around the lines $y = x$ and $y = -x$ and with respect to the pole

47. Four-leaved rose, symmetric around both coordinate axes, around both lines $y = \pm x$, and with respect to the pole

49. Three-leaved rose, symmetric around the x-axis, unchanged through any rotation around the origin of an integral multiple of $2\pi/3$

51. Five-leaved rose, symmetric around the y-axis, unchanged through any rotation around the origin of an integral multiple of $2\pi/5$

53. The only point of intersection has coordinates $(1, 0)$.

55. The points of intersection are $(\frac{1}{2}, \frac{1}{6}\pi)$, $(\frac{1}{2}, \frac{5}{6}\pi)$, $(-1, \frac{3}{2}\pi)$, and $(0, 0)$.

57. The points of intersection are $(0, 0)$, $(2, \pi)$,

$$\left(2\sqrt{2} - 2, \cos^{-1}\left(3 - 2\sqrt{2}\right)\right) \quad \text{and}$$

$$\left(2\sqrt{2} - 2, -\cos^{-1}\left(3 - 2\sqrt{2}\right)\right).$$

61. The polar equation can be written in the form $r = \pm a + b\sin\theta$. If $|a| = |b|$ and neither is zero, then the graph is a cardioid. If $|a| \neq |b|$ and neither a nor b is zero, then the graph is a limaçon. If either a or b is zero and the other is not, then the graph is a circle. If $a = b = 0$ then the graph consists of the pole alone.

SECTION 10.3 (PAGE 641)

1.

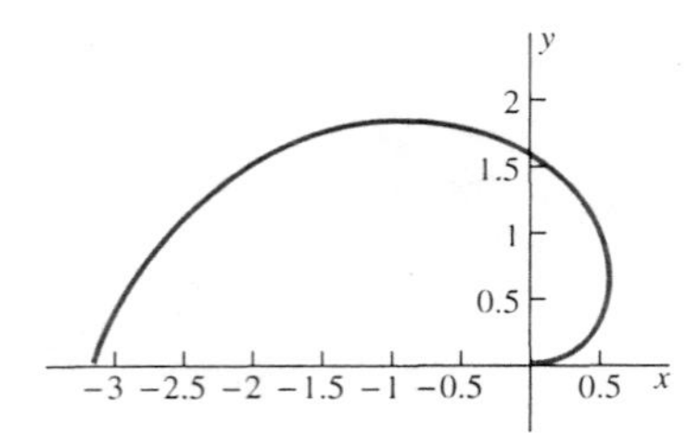

3.

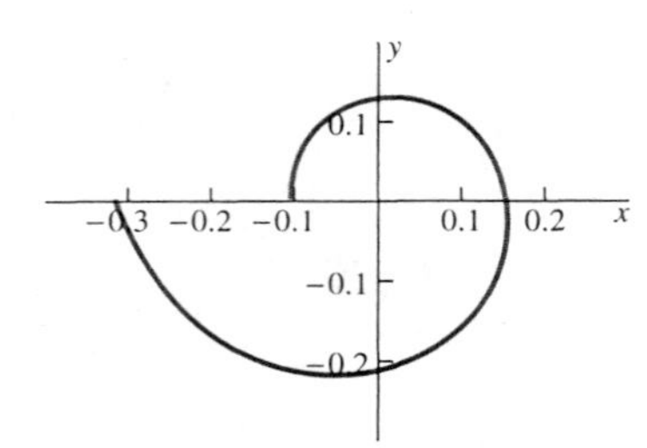

5.

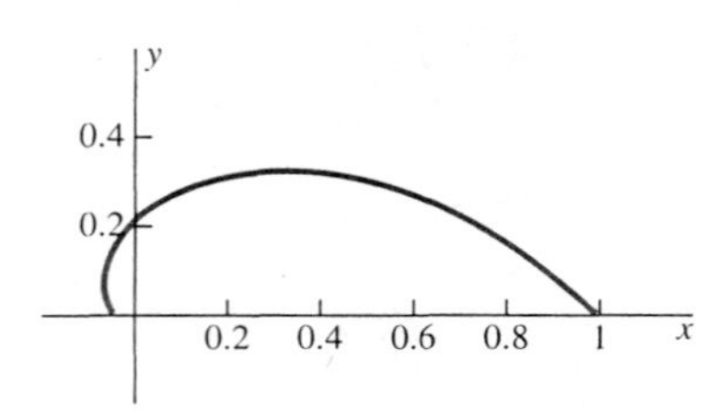

7. π **9.** $\frac{3}{2}\pi$ **11.** $\frac{9}{2}\pi$

13. 4π **15.** $\frac{19}{2}\pi$ **17.** $\frac{1}{2}\pi$

19. $\frac{1}{4}\pi$ **21.** 2 **23.** 4

25. $\dfrac{2\pi + 3\sqrt{3}}{6}$ **27.** $\dfrac{5\pi - 6\sqrt{3}}{24}$

29. $\dfrac{39\sqrt{3} - 10\pi}{6}$ **31.** $\dfrac{2 - \sqrt{2}}{2}$

33. $\dfrac{20\pi + 21\sqrt{3}}{6}$ **35.** $\dfrac{\pi - 2}{2}$

37. $\left(x - \frac{1}{2}\right)^2 + \left(y - \frac{1}{2}\right)^2 = \frac{1}{2}$; area $\frac{1}{2}\pi$

39. **a.** $A_1 = \dfrac{1}{2}\displaystyle\int_0^{2\pi} a^2\theta^2\, d\theta$;

b. $A_2 = \dfrac{1}{2}\displaystyle\int_{2\pi}^{4\pi} a^2\theta^2\, d\theta$;

c. $R_2 = A_2 - A_1$;

d. If $n \geqq 2$, then $A_n = \dfrac{1}{2}\displaystyle\int_{2(n-1)\pi}^{2n\pi} a^2\theta^2\, d\theta$.

41. **a.** $\frac{5}{2}\left(1 - e^{-2\pi/5}\right)^2$; **b.** $\frac{5}{2}e^{-2(n-1)\pi/5}\left(1 - e^{-2\pi/5}\right)^2$

43. Approximately 1.58069

SECTION 10.4 (PAGE 650)

1. $y = 2x - 3$ **3.** $y^2 = x^3$

5. $y = 2x^2 - 5x + 2$ **7.** $y = 4x^2,\ x > 0$

9. $9x^2 + 25y^2 = 225$ **11.** $9x^2 - 4y^2 = 16$

13. $x^2 + y^2 = 1$ **15.** $y = 1 - x,\ 0 \leqq x \leqq 1$

17. $9x = 4y + 7$; concave upward

19. $2\pi x + 4y = \pi^2$; concave downward

21. $\psi = \dfrac{\pi}{6}$ **23.** $\psi = \dfrac{\pi}{2}$

25. Horizontal tangents at $(1, -2)$ and $(1, 2)$; vertical tangent at $(0, 0)$ and *no* tangent line at $(3, 0)$.

27. Horizontal tangents at $\left(\frac{3}{4}, \pm\frac{3}{4}\sqrt{3}\right)$ and at $(0, 0)$; vertical tangent at $(2, 0)$.

29. $\dfrac{dy}{dx} = -2e^{3t}$ and $\dfrac{d^2y}{dx^2} = 6e^{4t}$.

31. $x = \dfrac{p}{m^2},\ y = \dfrac{2p}{m},\ -\infty < m < +\infty$

33. The slope of the line containing P_0 and P is

$$\frac{1 + \cos\theta}{\sin\theta},$$

and this is also the value of dy/dx at the point P.

35. The identities $\cos 3t = \cos^3 t - 3\sin^2 t \cos t$ and $\sin 3t = 3\sin t \cos^2 t - \sin^3 t$ will be very helpful.

41. $x = \dfrac{5t^2}{1 + t^5},\ y = \dfrac{5t^3}{1 + t^5},\ 0 \leqq t < +\infty$

43. No horizontal tangents; vertical tangents at $(-3, 2)$ and $(1, 0)$; inflection point at $(-1, 1)$.

45. Horizontal tangents at $(0, -2.3077)$, $(0, 1)$, and $(0, 2.1433)$ (numbers with decimal points are approximations); vertical tangents at $(-1.8559, 1.7321)$, $(2.4324, 1.7321)$, and $(1.5874, 0)$; inflection points at $(-5.1505, -3.1103)$, $(0, -2.3077)$, $(2.0370, -1.0443)$, $(1.5874, 0)$, $(0, 1)$, $(0, 2.1433)$, and $(4.2661, 2.8565)$. To see the graph, use a computer algebra system to plot the parametric equations $x = (t^5 - 5t^3 + 4)^{1/3}$, $y = t$ with the [suggested] range $-2.7 \leqq t \leqq 2.7$.

SECTION 10.5 (PAGE 657)

1. $\dfrac{22}{5}$ **3.** $\dfrac{4}{3}$

5. $\dfrac{1}{2}(e^{\pi} + 1)$ **7.** $\dfrac{358\pi}{35}$

9. $\dfrac{16\pi}{15}$ **11.** $\dfrac{74}{3}$

13. $\dfrac{\pi\sqrt{2}}{4}$ **15.** $(e^{2\pi} - 1)\sqrt{5}$

17. $\dfrac{8\pi}{3}\left(5^{3/2} - 2^{3/2}\right)$ **19.** $\dfrac{2\pi}{27}\left(13^{3/2} - 8\right)$

21. $16\pi^2$ **23.** $5\pi^2 a^3$

25. **a.** πab; **b.** $\dfrac{4}{3}\pi ab^2$

27. $\dfrac{1}{2}\left[2\pi\sqrt{1 + 4\pi^2} + \ln\left(2\pi + \sqrt{1 + 4\pi^2}\right)\right]$

29. $\dfrac{3}{8}\pi a^2$ **31.** $\dfrac{12}{5}\pi a^2$

33. $\dfrac{216\sqrt{3}}{5}$ **35.** $\dfrac{243\pi\sqrt{3}}{4}$

37. The length is $2\int_0^1 \frac{3\sqrt{t^8+4t^6-4t^5-4t^3+4t^2+1}}{(t^3+1)^2}\,dt$ $\approx 4.9174887217.$

39. $6\pi^3a^3$

41. $\frac{5}{6}\pi^3a^2$

43. $\int_0^{\pi}\sqrt{45+36\cos 6\theta}\,d\theta \approx 20.0473398308$

45. $\int_0^{2\pi}\sqrt{10-6\cos 4\theta}\,d\theta \approx 19.3768964411$

47. $\int_0^{2\pi}\sqrt{106+90\cos\theta}\,d\theta \approx 61.0035813739$

49. $\int_0^{3\pi}\frac{1}{3}\sqrt{29-20\cos(\frac{14}{3}\theta)}\,d\theta \approx 16.3428333739$

51. (a) Approximately 16.0570275666; (b) $\frac{16\pi}{15}$

53. $\int_0^{2\pi}\sqrt{25\cos^2 5t+9\sin^2 3t}\,dt \approx 24.6029616185$

55. Length: $\int_0^{2\pi}\sqrt{[x'(t)]^2+[y'(t)]^2}\,dt \approx 39.4035787129$

SECTION 10.6 (PAGE 676)

1. Opens to the right; equation $y^2 = 12x$
3. Opens downward; equation $(x-2)^2 = -8(y-3)$
5. Opens to the left; equation $(y-3)^2 = -8(x-3)$
7. Opens downward; equation $x^2 = -6(y+\frac{3}{2})$
9. Opens upward; equation $x^2 = 4(y+1)$
11. Opens to the right, vertex at $(0, 0)$, axis the x-axis, focus at $(3, 0)$, directrix $x = -3$
13. Opens to the left, vertex at $(0, 0)$, axis the x-axis, focus at $(-\frac{3}{2}, 0)$, directrix $x = \frac{3}{2}$
15. Opens upward, vertex at $(2, -1)$, axis $x = 2$, focus at $(2, 0)$, directrix $y = -2$
17. Opens downward, vertex at $(-\frac{1}{2}, -3)$, axis $x = -\frac{1}{2}$, focus at $(-\frac{1}{2}, \frac{13}{4})$, directrix $y = -\frac{11}{4}$

19. $\left(\frac{x}{4}\right)^2+\left(\frac{y}{5}\right)^2=1$

21. $\left(\frac{x}{13}\right)^2+\left(\frac{y}{17}\right)^2=1$

23. $\left(\frac{x}{4}\right)^2+\left(\frac{y}{\sqrt{7}}\right)^2=1$

25. $\frac{x^2}{100}+\frac{y^2}{75}=1$

27. $\frac{x^2}{16}+\frac{y^2}{12}=1$

29. $\left(\frac{x-2}{4}\right)^2+\left(\frac{y-3}{2}\right)^2=1$

31. $\left(\frac{x-1}{5}\right)^2+\left(\frac{y-1}{4}\right)^2=1$

33. $\frac{(x-1)^2}{81}+\frac{(y-2)^2}{72}=1$

35. Center $(0, 0)$, foci $(\pm 2\sqrt{5}, 0)$, axes 12 and 8

37. Center $(0, 4)$, foci $(0, 4\pm\sqrt{5})$, axes 6 and 4

39. $\frac{x^2}{1}-\frac{y^2}{15}=1$

41. $\left(\frac{x}{4}\right)^2-\left(\frac{y}{3}\right)^2=1$

43. $\left(\frac{y}{5}\right)^2-\left(\frac{x}{5}\right)^2=1$

45. $\frac{y^2}{9}-\frac{x^2}{27}=1$

47. $\frac{x^2}{4}-\frac{y^2}{12}=1$

49. $\frac{(x-2)^2}{9}-\frac{(y-2)^2}{27}=1$

51. $\left(\frac{y+2}{3}\right)^2-\left(\frac{x-1}{2}\right)^2=1$

53. Center $(1, 2)$, foci $(1\pm\sqrt{2}, 2)$, asymptotes $y-2 = \pm(x-1)$
55. Center $(0, 3)$, foci $(0, 3\pm 2\sqrt{3})$, asymptotes $y = 3\pm x\sqrt{3}$
57. Center $(-1, 1)$, foci $(-1\pm\sqrt{13}, 1)$, asymptotes $y-1 = \pm\frac{3}{2}(x+1)$
59. Parabola, opening to the left, vertex $(3, 0)$, axis the x-axis
61. Parabola, opening to the right, vertex $(-\frac{3}{2}, 0)$, axis the x-axis
63. Ellipse, center $(0, 2)$, vertices at $(0, 6)$ and $(0, -2)$
65. Minimize $(x-p)^2+y^2$ where $x = y^2/(4p)$.
69. About 16 h 38 min
71. Maximize $R(\alpha) = (v_0^2 \sin 2\alpha)/g$.
73. Approximately 14° 40′ 13″ and 75° 19′ 47″
75. Square the given equation twice to eliminate radicals, convert to polar form, rotate 45° by replacing θ with $\theta+(\pi/4)$, and finally return to Cartesian coordinates. You will recognize the equation as that of a parabola.
77. **a.** About 322 billion miles; **b.** about 120 billion miles
79. With focus $F(0, c)$ and directrix the line L: $y = c/e^2$ $(0 < e < 1)$, begin with $|PF| = e\cdot|PL|$, eliminate radicals, simplify, replace $a^2(1-e^2)$ with b^2, and convert the resulting Cartesian equation to "standard form."
81. Go to **www.augsburg.edu/depts/math/MATtours/ellipses.1.09.0.html**.
83. The only solution is $\frac{(x-1)^2}{4}+\frac{3y^2}{16}=1.$
85. (c) In this case there are no points on the graph.
89. $16x^2+50xy+16y^2 = 369$
91. If A is at $(-50, 0)$ and B is at $(50, 0)$, then the x-coordinate of the plane is approximately 41.3395 (in mi).
93. 2000 mi
95. Begin with $r = pe/(1-e\cos\theta)$ and first show that the area of the ellipse is

$$A = 2\int_0^{\pi}\frac{1}{2}r^2\,d\theta = a^2(1-e^2)^2\int_0^{\pi}\frac{1}{(1-e\cos\theta)^2}\,d\theta.$$

Then use the substitution discussed after Miscellaneous Problem 134 of Chapter 7.

CHAPTER 10 MISCELLANEOUS PROBLEMS (PAGE 679)

1. Circle, center $(1, 1)$, radius 2
3. Circle, center $(3, -1)$, radius 1
5. Parabola, vertex $(4, -2)$, focus $(4, -25)$, opening downward
7. Ellipse, center $(2, 0)$, vertices at $(0, 0)$, $(4, 0)$, $(2, 3)$, and $(2, -3)$, foci $(2, \pm\sqrt{5})$
9. Hyperbola, center $(-1, 1)$, foci $(-1, 1\pm\sqrt{3})$, vertices $(-1, 1\pm\sqrt{2})$
11. There are no points on the graph.
13. Hyperbola, center $(1, 0)$, vertices $(3, 0)$ and $(-1, 0)$, foci $(1\pm\sqrt{3}, 0)$

15. Circle, center $(4, 1)$, radius 1

17. The graph consists of the straight line $y = -x$ together with the isolated point $(2, 2)$; it is not a conic section.

19. Circle, center $(-1, 0)$, radius 1

21. The straight line with Cartesian equation $y = x + 1$

23. The horizontal line $y = 3$

25. A pair of tangent ovals through the origin; the figure is symmetric around the y-axis

27. A limaçon symmetric around the y-axis

29. Ellipse, center $(-\frac{4}{3}, 0)$, horizontal semimajor axis of length $\frac{8}{3}$, semiminor axis of length $\frac{4}{3}\sqrt{3}$, vertices $(-4, 0)$, $(0, \pm\frac{4}{3}\sqrt{3})$, and $(\frac{4}{3}, 0)$, foci $(-\frac{8}{3}, 0)$ and $(0, 0)$

31. $\dfrac{\pi - 2}{2}$ **33.** $\dfrac{39\sqrt{3} - 10\pi}{6}$

35. 2 **37.** $\dfrac{5\pi}{4}$

39. The straight line $y = x + 2$

41. The circle with center $(2, 1)$ and radius 1

43. The "semicubical parabola" with Cartesian equation $y^2 = (x - 1)^3$

45. $y = -\frac{4}{3}\left(x - 3\sqrt{2}\right)$ **47.** $4x + 2\pi y = \pi^2$

49. 24 **51.** 3π

53. $\dfrac{13\sqrt{13} - 8}{27}$ **55.** $\dfrac{43}{6}$

57. $1 + \dfrac{9\sqrt{5}}{10} \arcsin \dfrac{\sqrt{5}}{3} - \dfrac{\sqrt{31}}{8} - \dfrac{9\sqrt{5}}{10} \arcsin \dfrac{\sqrt{5}}{6}$

59. $\dfrac{471295\pi}{1024}$ **61.** $\dfrac{\pi\,(e^{\pi} + 1)\sqrt{5}}{2}$

63. Suppose that the circle rolls to the right through a central angle θ. Then $x = a\theta - b\sin\theta$, $y = a - b\cos\theta$.

65. If the epicycloid is shifted a units to the left, its equations will be

$$x = 2a\cos\theta - a\cos 2\theta - a, \qquad y = 2a\sin\theta - a\sin 2\theta.$$

Now compute and simplify $r^2 = x^2 + y^2$.

67. $6\pi^3 a^3$

69. $r = 2p\cos(\theta - \alpha)$

71. Maximum $2a$, minimum $2b$

73. $y = \dfrac{4hx(b - x)}{b^2}$

75. The ellipse has equation $\left(\dfrac{x}{a}\right)^2 + \left(\dfrac{y}{b}\right)^2 = 1.$

79. $e = 2$

81. $A = 9\displaystyle\int_0^{\pi/4} \frac{\sec^2\theta\,\tan^2\theta}{(1 + \tan^3\theta)^2}\,d\theta = \frac{3}{2}.$

83. If $B < \frac{5}{2}$, then the conic is an ellipse; if $B > \frac{5}{2}$, it is a hyperbola. If $B = \frac{5}{2}$ the graph is a degenerate parabola: two parallel lines. If the graph is normal to the y-axis at the point $(0, 4)$, then the graph is the ellipse with equation

$$\left(\frac{x}{5}\right)^2 + \left(\frac{y}{4}\right)^2 = 1.$$

SECTION 11.2 (PAGE 689)

1. $a_n = n^2$ for $n \geqq 1$.

3. $a_n = 3^{-n}$ for $n \geqq 1$.

5. $a_n = (3n - 1)^{-1}$ for $n \geqq 1$.

7. $a_n = 1 + (-1)^n$ for $n \geqq 1$.

9. $\dfrac{2}{5}$ **11.** 0 **13.** 1

15. Diverges **17.** 0 **19.** 0

21. 0 **23.** 1 **25.** 0

27. 0 **29.** 0 **31.** 0

33. e **35.** $\dfrac{1}{e^2}$ **37.** 2

39. 1 **41.** Diverges **43.** 1

45. 2 **47.** 1 **49.** π

51. To begin, suppose (without loss of generality) that $A > 0$.

53. Let $L = \lim_{n\to\infty} x_n$. Then $L = \lim_{n\to\infty} x_{n+1}$.

55. (b) $G_1 = G_2 = G_3 = 1$; $G_{n+1} = G_n + G_{n-2}$ for $n \geqq 3$. Check: $G_{25} = 5896$.

57. (b) 4

SECTION 11.3 (PAGE 699)

1. $\dfrac{3}{2}$

3. Diverges (the kth partial sum is k^2).

5. Diverges (geometric with ratio -2).

7. 6

9. Diverges (geometric with ratio 1.01).

11. Diverges by the nth-term test.

13. Diverges (geometric with ratio $-3/e$).

15. $2 + \sqrt{2}$

17. Diverges by the nth-term test.

19. $\dfrac{1}{12}$

21. $\dfrac{e}{\pi - e}$

23. Diverges (geometric with ratio $\frac{100}{99}$).

25. $\dfrac{65}{12}$ **27.** $\dfrac{247}{8}$ **29.** $\dfrac{1}{4}$

31. Diverges by the nth-term test.

33. Diverges (geometric with ratio $\tan 1 > 1$).

35. $\dfrac{\pi}{4 - \pi}$

37. Diverges: Show that $S_k \geqq \displaystyle\int_2^{k+1} \frac{1}{x\ln x}\,dx > \ln(\ln(k+1))$.

39. $\dfrac{47}{99}$ **41.** $\dfrac{41}{333}$ **43.** $\dfrac{314156}{99999}$

45. Converges to $\dfrac{x}{3 - x}$ if $-3 < x < 3$.

47. Converges to $\dfrac{x - 2}{5 - x}$ if $-1 < x < 5$.

49. Converges to $\dfrac{5x^2}{16 - 4x^2}$ if $-2 < x < 2$.

51. $\frac{1}{6}$

53. $\frac{1}{4}$ (Beaverbock's constant)

55. $\frac{3}{4}$ **57.** 2 **59.** $\frac{1}{3}$

61. Use the converse of part 2 of Theorem 2.

65. 4.5 s

67. $M_n \to 0$ as $n \to +\infty$.

69. Peter: $\frac{4}{7}$; Paul: $\frac{2}{7}$; Mary: $\frac{1}{7}$

71. $\frac{1}{12}$ of the incident light

SECTION 11.4 (PAGE 713)

1. $e^{-x} = 1 - x + \frac{x^2}{2!} - \frac{x^3}{3!} + \frac{x^4}{4!} - \frac{x^5}{5!} + \frac{x^6}{6!}e^{-z}$ for some number z between 0 and x.

3. $\cos x = 1 - \frac{x^2}{2!} + \frac{x^4}{4!} - \frac{x^5}{5!}\sin z$ for some number z between 0 and x.

5. $\sqrt{1+x} = 1 + \frac{x}{2} - \frac{x^2}{8} + \frac{x^3}{16} - \frac{5x^4}{128(1+z)^{7/2}}$ for some number z between 0 and x.

7. $\tan x = x + \frac{x^3}{3} + \frac{x^4}{4!}(16\sec^4 z \tan z + 8\sec^2 z \tan^3 z)$ for some number z between 0 and x.

9. $\arcsin x = x + \frac{x^3(1+2z^2)}{3!(1-z^2)^{5/2}}$ for some number z between 0 and x.

11. $e^x = e + e(x-1) + \frac{e}{2}(x-1)^2 + \frac{e}{6}(x-1)^3 + \frac{e}{24}(x-1)^4 + \frac{e^z}{120}(x-1)^5$ for some number z between 1 and x.

13. $\sin x = \frac{1}{2} + \frac{\sqrt{3}}{2}\left(x - \frac{\pi}{6}\right) - \frac{1}{4}\left(x - \frac{\pi}{6}\right)^2 - \frac{\sqrt{3}}{12}\left(x - \frac{\pi}{6}\right)^3 + \frac{\sin z}{24}\left(x - \frac{\pi}{6}\right)^4$ for some number z between $\pi/6$ and x.

15. $\frac{1}{(x-4)^2} = 1 - 2(x-5) + 3(x-5)^2 - 4(x-5)^3 + 5(x-5)^4 - 6(x-5)^5 + \frac{(x-5)^6}{720} \cdot \frac{5040}{(z-4)^8}$ for some number z between 5 and x.

17. $\cos x = -1 + \frac{(x-\pi)^2}{2} - \frac{(x-\pi)^4}{24} - \frac{\sin z}{120}(x-\pi)^5$ for some number z between π and x.

19. $x^{3/2} = 1 + \frac{3}{2}(x-1) + \frac{3}{8}(x-1)^2 - \frac{1}{16}(x-1)^3 + \frac{3}{128}(x-1)^4 - \frac{(x-1)^5}{120} \cdot \frac{45}{32z^{7/2}}$ for some number z between 1 and x.

21. $e^{-x} = \sum_{n=0}^{\infty} \frac{(-1)^n x^n}{n!}$. This representation is valid for all x.

23. $e^{-3x} = \sum_{n=0}^{\infty} \frac{(-1)^n 3^n x^n}{n!}$. This representation is valid for all x.

25. $\sin 2x = \sum_{n=0}^{\infty} \frac{(-1)^n (2x)^{2n+1}}{(2n+1)!}$. This representation is valid for all x.

27. $\sin(x^2) = \sum_{n=0}^{\infty} \frac{(-1)^n x^{4n+2}}{(2n+1)!}$. This representation is valid for all x.

29. $\ln(1+x) = \sum_{n=1}^{\infty} \frac{(-1)^{n+1} x^n}{n}$. This representation is valid if $-1 < x \leqq 1$.

31. $e^{-x} = \sum_{n=0}^{\infty} \frac{(-1)^n x^n}{n!}$. This representation is valid for all x.

33. $\ln x = \sum_{n=1}^{\infty} \frac{(-1)^{n+1}(x-1)^n}{n}$. This representation is valid if $0 < x \leqq 2$.

35. $\cos x = \frac{\sqrt{2}}{2} - \frac{\sqrt{2}}{2}\left(x - \frac{\pi}{4}\right) - \frac{\sqrt{2}}{2! \cdot 2}\left(x - \frac{\pi}{4}\right)^2 + \frac{\sqrt{2}}{3! \cdot 2}\left(x - \frac{\pi}{4}\right)^3 + \frac{\sqrt{2}}{4! \cdot 2}\left(x - \frac{\pi}{4}\right)^4 - \cdots$.

This representation is valid for all x.

37. $\frac{1}{x} = \sum_{n=0}^{\infty} (-1)^n (x-1)^n$. This representation is valid for $0 < x < 2$.

39. $\sin x = \frac{\sqrt{2}}{2} + \frac{\sqrt{2}}{2}\left(x - \frac{\pi}{4}\right) - \frac{\sqrt{2}}{2! \cdot 2}\left(x - \frac{\pi}{4}\right)^2 - \frac{\sqrt{2}}{3! \cdot 2} \times \left(x - \frac{\pi}{4}\right)^3 + \frac{\sqrt{2}}{4! \cdot 2}\left(x - \frac{\pi}{4}\right)^4 + \frac{\sqrt{2}}{5! \cdot 2}\left(x - \frac{\pi}{4}\right)^5 - \cdots$.

This representation is valid for all x.

45. Given $f(x) = e^{-x}$, its plot together with that of

$$P_3(x) = 1 - x + \frac{x^2}{2!} - \frac{x^3}{3!}$$

are shown next.

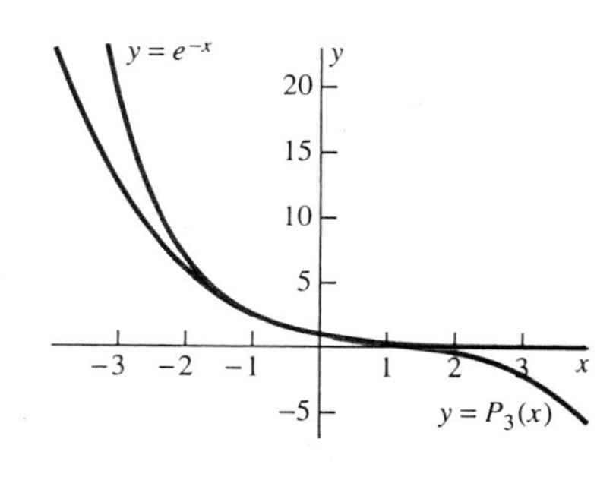

The graphs of $f(x) = e^{-x}$ and

$$P_6(x) = 1 - x + \frac{x^2}{2!} - \frac{x^3}{3!} + \frac{x^4}{4!} - \frac{x^5}{5!} + \frac{x^6}{6!}$$

are shown together next.

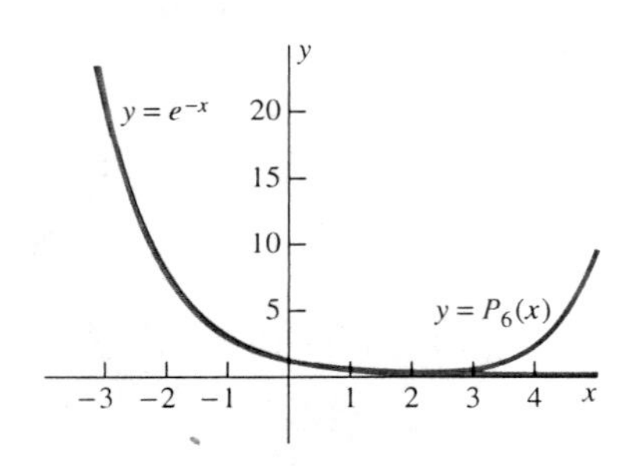

47. One of the Taylor polynomials for $f(x) = \cos x$ is $P_4(x) = 1 - \frac{x^2}{2!} + \frac{x^4}{4!}$. The graphs of f and P_4 are shown together, next.

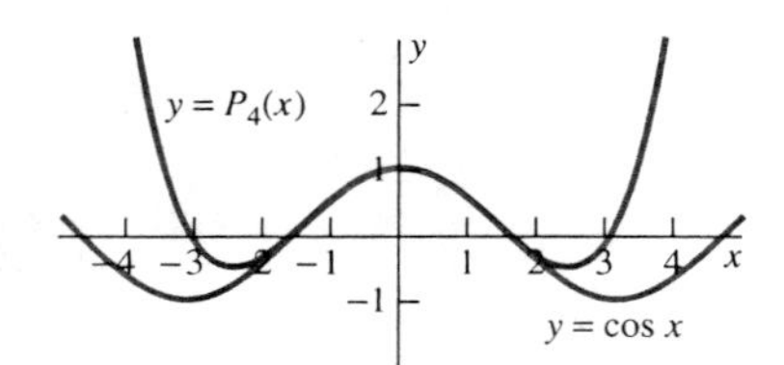

49. One of the Taylor polynomials for $f(x) = \frac{1}{1+x}$ is $P_4(x) = 1 - x + x^2 - x^3 + x^4$. The graphs of f and P_4 are shown together, next.

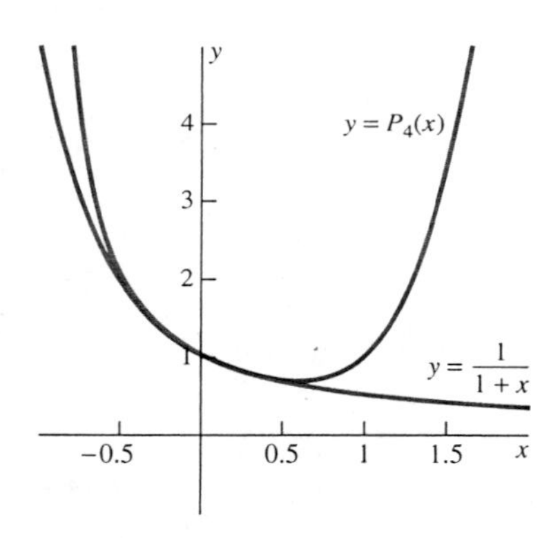

51. The graph of the Taylor polynomial

$$P_4(x) = 1 - \frac{x}{2!} + \frac{x^2}{4!} - \frac{x^3}{6!} + \frac{x^4}{8!}$$

of $f(x)$ and the graph of $g(x)$ are shown together, next.

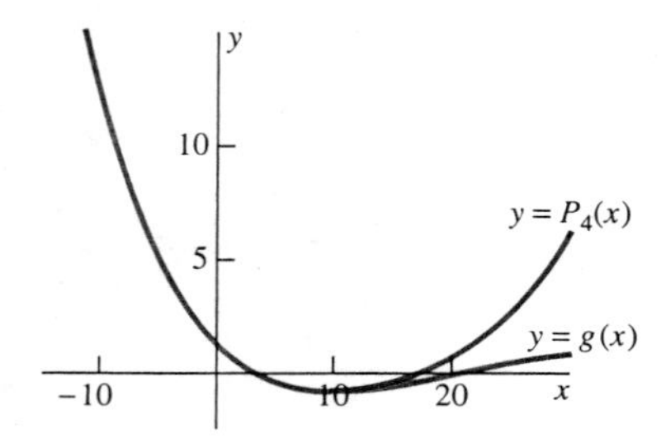

57. By Theorem 4 of Section 11.3, S is not a number. Hence attempts to do arithmetic with S will generally lead to false or meaningless results.

59. Results: With $x = 1$ in the Maclaurin series in Problem 56, we find that

$$a = \sum_{n=1}^{50} \frac{(-1)^{n+1}}{n} \approx 0.68324716057591818842565811649.$$

With $x = \frac{1}{3}$ in the second series in Problem 58, we find that

$$b = \sum_{\substack{n=1 \\ n \text{ odd}}}^{49} \frac{2}{n \cdot 3^n} \approx 0.69314718055994530941723210107.$$

Because $|a - \ln 2| \approx 0.009900019984$, whereas $|b - \ln 2| \approx 2.039 \times 10^{-26}$, it is clear that the second series of Problem 58 is *far* superior to the series of Problem 56 for the accurate approximation of $\ln 2$.

SECTION 11.5 (PAGE 720)

1. Diverges: $\displaystyle\int_0^\infty \frac{x}{x^2+1}\,dx = \left[\frac{1}{2}\ln(x^2+1)\right]_0^\infty = +\infty.$

3. Diverges: $\displaystyle\int_0^\infty (x+1)^{-1/2}\,dx = \left[2(x+1)^{1/2}\right]_0^\infty = +\infty.$

5. Converges: $\displaystyle\int_0^\infty \frac{1}{x^2+1}\,dx = \left[\arctan x\right]_0^\infty = \frac{\pi}{2} < +\infty.$

7. Diverges: $\displaystyle\int_2^\infty \frac{1}{x\ln x}\,dx = \left[\ln(\ln x)\right]_2^\infty = +\infty.$

9. Converges (to 1): $\displaystyle\int_0^\infty 2^{-x}\,dx = \left[-\frac{1}{2^x \ln 2}\right]_0^\infty = \frac{1}{\ln 2} < +\infty.$

11. Converges: $\displaystyle\int_0^\infty x^2 e^{-x}\,dx = -\left[(x^2 + 2x + 2)e^{-x}\right]_0^\infty = 2 < +\infty.$

13. Converges: $\displaystyle\int_1^\infty \frac{\ln x}{x^2}\,dx = \left[-\frac{1+\ln x}{x}\right]_1^\infty = \frac{1+0}{1} - \lim_{x\to\infty}\frac{1+\ln x}{x} = 1 < +\infty.$

15. Converges: $\displaystyle\int_0^\infty \frac{x}{x^4+1}\,dx = \left[\frac{1}{2}\arctan(x^2)\right]_0^\infty = \frac{\pi}{4} < +\infty.$

17. Diverges: $\displaystyle\int_1^\infty \frac{2x+5}{x^2+5x+17}\,dx = \left[\ln(x^2 + 5x + 17)\right]_1^\infty = +\infty.$

19. Converges: $\displaystyle\int_1^\infty \ln\left(1+\frac{1}{x^2}\right)dx = \frac{\pi}{2} - \ln 2 < +\infty.$

21. Diverges: $\displaystyle\int_1^\infty \frac{x}{4x^2+5}\,dx = \left[\frac{1}{8}\ln(4x^2+5)\right]_1^\infty = +\infty.$

23. Diverges: $\displaystyle\int_2^\infty \frac{1}{x\sqrt{\ln x}}\,dx = \int_2^\infty \frac{(\ln x)^{-1/2}}{x}\,dx = \left[2(\ln x)^{1/2}\right]_2^\infty = +\infty.$

25. Converges: $\displaystyle\int_1^{\infty} \frac{1}{4x^2+9}\,dx = \left[\frac{1}{6}\arctan\left(\frac{2x}{3}\right)\right]_1^{\infty}$
$\displaystyle = \frac{\pi}{12} - \frac{1}{6}\arctan\left(\frac{2}{3}\right) < +\infty.$

27. Converges: $\displaystyle\int_1^{\infty} \frac{x}{(x^2+1)^2}\,dx = \left[-\frac{1}{2(x^2+1)}\right]_1^{\infty} = \frac{1}{4} < +\infty.$

29. Converges: $\displaystyle\int_1^{\infty} \frac{\arctan x}{x^2+1}\,dx = \left[\frac{1}{2}(\arctan x)^2\right]_1^{\infty}$
$\displaystyle = \frac{3\pi^2}{32} < +\infty.$

31. This is not a positive-term series.

33. The terms of this series are not monotonically decreasing.

35. Diverges if $0 < p \leqq 1$, converges if $p > 1$.

37. Diverges if $p \leqq 1$, converges otherwise.

39. $n > 10000$

41. $n > 100$ **43.** $n > 160{,}000$

45. $n \geqq 15$ **47.** $p > 1$

49. Sloppy answer: Over 604,414 centuries. A more precise answer: A little over 922,460 centuries.

51. Apply Theorem 4 and Problem 52 of Section 11.2.

SECTION 11.6 (PAGE 727)

1. Converges: Dominated by the p-series with $p = 2$.

3. Diverges by limit-comparison with the harmonic series.

5. Converges: Dominated by the geometric series with ratio $\frac{1}{3}$.

7. Diverges by limit-comparison with the harmonic series.

9. Converges: Dominated by the p-series with $p = \frac{3}{2}$.

11. Converges: Dominated by the p-series with $p = \frac{3}{2}$.

13. Diverges by comparison with the harmonic series.

15. Converges: Dominated by the p-series with $p = 2$.

17. Converges: Dominated by a geometric series with ratio $\frac{2}{3}$.

19. Converges by comparison with the p-series with $p = 2$.

21. Converges: Dominated by the p-series with $p = \frac{3}{2}$ (among others).

23. Converges: Dominated by the p-series with $p = 2$.

25. Converges: Dominated by the geometric series with ratio $\dfrac{2}{e}$.

27. Diverges by limit-comparison with the p-series with $p = \frac{1}{2}$.

29. Diverges by limit-comparison with the p-series with $p = \frac{1}{2}$.

31. Converges by comparison with a geometric series with ratio $\frac{2}{3}$ and by limit comparison with a geometric series with ratio $\frac{1}{3}$.

33. Converges by comparison with the p-series with $p = 2$.

35. Diverges by limit-comparison with the harmonic series.

37. $S_{10} \approx 0.981793$ with error less than 0.094882.

39. $S_{10} \approx 0.528870$ with error less than 0.1.

41. $n = 10$; the sum is approximately 0.686503.

43. $n = 3$; the sum is approximately 0.100714.

45. Use the converse of Theorem 3 in Section 11.3.

47. Use the comparison test.

49. Apply the converse of Theorem 3 in Section 11.3 and the result in Problem 48.

51. Use the result in Problem 50 in Section 11.5.

SECTION 11.7 (PAGE 735)

1. Converges (to $\frac{1}{12}\pi^2$) by the alternating series test.

3. Diverges by the nth-term test for divergence.

5. Diverges by the nth-term test for divergence.

7. Diverges by the nth-term test for divergence.

9. Converges (to $-\frac{2}{9}$) by the alternating series test.

11. Converges by the alternating series test. (The sum is approximately -0.1782434556.)

13. Converges by the alternating series test. (The sum is approximately 0.711944418056.)

15. Converges by the alternating series test. (The sum is roughly -0.550796848134.)

17. Diverges by the nth-term test for divergence.

19. Diverges by the nth-term test for divergence.

21. Converges absolutely by the ratio test. (The sum is $\frac{1}{3}$.)

23. Converges by the alternating series test, but only conditionally by the integral test. (The sum is approximately 0.159868903742.)

25. Converges absolutely by the root test. (The sum is approximately 186.724948614024.)

27. Converges absolutely by the ratio test. (The sum is $e^{-10} \approx 0.00004539992976$.)

29. Diverges by the nth-term test for divergence.

31. Converges absolutely by the root test. (The sum is approximately 0.187967875056.)

33. Converges by the alternating series test, but only conditionally by the comparison test. (The sum is approximately 0.760209625219.)

35. Diverges by the nth-term test for divergence.

37. Diverges by the nth-term test for divergence.

39. Converges absolutely by the ratio test. (The sum is approximately 0.586781998767.)

41. Converges absolutely by the ratio test. (The sum is approximately 2.807109464185.)

43. 0.9044; 0.005; 0.90

45. 0.6319; 0.0002; 0.632

47. 0.6532; 0.08; 0.7

49. $n = 6$; 0.947 (the sum is $\frac{7}{720}\pi^4$)

51. $n = 5$; 0.6065

53. $n = 4$; 0.86603

55. The sequence of terms is not monotonically decreasing; the series diverges by comparison with the harmonic series.

57. Let $a_n = b_n = \dfrac{(-1)^n}{\sqrt{n}}$.

63. $1 + \frac{1}{3} - \frac{1}{2} + \frac{1}{5} - \frac{1}{4} + \frac{1}{7} + \frac{1}{9} - \frac{1}{6} + \frac{1}{11} + \frac{1}{13} - \frac{1}{8} + \frac{1}{15}$

65. It converges to zero.

SECTION 11.8 (PAGE 748)

1. $(-1, 1)$

3. $(-2, 2)$

5. $[0, 0]$

7. $\left[-\frac{1}{3}, \frac{1}{3}\right]$

9. $\left(-\frac{1}{2}, \frac{1}{2}\right)$

11. $[-2, 2]$

13. $(-3, 3)$

15. $\left(\frac{2}{5}, \frac{4}{5}\right)$

17. $\left[\frac{5}{2}, \frac{7}{2}\right]$

19. $[0, 0]$

21. $(-4, 2)$

23. $[2, 4]$

25. $[5, 5]$

27. $(-1, 1)$

29. $(-\infty, +\infty)$

31. $f(x) = x + x^2 + x^3 + x^4 + x^5 + \cdots;\quad R = 1$

33. $f(x) = \displaystyle\sum_{n=0}^{\infty} \frac{(-1)^n 3^n x^{n+2}}{n!};\quad R = +\infty$

35. $f(x) = \displaystyle\sum_{n=0}^{\infty} \frac{(-1)^n x^{4n+2}}{(2n+1)!};\quad R = +\infty$

37. $f(x) = 1 - \frac{1}{3}x - \frac{2}{3^2}\cdot\frac{x^2}{2!} - \frac{2\cdot 5}{3^3}\cdot\frac{x^3}{3!} - \frac{2\cdot 5\cdot 8}{3^4}\cdot\frac{x^4}{4!} - \frac{2\cdot 5\cdot 8\cdot 11}{3^5}\cdot\frac{x^5}{5!} - \cdots;\quad R = 1$

39. $f(x) = (1+x)^{-3} = 1 - 3x + 3\cdot 4\cdot\frac{x^2}{2!} - 3\cdot 4\cdot 5\cdot\frac{x^3}{3!} + 3\cdot 4\cdot 5\cdot 6\cdot\frac{x^4}{4!} - \cdots;\quad R = 1$

41. $f(x) = \displaystyle\sum_{n=0}^{\infty} \frac{(-1)^n x^n}{n+1};\quad R = 1$

43. $f(x) = \displaystyle\sum_{n=0}^{\infty} \frac{(-1)^n x^{6n+4}}{(2n+1)!\cdot(6n+4)};\quad R = +\infty$

45. $f(x) = \displaystyle\sum_{n=0}^{\infty} \frac{(-1)^n x^{3n+1}}{n!\cdot(3n+1)};\quad R = +\infty$

47. $f(x) = \displaystyle\sum_{n=1}^{\infty} \frac{(-1)^{n+1} x^{2n-1}}{n!\cdot(2n-1)};\quad R = +\infty$

49. $\dfrac{x}{(1-x)^2},\quad -1 < x < 1$

51. $\dfrac{x(1+x)}{(1-x)^3},\quad -1 < x < 1$

61. $f(x) = \displaystyle\sum_{n=0}^{\infty} \frac{(-1)^n x^{2n}}{(2n+1)!},\quad -\infty < x < +\infty$

SECTION 11.9 (PAGE 756)

1. $65^{1/3} = 4\cdot\left(1 + \frac{1}{64}\right)^{1/3} \approx 4 + \frac{4}{3}\cdot\frac{1}{64} \approx 4.021.$

3. $\sin(0.5) \approx \frac{1}{2} - \frac{1}{3!\cdot 2^3} \approx 0.479.$

5. 0.464

7. $\sin\left(\frac{\pi}{10}\right) \approx \frac{\pi}{10} - \frac{\pi^3}{3!\cdot 10^3} \approx 0.309.$

9. $\sin\left(\frac{\pi}{18}\right) \approx \frac{\pi}{18} - \frac{\pi^3}{3!\cdot 18^3} \approx 0.174.$

11. $\displaystyle\int_0^1 \frac{\sin x}{x}\,dx \approx 1 - \frac{1}{3!3} + \frac{1}{5!5} - \frac{1}{7!7} \approx 0.9641.$

13. $\displaystyle\int_0^{1/2} \frac{\arctan x}{x}\,dx \approx \frac{1}{2} - \frac{1}{2^3\cdot 3^2} + \frac{1}{2^5\cdot 5^2} - \frac{1}{2^7\cdot 7^2} \approx 0.4872.$

15. $\displaystyle\int_0^{0.1} \frac{\ln(1+x)}{x}\,dx \approx \frac{1}{10} - \frac{1}{4\cdot 10^2} + \frac{1}{9\cdot 10^3} \approx 0.0976.$

17. $\displaystyle\int_0^{1/2} \frac{1-e^{-x}}{x}\,dx \approx \frac{1}{2} - \frac{1}{2!\cdot 2\cdot 2^2} + \frac{1}{3!\cdot 3\cdot 2^3} - \frac{1}{4!\cdot 4\cdot 2^4} + \frac{1}{5!\cdot 5\cdot 2^5} \approx 0.4438.$

19. $0.7468241328124270 \approx 0.7468$

21. $0.5132555590033423 \approx 0.5133$

23. $-\frac{1}{2} - \frac{x}{6} - \frac{x^2}{24} - \cdots \to -\frac{1}{2}$ as $x \to +\infty$.

25. $\displaystyle\lim_{x\to 0} \frac{\frac{1}{2!} - \frac{x^2}{4!} + \frac{x^4}{6!} - \cdots}{1 + \frac{x}{2!} + \frac{x^2}{3!} + \cdots} = \frac{1}{2}.$

27. $\displaystyle\lim_{x\to 0} \frac{-\frac{x}{3!} + \frac{x^3}{5!} - \frac{x^5}{7!} + \cdots}{1 - \frac{x^2}{3!} + \frac{x^4}{5!} - \cdots} = \frac{0}{1} = 0.$

29. $\sin 80^\circ \approx 1 - \frac{1}{2!}\cdot\left(\frac{\pi}{18}\right)^2 \approx 0.9848.$

31. 0.681998

33. Six-place accuracy

35. Five-place accuracy

37. $e^{1/3} \approx 1.39$

39. **a.** $|R_3(x)| < 0.000002$; **b.** $|R_3(x)| < 0.000000003$

41. $V = 2\pi \displaystyle\int_0^{\pi} \frac{\sin^2 x}{x^2}\,dx = \frac{(2\pi)^2}{2!} - \frac{(2\pi)^4}{4!\cdot 3} + \frac{(2\pi)^6}{6!\cdot 5} - \cdots \approx 8.9105091465101038.$

43. $V = 2\pi \displaystyle\int_0^{2\pi} \frac{1-\cos x}{x}\,dx = \frac{(2\pi)^3}{2!\cdot 2} - \frac{(2\pi)^5}{4!\cdot 4} + \frac{(2\pi)^7}{6!\cdot 6} - \cdots \approx 15.316227983254.$

47. $a_0 + (a_1 - a_0)x + (a_2 - a_1)x^2 + (a_3 - a_2)x^3 + (a_4 - a_3)x^4 + \cdots = 1$

49. $a_0 + a_1 x + \left(a_2 - \frac{1}{2}a_0\right) x^2 + \left(a_3 - \frac{1}{2}a_1\right) x^3$
$+ \left(a_4 - \frac{1}{2}a_2 + \frac{1}{24}a_0\right) x^4 + \cdots = 1;$

$\sec x = 1 + \frac{1}{2}x^2 + \frac{5}{24}x^4 + \frac{61}{720}x^6 + \cdots.$

51. $1+x = a_0 + a_1 x + \left(a_2 - \frac{1}{2}a_1\right) x^2 + \left(a_3 - a_2 + \frac{1}{3}a_1\right) x^3 + \cdots.$

53. Apply Theorem 1 to determine R.

55. $\int_0^{1/2} \frac{1}{1+x^2+x^4}\,dx = \frac{1}{2} - \frac{1}{2^3 \cdot 3} + \frac{1}{2^7 \cdot 7} - \frac{1}{2^9 \cdot 9} + \frac{1}{2^{13} \cdot 13}$
$- \cdots \approx 0.4592398250.$

59. -1

SECTION 11.10 (PAGE 766)

1. $y(x) = a_0 \sum_{n=0}^{\infty} \frac{x^n}{n!} = a_0 e^x; \quad R = +\infty$

3. $y(x) = a_0 \sum_{n=0}^{\infty} \frac{(-1)^n}{n!}\left(\frac{3x}{2}\right)^n = a_0 e^{-3x/2}; \quad R = +\infty$

5. $y(x) = a_0 \left[1 + \frac{1}{1!} \cdot \frac{x^3}{3} + \frac{1}{2!}\left(\frac{x^3}{3}\right)^2 + \frac{1}{3!}\left(\frac{x^3}{3}\right)^3 + \cdots\right]$
$= a_0 \exp\left(\frac{x^3}{3}\right); \quad R = +\infty$

7. $y(x) = a_0 \sum_{n=0}^{\infty} 2^n x^n = a_0 \sum_{n=0}^{\infty} (2x)^n = \frac{a_0}{1-2x}; \quad R = \frac{1}{2}$

9. $y(x) = a_0 \sum_{n=0}^{\infty} (n+1)x^n = \frac{a_0}{(1-x)^2}; \quad R = 1$

11. $y(x) = a_0 \left(1 + \frac{x^2}{2!} + \frac{x^4}{4!} + \frac{x^6}{6!} + \cdots\right)$
$+ a_1 \left(x + \frac{x^3}{3!} + \frac{x^5}{5!} + \frac{x^7}{7!} + \cdots\right)$
$= a_0 \cosh x + a_1 \sinh x; \quad R = +\infty$

13. $y(x) = a_0 \left(1 - \frac{9x^2}{2!} + \frac{9^2 x^4}{4!} - \frac{9^3 x^6}{6!} + \cdots\right)$
$+ a_1 \left(x - \frac{9x^3}{3!} + \frac{9^2 x^5}{5!} - \frac{9^3 x^7}{7!} + \cdots\right)$
$= a_0 \cos 3x + \frac{a_1}{3} \sin 3x = c_1 \cos 3x + c_2 \sin 3x;$
$R = +\infty$

15. $a_0 = 0$ and $(n+1)a_n = 0$ if $n \geqq 1$, so $y(x) \equiv 0$.

17. $a_0 = a_1 = 0$ and $(n-1)a_{n-1} + a_n = 0$ for $n \geqq 2$, so $y(x) \equiv 0$.

19. $y(x) = \frac{3}{2} \sin 2x$

21. $y(x) = xe^x$

23. $c_1 = c_2 = 0$ and
$$c_n = -\frac{n-1}{n^2 - n + 1} c_{n-1}$$
if $n \geqq 2$, so $y(x) \equiv 0$.

25. $x + \frac{1}{3}x^3 + \frac{2}{15}x^5 + \frac{17}{315}x^7 + \frac{62}{2835}x^9 + \frac{1382}{155925}x^{11}$
$+ \frac{21844}{6081075}x^{13} + \frac{929569}{638512875}x^{15} + \cdots$

CHAPTER 11 MISCELLANEOUS PROBLEMS (PAGE 767)

1. 1
3. 10
5. 0
7. 0
9. The limit does not exist.
11. 0
13. $+\infty$ (or "Does not exist.")
15. 1
17. Converges by the alternating series test. (The sum is approximately 0.080357603217.)
19. Converges by the ratio test. (The sum is approximately 1.405253880284.)
21. Converges by the comparison test and Theorem 3 of Section 11.7. (The sum is approximately 0.230836643803.)
23. Diverges by the nth-term test for divergence.
25. Converges by the comparison test. (The sum is approximately 1.459973884376.)
27. Converges by the alternating series test. (The sum is approximately 0.378868816198.)
29. Diverges by the integral test.
31. Converges by the ratio test; the sum is e^{2x} and the radius of convergence is $+\infty$.
33. The interval of convergence is $[-2, 4)$.
35. The interval of convergence is $[-1, 1]$.
37. The series converges only if $x = 0$.
39. The series converges to $\cosh x$ on $(-\infty, +\infty)$.
41. Diverges for all x by the nth-term test for divergence.
43. Converges for all x to $\exp(e^x)$.
45. Let $a_n = b_n = (-1)^n \cdot n^{-1/2}$.
51. 1.084
53. 0.461
55. 0.797
65. $a_0 = 2$ and $a_n = 4$ for all $n \geqq 1$.

SECTION 12.1 (PAGE 777)

1. $\mathbf{v} = \overrightarrow{RS} = \langle 2, 3 \rangle$
3. $\mathbf{v} = \overrightarrow{RS} = \langle -10, -20 \rangle$
5. $\mathbf{w} = \mathbf{u} + \mathbf{v} = \langle 4, 2 \rangle$
7. $\mathbf{u} + \mathbf{v} = 5\mathbf{i} - 2\mathbf{j}$
9. $\sqrt{5}$, $2\sqrt{13}$, $4\sqrt{2}$, $\langle -2, 0 \rangle$, $\langle 9, -10 \rangle$

11. $2\sqrt{2}$, 10, $\sqrt{5}$, $\langle -5, -6\rangle$, $\langle 0, 2\rangle$

13. $\sqrt{10}$, $2\sqrt{29}$, $\sqrt{65}$, $3\mathbf{i} - 2\mathbf{j}$, $-\mathbf{i} + 19\mathbf{j}$

15. 4, 14, $\sqrt{65}$, $4\mathbf{i} - 7\mathbf{j}$, $12\mathbf{i} + 14\mathbf{j}$

17. $\mathbf{u} = -\frac{3}{5}\mathbf{i} - \frac{4}{5}\mathbf{j}$, $\mathbf{v} = \frac{3}{5}\mathbf{i} + \frac{4}{5}\mathbf{j}$

19. $\mathbf{u} = \frac{8}{17}\mathbf{i} + \frac{15}{17}\mathbf{j}$, $\mathbf{v} = -\frac{8}{17}\mathbf{i} - \frac{15}{17}\mathbf{j}$

21. $\mathbf{a} = \overrightarrow{PQ} = -4\mathbf{j}$

23. $\mathbf{a} = \overrightarrow{PQ} = 8\mathbf{i} - 14\mathbf{j}$

25. $\mathbf{a} \perp \mathbf{b}$

27. $\mathbf{a} \perp \mathbf{b}$

29. $\mathbf{i} = -4\mathbf{a} + 3\mathbf{b}$ and $\mathbf{j} = 3\mathbf{a} - 2\mathbf{b}$

31. $\mathbf{c} = -\frac{1}{2}\mathbf{a} + \frac{5}{2}\mathbf{b}$

33. **a.** $15\mathbf{i} - 21\mathbf{j}$; **b.** $\frac{5}{3}\mathbf{i} - \frac{7}{3}\mathbf{j}$

35. **a.** $\frac{5\sqrt{58}}{58}(7\mathbf{i} - 3\mathbf{j})$; **b.** $-\frac{5\sqrt{89}}{89}(8\mathbf{i} + 5\mathbf{j})$

37. $c = 0$ is the unique solution.

43. $T_1 = T_2 = 100$

45. $T_1 \approx 71.971$, $T_2 \approx 96.121$ (lb)

47. Compass bearing 86°13′, airspeed approximately 536.52 mi/h

49. Compass bearing 320°43′, airspeed approximately 502 mi/h

SECTION 12.2 (PAGE 786)

1. $\langle 5, 8, -11\rangle$; $\langle 2, 23, 0\rangle$; 4; $\sqrt{51}$; $\frac{1}{15}\sqrt{5}\langle 2, 5, -4\rangle$

3. $\langle 2, 3, 1\rangle$; $\langle 3, -1, 7\rangle$; 0; $\sqrt{5}$; $\frac{1}{3}\sqrt{3}\langle 1, 1, 1\rangle$

5. $\langle 4, -1, -3\rangle$; $\langle 6, -7, 12\rangle$; -1; $\sqrt{17}$; $\frac{1}{5}\sqrt{5}\langle 2, -1, 0\rangle$

7. $\theta \approx 81°$

9. $\theta = 90°$

11. $\theta \approx 98°$

13. $\text{comp}_{\mathbf{a}}\mathbf{b} = \frac{4}{15}\sqrt{5}$; $\text{comp}_{\mathbf{b}}\mathbf{a} = \frac{2}{7}\sqrt{14}$

15. $\text{comp}_{\mathbf{a}}\mathbf{b} = 0 = \text{comp}_{\mathbf{b}}\mathbf{a}$

17. $\text{comp}_{\mathbf{a}}\mathbf{b} = -\frac{1}{5}\sqrt{5}$; $\text{comp}_{\mathbf{b}}\mathbf{a} = -\frac{1}{10}\sqrt{10}$

19. $x^2 - 6x + y^2 - 2y + z^2 - 4z = 11$

21. $x^2 - 10x + y^2 - 8y + z^2 + 2z + 33 = 0$

23. $x^2 + y^2 + z^2 - 4z = 0$

25. Center $(-2, 3, 0)$, radius $\sqrt{13}$

27. Center $(0, 0, 3)$, radius 5

29. The xy-plane

31. The plane through $(0, 0, 10)$ parallel to the xy-plane

33. The union of the three coordinate planes

35. The single point $(0, 0, 0)$

37. The single point $(3, -4, 0)$

39. Parallel (and not perpendicular)

41. Parallel (and not perpendicular)

43. The points lie on one line.

45. All three angles have measure 60°.

47. $\angle A \approx 79°$, $\angle B \approx 64°$, $\angle C \approx 37°$

49. $\alpha \approx 74.206831°$, $\beta = \gamma \approx 47.124011°$

51. $\alpha \approx 64.895910°$, $\beta \approx 55.550098°$, $\gamma = 45°$

53. 3

55. Approximately 7323.385 cal

57. $W = mgh$

59. Begin with $|\mathbf{a} + \mathbf{b}|^2 = (\mathbf{a} + \mathbf{b}) \cdot (\mathbf{a} + \mathbf{b})$ and expand the right-hand side.

61. Any nonzero multiple of $\mathbf{w} = \langle -2, 7, 4\rangle$

65. $\alpha = \dfrac{b_2c_1 - b_1c_2}{a_1b_2 - a_2b_1}$, $\beta = \dfrac{a_1c_2 - a_2c_1}{a_1b_2 - a_2b_1}$

67. $2x + 9y - 5z = 23$; the plane that bisects AB and is perpendicular to that segment

69. The angle between any two edges is $\pi/3$.

SECTION 12.3 (PAGE 794)

1. $\langle 0, -14, 7\rangle$

3. $-10\mathbf{i} - 7\mathbf{j} + \mathbf{k}$

5. $\langle 0, 0, 22\rangle$

7. $\pm\frac{1}{13}\langle 12, -3, 4\rangle$

11. $\mathbf{a} \times (\mathbf{b} \times \mathbf{c}) = -\mathbf{k} \neq -\mathbf{i} + \mathbf{j} = (\mathbf{a} \times \mathbf{b}) \times \mathbf{c}$.

13. $\mathbf{b} \times \mathbf{c}$ is parallel to $\mathbf{a}$.

15. $\frac{1}{2}\sqrt{2546}$

17. **a.** 55; **b.** $\frac{55}{6}$

19. Coplanar

21. Not coplanar

23. The area is approximately 4395.6569291026 m^2.

25. The area is approximately 31271.643253 ft^2.

29. Begin with the observation that the area of the triangle in Fig. 12.3.13 is $\frac{1}{2}|\overrightarrow{PQ}| \cdot d$.

31. Begin with the observation that a vector perpendicular to both lines is $\mathbf{n} = \overrightarrow{P_1Q_1} \times \overrightarrow{P_2Q_2}$.

33. Use Eq. (12) and the result in Problem 32.

35. See the discussion following Eq. (3) in the text.

SECTION 12.4 (PAGE 801)

1. $x = t$, $y = 2t$, $z = 3t$, $-\infty < t < +\infty$

3. $x = 2t + 4$, $y = 13$, $z = -3t - 3$, $-\infty < t < +\infty$

5. $x = -6t$, $y = 3t$, $z = 5t$, $-\infty < t < +\infty$

7. $x = 3t + 3$, $y = 5$, $z = -3t + 7$, $-\infty < t < +\infty$

9. Parametric equations $x = t + 2$, $y = -t + 3$, $z = -2t - 4$, $-\infty < t < +\infty$; symmetric equations

$$x - 2 = -y + 3 = -\frac{z + 4}{2}.$$

11. Parametric equations $x = 1$, $y = 1$, $z = t + 1$, $-\infty < t < +\infty$; Cartesian equations $x = 1$, $y = 1$.

13. Parametric equations $x = 2t + 2$, $y = -t - 3$, $z = 3t + 4$, $-\infty < t < +\infty$; symmetric equations

$$\frac{x - 2}{2} = -(y + 3) = \frac{z - 4}{3}.$$

15. The lines meet at (and only at) the point $(2, -1, 3)$.

17. L_1 and L_2 are skew lines.

19. L_1 and L_2 are parallel and distinct.

21. $x + 2y + 3z = 0$

23. $x - z + 8 = 0$

25. $y = 7$

27. $7x + 11y = 114$

29. $3x + 4y - z = 0$

31. $2x - y - z = 0$

33. $2x - 7y + 17z = 78$

35. L and $\mathcal{P}$ are parallel and have no points in common.

37. They meet at (and only at) the point $(\frac{9}{2}, \frac{9}{4}, \frac{17}{4})$.

39. The angle between the planes is $\theta = \arccos(1/\sqrt{3})$.

41. The angle between the planes is $\theta = 0$ because the planes are parallel.

43. Parametric equations $x = 10,\ y = t,\ z = -10 - t,\ -\infty < t < +\infty$; Cartesian equations $x = 10,\ y = -10 - z$

45. There is no line of intersection because the planes are parallel.

47. Parametric equations $x = 3,\ y = 3 - t,\ z = 1 + t,\ -\infty < t < +\infty$; Cartesian equations $x = 3,\ z = 4 - y$.

49. $3x + 2y + z = 6$

51. $7x - 5y - 2z = 9$

53. $x - 2y + 4z = 3$

55. $\dfrac{10\sqrt{3}}{3}$

59. Part (b): $\dfrac{133\sqrt{501}}{501}$

SECTION 12.5 (PAGE 813)

1. Because $y^2 + z^2 = 1$ while x is arbitrary, the graph lies on the cylinder of radius 1 with axis the x-axis. A small part of the graph is shown in Fig. 12.5.17.

3. Because $x^2 + y^2 = t^2 = z^2$, the graph lies on the cone with axis the z-axis and equation $z^2 = x^2 + y^2$. A small part of the graph is shown in Fig. 12.5.16.

5. $\mathbf{r}'(1) = \mathbf{0} = \mathbf{r}''(1)$

7. $\mathbf{r}'(0) = 2\mathbf{i} - \mathbf{j}$ and $\mathbf{r}''(0) = 4\mathbf{i} + \mathbf{j}$

9. $\mathbf{r}'\left(\dfrac{3}{4}\right) = 6\pi\mathbf{i}$ and $\mathbf{r}''\left(\dfrac{3}{4}\right) = 12\pi^2\mathbf{j}$

11. $\mathbf{v}(t) = \langle 1, 2t, 3t^2\rangle,\ v(t) = \sqrt{1 + 4t^2 + 9t^4},\ \mathbf{a}(t) = \langle 0, 2, 6t\rangle$

13. $\mathbf{v}(t) = \langle 1, 3e^t, 4e^t\rangle,\ v(t) = \sqrt{1 + 25e^{2t}},\ \mathbf{a}(t) = \langle 0, 3e^t, 4e^t\rangle$

15. $\mathbf{v}(t) = \langle -3\sin t, 3\cos t, -4\rangle,$
$v(t) = \sqrt{9\sin^2 t + 9\cos^2 t + 16} = 5,$
$\mathbf{a}(t) = \langle -3\cos t, -3\sin t, 0\rangle$

17. $\left\langle \dfrac{2 - \sqrt{2}}{2}, \sqrt{2}\right\rangle$

19. $\dfrac{484}{15}\mathbf{i}$

21. 11

23. 0

25. $\mathbf{r}(t) = \langle 1, 0, t\rangle$

27. $\mathbf{r}(t) = \langle t^2, 10t, -2t^2\rangle$

29. $\mathbf{r}(t) = \langle 2, t^2, 5t - t^3\rangle$

31. $\mathbf{r}(t) = \left\langle \dfrac{1}{6}t^3 + 10, \dfrac{1}{12}t^4 + 10t, \dfrac{1}{20}t^5\right\rangle$

33. $\mathbf{r}(t) = \langle 1 - t - \cos t, 1 + t - \sin t, 5t\rangle$

35. $\mathbf{v}\left(\frac{7}{8}\pi\right) = \left\langle 3\sqrt{2}, 3\sqrt{2}, 8\right\rangle,\ v\left(\frac{7}{8}\pi\right) = 10$, and $\mathbf{a}\left(\frac{7}{8}\pi\right) = \left\langle -6\sqrt{2}, 6\sqrt{2}, 0\right\rangle$

37. $\mathbf{u}(t) \times \mathbf{v}'(t) + \mathbf{u}'(t) \times \mathbf{v}(t) = \langle 0, 40t, -15\rangle = D_t[\mathbf{u}(t) \times \mathbf{v}(t)]$.

41. 100 ft

43. $v_0 = \sqrt{32 \cdot 5280} = 32\sqrt{165} \approx 411.047442517284$ ft/s

47. (a) Range: $400\sqrt{3}$ ft, maximum height 100 ft. (b) Range: 800 ft, maximum height 200 ft; (c) Range: $400\sqrt{3}$ ft, maximum height 300 ft

49. $70\sqrt{10}$ m/s

51. Angle of inclination: approximately 41°50′33.739224″; initial velocity: approximately 133.6459515485 m/s.

53. First assume that $\mathbf{u}(t) = \langle u_1(t), u_2(t)\rangle$ and $\mathbf{v}(t) = \langle v_1(t), v_2(t)\rangle$. Your proof will be easy to generalize to vectors with three or more components.

55. First show that $D_t[\mathbf{v}(t) \cdot \mathbf{v}(t)] = 0$.

57. A central repulsive force with magnitude proportional to distance from the origin.

63. 5 ft north

65. **b.** 12 s; **c.** 2400 ft north, 144 ft east; **d.** 784 ft

SECTION 12.6 (PAGE 828)

1. 10π

3. $19(e - 1) \approx 32.647355$

5. $\dfrac{20 + 9\ln 3}{10} \approx 2.988751$

7. $\kappa(0) = 0$

9. $\kappa(0) = 1$

11. $\kappa\left(\dfrac{\pi}{4}\right) = \dfrac{40\sqrt{82}}{1681} \approx 0.215476$

13. $\left(-\frac{1}{2}\ln 2, \frac{1}{2}\sqrt{2}\right)$

15. Maximum at $(\pm 5, 0)$, minimum at $(0, \pm 3)$

17. $\mathbf{T}(-1) = \left\langle \dfrac{\sqrt{10}}{10}, \dfrac{3\sqrt{10}}{10}\right\rangle,\ \mathbf{N}(-1) = \left\langle \dfrac{3\sqrt{10}}{10}, -\dfrac{\sqrt{10}}{10}\right\rangle$

19. $\mathbf{T}(\pi/6) = \left\langle \dfrac{\sqrt{57}}{19}, -\dfrac{4\sqrt{19}}{19}\right\rangle,\ \mathbf{N}(\pi/6) = \left\langle -\dfrac{4\sqrt{19}}{19}, -\dfrac{\sqrt{57}}{19}\right\rangle$

21. $\mathbf{T}(3\pi/4) = \left\langle -\dfrac{\sqrt{2}}{2}, -\dfrac{\sqrt{2}}{2}\right\rangle,\ \mathbf{N}(3\pi/4) = \left\langle -\dfrac{\sqrt{2}}{2}, \dfrac{\sqrt{2}}{2}\right\rangle$

23. $a_T = \dfrac{18t}{\sqrt{9t^2 + 1}},\quad a_N = \dfrac{6}{\sqrt{9t^2 + 1}}$

25. $a_T = \dfrac{t}{\sqrt{t^2 + 1}},\quad a_N = \dfrac{t^2 + 2}{\sqrt{t^2 + 1}}$

27. $\kappa = \dfrac{1}{a}$

29. $x^2 + \left(y - \frac{1}{2}\right)^2 = \frac{1}{4}$

31. $(x - 2)^2 + (y - 2)^2 = 2$

33. $\kappa(t) \equiv \frac{1}{2}$

35. $\kappa(t) = \dfrac{\sqrt{2}}{3}e^{-t}$

37. $a_T = 0 = a_N$

39. $a_T = \dfrac{4t + 18t^3}{\sqrt{1 + 4t^2 + 9t^4}}, \quad a_N = \dfrac{\sqrt{4 + 36t^2 + 36t^4}}{\sqrt{1 + 4t^2 + 9t^4}}$

41. $a_T = \dfrac{t}{\sqrt{t^2 + 2}}, \quad a_N = \dfrac{\sqrt{t^4 + 5t^2 + 8}}{\sqrt{t^2 + 2}}$

43. $\mathbf{T}(0) = \left\langle \dfrac{\sqrt{2}}{2}, \dfrac{\sqrt{2}}{2}, 0 \right\rangle, \quad \mathbf{N}(0) = \langle 0, 0, -1 \rangle$

45. $\mathbf{T}(0) = \left\langle \dfrac{\sqrt{3}}{3}, \dfrac{\sqrt{3}}{3}, \dfrac{\sqrt{3}}{3} \right\rangle, \quad \mathbf{N}(0) = \left\langle -\dfrac{\sqrt{2}}{2}, \dfrac{\sqrt{2}}{2}, 0 \right\rangle$

47. $x(s) = 2 + \dfrac{4s}{13}, \quad y(s) = 1 - \dfrac{12s}{13}, \quad z(s) = 3 + \dfrac{3s}{13}$

49. $x(s) = 3\cos\dfrac{s}{5}, \quad y(s) = 3\sin\dfrac{s}{5}, \quad z(s) = \dfrac{4s}{5}$

51. Note that $D_t(\mathbf{v} \cdot \mathbf{v}) = 0$ (why?).

53. $\kappa(t) = \dfrac{1}{|t|}$

55. $y = 3x^5 - 8x^4 + 6x^3$

57. Approximately 36.651 mi/s; 24.130 mi/s

59. Approximately 0.672 mi/s; 0.602 mi/s

61. About 795 *below* the surface of the Earth

63. Approximately 1 h 42 min 2.588 s

65. Begin with Eq. (42), substitute Eqs. (37) and (41).

SECTION 12.7 (PAGE 837)

1. The plane with intercepts $x = \frac{20}{3}$, $y = 10$, and $z = 2$.

3. Circular cylinder, radius 3, axis the z-axis.

5. A hyperbolic cylinder with rulings parallel to the z-axis and meeting the xy-plane in the hyperbola with equation $xy = 4$.

7. Elliptic paraboloid, axis the z-axis, vertex at the origin, opening upward.

9. Circular paraboloid, axis the z-axis, vertex at $(0, 0, 4)$, opening downward.

11. Circular paraboloid, axis the z-axis, vertex at the origin, opening upward.

13. Both nappes of a circular cone, axis the z-axis, vertex at the origin.

15. Parabolic cylinder parallel to the y-axis, opening upward, lowest points those on the line $z = -2$, $x = 0$.

17. Elliptical cylinder parallel to the z-axis, centerline the z-axis.

19. Both nappes of an elliptical cone, axis the x-axis, vertex at the origin.

21. Paraboloid opening downward, axis the negative z-axis, vertex at the origin.

23. Hyperbolic paraboloid, saddle point at the origin; to see it, execute the *Mathematica* 3.0 command

```
ParametricPlot3D[ { 2*y*y - z*z, y, z },
                  { y, -1, 1 }, { z, -1, 1 } ];
```

25. Hyperboloid of one sheet, axis the z-axis.

27. Elliptical paraboloid, axis the nonnegative y-axis, vertex at the origin.

29. Hyperboloid of two sheets, axis the y-axis, center the origin, intercepts $(0, \pm 6, 0)$.

31. Equation $x = 2(y^2 + z^2)$; circular paraboloid opening along the positive x-axis.

33. Equation $x^2 + y^2 - z^2 = 1$; circular hyperboloid of one sheet with axis the z-axis.

35. Equation: $4x = y^2 + z^2$; circular paraboloid, axis the positive x-axis, vertex at the origin.

37. Equation: $z = \exp(-x^2 - y^2)$.

39. Equation: $z^2 = x^2 + y^2$; both nappes of a right circular cone with axis the z-axis and vertex at the origin.

41. The traces in horizontal planes are ellipses with centers on the z-axis and semiaxes 2 and 1.

43. The traces in the planes $x = a$ are circles if $|a| < 2$, single points if $|a| = 2$, empty if $|a| > 2$.

45. The trace in the plane $x = a$ is a parabola opening upward with vertex at $(a, 0, 4a^2)$.

47. The traces are generally parabolas; some open upward, some downward; rotate the surface of Fig. 12.7.22 around the z-axis 45° to see the surface.

55. Elliptic paraboloid; $z = 2u^2 + 4v^2$.

57. Hyperbolic paraboloid; $z = 5u^2 - 5v^2$.

59. Elliptical paraboloid; $z = 34u^2 + 17v^2$.

61. Hyperbolic paraboloid; $z = 169u^2 - 169v^2$.

63. Hyperboloid of one sheet; $-u^2 + v^2 + 2w^2 = 5$.

65. Hyperbolic cylinder; $-5u^2 + 5w^2 = 14$.

67. Elliptic cylinder; $2v^2 + 3w^2 = 5$.

69. Ellipsoid; $3u^2 + 3v^2 + 6w^2 = 23$.

SECTION 12.8 (PAGE 843)

1. $(0, 1, 2)$

3. $(-\sqrt{2}, \sqrt{2}, 3)$

5. $(1, \sqrt{3}, -5)$

7. $(0, 0, 2)$

9. $(-3, 0, 0)$

11. $(0, -\sqrt{3}, 1)$

Note that a given point does not have unique cylindrical or spherical coordinates. Indeed, there are infinitely many correct answers to Problem 13 through 22. If a computer programmed to implement Eqs. (3) and (6) converts your answer to correct rectangular coordinates, your answer is almost certainly correct.

13. Cylindrical: $(0, 0, 5)$; spherical: $(5, 0, 0)$

15. Cylindrical: $(\sqrt{2}, \pi/4, 0)$; spherical: $(\sqrt{2}, \pi/2, \pi/4)$

17. Cylindrical: $(\sqrt{2}, \pi/4, 1)$; spherical: $\left(\sqrt{3}, \cos^{-1}\dfrac{\sqrt{3}}{3}, \dfrac{\pi}{4}\right)$

19. Cylindrical: $\left(\sqrt{5}, \tan^{-1}\left(\dfrac{1}{2}\right), -2\right)$;

spherical: $\left(3, \cos^{-1}\left(-\dfrac{2}{3}\right), \tan^{-1}\left(\dfrac{1}{2}\right)\right)$

21. Cylindrical: $(5, \arctan\frac{4}{3}, 12)$;

spherical: $(13, \arcsin\frac{5}{13}, \arctan\frac{4}{3})$

23. Cylinder, radius 5, axis the z-axis

25. The vertical plane $y = x$

27. The circular cone $z^2 = 3x^2 + 3y^2$ with axis the z-axis and vertex at the origin

29. The xy-plane

31. The ellipsoid with center at the origin and intercepts $(\pm\sqrt{2}, 0, 0)$, $(0, \pm\sqrt{2}, 0)$, and $(0, 0, \pm 2)$

33. Circular cylinder, radius 2, axis the vertical line $x = 2$, $y = 0$

35. Two concentric circular cylinders with common axis the z-axis and radii 1 and 3

37. Two congruent circular paraboloids, each with axis the z-axis and vertex at the origin; one opens upward, the other downward

39. Cylindrical: $r^2 + z^2 = 25$; spherical: $\rho = 5$ (the same as the graph of $\rho = \pm 5$)

41. Cylindrical: $r \cos\theta + r \sin\theta + z = 1$;
spherical: $\rho \sin\phi \cos\theta + \rho \sin\phi \sin\theta + \rho \cos\phi = 1$

43. Cylindrical: $r^2 + z^2 = r\cos\theta + r\sin\theta + z$;
spherical: $\rho^2 = \rho \sin\phi \cos\theta + \rho \sin\phi \sin\theta + \rho\cos\phi$
(it's legal to cancel ρ from both sides of the last equation).

45. The part of the cylinder of radius 3 and centerline the z-axis that lies between the planes $z = -1$ and $z = 1$

47. The part of the spherical surface of radius 2 and center the origin that lies between the two horizontal planes $z = -1$ and $z = 1$

49. The solid is bounded above by the plane $z = 2$, below by the plane $z = -2$, outside by the cylinder of radius 3 with centerline the z-axis, and inside by the cylinder of radius 1 with centerline the z-axis.

51. The solid is the region between two concentric spherical surfaces centered at the origin, one of radius 3 and the other of radius 5.

53. $z = r^2$

55. **a.** $-\sqrt{4 - r^2} \leqq z \leqq \sqrt{4 - r^2}$, $1 \leqq r \leqq 2$, $0 \leqq \theta \leqq 2\pi$;
b. $\sec\phi \leqq \rho \leqq 2$, $\pi/3 \leqq \phi \leqq 2\pi/3$, $0 \leqq \theta \leqq 2\pi$

57. About 3821 mi (about 6149 km)

59. A little less than 41 mi (66 km)

61. $0 \leqq \rho \leqq \sqrt{R^2 + H^2}$, $0 \leqq \theta \leqq 2\pi$, $\phi = \arctan\left(\dfrac{R}{H}\right)$

63. **a.** $4a^2(x^2 + y^2) = (x^2 + y^2 + z^2 + a^2 - b^2)^2$;
b. $(r - a)^2 + z^2 = b^2$;
c. $2a\rho \sin\phi = \rho^2 + a^2 - b^2$

CHAPTER 12 MISCELLANEOUS PROBLEMS (PAGE 845)

1. $\frac{1}{2}\left(\overrightarrow{AP} + \overrightarrow{AQ}\right) = \frac{1}{2}\left(\overrightarrow{AM} - \overrightarrow{PM} + \overrightarrow{AM} + \overrightarrow{MQ}\right) = \overrightarrow{AM}.$

5. Note that $A = \frac{1}{2}\left|\overrightarrow{PQ} \times \overrightarrow{PR}\right|$.

7. Parametric equations

$$x = 1 + 2t, \quad y = -1 + 3t, \quad z = 2 - 3t, \quad -\infty < t < +\infty,$$

symmetric equations

$$\frac{x - 1}{2} = \frac{y + 1}{3} = \frac{z - 2}{-3}.$$

9. Both lines are parallel to $\mathbf{u} = \langle 6, 3, 2 \rangle$; the plane has Cartesian equation $13x - 22y - 6z = 23$.

11. $x - y + 2z = 3$

15. 3

19. The position vector $\mathbf{r}(t) = \langle -\sin t, \cos t \rangle$ traces the circle of radius 1 with center $(0, 0)$.

21. Two solutions: $\alpha \approx 0.033364$ (about 1°54′42″) and $\alpha \approx 1.291156$ (about 73°58′40″)

23. $\kappa(1) = \frac{1}{9}$, $a_T(1) = 2$, $a_N(1) = 1$

25. Begin with the observation that $\mathbf{v}_1 \times \mathbf{v}_2$ is normal to the plane.

27. $3x - 3y + z = 1$

33. $\rho = 2\cos\phi$

35. $\rho^2 = 2\cos 2\phi$

43. Minimal at every integral multiple of π, maximal at every odd integral multiple of $\pi/2$

45. $\mathbf{T} = \dfrac{1}{\sqrt{\pi^2 + 4}}\langle -\pi, 2 \rangle$, $\quad \mathbf{N} = \dfrac{1}{\sqrt{\pi^2 + 4}}\langle -2, -\pi \rangle$

49. $y(x) = \dfrac{15}{8}x - \dfrac{5}{4}x^3 + \dfrac{3}{8}x^5$

SECTION 13.2 (PAGE 857)

1. The entire xy-plane

3. The entire xy-plane except for the origin $(0, 0)$

5. All points of the xy-plane

7. All points on and within the unit circle

9. The entire xy-plane

11. The region *above* the straight line with equation $y = x$

13. All points of the xy-plane not on either coordinate axis

15. All points of the xy-plane not on either straight line $y = x$ or $y = -x$

17. All points in space other than the origin $(0, 0, 0)$

19. All points of space strictly above the paraboloid $z = x^2 + y^2$

21. The horizontal plane through $(0, 0, 10)$

23. The plane with equation $z = x + y$

25. A circular paraboloid with axis the nonnegative z-axis, opening upward, vertex at the origin

27. The upper half of the spherical surface with radius 2 and center $(0, 0, 0)$

29. The lower nappe of a circular cone with axis the z-axis and vertex at $(0, 0, 10)$

31. Straight lines of the form $x - y = c$ (where c is a constant)

33. Ellipses centered at the origin with major axes on the x-axis and minor axes on the y-axis

35. Curves with equations of the form $y = x^3 + C$ (C is a constant)

37. Circles centered at the point $(2, 0)$

39. Circles centered at the origin

41. Congruent circular paraboloids all with axis the z-axis and all opening upward

43. Spherical surfaces centered at the point $(2, 1, 3)$

45. The level surfaces of f are elliptical cylinders parallel to the z-axis and centered on the vertical line that meets the xy-plane at the point $(2, 1, 0)$. The ellipse in which each such cylinder meets the xy-plane has major axis parallel to the x-axis, minor axis parallel to the y-axis, and the major axis is twice the length of the minor axis.

47. Matches Fig. 13.2.32

49. Almost matches Fig. 13.2.30
51. Matches Fig. 13.2.28
53. Matches Fig. 13.2.41
55. Matches Fig. 13.2.42
57. Matches Fig. 13.2.44
59. If a and b are not both zero, then the surface has one pit and one peak.
61. Apparently n peaks and n pits alternately surround the origin.

SECTION 13.3 (PAGE 866)

1. 7 **3.** e **5.** $\frac{5}{3}$
7. 0 **9.** 1 **11.** $-\frac{3}{2}$
13. 1 **15.** -4 **17.** y; x
19. y^2; $2xy$ **21.** 0 **23.** $\frac{1}{3}$
25. 0
27. Does not exist; it is also correct to indicate that the limit is $+\infty$.
29. 0
31. All points (x, y) such that $y > -x$
33. All points (x, y) such that $x^2 + y^2 > 1$
35. Continuous at all points (x, y) other than $(0, 0)$
37. 0 **39.** 0 **41.** 0
43. Does not exist
45. Does not exist
47. Does not exist
49. Does not exist
55. Continuous for all (x, y)

SECTION 13.4 (PAGE 875)

1. $\frac{\partial f}{\partial x} = 4x^3 - 3x^2y + 2xy^2 - y^3$ and $\frac{\partial f}{\partial y} = -x^3 + 2x^2y - 3xy^2 + 4y^3$.
3. $\frac{\partial f}{\partial x} = e^x(\cos y - \sin y)$ and $\frac{\partial f}{\partial y} = -e^x(\cos y + \sin y)$.
5. $\frac{\partial f}{\partial x} = -\frac{2y}{(x-y)^2}$ and $\frac{\partial f}{\partial y} = \frac{2x}{(x-y)^2}$.
7. $\frac{\partial f}{\partial x} = \frac{2x}{x^2+y^2}$ and $\frac{\partial f}{\partial y} = \frac{2y}{x^2+y^2}$.
9. $\frac{\partial f}{\partial x} = yx^{y-1}$ and $\frac{\partial f}{\partial y} = x^y \ln x$.
11. $\frac{\partial f}{\partial x} = 2xy^3z^4$, $\frac{\partial f}{\partial y} = 3x^2y^2z^4$, and $\frac{\partial f}{\partial z} = 4x^2y^3z^3$.
13. $\frac{\partial f}{\partial x} = yze^{xyz}$, $\frac{\partial f}{\partial y} = xze^{xyz}$, and $\frac{\partial f}{\partial z} = xye^{xyz}$.
15. $\frac{\partial f}{\partial x} = 2xe^y \ln z$, $\frac{\partial f}{\partial y} = x^2e^y \ln z$, and $\frac{\partial f}{\partial z} = \frac{x^2e^y}{z}$.
17. $\frac{\partial f}{\partial r} = \frac{4rs^2}{(r^2+s^2)^2}$ and $\frac{\partial f}{\partial s} = -\frac{4r^2s}{(r^2+s^2)^2}$.
19. $\frac{\partial f}{\partial u} = we^u + e^v$, $\frac{\partial f}{\partial v} = ue^v + e^w$, and $\frac{\partial f}{\partial w} = e^u + ve^w$.
21. $z_x(x, y) = 2x - 4y$, $z_y(x, y) = -4x + 6y$, $z_{xy}(x, y) = -4$, $z_{yx}(x, y) = -4$.
23. $z_x(x, y) = 2x\exp(-y^2)$, $z_y(x, y) = -2x^2y\exp(-y^2)$, $z_{xy}(x, y) = -4xy\exp(-y^2)$, $z_{yx}(x, y) = -4xy\exp(-y^2)$.
25. $z_x(x, y) = \frac{1}{x+y} = z_y(x, y)$ and $z_{xy}(x, y) = -\frac{1}{(x+y)^2} = z_{yx}(x, y)$.
27. $z_x(x, y) = -3e^{-3x}\cos y$, $z_y(x, y) = -e^{-3x}\sin y$, $z_{xy}(x, y) = 3e^{-3x}\sin y$, $z_{yx}(x, y) = 3e^{-3x}\sin y$.
29. $z_x(x, y) = 2x\cosh\left(\frac{1}{y^2}\right)$, $z_y(x, y) = -\frac{2x^2}{y^3}\sinh\left(\frac{1}{y^2}\right)$, $z_{xy}(x, y) = -\frac{4x}{y^3}\sinh\left(\frac{1}{y^2}\right)$, $z_{yx}(x, y) = -\frac{4x}{y^3}\sinh\left(\frac{1}{y^2}\right)$.
31. $z = 6x + 8y - 25$
33. $z = -1$
35. $z = 27x - 12y - 38$
37. $z = 1 - x + y$
39. $z = 10x - 16y - 9$
41. One answer: $f(x, y) = x^2y^3$
43. $f_{xy}(x, y) = -2x\sin xy\cos xy \neq -2y\sin xy\cos xy = f_{yx}(x, y)$.
45. Matches Fig. 13.4.14
47. Matches Fig. 13.4.13
49. Matches Fig. 13.4.15
51. $f_{xy}(x, y) = mnx^{m-1}y^{n-1} = f_{yx}(x, y)$.
53. $f_{xx}(x, y, z) = y^2z^2e^{xyz}$,
$f_{xy}(x, y, z) = f_{yx}(x, y, z) = (xyz^2 + z)e^{xyz}$,
$f_{xz}(x, y, z) = f_{zx}(x, y, z) = (y + xy^2z)e^{xyz}$,
$f_{yz}(x, y, z) = f_{zy}(x, y, z) = (x + x^2yz)e^{xyz}$,
$f_{yy}(x, y, z) = x^2z^2e^{xyz}$,
$f_{zz}(x, y, z) = x^2y^2e^{xyz}$,
$f_{xyz}(x, y, z) = (1 + 3xyz + x^2y^2z^2)e^{xyz}$.
55. $u_t(x, t) = -n^2k\exp(-n^2kt)\sin nx$,
$u_x(x, t) = n\exp(-n^2kt)\cos nx$, and
$u_{xx}(x, t) = -n^2\exp(-n^2kt)\sin nx$.
57. Part (a): $y_t(x, t) = a\cos(x + at)$,
$y_x(x, t) = \cos(x + at)$,
$y_{tt}(x, t) = -a^2\sin(x + at)$,
$y_{xx}(x, t) = -\sin(x + at)$.
Part (b): $y_t(x, t) = -3a\sinh(3(x - at))$,
$y_x(x, t) = 3\sinh(3(x - at))$,
$y_{tt}(x, t) = 9a^2\cosh(3(x - at))$,
$y_{xx}(x, t) = 9\cosh(3(x - at))$.
Part (c): $y_t(x, t) = -ka\sin kx\sin kat$,
$y_x(x, t) = k\cos kx\cos kat$,
$y_{tt}(x, t) = -k^2a^2\sin kx\cos kat$,
$y_{xx}(x, t) = -k^2\sin kx\cos kat$.
59. $y_t(x, t) = af'(x + at) - ag'(x - at)$,
$y_{tt}(x, t) = a^2f''(x + at) + a^2g''(x - at)$,
$y_x(x, t) = f'(x + at) + g'(x - at)$,
$y_{xx}(x, t) = f''(x + at) + g''(x - at)$.

61. $u(0, t) = T_0 + a_0 e^0 \cos(\omega t - 0) = T_0 + a_0 \cos \omega t;$

$u_t(x, t) = -a_0\omega \exp\left(-x\sqrt{\omega/2k}\right) \sin\left(\omega t - x\sqrt{\omega/2k}\right),$

$u_x(x, t) = -a_0\left(\sqrt{\omega/2k}\right) \exp\left(-x\sqrt{\omega/2k}\right) \times \left[\cos\left(\omega t - x\sqrt{\omega/2k}\right) - \sin\left(\omega t - x\sqrt{\omega/2k}\right)\right],$

$u_{xx}(x, t) = -\dfrac{a_0\omega}{k} \exp\left(-x\sqrt{\omega/2k}\right) \sin\left(\omega t - x\sqrt{\omega/2k}\right).$

65. $(10, -7, -58)$

67. a. $\Delta V \approx -2570$ (cm^3); **b.** $\Delta V \approx 82.51$ (cm^3)

69. a. $f_{xx}(x, y) = -\sin x \sinh(\pi - y) = -f_{yy}(x, y)$.
b. $f_{xx}(x, y) = 4 \sinh 2x \sin 2y = -f_{yy}(x, y)$.
c. $f_{xx}(x, y) = -9 \sin 3x \sinh 3y = -f_{yy}(x, y)$.
d. $f_{xx}(x, y) = 16 \sinh 4(\pi - x) \sin 4y = -f_{yy}(x, y)$.

71. a. Initially descending at 45°; **b.** initially ascending at 45°

SECTION 13.5 (PAGE 886)

1. There are no horizontal tangent planes.

3. $(0, 0, 5)$

5. $(3, -1, -5)$

7. $(-2, 0, -4)$

9. $(-2, 0, -7)$ and $(-2, 1, -9)$

11. $(-1, 0, 2e^{-1})$, $(1, 0, 2e^{-1})$, and $(0, 0, 0)$

13. Lowest point $(1, 1, 1)$

15. Equally high highest points $(1, -1, 2)$ and $(1, 1, 2)$

17. Lowest point $(2, 3, -50)$

19. Equally low lowest points $(-4, 2, -16)$ and $(4, -2, -16)$

21. Highest point $(1, -2, e^5)$

23. -3 and 3

25. -1 and 4

27. -1 and 1

29. $(12, 4, 3)$

31. $(15, 5, 4)$

33. $\left(\sqrt{2}, \sqrt{2}, 1\right)$

35. 64000

37. $10 \times 10 \times 10$ in.

39. Base 10 by 10 in., height 6 in.

41. Base and top 15×10 in., front and back 15×5 in., sides 10×5 in.

43. 40 ft wide (in front), 20 ft deep, 10 ft high

45. $\dfrac{1131}{567}$

47. 11664 in.^{3}

49. $\dfrac{1}{2}$

51. $5^{1/6}(18\pi V^2)^{1/3}$

53. $\left(\dfrac{2}{3}, \dfrac{1}{3}\right)$

55. Base of front $2^{5/6}V^{1/3}$, height of front half that, depth of house $2^{1/3}V^{1/3}$

57. Maximum area: Make one square. Minimum area: Make three equal squares.

59. Maximum cross-sectional area $\frac{1}{12}L^2\sqrt{3}$

61. a. $x = \frac{45}{11}$, $y = \frac{48}{11}$; **b.** $x = \frac{37}{5}$, $y = \frac{98}{15}$

63. Raise 40 hogs and 40 head of cattle per unit of land, but no sheep.

69. The function g has no extrema, local or global.

SECTION 13.6 (PAGE 895)

1. $dw = (6x + 4y)\,dx + (4x - 6y^2)\,dy$

3. $dw = \dfrac{x\,dx + y\,dy}{\sqrt{1 + x^2 + y^2}}$

5. $dw = \dfrac{y\,dx - x\,dy}{x^2 + y^2}$

7. $dw = \dfrac{2x\,dx + 2y\,dy + 2z\,dz}{x^2 + y^2 + z^2}$

9. $dw = \tan yz\,dx + xz\sec^2 yz\,dy + xy\sec^2 yz\,dz$

11. $dw = -yze^{-xyz}\,dx - xze^{-xyz}\,dy - xye^{-xyz}\,dz$

13. $dw = 2u\exp(-v^2)\,du - 2u^2 v\exp(-v^2)\,dv$

15. $dw = \dfrac{x\,dx + y\,dy + z\,dz}{\sqrt{x^2 + y^2 + z^2}}$

17. 5.014

19. 0.0993

21. $\dfrac{16953}{1300} \approx 13.040769$

23. 1.06

25. 191.1

27. 1.4

29. $\dfrac{333}{26} \approx 12.807692$

31. 2.08

33. 2.5

35. $\dfrac{25\pi}{6} \approx 13.089969$

37. $300 + \dfrac{4375\pi\sqrt{3}}{36} \approx 961.281018$ ft^2, about 0.022068 acres

39. $\dfrac{17\pi}{1920} \approx 0.027816$

41. $125\sqrt{3} + \dfrac{250\pi}{9} \approx 303.772814$

43. a. Let $(x, y) \to (0, 0)$ along the lines $y = x$ and $y = 0$; **b.** you should find that $f_x(0, 0) = 0 = f_y(0, 0)$.

SECTION 13.7 (PAGE 904)

1. $\dfrac{dw}{dt} = -(2t + 1)\exp(-t^2 - t)$

3. $\dfrac{dw}{dt} = 6t^5 \cos t^6$

5. $\dfrac{\partial w}{\partial s} = \dfrac{2}{s + t} = \dfrac{\partial w}{\partial t}$

7. $\dfrac{\partial w}{\partial s} = 0$, $\dfrac{\partial w}{\partial t} = 5e^t$

9. $\dfrac{\partial r}{\partial x} = (y + z)\exp(yz + xz + xy)$,

$\dfrac{\partial r}{\partial y} = (x + z)\exp(yz + xz + xy)$, and

$\dfrac{\partial r}{\partial z} = (x + y)\exp(yz + xz + xy)$

11. Here we have

$$\frac{\partial r}{\partial x} = \frac{(2y + 3z)\sqrt{xy^2z^3}}{2x(x + 2y + 3z)^{3/2}} \cos\left(\frac{\sqrt{xy^2z^3}}{\sqrt{x + 2y + 3z}}\right),$$

$$\frac{\partial r}{\partial y} = \frac{(x + y + 3z)\sqrt{xy^2z^3}}{y(x + 2y + 3z)^{3/2}} \cos\left(\frac{\sqrt{xy^2z^3}}{\sqrt{x + 2y + 3z}}\right), \quad \text{and}$$

$$\frac{\partial r}{\partial z} = \frac{3(x + 2y + 2z)\sqrt{xy^2z^3}}{2z(x + 2y + 3z)^{3/2}} \cos\left(\frac{\sqrt{xy^2z^3}}{\sqrt{x + 2y + 3z}}\right).$$

13. The formulas are

$$\frac{\partial p}{\partial u} = \frac{\partial f}{\partial x} \cdot \frac{\partial x}{\partial u} + \frac{\partial f}{\partial y} \cdot \frac{\partial y}{\partial u},$$

$$\frac{\partial p}{\partial v} = \frac{\partial f}{\partial x} \cdot \frac{\partial x}{\partial v} + \frac{\partial f}{\partial y} \cdot \frac{\partial y}{\partial v}, \quad \text{and}$$

$$\frac{\partial p}{\partial w} = \frac{\partial f}{\partial x} \cdot \frac{\partial x}{\partial w} + \frac{\partial f}{\partial y} \cdot \frac{\partial y}{\partial w}.$$

15. Answer:

$$\frac{\partial p}{\partial x} = \frac{\partial f}{\partial u} \cdot \frac{\partial u}{\partial x} + \frac{\partial f}{\partial v} \cdot \frac{\partial v}{\partial x} + \frac{\partial f}{\partial w} \cdot \frac{\partial w}{\partial x},$$

$$\frac{\partial p}{\partial y} = \frac{\partial f}{\partial u} \cdot \frac{\partial u}{\partial y} + \frac{\partial f}{\partial v} \cdot \frac{\partial v}{\partial y} + \frac{\partial f}{\partial w} \cdot \frac{\partial w}{\partial y}, \quad \text{and}$$

$$\frac{\partial p}{\partial z} = \frac{\partial f}{\partial u} \cdot \frac{\partial u}{\partial z} + \frac{\partial f}{\partial v} \cdot \frac{\partial v}{\partial z} + \frac{\partial f}{\partial w} \cdot \frac{\partial w}{\partial z}.$$

17. $\frac{\partial p}{\partial x} = f'(w) \cdot \frac{\partial w}{\partial x}, \quad \frac{\partial p}{\partial y} = f'(w) \cdot \frac{\partial w}{\partial y},$

$\frac{\partial p}{\partial z} = f'(w) \cdot \frac{\partial w}{\partial z}, \quad \frac{\partial p}{\partial u} = f'(w) \cdot \frac{\partial w}{\partial u}, \quad$ and

$\frac{\partial p}{\partial v} = f'(w) \cdot \frac{\partial w}{\partial v}.$

19. $\frac{\partial z}{\partial x} = -\frac{z^{1/3}}{x^{1/3}}, \quad \frac{\partial z}{\partial y} = -\frac{z^{1/3}}{y^{1/3}}$

21. $\frac{\partial z}{\partial x} = -\frac{e^{xy} + xye^{xy} + yze^{zx} + yze^{xy}}{xye^{zx} + e^{xy}},$

$\frac{\partial z}{\partial y} = -\frac{x^2e^{xy} + e^{zx} + xze^{xy}}{xye^{zx} + e^{xy}}$

23. $\frac{\partial z}{\partial x} = -\frac{c^2x}{a^2z}, \quad \frac{\partial z}{\partial y} = -\frac{c^2y}{b^2z}$

25. $\frac{\partial w}{\partial x} = 6x, \quad \frac{\partial w}{\partial y} = 6y$

27. Answer:

$$\frac{\partial w}{\partial x} = \frac{2x^2y}{3(x^2+y^2)^{2/3}\left[(x^2+y^2)^{1/3} + (x^3+y^3)^{1/2}\right]} + \frac{3x^3y}{2(x^3+y^3)^{1/2}\left[(x^2+y^2)^{1/3} + (x^3+y^3)^{1/2}\right]} + y\ln\left((x^2+y^2)^{1/3} + (x^3+y^3)^{1/2}\right)$$

and

$$\frac{\partial w}{\partial y} = \frac{2xy^2}{3(x^2+y^2)^{2/3}\left[(x^2+y^2)^{1/3} + (x^3+y^3)^{1/2}\right]} + \frac{3xy^3}{2(x^3+y^3)^{1/2}\left[(x^2+y^2)^{1/3} + (x^3+y^3)^{1/2}\right]} + x\ln\left((x^2+y^2)^{1/3} + (x^3+y^3)^{1/2}\right).$$

29. $x + 2y + 2z = 9$

31. $z = x - y$

33. -2880 in.^{3}/h

35. $\frac{26\pi}{5} \approx 16.3363$ ft^3/min

37. Decreasing at $\frac{13}{3}$ L/min

57. It will break.

SECTION 13.8 (PAGE 915)

1. $\langle 3, -7\rangle$

3. $\langle 0, 0\rangle$

5. $\langle 0, 6, -4\rangle$

7. $\langle 1, 1, 1\rangle$

9. $\left\langle 2, -\frac{3}{2}, -2\right\rangle$

11. $8\sqrt{2}$

13. $\frac{12}{13}\sqrt{13}$

15. $-\frac{13}{20}$

17. $-\frac{1}{6}$

19. $-6\sqrt{2}$

21. $\sqrt{170}$ and $\langle 7, 11\rangle$

23. $\frac{2}{5}$ and $\langle 3, 4\rangle$

25. $14\sqrt{2}$ and $\langle 3, 5, -8\rangle$

27. $2\sqrt{14}$ and $\langle 1, 2, 3\rangle$

29. $3x + 4y = 25$

31. $29x - 4y = 70$

33. $x + y + z = 1$

39. Use the fact that $\nabla(-f(P)) = -\nabla f(P)$.

45. 14 deg/min

47. **a.** $\frac{34}{3}$ °C/ft; **b.** 13, in the direction $\langle 4, 3, 12\rangle$

49. **a.** $3x + 2y - 10z = 4$;
b. approximately 0.44 (true value: 0.448)

51. $x - 2y + z + 10 = 0$

55. Each such pyramid has volume 4.5.

57. Heading approximately 36° 52′11.6″; up at an angle of 45°

59. Heading approximately 203° 11′54.9″; up at an angle of approximately 75°17′ 8.327″

61. **a.** Descending, angle about 8° 2′58.1″;
b. descending, angle about 3° 37′56.7″

SECTION 13.9 (PAGE 924)

1. Maximum $\sqrt{5}$, minimum $-\sqrt{5}$

3. Maximum 4, minimum -4

5. Maximum 3, minimum -3

7. No maximum; minimum $\frac{18}{7}$

9. Maximum 7, minimum -7

11. Maximum 20, minimum -20

13. Maximum $\frac{81}{4}$, minimum 0

15. No maximum; minimum $\frac{25}{3}$

17. Maximum $1 + \sqrt{2}$, minimum $1 - \sqrt{2}$

19. $(12, 16)$

21. $(12, 4, 3)$

23. $(15, 5, 4)$

25. $\left(\sqrt{2}, \sqrt{2}, 1\right)$

27. Maximum: 64000

29. Minimum: 600 in.^{2}

31. \$18.00

33. Front 15 in. wide and 5 in. high, depth 10 in.

35. Two closest points: $(2, -2, 1)$ and $(-2, 2, 1)$

39. $(2, 3)$ and $(-2, -3)$

41. Highest point $\left(\frac{2}{5}\sqrt{5}, \frac{1}{5}\sqrt{5}, -4+\sqrt{5}\right)$,

lowest point $\left(-\frac{2}{5}\sqrt{5}, -\frac{1}{5}\sqrt{5}, -4-\sqrt{5}\right)$

43. Closest point

$$\left(\tfrac{3}{20}\left[-5+3\sqrt{5}\right], \tfrac{3}{10}\left[-5+3\sqrt{5}\right], \tfrac{3}{4}\left[3-\sqrt{5}\right]\right),$$

farthest point

$$\left(-\tfrac{3}{20}\left[5+3\sqrt{5}\right], -\tfrac{3}{10}\left[5+3\sqrt{5}\right], \tfrac{3}{4}\left[3+\sqrt{5}\right]\right).$$

47. Maximum area: $\frac{1}{4}\left(3-2\sqrt{2}\right)P^2 \approx (0.043)P^2$

51. $(0.410245, 0.347810)$ (coordinates approximate)

53. $(4, 6)$

55. Closest $(-0.604, -1.207, -1.811)$, farthest $(2.604, 5.207, 7.811)$ (coordinates approximate)

57. Closest point $\left(\frac{9}{13}\sqrt{13}, \frac{4}{13}\sqrt{13}\right)$,

farthest point $\left(-\frac{9}{13}\sqrt{13}, -\frac{4}{13}\sqrt{13}\right)$

59. $\dfrac{81}{4}\sqrt{2}$

63. The minimum is $\left(a^{2/3}+b^{2/3}\right)^{3/2}$

65. $(7, 43, -21)$ on L_1 and $(12, 41, -22)$ on L_2

SECTION 13.10 (PAGE 933)

1. Local (in fact, global) minimum at $(-1, 2)$

3. Saddle point at $\left(-\frac{1}{2}, -\frac{1}{2}\right)$

5. Local (in fact, global) minimum at $(-3, 4)$

7. Local maximum at $(-1, -1)$, saddle point at $(0, 0)$

9. No extrema

11. Local (in fact, global) minima at $(-1, -1)$ and $(1, 1)$, saddle point at $(0, 0)$

13. Saddle point at $(-1, 1)$, local minimum at $(3, -3)$

15. Local minimum at $(-5, 3)$, saddle point at $(0, -2)$

17. Local (in fact global) maxima at $(-1, -2)$ and $(1, 2)$, saddle point at $(0, 0)$

19. Saddle point at $(-1, 0)$, local minimum at $(2, 0)$

21. Saddle point at $(0, 0)$, local (in fact, global) maxima at $\left(-\frac{1}{2}\sqrt{2}, -\frac{1}{2}\sqrt{2}\right)$ and $\left(\frac{1}{2}\sqrt{2}, \frac{1}{2}\sqrt{2}\right)$, local (in fact, global) minima at $\left(\frac{1}{2}\sqrt{2}, -\frac{1}{2}\sqrt{2}\right)$ and $\left(-\frac{1}{2}\sqrt{2}, \frac{1}{2}\sqrt{2}\right)$

23. Global minimum at $(0, 0)$

25. Global maximum at $(0, 0)$

27. Global minimum value 3 at $(-1, -1)$ and $(1, 1)$, no extremum at $(0, 0)$

29. The global maximum value 900 occurs on the boundary of the domain.

31. If x and y are both even integers, then there is a saddle point at (x, y); if x and y are odd integers both of the form $4k+1$ or both of the form $4k+3$, then there is a global maximum at (x, y); if x and y are odd integers one of which is of the form $4k+1$ and the other of which is of the form $4k+3$, then there is a global minimum at (x, y).

33. Examine the behavior of $f(x, y)$ on lines of the form $y = mx$.

35. Local minimum at $(1.532, 0)$ (numbers with decimals are approximations), saddle point at $(0.347, 0)$, global minimum at $(-1.879, 0)$

37. Local (indeed, global) minimum at $(-1.879, 1.879)$ (numbers with decimals are approximations), saddle point at $(0.347, -0.347)$, local minimum at $(1.532, -1.532)$

39. Global minimum at $(3.625, -3.984)$ (numbers with decimals are approximations) and at $(3.625, 3.984)$, saddle point at $(0, 0)$

CHAPTER 13 MISCELLANEOUS PROBLEMS (PAGE 936)

1. You should obtain $r^2\sin^2\theta\cos^2\theta \to 0$ as $r \to 0$.

3. $g(x, y) \to \dfrac{1}{2} \neq g(0, 0)$ as $(x, y) \to (0, 0)$ along the line $y = x$.

5. $f(x, y) = x^2y^3 + e^x\sin y + y + C$ (where C is an arbitrary constant).

7. The origin and points on the circle formed by the intersection of the paraboloid and the horizontal plane $z = \frac{1}{2}$.

9. You should find that the normal to the cone at (a, b, c) (extended, if necessary) passes through the point $(0, 0, 2c)$.

11. You should find that

$$u_{xx}(x, y, t) = \frac{x^2-2kt}{16k^3\pi t^3}\exp\left(-\frac{x^2+y^2}{4kt}\right).$$

13. You should find that

$$\mathbf{r}_x \times \mathbf{r}_y = \langle -f_x(x, y), -f_y(x, y), 1\rangle = \nabla g(x, y, z)$$

where $g(x, y, z) = z - f(x, y)$.

15. The base of the shipping crate will be a square $2\cdot 3^{1/3} \approx 2.884449914$ feet on each side and the height of the crate will be $5\cdot 3^{1/3} \approx 7.21124785$ feet.

17. The estimate of the error is $2\ \Omega$.

19. The maximum error will be approximately 3%.

21. The six points $(\pm 4, 0, 0)$, $(0, \pm 2, 0)$, and $\left(0, 0, \pm\frac{4}{3}\right)$.

23. First rename **a**, **b**, and **c** (if necessary) so that **a**, **b**, **c** forms a right-handed triple, and thus $\mathbf{a}\times\mathbf{b} = \mathbf{c}$, etc.

25. Either $\langle -4, 3\rangle$ or $\langle 4, -3\rangle$. **27.** 1

29. The global minimum value of $f(x, y)$ is $0 = f(0, 0)$.

31. The semiaxes have lengths 1 and 2.

33. The minimum occurs when the triangle is totally degenerate: Its three vertices are all located at the same point of the circumference of the circle.

35. The closest and farthest points are (respectively)

$$\left(\frac{1}{3}\sqrt{6}, \frac{1}{6}\sqrt{6}\right) \quad\text{and}\quad \left(-\frac{1}{3}\sqrt{6}, -\frac{1}{6}\sqrt{6}\right).$$

37. Let n be a fixed positive integer and let $f(x_1, x_2, \dots, x_n) = x_1 + x_2 + \cdots + x_n$. Maximize this function subject to the constraint

$$g(x_1, x_2, \dots, x_n) = x_1^2 + x_2^2 + \cdots + x_n^2 - a^2 = 0$$

where a is a fixed but otherwise arbitrary nonnegative real number.

39. Maximum: 1; minimum: $-\dfrac{1}{2}$

41. Theorem 1 of Section 13.10 yields these results:

At $P(-1, -1)$: $A = 6$, $B = 0$, $C = 2$, $\Delta = 12$, $f(P) = -1$: Local minimum;

At $Q(0, 0)$: $A = 0$, $B = -3$, $C = 2$, $\Delta = -9$, $f(Q) = 0$: Saddle point;

At $R\left(-\sqrt{3}, 0\right)$: $A = 0$, $B = 6$, $C = 2$, $\Delta = -36$, $f(R) = 0$: Saddle point;

At $S\left(\sqrt{3}, 0\right)$: $A = 0$, $B = 6$, $C = 2$, $\Delta = -36$, $f(S) = 0$: Saddle point;

At $T(1, 1)$: $A = 6$, $B = 0$, $C = 2$, $\Delta = 12$, $f(T) = -1$: Local minimum.

There are no global extrema (examine $f(x, y)$ on the lines $y = \pm x$).

43. Saddle point at (0, 0), local (not global) minimum at (2, 2)

45. Local maximum at $(\frac{1}{2}, \frac{1}{3})$, saddle point at (0, 1), local maximum at every point of the x-axis for which $x < 0$ or $x > 1$, local minimum at every point of the x-axis for which $0 < x < 1$, and no global extrema.

47. Saddle point at (0, 0), global minimum at every point of the hyperbola with equation $xy = \ln 2$, and no other extrema.

49. Saddle points at $(-1, -1)$ and (1, 1); no extrema.

51. The coefficients m and b are the [generally] unique solutions of the equations

$$b\sum_{i=1}^{n} x_i + m\sum_{i=1}^{n} (x_i)^2 = \sum_{i=1}^{n} x_i y_i \quad \text{and}$$

$$b\sum_{i=1}^{n} 1 + m\sum_{i=1}^{n} x_i = \sum_{i=1}^{n} y_i.$$

SECTION 14.1 (PAGE 945)

1. a. 198; **b.** 480

3. 8

5. 88

7. $\frac{1}{2}\pi^2$

9. $L \leqq M \leqq U$

11. 80

13. -78

15. 128.25

17. -4.5

19. 1

21. $\dfrac{e-1}{2}$

23. $2e - 2$

25. $\dfrac{\pi^4 + 8\pi}{4}$

27. 1

29. $2 \ln 2$

31. Both values: -32

33. Both values: $\dfrac{4}{15}\left(9\sqrt{3} - 8\sqrt{2} + 1\right)$

35. $\displaystyle\int_0^1 \int_0^1 x^n y^n \, dx \, dy = \frac{1}{(n+1)^2}.$

37. Note that $0 \leqq f(x, y) \leqq \sin \frac{1}{2}\pi = 1$ if (x, y) is a point of R.

SECTION 14.2 (PAGE 953)

1. $\dfrac{5}{6}$

3. $\dfrac{1}{2}$

5. $\dfrac{1}{12}$

7. $\dfrac{1}{20}$

9. $-\dfrac{1}{18}$

11. $\dfrac{e-2}{2}$

13. $\dfrac{61}{3}$

15. 0

17. 0

19. π

21. 1

23. 2

25. $\dfrac{512}{21}$

27. $\dfrac{32}{3}$

29. $\dfrac{4}{3}$

31. 2

33. $\dfrac{\pi}{8}$

35. Approximately 7.9517471897

37. 0

39. 0

41. 0

43. 0

53. Midpoint approximation: 0.109696; exact value: $\dfrac{e-2}{4}$

SECTION 14.3 (PAGE 959)

1. $\dfrac{1}{6}$

3. $\dfrac{32}{3}$

5. $\dfrac{5}{6}$

7. $\dfrac{32}{3}$

9. $2 \ln 2$

11. 2

13. $2e$

15. $\dfrac{1}{3}$

17. $\dfrac{41}{60}$

19. $\dfrac{4}{15}$

21. $\dfrac{10}{3}$

23. 19

25. $\dfrac{4}{3}$

27. 6

29. 24

31. π

33. $\dfrac{\pi}{3}\left(32 - 12\sqrt{3}\right)$

35. $\dfrac{1}{6}abc$

37. $\dfrac{2}{3}$

39. The volume is $V = 8\displaystyle\int_0^a \int_0^{\sqrt{a^2 - x^2}} \sqrt{a^2 - x^2 - y^2} \, dy \, dx.$

41. $\dfrac{625\pi}{2}$

43. $\dfrac{3\sqrt{3} + 2\pi}{6} R^3$

45. $\dfrac{256}{15}$

47. Approximately 3.5729749639

49. 8π

51. 108π

53. The "hole volume" is approximately 26.7782.

SECTION 14.4 (PAGE 966)

1. $\displaystyle\int_0^{2\pi} \int_0^1 r \, dr \, d\theta = \pi$

3. $\dfrac{3}{2}\pi$

5. $\dfrac{4\pi - 3\sqrt{3}}{6}$

7. $\dfrac{2\pi - 3\sqrt{3}}{2}$

9. $\dfrac{16\pi}{3}$

11. $\dfrac{23\pi}{8}$

13. $\dfrac{\pi \ln 2}{4}$

15. $\dfrac{16\pi}{5}$

17. $\dfrac{\pi}{4}(1 - \cos 1) \approx 0.361046$

19. 2π

21. 4π

23. $\displaystyle\int_0^{2\pi} \int_0^a 2r\sqrt{a^2 - r^2} \, dr \, d\theta$

25. $\int_0^{2\pi}\int_0^a (h + r\cos\theta)\cdot r\,dr\,d\theta$

27. 2π

29. $\frac{1}{3}\pi\left(2-\sqrt{2}\right)a^3$

31. As stated, the volume is $\pi/2$. If the equation of the lemniscate should be $r^2 = 2\sin 2\theta$, then the volume is $\pi/4$.

35. $2\pi^2a^2b$ **37.** 8π **39.** 24π

41. $\frac{4}{3}\left(64-24\sqrt{3}\right)\pi$

43. Hexagonal hole: 9.83041 (numbers with decimals are approximations). Pentagonal hole: 9.03688. Heptagonal hole: 10.32347. 17-sided hole: 11.49809.

SECTION 14.5 (PAGE 975)

1. $(2, 3)$ **3.** $(1, 1)$

5. $m = 4$, $M_y = \frac{16}{3}$, $M_x = \frac{8}{3}$ **7.** $\left(\frac{3}{2}, \frac{6}{5}\right)$

9. $\left(0, -\frac{8}{5}\right)$ **11.** $\left(\frac{2}{5}, \frac{2}{5}\right)$

13. $\left(0, \frac{16}{7}\right)$ **15.** $\left(\frac{9}{14}, \frac{9}{14}\right)$

17. $\left(0, \frac{43}{35}\right)$ **19.** $\left(\frac{\pi}{2}, \frac{\pi}{8}\right)$

21. $\left(\frac{7a}{12}, \frac{7a}{12}\right)$ **23.** $\left(0, \frac{20}{7}\right)$

25. $\left(\frac{\pi^2-4}{\pi}, \frac{\pi}{8}\right)$ **27.** $\left(0, \frac{3a}{2\pi}\right)$

29. $\left(0, \frac{36\pi+33\sqrt{3}}{32\pi+12\sqrt{3}}\right)$ **31.** $I_0 = \frac{2\pi a^{n+4}}{n+4}$

33. $I_0 = \frac{3}{2}\pi k$ **35.** $I_0 = \frac{1}{9}$

37. $\hat{x} = \frac{2}{21}\sqrt{105}$, $\hat{y} = \frac{4}{3}\sqrt{5}$ **39.** $\hat{x} = \hat{y} = \frac{1}{10}a\sqrt{30}$

41. $\left(\frac{4a}{3\pi}, \frac{4a}{3\pi}\right)$ **43.** $\left(\frac{2r}{\pi}, \frac{2r}{\pi}\right)$

51. **a.** Centroid $\left(0, \frac{4a^2+3\pi ab+6b^2}{3\pi a+12b}\right)$;

b. volume $\frac{\pi a}{3}\cdot(4a^2+3\pi ab+6b^2)$

53. $I_0 = \frac{1}{12}m(a^2+b^2)$

55. $I_0 = \frac{484}{3}k$

57. Mass π, centroid $\left(0, \frac{5}{4}\right)$

59. Mass $\frac{\pi}{2}$, centroid $\left(\frac{5}{4}, \frac{4}{3\pi}\right)$

SECTION 14.6 (PAGE 985)

1. 18 **3.** 128 **5.** $\frac{1}{60}$

7. $-\frac{1}{6}$ **9.** 12 **11.** 6

13. $\frac{128}{5}$ **15.** $\frac{332}{105}$ **17.** $\frac{256}{15}$

19. $\frac{11}{30}$

21. Mass $\frac{128}{5}$; centroid $\left(0, \frac{20}{7}, \frac{10}{7}\right)$

23. $\left(0, \frac{8}{7}, \frac{12}{7}\right)$

25. $\left(0, \frac{44-9\pi}{72-9\pi}, \frac{9\pi-16}{72-9\pi}\right)$

27. $I_y = \frac{8}{7}$ **29.** $I_z = \frac{1}{30}$

31. Mass $m = \frac{1}{2}\pi h^2$, $M_{yz} = 0$, $M_{xz} = 0$, $M_{xy} = \frac{1}{3}\pi h^3$

33. $I_z = \frac{2}{3}a^5$ **35.** $I_z = \frac{38}{45}ka^7$

37. $I_z = \frac{1}{3}k$ **39.** $\left(\frac{9\pi}{64}, \frac{9\pi}{64}, \frac{3}{8}\right)$

41. 24π **43.** $\frac{1}{6}\pi$

45. $\left(\frac{2}{5}, \frac{3}{5}, \frac{12}{5}\right)$ **47.** $\bar{\delta} = \frac{3}{2}$

49. $\frac{1}{4}$ **51.** $\bar{d} = \frac{49}{10}$

53. $\frac{1}{24}\left[6\sqrt{3}-\pi+8\,\mathrm{arcsinh}\left(\frac{\sqrt{2}}{2}\right)-8\ln 2+16\ln\left(1+\sqrt{3}\right)\right]$

≈ 0.960592

SECTION 14.7 (PAGE 993)

1. 8π

3. $V = 2\int_0^{2\pi}\int_0^a\int_0^{\sqrt{a^2-r^2}} r\,dz\,dr\,d\theta$

5. $\frac{4\pi}{3}\left(8-3\sqrt{3}\right)$ **7.** $\frac{1}{2}\pi a^2h^2$

9. $I_z = \frac{1}{4}\pi a^4h^2$ **11.** $\frac{81\pi}{2}$; $(0, 0, 3)$

13. 24π **15.** $\frac{\pi}{6}\left(8\sqrt{2}-7\right)$

17. $I_x = \frac{1}{12}\delta\pi a^2h(3a^2+4h^2) = \frac{1}{12}m(3a^2+4h^2)$

19. $\frac{\pi}{3}$ **21.** $\left(0, 0, \frac{3}{8}a\right)$

23. $V = \int_0^{2\pi}\int_0^{\pi/4}\int_0^{\sec\phi} \rho^2\sin\phi\,d\rho\,d\phi\,d\theta = \frac{\pi}{3}$

25. $m = \frac{\pi}{3}(2-\sqrt{2})a^3$; $\bar{x} = \bar{y} = 0$, $\bar{z} = \frac{3}{16}(2+\sqrt{2})a$

27. $I_z = \frac{28}{15}\pi\delta a^5 = \frac{7}{5}ma^2$

29. This "pinched torus" has volume $V = 2\pi^2 a^3$.

31. $I_x = \frac{2}{15}(128 - 51\sqrt{3})\pi\delta a^5$

33. Mass $\frac{37}{48}\pi a^4$; $\bar{x} = \bar{y} = 0$, $\bar{z} = \frac{105}{74}a$

37. (a) $\frac{4}{3}\pi[1 - \exp(-a^3)]$

39. $\bar{d} = \frac{3}{4}a$

41. $I_z = \iint_S \delta(x^2 + y^2)\,dA$ where $dA = a^2 \sin\phi\, d\phi\, d\theta$

45. $\frac{3768}{11}\pi$

49. $x \approx 2.76449 \times 10^6$(meters); mantle thickness: about 3606 km

SECTION 14.8 (PAGE 1000)

1. $6\pi\sqrt{11}$

3. $\frac{\pi}{6}(17\sqrt{17} - 1)$

5. $3\sqrt{2} + \frac{1}{2}\ln(3 + 2\sqrt{2})$

7. $3\sqrt{14}$

9. $\frac{2\pi}{3}(2\sqrt{2} - 1)$

11. $\frac{\pi}{6}(65\sqrt{65} - 1)$

13. $A = \int_0^{2\pi}\int_0^h a\,dz\,d\theta$

15. $8a^2$

23. $A = \int_0^{2\pi}\int_0^h r\,dz\,d\theta$

25. **a.** $4 + \frac{7}{3}\operatorname{arcsinh}\left(\frac{2\sqrt{5}}{5}\right) - \frac{1}{3}\arctan\left(\frac{4}{3}\right) + \frac{7}{6}\ln 5$;

b. $\frac{2\sqrt{5}}{3} + \frac{5\sqrt{2}}{3}\operatorname{arcsinh}\left(\frac{\sqrt{6}}{3}\right) - \frac{1}{6}\arctan\left(\frac{72 - 25\sqrt{5}}{71}\right) - \frac{1}{6}\arctan\left(\frac{72 - 25\sqrt{5}}{29}\right) + \frac{1}{6}\arctan\left(\frac{25\sqrt{5} - 72}{29}\right) + \frac{1}{6}\arctan\left(\frac{25\sqrt{5} - 72}{71}\right)$

27. **a.** Approximately 5.123157; **b.** approximately 2.302311

29. Elliptic paraboloid

31. Hyperboloid of two sheets

33. Approximately 111.545775

SECTION 14.9 (PAGE 1007)

1. $x = \frac{u+v}{2}$, $y = \frac{u-v}{2}$; $\frac{\partial(x,y)}{\partial(u,v)} = -\frac{1}{2}$

3. Two solutions: $x = \pm(u/v)^{1/2}$, $y = \pm(uv)^{1/2}$ (choose the same sign); the Jacobian is $1/(2v)$ in each case.

5. $x = \frac{u+v}{2}$, $y = \pm\frac{\sqrt{u-v}}{2}$ (choose the sign so that $y \geqq 0$); the Jacobian is

$$\frac{\partial(x,y)}{\partial(u,v)} = -\frac{1}{4\sqrt{u-v}}.$$

7. $\frac{3}{5}$

9. $\ln 2$

11. $\frac{2-\sqrt{2}}{8}$

13. $\frac{39\pi}{2}$

15. 8

17. First use the substitution $x = u + v$, $y = u - v$. The value of the integral is

$$\frac{2\pi}{3}\left(1 - \frac{1}{e^3}\right)\sqrt{3} \approx 3.446991.$$

21. $I_z = \int_{\theta=0}^{2\pi}\int_{\phi=0}^{\pi}\int_{\rho=0}^{1}(\rho^2\sin^2\phi)(a^2\cos^2\theta + b^2\sin^2\theta)\delta abc\rho^2\sin\phi\,d\rho\,d\phi\,d\theta$

23. $(\bar{x}, \bar{y}) = \left(\frac{72\sqrt{3} - 40\sqrt{6}}{15\ln 2}, \frac{6\sqrt{6} - 8\sqrt{3}}{\ln 2}\right)$

25. $I_z = \frac{1}{5}M(a^2 + b^2)$; the other moments follow by symmetry.

27. 2.30026852

CHAPTER 14 MISCELLANEOUS PROBLEMS (PAGE 1010)

1. $\int_{x=0}^{1}\int_{y=0}^{x^3}\frac{1}{\sqrt{1+x^2}}\,dy\,dx = \frac{2-\sqrt{2}}{3}$

3. $\frac{e-1}{2e}$

5. $\frac{e^4-1}{4}$

7. $\frac{4}{3}$

9. 9π; $\left(0, 0, \frac{9}{16}\right)$

11. 4π

13. 4π

15. $\frac{\pi - 2}{16}$

17. $\frac{128}{15}$; $\left(\frac{32}{7}, 0\right)$

19. $k\pi$; $(1, 0)$

21. $\bar{y} = \frac{4b}{3\pi}$

23. $\left(0, \frac{8}{5}\right)$

25. $\frac{10\pi}{3}(\sqrt{5} - 2)$

27. $I_z = \frac{3}{10}Ma^2$

29. $I_x = \frac{4}{15}\pi abc(b^2 + c^2) = \frac{1}{5}M(b^2 + c^2)$

31. $I_z = \frac{128}{225}(15\pi - 26)$

33. $\frac{8\pi}{3}$

35. $\frac{1}{4}M(3a^2 + 4b^2)$ where $M = 2\pi^2\delta a^2 b$ is the mass of the torus.

37. $\bar{d} = \frac{1}{\pi a^2}\int_0^{\pi}\int_0^{2a\sin\theta} r^2\,dr\,d\theta$

39. $\bar{d} = \frac{3}{4\pi a^3}\int_0^{2\pi}\int_0^{\pi}\int_0^{a}\rho^3\sin\phi\,d\rho\,d\phi\,d\theta$

41. Use the spheres $\rho = 2\cos\phi$ and $\rho = 4\cos\phi$.

43. $\frac{\pi}{6}\left(37\sqrt{37} - 17\sqrt{17}\right)$

47. $4\sqrt{2}$

51. $I_0 = 3\delta$

53. $\frac{8}{15}\pi abc$

SECTION 15.1 (PAGE 1018)

1.

3.

5.

7.

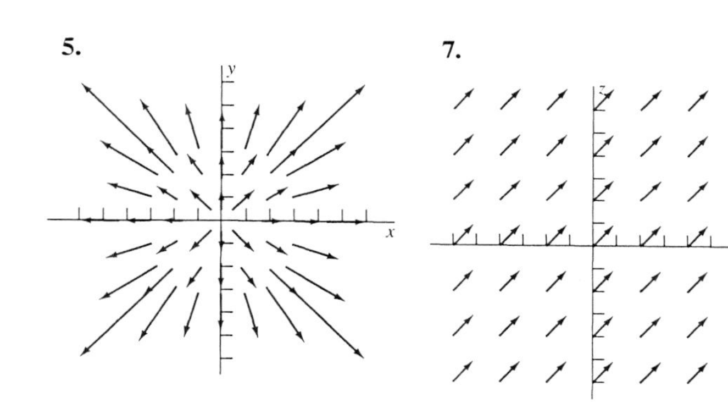

9.

11.

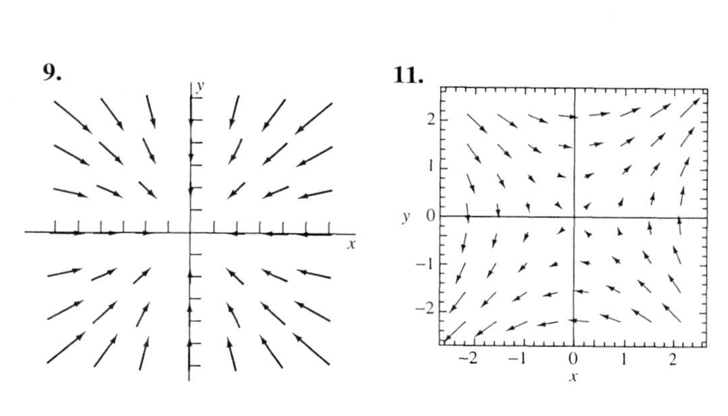

13.

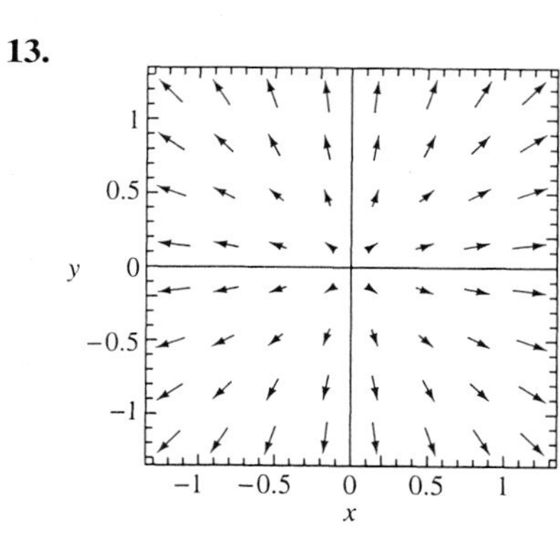

15. $\nabla \cdot \mathbf{F} = 3$, $\nabla \times \mathbf{F} = \mathbf{0}$

17. $\nabla \cdot \mathbf{F} = 0$, $\nabla \times \mathbf{F} = \mathbf{0}$

19. $\nabla \cdot \mathbf{F} = x^2 + y^2 + z^2$, $\nabla \times \mathbf{F} = \langle -2yz, -2xz, -2xy \rangle$

21. $\nabla \cdot \mathbf{F} = 0$, $\nabla \times \mathbf{F} = \langle 2y - 2z, 2z - 2x, 2x - 2y \rangle$

23. $\nabla \cdot \mathbf{F} = 3$, $\nabla \times \mathbf{F} = \langle x\cos xy - x\cos xz, y\cos yz - y\cos xy, z\cos xz - z\cos yz \rangle$

35. See the answer to Problem 15.

37. Use the results in Problems 28 and 35.

41. Use the results in Problems 28, 35, and 39.

SECTION 15.2 (PAGE 1028)

1. $\frac{310}{3}$, $\frac{248}{3}$, and 62

3. $3\sqrt{2}$ and 3

5. $\frac{49}{24}$ and $\frac{4}{3}$

7. $\frac{6}{5}$

9. 315

11. $\frac{19}{60}$

13. $\pi(1 + 2\pi)$

15. 28

17. $\frac{14\sqrt{14} - 1}{6}$

19. $\left(0, \frac{2a}{\pi}\right)$

21. Mass $10k\pi$; centroid $(0, 0, 4\pi)$

23. Mass $\frac{1}{2}ka^3$; centroid $\left(\frac{2}{3}a, \frac{2}{3}a, 0\right)$; $I_x = I_y = \frac{1}{2}ma^2$; $I_0 = ma^2$

25. $I_0 = 3k = \frac{1}{2}m$ where m is the mass of the wire

27. $\frac{4a}{\pi}$

29. $\frac{1}{2} + \frac{\sqrt{3}}{12}\operatorname{arctanh}\frac{\sqrt{3}}{2} \approx 0.690086$

31. $\frac{1}{2}$

33. a. $\frac{1}{2}k\ln 2$; **b.** $-\frac{1}{2}k\ln 2$

37. Note that $\mathbf{F}$ is normal to the sphere.

39. 15000 ft·lb

41. 20000 ft·lb

SECTION 15.3 (PAGE 1036)

1. $\phi(x, y) = x^2 + 3xy + y^2$

3. $\phi(x, y) = x^3 + 2xy^2 + 2y^3$

5. Not conservative

7. $\phi(x, y) = \frac{1}{4}x^4 + y\ln x + \frac{1}{3}y^3$

9. $\phi(x, y) = \sin x + x\ln y + e^y$

11. Not conservative

13. $\phi(x, y) = x^3y^3 + xy^4 + \frac{1}{5}y^5$

15. $\phi(x, y) = \frac{x^2}{y} + 2\sqrt{y} + \frac{y^2}{x^3}$

17. $\phi(x, y) = x^3 + 2xy^2 + 2y^3$

19. $\phi(x, y) = x^3y^3 + xy^4 + \frac{1}{5}y^5$

21. 6 **23.** $\dfrac{1}{e}$

25. $-\pi$ **27.** $\phi(x, y, z) = xyz$

29. $\phi(x, y, z) = xy\cos z - yze^x$

37. $W = 8.04442 \times 10^{10}$ N·m

SECTION 15.4 (PAGE 1045)

1. 0 **3.** 3 **5.** $\dfrac{3}{10}$

7. 2 **9.** 0 **11.** $\dfrac{16}{105}$

13. $\displaystyle\int_0^{2\pi} a^2\cos^2 t\,dt = \pi a^2$

15. $\dfrac{3\pi}{8}$ **17.** 30π **19.** $\dfrac{972}{5}$

21. 30π **23.** 45 **33.** $\dfrac{3}{2}$

39. **a.** $\dfrac{3}{4}\sqrt{3}$; **b.** $\dfrac{5}{8}\sqrt{10+2\sqrt{5}}$

SECTION 15.5 (PAGE 1055)

1. $\dfrac{\sqrt{3}}{3}$

3. $27\pi\sqrt{14}$

5. $\dfrac{\pi}{6}\left(-1+17\sqrt{17}\right)$

7. $\dfrac{81}{2}\pi\delta\sqrt{3} = \dfrac{9}{2}m$ where m is the mass of S

9. $\dfrac{10}{3}\pi\delta = \dfrac{5}{6}m$ where m is the mass of S

11. $\dfrac{520}{3}\pi\delta = \dfrac{26}{3}m$ where m is the mass of S

13. 36π **15.** 24π

17. 0 **19.** 6

21. 0 **23.** 1458π

25. $\left(\dfrac{1}{2}a, \dfrac{1}{2}a, \dfrac{1}{2}a\right)$

27. $\bar{x} = 0 = \bar{y}$,

$$\bar{z} = \frac{(24a^4+2a^2-1)\sqrt{1+4a^2}+1}{10\left[(1+4a^2)^{3/2}-1\right]};$$

$$I_z = \frac{1}{60}\pi\delta\left[(24a^4+2a^2-1)\sqrt{1+4a^2}+1\right]$$

29. $\left(\dfrac{4}{3\pi-6}, 0, \dfrac{\pi}{2\pi-4}\right)$

31. $\dfrac{460\sqrt{17}+13\,\text{arcsinh}\,4}{48}$

35. -1728π

37. $I_z \approx 5157.168115$

39. $I_z \approx 98546.934874$

SECTION 15.6 (PAGE 1063)

1. $\displaystyle\iiint_B \nabla\cdot\mathbf{F}\,dV = 4\pi = \iint_S \mathbf{F}\cdot\mathbf{n}\,dS$

3. 24 **5.** $\dfrac{1}{2}$

7. $\dfrac{2385\pi}{2}$ **9.** $\dfrac{1}{4}$

11. $\dfrac{703125\pi}{4}$ **13.** 16π

23. $\dfrac{482620+29403\ln 11}{48}\pi$

SECTION 15.7 (PAGE 1070)

1. -20π **3.** 0

5. -52π **7.** -8π

9. -2

11. $\phi(x, y, z) = 3xy - 2xz + yz$

13. $\phi(x, y, z) = 3xe^z + 17z + 5y\cos x$

CHAPTER 15 MISCELLANEOUS PROBLEMS (PAGE 1072)

1. $\dfrac{125}{3}$ **3.** $\dfrac{69}{8}$ **5.** $\dfrac{2148}{5}$

7. First assume (by way of contradiction) that there exists a function $\phi(x, y)$ such that $\nabla\phi = \langle x^2y, xy^2\rangle$.

9. $m = \dfrac{5\sqrt{5}-1}{3}$; $I_y = \dfrac{50\sqrt{5}+2}{15}$

11. $\dfrac{2816}{7}$

13. Both integrals are zero.

15. Begin with the observation that $\langle P, Q\rangle = \nabla\phi$ for some differentiable function ϕ.

17. $\dfrac{371\pi}{30}$ **19.** $60\pi + 12\pi = 72\pi$

29. **a.** $\dfrac{\mathbf{r}}{r}\phi'(r)$; **b.** $3\phi(r) + r\dfrac{d\phi}{dr}$; **c.** $\mathbf{0}$

APPENDIX A (PAGE A-5)

1. 14 **3.** $\dfrac{1}{2}$

5. 25 **7.** 27

9. $\dfrac{22}{7} - \pi$ (because $\pi < \frac{22}{7}$)

11. $3 - x$ **13.** $(-\infty, 2)$

15. $[7, +\infty)$ **17.** $\left(-\frac{5}{3}, +\infty\right)$

19. $(-4, 1)$ **21.** $\left(\frac{3}{2}, \frac{11}{2}\right]$

23. $(-1, 4)$ **25.** $\left(-\infty, \frac{1}{3}\right) \cup (1, +\infty)$

27. $\left[0, \frac{2}{5}\right] \cup \left[\frac{6}{5}, \frac{8}{5}\right]$ **29.** $\left(\frac{7}{3}, \frac{37}{15}\right]$

31. $\left(-\infty, \frac{1}{5}\right) \cup \left(\frac{1}{5}, +\infty\right)$

33. $(-\infty, -2) \cup (4, +\infty)$

35. $\left(-\infty, \frac{1}{2}\right] \cup \left[\frac{3}{2}, +\infty\right)$

37. $4 \leqq p \leqq 8$

39. $2 < I < 4$

APPENDIX B (PAGE A-12)

1. They lie on one line.

3. They do not lie on one line.

5. This parallelogram is a rectangle!

7. Right angle at A

9. Slope $\frac{2}{3}$, y-intercept 0

11. Slope 2, y-intercept 3

13. Slope $-\frac{2}{5}$, y-intercept $\frac{3}{5}$

15. $y = -5$

17. $y = 2x - 7$

19. $y = 6 - x$

21. $2x + y = 7$

23. $2y = 13 - x$

25. $\frac{4}{13}\sqrt{26}$

31. P_1M and MP_2 have the same slope $\frac{y_1 - y_2}{x_1 - x_2}$.

33. $K = \frac{125F + 57461}{225}$

35. 1136 gal/wk

37. $x = -\frac{11}{4}, \quad y = \frac{7}{2}$

39. $x = \frac{37}{6}, \quad y = -\frac{1}{2}$

41. $x = \frac{22}{5}, \quad y = -\frac{1}{5}$

43. $x = -\frac{7}{4}, \quad y = \frac{33}{8}$

45. $x = \frac{119}{12}, \quad y = -\frac{19}{4}$

APPENDIX C (PAGE A-17)

1. $\frac{2}{9}\pi$ (rad)

3. $\frac{7}{4}\pi$ (rad)

5. $-\frac{5}{6}\pi$ (rad)

7. 72°

9. 675°

11. If $x = -\frac{\pi}{3}$, then the values of the six trigonometric functions are given in the following table.

$\sin x$	$\cos x$	$\tan x$	$\sec x$	$\csc x$	$\cot x$
$-\frac{\sqrt{3}}{2}$	$\frac{1}{2}$	$-\sqrt{3}$	2	$-\frac{2\sqrt{3}}{3}$	$-\frac{\sqrt{3}}{3}$

13. If $x = \frac{7\pi}{6}$, then the values of the six trigonometric functions are given in the following table.

$\sin x$	$\cos x$	$\tan x$	$\sec x$	$\csc x$	$\cot x$
$-\frac{1}{2}$	$-\frac{\sqrt{3}}{2}$	$\frac{\sqrt{3}}{3}$	$-\frac{2\sqrt{3}}{3}$	-2	$\sqrt{3}$

15. $x = n\pi$ where n is an integer

17. $x = 2n\pi - \frac{\pi}{2}$ where n is an integer

19. $x = 2n\pi$ where n is an integer

21. $x = n\pi$ where n is an integer

23. $x = n\pi - \frac{\pi}{4}$ where n is an integer

25. The results are in the next table.

$\sin x$	$\cos x$	$\tan x$	$\sec x$	$\csc x$	$\cot x$
$-\frac{3}{5}$	$\frac{4}{5}$	$-\frac{3}{4}$	$\frac{5}{4}$	$-\frac{5}{3}$	$-\frac{4}{3}$

29. $\frac{1}{2}$

31. $-\frac{1}{2}$

33. $\frac{\sqrt{3}}{2}$

35. $-\frac{\sqrt{3}}{2}$

43. $\frac{\pi}{3}, \frac{2\pi}{3}$

45. $\frac{\pi}{2}$

47. $\frac{\pi}{8}, \frac{3\pi}{8}, \frac{5\pi}{8}, \frac{7\pi}{8}$

APPENDIX D (PAGE A-23)

1. Given $\epsilon > 0$, let $\delta = \epsilon$.

3. Given $\epsilon > 0$, let $\delta = \epsilon$.

5. Given $\epsilon > 0$, let δ be the minimum of 1 and $\epsilon/3$.

7. Given $\epsilon > 0$, let δ be the minimum of 1 and $\epsilon/6$.

9. Consider three cases: $a > 0$, $a < 0$, and $a = 0$.

13. Consider two cases: $L > 0$ and $L < 0$.

15. Given $a > 0$ and $\epsilon > 0$, let δ be the minimum of $a/2$ and $\epsilon\sqrt{2a}$.

APPENDIX G (PAGE A-36)

1. $\frac{1}{3}$

3. $\frac{8}{3}$

5. $\frac{1}{2}$

7. $\frac{52}{9}$

REFERENCES FOR FURTHER STUDY

References 2, 3, 7, and 10 may be consulted for historical topics pertinent to calculus. Reference 14 provides a more theoretical treatment of single-variable calculus topics than ours. References 4, 5, 8, and 15 include advanced topics in multivariable calculus. Reference 11 is a standard work on infinite series. References 1, 9, and 13 are differential equations textbooks. Reference 6 discusses topics in calculus together with computing and programming in BASIC. Those who would like to pursue the topic of fractals should look at reference 12.

1. Boyce, William E. and Richard C. DiPrima, *Elementary Differential Equations* (5th ed.). New York: John Wiley, 1991.
2. Boyer, Carl B., *A History of Mathematics* (2nd ed.). New York: John Wiley, 1991.
3. Boyer, Carl B., *The History of the Calculus and Its Conceptual Development.* New York: Dover Publications, 1959.
4. Buck, R. Creighton, *Advanced Calculus* (3rd ed.). New York: McGraw-Hill, 1977.
5. Courant, Richard and Fritz John, *Introduction to Calculus and Analysis.* Vols. I and II. New York: Springer-Verlag, 1982.
6. Edwards, C. H., Jr., *Calculus and the Personal Computer.* Englewood Cliffs, NJ: Prentice-Hall, 1986.
7. Edwards, C. H., Jr., *The Historical Development of the Calculus.* New York: Springer-Verlag, 1979.
8. Edwards, C. H., Jr., *Advanced Calculus of Several Variables.* New York: Academic Press, 1973.
9. Edwards, C. H., Jr. and David E. Penney, *Differential Equations with Boundary Value Problems: Computing and Modeling* (2nd ed.). Upper Saddle River, NJ: Prentice Hall, 2000.
10. Kline, Morris, *Mathematical Thought from Ancient to Modern Times.* Vols. I, II, and III. New York: Oxford University Press, 1972.
11. Knopp, Konrad, *Theory and Application of Infinite Series* (2nd ed.). New York: Hafner Press, 1990.
12. Peitgen, H.-O. and P. H. Richter, *The Beauty of Fractals.* New York: Springer-Verlag, 1986.
13. Simmons, George E., *Differential Equations with Applications and Historical Notes.* New York: McGraw-Hill, 1972.
14. Spivak, Michael E., *Calculus* (2nd ed.). Berkeley: Publish or Perish, 1980.
15. Taylor, Angus E. and W. Robert Mann, *Advanced Calculus* (3rd ed.). New York: John Wiley, 1983.

INDEX

Entries in bold type indicate the page (or pages) where a term is formally defined.

A

B

C

D

E

I

J

K

L

U

V

W

X

Y

Z

TABLE OF INTEGRALS

ELEMENTARY FORMS

1. $\int u\,dv = uv - \int v\,du$

2. $\int u^n\,du = \frac{1}{n+1}u^{n+1} + C$ if $n \neq -1$

3. $\int \frac{du}{u} = \ln|u| + C$

4. $\int e^u\,du = e^u + C$

5. $\int a^u\,du = \frac{a^u}{\ln a} + C$

6. $\int \sin u\,du = -\cos u + C$

7. $\int \cos u\,du = \sin u + C$

8. $\int \sec^2 u\,du = \tan u + C$

9. $\int \csc^2 u\,du = -\cot u + C$

10. $\int \sec u \tan u\,du = \sec u + C$

11. $\int \csc u \cot u\,du = -\csc u + C$

12. $\int \tan u\,du = \ln|\sec u| + C$

13. $\int \cot u\,du = -\ln|\csc u| + C$

14. $\int \sec u\,du = \ln|\sec u + \tan u| + C$

15. $\int \csc u\,du = -\ln|\csc u + \cot u| + C$

16. $\int \frac{du}{\sqrt{a^2-u^2}} = \sin^{-1}\frac{u}{a} + C$

17. $\int \frac{du}{a^2+u^2} = \frac{1}{a}\tan^{-1}\frac{u}{a} + C$

18. $\int \frac{du}{a^2-u^2} = \frac{1}{2a}\ln\left|\frac{u+a}{u-a}\right| + C$

19. $\int \frac{du}{u\sqrt{u^2-a^2}} = \frac{1}{a}\sec^{-1}\left|\frac{u}{a}\right| + C$

TRIGONOMETRIC FORMS

20. $\int \sin^2 u\,du = \frac{1}{2}u - \frac{1}{4}\sin 2u + C$

21. $\int \cos^2 u\,du = \frac{1}{2}u + \frac{1}{4}\sin 2u + C$

22. $\int \tan^2 u\,du = \tan u - u + C$

23. $\int \cot^2 u\,du = -\cot u - u + C$

24. $\int \sin^3 u\,du = -\frac{1}{3}(2 + \sin^2 u)\cos u + C$

25. $\int \cos^3 u\,du = \frac{1}{3}(2 + \cos^2 u)\sin u + C$

26. $\int \tan^3 u\,du = \frac{1}{2}\tan^2 u + \ln|\cos u| + C$

27. $\int \cot^3 u\,du = -\frac{1}{2}\cot^2 u - \ln|\sin u| + C$

28. $\int \sec^3 u\,du = \frac{1}{2}\sec u \tan u + \frac{1}{2}\ln|\sec u + \tan u| + C$

29. $\int \csc^3 u\,du = -\frac{1}{2}\csc u \cot u + \frac{1}{2}\ln|\csc u - \cot u| + C$

30. $\int \sin au \sin bu\,du = \frac{\sin(a-b)u}{2(a-b)} - \frac{\sin(a+b)u}{2(a+b)} + C$ if $a^2 \neq b^2$

31. $\int \cos au \cos bu\,du = \frac{\sin(a-b)u}{2(a-b)} + \frac{\sin(a+b)u}{2(a+b)} + C$ if $a^2 \neq b^2$

32. $\int \sin au \cos bu\,du = -\frac{\cos(a-b)u}{2(a-b)} - \frac{\cos(a+b)u}{2(a+b)} + C$ if $a^2 \neq b^2$

33. $\int \sin^n u\,du = -\frac{1}{n}\sin^{n-1} u \cos u + \frac{n-1}{n}\int \sin^{n-2} u\,du$

34. $\int \cos^n u\,du = \frac{1}{n}\cos^{n-1} u \sin u + \frac{n-1}{n}\int \cos^{n-2} u\,du$

35. $\int \tan^n u\,du = \frac{1}{n-1}\tan^{n-1} u - \int \tan^{n-2} u\,du$ if $n \neq 1$

36. $\int \cot^n u\,du = -\frac{1}{n-1}\cot^{n-1} u - \int \cot^{n-2} u\,du$ if $n \neq 1$

37. $\int \sec^n u\,du = \frac{1}{n-1}\sec^{n-2} u \tan u + \frac{n-2}{n-1}\int \sec^{n-2} u\,du$ if $n \neq 1$

38. $\int \csc^n u\,du = -\frac{1}{n-1}\csc^{n-2} u \cot u + \frac{n-2}{n-1}\int \csc^{n-2} u\,du$ if $n \neq 1$

39a. $\int \sin^n u \cos^m u\,du = -\frac{\sin^{n-1} u \cos^{m+1} u}{n+m} + \frac{n-1}{n+m}\int \sin^{n-2} u \cos^m u\,du$ if $n \neq -m$

39b. $\int \sin^n u \cos^m u\,du = -\frac{\sin^{n+1} u \cos^{m-1} u}{n+m} + \frac{m-1}{n+m}\int \sin^n u \cos^{m-2} u\,du$ if $m \neq -n$

40. $\int u \sin u\,du = \sin u - u\cos u + C$

41. $\int u \cos u\,du = \cos u + u \sin u + C$

42. $\int u^n \sin u\,du = -u^n \cos u + n\int u^{n-1}\cos u\,du$

43. $\int u^n \cos u\,du = u^n \sin u - n\int u^{n-1}\sin u\,du$

(Table of Integrals continues from previous page)

FORMS INVOLVING $\sqrt{u^2 \pm a^2}$

44 $$\int \sqrt{u^2 \pm a^2}\, du = \frac{u}{2}\sqrt{u^2 \pm a^2} \pm \frac{a^2}{2}\ln\left|u + \sqrt{u^2 \pm a^2}\right| + C$$

45 $$\int \frac{du}{\sqrt{u^2 \pm a^2}} = \ln\left|u + \sqrt{u^2 \pm a^2}\right| + C$$

46 $$\int \frac{\sqrt{u^2 + a^2}}{u}\, du = \sqrt{u^2 + a^2} - a \ln\left(\frac{a + \sqrt{u^2 + a^2}}{u}\right) + C$$

47 $$\int \frac{\sqrt{u^2 - a^2}}{u}\, du = \sqrt{u^2 - a^2} - a \sec^{-1}\frac{u}{a} + C$$

48 $$\int u^2\sqrt{u^2 \pm a^2}\, du = \frac{u}{8}(2u^2 \pm a^2)\sqrt{u^2 \pm a^2} - \frac{a^4}{8}\ln\left|u + \sqrt{u^2 \pm a^2}\right| + C$$

49 $$\int \frac{u^2\, du}{\sqrt{u^2 \pm a^2}} = \frac{u}{2}\sqrt{u^2 \pm a^2} \mp \frac{a^2}{2}\ln\left|u + \sqrt{u^2 \pm a^2}\right| + C$$

50 $$\int \frac{du}{u^2\sqrt{u^2 \pm a^2}} = \mp\frac{\sqrt{u^2 \pm a^2}}{a^2 u} + C$$

51 $$\int \frac{\sqrt{u^2 \pm a^2}}{u^2}\, du = -\frac{\sqrt{u^2 \pm a^2}}{u} + \ln\left|u + \sqrt{u^2 \pm a^2}\right| + C$$

52 $$\int \frac{du}{(u^2 \pm a^2)^{3/2}} = \pm\frac{u}{a^2\sqrt{u^2 \pm a^2}} + C$$

53 $$\int (u^2 \pm a^2)^{3/2}\, du = \frac{u}{8}(2u^2 \pm 5a^2)\sqrt{u^2 \pm a^2} + \frac{3a^4}{8}\ln\left|u + \sqrt{u^2 \pm a^2}\right| + C$$

FORMS INVOLVING $\sqrt{a^2 - u^2}$

54 $$\int \sqrt{a^2 - u^2}\, du = \frac{u}{2}\sqrt{a^2 - u^2} + \frac{a^2}{2}\sin^{-1}\frac{u}{a} + C$$

55 $$\int \frac{\sqrt{a^2 - u^2}}{u}\, du = \sqrt{a^2 - u^2} - a \ln\left|\frac{a + \sqrt{a^2 - u^2}}{u}\right| + C$$

56 $$\int \frac{u^2\, du}{\sqrt{a^2 - u^2}} = -\frac{u}{2}\sqrt{a^2 - u^2} + \frac{a^2}{2}\sin^{-1}\frac{u}{a} + C$$

57 $$\int u^2\sqrt{a^2 - u^2}\, du = \frac{u}{8}(2u^2 - a^2)\sqrt{a^2 - u^2} + \frac{a^4}{8}\sin^{-1}\frac{u}{a} + C$$

58 $$\int \frac{du}{u^2\sqrt{a^2 - u^2}} = -\frac{\sqrt{a^2 - u^2}}{a^2 u} + C$$

59 $$\int \frac{\sqrt{a^2 - u^2}}{u^2}\, du = -\frac{\sqrt{a^2 - u^2}}{u} - \sin^{-1}\frac{u}{a} + C$$

60 $$\int \frac{du}{u\sqrt{a^2 - u^2}} = -\frac{1}{a}\ln\left|\frac{a + \sqrt{a^2 - u^2}}{u}\right| + C$$

61 $$\int \frac{du}{(a^2 - u^2)^{3/2}} = \frac{u}{a^2\sqrt{a^2 - u^2}} + C$$

62 $$\int (a^2 - u^2)^{3/2}\, du = \frac{u}{8}(5a^2 - 2u^2)\sqrt{a^2 - u^2} + \frac{3a^4}{8}\sin^{-1}\frac{u}{a} + C$$

EXPONENTIAL AND LOGARITHMIC FORMS

63 $$\int ue^u\, du = (u - 1)e^u + C$$

64 $$\int u^n e^u\, du = u^n e^u - n\int u^{n-1}e^u\, du$$

65 $$\int \ln u\, du = u \ln u - u + C$$

66 $$\int u^n \ln u\, du = \frac{u^{n+1}}{n+1}\ln u - \frac{u^{n+1}}{(n+1)^2} + C$$

67 $$\int e^{au}\sin bu\, du = \frac{e^{au}}{a^2 + b^2}(a \sin bu - b \cos bu) + C$$

68 $$\int e^{au}\cos bu\, du = \frac{e^{au}}{a^2 + b^2}(a \cos bu + b \sin bu) + C$$

INVERSE TRIGONOMETRIC FORMS

69 $$\int \sin^{-1} u\, du = u \sin^{-1} u + \sqrt{1 - u^2} + C$$

70 $$\int \tan^{-1} u\, du = u \tan^{-1} u - \frac{1}{2}\ln(1 + u^2) + C$$

71 $$\int \sec^{-1} u\, du = u \sec^{-1} u - \ln\left|u + \sqrt{u^2 - 1}\right| + C$$

72 $$\int u \sin^{-1} u\, du = \frac{1}{4}(2u^2 - 1)\sin^{-1} u + \frac{u}{4}\sqrt{1 - u^2} + C$$

73 $$\int u \tan^{-1} u\, du = \frac{1}{2}(u^2 + 1)\tan^{-1} u - \frac{u}{2} + C$$

74 $$\int u \sec^{-1} u\, du = \frac{u^2}{2}\sec^{-1} u - \frac{1}{2}\sqrt{u^2 - 1} + C$$

75 $$\int u^n \sin^{-1} u\, du = \frac{u^{n+1}}{n+1}\sin^{-1} u - \frac{1}{n+1}\int \frac{u^{n+1}}{\sqrt{1 - u^2}}\, du \quad \text{if } n \neq -1$$

76 $$\int u^n \tan^{-1} u\, du = \frac{u^{n+1}}{n+1}\tan^{-1} u - \frac{1}{n+1}\int \frac{u^{n+1}}{1 + u^2}\, du \quad \text{if } n \neq -1$$

77 $$\int u^n \sec^{-1} u\, du = \frac{u^{n+1}}{n+1}\sec^{-1} u - \frac{1}{n+1}\int \frac{u^n}{\sqrt{u^2 - 1}}\, du \quad \text{if } n \neq -1$$

(Table of Integrals continues from previous page)

HYPERBOLIC FORMS

78 $\int \sinh u\,du = \cosh u + C$

79 $\int \cosh u\,du = \sinh u + C$

80 $\int \tanh u\,du = \ln(\cosh u) + C$

81 $\int \coth u\,du = \ln|\sinh u| + C$

82 $\int \operatorname{sech} u\,du = \tan^{-1}|\sinh u| + C$

83 $\int \operatorname{csch} u\,du = \ln\left|\tanh \frac{u}{2}\right| + C$

84 $\int \sinh^2 u\,du = \frac{1}{4}\sinh 2u - \frac{u}{2} + C$

85 $\int \cosh^2 u\,du = \frac{1}{4}\sinh 2u + \frac{u}{2} + C$

86 $\int \tanh^2 u\,du = u - \tanh u + C$

87 $\int \coth^2 u\,du = u - \coth u + C$

88 $\int \operatorname{sech}^2 u\,du = \tanh u + C$

89 $\int \operatorname{csch}^2 u\,du = -\coth u + C$

90 $\int \operatorname{sech} u \tanh u\,du = -\operatorname{sech} u + C$

91 $\int \operatorname{csch} u \coth u\,du = -\operatorname{csch} u + C$

MISCELLANEOUS ALGEBRAIC FORMS

92 $\int u(au+b)^{-1}\,du = \frac{u}{a} - \frac{b}{a^2}\ln|au+b| + C$

93 $\int u(au+b)^{-2}\,du = \frac{1}{a^2}\left(\ln|au+b| + \frac{b}{au+b}\right) + C$

94 $\int u(au+b)^n\,du = \frac{(au+b)^{n+1}}{a^2}\left(\frac{au+b}{n+2} - \frac{b}{n+1}\right) + C \quad \text{if } n \neq -1, -2$

95 $\int \frac{du}{(a^2 \pm u^2)^n} = \frac{1}{2a^2(n-1)}\left(\frac{u}{(a^2 \pm u^2)^{n-1}} + (2n-3)\int \frac{du}{(a^2 \pm u^2)^{n-1}}\right) \quad \text{if } n \neq 1$

96 $\int u\sqrt{au+b}\,du = \frac{2}{15a^2}(3au-2b)(au+b)^{3/2} + C$

97 $\int u^n\sqrt{au+b}\,du = \frac{2}{a(2n+3)}\left(u^n(au+b)^{3/2} - nb\int u^{n-1}\sqrt{au+b}\,du\right)$

98 $\int \frac{u\,du}{\sqrt{au+b}} = \frac{2}{3a^2}(au-2b)\sqrt{au+b} + C$

99 $\int \frac{u^n\,du}{\sqrt{au+b}} = \frac{2}{a(2n+1)}\left(u^n\sqrt{au+b} - nb\int \frac{u^{n-1}\,du}{\sqrt{au+b}}\right)$

100a $\int \frac{du}{u\sqrt{au+b}} = \frac{1}{\sqrt{b}}\ln\left|\frac{\sqrt{au+b}-\sqrt{b}}{\sqrt{au+b}+\sqrt{b}}\right| + C \quad \text{if } b > 0$

100b $\int \frac{du}{u\sqrt{au+b}} = \frac{2}{\sqrt{-b}}\tan^{-1}\sqrt{\frac{au+b}{-b}} + C \quad \text{if } b < 0$

101 $\int \frac{du}{u^n\sqrt{au+b}} = -\frac{\sqrt{au+b}}{b(n-1)u^{n-1}} - \frac{(2n-3)a}{(2n-2)b}\int \frac{du}{u^{n-1}\sqrt{au+b}} \quad \text{if } n \neq 1$

102 $\int \sqrt{2au-u^2}\,du = \frac{u-a}{2}\sqrt{2au-u^2} + \frac{a^2}{2}\sin^{-1}\frac{u-a}{a} + C$

103 $\int \frac{du}{\sqrt{2au-u^2}} = \sin^{-1}\frac{u-a}{a} + C$

104 $\int u^n\sqrt{2au-u^2}\,du = -\frac{u^{n-1}(2au-u^2)^{3/2}}{n+2} + \frac{(2n+1)a}{n+2}\int u^{n-1}\sqrt{2au-u^2}\,du$

105 $\int \frac{u^n\,du}{\sqrt{2au-u^2}} = -\frac{u^{n-1}}{n}\sqrt{2au-u^2} + \frac{(2n-1)a}{n}\int \frac{u^{n-1}\,du}{\sqrt{2au-u^2}}$

106 $\int \frac{\sqrt{2au-u^2}}{u}\,du = \sqrt{2au-u^2} + a\sin^{-1}\frac{u-a}{a} + C$

107 $\int \frac{\sqrt{2au-u^2}}{u^n}\,du = \frac{(2au-u^2)^{3/2}}{(3-2n)au^n} + \frac{n-3}{(2n-3)a}\int \frac{\sqrt{2au-u^2}}{u^{n-1}}\,du$

108 $\int \frac{du}{u^n\sqrt{2au-u^2}} = \frac{\sqrt{2au-u^2}}{a(1-2n)u^n} + \frac{n-1}{(2n-1)a}\int \frac{du}{u^{n-1}\sqrt{2au-u^2}}$

109 $\int \left(\sqrt{2au-u^2}\right)^n du = \frac{u-a}{n+1}(2au-u^2)^{n/2} + \frac{na^2}{n+1}\int \left(\sqrt{2au-u^2}\right)^{n-2} du$

110 $\int \frac{du}{\left(\sqrt{2au-u^2}\right)^n} = \frac{u-a}{(n-2)a^2}\left(\sqrt{2au-u^2}\right)^{2-n} + \frac{n-3}{(n-2)a^2}\int \frac{du}{\left(\sqrt{2au-u^2}\right)^{n-2}}$

DEFINITE INTEGRALS

111 $\int_0^\infty u^n e^{-u}\,du = \Gamma(n+1) = n! \quad (n \geqq 0)$

112 $\int_0^\infty e^{-au^2}\,du = \frac{1}{2}\sqrt{\frac{\pi}{a}} \quad (a > 0)$

113 $\int_0^{\pi/2} \sin^n u\,du = \int_0^{\pi/2} \cos^n u\,du = \begin{cases} \frac{1\cdot 3\cdot 5\cdot \cdots \cdot (n-1)}{2\cdot 4\cdot 6\cdot \cdots \cdot n}\cdot\frac{\pi}{2} & \text{if } n \text{ is an even integer and } n \geqq 2 \\ \frac{2\cdot 4\cdot 6\cdot \cdots \cdot (n-1)}{3\cdot 5\cdot 7\cdot \cdots \cdot n} & \text{if } n \text{ is an odd integer and } n \geqq 3 \end{cases}$